CONDENSED PYRAZINES

This is the Thirty-Fifth Volume in the Series

THE CHEMISTRY OF HETEROCYCLIC COMPOUNDS

THE CHEMISTRY OF HETEROCYCLIC COMPOUNDS

A SERIES OF MONOGRAPHS

ARNOLD WEISSBERGER and EDWARD C. TAYLOR

Editors

CONDENSED PYRAZINES

G. W. H. Cheeseman

DEPARTMENT OF CHEMISTRY, QUEEN ELIZABETH COLLEGE
UNIVERSITY OF LONDON

R. F. Cookson

JANSSEN PHARMACEUTICAL LTD.

AN INTERSCIENCE ® PUBLICATION

JOHN WILEY & SONS
NEW YORK · CHICHESTER · BRISBANE · TORONTO

An Interscience ® Publication

Library of Congress Cataloging in Publication Data:

Cheeseman, G. W. H., 1927–
Condensed pyrazines.

(The Chemistry of heterocyclic compounds; v. 35)
"An Interscience publication."
Includes indexes.
1. Pyridazine. 2. Quinoxaline. 3. Condensation products (Chemistry). I. Cookson, Ronald Frederick, 1943– joint author. II. Title

QD401.C493 547'.593 78—17533
ISBN 0-471-38204-3

Printed in the United States of America

10 9 8 7 6 5 4 3 2 1

The Chemistry of Heterocyclic Compounds

The chemistry of heterocyclic compounds is one of the most complex branches of organic chemistry. It is equally interesting for its theoretical implications, for the diversity of its synthetic procedures, and for the physiological and industrial significance of heterocyclic compounds.

A field of such importance and intrinsic difficulty should be made as readily accessible as possible, and the lack of a modern detailed and comprehensive presentation of heterocyclic chemistry is therefore keenly felt. It is the intention of the present series to fill this gap by expert presentations of the various branches of heterocyclic chemistry. The subdivisions have been designed to cover the field in its entirety by monographs which reflect the importance and the interrelations of the various compounds, and accommodate the specific interests of the authors.

In order to continue to make heterocyclic chemistry as readily accessible as possible new editions are planned for those areas where the respective volumes in the first edition have become obsolete by overwhelming progress. If, however, the changes are not too great so that the first editions can be brought up-to-date by supplementary volumes, supplements to the respective volumes will be published in the first edition.

ARNOLD WEISSBERGER

Research Laboratories
Eastman Kodak Company
Rochester, New York

EDWARD C. TAYLOR

Princeton University
Princeton, New Jersey

Preface

This book provides an account of the preparation, properties, and uses of the more important bicyclic and tricyclic ring systems incorporating the pyrazine ring. Twenty of the chapters survey the developments in quinoxaline chemistry since the publication of the Simpson monograph on condensed pyridazines and pyrazines in 1953. Continuity has been ensured by some small overlap with the previous monograph, and to facilitate cross-referencing the same basic organization of subject material has been retained.

The remaining 20 chapters incorporate reviews on selected 5,6–, 6,6–, 5,6,6–, and 6,6,6–ring systems. These reviews give comprehensive coverage to such important ring systems as the pyrrolopyrazines and the pyridopyrazines. The tricyclic heterocycles chosen for inclusion are those in which the third ring is fused to the pyrazine ring of a quinoxaline. Space limitations have dictated this somewhat arbitrary choice of material. Also excluded from this monograph is a discussion of the chemistry of pteridines, because a future volume in this series will be devoted to them, and of phenazines, because a monograph on them has already been published.

The chapters are organized for easy reference and incorporate tables by which information on specific compounds can be readily traced. Table entries on individual compounds are listed in order of their molecular formula. We suggest that a molecular formula check is the surest way of ascertaining if a particular compound is listed in the tabulation. Also, a molecular formula provides a convenient search term for locating specific information in *Chemical Abstracts*. It is clearly impossible to include the entries for polyfunctional compounds in each appropriate table so that in the case, for example, of a chloroamino compound the tabulation of both chloro- and amino compounds must be consulted.

The literature has been covered to the end of 1975, but additional material from papers published in 1976 and 1977 has been included. It is hoped that this book will help to stimulate further research as well as prove a useful source of reference for chemists with widely differing interests.

London
September 1977

G. W. H. Cheeseman
R. F. Cookson

Acknowledgments

We are grateful for help and encouragment from a variety of sources. Professor W. Pfleiderer provided hospitality for one of us at the University of Konstanz, where the initial planning of the book was done. Much of the research and writing was carried out in the libraries of the Chemical Society (London) and the Nicholas Research Institute, and we appreciated the facilities so readily made available to us. We should like to thank Aspro-Nicholas Ltd. and, more recently, Janssen Pharmaceutical Ltd. for allowing one of their employees to act as coauthor of this work. We are particularly indebted to colleagues at Queen Elizabeth College for their helpful comments. Special mention should also be made of C. W. Bird, D. J. Brown, and R. T. Parfitt, all of whom provided much appreciated advice and encouragement.

Above all, we should like to thank our wives, Ann and Mildred, for their help and support at every stage in the production of this book. We are most grateful that somehow they found time to undertake the arduous task of typing the manuscript.

London
September 1977

G. W. H. C.
R. F. C.

Contents

I. General Introduction to Quinoxaline Chemistry 1

II. Quinoxaline—The Parent Heterocycle 7

III. Quinoxalines Unsubstituted in the Heteroring 17

IV. Quinoxaline Mono- and Di-*N*-oxides 28

V. Quinoxalin-2-ones and Quinoxaline-2,3-diones 78

VI. Quinoxaline-2-thiones and Quinoxaline-2,3-dithiones 112

VII. Quinoxaline-2-carboxaldehydes and Quinoxaline-2,3-dicarboxaldehydes 123

VIII. Quinoxaline 2-Ketones and 2,3-Diketones 130

IX. Quinoxaline-2-carboxylic Acids and Quinoxaline-2,3-dicarboxylic Acids 137

X. 2-Halogenoquinoxalines and 2,3-Dihalogenoquinoxalines 162

XI. 2-Aminoquinoxalines and 2,3-Diaminoquinoxalines 179

XII. 2-Hydrazinoquinoxalines 194

XIII. 2-Alkoxy-, 2-Aryloxy-, 2,3-Dialkoxy-, 2-Alkoxy-3-aryloxy-, and 2,3-Diaryloxquinoxalines 199

XIV. 2-Alkyl- and 2,3-Dialkylquinoxalines 205

XV. 2-Aryl(heteroaryl)- and 2,3-Diaryl(diheteroaryl)quinoxalines 233

XVI. 2-Polyhydroxyalkylquinoxalines 241

XVII. Quinoxaline Quaternary Salts 247

XVIII. Reduced Quinoxalines 261

XIX. **Pyrrolo[1,2-*a*]pyrazines** 291

XX. **Pyrrolo[*b*]pyrazines** 348

XXI. **Imidazopyrazines** 360

XXII. **Pyrazolopyrazines** 398

XXIII. **Triazolo- and Tetrazolopyrazines** 410

XXIV. **Furo-, Oxazolo-, Isoxazolo-, and Oxadiazolopyrazines** . . . 431

XXV. **Thieno-, Thiazolo-, Isothiazolo-, and Thiadiazolo-pyrazines** 441

XXVI. **Miscellaneous 5,6-Systems** 456

XXVII. **Pyrido[1,2-*a*]pyrazines** 463

XXVIII. **Pyrido[2,3-*b*]pyrazines** 495

XXIX. **Pyrido[3,4-*b*]pyrazines** 544

XXX. **Pyrazinopyrimidines** 568

XXXI. **Pyrazinopyrazines** 575

XXXII. **Pyrazinotriazines and Pyrazinotetrazines** 583

XXXIII. **Pyrazinooxazines** 585

XXXIV. **Pyrazinothiazines** 591

XXXV. **Pyrroloquinoxalines** 598

XXXVI. **Imidazoquinoxalines** 654

XXXVII. **Pyrazoloquinoxalines** 691

XXXVIII. **Benzo[*f*]quinoxalines** 712

XXXIX. **Benzo[*g*]quinoxalines** 730

XL. **Pyridoquinoxalines** 743

Author Index 761

Subject Index 809

CHAPTER I

General Introduction to Quinoxaline Chemistry

I.	Nomenclature	1
II.	Synthesis	1
III.	Reactions of Quinoxalines with Electrophilic Reagents on Ring Nitrogen	2
IV.	Substitution Behavior of Quinoxaline Derivatives on Carbon	3
V.	Addition Reactions of Quinoxalines	4
VI.	Reactions of Substituted Quinoxalines	4
VII.	Tautomerism of Quinoxaline Derivatives	4
VIII.	Reactions of Quinoxalines Involving Ring Change	5
IX.	Biological Properties of Quinoxaline Derivatives	6
X.	Major Sources of Reference	6

I. Nomenclature

The approved numbering for the quinoxaline ring system is shown in structure **1**; positions 2 and 3 are sometimes designated α-positions. An alternative name for quinoxaline occasionally to be found in the literature is 1,4-diazanaphthalene.

5 4 6 3 (α) 7 2 (α) 8 1

1

II. Synthesis

The vast majority of quinoxalines are of synthetic origin, and with very few exceptions the synthetic method used is to condense an *o*-disubstituted benzene with a two-carbon synthon. Thus condensation of *o*-phenylenediamines with α-dicarbonyl compounds results in quinoxaline formation as shown in Scheme 1. By suitable choice of the α-dicarbonyl component, alkyl- and arylquinoxalines, quinoxalinones, and

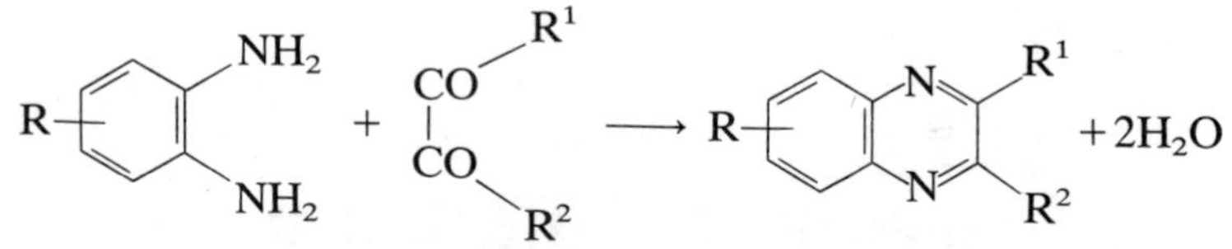

Scheme 1

quinoxalinecarboxylic acids have been prepared (Chapters XIV, XV, V, and IX, respectively). Other two-carbon synthons that have been reacted with *o*-phenylenediamines to form quinoxalines include α-halogenocarbonyl compounds, α,β-dihalides, and acetylene 1,2-dicarboxylic acid esters.

Major variants on this method are the use of *o*-nitrosoaminobenzenes (Chapter IX) and benzofuroxans (Chapter IV) as substrates for reaction with a two-carbon component as illustrated in Scheme 2. The *o*-nitrosoaminobenzene-based synthesis has the advantage that it leads to products

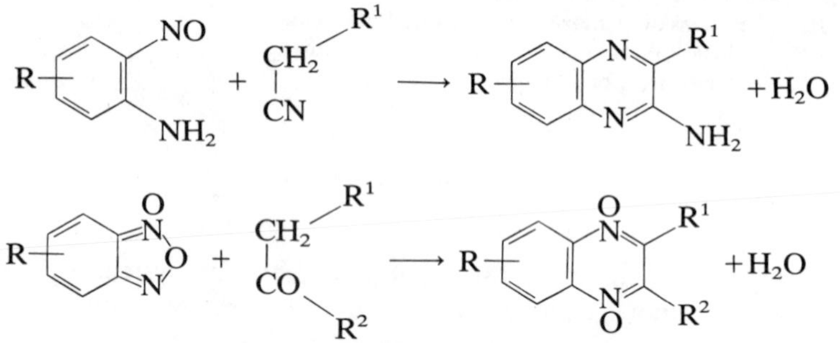

Scheme 2

of unambiguous structure, which is not the case where unsymmetric *o*-phenylenediamines or benzofuroxans are used. The synthesis of quinoxaline di-*N*-oxides from benzofuroxans is known as the Beirut reaction and has been exploited extensively in recent years.

III. Reactions of Quinoxalines with Electrophilic Reagents on Ring Nitrogen

Quinoxaline (1,4-diazanaphthalene) has a pK_a value of 0.6, and it is therefore less basic than either cinnoline (1,2-diazanaphthalene), quinazoline (1,3-diazanaphthalene), or phthalazine (2,3-diazanaphthalene) (Chapter II). Quinoxaline is reported to have a second pK_a of −5.52, and it is therefore only significantly diprotonated in a strongly acidic medium. Quinoxaline and its simple derivatives are readily converted into both mono- and di-*N*-oxides by oxidation with peracids (Chapter IV). As mentioned above, di-*N*-oxides are available from primary synthesis from benzofuroxans. Quinoxalines form monoquaternary salts when treated with the common quaternizing agents such as methyl sulfate and methyl *p*-toluenesulfonate (Chapter XVII). The quaternary salts of 2-alkylquinoxalines are unstable and on oxidation are converted into complex colored products (Chapter XVII). Quinoxaline quaternary salts have also been prepared by primary synthesis from *N*-substituted *o*-phenylenediamines and α-dicarbonyl compounds (Scheme 3) (Chapter XVII).

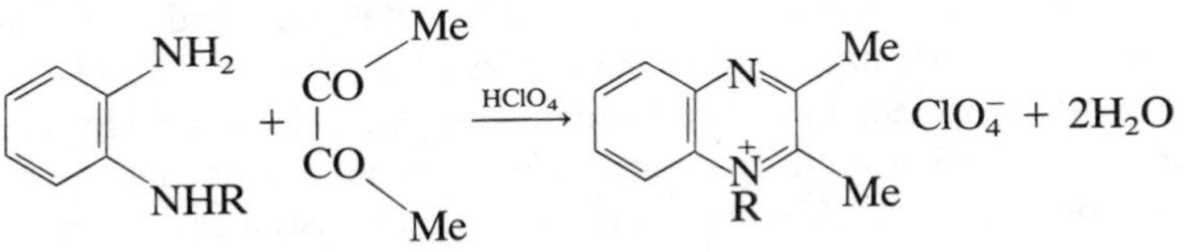

NH_2, NHR + Me–CO–CO–Me $\xrightarrow{HClO_4}$ $\overset{+}{N}$–R, Me, Me $ClO_4^- + 2H_2O$

Scheme 3

IV. Substitution Behavior of Quinoxaline Derivatives on Carbon

Quinoxaline itself and many of its simple derivatives do not readily undergo substitution on carbon when treated with electrophilic reagents; however, under forcing conditions the parent base is nitrated to give 5,6-dinitroquinoxaline as the major product (Chapter II). The benzene ring of quinoxalin-2-ones is activated to electrophilic substitution and nitration, and halogenation occurs smoothly in the 7-position when the reactions are carried out in acetic acid solution (Chapter V). The quinoxalinium cation is however susceptible to substitution at C-2 by a whole range of radical reagents. For example, acyl radicals ($RCO^{\cdot}$), generated under oxidizing conditions from aldehydes, react with quinoxaline to give 2-quinoxalinyl ketones (Chapter VIII). 2-Alkyl-, carboxamido-, and ethoxycarbonylquinoxalines have also been prepared by radical substitution (Chapter II). Homolytic δ-aminoalkylation of the quinoxalinium cation also occurs at the 2-position, but at high acidity, when a significant amount of diprotonated base is present, both 2- and 6-substitution occurs (Chapter II).

V. Addition Reactions of Quinoxalines

1,2-Dihydro-, 1,4-dihydro-, 1,2,3,4-tetrahydro-, and decahydroquinoxalines are known (Chapter XVIII). Thus reduction of quinoxaline with lithium aluminum hydride yields 1,2,3,4-tetrahydroquinoxaline (Chapter II). Quinoxaline also adds two molecular proportions of Grignard reagent to give a 2,3-disubstituted 1,2,3,4-tetrahydroquinoxaline (Chapter II). It also undergoes cycloaddition reactions with reagents such as diphenylcyclopropenone to form 1:1 molecular adducts (Chapter II).

VI. Reactions of Substituted Quinoxalines

2-Alkylquinoxalines show enhanced reactivity in terms of their ability to undergo condensation reactions with aldehydes and their ability to undergo Michael additions (Chapter XIV). Similarly 2-halogenoquinoxalines have been found to participate in a wide range of nucleophilic substitution reactions with oxygen, sulfur, nitrogen, and carbon nucleophiles. Chlorine in the 2-position is also readily removed by catalytic hydrogenation (Chapter X). Quinoxaline 2-carboxylic acids are very readily decarboxylated which renders their purification difficult but in some cases increases their utility as intermediates in other quinoxaline preparations; for example 2-chloroquinoxaline can be readily prepared from 3-chloroquinoxaline-2-carboxylic acid (Chapter X).

VII. Tautomerism of Quinoxaline Derivatives

2-Hydroxy- and 2-mercaptoquinoxalines exist in the quinoxalin-2-one (Chapter V) and quinoxaline-2-thione forms (Chapter VI), whereas 2-aminoquinoxaline exists as such rather than as an imine (Chapter XI) (Scheme 4).

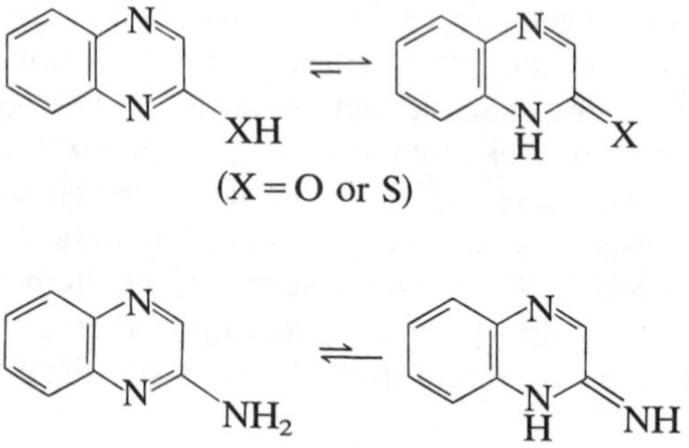

(X = O or S)

Scheme 4

2,3-Dihydroxy- and 2,3-dimercaptoquinoxaline similarly exist in 2,3-dione (Chapter V) and 2,3-dithione forms (Chapter VI), respectively. Literature statements that 2,3-dimethylquinoxaline (**2**) reacts in the tautomeric diene form (**3**) are incorrect (Chapter XIV), although in the case of the acyl derivatives (**4**) enamine forms (**5**) are preferred (Scheme 5) (Chapter V).

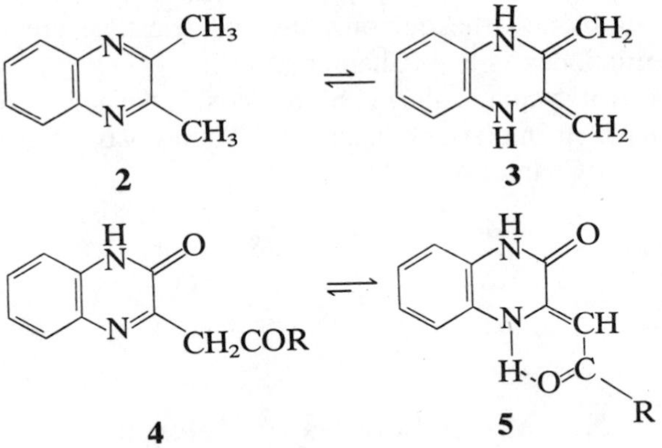

Scheme 5

VIII. Reactions of Quinoxalines Involving Ring Change

Relatively few reactions of quinoxaline derivatives occur with change of ring size. Isolated examples are noted in the following text. For example, ring contraction to benzimidazole derivatives occurs when 2,3-diphenylquinoxaline (Chapter XV) or 2-halogenoquinoxalines (Chapter X) are treated with potassium amide in liquid ammonia and quinoxalin-2-one is treated with hydrazine (Chapter V). It is also found that oxidation of 1-aminoquinoxalin-2-ones with lead tetraacetate give benzo-1,2,4-triazines (Scheme 6) (Chapter V).

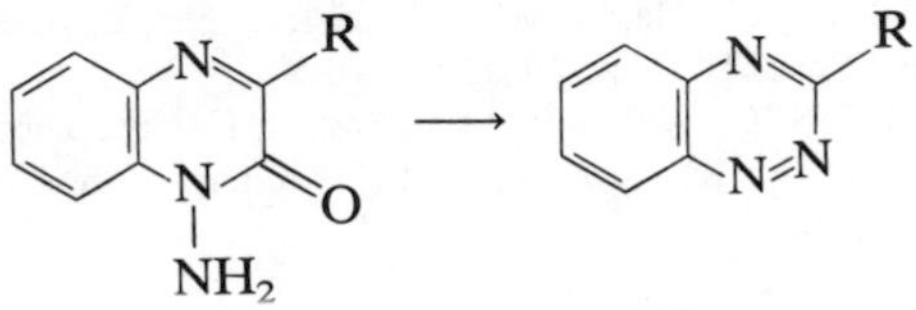

Scheme 6

IX. Biological Properties of Quinoxaline Derivatives

The main search for biologically active quinoxalines has centered around the preparation of quinoxaline *N*-oxides. 3-Substituted 2-methylquinoxaline 1,4-dioxides with high antibacterial activity have been prepared (Chapter IV). Quinoxaline 2-sulfonamide has had sustained use as a coccidiostat for poultry (Chapter XI). A series of naturally occurring quinoxaline antibiotics, the quinomycins and triostins, are known, but their therapeutic index is low (Chapter IX).

5,6,7,8-Tetrachloroquinoxaline (Chlorquinox) is the active compound in various fungicidal formulations (Chapter III) and Morestan (**6**) is used as an insecticide (Chapter VI).

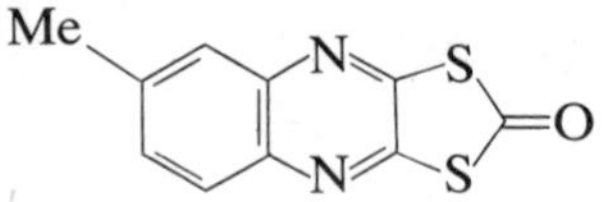

Me N S N S =O

6

X. Major Sources of Reference

The early literature on quinoxaline chemistry can be conveniently located either via Beilstein's *Handbuch der organischen Chemie* or in Meyer-Jacobson's *Lehrbuch der organischen Chemie.* The period 1917–1948 is covered by the previous monograph in this series by Simpson,[1] and it is the aim of the present volume to cover the quinoxaline literature in the period 1949–1975 and in addition to refer to major papers appearing in 1976. In an attempt to preserve continuity, quinoxaline chemistry is discussed as far as possible under the same chapter headings as used in Simpson's monograph. Much detailed information on quinoxalines has appeared in several review articles.[2–5]

1. J. C. E. Simpson, "Condensed Pyridazine and Pyrazine Rings," Interscience, New York, 1953.
2. Y. T. Pratt, "Heterocyclic compounds," Vol. 6, R. C. Elderfield, Ed., Wiley, New York, 1956, Chap. 10.
3. G. R. Ramage and J. K. Landquist, "Chemistry of Carbon Compounds," Vol. IVB, Elsevier, Amsterdam, 1959, Chap. 15.
4. G. W. H. Cheeseman, "Advances in Heterocyclic Chemistry," Vol. 2, A. R. Katritzky, Ed., Academic Press, New York, 1963, p. 203.
5. G. W. H. Cheeseman and E. S. G. Werstiuk, "Advances in Heterocyclic Chemistry," Vol. 22, A. J. Boulton and A. R. Katritzky, Eds., Academic Press, New York, 1978, p. 367.

CHAPTER II

Quinoxaline—The Parent Heterocycle

I. Methods of Preparation 7
II. Properties 7
1. Physical Properties 7
2. Chemical Properties 11
III. References 15

I. Methods of Preparation

Quinoxaline (**1**) has been prepared in 85–90% yield by reaction of *o*-phenylenediamine with glyoxal sodium bisulfite.[1] It has also been prepared in excellent yield from the diamine by treatment with 30% aqueous glyoxal in the presence of sodium carbonate (Scheme 1).[2]

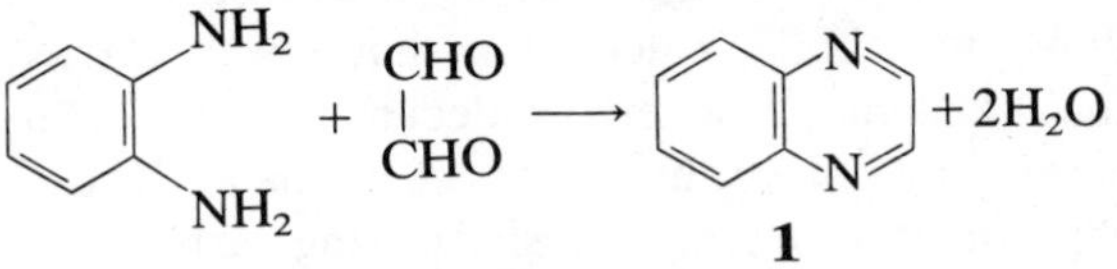

NH_2 NH_2 + CHO | CHO ⟶ (N, N) **1** $+ 2H_2O$

Scheme 1

II. Properties

1. Physical Properties

Quinoxaline is conveniently purified by distillation, and a fraction of b.p. 108–111°/12 mm has a m.p. of 29–30°.[1] Quinoxaline forms a 2:1 molecular complex with phloroglucinol of m.p. 131–132°.[3] The pK_a of quinoxaline in water at 20° is 0.56; it is therefore a considerably weaker base than the isomeric diazanaphthalenes, namely, cinnoline (pK_a 2.42), phthalazine (pK_a 3.47), and quinazoline (pK_a 1.95).[4] Quinoxaline has a second pK_a of −5.52.[72] The effect of substitution of Cl, OMe, SMe, $CONH_2$, and CO_2Et in the 2-position is base weakening. These substituents are less base weakening in the quinoxaline series than in the pyridine or in the quinoline series, probably because protonation occurs at N-4. 2-Substituents such as Me, NHCOMe, NH_2, NHMe, and NMe_2 are base strengthening and direct protonation to N-1. These substituents cause a greater enhancement of basic strength than in the pyridine or quinoline series. For example, 2-aminoquinoxaline has a pK_a of 3.96 (3.40 pK_a units greater than the parent base) whereas the pK_a values of 2-aminopyridine and pyridine differ by only 1.63 pK_a units.[5] 5-Amino-, 6-amino-, 5-hydroxy and 6-hydroxyquinoxalines are stronger bases than quinoxaline itself, and 5-aminoquinoxaline is exceptional in undergoing protonation at the amino group rather than at a ring nitrogen atom.[6] The basic center in 5-hydroxyquinoxaline is N-1. Reaction of 5-hydroxyquinoxaline with methyl iodide gives a methiodide which must be the 1-methiodide because of its ability to form a strongly bound nickel complex.[7] Quinoxaline has a dipole moment of 0.51 D in benzene,[8] whereas quinoline has a dipole moment of 2.18 D.

The first and second ionization potentials of quinoxaline measured by photon electron spectroscopy are 8.99 and 10.72 eV, respectively.[9] Since the highest-occupied π orbital and nonbonding orbitals of quinoxaline are very close in energy, it is not known unambiguously from which orbital the first electron is lost.[10] The heat of atomization of quinoxaline has been calculated by a self-consistent field molecular orbital treatment to be 79.739 eV, and similar calculations have been carried out out on related diazanaphthalenes.[11] Several molecular orbital calculations of the π-electron density in quinoxaline have been made. These calculations indicate that the highest electron density at ring carbon is at positions 5 and 8, the next at positions 6 and 7, and the lowest at positions 2 and 3; the related σ-charge densities have also been calculated (Fig. 1).[12] Various attempts have been made to correlate chemical reactivity and chemical shift data with calculated π-electron densities, but no very clear correlations have emerged.[13]

The infrared spectrum of quinoxaline, nine of its 2-substituted, five of its 5-substituted, and eight of its 6-substituted derivatives have been measured in chloroform.[14,15] Eight ring-stretching bands are found in the region 1620–1350 cm^{-1} at frequencies close to those observed in substituted quinolines. The positions of these bands are not very sensitive to substituent change, but the intensities are much more variable. Thus the

0.995 0.985 N 0.976 1.010 1.033

π-Electron densities

2.996 2.923 N 2.964 3.003 4.106

σ-Electron densities

Figure 1

intensity of the band near 1600 cm^{-1} rises as the substituents change from electron acceptors to electron donors. β-CH Bending modes give rise to absorption in the region 1300–1050 cm^{-1}, ring breathing modes to absorption near to 1000 cm^{-1}, and out-of-plane γ-CH bending modes to absorption in the region below 1000 cm^{-1}.

The ultraviolet absorption spectrum of quinoxaline in cyclohexane shows bands with vibrational fine structure at 340 (log ε 2.84), 312 (log ε 3.81), and 232 nm (log ε 4.51) which are attributed to n–π^* and π–π^* transitions.[16] In ethanol the vibrational fine structure disappears and the less intense n–π^* band appears as a shoulder on the long-wave π–π^* band.[17] However in methanol[18] and in water[19] the n–π^* band is obscured by the more intense π–π^* band. The weak n–π^* bands in the ultraviolet spectra of 6-chloro- and 6-bromoquinoxaline[20] and certain 2-substituted quinoxalines also show shifts to shorter wavelengths on change from a nonpolar to polar solvent, whereas the π–π^* bands are not greatly affected by change of solvent. Substitution in the 2-position of the quinoxaline nucleus produces bathochromic shifts in the π–π^* bands. This increases in the order Me$<$Cl$<$OMe$<$SMe$<$NMe$_2$.[21,22] The phosphorescence spectra of quinoxaline and other diazanaphthalenes have been examined, the diazanaphthalenes showing the same π–π^* phosphorescence band as naphthalene.[23]

The ^{1}H NMR spectrum of quinoxaline has been measured in acetone,[24] carbon tetrachloride,[24] dimethyl sulfoxide,[25] dichloromethane,[26] and trifluoroacetic acid.[26] The signal for H2 and H3 in carbon tetrachloride appears as a low-field singlet, and the aromatic ring protons appear as an $AA'BB'$ system. The low-field half of the $AA'BB'$ multiplet is assigned to the 5- and 8-protons and the high-field half, to protons 6 and 7. Some broadening of the signals from protons 6 and 7 is attributed to long-range coupling with protons 2 and 3. The computed chemical shifts for protons 2 and 3, 5 and 8, and 6 and 7 are 8.73, 8.06, and 7.67 δ, respectively. In the more polar solvent acetone, there is a general small low-field shift. Coupling constant values are J_{56} 8.4 Hz, J_{57} 1.4 Hz, J_{58} 0.6 Hz, and J_{67} 6.9 Hz. Analysis of the ^{1}H NMR spectra of a number of 2-, 5-, and 6-monosubstituted quinoxalines show the following coupling constant variations: J_{23} 1.7–1.9 Hz, J_{67} 5.0–8.3 Hz, J_{78} 8.4–10.3 Hz, J_{57} 1.4–2.7 Hz, J_{68} 0.7–2.9 Hz, and J_{58} 0.3–0.8 Hz.[25]

The ^{13}C NMR spectrum of the parent heterocycle in deuterochloroform shows resonances at δ 144.8, 142.8, 129.6, and 129.4 assigned to carbons 2 and 3, 4a and 8a, 5 and 8, and 6 and 7, respectively (Fig. 2).[12,27] The ^{14}N chemical shifts have been measured for a series of aromatic

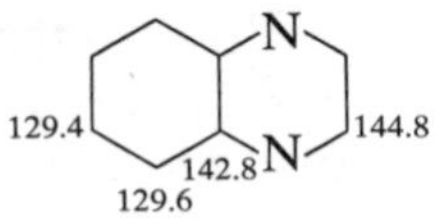

129.4 142.8 144.8 129.6

^{13}C Chemical shifts

Figure 2

heterocycles including quinoxaline, for which the ^{14}N resonance is at $+46 \pm 3$ ppm from nitromethane and the peak half-height width is 950 ± 50 Hz. The ^{14}N chemical shifts have been found to depend almost linearly on the π-electron density at nitrogen, as calculated by the Pariser–Parr–Pople method.[28]

The mass spectrum of quinoxaline shows fragment ions resulting from the loss of one and two molecules of hydrogen cyanide. Where structurally possible, loss of hydrogen cyanide from the molecular ion is found to be the major fragmentation process for many nitrogen heteroaromatic compounds.[29] Similarly in the case of 2-alkyl- and 2-arylquinoxalines, $(M—HCN)^+$ and $(M—RCN)^+$ ions are observed. A notable feature of the spectrum of 2-methyl-3-phenylquinoxaline is the formation of an intense $(M-1)^+$ ion.[30,31]

The polarographic reduction of quinoxaline and its derivatives has been studied by a number of workers.[32–38] Half-wave electrode potentials are pH dependent; and over the pH range of 1 to 10, $E_{\frac{1}{2}}$ has been reported to vary from -0.254 to -0.863 V.[32] The half-wave electrode potential in anhydrous dimethylformamide is -1.09 V for quinoxaline, -1.06 V for cinnoline, -1.41 V for phthalazine, and -1.22 V for quinazoline. The benzodiazines are more easily reduced than the corresponding diazines, thus pyrazine has a half-wave electrode potential of -1.57 V. Pyridine with a half-wave potential of -2.15 V is still more difficult to reduce. The reversible reduction potential of quinoxaline as determined by cyclic voltammetry is -1.097 V.[33] The observed energy differences between azine and radical anion are well correlated with the results of CNDO and SCF π-electron calculations.

Detailed study of the electrochemical reduction of quinoxaline in aqueous media has revealed that 1,2,3,4-tetrahydroquinoxaline is formed via a sequence involving 1,4-dihydroquinoxaline and acid-catalyzed rearrangement of the 1,4-dihydro derivative to 1,2-dihydroquinoxaline (Scheme 2).[34]

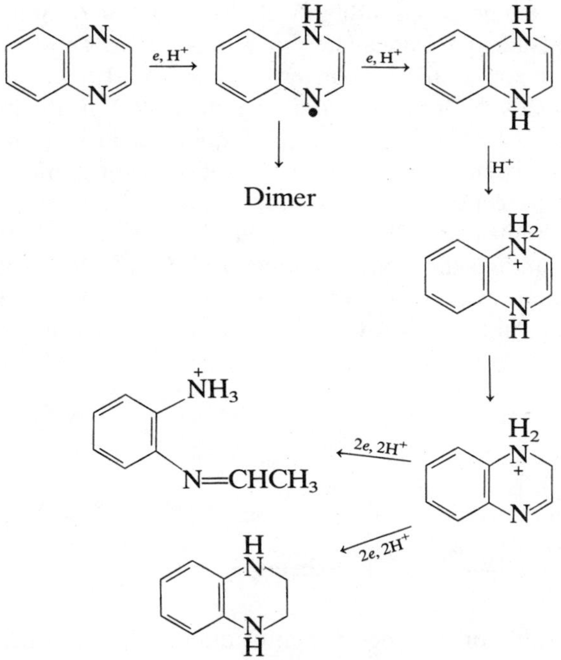

e, H^+ e, H^+ Dimer H^+ $\overset{+}{N}H_3$ $N{=}CHCH_3$ 2e, $2H^+$ 2e, $2H^+$

Scheme 2

2. Chemical Properties

Reduction of quinoxaline with sodium in THF at 20° yields a deep-purple solution from which 1,4-dihydroquinoxaline is isolated.[39] Reduction with either sodium in refluxing alcohol or lithium aluminum hydride in ether gives 1,2,3,4-tetrahydroquinoxaline.[39] Hydrogenation of quinoxaline over a 5% rhodium-on-alumina catalyst at 100° and 136 atm or over freshly prepared Raney nickel W-6 under similar conditions gives *meso*-(*cis*)-decahydroquinoxaline.[40] However hydrogenation of quinoxaline over a palladium-on-charcoal catalyst at 180° and 50 atm gives DL-(*trans*)-decahydroquinoxaline.[41]

Treatment of quinoxaline with one equivalent of peracetic acid in acetic acid yields quinoxaline 1-oxide, and with excess of peracetic acid quinoxaline 1,4-dioxide is formed.[42] Reaction of quinoxaline with 30% aqueous hydrogen peroxide in acetic acid, however, gives quinoxaline-2,3-dione as the main product.[43] Electrolytic oxidation of quinoxaline at a copper anode gives pyrazine-2,3-dicarboxylic acid in excellent yield.[44] The latter compound is also prepared in high yield by oxidation of quinoxaline with alkaline potassium permanganate[1]; a number of α-substituted quinoxalines have been similarly converted into substituted pyrazine-2,3-dicarboxylic acids.[45] On heating quinoxaline at 200° in the presence of a 5% palladium-on-carbon catalyst, 2,2′-biquinoxaline is formed.[46]

Quinoxaline itself is resistant to nitration with fuming nitric acid (d 1.52) and concentrated sulfuric acid at 100°,[47] but under prolonged treatment with nitric acid and oleum it has been converted into a mixture of 5-nitro- and 5,6-dinitroquinoxaline in 1.5 and 24% yield, respectively.[48,49] Later workers report* that nitration of quinoxaline gives a mixture of 5,6-dinitroquinoxaline and 4-nitrobenzotriazole (Scheme 3).[50]

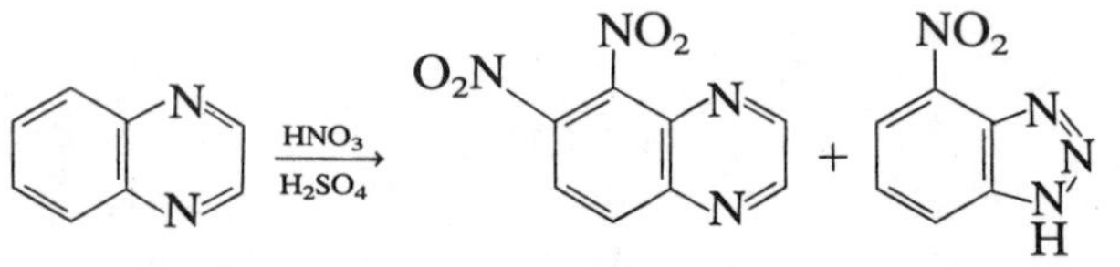

Scheme 3

A careful study of the phenylation of quinoxaline with benzoyl peroxide, various benzenediazonium salts, and N-nitrosoacetanilide showed that position 2 is the most reactive and that position 5 is more reactive than position 6.[51] With benzoyl peroxide in glacial acetic acid at 118°, the product formed contains 77% 2-phenylquinoxaline and 23% 5- and 6-phenylquinoxaline.[52] 2-Benzylquinoxaline is the only product isolated from the homolytic benzylation of quinoxaline with dibenzylmercury in acetic acid.[53] In a series of publications Italian workers have described the alkylation,[54] acylation,[55–57] α-oxyalkylation,[58] δ-aminoalkylation,[59] ethoxycarbonylation,[60] and amidation[61] of the quinoxalinium ion. For example, acylation of quinoxaline is achieved by the simultaneous addition of saturated solutions of ferrous sulfate and t-butyl hydroperoxide to a cooled (5–15°) mixture of aldehyde, quinoxaline, and $4M$ sulfuric acid.

* A possible route to 4-nitrobenzotriazole may be ring cleavage of the initially formed 5-nitroquinoxaline to 3-nitro-o-phenylenediamine and its subsequent ring closure to 4-nitrobenzotriazole by nitrous acid present in the nitrating medium.

Using a range of aliphatic and aromatic aldehydes the yields of 2-acylquinoxalines range from 45 to 73%. The reaction is formulated as shown in Scheme 4.

$$Fe^{2+} + ROOH \longrightarrow Fe^{3+} + OH^- + RO^\cdot \qquad (R = H \text{ or } t\text{-Bu})$$
$$RO^\cdot + R'CHO \longrightarrow ROH + R'CO^\cdot$$

(quinoxalinium N-H) + R'CO· ⟶ (2-COR' quinoxalinium N-H) + H·

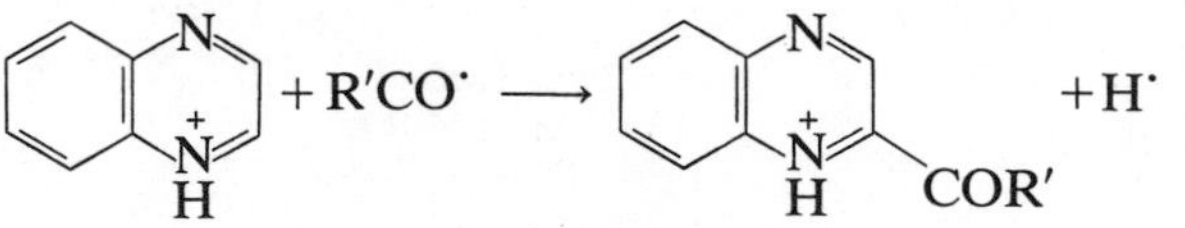

Scheme 4

Ethers have been used as reagents for α-oxyalkylation,[58] and the addition product of ethyl pyruvate and hydrogen peroxide

$$[CH_3C(OH)(OOH)CO_2Et]^{52}$$

and formamide[61] are used as the source of $EtO_2C^\cdot$ and $NH_2CO^\cdot$ radicals, respectively. In all cases substitution occurs at the 2-position although in some instances 2,3-disubstituted products are obtained.[58] Treatment of quinoxaline with *N*-chlorodi-*n*-butylamine and ferrous sulfate in 50% sulfuric acid gives exclusive 2-substitution, but in concentrated acid a mixture of 2- and 6-(4-*n*-butylaminobutyl)quinoxaline is obtained. Abnormal substitution at position 6 is explained by postulating radical attack on the diprotonated species.[59]

Substitution of quinoxaline takes place at C-2 when it is irradiated in ether, methanol, or ethanol.[62,63] Irradiation of quinoxaline in acidified methanol furnishes 2-methylquinoxaline, and the reaction is suggested to go through a pathway involving electron transfer from the solvent to an excited state of the protonated quinoxaline (Scheme 5).[64] Irradiation of quinoxaline under aerobic conditions in acidic aqueous solution leads to the formation of 5-hydroxyquinoxaline.[65]

Quinoxaline readily reacts with Grignard reagents. Addition of two molecular proportions of allylmagnesium bromide[66] and of 3-(dimethylamino)propylmagnesium chloride,[67] and hydrolysis of the initial adducts, gives the tetrahydroquinoxalines **2** and **3**, respectively. The 1:1 adducts **4**, **5**, and **6** are obtained from the reaction of quinoxaline with diphenylcyclopropenone,[68] diphenylcyclopropenethione,[69] and methyl phenylhydrazonochloroacetate in the presence of triethylamine,[70] respectively. In the latter case the presumed intermediate is the dipolar species ($MeO_2C\overset{+}{C}{=}N{-}\overset{-}{N}Ph$).

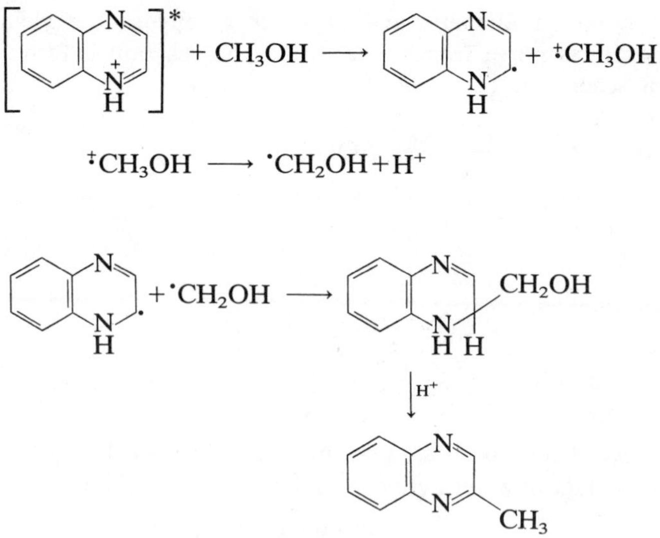

$[\text{quinoxalinium-H}^+]^* + CH_3OH \longrightarrow \text{(N-H radical)} + {}^{\ddagger}CH_3OH$

${}^{\ddagger}CH_3OH \longrightarrow {}^{\bullet}CH_2OH + H^+$

$\text{(N-H radical)} + {}^{\bullet}CH_2OH \longrightarrow$ CH₂OH adduct $\xrightarrow{H^+}$ 2-methylquinoxaline (CH_3)

Scheme 5

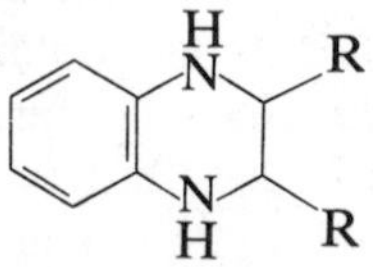

2 $R = CH_2CH{=}CH_2$
3 $R = CH_2CH_2CH_2NMe_2$

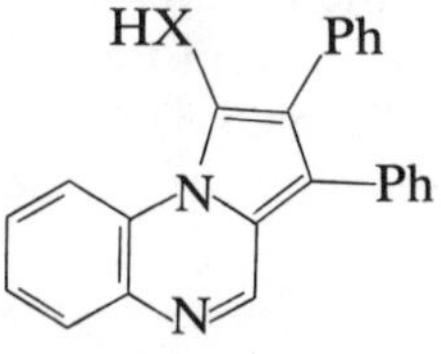

HX, Ph, Ph

4 X=O
5 X=S

MeO_2C, N, N—Ph

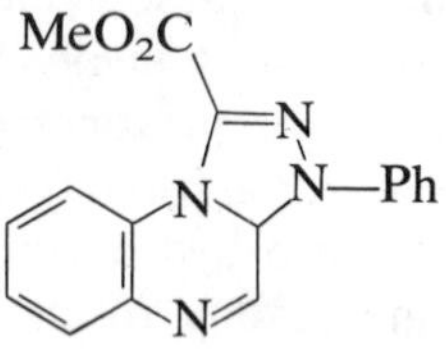

6

Reaction of quinoxaline with ethyl iodide in boiling acetonitrile gives ethylquinoxalinium iodide in 76% yield, and treatment of the parent base with methyl *p*-toluenesulfonate at room temperature gives methylquinoxalinium *p*-toluenesulfonate in quantitative yield.[71] There are apparently no reports of the isolation of quinoxaline bisquaternary salts.

III. References

1. R. G. Jones and K. C. McLaughlin, *Org. Synth.*, **30,** 86 (1950).
2. A. Zmujdzin, Pol. Pat. 69,644; *Chem. Abstr.*, **81,** 77966z (1974).
3. P. E. Verkade and M. van Leeuwen, *Rec. Trav. Chim.*, **70,** 142 (1951).
4. A. Albert, "Physical Methods in Heterocyclic Chemistry," Vol. 1, A. R. Katritzky, Ed. Academic Press, New York, 1963, Chap. 1.
5. M. M. Kaganskii, G. G. Dvoryantseva, and A. S. Elina, *Chem. Heterocycl. Comp.*, 369 (1973).
6. A. R. Osborn, K. Schofield, and L. N. Short, *J. Chem. Soc.*, 4191 (1956).
7. A. Albert and A. Hampton, *J. Chem. Soc.*, 505 (1954).
8. H. Lumbroso and G. Palamidessi, *Bull. Soc. Chim. Fr.* 3150 (1965).
9. M. J. S. Dewar and S. D. Worley, *J. Chem. Phys.*, **51,** 263 (1969).
10. S. D. Worley, *Chem. Rev.*, **71,** 295 (1971).
11. M. J. S. Dewar and T. Morita, *J. Am. Chem. Soc.*, **91,** 796 (1969).
12. R. J. Pugmire, D. M. Grant, M. J. Robins, and R. K. Robins, *J. Am. Chem. Soc.*, **91,** 6381 (1969).
13. W. W. Paudler and T. J. Kress, "Topics in Heterocyclic Chemistry," R. N. Castle, Ed., Interscience, New York, 1969, p. 86.
14. H. H. Perkampus and A. Roders, *Z. Naturforsch.*, **15b,** 1 (1960).
15. G. W. H. Cheeseman, A. R. Katritzky, and B. J. Ridgewell, *J. Chem. Soc.*, 3764 (1963).
16. S. F. Mason, *Chem. Soc. (London) Spec. Publ.*, **No. 3,** 139 (1955).
17. G. M. Badger and J. S. Walker, *J. Chem. Soc.*, 122 (1956).
18. F. Bohlmann, *Chem. Ber.*, **84,** 860 (1951).
19. A. Albert, D. J. Brown, and G. W. H. Cheeseman, *J. Chem. Soc.*, 474 (1951).
20. R. C. Hirt, T. F. King, and J. C. Cavagnol, *J. Chem. Phys.*, **25,** 574 (1956).
21. G. W. H. Cheeseman, *J. Chem. Soc.*, 108 (1958).
22. H. H. Perkampus, *Z. Naturforsch.*, **17a,** 614 (1962).
23. R. Müller and F. Dörr, *Z. Electrochem.*, **63,** 1150 (1959); *Chem. Abstr.*, **54,** 5243 (1960).
24. P. J. Black and M. L. Heffernan, *Austr. J. Chem.*, **18,** 707 (1965).
25. P. J. Brignell, A. R. Katritzky, R. E. Reavill, G. W. H. Cheeseman, and A. A. Sarsfield, *J. Chem. Soc. B*, 1241 (1967).
26. D. J. Blears and S. S. Danyluk, *Tetrahedron*, **23,** 2927 (1967).
27. L. F. Johnson and W. C. Jankowski, "Carbon-13 N.M.R. Spectra," Wiley-Interscience, New York, 1972.
28. M. Witanowski, L. Stefaniak, H. Januszewski, and G. A. Webb, *Tetrahedron*, **27,** 3129 (1971).
29. A. Karjalainen and H. Krieger, *Soumen Kemistilehti B*, **43,** 273 (1970).
30. S. N. Bannore, J. L. Bose, K. G. Das, and V. N. Gogte, *Indian J. Chem.*, **7,** 654 (1969).
31. V. Kovacik, M. Fedoronko, and I. Jezo, *Org. Mass Spectrom.*, **7,** 449 (1973).
32. M. P. Streier and J. C. Cavagnol, *J. Am. Chem. Soc.*, **79,** 4331 (1957).
33. K. B. Wiberg and T. P. Lewis, *J. Am. Chem. Soc.*, **92,** 7154 (1970).
34. O. Fischer and Tran Hong Thuy, *Collect. Czech. Chem. Commun.*, **41,** 1853 (1976).
35. M. Fedoronko and I. Jezo, *Collect. Czech. Chem. Commun.*, **37,** 1781 (1972).
36. B. J. Tabner and J. R. Yandle, *J. Chem. Soc. A*, 381 (1968).
37. S. Millefiori, *J. Heterocycl. Chem.*, **7,** 145 (1970).
38. D. van der Meer and D. Feil, *Rec. Trav. Chim.*, **87,** 746 (1968).

39. J. Hamer and R. E. Holliday, *J. Org. Chem.*, **28,** 2488 (1963).
40. H. Smith-Broadbent, E. L. Allred, L. Pendleton, and C. W. Whittle, *J. Am. Chem. Soc.*, **82,** 189 (1960).
41. S. Maffei and S. Pietra, *Gazz. Chim. Ital.*, **88,** 556 (1958); *Chem. Abstr.*, **53,** 20060 (1959).
42. J. K. Landquist, *J. Chem. Soc.*, 2816 (1953).
43. M. Asai, *Yakugaku Zasshi*, **79,** 260 (1959); *Chem. Abstr.*, **53,** 13160 (1959).
44. T. Kimura, S. Yamada, K. Yoshizue, and T. Nagoka, *Yakugaku Zasshi*, **77,** 891 (1957); *Chem. Abstr.*, **52,** 1181 (1958).
45. H. I. X. Mager and W. Berends, *Rec. Trav. Chim.*, **78,** 5 (1959).
46. H. Smith-Broadbent and R. C. Anderson, *J. Org. Chem.*, **27,** 2679 (1962).
47. H. Otomasu and S. Nakajima, *Chem. Pharm. Bull.* (Tokyo), **6,** 566 (1958).
48. M. J. S. Dewar and P. M. Maitlis, *J. Chem. Soc.*, 2518 (1957).
49. F. H. Case and J. A. Brennan, *J. Am. Chem. Soc.*, **81,** 6297 (1959).
50. R. Nasielski-Hinkens and M. Benedek-Vamos, *J. Chem. Soc., Perkin I*, 1229 (1975).
51. C. M. Atkinson and C. J. Sharpe, *J. Chem. Soc.*, 3040 (1959).
52. H. J. M. Dou and B. M. Lynch, *Bull. Soc. Chim. Fr.*, 3815 (1966).
53. K. C. Bass and P. Nababsing, *Org Prepr. and Proc. Int.*, **3,** 45 (1971).
54. G. P. Gardini and F. Minisci, *Ann. Chim.* (Rome), **60,** 746 (1970); *Chem. Abstr.*, **74,** 87128x (1971).
55. T. Caronna, G. P. Gardini, and F. Minisci, *Chem. Commun.*, 201 (1969).
56. G. P. Gardini and F. Minisci, *J. Chem. Soc.* (C), 929 (1970).
57. G. P. Gardini, *Tetrahedron Lett.*, 4113 (1972).
58. W. Buratti, G. P. Gardini, F. Minisci, F. Bertini, R. Galli, and P. Perchinunno, *Tetrahedron*, **27,** 3655 (1971).
59. A. Citterio, M. Ghirardini, and F. Minisci, *Tetrahedron Lett.*, 203 (1976); T. Caronna, A. Citterio, M. Ghirardini, and F. Minisci, *J. Heterocycl. Chem.*, **13,** 955 (1976).
60. R. Bernardi, T. Caronna, R. Galli, F. Minisci, and M. Perchinunno, *Tetrahedron Lett.*, 645 (1973).
61. F. Minisci, R. Galli, and A. Quilico, *Ger. Offen.*, 2,056,433; *Chem. Abstr.*, **75,** 49055j (1971).
62. T. T. Chen, W. Dörscheln, H. Göth, M. Hesse, and H. Schmid, *Helv. Chim. Acta*, **51,** 632 (1968).
63. A. Castellano, J. P. Catteau, A. Lablache-Combier, B. Planckaert, and G. Allen, *Tetrahedron*, **28,** 3511 (1972).
64. S. Wake, Y. Takayama, Y. Otsuji, and E. Imoto, *Bull. Chem. Soc. Jap.*, **47,** 1257 (1974).
65. J. Verbeek, W. Berends and H. C. A. van Beek, *Rec. Trav. Chim.*, **95,** 285 (1976).
66. H. Gilman, J. Eisch, and T. Soddy, *J. Am. Chem. Soc.*, **79,** 1249 (1957).
67. A. Marrer, U. Salzmann, and F. Hofer, *Helv. Chim. Acta*, **54,** 2507 (1971).
68. J. W. Lown and K. Matsumoto, *Can. J. Chem.*, **49,** 1165 (1971).
69. J. W. Lown and K. Matsumoto, *Can. J. Chem.*, **49,** 3119 (1971).
70. P. D. Croce, *J. Heterocycl. Chem.*, **12,** 1133 (1975).
71. R. F. Smith, W. J. Rebel, and T. N. Beach, *J. Org. Chem.*, **24,** 205 (1959).
72. T. Caronna, A. Citterio, T. Crolla, and F. Minisci, *J. Chem. Soc. Perkin I*, 865 (1977).

CHAPTER III

Quinoxalines Unsubstituted in the Heteroring

I. Methods of Preparation . . . 17
II. Properties . . . 18
  1. Physical Properties . . . 18
  2. Chemical Properties . . . 19
III. Uses . . . 21
IV. Tables . . . 22
  Table 1. Quinoxalines Unsubstituted in the Heteroring . . . 22
V. References . . . 25

This chapter is concerned with the chemistry of quinoxalines unsubstituted in the heteroring, and it updates the chapter to be found in Simpson's monograph.[1] For the most part substituents in positions 5, 6, 7, and 8 of the quinoxaline ring behave in a manner that is predictable by analogy with the chemistry of naphthalene. However this generalization is not entirely true, as is seen in the following discussion.

I. Methods of Preparation

In the main, quinoxalines unsubstituted in the heteroring are prepared by condensation of an arylenediamine with aqueous glyoxal (Scheme 1). Various chemical transformations of 5- and 6-monosubstituted quinoxalines involving modification of the substituent or which lead to the

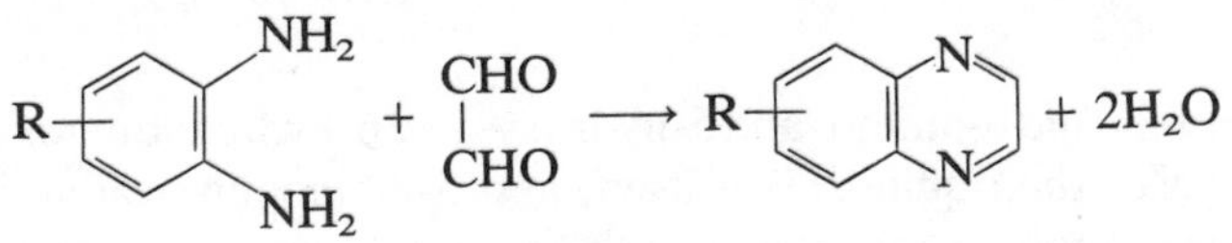

$$\text{R–}C_6H_3(NH_2)_2 + \begin{matrix} CHO \\ | \\ CHO \end{matrix} \longrightarrow \text{R–quinoxaline} + 2H_2O$$

Scheme 1

introduction of a second substituent in the benzenoid ring are mentioned later in this chapter.

Preparatively useful benzosubstitutions of the parent heterocycle are only of very limited value. Nitration of quinoxaline with a mixture of oleum and nitric acid at 90° gives a mixture of 1.5% 5-nitroquinoxaline and 24% 5,6-dinitroquinoxaline.[2,3] Although free-radical substitutions of quinoxaline occur mainly at the 2-position (see Chapter II), a significant amount of 6-(4-*n*-butylaminobutyl)quinoxaline is formed together with the expected 2-isomer on treatment of quinoxaline with *N*-chloro-di-*n*-butylamine and ferrous sulfate in concentrated sulfuric acid. The abnormal course of substitution is rationalized by the suggestion that it is the dication which is the species undergoing 6-substitution.[4] Phenylation of quinoxaline with *N*-nitrosoacetanilide gives a low yield of phenylquinoxalines containing 2-, 5-, and 6-phenylquinoxaline in the approximate ratio of 40:10:1.[5]

Quinoxaline *N*-oxides can act as precursors of benzosubstituted quinoxalines. For example, prolonged treatment of quinoxaline 1,4-dioxide with acetyl chloride gives some 6,7-dichloroquinoxaline, this probably arises from 6-chloroquinoxaline 1-oxide, which is the main product of reaction.[6]

II. Properties

1. Physical Properties

The infrared-absorbing properties of 5-ethoxycarbonylquinoxaline are abnormal in that split carbonyl absorption is observed at 1724 and 1741 cm^{-1}. Since a single carbonyl stretching absorption for 6-ethoxycarbonylquinoxaline is observed at 1727 cm^{-1}, it appears that the 5-substituted compound exists in the two rotomeric forms **1a** and **1b**.[7]

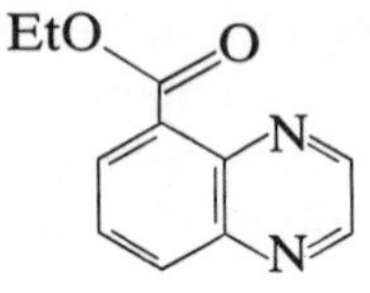

EtO O — N N — **1a**   O OEt — N N — **1b**

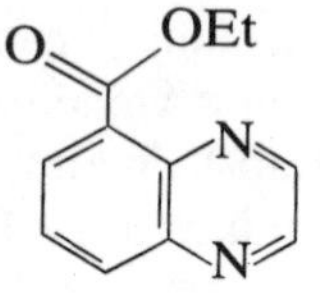

Both 5-amino- and 6-amino- and 5-hydroxy- and 6-hydroxyquinoxalines are stronger bases than quinoxaline itself, and 5-aminoquinoxaline is exceptional in undergoing protonation at the amino group rather than at a ring nitrogen. Thus the ultraviolet spectrum of the cation of the 5-amino compound is closely similar to that of quinoxaline.[8] The basic center in 5-hydroxyquinoxaline is thought to be nitrogen-1, since the methiodide of 5-hydroxyquinoxaline retains the ability of 5-hydroxyquinoxaline to form a strongly bound nickel complex.[9]

2. Chemical Properties

Oxidation of 5-substituted quinoxalines with peracetic acid gives mainly 1-oxides, although 5-methoxyquinoxaline is exceptional in forming a 1,4-dioxide.[10,11] In the case of 6-substituted quinoxalines, as the substituent becomes more electronegative, decreasing yields of 1,4-dioxide are obtained on peracetic acid oxidation but more of the corresponding 2,3-dioxo compounds are formed.[10,11] 5-Hydroxyquinoxaline is readily converted into the 1,4-oxide with *m*-chloroperbenzoic acid.[12]

Sodium borohydride–acetic acid treatment of 5- and 6-substituted quinoxalines yields 1,2,3,4-tetrahydroquinoxalines.[13] Alternatively, reduction is effected with hydrogen and a platinum catalyst.[14] The acetic acid–borohydride technique gives excellent yields of 5- and 6-amino-1,2,3,4-tetrahydroquinoxalines from the corresponding 5- and 6-nitroquinoxalines.[13] 6-Fluoro- and 6-chloroquinoxalines form radical anions on electrochemical reduction which are stable at room temperature, but the radical anions from 6-bromo- and 6-iodoquinoxaline fragment with loss of halide ion, and by subsequent reaction with solvent the parent heterocycle is formed.[15]

Nitration of 5-methoxyquinoxaline with potassium nitrate in concentrated sulfuric acid gives the 6,8-dinitro derivative, and the 6-methoxy compound is converted to 5-nitro-6-methoxyquinoxaline with these reagents.[16] Nitration of 6-*p*-toluenesulfonamidoquinoxaline in acetic acid also gives a 5-nitro derivative.[3] 6-Methylquinoxaline is brominated in the side chain when treated with *N*-bromosuccinimide in the presence of azobisisobutyronitrile, and 6-bromomethylquinoxaline is obtained.[14]

6-Substituted quinoxalines undergo some interesting reactions when treated with nucleophilic reagents. When 6-bromoquinoxaline is treated with potassium amide in liquid ammonia the main product is benzimidazole. Minor amounts of 2-amino- and 2,3-diaminoquinoxalines are also formed, but none of the expected 5- and 6-aminoquinoxalines are obtained (Scheme 2).[17] Similar treatment of 6-chloroquinoxaline gives a mixture of 2-amino-7-chloroquinoxaline and 6-chloro-2,3-diaminoquinoxaline.[18] It has been suggested that the difference in behavior between the 6-bromo and 6-chloro compounds may be explained by the

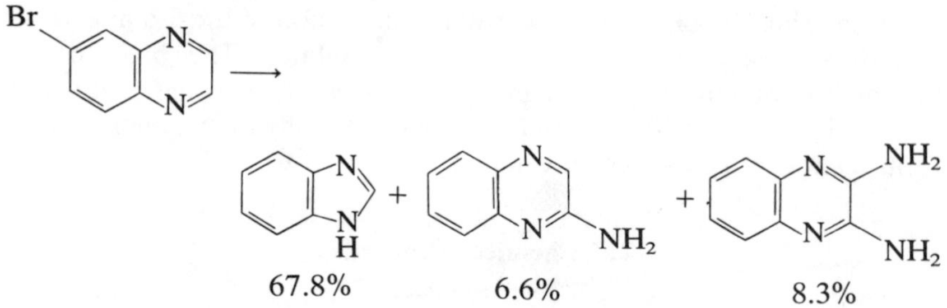

Br

67.8% 6.6% 8.3%

Scheme 2

ability of bromine, but not chlorine, to migrate to the 2-position.[78] 2-Bromoquinoxaline is known to give benzimidazole in good yield on treatment with potassium amide and liquid ammonia.

In contrast, 6-bromo-2,3-dimethylquinoxaline when subjected to the same reagents give a mixture of 5- and 6-amino-2,3-dimethylquinoxalines.[19] 6-Nitroquinoxalines are smoothly converted into 5-amino-6-nitroquinoxalines on reaction with hydroxylamine,[20] and 6-nitro-2,3-diphenylquinoxaline (**2**) yields a mixture of the 5-cyano-6-methoxyquinoxaline **3** and the isoxazole **4** on treatment with potassium cyanide in methanol.[21]

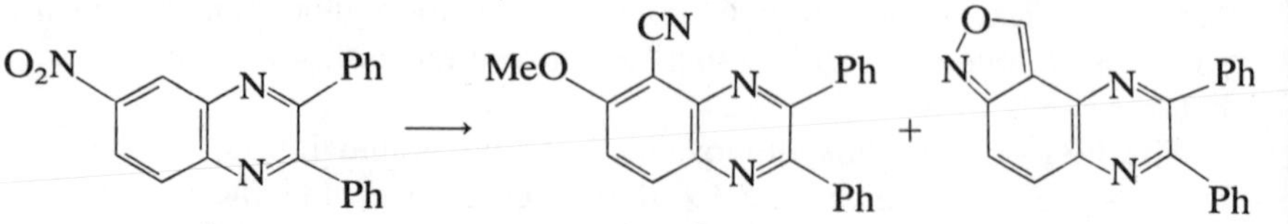

2 **3** **4**

Treatment of 5,6,8-trichloro-7-methylquinoxaline with alkali in aqueous ethanol gives 5,8-dichloro-7-methylquinoxaline in good yield, and 5,6,7,8-tetrachloroquinoxaline is converted under similar conditions into a mixture of 5,8-dichloro- and 5,6,8-trichloroquinoxalines.[22] 6-Azido-7-nitroquinoxalines (**5**), prepared by primary synthesis from suitably substituted *o*-phenylenediamines, have been converted thermally into quinoxalinofurazan oxides (**6**). An attempt to prepare 6-azido-2,3-dimethyl-7-nitroquinoxaline by reaction of the corresponding 6-chloro-7-nitroquinoxaline with sodium azide was not successful.[23,24]

Several groups have studied the synthesis and reactions of quinoxaline-5,8-diones. The parent compound is obtained by silver oxide oxidation of

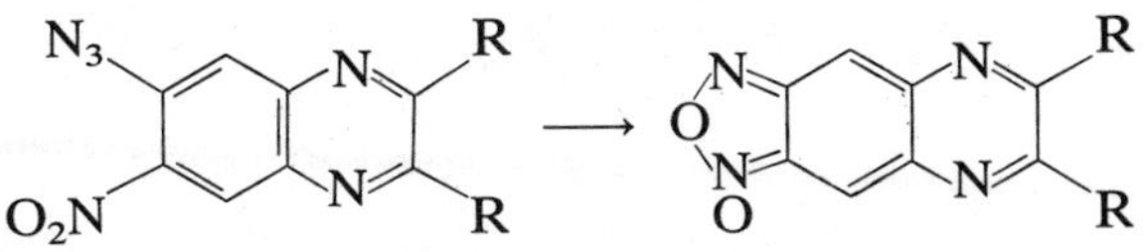

N3, O2N, N, N, R, R → N, O, N, O, N, N, R, R

5 **6**

5,8-dihydroxyquinoxaline.[25] 6-Methylquinoxaline-5,8-diones have been prepared by dichromate oxidation of the corresponding 5-amino-8-hydroxyquinoxalines[26] and 6-methoxyquinoxaline-5,8-dione from 5-amino-6-methoxyquinoxaline.[27] Quinoxaline-5,8-diones show expected reactivity as dienophiles in Diels–Alder reactions[28,29] and undergo 1,4-addition reactions with amines,[30,31] halogen acids,[31] and mercaptans[32] in analogous fashion to 1,4-naphthoquinone. The polarographic reduction of quinoxaline-5,8-diones is postulated to occur in two stages, as shown in Scheme 3.[31] The oxidative amination of 6-hydroxyquinoxalines yields aminosubstituted quinoxaline-5,6-diones.[77]

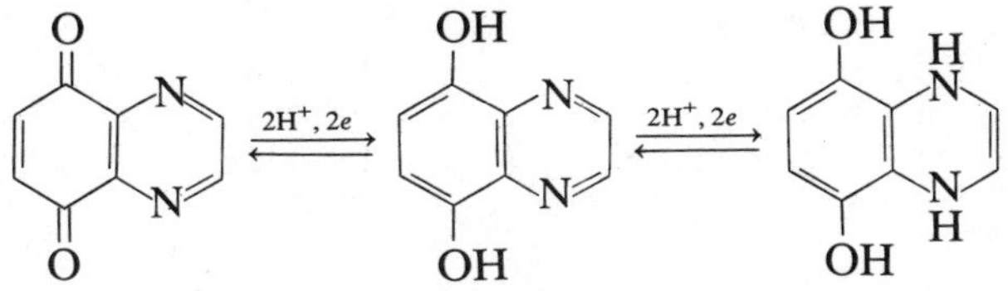

O, N, N, O ⇌ (2H⁺, 2*e*) OH, N, N, OH ⇌ (2H⁺, 2*e*) OH, H, N, N, H, OH

Scheme 3

III. Uses

The topical mosquito-repellent activity of a range of simple quinoxaline derivatives, including 6-chloro- and 6-methoxyquinoxalines, has been tested, but none of the compounds tested was superior to diethyltoluamide.[79]

The insecticidal and acaricidal properties of phosphoric, phosphonic, thionophosphoric and thionophosphomic esters of 6-hydroxyquinoxaline have been patented[33] and 5,6,7,8-tetrachloroquinoxaline (Chlorquinox) and related halogenated quinoxalines have been used in various fungicidal formulations.[34] Substituted 5,8-dimethoxyquinoxalines were synthesized and tested for antimalarial activity, but the biological results were negative.[35] The metal complex-forming behavior of 5-hydroxy- and 5,8-dihydroxyquinoxalines have been studied in several laboratories.[9,36,37]

IV. Tables

TABLE 1. QUINOXALINES UNSUBSTITUTED IN THE HETERORING

Molecular formula	Substituents or structure	m.p., °C	Ref.
$C_8H_2BrCl_3N_2$	5-Bromo-6,7,8-trichloro	199–201	38
	6-Bromo-5,7,8-trichloro	306–308	38
$C_8H_2Cl_3FN_2$	5,6,8-Trichloro-7-fluoro	155–157	38
$C_8H_2Cl_4N_2$	5,6,7,8-Tetrachloro	189.5–190.5	22, 38
$C_8H_2F_4N_2$	5,6,7,8-Tetrafluoro	90–91	39
$C_8H_3Cl_3N_2$	5,6,7-Trichloro	138–139	22, 38
	5,6,8-Trichloro	178.5–179.5	22, 38
$C_8H_4ClN_3O_2$	7-Chloro-5-nitro	174–175	40
$C_8H_4Cl_2N_2$	5,6-Dichloro	145.5–146.5	22
	5,7-Dichloro	159.5–160	22
	5,8-Dichloro	205–207, 210	10, 41
	6,7-Dichloro	208–210, 210	6, 10
$C_8H_4N_2O_2$	5,8-Dione	155–160 decomp. 171–177 decomp.	25, 30
$C_8H_4N_4O_2$	6,7-Dinitroso		24
$C_8H_4N_4O_4$	5,6-Dinitro	168–170, 172–173	2, 3
	6,7-Dinitro	193–194	38
$C_8H_4N_6O_2$	6-Azido-7-nitro	135–145 decomp.	23
$C_8H_5BrN_2$	6-Bromo	48–49, 49–50, b.p. 148–149/18 mm	10, 17, 42
$C_8H_5ClN_2$	5-Chloro	60–62, 61–62	10, 43
	6-Chloro	62.5–63, 63–64	42, 43
$C_8H_5FN_2$	6-Fluoro	35–36, b.p. 103–104/15 mm	38
$C_8H_5IN_2$	6-Iodo	114–115	10
$C_8H_5N_3O_2$	5-Nitro	95–96, 96–97	2, 3, 44
	6-Nitro	177	1
$C_8H_6BrN_3$	6-Amino-5-bromo		45
$C_8H_6ClN_3O_2S$	6-Chloro-7-sulfamoyl	254	46
$C_8H_6N_2O$	5-Hydroxy	100–101, 102–103, b.p. 184/7 mm	36, 37, 47, 48, 49, 50, 51
	Acetate	103–104, 106.5–108	47, 51
	6-Hydroxy	242 decomp., 252–254	47, 52, 53
	Acetate	80–81	47
$C_8H_6N_2O_2$	5,6-Dihydroxy	ca. 190 decomp.	47
	Diacetate	112	47
	5,7-Dihydroxy	ca. 250 decomp.	47
	Diacetate	113	47
	5,8-Dihydroxy	ca. 230, 237	25, 37, 47
	Diacetate	200–202, 212.5–213.5	25, 36, 47
	6,7-Dihydroxy	ca. 260 decomp.	47
	Diacetate	112	47

TABLE 1 (*contd.*)

Molecular formula	Substituents or structure	m.p., °C	Ref.
$C_8H_6N_4O_2$	5-Amino-6-nitro	236–237	20
	6-Amino-5-nitro	226–227	3
	6-Amino-7-nitro	249	23
$C_8H_7N_3$	5-Amino	91, 93–95	54, 55, 56
	Hydrate	87–88, 90.5	55, 56
	6-Amino	156–157, 157–159	3, 42, 55
$C_8H_8N_2O$	7,8-Dihydro-5(6*H*)-one		57
$C_8H_8N_4$	5,6-Diamino	144–145	20
	5,7-Diamino	170	58
	6,7-Diamino	>360	59
$C_8H_{10}N_2$	5,6,7,8-Tetrahydro	b.p. 85/3 mm	60
$C_9H_5Cl_3N_2$	5,6,8-Trichloro-7-methyl	199–201	22
$C_9H_5Cl_3N_2O$	5,6,8-Trichloro-7-methoxy	153–154	34, 38
$C_9H_5F_3N_2$	6-Trifluoromethyl	78	14
$C_9H_5N_3$	6-Cyano	176–178	10
$C_9H_6Cl_2N_2$	5,7-Dichloro-6-methyl	115–117	22
	5,8-Dichloro-6-methyl	166–167	22
$C_9H_6N_2O_2$	6,7-Methylenedioxy	156	14
	5-Carboxy	168–170	7
	Methyl ester	73–74	7
	Ethyl ester	215–216	7
	6-Carboxy	240–243 decomp., 266	7, 61
	Methyl ester	100	14
	Ethyl ester	66, 70–71	7, 52, 61
	6-Methyl-5,8-dione	171	26
$C_9H_6N_2O_3$	6-Methoxy-5,8-dione	235–236	27
$C_9H_6N_4O_5$	5-Methoxy-6,8-dinitro	204–206	16
$C_9H_7BrN_2$	6-Bromomethyl	100–105	14
	6-Bromo-7-methyl	127–128	10
$C_9H_7ClN_2$	6-Chloro-7-methyl	120–122	10
$C_9H_7ClN_2O$	6-Chloro-5-methoxy	115	12
$C_9H_7N_3O_2$	7-Methyl-5-nitro	134–135	40
$C_9H_7N_3O_3$	5-Methoxy-7-nitro	177–179	62
	6-Methoxy-5-nitro	203	16
$C_9H_8N_2$	5-Methyl	20–21, b.p. 120/15 mm	10
	Hydrate	49–50	10
	6-Methyl	b.p. 135/26 mm	38
$C_9H_8N_2O$	5-Methoxy	100–101	36, 37, 47
	6-Methoxy	60	42
$C_9H_9N_3$	5-Amino-7-methyl	103	55
	Hydrate	122	55
	7-Amino-5-methyl	158–160	55
	7-Amino-6-methyl	194–195	55
$C_9H_9N_3O$	5-Amino-6-methoxy	96	16
	5-Amino-8-methoxy	125	16

TABLE 1 (*contd.*)

Molecular formula	Substituents or structure	m.p., °C	Ref.
$C_9H_9N_3O$ (*contd.*)	7-Amino-5-methoxy	199–201	62
$C_9H_{11}N_3O_2$	5-Ethoxycarbonylamino		12
$C_9H_{12}N_2$	5,6,7,8-Tetrahydro-5-methyl	*a*	60
$C_{10}H_6N_2O_4$	6,7-Dicarboxy		
	N-Methylimide	211–212 decomp.	64
$C_{10}H_6N_2O_6$	5,8-Dihydroxy-6,7-dicarboxy		
	Dimethyl ester	183–184	63
	Diethyl ester	149	37, 63
	Diacetate of diethyl ester	205–206	63
$C_{10}H_7N_3O_4$	5-Amino-6,7-dicarboxy		
	N-Methylimide	272–275 decomp.	64
$C_{10}H_8Br_2N_2O_2$	6,7-Dibromo-5,8-dimethoxy	160.5–161.5	25
$C_{10}H_8Cl_2N_2O$	5,8-Dichloro-6-methoxy-7-methyl	140–142	22
$C_{10}H_8N_2O$	6-Acetyl	106–108	52
$C_{10}H_8N_2O_2$	6-carboxy-7-Methyl	238–244 decomp.	64
$C_{10}H_8N_2O_3$	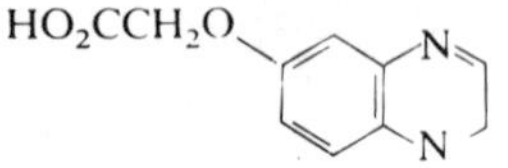 HO_2CCH_2O		
	Ethyl ester	99–100	52
$C_{10}H_9BrN_2O_2$	6-Bromo-5,8-dimethoxy	122–123	25
$C_{10}H_9N_3O$	5-Acetamido	131	2, 54
	6-Acetamido	196, 196–197	10, 38
$C_{10}H_9N_3O_2$	8-Acetamido-5-hydroxy	246–247	65
$C_{10}H_9N_3O_3$	7-Ethoxy-5-nitro	142–144	66
$C_{10}H_{10}N_2$	5,6-Dimethyl		
	Hydrochloride	177–178	67
	6,7-Dimethyl	95–98, 100–102	10, 64, 68
$C_{10}H_{10}N_2O$	5-Ethoxy	63–64, b.p. 165–166/ 18 mm	10, 69
	Methiodide	214–215 decomp.	69
	Ethiodide	146–148 decomp.	69
	6-Ethoxy	79.5–80	42
$C_{10}H_{10}N_2O_2$	5,6-Dimethoxy	69–70	70
	5,7-Dimethoxy	110	70
	5,8-Dimethoxy	146, 150, 152	14, 25, 37, 70
	6,7-Dimethoxy	151	37
$C_{10}H_{11}N_3O_2$	6-Amino-5,8-dimethoxy	164.5–165	35
$C_{10}H_{12}N_2O_2$	5-Acetoxy-5,6,7,8-tetrahydro		57
$C_{11}H_{10}Cl_2N_2O$	5,8-Dichloro-6-ethoxy-7-methyl	100–101	22
$C_{11}H_{11}N_3O$	5-Acetamido-7-methyl		12
$C_{11}H_{11}N_3O_2$	5-Acetamido-8-methoxy	149	16
$C_{12}H_{12}N_4O_2$	5,6-Diacetamido	231–232, 232–233	2, 3
	6,7-Diacetamido	279–280	59
$C_{12}H_{13}N_3O_2$	5-Acetamido-7-ethoxy	142–144	66

TABLE 1 (*contd.*)

Molecular formula	Substituents or structure	m.p., °C	Ref.
$C_{12}H_{13}N_3O_3$	6-Acetamido-5,8-dimethoxy	183–184	35
$C_{12}H_{14}N_2O_2$	5,8-Diethoxy	101	36, 71
$C_{13}H_{15}N_3O$	6-Morpholinomethyl	b.p. 135–145/0.1mm	14
$C_{14}H_9N_3O_4$	5,8-Dihydroxy-6-(*p*-nitrophenyl)	270–272	25
$C_{14}H_{10}N_2$	5-Phenyl	124	5, 72
	Picrate	148	5, 72
	6-Phenyl	112	5, 72
	Picrate	170	5, 72
$C_{14}H_{12}N_4O_2S$	5-Sulfanilamido	163–165, 168.5–169.5	55, 56
	6-Sulfanilamido	227–229, 231.5–232	55, 56
$C_{14}H_{14}N_2O_2$	5,8-Bis(allyloxy)	169–172	73
	6,7-Diallyl-5,8-dihydroxy	179–180	73
$C_{14}H_{20}N_4O_2$	5,8-Dimethoxy-6[*N*-(2-dimethyl-aminoethyl)amino]	64–65	35
$C_{15}H_{12}N_4O_4S$	5-Nitro-6-*p*-toluenesulfonamido	197–198	3
$C_{15}H_{13}N_3O_2S$	6-*p*-Toluenesulfonamido	182–183	3
$C_{15}H_{14}N_4O_2S$	5-Methyl-7-sulfanilamido	225–227	55
	6-Methyl-7-sulfanilamido	219–220	55
	7-Methyl-5-sulfanilamido	214–216	55
$C_{15}H_{22}N_4O_2$	5,8-Dimethoxy-6-[*N*-(3-dimethyl-aminopropyl)amino]	95–96	35
$C_{16}H_{15}N_3O_3S$	5-Methoxy-8-(*p*-toluene-sulfonylamino)	170–171	12
$C_{16}H_{15}N_5$	6-[(*p*-Dimethylaminophenyl)azo]	174–176	74
$C_{16}H_{22}N_2O_2$	6,7-Dibutoxy	65.5–67	75
$C_{16}H_{23}N_3$	6-(4-*n*-Butylaminobutyl)		76
$C_{17}H_{17}N_3O_4S$	5,8-Dimethoxy-6-(*N*-*p*-toluene-sulfonamido)	213–214	35

[a] Present in a fraction of b.p. 72°/3 mm to the extent of 63%.

V. References

1. J. C. E. Simpson, "Condensed Pyridazine and Pyrazine Rings," Interscience, New York, 1953, p. 228.
2. M. J. S. Dewar and P. M. Maitlis, *J. Chem. Soc.*, 2518 (1957).
3. F. H. Case and J. A. Brennan, *J. Am. Chem. Soc.*, **81,** 6297 (1959).
4. A. Citterio, M. Ghirardini, and F. Minisci, *Tetrahedron Lett.* 203 (1976).
5. C. M. Atkinson and C. J. Sharpe, *J. Chem. Soc.*, 3040 (1959).
6. Y. Ahmad, M. S. Habib, M. K. Qureshi, and M. A. Farooqi, *J. Org. Chem.*, **38,** 2176 (1973).
7. W. F. Gum and M. M. Joullie, *J. Org. Chem.*, **30,** 3982 (1965).

8. A. R. Osborne, K. Schofield, and L. N. Short, *J. Chem. Soc.*, 4191 (1956).
9. A. Albert and A. Hampton, *J. Chem. Soc.*, 505 (1954).
10. J. K. Landquist, *J. Chem. Soc.*, 2816 (1953).
11. J. A. Silk, *J. Chem. Soc.*, 2058 (1956).
12. R. H. B. Galt, U.S. Pat. 3,479,354 (1969); *Chem. Abstr.*, **72,** 79095*c* (1970). R. H. B. Galt and R. J. Pearce, Brit. Pat., 1,237,438 (1971); *Chem. Abstr.*, **75,** 110340*s* (1971).
13. K. V. Rao and D. Jackman, *J. Heterocycl. Chem.*, **10,** 213 (1973).
14. R. C. De Selms, R. J. Greaves, and W. R. Schleigh, *J. Heterocycl. Chem.*, **11,** 595 (1974).
15. K. Alwair and J. Grimshaw, *J. Chem. Soc., Perkin II*, 1811 (1973).
16. H. Otomasu and S. Nakajima, *Chem. Pharm. Bull.*, (Tokyo), **6,** 566 (1958).
17. W. Czuba and H. Poradowska, *Rocz. Chem.*, **48,** 1233 (1974).
18. H. Poradowska and W. Czuba, *Univ. Adama Mickiewicza Poznaniu, Wydz. Mat. Fiz. Chem. (Pr.) Ser. Chem.*, **18,** 299 (1975); *Chem. Abstr.*, **84,** 120749*g* (1976).
19. W. Czuba and H. Poradowska, *Rec. Trav. Chim.*, **93,** 162 (1974).
20. R. Nasielski and M. Benedek-Vamos, *J. Chem. Soc., Perkin I*, 1229 (1975).
21. H. Takahashi and H. Otomasu, *Chem. Pharm. Bull.*, (Tokyo), **18,** 22 (1970).
22. D. E. Burton, D. Hughes, G. T. Newbold, and J. A. Elvidge, *J. Chem. Soc. (C)*, 1274 (1968).
23. J. H. Boyer, R. S. Buriks, and U. Toggweiler, *J. Am. Chem. Soc.*, **82,** 2213 (1960).
24. J. H. Boyer and R. S. Buriks, *J. Am. Chem. Soc.*, **82,** 2216 (1960).
25. J. Adachi, *Nippon Kagaku Zasshi*, **76,** 311 (1955); *Chem. Abstr.*, **51,** 17936 (1957).
26. M. R. W. Levy and M. M. Joullie, *J. Heterocycl. Chem.*, **1,** 171 (1964).
27. Y. S. Tsizin and S. A. Chernyak, *Khim. Geterotsikl. Soedin.*, 714 (1975); *Chem. Abstr.*, **83,** 79198*r* (1975).
28. W. F. Gum and M. M. Joullie, *J. Org. Chem.*, **30,** 2583 (1965).
29. G. Kumar and A. P. Bhaduri, *Indian J. Chem.*, **13,** 1009 (1975).
30. K. H. Ford and M. M. Joullie, *J. Heterocycl. Chem.*, **3,** 529 (1966).
31. W. F. Gum and M. M. Joullie, *J. Org. Chem.*, **32,** 53 (1967).
32. T. H. Porter, A. Klaudy, and K. Folkers, *J. Med. Chem.*, **16,** 1310 (1973).
33. K. J. Schmidt and I. Hammann (to Farbenfabriken Bayer A.-G.), Brit. Pat. 1,160,493; *Chem. Abstr.*, **71,** 113084*q* (1969).
34. D. W. J. Lane and G. T. Newbold (to Fison Ltd.), Brit. Pat. 1,041,011 Chem. Abstr., **60,** 15891 (1964).
35. G. H. Fisher, H. R. Moreno, J. E. Oatis, and H. P. Schultz, *J. Med. Chem.*, **18,** 746 (1975).
36. E. S. Lane and C. Williams, *J. Chem. Soc.*, 2983 (1956).
37. S. Oguchi, *Bull. Chem. Soc. Jap.*, **41,** 980 (1968).
38. Fisons Pest Control Ltd., Belg. Pat. 631,044; *Chem. Abstr.*, **60,** 15891 (1964).
39. C. G. Allison, R. D. Chambers, J. A. H. MacBride, and W. K. R. Musgrave, *J. Fluorine Chem.*, **1,** 59 (1971).
40. H. Gillespie, F. Spano, and S. Graff, *J. Org. Chem.*, **25,** 942 (1960).
41. C. W. Bird, G. W. H. Cheeseman, and A. A. Sarsfield, *J. Chem. Soc.*, 4767 (1963).
42. M. P. Strier and J. C. Cavagnol, *J. Am. Chem. Soc.*, **80,** 1565 (1958).
43. G. Favini and M. Simonetta, *Gazz. Chim. Ital.*, **90,** 363 (1960); *Chem. Abstr.*, **55,** 11425 (1961).
44. H. P. Schultz, *J. Am. Chem. Soc.*, **72,** 3824 (1950).
45. J. C. Danilewicz, M. Snarey, and G. N. Thomas (to Pfizer Corp.), Ger. Offen. 2,309,160; *Chem. Abstr.*, **79,** 146425*v* (1973).
46. C. W. Whitehead and J. J. Traverso, *J. Org. Chem.*, **28,** 743 (1963).

47. F. E. King, N. G. Clark, and P. M. H. Davis, *J. Chem. Soc.*, 3012 (1949).
48. A. Albert and A. Hampton, *J. Chem. Soc.*, 4985 (1952).
49. S. K. Freeman and P. E. Spoerri, *J. Org. Chem.*, **16,** 438 (1951).
50. H. M. Irving, H. S. Rossotti, and G. Harris, *Analyst*, **80,** 83 (1955).
51. E. Sorkin and W. Roth, *Helv. Chim. Acta*, **34,** 427 (1951).
52. J. A. Silk, *J. Chem. Soc.*, 2058 (1956).
53. K. J. Schmidt and I. Ingeborg (to Farbenfabriken Bayer A.-G.), Brit. Pat. 1,160,493; *Chem. Abstr.*, **71,** 113084*q* (1969).
54. F. J. Wolf, R. H. Beutel, and J. R. Stevens, *J. Am. Chem. Soc.*, **70,** 2572 (1948).
55. B. C. Platt and T. M. Sharp, *J. Chem. Soc.*, 2129 (1948).
56. K. A. Jensen, *Acta Chem. Scand.*, **2,** 91 (1948).
57. B. D. Mookherjee, M. G. Beets, A. O. Pittet, M. E. Mason, E. T. Theimer, M. S. Tibbets, W. J. Evers, I. Katz, and R. A. Wilson (to Int. Flavors and Fragrances Inc.), Ger. Offen. 2,117,926; *Chem. Abstr.* **76,** 25313*m* (1972).
58. H. N. Rydon and K. Undheim, *J. Chem. Soc.*, 4685 (1962).
59. J. Harley-Mason and A. H. Laird, *Tetrahedron*, **7,** 70 (1959).
60. A. O. Pettit, R. Muralidhara, J. P. Walradt, and T. Kinlin, *J. Agr. Food. Chem.*, **22,** 273 (1974).
61. L. Birkhofer and A. Widmann, *Chem. Ber.*, **86,** 1295 (1953).
62. H. B. Gillespie, M. Engelman, and S. Graff, *J. Am. Chem. Soc.*, **78,** 2445 (1956).
63. S. Oguchi, *Nippon Kagaku Zasshi*, **86,** 435 (1965); *Chem. Abstr.*, **63,** 7011 (1965).
64. E. H. White and K. Matsuo, *J. Org. Chem.*, **32,** 1921 (1967).
65. M. H. Broyles and W. K. Easley, *J. Org. Chem.*, **25,** 2233 (1960).
66. H. B. Gillespie, M. Engelman, F. Spano, and S. Graff, *J. Am. Chem. Soc.*, **79,** 2245 (1957).
67. H. Euler and H. Hasselquist, *Arkiv Kemi*, **48,** 559 (1958); *Chem. Abstr.*, **53,** 9234 (1959).
68. R. M. Cresswell, W. R. D. Smith, and H. C. S. Wood, *J. Chem. Soc.*, 4882 (1961).
69. W. K. Easley and M. X. Sullivan, *J. Am. Chem. Soc.*, **74,** 4450 (1952).
70. F. E. King, N. G. Clark, and P. M. H. Davis, *J. Chem. Soc.*, 4009 (1949).
71. E. S. Lane and C. Williams, *J. Chem. Soc.*, 4106 (1954).
72. C. M. Atkinson and C. J. Sharpe, *J. Chem. Soc.*, 2858 (1959).
73. S. Kawai, M. Hatano, and E. Iida, *Nippon Kagaku Zasshi*, **80,** 806 (1959); *Chem. Abstr.*, **55,** 3598 (1961).
74. E. V. Brown and W. M. Fisher, *J. Med. Chem.*, **12,** 1113 (1969).
75. S. Kawai, Y. Okawa, Y. Yada, H. Hosoi, T. Murakoshi, and I. Yajima, *Nippon Kagaku Zasshi*, **80,** 551 (1959); *Chem. Abstr.*, **55,** 3598 (1961).
76. T. Caronna, A. Citterio, T. Crolla, M. Ghirardini, and F. Minisci, *J. Heterocycl. Chem.*, **13,** 955 (1976).
77. Y. S. Tsizin and S. A. Charnyak, *Khim. Geterotsikl. Soedin.*, 982 (1976).
78. H. C. van der Plas and J. W. Streef, "Aromatic and Heteroaromatic Chemistry," Vol. 5, Chemical Society London, Specialist Periodical Report, 1977, p. 196.
79. W. A. Skinner, H. T. Crawford, H. Tong, D. Skidmore, and H. I. Maibach, *J. Pharm. Sci.*, **65,** 1404 (1976).

CHAPTER IV

Quinoxaline Mono- and Di-*N*-oxides

I. Methods of Preparation 29
1. *N*-Oxidation of Quinoxalines 29
2. Quinoxaline Mono-*N*-oxides from *o*-Nitroanilines 33
3. Quinoxaline Di-*N*-oxides from Benzofuroxans 35
4. Quinoxaline Mono- and Di-*N*-oxides from *o*-Benzoquinone Dioxime . . 38
5. Miscellaneous Methods of Quinoxaline *N*-Oxide Formation 39
II. Properties 40
1. Physical Properties 40
2. Chemical Reactions 42
3. Reactions of 1,2-Dihydro-2-oxoquinoxaline 4-Oxides 51
III. Uses 55
IV. Tables 56
Table 1. Quinoxaline Mono-*N*-oxides 56
Table 2. Quinoxaline Di-*N*-oxides 63
V. References 72

In the period under review very many quinoxaline *N*-oxides have been prepared. The main stimulus for this activity has been the search for biologically active compounds, and quinoxaline di-*N*-oxides with useful antibacterial and growth-promoting properties have been synthesized. Although many quinoxaline *N*-oxides have been obtained by oxidation of the corresponding quinoxalines, alternative methods are now available. In a synthesis based on *o*-nitroaniline, one oxygen atom of the nitro group becomes incorporated in a quinoxaline mono-*N*-oxide; and in the benzofuroxan-based procedure, the two oxygen atoms of the starting material become incorporated in a quinoxaline di-*N*-oxide. Studies of the chemistry of 1,2-dihydro-2-oxoquinoxaline 4-oxides have revealed unexpected benzenoid substitution reactions and interesting rearrangement reactions. There are few examples of naturally occurring quinoxalines; it is therefore of some interest that cultures of *Streptomyces ambofaciens* NRRL 3455 produce 6-chloro-2-quinoxalinecarboxylic acid 1,4-dioxide (**1**). The acid and derived esters and amides have antibiotic activity.[1]

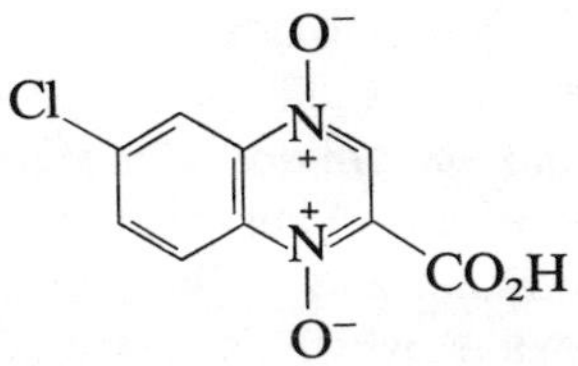

Cl O⁻ N CO₂H

1

I. Methods of Preparation

1. *N*-Oxidation of Quinoxalines

The most convenient reagent for oxidizing quinoxalines to their *N*-oxides is 30% aqueous hydrogen peroxide in acetic acid. However in the case of quinoxaline itself, treatment with this reagent gives mainly quinoxaline-2,3-dione (2,3-dihydroxyquinoxaline).[2] Conversion to the 1-oxide has been achieved by reaction with one equivalent of peracetic acid in acetic acid; and with excess of peracetic acid, quinoxaline 1,4-dioxide is formed.[3] Alternative reagents that have been used to oxidize quinoxalines include monoperphthalic acid,[4] *m*-chloroperbenzoic acid,[5] and permaleic acid.[6]

5-Substituted quinoxalines generally give 1-oxides which cannot be oxidized further to 1,4-dioxides; 5-methoxyquinoxaline is exceptional in this respect, being readily converted into the 1,4-dioxide.[3] Substituents in the 6-position have an unexpected effect. Thus quinoxalines with electron-donating groups (e.g., Me or MeO) in this position readily form di-*N*-oxides; but when electron-withdrawing groups (e.g., NO_2 and CN) are present, quinoxaline-2,3-diones are the major products of oxidation and only trace amounts of *N*-oxides are formed.[3,7] Oxidation of 6-chloroquinoxaline with 30% hydrogen peroxide in acetic acid gives a mixture of about 30% 6-chloroquinoxaline-2,3-dione and about 10% mono-*N*-oxide. Purification of the mono-*N*-oxide fraction gives 7-chloroquinoxaline 1-oxide, m.p. 151–152°.[8]

The effect of substitution at the 2- and 3-positions is now broadly understood. Oxidation of 2-methylquinoxaline with perphthalic acid gives a mixture of 38% of the 1-oxide, 25% of the 4-oxide, and 13% of the 1,4-dioxide. In the case of 2-ethylquinoxaline, the corresponding yields of 1-oxide, 4-oxide, and 1,4-dioxide are 23, 26, and 7%, respectively. 2-Isopropylquinoxaline is oxidized to a mixture of 38% of the 4-oxide and 4% of the 1,4-dioxide, and 2-*tert*-butylquinoxaline gives the 4-oxide in

63% yield. The lower yields of 1-oxide in this series clearly demonstrate the steric effect of the more bulky alkyl groups.[4] Electron-donating groups, such as amino and acetamido, in the 2-position direct oxidation preferentially to the 1-position. Thus oxidation of 2-aminoquinoxaline with permaleic acid in ethanol and in the presence of sodium hydrogen carbonate gives the 1-oxide, which is conveniently isolated as methyl quinoxalin-2-ylcarbamate 1-oxide (**2**) in 85% yield.[6] Oxidation of 2-acetamidoquinoxaline with peracetic acid in acetic acid yields a mixture of

NH_2 ⟶ NH_2, O^- —$ClCO_2Me$⟶ $NHCO_2Me$, O^-

2

the 1- and 4-oxides, with preferential formation of the 1-oxide and the 1,4-dioxide on further oxidation. With 30% hydrogen peroxide in acetic acid, a 20% yield of the 1,4-dioxide of 3-aminoquinoxalin-2-one is obtained.[9]

Electron-withdrawing substituents in the 2-position such as Cl, CO_2H, CO_2R, $CONH_2$, CN, SO_2Me, and Ph inhibit oxidation at the adjacent nitrogen, and thus oxidation results in the formation of 4-oxides.[4,10,11] Although treatment of 2-phenylquinoxaline 4-oxide with 30% hydrogen peroxide in acetic acid gives the 1,4-dioxide, in general, 1,4-dioxides cannot be formed by direct oxidation of quinoxalines substituted with an electron-withdrawing substituent at position 2.[12]

2,3-Dimethylquinoxaline can be oxidized with peracetic acid in acetic acid to a mono- or di-*N*-oxide by appropriate choice of reaction conditions.[13] 1,4-Dioxides have also been prepared with this reagent from other 2,3-dialkylquinoxalines, but similar treatment of 2-isopropyl-3-methylquinoxaline gives a considerable proportion of the 4-oxide, and 2,3-diisopropylquinoxaline is unchanged by hot peracetic or performic acids.[13] 2,3-Diphenylquinoxaline is converted by peracetic acid into a mixture of the 1-oxide and 1,4-dioxide; however with excess of aqueous hydrogen peroxide in acetic acid, in addition to the 1,4-dioxide, *N,N'*-dibenzoyl-*o*-phenylenediamine is also formed.[14]

The results of oxidizing other 2,3-disubstituted quinoxalines are broadly predictable from what is known about the oxidation of the monosubstituted compounds. Thus 1,4-dioxides can be prepared from 2-methyl- and 2-ethyl-3-phenylquinoxalines, but a considerable amount of 4-oxide is formed on oxidation of 2-isopropyl-3-phenylquinoxaline.[4] 4-Oxides are formed on oxidation of 2-methoxy-3-methyl(or phenyl)-quinoxalines or the corresponding 2-chloro compounds and 2-acetyl-3-methylquinoxaline.[13] Oxidation of 2,3-dimethoxyquinoxaline gives only 1% mono-*N*-oxide; the major product is quinoxaline-2,3-dione, and some 3-methoxyquinoxalin-2-one and 3*H*-benzimidazol-2-one are also formed.[4] Attempts to *N*-oxidize 2,3-diethoxy-, 2-chloro-3-methoxy-, and 2,3-dichloroquinoxaline also failed. A high yield (85%) of 1-oxide is formed on oxidation of 3-phenyl-2-quinoxalinecarboxamide.[4]

2,3-Dimethylquinoxalines with both electron-donating (Me, MeO, etc.) and electron-withdrawing (CN, MeCO, EtO_2C, etc.) substituents in the 6-position readily form di-*N*-oxides. 6,7-Disubstituted-2,3-dimethylquinoxalines also form di-*N*-oxides with the exception of the 6,7-dichloro derivative from which only a mono-*N*-oxide is obtained. Steric hindrance from groups in the 5-position is more pronounced than in the related 2,3-unsubstituted compounds. Peracetic acid oxidation of 2,3-dimethyl-5-nitroquinoxaline (**3**) gives the monooxide (**4**) and 4-nitro-2-methylbenzimidazole (**5**).[7,13]

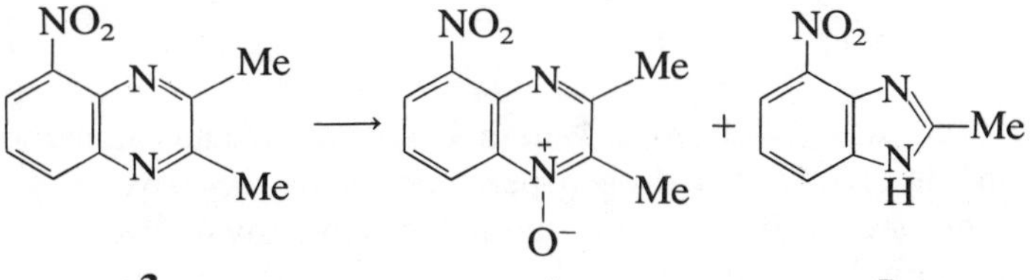

NO_2 Me Me N N → NO_2 Me Me N $\overset{+}{N}$ O^- + NO_2 N N H Me

3 **4** **5**

N N $SC_6H_4CH_3(p)$ $SC_6H_4CH_3(p)$

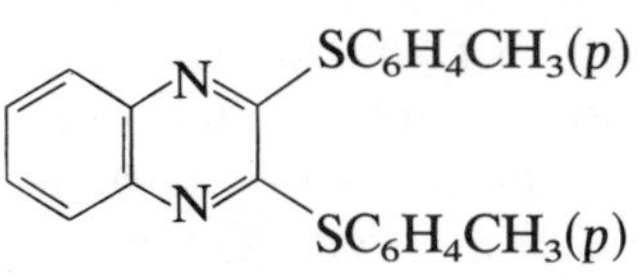

6

Oxidation of 2,3-bis-*p*-tolylthioquinoxaline (**6**) with 30% hydrogen peroxide in a mixture of acetic and formic acids gives quinoxaline-2,3-dione (**8**),[16] which is also isolated together with 2-methylsulfonylquinoxaline (**9**) and its 4-oxide (**10**) on oxidation of 2-methylthioquinoxaline (**7**).[11]

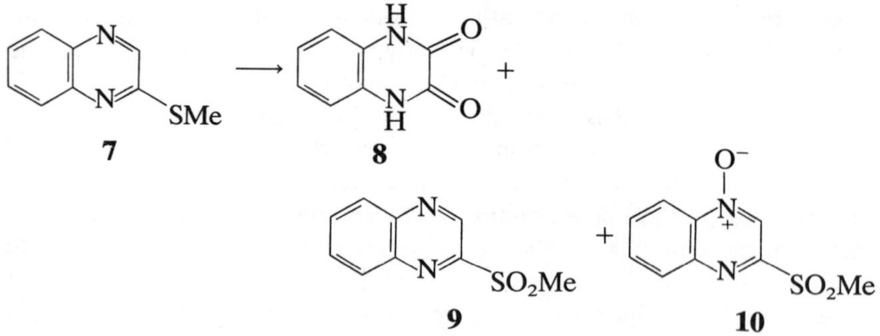

Peracetic acid treatment of 1,2-dihydro-1-methyl-2-oxoquinoxaline (**11**) gives the 3-oxygenated derivative (**13**), and similar treatment of 1,2-dihydro-1,3-dimethyl-2-oxoquinoxaline gives a small quantity of the 4-oxide.[17] Both compound **11** and its 6-nitro derivative **12** are converted in high yield to the corresponding quinoxalin-3-ones **13** and **14** on treatment with aqueous hydrogen peroxide in acetic acid at room temperature.[18]

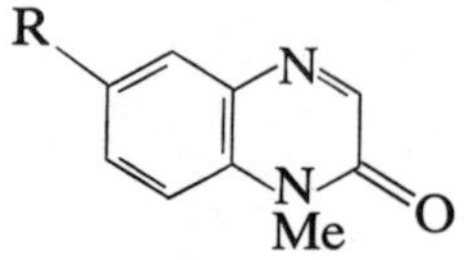

11 R=H
12 $R=NO_2$

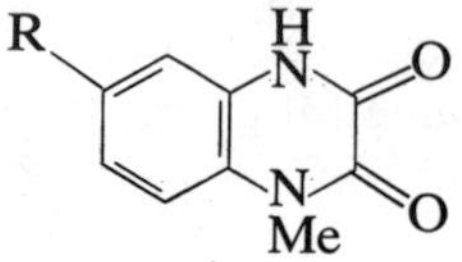

13 R=H
14 $R=NO_2$

Oxidation of 3,4-dihydro-4-methyl-3-oxoquinoxaline-2-carboxy-*N*-methylanilide (**15**) with hydrogen peroxide yields the 1-oxide;[19] but on removal of one or both of the methyl groups, giving **16**, **17**, or **18**, oxidation with hydrogen peroxide, peracetic acid, or perbenzoic acid results in the removal of the carboxyamide group and replacement with hydroxyl.[20] The mechanism proposed for this reaction is illustrated by reference to the conversion of **18** into quinoxaline-2,3-dione (Scheme 1).

It was subsequently shown that the introduction of methyl or methoxy groups into the 6- and 7-positions of the quinoxaline carboxyanilides resulted in increased yields of 1-oxides. This result is rationalized in terms

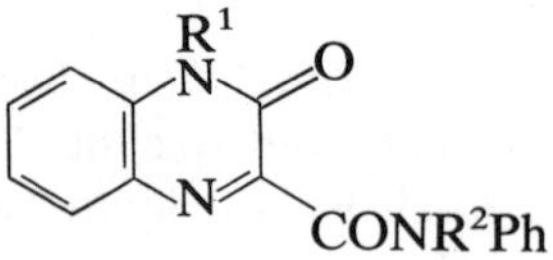

15 $R^1=R^2=Me$
16 $R^1=H$, $R^2=Me$
17 $R^1=Me$, $R^2=H$
18 $R^1=R^2=H$
19 $R^1=COMe$, $R^1=Me$

of the reduced electrophilic nature of carbon-2 in the substituted anilides.[21] 4-Acetyl-3,4-dihydro-3-oxoquinoxaline-2-carboxy-*N*-methylanilide (**19**) also forms an *N*-oxide normally.[20] Attempts to oxidize quinoxalin-2-one and 3-carboxyquinoxalin-2-one with hydrogen peroxide in acetic acid in each case give quinoxaline-2,3-dione.[10]

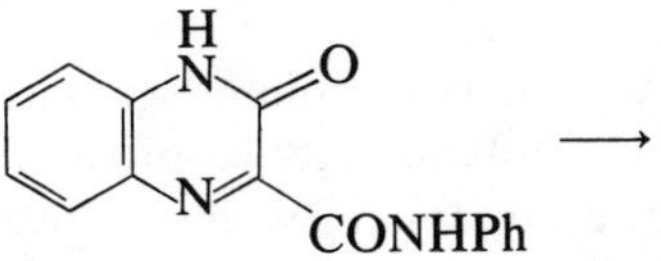

18

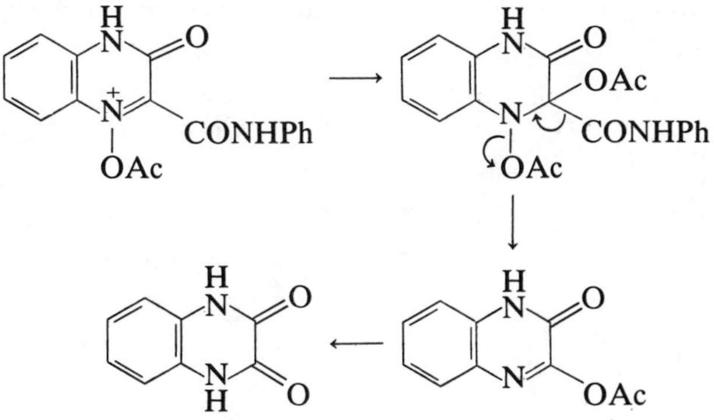

Scheme 1

2. Quinoxaline Mono-*N*-oxides from *o*-Nitroanilines

The preparation of quinoxaline *N*-oxides from *o*-nitroanilines was first reported by Tennant,[22–24] and work on similar lines has also been carried out by Italian[25,26] and Pakistani workers.[27,28] The method is illustrated in Scheme 2. The cyclization step is base promoted, and the activating groups may be CO_2R, COR, CN, $\overset{+}{N}C_5H_5$, etc. The activating group may not survive the cyclization reaction; for example, quinoxaline *N*-oxide

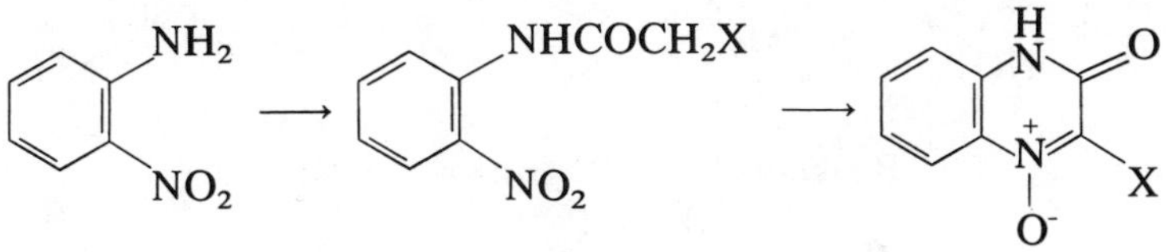

Scheme 2

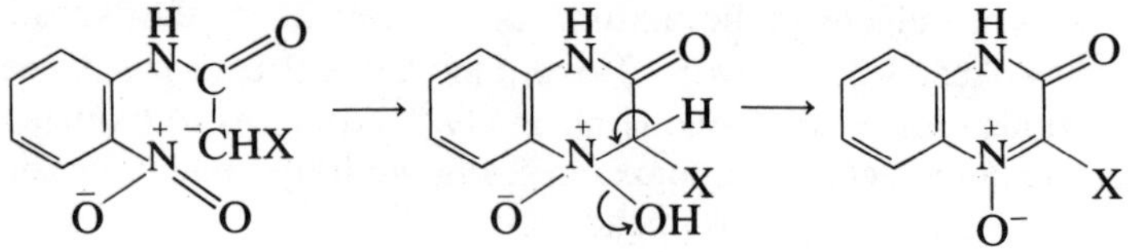

Scheme 3

formation from pyridinium salts is accompanied by concomitant opening of the pyridine ring.[26] The formation of the C—N bond may be pictured as carbanion attack on the nitro group, followed by loss of water to give the quinoxaline *N*-oxide (Scheme 3). A more complex example of the utilization of this type of reaction is shown in the formation of compound **20**.[29]

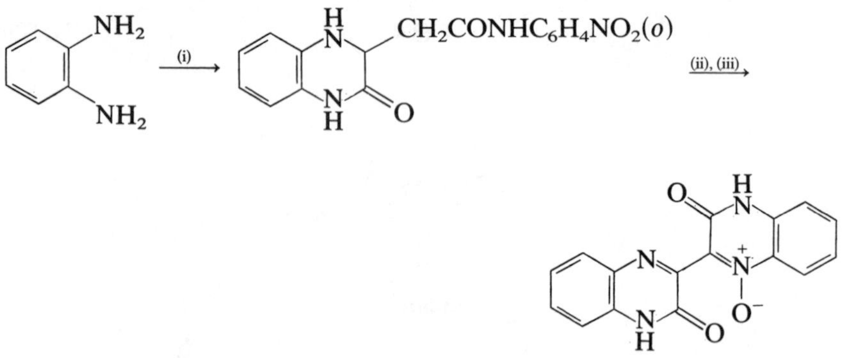

NH_2 NH_2 (i) $CH_2CONHC_6H_4NO_2(o)$ (ii), (iii)

20

Reagents (i) *o*-Nitrophenylmaleimide
(ii) Chloranil
(iii) 10% sodium hydroxide

A related synthesis of a quinoxaline *N*-oxide is illustrated by the reductive cyclization of the cyanoformyl derivative of *o*-nitroaniline (Scheme 4).[30]

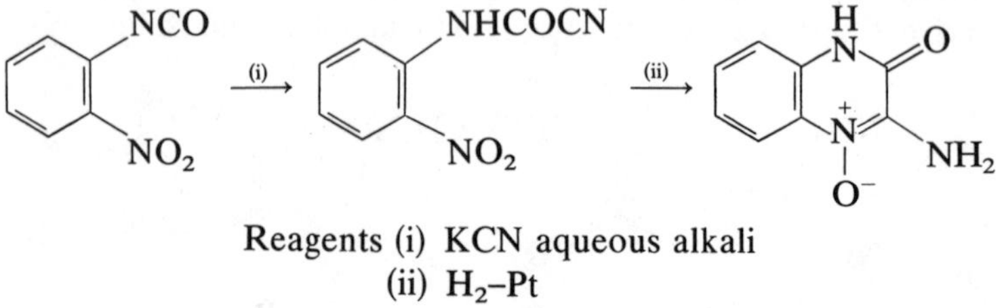

NCO NO_2 (i) NHCOCN NO_2 (ii) NH_2

Reagents (i) KCN aqueous alkali
(ii) H_2–Pt

Scheme 4

3. Quinoxaline Di-*N*-oxides from Benzofuroxans

This method of quinoxaline di-*N*-oxide formation is known as the Beirut reaction and is extremely useful, particularly in view of the limitations of direct oxidative procedures for preparing di-*N*-oxides. Probably several hundred quinoxaline di-*N*-oxides have been prepared in this way. Only a selection of those exemplified in the patent literature are included in the table at the end of this chapter. Valuable reviews on the conversion of benzofuroxans into quinoxaline di-*N*-oxides have been written by Haddadin and Issidorides[31] and by Ley and Seng.[32]

The first report of the preparation of quinoxaline di-*N*-oxides from benzofuroxans was by reaction with enamines.[33] Benzofuroxans are themselves conveniently prepared by oxidation of the appropriate *o*-nitroanilines with hypochlorite or by pyrolysis of the corresponding nitrophenyl azides. A variety of enamines react satisfactorily, but the less reactive morpholine enamines give higher yields and more easily isolatable products (Scheme 5).[34]

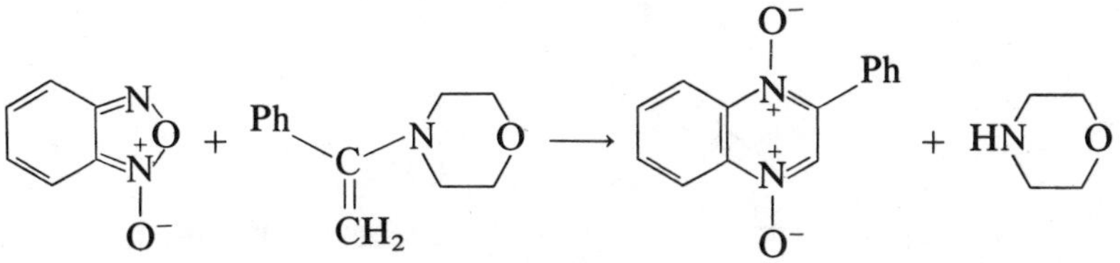

Scheme 5

When benzofuroxan is reacted with the enamine ($Me_2C{=}CHNMe_2$), the dihydroquinoxaline 1,4-dioxide **21** is obtained. In this case aromatization by elimination of amine is structurally precluded. It is suggested that dihydroquinoxalines are the common intermediates in these reactions.[35]

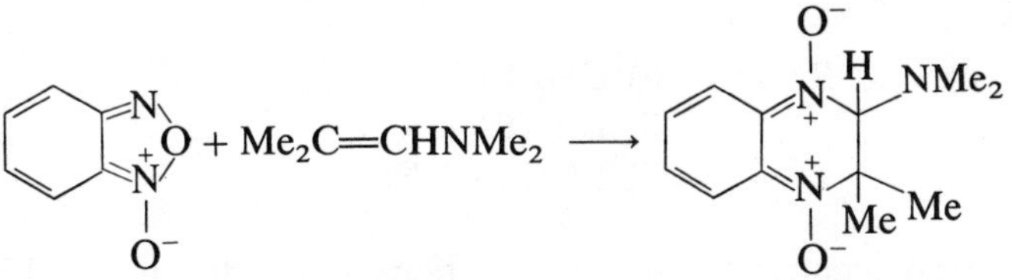

$$\text{benzofuroxan} + Me_2C{=}CHNMe_2 \longrightarrow \mathbf{21}$$

21

A variety of substrates of the type $R^1COCH_2R^2$ (where R^1 = alkyl or aryl and R^2 = H, alkyl, aryl, COR, CO_2R, SR, etc.) also react with

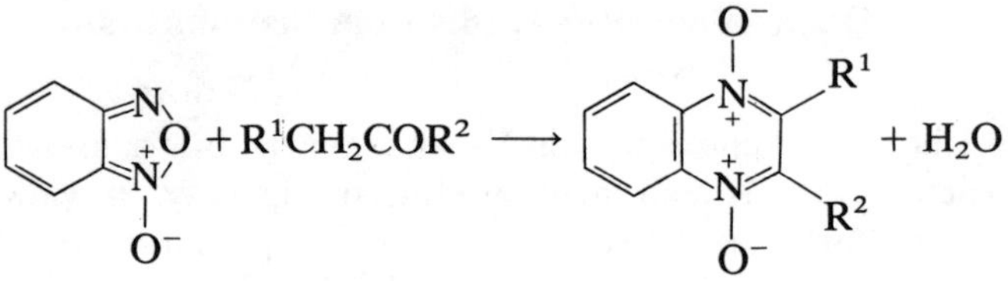

$$\text{benzofuroxan} + R^1CH_2COR^2 \longrightarrow \text{2-}R^1\text{-3-}R^2\text{-quinoxaline 1,4-dioxide} + H_2O$$

Scheme 6

benzofuroxans in the presence of base to give quinoxaline di-*N*-oxides (Scheme 6). In the simple case of reaction with methyl ethyl ketone in methanolic ammonia, 91% 2,3-dimethylquinoxaline 1,4-dioxide is obtained.[32] Other components found to react with benzofuroxan include malononitrile, ethyl cyanoacetate, and ynamines (Scheme 7).[32]

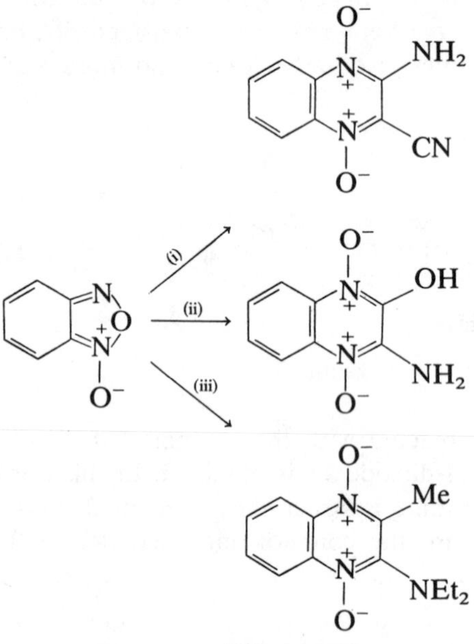

Scheme 7

Reagents (i) $CH_2(CN)_2$
(ii) EtO_2CCH_2CN
(iii) $MeC{\equiv}CNEt_2$

When unsymmetrically substituted benzofuroxans are used then in principle, product mixtures will be formed. However in practice a considerable regiospecificity of cyclization is often achieved. For example, in the reaction of 5-substituted benzofuroxans with benzoylacetonitrile, only products of structure **22** are formed.[36]

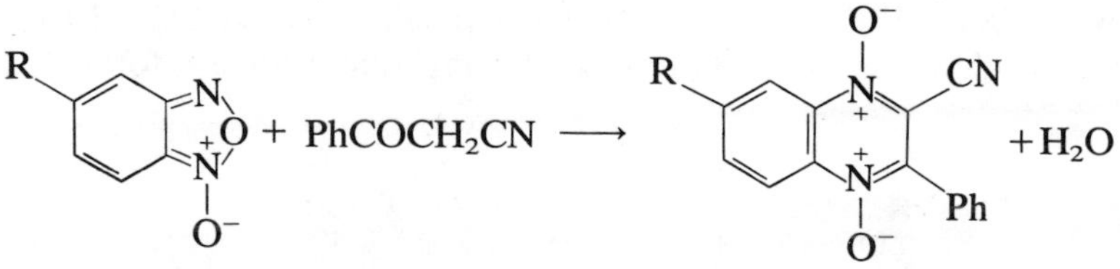

R = H, MeO, Cl, or Br

The main mechanistic features of the reaction are shown in Scheme 8, a key feature of which is the formation of an intermediate hydroxylamino-nitrone. However more detailed study is required as it is still not proved at which nitrogen of the benzofuroxan initial nucleophilic attack occurs. Experimental evidence has been obtained which suggests that reaction may occur via the tautomeric *o*-dinitrosobenzene form.[37]

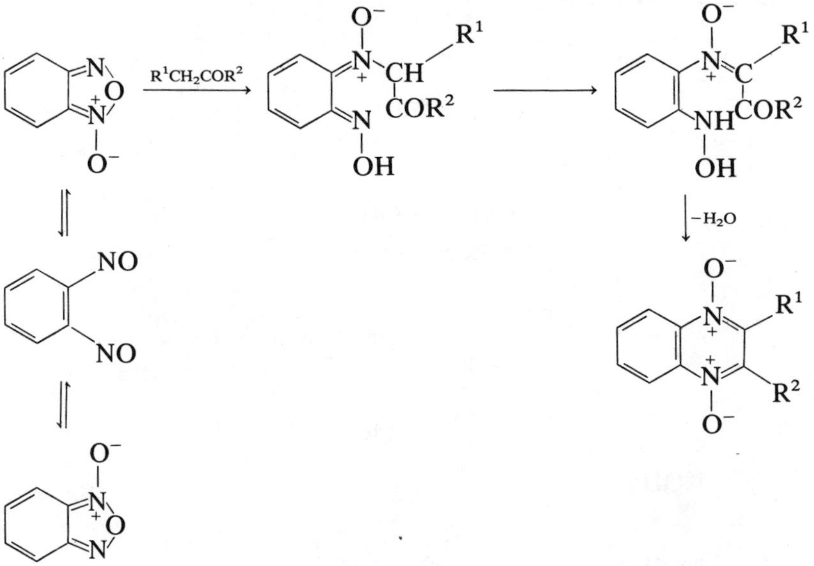

Scheme 8

Quinoxaline di-*N*-oxides without substituents in the pyrazine ring have been prepared by reaction of benzofuroxans with synthons of the type RCH=CHX, where R = COMe or 4-substituted benzoyl and X = ONa, or where R = H and X = NR_2 or OAc.[38,39] Two cases are known of mono-*N*-oxide rather than di-*N*-oxide formation using benzofuroxan. Thus reaction with benzofuran-3(2*H*)ones gives 3-(*o*-hydroxyaryl)quinoxaline

1-oxides (e.g., **23**);[40] and with the sodium salt of $EtO_2CCOCH_2CO_2Et$, the mono-*N*-oxide of 2,3-bis(ethoxycarbonyl)quinoxaline is formed.[41]

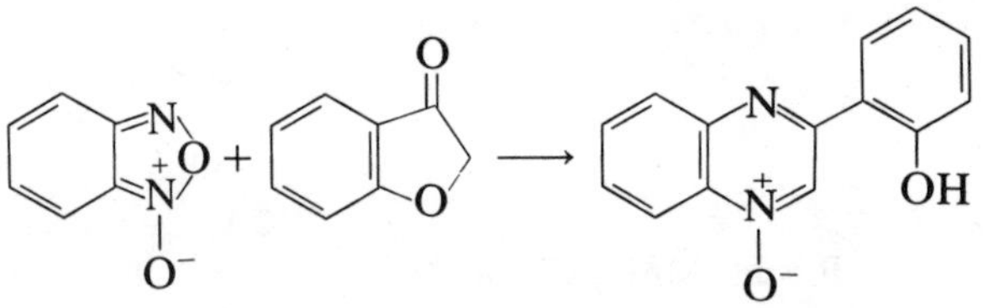

23

4. Quinoxaline Mono- and Di-*N*-oxides from *o*-Benzoquinone Dioxime

The reaction between *o*-benzoquinone dioxime and α-diketones and α-hydroxyketones yields quinoxaline di-*N*-oxides, though in relatively low yields (Scheme 9).[42]

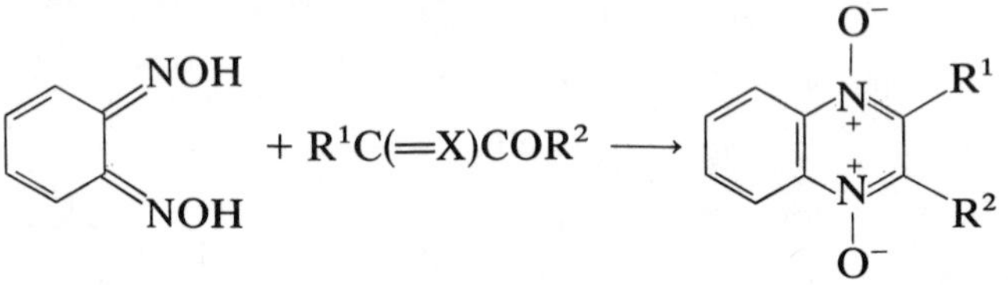

NOH + $R^1C(=X)COR^2$ ⟶

X = O or H,OH

Scheme 9

CH_2COMe Me

24

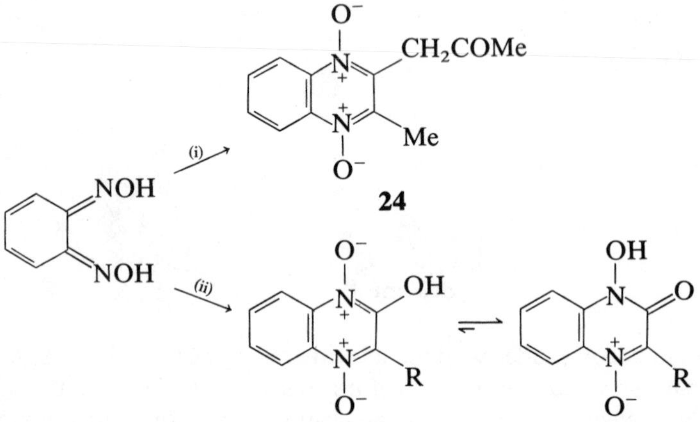

R = H or Me **25**

Reagents (i) MeCOCH=CHCOMe
(ii) $(CHO)_2$ or MeCOCHO

Quinoxaline di-*N*-oxide formation (e.g., **24**) has also been achieved by reaction of the dioxime with α,β-unsaturated ketones.[43] A further extension of this reaction is the preparation of 1-hydroxyquinoxalin-2-one 4-oxides (**25**) from *o*-quinone dioximes and glyoxal or pyruvaldehyde.[37,42]

5. Miscellaneous Methods of Quinoxaline *N*-Oxide Formation

Quinoxaline hydroxy-*N*-oxides (1-hydroxyquinoxalin-2-ones) have been prepared by alkaline hydrolysis of alloxazine *N*-oxides (Scheme 10).[44]

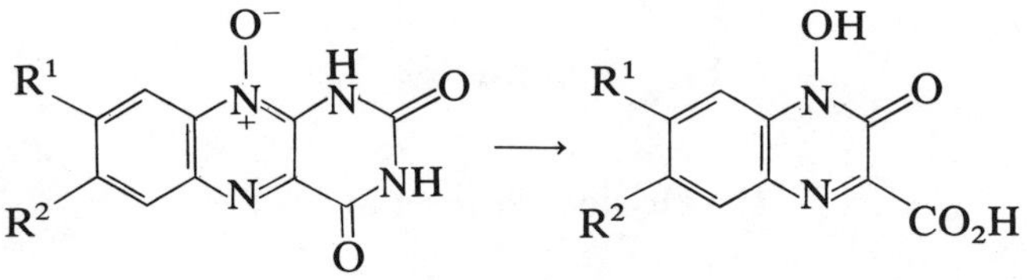

$R^1 = R^2 = H$
$R^1 = H, R^2 = Me$
$R^1 = R^2 = Me$
$R^1 = H, R^2 = Cl$

Scheme 10

Quinoxaline mono-*N*-oxides result from the partial reduction of the di-*N*-oxides. Among reagents that have been used for this purpose are hydrogen and a metal catalyst, phosphorus trichloride, and sodium dithionite. Thus 2-methylquinoxaline 1-oxide has been prepared by partial hydrogenation of the di-*N*-oxide;[45] quinoxaline-2-carboxyanilide 1-oxide has been prepared from the partial deoxygenation of the 1,4-dioxide with phosphorus trichloride;[20] and quinoxaline-2-carboxylic acid 1-oxide, by sodium dithionite reduction of the 1,4-dioxide.[46] While reduction with phosphorus trichloride occurs by attack at oxygen, dithionite anion attacks at the carbon alpha to the *N*-oxide function as shown in Scheme 11.[47] Selective monodeoxygenation of quinoxaline 1,4-dioxides has also been achieved with trimethyl phosphite.[153]

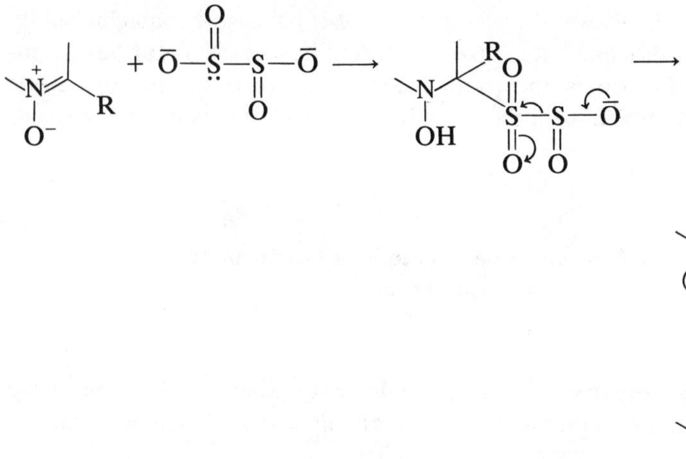

Scheme 11

II. Properties

1. Physical Properties

The infrared spectra of quinoxaline mono- and di-*N*-oxides show characteristic N—O stretching frequencies in the range of 1280–1380 cm^{-1}. The band usually appears in the range of 1320–1350 cm^{-1}.[31] Isomers such as **26** and **27**, produced by reaction of benzofuroxan with

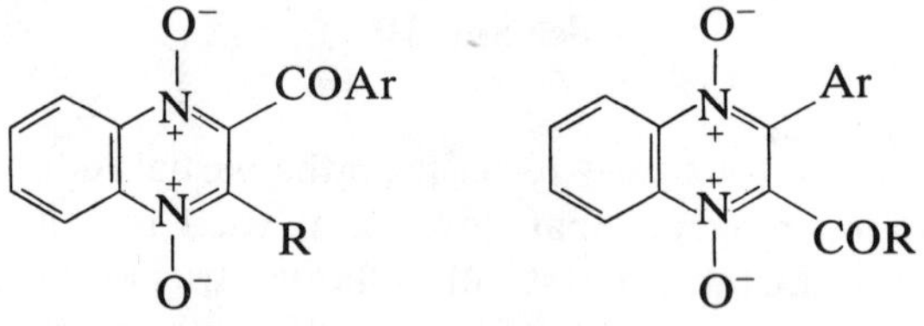

26 **27**

1,3-diketones of the type $ArCOCH_2COR$, are readily distinguishable because the former absorb around 1700–1710 cm^{-1} whereas the latter absorb around 1660–1680 cm^{-1}.[48]

Quinoxaline 1-oxide has a pK_a of 0.25 and is therefore a weaker base by about 0.3 pK_a units than the parent heterocycle. It undergoes protonation at N-4. The basic center in 3-aminoquinoxaline 1-oxide (pK_a 2.63) is also N-4; however the isomeric 2-amino-1-oxide (pK_a 1.09) undergoes protonation at the oxygen atom of the $\overset{+}{N}—\overset{-}{O}$ group. Quinoxaline 1,4-dioxide and 2-aminoquinoxaline 1,4-dioxide (pK_a 1.06) protonate to give the cations **28** and **29**, respectively.[49]

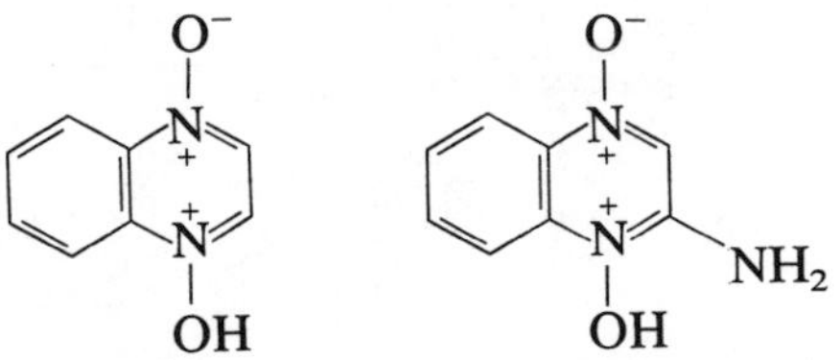

28 **29**

The ultraviolet spectrum of quinoxaline 1-oxide in a Nujol mull shows λ_{max} at 238, 252 sh, 325, 340 sh, and 348 nm. Both ultraviolet and infrared data have been used to assign structures to the complexes quinoxaline 1-oxide forms with transition metal chlorides.[50]

The longest ultraviolet absorption maximum for quinoxaline *N*-oxides is shifted to shorter wavelengths in solvents of increasing polarity, and this is attributed to hydrogen bonding.[51]

^1H NMR parameters for quinoxaline 1-oxide and 1,4-dioxide in $CDCl_3$ are shown in Fig. 1.[52] Thus compared to quinoxaline itself, the proton

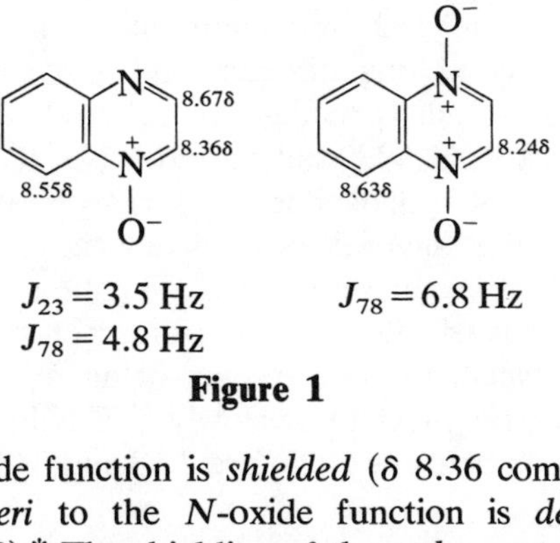

8.67δ 8.36δ 8.55δ 8.63δ 8.24δ

$J_{23} = 3.5$ Hz
$J_{78} = 4.8$ Hz

$J_{78} = 6.8$ Hz

Figure 1

ortho to the *N*-oxide function is *shielded* (δ 8.36 compared to δ 8.73)* and the proton *peri* to the *N*-oxide function is *deshielded* (δ 8.55 compared to δ 8.03).* The shielding of the *ortho* proton and deshielding of the *peri* proton in quinoxaline 1-oxides is a general phenomenon and has been suggested as a reliable criterion for structure determination.[53]

Quinoxaline di-*N*-oxides as a group show aromatic A_2B_2 multiplets in the range δ 8.3–8.7 for protons 5 and 8 and in the range δ 7.6–7.8 for protons 6 and 7. Hydrogen–fluorine long-range couplings have been

* Values in carbon tetrachloride.

observed in some 2-substituted 3-trifluoromethylquinoxalines and their 1- and 1,4-dioxides. It is found that J_{HF} is susceptible to changes in the hydrogen–fluorine internuclear distances, and this effect is explained by a through-space mechanism. In compounds **30** and **31**, J_{HF} is 0.0 and 3.0 Hz, although the number of intervening bonds is 4 and 5, respectively.[54]

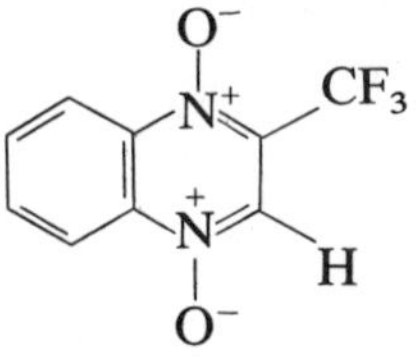

30 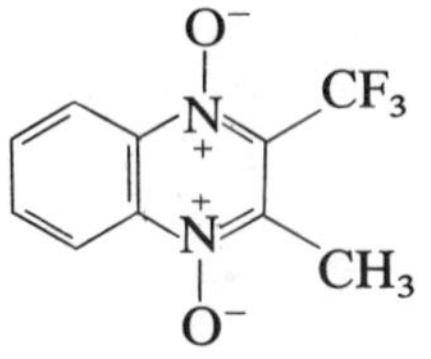 **31**

Alkylquinoxaline mono-*N*-oxides fragment with the loss of O$^{\bullet}$ and OH$^{\bullet}$ on electron impact. The presence of an alkyl group adjacent to the *N*-oxide function is not necessarily required for the one-step expulsion of OH$^{\bullet}$. In the case of quinoxaline di-*N*-oxides, the M-16 peak is larger than the M-17 peak.[55] An important feature of the mass spectra of several quinoxaline mono-*N*-oxides is the presence of an (M—CO)$^+$ peak which is indicative of molecular rearrangement on electron impact.[56]

Quinoxaline 1-oxide is more easily reduced polarographically than pyridine 1-oxide. In general, reduction occurs more easily with the introduction of a second ring nitrogen and a second aromatic ring.[57] Quinoxaline mono and di-*N*-oxides are reduced at exactly the same half-wave potential (−0.52 V),* and in both cases the subsequent waves correspond to those of quinoxaline itself ($E_{1/2} = -0.68$ and −1.52 V). It therefore appears that the di-*N*-oxide is reduced in a four-electron step to quinoxaline.[58] Kinetic data on the deoxygenation of quinoxaline di-*N*-oxide with phosphorus trichloride in chloroform, however, indicate stepwise removal of oxygen, the first oxygen being removed in a very fast reaction and the second, much more slowly.[59] The dipole moments of the 1- and 1,4-dioxides in benzene at 25° are 2.53 and 2.27 D, respectively.[60]

2. Chemical Reactions

Many of the reactions of quinoxaline *N*-oxides are typical of those of heteroaromatic *N*-oxides in general. Thus the *N*-oxide function is readily reduced, nuclear substitution products are obtained on treatment with

* At 25° w.r.t. the saturated calomel electrode.

acid anhydrides or acid chlorides, and the carbon alpha to the $\overset{+}{N}—\overset{-}{O}$ function is readily attacked by nucleophilic reagents.[61,62] The following discussion is divided into three main sections. In the first the reactions of quinoxaline mono- and di-*N*-oxide are discussed, the second illustrates the effect of the *N*-oxide function on substituent reactivity, and the third deals with the reactions of 1,2-dihydro-2-oxoquinoxaline 4-oxides.

Quinoxaline 1-oxide (**32**) can be reduced to the parent base by hydrogenation in the presence of a Raney nickel catalyst or by treatment with phosphorus tribromide. Reaction of the *N*-oxide with acetic anhydride gives initially a mixture of quinoxaline (**33**) and quinoxalin-2-one (**34**); but on more prolonged reaction a third product, 1-(2-quinoxalinyl)-quinoxalin-2-one (**35**), is formed.[63*]

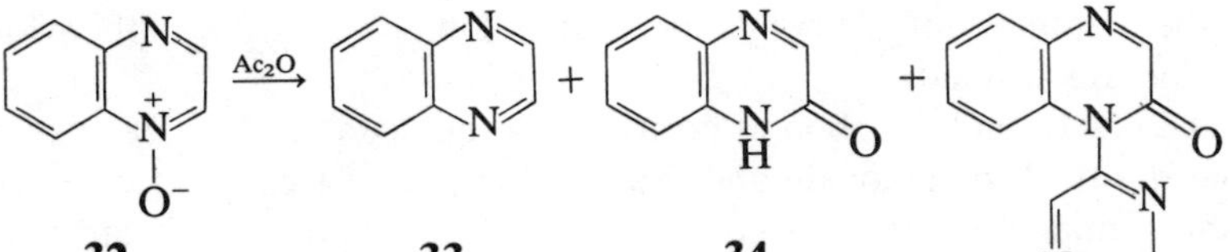

Ac_2O

32 **33** **34** **35**

A mixture of 2- and 6-chloroquinoxalines is obtained by reaction of quinoxaline 1-oxide with phosphoryl chloride, sulfuryl chloride, or acetyl chloride (Scheme 12).[63] Reaction with benzenesulfonyl chloride gives a

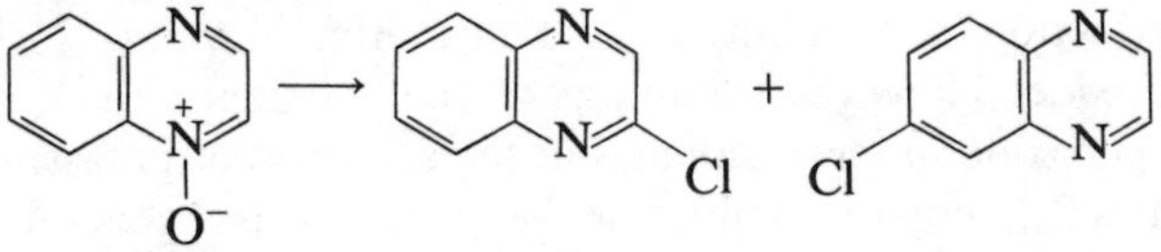

Scheme 12

mixture of 2-chloroquinoxaline and the benzenesulfonate of the 1-oxide.[64] Treatment of quinoxaline 1-oxide with benzoyl chloride and a nucleophilic reagent such as ethanol or aniline leads to a 2-substituted quinoxaline (Scheme 13), though only small yields of product are reported.[63] The main product formed on reaction of the 1-oxide with phenylmagnesium bromide is 2-phenylquinoxaline 1-oxide (17%). With

* An earlier report[45] stated that quinoxaline *N*-oxide is unchanged when heated with acetic anhydride.

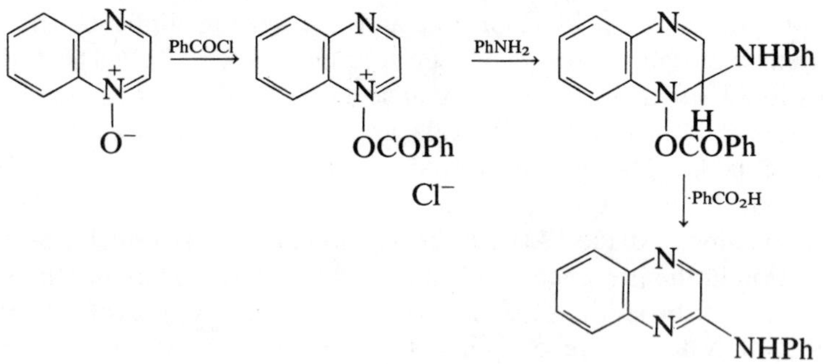

Scheme 13

phenyllithium, the major product is 2,2′-biquinoxaline (14%), although smaller amounts of 2-phenylquinoxaline 1-oxide and 2,3-diphenylquinoxaline are formed.[63]

The electrophilic carbon at position 2 in quinoxaline 1-oxide is susceptible to attack by alkoxide and cyanide ions. Brief treatment with potassium cyanide in methanol gives a mixture of quinoxaline 2-carbonitrile and quinoxaline 2-carboxamide. Increasing the reaction time increases the amount of amide formed, and 2-methoxyquinoxaline is also produced.[63] Treatment of the 1-oxide with potassium cyanide and potassium ferricyanide in water yields 2-cyanoquinoxaline 1-oxide.[65]

2-Phenylquinoxaline 4-oxide has been found to react with a similar range of reagents to quinoxaline 1-oxide;[12] thus deoxygenative chlorination with sulfuryl chloride gives 2-chloro-3-phenylquinoxaline in high yield. Oxidation of 2-phenylquinoxaline 4-oxide with 30% aqueous hydrogen peroxide in acetic acid[12] or formic acid[66] gives the 1,4-dioxide, and this on reduction with sulfurous acid in methanol yields 2-phenylquinoxaline 1-oxide. However treatment of the 4-oxide with 30% aqueous hydrogen peroxide in methanol and in the presence of potassium hydroxide furnishes 2-phenylbenzimidazole 3-oxide. This is a general reaction of 2-alkyl- and 2-alkoxyquinoxaline 4-oxides (Scheme 14).[67]

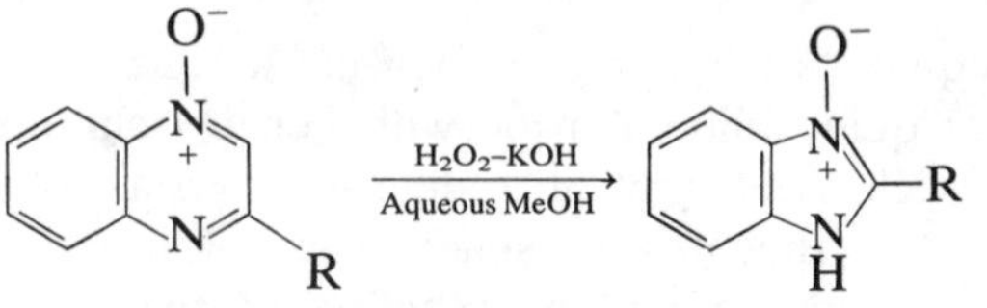

$R = Ph$, alkyl, or alkoxy

Scheme 14

Both quinoxaline 1-oxide[68] and 2-phenylquinoxaline 4-oxide[12] give anilinoquinoxalines (**36**) on reaction with phenyl isocyanate. With excess of reagent and at higher temperatures, quinoxaline 1-oxide yields the imidazoloquinoxaline (**37**). Treatment of 2-phenylquinoxaline 4-oxide and other 2-substituted quinoxaline 4-oxides with acetophenone in benzene and in the presence of sodamide gives 3-phenacyl derivatives.[69] Quinoxaline mono-*N*-oxides are believed to form quaternary salts by coordination at ring nitrogen rather than at exocyclic oxygen;[3,70] thus the methiodide of quinoxaline 1-oxide is thought to have structure **38**.[3]

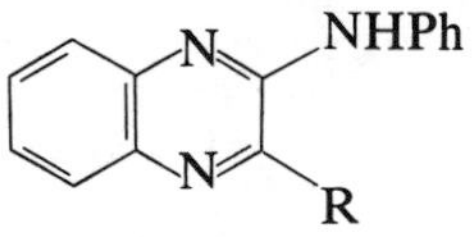

36 R = H or Ph 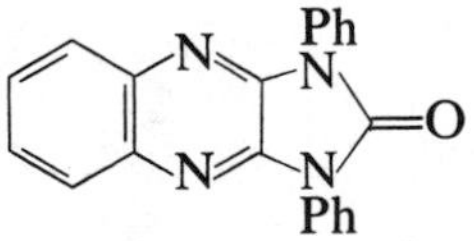 **37** 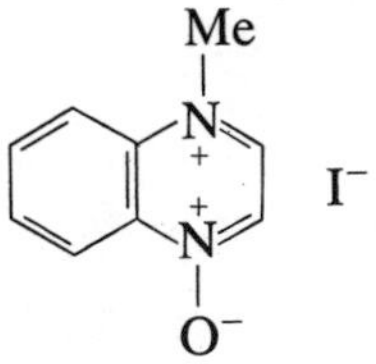 **38**

Quinoxaline 1-oxides are photochemically unstable. Irradiation of quinoxaline 1-oxide itself in dilute aqueous solution gives quinoxalin-2-one.[17] Irradiation of both the 1- and 4-oxides of 2-phenylquinoxaline gives a common product, the unstable benzoxadiazepine **39**; in each case some 2-phenylquinoxaline is also formed. Irradiation of the 4-oxide yields a third product, 3-phenylquinoxalin-2-one. A benzoxadiazepine is also obtained on irradiation of 2,3-diphenylquinoxaline 1-oxide.[71,72]

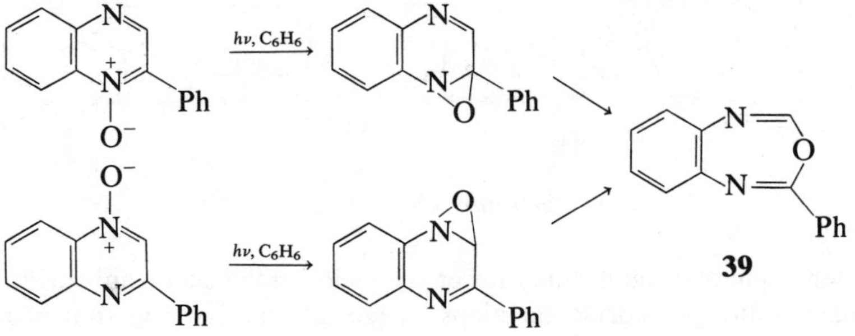

$h\nu$, C_6H_6

$h\nu$, C_6H_6

39

Reaction of quinoxaline 1,4-dioxide with boiling acetic anhydride gives about 50% 2-acetoxyquinoxaline 1-oxide (**40**) and some quinoxaline 1-oxide.[73] With acetyl chloride at room temperature, 6-chloroquinoxaline 1-oxide (**41**) is formed; and on heating and prolonged reaction times, 2-chloroquinoxaline 4-oxide (**42**) and 6,7-dichloroquinoxaline appear as additional products.[8] 6-Ethoxy-, 6-methoxy-, and 6-methylquinoxaline 1,4-dioxides show similar behavior, undergoing deoxygenative chlorination to give the corresponding chloro-*N*-oxides.

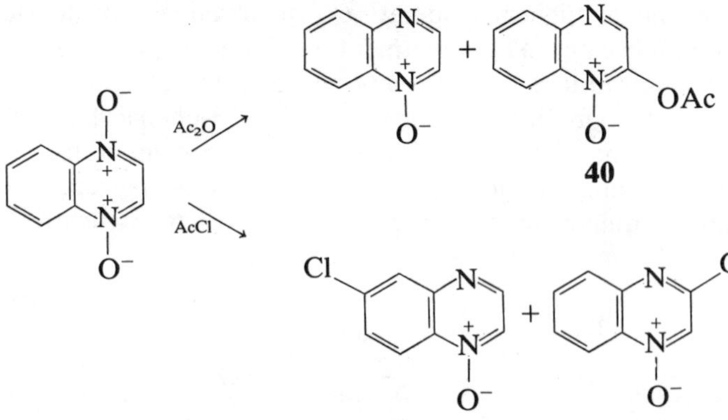

40

41

42

The reactions of the parent di-N-oxide with benzenesulfonyl chloride and benzoyl chloride are also illustrated in Scheme 15.[64] Small quantities

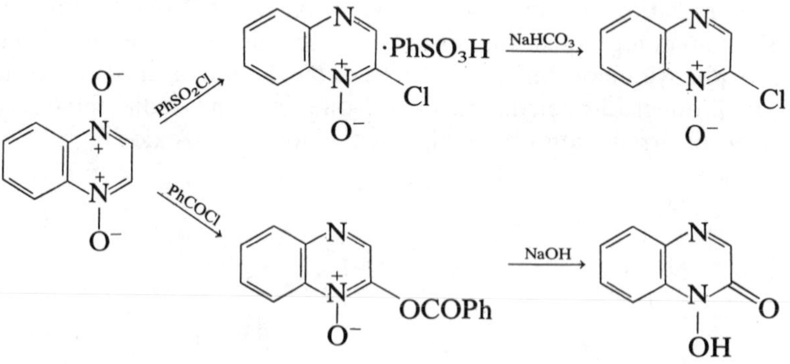

Scheme 15

of benzoquinone significantly lower the yields in the acetic anhydride and benzenesulfonyl chloride reactions, suggesting the participation of free-radical processes in the main reaction pathways. Quinoxaline 1,4-dioxide is photochemically unstable and in dilute aqueous solution is transformed to quinoxalin-2-one 4-oxide.[17] Irradiation in dilute hydrochloric acid yields mainly 2-chloroquinoxaline 1-oxide.[74]

Reaction of 2-methylquinoxaline 1-oxide with acetic anhydride gives 2-acetoxymethylquinoxaline,[45] and a similar reaction of 2,3-dimethylquinoxaline 1-oxide yields 2-acetoxymethyl-3-methylquinoxaline.[66] The acetoxymethyl compounds are readily hydrolyzed with alkali to the corresponding hydroxymethylquinoxalines. Treatment of

2,3-dimethylquinoxaline 1-oxide with acetyl chloride gives 2-chloromethyl-3-methylquinoxaline in high yield. Analogous reactions have been carried out on the 1,4-dioxide of 2,3-dimethylquinoxaline, which is smoothly converted into 2,3-bis(acetoxymethyl)quinoxaline[73] and 2,3-bis(chloromethyl)quinoxaline[66] on treatment with acetic anhydride and acetyl chloride, respectively. Neither a radical source (benzoyl peroxide) nor a radical scavenger (*p*-benzoquinone) has any appreciable effect on the yield of the bis(acetoxymethyl) compound.

There have been several observations on the reactivity of the methyl groups in 2-methyl- and 2,3-dimethylquinoxaline 1,4-dioxides. 2-Methylquinoxaline 1,4-dioxide is converted into 2-formylquinoxaline 1,4-dioxide by oxidation with selenium dioxide in refluxing benzene.[75] 2,3-Dimethylquinoxaline 1,4-dioxide reacts with formaldehyde only in the presence of alkaline reagents, and, depending on the amount of formaldehyde taken, either a mono- or dihydroxyethylquinoxaline is formed. No pure products were obtained from the attempted hydroxymethylation of 2-methylquinoxaline 1,4-dioxide.[76] Both the mono- and dimethyl di-*N*-oxides undergo aldol condensation with aromatic aldehydes in the presence of methanolic sodium methoxide, and the resulting carbinols are converted into styrylquinoxalines by the action of concentrated sulfuric acid.[77] Treatment with *p*-nitrosodimethylaniline in methanolic sodium hydroxide gives the nitrones **43**,[78] and coupled products are obtained

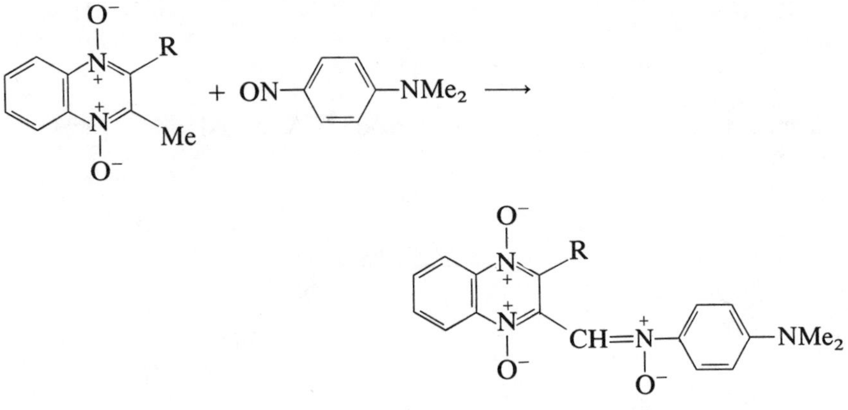

+ ON— —NMe_2 ⟶

CH=N— —NMe_2

43

R = H or Me

with aryldiazonium salts.[79] 2,3-Dimethylquinoxaline 1,4-dioxide has also been found to undergo two further reactions characteristic of compounds with active methyl groups. Thus treatment with piperidine and formaldehyde yields the bis(piperidinoethyl) derivative **44**, and with chloral the adduct **45** is formed.[16]

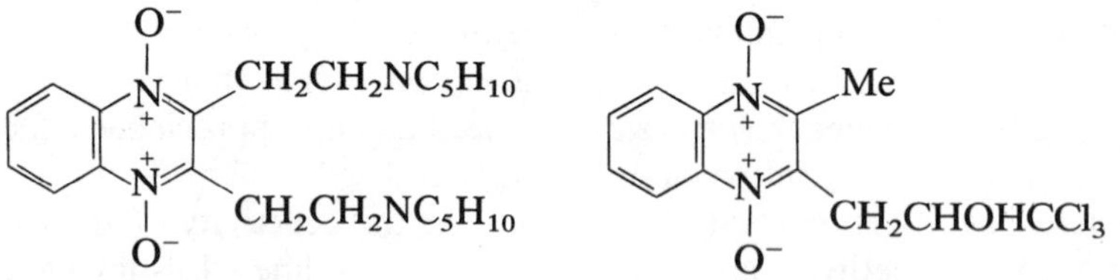

44 **45**

On heating with dimethyl sulfoxide, 2-hydroxymethylquinoxaline 1,4-dioxide undergoes a redox reaction in which 2-carboxyquinoxaline 4-oxide is formed. Similar treatment of 2,3-bis(hydroxymethyl)quinoxaline 1,4-dioxide (**46**) yields the internal hemiacetal **47** of 3-formyl-2-hydroxymethylquinoxaline 1-oxide. This transformation can also be

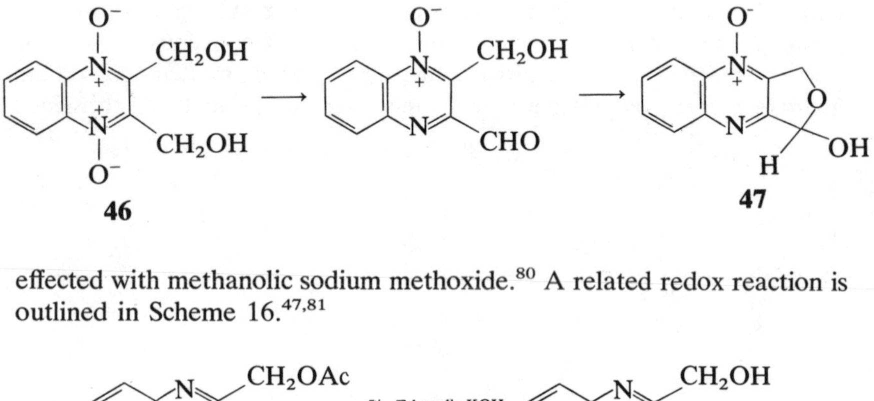

46 **47**

effected with methanolic sodium methoxide.[80] A related redox reaction is outlined in Scheme 16.[47,81]

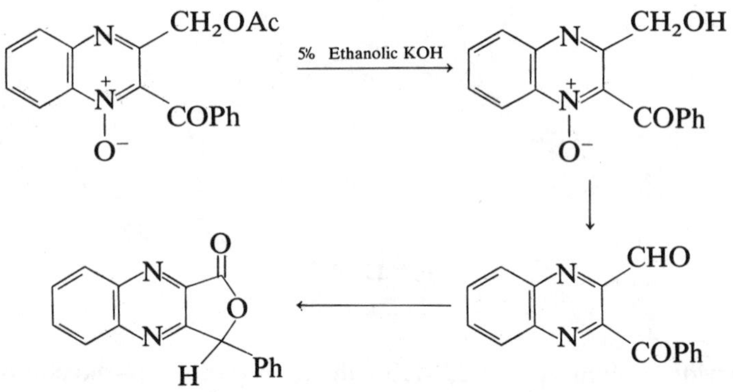

5% Ethanolic KOH

Scheme 16

A complex series of reactions occurs when di-*N*-oxides of type **48** are treated with primary amines, deoxygenation being accompanied by ring closure to give pyrroloquinoxalines of type **49** (see Chapter IX).[82]

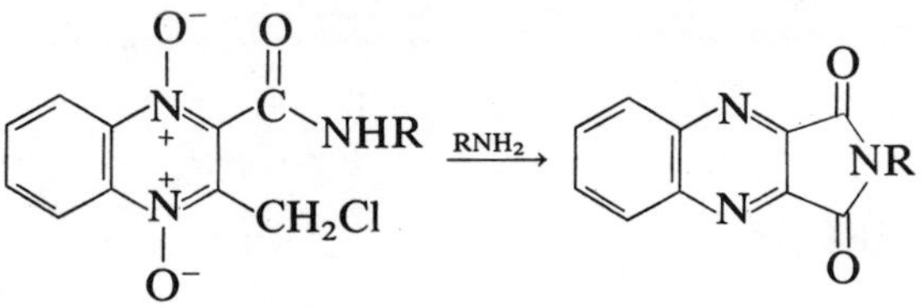

48 $\xrightarrow{RNH_2}$ **49**

The ease with which 2-chloroquinoxaline 1-oxide undergoes substitution reactions with a variety of nucleophilic reagents clearly demonstrates the activating effect of the *N*-oxide function. Thus whereas the 2-chloro-1-oxide is converted into 2-aminoquinoxaline 1-oxide by heating with aqueous ammonia at 60°, the corresponding conversion of 2-chloro- to 2-aminoquinoxaline is carried out by heating with ethanolic ammonia at 150–155°.[83] 2-Aminoquinoxaline 1-oxide undergoes an unexpected reaction with benzenesulfonyl chloride which results in the formation of 2-amino-3-benzenesulfonyloxyquinoxaline (**50**).[84]

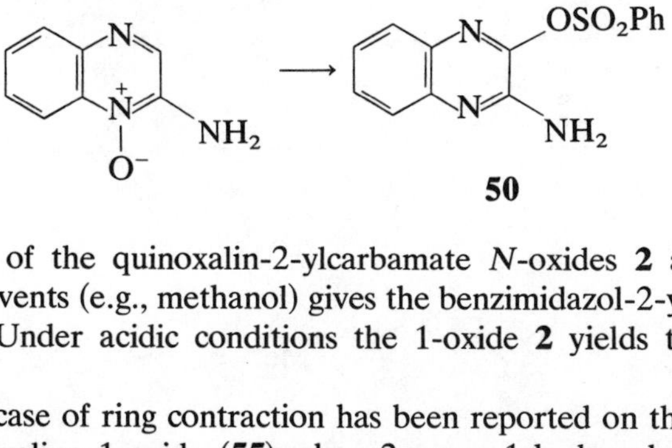

50

Photolysis of the quinoxalin-2-ylcarbamate *N*-oxides **2** and **51** in a variety of solvents (e.g., methanol) gives the benzimidazol-2-ylcarbamates **52** and **53**. Under acidic conditions the 1-oxide **2** yields the isonitrile **54**.[6]

A further case of ring contraction has been reported on thermolysis of 2-azidoquinoxaline 1-oxide (**55**) when 2-cyano-1-hydroxybenzimidazole (**56**) is formed.[85] Ring contraction also occurs on irradiation of quinoxaline di-*N*-oxides of the type **57**, the product of reaction being a 1,3-disubstituted benzimidazolone **58**.[86]

2-Methyl-3-methylthioquinoxaline 1,4-dioxide (**59**) is readily oxidized to the corresponding sulfoxide **60** or sulfon **61** with *m*-chloroperbenzoic acid. Treatment of either the sulfoxide or sulfone with hydrochloric or hydrobromic acid yields the corresponding 2-chloro- or

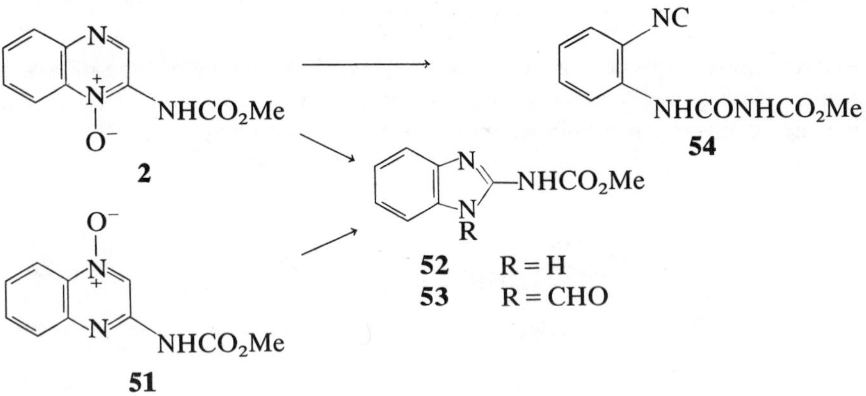

NC

$NHCO_2Me$ $NHCONHCO_2Me$

2 **54**

$NHCO_2Me$

52 R = H
53 R = CHO

$NHCO_2Me$

51

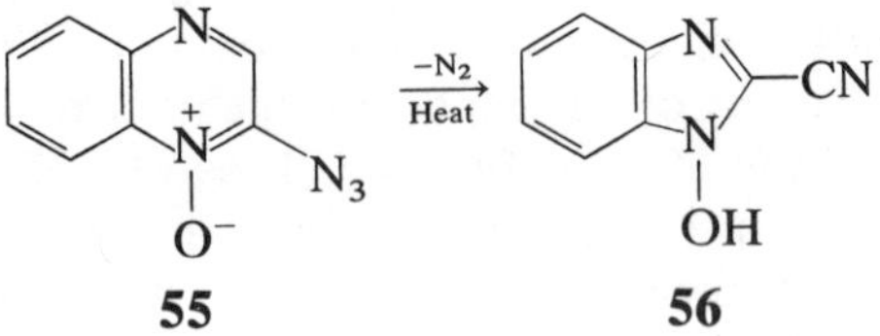

N_3 $\xrightarrow[\text{Heat}]{-N_2}$ CN

OH

55 **56**

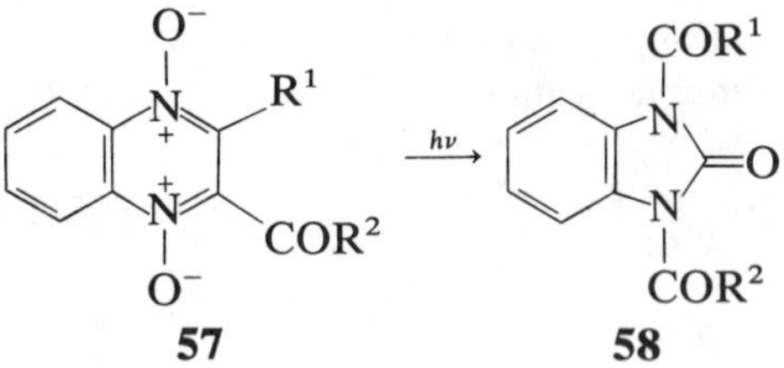

R^1 $\xrightarrow{h\nu}$ COR^1

COR^2 COR^2

57 **58**

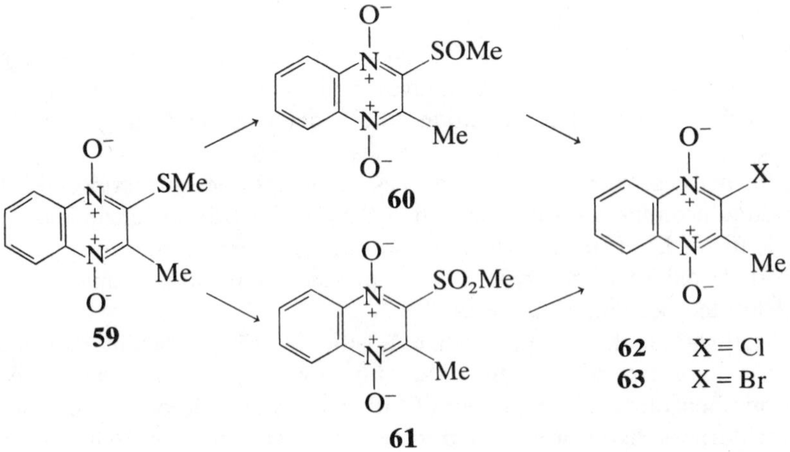

SOMe

Me

60

SMe X

Me

59 Me

SO_2Me

Me

61

62 X = Cl
63 X = Br

2-bromoquinoxaline 1,4-dioxides **62** and **63**. The latter compounds are not accessible by alternative methods.[87]

3. Reactions of 1,2-Dihydro-2-oxoquinoxaline 4-Oxides

N-oxides of this type are accessible from *o*-nitroaniline-based syntheses (see earlier section). 1,2-Dihydro-2-oxoquinoxaline 4-oxide (**64**) and its *N*-methyl derivative (**65**) condense with compounds containing an active methylene group in the presence of piperidine, yielding 3-alkylated quinoxalinones (**66**). The oxide **64**, when warmed with aqueous potassium cyanide, affords 3-cyano-1,2-dihydro-2-oxoquinoxaline, whereas oxide **65** gives 1,2-dihydro-3-hydroxy-2-oxo-1-methylquinoxaline.[23]

+ XCH_2R^2 ⟶ **66** (CH_2R^2, R^1)

64 $R^1 = H$
65 $R^1 = Me$

$R^2 = CO_2Et$, COMe, or CN
X = COMe or COPh

Attempts to deoxygenate 3-cyano-1,2-dihydro-2-oxoquinoxaline 4-oxide (**67**) failed, and in all cases 1,2-dihydro-2-oxoquinoxaline was produced. Treatment of compound **67** with aniline, *N*-methylaniline, or cyclohexylamine results in deoxygenation and displacement of the cyano group to give compounds of type **68**.[88]

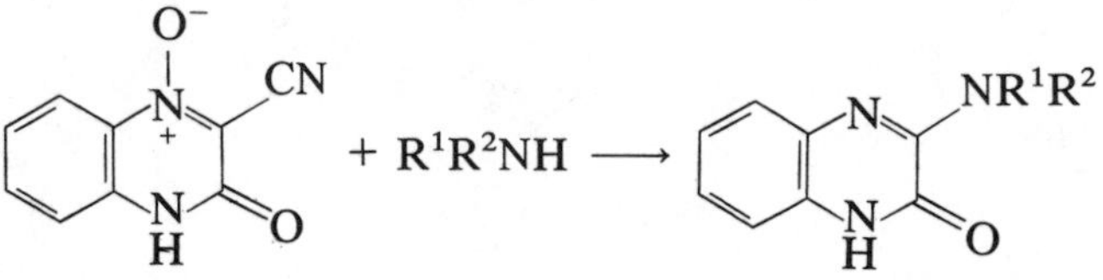

67 + R^1R^2NH ⟶ **68** (NR^1R^2)

Compounds **64** and **65** act as 1,3-dipoles and undergo cycloaddition with arylisocyanates and benzyne. Addition of arylisocyanates yields arylaminoquinoxalinones, and addition of benzyne gives *o*-hydroxyphenylquinoxalinones (Scheme 17).[89]

Attempted acylation of 1,2-dihydro-2-oxo-3-phenylquinoxaline 4-oxide (**69**) with acetyl chloride results in nuclear chlorination and loss of

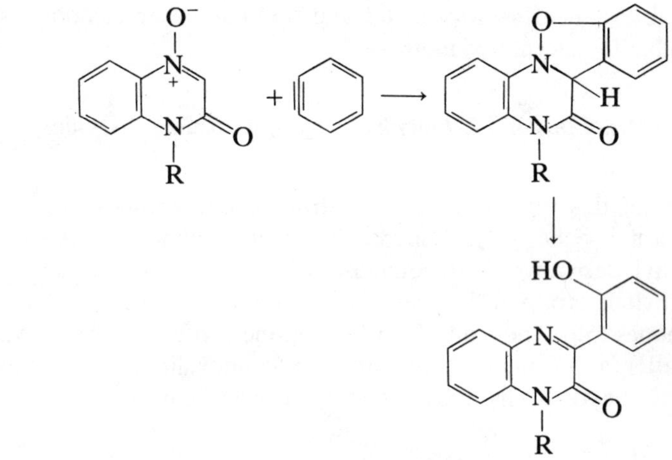

Scheme 17

the *N*-oxide function to give compound **70**. A similar transformation can be carried out with a 1:1 mixture of concentrated hydrochloric acid and glacial acetic acid.[90] Reaction with acetic anhydride does not give the anticipated 7-acetoxy derivative; instead, ring contraction occurs and the benzimidazolinones **71** and **72** are formed.[91] However when the *N*-methyl derivative **73** is treated with acetic anhydride, a *C*-acetoxy derivative, **74**, is formed with simultaneous loss of the *N*-oxide function.[91] It is probable that *N*-acetoxy derivatives are involved in these reactions, and when **73** is treated at room temperature with acetic anhydride in the presence of perchloric acid, the crystalline perchlorate **75** is obtained. Treatment of **75** with sodium acetate gives the 7-acetoxy derivative **76**

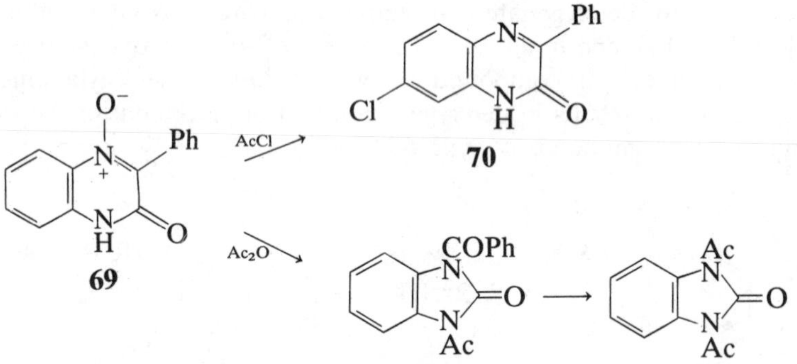

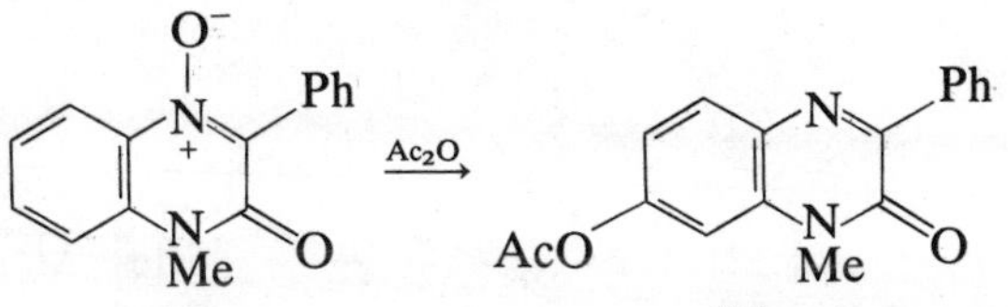

73 74

(X = OAc). Similarly, using lithium chloride, alcohols and amines, 7-chloro-, 7-alkoxy-, and 7-amino derivatives are formed.[92]

The mechanism suggested for these reactions is shown in Scheme 18.

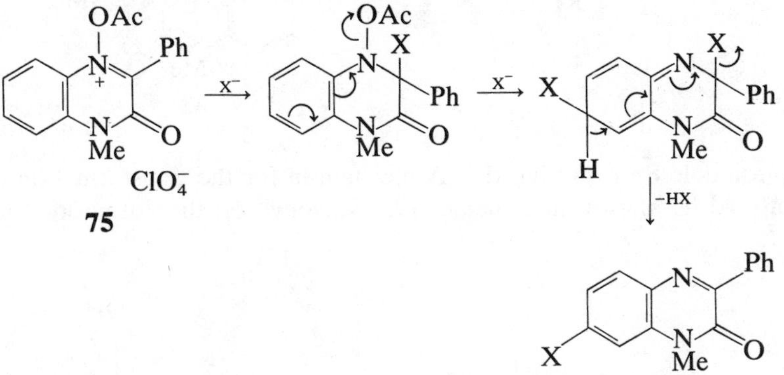

75 76

Scheme 18

It has also been found that when the 7-methylquinoxalin-2-one 4-oxides **77** and **78** are treated with acetic anhydride, the 7-acetoxymethyl derivatives **79** and **80** are obtained. A similar mechanism is invoked.[93]

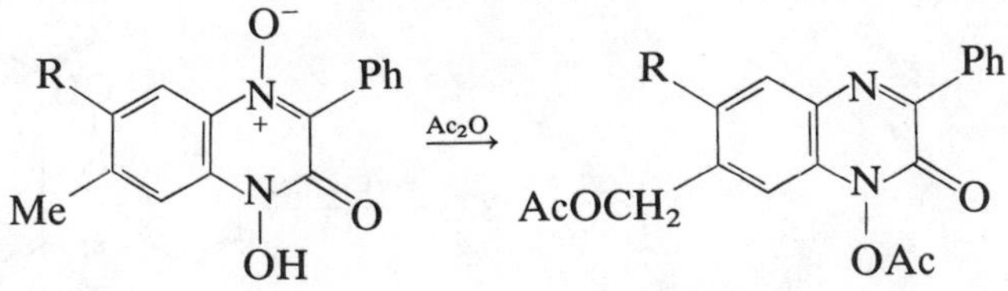

77 R = H
78 R = Me
79 R = H
80 R = Me

The *N*-oxide **81** rapidly evolves carbon dioxide when treated with concentrated sulfuric acid at 0°, and the *o*-methylaminophenylquinoxaline **82** is formed.[19,94] With ethanolic hydrogen chloride, the chlorinated spiroindole **83** is produced.[95] A mechanism for the transformation of **81** into **82** is shown in Scheme 19. Analogy for the formation of the

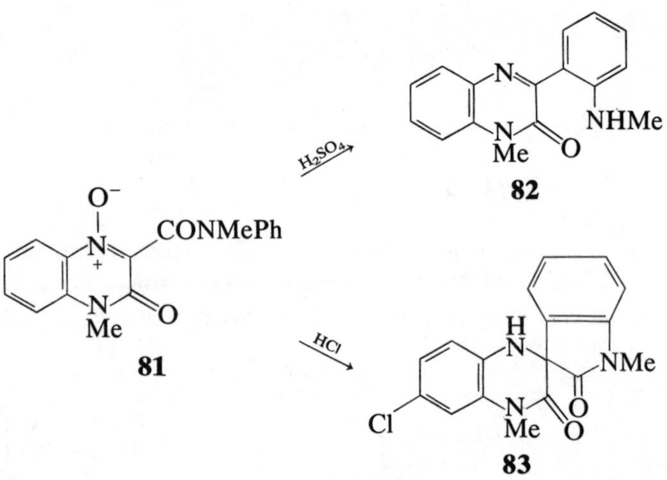

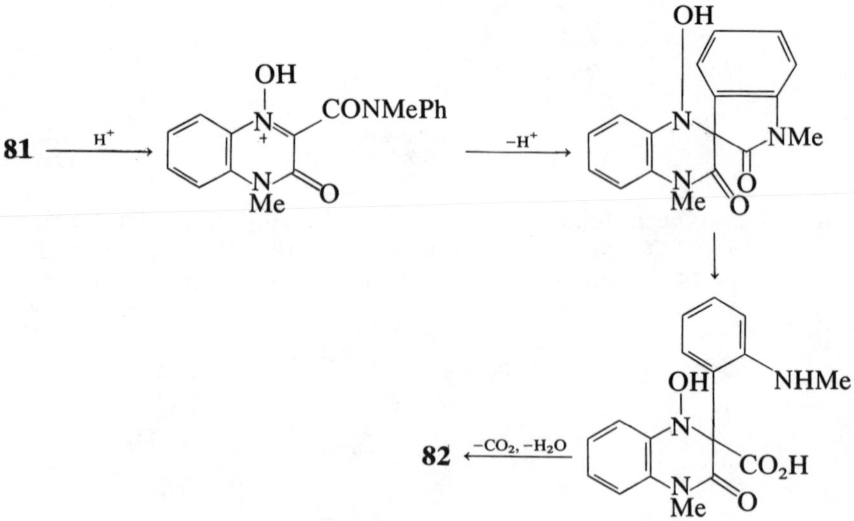

Scheme 19

chlorinated spiroindole may be derived from the transformation of the *N*-oxide **84** to the chloroquinoxalinone **85** on treatment with ethanolic hydrogen chloride.[10]

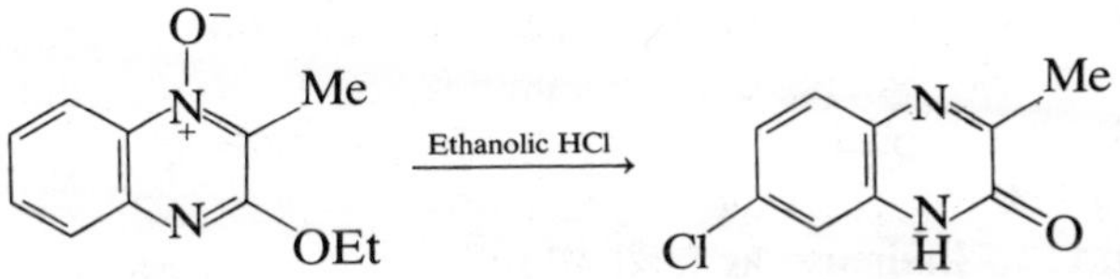

III. Uses

Various metal salts of 2-mercaptoquinoxaline 1-oxide such as the zinc salt have been patented for treating the bacteria and fungi on the scalp and have been incorporated in shampoos.[96] Quinoxaline 1,4-dioxides are active against a wide range of Gram-positive and Gram-negative bacteria, and they have been added to animal feeds both as bacteriocides and growth-promoting agents.

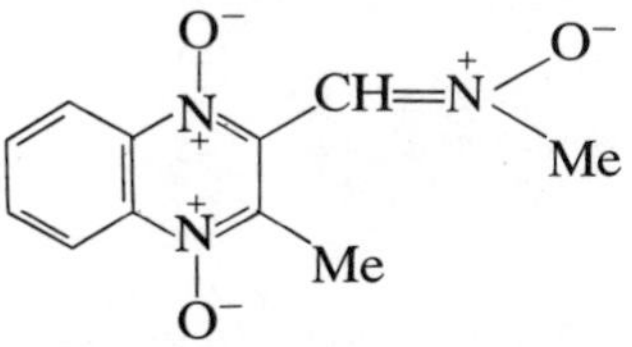

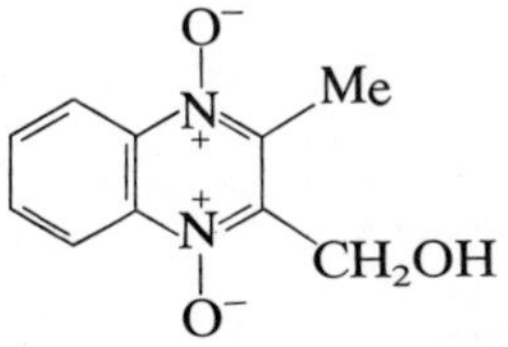

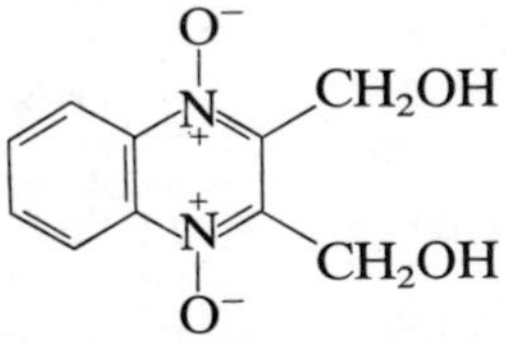

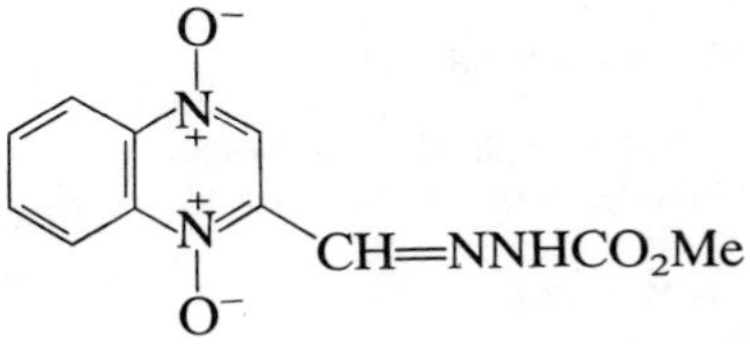

Quinoxaline 1,4-dioxide (Quindoxin) has higher antibacterial activity under *in vivo* than under *in vitro* conditions. It has been shown that its *in vitro* activity is enhanced 10 to 100-fold under anaerobic conditions. Derivatives of 3-methylquinoxaline 2-carboxaldehyde 1,4-dioxide such as the *N*-methylnitrone **86** also have higher *in vivo* activity than would be expected from *in vitro* tests.[97] The high *in vivo* activity of 2,3-dimethylquinoxaline 1,4-dioxide is due to the formation of active metabolites, the mono- and dihydroxymethyl derivatives **87** and **88**.[98] However the hydroxymethyl derivative of **86** did not show the expected enhanced activity.[97] Compound **88** is referred to in the literature as Dioxidine, and 2,3-bis(acetoxymethyl)quinoxaline 1,4-dioxide is known as Quinoxidine. The trade name for the 2-carboxaldehyde 1,4-dioxide derivative (**89**) is Carbadox.[99] Of the numerous quinoxaline 1,4-dioxide derivatives patented in the literature for antibacterial properties, most are 2-carboxylic acids, 2-carboxylic acid esters, 2-carboxamides, or 2-carboxaldehydes. Substituents in the 3-position include methyl, alkylthio, and iminomethyl groups and heterocyclic rings.[100,101]

In addition to their use as feed additives,[5,102] quinoxaline 1,4-dioxides have been patented for the treatment of diseases in aquatic animals[103] and as fungicides and herbicides.[6,104] There is also a report of the prophylactic effect of 6-chloro-2,3-dimethylquinoxaline 1,4-dioxide and 2,3-dimethylquinoxaline 1,4-dioxide on radiation injuries in mice.[105]

IV. Tables

TABLE 1. QUINOXALINE MONO-N-OXIDES

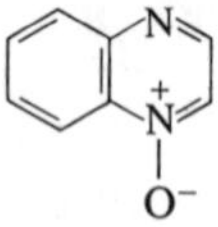

Molecular formula	Substituents or structure[a]	m.p. °C	Ref.
$C_8H_5ClN_2O$	2-Chloro 1-oxide	114, 115–116	17, 64, 74
	2-Chloro 4-oxide	147–148, 150–152, 151–153	3, 4, 8
	5-Chloro 1-oxide	177–179	3
	6-Chloro 1-oxide	136–138	8
	7-Chloro 1-oxide	151–152	8
$C_8H_5ClN_2O_2$	7-Chloro-2-hydroxy 1-oxide	248 decomp.	44
	6-Chloro-1,2-dihydro-2-oxo 4-oxide	275	25
$C_8H_5ClN_2O_3$	6-Chloro-1,2-dihydro-3-hydroxy-2-oxo 4-oxide	>360	27
$C_8H_5N_5O$	2-Azido 1-oxide	*e*	85
$C_8H_6N_2O$	Unsubstituted	122–123, 124–125	3, 45
	Picrate	184–185 decomp.	45
	Methiodide	188–189[b]	3
$C_8H_6N_2OS$	1,2-Dihydro-2-thioxo 4-oxide	183	25
	2-Mercapto 1-oxide		96, 106
$C_8H_6N_2O_2$	2-Hydroxy 1-oxide	208–209, 209–210, 211–212	45, 64, 83
	Benzoyl	162–163	64
	1,2-Dihydro-2-oxo 4-oxide	270 decomp., 273–274, 276	8, 11, 17, 22, 25, 45
$C_8H_6N_2O_3$	1,2-Dihydro-1-hydroxy-2-oxo 4-oxide	255–257	42
	3,4-Dihydro-2-hydroxy-3-oxo 1-oxide	290 decomp., 292	17, 22
	1-Acetoxy	204	22
$C_8H_7N_3O$	2-Amino 1-oxide	187–188, 191–192	9, 83
	Benzoyl	143–144	84
	Benzenesulfonyl	224–225	84
	Dichloroacetyl	159–160	84

TABLE 1 (*contd.*)

Molecular formula	Substituents or structure[a]	m.p., °C	Ref.
$C_8H_7N_3O$ (*contd.*)	Ethoxycarbonyl	127–129	6
	Methoxycarbonyl	178–179, 188–189	6, 84
	Trichloroacetyl	145–146	84
	Trifluoroacetyl	185–186	84
	2-Amino 4-oxide	276 decomp., 278 decomp.	9, 64
	Hydrochloride	242.5	107
	Ethoxycarbonyl	128–130	6
	Methoxycarbonyl	182–183	6
$C_8H_7N_3O_2$	3-Amino-2-hydroxy 1-oxide	282 decomp.	9
	3-Amino-1,2-dihydro-2-oxo 4-oxide	300 decomp., 313 decomp. 320, 350–355 decomp.	9, 22, 26, 108
$C_9H_4ClN_3O$	3-Chloro-2-cyano 1-oxide		25
$C_9H_4ClN_3O_2$	7-Chloro-2-cyano-3,4-dihydro-3-oxo 1-oxide	285 decomp., 287 decomp.	25, 27
$C_9H_5ClN_2O_4$	2-Carboxy-7-chloro-3,4-dihydro-3-oxo 1-oxide		
	Amide	275	25
$C_9H_5ClN_2O_4$	2-Carboxy-6-chloro-3-hydroxy 4-oxide	255	44
$C_9H_5N_3O$	2-Cyano 1-oxide	130, 135	4, 65, 109
	2-Cyano 4-oxide	163	4
$C_9H_5N_3O_2$	2-Cyano-3,4-dihydro-3-oxo 1-oxide	265, 278 decomp., 282	24, 25, 27
$C_9H_6Br_2N_2O$	2-Dibromomethyl 4-oxide	191–192 decomp.	83
$C_9H_6N_2O_2$	2-Formyl 1-oxide	129	4
	Oxime	183	4
	2-Formyl 4-oxide	169	4
	2,4-Dinitrophenylhydrazone	286.5	80
	Oxime	236	4
$C_9H_6N_2O_3$	2-Carboxy 1-oxide	164–165, 180 decomp., 181	4, 46, 109
	Methyl ester	104–105	109
	Amide	224, 227–228	4, 109
	Anilide	159	20
	N-Methylanilide	198–199	94
	2-Carboxy 4-oxide	180–182, 186.5–187	46, 80
	Amide	230–230.5, 234	4, 46
	Ethyl ester	116–117	46
	Hydrazide	216–217	46
	N-Methylanilide	150	20
$C_9H_6N_2O_4$	2-Carboxy-3,4-dihydro-3-oxo 1-oxide		
	Ethyl ester	193	24
	Amide	247, 259 decomp., >325	24, 25, 27
	2-Carboxy-3-hydroxy 4-oxide	175–176	44
	6-Bromo-7-methyl oxide[c]	167–168	3
$C_9H_7ClN_2O$	2-Chloro-3-methyl 4-oxide	94	4
	3-Chloro-7-methyl 1-oxide	142–143	8
	6-Chloro-7-methyl 1-oxide	166–168, 168–169	3, 8
$C_9H_7ClN_2O_2$	3-Chloro-7-methoxy 1-oxide	150–152	8
	6-Chloro-7-methoxy 1-oxide	190–192	8

TABLE 1 (*contd.*)

Molecular formula	Substituents or structure[a]	m.p., °C	Ref.
$C_9H_8N_2O$	2-Methyl 1-oxide	93–94	45
	2-Methyl 4-oxide	118	4
	5-Methyl 1-oxide		
	Monohydrate	131–132	3
$C_9H_8N_2O_2$	1,2-Dihydro-1-methyl-2-oxo 4-oxide	207–208, 208–209, 210–211	11, 17, 45
	1,2-Dihydro-3-methyl-2-oxo 4-oxide	222 decomp., 250	24, 25
	1,2-Dihydro-6-methyl-2-oxo 4-oxide	252	25
	2-Hydroxymethyl 4-oxide	155–156	45
	2-Methoxy 1-oxide	135–136	110
	2-Methoxy 4-oxide	105–107, 106–108	4, 18
$C_9H_8N_2O_3$	1,2-Dihydro-3-hydroxy-1-methyl-2-oxo 4-oxide	253 decomp., 253–255	17, 24
	Acetyl	177	24
	1,2-Dihydro-1-hydroxy-3-methyl-2-oxo 4-oxide	224–225, 231–232	42, 87
	1-Acetyl	142–143	87
	1-*m*-Chlorobenzoyl	161–162	87
	1,2-Dihydro-6-methoxy-2-oxo 4-oxide	275	25
$C_9H_8N_2O_3S$	2-Methylsulfonyl 4-oxide	197–199	11
$C_9H_8N_2O_4$	1,2-Dihydro-3-hydroxy-6-methoxy-2-oxo 4-oxide	325–327 decomp.	27
$C_9H_9N_3O$	2-Amino-3-methyl 4-oxide	208–209	9
$C_9H_9N_3O_2$	3-Amino-1,2-dihydro-1-methyl-2-oxo 4-oxide	314 decomp.	24
$C_{10}H_6BrF_3N_2O$	2-Bromomethyl-3-trifluoromethyl 1-oxide	130–131	54
$C_{10}H_6N_2O_5$	2,3-Dicarboxy 1-oxide		
	Diethyl ester	105–106	41
$C_{10}H_7F_3N_2O$	2-Methyl-3-trifluoromethyl 1-oxide	142–144	54
$C_{10}H_7N_3O_2$	2-Cyano-3,4-dihydro-4-methyl-3-oxo 1-oxide	210–211, 223	24, 27
	2-Cyano-3,4-dihydro-7-methyl-3-oxo 1-oxide	203 decomp.	25
	2-Cyano-3-methoxy 1-oxide	137	4
$C_{10}H_7N_3O_3$	2-Cyano-3,4-dihydro-7-methoxy-3-oxo 1-oxide	251, 266–267	25, 27
$C_{10}H_8Br_2N_2O$	2,3-Bis(bromomethyl) 1-oxide	167–168	16
$C_{10}H_8Cl_2N_2O$	6,7-Dichloro-2,3-dimethyl 1-oxide	238	13
$C_{10}H_8N_2O_3$	2-Acetoxy-1-oxide	110.5–111, 111–112	45, 73
	2-Acetyl-3,4-dihydro-3-oxo 1-oxide	188	22
	2-Carboxy-3-methyl oxide		
	Ethyl ester[c]	93–94	16
	Hydrazide[c]	224–225 decomp.	16
	2-Carboxy-3-methyl oxide[c]		149
	3-Formyl-2-hydroxymethyl 1-oxide[d]	212.5	80
	2-Formyl-3-methoxy 1-oxide	144	4
	Oxime	217 decomp.	4
$C_{10}H_8N_2O_4$	2-Carboxy-3,4-dihydro-4-methyl-3-oxo 1-oxide		
	Amide	256	24
	N-Methylanilide	189–190, 191	19, 24
	2-Carboxy-3,4-dihydro-7-methyl-3-oxo 1-oxide		
	Amide	235	25
	2-Carboxy-3-hydroxy-7-methyl 4-oxide	180	44
	2-Carboxy-3-methoxy-1-oxide		
	Amide	226	4
$C_{10}H_8N_2O_5$	2-Carboxy-3,4-dihydro-7-methoxy-3-oxo 1-oxide		
	Amide	245 decomp.	25

TABLE 1 (*contd.*)

Molecular formula	Substituents or structure[a]	m.p., °C	Ref.
$C_{10}H_9ClN_2O$	2-Chloromethyl-3-methyl oxide[c]	173–174	150
	2-Chloromethyl-3-methyl 4-oxide	181–182	111
	5-Chloro-2,3-dimethyl 1-oxide	114–116	13
$C_{10}H_9ClN_2O_2$	3-Chloro-7-ethoxy 1-oxide	143–145	8
	6-Chloro-7-ethoxy 1-oxide	188–189	8
$C_{10}H_9N_3O_2$	2-Acetamido 1-oxide	193–194	9
	2-Acetamido 4-oxide	239–240	9
	5-Acetamido oxide[c]	175–178	3
$C_{10}H_9N_3O_3$	2,3-Dimethyl-5-nitro 1-oxide	157–158	13
$C_{10}H_{10}N_2O$	2,3-Dimethyl 1-oxide	91, 93–94	4, 13, 112
	4-Methyl triiodide	144–145	13
	5,6-Dimethyl oxide[c]	152–153	113
	2-Ethyl 1-oxide	64.5–65.5	114
	2-Ethyl 4-oxide	84.5–85.5	114
$C_{10}H_{10}N_2O_2$	2-Ethoxy 4-oxide	92–93	4
	5-Ethoxy 1-oxide	114–116	3
	2-Ethyl-3,4-dihydro-3-oxo 1-oxide	212	115
	1,2-Dihydro-1,3-dimethyl-2-oxo 4-oxide	200–202, 207	17, 24, 69
	5-Hydroxy-2,3-dimethyl 1-oxide	143–144.5	7
	2-Methoxy-3-methyl 4-oxide	94.5	4, 69
$C_{10}H_{10}N_2O_3$	2,3-Dimethoxy 1-oxide		
	Monohydrate	157.5	4
	5,6-Dimethoxy 1-oxide	138–140	3
$C_{10}H_{10}N_2O_4$	6-Ethoxy-1,2-dihydro-3-hydroxy-2-oxo 4-oxide	285–286	27
$C_{11}H_8N_2O_4$	4-Acetyl-2-carboxy-3,4-dihydro-3-oxo 1-oxide	216–217	20
$C_{11}H_8N_2O_5$	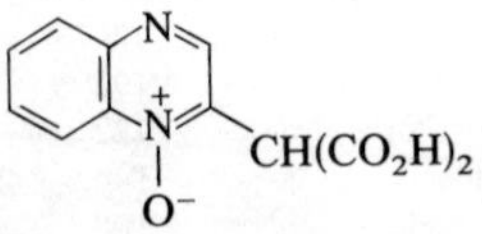 N, $\overset{+}{N}$, O^-, $CH(CO_2H)_2$		
	Diethyl ester	84.5–85	111
$C_{11}H_9N_3O$	6-Cyano-2,3-dimethyl 1-oxide	169–171	13
$C_{11}H_9N_3O_3$	2-Cyano-3,4-dihydro-7-ethoxy-3-oxo 1-oxide	246–248	27
$C_{11}H_{10}N_2O$	2-Methyl-3-vinyl 1-oxide	117–118	41
$C_{11}H_{10}N_2O_2$	2-Acetyl-3-methyl 4-oxide	92–94	13
$C_{11}H_{10}N_2O_3$	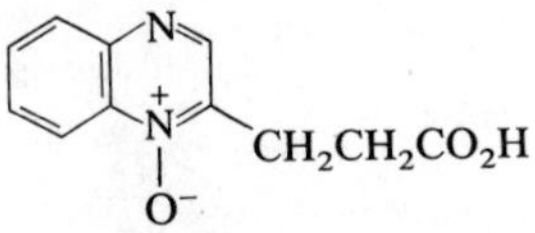 N, $\overset{+}{N}$, O^-, $CH_2CH_2CO_2H$	203–204 decomp.	116
	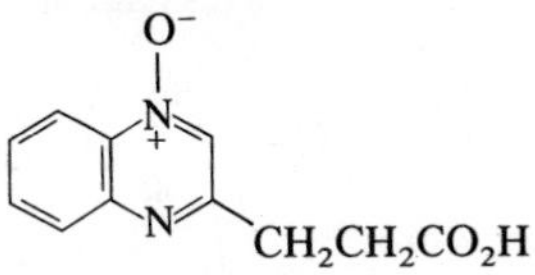 O^-, $\overset{+}{N}$, N, $CH_2CH_2CO_2H$	163–164	116
	Ethyl ester	73–74	116
	2-Acetoxymethyl 4-oxide	108–109	45
$C_{11}H_{10}N_2O_4$	2-Carboxy-3,4-dihydro-6,7-dimethyl-4-oxo 1-oxide		
	N-Methylanilide	266–267 decomp.	21

TABLE 1 (*contd.*)

Molecular formula	Substituents or structure[a]	m.p., °C	Ref.
$C_{11}H_{10}N_2O_4$ (*contd.*)	2-Carboxy-3-hydroxy-6,7-dimethyl 4-oxide	241–244	44
$C_{11}H_{10}N_2O_6$	2-Carboxy-3,4-dihydro-6,7-dimethoxy-3-oxo 1-oxide		
	N-Methylanilide	272–273 decomp.	21
$C_{11}H_{11}N_3O_2$	2-Acetamido-3-methyl 4-oxide	208–209	9
		119–120	109
$C_{11}H_{12}N_2O$	2-Ethyl-3-methyl 4-oxide	94	69
	2-Isopropyl 4-oxide	95–96	114
	2,3,5-Trimethyl 1-oxide	98–99	13
	2,6,7-Trimethyl oxide[c]	133–135	13
$C_{11}H_{12}N_2O_2$	2-Ethoxy-3-methyl 4-oxide	86	69
	2-Ethyl-3,4-dihydro-4-methyl-3-oxo 1-oxide	152	115
	1,2-Dihydro-3-propyl-2-oxo 4-oxide	210	115
	5-Methoxy-2,3-dimethyl 1-oxide	155–156	13
$C_{12}H_{12}N_2O_4$	2-Carboxy-3,4-dihydro-4,6,7-trimethyl-3-oxo 1-oxide		
	N-Methylanilide	263–264 decomp.	21
$C_{12}H_{12}N_2O_6$	2-Carboxy-3,4-dihydro-6,7-dimethoxy-4-methyl-3-oxo 1-oxide		
	N-Methylanilide	250–251	21
$C_{12}H_{14}N_2O$	2-*t*-Butyl 4-oxide	99–100	114
	2-Isopropyl-3-methyl 4-oxide	72–73	13, 69
$C_{12}H_{14}N_2O_2$	3-Butyl-1,2-dihydro-2-oxo 4-oxide	192	115
	1,2-Dihydro-1-methyl-2-oxo-3-propyl 4-oxide	122	115
$C_{13}H_{13}N_3O_3$	2-Diacetylamino-3-methyl oxide[c]	144.5–146	9
$C_{13}H_{14}N_2O_3$	2-(1′-Acetoxyethyl)-3-methyl oxide	100	16
$C_{13}H_{16}N_2O_2$	3-Butyl-1,2-dihydro-1-methyl-2-oxo 4-oxide	110	115
	2-*t*-Butyl-3-methyl 4-oxide	78.5	69
	3,4-Dihydro-3-oxo-2-pentyl 1-oxide	195	115
$C_{14}H_7Cl_2N_3O_4$	6,7-Dichloro-1,2-dihydro-3-(*p*-nitrophenyl)-2-oxo 4-oxide	>310	28
$C_{14}H_7Cl_3N_2O_2$	6,7-Dichloro-3(*p*-chlorophenyl)-1,2-dihydro-2-oxo 4-oxide	310 decomp.	90
$C_{14}H_8ClN_3O_4$	6-Chloro-1,2-dihydro-3-(*p*-nitrophenyl)-2-oxo 4-oxide	>325	28
$C_{14}H_8Cl_2N_2O_2$	6-Chloro-3-(*p*-chlorophenyl)-1,2-dihydro-2-oxo 4-oxide	333 decomp.	90
	6,7-Dichloro-1,2-dihydro-3-phenyl-2-oxo 4-oxide	305–306	28
$C_{14}H_8N_4O_6$	1,2-Dihydro-3-(2,4-dinitrophenyl)-2-oxo 4-oxide	250	25
$C_{14}H_9BrN_2O_3$	7-Bromo-1,2-dihydro-1-hydroxy-2-oxo-3-phenyl 4-oxide	231	36
$C_{14}H_9ClN_2O$	3-Chloro-2-phenyl 1-oxide	126–127	4, 66
	6-Chloro-2-phenyl 4-oxide	196–198	53
	7-Chloro-2-phenyl 4-oxide	155–156	53

TABLE 1 (*contd.*)

Molecular formula	Substituents or structure[a]	m.p., °C	Ref.
$C_{14}H_9ClN_2O_2$	3-(*p*-Chlorophenyl)-1,2-dihydro-2-oxo 4-oxide	316–318 decomp.	90
	6-Chloro-1,2-dihydro-2-oxo-3-phenyl 4-oxide	262–263	28
$C_{14}H_9ClN_2O_3$	7-Chloro-1,2-dihydro-1-hydroxy-2-oxo-3-phenyl 4-oxide	228	36
$C_{14}H_9N_3O_3$	2-*p*-Nitrophenyl 4-oxide	267–268	117
$C_{14}H_9N_3O_4$	1,2-Dihydro-3-(*o*-nitrophenyl)-2-oxo 4-oxide	244	25
	1,2-Dihydro-3-(*p*-nitrophenyl)-2-oxo 4-oxide	>325	28
$C_{14}H_{10}N_2O$	2-Phenyl 1-oxide	154	12
	2-Phenyl 4-oxide	137–138	13
$C_{14}H_{10}N_2O_2$	1,2-Dihydro-2-oxo-3-phenyl 4-oxide	278, 285, 307	4, 24, 28
	2-Hydroxy-3-phenyl 1-oxide	204	4
	3,4-Dihydro-3-oxo-2-phenyl 1-oxide	305–306	66
	2-(*o*-Hydroxyphenyl) 1-oxide	229–230	40
	2-(*p*-Hydroxyphenyl) 4-oxide	234–235	117
$C_{14}H_{10}N_2O_3$	1,2-Dihydro-1-hydroxy-2-oxo-3-phenyl 4-oxide	196	36
$C_{14}H_{11}ClN_2O_2$	6-Chloro-3-(*o*-hydroxyphenyl) 1-oxide	300–302 decomp.	40
$C_{14}H_{12}N_4O_3S$	2-(*p*-Aminobenzenesulphonamido) 1-oxide	218–219	83
	Acetyl	236–237	83
$C_{14}H_{14}N_2O_5$	2,3-Bis(acetoxymethyl) 1-oxide	125–126	112
$C_{14}H_{18}N_2O_2$	1,2-Dihydro-3-hexyl-2-oxo 4-oxide	169	115
	1,2-Dihydro-1-methyl-2-oxo-3-pentyl 4-oxide	105	115
$C_{15}H_9N_3O$	2-Cyano-3-phenyl 1-oxide	200.5	118
	2-Cyano-3-phenyl 4-oxide	183–184	118
$C_{15}H_{10}N_2O_2$	2-Formyl-3-phenyl 1-oxide	175–176	118
	Oxime	210–211 decomp.	118
$C_{15}H_{10}N_2O_3$	2-Benzoyl-3,4-dihydro-3-oxo 1-oxide	274	22
	2-Benzoyloxy 1-oxide	162–163	64
	2-Carboxy-3-phenyl 1-oxide		
	Amide	277	4, 118
	Hydrate of amide	236	118
	2-Carboxy-3-phenyl 4-oxide		
	Amide	223–224	118
$C_{15}H_{11}ClN_2O_2$	7-Chloro-1,2-dihydro-1-methyl-2-oxo-3-phenyl 4-oxide	191–192	90
$C_{15}H_{11}N_3O_5$	1,2-Dihydro-6-methoxy-3-(*p*-nitrophenyl)-2-oxo 4-oxide	278 decomp.	28
$C_{15}H_{12}N_2O$	2-Methyl-3-phenyl 1-oxide	106, 107	4, 69
	6-Methyl-2-phenyl 4-oxide	125–126	53
	7-Methyl-2-phenyl 4-oxide	124–125	53
	8-Methyl-2-phenyl 4-oxide	186–188	53
$C_{15}H_{12}N_2O_2$	2-Methoxy-3-phenyl 4-oxide	107	4
	1,2-Dihydro-1-methyl-2-oxo-3-phenyl 4-oxide	192–193, 196	24, 28
$C_{15}H_{12}N_2O_3$	1,2-Dihydro-1-hydroxy-7-methyl-2-oxo-3-phenyl 4-oxide	218	93
	1,2-Dihydro-6-methoxy-2-oxo-3-phenyl 4-oxide	297–300	28
$C_{15}H_{12}N_2O_4$	1,2-Dihydro-1-hydroxy-7-methoxy-3-phenyl-2-oxo 4-oxide	243	36
$C_{15}H_{20}N_2O_2$	3-Heptyl-1,2-dihydro-2-oxo 4-oxide	150	115
	3-Hexyl-1,2-dihydro-1-methyl-2-oxo 4-oxide	105	115
$C_{16}H_{12}N_2O_3$	2-Benzoyl-3,4-dihydro-4-methyl-3-oxo 1-oxide	193	22

TABLE 1 (*contd.*)

Molecular formula	Substituents or structure[a]	m.p., °C	Ref.
$C_{16}H_{12}N_2O_3$ (*contd.*)	2-Benzoyl-3-hydroxymethyl 1-oxide	164–165	47, 81
$C_{16}H_{13}N_3O_5$	6-Ethoxy-1,2-dihydro-3-(*p*-nitrophenyl)-2-oxo 4-oxide	>310	28
$C_{16}H_{14}N_2O$	2-Ethyl-3-phenyl 1-oxide	130.5	4
	2-Ethyl-3-phenyl 4-oxide	127–128	114
$C_{16}H_{14}N_2O_2$	2-Benzyloxy-3-methyl 4-oxide	72	69
$C_{16}H_{14}N_2O_3$	6-Ethoxy-1,2-dihydro-3-phenyl-2-oxo 4-oxide	290–292	28
	1,2-Dihydro-1-hydroxy-6,7-dimethyl-3-phenyl-2-oxo 4-oxide	225	93
$C_{16}H_{16}N_2O_7$	2-Acetoxymethyl-3-diacetoxymethyl 1-oxide	136–137	73
$C_{16}H_{22}N_2O_2$	3-Heptyl-1,2-dihydro-1-methyl-2-oxo 4-oxide	101	115
$C_{17}H_{13}ClN_2O$	2-(*p*-Chlorostyryl)-3-methyl 1-oxide	174	79
$C_{17}H_{13}N_3O_3$	2-(*p*-Nitrostyryl)-3-methyl 1-oxide	196	79
$C_{17}H_{14}N_2O$	2-Methyl-3-styryl 4-oxide	177	79
$C_{17}H_{14}N_2O_3$	1,2-Dihydro-1-methyl-2-oxo-3-phenacyl 4-oxide	171	69
$C_{17}H_{16}N_2O$	2-Isopropyl-3-phenyl 1-oxide	119.5	4
	2-Isopropyl-3-phenyl 4-oxide	137	4, 114
$C_{17}H_{16}N_2O_2$	1,2-Dihydro-1,6,7-trimethyl-2-oxo-3-phenyl 4-oxide	200	93
$C_{18}H_{14}N_2O_4$	2-Acetoxymethyl-3-benzoyl 4-oxide	177–178	47, 81
$C_{18}H_{16}N_2O_2$	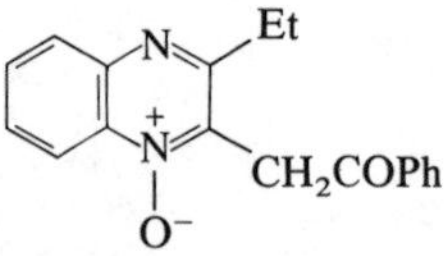 N, Et, $\overset{+}{N}$, CH_2COPh, O^-	149	69
$C_{18}H_{18}N_2O$	2-*t*-Butyl-3-phenyl 1-oxide		151
$C_{20}H_{14}N_2O$	2,3-Diphenyl 1-oxide	197, 207	12, 13, 70
$C_{21}H_{14}N_2O_2$	2-Benzoyl-3-phenyl 1-oxide	158	4
$C_{21}H_{16}N_2O$	6-Methyl-2,3-diphenyl oxide[c]	178–180	70
$C_{22}H_{16}N_2O_2$	2-Benzyl-3-benzoyl 1-oxide	139–140	119
$C_{22}H_{18}N_2O_3$	2,3-Bis(*p*-methoxyphenyl) 1-oxide	156	13
$C_{22}H_{34}N_2O$	2-Methyl-3-tridecyl 1-oxide	83–85	120

[a] Because of the possibility of isomerism, the position of the *N*-oxide function is specified where it is known for each table entry.

[b] 1-Methylquinoxalinium iodide 4-oxide.

[c] Structure not certain.

[d] Exists as the internal hemiacetal.

[e] Thermally unstable

TABLE 2. QUINOXALINE DI-*N*-OXIDES

Molecular formula	Substituents or structure	m.p., °C[a]	Ref.
$C_8H_4Br_2N_2O_3$	6,8-Dibromo-5-hydroxy	204–206 decomp.	5
$C_8H_4Cl_2N_2O_2$	6,7-Dichloro	206–208	3
$C_8H_4Cl_2N_2O_3$	6,8-Dichloro-5-hydroxy	195–197 decomp.	5
$C_8H_5BrN_2O_2$	6-Bromo	223–225	3
$C_8H_5ClN_2O_2$	6-Chloro	211–212	3
$C_8H_5ClN_2O_3$	6-Chloro-5-hydroxy	202–204 decomp.	5
	8-Chloro-5-hydroxy	206–207 decomp.	5
$C_8H_6N_2O_2$	Unsubstituted	241–243	3
$C_8H_6N_2O_3$	2-Hydroxy[b]	240–245, 255–257	42, 121
	5-Hydroxy	234 decomp.	5
	6-Hydroxy	247–250 decomp.	7
$C_8H_6N_2O_4$	2,3-Dihydroxy	261–262.5 decomp.	9
	5,8-Dihydroxy	233–234 decomp.	5
$C_8H_7N_3O_2$	2-Amino	242 decomp.	9
	Hydrochloride	248	107
	5-Amino		122
	6-Amino	245 decomp.	13
	Hydrochloride	208 decomp.	13
$C_8H_7N_3O_3$	3-Amino-2-hydroxy	300 decomp.	9
$C_8H_8N_4O_2$	2,3-Diamino		152
$C_9H_5ClN_2O_4$	6-Chloro-2-carboxy		1
	N-Dodecylamide		129
	N-Hydroxyamide		123, 124
$C_9H_5F_3N_2O_2$	2-Trifluoromethyl	190–192	54
$C_9H_6F_3N_3O_2$	2-Amino-3-trifluoromethyl		125
$C_9H_6N_2O_3$	2-Formyl	189–190	46, 75
	Dimethylacetal	195–195.5	78, 126
	Thiosemicarbazide	262 decomp.	78
$C_9H_6N_2O_4$	2-Carboxy	208–209, 221 decomp.	46, 78, 111
	Amide	285–287	130
	N-(4-Methyl-1-piperazinyl) methylamide	167–168	111
	Anilide	211	20
	N-Methylanilide	223	94
$C_9H_6N_4O_2$	2-Amino-3-cyano	232 decomp.	127
$C_9H_7BrN_2O_2$	2-Bromo-3-methyl	163–164	87
	6-Bromo-7-methyl	222–224	3
	2-Bromomethyl	160–161, 162, 163	16, 83, 128

TABLE 2 (*contd.*)

Molecular formula	Substituents or structure	m.p., °C[a]	Ref.
$C_9H_7ClN_2O_2$	2-Chloro-3-methyl	166–168	87
	6-Chloro-7-methyl	227	3
	7-Chloro-2-methyl	190–191	129
$C_9H_7ClN_2O_3$	6-Chloro-5-methoxy	204–206 decomp.	5
$C_9H_7N_3O_4$	3-Amino-2-carboxy		
	Amide	242–243	131
$C_9H_8N_2O_2$	2-Methyl	171, 180–181	4, 129
	5-Methyl	192–194	3
	6-Methyl	218–219	3
$C_9H_8N_2O_3$	2-Hydroxy-3-methyl	216–217 decomp.	9, 121
	2-Hydroxymethyl	191–192, 198–199	16, 121
	Acetyl	188–189	16
	2-Methoxy	171–172	110
	5-Methoxy	222	3
	6-Methoxy	227–228	3
$C_9H_8N_2O_4$	5-Hydroxy-8-methoxy	186–189	5
$C_9H_9N_3O_2$	2-Amino-3-methyl	215–216 decomp.	9
	2-Aminomethyl		
	Picrate	154.5	109
$C_9H_9N_3O_5S$	5-Hydroxy-8-methylsulfonylamino	211–214	5
$C_{10}H_4N_4O_2$	2,3-Dicyano	228 decomp.	65
$C_{10}H_5ClN_2O_6$	2,3-Dicarboxy-6-chloro		
	Diamide	300 decomp.	129
$C_{10}H_5F_3N_2O_3$	2-Formyl-3-trifluoromethyl		125
$C_{10}H_6BrF_3N_2O_2$	2-Bromomethyl-3-trifluoromethyl	153–154	54
$C_{10}H_6ClN_3O_2$	6-Chloro-2-cyano-3-methyl		132
	2-Chloromethyl-3-cyano	194	82
$C_{10}H_6F_4N_2O_2$	2-Fluoromethyl-3-trifluoromethyl		125
$C_{10}H_6N_2O_6$	2,3-Dicarboxy		152
	Diethyl ester		127
	Diamide	217 decomp.	129
$C_{10}H_7Br_2ClN_2O_2$	2,3-Bis(bromomethyl)-6-chloro	188.5–189	104
$C_{10}H_7ClN_2O_4$	2-Carboxy-6-chloro-3-methyl		
	Amide	232 decomp.	129, 134
	N-Dodecylamide		129
	2-Carboxy-7-chloro-3-methyl		
	Ethyl ester	178–179	129
	2-Carboxy-3-chloromethyl		
	Amide		135
	N-Methylamide	195–196	138
	N,N-Dimethylamide		135
$C_{10}H_7F_3N_2O_2$	2-Methyl-3-trifluoromethyl	135–135.5	54
$C_{10}H_7F_3N_2O_3$	2-Hydroxymethyl-3-trifluoromethyl	130–131	54
$C_{10}H_7N_3O_2$	2-Cyano-3-methyl	194	127, 132, 137

TABLE 2 (*contd.*)

Molecular formula	Substituents or structure	m.p., °C[a]	Ref.
$C_{10}H_8Br_2N_2O_2$	2,3-Bis(bromomethyl)	182.5–184, 193–194	16, 104. 128
	Hydrobromide	167	107
$C_{10}H_8Br_2N_2O_4$	6,7-Dibromo-5,8-dimethoxy	238.5–239.5	133
$C_{10}H_8Cl_2N_2O_2$	2,3-Bis(chloromethyl)		104
$C_{10}H_8N_2O_3$	2-Formyl-3-methyl	186–187	136
	Thiosemicarbazone	236 decomp.	78
	N-Methylnitrone		142
$C_{10}H_8N_2O_4$	2-Carboxy-3-methyl	167–169	16
	Ethyl ester	112–115, 132–134, 134–136	16, 41, 129, 139
	Amide	247 decomp.	16
	N-Hydroxyamide	210 decomp.	16
	N-Methylamide	214	142
	N-[3-(Dimethylamino)propyl]amide		134
	N,N-Dimethylamide		134
	N,N-Diethylamide		134, 142
	N-Cyclohexylamide		129, 145
	Anilide		145
	2-Formyl-3-hydroxymethyl[c]	215	97
	Oxime	185–186	97
	Semicarbazone	232	97
$C_{10}H_8N_2O_5$	2-Carboxy-3-hydroxymethyl		
	N-Methylamide		138, 143, 144
	N-Ethylamide		138, 143
	N,N-Diethylamide		138, 143
	Morpholide		143
$C_{10}H_9BrN_2O_2$	6-Bromo-2,3-dimethyl	186–188	13
	2-Bromomethyl-3-methyl	172–173	128
$C_{10}H_9ClN_2O_2$	6-Chloro-2,3-dimethyl	175–176	13, 129
$C_{10}H_9N_3O_3$	2-Acetamido	233 decomp.	9
	6-Acetamido	245–247	3
$C_{10}H_9N_3O_4$	8-Acetamido-5-hydroxy	230–231	5
	2,3-Dimethyl-6-nitro	192–194	13
$C_{10}H_{10}N_2O_2$	2-Ethyl	149, 150–152	13, 114
	2,3-Dimethyl	188–190, 192–193	4, 13, 32, 48, 121, 127
	Hydrochloride	179–180	107
	2,7-Dimethyl	183–184 decomp.	129
	6,7-Dimethyl	220	3
$C_{10}H_{10}N_2O_2S$	2-Methyl-3-methylthio	146–148	87
$C_{10}H_{10}N_2O_3$	6-Ethoxy	192–194	3
	5-Hydroxy-2,3-dimethyl	171–173	7
	6-Hydroxy-2,3-dimethyl	249–250 decomp.	7
	Nitro deriv.	258 decomp.	7

TABLE 2 (*contd.*)

Molecular formula	Substituents or structure	m.p., °C[a]	Ref.
$C_{10}H_{10}N_2O_3$ (*contd.*)	Bromo deriv.	180 decomp.	7
	2-(2′-Hydroxyethyl)	126–127	76
	2-Hydroxymethyl-3-methyl	172–173, 182–184	97, 98, 112
	Acetyl	117–118	97, 112
	6-Methoxy-2-methyl		15
	7-Methoxy-2-methyl	245 decomp.	129
$C_{10}H_{10}N_2O_3S$	2-Methyl-3-methylsulfinyl	201–202	87
$C_{10}H_{10}N_2O_4$	2,3-Bis(hydroxymethyl)	170–172	16
	Diacetyl	175–176, 176–177	16, 112, 128
	5,6-Dimethoxy	220–222	3
	6,7-Dimethoxy	264–265	3
$C_{10}H_{10}N_2O_4S$	2-Methyl-3-methylsulfonyl	153–154	87
$C_{10}H_{10}N_4O_3$	O^-, $\overset{+}{N}$, $\overset{+}{N}$, O^-; C(=NH)NHOH; Me	212	137
$C_{10}H_{11}BrN_4O_2S$	O^-, $\overset{+}{N}$, $\overset{+}{N}$, O^-; $CH_2SC(=\overset{+}{N}H_2)NH_2$ Br^-	190–191, 196	83, 109
$C_{10}H_{11}N_3O_2$	6-Amino-2,3-dimethyl	268 decomp.	13
	Hydrochloride	240 decomp.	13
	p-Toluenesulfonyl	276 decomp.	13
	2-Aminomethyl-3-methyl		
	Picrate	125 decomp.	109
	2-Dimethylamino	178–180	35
$C_{10}H_{11}N_3O_5S$	5-Methoxy-8-(methylsulfonylamino)		5
$C_{10}H_{11}N_5O_2$	O^-, $\overset{+}{N}$, $\overset{+}{N}$, O^-; $C(=NH)NHNH_2$; Me	191	137
$C_{11}H_8N_2O_4$	O^-, $\overset{+}{N}$, $\overset{+}{N}$, O^-; $CH=CHCO_2H$	219–220 decomp.	116

TABLE 2 (*contd.*)

Molecular formula	Substituents or structure	m.p., °C[a]	Ref.
$C_{11}H_8N_2O_6$	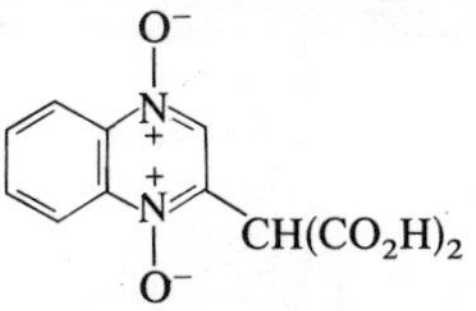 O^- / N^+ / N^+ / O^-, $CH(CO_2H)_2$		
	Diethyl ester	153.5–154.5	111
$C_{11}H_8N_2O_7$	2,3-Dicarboxy-6-methoxy		
	Diamide	222 decomp.	129
$C_{11}H_9F_3N_2O_2$	2-Ethyl-3-trifluoromethyl	107–108	54
	2,3-Dimethyl-6-trifluoromethyl	155–156	13
$C_{11}H_9F_3N_2O_3$	2-Methoxymethyl-3-trifluoromethyl	121–123	54
$C_{11}H_9F_3N_2O_3S$	2-Methylsulfinylmethyl-3-trifluoromethyl		125
$C_{11}H_9N_3O_2$	6-Cyano-2,3-dimethyl	216–218	13
$C_{11}H_{10}N_2O_3$	2-Acetyl-3-methyl	153–154	139
	6-Formyl-2,3-dimethyl	217–218	140
	N-2-Hydroxyethylnitrone	209 decomp.	140
	N-Methylnitrone	239 decomp.	140
	N-Phenylnitrone	211 decomp.	140
	Semicarbazone	281–283	140
$C_{11}H_{10}N_2O_4$	2-Carboxy-3,6-dimethyl		
	Amide	223 decomp.	129
	6-Carboxy-2,3-dimethyl	241 decomp. 243 decomp.	7, 41
	Ethyl ester	134–135	7
	Methyl ester	185–186	129
	Amide	266 decomp.	7
	N-Hydroxyamide hydrate	230–232	7
	N,N-Dimethylamide hydrate	215	7
	N-(3-Piperidinylpropyl)amide	172–173	7
	2-Carboxymethyl-3-methyl	188–190	41
	Ethyl ester	150–151	41, 145
	Amide	248–249	41
	Anilide		129
	2-Acetoxymethyl	188–189	16, 128
	2-Formyl-6(or 7)-methoxy-3-methyl		141
	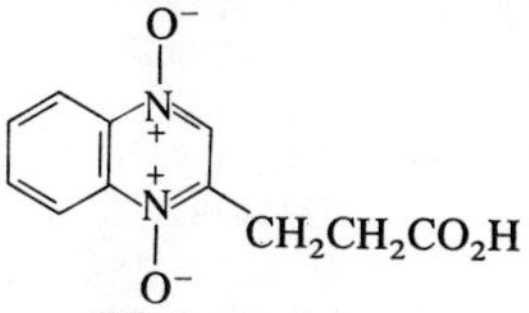 O^- / N^+ / N^+ / O^-, $CH_2CH_2CO_2H$		
	Ethyl ester	184–185	116

TABLE 2 (*contd.*)

Molecular formula	Substituents or structure	m.p., °C[a]	Ref.
$C_{11}H_{11}BrN_2O_2$	6-Bromo-2,3,7-trimethyl		
	Hemiethanolate	242–244	13
$C_{11}H_{11}ClN_2O_2$	6-Chloro-2-ethyl-3-methyl	142–144	129
	6-Chloro-2,3,7-trimethyl	250–252	13
$C_{11}H_{11}N_3O_4$	8-Acetamido-5-methoxy	182–185	5
	5-Ethoxycarbonylamino		122
	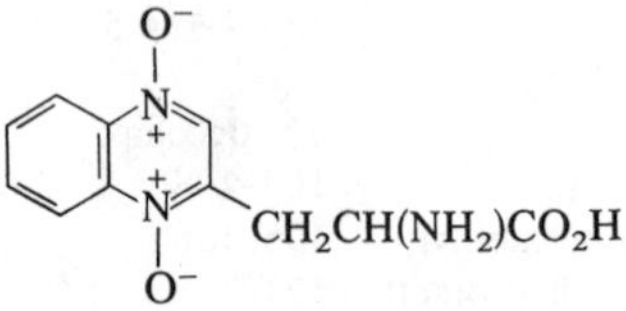 O^-, N^+, N^+, O^-, $CH_2CH(NH_2)CO_2H$	214 decomp.	109
	α-Acetyl	205.5 decomp.	109
$C_{11}H_{12}N_2O_2$	2-Ethyl-3-methyl	139–141, 141–142	13, 129
	2-Isopropyl	114.5–116	114
	2,3,5-Trimethyl		145
	2,3,6-Trimethyl	154–155, 155–156	13, 129
	2,3,7-Trimethyl	185–187	13
$C_{11}H_{12}N_2O_3$	7-Ethoxy-2-methyl	202 decomp.	129
	2-Ethoxymethyl	134	16
	2-(2-Hydroxyethyl)-3-methyl	185–186	76
	Acetyl	159–160	76
	5-Methoxy-2,3-dimethyl	206–208	13
	6-Methoxy-2,3-dimethyl	197–198	13, 129
$C_{11}H_{13}BrN_4O_2S$	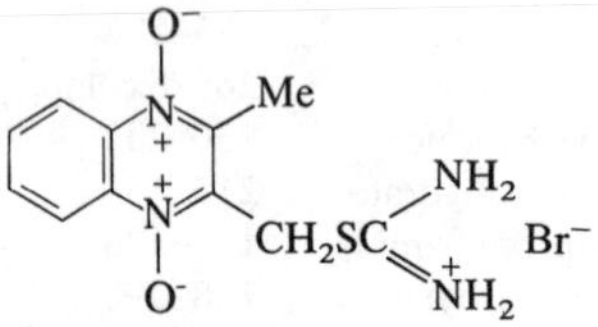 O^-, N^+, Me, N^+, O^-, $CH_2SC(NH_2)=\overset{+}{N}H_2$ Br^-	194–195 decomp.	109
$C_{11}H_{13}N_3O_2$	2-Dimethylamino-3-methyl	124–127	35
$C_{12}H_9F_3N_2O_3$	2-Acetyl-3-methyl-6 (or 7)-trifluoromethyl	152–153	34
	2-Acetyl-3-methyl-7 (or 6)-trifluoromethyl	167–168	34
$C_{12}H_{10}N_2O_5$	3-Acetoxymethyl-2-formyl	157–159	97
	Oxime	177–178	97
	N-Methylnitrone	169–170	97
	Semicarbazone	250	97
	Thiosemicarbazone	236–237	97
$C_{12}H_{10}N_2O_7$	2,3-Dicarboxy-6-ethoxy		
	Diamide	218 decomp.	129

TABLE 2 (*contd.*)

Molecular formula	Substituents or structure	m.p., °C[a]	Ref.
$C_{12}H_{11}Cl_3N_2O_2$	2-Methyl-3(3,3,3-trichloro-2-hydroxypropyl)	226 decomp.	16
$C_{12}H_{12}F_3N_3O_2$	2-[(Dimethylamino)methyl]-3-trifluoromethyl		125
$C_{12}H_{12}N_2O_3$	2-Acetyl-3,5(or 8)-dimethyl	169–170	34
	6-Acetyl-2,3-dimethyl	160–162	7
	Oxime	244–246	7
$C_{12}H_{12}N_2O_3S$	2-(Acetylthiomethyl)-3-methyl	150–151	109
$C_{12}H_{12}N_2O_4$	2-(2-Acetoxyethyl)	129–130	76
	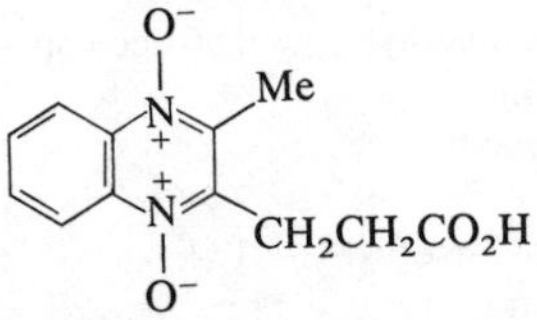	206–208 decomp.	41
$C_{12}H_{12}N_2O_5$	2-Carboxy-6-ethoxy-3-methyl		
	Amide		129
$C_{12}H_{13}ClN_4O_2$	6-Chloro-2-isopropyl-3-methyl	158	129
$C_{12}H_{13}N_3O_3$	6-Acetamido-2,3-dimethyl	259–260	13
$C_{12}H_{14}N_2O_2$	2,3-Diethyl	108–110	13
	2-Ethyl-3,6-dimethyl	150–152	129
	2-Isopropyl-3-methyl	184, 193–194	13, 129
	2-Methyl-3-propyl	108–110	13
	2,3,5,7-Tetramethyl	164–166	129
	2,3,6,7-Tetramethyl	247–249	13
$C_{12}H_{14}N_2O_3$	6-Ethoxy-2,3-dimethyl	160–162	13, 129
$C_{12}H_{14}N_2O_4$	6,7-Dimethoxy-2,3-dimethyl	275–277	13
	2,3-Bis(hydroxymethyl)	187.5–188.5	76
	Diacetyl	130–130.5	76
$C_{12}H_{17}N_3O_2$	2,3-Dihydro-2-dimethylamino-3,3-dimethyl	135–137	35
$C_{13}H_{14}N_2O_3$	2-Acetyl-3,6,7-trimethyl	200–202	34
$C_{13}H_{14}N_2O_4$	2-(1-Acetoxyethyl)-3-methyl	135	16
	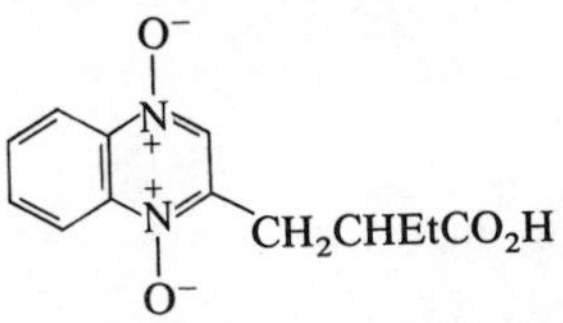	176–177 decomp.	146
$C_{13}H_{16}N_2O_2$	2-Ethyl-3-propyl		
	Hydrate	85–87	13
	2-Isopropyl-3,6-dimethyl	148	129
$C_{13}H_{16}N_2O_3$	6-Ethoxy-2-ethyl-3-methyl	167–168	129

TABLE 2 (*contd.*)

Molecular formula	Substituents or structure	m.p., °C[a]	Ref.
$C_{13}H_{16}N_2O_3$ (*contd.*)	6-Ethoxy-3-ethyl-2-methyl	174–175	129
	2-Isopropyl-6-methoxy-3-methyl	212	129
	2-(2-Ethoxyethyl)-3-methyl	117	13
$C_{13}H_{17}N_3O_2$	2-Ethyl-3-ethylmethylamino		127
$C_{14}H_8Cl_2N_2O_2$	6,7-Dichloro-2-phenyl	249–251	34
$C_{14}H_9ClN_2O_2$	6(or 7)-Chloro-2-phenyl	216–220	34
$C_{14}H_{10}N_2O_2$	2-Phenyl	202–203, 205–206, 209–210	12, 13, 33, 66, 129, 139, 147
$C_{14}H_{14}N_2O_4S_2$	2,3-Bis(acetylthiomethyl)	178–179	109
$C_{14}H_{14}N_2O_6$	2,3-Bis(acetoxymethyl)	176–177	16
$C_{14}H_{14}N_4O_3$	3-Cyano-2-morpholinomethyl	156 decomp.	82
$C_{14}H_{16}N_2O_2$	2-Cyclopentyl-3-methyl		145
$C_{14}H_{17}N_3O_3$	2-Methyl-3-morpholinomethyl		129
$C_{14}H_{18}N_2O_2$	2,3-Dipropyl	74–76	13
$C_{14}H_{18}N_2O_3$	6-Ethoxy-2-isopropyl-3-methyl	174	129
$C_{14}H_{19}N_3O_2$	2-Ethyl-3-diethylamino		127
$C_{14}H_{20}N_4O_2$	2,3-Bis(dimethylaminomethyl)	135–136 decomp.	16
$C_{15}H_8BrN_3O_2$	7-Bromo-2-cyano-3-phenyl	216	36
$C_{15}H_8ClN_3O_2$	7-Chloro-2-cyano-3-phenyl	218	36
$C_{15}H_9N_3O_2$	2-Cyano-3-phenyl	208	36
$C_{15}H_{10}N_2O_4$	2-Carboxy-3-phenyl		
	Ethyl ester	124	41
$C_{15}H_{11}ClN_2O_2$	6-Chloro-3-methyl-2-phenyl		129
$C_{15}H_{12}N_2O_2$	2-Methyl-3-phenyl	193–194, 195–197	4, 13, 41, 129
$C_{15}H_{12}N_2O_2S$	2-Methyl-3-phenylthio	153–154	87
$C_{15}H_{12}N_2O_3S$	2-Methyl-3-phenylsulfinyl	164–165	87
$C_{15}H_{12}N_2O_4S$	2-Methyl-3-phenylsulfonyl	180–181	87
$C_{15}H_{13}N_3O_5S$	5-Hydroxy-8-(*p*-toluene-sulfonamido)	215–218 decomp.	5
$C_{15}H_{16}N_2O_6$	O^-, N^+, N^+, O^-, $CH(CH_2OAc)_2$	125.5–127	76
$C_{15}H_{18}N_2O_4$	O^-, N^+, N^+, O^-, $CH_2CH(iso\text{-}Bu)CO_2H$	175–176	146
$C_{16}H_{11}N_3O_3$	2-Cyano-7-methoxy-3-phenyl	223	36
$C_{16}H_{12}N_2O_2$	2-Styryl	231 decomp.	77
$C_{16}H_{12}N_2O_3$	2-Benzoyl-3-methyl	223–224	48
$C_{16}H_{14}N_2O_2$	2-Ethyl-3-phenyl	129.5	4
	2-Benzyl-3-methyl	155	13

TABLE 2 (*contd.*)

Molecular formula	Substituents or structure	m.p., °C[a]	Ref.
$C_{16}H_{15}N_3O_5S$	5-Methoxy-8-(*p*-toluenesulfonamido	230–235 decomp.	5
$C_{16}H_{16}N_4O_4S$	6-(*p*-Aminobenzenesulfonamido)-2,3-dimethyl	265 decomp.	13
	Acetyl	290 decomp.	13
$C_{16}H_{17}N_3O_7$	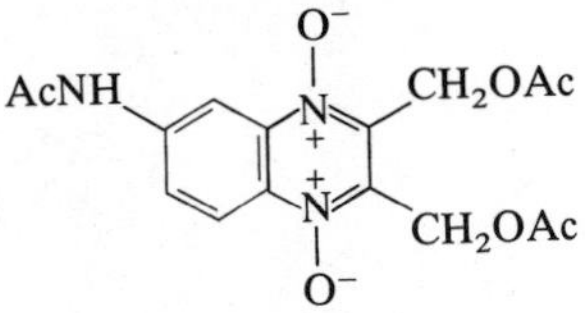 [structure: 6-AcNH-2,3-bis(CH₂OAc)quinoxaline 1,4-di-N-oxide]	208.5 decomp.	111
$C_{17}H_{14}N_2O_2$	2-Methyl-3-styryl	173–174	77
$C_{17}H_{14}N_2O_3$	2-Benzoyl-3-ethyl	199–200 decomp.	48[d]
$C_{17}H_{15}ClN_2O_2$	2-(*p*-Chlorophenethyl)-3-methyl	168–169	13
$C_{17}H_{16}N_2O_2$	2-Isopropyl-3-phenyl	163	4
$C_{17}H_{16}N_2O_3$	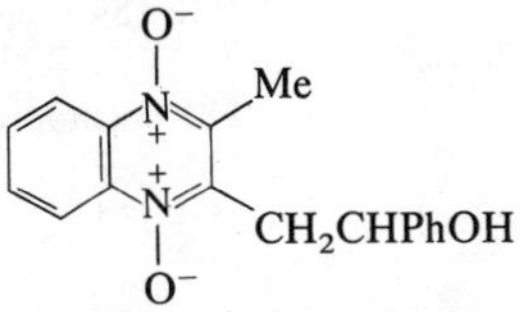 [structure: 2-Me-3-CH₂CHPhOH quinoxaline 1,4-di-N-oxide]	191–192	77
Acetyl		88.5–90	77
	2-Methyl-3-(2-phenoxyethyl)	146	13
$C_{17}H_{19}N_3O_4$	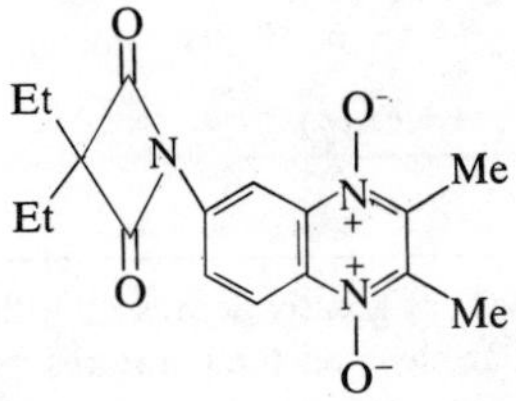 [structure: 6-(3,3-diethyl-2,4-dioxoazetidin-1-yl)-2,3-dimethylquinoxaline 1,4-di-N-oxide]	198–201	13
$C_{18}H_{16}N_2O_3$	2-*p*-Methoxystyryl-3-methyl	176–177	77
$C_{18}H_{16}N_2O_4$	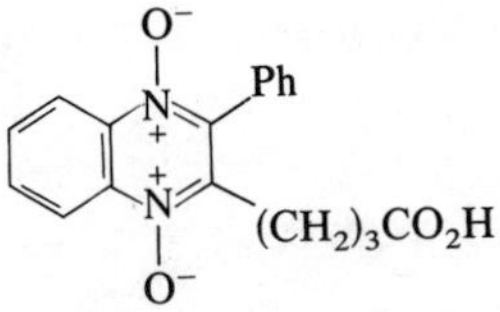 [structure: 2-Ph-3-(CH₂)₃CO₂H quinoxaline 1,4-di-N-oxide]	218–220	41
$C_{18}H_{18}N_2O_4$	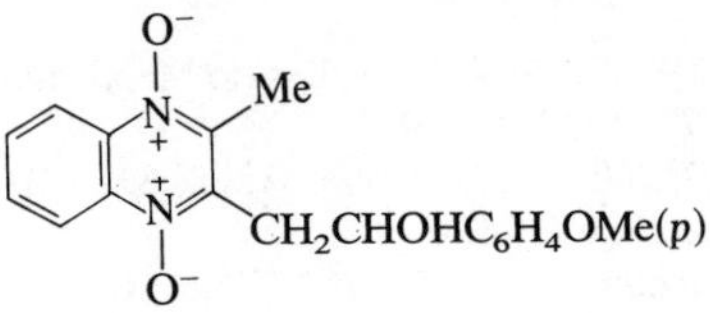 [structure: 2-Me-3-CH₂CHOHC₆H₄OMe(*p*) quinoxaline 1,4-di-N-oxide]	192–193	77

TABLE 2 (*contd.*)

Molecular formula	Substituents or structure	m.p., °C[a]	Ref.
$C_{19}H_{28}N_2O_2$	2-Decyl-3-methyl	111–113	129
$C_{20}H_{14}N_2O_2$	2,3-Diphenyl	210–211 decomp., 216	13, 14
$C_{20}H_{14}N_2O_2S$	2-Phenyl-3-phenylthio	205–207	148
$C_{20}H_{14}N_2O_4S$	2-Benzenesulfonyl-3-phenyl	231–233	148
$C_{20}H_{19}N_3O_3$	[structure: quinoxaline 1,4-dioxide, 6-EtO, 2-Me, 3-$CH_2CHPhCN$; N-oxides O^-]		129
$C_{20}H_{28}N_4O_2$	2,3-Bis(piperidinomethyl)	175–176 decomp.	16
$C_{21}H_{14}N_2O_3$	2-Benzoyl-3-phenyl	234	139
$C_{21}H_{29}N_2O_3$	2-Decyl-6-ethoxy-3-methyl		129
$C_{22}H_{18}N_2O_4$	2,3-Bis(*p*-methoxyphenyl)	183–184	13
$C_{22}H_{20}N_4O_2$	2,3-Bis(anilinomethyl)	152–154	16
$C_{22}H_{32}N_4O_2$	2,3-Bis(2′-piperidinoethyl) 2HBr·0.5H_2O	211 decomp.	16
$C_{22}H_{34}N_2O_2$	2-Methyl-3-tridecyl	104–107	120
$C_{24}H_{20}Cl_2N_2O_4$	2,3-Bis(*p*-chlorophenethyl)	150–151	13
$C_{24}H_{20}N_4O_8$	[structure: quinoxaline 1,4-dioxide (N^+–O^-), 2,3-bis($CH_2CHOHC_6H_4NO_2$(*p*))]	241.5 decomp.	77

[a] The melting points of quinoxaline dioxides are greatly depressed if the samples are heated slowly or are kept for any length of time at elevated temperatures.
[b] Exists as 1-hydroxyquinoxalin-2-one 4-oxide.
[c] Exists as the cyclic hemiacetal.
[d] Many related di-*N*-oxides are described in this reference.

V. References

1. E. P. Stapley, J. M. Mata, F. J. Wolf, and T. W. Miller (to Merck and Co., Inc.), U.S. Pat 3,692,633; *Chem. Abstr.*, **77,** 163041*t* (1972).
2. M. Asai, *Yakugaku Zasshi,* **79,** 260 (1959); *Chem. Abstr.*, **53,** 13160 (1959).
3. J. K. Landquist, *J. Chem. Soc.*, 2816 (1953).
4. E. Hayashi and C. Iijima, *Yakugaku Zasshi,* **84,** 163 (1964); *Chem. Abstr.*, **61,** 3108 (1964).

5. R. H. B. Galt (to I.C.I. Ltd.), U.S. Pat. 3,479,354; *Chem. Abstr.*, **72,** 79095*c* (1970).
6. R. A. Burrell, J. M. Cox, and E. G. Savins, *J. Chem. Soc., Perkin I*, 2707 (1973).
7. J. A. Silk, *J. Chem. Soc.*, 2058 (1956).
8. Y. Ahmad, M. S. Habib, M. I. Qureshi, and M. A. Farooqi, *J. Org. Chem.*, **38,** 2176 (1973).
9. A. S. Elina and L. G. Tsyrul'nikova, *J. Gen. Chem. USSR*, **33,** 1507 (1963).
10. G. T. Newbold and F. S. Spring, *J. Chem. Soc.*, 519 (1948).
11. G. W. H. Cheeseman, *J. Chem. Soc.*, 3236 (1957).
12. E. Hayashi and C. Iijima, *Yakugaki Zasshi*, **82,** 1093 (1962); *Chem. Abstr.*, **58,** 4551 (1963).
13. J. K. Landquist and G. J. Stacey, *J. Chem. Soc.*, 2822 (1953).
14. E. Grovenstein, W. Postman, and J. W. Taylor, *J. Org. Chem.*, **25,** 68 (1960).
15. A. S. Elina, G. N. Pershin, L. G. Tsyrul'nikova, and K. A. Belozerova, *Farmakol. Toksikol.*, **30,** 617 (1967); *Chem. Abstr.*, **68,** 11327*s* (1968).
16. J. K. Landquist and J. A. Silk, *J. Chem. Soc.*, 2052 (1956).
17. J. K. Landquist, *J. Chem. Soc.*, 2830 (1953).
18. G. W. H. Cheeseman, *J. Chem. Soc.*, 1246 (1961).
19. J. W. Clark-Lewis, *J. Chem. Soc.*, 439 (1957).
20. M. S. Habib and C. W. Rees, *J. Chem. Soc.*, 3386 (1960).
21. Y. Ahmad, M. S. Habib, M. Iqbal, M. I. Qureshi, and Ziauddin, *Can. J. Chem.*, **43,** 3424 (1965).
22. G. Tennant, *J. Chem. Soc.*, 2428 (1963).
23. G. Tennant, *J. Chem. Soc.*, 1986 (1964).
24. G. Tennant, *J. Chem. Soc.*, 2666 (1964).
25. R. Fusco and S. Rossi, *Gazz. Chim. Ital.*, **94,** 3 (1964); *Chem. Abstr.*, **61,** 4352 (1964).
26. R. Fusco, S. Rossi, and S. Maiorana, *Gazz. Chem. Ital.*, **95,** 1237 (1965); *Chem. Abstr.*, **64,** 9724 (1966).
27. Y. Ahmad, M. S. Habib, and Ziauddin, *Tetrahedron*, **20,** 1107 (1964).
28. Y. Ahmad, M. S. Habib, Ziauddin, and N. Bashir, *Tetrahedron*, **21,** 861 (1965).
29. V. D. Romanenko, N. E. Kul'chitskaya, and S. I. Burmistrov, *Zh. Org. Khim.*, **8,** 659 (1972); *Chem. Abstr.*, **77,** 34454*f* (1972).
30. C. W. Jefford and E. C. Taylor, *Chem. Ind.* (London), 1559 (1963).
31. M. J. Haddadin and C. H. Issidorides, *Heterocycles*, **4,** 769 (1976).
32. K. Ley and F. Seng, *Synthesis*, 415 (1975).
33. M. J. Haddadin and C. H. Issidorides, *Tetrahedron Lett.*, 3253 (1965).
34. N. A. Mufarrij, M. J. Haddadin, C. H. Issidorides, J. W. McFarland, and J. D. Johnston, *J. Chem. Soc., Perkin I*, 965 (1972).
35. J. W. McFarland, *J. Org. Chem.*, **36,** 1842 (1971).
36. J. C. Mason and G. Tennant, *J. Chem. Soc. Chem. Commun.*, 586 (1971).
37. E. Abushanab and N. A. Alteri, *J. Org. Chem.*, **40,** 157 (1975).
38. R. A. Bowie and G. Jones (to I.C.I. Ltd.) Brit. Pat. 1,301,944; *Chem. Abstr.*, **78,** 97705*e* (1973).
39. P. J. Diel and W. Schmid (to Ciba-Geigy A.-G.), Ger. Offen. 2,344,314; *Chem. Abstr.*, **80,** 146201*q* (1974).
40. J. J. Zamet, M. J. Haddadin, and C. H. Issidorides, *J. Chem. Soc., Perkin I*, 1687 (1974).
41. M. L. Edwards, R. E. Bambury, and H. W. Ritter, *J. Med. Chem.*, **19,** 330 (1976).
42. E. Abushanab, *J. Org. Chem.*, **35,** 4279 (1970).
43. Pfizer Inc., Brit. Pat. 1,271,194; *Chem. Abstr.*, **77,** 34575*w* (1972).

44. V. M. Berezovskii and Zh. I. Aksel'rod, *Dokl. Akad. Nauk. SSSR*, **168,** 577 (1966); *Chem. Abstr.*, **65,** 8911 (1966).
45. A. S. Elina, *J. Gen. Chem. USSR*, **32,** 2919 (1962).
46. A. S. Elina and O. Yu. Magidson, *J. Gen. Chem. USSR*, **25,** 145 (1955).
47. M. J. Haddadin, G. E. Zahr, T. N. Rawdah, N. C. Chelot, and C. H. Issidorides, *Tetrahedron*, **30,** 659 (1974).
48. M. J. Haddadin, M. U. Taha, A. A. Jarrar, and C. H. Issidorides, *Tetrahedron*, **32,** 719 (1976).
49. G. G. Dvoryantseva, M. M. Kaganskii, I. S. Musatova, and A. S. Elina, *Chem. Heterocycl. Comp.*, **10,** 1366 (1974).
50. D. E. Chasan, L. L. Pytlewski, C. Owens, and N. M. Karayannis, *J. Inorg. Nucl. Chem.*, **38,** 1799 (1976).
51. T. Kubota and H. Miyazaki, *Chem. Pharm. Bull.* (Tokyo), **9,** 948 (1961).
52. K. Tori, M. Ogata, and H. Kano, *Chem. Pharm. Bull.* (Tokyo), **11,** 681 (1963).
53. S. N. Bannore, J. L. Bose, A. A. Thakar, and M. S. Wadia, *Indian J. Chem.*, **13,** 609 (1975).
54. E. Abushanab, *J. Am. Chem. Soc.*, **93,** 6532 (1971).
55. A. Tatematsu, H. Yoshizumi, E. Hayashi, and H. Nakata, *Tetrahedron Lett.*, 2985 (1967).
56. A. Kubo, S. Sakai, S. Yamada, I. Yokoe, C. Kaneko, A. Tatematsu, H. Yoshizumi, E. Hayashi, and H. Nakata, *Chem. Pharm. Bull.* (Tokyo), **15,** 1079 (1967).
57. L. V. Varyukhina and Z. V. Pushkareva, *J. Gen. Chem. USSR*, **26,** 1953 (1956).
58. T. R. Emerson and C. W. Rees, *J. Chem. Soc.*, 1923 (1962).
59. T. R. Emerson and C. W. Rees, *J. Chem. Soc.*, 2319 (1964).
60. Z. V. Pushkareva, L. V. Varyukhina, and Z. Yu. Kokoshko, *Dokl. Akad. Nauk. SSSR*, **108,** 1098 (1956); *Chem. Abstr.*, **51,** 21 (1957).
61. E. Ochiai, "Aromatic Amine Oxides," Elsevier, Amsterdam, 1967.
62. A. R. Katritzky and J. M. Lagowski, "Chemistry of Heterocyclic *N*-Oxides," Academic Press, London and New York, 1971.
63. C. Iijima, *Yakugaku Zasshi*, **87,** 942 (1967).
64. A. S. Elina, *Chem. Heterocycl. Comp.*, **3,** 576 (1967).
65. Y. Kobayashi, I. Kumadaki, and H. Sato, *J. Org. Chem.*, **37,** 3588 (1972).
66. Y. Ahmad, M. S. Habib, Ziauddin, and B. Bakhtiari, *J. Org. Chem.*, **31,** 2613 (1966).
67. E. Hayashi and Y. Miura, *Yakugaku Zasshi*, **87,** 648 (1967); *Chem. Abstr.*, **67,** 90775*s* (1967).
68. E. Hayashi, C. Iijima, and K. Yamamoto, *Yakugaku Zasshi*, **87,** 164 (1967); *Chem. Abstr.*, **67,** 3067*w* (1967).
69. C. Iijima and E. Hayashi, *Yakugaku Zasshi*, **91,** 721 (1971); *Chem. Abstr.*, **75,** 129758*f* (1971).
70. Yu. S. Rozum and N. N. Lisovskaya, *J. Gen. Chem. USSR*, **29,** 231 (1959).
71. S. Kaneko, S. Yamada, I. Yokoe, and M. Ishikawa, *Tetrahedron Lett.*, 1873 (1967).
72. O. Buchardt and J. Feeney, *Acta Chem. Scand.*, **21,** 1399 (1965); O. Buchardt and B. Jensen, *Acta Chem. Scand.*, **22,** 877 (1968).
73. A. S. Elina, *Chem. Heterocycl. Comp.*, **4,** 403 (1968).
74. G. W. H. Cheeseman and E. S. G. Törz, *J. Chem. Soc.* (*C*), 157 (1966).
75. P. Benko, L. Buda, L. Pallos, I. Simonek, P. Foris, J. Kovacs, K. Magyar, and G. Tokar, *Hung. Teljes*, 8581 (1975); *Chem. Abstr.*, **82,** 57734*g* (1975).
76. A. S. Elina and L. G. Tsyrul'nikova, *Chem. Heterocycl. Comp.*, **2,** 316 (1966).
77. A. S. Elina and L. G. Tsyrul'nikova, *Chem. Heterocycl. Comp.*, **2,** 194 (1966).
78. A. S. Elina, L. G. Tsyrul'nikova, and I. S. Musatova, *Khim. Farm. Zh.*, **1**(5), 10 (1967); *Chem. Abstr.*, **68,** 78247*j* (1968).

79. N. E. Plevachuk and S. N. Baranov, *Chem. Heterocycl. Comp.*, **4,** 536 (1968).
80. A. S. Elina, L. G. Tsyrul'nikova, and G. P. Syrova, *Chem. Heterocycl. Comp.*, **5,** 115 (1969).
81. M. J. Haddadin and C. H. Issidorides, *Tetrahedron Lett.*, 4609 (1968).
82. U. Eholzer, H. Heitzer, F. Seng, and K. Ley, *Synthesis*, 296 (1974).
83. A. S. Elina, *Chem. Heterocycl. Comp.*, **3,** 740 (1967).
84. A. S. Elina, L. G. Tsyrul'nikova, E. M. Pereslini, and Yu. N. Sheinker, *Chem. Heterocycl. Comp.*, **2,** 72 (1966).
85. R. A. Abramovitch and B. W. Cue, *Heterocycles*, **1,** 227 (1973).
86. A. A. Jarrer, S. S. Halawi, and M. J. Haddadin, *Heterocycles*, **4,** 1077 (1976).
87. E. Abushanab, *J. Org. Chem.*, **38,** 3105 (1973).
88. Y. Ahmad, M. S. Habib, M. Iqbal, and Ziauddin, *Bull. Chem. Soc. Jap.*, **38,** 562 (1965).
89. J. C. Mason and G. Tennant, *J. Chem. Soc. Chem. Commun.*, 218 (1972).
90. Y. Ahmad, M. S. Habib, Ziauddin, and N. Bashir, *Bull. Chem. Soc. Jap.*, **38,** 1654 (1965).
91. Y. Ahmad, M. S. Habib, A. Mohammady, B. Bakhtiari, and S. S. Shamsi, *J. Org. Chem.*, **33,** 201 (1968).
92. D. B. Livingstone and G. Tennant, *Chem. Ind.* (London), 848 (1973).
93. J. C. Mason and G. Tennant, *J. Chem. Soc. Chem. Commun.*, 1550 (1971).
94. M. S. Habib and C. W. Rees, *J. Chem. Soc.*, 3371 (1960).
95. Y. Ahmad, M. S. Habib, M. Iqbal, M. I. Qureshi, and Ziauddin, *Bull. Chem. Soc. Jap.*, **38,** 1659 (1965).
96. M. L. Davis (to Colgate-Palmolive), U.S. Pat. 3,852,443; *Chem. Abstr.*, **83,** 48193*f* (1975).
97. M. L. Edwards, R. E. Bambury, and H. W. Ritter, *J. Med. Chem.*, **18,** 637 (1975).
98. J. Francis, J. K. Landquist, A. A. Levi, J. A. Silk, and J. M. Thorp, *Biochem. J.*, **63,** 455 (1956).
99. A. Monge, V. Huarte, A. Llamas, A. Gonzalez, and E. Martinez, *An. Quim.*, **71,** 248 (1975); *Chem. Abstr.*, **83,** 131548*w* (1975).
100. G. W. H. Cheeseman and E. S. G. Werstiuk, *Adv. Heterocycl. Chem.*, Vol. 22, A. J. Boulton and A. R. Katritzky Eds., Academic Press, 1978, p. 367.
101. J. Dukai, T. Pfliegel, M. Barath, and A. Kelemen–(to Chinoin Gyogyszer es Vegyeszeti Termekek Gyara Rt.), Ger. Offen. 2,423,093; *Chem. Abstr.*, **83,** 164229*w* (1975).
102. J. D. Johnston (to Chas. Pfizer and Co. Ltd.), U.S. Pat. 3,444,022; *Chem. Abstr.*, **67,** 111452*b* (1967).
103. H. Mulik, T. Takada, H. Ito, and T. Mikami, *Jap. Kokai* 75, 82, 217; *Chem. Abstr.*, **83,** 152385*f* (1975).
104. R. R. Shaffer (to E. I. du Pont de Nemours and Co.), U.S. Pat. 3,560,616; *Chem. Abstr.*, **75,** 47839*u* (1971).
105. T. J. Haley, A. M. Flesher, and N. Komescu, *Nature*, **184,** 198 (1959).
106. M. L. Douglas (to Colgate-Palmolive Co.), U.S. Pat. 3,733,323; *Chem. Abstr.*, **79,** 42557*y* (1973).
107. A. S. Elina, E. M. Peresleni, L. S. Tsyrul'nikova, and Yu. N. Sheinker, *Chem. Heterocycl. Comp.*, **5,** 540 (1969).
108. A. S. Elina, L. G. Tsyrul'nikova, and I. S. Musatova, *Khim.-Farm. Zh.*, **2**(5), 11 (1968); *Chem. Abstr.*, **69,** 96664*t* (1968).
109. A. S. Elina and L. G. Tsyrul'nikova, *Khim.-Farm. Zh.*, **1**(11), 36 (1967); *Chem. Abstr.*, **69,** 27384*q* (1968).
110. A. S. Elina, I. S. Musatova, E. M. Peresleni, and E. N. Padeiskaya, *Khim. Geterotsikl. Soedin.*, 278 (1976).

111. A. S. Elina, I. S. Musatova, and L. G. Tsyrul'nikova, *Khim.-Farm. Zh.*, **5**(8), 6 (1971); *Chem. Abstr.*, **75,** 151759*d* (1971).
112. A. S. Elina, *J. Gen. Chem. USSR*, **31,** 942 (1961).
113. H. von Euler and H. Hasselquist, *Arkiv Khemi*, **48,** 559 (1958); *Chem. Abstr.*, **53,** 9234 (1959).
114. E. Hayashi and Y. Miura, *Yakugaku Zasshi*, **87,** 643 (1967); *Chem. Abstr.*, **67,** 90774*r* (1967).
115. G. Tennant. *J. Chem. Soc.* (*C*), 2285 (1966).
116. A. S. Elina, *J. Gen. Chem. USSR*, **28,** 1437 (1958).
117. G. Y. Sarkis and S. Al-Azawe, *J. Chem. Eng. Data*, **18,** 102 (1973).
118. E. Hayashi and C. Iijima, *Yakugaku Zasshi*, **84,** 156 (1964); *Chem. Abstr.*, **61,** 3108 (1964).
119. E. J. Moriconi, R. E. Misner, and T. E. Brady, *J. Org. Chem.*, **34,** 1651 (1969).
120. D. E. Ames., G. Hall, and B. T. Warren, *J. Chem. Soc.* (*C*), 2617 (1968).
121. E. Abushanab (to Chas. Pfizer and Co. Inc.), Ger. Offen. 1,929,541; *Chem. Abstr.*, **72,** 66981*p* (1970).
122. R. H. B. Galt and R. J. Pearce (to I.C.I. Ltd.), Brit. Pat. 1,237,438; *Chem. Abstr.*, **75,** 110340*s* (1971).
123. E. O. Stapley, J. M. Mata, F. J. Wolf, and T. W. Miller (to Merck and Co. Inc.), U.S. Pat. 3,598,819; *Chem. Abstr.*, **75,** 110338*x* (1971).
124. T. W. Miller, A. K. Miller, and E. O. Stapley, *J. Antibiot.*, **22,** 293 (1969).
125. E. Abushanab (to Chas. Pfizer and Co. Inc.), Ger. Offen. 2,105,112; *Chem. Abstr.*, **75,** 110336*v* (1971).
126. F. Seng and K. Ley (to Farbenfabriken Bayer A.-G.), Ger. Offen. 1,934,887; *Chem. Abstr.* **74,** 76446y (1971).
127. K. Ley, F. Seng. U. Eholzer, R. Nast, and R. Schubart, *Angew. Chem. Int. Ed.*, **8,** 596 (1969).
128. A. S. Elina, P. M. Timkova, L. G. Tsyrul'nikova, and T. Ya. Filipenko, *Khim.-Farm. Zh.*, **10,** 44 (1976).
129. K. Ley, U. Eholzer, R. Nast, and F. Seng (to Farbenfabriken Bayer A.-G.), Brit. Pat. 1,187,991; Fr. Pat. 1,521,907; *Chem. Abstr.*, **71,** 70643*b* (1969).
130. Société des Usines Chimiques Rhône–Poulenc, Fr. M. 3717 *Chem. Abstr.*, **66,** 95087*h* (1967).
131. F. Seng, K. Ley, K. G. Metzger, and D. Fritsche (to Farbenfabriken Bayer A.-G.), Brit. Pat. 1,174,874; *Chem. Abstr.*, **72,** 66984*s* (1970).
132. Farbenfabriken Bayer A.-G., Fr. Addn. 95,298; *Chem. Abstr.*, **74,** 88059*n* (1971).
133. J. Adachi, *Nippon Kagaku Zasshi*, **76,** 311 (1955); *Chem. Abstr.*, **51,** 17936 (1957).
134. Abuel-Haj, J. Marwan, and T. H. Cronin (to Chas. Pfizer Inc.), Ger. Offen. 2,035,480; *Chem. Abstr.*, **74,** 112057*h* (1971).
135. K. Ley, U. Eholzer, R. Nast, K. G. Metzger, and D. Fritsche (to Farbenfabriken Bayer A.-G.), S. African Pat. 68 06 098; *Chem. Abstr.*, **72,** 3509*v* (1970).
136. J. D. Johnston (to Chas. Pfizer and Co. Inc.), U.S. Pat. 3,344,022; *Chem. Abstr.*, **67,** 111452*b* (1967).
137. K. Ley, U. Eholzer, K. G. Metzger, and D. Fritsche (to Farbenfabriken Bayer A.-G.), S. African Pat. 68 06 094; *Chem. Abstr.*, **72,** 3508*u* (1970).
138. K. Ley, U. Eholzer, R. Nast, K. G. Metzger, and D. Fritsche (to Farbenfabriken Bayer A.-G.), S. African Pat. 68 06 096; *Chem. Abstr.*, **71,** 91533*f* (1969).
139. C. H. Issidorides and M. J. Haddadin, *J. Org. Chem.*, **31,** 4067 (1966).
140. M. L. Edwards and R. E. Bambury, *J. Heterocycl. Chem.*, **12,** 835 (1975).
141. K. H. Kim (to Richardson-Merrell Inc.), Ger. Offen. 2,043,532; *Chem. Abstr.*, **74,** 125729*d* (1971).

142. K. Ley, U. Eholzer, K. G. Metzger, and D. Fritsche (to Farbenfabriken Bayer A.-G.), S. African Pat. 68 6099; *Chem. Abstr.*, **71,** 91528*h* (1969).
143. F. Seng, K. Ley, and K. G. Metzger (to Farbenfabriken Bayer A.-G.), Ger. Offen. 1,813,918; *Chem. Abstr.*, **73,** 45539*f* (1970).
144. K. Ley, U. Eholzer, and F. Seng (to Farbenfabriken Bayer A.-G.), Ger. Offen. 1,807,735; *Chem. Abstr.*, **73,** 56130*s* (1970).
145. C. H. Issidorides and M. J. Haddadin (to Research Corp.), Brit. Pat. 1,215,815; *Chem. Abstr.*, **74,** 141873*b* (1971).
146. A. S. Elina, *J. Gen. Chem. USSR,* **29,** 2728 (1959).
147. L. Marchetti and G. Tosi, *Ann. Chim.* (Rome), **57,** 1414 (1967); *Chem. Abstr.*, **69,** 2934*f* (1968).
148. D. P. Claypool, A. R. Sidani, and K. J. Flanagan, *J. Org. Chem.*, **37,** 2372 (1972).
149. J. R. Valenta, J. R. E. Hoover, and J. F. Pagano, *Antimicrob. Agents Chemother.*, 453 (1966); *Chem. Abstr.*, **67,** 80856*e* (1967).
150. B. A. Arbuzov and V. M. Zoroastrova, *Izv. Akad. Nauk. SSSR Ser. Khim.*, 254 (1966); *Chem. Abstr.*, **64,** 15915 (1966).
151. E. Hayashi and C. Iijima, *Yakugaku Zasshi,* **86,** 1109 (1966); *Chem. Abstr.*, **67,** 3066*v* (1967).
152. A. I. Brodskii, L. L. Gordienko, Yu. S. Rozum, T. Ya. Lavrenyuk, and N. P. Romazanovich, *Teor. Eksp. Khim.*, **6,** 784 (1970); *Chem. Abstr.*, **75,** 13157*e* (1971).
153. J. P. Dirlam and J. W. McFarland, *J. Org. Chem.*, **42,** 1360 (1977).

CHAPTER V

Quinoxalin-2-ones and Quinoxaline-2,3-diones

I. Quinoxalin-2-ones 79
  1. Methods of Preparation 79
    A. From *o*-Phenylenediamines 79
    B. From *N*-Substituted *o*-Nitroanilines 83
    C. From Quinoxaline Intermediates 84
    D. Quinoxalinones from Reactions in Which a Change of Ring Size Occurs 85
  2. Properties 86
    A. Physical Properties 86
    B. Chemical Properties 88
  3. Uses 93
II. Quinoxaline-2,3-diones 94
  1. Methods of Preparation 94
  2. Properties 95
  3. Uses 97
III. Tables 97
  Table 1. Quinoxalin-2-ones (2-Hydroxyquinoxalines) 97
  Table 2. Quinoxaline-2,3-diones (2,3-Dihydroxyquinoxalines) 104
IV. References 106

The chemistry of quinoxalin-2-ones and quinoxaline-2,3-diones* (2-hydroxy- and 2,3-dihydroxyquinoxalines) has been previously reviewed by Simpson.[1] Quinoxalin-2-ones are very easily prepared from condensation of *o*-phenylenediamines and α-keto acids; and because of their ease of preparation, they have been used as characterizing derivatives for α-keto acids. There are widespread references in the patent literature to the preparation of quinoxalinones for potential pharmaceutical and horticultural applications and also for the formation of polymers.

* Strictly speaking, these compounds are named quinoxalin-2(1*H*)-ones and quinoxaline-2,3(1*H*, 4*H*)-diones.

I. Quinoxalin-2-ones

1. Methods of Preparation

A. From o-Phenylenediamines

Quinoxalin-2-one (**3**) is readily prepared by the condensation of *o*-phenylenediamine and *n*-butyl glyoxylate (**2**). The latter compound is conveniently obtained from sodium periodate oxidation of *n*-butyl tartrate (**1**).[2–4]

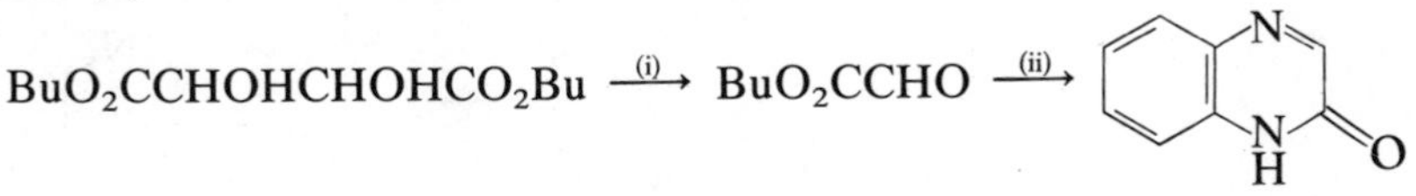

$$BuO_2CCHOHCHOHCO_2Bu \xrightarrow{(i)} BuO_2CCHO \xrightarrow{(ii)}$$ [quinoxalin-2-one]

1 **2** **3**

Reagents (i) Aqueous $NaIO_4$
(ii) *o*-Phenylenediamine

Simple extensions of this method involve the use of arylglyoxylic acids (e.g., phenylglyoxylic acid, $PhCOCO_2H$)[5] and a variety of pyruvic acids (RCH_2COCO_2H) and α-keto acids or their esters (Scheme 1).[6]

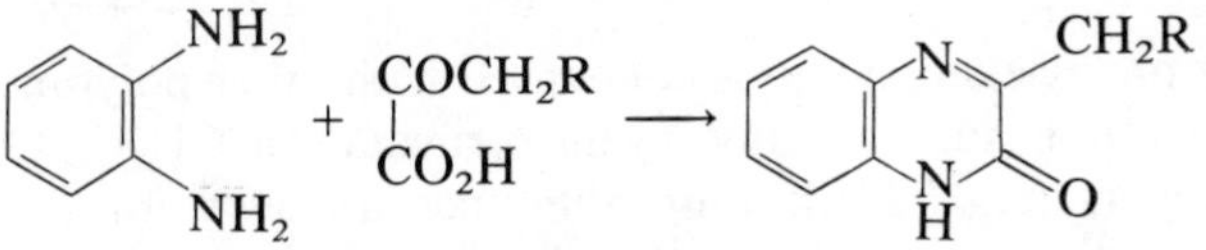

o-phenylenediamine (NH_2, NH_2) + $COCH_2R$ / CO_2H ⟶ 3-(CH_2R)quinoxalin-2-one (N, N–H, O)

Scheme 1

The use of unsymmetrical *o*-phenylenediamines in these reactions clearly leads to products of ambiguous structure; however the condensation reactions of *N*-substituted *o*-phenylenediamines (e.g., *N*-methyl-*o*-phenylenediamines) yields quinoxalinones of known structure (**4**). Since 1-methylquinoxalin-2-ones can also be prepared from quinoxalin-2-ones by methylation with methyl sulfate and alkali, these reactions have been useful in the assignment of structure to quinoxalin-2-ones prepared from routes leading to products of ambiguous structure.[7]

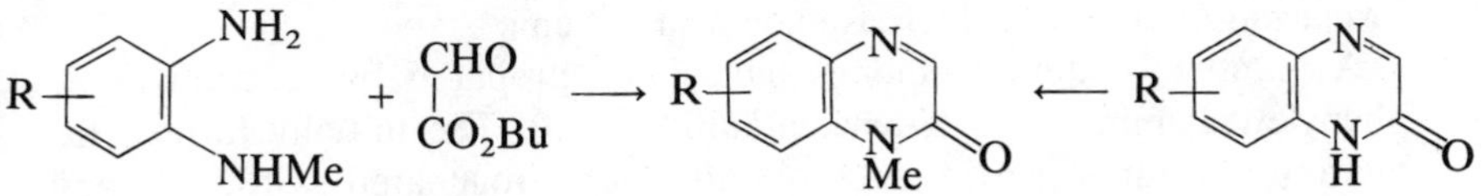

R–(NH_2, NHMe benzene) + CHO / CO_2Bu ⟶ R–(N, N–Me, O quinoxalinone) ⟵ R–(N, N–H, O quinoxalinone)

4

Condensation of *o*-phenylenediamines with mesoxalic ester (**5**) followed by hydrolysis and decarboxylation of the product yields quinoxalin-2-ones (e.g., **6**) unsubstituted in the heteroring.[8] An identical quinoxalinone (**9**) is obtained by condensation of *o*-phenylenediamine with either ethyl ethoxalylacetate (**7**)[9] or diethyl acetylenedicarboxylate (**8**).[10]

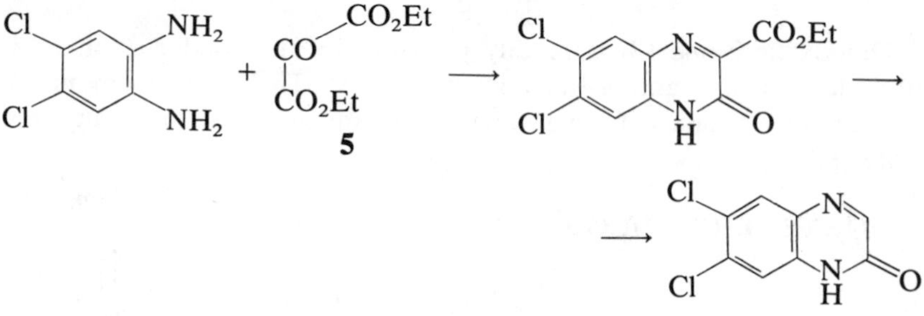

5 **6**

$EtO_2CCOCH_2CO_2Et$ **7** + $EtO_2CC{\equiv}CCO_2Et$ **8** → **9**

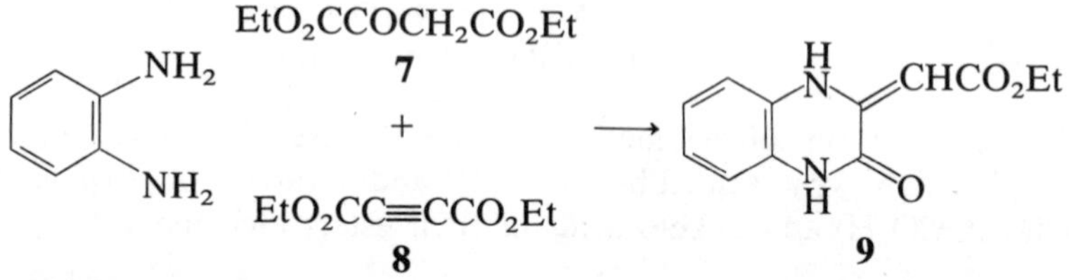

When *o*-phenylenediamine is condensed with other polyfunctional carbonyl compounds such as ethyl cyanopyruvate ($NCCH_2COCO_2Et$)[11] or α,γ-dioxoesters (e.g., **10**),[12] quinoxalinones are also formed, although cyclocondensation could in principle lead to products of increased ring size.

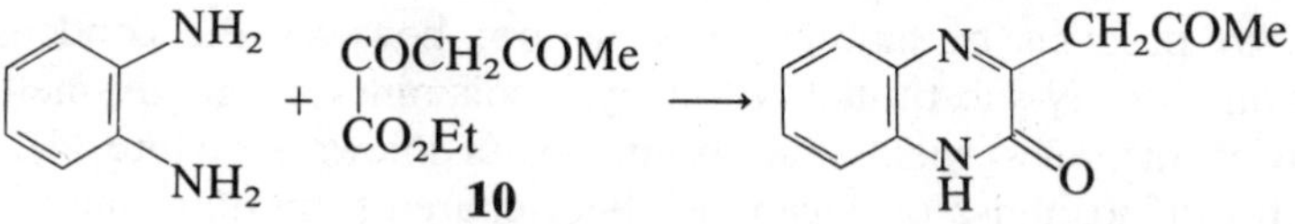

10

The reaction of *o*-phenylenediamine and α-oximinoacetoacetic acid ester (**11**) surprisingly gives mainly 3-methylquinoxalin-2-one (**12**) and diethyl carbonate as the main products. The suggested mechanism for the formation of these products is shown in Scheme 2.[13]

A number of quinoxalinones have been prepared by reaction of *o*-phenylenediamine with *N*-arylmaleimides (**13**). The initially formed dihydroquinoxalinones are subsequently dehydrogenated with chloranil (Scheme 3).[14] When *o*-phenylenediamines are condensed with alloxan

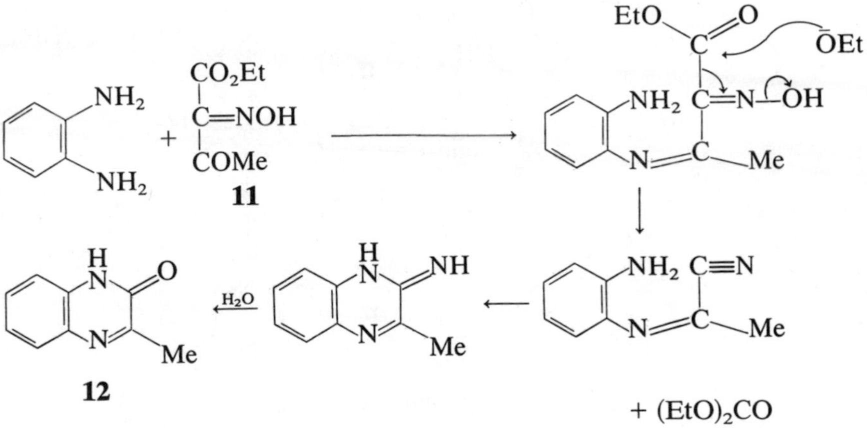

Scheme 2

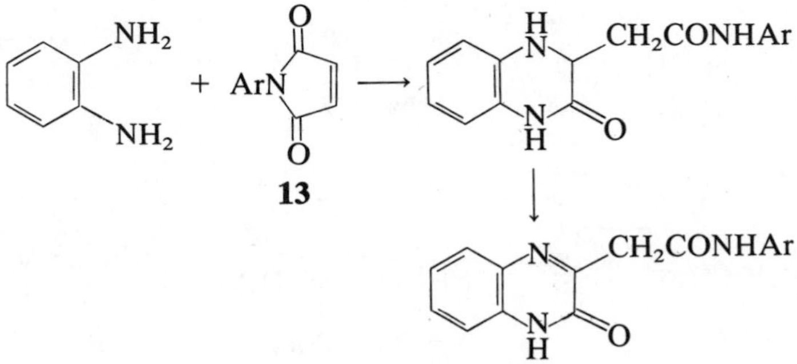

Scheme 3

($\mathbf{14}$) in neutral or slightly acidic aqueous solution, ureides are formed which, by hydrolysis and decarboxylation, may be transformed into 3-unsubstituted quinoxalinones (Scheme 4).[15]

The most convenient preparation of quinoxalin-2-one itself involves the reaction of *o*-phenylenediamine with chloroacetic acid and oxidation of the resulting dihydroquinoxalinone (Scheme 5).[16] This sequence is illustrative of a more general synthetic possibility, and both 3-substituted alkyl-[17] and arylquinoxalinones[18] have been prepared by this route starting from the appropriate α-halo acid.

Quinoxalinones have been prepared by reaction of the dichloroacetyl derivative of *o*-phenylenediamine with either morpholine or *N*-methylpiperazine in the presence of sulfur (Scheme 6).[19] Treatment of *o*-phenylenediamine with 1,2-dichloro-1,2-diethoxyethane gives a small

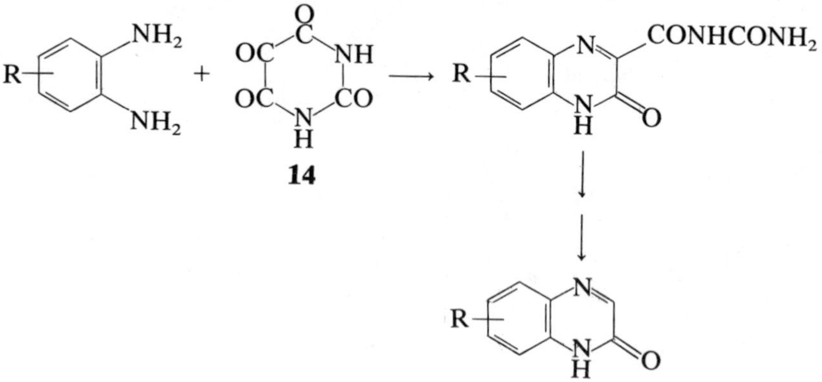

NH_2 NH_2 OC C(O) NH CO N H OC
14
$CONHCONH_2$

Scheme 4

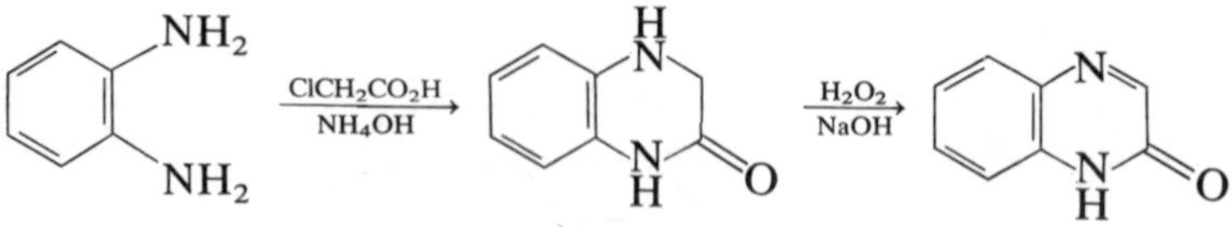

$ClCH_2CO_2H$, NH_4OH; H_2O_2, NaOH

Scheme 5

$NHCOCHCl_2$ → $NHCOCSNC_4H_8X$ → NC_4H_8X

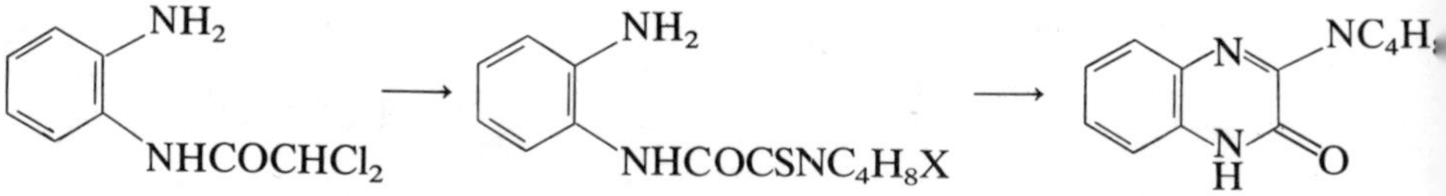

X = O or NMe

Scheme 6

quantity of quinoxalin-2-one together with quinoxaline and 2-(diethoxymethyl)benzimidazole.[20] 3-Phenylquinoxalin-2-one (**16**) is isolated in 84% yield when the α-azido acid **15** is heated with *o*-phenylenediamine.[21]

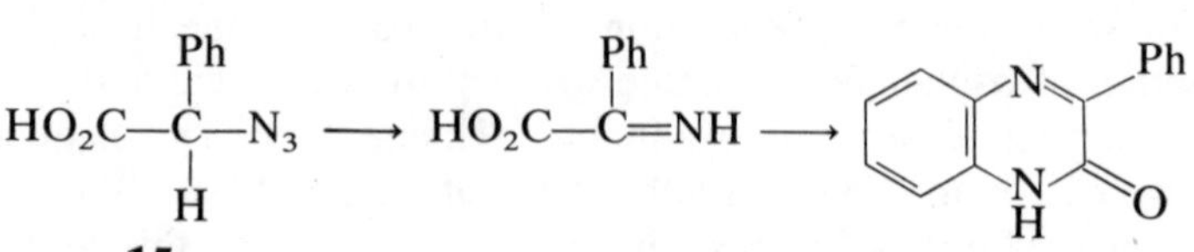

$HO_2C-C(Ph)(H)-N_3 \longrightarrow HO_2C-C(Ph)=NH \longrightarrow$

15 **16**

A series of quinoxalinones have been prepared by reaction of *o*-phenylenediamine with 2-trifluoromethyl-5-oxazolones (**17**). These intermediates are prepared by reaction of α-amino acids with trifluoroacetic anhydride (Scheme 7).[22]

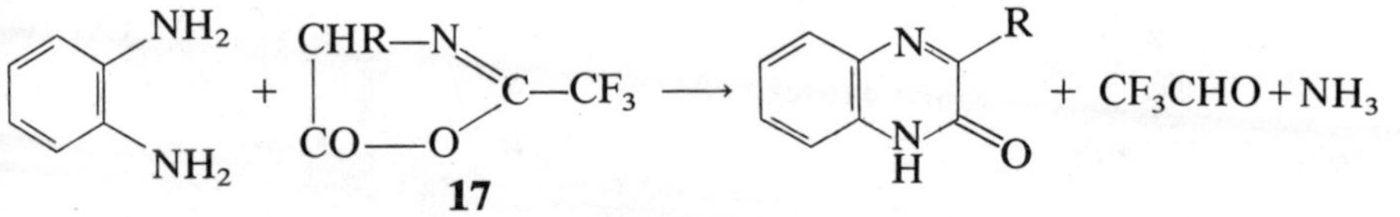

$$\text{o-C}_6\text{H}_4(\text{NH}_2)_2 + \mathbf{17}\ (\text{CHR—N=C—CF}_3,\ \text{CO—O}) \longrightarrow \text{3-R-quinoxalin-2(1H)-one} + CF_3CHO + NH_3$$

Scheme 7

2-Quinoxalinones result from the reaction of *o*-phenylenediamines and acetylenedicarboxylic acid, as exemplified by the synthesis of 3,6,7-trimethylquinoxalin-2-one (**18**).[23] The corresponding condensation reactions with acetylenedicarboxylic acid esters yield 3-alkoxycarbonylmethylenequinoxalin-2-ones, which on hydrolysis and rearrangement give 3-methylquinoxalin-2-ones (Scheme 8).[10,24]

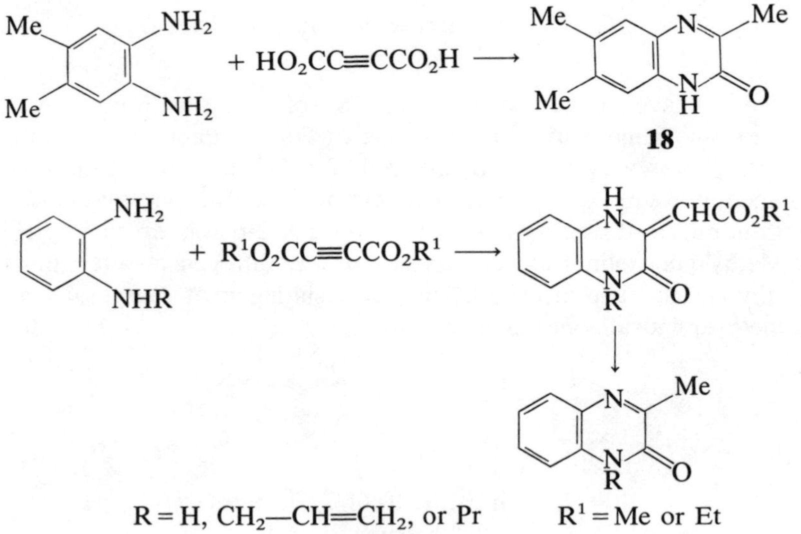

$$\text{4,5-Me}_2\text{C}_6\text{H}_2(\text{NH}_2)_2 + HO_2CC{\equiv}CCO_2H \longrightarrow \mathbf{18}$$

$$\text{o-H}_2\text{NC}_6\text{H}_4\text{NHR} + R^1O_2CC{\equiv}CCO_2R^1 \longrightarrow \text{3-(CHCO}_2\text{R}^1\text{)-1-R-quinoxalin-2-one} \longrightarrow \text{3-Me-1-R-quinoxalin-2-one}$$

$R = H, CH_2{—}CH{=}CH_2$, or Pr $\qquad R^1 = $ Me or Et

Scheme 8

B. *From N-Substituted o-Nitroanilines*

In the preceding quinoxalinone syntheses based on *o*-phenylenediamines, products of ambiguous structure are obtained when unsymmetrical diamines are used. However quinoxalinones of known orientation may be synthesized by cyclization of the appropriate *o*-nitro-*N*-substituted aminobenzene (e.g., **19**) (Scheme 9).[25] Intermediates of type **19** are alternatively obtained by reaction of an *o*-nitroaniline with an α-halo acid.

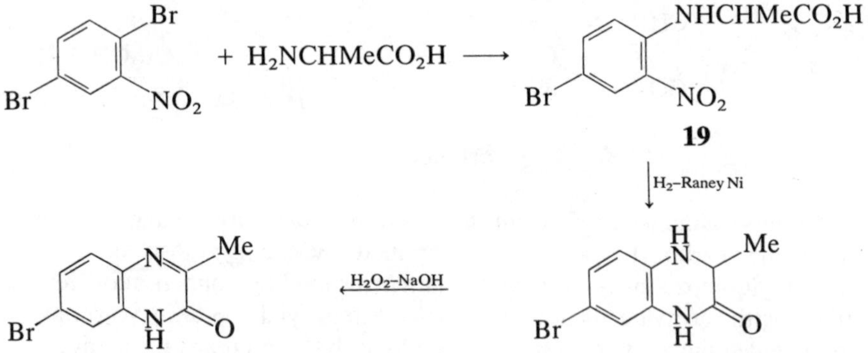

Br, Br, NO_2 + $H_2NCHMeCO_2H$ ⟶ $NHCHMeCO_2H$, Br, NO_2 **19**

H_2–Raney Ni

H_2O_2–NaOH

Scheme 9

C. *From Quinoxaline Intermediates*

There have been several reports of the hydrolysis of 2-aminoquinoxalines to quinoxalinones. For example, 2,3-diaminoquinoxaline is very readily hydrolyzed (2.5 *M* HCl, 100°, 5 min) to 3-aminoquinoxalin-2-one,[26] and treatment of 2-amino-3-phenylquinoxaline with nitrous acid furnishes 3-phenylquinoxalin-2-one in excellent yield.[27] 2-Methylquinoxalines are quaternized on the nitrogen remote from the methyl group, and hydrolysis of the resulting quaternary salts gives 3-methylquinoxalinones (Scheme 10).[28]

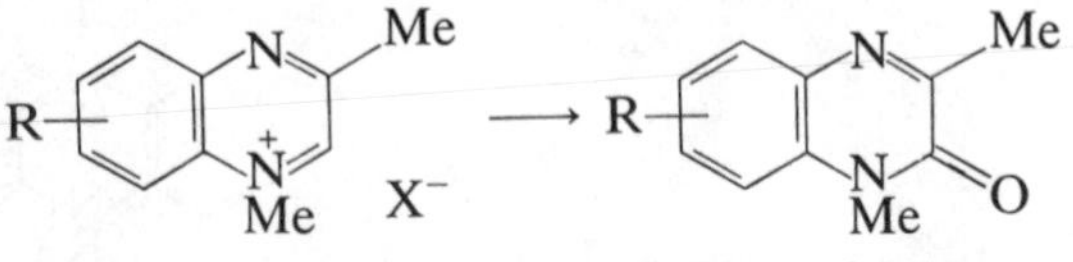

R = H, 6- or 7-Me, 6- or 7-Cl, or 6-NHAc

Scheme 10

An unusual reaction occurs when the quinoxalinylglyoxal oximes **20** are heated in refluxing dimethylaniline and 2-quinoxalinones are obtained as a result of the formation of a spiro intermediate (Scheme 11).[29]

Numerous quinoxalinones have been prepared from *N*-oxides. For example, treatment of 2-phenylquinoxaline 4-oxide (**21**) with acetic anhydride gives 3-phenylquinoxalin-2-one (**16**).[30] Illustrative of a general method for quinoxalinone synthesis is the ring closure of *o*-nitroanilides of type **22** to quinoxalin-2-one 4-oxides (**23**).[31] Related methods for quinoxalin-2-one 4-oxide synthesis are discussed in the previous chapter together with the range of reagents available for their reduction to quinoxalin-2-ones (**24**).

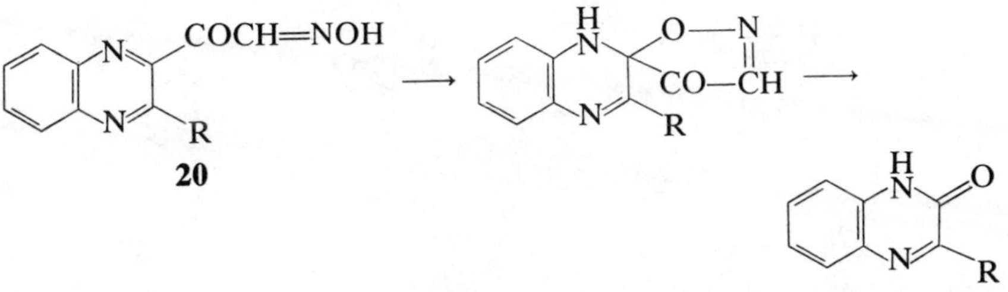

COCH=NOH

R

20

R = Me or Ph

Scheme 11

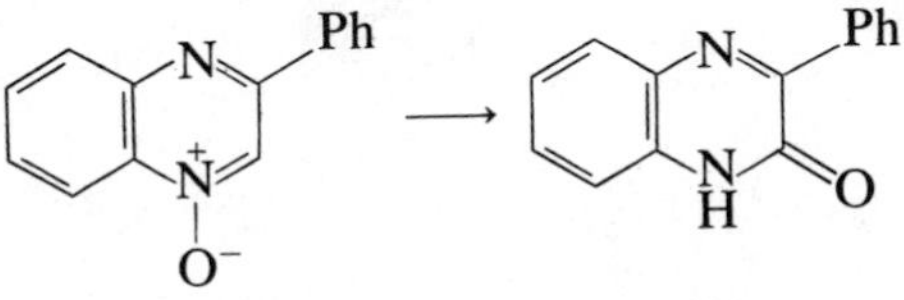

21 **16**

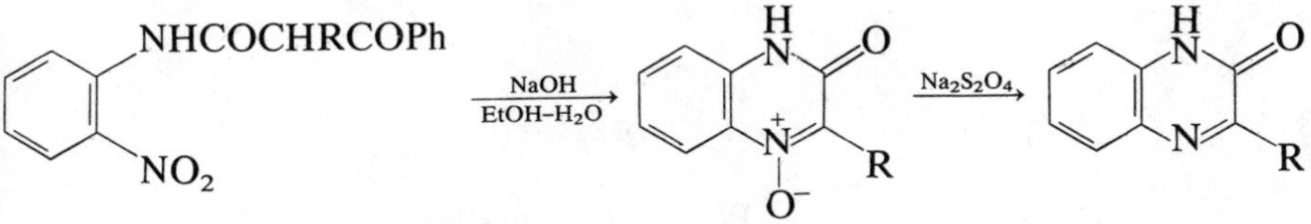

NHCOCHRCOPh

NO_2

NaOH / EtOH–H_2O

$Na_2S_2O_4$

22 **23** **24**

D. Quinoxalinones from Reactions in Which a Change of Ring Size Occurs

The quinoxalinone **26** is obtained in 34% yield when the diazetidone **25** is heated in boiling methanol. Rearrangement is accompanied by aerial oxidation of the intermediate 3,4-dihydroquinoxalinone and the formation of the ring-opened product **27**.[32,33] Irradiation of the ben-

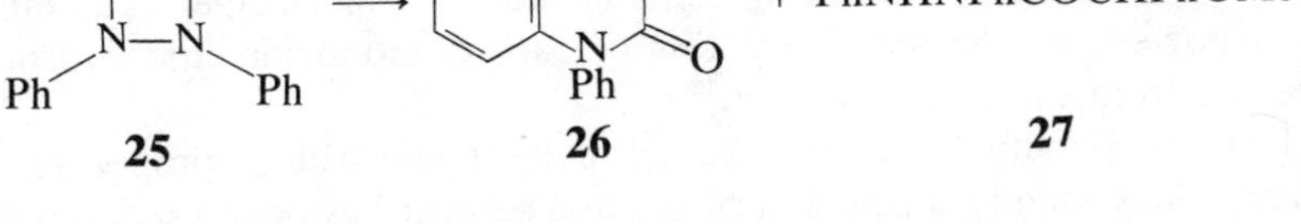

+ PhNHNPhCOCHPhOMe

25 **26** **27**

zodiazepine **28** (X=Cl) gives the dihydroquinoxalinone **29** (X=Cl), whereas the diazepine **28** (X=SMe) yields a mixture of the quinoxalinone

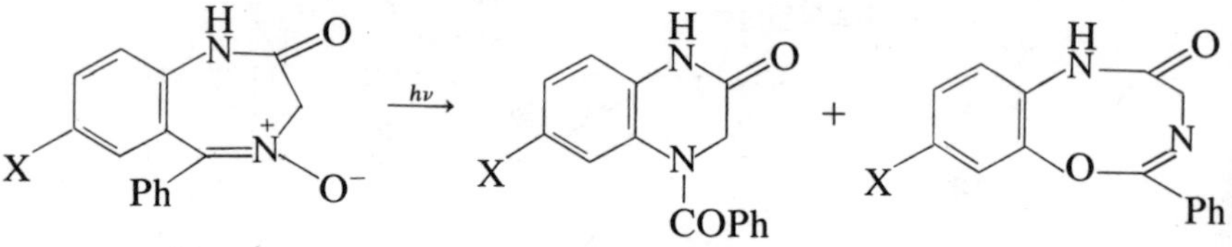

28 → (hν) **29** + **30**

29 (X=SMe) and the oxadiazocine **30** (X=SMe).[34] Other benzodiazepine to quinoxaline ring contractions have been reported.[35,35a,36] The quinoxalinone **32** is unexpectedly formed on hydrogenation of the benzotriazepinone **31**.[37]

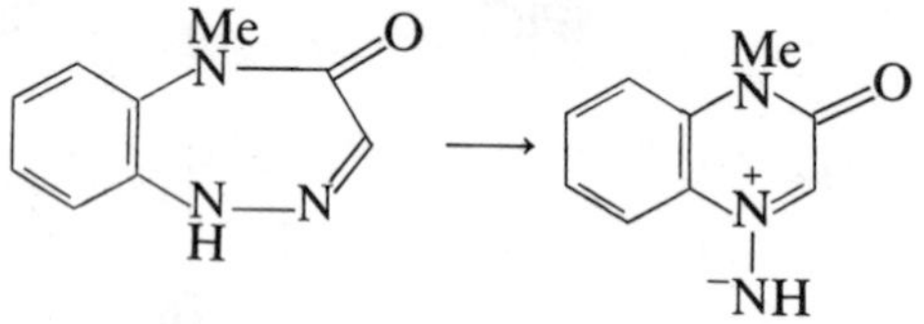

31 → **32**

2. Properties

A. *Physical Properties*

Quinoxalin-2-ones show carbonyl stretching absorption in the region of 1660–1690 cm^{-1} both in Nujol mulls and KBr discs.[38,39] The ultraviolet spectrum of quinoxalin-2-one shows maxima at 343, 287, 254, 250, and 228 nm in aqueous solution at pH 4.0 and is closely similar to that of its 1-methyl derivative. These data indicate that quinoxalin-2-one exist predominately in the cyclic amide form rather than as 2-hydroxyquinoxaline, and this conclusion is further supported by the closely similar pK_a values of quinoxalin-2-one and its 1-methyl derivative, which are -1.38 and -1.15, respectively. 2-Methoxyquinoxaline is by comparison an appreciably stronger base, with a pK_a value of 0.28.[40] X-Ray examination of crystals of quinoxalin-2-one confirms that it exists in the cyclic amide form.[199] Quinoxalin-2-one fragments in the mass spectrometer, as might be expected, by the successive loss of carbon monoxide and hydrogen cyanide from the molecular ion.[41]

Detailed examination of the ultraviolet absorbing properties of quinoxalin-2-one as a function of pH and solvent has been carried out,[42] and the fluorescence spectra of various quinoxalin-2-ones have been reported.[43] The fluorescence of quinoxalin-2-one has been explored for analytical purposes.[44]

The tautomerism of quinoxalinones with a variety of substituents in the 3-position has also been investigated.[9,45–47] The introduction of a 3-methyl group does not cause any change in tautomeric properties, but when the substituent is alkoxycarbonylmethyl (RO_2CCH_2), the tautomeric equilibrium is both solvent and temperature dependent. Ultraviolet measurements show that in trifluoracetic acid, the ester **33** exists predominantly as tautomer **33a** but that in chloroform, tautomer **33b** is favored.[45] In the solid state, the infrared spectrum indicates that tautomer **33b** is preferred.[9] Variable temperature ^1H NMR measurements in dimethyl sulfoxide solution show that a higher proportion of tautomer **33a** is present at higher temperatures. The dependence of tautomeric equilibrium on temperature is explained by preferential stabilization of tautomer **33b** by intramolecular hydrogen bonding (as shown in structure **33c**) and the disruption of hydrogen bonding at higher temperatures.[46]

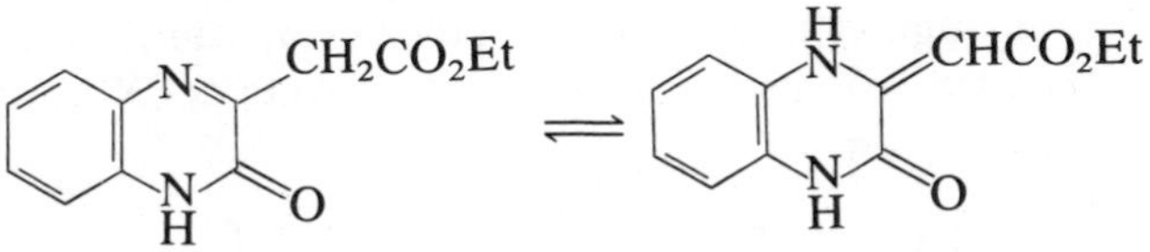

33a ⇌ **33b**

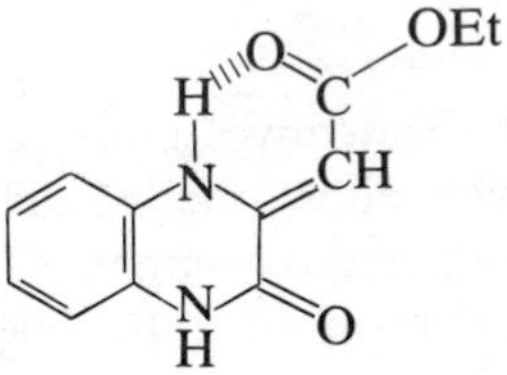

33c

The 1-methyl derivative of **33** has been shown by ^1H NMR studies to exist in deuterochloroform as a mixture of saturated and unsaturated esters (corresponding to **33a** and **33b**). Infrared and ^1H NMR measurements have been made on the 3-phenacylquinoxalin-2-one **34a, b** and a preference for the intramolecularly hydrogen-bonded enamine form **34b** is indicated.[47]

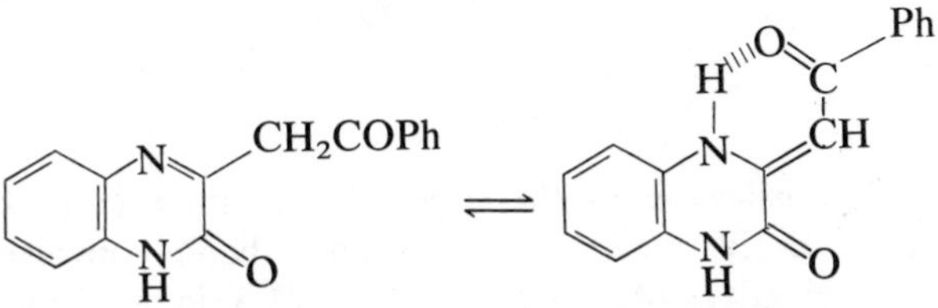

34a ⇌ **34b**

B. Chemical Properties

Polarographic reduction of 3-substituted quinoxalin-2-ones gives both 3,4-dihydro derivatives (**35**) and their 4,4′-dimers (**36**)[48,49] In the case

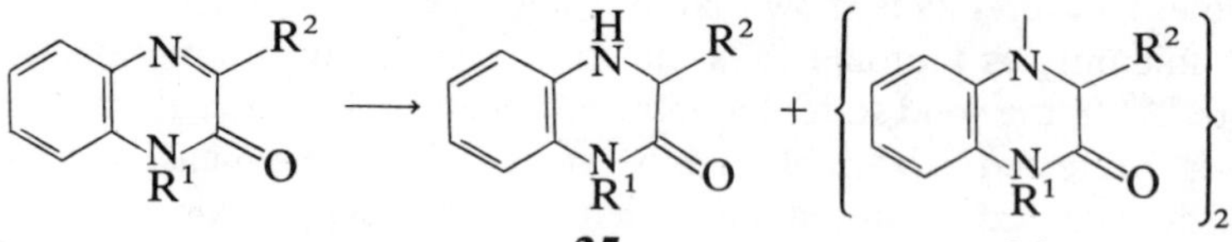

when $R^1 = R^2 = p\text{-MeOC}_6H_4CH_2$-, both reduction products are isolated, but when $R^1 = H$ or Me and $R^2 = Ph$, only the 3,4-dihydro derivative is obtained. Derivatives of the latter type are also formed by catalytic reduction over a platinum catalyst.

Methylation of quinoxalin-2-ones with ethereal diazomethane gives a mixture of *O*-methyl (**37**) and *C,N*-dimethyl derivatives (**38**). Mono-*N*-

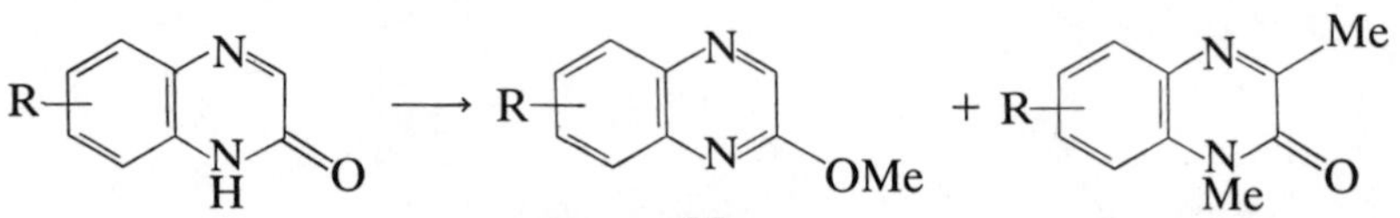

methyl derivatives (e.g., **39**) are conveniently prepared by treatment of the quinoxalinone anion with either methyl sulfate or methyl iodide.[7] Treatment of the anion with hydroxylamine *O*-sulfonic acid gives the corresponding *N*-amino derivative **40**.[50] Reaction of quinoxalin-2-ones

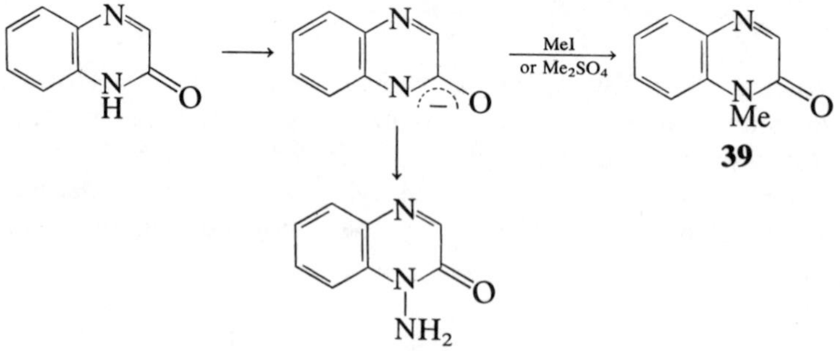

with hexamethyldisilazane in the presence of ammonium sulfate gives *O*-trimethylsilyl derivatives, and these have been converted into *N*-ribofuranosyl derivatives (**41**) (Scheme 12).[51] *O*-Trimethylsilyl derivatives

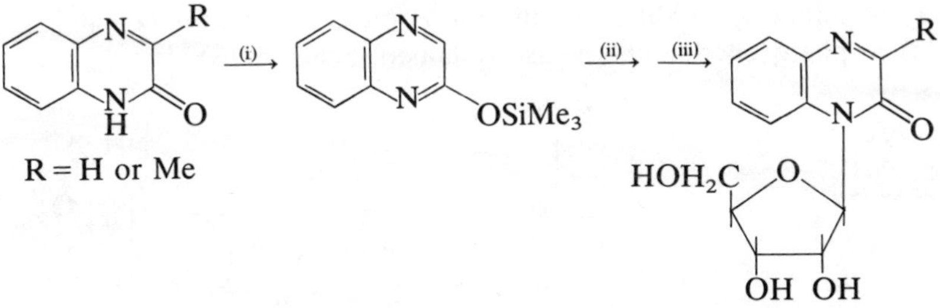

R = H or Me

41

Reagents (i) $Me_3SiNNSiMe_3$–$(NH_4)_2SO_4$
(ii) 1-Bromo-2,3,5-tri-*O*-benzoylribofuranose–$HgBr_2$/HgO or 1-acetyl-2,3,5-tri-*O*-benzoylribofuranose–$SnCl_4$
(iii) NaOMe–HOMe

Scheme 12

of quinoxalin-2-ones have also been prepared by using bistrimethylsilyltrifluoroacetamide as the silylating reagent.[52] *O*-Glucosyl derivatives are obtained when the silver or mercury salts of quinoxalin-2-ones are treated with α-bromoacetoglucose.[53]

Quinoxalin-2-ones are readily converted into 2-chloroquinoxalines on treatment with phosphoryl chloride[3] or thionyl chloride containing dimethylformamide,[54] and in view of the highly reactive nature of 2-chloroquinoxalines, quinoxalin-2-ones are much used intermediates in quinoxaline synthesis.

The nitration[7,55] and halogenation[56] of quinoxalin-2-ones in acetic acid occur at position 7, but in concentrated sulfuric acid, where the base is present in a protonated form, nitration occurs at position 6.[7,55] Hydroxylation of quinoxalin-2-one and its *N*-methyl derivative with hydrogen peroxide in acetic acid yields 3-hydroxy derivatives.[57] Phenylation also occurs at position 3 when *N*-methylquinoxalin-2-one is treated with benzenediazonium chloride.[7]

Quinoxalin-2-ones have been found to react with a number of nucleophilic reagents at position 3. For example, reaction of quinoxalin-2-one with morpholine or piperidine in the presence of sulfur gives the 3-substituted quinoxalin-2-ones **42**.[58] Reaction with hydrazine leads by

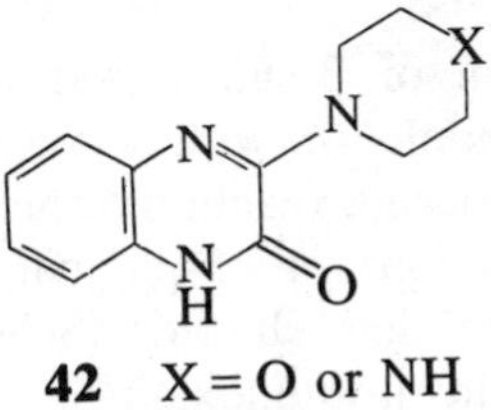

42 X = O or NH

way of initial 3,4-addition of the nucleophile to 2-methylbenzimidazole (**43**); *o*-phenylenediamine is also obtained (Scheme 13).[59]

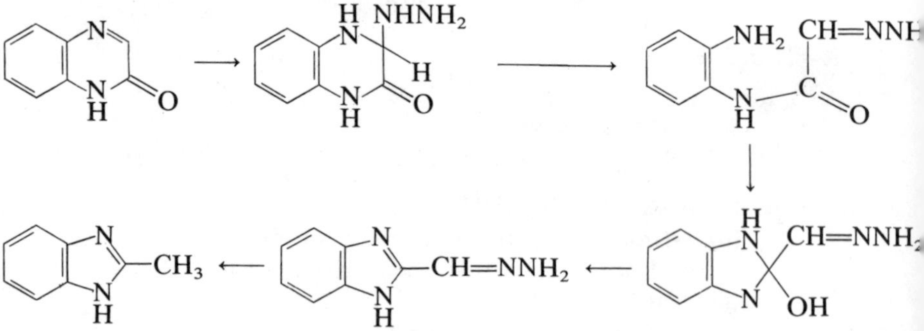

Scheme 13

Quinoxalin-2-one is also attacked in the 3-position by carbon nucleophiles such as Grignard reagents[60] and *N,N*-dimethylaniline[61] to give the products shown in Schemes 14 and 15. An extension of the latter

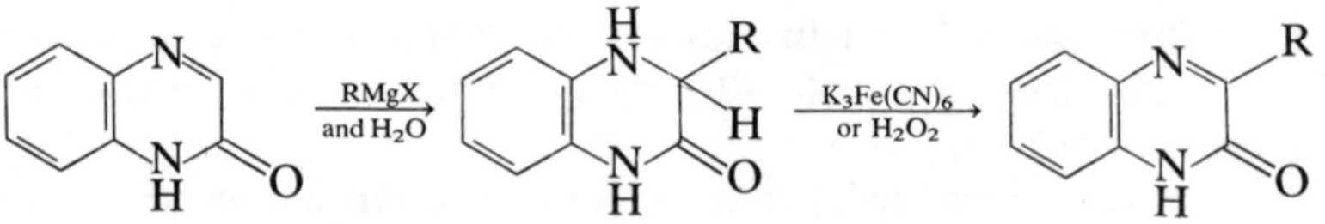

R = Me, Ph, or $(CH_2)_3NMe_2$
X = I, Br, or Cl

Scheme 14

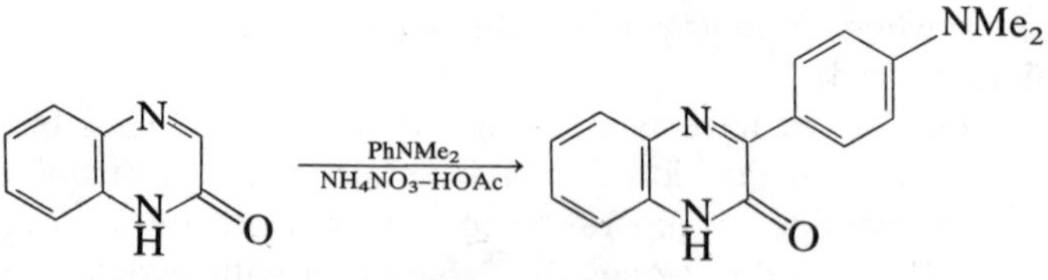

Scheme 15

reaction is the formation of 3-substituted derivatives on reaction of quinoxalin-2-one with indole or with 1-methyl- or 1-phenylpyrrole.[62] When the 3-position is blocked, reaction occurs at the carbonyl carbon as illustrated by the reaction of 3-phenylquinoxalin-2-one with 3-dimethylaminopropylmagnesium chloride (Scheme 16).[60] It is interesting to note that when 3,4-dihydroquinoxalin-2-ones such as compound **44**

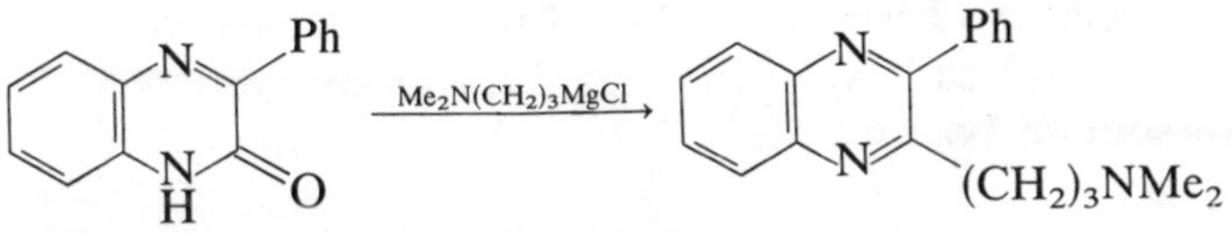

$Me_2N(CH_2)_3MgCl$

Ph, $(CH_2)_3NMe_2$

Scheme 16

are treated with ammonia or an amine in the presence of a titanium chloride–tetrahydrofuran complex, amidines (e.g., **45**) are formed.[63]

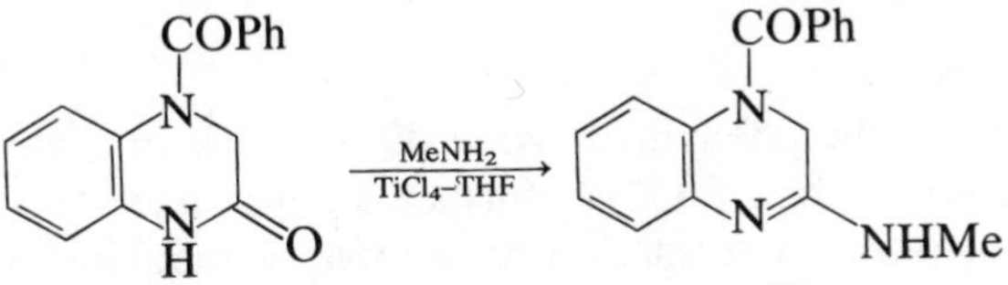

COPh, $MeNH_2$, $TiCl_4$–THF, NHMe

44 **45**

Various tricyclic derivatives have been prepared from quinoxalin-2-ones. Thus derivatives of the 4,6,6-system, azeto[1,2-*a*]quinoxaline, have been synthesized by the route illustrated in Scheme 17.[63a] β-Lactams of

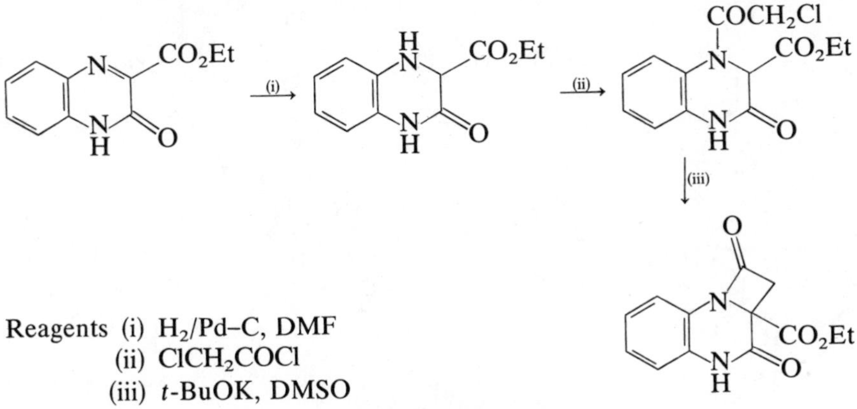

Reagents (i) H_2/Pd–C, DMF
(ii) $ClCH_2COCl$
(iii) *t*-BuOK, DMSO

Scheme 17

this type were not found to possess any antimicrobial activity. Reaction of the carboxyureide **46** with methyliodide and potassium carbonate gives a trimethyl derivative, the spiro compound **47**.[64]

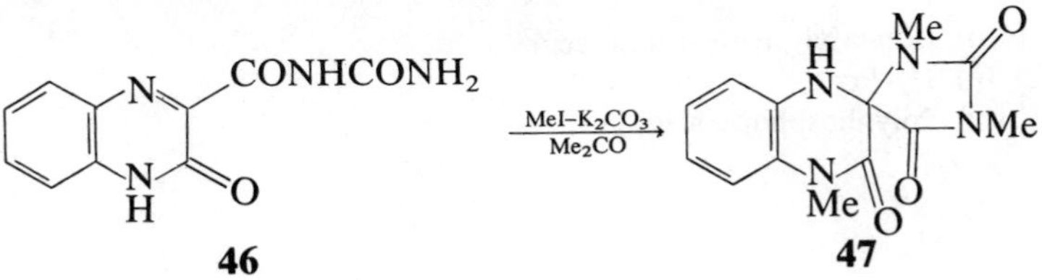

$CONHCONH_2$, $MeI–K_2CO_3$, Me_2CO

46 **47**

A very facile synthesis of tricyclic diamides of type **48** involves the treatment of 2-(2-carboxyethyl)- and 2-(3-carboxypropyl)quinoxalin-3-one with sodium borohydride followed by acidification.[65] Ring closure

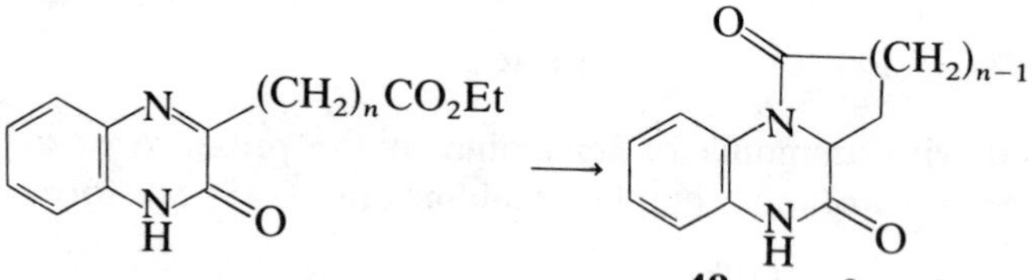

$(CH_2)_nCO_2Et$ → $(CH_2)_{n-1}$

48 $n = 2$ or 3

also occurs when the phenylhydrazone **49** is boiled in acetic acid to give the pyrazoloquinoxaline **50**.[66] A further extension of this type of ring closure is illustrated by formation of the pyridazinoquinoxaline **51**.[67]

CPh=NNHPh → **49** → **50**

CH_2CPh=NNHPh → **51**

33 $CHCO_2Et$ —(i)→ CO_2Et —(ii)→ NH_2, CO_2Et

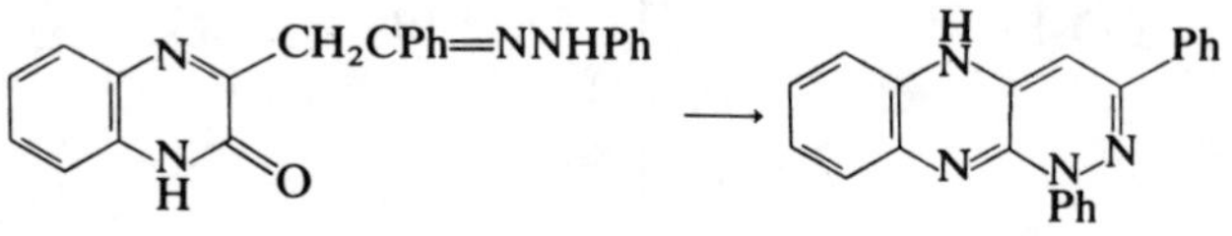

(iii) → **52** CO_2Et

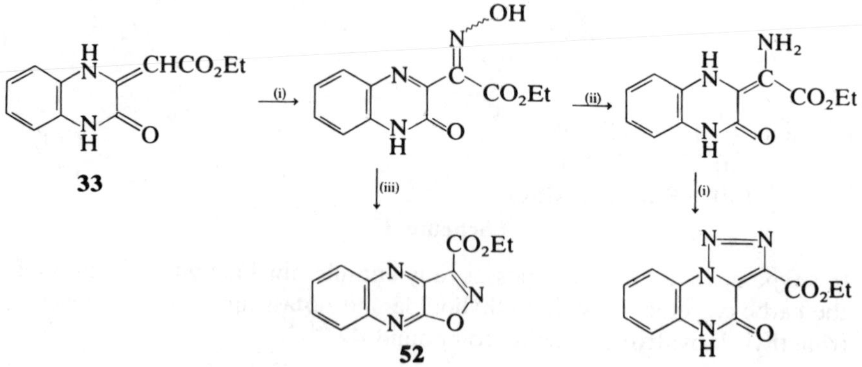

(i) → **53** CO_2Et

Reagents (i) Isopentyl nitrite–acetic acid
(ii) H_2–Pt
(iii) Polyphosphoric acid

2-Ethoxycarbonylmethylene-1,2,3,4-tetrahydroquinoxalin-2-one (**33**) has been used as an intermediate to synthesize the isoxazolo- and triazoloquinoxalines **52** and **53**.[68] On heating the quinoxalinylpropionic ester (**54**)* in diphenyl ether, ring closure to the furoquinoxaline (**55**) occurs.

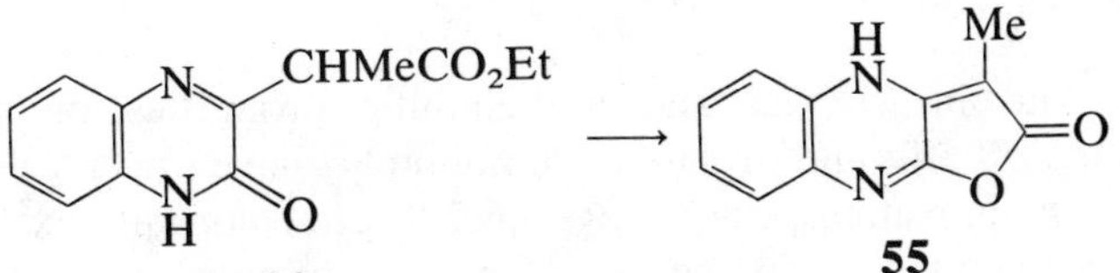

CHMeCO$_2$Et → Me, =O

54 **55**

3. Uses

There are numerous references in the literature to the preparation of quinoxalinones as characterizing derivatives for α-keto acids. Paper chromatographic,[69–73] gas–liquid chromatographic,[52] and fluorimetric methods[73] have been used for the microestimation of quinoxalinones. Quinoxalinone formation is also the basis of a method for amino acid identification,[74] as shown in Scheme 18. It has found application in

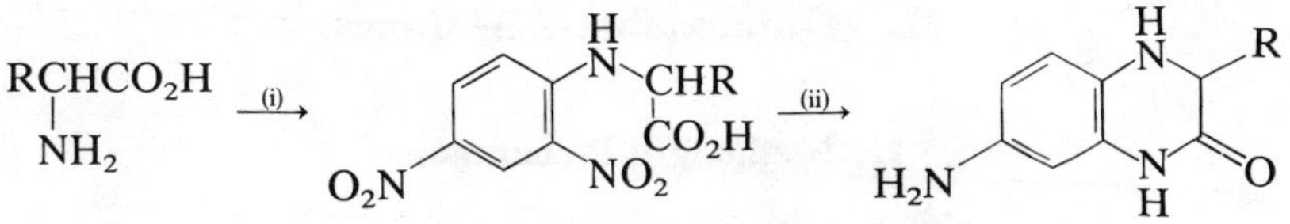

RCHCO$_2$H (NH$_2$) $\xrightarrow{(i)}$ O$_2$N–C$_6$H$_3$(NO$_2$)–NH–CHR–CO$_2$H $\xrightarrow{(ii)}$ H$_2$N-substituted 3-R-3,4-dihydroquinoxalin-2(1H)-one

Reagents (i) 1-Fluoro-2,4-dinitrobenzene
(ii) $(NH_4)_2S$

Scheme 18

various systematic studies of peptide sequencing.[75,76] 4-Acyl-3,4-dihydro-2-quinoxalinones have been reported to have a variety of useful pharmacological properties.[77] For example, the thienoyl derivative (**56**) is claimed to have antiinflammatory properties.[78] Various 3-benzylquinoxalin-2-ones have been found to have analgesic activity,[79] and the quinoxalinone **57** has been shown to have a tranquilizing and antidepressant effect.[80] 3-Methyl-7(or 6)-ethoxy-2(1*H*)-quinoxalinone strongly inhibits the regeneration of the amputated tail tip of *Xenopus*

*The propionic ester, unlike the corresponding acetic acid ester (**33**), exists entirely in the saturated ester form.

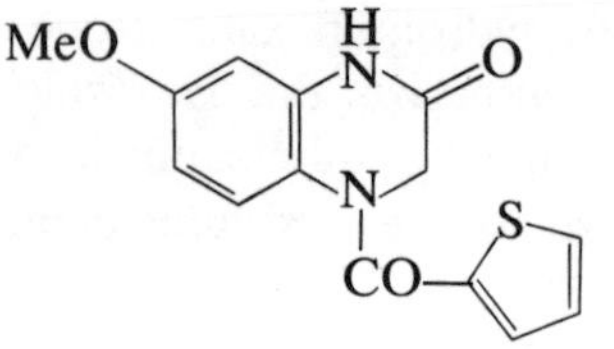

MeO, H, N, O, N, CO, S

56

N, $CH_2CH_2CONH_2$, N, H, O

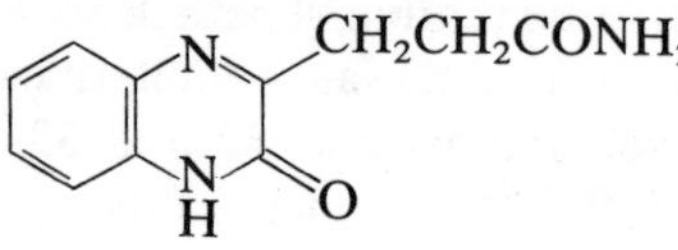

57

larvae.[81] The insecticidal and anthelmintic properties of the phosphorothionate **58** and related compounds have been reported.[82] Tetrachloroquinoxalines, including 5,6,7,8-tetrachloroquinoxalin-2-one, have been patented as fungicides,[83] and polyquinoxalinones have been found to be thermally stable polymers. For example, the polyquinoxalinone **59** has been prepared from *p*-phenylenediglyoxylic acid and 3,3′,4,4′-tetraminobiphenyl.[84]

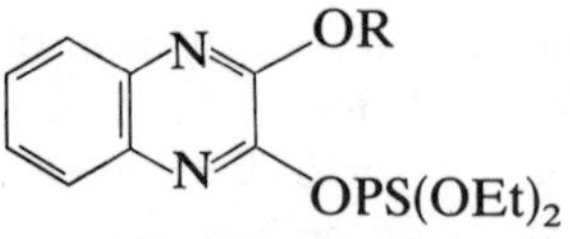

N, OR, N, $OPS(OEt)_2$

R = Me or Et

58

N, O, N, H, N, N, H, O, n

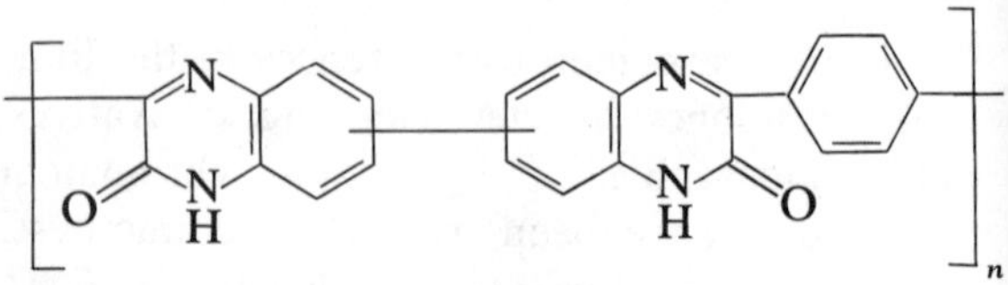

59

II. Quinoxaline-2,3-diones

1. Methods of Preparation

Quinoxaline-2,3-diones are conveniently prepared by boiling a mixture of the *o*-phenylenediamine with oxalic acid dihydrate in 4 *M* hydrochloric acid.[85] Alternatively a mixture of the diamine is refluxed with excess diethyl oxalate under an air condenser so that the ethyl alcohol formed by intermolecular condensation is allowed to escape (Scheme 19).[86] 6-Nitroquinoxaline-2,3-dione has been prepared by fusion of 4-nitro-*o*-phenylenediamine with oxalic acid at 160°; in this preparation diethyl

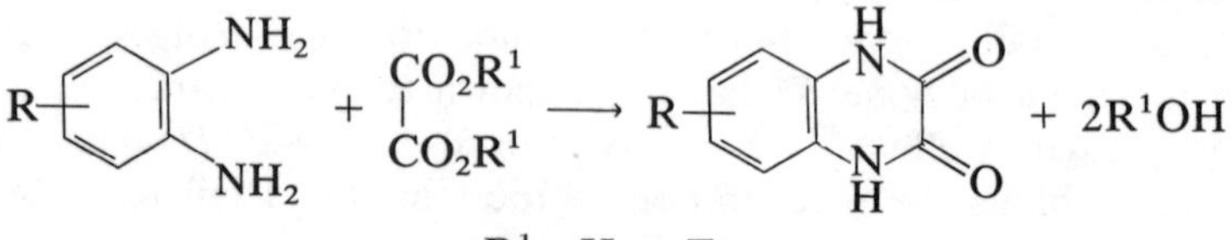

R, NH_2, NH_2 + CO_2R^1 | CO_2R^1 ⟶ R, H, N, O, N, H, O + $2R^1OH$

R^1 = H or Et

Scheme 19

oxalate could not be substituted for oxalic acid.[87] A simple extension of this method is the condensation of *N*-substituted *o*-phenylenediamines with ethyl oxalate to give 1-substituted or 1,4-disubstituted quinoxaline-2,3-diones.[88]

Quinoxaline-2,3-diones have been obtained by the hydrolysis of various quinoxaline derivatives. For example, hydrolysis of hexafluoroquinoxaline with sulfuric acid yields 5,6,7,8-tetrafluoroquinoxaline-2,3-dione.[89] Treatment of 3-hydrazinoquinoxaline-2-thione with 2.5*M* hydrochloric acid gives quinoxaline-2,3-dione,[90] which is also obtained on treatment of 3-aminoquinoxalin-2-one with nitrous acid.[91]

Oxidation of quinoxaline with aqueous nitric acid[91a] and 2,3-dimethylquinoxaline with 20% nitric acid yields 6-nitroquinoxaline-2,3-dione.[92] Quinoxalinediones are also isolated when quinoxalines are oxidized with peracids. Electronegative substituents such as nitro and cyano in the 6-position promote the formation of quinoxalinediones rather than *N*-oxides.[93]

Quinoxaline-2,3-diones have been obtained on heating benzodiazepinetriones; for example, when compound **60** is heated under reflux in xylene, the quinoxalinedione **61** is formed.[94] Derivatives of 6,7-dimethylquinoxaline-2,3-dione have been isolated from the products of bacterial degradation of riboflavin.[95]

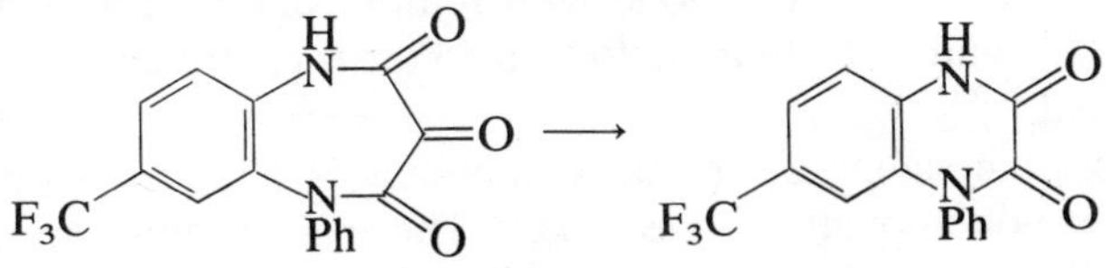

H N O =O N Ph O F_3C → H N O N Ph O F_3C

60 **61**

2. Properties

Quinoxaline-2,3-diones **62a** are known to exist as such rather than as hydroxyquinoxalinones **62b** or as dihydroxyquinoxalines **62c** from infrared and ultraviolet measurements.[38] A mixture of the dimethyl derivatives of the three possible tautomeric forms is obtained on treatment of a suspension of quinoxaline-2,3-dione in methanol with ethereal diazomethane.[88]

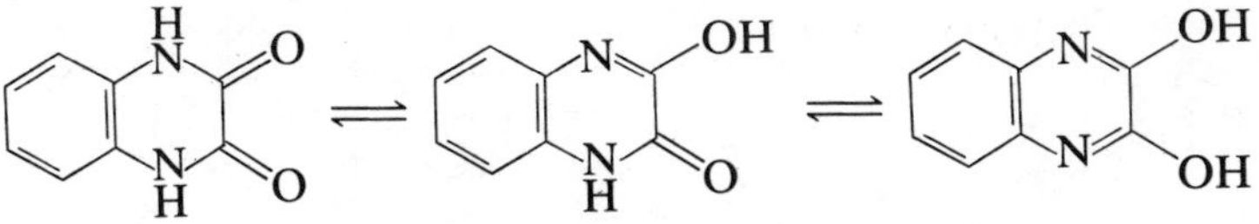

H N O N H O ⇌ N OH N H O ⇌ N OH N OH

62a **62b** **62c**

Diazomethane methylation of 5,6,7,8-tetrafluoroquinoxaline-2,3-dione is, however, reported to yield specifically the unsymmetrical *O,N*-dimenthyl derivative.[89] Both the *N,N*-dimethyl derivative (1,4-dimethylquinoxaline-2,3-dione)[88] and the mono-*N*-methyl derivative (1-methylquinoxaline-2,3-dione) have been obtained by methylation of quinoxaline-2,3-dione in alkaline solution with methylsulfate.[7]

A useful synthetic application of quinoxaline-2,3-diones is their conversion into 2,3-dichloroquinoxalines by treatment with a mixture of phosphoryl chloride and dimethylaniline,[88] with phosphorus pentachloride,[93] or thionyl chloride in the presence of dimethylformamide.[96] 2,3-Dibromoquinoxaline has been similarly prepared from quinoxaline-2,3-dione by reaction with a mixture of phosphoryl bromide and dimethylaniline.[97]

Electrophilic substitution of quinoxaline-2,3-dione and its 1,4-dimethyl derivative occurs at positions 6 and 7.[98] For example, treatment of a solution of quinoxaline-2,3-dione in concentrated sulfuric acid with one equivalent of potassium nitrate yields the 6-nitro derivative, whereas with two equivalents of potassium nitrate 6,7-dinitroquinoxaline-2,3-dione is formed. 6-Bromo and 6,7-dibromoquinoxaline-2,3-dione have been prepared by bromination of the 2,3-dione in sulfuric acid with bromine and silver sulfate, and chlorination is found to occur under similar conditions. Treatment of quinoxaline-2,3-dione with fuming sulfuric acid yields the 6-sulfonic acid,[99] and with chlorosulfonic acid the 6-chlorosulfonyl derivative is obtained.[100]

Reaction of quinoxaline-2,3-dione with boiling 50% aqueous hydrazine gives 3-hydrazinoquinoxalin-2-one (**63**).[59] The main product formed on reaction of the dione with *o*-phenylenediamine is 2,2′-dibenzimidazoyl (**65**), presumably as a result of ring closure of the intermediate diamide **64**.[101]

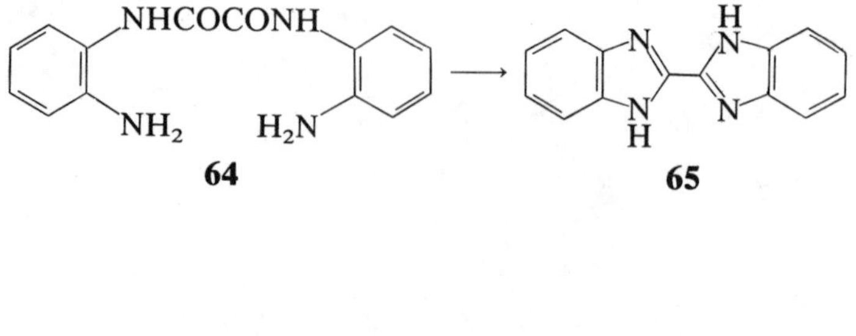

$NHNH_2$

63

NHCOCONH

NH_2 H_2N

64

65

3. Uses

The hypnotic activity of a number of 6-fluoroalkylquinoxaline-2,3-diones has been patented,[158] and the quinoxalinedione **66** is reported to

66

have hypotensive activity in dogs.[102] There is also a patent report of the insecticidal properties of 6-nitroquinoxaline-2,3-dione.[103] Complex Ni-containing pigments have been prepared from 6,7-diaminoquinoxaline-2,3-dione,[104] and heat-resistant ladder polymers are prepared from this diamino compound by heating with phosphoric acid at 210–230°.[105]

III. Tables

TABLE 1. QUINOXALIN-2-ONES (2-HYDROXYQUINOXALINES)

Molecular formula	Substituents or structure	m.p., °C	Ref.
$C_8H_2Cl_4N_2O$	5,6,7,8-Tetrachloro	319	106
$C_8H_4Cl_2N_2O$	6,7-Dichloro	343 decomp.	8
$C_8H_5BrN_2O$	7-Bromo	Decomp. before melting	56
$C_8H_5ClN_2O$	3-Chloro		
	1-Methyl deriv.	131–133	88
	6-Chloro	300–305, 312–313 decomp., 320–321	15, 107, 108, 110
	7-Chloro	251–252, 261–263 269–270 decomp.	56,107, 109, 110
	1-*p*-Chlorophenyl deriv.	229–230.5	111
$C_8H_5N_3O_3$	6-Nitro	294, 300–302	7, 87
	1-Methyl deriv	219–220	7
	7-Nitro	273, 274–276	2, 7, 87, 112
$C_8H_6N_2O$	Unsubstituted	262–264, 267–269, 268–270	2, 3, 16, 113
	1-Amino deriv.		50

TABLE 1 (*contd.*)

Molecular formula	Substituents or structure	m.p., °C	Ref.
$C_8H_6N_2O$ (*contd.*)	1-Methyl deriv.	122–123	3
	1-Carboxymethyl deriv.	225–226 decomp.	16
	1-β-D-Ribofuranosyl deriv.	165–167	51
$C_8H_7N_3O$	3-Amino	>360	26, 114, 115 116, 117
	N-Acetyl	>350	118
	N-Benzoyl	255, 263–267 decomp.	118, 200
	N-*o*-Chlorobenzoyl	230	118
	N-*p*-Chlorobenzoyl	281	118
	N-*m*-Nitrobenzoyl	272	118
	N-*p*-Nitrobenzoyl	288	118
	N-Ethoxycarbonyl	189–190	116
	O-Benzenesulfonyl	180–181	115a
	1-Methyl deriv.	273–275	88
	7-Amino	>360	2, 112
$C_8H_7N_3O_2$	7-Amino-5-hydroxy	>350	87
$C_8H_8N_4O$	3-Hydrazino	>360	59, 112
	Acetyl deriv.	>300	119
	Isonicotinoyl deriv.	>300	119
	1-Methyl deriv.	156–157	59
	Benzylidene deriv. of 1-methyl deriv.	229–233	59
$C_9H_3Cl_4N_2O$	5,6,7,8-Tetrachloro-3-methyl	324–325	106
$C_9H_5N_3O$	3-Cyano	284–287, 290, 300 decomp.	68, 120, 121
$C_9H_6N_2O_2$	3-Formyl	72–74	122
	Oxime	268–271 decomp.	120
	Phenylhydrazone	275–276	122
	Phenylhydrazone of 1-methyl deriv.	198	123
$C_9H_6N_2O_3$	3-Carboxy	268	124
	Ethyl ester	266	63a
	1-Methyl deriv.	172.5–173 decomp.	64, 124
	Ethyl ester of 1-methyl deriv.	125.5	124
	6-Carboxy		125
	7-Carboxy		125
$C_9H_7BrN_2O$	3-Bromomethyl	225 decomp.	126
	Pyridinium bromide	244–245	126
	1-Methyl deriv.	193 decomp.	65
	3-Methyl-6-bromo	252.5–254	25
	3-Methyl-7-bromo	291–293 decomp.	25
$C_9H_7ClN_2O$	3-Methyl-7-chloro	274–275	127
$C_9H_7N_3O_2$	3-Carbamoyl	307–308	68
$C_9H_7N_3O_3$	3-Methyl-6-nitro	284–285	128
	1-Methyl deriv.	222–223	7
	3-Methyl-7-nitro	252 decomp.	87

TABLE 1 (*contd.*)

Molecular formula	Substituents or structure	m.p., °C	Ref.
$C_9H_7N_3O_4$	5-Methoxy-7-nitro	276–278 decomp.	87
$C_9H_8N_2O$	3-Methyl	240–242, 244–245 250, 251–252	22, 23, 24, 29, 126, 129, 130, 131
	1-Methyl deriv.	85–86	88
	Picrate of 1-methyl deriv.	144–145	88
	1-Allyl deriv.	56–57	24
	1-Propyl deriv.	47–49	24
	1-Phenyl deriv.	201	132
	1-β-D-Ribofuranosyl deriv.	180–182	93
	5-Methyl	282–283	93
	6-Methyl		23, 125
$C_9H_8N_2O_2$	3-Methoxy	243–244 decomp.	88
	7-Methoxy	232–234	109
$C_9H_8N_2O_2$	5-Hydroxy-3-methyl	252-254	133
	8-Hydroxy-3-methyl	236–237	133
$C_9H_9N_3O$	6-Amino-3-methyl		
	Hemihydrate	296–298	128
	Benzenesulfonyl deriv. hydrate	267	128
	3-Methylamino		
	1-Methyl deriv.	158	64
	Acetyl deriv. of 1-methyl deriv.	205	64
$C_9H_9N_3O_2$	6-Amino-8-methoxy	198	87
	7-Amino-5-methoxy	318–320 decomp.	87
$C_9H_9N_3O_3$	3,4-Dihydro-3-methyl-7-nitro	225 decomp.	87
$C_9H_9N_3O_3S$	3-Methyl-7-sulfamoyl	325–327	127
$C_9H_9N_3O_4$	3,4-Dihydro-5-methoxy-7-nitro	262 decomp.	87
$C_9H_{10}N_4O$	5,7-Diamino-3-methyl	314 decomp.	134
	3-(1-Methylhydrazino)	215–220 decomp.	59
	1-Methyl deriv.	137–138	59
	Benzylidene deriv.	>185 decomp.	59
$C_{10}H_6N_4O_7$	3-Carboxymethyl-5,7-dinitro		
	Ethyl ester	265	134
	3-Carboxymethyl-6,8-dinitro		
	Ethyl ester	214 decomp.	134
$C_{10}H_7N_3O$	3-Cyanomethyl		11
$C_{10}H_8N_2O_2$	3-Acetyl	188	135
	1-Methyl deriv.	130	135
$C_{10}H_8N_2O_3$	3-Carboxymethyl	247–249.5	69
	Ethyl ester	204–205, 210	22, 129, 152
	1-Methyl deriv. of ethyl ester	144–146	9, 147
	Methyl ester	225	22, 24
	1-Allyl deriv. of methyl ester	109–110	24
	1-Methyl deriv. of methyl ester	164–166, 165–167	9, 147
	1-Propyl deriv. of methyl ester	129–131	24
	6-Carboxy-3-methyl		

TABLE 1 (*contd.*)

Molecular formula	Substituents or structure	m.p., °C	Ref.
$C_{10}H_8N_2O_3$ (*contd.*)	Hydrate	334–336 decomp.	137
	Ethyl ester	229–230	137
	7-Carboxy-3-methyl	329–332 decomp.	137
	Ethyl ester	199–200	137
$C_{10}H_9Cl_2N_3O_2$	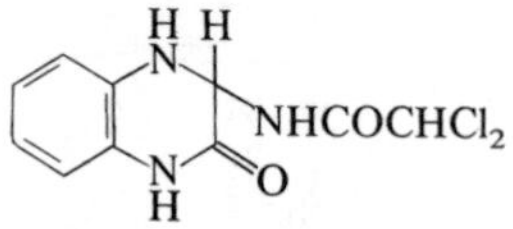 H H N NHCOCHCl$_2$ N H O	190–200 decomp.	136
$C_{10}H_9N_3O_2$	7-Acetamido	348–350	2
$C_{10}H_{10}N_2O$	3-Ethyl	191, 194–196, 196–197	31, 69, 129, 138
	Hydrate	198	129
	1-Methyl deriv.	107	31
	1-Benzyl deriv.	98–99	68
	3,5-Dimethyl	256.5–257.5	139
	3,6-Dimethyl	254–255, 209–210	23, 140, 141
	3,7-Dimethyl	238–239, 243–244	23, 140
	3,8-Dimethyl	256.5–257.5	139
	6,7-Dimethyl	291–292, 292–293	23, 86
$C_{10}H_{10}N_2O_2$	5-Methoxy-3-methyl	260–262	133
	8-Methoxy-3-methyl	220-222	133
$C_{10}H_{10}N_2O_3$	5,8-Dimethoxy	228–228.5	143
$C_{10}H_{10}N_4O$	5,7-Diamino-3-carboxymethyl		
	Ethyl ester	251 decomp.	134
$C_{10}H_{12}N_2O$	3,4-Dihydro-3,3-dimethyl		
	4-Ethoxycarbonyl deriv.	158–159	142
$C_{11}H_6Cl_4N_2O_3$	5,6,7,8-Tetrachloro-3-ethoxycarbonyl	244–245	106
$C_{11}H_{10}N_2O_2$	3-Formyl-6,7-dimethyl		
	p-Nitrophenylhydrazone	>350	23
	3-Acetonyl	257	12
	1-Methyl deriv.	187, 194.5–196, 200	12, 144, 145
$C_{11}H_{10}N_2O_3$	3-(1-Carboxyethyl)		
	Ethyl ester	160–162	129
	3-(2-Carboxyethyl)	255–257, 269–271 decomp., 275–277 decomp.	65, 69, 146
	Ethyl ester	196	22
	1-Methyl deriv.	180–182	65
	Amide		80
$C_{11}H_{12}N_2O$	3,6,7-Trimethyl	278–279	23
	3-Propyl	182–183, 185	6, 17, 31
	1-Methyl deriv.	88	31
	3-Isopropyl	228, 233–235	6, 17, 22, 69, 136

TABLE 1 (*contd.*)

Molecular formula	Substituents or structure	m.p., °C	Ref.
$C_{11}H_{14}N_2O$	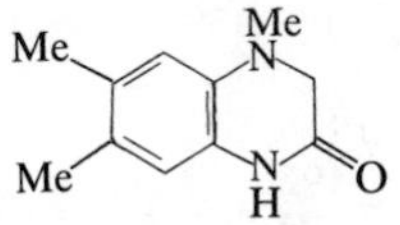	>325	148
$C_{12}H_7BrN_2O_2$	3-(5-Bromo-2-furyl)	124	149
$C_{12}H_7N_3O_4$	3-(5-Nitro-2-furyl)		150
$C_{12}H_8N_2O_2$	2-(2-Furyl)	103–104	151
$C_{12}H_{11}N_3O$	3-(3-Cyanopropyl)	200–201 decomp.	65
$C_{12}H_{12}N_2O_3$	3-(3-Carboxypropyl)	216–218	65
	1-Methyl deriv.	147–150	65
$C_{12}H_{13}N_3O_2$	3-Morpholino	204–205	19
$C_{12}H_{14}N_2O$	3-Butyl	153–154, 156	6, 17, 31
	1-Methyl deriv.	61	31
	3-Isobutyl	186–187, 188.5	6, 22
	3-*t*-Butyl	207–208, >300 decomp.	6, 153, 154
	DL-3-(2-Butyl)	181	6, 22
	(+)-3-(2-Butyl)	194	22
$C_{13}H_{10}N_2O_5$	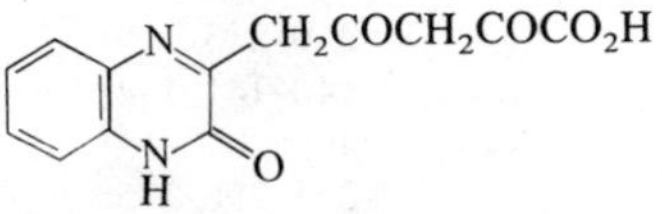 $CH_2COCH_2COCO_2H$		
	Ethyl ester	248 decomp.	172
	1-Phenyl deriv. of ethyl ester	225 decomp.	172
$C_{13}H_{14}N_2O$	3-Cyclopentyl	237–238 decomp.	155
$C_{13}H_{14}N_2O_3$	3-(4-Carboxybutyl)	179–180.5	65
$C_{13}H_{15}N_3O_2$	6-Methyl-3-morpholino		
	1-Benzyl deriv.	153–155	19
$C_{13}H_{16}N_2O$	3-Pentyl	151–152, 156	17, 31
	1-Methyl deriv.	69	31
$C_{13}H_{16}N_4O$	3-(4-Methylpiperazino)	199–200	19
$C_{13}H_{17}N_3O$	3-(3-Dimethylaminopropyl)	128–129	60
$C_{14}H_8Cl_2N_2O$	6,7-Dichloro-3-phenyl	291–292	171a
$C_{14}H_9BrN_2O$	6-Bromo-3-phenyl		171
$C_{14}H_9ClN_2O$	6-Chloro-3-phenyl	262–263	171a
	3-(*p*-Chlorophenyl)	284, 292	156, 157
$C_{14}H_9N_3O_3$	3-(*p*-Nitrophenyl)		
	1-Methyl deriv.	247–248	7
$C_{14}H_{10}N_2O$	3-Phenyl	243, 247, 248–249, 252	5, 21, 22, 27, 29, 30, 130 131, 156, 157, 157a 159, 170

TABLE 1 (*contd.*)

Molecular formula	Substituents or structure	m.p., °C	Ref.
$C_{14}H_{10}N_2O$ (*contd.*)	1-Methyl deriv.	138–139	7
	1-(2-Diethylaminoethyl) deriv. hydrochloride	176–178	5
	1-Phenyl deriv.	192	32, 33,
$C_{14}H_{10}N_2O_2$	3-(*o*-Hydroxyphenyl)	294–295, 306	160, 161, 162
	7-Hydroxy-3-phenyl		
	1-Methyl deriv.	298–300	163
	3-Phenoxy		
	1-Methyl deriv.	170–172	88
$C_{14}H_{10}N_2O_4$	3-(2,4,5-Trihydroxyphenyl)	>350	164
$C_{14}H_{11}N_3O$	3-(*o*-Aminophenyl		
	1-Methyl deriv.	185–186	165
	Acetyl deriv. of 1-methyl deriv.	202–204	165
	3-(*p*-Aminophenyl)		
	1-Methyl deriv.	158–159	7
$C_{14}H_{16}N_2O$	3-Cyclohexyl	257–258	155
$C_{14}H_{16}N_2OS$	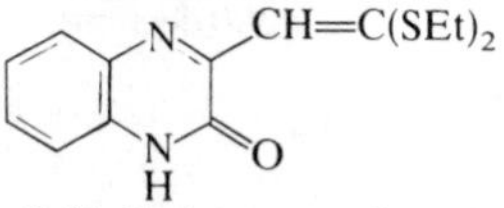 N $CH{=}C(SEt)_2$ N O H	200–203 decomp.	166
$C_{14}H_{16}N_2O_3$	3-(5-Carboxypentyl)	210–213	65
$C_{14}H_{18}N_2O$	3-Hexyl	140–141, 144	17, 31
	1-Methyl deriv.	46	31
$C_{15}H_{10}N_2O_2$	3-Benzoyl	255–258, 275	66, 135
	Phenylhydrazone	248–250	66
	1-Methyl deriv.	159	135
$C_{15}H_{11}ClN_2O$	3-*m*-Chlorobenzyl	228–230	167
	3-*p*-Chlorobenzyl	231	22
$C_{15}H_{11}ClN_2O_2$	4-Benzoyl-6-chloro-3,4-dihydro	255–257	34, 35
$C_{15}H_{12}N_2O$	3-Benzyl	197–198, 198–199, 201–203	22, 66, 68, 69, 146, 168
$C_{15}H_{12}N_2O_2$	3-(*o*-Methoxyphenyl)		169
	3-(*m*-Methoxyphenyl)		169
	3-(*p*-Methoxyphenyl)	269, 275	22, 157, 170
	7-Methoxy-3-phenyl	234–235	171, 171a
	1-Methyl deriv.	94–96	163
	1-*p*-Methoxyphenyl deriv.	208.5–209.5	180
$C_{15}H_{13}N_3O$	3-*N*-Methylanilino		
	1-Methyl deriv.	145–146	165
	3-(*o*-Methylaminophenyl)		
	1-Methyl deriv.	135	165
	N-nitroso deriv. of 1-methyl deriv.	199 decomp.	165
$C_{15}H_{18}N_2O_3$	3-(6-Carboxyhexyl)	150–153	65
$C_{15}H_{20}N_2O$	3-(*n*-Heptyl)	141	31
	1-Methyl deriv.	44	31

TABLE 1 (*contd.*)

Molecular formula	Substituents or structure	m.p., °C	Ref.
$C_{16}H_{11}ClN_2O_2$	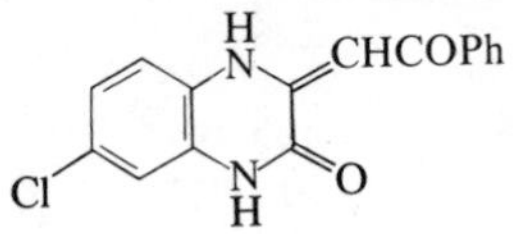 H N CHCOPh Cl N H O	323	47
$C_{16}H_{12}N_2O$	3-Styryl	253–256	173, 174
$C_{16}H_{12}N_2O_2$	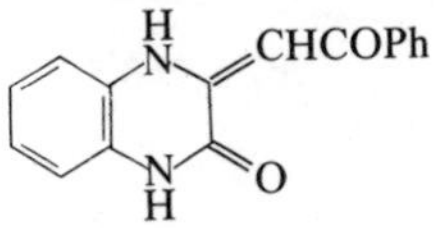 H N CHCOPh N H O	268–269	47
	1-Methyl deriv.	181, 187–188	175, 176
$C_{16}H_{12}N_2O_3$	7-Acetoxy-3-phenyl		
	1-Methyl deriv.	132–134	163
	3-(3,4-Methylenedioxybenzyl)	212, 220	177, 186
	3-*p*-Anisoyl	244–246	178
	Phenylhydrazone	241–242	178
	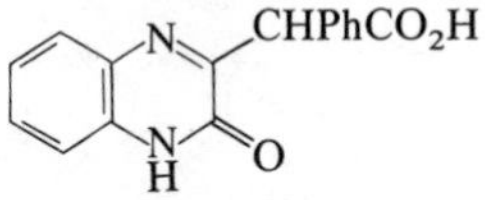 N CHPhCO₂H N H O		
	Ethyl ester	191–192, 196–198	68, 168
$C_{16}H_{14}N_2O_2$	6-Ethoxy-3-phenyl	263–264	171a
	3-(2-Hydroxy-4-ethylphenyl)	260	179
	3-(4-Methoxybenzyl)[a]	198	186
	1-(4-Methoxybenzyl) deriv.		48, 49
$C_{16}H_{14}N_2O_2S$	4-Benzoyl-3,4-dihydro-6-methylthio	242–243	34
$C_{16}H_{14}N_2O_3$	3-(4-Hydroxy-3-methoxybenzyl)	160	167
	7-Methoxy-3-(*p*-methoxyphenyl)		
	1-Phenyl deriv.	215–217	180
	3-(2,4-Dimethoxyphenyl)	262	181
$C_{16}H_{14}N_4O_2$	3-(1-Methyl-2-benzoylhydrazino)		
	1-Methyl deriv.	204–205	59
$C_{17}H_{16}N_2O$	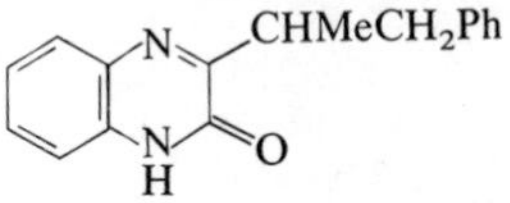 N CHMeCH₂Ph N H O	174–175	68
$C_{17}H_{16}N_2O_4$	3-(3,4,5-Trimethoxyphenyl)	233, 239	182, 183
$C_{18}H_{12}N_2O_3$	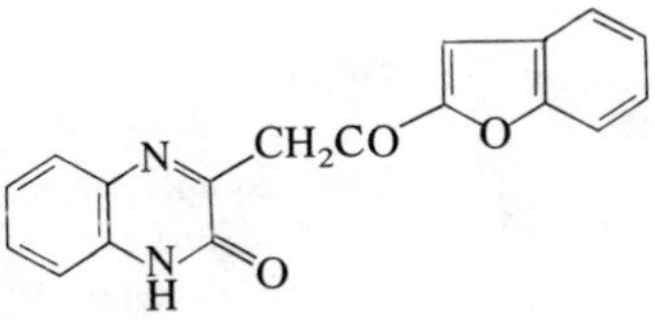 N CH₂CO O N H O	290–292	184

TABLE 1 (*contd.*)

Molecular formula	Substituents or structure	m.p., °C	Ref.
$C_{19}H_{14}N_2O$	3-(1-Naphthylmethyl)	228	167
$C_{19}H_{19}N_3O_2$	3-Benzoyl-7-diethylamino		
	1-Phenyl deriv.	192–193	185
$C_{20}H_{16}N_2O$	3,4-Dihydro-3,3-diphenyl	211–212	142
$C_{21}H_{16}N_2O$	3-Diphenylmethyl	312	186
$C_{22}H_{18}N_2O_2$	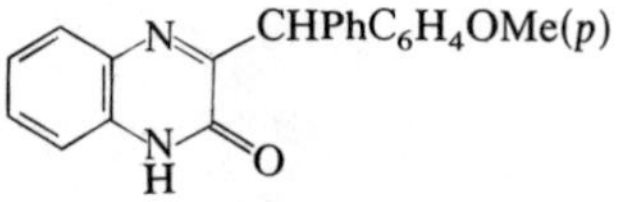 3-[$CHPhC_6H_4OMe(p)$]quinoxalin-2(1H)-one (structure)	267.5–268.5	186

[a] Many related substituted benzylquinoxalinones have been prepared.

TABLE 2. QUINOXALINE-2,3-DIONES (2,3-DIHYDROXYQUINOXALINES)

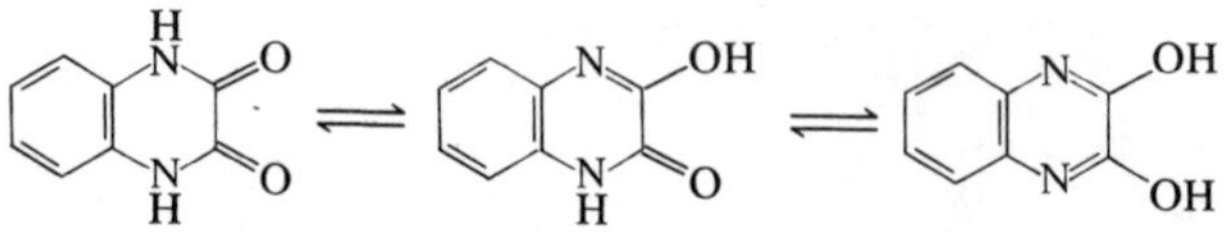

Molecular formula	Substituents	m.p., °C[a]	Ref.
$C_8H_2Cl_4N_2O_2$	5,6,7,8-Tetrachloro	>360	106
$C_8H_2F_4N_2O_2$	5,6,7,8-Tetrafluoro[b]	300 decomp.	89
$C_8H_4Br_2N_2O_2$	6,7-Dibromo	>360	98
	1,4-Dimethyl deriv.	311–312	98
$C_8H_4Cl_2N_2O_2$	5,7-Dichloro	320	187
	5,8-Dichloro		93
	6,7-Dichloro[c]	>360	93, 98
	1,4-Dimethyl deriv.	298–300	98
$C_8H_4N_4O_6$	5,7-Dinitro	270	87
	6,7-Dinitro	>360	98
	1,4-Dimethyl deriv.	279–280	98
$C_8H_5BrN_2O_2$	6-Bromo		93
	1-Methyl deriv.	339–340	98
	1,4-Dimethyl deriv.	214–216	98
$C_8H_5ClN_2O_2$	5-Chloro		93
	6-Chloro		93
	1,4-Dimethyl deriv.	191–192	98
$C_8H_5ClN_2O_4S$	6-Chlorosulfonyl		100
$C_8H_5N_3O_4$	5-Nitro	296 decomp.	188
	6-Nitro	345–346, 355	87, 93, 98
	1-Methyl deriv.	350–352 decomp.	7

TABLE 2 (*contd.*)

Molecular formula	Substituents	m.p., °C[a]	Ref.
$C_8H_5N_3O_4$ (*contd.*)	4-Methyl deriv.	344, >360	93, 98
	1,4-Dimethyl deriv.	249–250, 263–264	98, 189
$C_8H_6N_2O_2$	Unsubstituted[d]	>360	85
	1-Methyl deriv.	278–280, 285–286	7, 37
	1,4-Dimethyl deriv.	252–253	88
	1-Amino-4-methyl deriv.	280–282	37
$C_8H_6N_2O_5S$	6-Sulfo		99, 100
$C_8H_6N_4O_4$	5-Amino-7-nitro	>300	87
$C_8H_7N_3O_2$	5-Amino	339–340 decomp.	188
	6-Amino		
	Hydrochloride	365 decomp.	188
$C_8H_7N_3O_3$	7-Amino-5-hydroxy	>300	87
$C_8H_8N_4O_2$	6,7-Diamino		105
	6-Hydrazino		190
$C_9H_5Cl_2N_3O_4$	5,8-Dichloro-6-methyl-7-nitro	>340 decomp.	191
$C_9H_7F_3N_2O_2$	6-Trifluoromethyl		
	4-Phenyl deriv.		94
$C_9H_5N_3O_2$	6-Cyano	>360	93
	Monomethyl deriv.	353–354	93
$C_9H_6N_2O_4$	5-Carboxy		192
	6-Carboxy		125
$C_9H_7BrN_2O_2$	6-Bromo-7-methyl		93
$C_9H_7ClN_2O_2$	6-Chloro-7-methyl		93
$C_9H_7N_3O_3$	7-Amino-5-methoxy		87
$C_9H_7N_3O_5$	5-Methoxy-7-nitro		
	Hemihydrate	285	87
$C_9H_8N_2O_2$	6-Methyl		93, 125
$C_9H_8N_2O_3$	5-Methoxy	272–274	193
$C_{10}H_6N_2O_6$	6,7-Dicarboxy		125
$C_{10}H_8N_2O_4$	7-Carboxy-6-methyl		125
$C_{10}H_{10}N_2O_2$	6,7-Dimethyl	>325	125, 148, 194
	1,4-Dimethyl deriv.	305	148
$C_{10}H_{10}N_2O_4$	5,8-Dimethoxy	325–329, 332	195, 196
	6,7-Dimethoxy	345–346	196
$C_{12}H_{10}N_2O_6$	5,6-Dicarbomethoxy	264–265 decomp.	197
	6,7-Dicarbomethoxy	285–286	197
$C_{12}H_{14}N_2O_4$	5,8-Diethoxy	269–270, 271	193, 198
$C_{16}H_{22}N_2O_4$	5,8-Dibutoxy	217–218	195
	6,7-Dibutoxy	304	195

[a] The melting points of many of these compounds are very high and therefore not suitable for characterization purposes.
[b] *O,N*-Dimethyl deriv. m.p. 134–136°.
[c] *O,N*-Dimethyl deriv. m.p. 228–230°.
[d] *O,N*-Dimethyl deriv. m.p. 122–123°.

IV. References

1. J. C. E. Simpson, "Condensed Pyridazine and Pyrazine Rings," Interscience, New York, 1953, p. 235.
2. C. M. Atkinson, C. W. Brown, and J. C. E. Simpson, *J. Chem. Soc.*, 26 (1956).
3. G. W. H. Cheeseman, *J. Chem. Soc.*, 3236 (1957).
4. E. J. Wolf, K. Pfister, R. H. Beutel, R. M. Wilson, C. A. Robinson, and J. R. Stevens, *J. Am. Chem. Soc.*, **71,** 6 (1949).
5. J. Drury and A. Hüne, *Helv. Chim. Acta*, **35,** 2301 (1952).
6. D. C. Morrison, *J. Am. Chem. Soc.*, **76,** 4483 (1954).
7. G. W. H. Cheeseman, *J. Chem. Soc.*, 1246 (1961).
8. R. M. Acheson, *J. Chem. Soc.*, 4731 (1956).
9. D. D. Chapman, *J. Chem. Soc.* (*C*), 806 (1966).
10. Y. Iwanami, *J. Chem. Soc. Jap.*, **82,** 788 (1961).
11. N. S. Vul'fson, L. I. Lu Kashina, and S. L. Davydova, *Khim. Tekhnol. i Primenenie Proizv. Piridina i Khinolina, Materialy Soveshch., Inst. Khim. Akad. Nauk. Latv. SSR Riga*, 243 (1957); *Chem. Abstr.*, **57,** 16604 (1962); *Chem. Abstr.*, **55,** 18748 (1961).
12. S. Fatutta and A. Stener, *Gazz. Chim. Ital.*, **88,** 89 (1958); *Chem. Abstr.*, **53,** 2244 (1959).
13. H. Dahn, J. P. Leresche, and H. P. Schlunke, *Helv. Chim. Acta*, **49,** 26 (1966).
14. V. D. Romanenko, N. E. Kul'chitskaya, and S. I. Burmistrov, *Chem. Heterocycl. Comp.*, **9,** 244 (1973).
15. H. G. Petering and G. J. van Giessen, *J. Org. Chem.*, **26,** 2818 (1961).
16. F. Cuiban, M. Ionesco, H. Bala, and M. Steresco, *Bull. Soc. Chim. France*, 356 (1963).
17. M. Goldweber and H. P. Schultz, *J. Am. Chem. Soc.*, **76,** 287 (1954).
18. H. Zellner (to Donau-Pharmazie G.m.b.H), Ger. Offen. 1,804,328; *Chem. Abstr.*, **71,** 70642*a* (1969).
19. Ciba Ltd., Brit. Pat. 1,021,195; *Chem. Abstr.*, **65,** 13738 (1966).
20. H. Baganz and P. Pflug, *Chem. Ber.*, **89,** 689 (1956).
21. J. H. Boyer and D. Straw, *J. Am. Chem. Soc.*, **75,** 1642 (1953).
22. F. Weygand, W. Steglich, and H. Tanner, *Annalen*, **658,** 128 (1962).
23. N. G. Koshel and I. Ya. Postovskii, *Chem. Heterocycl. Comp.*, **6,** 633 (1970).
24. H. Suschitzky, B. J. Wakefield, and R. A. Whittaker, *J. Chem. Soc., Perkin I*, 401 (1975).
25. J. Van Dusen and H. P. Schultz, *J. Org. Chem.*, **21,** 1326 (1956).
26. E. Schipper and A. R. Day, *J. Am. Chem. Soc.*, **73,** 5672 (1951).
27. F. Krönke and H. Leister, *Chem. Ber.*, **91,** 1479 (1958).
28. M. T. LeBris and D. Berkovitch, *Bull. Soc. Chim. Fr.*, 1285 (1973).
29. V. V. Titov and L. F. Kozhokina, *Tetrahedron Lett.*, 1105 (1973).
30. E. Hayashi and C. Iijima, *Yakagaku Zasshi*, **82,** 1093 (1962); *Chem. Abstr.*, **58,** 4551 (1963).
31. G. Tennant, *J. Chem. Soc.* (*C*), 2285 (1966).
32. W. Fischer and E. Fahr, *Angew. Chem. Int. Ed.*, **6,** 630 (1967).
33. D. F. Morrow and L. A. Regan, *J. Org. Chem.*, **36,** 27 (1971).
34. R. Y. Ning, G. F. Field, and L. H. Sternbach, *J. Heterocycl. Chem.*, **7,** 475 (1970).
35. S. C. Bell and S. J. Childress, *J. Org. Chem.*, **29,** 506 (1964).

35a. G. F. Field and L. H. Sternbach (to Hoffmann–LaRoche Inc.), U.S. Pat. 3,697,545; *Chem. Abstr.*, **78,** 16253*f* (1973).

36. M. Matsumoto, Y. Matsumura, A. Iio, and T. Yonezawa, *Bull. Chem. Soc. Jap.*, **43,** 1496 (1970).
37. S. Rossi, O. Pirola, and F. Selva, *Tetrahedron,* **24,** 6395 (1968).
38. G. W. H. Cheeseman, A. R. Katritzky, and S. Oksne, *J. Chem. Soc.*, 3983 (1961).
39. J. Derkosch, *Monatsh.*, **92,** 1107 (1961).
40. G. W. H. Cheeseman, *J. Chem. Soc.*, 108 (1958).
41. V. Kovacik, M. Federonko and I. Jezo, *Org. Mass Spectrom.*, **7,** 449 (1973).
42. E. Tombacz, *Acta Univ. Szegediensis, Acta Phys. Chem.*, **3,** 56 (1957); *ibid.* **6,** 43 (1960); *Chem. Abstr.*, **53,** 1363 (1959); *ibid.* **55,** 16138 (1961).
43. I. Kumashiro, *Nippon Kagaku Zasshi,* **82,** 1224 (1961); *Chem. Abstr.*, **58,** 4047 (1963).
44. G. Milch, *Magy Kem. Lapja,* **28,** 426 (1973); *Chem. Abstr.*, **80,** 30756*v* (1974).
45. R. Mondelli and L. Merlini, *Tetrahedron,* **22,** 3253 (1966).
46. H. Sterk and Th. Kappa, *Monatsh.*, **100,** 1274 (1969).
47. Y. Iwanami, T. Seki, and T. Inagaki, *Bull. Chem. Soc. Jap.*, **44,** 1316 (1971).
48. P. Pflegel and G. Wagner, *Z. Chem.*, **8,** 179 (1968).
49. P. Pflegel and G. Wagner, *Pharmazie,* **34,** 308 (1969); *Chem. Abstr.*, **71,** 108423*v* (1969).
50. B. Adger, C. W. Rees, A. A. Sale, and R. C. Storr, *Chem. Commun.*, 695 (1971).
51. W. Pfleiderer and M. Schranner, *Chem. Ber.*, **104,** 1915 (1971).
52. U. Langenbeck, H. U. Moehring, and K. P. Dieckmann, *J. Chromatogr.*, **115,** 65 (1975).
53. F. Reisser and W. Pfleiderer, *Chem. Ber.*, **99,** 547 (1966).
54. F. Eiden and G. Bachmann, *Arch. Pharm.*, **306,** 401 (1973).
55. H. Otomasu and K. Yoshida, *Chem. Pharm. Bull.* (Tokyo), **8,** 475 (1960).
56. P. Linda and G. Marino, *Ric. Sci. Rend. Sez. A.*, **3,** 225 (1963); *Chem. Abstr.*, **59,** 7523 (1963).
57. R. A. Baxter, G. Dunn, J. J. Gallagher, G. T. Newbold, W. Sharp, W. Sweeny, and F. S. Spring, *J. Roy. Tech. Coll.* (Glasgow), **5,** 26 (1950); *Chem. Abstr.*, **44,** 8355 (1950).
58. K. Eichenberger and P. Schmidt, *Chem. Biol. Pteridines,* Proc. 4th Int. Symp. 1969, p. 99; *Chem. Abstr.*, **76,** 3802*t* (1972).
59. G. W. H. Cheeseman and M. Rafiq, *J. Chem. Soc.* (*C*), 452 (1971).
60. A. Marxer, U. Salzmann, and F. Hofer, *Helv. Chim. Acta,* **54,** 2507 (1971).
61. O. N. Chupakhin, E. O. Sidorov, and I. Ya. Postovskii, *Khim. Geterotsikl. Soedin.*, 993 (1974); *Chem. Abstr.*, **81,** 120578*p* (1974).
62. O. N. Chupakhin, E. O. Sidorov, and I. Ya. Postovskii, *Khim. Geterotsikl. Soedin.*, 1433 (1975); *Chem. Abstr.*, **84,** 43979*c* (1976).
63. R. I. Fryer, J. V. Earley, G. F. Field, W. Zally, and L. H. Sternbach, *J. Org. Chem.*, **34,** 1143 (1969).
63a. R. F. Abdulla and K. H. Fuhr, *J. Heterocycl. Chem.*, **13,** 427 (1976).
64. J. W. Clark-Lewis, *J. Chem. Soc.*, 422 (1957).
65. E. C. Taylor, A. McKillop, and R. E. Ross, *J. Am. Chem. Soc.*, **87,** 1990 (1965).
66. V. D. Romanenko and S. I. Burmistrov, *Chem. Heterocycl. Comp.*, **9,** 785 (1973).
67. Yu. S. Andreichikov, G. D. Plakhina, and R. F. Sarava, *Khim. Geterotsikl. Soedin.*, 1690 (1974); *Chem. Abstr.*, **82,** 97992*d* (1975).
68. D. D. Chapman, *J. Org. Chem.*, **37,** 2498 (1972).
69. R. Pohloudek-Fabini and E. Papke, *Pharmazie,* **18,** 273 (1963); *Chem. Abstr.*, **59,** 8742 (1963).
70. K. H. Nielsen, *J. Chromatogr.*, **10,** 463 (1963).

71. M. S. Lanning and S. S. Cohen, *J. Biol. Chem.*, **189,** 109 (1951).
72. D. J. D. Hockenhull and G. D. Floodgate, *Biochem. J.*, **52,** 38 (1952).
73. J. E. Spikner and J. C. Towne, *Anal. Chem.*, **34,** 1468 (1962).
74. E. Scoffone, E. Vianello, and A. Lorenzini, *Gazz. Chim. Ital.*, **87,** 354 (1957); *Chem. Abstr.*, **51,** 16490 (1957).
75. L. Scarso, E. Scoffone, and D. Chellemi, *Gazz. Chim. Ital.*, **87,** 1348 (1957); *Chem. Abstr.*, **52,** 13813 (1958).
76. P. de la Llosa, M. Jutisz, and E. Scoffone, *Bull. Soc. Chim. Fr.*, 1621 (1960).
77. H. Yamamoto, M. Nakao, and I. Maruyama (to Sumitomo Chemical Co. Ltd.), U.S. Pat. 3,479,348; *Chem. Abstr.*, **72,** 90512*g* (1970).
78. H. Yamamoto (to Sumitomo Chemical Co. Ltd.), Jap. Pat. 69,17,136; *Chem. Abstr.*, **71,** 124505*d* (1969).
79. H. Zellner, M. Pailer, and G. Pruckmayr (to Donau Pharmazie G.m.b.H), Austrian Pat. 228,204; *Chem. Abstr.*, **59,** 11514 (1963).
80. T. O. Yellin (to Abbot Laboratories), U.S. Pat. 3,635,971; *Chem. Abstr.*, **76,** 99708*r* (1972).
81. F. E. Lehmann, *Helv. Physiol. Pharmacol. Acta*, **15,** 431 (1957); *Chem. Abstr.*, **52,** 10396 (1958).
82. R. J. Magee (to American Cyanamid Co.), U.S. Pat. 3,634,425; *Chem. Abstr.*, **76,** 99709*s* (1972).
83. Fisons Pest Control Ltd., Belg. Pat. 631,044; *Chem. Abstr.*, **60,** 15891 (1964).
84. J. Higgins and C. S. Menon, *J. Polym. Sci. B*, **10,** 129 (1972).
85. H. I. X. Mager and W. Berends, *Rec. Trav. Chim*, **77,** 842 (1958).
86. D. W. Woolley and J. M. Stewart, *J. Med. Chem.*, **6,** 599 (1963).
87. L. Horner, U. Schwenk, and E. Junghanns, *Justus Liebigs Ann. Chem.*, **579,** 212 (1953).
88. G. W. H. Cheeseman, *J. Chem. Soc.*, 1804 (1955).
89. C. G. Allison, R. D. Chambers, J. A. H. MacBride, and W. R. K. Musgrave, *J. Fluorine Chem.*, **1,** 59 (1971/2).
90. K. Asano and S. Asai, *Yakugaku Zasshi*, **79,** 567 (1959); *Chem. Abstr.*, **53,** 21979 (1959).
91. D. Shiho and S. Tagami, *Chem. Pharm. Bull.* (Tokyo), **5,** 45 (1957); *Chem. Abstr.*, **52,** 1182 (1958).
91a. M. Asai, *Yakugaku Zasshi*, **79,** 260 (1959); *Chem. Abstr.*, **53,** 13160 (1959).
92. A. S. Elina, L. G. Tsyrul'nikova, and M. I. Medvedeva, *J. Org. Chem. USSR*, **1,** 143 (1965).
93. J. K. Landquist, *J. Chem. Soc.*, 2816 (1953).
94. A. Bauer and K. H. Weber, *Justus Liebigs Ann. Chem.*, **762,** 73 (1972).
95. H. Todd Miles, P. Z. Smyrniotis, and E. R. Stadtman, *J. Am. Chem. Soc.*, **81,** 1946 (1959).
96. A. P. Komin and M. Cormack, *J. Heterocycl. Chem.*, **13,** 13 (1976).
97. C. Sosnowski and L. Wojciechowski, Pol. Pat. 61,711; *Chem. Abstr.*, **74,** 125728*c* (1971).
98. G. W. H. Cheeseman, *J. Chem. Soc.*, 1170 (1962).
99. Sumitomo Chem. Co. Ltd., Belg. Pat. 635,579; *Chem. Abstr.*, **61,** 15288 (1964).
100. H. Sugiyama, T. Ikeda, and S. Koike (to Sumitomo Chem. Co. Ltd.), Jap. Pat. 26 975 (1964); *Chem. Abstr.*, **62,** 11833 (1965).
101. E. S. Lane, *J. Chem. Soc.*, 1079 (1955).
102. E. E. Gilbert (to Allied Chem. Corp.), U.S. Pat. 3,772,273; *Chem. Abstr.*, **80,** 59959*z* (1974).

103. P. Rapos, E. Beska, J. Synak, V. Matous, and J. Pirkl, Czech. Pat. 146,816; *Chem. Abstr.*, **79,** 39342*u* (1973).
104. L. L. Kaul (to Sandoz Ltd.), Ger. Offen. 2,263,235; *Chem. Abstr.*, **79,** 127463*q* (1973).
105. W. G. Jackson and W. Schroeder (to Burdick and Jackson Laboratories Inc.), U.S. Pat. 3,484,387; *Chem. Abstr.*, **73,** 99417*h* (1970).
106. D. E. Burton, A. J. Lambie, D. W. J. Lane, G. T. Newbold, and A. Percival, *J. Chem. Soc.* (*C*), 1268 (1968).
107. H. Moreno and H. P. Schultz, *J. Org. Chem.*, **36,** 1158 (1971).
108. Y. Ahmad, M. S. Habib, and Ziauddin, *Tetrahedron,* **20,** 1107 (1964).
109. R. L. Wear and C. S. Hamilton, *J. Am. Chem. Soc.*, **72,** 2893 (1950).
110. Y. Ahmad, M. S. Habib, M. Iqbal, M. I. Qureshi, and Ziauddin, *Bull. Chem. Soc. Jap.*, **38,** 1659 (1965).
111. Y. Mori and J. Tsuji, *Tetrahedron,* **37,** 3811 (1971).
112. K. Asano and S. Asai, *Yakugaku Zasshi,* **79,** 658 (1959); *Chem. Abstr.*, **53,** 21979 (1959).
113. J. H. Gardner and J. R. Stevens, *J. Am. Chem. Soc.*, **71,** 1868 (1949).
114. H. Saikachi and S. Tagami, *Chem. Pharm. Bull.* (Tokyo), **9,** 941 (1061).
115. A. S. Elina, L. G. Tsyrul'nikova, and I. S. Musatova, *Khim. Farm. Zh.*, **2,** 11 (1968); *Chem. Abstr.*, **69,** 96664*t* (1968).
115a. A. S. Elina, L. G. Tsyrul'nikova, E. M. Peresleni, and Yu. N. Sheinker, *Chem. Heterocycl. Comp.*, **2,** 72 (1966).
116. R. A. Burrell, J. M. Cox, and E. G. Savins, *J. Chem. Soc., Perkin I,* 2707 (1973).
117. G. Tennant, *J. Chem. Soc.*, 2428 (1963).
118. D. Shiho and S. Tagami, *Chem. Pharm. Bull.* (Tokyo), **5,** 45 (1957).
119. K. Asano and S. Asai, *Yakugaku Zasshi,* **79,** 661 (1959); *Chem. Abstr.*, **53,** 21979 (1959).
120. J. Klicnar, M. Hajek, and I. Dobas, *Collect. Czech. Chem. Commun.*, **30,** 3092 (1965).
121. R. W. Begland and D. R. Hartter, *J. Org. Chem.*, **37,** 4136 (1972).
122. S. N. Baranov and N. E. Plevachuk-Tarnavskaya, *Ukr. Khim. Zh.*, **29,** 82 (1963); *Chem. Abstr.*, **59,** 6406 (1963).
123. G. Henseke and K. Dittrich, *Chem. Ber.*, **92,** 1550 (1959).
124. M. S. Habib and C. W. Rees, *J. Chem. Soc.*, 3371 (1960).
125. Y.-W. Chang (to E. I. du Pont de Nemours and Co.), Fr. Pat. 1,386,355; *Chem. Abstr.*, **62,** 14702 (1965).
126. C. L. Leese and H. N. Rydon, *J. Chem. Soc.*, 303 (1955).
127. J. Kavalek and J. Socha, *Collect. Czech. Chem. Commun.*, **33,** 619 (1968).
128. I. Baxter and D. W. Cameron, *J. Chem. Soc.* (*C*), 2471 (1968).
129. Y. J. L'Italien and C. K. Banks, *J. Am. Chem. Soc.*, **73,** 3246 (1951).
130. B. Görlich, *Chem. Ber.*, **89,** 2145 (1956).
131. J. A. Barltrop, C. G. Richards, D. M. Russell, and G. Rybach, *J. Chem. Soc.*, 1132 (1959).
132. M. D. Nair and S. R. Mehta, *Indian J. Chem.*, **6,** 490 (1968).
133. I. Ya. Postovskii and N. G. Koshel, *Chem. Heterocycl. Comp.*, **6,** 913 (1970).
134. H. N. Rydon and K. Undheim, *J. Chem. Soc.*, 4685 (1962).
135. G. Tennant, *J. Chem. Soc.*, 2428 (1963).
136. W. Steglich, H. Tanner, and R. Hurnaus, *Chem. Ber.*, **100,** 1824 (1967).
137. W. Blackburn, M. Danzig, H. Hubinger, D. Soisson, and H. P. Schultz, *J. Org. Chem.*, **26,** 2805 (1961).

138. A. Ya. Strakov and D. Brutane, *Latv. PSR Zinat. Akad. Vestis Khim. Ser.*, 591 (1967); *Chem. Abstr.*, **68,** 49499*n* (1968).
139. G. Kyryacos and H. P. Schultz, *J. Am. Chem. Soc.*, **75,** 3597 (1953).
140. B. Marks and H. P. Schultz, *J. Am. Chem. Soc.*, **73,** 1368 (1951).
141. B. Munk and H. P. Schultz, *J. Am. Chem. Soc.*, **74,** 3433 (1952).
142. C. W. Bird, *J. Chem. Soc.*, 674 (1963).
143. C. W. Rees and D. E. West, *J. Chem. Soc.* (*C*), 583 (1970).
144. G. Tennant, *J. Chem. Soc.*, 1986 (1964).
145. G. B. Barlin and W. Pfleiderer, *Chem. Ber.*, **102,** 4032 (1969).
146. F. Weygand and K. Burger, *Chem. Ber.*, **99,** 2880 (1966).
147. J. K. Landquist, *J. Chem. Soc.*, 2830 (1953).
148. Y. Ahmad, M. S. Habib, M. Iqbal, M. I. Qureshi, and Ziauddin, *Can. J. Chem.*, **43,** 3424 (1965).
149. N. O. Saldabols, A. Ya. Medne, and S. A. Hiller, *Chem. Heterocycl. Comp.*, **3,** 444 (1967).
150. N. O. Saldabols, A. Cimanis, and S. A. Hiller, USSR Pat. 410,015 (1974); *Chem. Abstr.*, **80,** 120992y (1974).
151. A. Gomez-Sanchez, M. Y. Antinola, and F. G. Gonzalez, *An. Real Soc. Espan. Fis. y Quim.* (Madrid), **50B,** 431 (1954); *Chem. Abstr.*, **52,** 11078 (1958).
152. P. H. Gore and G. K. Hughes, *J. Am. Chem. Soc.*, **77,** 5738 (1955).
153. F. Krönke, *Chem. Ber.*, **80,** 298 (1947).
154. F. Piozzi, *Chim. Ind.* (Milan), **45,** 347 (1963); *Chem. Abstr.*, **61,** 659 (1964).
155. J. D. Fissekis, C. G. Skinner, and W. Shive, *J. Org. Chem.*, **24,** 1722 (1959).
156. Y. Ahmad, M. S. Habib, Ziauddin, and N. Bashir, *Bull. Chem. Soc. Jap.*, **38,** 1654 (1965).
157. B. Eistert and J. Grammel, *Chem. Ber.*, **104,** 1942 (1971).
157a. F. Micheel, R. Austrup, and A. Striebeck, *Chem. Ber.*, **94,** 132 (1961).
158. R. L. St. Clair and T. B. Thibault (to Eli Lilly and Co.), Ger. Offen. 2,459,453; *Chem. Abstr.*, **83,** 164237*x* (1975).
159. E. Grovenstein, W. Postman, and J. W. Taylor, *J. Org. Chem.*, **25,** 68 (1960).
160. J. J. Zamet, M. J. Haddadin, and C. H. Issidorides, *J. Chem. Soc., Perkin I*, 1687 (1974).
161. W. Logemann, G. Cavagna, and G. Tosolini, *Chem. Ber.*, **96,** 2248 (1963).
162. R. Howe, B. S. Rao, and H. Heyneker, *J. Chem. Soc.* (*C*), 2510 (1967).
163. Y. Ahmad, M. S. Habib, A. Mohammady, B. Bakhtiari, and S. A. Shamsi, *J. Org. Chem.*, **33,** 201 (1968).
164. B. J. Ralph and A. Robertson, *J. Chem. Soc.*, 3380 (1950).
165. J. W. Clark-Lewis, *J. Chem. Soc.*, 439 (1957).
166. H.-D. Stachel, *Chem. Ber.*, **95,** 2172 (1962).
167. S. N. Baranov and N. E. Tarnavskaya, *Ukr. Khim. Zh.*, **25,** 620 (1959); *Chem. Abstr.*, **54,** 11041 (1960).
168. J. Klicnar, M. Hajek, J. Hoffmann, and M. Vecera, *Collect. Czech. Chem. Commun.*, **36,** 262 (1971).
169. A. Nardi, A. Tajana, and E. Massarani, *Ann. Chim.* (Rome), **59,** 1075 (1969); *Chem. Abstr.*, **72,** 89963y (1970).
170. K. Burger, J. Fehn, and E. Müller, *Chem. Ber.*, **106,** 1 (1973).
171. G. Tennant and J. C. Mason, *J. Chem. Soc.* (*D*), 586 (1971).
171*a*. Y. Ahmad, M. S. Habib, Ziauddin, and N. Bazir, *Tetrahedron*, **21,** 861 (1965).
172. E. Bickert and L. Enslein, *Chem. Ber.*, **92,** 1851 (1961).
173. S. Bodforss, *Justus Liebigs Ann., Chem.*, **609,** 103 (1957).

174. W. Reid and P. Stahlhofen, *Chem. Ber.*, **90,** 828 (1957).
175. C. Iijima and E. Hayashi, *Yakugaku Zasshi*, **91,** 721 (1971); *Chem. Abstr.*, **75,** 129758*f* (1971).
176. G. S. Predvoditeleva, T. V. Kartseva, M. N. Shchukina, T. N. Zykova, and G. N. Pershin, *Khim. Farm. Zh.*, **2,** 19 (1968); *Chem. Abstr.*, **70,** 77912*b* (1969).
177. L. Farkas and V. Szantho, *Acta Chim. Acad. Hung.*, **19,** 217 (1959); *Chem. Abstr.*, **54,** 3398 (1960).
178. H. Dahn and J. Nussbaum, *Helv. Chim. Acta*, **52,** 1661 (1969).
179. J. N. Chatterjea, S. N. P. Gupta, and V. N. Mehrotra, *J. Indian Chem. Soc.*, **42,** 205 (1965); *Chem. Abstr.*, **63,** 9899 (1965).
180. D. F. Morrow and L. A. Regan, *J. Org. Chem.*, **36,** 27 (1971).
181. J. W. Clark-Lewis, J. A. Edgar, and K. Moody, *J. Chem. Soc.* (*C*), 1221 (1966).
182. J. Michalsky and L. Sadilek, *Monatsh.*, **90,** 171 (1959); *Chem. Abstr.*, **54,** 4481 (1960).
183. G. P. Schiemenz and H. Engelhard, *Chem. Ber.*, **92,** 1336 (1959).
184. S. Fatutta, *Gazz. Chim. Ital.* **89,** 964 (1959); *Chem. Abstr.*, **54,** 22554 (1960).
185. P. M. Mader, *J. Org. Chem.*, **27,** 2217 (1962).
186. M. Pailer, G. Pruckmayr, H. Zellner, and G. Zellner, *Monatsh.*, **93,** 1005 (1963).
187. S. H. Dandegaonker and C. K. Mesta, *J. Med. Chem.*, **8,** 884 (1965).
188. H. I. X. Mager and W. Berends, *Rec. Trav. Chim.*, **78,** 5 (1959).
189. A. V. Et'tsov, Kh. L. Muravich-Aleksandr, and L. M. Roitshtein, *J. Org. Chem. USSR*, **3,** 196 (1967).
190. L. Wojciechowski, *Soc. Sci. Lodz. Acta Chim.*, **17,** 207 (1972); *Chem. Abstr.*, **78,** 58351*s* (1973).
191. D. E. Burton, D. Hughes, G. T. Newbold, and J. A. Elvidge, *J. Chem. Soc.* (*C*), 1274 (1968).
192. Y.-W. Chang (to E. I. du Pont de Nemours and Co.), Fr. Pat. 1,385,595; *Chem. Abstr.*, **62,** 14703 (1965).
193. E. S. Lane and C. Williams, *J. Chem. Soc.*, 2983 (1956).
194. K. Hotta, I. Ishiguro, and S. Imaizumi, *Bitamin*, **14,** 760 (1958); *Chem. Abstr.*, **55,** 27513 (1961).
195. S. Kawai, Y. Okawa, Y. Yada, H. Hosi, T. Murakoshi, and I. Yajima, *Nippon Kagaku Zasshi*, **80,** 551 (1959); *Chem. Abstr.*, **55,** 3598 (1961).
196. S. Oguchi, *Nippon Kagaku Zasshi*, **86,** 319 (1965); *Chem. Abstr.*, **63,** 4296 (1965).
197. R. L. Williams and S. W. Shalaby, *J. Heterocycl. Chem.*, **10,** 891 (1973).
198. Y. Sakakibara, *Nippon Kagaku Zasshi*, **83,** 172 (1962); *Chem. Abstr.*, **58,** 11357 (1963).
199. A. Stepien, M. J. Grabowski, M. Cygler, and E. Wajsman, *Acta Cryst.*, **B32,** 2048 (1976).
200. M. Sato and C. H. Stammer, *J. Heterocycl. Chem.*, **14,** 149 (1977).

CHAPTER VI

Quinoxaline-2-thiones and Quinoxaline-2,3-dithiones

I. Quinoxaline-2-thiones (Quinoxaline-2-thiols or 2-Mercaptoquinoxalines) 112
   1. Methods of Preparation 112
   2. Properties 114
   3. Uses 116
II. Quinoxaline-2,3-dithiones (Quinoxaline-2,3-dithiols or 2,3-Dimercaptoquinoxalines) 116
   1. Methods of Preparation 116
   2. Properties 116
   3. Uses 118
III. Tables 119
   Table 1. Quinoxaline-2-thiones (2-Mercaptoquinoxalines) 119
   Table 2. Quinoxaline-2,3-dithiones (2,3-Dimercaptoquinoxalines) 120
IV. References 121

I. Quinoxaline-2-thiones (Quinoxaline-2-thiols or 2-Mercaptoquinoxalines)

1. Methods of Preparation

Quinoxaline-2-thione (**2**) is prepared from 2-chloroquinoxaline by treatment with thiourea in methanol, followed by hydrolysis of the resulting isothiouronium salt (**1**) with 2.5 *M* sodium hydroxide.[1] A number of other quinoxaline-2-thiones have been prepared by this method, in most cases without the isolation of the intermediate isothiouronium salt.[2,3] 3-Hydrazinoquinoxaline-2-thiones are obtained from the corresponding 2-chloro-3-hydrazinoquinoxalines by treatment with an aqueous solution of alkali sulfide or alkali hydrogen sulphide.[4] 3-Aminoquinoxaline-2-thione has been prepared similarly by treating 2-amino-3-chloroquinoxaline with methanolic potassium hydrogen sulfide in an

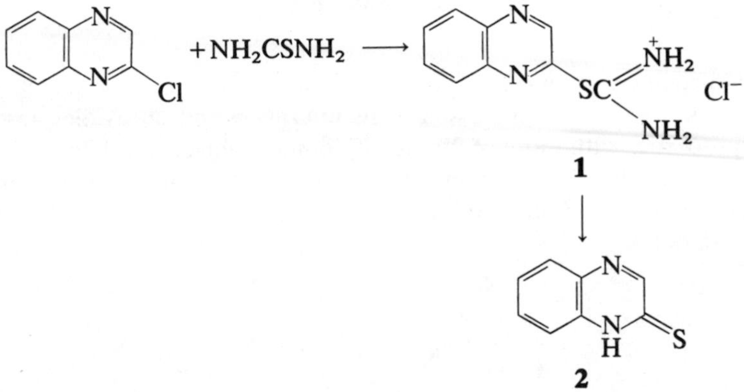

+ NH_2CSNH_2 →

1

↓

2

autoclave at 150°, alternatively by reacting 3-aminoquinoxalin-2-one with phosphorus pentasulfide in xylene.[5]

Various 3-substituted aminoquinoxaline-2-thiones have been prepared from quinoxaline-2,3-dithione. For example, treatment of the dithione with methylamine yields 3-methylaminoquinoxaline-2-thione,[6] and the corresponding reaction with hydrazine hydrate yields 3-hydrazinoquinoxaline-2-thione.[7] The latter compound is also obtained by treatment of 2,3-dichloroquinoxaline with thiourea and then with hydrazine hydrate.[8] When 2-chloroquinoxaline 1-oxide is heated with thiourea in methanol at room temperature, an unidentified solid m.p. 184–185° is obtained; alkaline hydrolysis of methanol-soluble products gives quinoxaline-2-thione in 41% yield.[9] A method of some general applicability is illustrated by the preparation of the thioxo compound **4** from *o*-anilinoaniline (**3**).[10]

3 →(i)→ →(ii)→ →(iii)→ 4

Reagents (i) $Cl_2CHCOCl$
(ii) Morpholine–S
(iii) EtOH–HCl

2. Properties

Quinoxaline-2-thione is a very weak base of pK_a −1.24 and has an acidic pK_a of 7.16. Comparison of its ionization and ultraviolet absorption properties with those of 2-methylthioquinoxaline (pK_a 0.26) and 1,2-dihydro-1-methyl-2-thioxoquinoxaline (pK_a −1.01) shows that quinoxaline-2-thione exists in solution in the thioamide form (**2a**) rather than as the thiol (**2b**).[11]

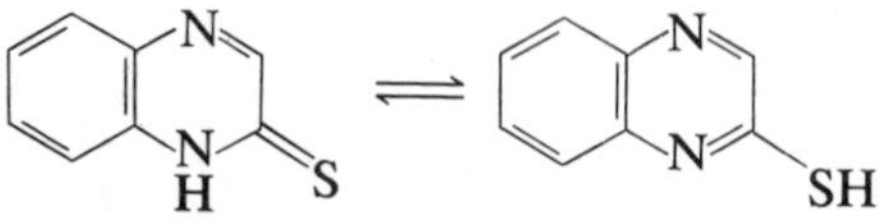

2a ⇌ **2b**

Quinoxaline-2-thione is readily converted into 2-hydrazinoquinoxaline on treatment with hydrazine hydrate, and 3-hydrazinoquinoxaline-2-thione is readily hydrolyzed (2.5 *M* HCl at 70°) to quinoxaline-2,3-dione.[12] The sodium salt of quinoxaline-2-thione has been *S*-alkylated with a wide range of alkyl halides to give 2-quinoxalinyl sulfides,[3,13] but attempts to carry out a parallel reaction with *p*-nitrochlorobenzene failed.[1] *p*-Nitrophenyl-2-quinoxalinyl sulfide (**5**) can however be prepared by reaction of 2-chloroquinoxaline with the sodium salt of *p*-nitrobenzenethiol.[1,14]

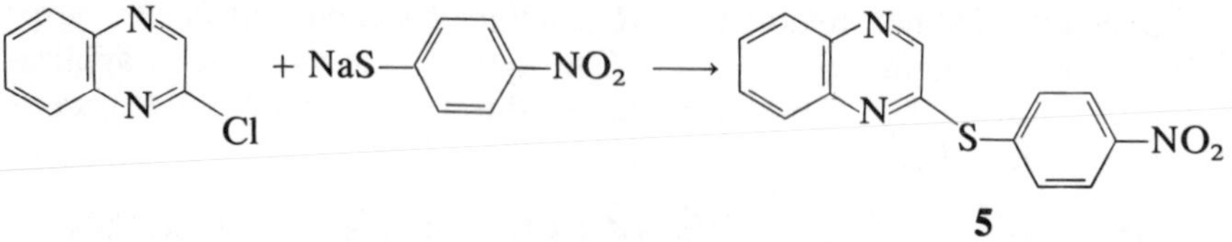

2-chloroquinoxaline + NaS–C_6H_4–NO_2 ⟶ **5**

The reaction of 3-aminoquinoxaline-2-thiones (**6**) with α-halo ketones such as chloroacetone and ω-bromoacetophenone yields quinoxalinothiazines (**7**), and hydrolysis of the thiazines in an alkaline medium gives 3-aminoquinoxalin-2-ones (**8**).[15]

A number of 3-chloro-, 3-methoxy-, and 3-methylthioquinoxaline-2-thiones have been reacted with acylating agents such as alkyl chloroformates, alkyl chlorothioformates, and thiophosgene in the search for compounds for use as pesticides. For example, treatment of 3-chloroquinoxaline-2-thione with ethyl chloroformate gives the ester **9**.

Oxidation of a solution of 2-methylthioquinoxaline in glacial acetic acid with 30% hydrogen peroxide at room temperature gives mainly 2-methylsulfonylquinoxaline, but at 55° the major oxidation product is 2-methylsulfonylquinoxaline 4-oxide.[3] Both 2-methylsulfonylquinoxaline

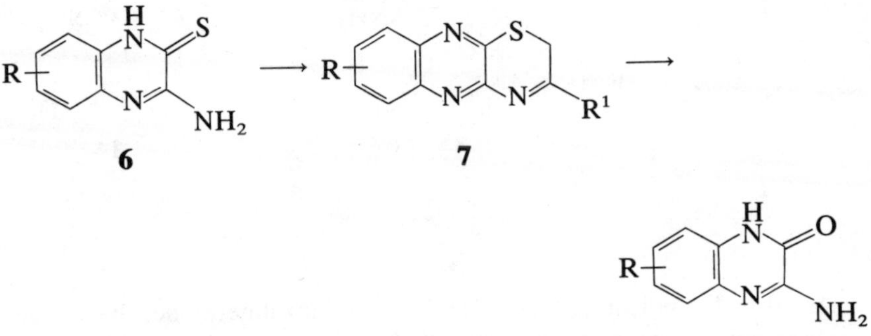

R-quinoxaline-2-thione, 3-amino (**6**) → (**7**) → (**8**)

6 **7** **8**

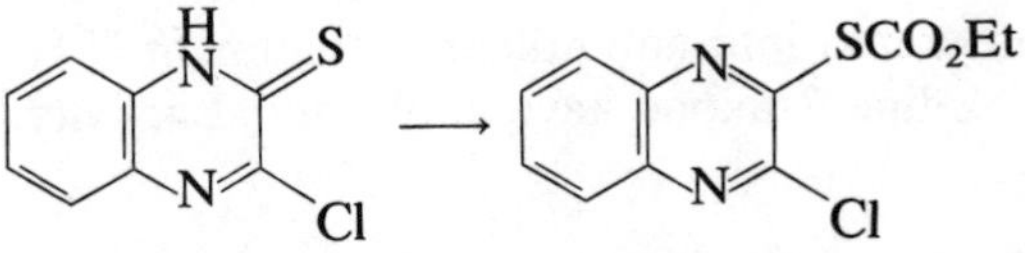

9

and its 4-oxide are readily hydrolyzed in 2 *M* aqueous sodium hydroxide (95°, 15 min) to the corresponding quinoxalinones.[3] Quinoxalinyl ylides (**10**) are formed by reaction of 2-methylsulfonylquinoxaline with $Ph_2Me\overset{+}{S}BF_4^-$ or $Ph_2Me\overset{+}{S}ClO_4^-$. These form oxiranes **11** on treatment with carbonyl compounds. Reaction of the oxiranes with $LiNEt_2$ gives quinoxalinyl ketones (**12**) of the type shown.[17]

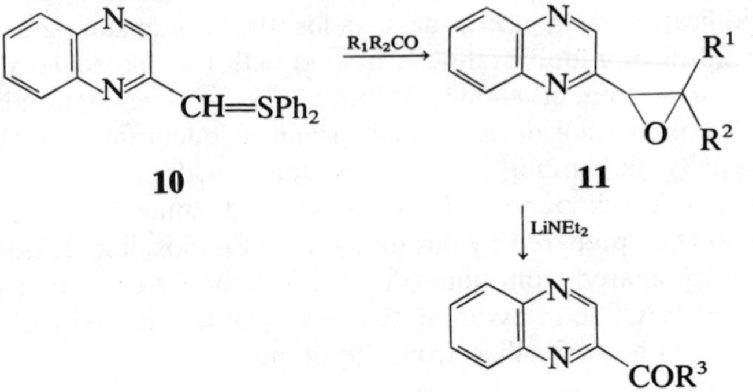

$\xrightarrow{R_1R_2CO}$

10 **11**

$\downarrow LiNEt_2$

12

R^3 = Et, Pr, or $CHMe_2$

The conversion of 3-aminoquinoxaline-2-thione into 3-amino-2-quinoxalinesulfonamide (**14**) has been effected via the corresponding sulfenamide (**13**).[18]

3. Uses

As already indicated, many acylated quinoxalinethiones have been synthesized as potential pesticides.[16] Quinoxaline-2-thiones have been tested for antitubercular activity[19] and 3-hydrazinoquinoxaline-2-thione has been incorporated into an antileprosy ointment.[20] Hydrazones of 3-hydrazinoquinoxaline-2-thione have antibacterial activity.[7]

II. Quinoxaline-2,3-dithiones (Quinoxaline-2,3-dithiols or 2,3-Dimercaptoquinoxalines)

1. Methods of Preparation

Many quinoxaline-2,3-dithiones have been prepared from the corresponding 2,3-dichloroquinoxalines by heating under reflux with excess of aqueous potassium hydrogen sulfide solution. The alkali-soluble product on acidification with acetic acid yields the quinoxaline-2,3-dithione.[21] When 2,3-dichloroquinoxaline is treated with two molecular proportions of thiourea in ethanolic solution, *S*-2-(3-mercaptoquinoxalinyl)-isothiouronium chloride is formed in almost quantitative yield. This on treatment with boiling aqueous sodium hydroxide solution yields quinoxaline-2,3-dithione (**15**).[22] Substituted quinoxaline-2,3-dithiones have also been prepared by this method.[2,23] Quinoxaline-2,3-dithione has also been prepared from quinoxaline-2,3-dione by treatment with phosphorus pentasulfide in pyridine, thus avoiding the necessity of first forming the dichloroquinoxaline from the dione.[2]

2. Properties

The acidic dissociation constants of quinoxaline-2,3-dithione are 6.84 and 9.95.[24] The dithione is readily alkylated in sodium hydroxide solution with either methyl sulfate or ethyl iodide to give 2,3-bis(alkylthio)-

quinoxalines. When acetobromoglucose is used as the alkylating reagent, either the monoglucosyl derivative (**16**) or diglucosyl derivative (**17**) can be obtained depending on the molecular proportions of reagents used. Deacetylation of **17** gives 2,3-bis(1-β-glucosylmercapto)quinoxaline (**18**).[25]

1,3-Dithiolanes have been prepared by the reaction of quinoxaline-2,3-dithiones with geminal dihalogeno compounds, for example, compound **19** is formed from the reaction of the dithione and methylene dibromide.[26]

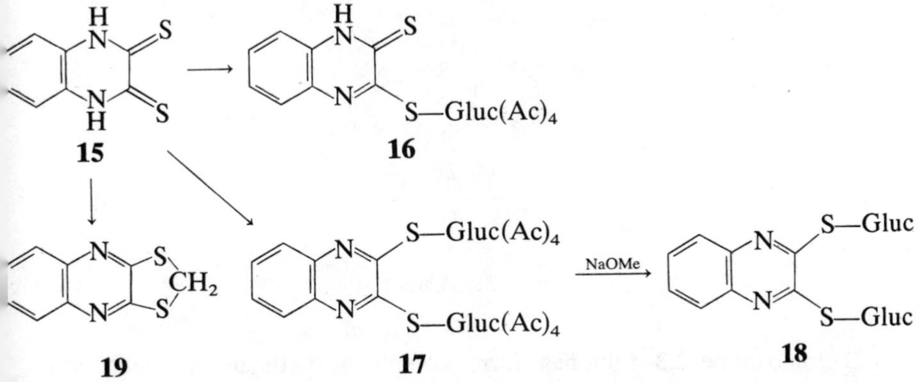

S—Gluc(Ac)$_4$

15 **16**

CH_2 S—Gluc(Ac)$_4$ NaOMe S—Gluc

S—Gluc(Ac)$_4$ S—Gluc

19 **17** **18**

Reaction of quinoxaline-2,3-dithiones with acylating reagents gives exclusively *S*-acylated products. For example, treatment of quinoxaline-2,3-dithione in 5% sodium hydroxide solution with ethyl chloroformate gives 2,3-bis(carbethoxythio)quinoxaline.[27] Similarly reaction of the sodium salt of quinoxaline-2,3-dithione with $(EtO)_2POCl$ gives the thiophosphoric ester **20**. The cyclic ester **21** is similarly obtained from reaction with $EtOPSCl_2$.[28]

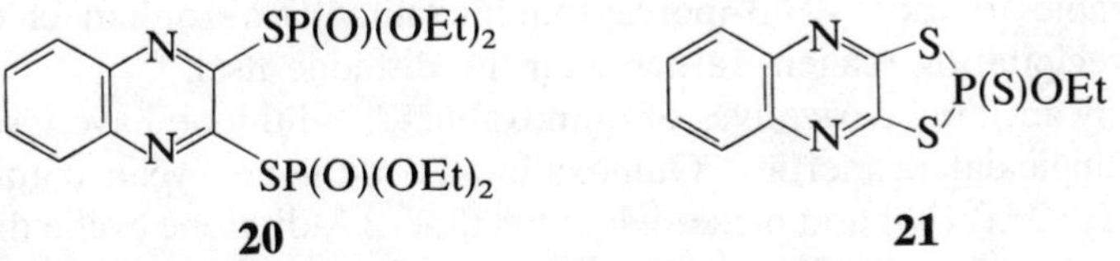

$SP(O)(OEt)_2$ P(S)OEt

$SP(O)(OEt)_2$

20 **21**

Cyclic derivatives have also been prepared from quinoxaline-2,3-dithiones by reaction with phosgene ($COCl_2$), thiophosgene ($CSCl_2$), and thionyl chloride ($SOCl_2$). 2,3-Disubstituted quinoxalines result from the reaction of the dithiones with isocyanates, isothiocyanates, and imidoyl chlorides (e.g., PhCCl=NMe).[29] Derivatives of quinoxaline-2,3-dithiones incorporating arsenic, antimony, and tin have been prepared; for example, with Bu_3SnCl the 2,3-disubstituted quinoxaline **22** is obtained, and with Et_2SnCl_2 the cyclic derivative **23** is formed.[30] Quinoxaline-2,3-dithiones undergo condensation reactions with 2,3-dichloroquinoxalines to give pentacyclic products (**24**) of the type shown.[31]

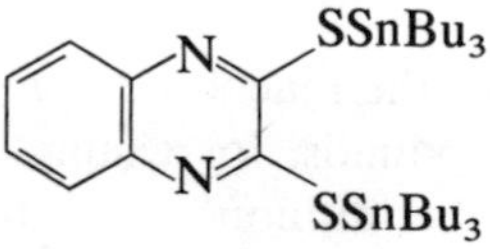

3. Uses

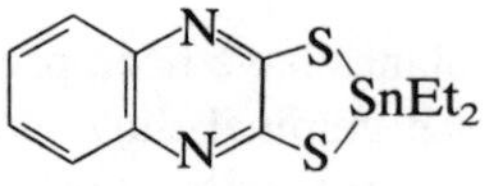

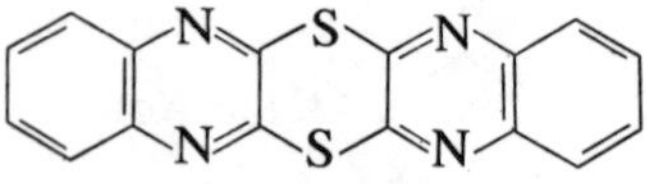

Quinoxaline-2,3-dithiones form complexes with many metal ions. Complexes of the type $[Bu_4N]_2[M(QDT)_2]$, where for example M = Cu, Ni, Pd, Pt, Zn, or Co and H_2QDT = quinoxaline-2,3-dithione have been isolated and their spectroscopic and magnetic properties examined.[32,33] Interest in the colored complexes formed between quinoxaline-2,3-dithione and metallic ions has led to the development of spectroscopic procedures for the determination of a number of metals including cobalt, nickel, and copper.[34] A gravimetric procedure for the determination of nickel involves the precipitation of $(C_8H_5N_2S_2)_2Ni$; in this case it is found preferable to use *S*-2-(3-mercaptoquinoxalinyl)thiouronium chloride as the precipitating reagent rather than the dithione itself.[35]

Many acylated derivatives of quinoxaline-2,3-dithione have insecticidal and fungicidal properties. Quinoxaline-2,3-dithione cyclic trithiocarbonate (Eradex) (**25**) and 6-methylquinoxaline-2,3-dithione cyclic dithiocarbonate (Morestan) (**26**) are particularly effective.[36] 2,3-Bis(carbethoxythio)quinoxaline is reported to have tuberculostatic action.[27]

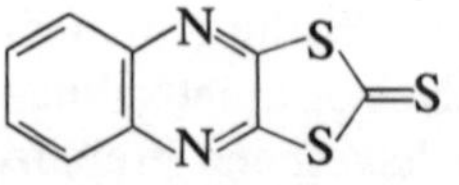

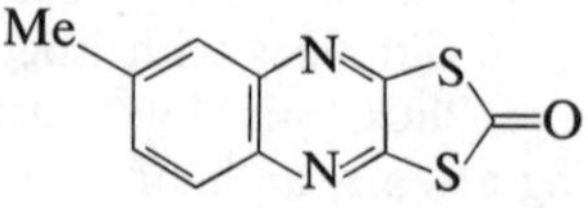

III. Tables

TABLE 1. QUINOXALINE-2-THIONES (2-MERCAPTOQUINOXALINES)

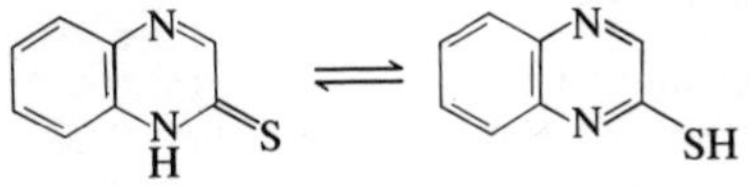

Molecular formula	Substituents or structure	m.p., °C	Ref.
$C_8H_5ClN_2S$	3-Chloro		16
	S-Methoxycarbonyl	167	16
	S-Ethoxycarbonyl	148–150	16
	S-Methyl	102–103	37
$C_8H_5N_3O_2S$	7-Nitro		
	S-*p*-Nitrophenyl	196	1
	SS-Dioxide of *S*-*p*-nitrophenyl	>240	1
$C_8H_6N_2S$	Unsubstituted	204–205, 209	1, 3, 11, 13
	S-methyl	46–47, b.p. 134/1 mm	3, 38
	Picrate of *S*-methyl	127, 128	38
	S,S-Dioxide of *S*-methyl	124–125, 126–127	3, 38
	1-Methyl	123–125	11
	S-Ethyl	47	13
	S-Allyl	b.p. 150/4 mm	13
	S-Butyl	b.p. 162/5 mm	13
	S-Cyclohexyl	66	13
	S-Benzyl	44	13
	S-Phenyl	87	13
	S-*p*-Nitrophenyl	151–152	1
	S,S-Dioxide of *S*-*p*-nitrophenyl	202–203	1
	S-1-β-D-Glucosyl[a]	202–203	25
	S-(Tetraacetyl-β-D-glucosyl)[b]	140.5–141.5	25
	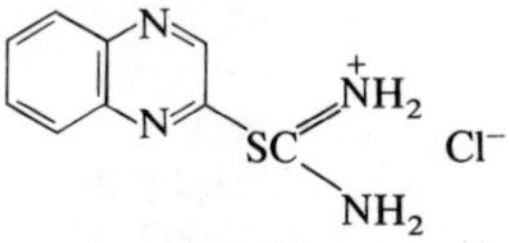	159–160	1
	S-2-Quinoxalinyl	159	1
	SS-Dioxide of *S*-2-quinoxalinyl	>360	1
$C_8H_7N_3S$	3-Amino	>300	5, 15
	S-Methyl	170	5
$C_8H_8N_4S$	3-Hydrazino		4, 7, 8, 12, 37
	S-Methyl	157	37
$C_9H_8N_2OS$	3-Methoxy		16
	S-Ethoxycarbonyl	82–84	16

TABLE 1 (*contd.*)

Molecular formula	Substituents or structure	m.p., °C	Ref.
$C_9H_8N_2S$	3-Methyl	240–245 decomp.	2
	S-Methyl	51.5–53	2
	S-1-β-D-glucosyl[c]	215–216	25
	S-(Tetraacetyl-1-β-D-glucosyl)[d]	171–172	25
$C_9H_9N_3S$	3-Methylamino	221–223	6
	S-Methyl	87	6
$C_{10}H_{11}N_3S$	3-Amino-6,7-dimethyl	299–301	15
$C_{11}H_{12}N_2S$	3-Isopropyl	215–225 decomp., sinters 195	2
	S-Methyl	66–67	2
$C_{12}H_{13}N_3OS$	3-Morpholino	134	6

[a] $[\alpha]_D^{20} -76.8°$ (*c* 2.5; DMF).
[b] $[\alpha]_D^{20} -13.6°$ (*c* 5.0; $CHCl_3$).
[c] $[\alpha]_D^{20} -67.1°$ (*c* 2.5; DMF).
[d] $[\alpha]_D^{20} -3.5°$ (*c* 5.0; DMF).

TABLE 2. QUINOXALINE-2,3-DITHIONES (2,3-DIMERCAPTOQUINOXALINES)

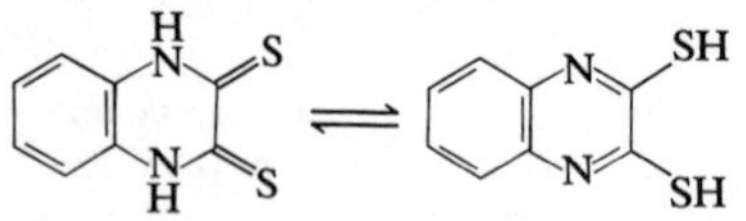

Molecular formula	Substituents or structure	m.p., °C	Ref.
$C_8H_4Cl_2N_2S_2$	6,7-Dichloro		28
$C_8H_5ClN_2S_2$	6-Chloro	336–338	21, 31
$C_8H_5N_3O_2S_2$	6-Nitro	>360	31
$C_8H_6N_2O_2S_2$	5,8-Dihydroxy	>360	39
	Diacetate	318–320	39
	6,7-Dihydroxy	>360	39
$C_8H_6N_2S_2$	Unsubstituted	290–295 decomp., >300	2, 19, 22
	S,S-Bisethoxycarbonyl	97, 102–103	29, 40
	S-Monomethyl	227	37
	S-(Tetraacetyl-1-β-D-glucosyl)[a]	109–110	25
	S,S-Dimethyl	134–135	21
	S,S-Diethyl	50–51	2
	S,S-Bis(1-β-D-glucosyl)[b]	230–231	25
	S,S-Bis(tetraacetyl-1-β-D-glucosyl)[c]	169–170	25
	S,S-Di-*p*-tolyl	140–142	41

TABLE 2 (*contd.*)

Molecular formula	Substituents or structure	m.p., °C	Ref.
$C_8H_6N_2S_2$ (*contd.*)	[structure: 2-thioxo-1H-quinoxalin-3-yl isothiouronium chloride, $SC(NH_2)=\overset{+}{N}H_2$ Cl^-]	273 decomp.	22
$C_9H_5F_3N_2S_2$	6-Trifluoro		28
$C_9H_8N_2OS_2$	6-Methoxy		28
$C_9H_8N_2S_2$	6-Methyl	287–289 decomp.	21
$C_{10}H_{10}N_2O_2S_2$	5,8-Dimethoxy	260–261 decomp.	39
	6,7-Dimethoxy	>350	39
$C_{10}H_{10}N_2S_2$	6,7-Dimethyl		28

[a] $[\alpha]_D^{20} - 12.7$ (*c* 5.0; $CHCl_3$).
[b] $[\alpha]_D^{20} - 139.3$ (*c* 2.5; DMF).
[c] $[\alpha]_D^{20} - 25.7$ (*c* 5.0; $CHCl_3$).

IV. References

1. F. J. Wolf, R. M. Wilson, and M. Tishler, *J. Am. Chem. Soc.*, **76,** 2266 (1954).
2. D. C. Morrison and A. Furst, *J. Org. Chem.*, **21,** 470 (1956).
3. G. W. H. Cheeseman, *J. Chem. Soc.*, 3236 (1957).
4. K. Sasse and G. Damagk (to Farbenfabriken Bayer A.-G.), Ger. Pat. 1,104,962; *Chem. Abstr.*, **57,** 8591 (1962).
5. H. Saikachi and S. Tagami, *Chem. Pharm. Bull.* (Tokyo), **9,** 941 (1961).
6. K. Sasse, R. Wegler, and G. Unterstenhoefer (to Farbenfabriken Bayer A.-G.), Ger. Pat. 1,117,586; *Chem. Abstr.*, **57,** 4684 (1962).
7. K. Asano, S. Asai, and N. Inoue, *Yakugaku Zasshi*, **79,** 24 (1959); *Chem. Abstr.*, **53,** 10242 (1959).
8. K. Asano and S. Asai (to First Industrial Drug Manufg. Co.), Jap. Pat. 2932 (1961); *Chem. Abstr.*, **57,** 3456 (1962).
9. A. S. Elina, *Chem. Heterocycl. Comp.*, **3,** 740 (1967).
10. Ciba Ltd., Fr. Pat. 1,361,010; *Chem. Abstr.*, **61,** 13329 (1964).
11. A. Albert and G. B. Barlin, *J. Chem. Soc.*, 3129 (1962).
12. K. Asano and S. Asai, *Yakugaku Zasshi*, **79,** 567 (1959); *Chem. Abstr.*, **53,** 21979 (1959).
13. F. Cuiban, M. Ionescu, H. Bala, and M. Steresco, *Bull. Soc. Chim. Fr.* 356 (1963).
14. J. Tulecki, J. Kalinowska-Torz, and J. Powalska, *Arch. Pharm.* (Poznan), **10,** 13 (1973); *Chem. Abstr.*, **79,** 42448*p* (1973).
15. I. Ya. Postovskii and N. G. Koshel, *Chem. Heterocycl. Comp.*, **7,** 796 (1971).
16. K. Sasse, R. Wegler, and G. Unterstenhoefer (to Farbenfabriken Bayer A.-G.), Belg. Pat. 623,228; *Chem. Abstr.*, **60,** 14522 (1964); Ger. Pat. 1,109,695; *Chem. Abstr.*, **56,** 8726 (1962).
17. E. C. Taylor, M. L. Chittenden, and S. F. Martin, *Heterocycles*, **1,** 59 (1973).

18. M. Hattori, M. Yoneda, and M. Gota, *Bull. Chem. Soc. Jap.* **46,** 1890 (1973).
19. K. Asano, *Yakugaku Zasshi* **78,** 729 (1958); *Chem. Abstr.*, **52,** 18428 (1958).
20. T. Hirako, S. Tsutsumi, and K. Asano (to Daiichi Seiyaku Co. Ltd.), Brit. Pat. 1,038,505; *Chem. Abstr.*, **65,** 15170 (1966).
21. K. Asano and S. Asai (to Daiichi Industrial Drug. Manufg. Co.), Jap. Pat. 3375 (1959); *Chem. Abstr.*, **54,** 14278 (1960); K. Sasse, R. Wegler, and G. Unterstenhoefer (to Farbenfabriken Bayer A.-G.), Ger. Pat. 1,088,965; *Chem. Abstr.*, **57,** 12507 (1962).
22. E. Taeger and Z. El-Hawehi, *J. Prakt. Chem.*, **18,** 255 (1962).
23. L. Wojciechowski, Pol. Pat. 65,740; *Chem. Abstr.*, **77,** 140150*p* (1972).
24. J. A. W. Dalziel and A. K. Slawinski, *Talanta,* **19,** 1240 (1972).
25. G. Wagner and H. Frenzel, *Arch. Pharm.* **300,** 433 (1967).
26. K. Sasse, R. Wegler, and G. Unterstenhoefer (to Farbenfábriken Bayer A.-G.), Ger. Pat. 1,102,749; *Chem. Abstr.*, **57,** 4681 (1962).
27. K. Asano and S. Asai (to First Industrial Drug Manufg. Co.), Jap. Pat. 1872 (1961); *Chem. Abstr.*, **56,** 8724 (1962).
28. K. Sasse, R. Wegler, and G. Unterstenhoefer (to Farbenfabriken Bayer A.-G.), Ger. Pat. 1,115,738; *Chem. Abstr.*, **56,** 14305 (1962).
29. K. Sasse, R. Wegler, and G. Unterstenhoefer (to Farbenfabriken Bayer A.-G.), Ger. Pat. 1,088,965; *Chem. Abstr.*, **57,** 12507 (1962); Belg. Pat. 612,972; *Chem. Abstr.*, **58,** 1475 (1963).
30. K. Sasse, R. Wegler, and G. Unterstenhoefer (to Farbenfabriken Bayer A.-G.), Brit. Pat. 919,248; *Chem. Abstr.*, **60,** 1775 (1964).
31. L. Wojciechowski, Pol. Pat. 64,713; *Chem. Abstr.*, **79,** 6780*g* (1973); *Rocz. Chem.*, **44,** 2461 (1970); *Chem. Abstr.*, **75,** 35956*s* (1971).
32. K. K. Ganguli, G. O. Carlisle, H. J. Hu, L. J. Theriot, and I. Bernal, *J. Inorg. Nucl. Chem.*, **33,** 3579 (1971).
33. L. J. Theriot, K. K. Ganguli, S. Kavarnos, and I. Bernal, *J. Inorg. Nucl. Chem.*, **31,** 3133 (1969).
34. R. W. Burke and E. R. Deardorff, *Talanta,* **17,** 255 (1970).
35. J. A. W. Dalziel and A. K. Slawinski, *Talanta,* **15,** 1385 (1968).
36. K. Sasse, R. Wegler, G. Unterstenhoefer, and F. Grewe, *Angew. Chem.*, **72,** 973 (1960).
37. K. Asano and S. Asai, *Yakugaku Zasshi* **79,** 661 (1959); *Chem. Abstr.*, **53,** 21979 (1959).
38. E. Hayashi and T. Miyagishima, *Yakugaku Zasshi,* **87,** 826 (1967); *Chem. Abstr.*, **67,** 116866*z* (1967).
39. S. Oguchi, *Nippon Kagaku Zasshi,* **86,** 319 (1965); *Chem. Abstr.*, **63,** 4296 (1965).
40. K. Asano and S. Asai, Jap. Pat. 1,872 (1961); *Chem. Abstr.*, **56,** 8724 (1962).
41. J. K. Landquist and J. A. Silk, *J. Chem. Soc.*, 2052 (1956).

CHAPTER VII

Quinoxaline-2-carboxaldehydes and Quinoxaline-2,3-dicarboxaldehydes

I. Methods of Preparation 123
II. Properties 125
III. Tables 127
Table 1. Quinoxaline-2-carboxaldehydes and -2,3-dicarboxaldehydes 127
IV. References 128

A review of the earlier literature on quinoxaline-2-carboxaldehydes is available in Simpson's monograph.[1]

I. Methods of Preparation

Quinoxaline-2-carboxaldehydes are conveniently prepared either from the corresponding methyl- or polyhydroxyalkylquinoxalines. The oxidation of methylquinoxalines to quinoxalinecarboxaldehydes is usually carried out with selenium dioxide,[2] and sodium metaperiodate ($NaIO_4$) is the reagent recommended for the oxidation of tetrahydroxybutylquinoxalines.[3] An indirect method for the conversion of methylquinoxalines into quinoxalinecarboxaldehydes is illustrated in Scheme 1.

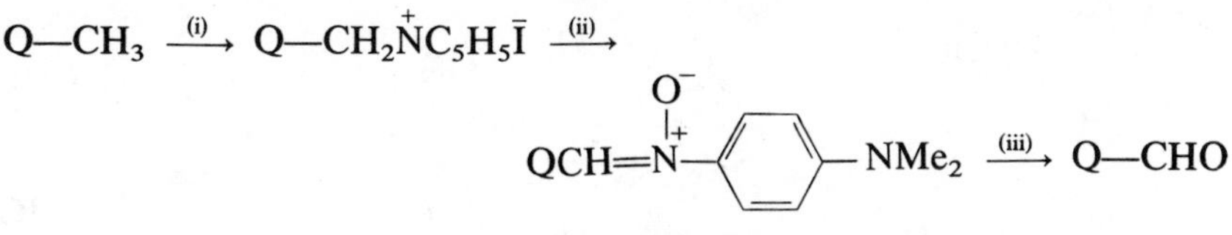

$$Q{-}CH_3 \xrightarrow{\text{(i)}} Q{-}CH_2\overset{+}{N}C_5H_5\overset{-}{I} \xrightarrow{\text{(ii)}}$$

$$QCH{=}\overset{+}{N}(O^-){-}C_6H_4{-}NMe_2 \xrightarrow{\text{(iii)}} Q{-}CHO$$

Q = 2-Quinoxalinyl

Reagents (i) C_5H_5N and I_2
(ii) p-$ONC_6H_4NMe_2$
(iii) Aqueous HCl

Scheme 1

This has been used by Henseke and his co-workers to prepare a number of quinoxalinecarboxaldehydes[4,5] and appears to be a general reaction of methylazines and diazines.

Quinoxaline on treatment with trioxane, hydrogen peroxide, and ferrous sulfate in acid solution yields 2-trioxanylquinoxaline (**1**), and this on acid hydrolysis gives quinoxaline-2-carboxaldehyde in an overall yield of 17% (Scheme 2).[32]

$$Fe^{2+} + HOOH \longrightarrow Fe^{3+} + {}^{\bullet}OH + OH^-$$

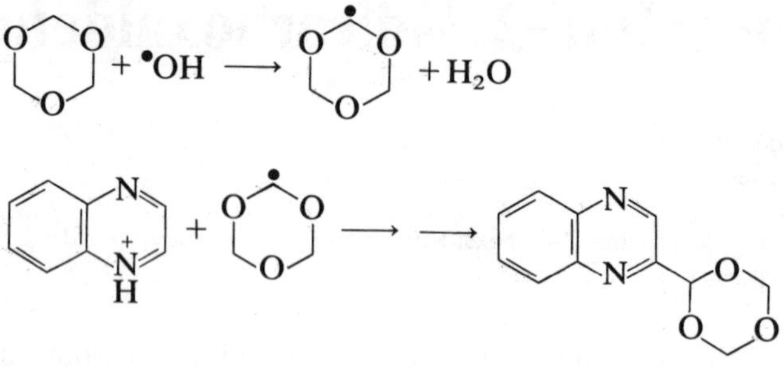

Scheme 2

Dimethylsulfoxide oxidation of α-bromomethylquinoxalines yields quinoxalinecarboxaldehydes, and 2,3-bis(bromomethyl)quinoxaline (**2**) is oxidized by this reagent to the 2,3-dicarboxaldehyde **4**.[6] This preparation yields 60% of the cyclic monohydrate **3**, which is converted into the dialdehyde by vacuum sublimation. The cyclic monohydrate is also formed by treatment of 2,3-bis(hydroxymethyl)quinoxaline 1,4-dioxide (**5**) with alkali.[7]

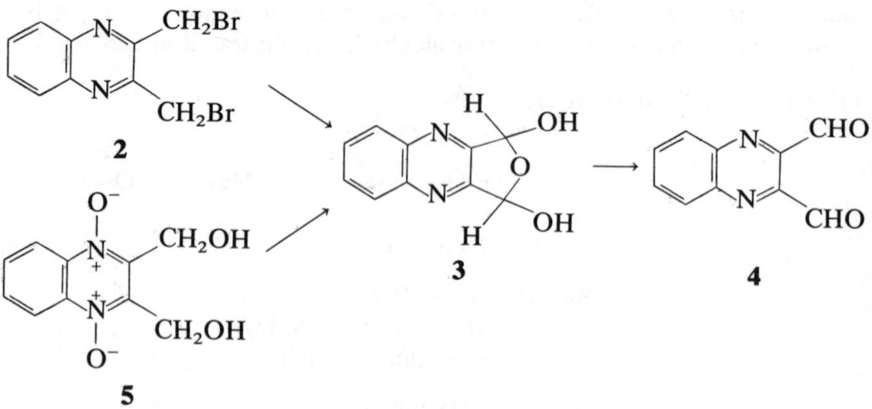

The ^1H NMR spectrum of the hydrate **3** reveals the presence of two types of methine proton, and this is accounted for by the production of a cis–trans mixture of diols.[8] Alkali treatment of the mono-*N*-oxide of 2,3-bis(hydroxymethyl)quinoxaline (**6**) similarly gives the cyclic hemiacetal (**7**) of 3-hydroxymethylquinoxaline-2-carboxaldehyde.[7]

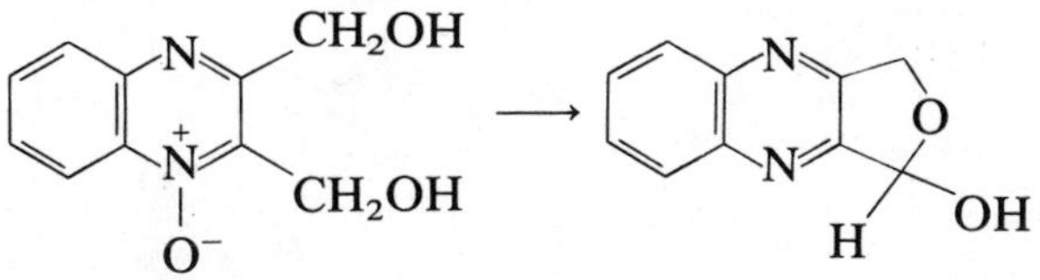

6 → **7**

A successful preparation of 3-methylquinoxaline-2-carboxaldehyde (**9**) by the selective reduction of the ester **8** is reported.[9]

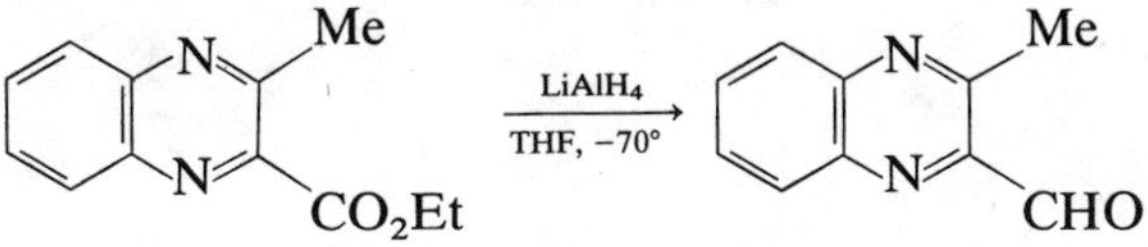

$\xrightarrow[\text{THF, } -70^\circ]{LiAlH_4}$ **9**

II. Properties

Quinoxaline-2-carboxaldehyde has been converted into the 2-carboxylic acid by oxidation with potassium permanganate in acetone[10] and reduced to the 2-hydroxymethyl compound by treatment with formalin and potassium hydroxide.[2] It also undergoes other typical reactions of aromatic aldehydes such as benzoin formation on reaction with potassium cyanide[11,12] and condensation reactions with malonic acid and its diethyl ester[13] and Schiff base formation.[33] Acid-catalyzed reaction of quinoxaline-2-carboxaldehyde with ethylene glycol gives the cyclic acetal[9]; the diethylacetal has been prepared by reaction of 2-dibromomethylquinoxaline with sodium ethoxide.[14] An indirect preparation of the oxime **11** is achieved by treatment of 2-nitromethylquinoxaline (**10**) with diazomethane followed by thermolysis of the resulting nitronic ester.[15]

An interesting property of quinoxaline-2-carboxaldehyde phenylhydrazone (**12**) is its ability to undergo intramolecular cyclization to the pyrazoloquinoxaline **13**.[16] A series of arylhydrazones (**15**) have been

CH_2NO_2 **10** $+ CH_2N_2$ ⟶ $CH{=}NO_2CH_3$ ⟶ (Δ) $CH{=}NOH$ **11**

$CH{=}NNHPh$ **12** ⟶ Ph, N, N **13**

Me, Me **14** $\xrightarrow{ArN_2^+}$ Me, $CH{=}N{-}NHAr$ **15** ⇌ Me, $CH_2{-}N{=}NAr$ **16**

Ar = Ph, *p*-$EtO_2CC_6H_4$, or *p*-$HO_2CC_6H_4$

prepared by the action of aryldiazonium salts on 2,3-dimethylquinoxaline (**14**). The arylhydrazones clearly have tautomeric potential and may alternatively exist in the azo form (**16**).[17]

Quinoxaline-2-carboxaldehyde thiosemicarbazones have been screened for antitubercular[18] and antitumour activity[19] but failed to show significant activity.

III. Tables

TABLE 1. QUINOXALINE 2-CARBOXALDEHYDES AND -2,3-DICARBOXALDEHYDES

Molecular formula	Substituents	m.p., °C	Ref.
$C_9H_5ClN_2O$	3-Chloro-2-carboxaldehyde	153–154	20
	Phenylhydrazone	163–164	20
	p-Nitrophenylhydrazone	289–290	20
	2,4-Dinitrophenylhydrazone	275–277	20
	2,4,6-Trinitrophenylhydrazone	267–270	20
	N-Acetylphenylhydrazone	216–218	20
	p-Dimethylaminophenylnitrone	124–125	20
	6-Chloro-2-carboxaldehyde	206	5
	Phenylhydrazone	223–224 decomp.	5
	7-Chloro-2-carboxaldehyde	163–164	5
	Phenylhydrazone	209–211 decomp.	5
	6-Nitro-2-carboxaldehyde		
	Hydrate	152–154	21
	2,4-Dinitrophenylhydrazone	238–240	21
$C_9H_6N_2O$	2-Carboxaldehyde	107–108, 110	2, 3, 10, 14, 22, 32
	Phenylhydrazone	229–230	14
	Thiosemicarbazone	238–239 decomp.	10
	Oxime	202, 204–205	15, 22,
	Diethyl acetal	b.p. 124.5–125.5/1 mm	14
	Ethylene acetal	69–70	9
	4′-Carboxyanil	288	3
	4′-Methylanil	120–121	3
$C_9H_6N_2O_2$	3-Hydroxy-2-carboxaldehyde	72–74	23
	Phenylhydrazone	290 decomp.	25
	Thiosemicarbazone	262–263 decomp.	19
	Oxime	268–271 decomp.	24
$C_{10}H_4Br_2Cl_2N_2O$	3-(Dibromomethyl)-6,7-dichloro-2-carboxaldehyde	162–164	6
$C_{10}H_4Cl_2N_2O_2$	6,7-Dichloro-2,3-dicarboxaldehyde	162–164	6
$C_{10}H_5ClN_2O_2$	6-Chloro-2,3-dicarboxaldehyde	147.5–148	6
$C_{10}H_6Br_2N_2O$	3-(Dibromomethyl)-2-carboxaldehyde	171.5–172	6
	2,4-Dinitrophenylhydrazone	246–247	6
$C_{10}H_6N_2O_2$	2,3-Dicarboxaldehyde	176, 177–178 decomp.	6, 7
	Hydrate	171–172	26
	Bisphenylhydrazone	188 decomp.	4

TABLE 1 (*contd.*)

Molecular formula	Substituents	m.p., °C	Ref.
$C_{10}H_7BrN_2O$	Bis-2,4-dinitrophenylhydra-zone	279 decomp., 283 decomp.	4,27
	3-Bromomethyl-2-carboxalde-hyde	158–159	6
$C_{10}H_7IN_2O$	3-Iodomethyl-2-carboxaldehyde	138.5–139.5 sublimes	6
$C_{10}H_8N_2O$	3-Methyl-2-carboxaldehyde	138, 143.5–144	6, 9, 27, 28
	Hydrobromide	190–191	6
	p-Nitrophenylhydrazone	250	27
	Thiosemicarbazone	251–252 decomp.	29
	6(or 7)-Methyl-2-carbox-aldehyde	114 decomp.	4
	Phenylhydrazone	200–204 decomp.	4
	Thiosemicarbazone	238 decomp.	4
	Oxime	212–214	30
$C_{10}H_8N_2O_2$	3-Hydroxymethyl-2-carbox-aldehyde[a]	158–159	7
	2,4-Dinitrophenylhydrazone	204 decomp.	7
	3-Methoxy-2-carboxaldehyde	98	19, 22
	Thiosemicarbazone	253–254 decomp.	19
	Oxime	207 decomp.	22
$C_{12}H_{10}Br_2N_2O$	3-Dibromomethyl-6,7-dimethyl-2-carboxaldehyde	135.5–137.5	6
$C_{15}H_{10}N_2O$	3-Phenyl-2-carboxaldehyde	133.5	31

[a] Exists as the cyclic hemiacetal.

IV. References

1. J. C. E. Simpson, "Condensed Pyridazine and Pyrazine Rings," Interscience, New York, 1953, p. 246.
2. J. K. Landquist and J. A. Silk, *J. Chem. Soc.*, 2052 (1956).
3. C. L. Leese and H. N. Rydon, *J. Chem. Soc.*, 303 (1955).
4. G. Henseke and K.-J. Bähner, *Chem. Ber.*, **91,** 1605 (1958).
5. G. Henseke and R. Jacobi, *Justus Liebigs Ann. Chem.*, **684,** 146 (1965).
6. E. J. Moriconi and A. J. Fritsch, *J. Org. Chem.*, **30,** 1542 (1965).
7. A. S. Elina, L. G. Tsyrul'nikova, and G. P. Syrova, *Chem. Heterocycl. Comp.*, **5,** 115 (1969).
8. E. J. Moriconi, J. W. Young, T. E. Brady, and A. J. Fritsch, *Can. J. Chem.*, **44,** 1596 (1966).
9. M. P. Mertes and A. J. Lin, *J. Med. Chem.*, **13,** 77 (1970).
10. A. S. Elina and O. Yu. Magidson, *J. Gen. Chem. USSR*, **25,** 145 (1955).
11. E. Lippmann, P. di Dio, W. Schweitzer, and M. Mühlstädt, *Z. Chem.*, **6,** 311 (1966).

12. H.-J. Binte, G. Henseke, W. Bauer, and K. Köhnke, *Z. Chem.*, **8,** 104 (1968).
13. E. C. Taylor and A. McKillop, *J. Am. Chem. Soc.*, **87,** 1984 (1965).
14. A. S. Elina, *J. Gen. Chem. USSR*, **29,** 2728 (1959).
15. P. E. Fanta, R. M. W. Rickett, and D. S. James, *J. Org. Chem.*, **26,** 938 (1961).
16. G. Henseke and W. Lemke, *Chem. Ber.*, **91,** 113 (1958).
17. N. E. Plevachuk and S. N. Baranov, *Chem. Heterocycl. Comp.*, **4,** 536 (1968).
18. K. Asano, *Yakugaku Zasshi*, **78,** 729 (1958); *Chem. Abstr.*, **52,** 18428 (1958).
19. F. A. French and E. J. Blanz, *J. Med. Chem.*, **9,** 585 (1966).
20. V. D. Romanenko and S. I. Burmistrov, *Chem. Heterocycl. Comp.*, **9,** 785 (1973).
21. J. Klicnar and F. Kosek, *Collect. Czech. Chem. Commun.*, **30,** 3102 (1965).
22. E. Hayashi, C. Iijima, and Y. Nagasawa, *Yakugaku Zasshi*, **84,** 163 (1964); *Chem. Abstr.*, **61,** 3108 (1964).
23. S. N. Baranov and N. E. Plevachuk-Tarnaskaya, *Ukr. Khim. Zh.*, **29,** 82 (1963); *Chem. Abstr.*, **59,** 6406 (1963).
24. J. Klicnar, M. Hajek, and I. Dobas, *Collect. Czech. Chem. Commun.*, **30,** 3092 (1965).
25. R. Kuhn and K. Dury, *Justus Liebigs Ann. Chem.*, **571,** 44 (1951).
26. V. Kovacik, M. Fedoronko, and I. Jezo, *Org. Mass Spectrom.* **7,** 449 (1973).
27. M. Seyhan, *Chem. Ber.*, **84,** 477 (1951).
28. J. Francis, J. K. Landquist, A. A. Levi, J. K. Silk, and J. M. Thorpe, *Biochem. J.*, **63,** 455 (1956).
29. S. Kushner, H. Dalalian, J. L. Sanjurjo, F. L. Bach, S. R. Safir, V. K. Smith, and J. H. Williams, *J. Am. Chem. Soc.*, **74,** 3617 (1952).
30. P. E. Fanta and C.-S. Wang, *J. Heterocycl. Chem.*, **3,** 525 (1966).
31. E. Hayashi and C. Iijima, *Yakugaku Zasshi*, **82,** 1093 (1962); *Chem. Abstr.*, **58,** 4552 (1963).
32. G. P. Gardini, *Tetrahedron Lett.*, 4113 (1972).
33. J. R. Potoski and M. E. Freed (to American Home Products Corp.), U.S. Pat. 3,632,581; *Chem. Abstr.*, **76,** 113252*w* (1972).

CHAPTER VIII

Quinoxaline 2-Ketones and Quinoxaline 2,3-Diketones

I. Methods of Preparation 130
II. Properties 132
III. Tables 134
Table 1. Quinoxaline 2-Ketones and 2,3-Diketones 134
IV. References 136

In this chapter ketones in which the carbonyl group is directly linked to either position 2 or positions 2 and 3 of the quinoxaline nucleus are discussed.

I. Methods of Preparation

Quinoxalinyl ketones have been prepared by application of the classical quinoxaline synthesis, an *o*-phenylenediamine being condensed with an appropriate 1,2,3-tricarbonyl compound.[1] Reactions of this type do not provide an unambiguous synthesis of quinoxalinyl ketones, since an alternative mode of condensation is possible leading to a benzo-1,5-diazepinone. The latter possibility is eliminated in the example cited in Scheme 1 because ^{1}H NMR examination of the condensation product

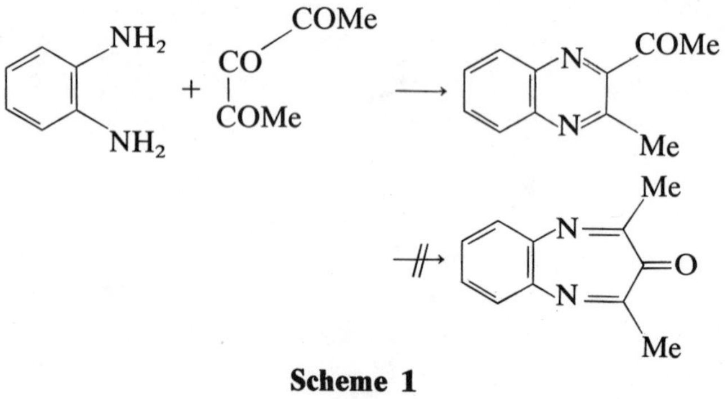

NH_2 NH_2 + CO(COMe)COMe ⟶ (quinoxaline with COMe, Me)

⟶/⟶ (benzodiazepinone with Me, =O, Me)

Scheme 1

reveals the presence of two methyl groups in different chemical environments.[2] Quinoxalinyl ketones (e.g., **1**) have also been prepared from the reaction of *o*-phenylenediamine with cyclobutenediones. The suggested mode of reaction is as shown in Scheme 2.[3]

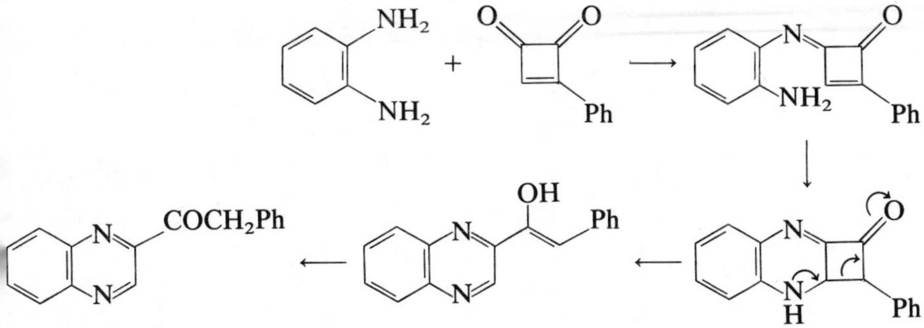

1

Scheme 2

Diazomethane methylation of quinoxaline-2-carboxaldehyde yields 2-acetylquinoxaline in excellent yield,[4] and a number of 2-benzoylquinoxalines have been prepared by chromium trioxide oxidation of the corresponding 2-benzylquinoxalines.[5]

Reaction of the quinoxaline ester **2** with acetone and sodium hydride yields the acetoacetylquinoxaline **3** in excellent yield.[15]

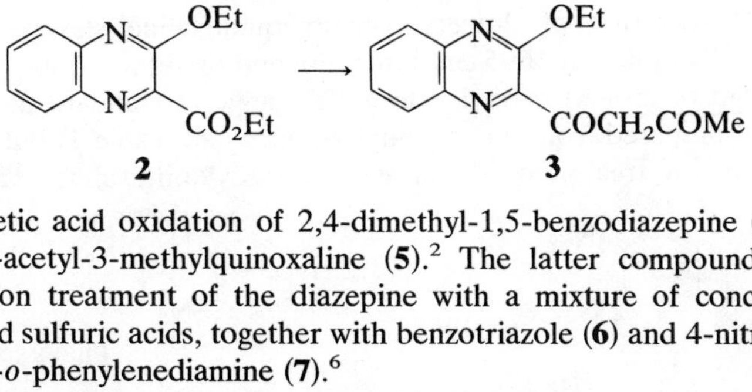

2 **3**

Peracetic acid oxidation of 2,4-dimethyl-1,5-benzodiazepine (**4**) gives some 2-acetyl-3-methylquinoxaline (**5**).[2] The latter compound is also formed on treatment of the diazepine with a mixture of concentrated nitric and sulfuric acids, together with benzotriazole (**6**) and 4-nitro-*N,N'*-diacetyl-*o*-phenylenediamine (**7**).[6]

Oxidation of the sulfone **8**, prepared from *o*-phenylenediamine and the appropriate diketone, with hydrogen peroxide in aqueous ethanolic sodium hydroxide furnishes 2-benzoyl-3-benzylquinoxaline (**9**) in excellent yield.[7]

A general procedure for the acylation of quinoxaline involves the simultaneous addition of solutions of ferrous sulfate and *tert*-butyl hydroperoxide to a stirred and cooled mixture of quinoxaline and aldehyde in 4 *M* sulfuric acid (see Chapter II). Under these conditions acyl radicals

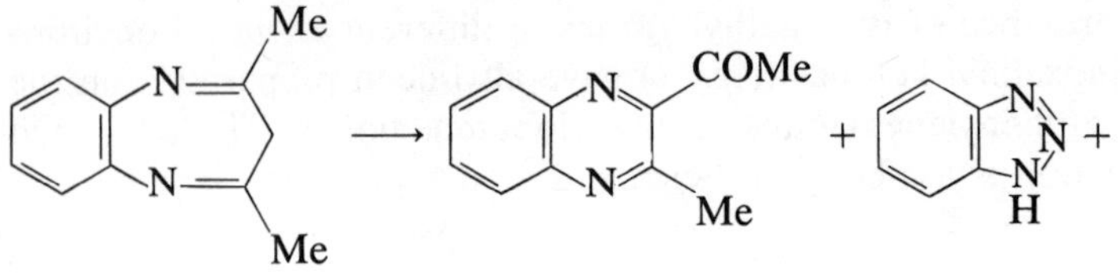

4 → 5 + 6 +

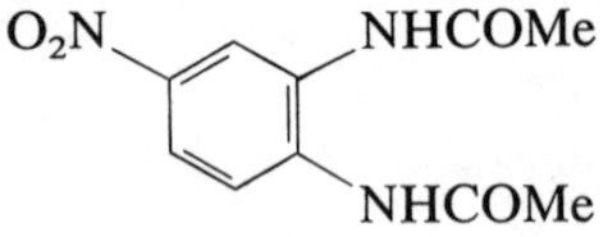

7

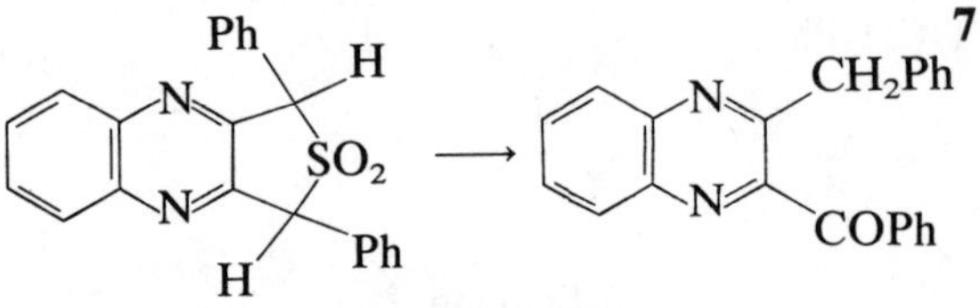

8 → 9

(RCO$^{\cdot}$) are obtained by oxidation of the aldehyde, and substitution occurs at the 2-position.[18]

II. Properties

The IR spectrum of 2-acetyl-3-methylquinoxaline shows carbonyl stretching absorption at 1695 cm^{-1} in Nujol and therefore in the region to be expected of an aryl methyl ketone.[2] A range of standard derivatives have been prepared from quinoxalinyl ketones (see Table 1) but cyclization occurs on treatment of various 2-benzoylquinoxalines (**10**) with

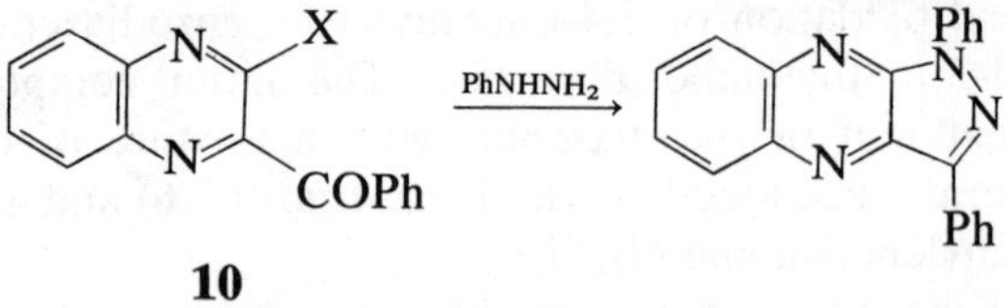

10 $\xrightarrow{PhNHNH_2}$

10

X = H, CO_2H, or $CONH_2$

10 $\xrightarrow{NH_2OH}$

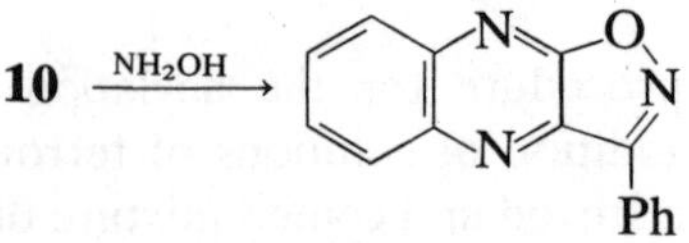

X = CO_2H or $CONH_2$

phenylhydrazine and hydroxylamine.[8] Attempted Beckmann rearrangement of the ketoxime **11** results in formation of the quinoquinoxaline **12**.[9]

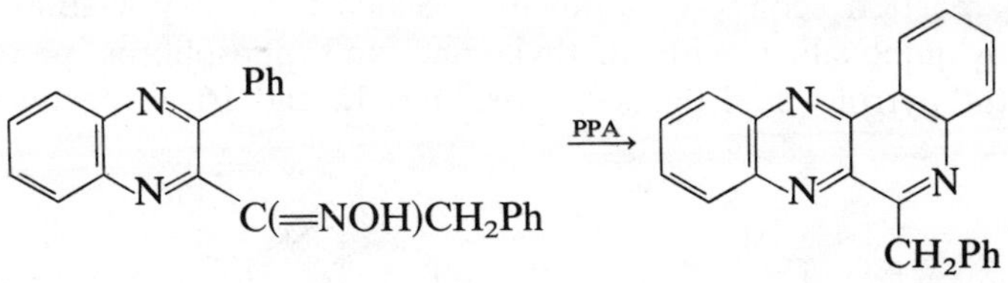

PPA

C(=NOH)CH_2Ph

CH_2Ph

11 **12**

Chromium trioxide oxidation of 2-phenacetyl-3-phenylquinoxaline (**13**) gives a mixture of 3-phenylquinoxaline-2-carboxylic acid, 3-phenylquinoxalin-2-one, and benzoic acid.[9]

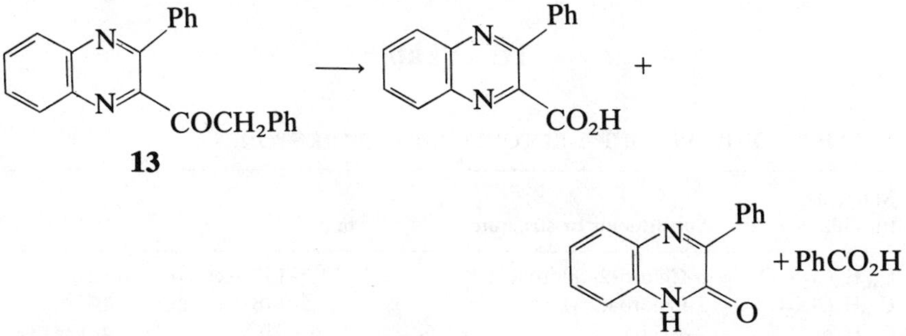

COCH_2Ph

13

CO_2H

+ $PhCO_2H$

Treatment of a number of 2-acetylquinoxalines with nitrous acid gives the expected glyoxal oximes (e.g., **14**).[10]

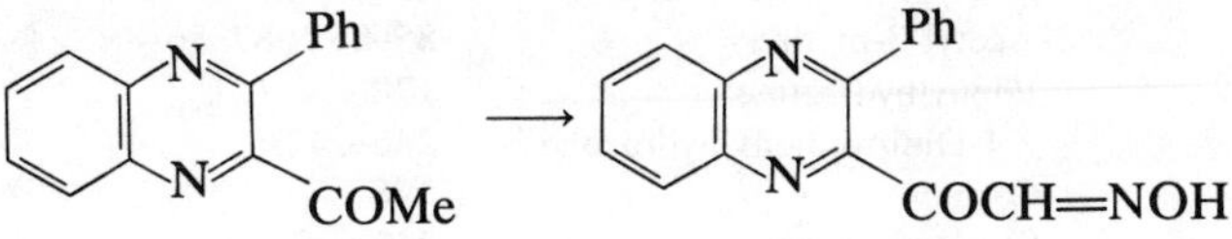

COMe

COCH=NOH

14

COPh

CH_2Ph

Ac_2O

COPh

CH(OAc)Ph

KOH–MeOH

COPh

COPh

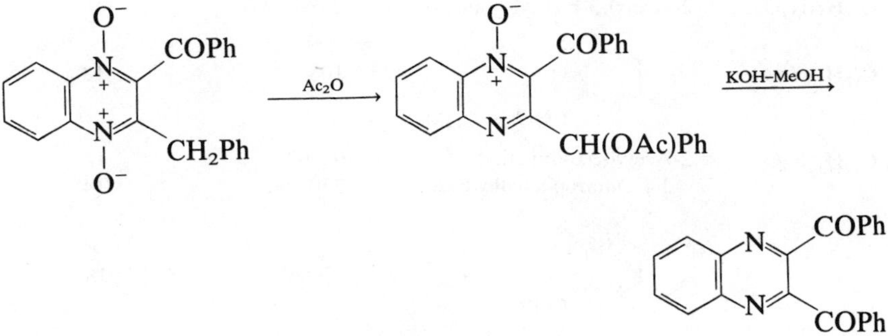

Scheme 3

2,3-Dibenzoylquinoxaline has been prepared by chromium trioxide oxidation of either 2,3-dibenzyl-[5] or 2-benzoyl-3-benzylquinoxaline,[7] or by the reaction sequence outlined in Scheme 3.[11] Cyclization of 2,3-dibenzoylquinoxaline with methylamine and phosphorus pentasulfide yields the pyrrolo- and thienoquinoxalines **15** and **16**, respectively.[11]

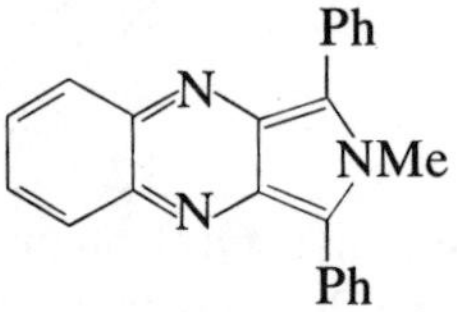

15 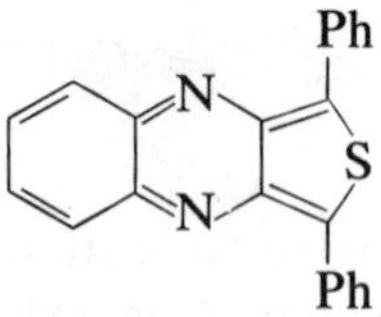 **16**

III. Tables

TABLE 1. QUINOXALINE 2-KETONES AND 2,3-DIKETONES

Molecular formula	Substituents or structure	m.p., °C	Ref.
$C_{10}H_6Cl_2N_2O$	6-Chloro-2-chloroacetyl	151.5–152 decomp.	12
$C_{10}H_7ClN_2O$	2-Chloroacetyl	147.5–148 decomp.	13
$C_{10}H_8N_2O$	2-Acetyl	77, 79–79.5	4, 13, 18
	Phenylhydrazone	222 decomp.	4
	Benzylidene deriv.	159–160	4
$C_{11}H_9N_3O_3$	2-Acetyl-3-methyl-7-nitro	138.5–139	14
$C_{11}H_{10}N_2O$	2-Acetyl-3-methyl	85–86.5, 87, 86–88	1, 2, 6, 14
	Phenylhydrazone	178	2
	2,4-Dinitrophenylhydrazone	246–247	2
	Oxime	196	2
	Semicarbazone	247	2
	2-Propionyl	104–105	18
$C_{11}H_{11}N_3O$	2-Acetyl-3-methyl-6-amino	233–234	14
$C_{12}H_{10}N_2O$	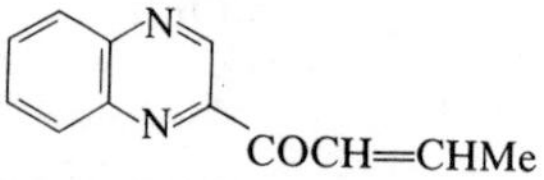 [quinoxaline structure] COCH=CHMe	105	18
$C_{12}H_{12}N_2O$	2-Acetyl-3,6-dimethyl	91–92	14
	2,4-Dinitrophenylhydrazone	230–232	14
	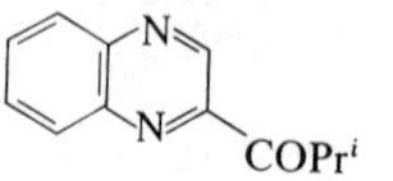 [quinoxaline structure] $COPr^i$	55–56	18
$C_{12}H_{12}N_2O_2$	2-Acetyl-3-ethoxy		20

TABLE 1 (*contd.*)

Molecular formula	Substituents or structure	m.p., °C	Ref.
$C_{13}H_8N_2O_2$	N, N, CO, O	124–125	18
$C_{13}H_{14}N_2O$	N, N, $COBu^t$	75–76	18
$C_{14}H_{14}N_2O_3$	2-Acetoacetyl-3-ethoxy	83–84	15
$C_{15}H_9ClN_2O$	2-Benzoyl-3-chloro	138–141	16
	2-*p*-Chlorobenzoyl	105	18
$C_{15}H_9N_3OS$	N, N, S, $COCH_2CN$	172	19
$C_{15}H_{10}N_2O$	2-Benzoyl	80–81	8, 18
	Oxime	208–210	8
$C_{16}H_{10}N_2O_3$	2-Benzoyl-3-carboxy		
	Monohydrate	92–94 decomp.	8
	Amide	198–199	8
$C_{16}H_{11}ClN_2O_2$	2-*p*-Anisoyl-3-chloro	114–115	5
$C_{16}H_{12}N_2O$	2-Acetyl-3-phenyl	110, 111–113	1, 2
	2,4-Dinitrophenylhydrazone	223	2
	2-Benzoyl-3-methyl	89–91	1
	2-Phenacetyl	83–84	3
	Oxime	210–212	3
	Bromo deriv.	115.5–116.5	3
$C_{16}H_{12}N_2O_2$	2-*p*-Anisoyl	112, 113	5, 18
	Oxime	227–228	5
$C_{17}H_{11}N_3O$	N, N, Ph, $COCH_2CN$	180	19
$C_{17}H_{12}N_2O_4$	2-*p*-Anisoyl-3-carboxy	155–156 decomp.	5
	p-Bromophenacyl ester	145–146	8
	Amide	223–225	17
	Amide oxime	193–194	8
	Nitrile	177–179	5
$C_{17}H_{14}N_2O$	3-Methyl-2-phenacetyl	102–103	21
$C_{17}H_{14}N_2O_2$	2-*p*-Anisoyl-3-methyl	112–113	5
$C_{17}H_{14}N_2O_3$	2-(3,4-Dimethoxy)benzoyl	124	18
$C_{18}H_{13}N_3O_2$	N, N, C_6H_4OMe-*p*, $COCH_2CN$	220	19

TABLE 1 (*contd.*)

Molecular formula	Substituents or structure	m.p., °C	Ref.
$C_{20}H_{14}N_2O$	2-Benzoyl-3-phenyl	154–156	1
$C_{22}H_{14}N_2O_2$	2,3-Dibenzoyl	169–170, 170–171	5, 7, 11
	Hydrazine deriv.	237–238, 239–240.5	7, 11
$C_{22}H_{16}N_2O$	2-Benzoyl-3-benzyl	96–96.5, 103–105	5, 7
	Oxime	200	5
	2-Phenacetyl-3-phenyl	129.5–130	9
	2,4-Dinitrophenylhydrazone	216–217	9
	Oxime(s).	65–67 and 130–131	9
$C_{24}H_{18}N_2O_4$	2,3-Di-*p*-anisoyl	202–204	5
$C_{24}H_{20}N_2O_3$	2-*p*-Anisoyl-3-*p*-methoxybenzyl	114–115	5

IV. References

1. M. Schellenberg and R. Steinmetz, *Helv. Chim. Acta*, **52,** 431 (1969).
2. J. A. Barltrop, C. G. Richards, D. M. Russell, and G. Ryback, *J. Chem. Soc.*, 1132 (1959).
3. E. J. Smutney, M. C. Caserio, and J. D. Roberts, *J. Am. Chem. Soc.*, **82,** 1793 (1960).
4. G. Henseke and K.-J. Bähner, *Chem. Ber.*, **91,** 1605 (1958).
5. H. Dahn and J. Nussbaum, *Helv. Chim. Acta*, **52,** 1661 (1969).
6. K. V. Levshina, L. P. Glazyrina, and T. S. Safonova, *Chem. Heterocycl. Comp.*, **6,** 1059 (1970).
7. E. J. Moriconi, R. E. Misner, and T. E. Brady, *J. Org. Chem.*, **34,** 1651 (1969).
8. H. Dahn and H. Moll, *Helv. Chim. Acta*, **49,** 2426 (1966).
9. A. T. Blomquist and E. A. LaLancette, *J. Am. Chem. Soc.*, **84,** 220 (1962).
10. V. V. Titov and L. F. Kozhokina, *Khim. Geterotsikl. Soedin.*, 1289 (1972); *Chem. Abstr.*, **78,** 4217*j* (1973).
11. M. J. Haddadin, N. C. Chelhot, and M. Pieridou, *J. Org. Chem.*, **39,** 3278 (1974).
12. H. R. Moreno and H. P. Schultz, *J. Med. Chem.*, **13,** 1005 (1970).
13. H. R. Moreno and H. P. Schultz, *J. Med. Chem.*, **13,** 119 (1970).
14. V. V. Titov and L. F. Kozhokina, *Chem. Heterocycl. Comp.*, **7,** 1328 (1971).
15. F. Eiden and G. Bachmann, *Arch. Pharm.*, **306,** 401 (1973); *Chem. Abstr.*, **79,** 78740*s* (1973).
16. V. D. Romanenko and S. I. Burmistrov, *Chem. Heterocycl. Comp.*, **9,** 785 (1973).
17. H. Dahn and H. Moll, *Helv. Chim. Acta*, **47,** 1860 (1964).
18. G. P. Gardini and F. Minisci, *J. Chem. Soc. (C)*, 929 (1970).
19. P. H. J. Ooms, J. W. Scheeren, and R. J. F. Nivard, *Synthesis*, 639 (1975).
20. F. Eiden and G. Bachmann, *Arch. Pharm.*, **305,** 580 (1972); *Chem. Abstr.*, **77,** 126564*v* (1972).
21. W. Reid and H. Kohl, *Synthesis*, 542 (1971).

CHAPTER IX

Quinoxaline-2-carboxylic Acids and Quinoxaline-2,3-dicarboxylic Acids

I. Quinoxaline-2-carboxylic Acids 137
1. Methods of Preparation 137
A. From *o*-Phenylenediamines 138
B. From *o*-Nitrosoaminobenzenes and *o*-Aminoazobenzenes 144
C. From Oxidation of 2-Alkylquinoxalines and 2-Polyhydroxyalkylquinoxalines 145
D. Miscellaneous Methods 145
2. Properties 145
3. Quinoxaline Antibiotics Containing Quinoxaline-2-carboxylic Acid Residues 148
4. Uses 148
II. Quinoxaline-2,3-dicarboxylic Acids 149
1. Methods of Preparation 149
2. Properties 150
III. Tables 152
Table 1. Quinoxaline-2-carboxylic Acids 152
Table 2. Quinoxaline-2,3-dicarboxylic Acids 157
IV. References 158

The earlier literature on quinoxaline-2-carboxylic acids and quinoxaline-2,3-dicarboxylic acids has been reviewed in Simpson's monograph.[1]

I. Quinoxaline-2-carboxylic Acids

1. Methods of Preparation

Most preparations of quinoxaline-2-carboxylic acids and their derivatives are based on the reaction of an *o*-phenylenediamine with a suitable tricarbonyl compound. The inherent difficulty with this synthetic approach is the formation of isomeric mixtures when unsymmetrical diamines are used. This problem is circumvented by the use of *o*-nitrosoamino and *o*-aminoazo compounds in place of an *o*-phenylenediamine; however these types of intermediates are less readily available. Quinoxalinecarboxylic acids have also been prepared by oxidative methods; they are formed by oxidative degradation of alkylquinoxalines and of polyhydroxyalkylquinoxalines.

A. *From o-Phenylenediamines*

Quinoxaline-2-carboxylic acids substituted in the 3-position have been widely prepared by reaction of an *o*-phenylenediamine with mesoxalic acid or its diethyl ester. Mesoxalic acid ester condensation proceeds as shown in Scheme 1. The resulting quinoxalinone carboxylate can be

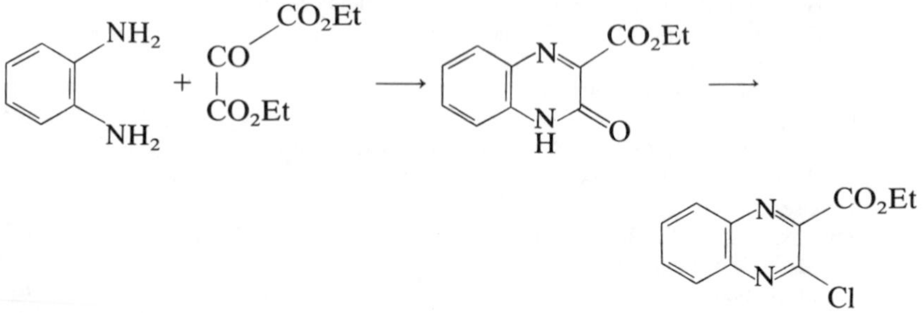

NH_2 NH_2 + $CO(CO_2Et)_2$ → N, N–H, CO_2Et, O → N, N, CO_2Et, Cl

Scheme 1

converted into a 3-chloro derivative by reaction with phosphoryl chloride. Quinoxaline-2-carboxylic acid esters are readily hydrolysed into the corresponding acids by mild alkali treatment.[1a] The use of an excess of mesoxalic acid ester is recommended so that reaction of the quinoxalinone ester with a second molecule of *o*-phenylenediamine to give an *o*-aminoanilide is avoided.[2]

Although the condensation of mesoxalic acid with *o*-phenylenediamine and *N*-methyl-*o*-phenylenediamine hydrochloride proceeds in an analogous manner, when sodium mesoxalate is reacted with *o*-phenylenediamine and the reaction mixture is acidified, the reduced acid **3** and benzimidazole-2-carboxylic acid (**4**) are formed. The primary products of reaction are thought to be the expected quinoxalinonecarboxylic acid (**1**) and the dihydrobenzimidazoledicarboxylic acid (**2**). Reduction of the quinoxalinonecarboxylic acid is thought to involve the dihydro-dicarboxylic acid rather than its decarboxylation product, the corresponding dihydrobenzimidazole-2-carboxylic acid.[3]

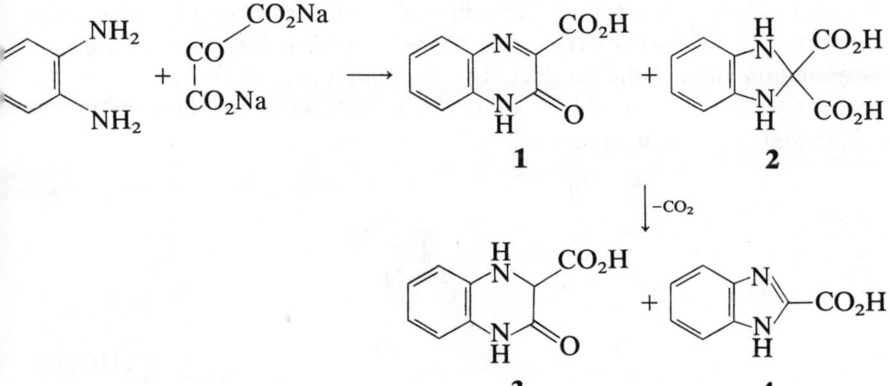

The 2-hydroxy-3-oxo-2,3-dihydrofuran **5** reacts with *o*-phenylenediamine to give the quinoxaline-2-carboxylic acid ester **7**. In this case the reaction may be envisaged as proceeding through the ring-opened tautomer (**6**) of the dihydrofuran.[4]

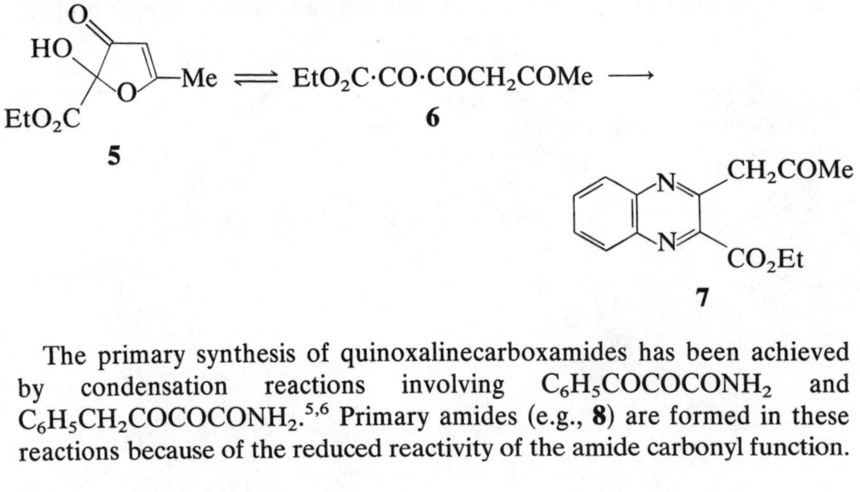

$EtO_2C{\cdot}CO{\cdot}COCH_2COMe$ (**6**)

The primary synthesis of quinoxalinecarboxamides has been achieved by condensation reactions involving $C_6H_5COCOCONH_2$ and $C_6H_5CH_2COCOCONH_2$.[5,6] Primary amides (e.g., **8**) are formed in these reactions because of the reduced reactivity of the amide carbonyl function.

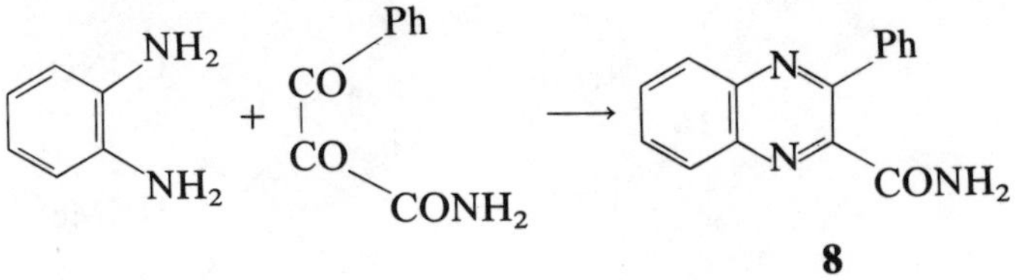

A two-step synthesis of the quinoxalinecarboxamide **11** results from reaction of the dioxobutyrolactone **9** with *o*-phenylenediamine followed by ring opening of the product (**10**) with ammonia.[7] Successive treatment of the lactone **10** with dilute sodium hydroxide and potassium permanganate yields 3-benzoylquinoxaline-2-carboxylic acid.[7a]

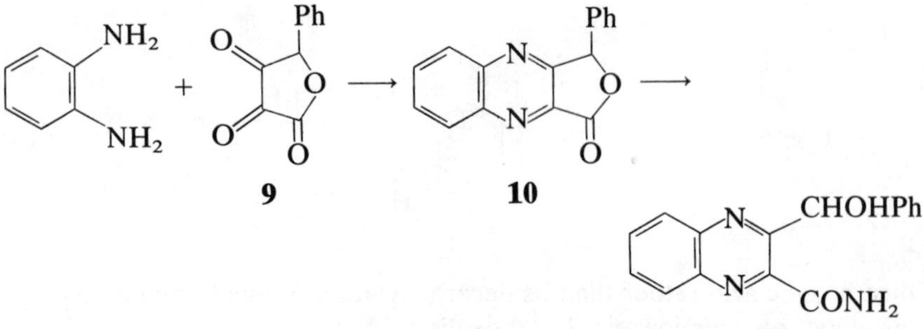

9 **10** **11**

The products obtained by condensation of *o*-phenylenediamines and alloxan and its derivatives are dependent on the pH of the reaction medium. Thus, when *o*-phenylenediamine hydrochloride is allowed to react with alloxan (**12**) in aqueous solution, alloxazine (**13**) is formed.[8]

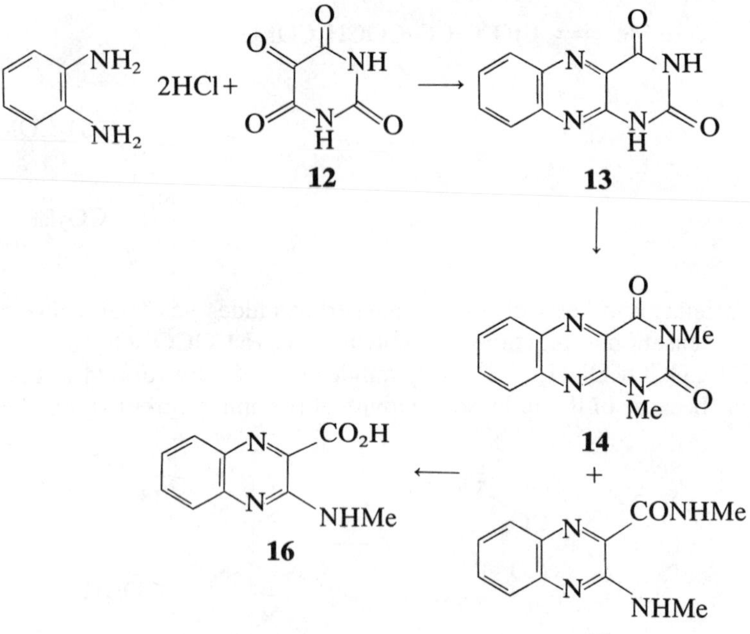

12 **13** **14** **15** **16**

Similarly, condensation of *o*-phenylenediamine hydrochloride with 1,3-dimethylalloxan gives dimethylalloxazine (**14**). The latter compound is also obtained by methylation of alloxazine with dimethyl sulfate and alkali together with the methylamide of 3-methylaminoquinoxaline-2-carboxylic acid (**15**). Prolonged treatment of compounds **14** and **15** with 1 *M* sodium hydroxide at 100° yields 3-methylaminoquinoxaline-2-carboxylic acid (**16**).[8]

8-Chloroalloxazine (**17**) is the sole product of condensation of 4-chloro-*o*-phenylenediamine and alloxan in 1 *M* hydrochloric acid, or in glacial acetic in the presence of boric acid. There is no evidence for the formation of 7-chloroalloxazine under these conditions. At a lower concentration of hydrochloric acid some of the 6-chloroquinoxaline-3-carboxyureide **18** is also formed.[9]

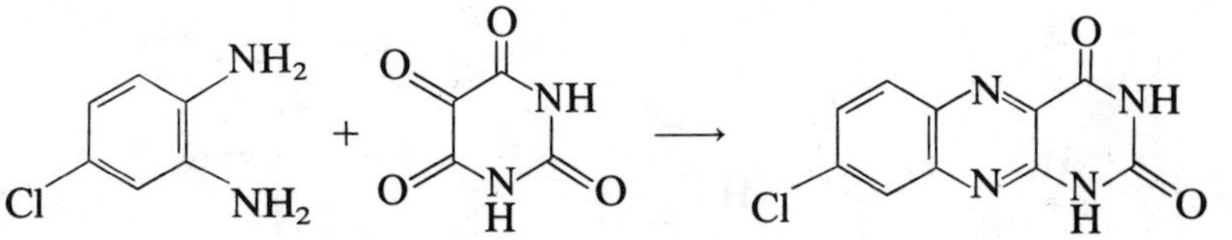

17

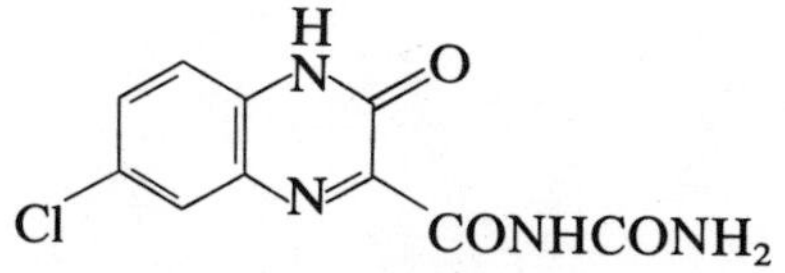

18

When *o*-phenylenediamine is allowed to react with alloxan in aqueous solution, the quinoxaline ureide **19** is produced.[10,11] The structure of the product formed in this reaction has been the subject of controversy. It was at first suggested to be the anil **20** since a superficially similar product, now known to be the spiro compound **22**, is obtained by reaction of *o*-aminodimethylaniline and alloxan in aqueous ethanol. Compound **22** is believed to be formed via the intermediate anil **21**, which undergoes a unique cyclization in which an *N*-methyl group is involved.[12,13]

The reaction of *o*-phenylenediamine with 1,3-dimethylalloxan (**23**) yields the quinoxalinonecarboxy-*N*-methylamide **25**; the latter compound is apparently formed by condensation of the diamine with mesoxalic acid monomethylamide (**24**) resulting from primary degradation of the alloxan.[14]

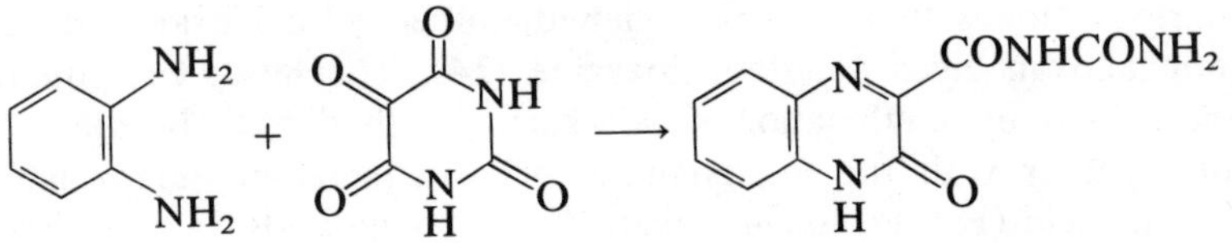

19

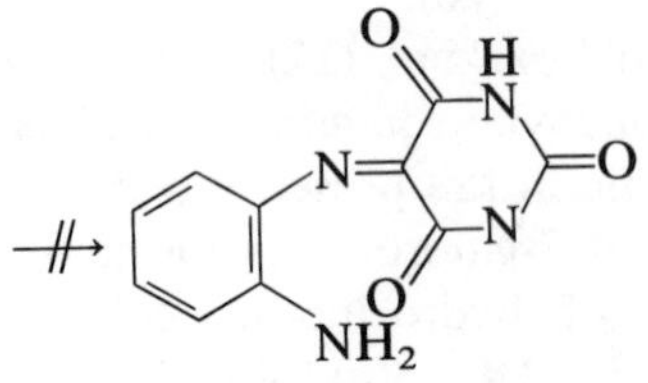

20

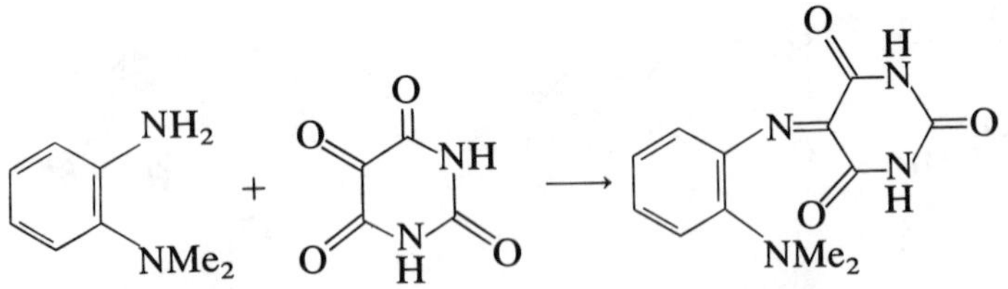

21

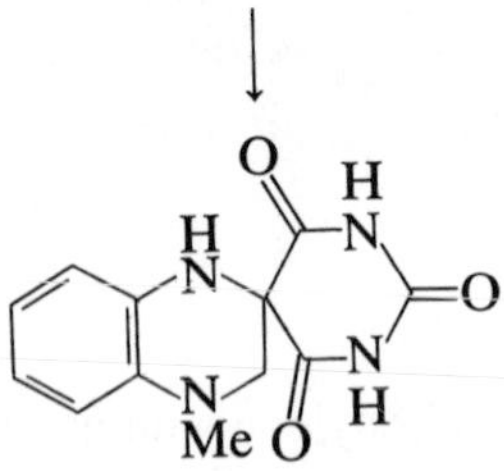

22

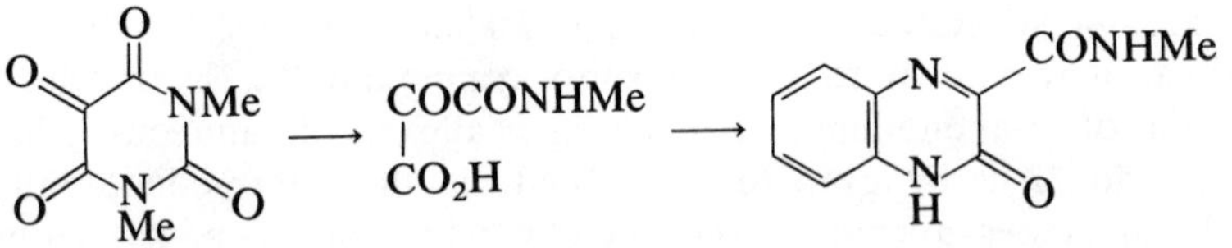

23

24

25

The ready availability of alloxazines from reaction of *o*-phenylenediamine hydrochlorides and alloxan makes them attractive starting materials for the preparation of quinoxaline-2-carboxylic acids. When the pyrimidine ring of the alloxazine is cleaved with ammonia, a 3-

aminoquinoxaline-2-carboxylic acid is formed (e.g., **26**)[15] whereas cleavage with hydroxide ion under vigorous conditions gives the corresponding quinoxalinonecarboxylic acid (e.g., **27**).[16] Isoalloxazines (e.g., **28**), formed by condensation of *N*-substituted *o*-phenylenediamine hydrochlorides with alloxan, are similarly cleaved by hydroxide ion.[17] The milder conditions used to give compound **29** presumably reflect the greater ease of hydrolysis of the intermediate imine compared to the amine in the previous example.

Cl, N, NH, O, N, N, H, O → NH_4OH, 165° → Cl, N, CO_2H, N, NH_2

26

N, NH, O, H_2N, N, N, H, O → 2*M* NaOH, 150° → N, CO_2H, H_2N, N, H, O

27

Me, NH_2, Me, NHMe ·HCl + O, O, NH, O, N, H, O → 0.25*M* $Ba(OH)_2$, 100°

Me, N, O, NH, Me, N, N, Me, O

28

↓

Me, N, CO_2H, Me, N, Me, O

29

Reaction of *N,N'*-dibenzenesulfonyl-*o*-phenylenediamine with methyl 2,3-dibromopropionate gives the tetrahydroquinoxaline **30**, and this is converted into methyl 2-quinoxalinecarboxylate (**31**) on treatment with sodium hydride in tetrahydrofuran.[18]

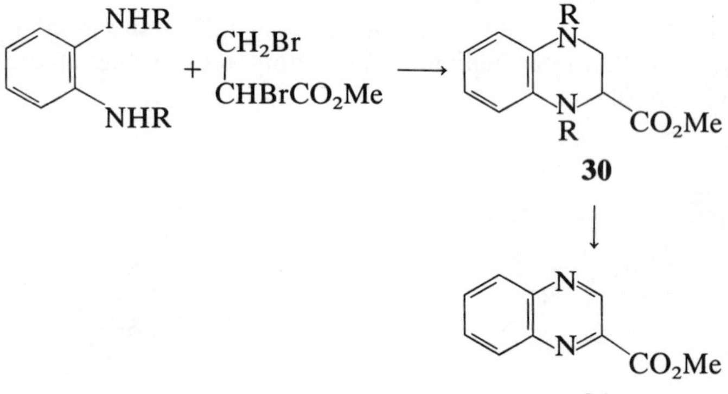

NHR, NHR + CH_2Br–$CHBrCO_2Me$ ⟶ **30** (R–N, N–R, CO_2Me) ⟶ **31** (CO_2Me)

$R = SO_2Ph$

B. *From o-Nitrosoaminobenzenes and o-Aminoazobenzenes*

An unambiguous synthesis of 3-aminoquinoxaline-2-carboxylic acids and the corresponding carboxamides (Scheme 2) is available from the sodium ethoxide-promoted condensation of *o*-nitrosoaminobenzenes and cyanoacetic acid or cyanoacetamide.[19]

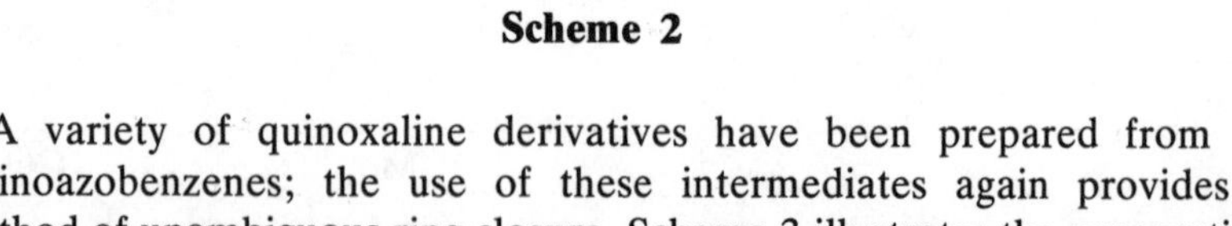

R, NO, H_2N, NH_2 + COX–CH_2–CN ⟶ R, COX, H_2N, NH_2

R = H or Me
X = OH or NH_2

Scheme 2

A variety of quinoxaline derivatives have been prepared from *o*-aminoazobenzenes; the use of these intermediates again provides a method of unambiguous ring closure. Scheme 3 illustrates the preparation of the quinoxaline ester **32**; similar ring closures take place with other carbonyl compounds, for example, acetophenone, with an active methylene group.[20]

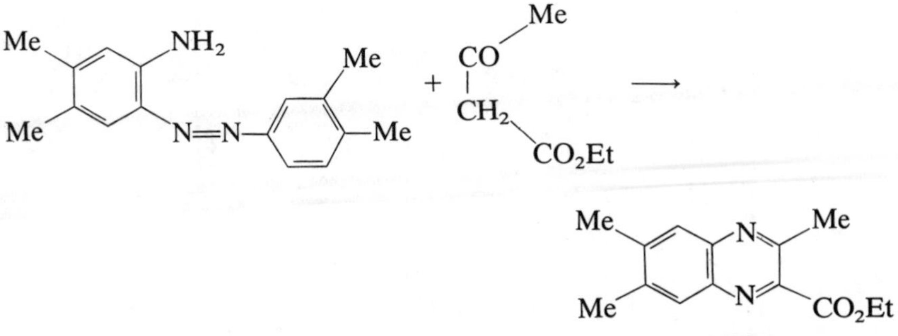

Scheme 3

C. *From Oxidation of 2-Alkylquinoxalines and 2-Polyhydroxyalkylquinoxalines*

2-Methylquinoxalines are converted into the corresponding carboxylic acids by condensation with an aromatic aldehyde and oxidation of the resulting styryl derivative with potassium permanganate.[21] Quinoxaline-2-carboxylic acids are also conveniently prepared by oxidation of the readily available D-arabinotetrahydroxybutylquinoxalines with sodium peroxide in water[22] or in a heterogeneous benzene–water system.[23]

D. *Miscellaneous Methods*

Reaction of quinoxalin-2(1*H*)-one 4-oxide (**33**) with potassium cyanide gives the cyanoquinoxalinone **35**. The latter compound may be converted into the amide **36** by treatment with hydrogen chloride in acetic acid. The amide can be converted into the corresponding acid by reaction with warm aqueous acetic and sulfuric acids. A smilar sequence of reactions is not possible with the *N*-methyl compound **34** because in this case aqueous potassium cyanide treatment gives the quinoxalinedione **37**.[24]

2. Properties

Two groups of workers have measured the pK_a value of quinoxaline-2-carboxylic acid and reported values of 2.87[25] and 2.80.[26] 3-Chloro- and 3-oxo groups are acid strengthening as shown by the measured pK_a values of 1.83 and 2.58, respectively.

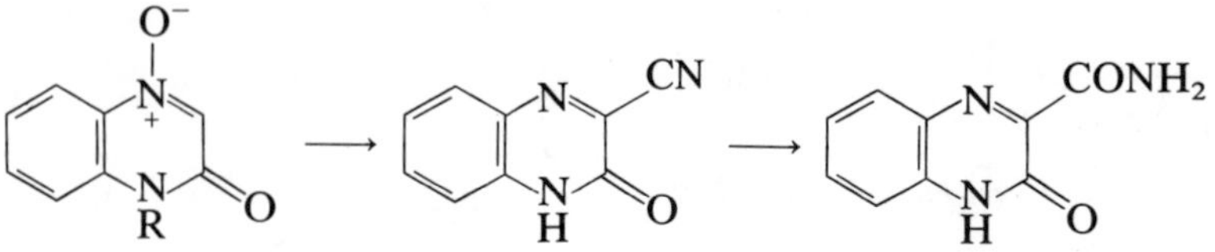

Quinoxaline-2-carboxylic acids are readily decarboxylated on heating, and ease of decarboxylation in some cases makes purification by crystallization difficult. Although a standard range of derivatives is known for the commoner acids, these have often been obtained by primary synthesis rather than from the carboxylic acids themselves. Quinoxaline-2-carboxylic acid is converted into 2-quinoxalinoyl chloride on treatment with thionyl chloride. Successive treatment of the acid chloride with diazomethane and hydrogen chloride gives 2-chloroacetylquinoxaline (**38**) in excellent yield.[22]

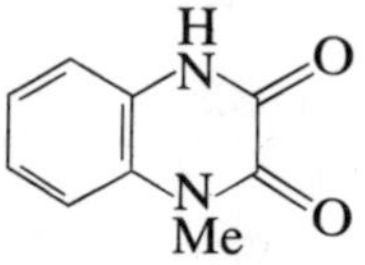

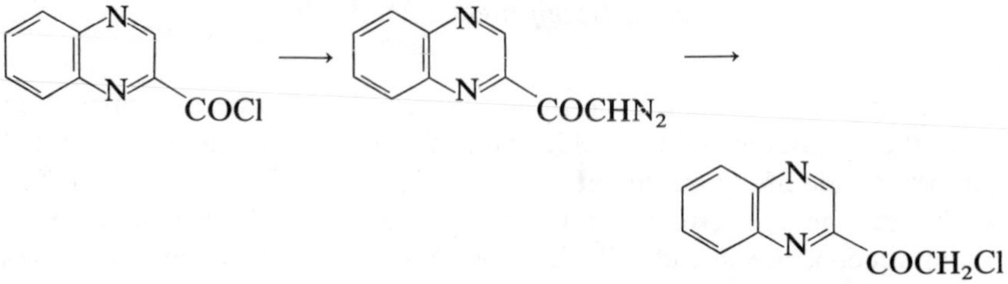

The reaction of 2-cyanoquinoxaline with the anion derived from acetophenone gives some 2-(2-quinoxalinyl)acetophenone (**39**) and some of the imine (**40**). The anion derived from benzylcyanide displaces cyanide ion from 2-cyanoquinoxaline to give the disubstituted acetonitrile **41**.[27]

3(4*H*)-Quinoxalinone-2-carboxylic acid is readily reduced by a variety of reagents (e.g., Zn/HOAc[28] and sodium dithionite[29]) to a 1,2-dihydro derivative. The ureides of 3(4*H*)-quinoxalinone-2-carboxylic acid and of

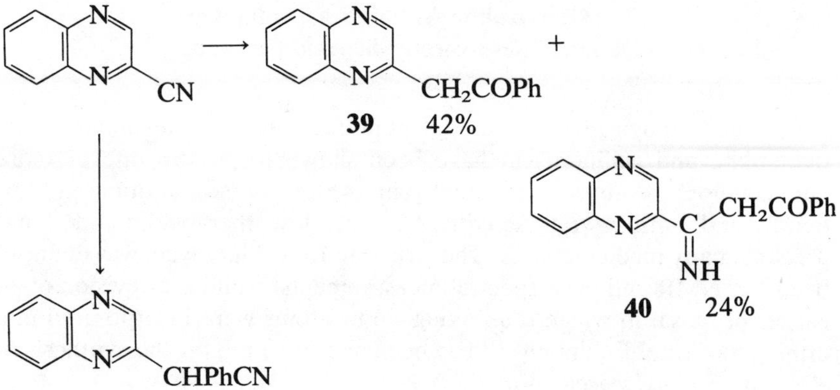

CN

CH$_2$COPh

39 42%

CH$_2$COPh

NH

40 24%

CHPhCN

41 34%

its 4-methyl derivative cyclize under the influence of sodium carbonate to give the tetrahydrospiroquinoxalines **42** and **43**.[30] Spiro derivatives of analogous structure are obtained when the methyl ester of 3(4*H*)-quinoxalinone-2-carboxylic acid is treated with either urea or guanidine in the presence of sodium methoxide.[31]

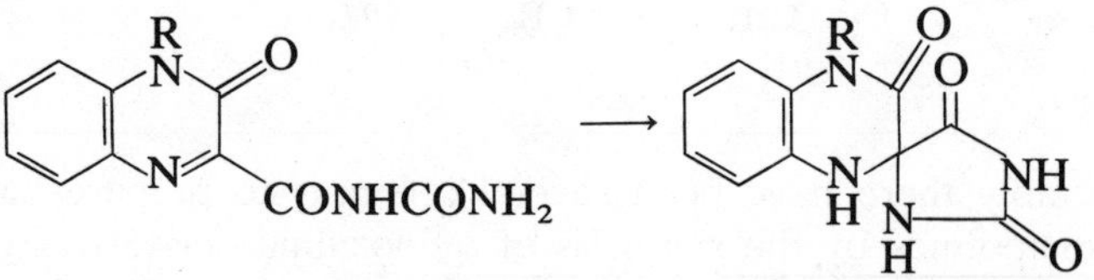

CONHCONH$_2$

42 R = H
43 R = Me

Methylation of the ureide of 3(4*H*)-quinoxalinone-2-carboxylic acid with methyl iodide and potassium carbonate gives the trimethyl derivative **44**, which is also formed by successive treatment of the 4-methyl derivative of the acid with thionyl chloride and *N*,*N*′-dimethylurea.[30]

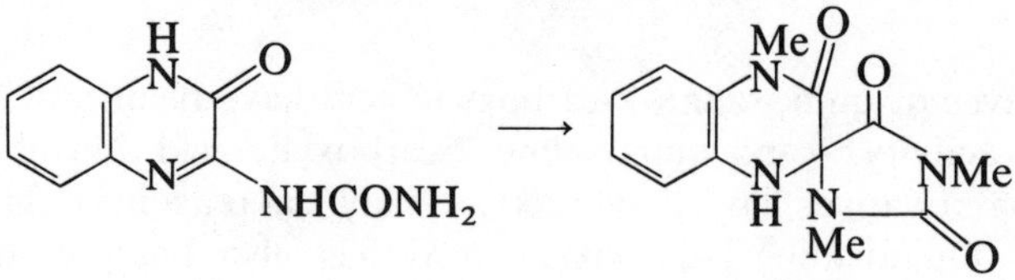

NHCONH$_2$

44

3. Quinoxaline Antibiotics Containing Quinoxaline-2-carboxylic Acid Residues

Several biologically active polypeptides such as levomycin, actinoleukin, and echinomycin have been shown to possess one or more quinoxalinoyl residues.[32] Echinomycin (**45**) possesses antitumour, antiviral, and antibacterial activities, but its low therapeutic index has precluded any medicinal uses. The structure for echinomycin was deduced from ^{13}C NMR and mass spectral measurements[32a] and is a revision of an earlier proposal in which the linking sulfur atoms were incorporated in a dithiapyran ring. Echinomycin has been shown to inhibit the synthesis of RNA by *Staphylococcus aureus.*[32]

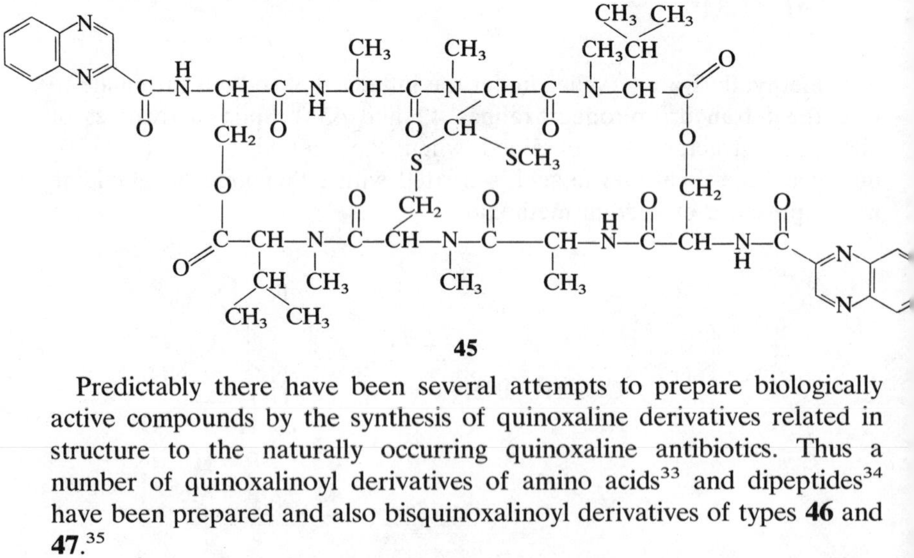

45

Predictably there have been several attempts to prepare biologically active compounds by the synthesis of quinoxaline derivatives related in structure to the naturally occurring quinoxaline antibiotics. Thus a number of quinoxalinoyl derivatives of amino acids[33] and dipeptides[34] have been prepared and also bisquinoxalinoyl derivatives of types **46** and **47**.[35]

$RCONH(CH_2)_xNHCOR$ **46**

$RCONHCH_2CONH(CH_2)_xNHCOCH_2NHCOR$ **47**

R = 2-Quinoxalinyl

4. Uses

Derivatives of quinoxaline-2-carboxylic acid have been tested for anthelminitic activity,[36] and quinoxaline-2-carboxylic acid and its 3-chloro and 3-oxo derivatives have been used as analytical reagents in gravimetric analysis.[37] Quinoxaline-2-carboxylic acid has also been used for the amperometric determination of Cu, Zn, Co, and Ni.[38]

II. Quinoxaline-2,3-dicarboxylic Acids

1. Methods of Preparation

Quinoxaline-2,3-dicarboxylic acids are directly synthesized by condensation of an *o*-phenylenediamine with dihydroxytartaric acid (Scheme 4).[39,40] Condensation of *o*-phenylenediamines with the esters of dioxosuccinic acid ($RO_2C \cdot CO \cdot CO \cdot CO_2R$) similarly gives the corresponding

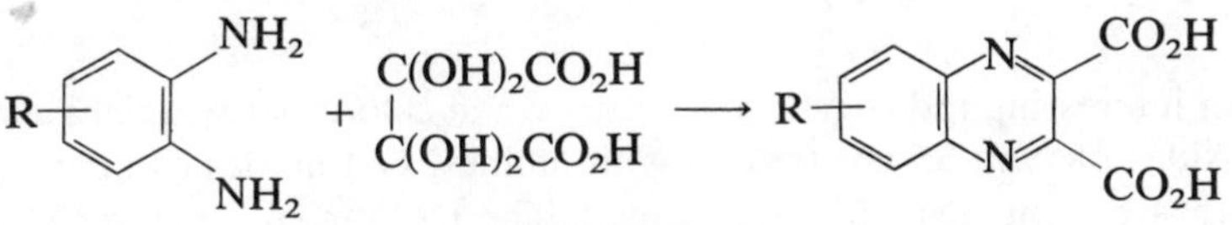

Scheme 4

quinoxalinedicarboxylic acid diesters.[41] An alternative approach involves the oxidative ring cleavage of phenazines; thus potassium permanganate oxidation of phenazine (**48**) yields quinoxaline-2,3-dicarboxylic acid (**49**) in 80% yield.[42] Further oxidation of the diacid yields pyrazinetetracarboxylic acid (**50**).[43]

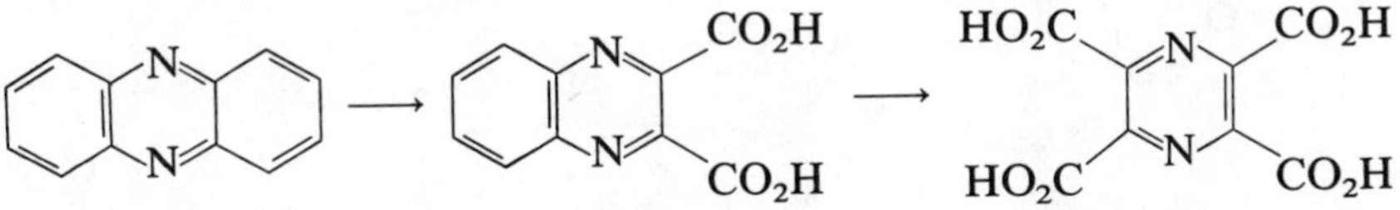

48 **49** **50**

Treatment of the di-*N*-oxide **51** with dimethyl sulfoxide gives the mono-*N*-oxide **52** as the major product; and this, on oxidation with hydrogen peroxide and alkali, yields quinoxaline-2,3-dicarboxylic acid in excellent yield.[44]

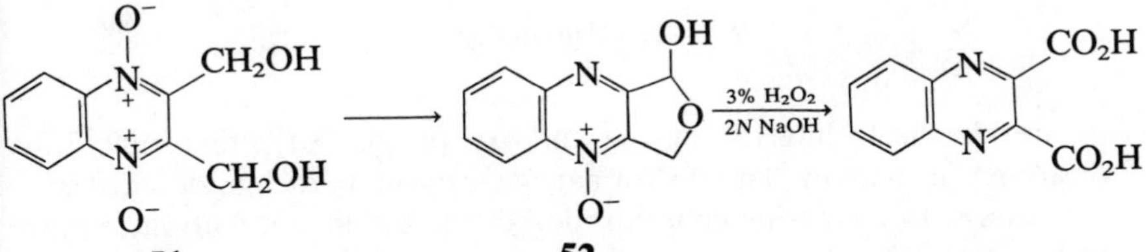

3% H_2O_2, 2*N* NaOH

51 **52**

N-Aryl-2,3-quinoxalinedicarboximides (**54**) are formed instead of the expected sydnones when 3-arylnitrosoamino-1-arylpyrrolidine-2,5-diones (**53**) are treated with acetic anhydride.[45] Ring opening of the imides with ammonia, aniline, and alkali gives the expected quinoxaline-2,3-dicarboxylic acid derivatives.

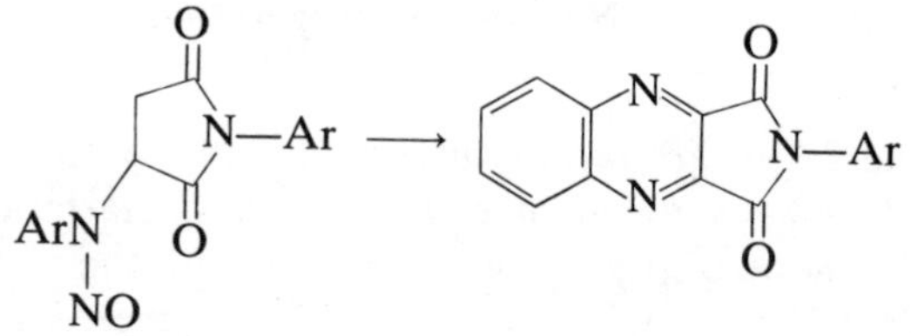

53 **54**

An interesting redox reaction occurs when chloromethylquinoxaline di-*N*-oxides of type **55** are treated with amines, and imides of quinoxaline-2,3-dicarboxylic acid (**57**) are formed. The mechanism suggested for this reaction is shown in Scheme 5; it involves the imino-imide **56**.[46]

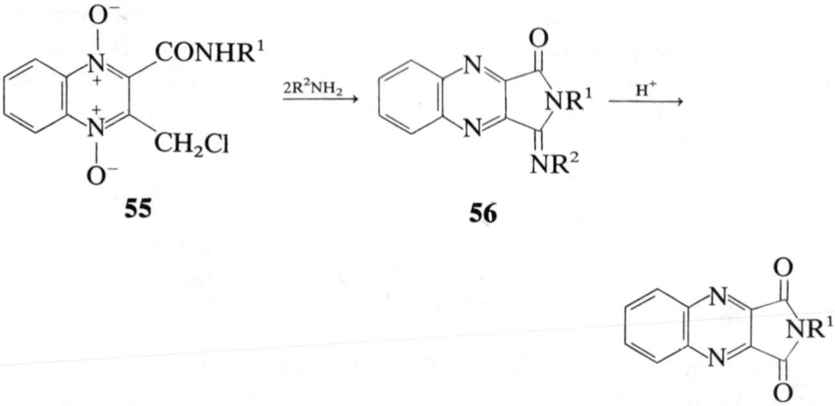

55 **56**

$R^1 = R^2 =$ alkyl or cyclohexyl **57**

2. Properties

The chemical shift (δ) of the carboxy protons of quinoxaline-2,3-dicarboxylic acid in dimethylformamide–carbon tetrachloride mixtures *increases* with increasing concentration of the diacid and with increasing carbon tetrachloride content in the solvent, and *decreases* with increasing temperature. However the chemical shift of the carboxy protons in quinoxaline-2,3-dicarboxylic acid 1,4-dioxide is concentration independent. This suggests that the diacid is intermolecularly hydrogen bonded and that the diacid di-*N*-oxide is intramolecularly hydrogen bonded.[47]

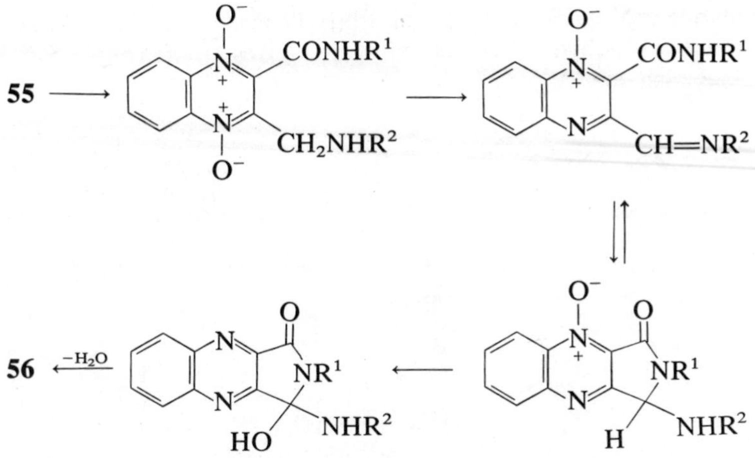

Scheme 5

Gas phase pyrolysis of quinoxaline-2,3-dicarboxylic acid anhydride (**58**) yields *o*-phthalonitrile (**60**) in 72% yield. The intermediacy of the quinoxalyne intermediate (**59**) is postulated.[48] The anhydride undergoes

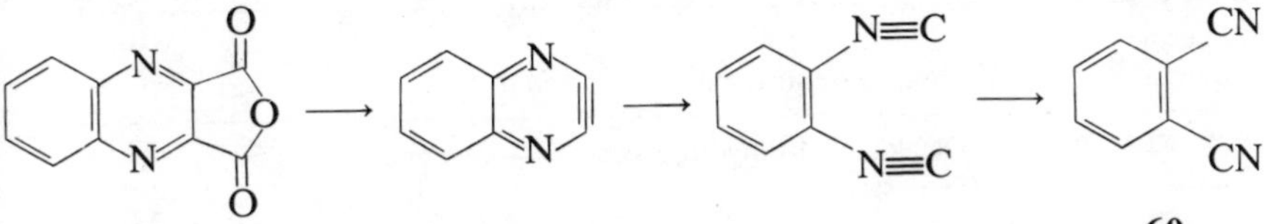

58 **59** **60**

standard ring-opening reactions with amines[49]; the product (**61**) obtained on ring-opening with hydroxylamine has been converted into 3-aminoquinoxaline-2-carboxylic acid as shown in Scheme 6.[50]

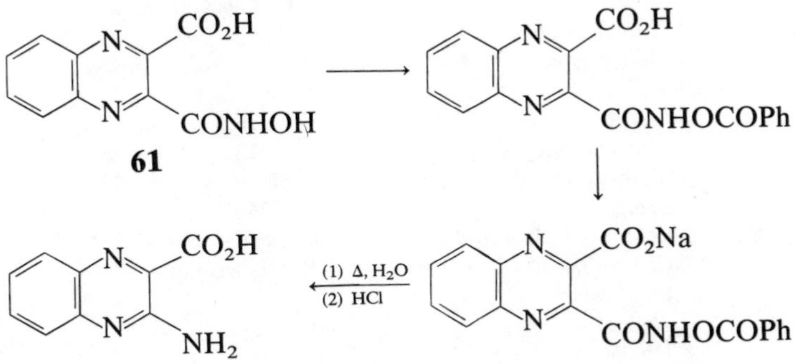

61

Scheme 6

Treatment of quinoxaline-2,3-dicarboxylic acid anhydrides, diesters, or dinitriles with hydrazine monohydrate gives pyridazino[4,5-*b*]quinoxalines (Scheme 7).[51]

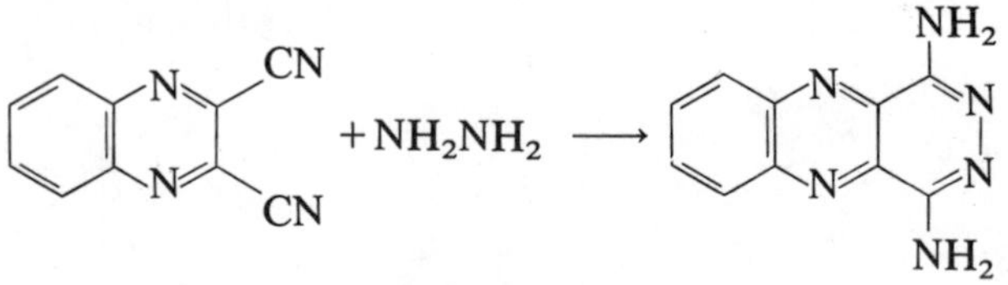

$$\text{Quinoxaline-2,3-dicarbonitrile} + NH_2NH_2 \longrightarrow \text{1,4-diaminopyridazino[4,5-}b\text{]quinoxaline}$$

Scheme 7

III. Tables

TABLE 1. QUINOXALINE-2-CARBOXYLIC ACIDS

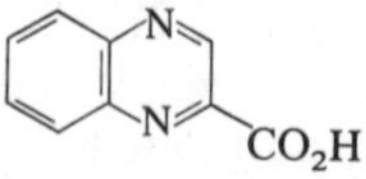

(Structure: quinoxaline with CO_2H at position 2)

Molecular formula	Substituents or structure	m.p., °C	Ref.
$C_9H_4Cl_2N_2O_3$	6,7-Dichloro-3,4-dihydro-3-oxo	340 decomp.	52
	Ethyl ester	230	52
$C_9H_5ClN_2O_2$	3-Chloro	146 decomp., 151–152	21, 37, 53
	Acid chloride	127	54
	Ethyl ester	38–40, 41.5–42	21, 54
	Amide	214–215	21
	Dimethylamide	98	54
	Ureide	242	54
	6-Chloro	223–224 decomp.	23
	Acid chloride	103–103.5	23
	Methyl ester	147.5–148.5	23
	Amide	300–301	55
	7-Chloro	225.5–226.5 decomp.	23
	Acid chloride	122.5–123.5	23
	Methyl ester	153–154	23
	Amide	259.5–260.5	55
$C_9H_5ClN_2O_3$	6-Chloro-3,4-dihydro-3-oxo	194–196 decomp.	2
	7-Chloro-3,4-dihydro-3-oxo	308–310 decomp.	2
	Ethyl ester	206–207	2
	Amide	310–311	2
	Ureide	249–250	9
	Anilide	>340	2
$C_9H_5ClN_2O_4$	3-Chloro-5,8-dihydroxy	228	1a

TABLE 1. (*contd.*)

Molecular formula	Substituents or structure	m.p., °C	Ref.
$C_9H_5N_3O_4$	6-Nitro	216–218	56
	Ethyl ester	223–226	56
	Amide	318–321	56
$C_9H_5N_3O_5$	3,4-Dihydro-6- and 7-nitro-3-oxo		57
$C_9H_6ClN_3O_2$	3-Amino-7-chloro	191–192 decomp.	15
	Methyl ester	224.5–225.5	15
$C_9H_6N_2O_2$	Unsubstituted	212.5–213 decomp.[a]	7, 21, 22, 37, 58, 59, 60, 61, 62, 63
	Acid chloride	112–113, 114–116	34, 35
	Methyl ester	111–112	18, 41, 60
	Ethyl ester	81.5–83, 83–84	21, 58
	Amide	191, 197–198, 200, 203.5–204	33, 42, 65, 66, 67
	N-Hydroxyamide	190 decomp.	42
	Hydrazide	208, 212	42, 65
	N-Methylanilide	128	64
	Nitrile	120	66, 67
$C_9H_6N_2O_3$	3,4-Dihydro-3-oxo	268 decomp.[b]	8, 10, 11, 29, 37, 64, 68, 69
	Methyl ester	222	31, 70
	Amide	306–308	24, 30, 64
	N-Methylamide	304 decomp. 310–311 decomp.	8, 14, 30
	Ureide	240 decomp. 250 decomp.	9, 10, 11
	Anilide	343–345 decomp.	30, 69
	N-Methylanilide	242	69
	N-Methyl-*p*-nitroanilide	226	64
	p-Toluidide	337–339	77
	Nitrile	298	24, 71
	Ethyl ester of 3-(β-D-glucopyranosyloxy)-tetraacetate	154.5–155.5	72
$C_9H_6N_2O_5$	3,4-Dihydro-5,8-dihydroxy-3-oxo	223	1a
$C_9H_7N_3O_2$	3-Amino	210 decomp.	50
	Hydrazide	196–198	73
	Anilide	213	29
	N-Methylanilide	157	29
$C_9H_7N_3O_3$	6-Amino-3,4-dihydro-3-oxo	292 sublimes	16
$C_9H_8N_2O_3$	1,2,3,4-Tetrahydro-3-oxo		
	Hydrate	152 decomp., 154–155 decomp.	3, 75
	Ethyl ester	145–146	75, 76
	Anilide	208	29
$C_9H_8N_4O_2$	3,6-Diamino	255 decomp.	19
	Amide	295 decomp.	19

TABLE 1. (*contd.*)

Molecular formula	Substituents or structure	m.p., °C	Ref.
$C_9H_{11}N_3O_2$	3-Amino-5,6,7,8-tetrahydro	174.5	78
	Methyl ester	154–155	78
$C_{10}H_5F_3N_2O_2$	3-Trifluoromethyl		
	Ethyl ester	51–52	79
$C_{10}H_7ClN_2O_3$	6-Chloro-3,4-dihydro-4-methyl-3-oxo	186–187 decomp.	2
	Ethyl ester	134–135	2
	N-Methylanilide	192–193	2
	7-Chloro-3,4-dihydro-4-methyl-3-oxo	198–199 decomp.	2
	Ethyl ester	125–126	2
	Amide	325–326 decomp.	2
	N-Methylanilide	164–166	2
$C_{10}H_8N_2O_2$	3-Methyl	154.5[c]	59
	Ethyl ester	73–73.5	82, 83, 85
	t-Butyl ester	70–71	81
	Amide	194–196	80
	N-Hydroxyamide hydrate	178–179	80
	Hydrazide	172–174	80
	Nitrile	151–152	66
	6(or 7)-Methyl	166 decomp.	86
$C_{10}H_8N_2O_3$	3,4-Dihydro-4-methyl-3-oxo	173–174 decomp.	8, 10, 64
	Methyl ester	126	10, 31
	Ethyl ester	125.5	64
	Amide	254, 256	24, 30, 64
	N-Methylamide	165–166, 168–170	8, 30
	N,N-Dimethylamide	115, 129	30, 64
	N,N-Diphenylamide	209	64
	Ureide	247 decomp.	10
	Anilide	193–195	29
	N-Methylanilide	162–163, 169	30, 64
	N-Methyl-2,4-dimethylanilide	213	64
	N-Methyl-2,6-dimethylanilide	264	64
	N-Methyl-*p*-nitroanilide	198	64
	p-Toluidide	184–185	77
	Nitrile	211	24, 87
	3-Methoxy	132 decomp., 140–142 decomp.	10, 30
	Methyl ester	107	10
	Amide	164	67
	Ureide	225	10
	Nitrile hydrate	150	67
$C_{10}H_9N_3O_2$	3-Methylamino	158–159 decomp.	8
	Methyl ester	154–156	8
	Methylamide	171	8

TABLE 1. (*contd.*)

Molecular formula	Substituents or structure	m.p., °C	Ref.
$C_{10}H_{10}N_2O_3$	1,2,3,4-Tetrahydro-4-methyl-3-oxo		
	N,N-Diphenylamide	179	29
	Anilide	161	29
	N-Methylanilide	188	29, 88
$C_{10}H_{10}N_4O_2$	3,6-Diamino-7-methyl[d]	248 decomp.	19
	Amide	245	19
$C_{11}H_9BrN_2O_3$	7-Bromo-4-ethyl-3,4-dihydro-3-oxo		
	Ureide	230–232	13
$C_{11}H_9ClN_2O_4$	3-Chloro-5,8-dimethoxy	168	1a
	Ethyl ester	154	1a
	3-Chloro-6,7-dimethoxy		
	Ethyl ester	140–140.5	89
$C_{11}H_{10}N_2O_2$	2,7-Dimethyl	156.5–157.5	90
	3-Ethyl		
	Nitrile	93–94	66
$C_{11}H_{10}N_2O_3$	3,4-Dihydro-6,7-dimethyl-3-oxo	309–310 decomp.	77
	Ethyl ester	199	75, 77, 91
	Anilide	330–331	77
	N-Methylanilide	293–294 decomp.	77
	4-Ethyl-3,4-dihydro-3-oxo	177 decomp.	13
	Ureide	235	13
$C_{11}H_{10}N_2O_4$	6,7-Dimethoxy	260–262	89
	Ethyl ester	159.5–160.5	89
	Amide	232	89
	N-(2-Diethylaminoethyl)amide	116–120	89
$C_{11}H_{10}N_2O_5$	3,4-Dihydro-5,8-dimethoxy-3-oxo	213, 215 decomp.	1a, 92
	Ethyl ester	242, 259–260	1a, 92
	3,4-Dihydro-6,7-dimethoxy-3-oxo	291–293 decomp.	77
	Ethyl ester	251–252, 255	77, 89
	Amide	283 decomp.	77
	Anilide	288–289 decomp.	77
	N-Methylanilide	238–239	77
	Hydrazide	234–237	89
$C_{11}H_{11}N_3O_2$	3-Amino-6,7-dimethyl	221.5–222	17
$C_{11}H_{11}N_3O_4$	3,4-Dihydro-6-(2-hydroxyethylamino)-3-oxo	151 decomp.	16
$C_{12}H_{12}N_2O_2$	3,6,7-Trimethyl		
	Ethyl ester	90	20
$C_{12}H_{12}N_2O_3$	3,4-Dihydro-4,6,7-trimethyl-3-oxo	204–205 decomp., 210–211 decomp., 212–214, 213–215	31, 77, 91, 93, 109

TABLE 1. (*contd.*)

Molecular formula	Substituents or structure	m.p., °C	Ref.
$C_{12}H_{12}N_2O_3$ (*contd.*)	Methyl ester	169–170	31
	Ethyl ester	125–126	91
	Anilide	243–244	77
	Hydrazide	273–276	93
	Ureide	245–255	93, 109
	3-(2-Oxopropyl)	250–251 decomp.	
	Ethyl ester	86–87.5	4
	N,N-Dimethylamide	98–101	94
$C_{12}H_{12}N_2O_5$	3,4-Dihydro-6,7-dimethoxy-4-methyl-3-oxo	233–235 decomp.	77
	Ethyl ester	141–143	77
	Amide	304–305 decomp.	77
	Anilide	326 decomp.	77
	N-Methylanilide	181–182	77
$C_{13}H_{14}N_2O_5$	3,4-Dihydro-6,7-diethoxy-3-oxo		
	Ethyl ester	181–183	89
$C_{13}H_{16}N_4O_3$	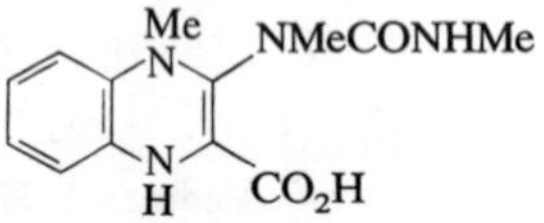 Me N NMeCONHMe N H CO_2H	192–193	95
$C_{14}H_8N_4O_4$	5-Nitro-3-(2-pyridyl)		
	Ethyl ester	140–141	96
$C_{14}H_9N_3O_2$	3-(3-Pyridyl)		
	Nitrile	193–194	97
	3-(4-Pyridyl)		
	Nitrile	228–229	97
$C_{14}H_{10}N_4O_2$	5-Amino-3-(2-pyridyl)		
	Ethyl ester	158–160	96
$C_{14}H_{15}N_3O_2$	3-*N*-Piperidyl		
	Piperidide	108	54
$C_{15}H_{10}N_2O_2$	3-Phenyl		98
	Ethyl ester	62, 65–66	79, 99
	Amide	198–199, 200–201	5, 100
	Nitrile	163	100
$C_{15}H_{10}N_2O_3$	3,4-Dihydro-3-oxo-4-phenyl	158–160 decomp.	10
	Ureide	244	10
$C_{16}H_{10}N_2O_3$	3-Benzoyl		
	Monohydrate	92–94 decomp.	7a
	Amide	198–199	7a
$C_{16}H_{11}ClN_2O_2$	3-*o*-Chlorobenzyl		
	Amide	193	6
$C_{16}H_{12}N_2O_2$	3-Benzyl	145 decomp.	6
	Amide	163	6
$C_{16}H_{12}N_2O_3$	3-(α-Hydroxybenzyl)		

TABLE 1. (*contd.*)

Molecular formula	Substituents or structure	Ref.	m.p., °C
$C_{16}H_{12}N_2O_3$ (*contd.*)	Amide	152–153	7
	o-Aminoanilide	193–194 decomp.	7
	Phenylhydrazide	188 decomp.	7
	Lactone	184	7
$C_{16}H_{19}N_3O_4$	6,7-Dimethoxy-3-piperidino		
	Ethyl ester	102–105	89
$C_{16}H_{20}N_2O_7$	3,4-Dihydro-6,7-dimethyl-3-oxo-4-D-ribityl	183–183.5 decomp.	101
	Na salt monohydrate	242–243 decomp.	101
	Methyl ester	185–186	31
$C_{17}H_{14}N_2O_3$	3-(*p*-Methoxybenzyl)	143–144 decomp.	7a
	p-Bromophenacyl ester	110–111	7a
	Nitrile	112–113	102
$C_{17}H_{12}N_2O_4$	3-*p*-Anisoyl	156–157 decomp.	102
	p-Bromophenacyl ester	145–146	7a
	Amide	223–225	7
	Nitrile	177–179	102
$C_{17}H_{14}N_2O_3$	3-*p*-Methoxybenzyl	143–144 decomp.	7a
	p-Bromophenacyl ester	110–111	7a
	Amide	165	6

[a] Quoted decomposition points range from 201–201.5° to 215–220°.
[b] Quoted decomposition points range from 260° to 268°.
[c] Quoted in reference 84 as 172°.
[d] Analysed with 1 molecule of acetic acid.

TABLE 2. QUINOXALINE-2,3-DICARBOXYLIC ACIDS

Molecular formula	Substituents	m.p., °C	Ref.
$C_{10}H_2Cl_4N_2O_4$	5,6,7,8-Tetrachloro		
	Diethyl ester	143–144	41
$C_{10}H_5N_3O_6$	6-Nitro	298–300	103
	Diethyl ester	81–82	41
$C_{10}H_6N_2O_4$	Unsubstituted	185–190, 187 decomp. 189.5, 190 decomp.	42, 44, 104, 105
	Monomethyl ester	151 decomp.	42
	Dimethyl ester	131.5–133	106
	Diethyl ester	81.5–83, 82.5–83.5	41, 44, 58, 104
	Monoanilide	173–174	45

TABLE 2. (*contd.*)

Molecular formula	Substituents	m.p., °C	Ref.
$C_{10}H_6N_2O_4$ (*contd.*)	Dianilide	251–253	45
	Monohydroxamate	213–216 decomp.	50
	Monophenylhydrazide	162.5 decomp.	106
	Anhydride	254 decomp., 255 decomp.	42, 105
	N-Propylimide	246	46
	N-Phenylimide	>300	45
	Dinitrile	218–220	71
$C_{12}H_{10}N_2O_4$	6,7-Dimethyl	180–182	51
	Dimethyl ester	118–120	51
	Diamide	286–288	51
	Dinitrile	192–194	51
	Anhydride	248–250	51
$C_{12}H_{10}N_2O_6$	5,8-Dimethoxy	186 decomp.	40
	Diethyl ester	164–165.5, 165	40, 107
	Monohydrazide	298 decomp.	40
	Anhydride	233–236 decomp.	40
	6,7-Dimethoxy	238–240	107, 108
	Diethyl ester	146	108
	Dibutyl ester	97–98	108
$C_{14}H_{14}N_2O_6$	5,8-Diethoxy	195–196 decomp.	39
	Monomethyl ester	174.5	74
	Dimethyl ester	141–142	74
	Monoethyl ester	134–135	74
	Diethyl ester	91–91.5	39
	Monopropyl ester	87–89	74
	Dipropyl ester	72.5–74	74
	Monoisopropyl ester	168	74
	Dibutyl ester	80–81	74
	Anhydride	218–219 decomp.	74
$C_{16}H_{18}N_2O_6$	5,8-Dipropoxy	157 decomp.	39
	Diethyl ester	68.5–69	39
$C_{18}H_{22}N_2O_6$	6,7-Dibutoxy	148–150 decomp.	108
	Diethyl ester	95–96, b.p. 234°/7 mm	108

IV. References

1. J. C. E. Simpson, "Condensed Pyridazine and Pyrazine Rings," Interscience, New York, 1953, p. 250.

1a. A. B. Sen and O. P. Madan, *J. Indian Chem. Soc.*, **38,** 225 (1961).

2. Y. Ahmad, M. S. Habib, M. Iqbal, M. I. Qureshi, and Ziauddin, *Bull. Chem. Soc. Jap.*, **38,** 1659 (1965).

3. E. C. Taylor and M. J. Thompson, *J. Org. Chem.*, **26,** 3511 (1961).

4. H. Taki and T. Mukaiyama, *Bull. Chem. Soc. Jap.*, **43,** 3607 (1970).
5. I. J. Pachter and P. E. Nemeth, *J. Org. Chem.*, **28,** 1203 (1963).
6. H. Dahn and G. Rotzler, *Helv. Chim. Acta*, **43,** 1555 (1960).
7. H. Dahn and H. Moll, *Helv. Chim. Acta*, **47,** 1860 (1964).
7a. H. Dahn and H. Moll, *Helv. Chim. Acta*, **49,** 2426 (1966).
8. H. Bredereck and W. Pfleiderer, *Chem. Ber.*, **87,** 1119 (1954).
9. H. G. Petering and G. J. van Giessen, *J. Org. Chem.*, **26,** 2818 (1961).
10. F. E. King and J. W. Clark-Lewis, *J. Chem. Soc.*, 3379 (1951).
11. R. B. Barlow, H. R. Ing, and I. M. Lewis, *J. Chem. Soc.*, 3242 (1951).
12. F. E. King and J. W. Clark-Lewis, *J. Chem. Soc.*, 172 (1953).
13. J. W. Clark-Lewis, J. A. Edgar, J. S. Shannon, and M. J. Thompson, *Aust. J. Chem.*, **17,** 877 (1964).
14. W. Pfleiderer, *Chem. Ber.*, **88,** 1625 (1955).
15. E. J. Cragoe (to Merck and Co. Ltd.), Belg. Pat. 639,386; *Chem. Abstr.* **62,** 14698 (1965); U.S. Pat. 3,313,813; *Chem. Abstr.*, **69,** 36172*w* (1968).
16. V. M. Berzovskii, N. A. Polyakova, and L. S. Tyl'chinskaya, *Chem. Heterocycl. Comp.*, **3,** 580 (1967).
17. W. S. McNutt, *J. Biol. Chem.*, **219,** 365 (1956).
18. E. Negishi and A. R. Day, *J. Org. Chem.*, **30,** 43 (1965).
19. T. S. Osdene and G. M. Timmis, *J. Chem. Soc.*, 2027 (1955).
20. V. M. Berezovskii and A. M. Yurkevich, *J. Gen. Chem. USSR*, **31,** 3526 (1961).
21. A. S. Elina and O. Yu. Magidson, *J. Gen. Chem. USSR*, **25,** 145 (1955).
22. H. R. Moreno and H. P. Schultz, *J. Med. Chem.*, **13,** 119 (1970).
23. H. R. Moreno and H. P. Schultz, *J. Med. Chem.*, **13,** 1005 (1970).
24. G. Tennant, *J. Chem. Soc.*, 1986 (1964).
25. W. F. Gum and M. M. Joullie, *J. Org. Chem.*, **30,** 3982 (1965).
26. N. K. Dutt, G. Sanyal, and K. Nag, *J. Indian Chem. Soc.*, **45,** 334 (1968).
27. E. Hayashi and S. Suzuki, *Yagugaku Zasshi*, **93,** 881 (1973); *Chem. Abstr.*, **79,** 92156*n* (1973).
28. E. C. Taylor, M. J. Thompson, and W. Pfleiderer, "Pteridine Chemistry," Proceedings of the Third International Symposium, Stuttgart 1962 (Publ. 1964). Pergamon Press, p. 196.
29. M. S. Habib and C. W. Rees, *J. Chem. Soc.*, 3384 (1960).
30. J. W. Clark-Lewis, *J. Chem. Soc.*, 422 (1957).
31. R. M. Cresswell, A. C. Hill, and H. C. S. Wood, *J. Chem. Soc.*, 698 (1959).
32. K. Sato, O. Shiratori, and K. Katagiri, *J. Antibiot. Ser.* A, **20,** 270(1967).
32a. A. Dell, D. H. Williams, H. R. Morris, G. A. Smith, J. Feeney, and G. C. K. Roberts, *J. Am. Chem. Soc.*, **97,** 2497 (1975).
33. S. Gerchakov, P. J. Whitman, and H. P. Schultz, *J. Med. Chem.*, **9,** 266 (1966).
34. S. Gerchakov and H. P. Schultz, *J. Med. Chem.*, **12,** 141 (1969).
35. H. C. Koppel, I. L. Honigberg, R. H. Springer, and C. C. Cheng, *J. Org. Chem.*, **28,** 1119 (1963).
36. J. Danek and B. Sevcik, *Vet. Med.* (Prague), **14,** 607 (1969); *Chem. Abstr.*, **73,** 12826*u* (1970).
37. N. K. Dutt, G. S. Sanyal, and K. Nag, *Anal. Chim. Acta*, **41,** 331 (1968); *Chem. Abstr.*, **68,** 84071*p* (1968).
38. N. K. Dutt and G. S. Sanyal, *Fesenius Z. Anal. Chem.*, **258,** 107 (1972); *Chem. Abstr.*, **76,** 107593*s* (1972).
39. S. Kawai, M. Torigoe, S. Fujiki, K. Shibata, A. Otaki, Y. Sakakibara, and S. Oguchi, *Nippon Kagaku Zasshi*, **80,** 788 (1959); *Chem. Abstr.*, **55,** 3597 (1961).

40. J. Adachi, *Nippon Kagaku Zasshi*, **76,** 311 (1955); *Chem. Abstr.*, **51,** 17936 (1957).
41. M. G. Gal'pern and E. A. Luk'yanets, *Chem. Heterocycl. Comp.*, **7,** 257 (1971).
42. I. Yoshioka and H. Otomasu, *Chem. Pharm. Bull.* (Tokyo), **5,** 277 (1957).
43. H. I. X. Mager and W. Berends, *Rec. Trav. Chim.*, **78,** 5 (1959).
44. A. S. Elina, L. G. Tsyrul'nikova, and G. P. Syrova, *Chem. Heterocycl. Comp.*, **5,** 115 (1969).
45. S. I. Burmistrov, N. E. Kul'chitskaya, and V. D. Romanenko, *J. Org. Chem. USSR*, **8,** 1107 (1972).
46. U. Eholzer, H. Heitzer, F. Seng, and K. Ley, *Synthesis*, 296 (1974).
47. L. L. Gordienko, Yu. S. Rozum, N. P. Romazanovich, and T. Ya. Lavrenyuk, *Chem. Heterocycl. Comp.*, **9,** 646 (1973).
48. M. P. Cava and L. Bravo, *J. Chem. Soc. Chem. Commun.*, 1538 (1968).
49. A. Cesari, *Ann. Inst. Super. Sanità*, **1,** 555 (1965); *Chem. Abstr.*, **65,** 10588 (1966).
50. C. D. Hurd and V. G. Bethune, *J. Org. Chem.*, **35,** 1471 (1970).
51. T. G. Koksharova, V. N. Konyukhov, Z. N. Pushhareva, and T. A. Pryakhina, *Khim. Geterotsikl. Soedin.*, 274 (1972); *Chem. Abstr.*, **76,** 153715*f* (1972).
52. R. M. Acheson, *J. Chem. Soc.*, 4731 (1956).
53. V. D. Romanenko and S. I. Burmistrov, *Chem. Heterocycl. Comp.*, **9,** 785 (1973).
54. F. Eiden and G. Bachmann, *Arch. Pharm.*, **306,** 401 (1973).
55. H. R. Moreno and H. P. Schultz, *J. Org. Chem.*, **36,** 1158 (1971).
56. J. Klicnar and F. Kosek, *Collect. Czech. Chem. Commun.*, **30,** 3102 (1965).
57. J. Klicnar and M. Adamek, *Sbornik Ved. Praci, Vysoka Skola Chem. Technol., Pardubici*, 81 (1960); *Chem. Abstr.*, **55,** 8418 (1961).
58. A. S. Elina, *J. Org. Chem. USSR*, **1,** 143 (1965).
59. A. S. Elina *J. Gen. Chem. USSR*, **34,** 2089 (1964).
60. W. Keller-Schierlien and V. Prelog, *Helv. Chim. Acta*, **40,** 205 (1957).
61. P. E. Fanta, R. A. Stein, and R. M. W. Rickett, *J. Am. Chem. Soc.*, **80,** 4577 (1958).
62. E. Grovenstein, W. Postman, and J. W. Taylor, *J. Org. Chem.*, **25,** 68 (1960).
63. E. J. Smutney, M. C. Caserio, and J. D. Roberts, *J. Am. Chem. Soc.*, **82,** 1793 (1960).
64. M. S. Habib and C. W. Rees, *J. Chem. Soc.*, 3371 (1960).
65. K. Asano and S. Asai, *Yakugaku Zasshi*, **78,** 450 (1958); *Chem. Abstr.*, **52,** 18428 (1958).
66. E. Hayashi and T. Miyagashima, *Yakugaku Zasshi*, **87,** 1103 (1967); *Chem. Abstr.*, **68,** 49560*a* (1968).
67. E. Hayashi, C. Iijima, and Y. Nagasawa, *Yakugaku Zasshi*, **84,** 163 (1964); *Chem. Abstr.*, **61,** 3108 (1964).
68. V. M. Berezovskii and Zh. I. Aksel'rod, *Dokl. Akad. Nauk SSSR*, **168,** 577 (1966); *Chem. Abstr.*, **65,** 8911 (1966).
69. M. S. Habib and C. W. Rees, *J. Chem. Soc.*, 3386 (1960).
70. H. Reinheckel, *Monatsh.*, **99,** 2203 (1968).
71. R. W. Begland and D. R. Hartter, *J. Org. Chem.*, **37,** 4136 (1972).
72. F. Reisser and W. Pfleiderer, *Chem. Ber.*, **99,** 547 (1966).
73. A. Dornow and E. Hinz, *Chem. Ber.*, **91,** 1834 (1958).
74. Y. Sakakibara, *Nippon Kagaku Zasshi*, **83,** 172 (1962); *Chem. Abstr.* **58,** 11357 (1963).
75. Y. Ahmad, M. S. Habib, M. Iqbal, and M. I. Qureshi, *J. Chem. Soc.*, 4056 (1964).
76. V. G. Granik and R. G. Glushkov, *J. Org. Chem. USSR*, **4,** 707 (1968).
77. Y. Ahmad, M. S. Habib, M. Iqbal, and M. I. Qureshi, *J. Chem. Soc.*, 4053 (1964).

78. J. B. Bicking, C. M. Robb, S. F. Kwong, and E. J. Cragoe, *J. Med. Chem.*, **10,** 598 (1967).
79. H. J. Bestmann and O. Klein, *Justus Liebigs Ann. Chem.*, **676,** 97 (1964).
80. J. K. Landquist and J. A. Silk, *J. Chem. Soc.*, 2052 (1956).
81. H. Dahn, H. Gowal, and H. P. Schunke, *Helv. Chim. Acta*, **53,** 1598 (1970).
82. H. Dahn and H. Hauth, *Helv. Chim. Acta*, **42,** 1214 (1959).
83. F. Wegland, H. J. Bestmann, and H. Fritsche, *Chem. Ber.*, **93,** 2340 (1960).
84. M. Seyhan, *Chem. Ber.*, **84,** 477 (1951).
85. G. Sarkis and S. Al-Azawe, *J. Chem. Eng. Data*, **18,** 102 (1973).
86. G. Henseke and K.-J. Bähner, *Chem. Ber.*, **91,** 1605 (1958).
87. G. Tennant, *J. Chem. Soc.*, 2666 (1964).
88. J. W. Clark-Lewis, *J. Chem. Soc.*, 439 (1957).
89. Z. Budesinsky and A. Valenta, *Collect. Czech. Chem. Commun.*, **36,** 2527 (1971).
90. V. V. Titov and L. F. Kozhokina, *Khim. Geterotsikl. Soedin.*, 1700 (1972); *Chem. Abstr.*, **78,** 72064*g* (1973).
91. J. M. Wellman and M. Tishler, *J. Am. Chem. Soc.*, **69,** 714 (1947).
92. C. W. Rees and D. E. West, *J. Chem. Soc.* (*C*), 583 (1970).
93. K. H. Dudley and P. Hemmerich, *J. Org. Chem.*, **32,** 3049 (1967).
94. W. H. Dekker, *Rec. Trav. Chim.*, **91,** 1338 (1972).
95. P. Hemmerich, B. Prijs, and H. Erlenmeyer, *Helv. Chim. Acta*, **43,** 372 (1960).
96. F. R. Pfeiffer and F. H. Case, *J. Org. Chem.*, **31,** 3384 (1966).
97. F. Krönke and F. Gross, *Chem. Ber.*, **92,** 22 (1959).
98. A. T. Blomquist and E. A. LaLancette, *J. Am. Chem. Soc.*, **84,** 220 (1962).
99. H. Böhme and H. Schneider, *Chem. Ber.*, **91,** 988 (1958).
100. E. Hayashi and C. Iijima, *Yakugaku Zasshi*, **82,** 1093 (1962); *Chem. Abstr.*, **58,** 4551 (1963).
101. A. R. Surrey and F. C. Nachod, *J. Am. Chem. Soc.*, **73,** 2336 (1951).
102. H. Dahn and J. Nussbaum, *Helv. Chim. Acta*, **52,** 1661 (1969).
103. J. Klicnar, F. Kosek, and S. Panusova, *Collect. Czech. Chem. Commun.*, **29,** 206 (1964).
104. H. Saikachi and J. Matsuo, *Yakugaku Zasshi*, **84,** 416 (1964); *Chem. Abstr.*, **61,** 5647 (1964).
105. M. L. Tamayo, R. Perez Ossorio, and M. Sanz Burata, *Ann. Real Soc. Espan. Fis. y Quim.* (*Madrid*), **50B,** 865 (1954); *Chem. Abstr.*, **50,** 361 (1956).
106. M. Sanz Burata, *Rev. Real Acad. Ciene* (*Madrid*), **49,** 23 (1955); *Chem. Abstr.*, **50,** 7112 (1956).
107. S. Oguchi *Bull. Chem. Soc. Jap.*, **41,** 980 (1968).
108. S. Oguchi, *Nippon Kagaku Zasshi*, **86,** 246 (1965); *Chem. Abstr.*, **63,** 4295 (1965).
109. V. M. Berezovskii and G. D. Glebova, *J. Gen. Chem.* U.S.S.R., **34,** 1007 (1964).

CHAPTER X

2-Halogenoquinoxalines and 2,3-Dihalogenoquinoxalines

I. 2-Halogenoquinoxalines . . . 162
1. Methods of Preparation . . . 162
2. Properties . . . 163
3. Uses . . . 168
II. 2,3-Dihalogenoquinoxalines . . . 168
1. Methods of Preparation . . . 168
2. Properties . . . 169
3. Uses . . . 171
III. Tables . . . 171
Table 1. 2(or 3)-Halogenoquinoxalines . . . 171
Table 2. 2,3-Dihalogenoquinoxalines . . . 174
IV. References . . . 175

The chemistry of 2-chloro- and 2,3-dichloroquinoxalines has been previously reviewed in Simpson's monograph.[1] The present review further demonstrates the extensive use which these derivatives have found in the preparation of other quinoxalines. This is a result of the ease with which nucleophilic displacement of halogen from the 2- and 3-positions can be carried out.

I. 2-Halogenoquinoxalines

1. Methods of Preparation

2-Chloroquinoxalines are usually prepared by chlorination of quinoxalin-2-ones. Reagents that have been employed for chlorination include phosphoryl chloride,[2] mixtures of phosphoryl chloride and phosphorus pentachloride,[3] and phosphoryl chloride and pyridine.[4] The reaction is normally uncomplicated by side reactions, but the use of phosphorus pentachloride alone as the chlorinating reagent is best avoided because of the possibility of nuclear substitution.[1] 2-Chloroquinoxalines have been purified by crystallization, distillation, steam distillation, or sublimation.[5]

2-Chloroquinoxalines have also been prepared from quinoxaline mono-*N*-oxides.[6–8] For example, treatment of 5-methylquinoxaline 1-oxide with phosphoryl chloride gives 2-chloro-5-methylquinoxaline,[6] and reaction of 2-phenylquinoxaline 4-oxide with acetyl chloride at 100° yields 2-chloro-3-phenylquinoxaline in excellent yield.[8]

2-Bromoquinoxaline has been prepared by heating quinoxalin-2-one with phosphoryl bromide at 120°,[9] and a series of 2-bromo-3-alkylquinoxalines have been prepared in analogous fashion.[10] 2-Iodoquinoxaline is prepared by the action of sodium iodide and hydriodic acid on 2-chloroquinoxaline[9] and 2-fluoroquinoxaline by application of the Balz-Schiemann reaction to 2-aminoquinoxaline.[11]

Many 2-chloro-3-substituted quinoxalines have been prepared from 2,3-dichloroquinoxaline by partial displacement of chlorine. This has been achieved with nitrogen nucleophiles such as ammonia,[12] hydrazine,[13] and primary[14] and secondary amines;[15,16] with oxygen nucleophiles such as alkoxide and aryloxide ions;[14] with the corresponding sulfur nucleophiles;[17] and with carbon nucleophiles such as the anions derived from malononitrile and ethyl cyanoacetate.[18] Further substitution of chlorine is often feasible. However in the majority of cases it is the electron-donating character, and therefore the deactivating effect of the first-introduced substituent, that ensures that displacement of a single chlorine from 2,3-dichloroquinoxalines is an effective synthetic procedure. Illustrative of the preparative conditions used are the reactions of 2,3-dichloroquinoxaline with secondary amines such as pyrrolidine, piperidine, and morpholine. When the reactants are mixed in ethanol at room temperature, it is possible to replace one chlorine atom of the dichloroquinoxaline; however when the reactants are heated under reflux in ethanol, 2,3-dicycloalkylaminoquinoxalines are formed.[16] In the case of the reaction of 2,3,6-trichloroquinoxaline with aminoalkylamines (e.g., $Et_2NCH_2CH_2NH_2$) it is found that the 3-chlorine atom is more reactive than the 2-chlorine atom.[19] However with 6-nitro-2,3-dichloroquinoxaline it has been shown that the 6-nitro group selectively activates the 2-chlorine to displacement with aziridine (see Chapter XI).

2. Properties

Kinetic studies have been made in order to compare the mobility of halogen in the various chlorodiazines and chloroazines and their benzo derivatives. Chloride ion is displaced by ethoxide ion in 2-chloroquinoxaline rather more readily than in 2-chloroquinazoline and in 1-chlorophthalazine. The chlorodiazines do not vary greatly in reactivity, as indicated by their respective rate constants 8.28×10^{-3}, 2.98×10^{-3}, and 1.86×10^{-3} l./(mole)(sec) in ethanol at 20°; but they are much more reactive than 2-chloroquinoline, for which the corresponding rate constant is 6.3×10^{-7}.[20] It is not possible to generalize about the relative reactivity of the benzochlorodiazines because this may be influenced by choice of nucleophile, solvent, or temperature. For example, the rate constant data for the displacement of chloride ion with piperidine are 6.36×10^{-5} for 2-chloroquinoxaline, 4.79×10^{-4} for 2-chloroquinazoline, and 2.0×10^{-5} for 1-chlorophthalazine.[20] A comparative study of halogen mobility in 2-chloroquinoxaline and 2-fluoroquinoxaline has shown that the fluoro compound is much more reactive. For displacement of halide with methoxide ion, K_F/K_{Cl} is 875 at −10°; K_F/K_{Cl} values are temperature dependent and increase with decreasing temperature. The high ratio of fluoro to chloro reactivity indicates that bond making rather than bond breaking is important in determining the overall rate of displacement.[21] 2-Fluoroquinoxaline has been found to hydrolyze 240 times faster than 2-fluoropyrazine in 0.01 *M* sodium hydroxide at 26°C.[10] This observation is illustrative of the general trend toward increased reactivity resulting from benzene ring annulation. A study of micellar effects on the reaction of 2-chloroquinoxaline and hydroxide ion in water has shown that reaction is catalyzed by cationic micelles of hexadecyltrimethylammonium bromide and retarded by anionic micelles of sodium dodecyl sulfate.[22]

A study of the reaction of 2-halogenoquinoxalines and 3-substituted 2-halogenoquinoxalines with potassium amide in liquid ammonia has shown that, in addition to the expected 2-aminoquinoxalines, benzimidazole may also be formed (Scheme 1). With 2-chloroquinoxaline the main product is

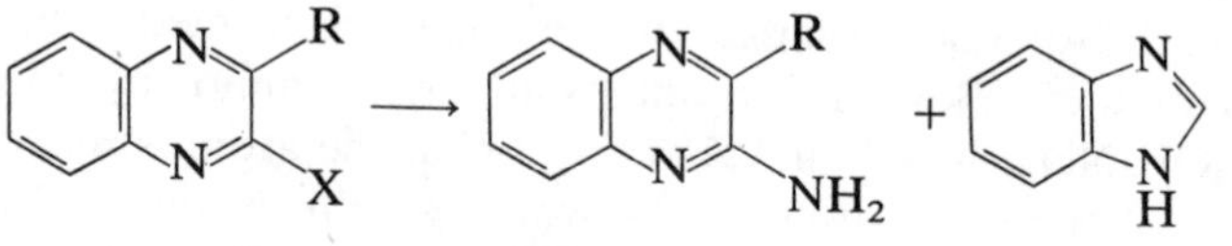

Scheme 1

benzimidazole, and only some 2-aminoquinoxaline is formed. 2-Bromo- and 2-iodoquinoxaline are converted under these conditions almost exclusively into benzimidazole, whereas with 2-fluoroquinoxaline only a trace of benzimidazole is formed. The formation of benzimidazole from 2-chloro 3-substituted quinoxalines indicates that ring contraction involves the elimination of the three-carbon atom of the quinoxaline ring.

Experiments with ^{14}C- and ^{15}N-labeled 2-chloroquinoxaline show that an addition–elimination sequence is operative and therefore that the ion (**1**) rather than didehydroquinoxaline (**2**) is the intermediate involved.[23,24]

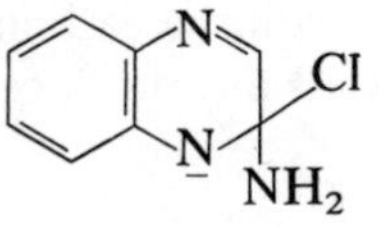

1 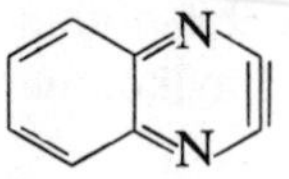 **2**

The chlorine of 2-chloroquinoxalines has been displaced with a wide range of nitrogen-, oxygen-, and sulfur-containing nucleophiles. Thus reaction with ammonia gives the corresponding 2-aminoquinoxaline;[25] with hydrazine, the 2-hydrazinoquinoxaline;[25,26] with sulfanilamide, the 2-sulfanilamidoquinoxaline;[102] and with the arylamines, the corresponding arylaminoquinoxalines[25,27] are obtained. The best preparative procedure for the latter reaction is to heat the reactants under reflux in dilute hydrochloric acid using a slight excess of the amine.[28,29] When 2-chloroquinoxaline (**3**) is boiled with excess of aniline, a mixture of 2-anilinoquinoxaline (**4**), indolo[2,3-*b*]quinoxaline (**5**), and 2-*p*-aminophenyl-3-anilinoquinoxaline (**6**) is obtained.[28]

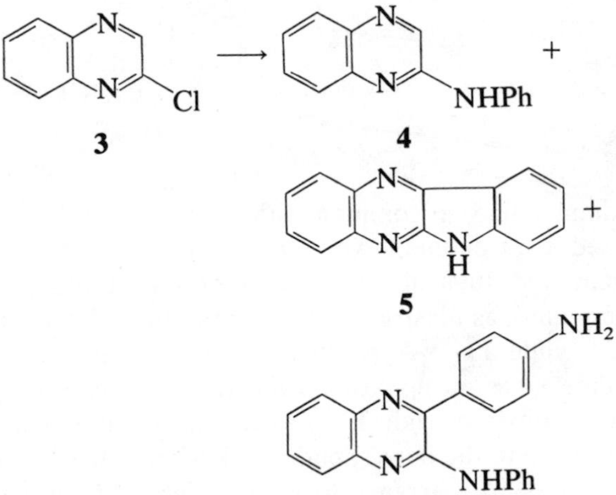

3 → **4** + **5** + **6**

Similar reactions occur when 2-chloroquinoxaline is treated with aryloxide ions. Thus treatment of 2-chloroquinoxaline with sodium phenoxide in excess of phenol gives a mixture of 2-phenoxyquinoxaline (**7**) and benzofuro[2,3-*b*]quinoxaline (**8**).[30] However, displacement of chloride ion with alkoxide ions is uncomplicated by side reactions.

Analogous displacements have been carried out with thioalkoxide and thiophenoxide ions.[31] Chloride ion has also been displaced from 2-chloroquinoxalines with various carbon-containing nucleophiles derived from active methylene compounds. For example, with benzyl cyanide, the substituted acetonitrile **9** is obtained;[32] and with the carbanions derived from ketones, 2-quinoxalinylketones of type **10** are produced.[33]

Cl → (PhONa–PhOH) → OPh **7** + **8**

↓ $PhCH_2CN$–$NaNH_2$

CH(CN)Ph **9**

Cl → ($R^1CH_2COR^2$–$NaNH_2$) → CHR^1COR^2 **10**

2-Chloroquinoxaline, in common with related chloroheterocycles, may be alkenylated by treatment with two equivalents of the appropriate Wittig reagent and subsequent reaction of the resulting ylide with a carbonyl compound, as illustrated by the formation of 2-styrylquinoxaline (**12**) from the ylide **11**.[34] A parallel sequence of reactions involving a sulfonium ylide gives an epoxyalkyl derivative. Thus the epoxide **14** is formed via the sulfonium ylide **13**.[35] Since phosphonium ylides such as **11** may be hydrolyzed to the corresponding alkyl derivative[34] and epoxyalkyl derivatives such as **14** rearrange to ketones (e.g., **15**) on treatment with strong base,[35] these reactions represent a general procedure for the alkylation and acylation of heterocyclic compounds using the readily available chloro (or bromo) compounds.

2-Chloroquinoxalines bearing appropriate 3-substituents have been converted into tricyclic derivatives as illustrated by the transformations shown in Scheme 2.

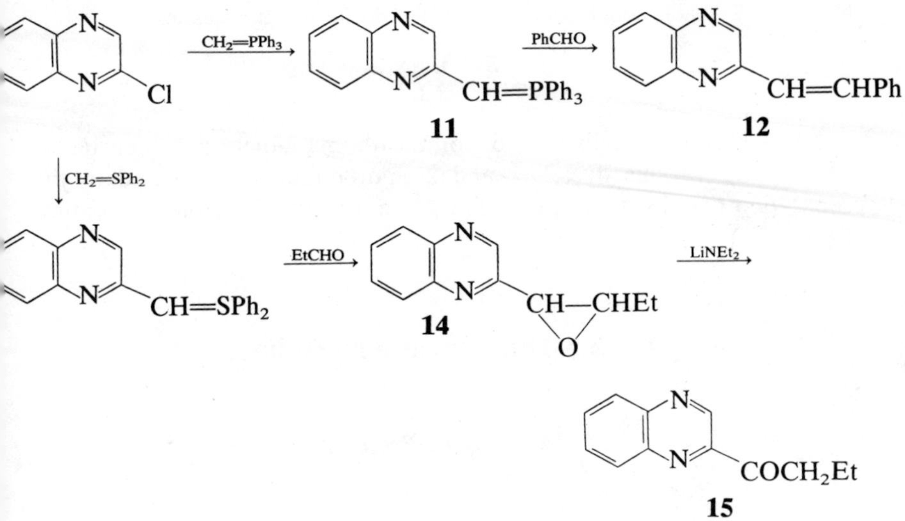

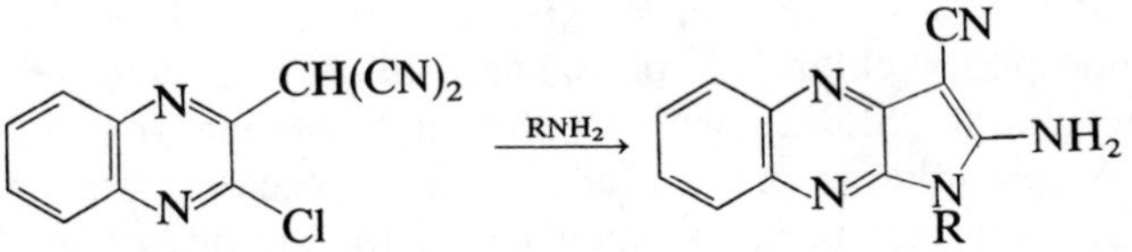

$CH_2{=}PPh_3$ → $CH{=}PPh_3$ **11** → PhCHO → $CH{=}CHPh$ **12**

$CH_2{=}SPh_2$ → $CH{=}SPh_2$ → EtCHO → **14** → $LiNEt_2$ → $COCH_2Et$ **15**

$CH(CN)_2$, Cl → RNH_2 → CN, NH_2, R (Ref. 36)

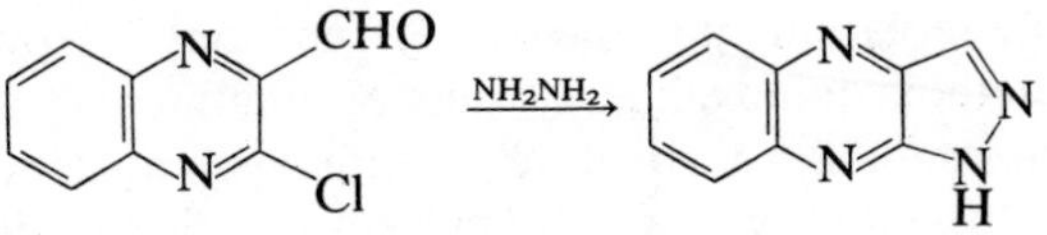

CHO, Cl → NH_2NH_2 → (Ref. 5)

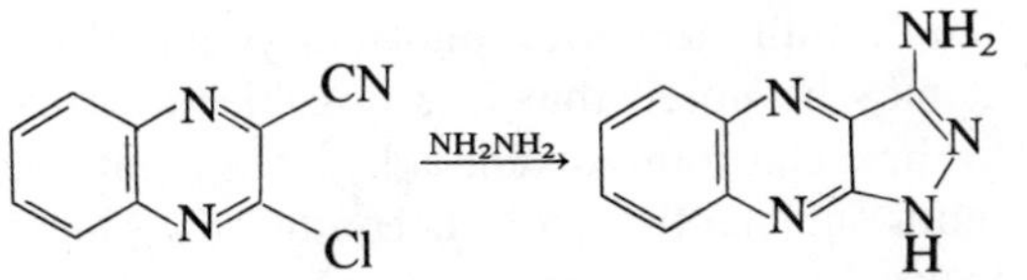

CN, Cl → NH_2NH_2 → NH_2 (Ref. 37)

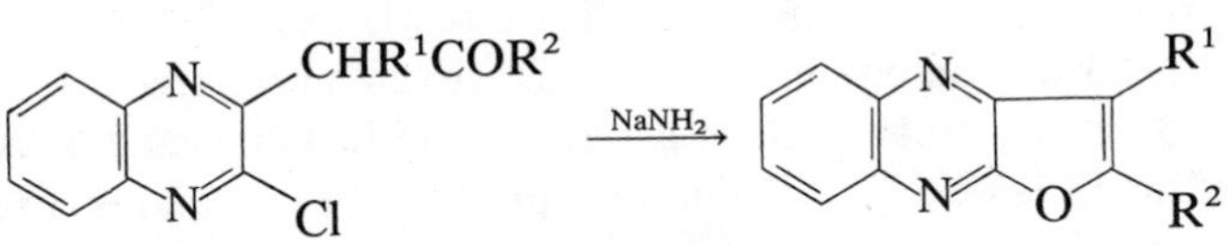

CHR^1COR^2, Cl → $NaNH_2$ → R^1, R^2 (Ref. 33)

Scheme 2

3. Uses

The 2-chloroquinoxaline 3-, 6-, or 7-carbonyl moiety has been incorporated into various dyes,[38,39] and 2-chloroquinoxaline is used in the preparation of substituted piperidines[40] and indoles[41] for pharmaceutical testing.

II. 2,3-Dihalogenoquinoxalines

1. Methods of Preparation

Compounds of this type are usually prepared from the corresponding quinoxaline-2,3-diones. A variety of chlorinating reagents have been used, including phosphoryl chloride,[42] phosgene,[43] thionyl chloride,[25] and mixtures of phosphoryl chloride with *N,N*-dimethylaniline[6] or *N,N*-diethylaniline[44] or phosphorus pentachloride.[45] The reactions are normally carried out under reflux in an open vessel for periods of about 1–4 hr. In the case of the conversion of 5-nitroquinoxaline-2,3-dione into the corresponding dichloro compound, it has been specifically reported that it is necessary to use phosphorus pentachloride rather than phosphoryl chloride for complete replacement of the oxygen functions.[47] However in the case of 6-methoxyquinoxaline-2,3-dione, treatment with phosphorous pentachloride yields 2,3,*x*-trichloro-6-methoxyquinoxaline, and phosphoryl chloride chlorination yields 2,3-dichloro-6-methoxyquinoxaline.[19] The normal procedure in using phosphorus pentachloride is to heat a mixture of the chlorinating reagent and the quinoxalinedione in a distillation flask at about 160° until no more phosphoryl chloride is distilled over.[46] In passing it may be noted that it is possible to convert 2,3-dichloroquinoxaline into hexachloroquinoxaline by treatment with phosphorus pentachloride under sufficiently vigorous conditions. Thus an excellent yield of the hexachloro compound is obtained when the reaction is carried out in an autoclave at 300° for 17 hr.[48] 2,3-Dichloroquinoxaline is formed on treatment of quinoxaline 1,4-dioxide with phosphoryl chloride,[6] and 2,3-dibromoquinoxaline has been prepared from quinoxaline-2,3-dione by treatment with phosphoryl bromide and *N,N*-dimethylaniline at 115–120°.[49] 2,3-Difluoroquinoxaline is readily prepared from 2,3-dichloroquinoxaline.[48]

2. Properties

There have been two reports of the oxidation of 2,3-dichloroquinoxaline with potassium permanganate to give 2,3-dichloropyrazine-5,6-dicarboxylic acid. In one case the primary product of oxidation was converted into 2,3-dioxopyrazine-5,6-dicarboxylic acid;[50] in the other it was converted into 2,3-dichloro-5,6-dimethoxycarbonylpyrazine.[51]

Many nucleophilic displacement reactions have been carried out on 2,3-dichloroquinoxaline and its benzosubstituted derivatives. Those leading to partial displacement of chlorine have been dealt with earlier, in the section dealing with the preparation of 3-substituted 2-chloroquinoxalines. The remainder are conveniently classified into displacement reactions leading to 2,3-disubstituted quinoxalines and those leading to polycyclic systems. Both chlorine atoms of 2,3-dichloroquinoxaline have been displaced with ammonia, and a series of benzosubstituted 2,3-diaminoquinoxalines have been prepared in an analogous fashion. Optimum yields of 2,3-diaminoquinoxalines are obtained by heating the dichloro compound with liquid ammonia in an autoclave at 90°.[52] A series of 2,3-dianilinoquinoxalines have been prepared by heating 2,3-dichloroquinoxaline with excess of the aromatic amine at 160°.[28] The dianilinoquinoxalines have melting points ranging from about 120° to 190°. The higher melting points quoted in the literature for 2,3-dianilinoquinoxaline[53] and 2,3-bis(*m*-toluidino)quinoxaline[53] are probably those of the corresponding acetates since the products were crystallized from acetic acid. The amino group of a sulfonamide is sufficiently nucleophilic to enable quinoxaline bis-sulfonamides to be prepared from 2,3-dichloroquinoxaline.[54]

Both chlorine atoms of 2,3-dichloroquinoxalines are readily displaced by alkoxide and aryloxide ions to give the corresponding dialkoxy- and diaryloxyquinoxalines.[46] Violent reaction, accompanied by pyrotechnic effects, is reported to occur between 2,3-dichloroquinoxaline and the potassium salts of *o*-nitrophenol and 2,4-dinitrophenol.[46] Quinoxaline-2,3-dithiones have been prepared from 2,3-dichloroquinoxalines by reaction with thiourea and alkaline hydrolysis of the resulting thiouronium salts.[55] However when 2,3-dichloroquinoxaline and thiourea are allowed to react in ethanol in 1:1 molecular proportion, 1,4-dithia-1,4-dihydrophenazino[2,3-*b*]quinoxaline (**16**) is obtained in excellent yield.[23]

2,3-Bis(ethylthio)quinoxaline results from reaction of 2,3-dichloroquinoxaline and sodium ethylmercaptide.[55] Displacement of chloride ion

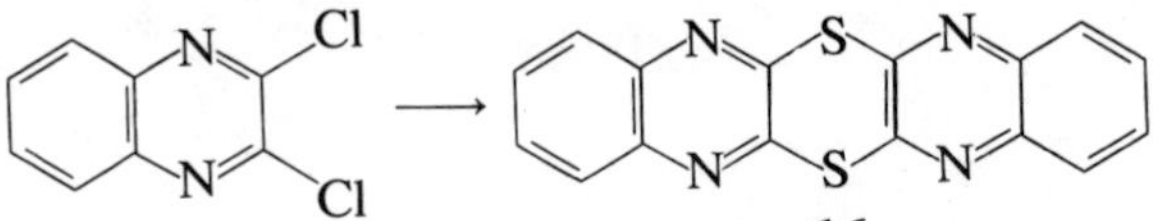

16

by fluoride ion in both 2,3-dichloroquinoxaline and perchloroquinoxaline gives 2,3-difluoroquinoxaline and perfluoroquinoxaline, respectively, in high yield. In the latter reaction some 5-chloro-2,3,6,7,8-pentafluoroquinoxaline is obtained as a by-product.[48] Attempted steam distillation of 2,3-difluoroquinoxaline gives quinoxaline-2,3-dione in quantitative yield.[56]

2,3-Dichloroquinoxaline has been converted into a number of polycyclic derivatives by reaction with *o*-disubstituted benzenes. Thus reaction with *o*-phenylenediamines gives tetraazanaphthacenes (**17**),[57] and the corresponding reaction with 2,3-diaminoquinoxaline gives fluorubin derivatives (**18**).[58] Quinoxalino[2,3-*b*]benzothiazine derivatives (**19**) result from the reaction of dichloroquinoxaline and *o*-aminobenzenethiol and *o*-aminoalkylbenzenethiols.[28,47,59] The condensation of 2,3-dichloroquinoxaline with catechol to give [1,4]benzodioxino[2,3-*b*]quinoxaline (**20**) is carried out in high yield by heating the reactants in diethyl ketone under reflux and in the presence of potassium carbonate.[60]

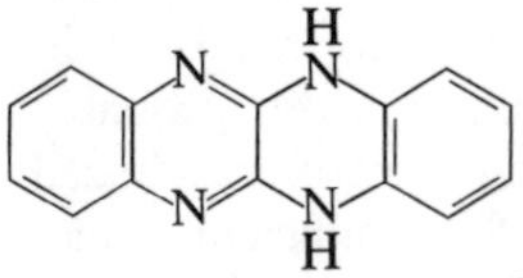

17 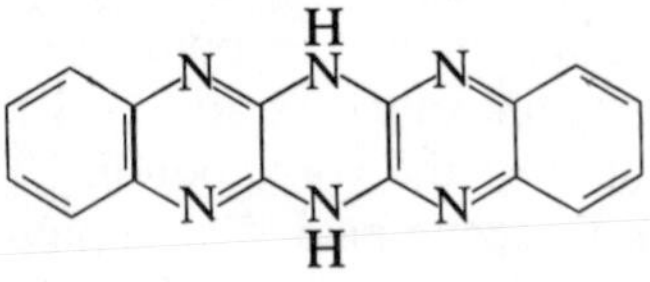 **18**

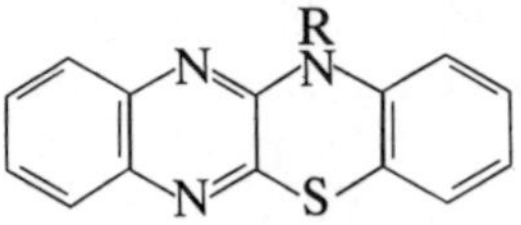

19 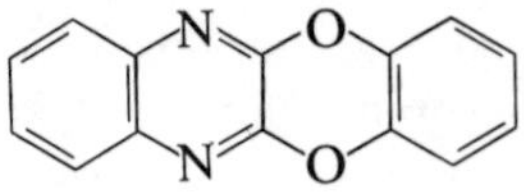 **20**

2,3-Dichloroquinoxaline undergoes an interesting cyclization reaction with acetone thiosemicarbazone to give the thiazoloquinoxaline **21**.[23]

$+ Me_2C{=}NNHCSNH_2 \longrightarrow$

21

3. Uses

There are numerous patents dealing with the biological activity of 2,3-dichloroquinoxalines, and a selection of these claims is summarized below. Thus derivatives with a 6-sulfonamido function have been patented as microbiocidal agents,[61] and 2,3,7-trichloro-6-methylaminosulfonylquinoxaline is patented as an anticancer drug.[62] Various 2,3-dichloroquinoxaline-6-sulfonates have antibacterial activity,[63] and there are patent claims for the fungicidal and bactericidal activity of nitro-2,3-dichloroquinoxalines.[64] The quinoxalinecarboxamide **22** is claimed to inhibit gastric acidity.[65] Perhaps the most widespread use of 2,3-dichloroquinoxalines is in their incorporation in reactive dyes for cellulose fibers. This incorporation is achieved by the use of the appropriate 2,3-dichloroquinoxalinecarbonyl chloride to give dyes of type **23**.[66] Dyes have also

$HOCH_2CH_2NHCO$–(2,3-dichloroquinoxalinyl)

22

Dye NHCO–(2,3-dichloroquinoxalinyl)

23

been prepared from 2,3-dichloroquinoxaline-6-sulfonyl chloride.[67] Polycyclic derivatives of 2,3-dichloroquinoxaline such as fluorubin (**18**) have been used in the mass dyeing of polyamides.[47]

III. Tables

TABLE 1. 2(OR 3)-HALOGENOQUINOXALINES

(2-X-quinoxaline)

Molecular formula	Substituents or structure	m.p., °C	Ref.
$C_8HCl_5N_2$	2,5,6,7,8-Pentachloro	168–169	68
$C_8H_4ClN_3O_2$	2-Chloro-6-nitro	202, 208–209	69, 70, 71
	2-Chloro-7-nitro	185–186, 188–190	69, 70, 71, 72, 73
$C_8H_4Cl_2N_2$	2,6-Dichloro	153–155, 159–160	74, 75
	2,7-Dichloro	141, 146–146.5	75, 76, 77

TABLE 1. (*contd.*)

Molecular formula	Substituents or structure	m.p., °C	Ref.
$C_8H_5BrN_2$	2-Bromo	58–59	9
$C_8H_5ClN_2$	2-Chloro	46–48, 49–49.5, b.p. 100/1.4 mm	2, 6, 78
$C_8H_5ClN_2O_2$	2-Chloro-5,8-dihydroxy	211	79
$C_8H_5Cl_2N_3$	2-Amino-3,6(or 7)-dichloro	220	14
$C_8H_5FN_2$	2-Fluoro	b.p. 96–97.5/11 mm,[a] 180–182/25 mm	9, 11, 21
$C_8H_5IN_2$	2-Iodo	104–105	9
$C_8H_6ClN_3$	2-Amino-3-chloro	139, 144–146	12, 14, 52
	7-Amino-2-chloro	200–201	72
$C_9H_3F_5N_2O$	2,5,6,7,8-Pentafluoro-3-methoxy	70–72	48
$C_9H_4Br_2N_2O$	2-Bromo-6-bromocarbonyl		39
	2-Bromo-7-bromocarbonyl		39
$C_9H_4ClF_3N_2$	2-Chloro-7-trifluoromethyl	117–118	81
$C_9H_4ClN_3$	2-Chloro-3-cyano		37
$C_9H_4Cl_2N_2O$	2-Chloro-6-chlorocarbonyl	127–129	82
	2-Chloro-7-chlorocarbonyl	120–122	82
$C_9H_5Br_2ClN_2$	2-Chloro-3-dibromomethyl		83
$C_9H_6ClN_3O$	2-Carbamoyl-3-chloro	207–209	84
	N-Benzyl deriv.	145–147	84
$C_9H_6ClN_3O_2$	2-Chloro-3-nitromethyl	115 decomp.	18
	2-Chloro-3-methyl-6-nitro	136	85
	2-Chloro-3-methyl-7-nitro	153	85
$C_9H_7BrN_2$	2-Bromo-3-methyl	102.5–103	10
$C_9H_7ClN_2$	2-Chloro-3-methyl	79–81, 85–86, 90–92	5, 86, 87
	2-Chloro-5-methyl	95	6, 88
	2-Chloro-7-methyl	76	88
$C_9H_7ClN_2O$	2-Chloro-3-methoxy	79–80	44
	2-Chloro-7-methoxy	100.5–102	76
	3-Chloro-1,2-dihydro-1-methyl-2-oxo	131–133	44
$C_9H_7ClN_2S$	2-Chloro-3-methylthio	102–103	17
$C_9H_8ClN_3$	2-Amino-3-chloro-6-(or 7)methyl	140–142	14
	2-Chloro-3-methylamino	92–93	14
$C_9H_8ClN_3O$	2-Amino-3-chloro-6-(or 7)-methoxy	159–161	14
$C_{10}H_8ClN_3$	2-(1-Aziridinyl)-3-chloro	101–102	15
$C_{10}H_9BrN_2$	2-Bromo-3-ethyl	67.5–68	10
$C_{10}H_9ClN_2$	2-Chloro-3-ethyl	38–40, b.p. 105/5 mm	86, 89
	2-Chloro-3,6-dimethyl	76–77	87
	2-Chloro-3,7-dimethyl	86–87	87
	2-Chloro-6,7-dimethyl	91–92	26
$C_{10}H_9ClN_2O$	2-Chloro-5-methoxy-3-methyl	94–96	90
	2-Chloro-8-methoxy-3-methyl	124–126	90

TABLE 1. (*contd.*)

Molecular formula	Substituents or structure	m.p., °C	Ref.
$C_{10}H_9ClN_2O_2$	2-Chloro-5,8-dimethoxy	128	79
$C_{11}H_5ClN_4$	2-Chloro-3-dicyanomethyl	217 decomp.	18
$C_{11}H_{11}BrN_2$	2-Bromo-3-propyl	79.5–81	10
$C_{11}H_{11}ClN_2$	2-Chloro-3,6,7-trimethyl	142–143	26
	2-Chloro-3-propyl	70–71	10
	2-Chloro-3-isopropyl	61, 69–70, b.p. 122/7 mm	55, 89
$C_{12}H_8ClN_3O_2$	2-Carbomethoxycyanomethyl-3-chloro	198	18
$C_{12}H_{11}ClN_2O_2$	2-Carbethoxymethyl-3-chloro	79–79.5	4
$C_{12}H_{12}ClN_3$	2-Chloro-3-(1-pyrrolidinyl)	86, 93–94	15, 16
$C_{12}H_{12}ClN_3O$	2-Chloro-3-morpholino	78, 84	15, 16
$C_{12}H_{13}BrN_2$	2-Bromo-3-butyl	42.5–43	10
$C_{12}H_{14}ClN_3$	2-Chloro-3-diethylamino	b.p. 132/0.5 mm	91
$C_{13}H_8ClN_3$	2-Chloro-3-(2-pyridyl)	115–117	92
$C_{13}H_{10}ClN_3O_2$	2-Carbethoxycyanomethyl-3-chloro	174–175	18
$C_{13}H_{14}ClN_3$	2-Chloro-3-piperidino	67, 68.5	15, 16
$C_{14}H_8ClN_3O_2$	2-Chloro-3-*p*-nitrophenoxy	151–154	14
$C_{14}H_8Cl_2N_2$	2,7-Dichloro-3-phenyl		93
$C_{14}H_8Cl_2N_2O$	2-Chloro-3-*p*-chlorophenoxy	131–133	14
$C_{14}H_9ClN_2$	2-Chloro-3-phenyl	127–128, 130–131	7, 8, 9, 94
$C_{14}H_9ClN_2O$	2-Chloro-3-phenoxy	107–109	14
$C_{14}H_9ClN_4O_2$	2-Chloro-3-*p*-nitroanilino	210–214	14
$C_{14}H_9Cl_2N_3$	2-Chloro-3-*p*-chloroanilino	132–133	14
$C_{14}H_{10}ClN_3$	2-Anilino-3-chloro	73–76	14
$C_{14}H_{16}ClN_3$	2-Chloro-3-(hexahydro-1*H*-azepin-1-yl)	68–68.5	15
$C_{15}H_9ClN_2O$	2-Benzoyl-3-chloro	138–141	5
$C_{15}H_{11}ClN_2$	2-Benzyl-3-chloro	86–87, 86–88	5, 95
$C_{15}H_{11}ClN_2O$	2-Chloro-3-*p*-methylphenoxy	108–110	14
$C_{15}H_{12}ClN_3$	2-Chloro-3-*p*-methylanilino	103–105	14
	2-Chloro-3-*N*-methylanilino	44–48	14
$C_{15}H_{12}ClN_3O_2S$	2-Chloro-3-(*p*-toluenesulfonamido)	162–164	14
	2-Chloro-3-(*o*-methylbenzenesulfonamido)	251–253	54
$C_{15}H_{14}ClN_3O_2$	2-(Carbo-*t*-butoxycyanomethyl)-3-chloro	168–170 decomp.	18
$C_{16}H_{13}ClN_2O$	2-Chloro-3-*p*-methoxybenzyl	101	96
$C_{18}H_{15}ClN_2O_2$	 N, N, $CHPhCO_2Et$, Cl	87.5–88.5	4

[a] n_D^{20} 1.5897; $n_D^{22.5}$ 1.5883.

TABLE 2. 2,3-DIHALOGENOQUINOXALINES

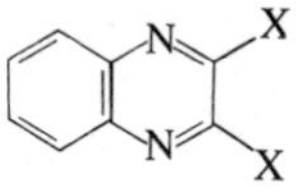

Molecular formula	Substituents or structure	m.p., °C	Ref.
$C_8ClF_5N_2$	5-Chloro-2,3,6,7,8-pentafluoro	119–120	48
$C_8Cl_2F_4N_2$	2,3-Dichloro-5, 6, 7, 8-tetrafluoro	80–82	48
$C_8Cl_6N_2$	Hexachloro	200–201, 207–209	48, 68, 97
$C_8F_6N_2$	Hexafluoro	142–144, b.p. 196–198	48
$C_8H_2Br_2Cl_2N_2$	6,7-Dibromo-2,3-dichloro	182–183	98
$C_8H_2Cl_2N_4O_4$	2,3-Dichloro-5,7-dinitro	215	99
	2,3-Dichloro-6,7-dinitro	213–214, 216–218	98, 100
$C_8H_2Cl_4N_2$	2,3,5,7-Tetrachloro	110–111, 114–115	45, 54
	2,3,5,8-Tetrachloro	159–163, 160–161	6, 45
	2,3,6,7-Tetrachloro	170–170.5, 173–174	6, 98
$C_8H_3BrCl_2N_2$	6-Bromo-2,3-dichloro	132	6
$C_8H_3Cl_2N_3O_2$	2,3-Dichloro-5-nitro	154, 156	46, 99
	2,3-Dichloro-6-nitro	150, 153–154	6, 46, 98, 99, 100, 101
$C_8H_3Cl_3N_2$	2,3,5-Trichloro	142–143	6
	2,3,6-Trichloro	141–142, 144	6, 98
	2,3,7-Trichloro	147	27
$C_8H_3Cl_3N_2O_2S$	6-Chlorosulfonyl	86–87	67
$C_8H_4Br_2N_2$	2,3-Dibromo	165–170	49
$C_8H_4Cl_2N_2$	2,3-Dichloro	151–153	25, 44
$C_8H_4Cl_2N_2O$	2,3-Dichloro-5-hydroxy	144	80
	5-Acetyl deriv.	90–94	80
$C_8H_4Cl_2N_2O_2$	2,3-Dichloro-5,8-dihydroxy	151–152	80
	5,8-Diacetyl deriv.	215, 218	42, 80
	2,3-Dichloro-6,7-dihydroxy	>360	42
	6,7-Diacetyl deriv.	183	42
$C_8H_4F_2N_2$	2,3-Difluoro	94–95	48
$C_9H_3Cl_3N_2O$	2,3-Dichloro-5-chlorocarbonyl	113–115	82
	2,3-Dichloro-6-chlorocarbonyl	111–113	82
$C_9H_5BrCl_2N_2$	6-Bromo-2,3-dichloro-7-methyl	160–161	6
$C_9H_5Cl_3N_2$	2,3,6-Trichloro-7-methyl	172–173	6
$C_9H_5Cl_3N_2O_2S$	2,3-Dichloro-6-chlorosulfonyl-7-methyl	103	67
$C_9H_6Cl_2N_2$	2,3-Dichloro-6-methyl	112	6
$C_9H_6Cl_2N_2O$	2,3-Dichloro-5-methoxy	144.5–145	80
	2,3-Dichloro-6-methoxy	159–160	19, 46
$C_9H_6Cl_3N_3O_2S$	2,3,7-Trichloro-6-methylaminosulfonyl	244–246	62
$C_{10}H_2Cl_2N_2O_3$	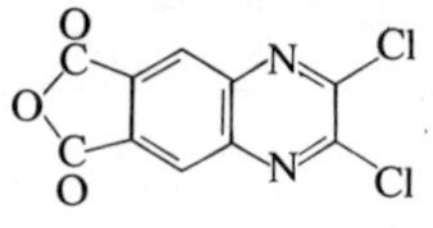	284–285	82

TABLE 2. (*contd.*)

Molecular formula	Substituents or structure	m.p., °C	Ref.
$C_{10}H_5Cl_3N_2O$	2,3-Dichloro-6-chloro-carbonyl-7-methyl	111–112	82
$C_{10}H_6Cl_4FN_3O_2S_2$	2,3-Dichloro-6-($Cl_2FCSNMeSO_2$)quinoxaline (structure)	126–128	61
$C_{10}H_6Cl_5N_3O_2S_2$	2,3-Dichloro-6-($Cl_3CSNMeSO_2$)quinoxaline (structure)	137–139	61
$C_{10}H_8Cl_2N_2O_2$	2,3-Dichloro-5,8-dimethoxy	214.5–215.5, 215–216	57, 58
	2,3-Dichloro-6,7-dimethoxy	178	42
$C_{12}H_{12}Cl_2N_2O_2$	2,3-Dichloro-5,8-diethoxy	199–199.5	80
$C_{15}H_8Cl_4FN_3O_2S_2$	2,3-Dichloro-6-($Cl_2FCSNPhSO_2$)quinoxaline (structure)	200–204	61
$C_{15}H_8Cl_5N_3O_2S_2$	2,3-Dichloro-6-($Cl_3CSNPhSO_2$)quinoxaline (structure)	212–214	61
$C_{16}H_{20}Cl_2N_2O_2$	5,8-Dibutoxy-2,3-dichloro	104.5–105.5	58
	6,7-Dibutoxy-2,3-dichloro		58

IV. References

1. J. C. E. Simpson, "Condensed Pyridazine and Pyrazine Rings," Interscience, New York, 1953, p. 258.
2. G. W. H. Cheeseman, *J. Chem. Soc.*, 3236 (1957).
3. F. J. Wolf (to Merck and Co.), U.S. Pat. 2,537,870; *Chem. Abstr.*, **45,** 4274 (1951).
4. J. Klicnar, M. Hajek, J. Hoffman, and M. Vecera, *Collect. Czech. Chem. Commun.*, **36,** 262 (1971).
5. V. D. Romanenko and S. I. Burmistrov, *Chem. Heterocycl. Comp.*, **9,** 785 (1973).
6. J. K. Landquist, *J. Chem. Soc.*, 2816 (1953).
7. E. Hayashi and C. Iijima, *Yakugaku Zasshi*, **82,** 1093 (1962); *Chem. Abstr.*, **58,** 4551 (1963).
8. Y. Ahmad, M. S. Habib, Ziauddin, and B. Bakhtiari, *J. Org. Chem.*, **31,** 2613 (1966).
9. P. J. Lont and H. C. van der Plas, *Rec. Trav. Chim.*, **91,** 850 (1972).
10. H. Reinheckel, *Monatsh.*, **99,** 2215 (1968).
11. H. Rutner and P. E. Spoerri, *J. Heterocycl. Chem.*, **3,** 435 (1966).
12. H. Saikachi and S. Tagami, *Chem. Pharm. Bull.* (Tokyo), **9,** 941 (1961).

13. K. Sasse and G. Demagk (to Farbenfabriken–Bayer A.-G), Ger. Pat. 1,104,962; *Chem. Abstr.*, **57,** 8591 (1962).
14. W. Deuschel and G. Reidel (to Badische Anilin- und Soda Fabrik A.-G.), Ger. Pat. 1,135,471; *Chem. Abstr.*, **58,** 537 (1963).
15. W. R. Vaughan and M. S. Habib, *J. Org. Chem.*, **27,** 324 (1962).
16. I. N. Goncharova and I. Ya. Postovskii, *J. Gen. Chem. USSR*, **32,** 3271 (1962).
17. K. Asano and S. Asai, *Yakugaku Zasshi*, **79,** 661 (1959); *Chem. Abstr.*, **53,** 21979 (1959).
18. E. F. Pratt and J. C. Keresztesy, *J. Org. Chem.*, **32,** 49 (1967).
19. F. H. S. Curd, D. G. Davey, and G. J. Stacey, *J. Chem. Soc.*, 1271 (1949).
20. N. B. Chapman and D. Q. Russell-Hill, *J. Chem. Soc.*, 1563 (1956).
21. G. B. Bressan, I. Giardi, G. Illuminati, P. Linda, and G. Sleiter, *J. Chem. Soc. (B)*, 225 (1971)..
22. V. Flamini, P. Linda, and G. Savelli, *J. Chem. Soc., Perkin II*, 421 (1975).
23. I. M. Ismail, R. Jacobi, and W. Sauer, *Z. Chem.*, **17,** 15 (1977).
24. P. J. Lont, H. C. van der Plas, and A. J. Verbeek, *Rec. Trav. Chim.*, **91,** 949 (1972).
25. A. P. Komin and M. Carmack, *J. Heterocycl. Chem.*, **13,** 13 (1976).
26. N. G. Koshel, E. G. Kovalev, and I. Ya. Postovskii, *Chem. Heterocycl. Comp.*, **6,** 791 (1970).
27. R. M. Acheson, *J. Chem. Soc.*, 4731 (1956).
28. S. D. Carter, and G. W. H. Cheeseman, Tetrahedron, **34,** 981 (1978).
29. J. Drumheller and H. P. Schultz, *J. Am. Chem. Soc.*, **77,** 6637 (1955).
30. R. K. Anderson and G. W. H. Cheeseman, *J. Chem. Soc., Perkin I*, 129 (1974).
31. F. Cuiban, M. Ionesco, H. Bala, and M. Steresco, *Bull. Chim. Soc. Fr.*, 356 (1963); J. Tulecki, J. Kalinowska-Torz, and J. Powalska, *Ann. Pharm.* (Poznan), **10,** 13 (1973); *Chem. Abstr.*, **79,** 42488*p* (1973).
32. R. N. Castle, A. Aldous, and C. Moore, *J. Org. Chem.*, **21,** 139 (1956); Y. Mizuno, K. Adachi, and K. Ikeda, *Chem. Pharm. Bull.* (Tokyo), **2,** 225 (1954); *Chem. Abstr.*, **50,** 1034 (1956).
33. C. Iijima and E. Hayashi, *Yakugaku Zasshi*, **92,** 729 (1972); *Chem. Abstr.*, **77,** 88434*t* (1972).
34. E. C. Taylor and S. F. Martin, *J. Am. Chem. Soc.*, **96,** 8095 (1974).
35. E. C. Taylor, M. L. Chittenden, and S. F. Martin, *Heterocycles*, **1,** 59 (1973).
36. H. Otomasu, S. Omiya, T. Sekiguchi, and H. Takahashi, *Chem. Pharm. Bull.*, (Tokyo), **18,** 2065 (1970).
37. G. Henseke and W. Sauer, Ger. (East) Pat. 83,361; *Chem. Abstr.*, **78,** 58459*h* (1973).
38. J. Benz (to Sandoz Ltd.), Belg. Pat. 659,696; *Chem. Abstr.*, **64,** 198432 (1966).
39. J. E. Cole and W. E. Gumprecht (to E. I. du Pont de Nemours and Co.) U.S. Pat. 3,326,888; *Chem. Abstr.*, **67,** 65405*p* (1967).
40. A. S. Tomcufcik, P. E. Fabio, and A. M. Hoffman (to American Cyanamid), Brit. Pat. 1,077,173; *Chem. Abstr.*, **68,** 59441*b* (1968).
41. G. R. Burchall, W. Hepworth, and S. C. Smith (to Imperial Chemical Industries), Ger. Offen. 2,253,927; *Chem. Abstr.*, **79,** 31865*w* (1973).
42. S. Oguchi, *Nippon Kagaku Zasshi*, **86,** 319 (1965); *Chem. Abstr.*, **63,** 4295 (1965).
43. H. Weidinger and G. Wellenreuther (Badische Anilin- und Soda Fabrik A.-G.), Brit. Pat. 927,974; *Chem. Abstr.*, **60,** 2987 (1964).
44. G. W. H. Cheeseman, *J. Chem. Soc.*, 1804 (1955).
45. K. Sasse, R. Wegler, H. Scheinpflug, and H. Jung, (to Farbenfabriken Bayer A.-G.), Ger. Pat. 1,194,631; *Chem. Abstr.*, **63,** 8381 (1965).

46. H. I. X. Mager and W. Berends, *Rec. Trav. Chim.*, **78,** 5 (1959).
47. H. Altermatt (to Ciba-Geigy A.-G.), Ger. Pat. 1,469,868; *Chem. Abstr.* **83,** 195097*f* (1975).
48. C. G. Allison, R. D. Chambers, J. A. H. McBride, and W. K. R. Musgrave, *J. Fluorine Chem.*, **1,** 59 (1971/1972).
49. C. Sosnowski and L. Wojciechowski (to Instytut Przemyslu Organicznego), Pol. Pat. 61,711; *Chem. Abstr.*, **74,** 125728*c* (1971).
50. H. I. X. Mager and W. Berends, *Rec. Trav. Chim.*, **77,** 843 (1958).
51. A. S. Elina and I. S. Musatova, *Chem. Heterocycl. Comp.*, **9,** 1403 (1973).
52. W. Deuschel and W. Vilsmeier (to Badische Anilin- und Soda Fabrik A.-G.), Belg. Pat. 612,092; *Chem. Abstr.*, **57,** 16634 (1962).
53. D. Lockhart and E. E. Turner, *J. Chem. Soc.*, 424 (1937).
54. S. H. Dandegaonker and C. K. Mesta, *J. Med. Chem.*, **8,** 884 (1965).
55. D. C. Morrison and A. Furst, *J. Org. Chem.*, **21,** 470 (1956).
56. J. Homer, *J. Heterocycl. Chem.*, **3,** 244 (1966).
57. S. Noguchi, *Nippon Kagaku Zasshi,* **80,** 569 (1959); *Chem. Abstr.*, **55,** 4516 (1961).
58. S. Kawai, Y. Okawa, Y. Yada, H. Hosoi, T. Murakoshi, and I. Yajima, *Nippon Kaguku Zasshi,* **80,** 551 (1959); *Chem. Abstr.*, **55,** 3598 (1961).
59. W. Schindler and H. J. Peterli (to Geigy Chemical Corp.), U.S. Pat. 3,010,961; *Chem. Abstr.*, **56,** 8729 (1962).
60. D. E. Ames and R. J. Ward, *J. Chem. Soc., Perkin I,* 534 (1975).
61. W. Paulus, E. Keuhle, O. Pauli, K. Sasse, and E. Klauke (to Farbenfabriken Bayer A.-G.), Brit. Pat. 1,120,339; *Chem. Abstr.*, **69,** 96774*d* (1968).
62. T. Miyagi and H. Yamamoto (to Sumitomo Chemical Co. Ltd.), Jap. Pat. 17,747 (1967); *Chem. Abstr.*, **69,** 10475*x* (1968).
63. S. Koike, J. Hattori, and T. Ozaki (to Sumitomo Chemical Co. Ltd.), Jap. Pat. 27,255 (1964); *Chem. Abstr.*, **63,** 7023 (1965).
64. J. Hattori, H. Sugiyama, K. Yoshioka, and S. Koike (to Sumitomo Chemical Co. Ltd.), U.S. Pat. 3,186,905; *Chem. Abstr.*, **63,** 6264 (1965).
65. W. A. Bolhofer and J. J. Baldwin (to Merck and Co. Inc.), U.S. Pat. 3,691,166; (*Chem. Abstr.*, **77,** 152222*b* (1972).
66. Farbenfabriken Bayer A.-G., Brit. Pat. 995,796; *Chem. Abstr.*, **64,** 8382 (1966).
67. E. Siegel and K. Sasse (to Farbenfabriken Bayer A.-G.), Ger. Pat. 1,218,092; *Chem. Abstr.*, **65,** 9063 (1966).
68. D. E. Burton, A. J. Lambie, D. W. J. Lane, G. T. Newbold, and A. Percival, *J. Chem. Soc.* (*C*), 1268 (1968).
69. G. W. H. Cheeseman, *J. Chem. Soc.*, 1246 (1961).
70. L. Horner, U. Schwenk, and E. Junghams, *Justus Liebigs Ann. Chem.*, **579,** 212 (1953).
71. H. Otomasu and K. Yoshida, *Chem. Pharm. Bull.* (Tokyo), **8,** 475 (1960).
72. C. M. Atkinson, C. W. Brown, and J. C. E. Simpson, *J. Chem. Soc.*, 26 (1956).
73. K. Asano and S. Asai, *Yakugaku Zasshi,* **79,** 658 (1959); *Chem. Abstr.*, **53,** 21979 (1959).
74. H. G. Petering and G. J. Van Geissen, *J. Org. Chem.*, **26,** 2818 (1961).
75. A. F. Crowther, F. H. S. Curd, D. G. Davey, and G. J. Stacey, *J. Chem. Soc.*, 1260 (1949).
76. R. L. Wear and C. S. Hamilton, *J. Am. Chem. Soc.*, **72,** 2893 (1950).
77. P. Linda and G. Marino, *Ric. Sci. Rend. Sez. A,* **3,** 225 (1963); *Chem. Abstr.*, **59,** 7523 (1963).
78. J. H. Gardner and J. R. Stevens, *J. Am. Chem. Soc.*, **71,** 1868 (1949).

79. A. B. Sen and O. P. Madan, *J. Indian Chem. Soc.*, **38,** 225 (1961).
80. E. S. Lane and C. Williams, *J. Chem. Soc.*, 2983 (1956).
81. M. L. Belli, S. Fatutta, M. Forchiassin, G. Illuminati, P. Linda, G. Marino, and E. Zinato, *Ric. Sci. Rend. Sez. A*, **3,** 530 (1963); *Chem. Abstr.*, **59,** 8743 (1963).
82. Y.-W. Chang (to E. I. du Pont de Nemours and Co.), Fr. Pat. 1,386,355; *Chem. Abstr.*, **62,** 14702 (1965); Fr. Pat. 1,386,595; *Chem. Abstr.*, **62,** 14703 (1965).
83. W. Sauer and G. Henseke, *Z. Chem.*, **10,** 381 (1970).
84. S. Kushner, H. Dalalian, J. L. Sanjurjo, F. L. Bach, S. R. Safir, V. K. Smith, and J. H. Williams, *J. Am. Chem. Soc.*, **74,** 3617 (1952).
85. H. Otomasu, *Chem. Pharm. Bull.* (Tokyo), **8,** 475 (1960).
86. Y. J. L'Italien and C. K. Banks, *J. Am. Chem. Soc.*, **73,** 3246 (1951).
87. M. Munk and H. P. Schultz, *J. Am. Chem. Soc.*, **74,** 3433 (1952).
88. B. C. Platt and T. M. Sharp, *J. Am. Chem. Soc.*, **71,** 3247 (1949).
89. E. Hayashi and Y. Miura, *Yakugaku Zasshi*, **87,** 643 (1967); *Chem. Abstr.*, **67,** 90774r (1967).
90. I. Ya. Postovskii and N. G. Koshel, *Chem. Heterocycl. Comp.*, **6,** 913 (1970).
91. J. R. Geigy A.-G., Brit. Pat. 822,069; *Chem. Abstr.*, **55,** 2005 (1961).
92. F. R. Pfeiffer and F. H. Case, *J. Org. Chem.*, **31,** 3384 (1966).
93. R. F. Abdulla, *Tetrahedron Lett.*, 3559 (1974).
94. D. Shiho and S. Tagami, *J. Am. Chem. Soc.*, **82,** 4044 (1960).
95. D. D. Chapman, *J. Org. Chem.*, **37,** 2498 (1972).
96. H. Dahn and J. Nussbaum, *Helv. Chim. Acta*, **52,** 1661 (1969).
97. A. Roedig, H.-H. Bauer, G. Bonse, and R. Ganns, *Chem. Ber.*, **107,** 558 (1974).
98. G. W. H. Cheeseman, *J. Chem. Soc.*, 1170 (1962).
99. Sumitomo Chemical Co. Ltd., Belg. Pat. 633,824; *Chem. Abstr.*, **61,** 7030 (1964).
100. A. S. Elina, L. G. Tsyrul'nikova, and M. I. Medvedeva *J. Org. Chem. USSR*, **1,** 143 (1965).
101. M. Asai, *Yakugaku Zasshi*, **79,** 260 (1959); *Chem. Abstr.*, **53,** 13160 (1959).
102. Z. Budesinsky, V. Valenta, V. Vosatka, and A. Simek, Czech.. Pat. 156,880; *Chem. Abstr.*, **83,** 179108*e* (1975).

CHAPTER XI

2-Aminoquinoxalines and 2,3-Diaminoquinoxalines

I. 2-Aminoquinoxalines and 2-Substituted Aminoquinoxalines 179
 1. Methods of Preparation 179
  A. From Quinoxalines by Nucleophilic Displacement 179
  B. From Quinoxalines by Other Methods 181
  C. From *o*-Phenylenediamines and *o*-Nitrosoaminobenzenes 181
 2. Properties 183
 3. Uses 184
II. 2,3-Diaminoquinoxalines and 2,3-Disubstituted Aminoquinoxalines 184
 1. Methods of Preparation 184
 2. Properties 185
 3. Uses 187
III. Tables 187
 Table 1. 2-Aminoquinoxalines and 2(or 3)-Substituted Aminoquinoxalines 187
 Table 2. 2,3-Diamino- and 2,3-Disubstituted Aminoquinoxalines 190
IV. References 191

The chemistry of 2-amino- and 2,3-diaminoquinoxalines has been previously reviewed by Simpson.[1]

I. 2-Aminoquinoxalines and 2-Substituted Aminoquinoxalines

1. Methods of Preparation

A. *From Quinoxalines by Nucleophilic Displacement*

The most common substrates for the preparation of aminoquinoxalines are the corresponding chloro compounds. 2-Aminoquinoxaline itself has been prepared by treatment of 2-chloroquinoxaline with methanolic ammonia at 120° for 8 hr.[2] The amination of 3-substituted 2-chloroquinoxalines is usually carried out by heating with either methanolic or ethanolic ammonia for several hours at temperatures in the range of 150–180°. The relatively severe conditions used in these reactions give scope for the monodisplacement of chlorine in 2,3-dichloroquinoxalines. Thus 2,3-dichloroquinoxaline has been converted into 2-amino-3-chloroquinoxaline by treatment with ethanolic ammonia at 70–80° in an autoclave,[3] alternatively by passing ammonia through a solution of the dichloro compound in *N*-methylpyrrolidone at 140°.[4]

2-Substituted aminoquinoxalines have been prepared in an analogous fashion by displacement of chlorine from 2-chloroquinoxalines with amines of a wide range of structural type. The reaction conditions employed are in general less severe than in the case of ammonia. While reactions with the lower molecular weight, more volatile amines are conveniently carried out in closed vessels,[5] reaction with amines such as morpholine and piperidine are carried out in open vessels at temperatures of about 100°.[6] Indeed three 2-chloroquinoxalines have been found to react with piperidine at room temperature.[7] Reactions with ethylenediamine derivatives have been carried out at temperatures in the range of 110–140°.[8,9] The preferred conditions for preparation of anilinoquinoxalines from 2-chloroquinoxalines appear to be refluxing the aniline with chloroquinoxaline in slightly acid aqueous suspension; as discussed in Chapter X, more complex reactions can occur under more vigorous conditions.[12,44]

2,3-Dichloro-6-nitroquinoxaline reacts with aziridine(ethyleneimine) in benzene solution at room temperature, in the presence of triethylamine as acid scavenger, to give the 2-aziridinyl derivative, the 6-nitro group thus selectively activating the 2-chloro substituent.[10] Chlorine activation is also apparent in the substitution reactions of 2-chloroquinoxaline 1-oxide, which is converted into the 2-amino 1-oxide by treatment with aqueous or ethanolic ammonia at 60-80° [11] and thus under much milder conditions than those used for the conversion of 2-chloroquinoxaline into 2-aminoquinoxaline.

As alternatives to the chloro compounds for substrates in nucleophilic displacement reactions, phenoxy compounds and cyclic amides and cyclic thioamides have been used. It appears that 2-phenoxyquinoxaline is significantly less reactive than the corresponding cinnoline or phthalazine derivatives and only gives a negligible yield of 2-aminoquinoxaline when heated with ammonium acetate.[13] However the phenoxyquinoxaline **1** is converted in high yield into the aminoquinoxaline **2** on treatment with this reagent.[14]

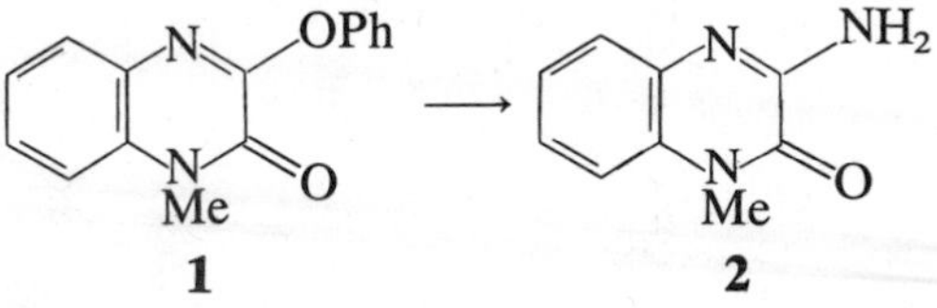

3-Aminoquinoxalin-2-ones have been prepared by heating quinoxaline-2,3-diones with ammonia or primary amines under pressures of 500–3500 psi.[15] The corresponding preparations of 3-aminoquinoxaline-2-thiones from quinoxaline-2,3-dithiones are carried out under much milder conditions and without the use of pressure vessels.[16] A comparative study of the replacement of the methylsulfinyl group from a variety of six-membered ring nitrogen heterocycles with butylamine has been made and reaction of 2-methylsulfinylquinoxaline with butylamine shown to give 2-butylaminoquinoxaline in good yield.[17]

B. *From Quinoxalines by Other Methods*

Quinoxaline-2-carboxamides have been used in the preparation of 2-aminoquinoxalines; these derivatives are converted into the corresponding amines by treatment with sodium hypohalite under normal Hofmann-type conditions.[18] 2-Aminoquinoxalines have been prepared by the reduction of aminoquinoxaline *N*-oxides with hydrogen and a Raney Ni catalyst,[19] and 2-anilinoquinoxaline has been obtained from hydrolysis of 1,3-diphenylbis(2-quinoxalinyl)urea. The latter compound is derived from quinoxaline 1-oxide and phenyl isocyanate.[20]

C. *From o-Phenylenediamines and o-Nitrosoaminobenzenes*

2-Aminoquinoxalines have been prepared by treatment of alloxazines (e.g., **3**) with aqueous sulfuric acid (Scheme 1).[13,22] An alternative procedure involves the decarboxylation of the 3-aminoquinoxaline-2-carboxylic acid produced by ammonia cleavage of the alloxazine.[21] The alloxazines are themselves readily available by condensation of *o*-phenylenediamines with alloxan. *o*-Phenylenediamines give *N*-cyanomethyl derivatives on treatment with formaldehyde and hydrogen cyanide. Cyclization of the cyanomethyl derivatives using methanolic potassium hydroxide and oxidation of the resulting dihydro compound with hydrogen peroxide gives a 2-aminoquinoxaline (Scheme 2).[23]

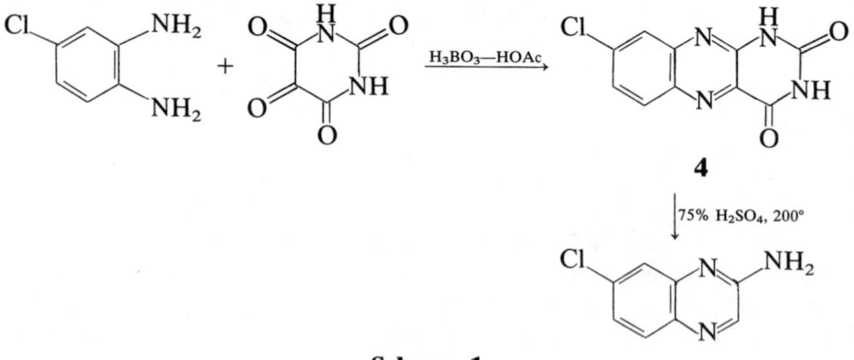

H_3BO_3—HOAc

4

75% H_2SO_4, 200°

Scheme 1

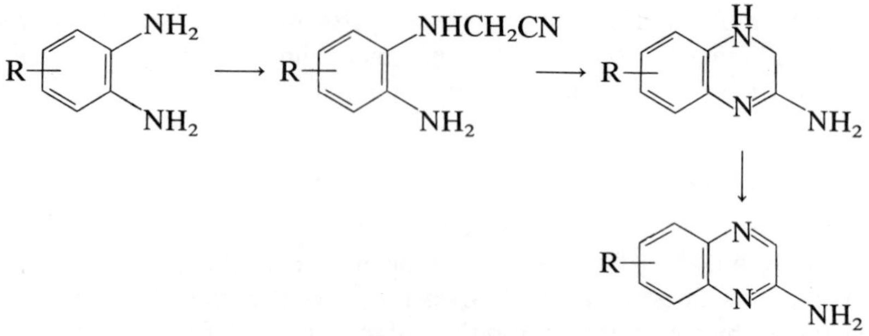

Scheme 2

Condensation of *o*-phenylenediamine with the *p*-dimethylaminoanils of aroylcyanides in hot glacial acetic acid yields 2-amino-3-aryl(heteroaryl)quinoxalines (Scheme 3). The required anils may be prepared in situ from the appropriate benzylpyridinium bromide ($ArCH_2\overset{+}{N}C_5H_5Br^-$), *p*-nitrosodimethylaniline, and sodium cyanide.[24,25] A related method of aminoquinoxaline synthesis involves the condensation of *o*-nitrosoaminobenzenes with a benzyl cyanide and gives products of unambiguous structure (Scheme 4).[26]

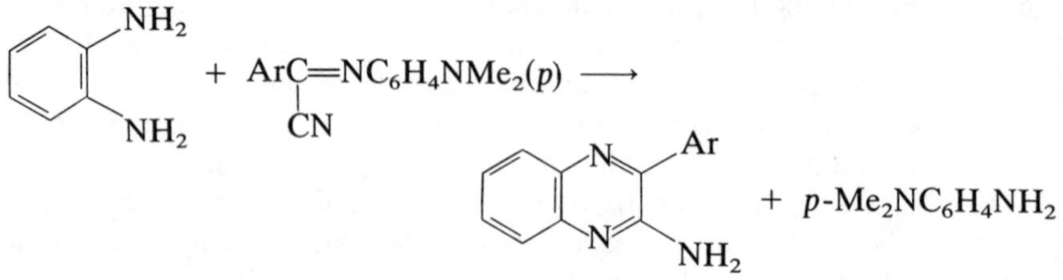

$+ ArC(CN){=}NC_6H_4NMe_2(p) \longrightarrow$

$+ p\text{-}Me_2NC_6H_4NH_2$

Scheme 3

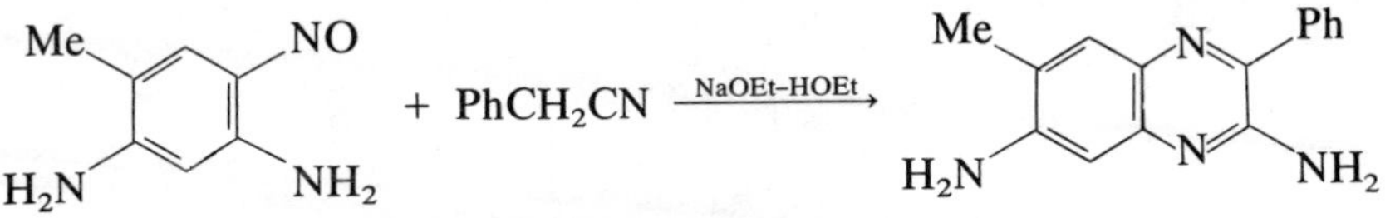

Scheme 4

2. Properties

2-Aminoquinoxaline is a relatively weak base having a pK_a value of 3.96 in water at 25°. Comparison of its ultraviolet spectrum with those of 2-methylamino- and 2-dimethylaminoquinoxaline, together with the similar basic strengths of the three quinoxalines, suggests that 2-aminoquinoxaline exists predominantly in the amino rather than imino form.[27] 2-Aminoquinoxalines are more readily hydrolyzed than is normally the case with aromatic amines. For example, 2-amino-3-methylquinoxaline is slowly hydrolyzed upon heating on the steam bath with 2 *M* sodium hydroxide.[28] The conversion of 2-aminoquinoxalines into quinoxalin-2-ones is more effectively carried out by treatment with sodium nitrite in concentrated sulfuric acid.[24] 2-Aminoquinoxalines have been converted into 2-aminoquinoxaline 1-oxides by oxidation with permaleic acid.[29] Attempts to quaternize 2-aminoquinoxaline are largely unsuccessful. However 2-acetamidoquinoxaline reacts smoothly with methyl *p*-toluenesulfonate to give a quaternary salt which is isolated as the methiodide **4**.[13] The structure of the methiodide follows from its hydrolysis by aqueous alkali to the quinoxalinone **5**. 2-Amino-3-methylquinoxaline is converted by excess of boiling acetic anhydride into a diacetyl derivative; monoacetylation is achieved by heating in toluene with the calculated amount of acetic anhydride.[19] Interaction of aminoquinoxalines with sulfonyl chlorides gives the expected sulfonamides.[30] 2-Aminoquinoxaline reacts with hydroxylamine hydrochloride to give 2-(hydroxyimino)-1,2-dihydroquinoxaline (**6**).[31] Refluxing the latter compound with acetic anhydride gives 2-acetamidoquinoxaline.

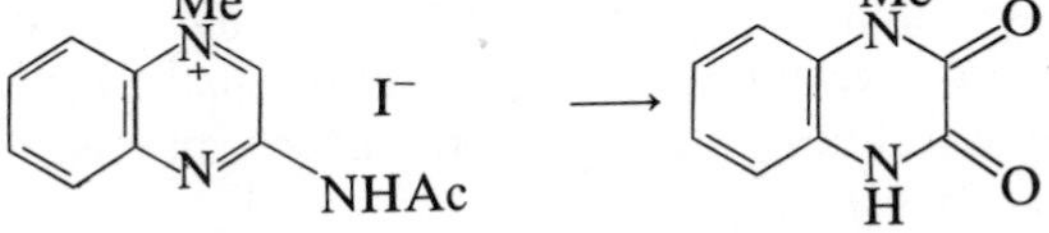

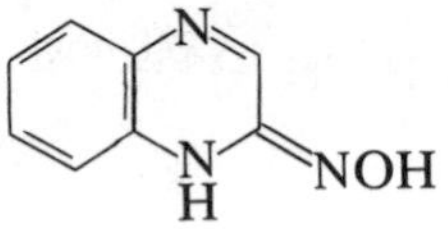

6

3. Uses

2-Sulfonamidoquinoxaline has been in veterinary use for many years for the control of coccidiosis in poultry.[32] There have been detailed investigations of its pharmacology,[33,34] and a variety of chromatographic and other techniques are available for its determination.[35–37] 2-Sulfonamidoquinoxaline has systemic activity against mosquitoes and has been shown to produce hyperglycemia in mice.[38] 2-Aminoquinoxaline has been screened for anthelmintic activity;[39] and 2-diethylamino-, 2-chloro-3-diethylamino-, and 2,3-bis(ethylamino)quinoxaline have been tested as plant growth regulators.[40] 2-Phenyl-3-piperidinoquinoxaline and some of its derivatives are selective herbicides.[41]

II. 2,3-Diaminoquinoxalines and 2,3-Substituted Aminoquinoxalines

1. Methods of Preparation

A number of 2,3-diaminoquinoxalines have been prepared by heating a 2,3-dichloroquinoxaline with anhydrous ammonia in an autoclave at 90°.[4,42,57] 2,3-Bis(alkylamino)-,[43,57] 2,3-bis(cycloalkylamino)-,[45,46] and 2,3-dianilinoquinoxalines[44] have also been prepared from 2,3-dichloroquinoxalines.

2,3-Diaminoquinoxalines are obtained from the reaction of *o*-phenylenediamines with cyanogen in methanol containing a little aqueous sodium hydroxide solution.[47] A variant on this procedure is first to prepare the oxaldiimido ester **7** from cyanogen and methanolic sodium methoxide and then to react the isolated ester with an *o*-phenylenediamine (Scheme 5).[48,49] The latter method has been used with a wide range of nuclear-substituted diamines.

o-Phenylenediamine reacts with diiminosuccindinitrile (**8**) under neutral conditions to give 2,3-diaminoquinoxaline. Under acidic conditions the primary adduct **9** decomposes to give 2,3-dicyanoquinoxaline

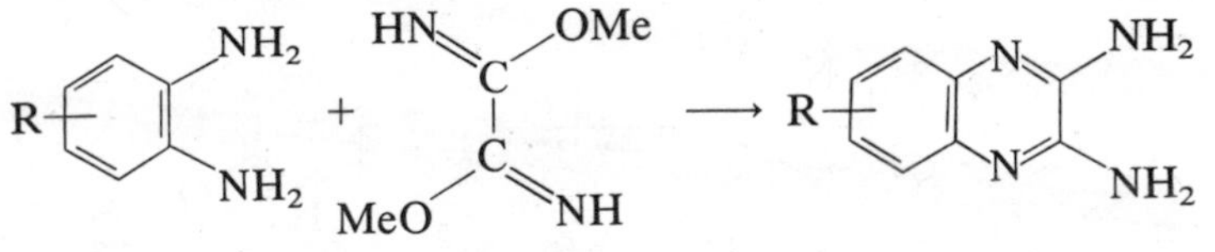

7

Scheme 5

and, in the presence of one equivalent of *p*-toluenesulfonic acid, 2-amino-3-cyanoquinoxaline is formed (Scheme 6).[51]

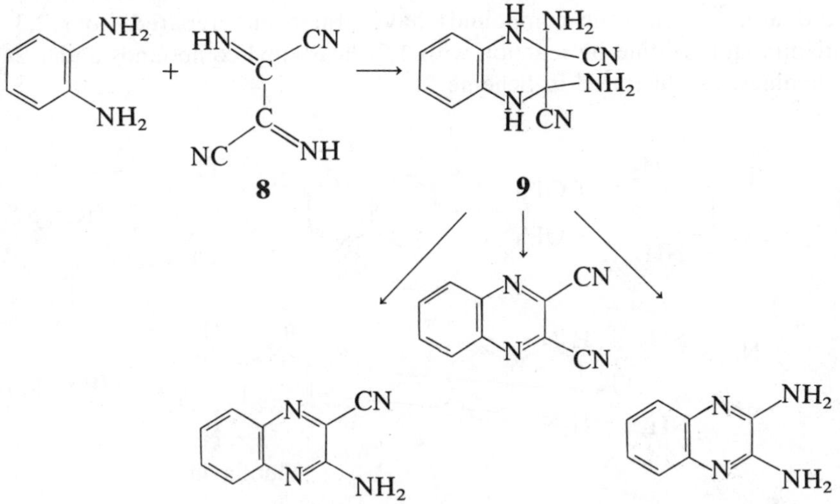

8

9

Scheme 6

2. Properties

There is evidence that in the acid hydrolysis of the 2,3-dicycloalkylaminoquinoxalines to the corresponding substituted quinoxalinones, the ease of hydrolysis is in the order morpholino > piperidino > pyrrolidinyl. Thus 2-morpholino-3-pyrrolidinylquinoxaline (**10**) can be selectively hydrolyzed to 3-pyrrolidinylquinoxalin-2-one (**11**). The hydrolysis of 2,3-dicycloaminoquinoxalines proceeds much more readily than is the case with the corresponding 2,4-disubstituted quinazolines, presumably because of the steric hindrance in the quinoxaline derivatives.[45]

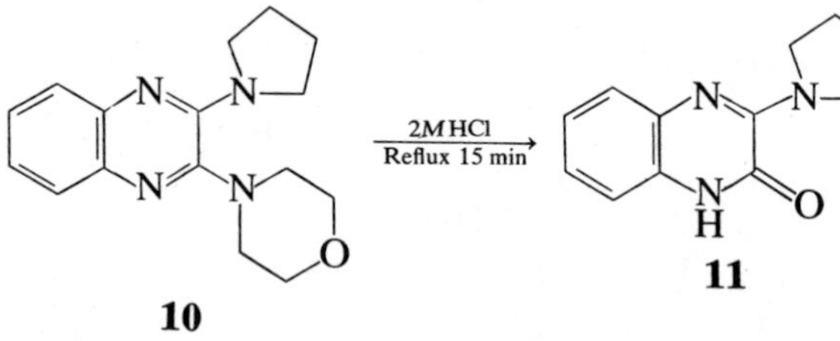

2M HCl
Reflux 15 min

10

11

Monoacylation of 2,3-diaminoquinoxalines has been achieved by reaction with an acid anhydride in THF.[52] Under more vigorous conditions reaction of 2,3-diaminoquinoxalines with acid anhydrides results in ring closure to an imidazo[*b*]quinoxaline.[47,49,50] Similar ring closures have been carried out with aldehydes, acyl halides, formic acid, orthoesters, and urea.[47] Polycyclic compounds have also been prepared from 2,3-diaminoquinoxaline by reaction with 1,2-dicarbonyl compounds and 1,2-diamines, as illustrated in Scheme 7.

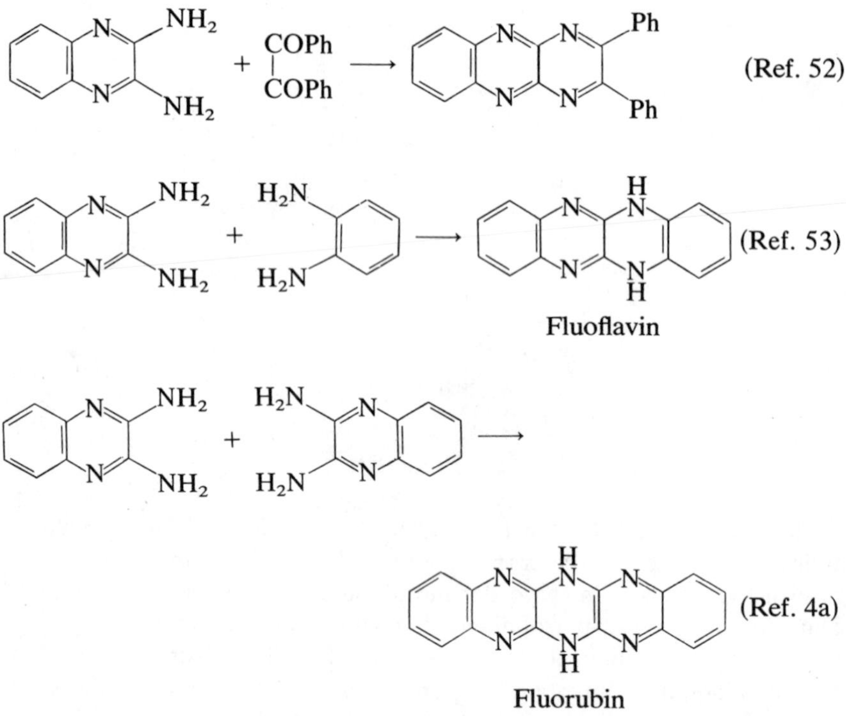

(Ref. 52)

(Ref. 53)

Fluoflavin

(Ref. 4a)

Fluorubin

Scheme 7

3. Uses

The imidazoquinoxalines derived from 2,3-diaminoquinoxalines have application as fungicides.[49] Monoacetylated 2,3-diaminoquinoxalines[52] also have fungicidal properties. The fluorubins are reddish-yellow pigments with an orange or red fluorescence. It has been suggested that they are suitable substances for identification and signaling in sea rescue work.[55]

III. Tables

TABLE 1. 2-AMINOQUINOXALINES AND 2(OR 3)-SUBSTITUTED AMINOQUINOXALINES

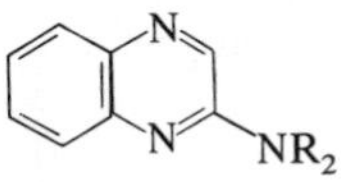

Molecular formula	Substituents or structure	m.p., °C	Ref.
$C_8H_6ClN_3$	2-Amino-3-chloro	141–143	3, 4, 56, 57
	2-Amino-6-chloro	213–217 decomp., 220–221	18, 58
	2-Amino-7-chloro	219–220.5	18
$C_8H_7N_3$	2-Amino	151–153, 155–157	2, 13, 22, 23
	Acetyl	193–194	1
	Methiodide of acetyl deriv.	343–344 decomp.	13
	Benzoyl	147–148	59
	Benzenesulfonyl	176–177	59
	Dichloroacetyl	135–136	59
	Ethoxycarbonyl	159–160	29
	Methoxycarbonyl	188–189, 194–195	29, 59
	Trichloroacetyl	107–108	59
	Trifluoroacetyl	155–156	59
	Methiodide	188	13
$C_8H_8N_4$	3,6-Diamino	204 decomp.	26
	Diacetyl	302	26
	Picrate	200	26
$C_8H_8N_4O_2S$	2-Amino-3-sulfonamido		60
$C_9H_6ClN_3O_2$	3-Amino-2-carboxy-7-chloro	191–192 decomp.	61
	Methyl ester	224.5–225.5	61
$C_9H_6N_4$	2-Amino-3-cyano	196–200 decomp.	51
$C_9H_9N_3$	2-Amino-3-methyl	163–165, 166–167	28, 62, 63, 64, 65
	Monoacetyl	145–146	19
	Diacetyl	87.5–88	19

TABLE 1 (*contd.*)

Molecular formula	Substituents or structure	m.p., °C	Ref.
$C_9H_9N_3$ (*contd.*)	2-Amino-5-methyl	202–203	65, 66
	2-Amino-6-methyl	181–182	66
	2-Amino-7-methyl	178–180	66
	2-Amino-8-methyl	129, 135.5–136	23, 66
	2-Methylamino	129–131	6
$C_9H_9N_3O$	2-Amino-3-methoxy	149, 151–152	14, 63
	Acetyl	163	63
	3-Amino-1,2-dihydro-1-methyl-		14
	2-oxo	273–274	14
$C_9H_{10}N_4$	3,6-Diamino-7-methyl	221	26
	Diacetyl	310	26
	Picrate	288 decomp.	26
$C_{10}H_7ClN_4O_2$	2-(1-Aziridinyl)-3-chloro-6-nitro		10
$C_{10}H_{10}ClN_3O$	2-Chloro-3-(2-hydroxyethylamino)		67
$C_{10}H_{11}N_3$	2-Amino-3-ethyl	119	63
	Acetyl	147.5	63
	2-Dimethylamino	94–95	6
$C_{10}H_{11}N_3O$	2-Amino-3-ethoxy	130	63
	Acetyl	150.5	63
$C_{10}H_{11}N_3O_2$	2-Amino-6,7-dimethoxy		68
	Methoxycarbonyl	188–193	68
$C_{10}H_{12}N_4$	2-(2-Aminoethyl)amino	109–111	69
	Dihydrochloride	222–224	69
$C_{11}H_{11}N_3O$	2-(1-Aziridinyl)-3-methoxy	78–79	70
$C_{11}H_{13}N_3$	2-Amino-3-isopropyl	160	63
	Acetyl	177	63
$C_{12}H_{13}N_3O$	2-Morpholino	87–89	6
$C_{12}H_{14}ClN_3$	2-Chloro-3-diethylamino	b.p. 132/0.5 mm	40
$C_{12}H_{15}N_3$	2-Amino-3-*t*-butyl	103	63
	Acetyl	165	63
	2-Butylamino		
	Picrate	184	17
	2-Diethylamino	62–64	40
$C_{13}H_{10}N_4$	2-Amino-3-(2-pyridyl)	178–179	25
$C_{13}H_{14}ClN_3$	7-Chloro-2-piperidino	114–115	7
$C_{13}H_{15}N_3$	2-Piperidino	59–60.5, 62–63	6, 7
	2-Methyl-3-(1-pyrrolidinyl)	108–109	46
$C_{13}H_{15}N_3O$	2-Methyl-3-morpholino	91–92	46
$C_{14}H_{10}ClN_3$	2-Amino-3-(*p*-chlorophenyl)	185	24
$C_{14}H_{10}N_4O_2$	2-Amino-3-(*p*-nitrophenyl)	283.5	24
	2-(*o*-Nitroanilino)	183–185	44
$C_{14}H_{11}ClN_4$	3,6-Diamino-2-(*o*-chlorophenyl)	230–231	26
	3,6-Diamino-2-(*p*-chlorophenyl)	212–213	26
$C_{14}H_{11}N_3$	2-Anilino	137	20
	2-Amino-3-phenyl	160, 163–164	16, 24, 63, 64
	Acetyl	180	24

TABLE 1 (*contd.*)

Molecular formula	Substituents or structure	m.p., °C	Ref.
$C_{14}H_{11}N_3O_3S$	2-Amino-3-phenylsulfonyloxy	180–181	71
$C_{14}H_{17}N_3$	2-Methyl-3-piperidino	55–56	46
$C_{14}H_{17}N_3O$	7-Methoxy-2-piperidino	66–68	7
$C_{14}H_{18}Cl_2N_4$	6,7-Dichloro-2-(2′-diethyl-aminoethylamino	b.p. 168–173/0.03 mm	8
	Methiodide	196–197	8
$C_{14}H_{20}N_4$	2-(2′-Diethylaminoethylamino)	b.p. 140/0.02 mm	8
	Dipicrate	185	8
$C_{15}H_{12}ClN_3$	2-(*o*-Chloroanilino)-3-methyl	114–115	12
$C_{15}H_{11}N_3O_2$	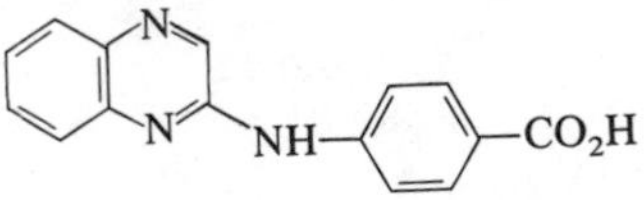 N, N, NH, CO_2H	344–345 decomp.	8
$C_{15}H_{13}ClN_4$	3,6-Diamino-2-(*o*-chlorophenyl)-7-methyl	236	26
$C_{15}H_{13}N_3$	2-Amino-3-benzyl	155–157	72
	2-Benzylamino	62–64, 73–75	6, 73
	Hydrochloride	255–256	73
	2-Methylamino-3-phenyl	132	5
	2-*m*-Toluidino	105–107	44
	2-*p*-Toluidino	155–157	44
$C_{15}H_{13}N_3O$	2-*m*-Anisidino	129–131	44
$C_{15}H_{14}N_4$	3,6-Diamino-7-methyl-2-phenyl	174	26
$C_{15}H_{19}N_3$	2-(Hexahydro-1*H*-azepin-1-yl)-3-methyl	60	46
$C_{15}H_{20}Cl_2N_4$	6,7-Dichloro-2-(3′-diethyl-aminopropylamino	84–86	8
	Picrate	182	8
	Methiodide	212 decomp.	8
$C_{15}H_{21}N_5O_3$	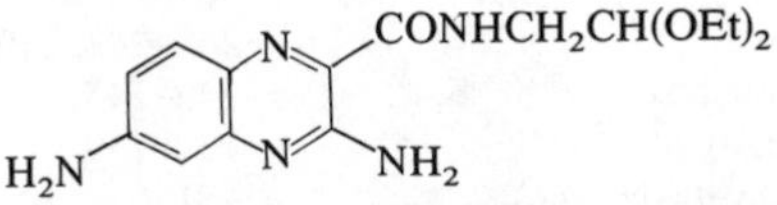 N, N, $CONHCH_2CH(OEt)_2$, H_2N, NH_2		74
$C_{15}H_{22}N_4$	2-(3′-Diethylaminopropylamino)	200–205/0.1 mm	8
	Dipicrate	164	8
$C_{16}H_{15}N_3$	2-(3,5-Dimethylanilino)	133–135	44
$C_{16}H_{22}N_4O$	2-(3′-Diethylaminopropylamino)-7-methoxy	Oil	7
	Methiodide	177–178	7
$C_{16}H_{24}N_6O$	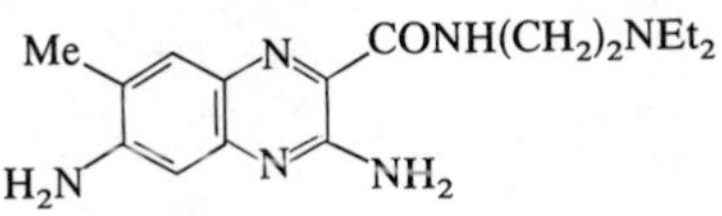 Me, N, N, $CONH(CH_2)_2NEt_2$, H_2N, NH_2	152–153	75

TABLE 1 (*contd.*)

Molecular formula	Substituents or structure	m.p., °C	Ref.
$C_{16}H_{24}N_6O$ (*contd.*)	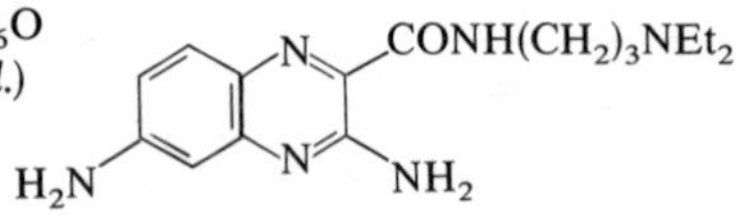 N, N, $CONH(CH_2)_3NEt_2$, H_2N, NH_2	131–132	75
$C_{17}H_{12}N_4$	2-Amino-3-(2-quinolyl)	215.5–217	25
$C_{18}H_{16}N_6$	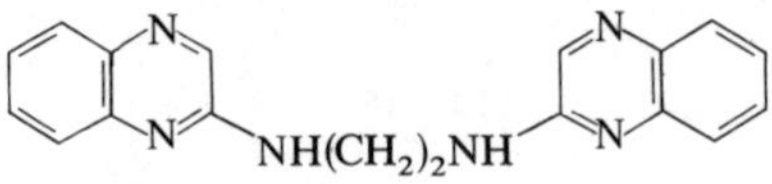 N, N, $NH(CH_2)_2NH$, N, N	262–264	69
	Dihydrochloride	280–282	69
$C_{19}H_{19}N_3$	2-Phenyl-3-piperidino	82–84	77
$C_{19}H_{22}N_4$	2-*N*-Benzyl-*N*-(2′-dimethyl-aminoethylamino)		73
	Hydrochloride	217–218	73
$C_{20}H_{15}N_3$	2-Anilino-3-phenyl	112–113	76

TABLE 2. 2,3-DIAMINO- AND 2,3-DISUBSTITUTED AMINOQUINOXALINES

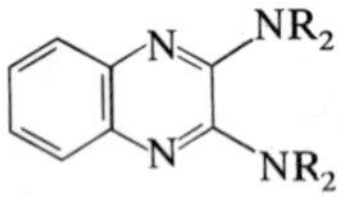

N, N, NR_2, NR_2

Molecular formula	Substituents	m.p., °C	Ref.
$C_8H_7ClN_4$	2,3-Diamino-6-chloro	280–282, 286	47, 48
$C_8H_7N_5O_2$	2,3-Diamino-6-nitro	>360	48
$C_8H_8N_4$	2,3-Diamino	331, >360	47, 48, 51, 78
$C_9H_{10}N_4$	2,3-Diamino-6-methyl	246–248, 249	47, 48
$C_9H_{10}N_4O$	2,3-Diamino-6-methoxy	243–245	48
$C_{12}H_{16}N_4$	2,3-Bis(ethylamino)	157–159	40
$C_{12}H_{16}N_4O_2$	2,3-Bis(2′-hydroxyethylamino)	181	45
$C_{14}H_{16}N_4$	2-(1-Aziridinyl)-3-(1-pyrrolidinyl)	56–57	46
$C_{14}H_{16}N_4O$	2-(1-Aziridinyl)-3-morpholino	125	46
$C_{15}H_{18}N_4$	2-(1-Aziridinyl)-3-piperidino	102	46
$C_{16}H_{20}N_4$	2,3-Di-1-pyrrolidinyl	124, 127	45, 46
	2-(1-Aziridinyl)-3-(hexahydro-1*H*-azepin-1-yl)	58	46
$C_{16}H_{20}N_4O$	2-Morpholino-3-(1-pyrrolidinyl)	92, 104	45, 46
$C_{16}H_{20}N_4O_2$	2,3-Dimorpholino	210, 225	45, 46
$C_{17}H_{22}N_4$	2-Piperidino-3-(1-pyrrolidinyl)	87, 88	45, 46
$C_{17}H_{22}N_4O$	2-Morpholino-3-piperidino	144	45, 46

TABLE 2 (*contd.*)

Molecular formula	Substituents	m.p., °C	Ref.
$C_{18}H_{24}N_4$	2,3-Dipiperidino	142, 148	45, 46
	2-(Hexahydro-1*H*-azepin-1-yl)-3-(1-pyrrolidinyl)	98	46
$C_{18}H_{24}N_4O$	2-(Hexahydro-1*H*-azepin-1-yl)-3-morpholino	104–105	46
$C_{19}H_{26}N_4$	2-(Hexahydro-1*H*-azepin-1-yl)-3-piperidino	79	46
$C_{20}H_{16}N_4$	2,3-Dianilino	143–144	44
$C_{20}H_{28}N_4$	2,3-Di(hexahydro-1*H*-azepin-1-yl)	57	46
$C_{22}H_{20}N_4$	2,3-Di-*m*-toluidino	125–127	44
	2,3-Di-*p*-toluidino	147–149	44
$C_{22}H_{20}N_4O_2$	2,3-Di-*m*-anisidino	190–192	44
$C_{24}H_{24}N_4$	2,3-Bis(3,5-dimethylanilino)	148–150	44

IV. References

1. J. C. E. Simpson, "Condensed Pyridazine and Pyrazine Rings," Interscience, New York, 1953, p. 263.
2. F. J. Wolf (to Merck and Co.) U.S. Pat. 2,537,870; *Chem. Abstr.*, **45,** 4274 (1951).
3. H. Saikachi and S. Tagami, *Chem. Pharm. Bull.* (Tokyo), **9,** 941 (1961).
4. W. Deuschel and G. Riedel (to Badische Anilin- und Soda-Fabrik A.-G.), Ger. Pat. 1,135,471; *Chem. Abstr.*, **58,** 537 (1963).
5. S. Tagami and D. Shiho, *Yakugaku Zasshi*, **84,** 1085 (1964); *Chem. Abstr.*, **62,** 5278 (1965).
6. G. W. H. Cheeseman, *J. Chem. Soc.*, 3236 (1957).
7. R. L. Wear and C. S. Hamilton, *J. Am. Chem. Soc.*, **72,** 2893 (1950).
8. R. M. Acheson, *J. Chem. Soc.*, 4731 (1956).
9. N. B. Chapman, K. Clarke, and K. Wilson, *J. Chem. Soc.*, 2256 (1963).
10. S. Inaba, C. Saito, K. Nishikawa, and A. Misaki (to Somitomo Chemical Co., Ltd.), Jap. Pat. 5706 (1965); *Chem. Abstr.*, **62,** 16273 (1965).
11. A. S. Elina, *Chem. Heterocycl. Comp.*, **3,** 740 (1967).
12. Y. J. L'Italien and C. K. Banks, *J. Am. Chem. Soc.*, **73,** 3246 (1951).
13. C. M. Atkinson, C. W. Brown, and J. C. E. Simpson, *J. Chem. Soc.*, 26 (1956).
14. G. W. H. Cheeseman, *J. Chem. Soc.*, 1804 (1955).
15. R. C. Harris (to Eastman Kodak Co.), U.S. Pat. 3,446,809; *Chem. Abstr.*, **71,** 51237*f* (1969).
16. K. Sasse, R. Wegler, and G. Unterstenhoeffer (to Farbenfabriken Bayer A.-G.), Ger. Pat. 1,117,586; *Chem. Abstr.*, **57,** 4684 (1962); I. J. Pachter and P. E. Nemeth, *J. Org. Chem.*, **28,** 1203 (1963).
17. G. B. Barlin and W. V. Brown, *J. Chem. Soc.* (*C*), 921 (1969).
18. H. R. Moreno and H. P. Schultz, *J. Org. Chem.*, **36,** 1158 (1971).
19. A. S. Elina and L. G. Tsyrul'nikova, *J. Gen. Chem. USSR*, **33,** 1507 (1963).

20. C. Iijima, *Yakugaku Zasshi*, **87,** 164 (1967); *Chem. Abstr.*, **67,** 3067*w* (1967).
21. J. Weijlard and M. Tishler (to Merck and Co. Inc.), U.S. Pat. 2,479,443; *Chem. Abstr.*, **44,** 1146 (1950).
22. H. G. Petering and G. J. van Giessen, *J. Org. Chem.*, **26,** 2818 (1961).
23. K. Pfister, A. P. Sullivan, J. Weijlard, and M. Tishler, *J. Am. Chem. Soc.*, **73,** 4955 (1951).
24. F. Krönhke and H. Leister, *Chem. Ber.*, **91,** 1479 (1958).
25. F. Krönhke and K. F. Gross, *Chem. Ber.*, **92,** 22 (1959).
26. T. S. Osdene and G. M. Timmis, *J. Chem. Soc.*, 2027 (1955).
27. G. W. H. Cheeseman, *J. Chem. Soc.*, 108 (1958).
28. J. A. Barltrop, C. G. Richards, D. M. Russell, and G. Ryback, *J. Chem. Soc.*, 1132 (1959).
29. R. A. Burrell, J. M. Cox, and E. G. Savins, *J. Chem. Soc.*, *Perkin I*, 2707 (1973).
30. American Cyanamid Co., Brit. Pat. 615,328; *Chem. Abstr.*, **43,** 7516 (1949).
31. K. Harsanyi, C. Gonezi, and D. Korbonits, *Justus Liebigs Ann. Chem.*, 190 (1973).
32. I. A. Koblova and A. B. Piskov, *Khim. Sel. Khoz.*, **10,** 40 (1972); *Chem. Abstr.*, **76,** 135459*w* (1972).
33. G. Aguggini and M. G. Clement, *Boll. Soc. Ital. Biol. Sper.*, **47,** 483 (1971); *Chem. Abstr.*, **76,** 149134*y* (1972).
34. K. Fried and J. Jantosovic, *Veterinaria* (Prague), **14,** 97 (1972); *Chem. Abstr.*, **77,** 135713*v* (1972).
35. A. Begliomini and A. Fravolini, *Arch. Vet. Italy*, **22,** 273 (1971); *Chem. Abstr.*, **76,** 152115*y* (1972).
36. R. J. Daun, *J. Ass. Offic. Anal. Chem.*, **54,** 1277 (1971); *Chem. Abstr.*, **76,** 12875*w* (1972).
37. T. C. Kram, *J. Pharm. Sci.*, **61,** 254 (1972); *Chem. Abstr.*, **76,** 103820*r* (1972).
38. W. N. Beesley, *Ann. Trop. Med. Parasitol.*, **65,** 417 (1971); *Chem. Abstr.*, **76,** 31037*r* (1972).
39. K. B. Kerr and J. W. Cavett, *Exp. Parasitol.*, **1,** 161 (1952); *Chem. Abstr.*, **46,** 7708 (1952).
40. J. R. Geigy A.-G., Brit. Pat. 822,069; *Chem. Abstr.*, **55,** 2005 (1961).
41. Q. F. Soper (to Eli Lilly and Co.), U.S. Pat. 3,647,793; *Chem. Abstr.*, **77,** 30339*z* (1972).
42. A. P. Komin and M. Carmack, *J. Heterocycl. Chem.*, **13,** 13 (1976).
43. L. G. S. Brooker and E. J. Van Lare (to Eastman Kodak Co.), U.S. Pat. 3,431,111; *Chem. Abstr.*, **72,** 68222*r* (1970).
44. S. D. Carter and G. W. H. Cheeseman, *Tetrahedron*, **34,** 981 (1978).
45. I. N. Goncharova and I. Ya. Postovskii, *J. Gen. Chem. USSR*, **32,** 3271 (1962).
46. W. R. Vaughan and M. S. Habib, *J. Org. Chem.*, **27,** 324 (1962).
47. E. Schipper and A. R. Day, *J. Am. Chem. Soc.*, **73,** 5672 (1951).
48. H. Weidinger and J. Kranz, *Chem. Ber.*, **97,** 1599 (1964).
49. D. J. Sam and M. A. Wuonola (to E. I. du Pont de Nemours and Co.), Ger. Offen. 2,339,012; *Chem. Abstr.*, **80,** 108578*k* (1974).
50. W. Lunkenheimer, K. H. Beuchel, and H. Kaspers (to Bayer A.-G.), Ger. Offen. 2,249,350; *Chem. Abstr.*, **81,** 13557*j* (1974).
51. R. W. Begland and D. R. Hartter, *J. Org. Chem.*, **37,** 4136 (1972).
52. D. J. Sam (to E. I. du Pont de Nemours and Co.), Ger. Offen. 2,339,023; *Chem. Abstr.*, **80,** 133479*m* (1974).
53. E. B. Nyquist and M. M. Joullie, *J. Chem. Soc.* (*C*), 947 (1968).

54. H. M. Woodburn and W. E. Hoffman, *J. Org. Chem.*, **23,** 262 (1958).
55. J. L. Switzer and R. C. Switzer (to J. L. Switzer and R. C. Switzer), U.S. Pat. 2,495,202; *Chem. Abstr*; **44,** 5501 (1950).
56. W. Deuschel and G. Riedel (to Badische Anilin- und Soda-Fabrik A.-G.), Ger. Pat. 1,142,981; *Chem. Abstr.*, **59,** 2831 (1963).
57. W. Deuschel, W. Vilsmeier, and G. Riedel (Badische Anilin- und Soda-Fabrik A.-G.), Belg. Pat. 612,092; *Chem. Abstr.*, **57,** 16634 (1962).
58. J. K. Horner and D. W. Henry, *J. Med. Chem.*, **11,** 946 (1968).
59. A. S. Elina, L. G. Tsyrul'nikova, E. M. Peresleni, and Yu. N. Sheinker, *Chem. Heterocycl. Comp.*, **2,** 72 (1966).
60. M. Hattori, M. Yoneda, and M. Goto, *Bull. Chem. Soc. Jap.*, **46,** 1890 (1973).
61. E. J. Cragoe (to Merck and Co.), Belg. Pat. 639,386; *Chem. Abstr.*, **62,** 14700 (1965).
62. F. J. Wolf, R. H. Beutel, and J. R. Stevens, *J. Am. Chem. Soc.*, **70,** 2572 (1948).
63. C. Iijima, T. Morikawa, and E. Hayashi, *Yakugaku Zasshi*, **95,** 784 (1975); *Chem. Abstr.*, **84,** 4901*b* (1976).
64. P. J. Lont and H. C. van der Plas, *Rec. Trav. Chim.*, **91,** 850 (1972).
65. B. C. Platt and T. M. Sharp (to Wellcome Foundation Ltd.), Brit. Pat. 621,289; *Chem. Abstr.*, **43,** 6661 (1949).
66. B. C. Platt and T. M. Sharp, *J. Am. Chem. Soc.*, **71,** 3247 (1949).
67. T. Shimada and R. Oda, *Kogyo Kagaku Zasshi*, **67,** 576 (1964); *Chem. Abstr.*, **61,** 9606 (1964).
68. Z. Budesinsky and A. Valenta, *Collect. Czech. Chem. Commun.*, **36,** 2527 (1971).
69. Z. Kolodynska and S. Biniecki, *Acta Polon. Pharm.*, **20,** 285 (1963); *Chem. Abstr.*, **62,** 559 (1965).
70. H. W. Heine and A. C. Brooker, *J. Org. Chem.*, **27,**. 2943 (1962).
71. A. S. Elina, L. G. Tsyrul'nikova, E. M. Peresleni, and Yu. N. Sheinker, *Khim. Geterotsikl. Soedin. Akad. Nauk. Latv. SSR*, 101 (1966); *Chem. Abstr.*, **65,** 712 (1966).
72. D. D. Chapman, *J. Org. Chem.*, **37,** 2498 (1972).
73. J. H. Gardner and J. R. Stevens, *J. Am. Chem. Soc.*, **71,** 1868 (1949).
74. A. A. Santilli and T. S. Osdene (to American Home Products Corp.), U.S. Pat. 3,192,212; *Chem. Abstr.*, **63,** 11589 (1965).
75. T. S. Osdene (to American Home Products Corp.), U.S. Pat. 3,185,688; *Chem. Abstr.*, **63,** 11589 (1965).
76. E. Hayashi and C. Iijima, *Yakugaku Zasshi*, **82,** 1093 (1962); *Chem. Abstr.*, **58,** 4551 (1963).
77. Q. F. Soper (to Eli Lilly and Co.), U.S. Pat. 3,582,315; *Chem. Abstr.*, **75,** 62476*s* (1971).
78. D. Shiho and S. Tagami, *Chem. Pharm. Bull.* (Tokyo), **5,** 45 (1957).

CHAPTER XII

2-Hydrazinoquinoxalines

I. Methods of Preparation . . . 194
II. Properties . . . 195
III. Uses . . . 196
IV. Tables . . . 196
Table 1. 2(or 3)-Hydrazinoquinoxalines . . . 196
V. References . . . 197

I. Methods of Preparation

2-Hydrazinoquinoxalines are generally prepared from the corresponding chloroquinoxalines by treatment with hydrazine hydrate in refluxing ethanol.[1] It has been found that displacement reactions with hydrazine can be carried out under milder conditions than the parallel reactions with ammonia or amines,[2] and that when methylhydrazine is reacted with chloroquinoxalines, a 1:1 disubstituted hydrazine is formed, rather than a 1:2 disubstituted product.[3]

2-Hydrazinoquinoxalines have also been prepared by the action of hydrazine hydrate on quinoxalin-2-ones[1,4] and quinoxaline-2-thiones.[5] 2,3-Dihydrazinoquinoxaline is obtained by the action of hydrazine hydrate on 2,3-dichloroquinoxaline,[6] a 2-chloro-3-cycloalkylaminoquinoxaline, or 2-hydrazino-3-piperidinoquinoxaline.[7]

An alternative method of synthesis is illustrated by the preparation of the 1:1 disubstituted hydrazine **1**.[8]

Ph, NHMe $\xrightarrow{HNO_2}$ Ph, NMe–NO $\xrightarrow{LiAlH_4}$ Ph, NMe–NH_2

1

II. Properties

2-Hydrazino-3-methoxyquinoxaline has been hydrolyzed to quinoxaline-2,3-dione by treatment with boiling 48% HBr or 2% HCl at 140–150°.[1] A series of 2-hydrazino-3-alkoxyquinoxalines gives 2-(2-acetylhydrazino)-3-alkoxyquinoxalines by reaction with 1-acetyl-3,5-dimethylpyrazole,[9] and benzoylation of the methylhydrazino compound **2** similarly yields the benzoyl derivative **3**.[4]

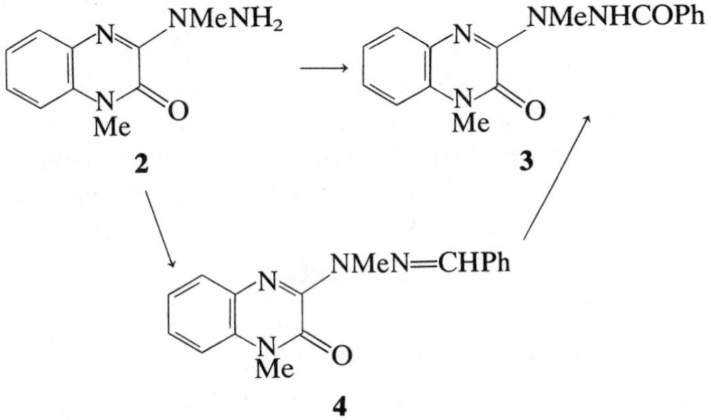

NMeNH$_2$ N N Me O **2** → NMeNHCOPh N N Me O **3**

NMeN=CHPh N N Me O **4**

2-Hydrazinoquinoxalines are readily converted into hydrazones on treatment with aldehydes or ketones.[8] Attempted formation of the benzylidene derivative **4** gives the benzoyl compound **3**, presumably as a result of the ready aerial oxidation of the benzylidene derivative.[4]

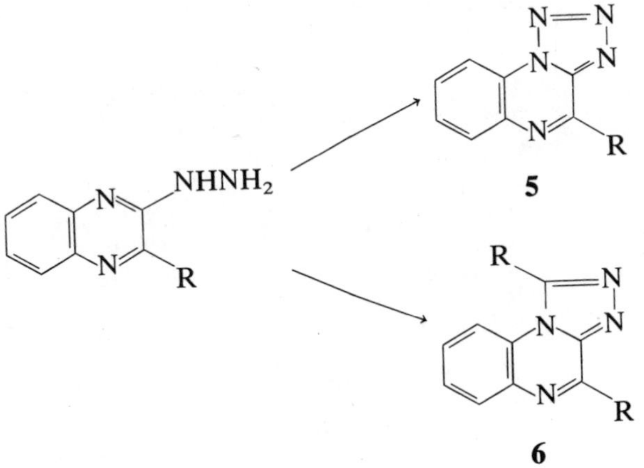

NHNH$_2$ N N R → N=N N N N R **5**

→ R N N N N R **6**

A number of ring closure reactions have been carried out on 2-hydrazinoquinoxalines. Thus treatment with nitrous acid results in tetrazolo[1,5-*a*]quinoxaline (**5**) formation,[4,10] and *s*-triazoloquinoxalines (**6**) are formed on pyrolysis of the hydrazones derived from 2-hydrazinoquinoxalines.[11,12] *s*-Triazoloquinoxalines have also been obtained by treatment of 2-hydrazinoquinoxalines with carboxylic acids, acid chlorides, acid anhydrides, and orthoesters[11]; with cyanogen bromide[3,13]; and with phosgene and thiophosgene.[3]

III. Uses

2-Hydrazinoquinoxalines have been investigated as potential antileprous drugs,[14] as hypotensives,[15] and as drugs for the treatment of myeloid leukemia.[16] None of the hydrazinoquinoxalines had hypotensive activity comparable to either that of 1-hydrazino- or 1,4-dihydrazinophthalazine.

IV. Table

TABLE 1. 2(OR 3)-HYDRAZINOQUINOXALINES

[Structure: quinoxaline with NHR^1NHR^2 at position 2]

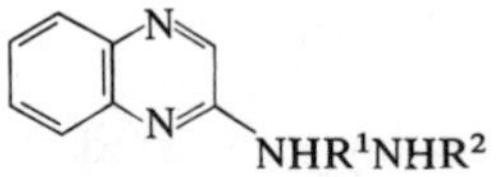

Molecular formula	Substituents or structure	m.p., °C	Ref.
$C_8H_8N_4$	Unsubstituted	167	4, 5, 11
$C_9H_{10}N_4$	3-Methyl	172	2, 6, 11
	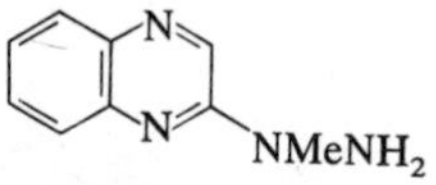 [Structure: quinoxaline with $NMeNH_2$ at position 2]	108–110	3
	p-Nitrobenzylidene	245–247	3
	Picrate	157	3
$C_9H_{10}N_4O$	3-Methoxy	168	1, 9
$C_{10}H_{12}N_4$	3-Ethyl	192–193	2
	6,7-Dimethyl	209–210	10
	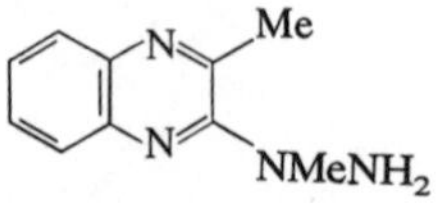 [Structure: 3-Me-quinoxaline with $NMeNH_2$ at position 2]	65–67	3
	p-Nitrobenzylidene	254–256	3
	Picrate	167–168	3

TABLE 1 (*contd.*)

Molecular formula	Substituents or structure	m.p., °C	Ref.
$C_{10}H_{12}N_4O$	3-Ethoxy	149	9
	5-Methoxy-3-methyl	198–201	17
	8-Methoxy-3-methyl	152–155	17
$C_{11}H_{14}N_4$	3,6,7-Trimethyl	210–211	10
	3-Propyl	145–146	2
$C_{12}H_{14}N_4O$	3-Isopropoxy	117–119	9
$C_{12}H_{14}N_4O_2$	[quinoxaline: 2-OEt, 3-NHNHAc]	183	9
$C_{12}H_{16}N_4$	3-Butyl	118–119	2
	3-*sec*-Butyl	112	11
$C_{12}H_{16}N_4O$	3-Butoxy	99–101	9
	3-Isobutoxy	105–107	9
	3-*sec*-Butoxy	83–85	9
$C_{13}H_{17}N_5$	3-Piperidino	162	7
$C_{13}H_{18}N_4$	2-Amyl	137	6
$C_{14}H_{12}N_4$	3-Phenyl	140	11
$C_{14}H_{18}N_4O_2$	[quinoxaline: 2-OBu^i, 3-NHNHAc]	166	9
$C_{15}H_{14}N_4$	[quinoxaline: 2-Ph, 3-NMeNH$_2$]	103–105	3
	Hydrochloride	245 decomp.	8
	Picrate	225 decomp.	3
	Benzylidene deriv.	181.5	8
	p-Chlorobenzylidene deriv.	115	8
	p-Nitrobenzylidene deriv.	171, 174–175	3, 8

V. References

1. K. Asano and S. Asai, *Yakugaku Zasshi*, **79,** 658 (1959); *Chem. Abstr.*, **53,** 21979 (1959).
2. H. Reinheckel, *Monatsh.* **99,** 2215 (1968).
3. K. T. Potts, S. K. Roy, S. W. Schneller, and R. M. Huseby, *J. Org. Chem.*, **33,** 2559 (1968).
4. G. W. H. Cheeseman and M. Rafiq, *J. Chem. Soc.* (*C*), 452 (1971).

5. K. Asano and S. Asai, *Yakugaku Zasshi* **79,** 661 (1959); *Chem. Abstr.*, **53,** 21979 (1959).
6. K. Asano, *Yakugaku Zasshi*, **78,** 729 (1958); *Chem. Abstr.*, **52,** 18428 (1958).
7. I. N. Goncharova and I. Ya. Postovskii, *J. Gen. Chem. USSR*, **32,** 3271 (1962).
8. S. Tagami and D. Shiho, *Yakugaku Zasshi*, **84,** 1085 (1964); *Chem. Abstr.*, **62,** 5278 (1965).
9. H. Saikachi and S. Tagami, *Yakugaku Zasshi*, **82,** 1246 (1962); *Chem. Abstr.*, **59,** 1634 (1963).
10. N. G. Koshel, E. G. Kovalev, and I. Ya. Postovskii, *Chem. Heterocycl. Comp.*, **6,** 791 (1970).
11. D. Shiho and S. Tagami, *J. Am. Chem. Soc.*, **82,** 4044 (1960).
12. S. Tagami, K. Sasayama, and D. Shiho, *Yakugaku Zasshi*, **86,** 622 (1966); *Chem. Abstr.*, **65,** 12185 (1966).
13. S. Yurugi, A. Miyake, and M. Tomimoto, *Takeda Kenkyusho Ho*, **32,** 111 (1973); *Chem. Abstr.*, **80,** 37073*h* (1974).
14. S. Tsutsumi, Y. Sakamoto, K. Nakamura, S. Hisai, and E. Kashima, *Repura*, **39,** 33 (1970); *Chem. Abstr.*, **76,** 135847*w* (1972).
15. S. Kumada, N. Watanabe, K. Yamamoto, and H. Zenno, *Yakugaku Kenkyo*, **30,** 635 (1958); *Chem. Abstr.*, **53,** 20554 (1959).
16. B. L. Freedlander and A. Furst, *Proc. Soc. Exp. Biol. Med.*, **81,** 638 (1952); *Chem. Abstr.*, **47,** 2877 (1953).
17. I. Ya. Postovskii and N. G. Koshel, *Chem. Heterocycl. Comp.*, **6,** 913 (1970).

CHAPTER XIII

2-Alkoxy-, 2-Aryloxy-, 2,3-Dialkoxy-, 2-Alkoxy-3-aryloxy-, and 2,3-Diaryloxyquinoxalines

I. 2-Alkoxy- and 2-Aryloxyquinoxalines 199
1. Methods of Preparation 199
2. Properties 200
II. 2,3-Dialkoxy- and 2,3-Diaryloxyquinoxalines 201
1. Methods of Preparation 201
2. Properties 201
3. Uses 202
III. Tables 202
Table 1. 2(or 3)-Alkoxy- and 2(or 3)-Aryloxyquinoxalines 202
Table 2. 2,3-Dialkoxy-, 2-Alkoxy-3-aryloxy-, and 2,3-Diaryloxyquinoxalines 203
IV. References 204

Compounds of this type have been reviewed previously by Simpson.[1]

I. 2-Alkoxy- and 2-Aryloxyquinoxalines

1. Methods of Preparation

2-Alkoxyquinoxalines are conveniently prepared from the corresponding 2-chloroquinoxalines by treatment with an alcoholic sodium alkoxide solution. Thus 2-methoxyquinoxalines are readily prepared from 2-chloroquinoxalines by reaction with methanolic sodium methoxide; they are also formed on methylation of quinoxalin-2-ones with diazomethane.[2,3] In the latter reaction a mixture of both *O*- and *N*-methyl derivatives is invariably formed (Scheme 1).

2-Chloroquinoxalines are converted into 2-phenoxyquinoxalines by treatment with phenol in which potassium hydroxide has been dissolved.[2,4,5] When 2-chloroquinoxaline is heated with a sodium aryloxide in

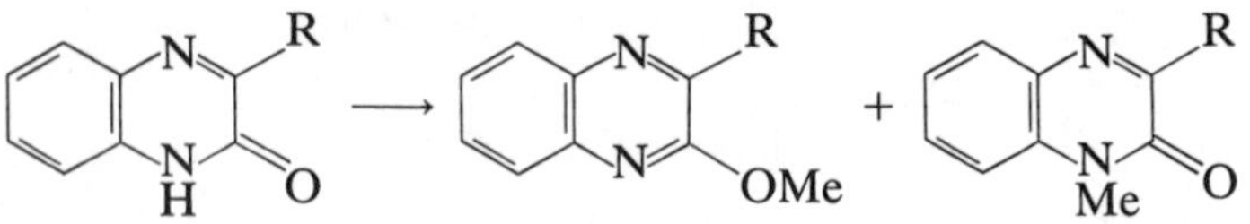

R = H, Me, OMe, or NH_2

Scheme 1

excess of the corresponding phenol, in addition to the expected 2-aryloxyquinoxaline (**1**), some of the corresponding benzofuro[2,3-*b*]-quinoxaline (**2**) is obtained.[6]

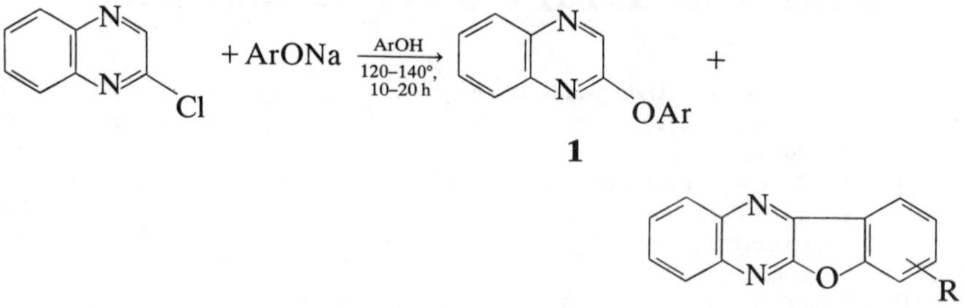

ArOH, 120–140°, 10–20 h

1

2

R = Me or MeO

2-Alkoxy-3-chloroquinoxalines have been prepared from 2,3-dichloroquinoxaline by treatment with one molecular proportion of sodium alkoxide,[2,7] and 2-aryloxy-3-chloroquinoxalines have been obtained from the dichloro compound by treatment with a phenol in boiling dimethylformamide in the presence of anhydrous sodium carbonate.[8]

2. Properties

2-Methoxyquinoxaline has been shown to undergo protonation[9] at position 4 and on treatment with hydrogen peroxide in glacial acetic acid to yield the 4-*N*-oxide.[3] Both 2-methoxy- and 2-phenoxyquinoxalines are hydrolyzed on treatment with hydrochloric acid to the corresponding quinoxalinones.[2] 2-Phenoxyquinoxaline is significantly less reactive to ammonium acetate than the corresponding cinnoline and phthalazine derivatives.[4] A number of 2-aryloxyquinoxalines have been found to undergo cyclization to benzofuro[2,3-*b*]quinoxalines in the presence of polyphosphoric acid (Scheme 2).[6]

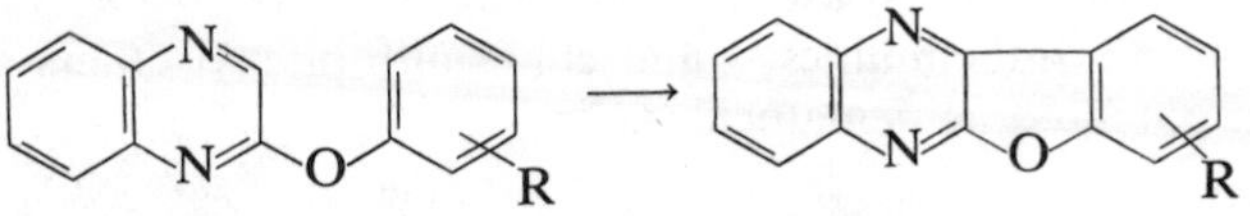

R = Me, diMe, or MeO

Scheme 2

II. 2,3-Dialkoxy- and 2,3-Diaryloxyquinoxalines

1. Methods of Preparation

2,3-Dialkoxyquinoxalines have been prepared by the reaction of sodium or potassium[10] alkoxides on 2,3-dichloroquinoxalines, or by primary synthesis from *o*-phenylenediamine and dithionoesters (Scheme 3).[21] 2,3-Dimethoxyquinoxaline is obtained as a minor product in the methylation of quinoxaline-2,3-dione with diazomethane.[2]

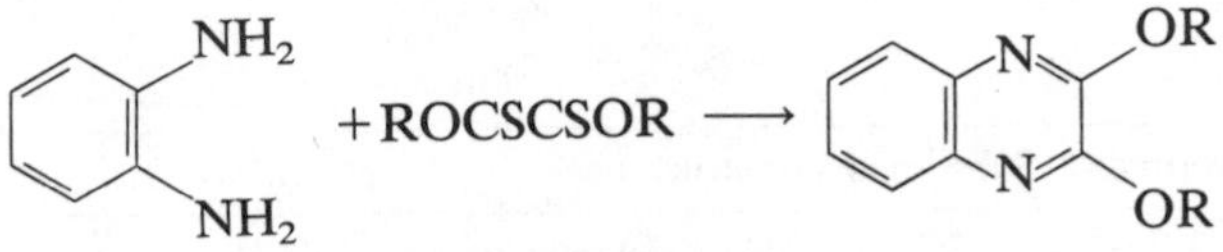

NH_2 NH_2 + ROCSCSOR ⟶ OR OR

Scheme 3

2,3-Diphenoxyquinoxaline and 2,3-bis(*p*-nitrophenoxy)quinoxaline have been prepared by dissolving potassium hydroxide in excess of the phenol and heating the resulting mixture with 2,3-dichloroquinoxaline. There is a specific warning of violent reactions when analogous preparations were attempted of the ethers derived from *o*-nitrophenol and 2,4-dinitrophenol.[11]

2. Properties

Both 2,3-dialkoxy- and 2,3-diaryloxyquinoxalines are reported to be relatively stable to oxidation to potassium permanganate.[11] 5-Amino-2,3-dimethoxyquinoxaline is readily hydrolyzed to 5-aminoquinoxaline-2,3-dione on heating with 6 *M* hydrochloric acid.[11]

3. Uses

Two alkoxyquinoxalinyl phosphorothioates (**3**) have been evaluated as insecticides and anthelmintics,[12] and chloronitrophenyl ethers such as **4** have been patented as nematocides.[13]

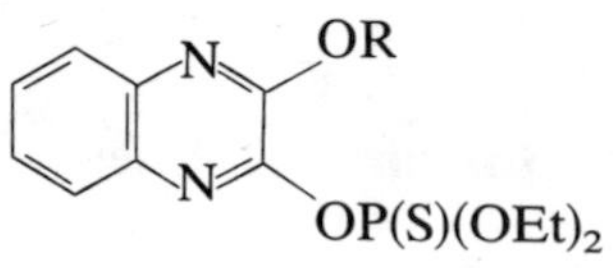

3 R = Me or Et

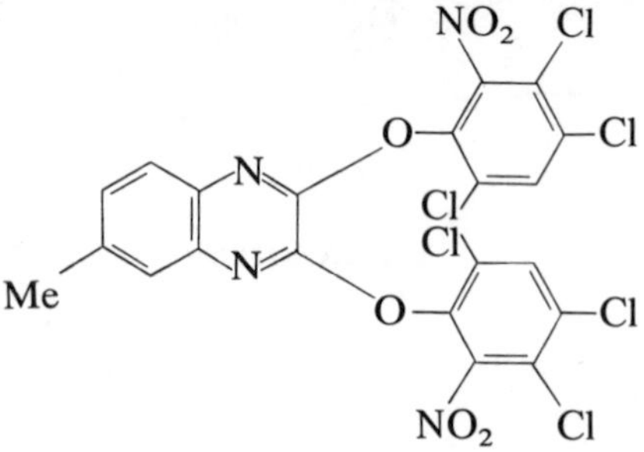

4

III. Tables

TABLE 1. 2(OR 3)-ALKOXY- AND 2(OR 3)-ARYLOXYQUINOXALINES

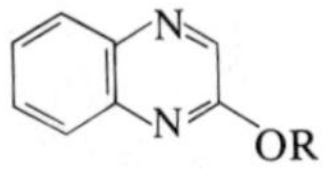

Molecular formula	Substituents or structure	m.p., °C	Ref.
$C_9H_4Cl_4N_2O$	5,6,7,8-Tetrachloro-2-methoxy	180–182	14
$C_9H_7ClN_2O$	2-Chloro-3-methoxy	74–75, 79–80	2, 7
$C_9H_7N_3O_3$	2-Methoxy-6-nitro	171–172	3
$C_9H_8N_2O$	2-Methoxy	31.5–33, b.p. 101–102/1.5 mm	3, 15
	Picrate	141–142	2
$C_9H_{10}N_4O$	2-Hydrazino-3-methoxy	168	16
$C_{10}H_6Cl_4N_2O$	5,6,7,8-Tetrachloro-2-ethoxy	171–173	14
$C_{10}H_9ClN_2O$	2-Chloro-3-ethoxy	71–73	7
$C_{10}H_9N_3O_3$	2-Methoxy-3-methyl-7-nitro	215	17
$C_{10}H_{10}N_2O$	2-Methoxy-3-methyl	68-69	2, 18
	Picrate	158	18
$C_{11}H_{11}ClN_2O$	2-Chloro-3-isopropoxy	56–57	7
$C_{11}H_{12}N_2O$	2-Ethoxy-3-methyl	48–49	18
	Picrate	119	18
$C_{12}H_{13}ClN_2O$	2-Butoxy-3-chloro	b.p. 151–155/1 mm	7
	2-*sec*-Butoxy-3-chloro	40–42	7
	2-Chloro-3-isobutoxy	b.p. 130–133/2 mm	7
$C_{13}H_{17}N_2O_4PS$	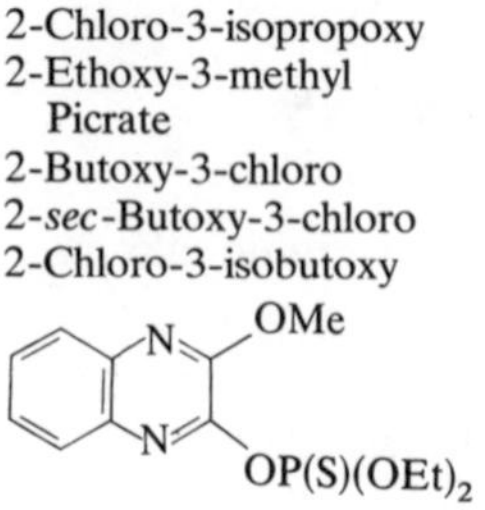 2-OMe, 3-OP(S)(OEt)$_2$ quinoxaline (structure)		12

TABLE 1 (*contd.*)

Molecular formula	Substituents or structure	m.p., °C	Ref.
$C_{14}H_8ClN_3O_3$	2-Chloro-3-(4-nitrophenoxy)	151–154	8
$C_{14}H_8Cl_2N_2O$	2-Chloro-3-(4-chlorophenoxy)	131–133	8
$C_{14}H_9ClN_2O$	2-Chloro-3-phenoxy	107–109	8
$C_{14}H_9N_3O_3$	7-Nitro-2-phenoxy	105–106	4
$C_{14}H_{10}N_2O$	2-Phenoxy	99–100, 100–101	4, 6
$C_{14}H_{11}N_3O$	7-Amino-2-phenoxy	126–128	4
$C_{14}H_{19}N_2O_4PS$	[quinoxaline structure: 2-OEt, 3-OP(S)(OEt)$_2$]		12
$C_{15}H_{11}ClN_2O$	2-Chloro-3-(4-methylphenoxy)	108–110	8
$C_{15}H_{12}N_2O$	2-Benzyloxy	49–51	6
	2-Methyl-3-phenoxy	101–102	6
	2-(2-Methylphenoxy)	87–88	6
	2-(3-Methylphenoxy)	100–102	6
	2-(4-Methylphenoxy)	90–92	6
$C_{15}H_{12}N_2O_2$	2-(3-Methoxyphenoxy)	120	6
$C_{16}H_{14}N_2O$	2-(2,3-Dimethylphenoxy)	124–125	6
	2-(2,5-Dimethylphenoxy)	82	6
	2-(3,5-Dimethylphenoxy)	72–73	6
$C_{18}H_{12}N_2O$	2-(1-Naphthalenyloxy)	118	6
	2-(2-Naphthalenyloxy)	145–147	6
$C_{20}H_{14}N_2O$	2-Phenoxy-3-phenyl	150, 152–153	5, 6

TABLE 2. 2,3-DIALKOXY-, 2-ALKOXY-3-ARYLOXY-, AND 2,3-DIARYLOXY QUINOXALINES

[quinoxaline structure: 2-OR, 3-OR]

Molecular formula	Substituents	m.p., °C	Ref.
$C_{10}H_6F_4N_2O_2$	5,6,7,8-Tetrafluoro-2,3-dimethoxy	146–148	19
$C_{10}H_9N_3O_4$	2,3-Dimethoxy-6-nitro	175–175.5	11
$C_{10}H_{10}N_2O_2$	2,3-Dimethoxy	90–92, 92–93	10, 11, 20
$C_{10}H_{11}N_3O_2$	5-Amino-2,3-dimethoxy	96–97	11
	6-Amino-2,3-dimethoxy	130–130.5	11
$C_{11}H_9F_3N_2O_3$	5,6,8-Trifluoro-2,3,7-trimethoxy	132–134	19
$C_{11}H_{12}N_2O_3$	2,3,6-Trimethoxy	127	11
$C_{12}H_{14}N_2O_2$	2,3-Diethoxy	72–74, 77–78	10, 11, 20
$C_{14}H_{18}N_2O_2$	2,3-Dipropoxy	48, 53–54	10, 11
	2,3-Diisopropoxy	93–94	10
$C_{15}H_{12}N_2O_2$	2-Methoxy-3-phenoxy	134–135.5	2

TABLE 1 (*contd.*)

Molecular formula	Substituents or structure	m.p., °C	Ref.
$C_{16}H_{14}N_2O_3$	2-Ethoxy-3-(2-hydroxyphenoxy)	143–144	20
$C_{16}H_{22}N_2O_2$	2,3-Dibutoxy	50–51	10
	2,3-Diisobutoxy	liq., n_D^{25} 1.5370	10
$C_{20}H_{12}N_4O_6$	2,3-Bis(4-nitrophenoxy)	215–216	11
$C_{20}H_{14}N_2O_2$	2,3-Diphenoxy	166	11
$C_{22}H_{18}N_2O_2$	2,3-Bis(3-methylphenoxy)	142–143	6

IV. References

1. J. C. E. Simpson, "Condensed Pyridazine and Pyrazine Rings," Interscience, London, 1953, p. 270.
2. G. W. H. Cheeseman, *J. Chem. Soc.*, 1804 (1955).
3. G. W. H. Cheeseman, *J. Chem. Soc.*, 1246 (1961).
4. C. M. Atkinson, C. W. Brown, and J. C. E. Simpson, *J. Chem. Soc.*, 26 (1956).
5. E. Hayashi and C. Iijima, *Yakugaku Zasshi*, **82,** 1093 (1962); *Chem. Abstr.*, **58,** 4551 (1963).
6. R. K. Anderson and G. W. H. Cheeseman, *J. Chem. Soc., Perkin I*, 129 (1974).
7. H. Saikachi and S. Tagami, *Yakugaku Zasshi*, **82,** 1246 (1962); *Chem. Abstr.*, **59,** 1634 (1963).
8. W. Deuschel and G. Riedel (to Badische Anilin- und Soda-Fabrik A.-G.), Ger. Pat. 1,135,471; *Chem. Abstr.*, **58,** 537 (1963).
9. M. M. Kaganskii, G. G. Dvoryantseva, and A. S. Elina, *Chem. Heterocycl. Comp.*, **9,** 369 (1973).
10. R. Patton and H. P. Schultz *J. Am. Chem. Soc.*, **73,** 5899 (1951).
11. H. I. X. Mager and W. Berends, *Rec. Trav. Chim.*, **78,** 5 (1959).
12. R. J. Magee (to American Cyanamid Co.) U.S. Pat. 3,634,425; *Chem. Abstr.*, **76,** 99709*s* (1972).
13. D. W. Gier and D. M. Wasleski (to Chemagro Corp.), U.S. Pat. 3,655,893; *Chem. Abstr.*, **77,** 44340*e* (1972).
14. D. E. Burton, A. J. Lambie, D. W. J. Lane, G. T. Newbold, and A. Percival, *J. Chem. Soc. (C)*, 1268 (1968).
15. G. W. H. Cheeseman, *J. Chem. Soc.*, 3236 (1957).
16. K. Asano and S. Asai, *Yakugaku Zasshi*, **79,** 658 (1959); *Chem. Abstr.*, **53,** 21979 (1959).
17. L. Horner, U. Schwenk, and E. Junghanns, *Justus Liebigs Ann. Chem.*, **579,** 212 (1953).
18. K. Adachi, *Yakugaku Zasshi*, **75,** 1426 (1955); *Chem. Abstr.*, **50,** 10105 (1956).
19. C. G. Allison, R. D. Chambers, J. A. H. McBride, and W. K. R. Musgrave, *J. Fluorine Chem.*, **1,** 59 (1971/2).
20. D. E. Ames and R. L. Ward, *J. Chem. Soc., Perkin I*, 534 (1975).
21. H. Hoppe and K. Hartke, *Arch. Pharm.* (Weinheim), **308,** 526 (1975); *Chem. Abstr.*, **83,** 178997g (1975).

CHAPTER XIV

2-Alkyl- and 2,3-Dialkylquinoxalines

I. Methods of Preparation 205
II. Properties 210
1. Physical Properties 210
2. Chemical Properties 211
III. Uses 221
IV. Tables 221
Table 1. 2-Alkylquinoxalines and Their Derivatives, Unsubstituted in the 3-Position 221
Table 2. 2,3-Dialkyl- and 2-Alkyl-3-arylquinoxalines and Their Derivatives 223
V. References 229

The Simpson monograph includes a chapter on 2-methyl- and 2,3-dimethylquinoxalines.[1]

I. Methods of Preparation

2-Methylquinoxaline has been prepared in 92% yield by the reaction of *o*-phenylenediamine and pyruvaldehyde sodium bisulfite[2] and in 76% yield by the reaction of the diamine with hydroxyiminoacetone.[3] The condensation of *o*-phenylenediamine with hydroxyimino ketones in general proceeds according to Scheme 1. The product, in some cases, is a

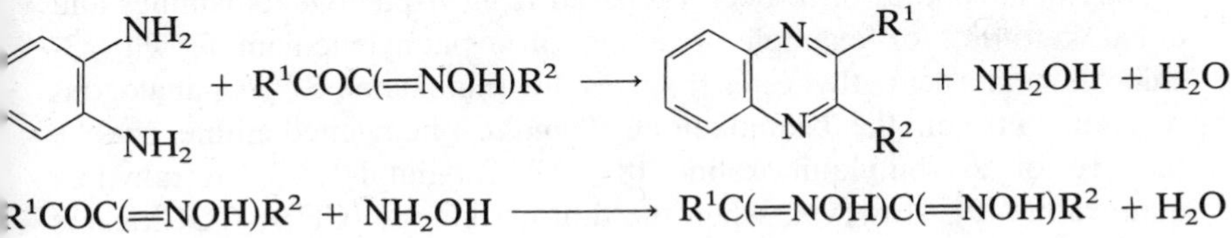

$$o\text{-}C_6H_4(NH_2)_2 + R^1COC(=NOH)R^2 \longrightarrow \text{2-}R^1\text{-3-}R^2\text{-quinoxaline} + NH_2OH + H_2O$$

$$R^1COC(=NOH)R^2 + NH_2OH \longrightarrow R^1C(=NOH)C(=NOH)R^2 + H_2O$$

Scheme 1

molecular compound of the quinoxaline with the dioxime, which is separable into its components by treatment with alkali. This reaction has been used by Landquist and his colleagues to prepare a large number of 2,3-dialkylquinoxalines.[4,5] Dioximes will themselves condense with *o*-phenylenediamine under suitable conditions; thus reaction of *o*-phenylenediamine and dimethylglyoxime in 2 *M* hydrochloric acid, followed by basification, yields 2,3-dimethylquinoxaline in 89% yield.[3]

Some hindered α-dicarbonyl compounds such as the α-diketone **1** fail to form quinoxalines. However carbonyl activation is achieved by nitro group substitution, and the diketones **2** and **3** are converted into the corresponding quinoxalines on reaction with *o*-phenylenediamine.[6]

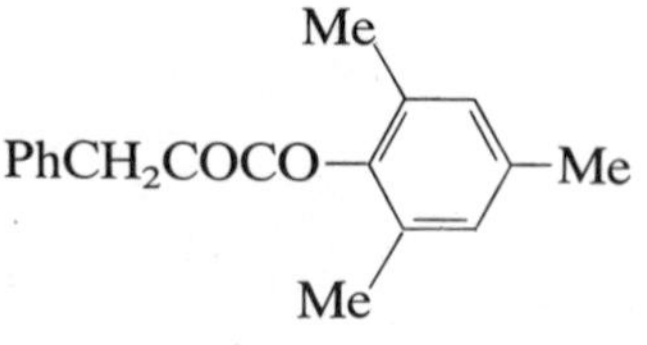

Me, $PhCH_2COCO$, Me, Me

1

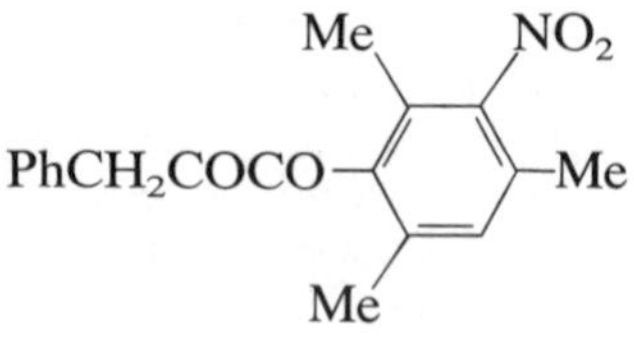

Me, NO_2, $PhCH_2COCO$, Me, Me

2

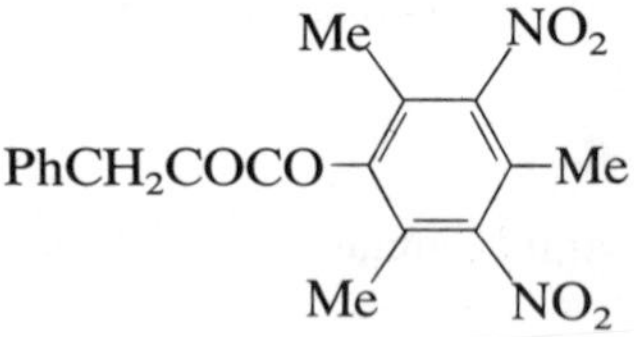

Me, NO_2, $PhCH_2COCO$, Me, Me, NO_2

3

An interesting variant on the general *o*-phenylenediamine-α-dicarbonyl procedure for forming quinoxalines is illustrated by the preparation of 2,3-di-*t*-butylquinoxaline (**4**)[7] from *o*-phenylenediamine and 3,3,6,6-tetramethyl-1-thiacycloheptane-4,5-dione X-Ray crystallography has shown that in this sterically hindered compound (**4**) both the C_2–C_3 and C–*t*-Bu bonds are elongated by about 0.05 Å and the exocyclic C_2–C_3–C–*t*-Bu bond angle is 129.7° rather than 120°.[8,9]

Alkylquinoxalines have been prepared from *o*-phenylenediamines and α-haloketones, for example, reaction of *o*-phenylenediamine with 3-chlorolevulinic acid ethyl ester (**5**) gives the quinoxaline **6**.[10] An analogous reaction between the bromoketone **7** and *o*-phenylenediamine gives a mixture of 2-*t*-butylquinoxaline (**8**) and 2-*t*-butyl-1,2,3,4-tetrahydroquinoxaline (**9**). However, when the dibromoketone ($CHBr_2COCMe_3$) is used, only 2-*t*-butylquinoxaline is obtained.[11]

Deactivated Rainey Ni

4

5

6

7

8

9

Aminoalkylquinoxalines (**17**) have been derived from α-amino acids (**10**) as shown in Scheme 2, the amino group is protected by formation of a phthalimide derivative (**11**). The intermediate chloromethyl ketone (**12**) is converted via a pyridinium salt (**13**) to a nitrone (**14**) which on hydrolysis yields the α-dicarbonyl compound (**15**) required for diamine condensation. Finally the phthalimide residue is removed from the condensation product (**16**) by hydrazine treatment.[12,13]

Alkylquinoxalines have also been prepared by various transformations of preformed quinoxalines. Thus 2-carboxy groups are usually readily decarboxylated and 2-chloro groups removed by catalytic hydrogenation. Illustrative of this approach is the reduction of 2-chloro-3-methylquinoxaline to 2-methylquinoxaline.[14]

2-Chloroquinoxalines react with substituted acetonitriles in the presence of sodium amide to give cyanoalkylquinoxalines (e.g., **18**).[15,16]

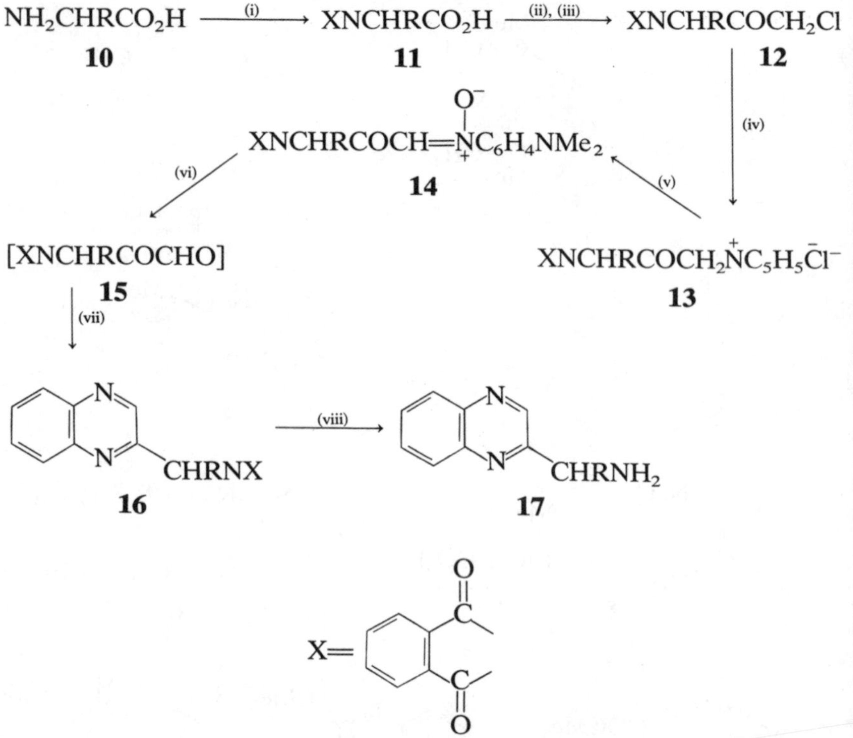

NH_2CHRCO_2H (10) —(i)→ $XNCHRCO_2H$ (11) —(ii), (iii)→ $XNCHRCOCH_2Cl$ (12) —(iv)→ $XNCHRCOCH_2\overset{+}{N}C_5H_5Cl^-$ (13) —(v)→ $XNCHRCOCH{=}\overset{+}{N}(O^-)C_6H_4NMe_2$ (14) —(vi)→ [XNCHRCOCHO] (15) —(vii)→ 2-(CHRNX)quinoxaline (16) —(viii)→ 2-($CHRNH_2$)quinoxaline (17)

X= phthaloyl (1,2-$C_6H_4(CO)_2$)

Reagents (i) Phthalic anhydride
(ii) CH_2N_2
(iii) HCl
(iv) Pyridine
(v) p-$ONC_6H_4NMe_2$
(vi) Aqueous HCl
(vii) o-Phenylenediamine
(viii) NH_2NH_2

Scheme 2

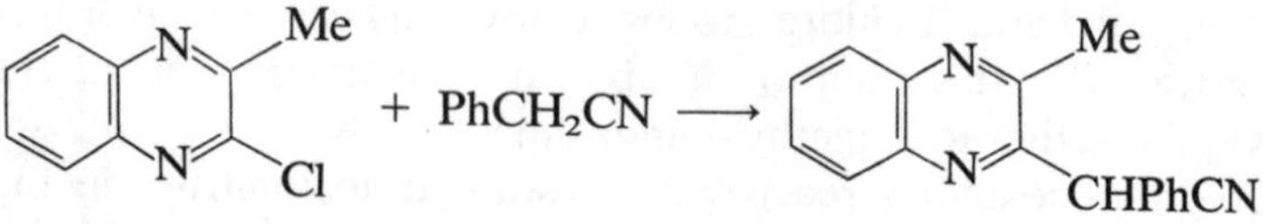

2-Chloro-3-methylquinoxaline + $PhCH_2CN$ ⟶ 2-(CHPhCN)-3-methylquinoxaline (18)

Alkylquinoxalines have also been prepared by homolytic substitution. Thus reaction of quinoxaline with a mixture of the appropriate carboxylic acid and its sodium salt in the presence of aqueous ammonium peroxidisulfate gives the 2-alkylated quinoxaline. Alkylation can also be achieved by the use of the acid and salt in the presence of trifluoroacetic acid and lead(IV) acetate in benzene (Scheme 3).[17,18] 2-Benzylquinoxaline is also conveniently prepared by homolytic substitution employing dibenzylmercury in acetic acid as the radical source.[19]

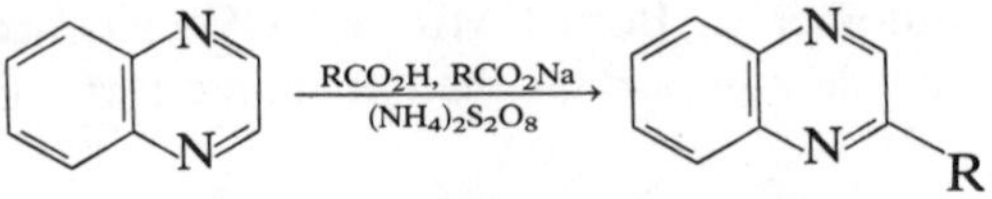

RCO_2H, RCO_2Na, $(NH_4)_2S_2O_8$

R = *n*-Pr, isoPr, or *t*-Bu

Scheme 3

Landquist and Stacey have also investigated the possibility of obtaining alkylquinoxalines from the reaction of Grignard reagents with quinoxaline-2,3-dione (**19**) and its 1,4-dimethyl derivative (**22**). The dione and ethylmagnesium bromide give a complex mixture of products from which some 2,3-diethylquinoxaline (**20**) and 3-ethylquinoxalin-2-one (**21**) are obtained. A product analyzing for the trihydrate of **23** is isolated from the 1,4-dimethyl compound and ethylmagnesium bromide. Hydrogenation of 2-styrylquinoxalines over Raney nickel gives 2-aralkylquinoxalines, or 2-aralkyl-1,2,3,4-tetrahydroquinoxalines if the 3-position is unoccupied.[4]

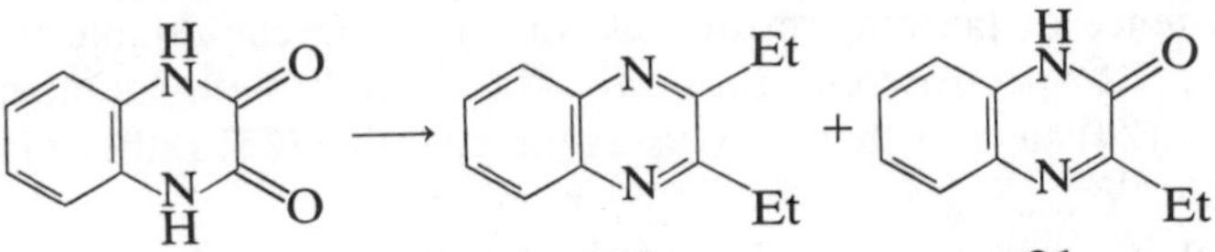

19 → **20** + **21**

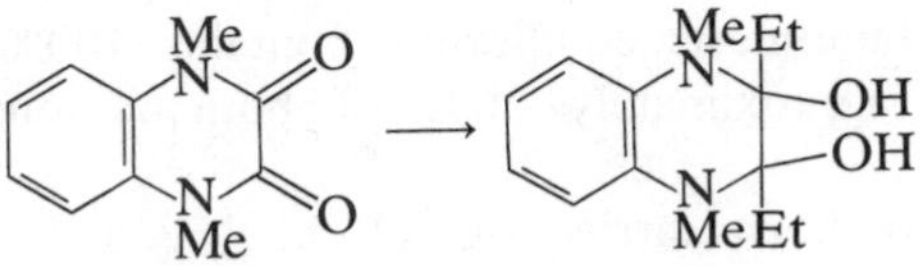

22 → **23**

II. Properties

1. Physical Properties

^1H NMR and UV studies have been made of the tautomerism of compounds of type **A** and **C** (Scheme 4). These measurements reveal a noticeable contribution of structures **B** and **D** to some of the tautomeric equilibria. It is often difficult to distinguish between tautomers of types **D** and **E**; however in the case of $R = CH_3$, the observation of allylic coupling is indicative of tautomer **E**. Both NMR and UV evidence suggests that enolic forms (**E**) are less important than enaminic forms (**D**).[20]

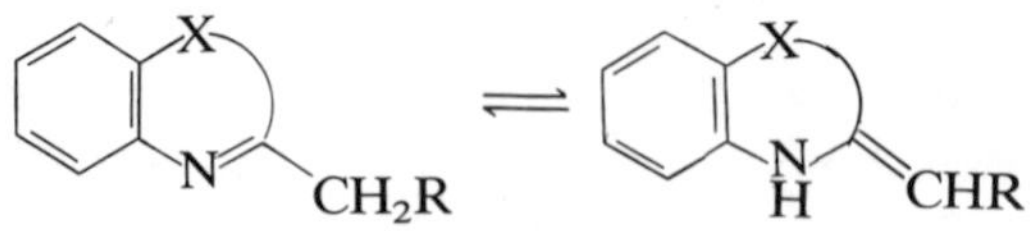

CH_2R **A** ⇌ CHR **B**

X = —N=CR— or —NH—CO— R = H, CN, or CO_2R^1

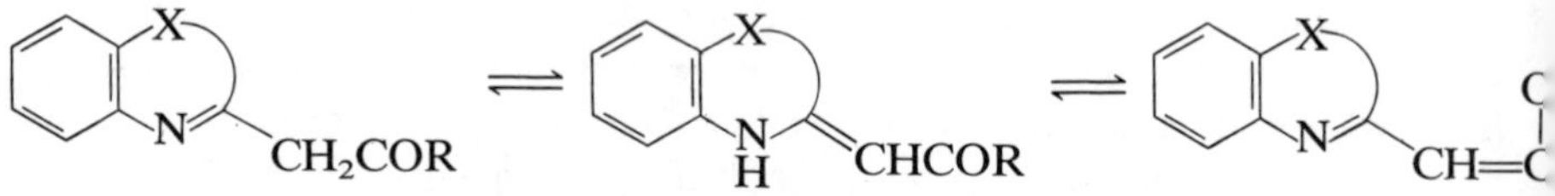

CH_2COR **C** ⇌ $CHCOR$ **D** ⇌ CH=C **E**

X = —N=CR— or —NH—CO— R = Me or CO_2R^1

Scheme 4

In the case of 2-methylquinoxalines (system **A**) there is no evidence for the existence of tautomeric form **B** with an exocyclic double bond. The UV and IR spectroscopic properties of ethyl 3-methylquinoxalin-2-yl pyruvate (**24**) suggest that it exists as the enamine (**25**) rather than as the enol (**26**).[21,22]

Variable-temperature studies of the NMR and IR absorption properties of compounds such as 3,4-dihydro-3-oxo-2-carbethoxymethylquinoxaline (**27**), which is in tautomeric equilibrium with the enamine **28,** show that the enamine is less favored at high temperature. Thus in DMSO at 40°, the tautomeric equilibrium contains 100% of the enamine, but at 140° approximately 50% of both tautomers are present.[23]

Systematic studies have been carried out of the UV and ^1H NMR spectra of a wide range of 5- and 6-substituted 2,3-dimethylquinoxalines and also of some 5,8- and 6,7-disubstituted 2,3-dimethylquinoxalines.[24] The ionization properties of these compounds have also been studied.[25–28] 2,3-Dimethylquinoxaline has a first ionization constant of 2.08 pK_a units and a second ionization constant of $-3.84\ pK_a$ units. The variations in first and second ionization constants for a series of 6-substituted 2,3-dimethylquinoxalines are tabulated in Figure 1. These are broadly predictable from the known electron-donating or electron-attracting properties of the groups. Thus electron-donating groups are found to enhance basic strength, whereas electron-withdrawing groups have the opposite effect.

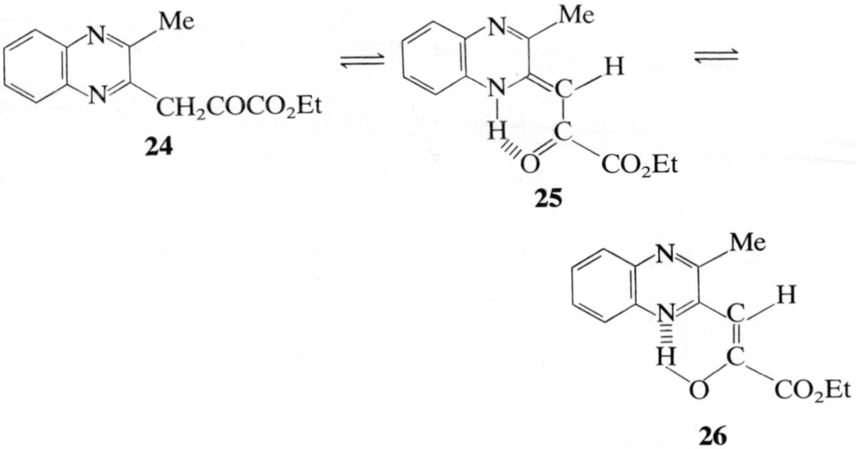

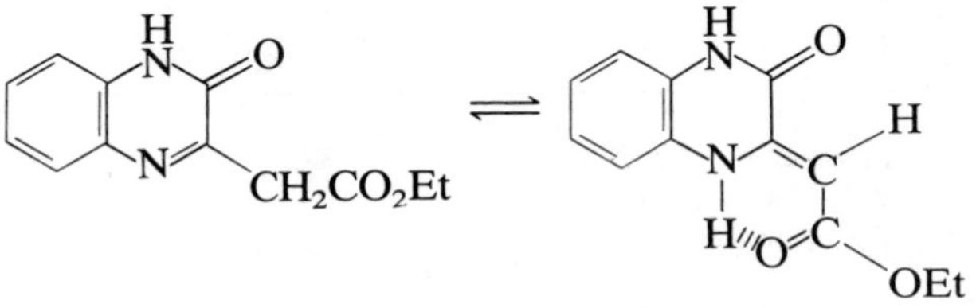

The first ionization constants of a number of 5-substituted 2,3-dimethylquinoxalines are given in Figure 2.

2. Chemical Properties

The following discussion of the reactivity of 2-alkyl- and 2,3-dialkylquinoxalines is mainly illustrated by the reactions of 2-methyl- and 2,3-dimethylquinoxaline. The pattern of reactivity which emerges is

FIGURE 1. IONIZATION CONSTANTS OF 6-SUBSTITUTED 2,3-DIMETHYLQUINOXALINES

Substituent	First ionization constant[a]	Second ionization constant[a]
H	2.08	−3.84
NH_2	3.44	0.22[b]
EtO	2.48	−3.76
MeO	2.45	−3.47
Me	2.41	−3.45
t-Bu	2.36	−3.37
F	1.48	−4.16
Cl	1.29	−4.21
Br	1.22	−4.38
I	1.34	−4.18
CO_2H	1.05[c]	−4.30
CN	0.41	−4.92
NO_2	0.07	−5.02

[a] In water at 25°.
[b] Third ionization constant is −5.10.
[c] For proton gain.

FIGURE 2. IONIZATION CONSTANTS OF 5-SUBSTITUTED 2,3-DIMETHYLQUINOXALINES

Substituent	First ionization constant[a]
NH_2	3.27[b]
MeO	2.09
Me	2.46
Cl	0.96
Br	2.16
NO_2	0.12

[a] In water at 25°
[b] Second ionization constant is 0.16

broadly similar to that of other methylazines and methyldiazines. Thus 2-methyl- and 2,3-dimethylquinoxaline react with a variety of electrophilic reagents such as aldehydes, diazonium compounds, and bromine. They undergo oxidation to the corresponding aldehydes and under basic conditions form anions which may be acylated or alkylated. An additional feature of the reactions of methylquinoxalines is their addition reactions with *N*-phenylmaleimide, maleic anhydride, and acetylenic esters.

Reaction of 2-methylquinoxaline with formaldehyde gives 26% of the monohydroxymethylene derivative **29** and 6% of the bishydroxymethylene derivative **30**. Similar reaction of 2,3-dimethylquinoxaline gives 17% of the symmetric bishydroxymethylene derivative **31**, 4% of the unsymmetric bishydroxymethylene derivative **32,** and in addition 5% of the trishydroxymethylene derivative **33**.[29]

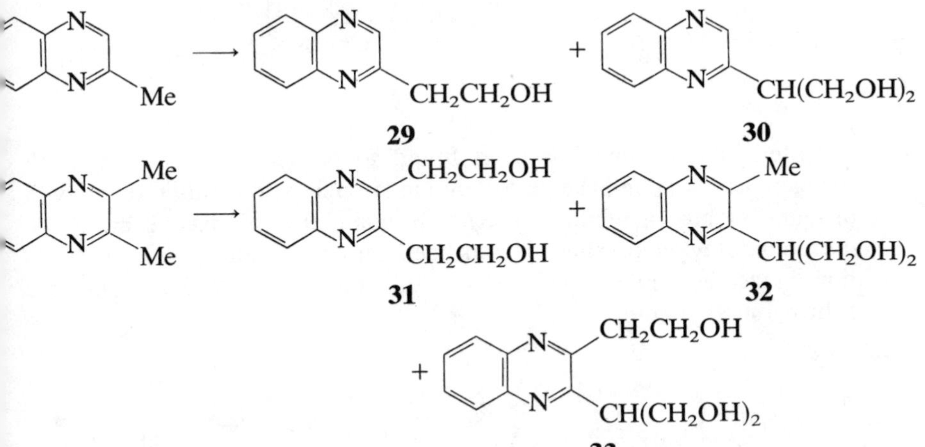

Me → CH_2CH_2OH (**29**) + $CH(CH_2OH)_2$ (**30**)

Me, Me → CH_2CH_2OH, CH_2CH_2OH (**31**) + Me, $CH(CH_2OH)_2$ (**32**)

+ CH_2CH_2OH, $CH(CH_2OH)_2$ (**33**)

With chloral, 2-methylquinoxaline and 2-methyl-3-phenylquinoxaline form monoadducts. Both mono- and diadducts have been obtained from 2,3-dimethylquinoxaline. Products of this type are readily converted by alkaline hydrolysis to the corresponding α,β-unsaturated acids, for example, the conversion of the adduct **34** to the acid **35**.[2,21,30]

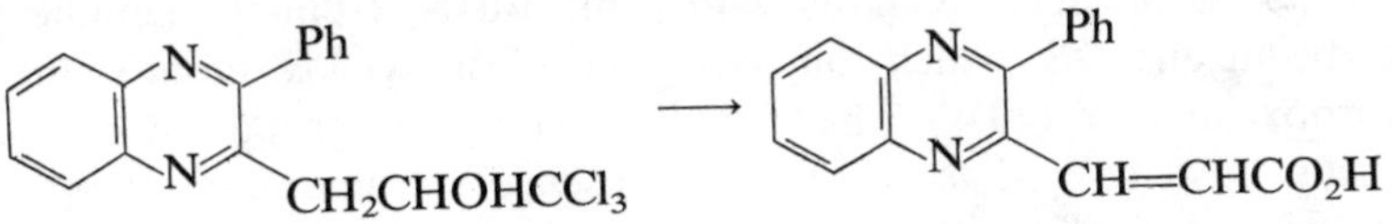

Ph, $CH_2CHOHCCl_3$ (**34**) → Ph, $CH{=}CHCO_2H$ (**35**)

There are numerous reports of the formation of styrylquinoxalines* from the reaction of methylquinoxalines with aromatic aldehydes. In nearly all cases reaction is carried out in boiling acetic anhydride; in one

* Styrylquinoxalines are listed in Tables 1 and 2 as arylidene derivatives of the parent methylquinoxaline.

case 10% potassium methoxide in methanol was used as the condensation medium.[4,31] 2,3-Dimethylquinoxalines form both mono- and distyrylquinoxalines depending on the conditions. 2,3-Dimethylquinoxaline itself couples with aryldiazonium salts in both acid and alkaline medium; the structure **36** of the compounds obtained was confirmed by independent synthesis from 3-methylquinoxaline-2-carboxaldehyde and the appropriate phenylhydrazine.[32]

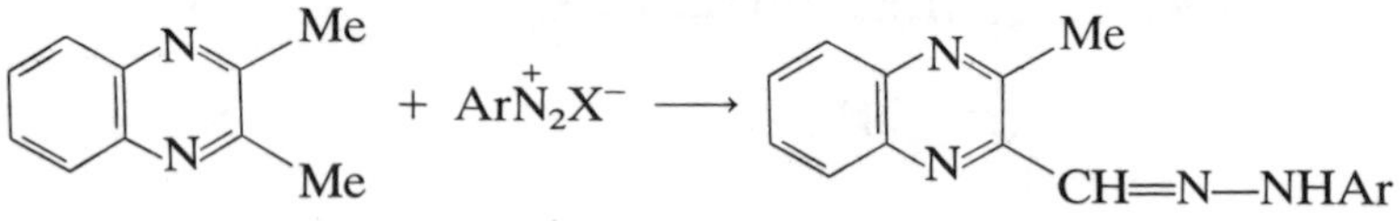

Me, Me + $Ar\overset{+}{N}_2X^-$ ⟶ Me, CH=N—NHAr

36

2-Methylquinoxaline has been found to participate in the Mannich reaction, as shown in Scheme 5,[33] and in strongly acid medium reacts with bromine to give a mixture of 37% bromomethylquinoxaline and 27% dibromomethylquinoxaline. It had been shown previously that bromination in acetic acid in the presence of sodium acetate gives only the tribromomethyl compound.[34,35]

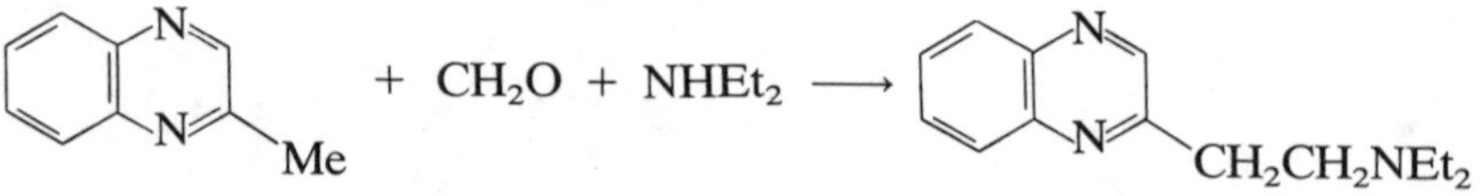

Me + CH_2O + $NHEt_2$ ⟶ $CH_2CH_2NEt_2$

Scheme 5

Bromination of 2,3-dibenzylquinoxaline with excess of *N*-bromosuccinimide in carbon tetrachloride yields a mixture of *meso*- and *dl*-2,3-bis(α-bromobenzyl)quinoxalines.[36] When 2,3-dimethylquinoxaline is treated in pyridine solution with powdered iodine, the bispyridinium iodide **37** is formed. Reaction with *p*-nitroso-*N,N*-dimethylaniline gives the bisnitrone **38** which on acid hydrolysis yields quinoxaline-2,3-dicarboxaldehyde (**40**).[37] The activated methylene groups of compound **37** condense with diacetyl in the presence of piperidine to give the phenazine derivative **39**.

2,3-Bis(bromomethyl)quinoxaline (**41**), prepared by direct synthesis from *o*-phenylenediamine and 1,4-dibromo-2,3-butanedione ($BrCH_2COCOCH_2Br$),[38] can be similarly converted into the bispyridinium bromide **42**, and this on treatment with *p*-nitroso-*N,N*-dimethylaniline and sodium cyanide yields the biscyanoanil **44**.[38] Reaction with secondary amines such as diethylamine yields pyrroline derivatives (e.g., **43,** see chapter XXXV).[143]

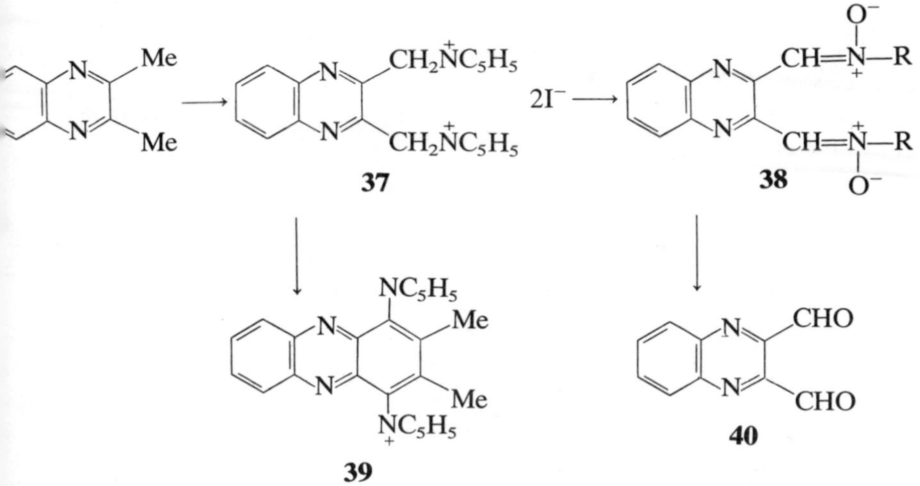

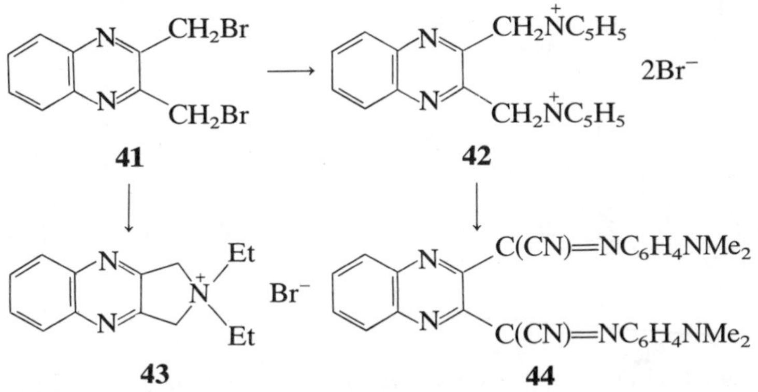

Selenium dioxide oxidation of 2-methylquinoxaline yields quinoxaline-2-carboxaldehyde,[5] and 3-methylquinoxaline-2-carboxaldehyde has been similarly prepared from 2,3-dimethylquinoxaline.[39] Dimethylsulfoxide oxidation of 2,3-bis(bromomethyl)quinoxaline yields the furoquinoxaline **45**, and this on sublimation gives quinoxaline-2,3-dicarboxaldehyde. 3-Dibromomethylquinoxaline-2-carboxaldehyde (**46**) is a by-product of the oxidation reaction.[39] Direct selenium dioxide oxidation of 2,3-dimethylquinoxaline appears to be a less satisfactory method of preparing the dialdehyde.[37]

Oxidation of 2,3-dimethylquinoxaline with 20% aqueous nitric acid gives a mixture of products from which quinoxaline-2,3-dicarboxylic acid (**47**), 6-nitroquinoxaline-2,3-dione (**48**), and 6,7-dinitroquinoxaline-2,3-dione (**49**) are isolated. The probable sequence of reactions is as shown.[40]

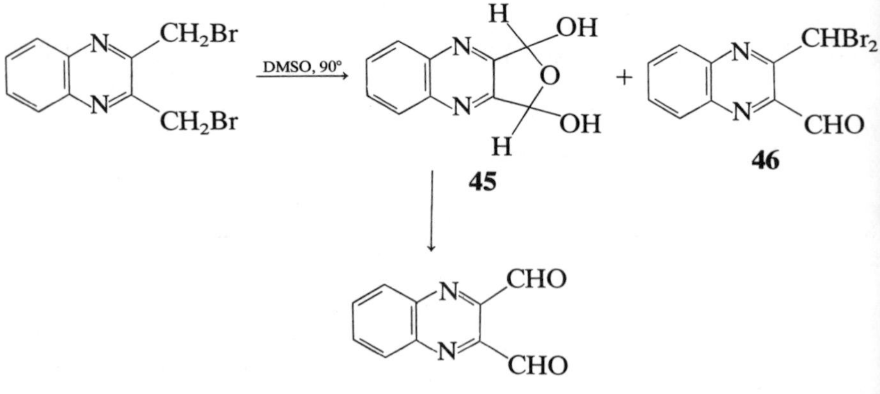

2,3-Dimethylquinoxaline is however resistant to attempted nitration at 100° with mixed fuming nitric acid and concentrated sulfuric acid.[41]

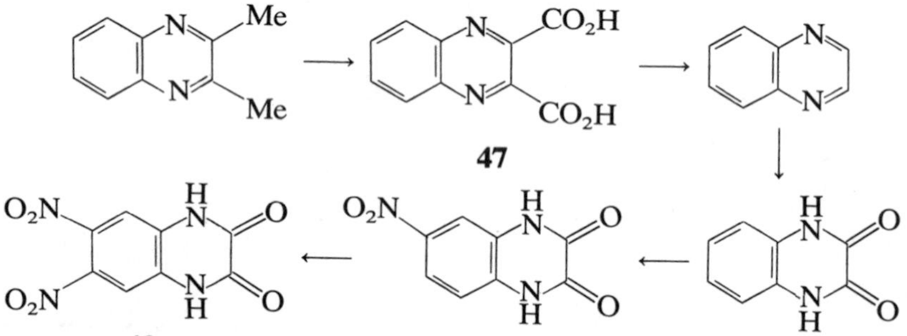

Potassium permanganate oxidation of 2-methylquinoxaline under carefully controlled conditions yields 70–75% 2-methyl-5,6-pyrazinedicarboxylic acid.[42] The further oxidation of 2-methylquinoxaline to 2,5,6-pyrazinetricarboxylic acid has also been described.[43]

The H–D exchange rates of a series of α-methylazines, including 2-methylquinoxaline, have been determined.[44] The anion of 2-methylquinoxaline generated in liquid ammonia with potassamide reacts with n-propyl nitrate to give 2-nitromethylquinoxaline (**50**) in 58% yield.[45,46] 2-Dichloronitromethylquinoxaline (**51**) is formed when 2-nitromethylquinoxaline is treated with sodium hypochlorite.[47] Treatment of 2,3-dimethylquinoxaline with sodamide in hexamethylphosphoramide (HMPT) similarly gives a dianion which on reaction with methyl iodide is converted into 2,3-diethylquinoxaline in 75% yield.[48]

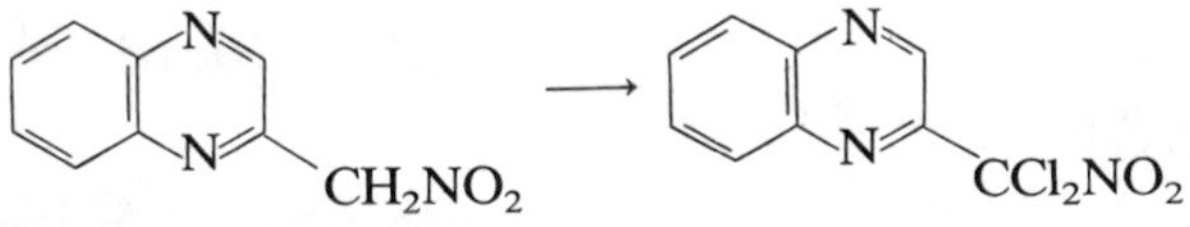

50 → 51

Alkyl and arylmagnesium halides react with 2-methylquinoxaline by addition of one mole of reactant to the 3,4-bond. After hydrolysis the 2-alkyl- or 2-aryl-1,2-dihydro-3-methylquinoxalines (**52**) are obtained. When ethylmagnesium bromide is used a dimeric by-product (**53**) is also isolated.[49,50] Reaction of 2,3-dimethylquinoxaline with benzonitrile and lithium amide gives 1-amino-1-phenyl-2-(3-methyl-2-quinoxalinyl)-ethylene (**54**).[51] The mono- and dilithium salts of 2,3-dimethylquinoxaline have been generated from the quinoxaline by reaction with one or two equivalents of lithium diisopropylamide ($LiNPr^i_2$), respectively. These salts have been reacted with a variety of electrophilic reagents such as alkyl halides, aryl ketones, esters, and nitriles.[145]

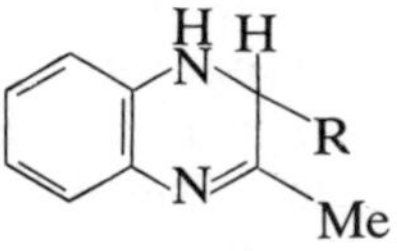

52

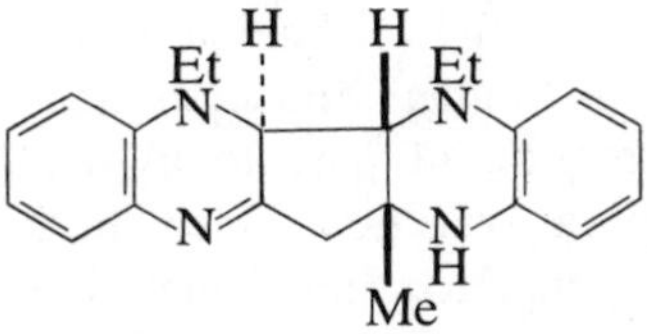

53

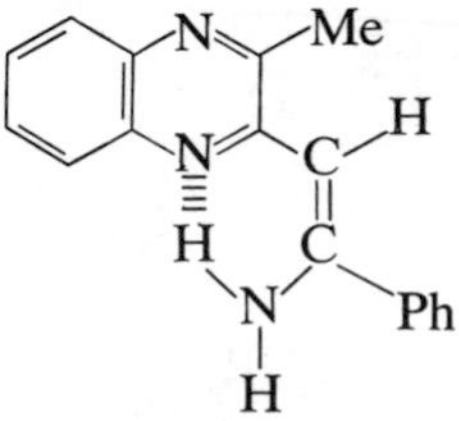

54

2-Methyl- and 2,3-dimethylquinoxalines have been acylated with benzoate, trifluoroacetate, oxalate, and phthalate esters using sodium hydride as the condensing agent. Both mono- and diacyl derivatives are obtained from 2,3-dimethylquinoxaline; with excess of methyl benzoate and diethyl oxalate the products **55** and **56** are obtained, respectively. Compound **57** is produced by condensation of the dimethyl compound with diethyl phthalate.[52]

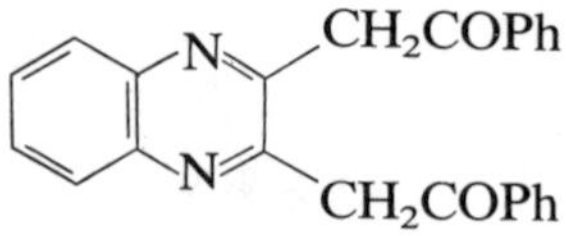

The product formed when 2,3-dimethylquinoxaline reacts with *N*-phenylmaleimide has been reformulated as compound **58**[53,54] and that from the quinoxaline and maleic anhydride, as compound **60**. In both reactions the initial step appears to be one of Michael addition; but whereas the Michael adduct **58** survives, the succinic anhydride ring in **59** opens to give, after hydrogen migration, the pyrroloquinoxaline **60**.[21] The formation of pyrroloquinoxalines by this route appears to be a fairly general reaction since both 2-methylquinoxaline[21] and 2-methyl-3-phenylquinoxaline[30] have been shown to add to maleic anhydride, and 2,3-dimethylquinoxaline has been shown to add to monosubstituted maleic anhydrides.[55]

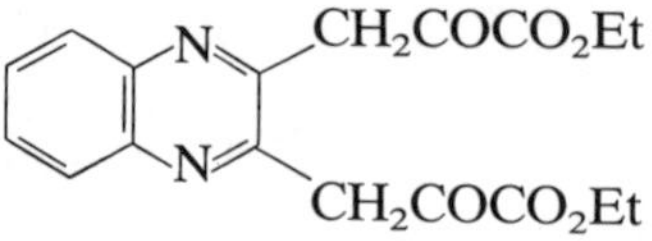

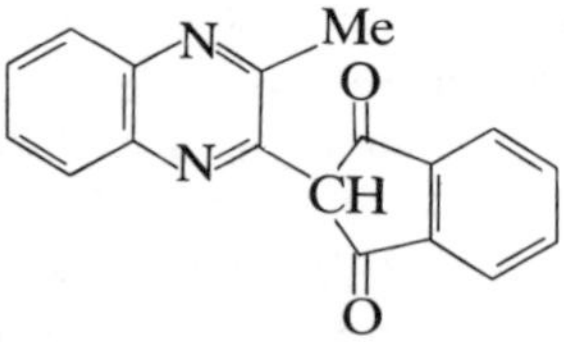

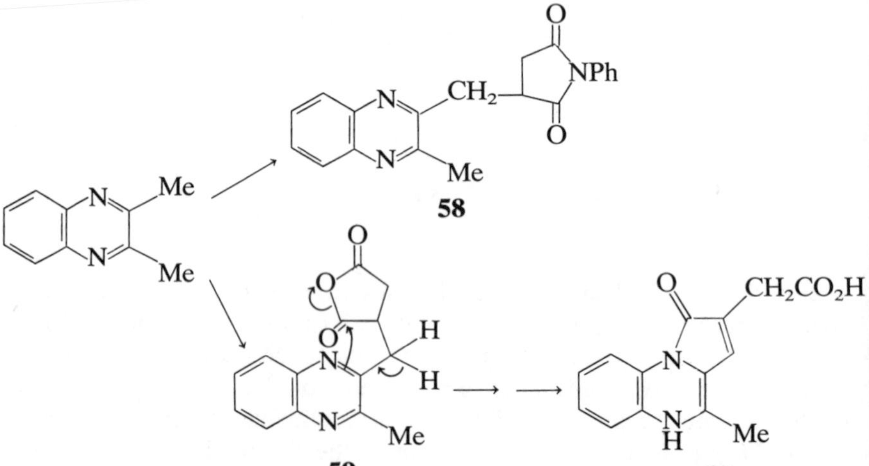

2-Methyl- and 2,3-dimethylquinoxaline also undergo interesting addition reactions with dimethyl acetylenedicarboxylate. Addition of 2-methylquinoxaline to two molecules of the ester results in the formation of the isomeric azepino[1,2-*a*]quinoxalines **61** and **62**. Addition is accompanied by the migration of an ester group to the extranuclear carbon of the quinoxaline.[56] These adducts are structurally analogous to those previously isolated from 6-bromo-2-methylquinoline and the acetylenic ester. The azepino[1,2-*a*]quinoxaline **63** is obtained from the reaction of 2,3-dimethylquinoxaline with dimethyl acetylenedicarboxylate, but in this case the second product is a dipyrido[1,2-*a*;2,1-*c*]quinoxaline (either compound **64** or **65**). The pyridone ring is formed by incorporation of an ester carbonyl carbon into the six-membered ring.

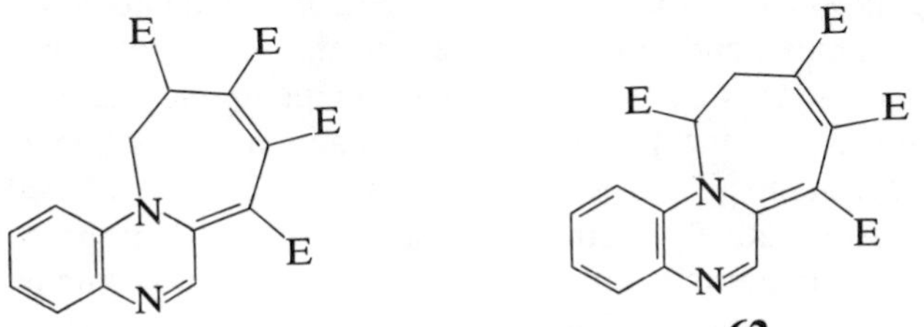

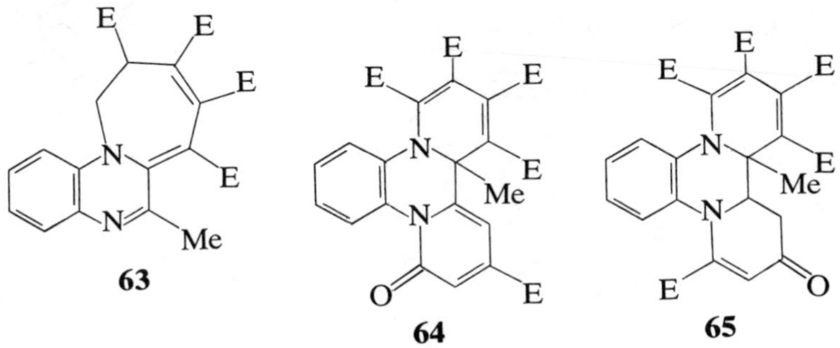

61 **62**

63 **64** **65**

$E = CO_2Me$

2,3-Dimethylquinoxaline and benzoquinone undergo reaction in boiling toluene to give a 2:1 complex of the quinoxaline and hydroquinone. This was originally incorrectly formulated as a Diels-Alder adduct, as were the addition products of the quinoxaline with *N*-phenylmaleimide and maleic anhydride. Identical material is obtained by crystallizing hydroquinone from toluene containing an excess of 2,3-dimethylquinoxaline.[53] With tetrachloro-*o*-benzoquinone, 2,3-dimethylquinoxaline forms the dioxole **66**. A similar product is formed from 2-methylquinoxaline, but a second, colored product (**67**) is also formed.[57] The chromophore in **67** is similar to that in methylquinoxaline orange (**68**), a by-product in the preparation of 2-methylquinoxaline from *o*-phenylenediamine and isonitrosoacetone in acid solution. Methylquinoxaline orange is also formed when 2-methylquinoxaline is heated at 200°, preferably in the presence of a palladium on carbon catalyst or when methylquinoxaline is heated under reflux with phosphoryl chloride.[49] Alternatively methylquinoxaline orange is prepared by mild reaction of 2,3-dimethylquinoxaline with 2-chloroquinoxaline in the presence of an acid catalyst. The latter reaction is a general one for 2-methylquinoxalines and for other types of methylbenzodiazines.[58]

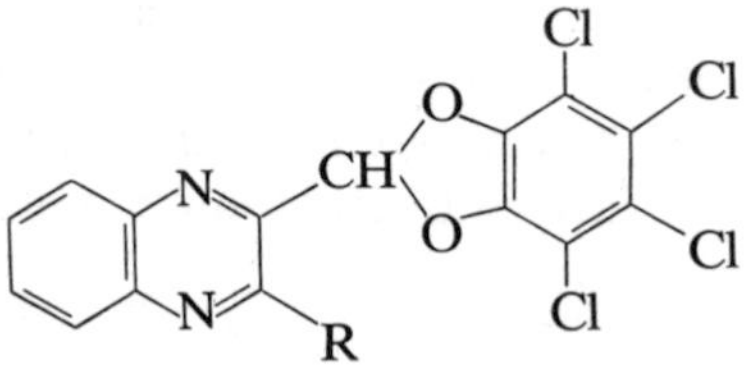

66 R = H or Me

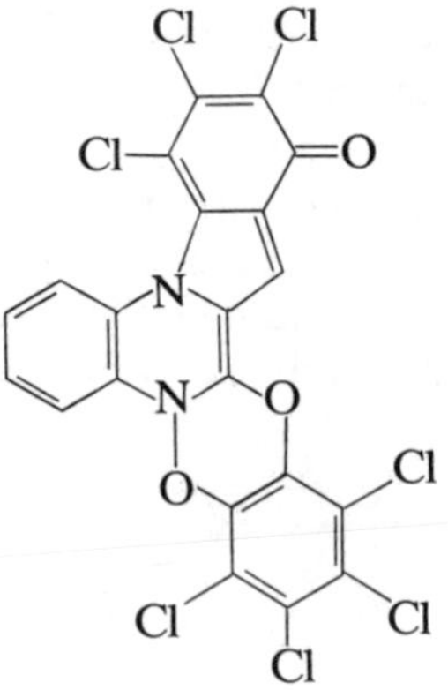

67

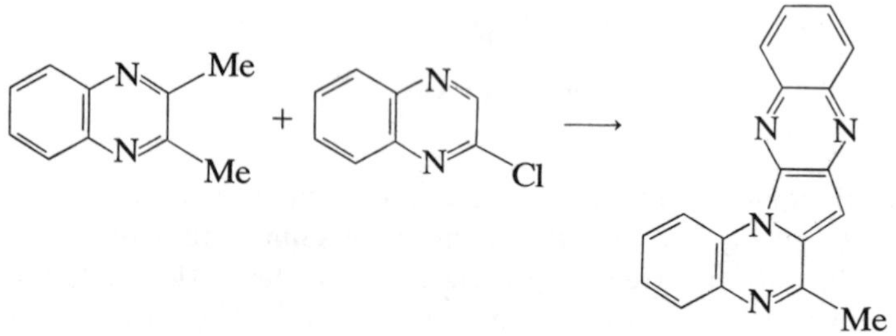

68

III. Uses

There are various references to the biological activity of 2-alkyl- and 2,3-dialkylquinoxalines and their derivatives. For example, 2-(dihalonitromethyl)quinoxalines have antimicrobial activity and are useful as germicides.[47] 2,3-Bis(thiomethyl)quinoxalines (e.g., **69**, R = nicotinoyl) have antiviral activity,[59] and compounds such as 2-(*p*-diethylaminoacetamidophenyl)-3-methylquinoxaline (**70**) have psychopharmacological effects.[60] Various 2,3-bis(halomethyl)quinoxalines, for example, 6-chloro-2,3-bis(chloromethyl)quinoxaline, have been patented for use as foliar fungicides.[61,62] 6-Chloro-5-methoxy-2,3-dimethylquinoxaline and related compounds are catalysts for photographic silver dye bleach baths.[63,64]

CH_2SR CH_2SR

69

Me C_6H_4NCOMe NEt_2

70

IV. Tables

TABLE 1. 2-ALKYLQUINOXALINES AND THEIR DERIVATIVES UNSUBSTITUTED IN THE 3- POSITION

R

Molecular formula	Substituents or structure	m.p., °C	Ref.
$C_9HCl_7N_2$	5,6,7,8-Tetrachloro-2-trichloromethyl	186–187	65
$C_9H_4Cl_4N_2$	5,6,7,8-Tetrachloro-2-methyl	174–175	65
$C_9H_5Cl_2N_3O_2$	2-Dichloronitromethyl		47
$C_9H_6Br_2N_2$	2-Dibromomethyl	120–121	34
$C_9H_6N_4O_4$	2-Methyl-6,8-dinitro	166–168	66
$C_9H_7BrN_2$	2-Bromomethyl	67–68	34
$C_9H_7ClN_2$	6-Chloro-2-methyl	131, 133–134	67, 68
	7-Chloro-2-methyl	91	67
$C_9H_7N_3O_2$	2-Nitromethyl	121, 122–123	45, 46, 69
	2-Methyl-6-nitro	170.5–171	70
	Benzylidene deriv.	200–203	71
$C_9H_8N_2$	2-Methyl	b.p. 110–111/15 mm, 125–127/11 mm	2, 3, 4, 14

TABLE 1 (*contd.*)

Molecular formula	Substituents or structure	m.p., °C	Ref.
$C_9H_8N_2$ (*contd.*)	Benzylidene deriv.	105, 107	4, 31, 72
	o-Chlorobenzylidene deriv.	116.5–118.5	73
	m-Chlorobenzylidene deriv.	159–161	73
	3,4-Dimethoxybenzylidene deriv.	154–156	73
	p-Dimethylaminobenzylidene deriv.	166–167	74
	3,4-Methylenedioxybenzylidene deriv.	100–101	73
	p-Methoxybenzylidene deriv.	129	31
	p-Nitrobenzylidene deriv.	107	31
$C_9H_8N_2O$	2-Hydroxymethyl	79–81[a]	5, 75
	Acetyl	49–51	5
$C_9H_9N_3$	2-Aminomethyl		
	Hydrochloride	205–207 decomp.	12
	5-Amino-2-methyl	83–84	28
	Acetyl	108–109	28
	6-Amino-2-methyl	162–163, 164–164.5	28, 71
	Dihydrate	149–152	76
	Hydrochloride	218–222 decomp.	71
	Picrate	212–214	71
	Acetyl	205.5–206	28
	7-Amino-2-methyl	177–177.5, 177.5–178	28, 71
	Hydrochloride	215–220 decomp.	71
	Picrate	232–235 decomp.	71
	Acetyl	162–163	28
	8-Amino-2-methyl	113–115	28
	Acetyl	167–168	28
$C_{10}H_7N_3$	2-Cyanomethyl	116–117	77
$C_{10}H_{10}N_2$	2,7-Dimethyl	74–75	78
	2-Ethyl	b.p. 97–100/3 mm, 113–120/8 mm	4, 79
	Picrate	123 decomp.	79
$C_{10}H_{10}N_2O$	2-(2-Hydroxyethyl)	78–79.5	29
$C_{10}H_{11}N_3$	2-(1-Aminoethyl)	194–196	12
$C_{11}H_8Cl_4N_2$	5,6,7,8-Tetrachloro-2-propyl	121–122	65
$C_{11}H_8N_2O_3$	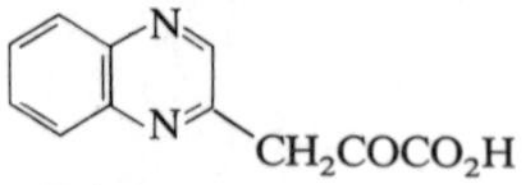 Quinoxalin-2-yl–CH_2COCO_2H (structure)		
	Ethyl ester	162–163	80
$C_{11}H_{12}N_2$	2-Isopropyl	b.p. 97–103/5 mm	79
	Picrate	113–114	79
	2,6,7-Trimethyl	116	4
$C_{11}H_{12}N_2O_2$	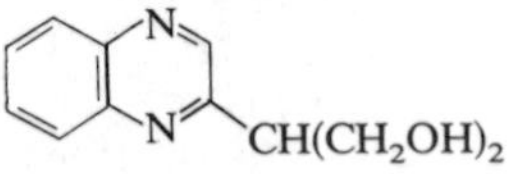 Quinoxalin-2-yl–$CH(CH_2OH)_2$ (structure)	103.5–105	29

TABLE 1 (*contd.*)

Molecular formula	Substituents or structure	m.p., °C	Ref.
$C_{12}H_{12}N_2$	2-(1-Butenyl)	b.p. 88–89/0.08 mm	72
	$HgCl_2$ salt	143–144 decomp.	72
$C_{12}H_{12}N_2O$	2-(2-Tetrahydrofuryl)		
	Picrate	107–109	11
$C_{12}H_{14}N_2$	2-*sec*.-Butyl		18
	2-*t*-Butyl	38–39.5, b.p. 70–75/0.05 mm, 118–124/7.5 mm.	11, 79
	Picrate	133	79
$C_{12}H_{14}N_2O$	2-(1-Ethoxyethyl)	80–85/0.01 mm	11
$C_{12}H_{14}N_2O_2$	2-(*p*-Dioxanyl)	64–65	11
	Picrate	125–127 decomp.	11
$C_{14}H_{19}N_3$	2-(2-Diethylaminoethyl)	b.p. 168–172/9 mm	33
	Picrate	141–143	33
$C_{15}H_{11}ClN_2$	2-*o*-Chlorobenzyl	55	81
$C_{15}H_{12}N_2$	2-Benzyl	38.5; b.p. 80/0.01 mm, 206–210/12 mm	19, 81
	Picrate	116	19

[a] Unstable on exposure to air.

TABLE 2. 2,3-DIALKYL- AND 2-ALKYL-3-ARYLQUINOXALINES AND THEIR DERIVATIVES

[Structure: quinoxaline with R^2 at C-2 and R^1 at C-3]

Molecular formula	Substituents or structure	m.p., °C	Ref.
$C_{10}H_2Cl_8N_2$	2,3-Bis(dichloromethyl)-5,6,7,8-tetrachloro	205–206	65
$C_{10}H_4Br_4Cl_2N_2$	2,3-Bis(dibromomethyl)-6,7-dichloro	198.5–200.5	39
$C_{10}H_4Cl_6N_2$	5,6,7,8-Tetrachloro-2-dichloromethyl-3-methyl	203–204	65
$C_{10}H_4F_6N_2$	2,3-Bis(trifluoromethyl)	117–118	82
$C_{10}H_5Br_4ClN_2$	2,3-Bis(dibromomethyl)-6-chloro	218–219.5	39
$C_{10}H_6Br_2Cl_2N_2$	2,3-Bis(bromomethyl)-6,7-dichloro	171–172.5	39
$C_{10}H_6Br_4N_2$	2,3-Bis(dibromomethyl)	224	83
$C_{10}H_6Cl_2I_2N_2$	6,7-Dichloro-2,3-bis(iodomethyl)	198–200	39
$C_{10}H_6Cl_4N_2$	5,6,7,8-Tetrachloro-2,3-dimethyl	197–198	65
$C_{10}H_7Br_2ClN_2$	2,3-Bis(bromomethyl)-6-chloro	150.5–152	39
$C_{10}H_7ClI_2N_2$	6-Chloro-2,3-bis(iodomethyl)	161.5–162.5	39

TABLE 2 (*contd.*)

Molecular formula	Substituents or structure	m.p., °C	Ref.
$C_{10}H_7F_3N_2$	2-Methyl-3-trifluoromethyl	80–82, 83–84	84, 85
$C_{10}H_8Br_2N_2$	2,3-Bis(bromomethyl)	150–151, 152–153	39, 86, 87, 88
$C_{10}H_8Cl_2N_2$	2,3-Bis(chloromethyl)	151–152	89
	5,8-Dichloro-2,3-dimethyl	146–148	4
	6,7-Dichloro-2,3-dimethyl	191–192, 194–196	4, 90
	Monobenzylidene deriv.	184–187	90
	Dibenzylidene deriv.	205–207	90
$C_{10}H_8I_2N_2$	2,3-Bis(iodomethyl)	145–145.5, 152–153	39, 83
$C_{10}H_8N_4O_2$	2,3-Dimethyl-6,7-dinitroso		91
$C_{10}H_8N_4O_4$	2,3-Dimethyl-5,7-dinitro	150	66
$C_{10}H_9BrN_2$	2-Bromomethyl-3-methyl	119–120	92
	5-Bromo-2,3-dimethyl	72.5–73.5	142
	6-Bromo-2,3-dimethyl	84–85, 86–87	4, 142
$C_{10}H_9ClN_2$	2-Chloromethyl-3-methyl	126–127	89, 93
	5-Chloro-2,3-dimethyl	78–80, b.p. 158/10 mm	4, 142
	6-Chloro-2,3-dimethyl	91–92	4, 142
$C_{10}H_9FN_2$	5-Fluoro-2,3-dimethyl	116–117	142
	6-Fluoro-2,3-dimethyl	110.5–111	142
$C_{10}H_9IN_2$	6-Iodo-2,3-dimethyl	75–75.5	4
$C_{10}H_9N_3O_2$	5-Nitro-2,3-dimethyl	131–131.5	4
	6-Nitro-2,3-dimethyl	133–134, 133.5–135	4, 76
	Dibenzylidene deriv.	219–220	94
$C_{10}H_{10}N_2$	2,3-Dimethyl[a]	104–105, 105–107	3, 95
	Monobenzylidene deriv.	136–136.5, 138–139	90, 96
	Dibenzylidene deriv.	191–192, 195–197	90, 96, 97
	Bis(*p*-hydroxybenzylidene) deriv.	220–225 decomp.	98
	Bis(*p*-methoxybenzylidene deriv.	163–164	99
	Bis(3,4-dimethoxybenzylidene) deriv.	196–197, 208	35, 99
$C_{10}H_{10}N_2O$	5-Hydroxy-2,3-dimethyl	146–147	100, 144
	6-Hydroxy-2,3-dimethyl	247–249	100
	2-Hydroxymethyl-3-methyl	108–110	89
$C_{10}H_{10}N_2O_2$	2,3-Bis(hydroxymethyl)	132–133	89
	5,8-Dihydroxy-2,3-dimethyl	228.5–230	101, 102
$C_{10}H_{11}N_3$	5-Amino-2,3-dimethyl	162–163	28, 147
	Acetyl	196–197.5	28
	6-Amino-2,3-dimethyl	186–187, 192	28, 103, 147
	Acetyl	192–193, 193–193.5	4, 28
	p-Acetamidobenzenesulfonyl	310–312	4
$C_{10}H_{12}N_4$	5,7-Diamino-2,3-dimethyl	227	66
$C_{11}H_8Br_4N_4$	2,3-Bis(dibromomethyl)-6-methyl	200–200.5	39
$C_{11}H_9F_3N_2$	2,3-Dimethyl-6-trifluoromethyl	93	4
$C_{11}H_9N_3$	2-Cyanomethyl-3-methyl	131–132	77
	6-Cyano-2,3-dimethyl	199–200	4

TABLE 2 (*contd.*)

Molecular formula	Substituents or structure	m.p., °C	Ref.
$C_{11}H_9N_3O_4$	2-Carboxymethyl-3-methyl-6-nitro		
	Ethyl ester	170–174	10
$C_{11}H_{10}Br_2N_2$	2,3-Bis(bromomethyl)-6-methyl	126–127.5	39
	2-(1,2-Dibromoethyl)-3-methyl	130.5–131.5	29
$C_{11}H_{10}I_2N_2$	2,3-Bis(iodomethyl)-6-methyl	149.5–151.5	39
$C_{11}H_{10}N_2O_2$	2-Carboxymethyl-3-methyl		
	Ethyl ester	68, b.p. 135/0.5 mm	10
	6-Carboxy-2,3-dimethyl		25
	Ethyl ester	102–104	100
	2-Diethylaminoethyl ester	43–46	100
$C_{11}H_{11}BrN_2$	6-Bromo-2,3,7-trimethyl	145	4
$C_{11}H_{11}ClN_2$	6-Chloro-2,3,7-trimethyl	154	4
$C_{11}H_{11}N_3O_3$	5-Methoxy-2,3-dimethyl-7-nitro	218–220	104
	6-Methoxy-2,3-dimethyl-5-nitro	145–146	105
	7-Methoxy-2,3-dimethyl-5-nitro	155–156	104
$C_{11}H_{12}N_2$	2-Ethyl-3-methyl	52–53, 53–54	49, 106, 113
	Picrate	159–160	106
	2,3,5-Trimethyl	72–73, b.p. 134–137/10 mm	4
	2,3,6-Trimethyl	91, b.p. 270–271	10, 107
$C_{11}H_{12}N_2O$	6-Hydroxymethyl-2,3-dimethyl	113–114	100
	5-Methoxy-2,3-dimethyl	118	4
	6-Methoxy-2,3-dimethyl	99–100	108
	2-(1-Hydroxyethyl)-3-methyl	80.5–81	5, 109
	Acetyl	56–57	5
	Benzoyl	144	109
$C_{11}H_{12}N_2S$	2,3-Dimethyl-6-methylthio	76–78	110
$C_{11}H_{13}N_3O$	5-Amino-6-methoxy-2,3-dimethyl	135–136	105
	7-Amino-5-methoxy-2,3-dimethyl	226–229	104
$C_{12}H_8N_2O_5$	N, N, CO_2H, CH_2COCO_2H		
	Diethyl ester		130
$C_{12}H_{10}Br_4N_2$	2,3-Bis(1,2-dibromoethyl)	176–177	29
	2,3-Bis(dibromomethyl)-6,7-dimethyl	201–202	39
$C_{12}H_{10}N_2O_3$	N, N, Me, CH_2COCO_2H		
	Ethyl ester	127–129	21
$C_{12}H_{10}N_2O_4$	7-Carboxy-2-carboxymethyl-3-methyl		
	2-Ethyl ester	186–188	10

TABLE 2 (*contd.*)

Molecular formula	Substituents or structure	m.p., °C	Ref.
$C_{12}H_{10}N_2O_4$ (*contd.*)	5,6-Dicarboxy-2,3-dimethyl		
	Dimethyl ester	153–154	116
	6,7-Dicarboxy-2,3-dimethyl		
	Dimethyl ester	159–161	116
$C_{12}H_{12}Br_2N_2$	2,3-Bis(1-bromoethyl)	110–111	83
	2,3-Bis(bromomethyl)-6,7-dimethyl	158.5–159.5	39
$C_{12}H_{12}I_2N_2$	2,3-Bis(iodomethyl)-6,7-dimethyl	176–178	39
$C_{12}H_{12}N_2$	2,3-Dimethyl-6-vinyl		111
$C_{12}H_{12}N_2O$	6-Acetyl-2,3-dimethyl	116–118	100
$C_{12}H_{12}N_2O_2$	2-Carboxymethyl-3,6-dimethyl		
	Ethyl ester	48–50, b.p. 130–135/0.8 mm	10
$C_{12}H_{14}N_2$	2,3-Diethyl	50–52	48, 112
	2-Isopropyl-3-methyl	37	4
	2,3,6,7-Tetramethyl	189–190	4
	2-Methyl-3-propyl	63–64	113
$C_{12}H_{14}N_2O$	5-Ethoxy-2,3-dimethyl	90–92	4
	6-Ethoxy-2,3-dimethyl	107–109	4
	2-Ethoxymethyl-3-methyl	b.p. 126–128/2 mm	5
$C_{12}H_{14}N_2O_2$	2,3-Bis(2-hydroxyethyl)	128.5–130	29
	5,6-Dimethoxy-2,3-dimethyl	105–107	4
	5,8-Dimethoxy-2,3-dimethyl	170	102
	6,7-Dimethoxy-2,3-dimethyl	173–174, 176–178	4, 146
	Me; $CH(CH_2OH)_2$	139.5–140.5	29
$C_{13}H_9N_3O_3$	2-Methyl-3-(5-nitro-2-furanyl)		114
$C_{13}H_{14}N_2O_2$	2-Carboxymethyl-3,6,7-trimethyl		
	Ethyl ester	81–83	10
$C_{13}H_{16}N_2$	2-Ethyl-3-propyl	44, 45	4, 115
$C_{13}H_{16}N_2O$	2-(2-Ethoxyethyl)-3-methyl	64	4
$C_{13}H_{16}N_2O_3$	CH_2CH_2OH; $CH(CH_2OH)_2$	121.5–122.5	29
$C_{14}H_{14}Cl_4N_2$	5,6,7,8-Tetrachloro-2,3-dipropyl	98–99	65
$C_{14}H_{14}N_2O_4$	2,3-Bis(acetoxymethyl)	88–89	88
$C_{14}H_{17}N_3O_2$	8-Acetamido-5-methoxy-2,3,7-trimethyl	221–222	101

TABLE 2 (*contd.*)

Molecular formula	Substituents or structure	m.p., °C	Ref.
$C_{14}H_{18}N_2$	6-*t*-Butyl-2,3-dimethyl	59–59.5	27
	2,3-Diisopropyl	74	4
	2,3-Dipropyl	43	117
$C_{14}H_{18}N_2O$	2-Isopropyl-3-(1-methoxyethyl)	60.5–61	118
$C_{14}H_{18}N_2O_2$	5,8-Diethoxy-2,3-dimethyl	129–130	119
$C_{15}H_{10}BrN_3O_2$	2-Bromomethyl-3-*p*-nitrophenyl	200–201	120
$C_{15}H_{10}Cl_2N_2$	2-(2,4-Dichlorophenyl)-3-methyl	130–131	121
	Benzylidene deriv.	188–189	121
	p-Chlorobenzylidene deriv.	207–208	121
	3,4-Methylenedioxybenzylidene deriv.	208–209	121
	p-Nitrobenzylidene deriv.	231–233	121
	2-(3,4-Dichlorophenyl)-3-methyl	115–116	121
	Benzylidene deriv.	154–155	121
	p-Chlorobenzylidene deriv.	198–199	121
	3,4-Methylenedioxybenzylidene deriv.	198–200	121
	p-Nitrobenzylidene deriv.	293–294	121
	2-(2,5-Dichlorophenyl)-3-methyl	131–132	121
	Benzylidene deriv.	135–136	121
	p-Chlorobenzylidene deriv.	215–216	121
	3,4-Methylenedioxybenzylidene deriv.	181–182	121
	p-Nitrobenzylidene deriv.	252–254	121
$C_{15}H_{11}BrN_2$	2-Bromomethyl-3-phenyl	121–122	122
$C_{15}H_{11}ClN_2$	2-Chloromethyl-3-phenyl	118–120	30
	2-*p*-chlorophenyl-3-methyl	103–104	121
	Benzylidene deriv.	178–179	121
	p-Chlorobenzylidene deriv.	222–223	121
	3,4-Methylenedioxybenzylidene deriv.	188–189	121
	p-Nitrobenzylidene deriv.	268–269	121
$C_{15}H_{11}N_3O_2$	2-Methyl-3-(*p*-nitrophenyl)	134–136	60
$C_{15}H_{12}N_2$	2-Methyl-3-phenyl[a]	55–57, 63	30, 123, 124
	Benzylidene deriv.	69–70	125
$C_{15}H_{13}N_3$	2-(*p*-Aminophenyl)-3-methyl	160–162	60
	Acetyl	234	60
	Chloroacetyl	192	60
	Dichloroacetyl	175–177	60
	Diethylaminoacetyl	85	60
	3,4,5-Trimethoxybenzoyl	200	60
$C_{16}H_{13}FN_2$	6-(*p*-Fluorophenyl)-2,3-dimethyl	117	126
$C_{16}H_{14}N_2$	2-Benzyl-3-methyl	56–57	127, 128
	3-Benzylidene deriv.	142–143	129

TABLE 2 (*contd.*)

Molecular formula	Substituents or structure	m.p., °C	Ref.
$C_{16}H_{14}N_2$ (*contd.*)	2-Ethyl-3-phenyl	45–46	123
	2-Methyl-3-*p*-tolyl	60–61, 63–65	49, 121
	Benzylidene deriv.	128–129	121
	p-Chlorobenzylidene deriv.	190–191	121
	3,4-Methylenedioxybenzylidene deriv.	180–182	121
	p-Nitrobenzylidene deriv.	246–247	121
$C_{16}H_{14}N_2O$	2-*p*-Methoxyphenyl-3-methyl	100–102, 101–102	49, 121
	Benzylidene deriv.	136–137	121
	p-Chlorobenzylidene deriv.	183–185	121
	3,4-Methylenedioxybenzylidene deriv.	178–179	121
	p-Nitrobenzylidene deriv.	234–235	121
$C_{16}H_{22}N_2$	2,3-Di-*t*-Butyl	53–54	7
$C_{17}H_{13}N_3$	2-Cyanophenylmethyl-3-methyl	220–222	15
$C_{17}H_{16}N_2$	2-Isopropyl-3-phenyl	98	123
$C_{17}H_{16}N_2O$	3-*p*-Methoxybenzyl-2-methyl	81–82, 84–85	127, 131
	2-Methyl-3-(2-phenoxyethyl)	156–156.5	4
$C_{18}H_{18}N_2$	2-*t*-Butyl-3-phenyl	108–109	123
$C_{19}H_{20}N_2$	2-Methyl-3-(2,4,6-trimethylbenzyl)	137–138	128
$C_{20}H_{28}N_4$	2,3-Bis(piperidinomethyl)	101–102	5
$C_{21}H_{14}ClN_3O_2$	2-(2-chlorobenzyl)-3-(*o*-nitrophenyl)	92	133
	2-(4-chlorobenzyl)-3-(*o*-nitrophenyl)	147–149	133
$C_{21}H_{15}N_3O_2$	2-Benzyl-3-(*o*-nitrophenyl)	126–127	133
	2-Benzyl-3-(*p*-nitrophenyl)	167	120
$C_{21}H_{16}N_2$	2-Benzyl-3-phenyl	96–98, 98–99	123, 134, 135
$C_{22}H_{15}N_3O_2$	2-(4-methylbenzyl)-3-(*o*-nitrophenyl)	100	133
$C_{22}H_{16}N_2O$	2-Benzoyl-3-benzyl	96–96.5	36
$C_{22}H_{18}N_2$	2,3-Dibenzyl	117–118, 118–118.5	36, 136, 137
	2-α-Methylbenzyl-3-phenyl	116–116.5	138
$C_{22}H_{18}N_2O$	2-Benzyl-3-(*o*-methoxyphenyl)	105–107	139
	2-(*p*-Methoxybenzyl)-3-phenyl	119–120	127
$C_{22}H_{20}N_4$	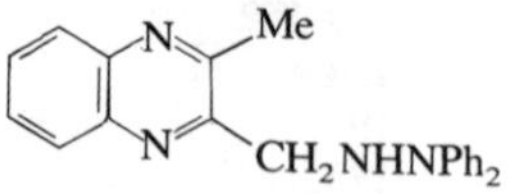 N, N, Me, CH_2NHNPh_2		132
$C_{24}H_{20}N_4O_4$	2-Benzyl-3-(3,5-dinitromesityl)	214	6
$C_{24}H_{21}N_3O_2$	2-Benzyl-3-(3-nitromesityl)	163	6
	2-(2,4,6-Trimethylbenzyl)-3-(*o*-nitrophenyl)	165	133
$C_{24}H_{22}N_2$	2,3-Bis(α-methylbenzyl)*meso*–*dl* mixture	89–90.5	36
$C_{24}H_{22}N_2O_2$	2,3-Bis(*p*-methoxybenzyl)	89–90, 90–91	140, 141

[a] The quinoxaline has been condensed with a number of more complex aromatic aldehydes; see, for example, G. Drefahl and G. Plötner, *Chem. Ber.*, **95,** 2782 (1962).

V. References

1. J. C. E. Simpson, "Condensed Pyridazine and Pyrazine Rings," Interscience, New York, 1953, p. 273.
2. R. G. Jones, E. C. Kornfeld, and K. C. McLaughlin, *J. Am. Chem. Soc.*, **72,** 3539 (1950).
3. C. L. Leese and H. N. Rydon, *J. Chem. Soc.*, 303 (1955).
4. J. K. Landquist and G. J. Stacey, *J. Chem. Soc.*, 2822 (1953).
5. J. K. Landquist and J. A. Silk, *J. Chem. Soc.*, 2052 (1956).
6. R. P. Barnes and L. A. Gist, *J. Am. Chem. Soc.*, **72,** 2509 (1950).
7. Ae. de Groot and H. Wynberg, *J. Org. Chem.*, **31,** 3954 (1966).
8. G. J. Visser and A. Vos, *Acta Crystallogr. Sec. B.*, **27,** 1793 (1971); *Chem. Abstr.*, **75,** 144941*d* (1971).
9. G. J. Visser, A. Vos, Ae. de Groot, and H. Wynberg, *J. Am. Chem. Soc.*, **90,** 3253 (1968).
10. G. Westphal, *Z. Chem.*, **9,** 145 (1969).
11. T. T. Chen, W. Dörscheln, H. Göth, M. Hesse, and H. Schmid, *Helv. Chim. Acta*, **51,** 632 (1968).
12. J. Borkovec, J. Michalsky, E. Rabusic, and J. Hadacek, *Chem. Listy*, **48,** 717 (1954); *Chem. Abstr.*, **49,** 9662 (1955).
13. J. Borkovec, J. Michalsky, and M. Ambroz, *Chem. Listy*, **48,** 865 (1954); *Chem. Abstr.*, **49,** 9662 (1955).
14. E. Hayashi, C. Iijima, and Y. Nagasawa, *Yakugaku Zasshi*, **84,** 163 (1964); *Chem. Abstr.*, **61,** 3108 (1964).
15. Y. Mizuno, K. Adachi, and K. Ikeda, *Chem. Pharm. Bull.* (Tokyo), **2,** 225 (1954)
16. R. N. Castle, A. Aldous, and C. Moore, *J. Org. Chem.*, **21,** 139 (1956).
17. F. Bertini, T. Caronna, R. Galli, F. Minisci, and O. Porta, *Chim. Ind.* (Milan), **54,** 425 (1972); *Chem. Abstr.*, **77,** 34381*e* (1972).
18. G. P. Gardini and F. Minisci, *Ann. Chim.* (Rome), **60,** 746 (1970); *Chem. Abstr.*, **74,** 87128*x* (1971).
19. K. C. Bass and P. Nababsing, *Org. Prep. Proc. Int.*, **3,** 45 (1971).
20. R. Mondelli and L. Merlini, *Tetrahedron*, **22,** 3253 (1966).
21. E. C. Taylor and E. S. Hand, *J. Am. Chem. Soc.*, **85,** 770 (1963).
22. B. Golankiewicz and K. Golankiewicz, *Bull. Acad. Polon. Sci., Ser. Sci. Chim.*, **14,** 199 (1966); *Chem. Abstr.*, **65,** 10466 (1966).
23. H. Sterk and Th. Kappe, *Monatsh.*, **100,** 1274 (1969)
24. P. Vetesnik, V. Bekarek, V. Beranek, and J. Kavalek, *Sb. Ved. Pr. Vys. Sk. Chemickotechnol. Pardubice*, **20,** 3 (1969); *Chem. Abstr.*, **72,** 84535*a* (1970).
25. P. Vetesnik, V. Beranek, and J. Kavalek, *Sb. Ved. Pr. Vys. Sk Chemickotechnol. Pardubice*, **14,** 37 (1966); *Chem. Abstr.*, **67,** 63717*k* (1967).
26. P. Vetesnik, J. Kavalek, V. Beranek, and O. Exner, *Collect. Czech. Chem. Commun.*, **33,** 566 (1968).
27. P. Vetesnik, J. Kavalek, and V. Beranek, *Collect. Czech. Chem. Commun.*, **36,** 2486 (1971).
28. P. Vetesnik, *Collect. Czech. Chem. Commun.*, **33,** 556 (1968).
29. A. S. Elina and L. G. Tsyrul'nikova, *J. Gen. Chem. USSR*, **34,** 2089 (1964).
30. E. C. Taylor and G. W. H. Cheeseman, *J. Am. Chem. Soc.*, **86,** 1830 (1964).
31. W. Reid and S. Hinsberg, *Justus Liebigs Ann. Chem.*, **600,** 47 (1956).
32. N. E. Plevachuk and S. N. Baranov, *Chem. Heterocycl. Comp.*, **4,** 536 (1968).
33. R. L. Wear and C. S. Hamilton, *J. Am. Chem. Soc.*, **72,** 2893 (1950).
34. A. S. Elina, *J. Gen. Chem. USSR*, **29,** 2728 (1959).

35. G. M. Bennett and G. H. Willis, *J. Chem. Soc.*, 1960 (1928).
36. E. J. Moriconi, R. E. Misner, and T. E. Brady, *J. Org. Chem.*, **34,** 1651 (1969).
37. G. Henseke and K.-J. Bähner, *Chem. Ber.*, **91,** 1605 (1958).
38. O. Westphal and K. Jann, *Justus Liebigs Ann. Chem.*, **605,** 8 (1957).
39. E. J. Moriconi and A. J. Fritsch, *J. Org. Chem.*, **30,** 1542 (1965).
40. A. S. Elina, L. S. Tsyrul'nikova, and I. Medvedeva, *J. Org. Chem. USSR*, **1,** 143 (1965).
41. H. Otomasu and S. Nakajima, *Chem. Pharm. Bull.* (Tokyo), **6,** 566 (1958).
42. R. G. Jones and K. C. McLaughlin, *Org. Synth.*, Coll. Vol. 4, p. 827.
43. K. H. Schaaf and P. E. Spoerri, *J. Am. Chem. Soc.*, **71,** 2043 (1949).
44. C. Weiss, F. Hoeppner, S. Becker, and W. Blaschke, *Tetrahedron*, **29,** 3071 (1973).
45. H. Feuer and J. P. Lawrence, *J. Am. Chem. Soc.*, **91,** 1856 (1969).
46. H. Feuer and J. P. Lawrence, *J. Org. Chem.*, **37,** 3662 (1972).
47. W. F. Gum and C. T. Goralski (to Dow Chemical Co.), U.S. Pat. 3,703,515; *Chem. Abstr.*, **78,** 43525*c* (1973).
48. H. Normant and T. Cuvigny, *Bull. Soc. Chim. Fr.*, 1881 (1965).
49. W. Schwaiger and J. P. Ward, *Rec. Trav. Chim.*, **91,** 1175 (1972).
50. T. C. Van Soest, *J. Cryst. Mol. Struct.*, **3,** 323 (1973); *Chem. Abstr.*, **80,** 75291*f* (1974).
51. O. Compagnon and P. L. Compagnon, *Bull. Soc. Chim. Fr.*, 3381, 3385 (1973).
52. J. F. Wolfe, D. E. Portlock, and D. J. Feuerbach, *J. Org. Chem.*, **39,** 2006 (1974).
53. E. C. Taylor and E. S. Hand, *J. Org. Chem.*, **27,** 3734 (1962).
54. C. W. Bird and G. W. H. Cheeseman, *J. Chem. Soc.*, 3037 (1962).
55. G. W. H. Cheeseman and B. Tuck, *J. Chem. Soc.*, 3678 (1965).
56. R. M. Acheson and M. W. Foxton, *J. Chem. Soc.* (*C*), 378 (1968).
57. J. W. Lown, R. Westwood, and A. S. K. Aidoo, *Can. J. Chem.*, **48,** 327 (1970).
58. G. W. H. Cheeseman and B. Tuck, *Tetrahedron Lett.*, 4851 (1968).
59. C. Egli (to Ciba-Geigy A.-G.), Ger. Offen. 2,132,175; *Chem. Abstr.*, **76,** 113251*v* (1972).
60. V. Hack and J. Kolinsky, *Pharm. Zentralhalle*, **105,** 291 (1966); *Chem. Abstr.*, **67,** 64358*p* (1967).
61. C. W. Huffman, J. J. Krajewski, P. J. Kotz, J. T. Traxler, and S. S. Ristich, *J. Agr. Food Chem.*. **1,** 298 (1971).
62. R. R. Schaffer (to E. I. du Pont de Nemours and Co.), U.S. Pat. 3,560,616; *Chem. Abstr.*, **75,** 47839*u* (1971).
63. H. P. Schlunke and K. Ronco (to Ciba Ltd.), Ger. Offen. 2,010,280; *Chem. Abstr.*. **74,** 133012*e* (1971).
64. H. P. Schlunke and C. Egli (to Ciba-Geigy A.-G.), Ger. Offen. 2,144,297; *Chem. Abstr.*, **77,** 54870*s* (1972).
65. D. E. Burton, A. J. Lambie, D. W. J. Lane, G. T. Newbold, and A. Percival, *J. Chem. Soc.* (*C*), 1268 (1968).
66. H. N. Rydon and K. Undheim, *J. Chem. Soc.*, 4685 (1962).
67. G. Henseke and R. Jacobi, *Justus Liebigs Ann. Chem.* **684,** 146 (1965).
68. H. Moreno and H. P. Schultz, *J. Med. Chem.*, **13,** 1005 (1970).
69. P. E. Fanta, R. A. Stein, and R. M. W. Rickett, *J. Am. Chem. Soc.*, **80,** 4577 (1958).
70. J. Klicnar and F. Kosek, *Collect. Czech. Chem. Commun.*, **30,** 3087 (1965).
71. J. Klicnar and F. Kosek, *Collect. Czech. Chem. Commun.*, **30,** 3102 (1965).
72. E. C. Taylor and S. F. Martin, *J. Am. Chem. Soc.*, **96,** 8095 (1974).
73. A. Jonsson, P. Stone, and N. Willman, *Acta Pharm. Suecica*, **5,** 537 (1968).
74. E. D. Parker and A. Furst, *J. Org. Chem.*, **23,** 201 (1958).

75. A. S. Elina, *J. Gen. Chem. USSR*, **32,** 2919 (1962).
76. J. Klicnar and P. Vetesnik, Czech. Pat. 104,455; *Chem. Abstr.*, **59,** 10079 (1963).
77. K. Golankiewicz, *Bull. Acad. Polon. Sci., Ser. Sci. Chim.*, **12,** 307 (1964); *Chem. Abstr.*, **61,** 14673 (1964).
78. V. V. Titov and L. F. Kozhokina *Khim. Geterotsikl. Soedin.*, 1700 (1972); *Chem. Abstr.*, **78,** 72064g (1973).
79. E. Hayashi and Y. Miura, *Yakugaku Zasshi*, **87,** 643 (1967); *Chem. Abstr.*, **67,** 90774*r* (1967).
80. A. M. Stock, W. E. Donahue, and E. D. Amstutz, *J. Org. Chem.*, **23,** 1840 (1958).
81. H. Dahn and G. Rotzler, *Helv. Chim. Acta*, **43,** 1555 (1960).
82. L. O. Moore and J. W. Clark, *J. Org. Chem.*, **30,** 2472 (1965).
83. J. Wegmann and H. Dahn, *Helv. Chim. Acta*, **29,** 101 (1946).
84. E. Abushanab, *J. Am. Chem. Soc.*, **93,** 6532 (1971).
85. R. Belcher, A. Sykes, and J. C. Tatlow, *J. Chem. Soc.*, 2393 (1957).
86. J. Wegmann and H. Dahn, *Helv. Chim. Acta*, **29,** 95 (1946).
87. B. A. Arbuzov and V. M. Zoroastrova, *Izv. Akad. Nauk. SSSR Otd. Khim. Nauk.*, 806 (1954); *Chem. Abstr.*, **49,** 13222 (1955).
88. V. Kovacik, M. Fedoronko, and I. Jezo, *Org. Mass Spectrom.*, **7,** 449 (1973).
89. Y. Ahmad, M. S. Habib, Ziauddin, and B. Bakhtiari, *J. Org. Chem.*, **31,** 2613 (1966).
90. J. Kavalek *Collect. Czech. Chem. Commun.*, **34,** 1819 (1969).
91. J. H. Boyer and R. S. Buriks, *J. Am. Chem. Soc.*, **82,** 2216 (1960).
92. H. Simon, G. Heubach, W. Bitterlich, and H. Gleinig, *Chem. Ber.*, **98,** 3692 (1965).
93. B. A. Arbuzov and V. M. Zoroastrova, *Izv. Akad. Nauk. SSSR Ser. Khim.*, 254 (1966); *Chem. Abstr.*, **64,** 15915 (1966).
94. J. Klicnar, F. Kosek, and S. Panusova, *Collect. Czech. Chem. Commun.*, **29,** 206 (1964).
95. C. A. C. Haley and P. Maitland, *J. Chem. Soc.*, 3155 (1951).
96. A. S. Elina and L. G. Tsyrul'nikova, *Chem. Heterocycl. Comp.*, **2,** 194 (1966).
97. N. A. Sörenson, E. Samuelson, and Fr. Oxaal, *Acta Chem. Scand.*, **1,** 458 (1947).
98. J. L. Towle, *Iowa State Coll. J. Sci.*, **26,** 308 (1952); *Chem. Abstr.*, **47,** 8062 (1953).
99. H. Gilman, J. L. Towle, and R. K. Ingram, *J. Org. Chem.*, **21,** 595 (1956).
100. J. A. Silk, *J. Chem. Soc.*, 2058 (1956).
101. M. R. W. Levey and M. M. Joullie, *J. Heterocycl. Chem.*, **1,** 171 (1964).
102. S. Kawai, S. Tanake, and K. Ichikawa, *J. Chem. Soc. Jap.*, **75,** 40 (1954); *Chem. Abstr.*, **49,** 10314 (1955).
103. H. Gilman and H. Smith Broadbent, *J. Am. Chem. Soc.*, **70,** 2619 (1948).
104. H. B. Gillespie, M. Engleman, and S. Graff, *J. Am. Chem. Soc.*, **78,** 2445 (1956).
105. H. B. Gillespie, M. Engelman, F. Spano, and S. Graff, *J. Am. Chem. Soc.*, **79,** 2245 (1957).
106. F. Asinger, M. Theil, and W. Schaefer, *Justus Liebigs Ann. Chem.*, **637,** 146 (1960).
107. H. von Pechmann, *Chem. Ber.*, **21,** 1411 (1888).
108. R. W. Bost and E. E. Towell, *J. Am. Chem. Soc.*, **70,** 903 (1948).
109. J. A. Barltrop, C. G. Richards, and D. M. Russell, *J. Chem. Soc.*, 1423 (1959).
110. E. Sawicki, B. Chastain, H. Bryant, and A. Carr, *J. Org. Chem.*, **22,** 625 (1957).
111. G. Manecke and U. Rotter, *Makromol. Chem.*, **171,** 49 (1973); *Chem. Abstr.*, **80,** 60236*t* (1974).
112. J. P. Freeman and W. D. Emmons, *J. Am. Chem. Soc.*, **79,** 1712 (1957).
113. I. M. Heilbron, E. R. H. Jones, P. Smith, and B. C. L. Weedon, *J. Chem. Soc.*, 54 (1946).

114. N. Saldabols and S. Hillers, *Khim. Geterotsikl. Soedin. Sb.* 2: *Kislorodsoderzhashchie Geterotsikly*, 122 (1970); *Chem. Abstr.*, **76,** 140374*u* (1972).
115. E. A. Braude and C. J. Timmons, *J. Chem. Soc.*, 3144 (1953).
116. R. L. Williams and S. W. Shalaby, *J. Heterocycl. Chem.*, **10,** 891 (1973).
117. R. A. Ogg and F. W. Bergstrom, *J. Am. Chem. Soc.*, **53,** 1846 (1931).
118. R. Adams, C. C. J. Culvenor, C. N. Robinson, and H. A. Stingl, *Austr. J. Chem.*, **12,** 706 (1959).
119. Y. Sakakibura, *Nippon Kagaku Zasshi*, **83,** 172 (1962); *Chem. Abstr.*, **58,** 11357 (1963).
120. V. Petrow, O. Stephenson, and B. Sturgeon, *J. Chem. Soc.*, 4066 (1953).
121. C. S. Mahajanshetti and S. Siddappa, *Indian J. Chem.*, **1,** 541 (1963); *Chem. Abstr.*, **60,** 8029 (1964).
122. J. Wegmann and H. Dahn, *Helv. Chim. Acta*, **29,** 1247 (1946).
123. E. Hayashi and C. Iijimà, *Yakugaku Zasshi*, **86,** 571 (1966); *Chem. Abstr.*, **65,** 15375 (1966).
124. M. I. Rybinskaya, L. V. Rybin, and A. N. Nesmeyanov, *Izv. Akad. Nauk. SSSR Otd. Khim. Nauk.*, 899 (1963); *Chem Abstr.*, **59,** 7413 (1963).
125. P. Ruggli, P. Weis, and H. Rupe, *Helv. Chim. Acta*, **29,** 1788 (1946).
126. P. Miles and H. Suschitzky, *Tetrahedron*, **18,** 1369 (1962).
127. H. Moureu, *Ann. Chim.*, **14,** 352 (1930).
128. G. Bianchi, C. De Micheli, R. Gandolfi, P. Grünanger, P. V. Finzi, and O. Vajna de Pava, *J. Chem. Soc., Perkin I*, 1148 (1973).
129. A. M. Cragg, F. M. Dean, and G. Winfield, *J. Chem. Soc.*, 2431 (1959).
130. F. Eiden, H. Mueller, and G. Bachmann, *Arch. Pharm.* (Weinheim) **305,** 2 (1972).
131. H. Dahn and J. Nussbaum, *Helv. Chim. Acta*, **52,** 166 (1969).
132. F. G. Valitova and L. P. Petrova, *Izv. Akad. Nauk. SSSR Ser. Khim.*, 695 (1973); *Chem. Abstr.*, **79,** 5313*v* (1973).
133. R. P. Barnes, J. H. Graham, and M. A. Salim Qureshi, *J. Org. Chem.*, **28,** 2890 (1963).
134. H. O. House, *J. Am. Chem. Soc.*, **78,** 2298 (1956).
135. H. O. House, *J. Org. Chem.*, **21,** 1306 (1956).
136. P. Ruggli and P. Zeller, *Helv. Chim. Acta*, **28,** 741 (1945).
137. H. F. Fiesselman and H. J. Lindner, *Chem. Ber.*, **89,** 1799 (1956).
138. C. L. Stevens and R. G. Hiskey, *J. Org. Chem.*, **24,** 32 (1959).
139. R. Bognar and G. Litkei *Acta Univ. Debrecen. Ludovico Kossuth Nom.*, **8,** 25 (1962) *Chem. Abstr.*, **59,** 13924 (1963).
140. I. Hagedorn, U. Eholzer, and A. Lüttringhaus, *Chem. Ber.*, **93,** 1584 (1960).
141. I. Hagedorn and H. Tönjes, *Pharmazie*, **12,** 567 (1957); *Chem. Abstr.*, **52,** 6362 (1958).
142. J. Kavalek, M. Dolecek, and P. Vetesnik, *Sb. Ved. Pr. Vys. Sk. Chemickotechnol. Pardubice*, **16,** 9 (1967); *Chem. Abstr.*, **71,** 60901*v* (1969).
143. W. Hahn and J. Lesiak, Pol. Pat. 71,049; *Chem. Abstr.*, **83,** 28279*k* (1975).
144. Y. Kidani, K. Inagaki, and H. Koike, *Bunseki Kagaku*, **24,** 221 (1975); *Chem. Abstr.*, **83,** 66345*m* (1975).
145. E. M. Kaiser and J. D. Petty, *J. Organomet. Chem.*, **108,** 139 (1976).
146. D. Schelz and M. Preister, *Helv. Chim. Acta*, **58,** 2536 (1975).
147. M.-T. Le Bris, *Bull. Soc. Chim. Fr.*, 921 (1976).

CHAPTER XV

2-Aryl(heteroaryl)- and 2,3-Diaryl-(diheteroaryl)quinoxalines

I.	Methods of Preparation	233
II.	Properties	235
III.	Uses	237
IV.	Tables	237
	Table 1. 2-Aryl(heteroaryl)- and 2,3-Diaryl(diheteroaryl)quinoxalines	237
V.	References	238

The compounds discussed in this chapter include some of the most readily prepared quinoxaline derivatives; it was thought logical to group together both aryl and heteroarylquinoxalines because of their similar modes of preparation and properties.

I. Methods of Preparation

2-Arylquinoxalines result from the condensation of arylglyoxals with *o*-phenylenediamines (Scheme 1). Arylglyoxals are conveniently prepared

$$ArCOCH_3 \longrightarrow ArCOCHO \longrightarrow$$ [2-arylquinoxaline: quinoxaline ring with Ar at C-2]

Scheme 1

by selenium dioxide oxidation of the corresponding methyl ketones.[1] Alternatively condensation of an α-bromoketone with *o*-phenylenediamine gives a dihydroquinoxaline which is usually so readily oxidized under the reaction conditions that the fully aromatic quinoxaline is isolated from the reaction mixture (Scheme 2).[1]

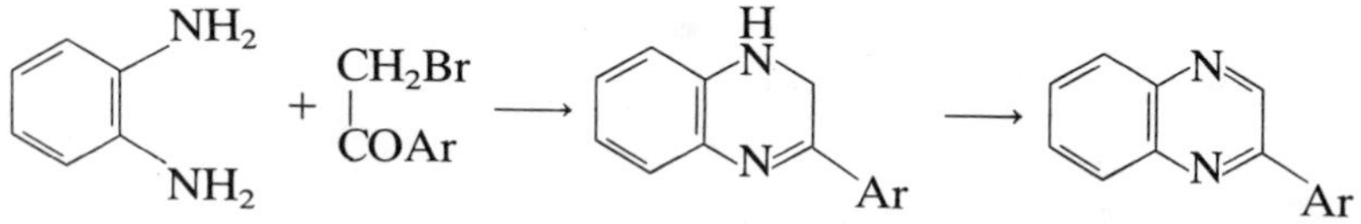

Scheme 2

2,3-Diarylquinoxalines are similarly prepared by reaction of *o*-phenylenediamine with benzils. The required benzils have been prepared traditionally by oxidation of the corresponding benzoins, which in turn have been prepared by treatment of the appropriate aldehydes with potassium cyanide. In a more recent procedure, benzils have been obtained in good yield by oxidation of para-substituted diphenyloxazoles with bromine in acetic acid. The oxazoles themselves are readily prepared by cyclization of the benzoin acetate with urea (Scheme 3).[2]

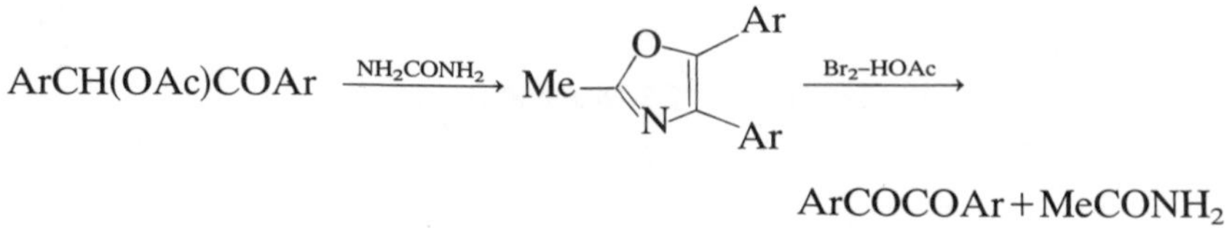

ArCH(OAc)COAr $\xrightarrow{NH_2CONH_2}$ Me–oxazole(Ar)(Ar) $\xrightarrow{Br_2\text{–}HOAc}$ ArCOCOAr + $MeCONH_2$

Scheme 3

2-Phenylquinoxalines are unexpectedly isolated together with 4-phenylcinnolines from cyclodehydration of *cis*-arylglyoxal 2-phenylhydrazones with a fused mixture of aluminum chloride and sodium chloride. Quinoxaline formation is thought to involve a diazetidine intermediate (Scheme 4).[3]

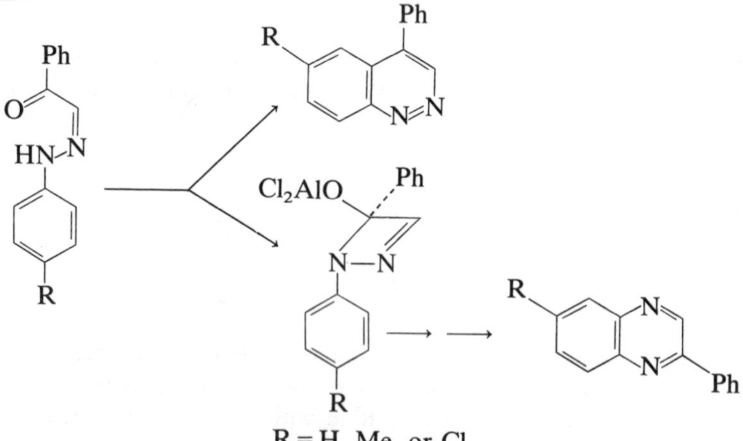

R = H, Me, or Cl

Scheme 4

2,3-Diphenylquinoxaline is formed on reaction of 2,3-diphenylindole with iodine azide (IN_3).[43]

II. Properties

The ultraviolet spectrum of 2-phenylquinoxaline shows maxima at 262 nm (ε 28,200) and 335 nm (ε 12,000). In the ^1H NMR spectrum a singlet at δ 9.28 is assigned to the proton at C-3. Two two-proton quartets ($J = 3$ and 8 Hz) appearing at δ 8.18 and 8.05 are assigned to the protons at C-5 and C-8 and at C-2′ and C-6′, respectively, although these assignments could be reversed. The remaining five protons appear as a multiplet at δ 7.50.[3] In the ^1H NMR spectrum of 2-mesitylquinoxaline in carbon tetrachloride the chemical shift difference between the *ortho*- and *para*-methyl groups is 0.25 p.p.m.; however in concentrated sulfuric acid, where the mesitylquinoxaline has been assumed to exist as the dication **1**, the *ortho*- and *para*-methyl signals are coincident.[4]

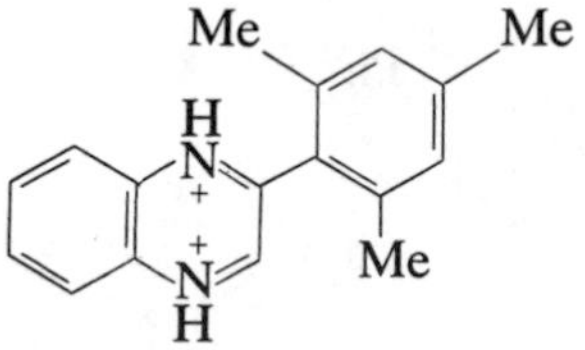

Me Me H N + + N H Me

1

The mass spectrum of 2,3-diphenylquinoxaline shows M^+, M^{2+}, and M^{3+} ions and, in addition, ions resulting from the cleavage of a hydrogen radical from these species.[5]

2-Phenylquinoxaline undergoes amination at position 4 when treated with *O*-mesitylenesulfonylhydroxylamine. The product **2** is converted into the *N*-benzoylimino derivative **3** on reaction with benzoyl chloride.[40] Treatment of **2** with base and an acetylenic ester results in 1,3-dipolar cycloaddition and the formation of a tricyclic adduct.[41]

2,3-Diphenylquinoxaline is reduced by sodium to a dianion which was characterized by reaction with ethyl chloroformate, methyl iodide, and 1,3-diiodopropane. Thus reaction with ethyl chloroformate yields the dihydroquinoxaline **4**. Reaction with lithium results in reduction and cyclodehydrogenation. Hydrolysis of the resulting dianion, and concomitant oxidation gives dibenzo[*a, c*]phenazine (**5**).[6] Treatment of 2,3-diphenylquinoxaline with potassium amide in liquid ammonia gives 2-phenylbenzimidazole and not, as originally postulated, 2,2-diphenyl-3-amino-1,2-dihydroquinoxaline (Scheme 5).[7]

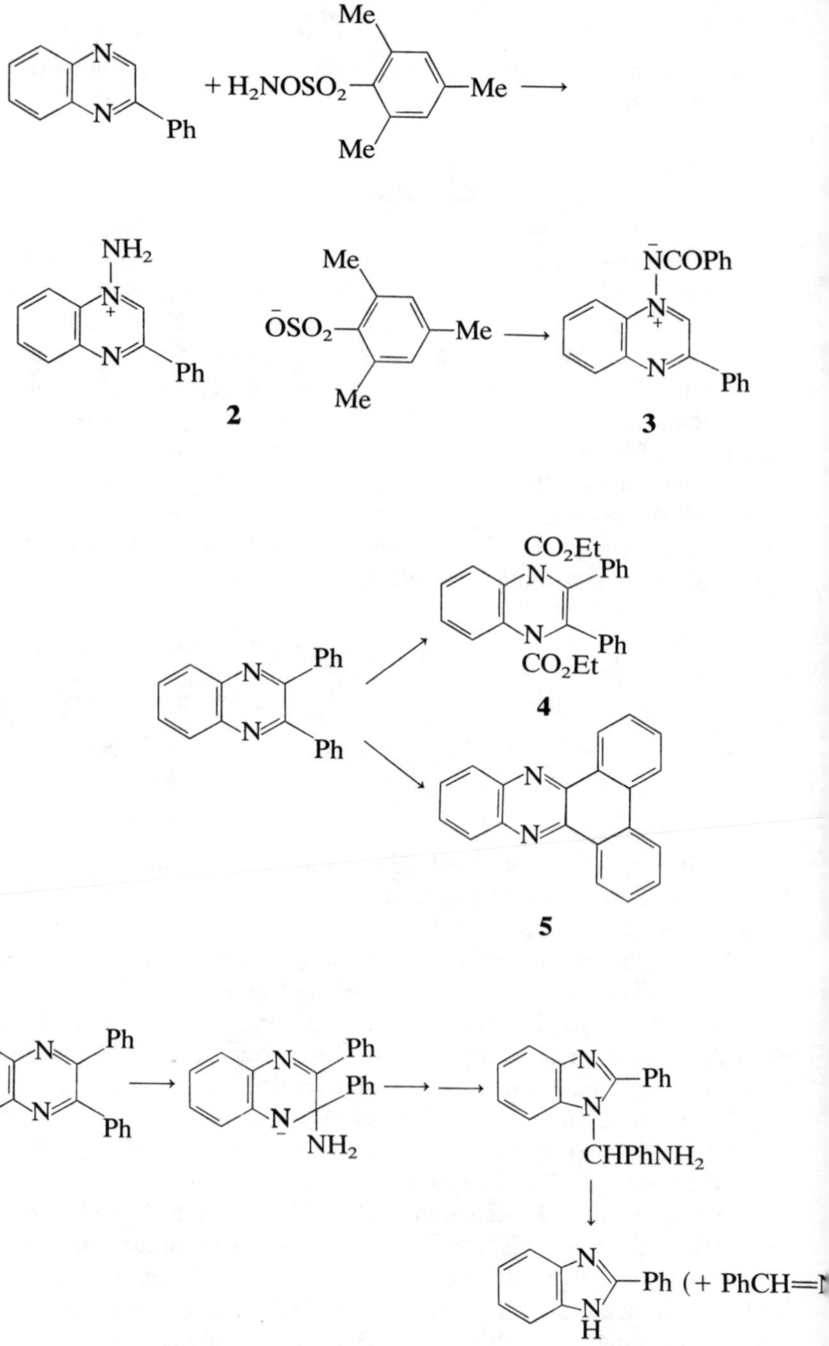

2

3

4

5

(+ PhCH=N

Scheme 5

Cobalt(II), nickel(II), iron(II), and copper(II) halide complexes of 2,3-di-(2-pyridyl)quinoxaline and related compounds have been described.[8,9]

III. Uses

Diarylquinoxalines have been prepared in the search for antimalarial[10] and antifertility[11] agents. They have been incorporated into light-sensitive copying compositions,[12] and 2,3-diphenylquinoxaline has been used as a photosensitizer for the decarboxylation of carboxylic acids.[13] Various phenylated quinoxaline polymers have been prepared,[36] and polymers have also been obtained from 6-acryloylamino-2,3-diphenylquinoxaline (**6**).[42]

$CH_2{=}CHCONH$ — N — Ph — N — Ph

6

IV. Tables

TABLE 1. 2-ARYL(HETEROARYL)- AND 2,3-DIARYL(DIHETEROARYL)QUINOXALINES

N R N R

Molecular formula	Substituents or structure	m.p., °C	Ref.
$C_{12}H_7N_3O_3$	2-(2-Furyl)-8-nitro	143.5–144.5	39
$C_{12}H_8N_2O$	2-(2-Furyl)	101, 101–102	1, 14
$C_{12}H_8N_2O_2$	2-(2-Furyl)-5-hydroxy	169.5–170	39
	2-(2-Furyl)-8-hydroxy	211.5–212.5	39
$C_{12}H_8N_2S$	2-(2-Thienyl)	117–119, 120–121	1, 15
$C_{12}H_9N_3O$	8-Amino-2-(2-furyl)	133–134	39
$C_{13}H_9N_3$	2-(2-Pyridyl)	112–114, 113–114	1, 16
	2-(3-Pyridyl)	243–244	1
	2-(4-Pyridyl)	255–256	1
$C_{14}H_6Cl_4N_2$	5,6,7,8-Tetrachloro-2-phenyl	276–278	17
$C_{14}H_6Cl_4N_2O$	5,6,7,8-Tetrachloro-2-(*p*-hydroxyphenyl)	313	17
$C_{14}H_9N_3O_2$	2-(*p*-Nitrophenyl)	187–188, 189–190	1, 18
$C_{14}H_{10}N_2$	2-Phenyl	76–77, 78	19, 20, 21

TABLE 1 (*contd.*)

Molecular formula	Substituents or structure	m.p., °C	Ref.
$C_{14}H_{10}N_2O$	2-(*p*-Hydroxyphenyl)	204, 209–210	1, 22
$C_{15}H_9ClN_2$	6-Chloro-2-phenyl		38
	7-Chloro-2-phenyl		38
$C_{15}H_{12}N_2$	5-Methyl-2-phenyl	120	38
	6-Methyl-2-phenyl		38
	7-Methyl-2-phenyl		38
	8-Methyl-2-phenyl	104	38
$C_{16}H_{11}N_3$	2-(3-Indolyl)	202–203, 208–209	1, 23
$C_{16}H_{14}N_2$	6,7-Dimethyl-2-phenyl	128–129	24
$C_{17}H_{14}N_2$	2-(5-Indanyl)	116–117	1
$C_{17}H_{16}N_2$	2-Mesityl	59, b.p. 162–164/0.55 mm	4
$C_{18}H_{12}N_4$	2,3-Di(2-pyridyl)	183–184	25
$C_{19}H_{14}N_4$	6-Methyl-2,3-di(2-pyridyl)	138	26
$C_{20}H_{10}Cl_4N_2$	5,6,7,8-Tetrachloro-2,3-diphenyl	250–252	17
$C_{20}H_{12}Cl_2N_2$	5,6-Dichloro-2,3-diphenyl	163–164	27
	5,7-Dichloro-2,3-diphenyl	154.5–155.5	27
	2,3-Bis(*m*-chlorophenyl)	186–186.5	27
$C_{20}H_{12}N_4O_4$	2,3-Bis(*p*-nitrophenyl)	201–202	2
$C_{20}H_{13}ClN_2$	5-Chloro-2,3-diphenyl	158–159	27
$C_{20}H_{14}N_2$	2,3-Diphenyl	125	28, 29
$C_{20}H_{16}N_4$	2,3-Di(6-methyl-2-pyridyl)	115	30
	6,7-Dimethyl-2,3-di(2-pyridyl)	189	26
$C_{21}H_{14}N_2$	2-(2-Fluorenyl)	193–194, 200–201	1, 31
$C_{21}H_{15}N_3O_3$	5-Methoxy-7-nitro-2,3-diphenyl Hemiethanolate	212	32
$C_{21}H_{18}N_4$	6-Methyl-2,3-di(6-methyl-2-pyridyl)	162	8
$C_{22}H_{18}N_2$	2,3-Di(*p*-tolyl)	144–145	33
$C_{22}H_{20}N_4$	6,7-Dimethyl-2,3-di(6-methyl-2-pyridyl)	171	8
$C_{24}H_{18}N_2$	2,3-Bis(4-vinylphenyl)	149–152	33
$C_{24}H_{22}N_2O_2$	5,8-Diethoxy-2,3-diphenyl	160–161	34
$C_{26}H_{23}N_3$	2,3-Diphenyl-6-(*N*-piperidino)	170–170.5	37
$C_{31}H_{26}N_2O$	[structure: quinoxaline with N, N, Ph, COPh, Bu^t]	153–154	35

V. References

1. G. Y. Sarkis and S. Al-Azawe, *J. Chem. Eng. Data*, **18,** 102 (1973).
2. T. van Es and O. G. Backeberg, *J. Chem. Soc.*, 1371 (1963).
3. S. N. Bannore and J. L. Bose, *Indian J. Chem.*, **11,** 631 (1973).

4. G. Haefelinger, R. G. Weissenhorn, F. Hack, and G. Westermayer, *Angew. Chem. Int. Ed.*, **11,** 725 (1972).
5. V. Kovacik, M. Fedoronko, and I. Jezo, *Org. Mass Spectrom.*, **7,** 449 (1973).
6. J. G. Smith and E. M. Levi, *J. Organomet. Chem.*, **36,** 215 (1972).
7. E. C. Taylor and A. McKillop, *J. Org. Chem.*, **30,** 2858 (1965).
8. D. F. Colton and W. J. Geary, *J. Chem. Soc., Dalton.*, 547 (1972).
9. H. A. Goodwin and F. Lions, *J. Am. Chem. Soc.*, **81,** 6415 (1959).
10. G. S. Bajwa, K. E. Hartman, and M. M. Joullie, *J. Med. Chem.*, **16,** 134 (1973).
11. R. N. Iyer and R. Gopalchari, *Indian J. Chem.*, **11,** 234 (1973).
12. S. Bauer, H. Sikora, and W. Frass (to Kalle A.-G.), Ger. Offen. 2,064,380; *Chem. Abstr.*, **78,** 65232g (1973).
13. D. R. G. Brimage, R. S. Davidson, and P. R. Steiner, *J. Chem. Soc., Perkin I*, 526 (1973).
14. S. Fattuta, *Gazz. Chim. Ital.*, **88,** 1122 (1958); *Chem. Abstr.*, **53,** 18936 (1959).
15. J. Druey and A. Hüni, *Helv. Chim. Acta*, **35,** 2301 (1952).
16. F. R. Pfeiffer and F. H. Case, *J. Org. Chem.*, **31,** 3384 (1966).
17. D. E. Burton, A. J. Lambie, D. W. J. Lane, G. T. Newbold, and A. Percival, *J. Chem. Soc.* (*C*), 1268 (1968).
18. N. P. Buu-Hoi and N. H. Khoi, *Bull. Soc. Chim. Fr.*, **15,** 753 (1950).
19. J. Figueras, *J. Org. Chem.*, **31,** 803 (1966).
20. A. T. Blomquist and E. A. LaLancette, *J. Am. Chem. Soc.*, **84,** 220 (1962).
21. E. Hayashi and C. Iijima, *Yakugaku Zasshi*, **82,** 1093 (1962); *Chem. Abstr.*, **58,** 4551 (1963).
22. G. Fodor, D. Beke, and O. Kovacs, *Acta Chim. Hung.*, **1,** 149 (1951); *Chem. Abstr.*, **46,** 3514 (1952).
23. Y. Kishi, T. Goto, S. Inoue, S. Sigiùra, and H. Kishimoto, *Tetrahedron Lett.*, 3445 (1966).
24. V. M. Berezovskii and A. M. Yurkevich, *J. Gen. Chem. USSR*, **31,** 3526 (1961).
25. W. J. Geary, *J. Chem. Soc.* (*A*), 71 (1969).
26. D. F. Colton and W. J. Geary, *J. Chem. Soc.* (*A*), 2457 (1971).
27. G. W. H. Cheeseman and E. S. G. Törzs, *J. Chem. Soc.* (*C*), 157 (1966).
28. E. Hayashi and C. Iijima, *Yakugaku Zasshi*, **86,** 571 (1966); *Chem. Abstr.*, **65,** 15375 (1966).
29. R. W. Bost and E. E. Towell, *J. Am. Chem. Soc.*, **70,** 904 (1948).
30. W. J. Geary, *J. Chem. Soc.* (*A*), 2118 (1969).
31. F. Gualtieri, F. M. Riccieri, and M. L. Stein, *Farm. Ed. Sci.* (Pavia), **17,** 430 (1962); *Chem. Abstr.*, **58,** 4553 (1963).
32. L. Horner, U. Schwenk, and E. Junghanns, *Justus Liebigs Ann. Chem.*, **579,** 212 (1953).
33. G. Drefahl and W. Winnefeld, *J. Prakt. Chem.*, **28,** 236 (1965).
34. Y. Sakakibara, *Nippon Kagaku Zasshi*, **83,** 172 (1962); *Chem. Abstr.*, **58,** 11357 (1963).
35. H. Crawford, *J. Org. Chem.*, **36,** 3533 (1971).
36. J. A. Harvey and M. A. Ogliaruso, *Org. Prep. Proc. Int.*, 146 (1976).
37. A. M. Jefferson, Ph.D. Thesis, University of Salford, 1976.
38. S. N. Bannore, J. L. Bose, A. A. Thakar, and M. S. Wadia, *Indian J. Chem.*, **13,** 609 (1975).
39. N. Saldabols, A. Cimanis, S. Hillers, E. Liepins, and I. S. Yankovskaya, *Khim. Geterotsikl. Soedin.*, 306 (1976); *Chem. Abstr.*, **85,** 32951g (1976).
40. Y. Tamura, Y. Miki, J. Minamika, and M. Ikeda, *J. Heterocycl. Chem.*, **11,** 675 (1974).

41. Y. Tamura, Y. Miki, and M. Ikeda, *J. Heterocycl. Chem.*, **12,** 119 (1975).
42. D. Braun and G. Quarg, *Makromol. Chem.*, **176,** 593 (1975); *Chem. Abstr.*, **83,** 43798*x* (1975).
43. M. Ikeda, F. Tabusa, Y. Nishimura, S. Kwon, and Y. Tamura, *Tetrahedron Lett.* 2347 (1976).

CHAPTER XVI

2-Polyhydroxyalkylquinoxalines

I. Methods of Preparation 241
II. Properties 242
III. Tables 245
Table 1. 2-Polyhydroxyalkylquinoxalines 245
IV. References 246

The chemistry of this class of compound has been reviewed previously by Simpson.[1] 2-Polyhydroxyalkylquinoxalines have been prepared from many common sugars and are thus of interest as characterizing derivatives. Since carbon atoms 1 and 2 of the sugar molecule become incorporated in the quinoxaline ring on reaction with *o*-phenylenediamine, sugars epimeric at C-2 give a common derivative. It also follows that a common derivative, 2-(D-arabinotetrahydroxybutyl)quinoxaline, is obtained from D-glucose and D-fructose. In cases where the solubility of polyhydroxyalkylquinoxalines in water and ethanol makes isolation difficult, they can be converted either into the more easily purified fully acetylated derivatives or into flavazoles. Study of the optical rotatory dispersion (O.R.D.) of polyhydroxyalkylquinoxalines has shown that it is possible to correlate this property with the chirality of carbon atoms 1 and 2 of the side chain and thus to the stereochemistry of carbons 3 and 4 of the parent sugar.[2]

I. Methods of Preparation

It should be noted that reaction of aldoses with *o*-phenylenediamine may produce either quinoxaline or benzimidazole derivatives. In order to obtain a polyhydroxyalkylquinoxaline and not a polyhydroxyalkylbenzimidazole from either D-galactose or L-arabinose it is necessary to carry out reaction with *o*-phenylenediamine in the presence of hydrazine and boric acid.[1] More recently L-arabinose has been converted into 2-(L-erythrotrihydroxypropyl)quinoxaline (**1**) by reaction with the diamine in pyridine containing hydrazine hydrate and acetic acid. L-Sorbose is converted into 2-(L-xylotetrahydroxybutyl)quinoxaline under similar conditions.[2]

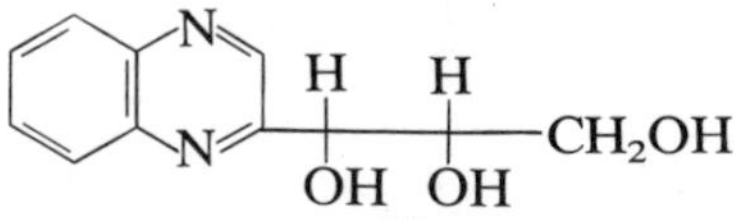

H H
CH$_2$OH
OH OH
1

2-(D-Arabinotetrahydroxybutyl)quinoxaline (**3**) has been prepared in 90% yield by heating a mixture of *o*-phenylenediamine, *p*-tolyl-D-isoglucosamine (**2**), and hydrazine hydrate in dilute acetic acid on the steam bath.[3] The product is isolated as a monohydrate, m.p. 190°, and

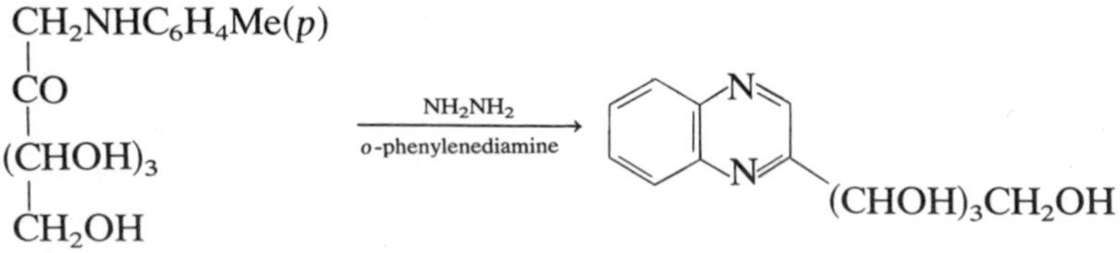

$CH_2NHC_6H_4Me(p)$
CO
$(CHOH)_3$
CH_2OH
2

$\xrightarrow[o\text{-phenylenediamine}]{NH_2NH_2}$

$(CHOH)_3CH_2OH$
3

this can be obtained more simply, though in much lower yield (20%), by heating a mixture of the diamine and sucrose in dilute acetic acid while passing air through the boiling solution.[4]

Osone hydrazones (e.g., **4**) derived from D-galactose, L-sorbose,[5] and D-fructose[6] react with *o*-phenylenediamine and its benzene-substituted derivatives to give the corresponding 2-tetrahydroxybutylquinoxaline.

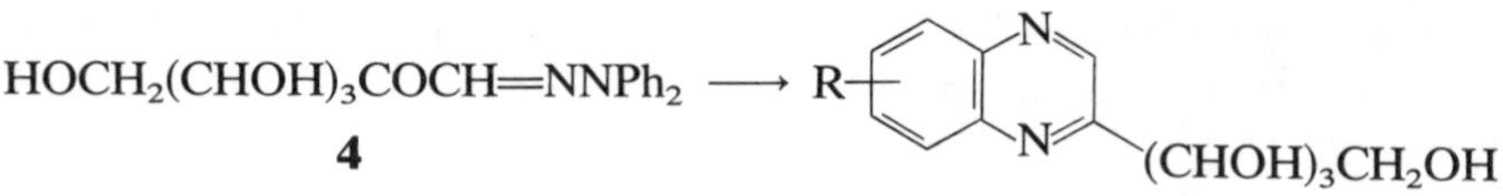

$HOCH_2(CHOH)_3COCH{=}NNPh_2 \longrightarrow$ R
4

$(CHOH)_3CH_2OH$

R = H, Me, NO_2, or Cl

II. Properties

It has been suggested that 2-(D-arabinotetrahydroxybutyl)quinoxaline (**3**) may have the cyclic structure **5**, but this possibility has been disproved by a ^{1}H NMR study which shows that the benzenoid protons appear as an almost symmetric four-proton multiplet of the A_2B_2 type which is

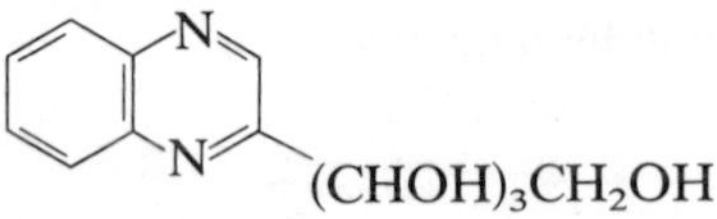

3 (CHOH)$_3$CH$_2$OH

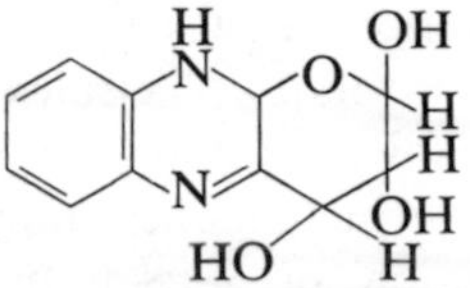

5

consistent with structure **3** but not with structure **5**. The ^{1}H NMR of the tetraacetyl derivative of **3** is indicative of a preferred planar zigzag conformation **6** for the side chain. Thus $J_{1'2'} = 3.0$ Hz and $J_{2'3'} = 8.5$ Hz, consistent with dihedral angles of approximately 60° and 180°, respectively.[6]

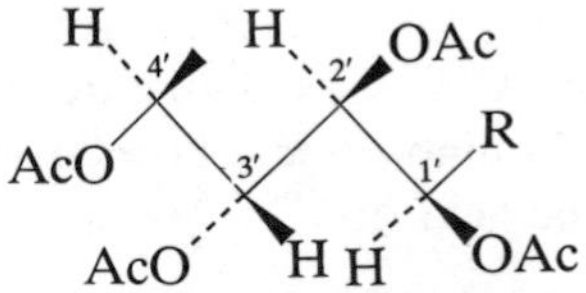

6 R = 2-Quinoxalinyl

Studies of the optical rotatory dispersion of polyhydroxyalkylquinoxalines have been made.[2,8] In methanol or in dimethylsulfoxide, negative Cotton effects centered at 315 nm are associated with R chirality at C-1′. The same correlation holds for the *O*-acetyl derivatives of polyhydroxyalkylquinoxalines. Compounds with threo relationship at C-1′ and C-2′ (**A, B, C**) have higher amplitude Cotton effects than erythro compounds (**D, E**). Quinoxalines with threo relationship at C-1′ and C-2′ also have higher rotatory power at long wavelength. Acetylation destroys the difference between threo and erythro compounds.[2]

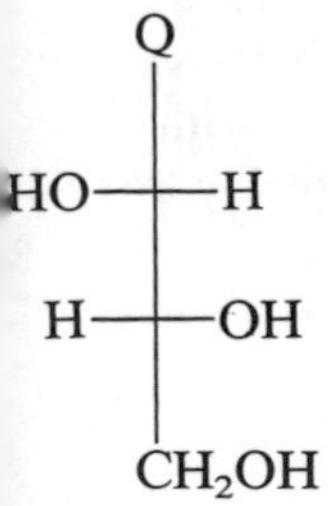

A Threo

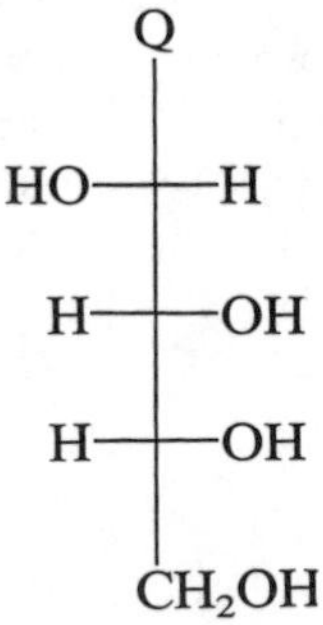

B Arabino

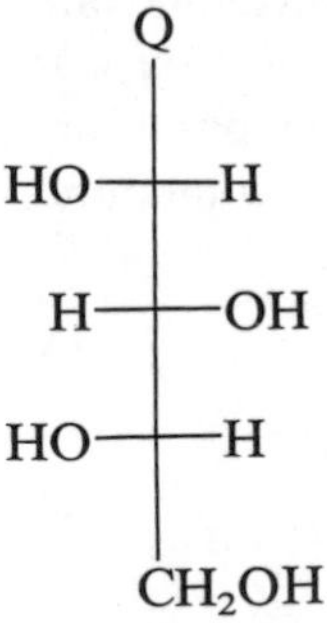

C Xylo

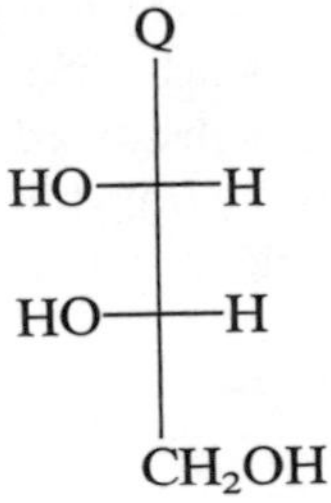

D Erythro

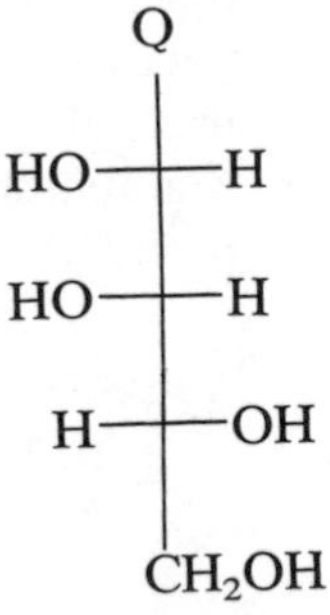

E Lyxo

Reference has already been made to the acetylation of polyhydroxyalkylquinoxalines. Tetrahydroxybutylquinoxalines are acetylated with acetic anhydride and pyridine.[2,7] Deacetylation is carried out by treatment with methanolic ammonia.[2]

Oxidation of 2-(D-arabinotetrahydroxybutyl)quinoxaline with either hydrogen peroxide and alkali or sodium peroxide yields quinoxaline-2-carboxylic acid. Sodium peroxide is the reagent of choice, and the 2-carboxylic acid is conveniently isolated as its sodium salt.[9] Further oxidation occurs when the tetrahydroxybutyl compound is treated with potassium permanganate in dilute potassium hydroxide and pyrazine 2,3,5-tricarboxylic acid is obtained in 87% yield.[3]

Sulfuric acid dehydration of 2-polyhydroxyalkylquinoxalines is now known to yield 2-(2-furyl)quinoxalines rather than pyrido[1,2-*a*]-quinoxalines (glucazidones).[10,11] The structure of product **7** from 2-(D-arabinotetrahydroxybutyl)quinoxaline follows from its independent synthesis from 2-furanglyoxylic acid (**9**).[10]

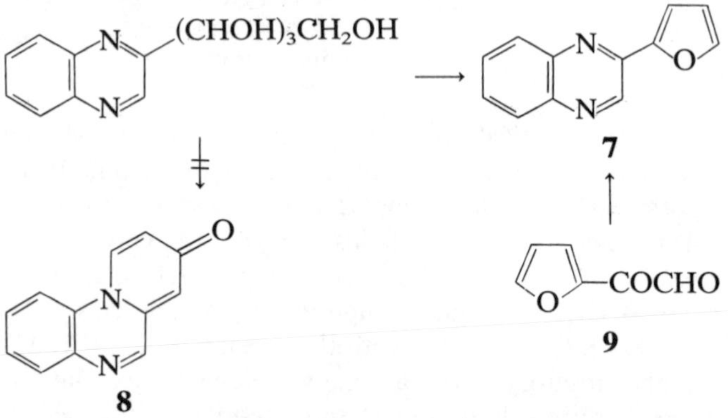

$(CHOH)_3CH_2OH$

7

8

—COCHO

9

Pyrazolo[3,4-*b*]quinoxalines (flavazoles) (e.g., **10**) are formed by the reaction of 2-polyhydroxyalkylquinoxalines with hydrazines. Thus with phenylhydrazine, 1-phenyl derivatives are formed; and with hydrazine itself, the 1-unsubstituted compounds are obtained.[8,12] Starting with a 2-D-arabinotetrahydroxybutyl compound, a 3-(D-erythrotrihydroxypropyl)flavazole is obtained.*

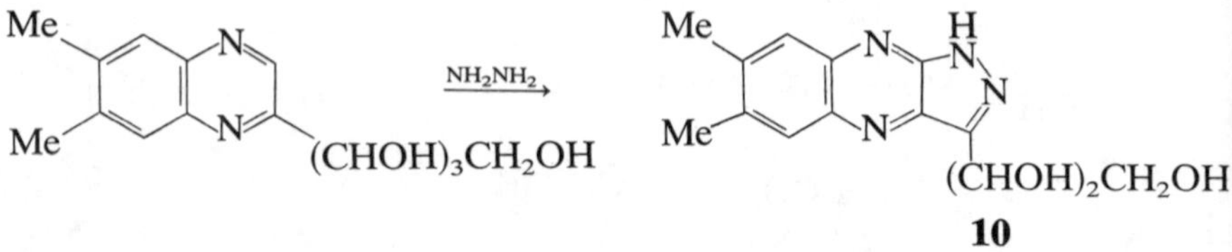

Me Me $(CHOH)_3CH_2OH$ $\xrightarrow{NH_2NH_2}$ Me Me $(CHOH)_2CH_2OH$

10

* A systematic review of the chemistry of pyrazolo[3,4-*b*]quinoxalines is given in Chapter XXXVII.

III. Tables

TABLE 1. 2-POLYHYDROXYALKYLQUINOXALINES

Molecular formula	Substituents or structure	m.p., °C	Ref.
$C_{11}H_{12}N_2O_3$	2-(L-Erythrotrihydroxypropyl)	121.5–122	2
$C_{12}H_{13}ClN_2O_4$	6- and 7-Chloro-2-(D-arabinotetrahydroxybutyl)[a]	180–181.5	13
$C_{12}H_{13}N_3O_6$	6(or 7)-Nitro-2-(D-arabinotetrahydroxybutyl)[b]	196 decomp.	7
	Tetraacetyl[c]	127	7
	6(or 7)-Nitro-2-(D-lyxotetrahydroxybutyl)[d]	155 decomp.	5
	6(or 7)-Nitro-2-(D-xylotetrahydroxybutyl)[e]	168 decomp.	5
$C_{12}H_{14}N_2O_4$	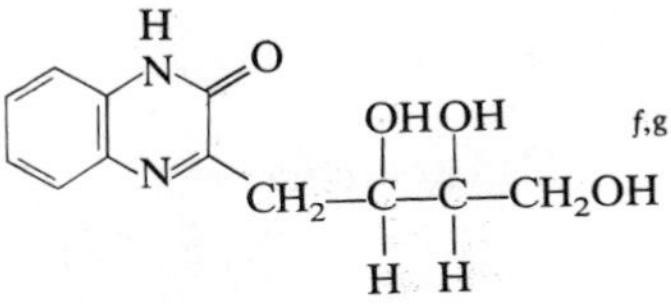 2(1H)-quinoxalinone, 3-CH_2—C(OH)(H)—C(OH)(H)—CH_2OH [f,g]	173–175, 174	14, 15
	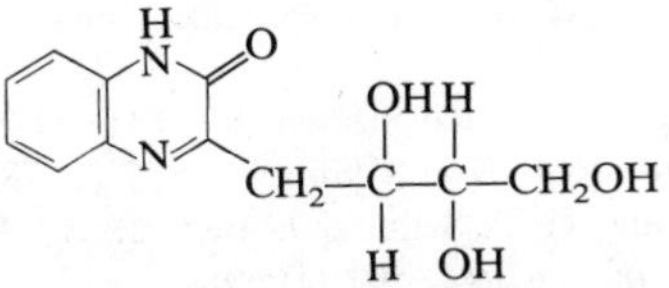 2(1H)-quinoxalinone, 3-CH_2—C(OH)(H)—C(H)(OH)—CH_2OH [h]	160–161	14
	2-(D-Arabinotetrahydroxybutyl)	152 decomp.	16
	Monohydrate	189–190, 192[i]	3, 4, 6, 7
	Tetraacetyl	119–120	6
	2-(D-Lyxotetrahydroxybutyl)	138	5
	2-(L-Xylotetrahydroxybutyl)	149–150, 152	2, 17
$C_{12}H_{15}N_3O_4$	6-Amino-2-(D-arabinotetrahydroxybutyl)[j]	216 decomp.	18
	7-Amino-2-(D-arabinotetrahydroxybutyl)[k]	210 decomp.	18
$C_{13}H_{16}N_2O_4$	6(or 7)-Methyl-2-(D-arabinotetrahydroxybutyl)[l]	176 decomp.	19
	Tetraacetyl[m]	70	19
	6(or 7)-Methyl-2-(D-lyxotetrahydroxybutyl)	145 decomp.	5

[a] Mixture had $[\alpha]_D^{25} - 129.2°$ (c 2.0, DMF).
[b] $[\alpha]_D^{20} - 160°$ (c 1.0, C_5H_5N).
[c] $[\alpha]_D^{18} - 30°$ (c 0.5, $CHCl_3$).
[d] $[\alpha]_D^{20} - 51°$ (c 0.6, C_5H_5N).
[e] $[\alpha]_D^{20} - 177°$ (c 0.5, C_5H_5N).
[f] $[\alpha]_D^{20} - 55.3°$ (c 0.5, MeOH).
[g] $[\alpha]_D^{19} - 58.4°$ (c 0.48, MeOH).
[h] $[\alpha]_D^{20} + 39.4°$ (c 1.01, MeOH).
[i] $[\alpha]_D^{20} - 150°$ (c 1.0, C_5H_5N).
[j] $[\alpha]_D^{23} - 166°$ (c 0.5, C_5H_5N).
[k] $[\alpha]_D^{23} - 175°$ (c 0.5, C_5H_5N).
[l] $[\alpha]_D^{20} - 181°$ (c 1.0, C_5H_5N).
[m] $[\alpha]_D^{20} - 37°$ (c 1.0, $CHCl_3$).

TABLE 1 (*contd.*)

Molecular formula	Substituents or structure	m.p., °C	Ref.
$C_{14}H_{16}N_2O_4$	[structure: 3-substituted quinoxalin-2(1H)-one, CH₂–CH(O–CHMe–O–CH₂)–CH(OH), with note n]	220–221	15
	Monoacetyl	188–190	15
$C_{18}H_{24}N_2O_9$	2-[2-(β-D-Galactosido)-D-arabinotetra-hydroxybutyl]	110 decomp.	20

[n] $[\alpha]_D^{20} -66°$ (*c* 0.5, DMF).

IV. References

1. J. C. E. Simpson, "Condensed Pyridazine and Pyrazine Rings," Interscience, New York, 1953, p. 300.
2. W. S. Chilton and R. C. Krahn, *J. Am. Chem. Soc.*, **90,** 1318 (1968).
3. H. I. X. Mager and W. Berends, *Rec. Trav. Chim.*, **77,** 827 (1958).
4. S. Gerchakov, P. J. Whitman, and H. P. Schultz, *J. Med. Chem.*, **9,** 266 (1966).
5. G. Henseke and C. Bauer, *Chem. Ber.*, **92,** 501 (1959).
6. H. Horton and M. J. Miller, *J. Org. Chem.*, **30,** 2457 (1965).
7. G. Henseke and M. Winter, *Chem. Ber.*, **89,** 956 (1956).
8. W. S. Chilton and R. C. Krahn, *J. Am. Chem. Soc.*, **89,** 4129 (1967).
9. H. R. Moreno and H. P. Schultz, *J. Med. Chem.*, **13,** 119 (1970).
10. A. Gomez-Sanchez, M. Yruela Antinolo, and F. Garcia, *An. Real. Soc. Espan. Fis. y Quim.* (Madrid), **50B,** 431 (1954); *Chem. Abstr.*, **52,** 11078 (1958).
11. A. Gomez-Sanchez and M. Yruela Antinolo, *An. Real. Soc. Espan. Fis. y Quim.* (Madrid), **51B,** 423 (1955); *Chem. Abstr.*, **50,** 10108 (1956).
12. N. P. Buu-Hoi, J. N. Vallat, G. Saint-Ruf, and G. Lambelin, *Chim. Ther.*, **6,** 245 (1971); *Chem. Abstr.*, **76,** 3794*s* (1972).
13. H. R. Moreno and H. P. Schultz, *J. Med. Chem.*, **13,** 1005 (1970).
14. R. Kuhn, D. Wieser, and H. Fischer, *Justus Liebigs Ann. Chem.*, **628,** 207 (1959).
15. G. Baschang and H. Fritz, *Helv. Chim. Acta*, **52,** 300 (1969).
16. N. K. Dutt, G. S. Sanayal, and K. Nag, *Anal. Chim. Acta*, **41,** 331 (1968).
17. G. Henseke and W. Liebenow, *Chem. Ber.*, **87,** 1068 (1954).
18. G. Henseke and R. Jacobi, *Justus Liebigs Ann. Chem.*, **684,** 146 (1965).
19. G. Henseke and K. H. Bähner, *Chem. Ber.*, **91,** 1605 (1958).
20. G. Henseke and E. Brose, *Chem. Ber.*, **91,** 2273 (1958).

CHAPTER XVII

Quinoxaline Quaternary Salts

I. Methods of Preparation 247
1. By Quaternization of Quinoxalines 247
2. From *N*-Substituted *o*-phenylenediamines 248
3. Miscellaneous 249
II. Properties 250
III. Uses 254
IV. Tables 255
Table 1. Quinoxaline Quaternary Salts. 255
V. References 259

Quinoxaline quaternary salts are an interesting group of compounds which have attracted a good deal of attention since the previous review was written.[1]

I. Methods of Preparation

1. By Quaternization of Quinoxalines

Quinoxalines have been quaternized with a number of reagents such as alkyl iodides,[2] alkyl sulfates,[2] methyl *p*-toluenesulfonate,[3,4] phenacyl bromides,[2] and bromoesters.[5] The conditions for carrying out quaternization of quinoxalines are sometimes quite critical. For example, the optimum conditions for reacting 2-methylquinoxaline with methyl sulfate involve the heterogeneous reaction of an aqueous solution of the base with a carbon tetrachloride solution of the methylating reagent in an atmosphere of nitrogen. Treatment of the water-soluble product with perchlorate ion gives 1,3-dimethylquinoxalinium perchlorate.[6] In general, quaternization of a 2-substituted quinoxaline occurs on the ring nitrogen remote from the substituent.[4,6] The necessity for carrying out the quaternization of 2-methylquinoxaline in an inert atmosphere follows from the observation that 1,3-dimethylquinoxalinium salts readily undergo oxidative dimerization. For example, the quinoxalinophenazinium salt **1** is obtained when 2-methylquinoxaline is treated with methyl sulfate in acetonitrile solution.[7] When quinoxaline 1-oxides are quaternized, 4-substituted quinoxalinium salts are formed.[8,9]

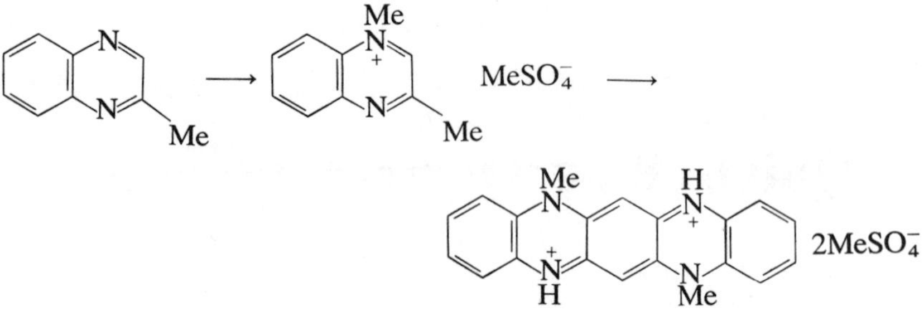

$MeSO_4^-$ ⟶ $2MeSO_4^-$

1

2. From *N*-Substituted *o*-Phenylenediamines

Quinoxalinium salts have been prepared from *N*-substituted *o*-phenylenediamines by condensation with either 2-halogenoketones or 1,2-diketones. In the first case ring closure is accompanied by oxidation of the first-formed 1,2-dihydroquinoxaline to the quinoxalinium salt (Scheme 1).[10] The formation of 1-substituted 2,3-dimethylquinoxalinium

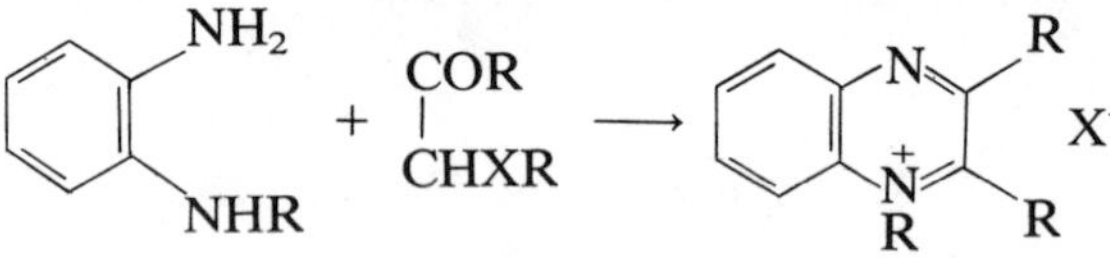

Scheme 1

perchlorates (**2**) by reaction of the substituted diamines with diacetyl is carried out in strongly acidic solution[11] since under neutral conditions unstable enamines (**3**) are formed.

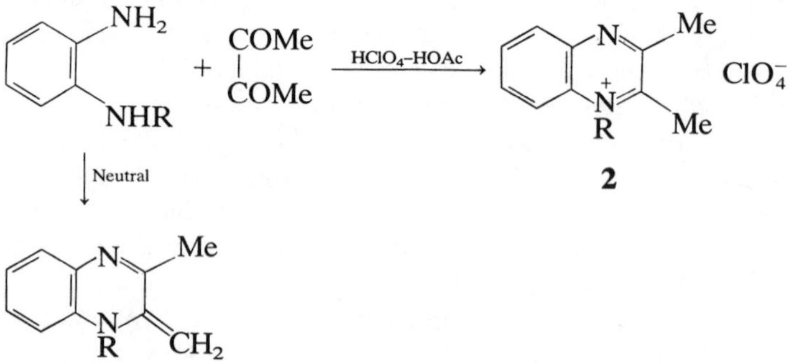

$HClO_4$–HOAc; ClO_4^-; Neutral

2

3

The preparation of 6- and 7-amino-1,2,3-trimethylquinoxalinium chlorides is carried out in an analogous manner.[46,47] Quaternization of 6-amino-2,3-dimethylquinoxaline with methyl sulfate occurs on the exocyclic nitrogen atom. Thus basification of the product yields 2,3-dimethyl-6-methylaminoquinoxaline.[46] Isonitrosoketones, rather than the corresponding 1,2-diketones, are sometimes used in quinoxalinium salt formation, as illustrated by the preparation of compound **4**.[10]

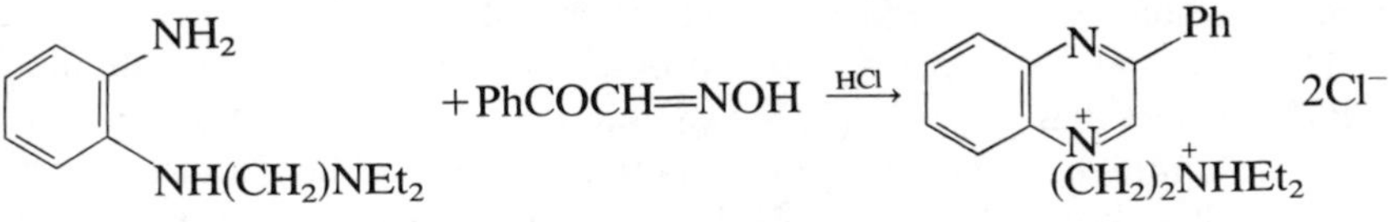

NH_2 / $NH(CH_2)NEt_2$ + PhCOCH=NOH $\xrightarrow{HCl}$ Ph, $(CH_2)_2\overset{+}{N}HEt_2$ $2Cl^-$

4

3. Miscellaneous

The quinoxalinium imine **6** is formed on treatment of the triazepinone **5** with hydrochloric acid,[12] and a product (**7**) of similar structure is formed when quinoxaline is heated with ethyl azidoformate.[13]

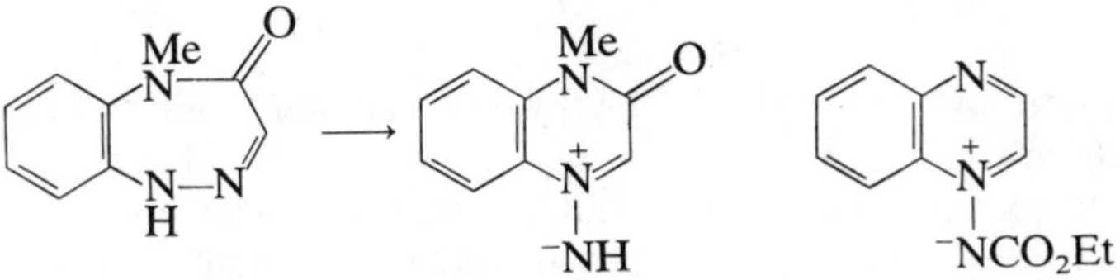

Me, O, N–N, H → Me, O, ^-NH ; $^-NCO_2Et$

5 **6** **7**

Reaction of 3,4-dihydro-3-oxo-2-phenylquinoxaline 1-oxides **8** with acetic anhydride in the presence of perchloric acid at room temperature gives high yields (80–96%) of the yellow to red *N*-acetoxyquinoxalinium perchlorates **9**. The salts ($R^1 = H$ or MeO; $R^2 = Me$) decompose rapidly at room temperature, but the remaining quinoxalinium salts are relatively stable.[14]

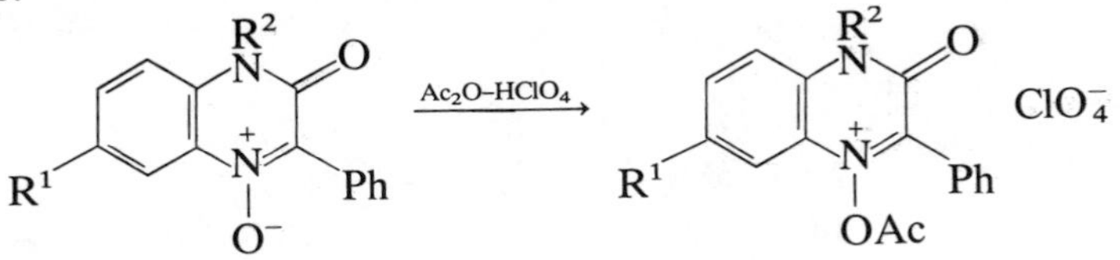

R^2, O, R^1, Ph, O^- $\xrightarrow{Ac_2O–HClO_4}$ R^2, O, R^1, Ph, OAc ClO_4^-

8 **9**

$R^1 = H$, Cl, or MeO
$R^2 = H$ or Me

II. Properties

^{14}N–H spin couplings have been observed in a number of methylpyrazinium iodides. Thus the ^{1}H NMR spectrum of 1-methylquinoxalinium iodide measured in D_2O at 96° shows a H–2, H–3 coupling of 3.0 Hz and a N–1, H–3 coupling of 2.1 Hz. No N–1, H–2 coupling is observed under these conditions; the resonance of H–2 appears as a sharp doublet when the methyl protons are irradiated.[15]

The UV and ^{1}H NMR spectral properties of 1-methylquinoxalinium iodide suggest that at pH 9–12, the pseudo base **10** is in equilibrium with the covalent hydrate **11** formed by addition of a molecule of water across the C–3, N–4 bond. At higher pH the anion **12** is formed, and this is less susceptible to covalent hydration across the C–3, N–4 bond.[16]

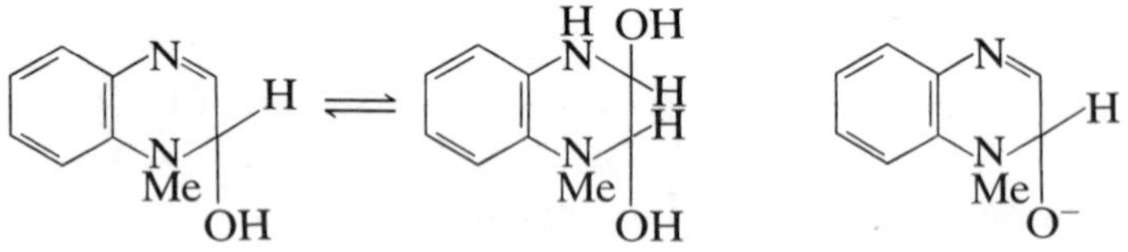

10 **11** **12**

Reduction of 1-methyl- or 1-ethylquinoxalinium salts with sodium borohydride yields 1-alkyltetrahydroquinoxalines,[17] and treatment of 1,3-dimethylquinoxalinium salts with dilute ammonia gives the corresponding 1,3-dimethylquinoxalinones (**13**).[6] On oxidation with silver oxide, diacylated *o*-phenylenediamines (**14**) are formed.[6]

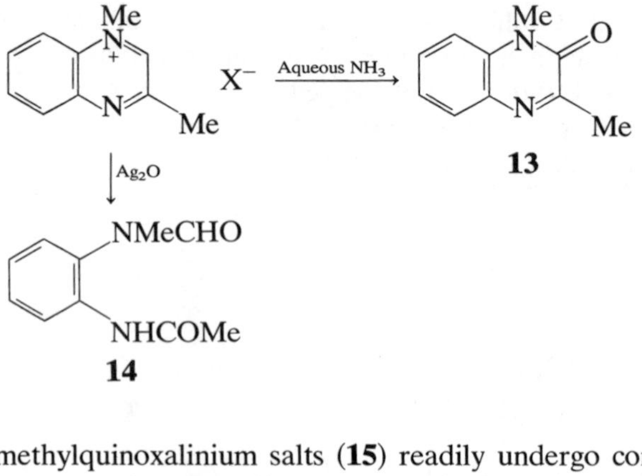

Aqueous NH_3 Ag_2O

13 **14**

1,2,3-Trimethylquinoxalinium salts (**15**) readily undergo condensation with *p*-nitroso-*N,N*-dimethylaniline and substitution with aryldiazonium ions to give the products **16** and **17** shown.[18] On treatment of 1,2,3-trimethylquinoxalinium salts with sodium acetate a mixture of two products, **19** and **21**, is obtained. Compound **19** is formed by oxidative dimerization of the intermediate **18** resulting from deprotonation of the 2-methyl group. The quinoxalinophenazine **21** arises from a reaction sequence involving deprotonation of the 3-methyl group and addition.[19] Treatment of 1,2,3-trimethylquinoxalinium salts with acetone and ammonia gives compound **22,** the product of addition of the intermediate **18** with the imine **20**.[20]

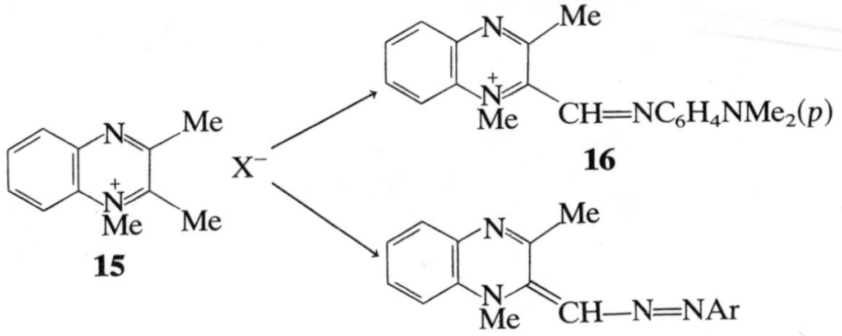

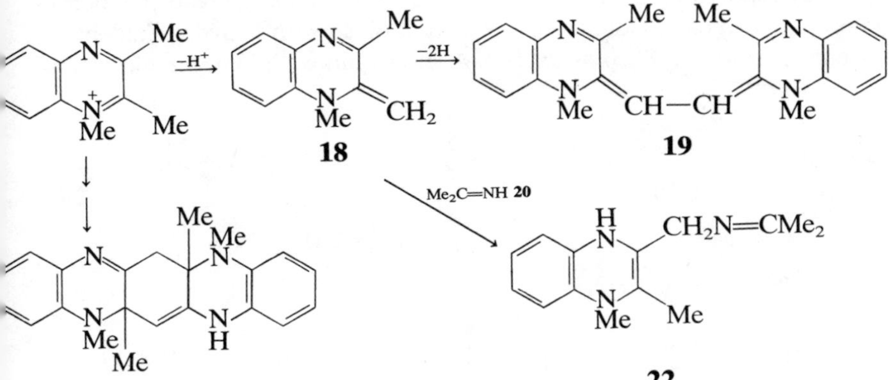

A series of dyestuffs has been prepared from 1-alkyl-2,3-dimethylquinoxalinium salts by reaction with 2,3-dichloro-1,4-naphthoquinone. The initial products of condensation such as **23** are cyclized to give naphthophenazinones (e.g., **24**) in base[21] but undergo acid-catalyzed cyclization to give pentacyclic derivatives such as **25**.[45]

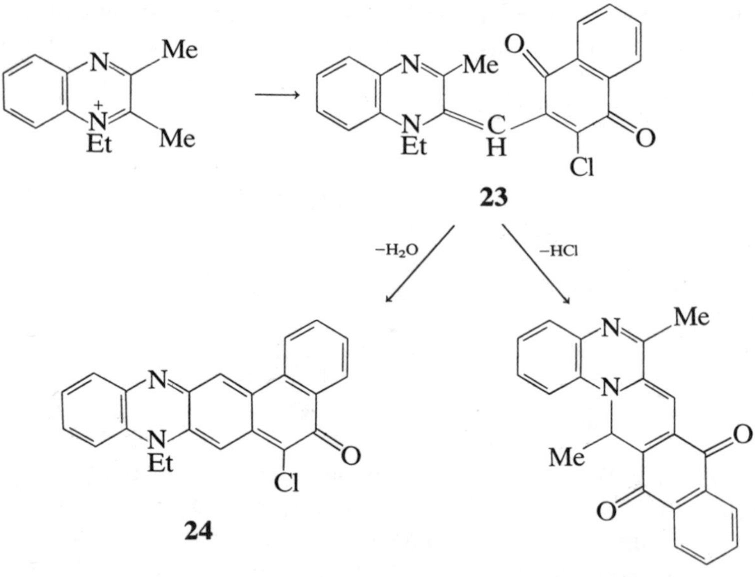

23

24

25

The cyanine salt **27** is obtained by treatment of the quinoxalinium perchlorate **26** in pyridine with triethylorthoformate,[22] and as expected the corresponding methylsulfate (**28**) is found to undergo ready condensation with aromatic aldehydes.[23] Reaction of **28** with *o*-hydroxyaldehydes yields spiropyrans such as **29**.[24] 1-Acetoxyquinoxalinium perchlorates of

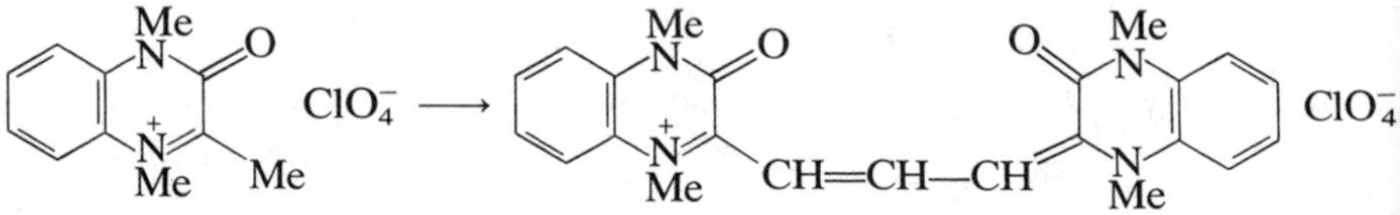

26

27

type **30** undergo ready substitution reactions with a variety of nucleophiles at position 6; for example, with sodium acetate, 6-acetoxy derivatives (**32**) are formed, possibly through intermediates of type **31**.[14]

The zwitterion **34**, formed by treating the quinoxalinium salt **33** with base, undergoes addition reactions when treated with water, dimethyl acetylenedicarboxylate, and phenyl isocyanate.[25] Base-promoted reaction with nitromethane results in pyrrolo[1,2-*a*]quinoxaline formation.[26] The salt derived from quinoxaline and ethyl bromoacetate similarly gives a

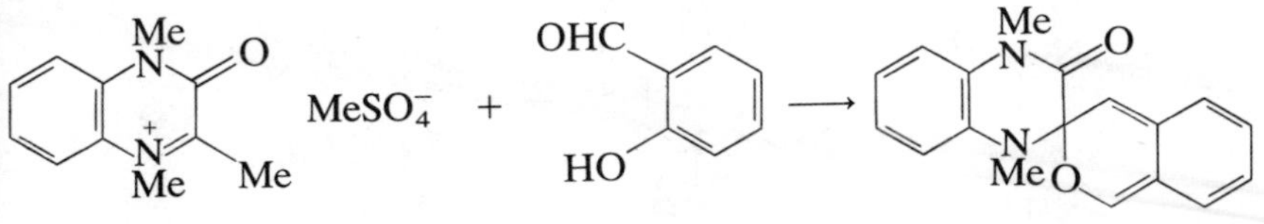

28 29 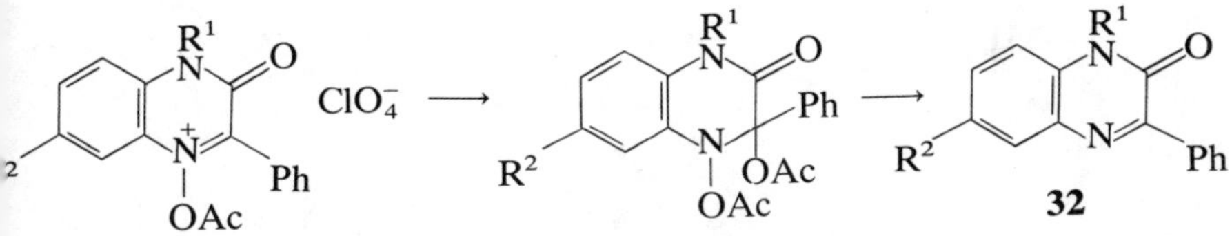 30 31 32 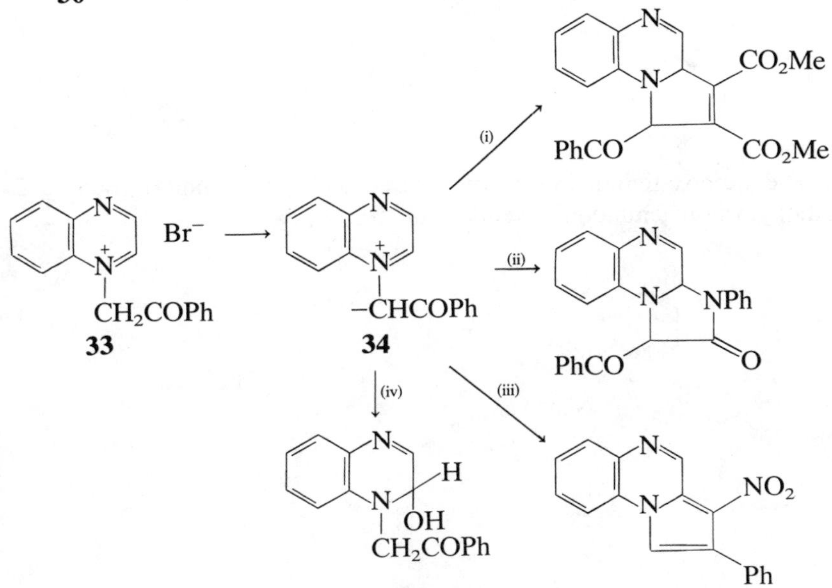 33 34

Reagents (i) $MeO_2CC{\equiv}CCO_2Me$
(ii) PhNCO
(iii) $MeNO_2$
(iv) H_2O

zwitterion on treatment with triethylamine which also undergoes addition reactions with dimethyl acetylenedicarboxylate and phenyl isocyanate.[5] 1,3-Dipolar properties for the quinoxalinium imine (**35a,b**) have also been demonstrated; thus addition of dimethyl acetylenedicarboxylate gives an adduct (**36**) which has been transformed to the pyrazoloquinoxaline **37**.[12]

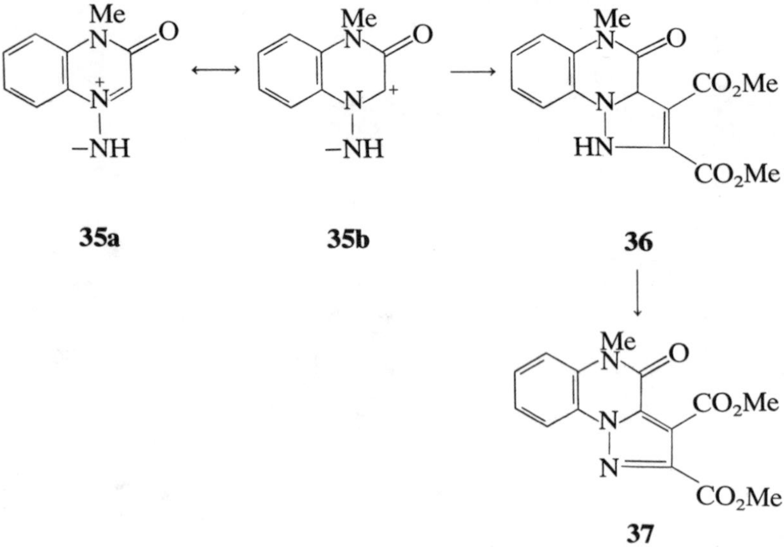

35a 35b 36

37

The quinoxalinium inner salt **7** undergoes photoisomerization to 2-ethoxycarbonylaminoquinoxaline (**38**).[13]

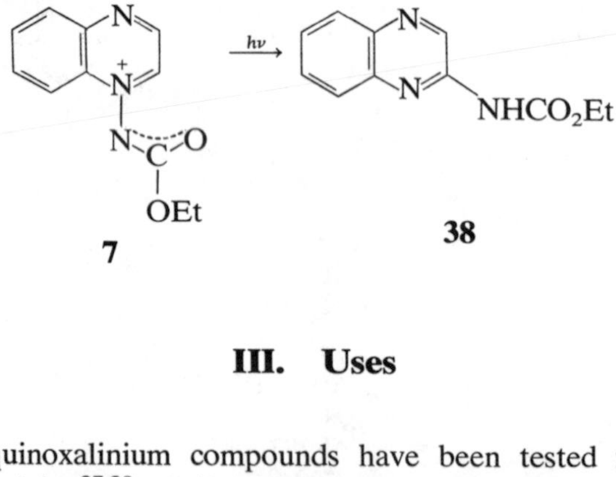

7 38

III. Uses

Various quinoxalinium compounds have been tested for their antitumour activity.[27,28] 1-(2-Diethylaminoethyl)-3-phenylquinoxalinium chloride hydrochloride (**4**) and related compounds have been tested for their action on the heart.[29] Other quinoxalinium salts have been evaluated for their antiviral[30] and insecticidal properties.[31] Cyanine dyes incorporating a quinoxalinium chromophore have been synthesized,[32] and quinoxalinium salts have been used to induce the photopolymerization of ethylenic derivatives.[33]

IV. Tables

TABLE 1. QUINOXALINE QUATERNARY SALTS[a]

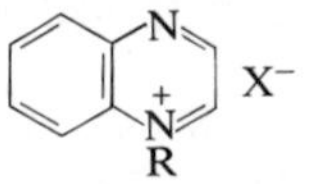

N + N R X^-

Molecular formula	Substituents or structure	m.p., °C	Ref.
$C_9H_8ClIN_2$	6(or 7)-Chloro-1-methyl iodide	190	2
$C_9H_9IN_2O$	5-Hydroxy-1-methyl iodide	177 decomp.	34
$C_9H_9IN_2O$	1-Methyl 4-oxide iodide	188–189	8
$C_9H_9N_3O$	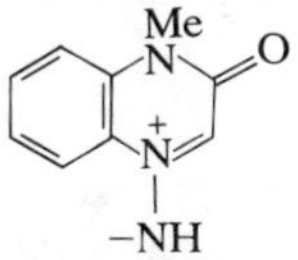 Me N O + N –NH	250–253 decomp.	12
$C_9H_{10}IN_3$	2(or 3)-Amino-1-methyl iodide	188	4
$C_9H_{11}IN_4$	1-Methyl-3,6-diamino iodide		
	Hydrate	233–235 decomp.	35
$C_{10}H_{10}ClIN_2$	6(or 7)-Chloro-1-ethyl iodide	171	2
$C_{10}H_{10}Cl_2N_2O_4$	6(or 7)-Chloro-1,3-dimethyl perchlorate	162–163	6
$C_{10}H_{11}ClN_2O_4$	1,3-Dimethyl perchlorate	207.5	6
$C_{10}H_{11}ClN_2O_5$	3,4-Dihydro-1,2-dimethyl-3-oxo perchlorate		22
$C_{10}H_{11}IN_2$	1-Ethyl iodide	141–143 decomp.	17
	1,6(or 1,7)-Dimethyl iodide	174–175	2
$C_{10}H_{12}N_2O_4S$	1-Methyl methyl sulfate	144–145	2
$C_{11}H_{11}BrN_2$	1-Allyl bromide	111	3
$C_{11}H_{11}Cl_3N_2O_4$	6,7-Dichloro-1,2,3-trimethyl perchlorate	228–229 decomp.	21
$C_{11}H_{11}N_3O_2$	1-Ethoxycarbonylamino hydroxide inner salt		13
$C_{11}H_{12}IN_3O$	3-Acetamido-1-methyl iodide	343–344 decomp.	4
$C_{11}H_{13}ClN_2O_4$	1,2,3-Trimethyl perchlorate	197 decomp.	18, 21
	1,3,6(or 7)-Trimethyl perchlorate	171	6
$C_{11}H_{13}FN_2O_3S$	1,2,3-Trimethyl fluorosulfate	142–146 decomp.	21
$C_{11}H_{13}IN_2$	1-Propyl iodide	145	3
$C_{11}H_{13}IN_2O$	5(or 8)-Ethoxy-1-methyl iodide	214–215 decomp.	36
$C_{11}H_{13}I_3N_2O$	1,2,3-Trimethyl 4-oxide triiodide	144–145	48
$C_{11}H_{14}ClN_3$	6-Amino-1,2,3-trimethyl chloride	270 decomp.	46
	7-Amino-1,2,3-trimethyl chloride	260 decomp.	47
$C_{11}H_{14}IN_3$	3-Dimethylamino-1-methyl iodide	245–246	37
$C_{11}H_{14}N_2O_4S$	3,4-Dihydro-1,2-dimethyl-3-oxo methylsulfate	145–146	23
$C_{11}H_{15}IN_2O_2$	4-Carboxy-1,2,3,4-tetrahydro-1,1-dimethyl iodide		
	Ethyl ester	172–173 decomp.	39

TABLE 1 (*contd.*)

Molecular formula	Substituents or structure	m.p., °C	Ref.
$C_{11}H_{17}IN_2$	1,2,3,4-Tetrahydro-1,1,4-trimethyl iodide	215–216	17, 38
$C_{12}H_{13}BrN_2O_2$	1-(2-Ethoxy-2-oxoethyl)bromide		5
$C_{12}H_{15}BF_4N_2$	1-Ethyl-2,3-dimethyl tetrafluoroborate	172–174	21
$C_{12}H_{15}IN_2O$	5(or 8)-Ethoxy-1-ethyl iodide	146–148 decomp.	36
$C_{12}H_{16}N_2O_4S$	1-Ethyl ethyl sulfate	171	2
	1,2,3-Trimethyl methyl sulfate	104.5, 150–153	10, 18
$C_{12}H_{16}N_2O_5S$	3,4-Dihydro-1,2,4-trimethyl-3-oxo methylsulfate		24
$C_{12}H_{17}IN_2O$	4-Acetyl-1,2,3,4-tetrahydro-1,1-dimethyl iodide	184–185 decomp.	39
$C_{12}H_{17}N_3SO_4$	6(or 7)-Methylamino-2,3-dimethyl methyl sulfate	251	46
$C_{13}H_{13}BrN_2O_2$	1-(4-Methoxy-4-oxo-2-butenyl) bromide		5
$C_{13}H_{14}ClN_3O_4$	1-(2-Cyanoethyl)-2,3-dimethyl perchlorate	174–176	11
$C_{13}H_{15}IN_4O_2$	1-Methyl-3,6-diacetamido iodide	301–303 decomp.	35
	Dihydrate	233–236	35
$C_{13}H_{16}N_3ClO$	6-Acetamido-1,2,3-trimethyl chloride		46
	7-Acetamido-1,2,3-trimethyl chloride		47
$C_{13}H_{17}ClN_2O_4$	1,2,3,6,7-Pentamethyl perchlorate	242–245 decomp.	21
$C_{13}H_{17}N_3O_5S$	7-Acetamido-1,3-dimethyl methyl sulfate	207 decomp.	6
$C_{13}H_{19}IN_2O_2$	4-Ethoxycarbonyl-1,2,3,4-tetrahydro-1,1-dimethyl iodide	172–173 decomp.	39
$C_{13}H_{22}N_2O_3$	1,2,3,4-Tetrahydro-6,7-dimethoxy-1,1,4-trimethyl hydroxide	185 decomp.	41
$C_{14}H_{17}ClN_2O_6$	1-(2-Acetoxyethyl)-2,3-dimethyl perchlorate	184–185	11
$C_{14}H_{19}BF_4N_2$	1-Ethyl-2,3,6,7-tetramethyl tetrafluoroborate	235–240	21
$C_{14}H_{19}BF_4N_2O_2$	1-Ethyl-6,7-dimethoxy-2,3-dimethyl tetrafluoroborate	243–245	45
$C_{14}H_{19}N_3O_4S$	5-Acetamido-1,2,3-trimethyl methyl sulfate		47
$C_{14}H_{20}Cl_4N_2$	1,1,4-Tris(2-chloroethyl)-1,2,3,4-tetrahydro chloride	145–146	40
$C_{14}H_{22}Cl_2N_2$	4-(2-Chloroethyl)-1,1-diethyl-1,2,3,4-tetrahydro chloride	70–72	40
$C_{14}H_{24}N_2O_3$	1,2,3,4-Tetrahydro-6,7-dimethoxy-1,1,2,4-tetramethyl hydroxide	200 decomp.	41
$C_{15}H_{13}BrN_2O$	1-(2-Oxo-2-phenylethyl) bromide		25
$C_{15}H_{13}IN_2$	1-Methyl-2(or 3)-phenyl iodide	221 decomp.	42
$C_{16}H_{10}N_2O_6$	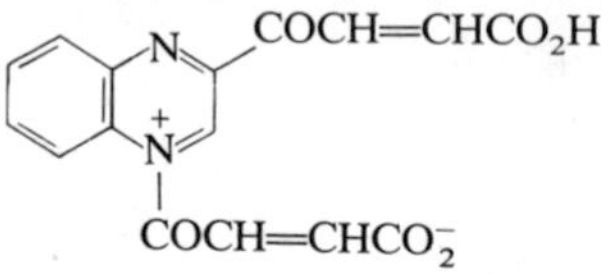 N, $\overset{+}{N}$, COCH=CHCO$_2$H, COCH=CHCO$_2^-$		25

TABLE 1 (*contd.*)

Molecular formula	Substituents or structure	m.p., °C	Ref.
$C_{16}H_{12}BrClN_2O$	1-(*p*-Chlorophenacyl) bromide	202.5	2
$C_{16}H_{12}BrN_3O_3$	1-(*m*-Nitrophenacyl) bromide	181	2
$C_{16}H_{12}Br_2N_2O$	1-(*p*-Bromophenacyl) bromide	206	2
$C_{16}H_{13}BrN_2O$	1-Phenacyl bromide	177	2
$C_{16}H_{12}Cl_2N_2O_7$	1-Acetoxy-7-chloro-3,4-dihydro-3-oxo-2-phenyl perchlorate		14
$C_{16}H_{13}ClN_2O_7$	1-Acetoxy-3,4-dihydro-3-oxo-2-phenyl perchlorate		14
$C_{16}H_{15}ClN_2O_4$	1-Phenyl-2,3-dimethyl perchlorate	230–233	11
$C_{16}H_{15}ClN_2O_4S$	6(or 7)-Chloro-1-methyl-3-phenyl methyl sulfate	232–234 decomp.	10 10
$C_{16}H_{15}IN_2$	1,6-Dimethyl-2(or 3)-phenyl iodide	226 decomp.	42
	1,7-Dimethyl-2(or 3)-phenyl iodide	222 decomp.	42
	1-Ethyl-3-phenyl iodide	188–190 decomp.	10
	1-Phenylethyl iodide	146–148	2
$C_{16}H_{16}N_2O_3S$	1-Methyl *p*-toluenesulfonate	150, 150–152	3, 17
$C_{16}H_{16}N_2O_4S$	1-Methyl-3-phenyl methyl sulfate	156–158	10
$C_{16}H_{21}ClN_2O_4$	1-Cyclohexyl-2,3-dimethyl perchlorate	159–160	11
$C_{17}H_{14}BrClN_2O$	1-(*p*-Chlorophenacyl)-6(or 7)-methyl bromide	226–227	2
$C_{17}H_{14}BrFN_2O$	1-(*p*-Fluorophenacyl)-6(or 7)-methyl bromide	192–193	2
$C_{17}H_{14}Br_2N_2O$	1-(*p*-Bromophenacyl)-6(or 7)-methyl bromide	226	2
$C_{17}H_{14}Cl_2N_2O_7$	1-Acetoxy-7-chloro-3,4-dihydro-4-methyl-3-oxo-2-phenyl perchlorate		14
$C_{17}H_{15}BrN_2O_2$	1-(*p*-Methoxyphenacyl) bromide	182.5	2
$C_{17}H_{15}ClN_2O_7$	1-Acetoxy-3,4-dihydro-4-methyl-3-oxo-2-phenyl perchlorate[b]		14
$C_{17}H_{15}ClN_2O_8$	1-Acetoxy-3,4-dihydro-7-methoxy-3-oxo-2-phenyl perchlorate		14
$C_{17}H_{16}ClN_3O_6$	1-(*p*-Nitrobenzyl)-2,3-dimethyl perchlorate	218–221	11
$C_{17}H_{16}Cl_2N_4O_4$	[structure: quinoxalinium ring, N-Me (+), 2-Me, 3-$CH_2N{=}NC_6H_4Cl$*p*] ClO_4^-	213 decomp.	18
$C_{17}H_{21}IN_2$	4-Benzyl-1,2,3,4-tetrahydro-1,1-dimethyl iodide	175–176	38
$C_{17}H_{23}ClN_2O_4$	1-Cyclohexyl-2,3,6-trimethyl perchlorate	183–185	11
	1-Cyclohexyl-2,3,7-trimethyl perchlorate	190–193	11
$C_{18}H_{17}BrN_2O$	6(or 7)-Methyl-1-(*p*-methylphenacyl) bromide	192–193	2
$C_{18}H_{17}BrN_2O_2$	1-(*p*-Methoxyphenacyl)-6(or 7)-methyl bromide	198	2

TABLE 1 (*contd.*)

Molecular formula	Substituents or structure	m.p., °C	Ref.
$C_{18}H_{17}ClN_2O_8$	1-Acetoxy-3,4-dihydro-7-methoxy-4-methyl-3-oxo-2-phenyl perchlorate[b]		14
$C_{18}H_{21}Cl_2N_3$	1-(2-Dimethylaminoethyl)-3-phenyl chloride hydrochloride	191–193 decomp.	10
$C_{18}H_{33}N_3O_3$	4-(2-Diethylaminoethyl)-1,2,3,4-tetrahydro-6,7-dimethoxy-1,1-dimethyl hydroxide	199–204 decomp.	41
$C_{19}H_{22}N_2O_3S$	1-Ethyl-2,3-dimethyl p-toluene sulfonate	180–182	10
$C_{20}H_{15}BrN_2O$	1-(2-Naphthoylmethyl) bromide Hydrate	182	2
$C_{20}H_{15}ClN_2$	1,3-Diphenyl chloride	210–213 decomp.	10
$C_{20}H_{24}N_4O_4S$	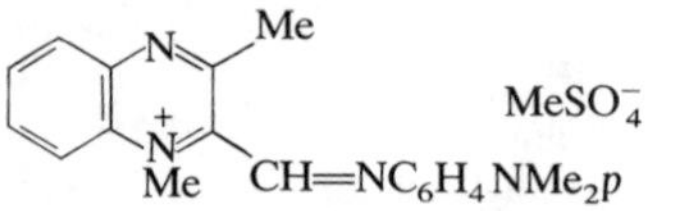 [structure: 1-Me quinoxalinium, 3-Me, 2-$CH{=}NC_6H_4NMe_2p$] $MeSO_4^-$	142 decomp.	18
$C_{20}H_{25}Cl_2N_3$	1-(2-Diethylaminoethyl)-3-phenyl chloride hydrochloride	184–186 decomp.	10
$C_{20}H_{27}ClN_2O_3$	4-(p-Chlorophenethyl)-1,2,3,4-tetrahydro-6,7-dimethoxy hydroxide	178–179 decomp.	41
$C_{21}H_{17}ClN_2O_5$	1-Methyl-2,3-diphenyl 4-oxide perchlorate	177	9, 43
$C_{21}H_{30}N_2O_4$	1,2,3,4-Tetrahydro-6,7-dimethoxy-4-(p-methoxyphenethyl)-1,1-dimethyl hydroxide	155–157 decomp.	41
$C_{22}H_{19}ClN_2O_5$	1,6(or 7)-Dimethyl-2,3-diphenyl 4-oxide perchlorate	220	9, 43
$C_{22}H_{20}N_2O_3S$	1-Methyl-3-phenyl p-toluenesulfonate	209–211	10
$C_{22}H_{22}N_2O_5S$	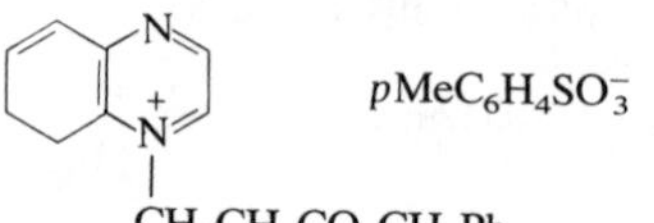 [structure: 5,6,7,8-tetrahydroquinoxalinium, N-$CH_2CH_2CO_2CH_2Ph$] $pMeC_6H_4SO_3^-$		31
$C_{23}H_{22}N_2O_3S$	1-Ethyl-3-phenyl p-toluenesulfonate	187–188	10
$C_{24}H_{25}Cl_2N_3$	1-(2-Dimethylaminoethyl)-2,3-diphenyl chloride hydrochloride Hemiethanolate	185–186 decomp.	10
$C_{26}H_{19}ClN_2O_2$	2,3-Bis(p-hydroxyphenyl)-1-phenyl chloride	>300 decomp.	44
$C_{26}H_{29}Cl_2N_3$	1-(2-Diethylaminoethyl)-2,3-diphenyl chloride hydrochloride	181–182 decomp.	10
$C_{28}H_{23}ClN_2O_6$	2,3-Bis(p-methoxyphenyl)-1-phenyl perchlorate	260 decomp.	44
$C_{28}H_{24}ClN_3O_4$	2(or 3)-(p-Dimethylaminophenyl)-1,3-(or 2)-diphenyl perchlorate	263	44

TABLE 1 (*contd.*)

Molecular formula	Substituents or structure	m.p., °C	Ref.
$C_{30}H_{27}ClN_2O_8$	2,3-Bis(3,4-dimethoxyphenyl)-1-phenyl perchlorate	155 decomp.	44
$C_{30}H_{29}ClN_4O_4$	2,3-Bis(*p*-dimethylaminophenyl)-1-phenyl perchlorate		44
$C_{32}H_{35}Cl_3N_4O_{12}$	2,3-Bis(*p*-dimethylaminophenyl)-1-phenyl perchlorate dimethoperchlorate		44

[a] It is important to note that in this tabulation by molecular formula, the quaternary salts of a given quinoxaline will appear as separate entries. For example, the methyl sulfate and methyl *p*-toluenesulfonate of the parent base are listed under $C_{10}H_{12}N_2O_4S$ and $C_{16}H_{16}N_2O_3S$, respectively.

[b] Unstable at room temperature.

V. References

1. J. C. E. Simpson, "Condensed Pyrazine and Pyridazine Rings, Interscience," New York, 1953, p. 286.
2. W. K. Easley and C. T. Bahner, *J. Am. Chem. Soc.*, **72,** 3803 (1950).
3. C. T. Bahner, L. R. Barclay, G. Biggerstaff, D. E. Bilancio, G. W. Blanc, M. Close, M. M. Isenberg, and E. Page, *J. Am. Chem. Soc.*, **75,** 4838 (1953).
4. C. M. Atkinson, C. W. Brown, and J. C. E. Simpson, *J. Chem. Soc.*, 26 (1956).
5. M. Ungureanu, I. Druta, and I. Zugravescu, *An. Stiint Univ. "Al. I. Cuza" Iasi, Sect. Ic*, **19,** 167 (1973); *Chem. Abstr.*, **80,** 108478*c* (1974).
6. H. Wahl, M.-T. Le Bris, and D. Berkovitch, *Bull. Soc. Chim. Fr.*, 1285 (1973).
7. H. Wahl, M.-T. Le Bris, and D. Berkovitch, *Bull. Soc. Chim. Fr.*, 1289 (1973).
8. J. K. Landquist, *J. Chem. Soc.*, 2816 (1953).
9. Yu. S. Rozum and N. N. Lisovskaya, *J. Gen. Chem. USSR*, **29,** 231 (1959).
10. J. Druey and A. Hüni, *Helv. Chim. Acta*, **35,** 2301 (1952).
11. D. Schelz and M. Priester, *Helv. Chim. Acta* **58,** 317 (1975).
12. S. Rossi, O. Pirola, and F. Selva, *Tetrahedron*, **24,** 6395 (1968).
13. B. Agai, K. Lempert, and J. Moller, *Acta Chim.* (Budapest), **80,** 465 (1974); *Chem. Abstr.*, **81,** 63582*j* (1974).
14. D. B. Livingstone and G. Tennant, *Chem. Ind.* (London), 848 (1973).
15. T. Goto, M. Isobe, M. Ohtsura, and K. Tori, *Tetrahedron Lett.*, 1511 (1968).
16. J. W. Bunting and W. G. Meathrel, *Can. J. Chem.*, **50,** 917 (1972).
17. R. F. Smith, W. J. Rebel, and T. N. Beach, *J. Org. Chem.*, **24,** 205 (1959).
18. M.-T. Le Bris, *Bull. Chim. Soc. Fr.*, 563 (1970).
19. M.-T. Le Bris, *Bull. Chim. Soc. Fr.*, 2277 (1970).
20. M.-T. Le Bris, *Bull. Chim. Soc. Fr.*, 1865 (1972).
21. D. Schelz, *Helv. Chim. Acta*, **57,** 1075 (1974).
22. L. Mandasescu, L. Stoicescu-Crivat, I. Gabe, S. Lica, and M. Stefanescu, *Acad. Rep. Populare Romine, Filiala Iasi, Studii Cercetari Stiint Chim.*, **13,** 115 (1962); *Chem. Abstr.*, **59,** 4069 (1963).

23. L. Mandasescu, L. Stoicescu-Crivat, and I. Gabe, *Acad. Rep. Populare Romine, Filiala Iasi, Studii Cercetari Stiint Chim.*, **11,** 311 (1960); *Chem. Abstr.*, **56,** 11751 (1962).
24. A. Oberlinner, H. Baumann, and K. Crychtol (BASF A.-G.), Ger. Offen. 2,230,225; *Chem. Abstr.*, **80,** 83061*w* (1974).
25. M. Ungareanu, I. Druta, M. Petrovanu, and I. Zugravescu, *An. Stiint Univ. "Al. I. Cuza" Iasi, Sect. Ic*, **18,** 49 (1972); *Chem. Abstr.*, **78,** 29712*e* (1973).
26. W. Kiel and F. Kröhnke, *Chem. Ber.*, **105,** 3709 (1972).
27. J. Leiter, A. D. Paradis, and V. S. Waravdekar, *J. Nat. Cancer Inst.*, **14,** 177 (1953); *Chem. Abstr.*, **47,** 11555 (1953).
28. M. G. Kelly, N. H. Smith, and J. Leiter, *J. Nat. Cancer Inst.*, **20,** 1113 (1958); *Chem. Abstr.*, **52,** 17492 (1958).
29. J. Tripod, W. Schuler and R. Meier, *Naunyn-Schmiedebergs Arch., Exp. Pathol. Pharmakol.*, **220,** 268 (1953); *Chem. Abstr.*, **48,** 2917 (1954).
30. E. Weston Hurst, J. K. Landquist, P. Melvin, J. M. Peters, N. Senior, J. A. Silk, and G. J. Stacey, *Brit. J. Pharmacol.*, **8,** 297 (1953).
31. S. A. Heininger (to Monsanto Chem. Co.), U.S. Pat. 2,857,310; *Chem. Abstr.*, **53,** 5291 (1959).
32. A. I. Kiprianov and G. A. Lezenko, *Zh. Org. Khim.*, **9,** 2587 (1973); *Chem. Abstr.*, **80,** 97324*w* (1974).
33. N. Baumann and H. P. Schlunke (to Ciba-Geigy A.-G.), Ger. Offen. 2,360,350; *Chem. Abstr.*, **81,** 170266*f* (1974).
34. A. Albert and A. Hampton, *J. Chem. Soc.*, 505 (1954).
35. S. S. Berg and E. W. Parnell, *J. Chem. Soc.*, 5275 (1961).
36. W. K. Easley and M. X. Sullivan, *J. Am. Chem. Soc.*, **74,** 4450 (1952).
37. G. B. Barlin and A. C. Young, *J. Chem. Soc. (B)*, 2323 (1971).
38. A. S. Elina and I. S. Musatova, *Chem. Heterocycl. Comp.*, **1,** 192 (1965).
39. J. S. Morley, *J. Chem. Soc.*, 4002 (1952).
40. J. Degutis and V. Breska, *Liet. TSR Mokslu Akad. Darb. Ser. B.* 105 (1968); *Chem. Abstr.*, **71,** 38911*b* (1969).
41. H. Zellner and G. Zellner, *Helv. Chim. Acta*, **49,** 913 (1966).
42. A. Dornow and N. Sassenberg, *Justus Liebigs Ann. Chem.*, **594,** 185 (1955).
43. Yu. S. Rozum, *J. Gen. Chem. USSR* **30,** 1657 (1960).
44. S. Bodforss, *Justus Liebigs Ann. Chem.*, **633,** 66 (1960).
45. D. Schelz and M. Priester, *Helv. Chim. Acta*, **58,** 2536 (1975).
46. M.-T. Le Bris, *Bull. Chim. Soc. Fr.*, 921 (1976).
47. M.-T. Le Bris, *Bull. Chim. Soc. Fr.*, 925 (1976).
48. J. K. Landquist and G. J. Stacey, *J. Chem. Soc.*, 2822 (1953).

CHAPTER XVIII

Reduced Quinoxalines

I. 1,2- and 1,4-Dihydroquinoxalines 261
II. Tetrahydroquinoxalines 266
1. Methods of Preparation 266
2. Properties 270
A. Physical Properties 270
B. Chemical Properties 272
3. Uses 274
III. Hexahydroquinoxalines 275
IV. Decahydroquinoxalines 275
V. Tables 276
Table 1. 1,2- and 1,4-Dihydroquinoxalines 276
Table 2. 1,2,3,4-Tetrahydroquinoxalines 278
Table 3. Decahydroquinoxalines 287
VI. References 287

This chapter updates the earlier review in Simpson's monograph except that consideration of 1,2-dihydro-2-oxoquinoxalines and 1,2,3,4-tetrahydro-2,3-dioxoquinoxalines has been transferred to Chapter V, which deals with quinoxalin-2-ones and quinoxaline-2,3-diones.[1]

I. 1,2- and 1,4-Dihydroquinoxalines

Reduction of quinoxaline with sodium in tetrahydrofuran,[2] or electrochemically in an alkaline medium,[3] gives 1,4-dihydroquinoxaline. This is described as a white, insoluble compound of m.p. 158–159°, and it is thought to be a 1,4- rather than a 1,2-dihydro compound because reaction with methyl magnesium iodide shows that it contains approximately 1.8 atoms of active hydrogen per molecule.

A systematic study of the electrochemical reduction of 2-methyl-, 2,3-dimethyl-, 2-phenyl-, 2,3-diphenyl-, and 2-methyl-3-phenylquinoxaline in alkaline or neutral media has been made. For example, reduction of 2-methylquinoxaline in either an alkaline or neutral medium gives 1,2-dihydro-3-methylquinoxaline; at pH 9 and a higher reduction potential, some 1,2,3,4-tetrahydro derivative is formed. The 1,2-dihydro derivative readily oxidizes to starting material on exposure to air. The primary products of electrochemical reduction are thought to be 1,4-dihydroquinoxalines; these subsequently rearrange to either the 1,2- or 3,4-dihydro derivatives. The driving force for rearrangement is the formation of a more conjugated system. Further reduction then gives a 1,2,3,4-tetrahydroquinoxaline (Scheme 1).[3]

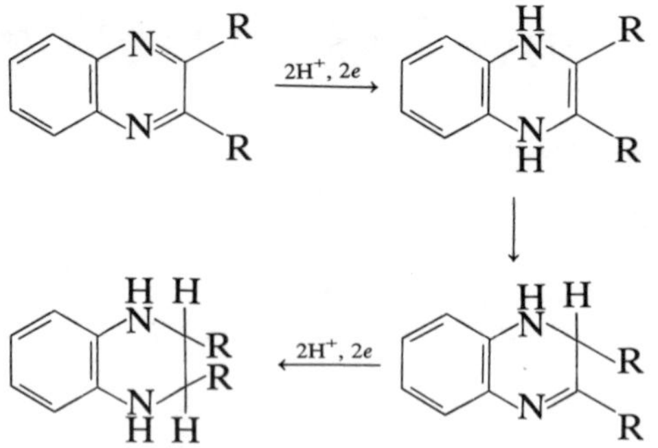

Scheme 1

Reduction of 2-phenylquinoxaline with zinc amalgam, in aqueous methanolic buffer at pH 9.2, yields an orange, crystalline 1,4-dihydro derivative. The 1,4-dihydro compound undergoes tautomeric rearrangement to the thermodynamically more stable 1,2-dihydro-3-phenylquinoxaline.[4] Reduction of 2-acetyl-3-methylquinoxaline with hydrogen in the presence of a palladium-on-carbon catalyst also gives a 1,4-dihydro derivative.[5] The kinetics of the reduction of azo compounds with dihydro compounds of this type have been measured; reduction may stop at the hydrazo stage, and alternatively cleavage products may be formed.[4,6]

1,2-Dihydroquinoxalines are formed in the photochemical reactions of quinoxalines with diethyl ether, tetrahydrofuran, and dioxan in the presence of benzophenone. Thus reaction of 2,3-dimethylquinoxaline and tetrahydrofuran yields the product **1** of 1,2-addition. With 2-*t*-butylquinoxaline a mixture of the diastereoisomers resulting from addition to the 3,4-bond is formed; but with quinoxaline itself, 2-substituted quinoxalines are produced. These are thought to arise from autoxidation of the intermediate 1,2-adducts.[7]

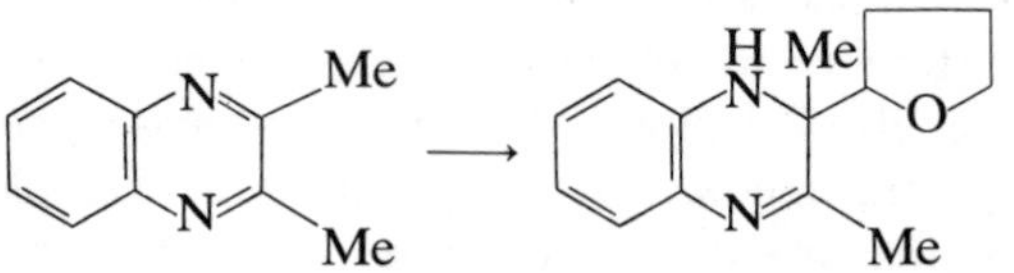

1

2-Methylquinoxaline reacts with both alkyl and aryl magnesium compounds to give, after hydrolysis, 2-alkyl- or 2-aryl-1,2-dihydro-3-methylquinoxalines. When the dihydro compound **2**, derived from *p*-tolylmagnesium bromide, is heated in ethanol, isomerization occurs, and 1,2-dihydro-2-methyl-3-*p*-tolylquinoxaline (**3**) is formed.[103]

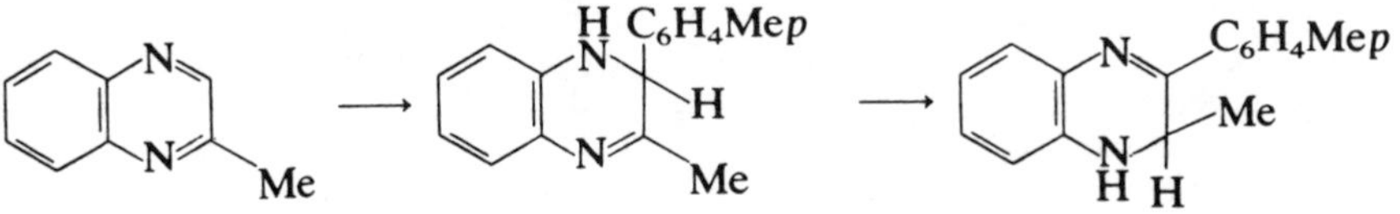

2

3

1,2-Dihydro-3-phenylquinoxaline (**4**) and related 3-aryl-1,2-dihydroquinoxalines have been synthesized by reaction of *o*-phenylenediamine with 2-haloacetophenones.[8] The IR spectrum of compound **4** shows a sharp absorption in the region 3300–3400 cm^{-1} due to

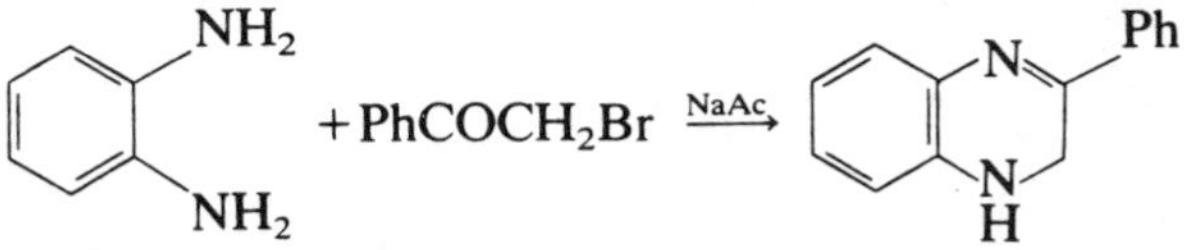

$+ PhCOCH_2Br \xrightarrow{NaAc}$

4

N–H stretching. The ^1H NMR spectrum, in deuterated acetone, shows a methylene singlet at δ 4.44 and a broad band at δ 5.3 due to N–H absorption, which disappears on addition of D_2O. 3-Aryl-1,2-dihydroquinoxalines give monoacetyl derivatives on treatment with acetic anhydride and are oxidized to the corresponding quinoxalines with benzoquinone, ferric chloride, or *m*-nitrobenzoic acid. Other 1,2-dihydroquinoxalines have been prepared from *o*-phenylenediamine as illustrated in Scheme 2.

An unexpected property of 2-amino-3,4-dihydroquinoxaline is its ability to form a molecular complex with 2-aminoquinoxaline. Reaction of *o*-phenylenediamine with mesityl oxide does not give a dihydroquinoxaline, as originally suggested; the product formed is the diazepine **5**.[12]

1,2-Dihydroquinoxalines are obtained by the reaction of benzils with *N*-phenyl-*o*-phenylenediamine;[13] thus reaction of the diamine with benzil gives 1,2-dihydro-1,2,3-triphenyl-2-quinoxalinol (**6**).[14] The quinoxalinol is very readily transformed to the corresponding 2-alkoxyderivatives (**7**) on treatment with alcohols.

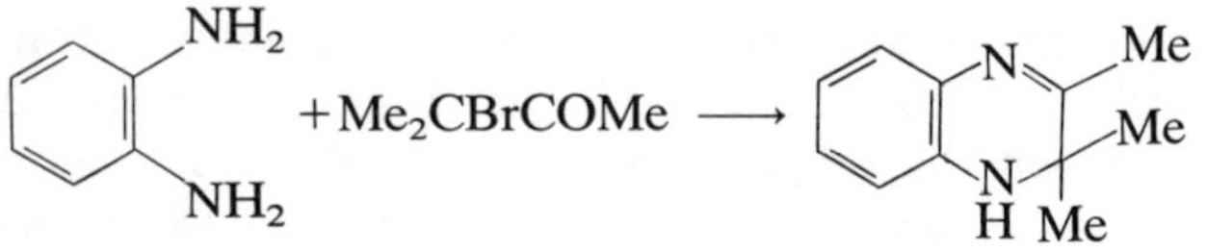

$+Me_2CBrCOMe \longrightarrow$ (Ref. 9)

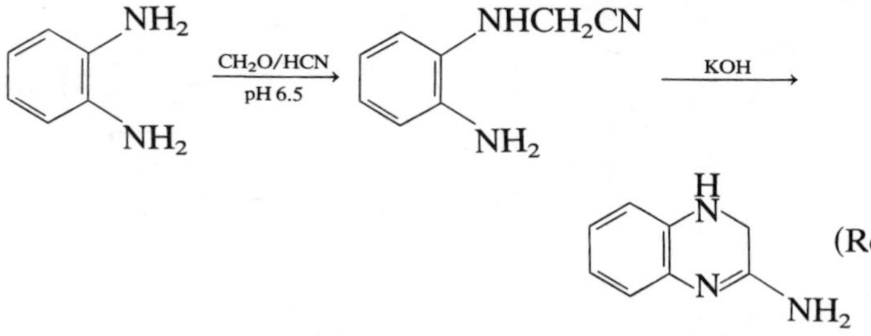

$\xrightarrow[pH\ 6.5]{CH_2O/HCN}$ $NHCH_2CN$ $\xrightarrow{KOH}$ (Ref. 10)

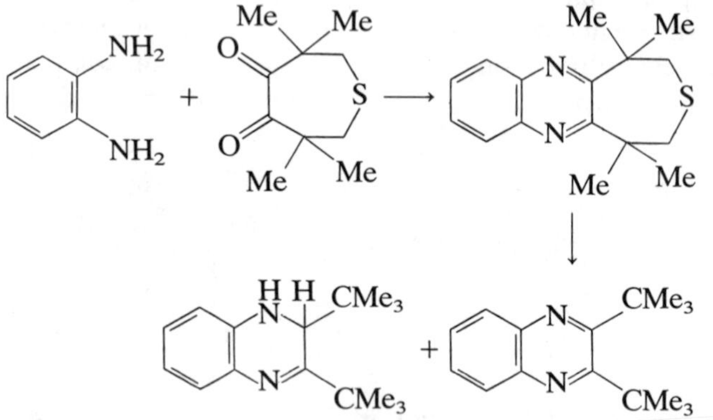

(Ref. 11)

Scheme 2

$+Me_2C{=}CHCOMe \longrightarrow$

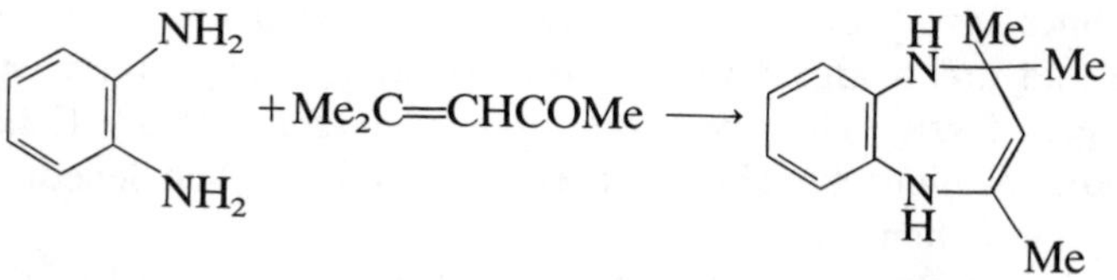

5

Irradiation of the diazepine *N*-oxide **8** gives a mixture of the 1,2-dihydroquinoxaline (**10**) and the oxadiazocine **11** through an intermediate oxaziridine (**9**). The dihydroquinoxaline undergoes acid hydrolysis to the quinoxalinone **12**.[15] It was subsequently shown that quinoxalinones of type **12** undergo ready reaction with ammonia or primary or secondary amines, in the presence of titanium tetrachloride, to give 3-amino- or 3-substituted amino-1,2-dihydroquinoxalines.[16]

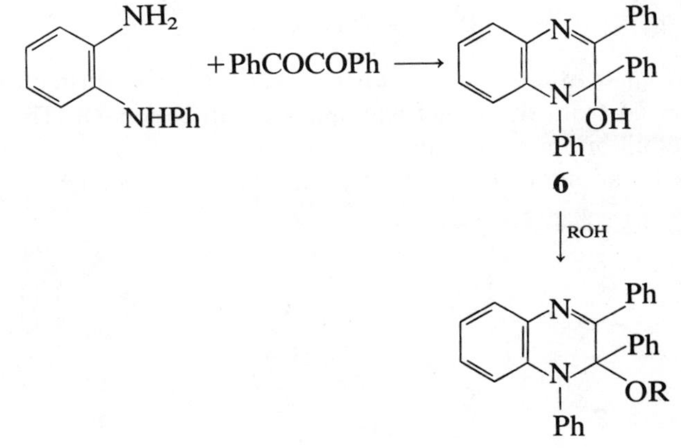

NH_2

NHPh

+ PhCOCOPh ⟶

N Ph Ph N OH Ph

6

ROH

N Ph Ph N OR Ph

7

R = Me, Et, or Bu

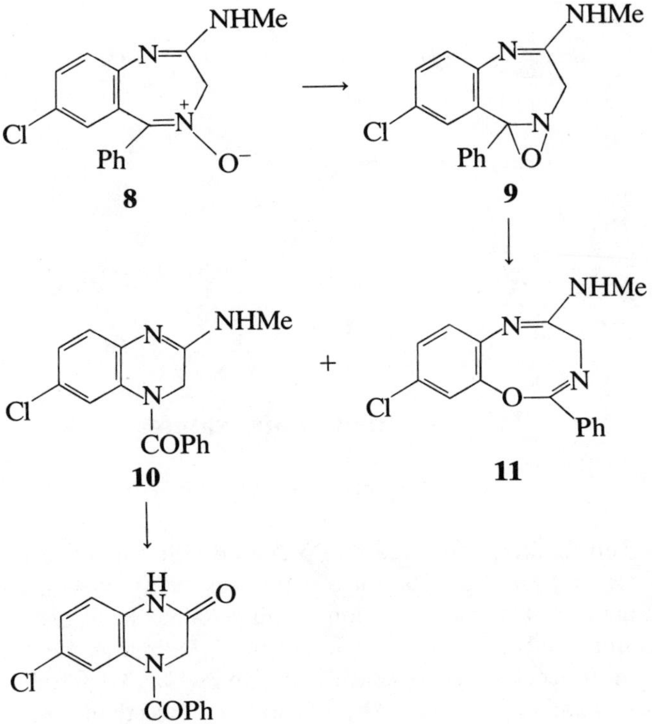

NHMe N Cl Ph $\overset{+}{N}$ O^-

8

⟶

NHMe N Cl Ph N O

9

NHMe N Cl N COPh

10

+

NHMe N N Cl O Ph

11

H N O Cl N COPh

12

Equilibrium constants have been measured for the formation of the pseudobase **14** from the 1-methylquinoxalinium cation **13**. The pseudobase is in equilibrium with a considerable amount of the covalent hydrate **15** formed by addition of water across the C-3, N-4 bond.[17]

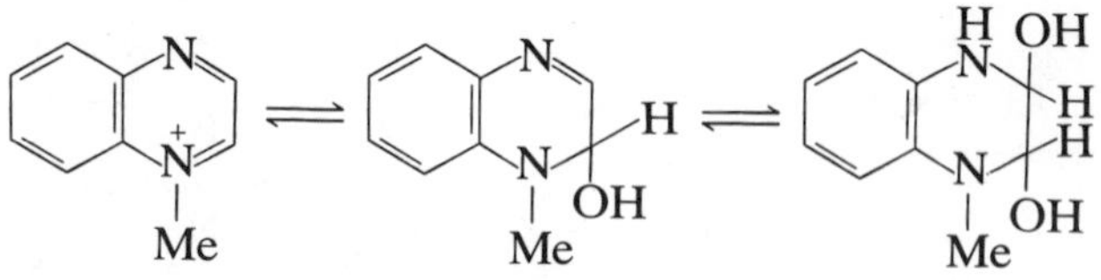

13 **14** **15**

Treatment of the methosulfate of 2,3-dimethylquinoxaline (**16**) with ammonia gives 1,2-dihydro-1,3-dimethyl-2-methylenequinoxaline (**17**) as an unstable oil; reaction with *p*-chlorophenyldiazonium chloride yields the 1,2-dihydroquinoxaline **18**.[18]

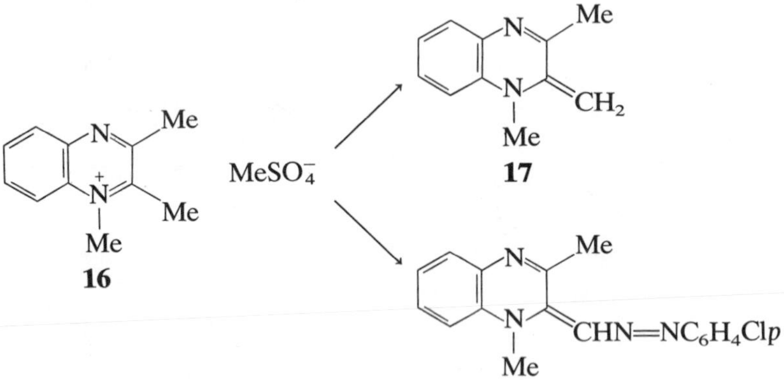

$MeSO_4^-$ **16** **17** **18**

II. Tetrahydroquinoxalines

1. Methods of Preparation

1,2,3,4-Tetrahydroquinoxalines are prepared either by the reduction of quinoxalines or by the cyclization of ortho-disubstituted benzenes. The lithium aluminum hydride reduction of quinoxaline itself gives 1,2,3,4-tetrahydroquinoxaline,[19] and reduction of the 2,3-dimethyl derivative has been shown to occur stereospecifically to give *cis*-1,2,3,4-tetrahydro-2,3-dimethylquinoxaline (m.p. 113–114°).[20] Sodium and ethanol reduction of 2,3-dimethylquinoxaline yields a mixture of *cis*- and *trans*-tetrahydro compounds. *dl-trans*-1,2,3,4-Tetrahydro-2,3-dimethylquinoxaline (m.p. 104–105°) is resolved into its optically active enantiomorphs (m.p. 94.5°) by fractional crystallization of the base oxalates.[21]

Several examples of the successful lithium aluminum hydride reduction of 1,2,3,4-tetrahydro-2-oxoquinoxalines to the corresponding tetrahydroquinoxalines have been recorded in the literature.[22–27] A moderate yield of 1,2,3,4-tetrahydroquinoxaline is obtained by reduction of 1,2,3,4-tetrahydro-2,3-dioxoquinoxaline with lithium aluminum hydride in *N*-ethylmorpholine.[28] The anils **19** give the tetrahydroquinoxalines **20** on treatment with sodium borohydride.[29–31] This reagent has also been used to reduce a number of 2,3-disubstituted quinoxaline di-*N*-oxides to the corresponding *cis*-tetrahydro derivatives. The trans isomers can be prepared by first reducing the *N*-oxides with sodium dithionite to the parent quinoxalines, and then the further reduction of the quinoxalines so formed with sodium and ethanol.[32]

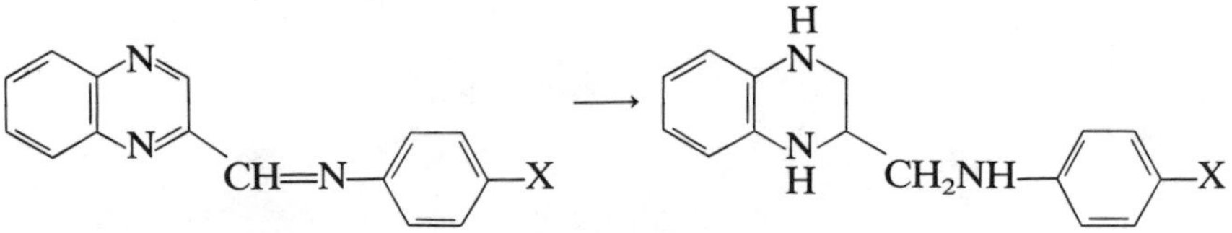

19 $\longrightarrow$ **20**

X = CO_2Et, Me, Cl, or CN

The selective reduction of 5- and 6-nitroquinoxalines to the corresponding 1,2,3,4-tetrahydro derivatives has been achieved by reaction with sodium borohydride in acetic acid at 5°. 6-Cyano-, 6-ethoxycarbonyl-, and 6-trifluoromethylquinoxaline have been reduced in a similar manner.[33]

There are numerous papers describing the reduction of quinoxalines to 1,2,3,4-tetrahydroquinoxalines with hydrogen and a metal catalyst.[34–45] Reduction at room temperature and pressures not exceeding 3 atm is normally employed; at higher temperatures and pressures decahydroquinoxalines may be obtained. The choice of solvent can sometimes be critical; thus a much increased yield of 1,2,3,4-tetrahydro-2-methylquinoxaline (10% → 93%) is obtained from the platinum-catalyzed hydrogenation of 2-methylquinoxaline when glacial acetic acid, rather than ethanol, is used as solvent.[41] Reduction of quinoxaline and 2-methylquinoxaline by formic acid in formamide or *N,N*-dimethylformamide yields the corresponding 1,4-diformyl-1,2,3,4-tetrahydro compound.[44,46] Higher-melting by-products are also isolated. 2-Oxoquinoxalines are similarly converted into 4-formyl-1,2,3,4-tetrahydro-2-oxoquinoxalines.[46]

1,2,3,4-Tetrahydro-2,3-diphenylquinoxaline results from reaction of quinoxaline with two molecular proportions of phenyllithium. With one molecular proportion of reagent, a 1:1 molecular complex of 1,2,3,4-tetrahydro-2,3-diphenylquinoxaline and quinoxaline is formed. Similar results are obtained with methyllithium; in each case products with trans stereochemistry are formed.[47] Allylmagnesium bromide may be similarly added in high yield to quinoxaline.[48]

Tetrahydroquinoxalines have also been prepared by cyclization of ortho-disubstituted benzenes. Thus high-temperature (200–210°) reaction of catechol with ethylenediamine gives 1,2,3,4-tetrahydroquinoxaline;[49] tetrahydroquinoxalines have also been prepared from hexafluorobenzene.[50,51] The reaction of *N,N'*-ditosyl-*o*-phenylenediamine with a variety of 1,2-dibromo compounds[52–54] gives 1,2,3,4-tetrahydroquinoxalines, as illustrated in Scheme 3.

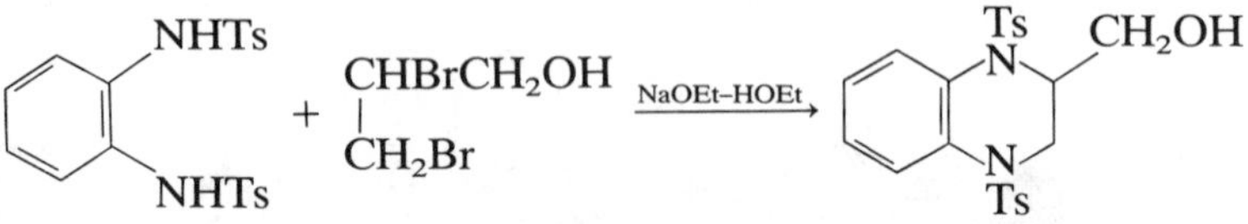

Scheme 3

A number of tetrahydroquinoxalines (**23**) have been derived from *o*-amino-*N*-(2-hydroxyethyl)anilines (**21**) by cyclization with concentrated hydrobromic or hydrochloric acid[55–57] and alternatively by reduction of *N*-2'-chloroethyl-*o*-nitroanilines (**22**).[58,59] The starting materials **21** and **22** are generally obtained from the appropriate *o*-chloronitrobenzene.

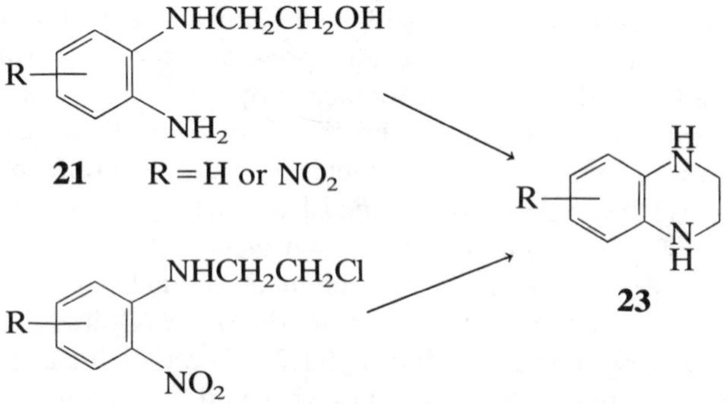

21 R = H or NO_2

23

22 R = Br, Cl, NHAc, NO_2, OMe, or SMe

Tetrahydroquinoxalines (e.g., **24**) are also formed by the reaction of *N*,*N'*-dimethyl-*o*-phenylenediamine with 1,4-dihalogenobut-2-enes. Hydrogenation of compound **24** gives the 2-ethyl derivative.[60] 1,2,3,4-Tetrahydro-1,4-dimethylquinoxaline (**25**) is obtained by an intramolecular ring-closure reaction; in this case a benzyne-type intermediate is involved.[61] Substituted 1,4-dimethyl-1,2,3,4-tetrahydroquinoxalines have been prepared by the cyclocondensation of the appropriate polychlorobenzene and *N*,*N'*-dimethylethylenediamine. In some cases it is advantageous to carry out the reaction in the presence of sodamide and sodium *t*-butoxide rather than in the presence of sodamide alone.[62]

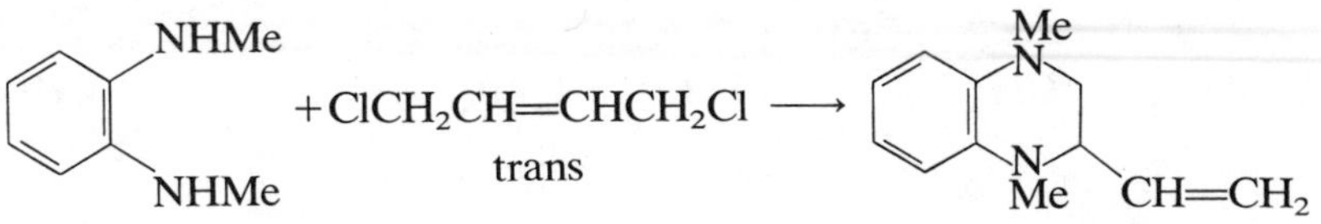

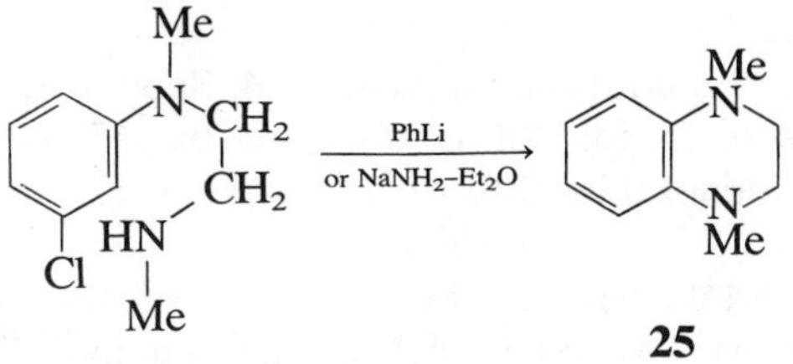

A novel method of pyrazine ring formation involves the electrolysis of an activated dialkene and is illustrated by the conversion of *o*-bis(β-diethoxycarbonylvinylamino)benzene (**26**) to the tetrahydroquinoxaline **27**. The latter compound decomposes on heating at about 160° to quinoxaline and diethyl malonate.[63]

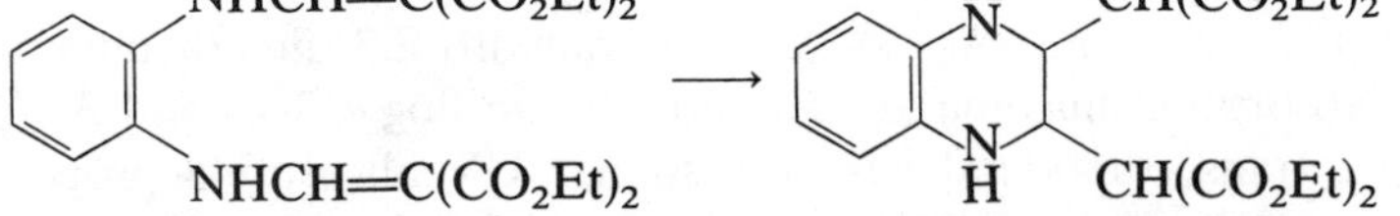

1,2,3,4-Tetrahydroquinoxalines **29** have also been obtained from the ring contraction of reduced 1,4-benzodiazepines by treatment with thionyl chloride or phosphoryl chloride. The mechanism proposed for this

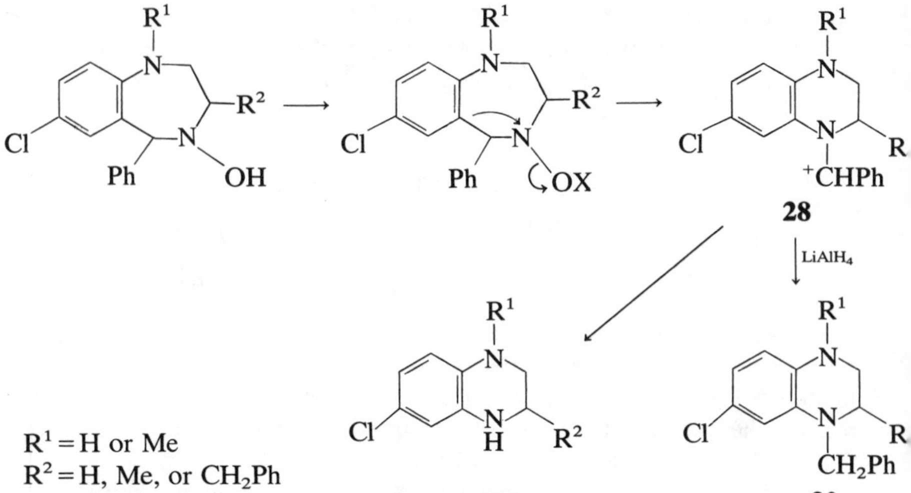

28

$LiAlH_4$

R^1 = H or Me
R^2 = H, Me, or CH_2Ph
X = $POCl_2$ or SOCl

29 30

Scheme 4

ring transformation is illustrated in Scheme 4. Support for the intermediacy of the carbonium ion **28** is obtained by the isolation of a 4-benzyltetrahydroquinoxaline (**30**) on reductive work-up of the reaction mixture.[64]

1-Morpholinocyclohexene has been used as starting material for the synthesis of 1-alkyl-2-amino-1,5,6,7-tetrahydroquinoxaline 3-carboxylates (**31**) as shown in Scheme 5.[65]

2. Properties

A. *Physical Properties*

The UV spectrum of *cis*-1,2,3,4-tetrahydro-2,3-dimethylquinoxaline shows absorption maxima at 256 and 310 nm (log ε 3.70 and 3.60); the isomeric trans compound has very similar UV absorption properties.[66] Reference IR[67,68] and NMR[69] spectra for 1,2,3,4-tetrahydroquinoxaline are available. The ^1H NMR spectrum shows absorption at δ 6.79, 5.93, and ca. 3.63 due to hydrogens at positions 2 and 3, 1 and 4, and 5, 6, 7, and 8, respectively. The thermodynamic acidity constants in water of tetrahydroquinoxaline are 2.11 and 4.84; data on the ionization properties of substituted tetrahydroquinoxalines are also available.[70,71]

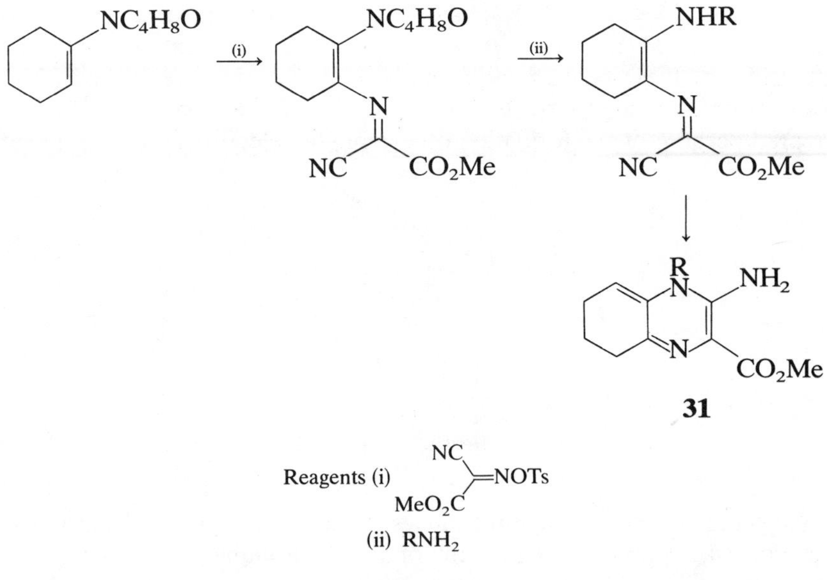

Reagents (i) NC(MeO$_2$C)C=NOTs
(ii) RNH$_2$

Scheme 5

On the basis of ^{1}H NMR measurements it has been concluded that 2,3-disubstituted 1,2,3,4-tetrahydroquinoxalines have a half-chair conformation. The chemical shift difference (0.5 ppm) between the 2- and 3-protons of *cis* and *trans*-1,2,3,4-tetrahydro-2,3-dimethylquinoxalines is ascribed to the magnetic anisotropy of the C–N bond. It is concluded that in both isomers interconversion of conformers is rapid down to $-87°$ and that in the trans isomer (**32a** and **32b**), the signal corresponds to hydrogens which are mainly axially oriented (**32a**). In the cis compound the signal is an averaged signal due to the rapidly interconverting axial and equatorial protons.[66] Comparison of the cis coupling constants, 3.35 and

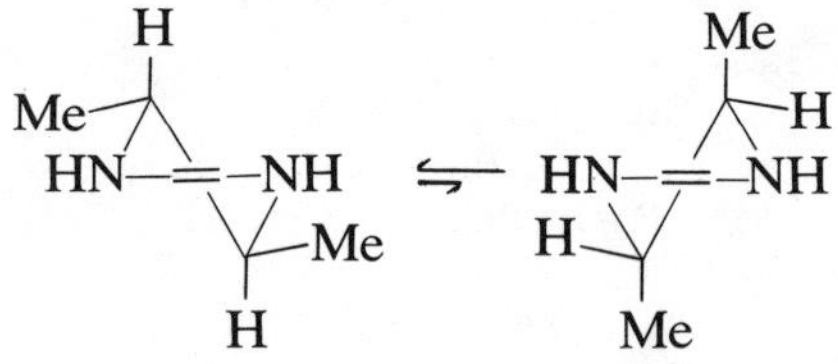

32a **32b**

8.4 Hz, for the 2- and 3-hydrogens of the cis monobenzoyl- and dibenzoyl-2,3-dimethyltetrahydroquinoxalines indicates that benzoylation of nitrogen results in a flattening of the heterocyclic ring.[72]

The mass spectrum of 1,2,3,4-tetrahydroquinoxaline shows a much more intense $(M-2)^{2+}$ peak at *m/e* 66 than M^{2+} peak at *m/e* 67. The intensities of these peaks relative to the molecular ion are 39 and 21%, respectively. It is suggested that loss of two hydrogens from the M^{2+} ion is favored because of the formation of a stable 6π-electron system (Scheme 6).[73]

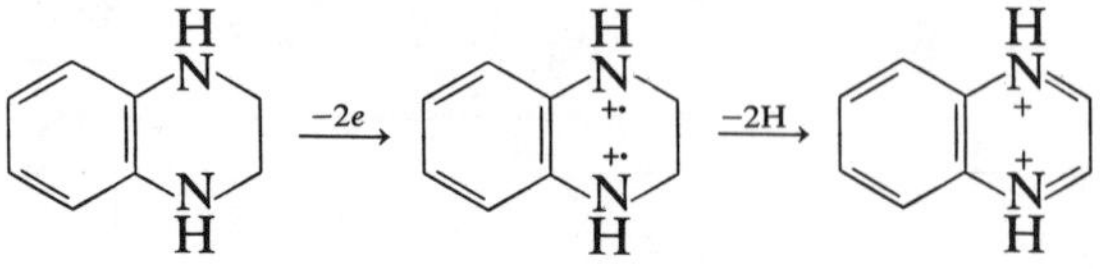

Scheme 6

B. *Chemical Properties*

Alkylation of tetrahydroquinoxalines cannot be stopped at the monoalkylation stage; however treatment of tetrahydroquinoxalines with benzenesulfonyl chloride, at a low temperature, gives monophenylsulfonyl derivatives. These on alkylation and subsequent acid hydrolysis of the protecting group yield 1-alkyl-1,2,3,4-tetrahydroquinoxalines.[74] An alternative approach involves the alkylation of tetrahydroquinoxalin-2-ones and subsequent reductive removal of the carbonyl function (Scheme 7).[22,75] The recorded difficulty in purification of the intermediate monoalkylated quinoxalinones may be due to alkylation both at amido and amino nitrogen. Subsequent lithium aluminum hydride reduction of either the N-1 or N-4 monoalkyl derivative however gives a common product.

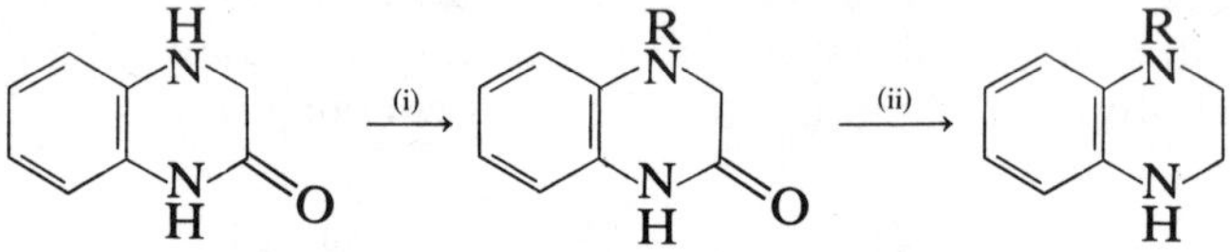

Reagents (i) RI–Na_2CO_3
(ii) $LiAlH_4$
R = Et, Pr, Bu, or CH_2Ph

Scheme 7

The best method for preparing 1-methyl- and 1-ethyltetrahydroquinoxaline (**35**) is by the sodium borohydride reduction of methylquinoxalinium *p*-toluenesulfonate (**33**) or ethylquinoxalinium iodide (**34**), respectively.[22] Further alkylation of 1-methyl-1,2,3,4-tetrahydroquinoxaline with methyl iodide, in the presence of sodium carbonate, gives the 1,1,4-trimethylquinoxalinium iodide **36**. Hofmann degradation of quaternary salts such as **36** yields alkylated *o*-phenylenediamine derivatives.[76] Monoacyl derivatives of tetrahydroquinoxaline may be readily obtained by direct reaction of the base with acetic anhydride, ethyl chloroformate, and benzoyl chloride. With equimolecular proportions of reactants the yields of monoacyl derivatives exceed 70%, provided that a relatively high pH is maintained. Below pH 5, the yields drop markedly and disubstitution predominates. This result is interpreted by the assumption that the free base acylates more rapidly than its monoacyl derivative. However at a lower pH the free base is present in a protonated form to a significant extent, and this is less readily acylated than its monoacyl derivative.[71]

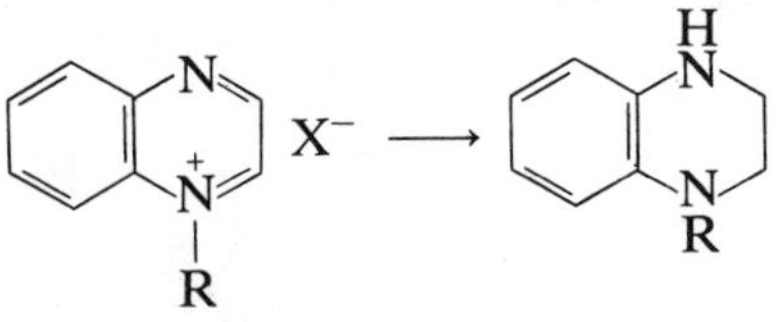

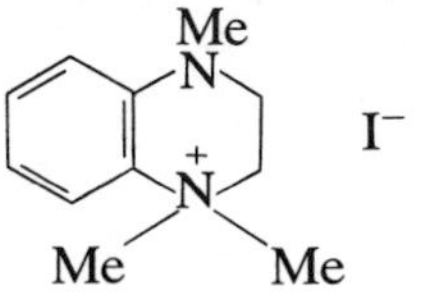

33 R = Me, X = $p\text{MeC}_6\text{H}_4\text{SO}_3^-$ **35** **36**

34 R = Et, X = I

The electronic effects of various substituents in the 6-position are pronounced enough to allow selective monoacylation. Thus 1,2,3,4-tetrahydro-6-nitroquinoxaline is exclusively monoacylated with acetic anhydride or trifluoracetic anhydride at the 4-position. The corresponding 1,4-diacylated derivatives are conveniently prepared by nitration of 1,4-diacetyl- and 1,4-bistrifluoroacetyl-1,2,3,4-tetrahydroquinoxaline, respectively.[45]

The ditosyl derivative of 2-hydroxymethyl-1,2,3,4-tetrahydroquinoxaline (**37**) undergoes an interesting ring expansion reaction to give the diazepine **39** on oxidation with dicyclohexylcarbodiimide (DCC) in dimethylsulfoxide. The intermediate nitrenium ion **38** is postulated.[27]

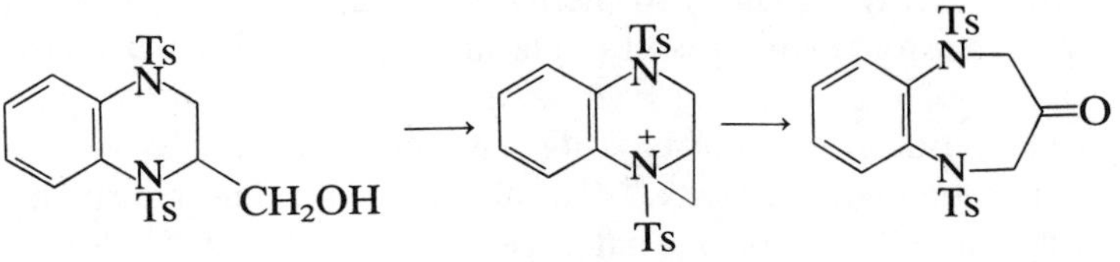

37 **38** **39**

Tetrahydroquinoxalines such as the ester (**20,** R = H) have been studied as model compounds for tetrahydrofolic acid. The latter has a vital role in one-carbon metabolism. On treating the ester with formaldehyde, in aqueous dioxan, the fused imidazoline **40** is isolated. Compound **40** undergoes acid-catalyzed rearrangement to the diazepine **42.** Using a suitably blocked ester (**20**, R = Me), it has been possible to isolate the formaldehyde condensation product **41**, formed by interaction with nitrogens 4 and 10. It is suggested that a compound of this type is formed prior to the imidazoline **40**.[31]

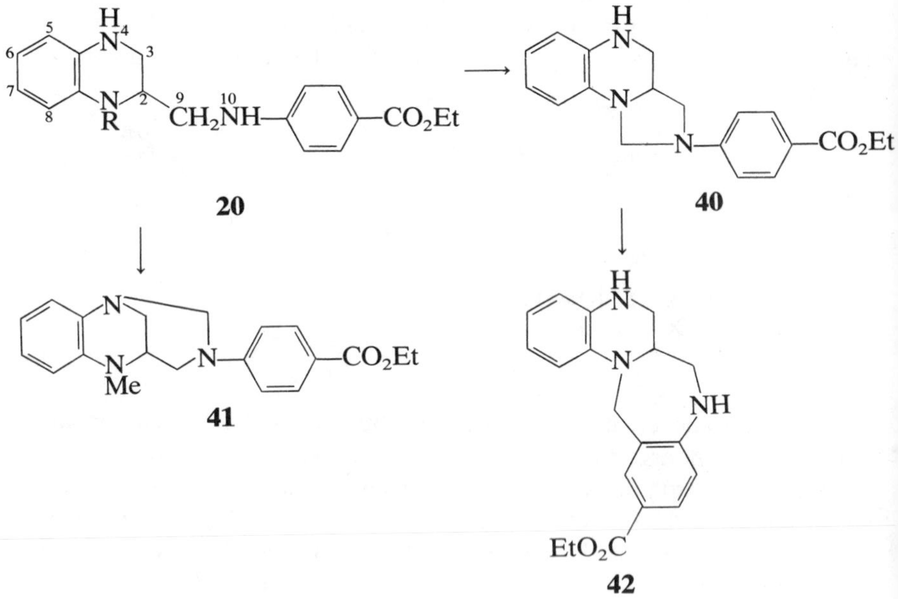

20

40

41

42

3. Uses

There are several indications, mainly in the patent literature, of possible uses of tetrahydroquinoxalines. Derivatives related in structure to Hetrazan (1-diethylcarbamoyl-4-methylpiperazine) have been prepared as potential drugs for the treatment of filarial infections.[71] Other tetrahydroquinoxalines of type **43** have been screened for antineoplastic activity[77] and bisindolyl derivatives of type **44** have been claimed to have a wide spectrum of biological activity.[78] Tetrahydroquinoxaline phosphates and thiophosphates have been patented as insecticides,[79] and other tetrahydroquinoxalines have been patented for protecting rubber against ozone degradation.[80–82] Tetrahydroquinoxalines have also been prepared in the search for new dyestuffs.[83,84]

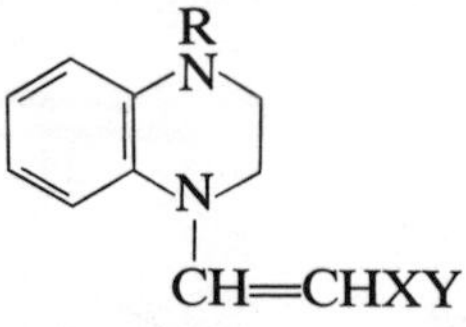

43

R = H or $COCH_2CH_2Cl$
X = CN, Y = CN or CO_2Et

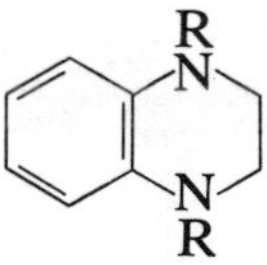

44

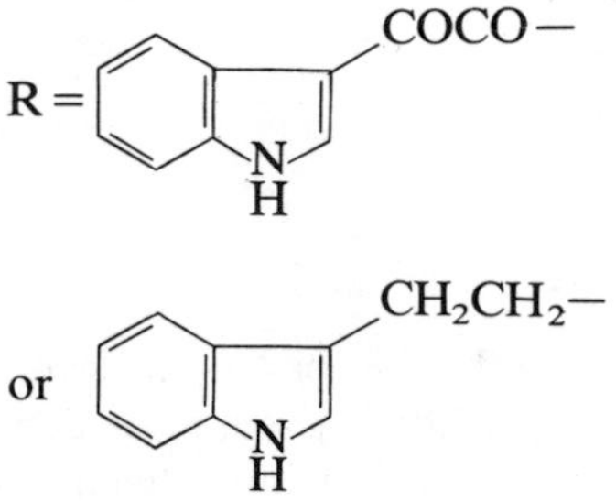

III. Hexahydroquinoxalines

trans-4a,5,6,7,8,8a-Hexahydro-2,3-diphenylquinoxaline (**45**), m.p. 173–175°, has been separated from the stereoisomeric mixture obtained from condensing a mixture of *cis*- and *trans*-1,2-diaminocyclohexane with benzil.[85] It undergoes an interesting photochemical transformation on irradiation in either ethanol or cyclohexane solution. Thus in ethanol a mixture of the imidazo[1,2-*a*]azepines **46** and **47** is formed.[86]

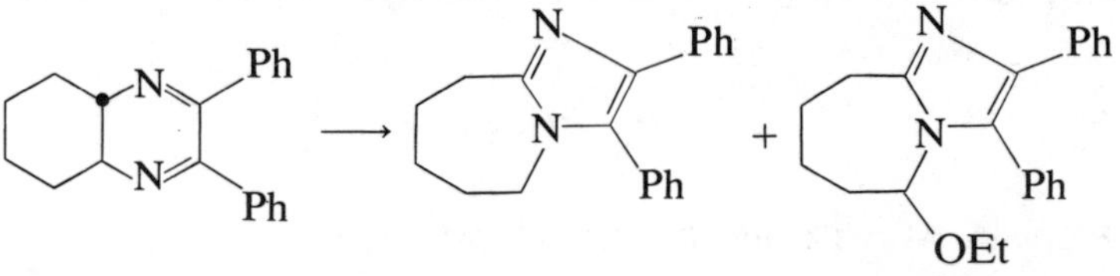

45 **46**, 9% **47**, 62%

IV. Decahydroquinoxalines

Synthetic routes to *cis*- and *trans*-decahydroquinoxaline have been carefully evaluated; the best route to *cis*-decahydroquinoxaline is shown in Scheme 8.[87] Condensation of 1,2-cyclohexanedione with glycine amide gives 5,6,7,8-tetrahydroquinoxalin-2-one (**48**), and this on reduction with hydrogen in the presence of a platinum or Raney nickel catalyst gives *cis*-decahydroquinoxalin-2-one (**49**) in almost quantitative yield.[88] Lithium aluminum hydride reduction of compound **46** then furnishes *cis*-decahydroquinoxaline (**50**).[87]

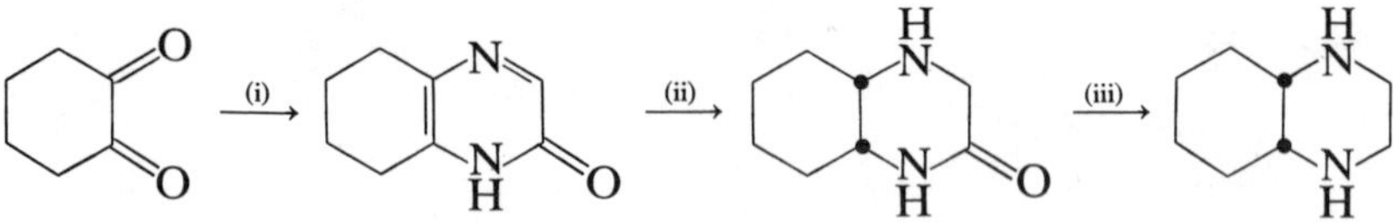

48 **49** **50**

Reagents (i) $NH_2CH_2CONH_2$
(ii) H_2–Pt (iii) $LiAlH_4$

Scheme 8

Reduction of 5,6,7,8-tetrahydroquinoxalin-2-one (**48**) with sodium and amyl alcohol gives *trans*-decahydroquinoxaline in 58% yield.[88] The trans derivative has also been prepared by cyclization of 2-(β-aminoethylamino)cyclohexanol,[89] and by platinum and hydrogen reduction of an ethanolic hydrogen chloride solution of 1,2,3,4-tetrahydroquinoxaline.[90] Reduction of quinoxaline itself over a rhodium-on-alumina catalyst (100°/2000 psi) gives *cis*-decahydroquinoxaline.[91] Substituted decahydroquinoxalines (see Table 3) have also been prepared by the hydrogenation of the corresponding quinoxalines; depending on the severity of the reaction conditions either deca- or tetrahydro derivatives are obtained.[37,38] The resolution of *dl-trans*-decahydroquinoxaline is accomplished by the use of first dibenzoyl-*d*-tartaric acid and then of dibenzoyl-*l*-tartaric acid.[87]

V. Tables

TABLE 1. 1,2- AND 1,4-DIHYDROQUINOXALINES

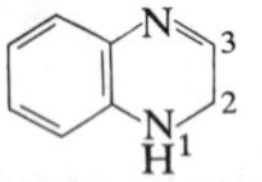

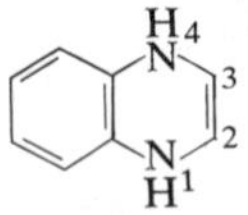

Molecular formula	Substituents or structure	m.p., °C	Ref.
$C_8H_8N_2$	1,4-Dihydro	158–159	2
$C_8H_9N_3$	3-Amino-1,2-dihydro	142–144	10
$C_9H_{10}N_2$	1,2-Dihydro-3-methyl	90	3

TABLE 1 (*contd.*)

Molecular formula	Substituents or structure	m.p., °C	Ref.
$C_{10}H_6N_2O_3$	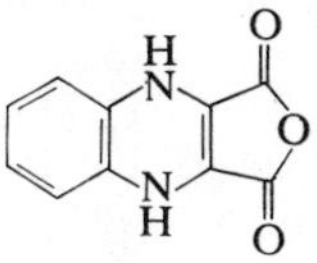	230	40
$C_{10}H_7N_3O_2$	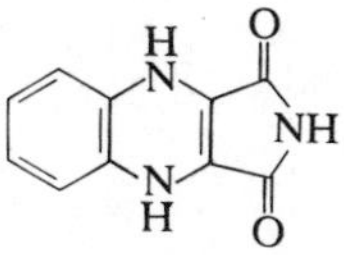	244 decomp.	40
	N-Acetylimide	202–206 decomp.	40
	1,4-Diacetyl-*N*-acetylimide	180 decomp.	40
$C_{10}H_{10}Br_2N_2O_4$	6,7-Dibromo-1,4-dihydro-1,4-dihydroxy-5,8-dimethoxy		92
$C_{10}H_{12}N_2$	1,2-Dihydro-2,3-dimethyl	136	3
$C_{11}H_{12}N_2$	1,2-Dihydro-1,3-dimethyl-2-methylene[a]	b.p. 146–148°/1 mm	18
$C_{11}H_{12}N_2O$	2-Acetyl-1,4-dihydro-3-methyl	149–151	5, 6
$C_{11}H_{14}N_2$	2-Ethyl-1,2-dihydro-3-methyl	110-113	103
	1,2-Dihydro-2,2,3-trimethyl	142–145	9
$C_{14}H_{11}BrN_2$	3-*p*-Bromophenyl-1,2-dihydro	121–122	8
$C_{14}H_{11}N_3O_2$	1,2-Dihydro-3-*m*-nitrophenyl	112-114	8
	Acetyl	194.5–195.5	8
	1,2-Dihydro-3-*p*-nitrophenyl	144–146	8
	Acetyl	141–142	8
$C_{14}H_{12}N_2$	1,2-Dihydro-2-phenyl	166	3
	1,4-Dihydro-2-phenyl[b]		3, 4
	1,2-Dihydro-3-phenyl	141–142	8
	Acetyl	103–104	8
$C_{14}H_{18}N_2O$	1,2-Dihydro-2,3-dimethyl-2-(2-tetrahydrofuryl)	85–86	7
	Picrate	150–153 decomp.	7
$C_{15}H_{14}N_2$	1,2-Dihydro-2-methyl-3-phenyl	128	3
	Acetyl	113	3
	1,2-Dihydro-3-methyl-2-phenyl	161, 163–165	3, 103
$C_{15}H_{14}N_2O$	1,2-Dihydro-3-*p*-methoxyphenyl	138–139	8
	Acetyl	142.5–143	8
$C_{16}H_{14}ClN_3O$	1-Benzoyl-7-chloro-1,2-dihydro-3-methylamino	220–230 decomp.	15
	Acetyl	123–126	15
	1-(*o*-Chlorobenzoyl)-1,2-dihydro-3-methylamino[c]	184–187	16
$C_{16}H_{14}N_2O$	2-Acetyl-1,4-dihydro-3-phenyl	181–182	5, 6
	2-Benzoyl-1,4-dihydro-3-methyl		6
$C_{16}H_{14}N_2O_2$	1-Acetyl-1,2-dihydro-2-hydroxy-3-phenyl	153–154	8

TABLE 1 (*contd.*)

Molecular formula	Substituents or structure	m.p., °C	Ref.
$C_{16}H_{15}N_3O$	1-Benzoyl-1,2-dihydro-3-methylamino	209–210 decomp.	16
$C_{16}H_{16}N_2$	1,2-Dihydro-2-methyl-3-*p*-tolyl	154–155	103
	1,2-Dihydro-3-methyl-2-*p*-tolyl	127–129	103
$C_{16}H_{16}N_2O$	1,2-Dihydro-2-*p*-methoxyphenyl-3-methyl	141–142	103
$C_{16}H_{22}N_2O$	3-*t*-Butyl-1,2-dihydro-2-(2-tetrahydrofuryl)[d]		7
$C_{16}H_{24}N_2$	2,3-Di-*t*-butyl-1,2-dihydro	110–116	11
$C_{20}H_{16}N_2$	1,4-Dihydro-2,3-diphenyl	101	3
$C_{22}H_{16}N_2O$	[structure: 1,2-dihydroquinoxaline, 3-Ph, 2-=CHCOPh, N1-H]	169–171	99
$C_{22}H_{18}N_2O$	[structure: 1,2-dihydroquinoxaline, 3-Ph, 2-CH$_2$COPh, N1-H]	124–125	99
$C_{24}H_{18}N_4$	1,2-Dihydro-1-phenyl-2,3-di-2-pyridyl Diperchlorate		13
$C_{24}H_{18}N_4O$	1,2-Dihydro-2-hydroxy-1-phenyl-2,3-di-2-pyridyl	190	13
$C_{26}H_{20}N_2O$	1,2-Dihydro-2-hydroxy-1,2,3-triphenyl	143–147	14
$C_{27}H_{22}N_2O$	1,2-Dihydro-2-methoxy-1,2,3-triphenyl	143–147	14
$C_{28}H_{24}N_2O$	1,2-Dihydro-2-ethoxy-1,2,3-triphenyl	126–129	14
$C_{28}H_{24}N_2O$	1,2-Dihydro-2,3-bis(*p*-methoxyphenyl)-1-phenyl	157	13 13
$C_{28}H_{25}N_3$	1,2-Dihydro-3-(*p*-dimethylamino-phenyl)-1,2-diphenyl	165	13 13
$C_{30}H_{28}N_2O$	2-Butoxy-1,2-dihydro-1,2,3-triphenyl	118–121	14

[a] Yellow oil which turns red in air.
[b] Immediately oxidized in air.
[c] A number of related compounds are known.
[d] Two diastereoisomers m.p. 104–106° and 128–129° have been isolated.

TABLE 2. 1,2,3,4-TETRAHYDROQUINOXALINES

[structure: 1,2,3,4-tetrahydroquinoxaline with positions N1(H), 2, 3, N4(H)]

Molecular formula	Substituents or structure	m.p., °C	Ref.
$C_8H_6F_4N_2$	5,6,7,8-Tetrafluoro	148–150	50, 51
	1,4-Dinitroso	95–97 decomp.	50

TABLE 2 (*contd.*)

Molecular formula	Substituents or structure	m.p., °C	Ref.
$C_8H_8D_2N_2$	*cis*-2,3-Dideuterio	96	72
	1-Benzoyl	152	72
	1,4-Dibenzoyl	203	72
	1,4-Diformyl	123	72
$C_8H_8N_2O$	2-Oxo	136–138	10
$C_8H_9BrN_2$	6-Bromo	118	58
$C_8H_9ClN_2$	6-Chloro	114	58, 64, 74
	1-Acetyl	80–90	45
	1,4-Dibenzoyl	168.5–169	74
$C_8H_9N_3O_2$	5-Nitro	127–128	33
	6-Nitro	114–115, 116	33, 45
	4-Acetyl	204–205	45, 58
	4-Formyl	177	58
	4-Trifluoroacetyl	182.5–183.5	45
	1,4-Diformyl	176	58
	1,4-Bis(trifluoroacetyl)	120–121	45
$C_8H_{10}N_2$	Unsubstituted	96–97, 98.5–99	19, 28, 49, 55, 74, 90
	1-Acetyl	109–110	71
	1-Benzoyl	152–153	71
	1-Benzyloxycarbonyl	117–118	76
	1-Benzyl	b.p. 195–205/2 mm	76
	Hydrochloride of 1-benzyl	170–171	76
	Methiodide	175–176	76
	Hydrochloride of 1-ethoxycarbonyl	149–150	71
	1-Phenylsulfonyl	138–139	74
	1-Tosyl	128–129	75
	1-Acetyl-4-benzyl	60.5–62.5	74
	1-Acetyl-4-phenylsulfonyl	111.5–112	74
	1,4-Diacetyl	143–144, 147–147.5	71, 74
	1,4-Dibenzoyl	204–205, 206–207	71, 74
	1,4-Diacryloyl	172	83
	1,4-Dibenzyl	b.p. 210–230/2 mm	76
	Hydrochloride of 1,4-dibenzyl	175–177	76
	1,4-Diethoxycarbonyl	42–44	74
	1,4-Diformyl	119–122, 125–126, 126–128	41, 46, 61
	1,4-Diphenylsulfonyl	180–181	74
	1,4-Ditosyl	162	52
	1,4-Bis(trifluoroacetyl)	118–118.5	45
$C_8H_{11}N_3$	5-Amino		
	Dihydrochloride	206–212	57
$C_9H_9F_3N_2$	6-Trifluoroacetyl	123–124, 126–127	33, 45
$C_9H_9N_3$	6-Cyano		
	Hydrochloride	236–237	33

TABLE 2 (*contd.*)

Molecular formula	Substituents or structure	m.p., °C	Ref.
$C_9H_9N_3O_3$	3-Methyl-6-nitro-2-oxo		
	4-Formyl	253–254	46
$C_9H_{10}N_2O$	2-Formyl		
	Ethylene acetal	61–63	27
	Benzoyl deriv. of ethylene acetal	135	27
	Dibenzoyl deriv. of ethylene acetal	166–167	27
	3-Methyl-2-oxo		
	4-Formyl	231–233	46
$C_9H_{10}N_2O_2$	2-Carboxy	166–168	39
	Ethyl ester	77	39
	Amide	111	39
	Ditosyl deriv. of methyl ester	119–121	53
	6-Carboxy		
	Hydrochloride	204–230 decomp.	45
	Methyl ester	88	58
	Hydrochloride of methyl ester	215 decomp.	58
	Ethyl ester hydrochloride	193–194	33
$C_9H_{11}BrN_2$	7-Bromo-2-methyl		
	S Form[a]	131–132	24
	1,4-Diacetyl[b]	131–132.5	24
	RS Form	156.5–158	24
	1,4-Diacetyl	129–130	24
$C_9H_{11}ClN_2$	6-Chloro-1-methyl	51–53	64
	4-Benzoyl	128–130	64
	4-Phenylcarbamoyl	190–192	64
	6-Chloro-2-methyl	145–146	93
	4-Phenylsulfonyl	143	93
$C_9H_{12}N_2$	1-Methyl	b.p. 108.5/2 mm	22, 74
	Hydrochloride	173–174	71
	4-Acetyl	Oil	71
	Hydrochloride of 4-acetyl	154–155	71
	Methiodide of 4-acetyl	184–185 decomp.	71
	4-Benzoyl	109–110	74
	4-Ethoxycarbonyl	Oil	71
	Hydrochloride of 4-ethoxy-carbonyl	133–134	71
	Methiodide of 4-ethoxy-carbonyl	172–173 decomp.	71
	Picrate	123–126.5	74
	2-Methyl		
	R Form[c]	90.5–91	24
	1,4-Diacetyl	143–144	24
	S Form[d]	90–90.5	24
	RS Form	70–71	34, 37, 41
	1,4-Diacetyl	138–139, 141–143	24, 37

TABLE 2 (*contd.*)

Molecular formula	Substituents or structure	m.p., °C	Ref.
$C_9H_{12}N_2$ (*contd.*)	1,4-Diformyl	44–45	46
	6-Methyl	104.5–105.5	74
	1-Acetyl	71–73	45
	1,4-Diacetyl	105.2–106.2	74
	1,4-Dibenzoyl	141.5–142	74
	1,4-Diphenylsulfonyl	124–125	74
	1,4-Bis(trifluoroacetyl)	98–99	45
	Picrate	148–148.5	74
$C_9H_{12}N_2O$	2-Hydroxymethyl	140–141	54
	1,4-Ditosyl	193	27, 54
	Tritosyl	186–188	27
	Picrate	179–180 decomp.	54
	6-Methoxy	81	58, 74
	1,4-Dibenzoyl	138.5–138.8	74
	1,4-Bis(trifluoroacetyl)	110–110.5	45
	Picrate	134–134.5	74
$C_9H_{12}N_2S$	6-Methylthio	65	58
	Hydrochloride	185	58
	Picrate	158 decomp.	58
$C_9H_{13}N_3$	6-Amino-3-methyl		
	Hydrochloride	198–205 decomp.	94
	6-Amino-7-methyl		
	Trihydrobromide	238–240	56
$C_{10}H_{10}N_2O_4$	2,3-Dicarboxy	213–216	39
	Monoethyl ester	173	39
	Monoamide	208–210	39
	Diethyl ester	84–85	39
	Diamide	236–237	39
$C_{10}H_{13}BrN_2$	2-(2-Bromoethyl)		
	Dihydrobromide	>300	100
$C_{10}H_{13}ClN_2$	1-(2-Chloroethyl)		
	Hydrochloride	154–160 decomp.	59
	4-Acetyl	109–110	59
	6-Chloro-1,3-dimethyl	35–37	64
	p-Chlorobenzoyl	124–128	64
	5-Chloro-1,4-dimethyl	Oil	62
	6-Chloro-1,4-dimethyl	62–63	62
$C_{10}H_{13}N_3O$	6-Acetamido	Unstable	58
	1,4-Diacetyl	225	58
$C_{10}H_{13}N_3O_2$	2-Amino-3-carboxy-1-methyl[e]		
	Methyl ester	212–213	65
$C_{10}H_{14}N_2$	1-Ethyl	b.p. 170–173/21 mm, 88–90/1 mm	22, 74
	4-Benzoyl	123–124	74
	4-Phenylsulfonyl	118.5–119.5	74
	Picrate	111.5–112	74

TABLE 2 (*contd.*)

Molecular formula	Substituents or structure	m.p., °C	Ref.
$C_{10}H_{14}N_2$ (*contd.*)	Oxalate	130–131	74
	1,4-Dimethyl	b.p. 92–94/1 mm, 180–190/16 mm	61, 62, 76
	Picrate	122–124, 128–129	61, 76
	cis-2,3-Dimethyl	112–113, 113–114	20, 66
	1-Benzoyl	136	72
	1,4-Diacetyl	147	37
	1,4-Dibenzoyl	221	72
	1,4-Diformyl	98	72
	dl-trans-2,3-Dimethyl	99–100, 103–104, 104–105	21, 47, 66
	1,4-Dibenzoyl	170	72
	1,4-Diformyl	186	72
	d-trans-2,3-Dimethyl	94.5	21
	l-trans-2,3-Dimethyl	94.5	21
	dl-2,6-Dimethyl	115–115.5	34
	dl-2,7-Dimethyl	117–118, 118–118.5	34, 101
	6,7-Dimethyl	154	45
	1,4-Bis(trifluoroacetyl)	84, 103	45
$C_{10}H_{14}N_2O$	2-(2-Hydroxyethyl)	97–98	100
$C_{10}H_{14}N_2O_2$	6,7-Dimethoxy	134–134.5	75, 95
	1-Phenylsulfonyl	170–172	75
$C_{11}H_{11}N_3O_2$	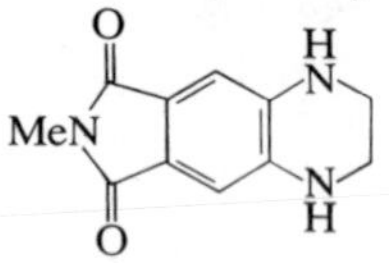 MeN	286, with sintering at 294–296	36
$C_{11}H_{16}N_2$	1-Isopropyl	b.p. 107.5/1.5 mm	74
	4-Benzoyl	114–115	74
	4-Phenylsulfonyl	142.5–143.5	74
	Picrate	131–132	74
	1-Propyl	b.p. 113.5/1.5 mm 176–179/20 mm	22,74
	4-Benzoyl	88–89	74
	4-Phenylsulfonyl	119.5–120	74
	Picrate	135–136	74
$C_{11}H_{16}N_2O$	6-Methoxy-1,4-dimethyl	Oil	62
$C_{11}H_{16}N_2O_2$	6,7-Dimethoxy-1-methyl	84–85	75
	6,7-Dimethoxy-3-methyl	b.p. 172–174/ 0.2 mm[f]	75
$C_{12}H_{10}N_4$	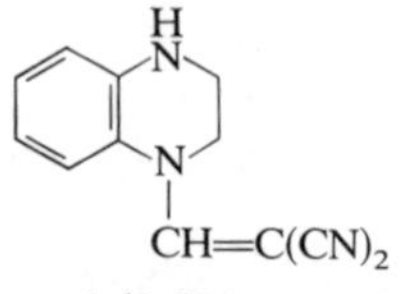 $CH{=}C(CN)_2$	178–179	77
	4-(3-Chloropropionyl)	140–142	77

TABLE 2 (*contd.*)

Molecular formula	Substituents or structure	m.p., °C	Ref.
$C_{12}H_{11}N_3O_2$			
	Ethyl ester	121–122	77
	4-(3-Chloropropionyl)	129–130	77
$C_{12}H_{16}N_2$	1,4-Dimethyl-2-vinyl	36	60
$C_{12}H_{16}N_2O$	2-(2-Tetrahydrofuryl)	Oil[g]	7
$C_{12}H_{17}N_3O_3$	2-Amino-3-carboxy-1-(2-methoxyethyl)[e]		
	Methyl ester	161–162	65
$C_{12}H_{18}N_2$	1-Butyl	b.p. 107.5/1 mm, 187–191/21 mm	22, 74
	4-Benzoyl	87–88	74
	4-Phenylsulfonyl	95–95.5	74
	Picrate	130–131.5	74
	Oxalate	142.5–143.5	74
	2-*t*-Butyl	86	7
	1,4-Dibenzoyl	191	72
	1,4-Diformyl	99	72
	1,4-Dimethyl-2-ethyl		
	Picrate	126	60
	1,2,2,4-Tetramethyl	b.p. 156/16 mm[h]	25
	Picrate	152	25
$C_{12}H_{18}N_2O$	2-(1-Ethoxyethyl)	Oil[i]	7
	Picrate	117–117.5	98
$C_{12}H_{18}N_2O_2$	6,7-Dimethoxy-1,3-dimethyl	132–135/0.01 mm[j]	75
	6,7-Dimethoxy-2,3-dimethyl	127–128	75
$C_{12}H_{19}N_3$	2-(2-Dimethylaminoethyl)	b.p. 120–124/0.2 mm	43
	1,4-Bis(ethylcarbamoyl)	127–129	43
	1,4-Bis(phenylcarbamoyl)	191–192	43
$C_{13}H_{15}N_3O_2$		170–172	36
$C_{13}H_{20}N_4$	2-(1-Piperazinylmethyl)	140–142	23
$C_{14}H_{14}N_2$	2-Phenyl	77–78	3, 8
$C_{14}H_{14}N_2O_8$			
	Tetraethyl ester	94.5	63

TABLE 2 (*contd.*)

Molecular formula	Substituents or structure	m.p., °C	Ref.
$C_{14}H_{16}N_4$	1,4-Bis(2-cyanoethyl)	88.5–89	49
$C_{14}H_{18}N_2$	2,3-Diallyl	b.p. 142–143/ 1.8 mm	48
$C_{14}H_{18}N_2O_4$	1,4-Bis(2-carboxyethyl)	136–137	49
$C_{14}H_{20}N_2$	2-Isopropenyl-1,2,4-trimethyl	39	60
$C_{14}H_{22}ClN_3O_2S$	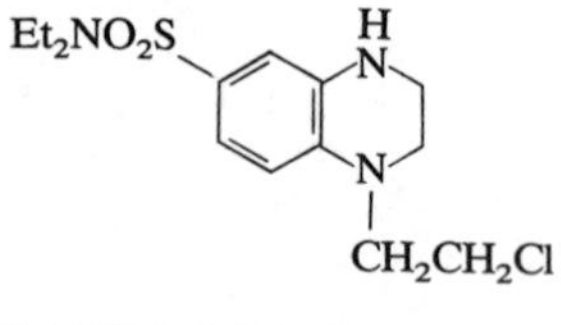 Et_2NO_2S, H, N, N, CH_2CH_2Cl	98–99	96
$C_{14}H_{22}N_2O_2$	2,3-Diethyl-1,4-dimethyl-2,3-dihydroxy		
	Trihydrate	54–55	35
$C_{14}H_{22}N_4$	2-(4-Methyl-1-piperazinyl)-methyl	125–127	23
$C_{14}H_{23}N_3$	2-(2-Diethylaminoethyl)	b.p. 140–142/ 0.2 mm	43
	1,4–Bis(ethylcarbamoyl)	138–140	43
	1,4-Bis(phenylcarbamoyl)	148–150	43
$C_{15}H_{11}ClN_2O_2$	4-Benzoyl-6-chloro-2-oxo	255–257, 259–263	15, 104
$C_{15}H_{16}N_2$	1-Benzyl	b.p. 178–179/1.5 mm, 242–250/19 mm	22, 74
	4-Benzoyl	123.3–123.8	74
	4-Phenylsulfonyl	134–135	74
	Picrate	150–151.5	74
	cis-2-Methyl-3-phenyl	b.p. 172/15 mm	3
	1-Acetyl	122–123	3
$C_{15}H_{16}ClN_3$	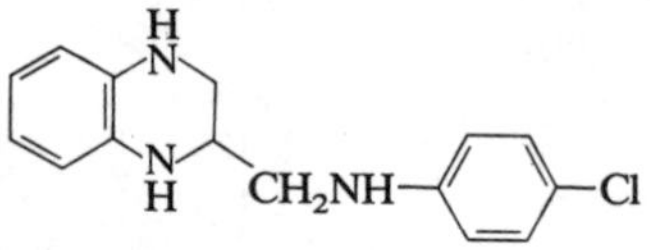 H, N, N, H, CH_2NH—, —Cl	106–108	30
$C_{16}H_{14}N_2O_4$	*trans*-2,3-Dicarboxy-1-phenyl		
	Dimethyl ester	114–116	102
$C_{16}H_{16}N_4$	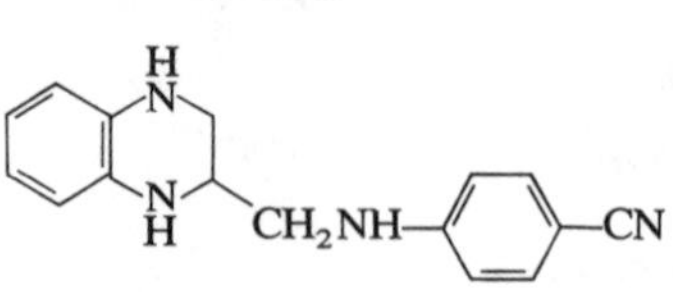 H, N, N, H, CH_2NH—, —CN	152–153	31
$C_{16}H_{17}ClN_2$	3-Benzyl-6-chloro-1-methyl	67–69	64
	Hydrochloride	135–145 decomp.	64
	4-Benzyl-6-chloro-1-methyl	122–124	64
	2-(*p*-Chlorophenethyl)	103	35

TABLE 2 (*contd.*)

Molecular formula	Substituents or structure	m.p., °C	Ref.
$C_{16}H_{17}N_3O_2$	H, N, N, H, CH_2NH, CO_2H		
	Ethyl ester	134–135	29
$C_{16}H_{18}N_2$	2-Phenethyl	68.5–69.5	35
$C_{16}H_{18}N_2O$	H, N, Ph, N, CH_2CH_2OH		
	Hydrochloride	170–172	26
$C_{16}H_{19}N_3$	H, N, N, H, CH_2NH, Me	111–112	29
$C_{16}H_{22}N_2O_2$	2,3-Bis(tetrahydro-2-furyl)	246–247	7
$C_{16}H_{22}N_2O_6$	$COCH(OMe)_2$, H, N, N, H, $COCH(OMe)_2$	107	84
$C_{16}H_{26}N_2$	2,3-Di-*t*-butyl	79–80	11
$C_{16}H_{26}N_2O_6P_2$	H, N, $PO(OEt)_2$, N, H, $PO(OEt)_2$	168–170	97
$C_{16}H_{27}N_3O_2$	1-(2-Diethylaminoethyl)-6,7-dimethoxy	b.p. 170/0.01 mm	75
$C_{17}H_{19}N_3O_2$	H, N, N, Me, CH_2NH, CO_2H		
	Ethyl ester	106–107	31
	H, N, N, H, CH_2NMe, CO_2H		
	Ethyl ester	92–94	31

TABLE 2 (*contd.*)

Molecular formula	Substituents or structure	m.p., °C	Ref.
$C_{17}H_{19}N_3O_2$ (*contd.*)	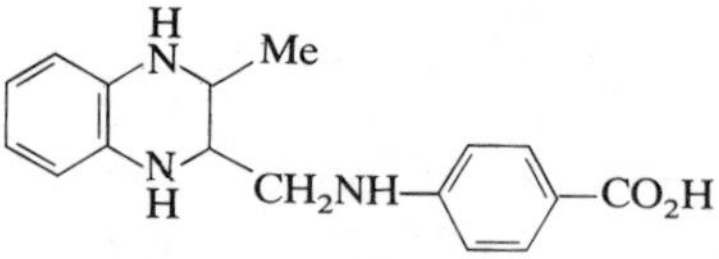 H N Me N H CH_2NH—⟨⟩—CO_2H		
	1,4-Ditosyl deriv. of ethyl ester	196–197	27
$C_{18}H_{31}N_3O_2$	4-(2-Diethylaminoethyl)-6,7-dimethoxy-1-ethyl	Oil[k]	95
$C_{19}H_{25}N_3$	2-(2-Benzylethylaminoethyl)	178–185/0.2 mm	43
$C_{20}H_{18}N_2$	*cis*-2,3-Diphenyl	141–142	3, 37, 66
	trans-2,3-Diphenyl	105	47, 66
$C_{20}H_{26}ClN_3$	3-(*p*-Chlorophenyl)-1-(2′-diethylaminoethyl)		
	Dihydrochloride	202–205 decomp.	26
$C_{20}H_{26}N_4$	2-(4-Benzyl-1-piperazinyl)-methyl	95–97	23
$C_{20}H_{27}N_3$	1-(2-Diethylaminoethyl)-3-phenyl		
	Dihydrochloride	212–214 decomp.	26
$C_{22}H_{22}N_2O_2$	5,8-Dimethoxy-2,3-diphenyl		
	1,4-Diacetyl		44
	2,3-Bis(*o*-methoxyphenyl)	114	47
$C_{22}H_{40}N_4O_2$	1,4-Bis(2-diethylaminoethyl)-6,7-dimethoxy	b.p. 185–187/0.04 mm, 200/0.5 mm	75, 95
$C_{24}H_{44}N_4O_2$	1,4-Diethyl-2,3-bis(2-diethylaminomethyl)-6,7-dimethoxy	Oil[l]	95
$C_{26}H_{31}N_3$	1-(2-Diethylaminoethyl)-2,3-diphenyl		
	Dihydrochloride	220–224 decomp.	26

[a] $[\alpha]_D^{24}$ +13.0° (*c* 1.0, THF).
[b] $[\alpha]_D^{24}$ +29.9° (*c* 1.1, THF).
[c] $[\alpha]_D^{25}$ −60.3° (*c* 1.0, THF).
[d] $[\alpha]_D^{24}$ +60.2° (*c* 1.0, THF).
[e] 1,5,6,7-Tetrahydro deriv.
[f] n_D 1.599.
[g] Distilled at 0.01 mm, bath temp. 120°.
[h] n_D^{25} 1.5791.
[i] Distilled at 0.01 mm, bath temp. 100–120°.
[j] n_D 1.584.
[k] Distilled at 0.01 mm, bath temp. 160°.
[l] Distilled at 0.01 mm, bath temp. 160–165°.

TABLE 3. DECAHYDROQUINOXALINES

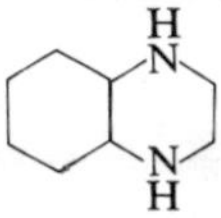

Molecular formula	Substituents	m.p., °C	Ref.
$C_8H_{12}N_2O_2$	2,3-Dioxo		
	Cis	248–249	87
	Trans	301–303	87
$C_8H_{14}N_2O$	2-Oxo		
	Cis	135–136	87
	Nitroso	165–167	87
	Trans	199–199.5	87, 88
	Nitroso	194–195	87
$C_8H_{16}N_2$	Unsubstituted		
	Cis	56–57, b.p. 85–87/0.25 mm	87, 91
	Dihydrochloride	298–301	87, 91
	Dinitroso	86–88	87, 91
	Dipicrate	274–275 decomp.	87, 91
	Diacetyl	144–146	91
	dl-Trans	146–148, 150–151, 152.5–153	87, 88, 89, 90, 91
	Dihydrochloride	365 decomp.	89
	Dinitroso	110–111	90
	Dipicrate	301–303	90
	d-Trans	175–175.5[a]	87
	l-Trans	175–175.5[b]	87
$C_9H_{16}N_2O_2$	2-Carboxy		
	Hydrate	249 decomp.	38
$C_9H_{18}N_2$	1-Methyl	b.p. 88–90/8 mm	89
	4-Ethoxycarbonyl	b.p. 103–104/0.6 mm	89
	4-*p*-Chlorobenzhydryl	b.p. 196/0.6 mm	89
	2-Methyl	80, b.p. 115–117/20 mm	37
$C_{10}H_{16}N_2O_4$	2,3-Dicarboxy		
	Hydrochloride[c]	230–231 decomp.	38
$C_{20}H_{36}N_2$	2,3-Dicyclohexyl	137.5	37

[a] $[\alpha]_D^{26} + 10.4°$ (*c* 10, $CHCl_3$).
[b] $[\alpha]_D^{26} - 9.7°$ (*c* 10.2, $CHCl_3$).
[c] Probably trans configuration.

VI. References

1. J. C. E. Simpson, "Condensed Pyridazine and Pyrazine Rings," Interscience, New York, 1953, p. 310.
2. J. Hamer and R. E. Holliday, *J. Org. Chem.*, **28,** 2488 (1963).

3. J. Pinson and J. Armand, *Collect. Czech. Chem. Commun.*, **36,** 585 (1971).
4. M. Schellenberg, *Helv. Chim. Acta*, **53,** 1151 (1970).
5. J. A. Barltrop, C. G. Richards, and D. M. Russell, *J. Chem. Soc.*, 1423 (1959).
6. M. Schellenberg and R. Steinmetz, *Helv. Chim. Acta*, **52,** 431 (1969).
7. T. T. Chen, W. Doerscheln, H. Goeth, M. Hesse, and H. Schmid, *Helv. Chim. Acta*, **51,** 632 (1968).
8. J. Figueras, *J. Org. Chem.*, **31,** 803 (1966).
9. J. Mirza, W. Pfleiderer, A. D. Brewer, A. Stuart, and H. C. S. Wood, *J. Chem. Soc.* (*C*), 437 (1970).
10. K. Pfister, A. P. Sullivan, J. Weijlard, and M. Tishler, *J. Am. Chem. Soc.*, **73,** 4955 (1951).
11. Ae. De Groot and H. Wynberg, *J. Org. Chem.*, **31,** 3954 (1966).
12. W. Reid and P. Stahlhofen, *Chem. Ber.*, **90,** 815 (1957).
13. S. Bodforss, *Justus Liebigs Ann. Chem.*, **633,** 66 (1960).
14. P. M. Mader, *J. Org. Chem.*, **26,** 3605 (1961).
15. G. F. Field and L. H. Sternbach, *J. Org. Chem.*, **33,** 4438 (1968).
16. R. I. Fryer, J. V. Earley, G. F. Field, W. Zally, and L. H. Sternbach, *J. Org. Chem.*, **34,** 1143 (1969).
17. J. W. Bunting and W. G. Meathrel, *Can. J. Chem.*, **50,** 917 (1972).
18. M. T. Le Bris, *Bull. Soc. Chim. Fr.*, 563 (1970).
19. F. Bohlmann, *Chem. Ber.*, **85,** 390 (1952).
20. R. C. De Selms and H. S. Mosher, *J. Am. Chem. Soc.*, **82,** 3762 (1960).
21. C. S. Gibson, *J. Chem. Soc.*, 342 (1927).
22. R. F. Smith, W. J. Rebel, and T. N. Beach, *J. Org. Chem.*, **24,** 205 (1959).
23. A. Jonsson, P. Stene, and N. Willman, *Acta Pharm. Suecica*, **5,** 237 (1968); *Chem. Abstr.*, **69,** 86957*b* (1968).
24. G. H. Fisher, P. J. Whitman, and H. P. Schultz, *J. Org. Chem.*, **35,** 2240 (1970).
25. J. W. Clark-Lewis, J. A. Edgar, J. S. Shannon, and M. J. Thompson, *Austr. J. Chem.*, **17,** 877 (1964).
26. J. Druey and A. Hüni, *Helv. Chim. Acta*, **35,** 2301 (1952).
27. M. P. Mertes and A. J. Lin, *J. Med. Chem.*, **13,** 77 (1970).
28. J. Cymerman-Craig, W. P. Rogers, and M. E. Tate, *Austr. J. Chem.*, **9,** 397 (1956).
29. S. J. Benkovic, P. A. Benkovic, and D. R. Comfort, *J. Am. Chem. Soc.*, **91,** 5270 (1969).
30. S. J. Benkovic, P. A. Benkovic, and R. Chrzanowski, *J. Am. Chem. Soc.*, **92,** 523 (1970).
31. T. H. Barrows, P. R. Farina, R. L. Chrzanowski, P. A. Benkovic, and S. J. Benkovic, *J. Am. Chem. Soc.*, **98,** 3678 (1976).
32. M. J. Haddadin, H. N. Alkaysi, and S. E. Saheb, *Tetrahedron*, **26,** 115 (1970).
33. K. V. Rao and D. Jackman, *J. Heterocycl. Chem.*, **10,** 213 (1973).
34. M. Munk and H. P. Schultz, *J. Am. Chem. Soc.*, **74,** 3433 (1952).
35. J. K. Landquist and G. J. Stacey, *J. Chem. Soc.*, 2822 (1953).
36. E. H. White and K. Matsuo, *J. Org. Chem.*, **32,** 1921 (1967).
37. S. Maffei and S. Pietra, *Gazz. Chim. Ital.*, **88,** 556 (1958); *Chem. Abstr.*, **53,** 20060 (1959).
38. G. F. Bettinetti, *Atti Accad. Nazl. Lincei, Rend. Classe Sci. Fis. Mat. Nat.*, **31,** 429 (1961); *Chem. Abstr.*, **57,** 9853 (1962).
39. G. F. Bettinetti, *Ann. Chim.* (Rome), **51,** 920 (1961); *Chem. Abstr.*, **57,** 5914 (1962).
40. G. F. Bettinetti and E. Tisselli, *Ann. Chim.* (Rome), **51,** 1102 (1961); *Chem. Abstr.*, **57,** 11194 (1962).

41. P. Schuyler, F. D. Popp, A. C. Noble, D. W. Alwani, and B. R. Masters, *J. Med. Chem.*, **9,** 704 (1966).
42. F. D. Popp, E. Cullen, R. B. Davis, and W. Kirsch. *J. Med. Chem.*, **5,** 398 (1962).
43. P. F. Wiley, *J. Am. Chem. Soc.*, **76,** 4924 (1954).
44. S. Kawai, S. Tanaka, and K. Ichikawa, *J. Chem. Soc. Jap.*, **75,** 40 (1954); *Chem. Abstr.*, **49,** 10314 (1955).
45. R. C. De Selms, R. J. Greaves, and W. R. Schleigh, *J. Heterocycl. Chem.*, **11,** 595 (1974).
46. I. Baxter and D. W. Cameron, *J. Chem. Soc.* (*C*), 2471 (1968).
47. E. S. Lane and C. Williams, *J. Chem. Soc.*, 4106 (1954).
48. H. Gilman, J. Eisch, and T. Soddy, *J. Am. Chem. Soc.*, **79,** 1245 (1957).
49. F. G. Mann and B. B. Smith, *J. Chem. Soc.*, 1906 (1951).
50. J. Burdon, V. A. Damodaran, and J. C. Tatlow, *J. Chem. Soc.*, 763 (1964).
51. F. I. Abezgauz and S. V. Sokolov, *J. Gen. Chem. USSR*, **38,** 2418 (1968).
52. H. Stetter, *Chem. Ber.*, **86,** 197 (1953).
53. E. Negishi and A. R. Day, *J. Org. Chem.*, **30,** 43 (1965).
54. R. M. Acheson, *J. Chem. Soc.*, 4731 (1956).
55. G. R. Ramage and G. Trappe, *J. Chem. Soc.*, 4406 (1952).
56. J. Perello P. and H. Zunza F., *Bol. Soc. Chil. Quim.*, **12,** 3 (1962); *Chem. Abstr.*, **59,** 3925 (1963).
57. W. Knobloch and G. Lietz, *J. Prakt. Chem.*, **36,** 113 (1967).
58. P. Clarke and A. Moorhouse, *J. Chem. Soc.*, 4763 (1963).
59. V. Beksha and Yu. Degutis, *J. Org. Chem. USSR*, **1,** 1873 (1965).
60. W. Schroth and B. Streckenbach, *Z. Chem.*, **3,** 465 (1963).
61. H. König and R. Huisgen, *Chem. Ber.*, **92,** 429 (1959).
62. L. Lalloz and P. Caubere, *Synthesis*, 657 (1975).
63. J. D. Anderson and M. M. Baizer, *Tetrahedron Lett.*, 511 (1966).
64. A. Walser, G. Silverman, R. I. Fryer, and L. H. Sternbach, *J. Org. Chem.*, **36,** 1248 (1971).
65. D. Clerin, A. Lacroix, and J.-P. Fleury, *Tetrahedron Lett.*, 2899 (1976).
66. R. A. Archer and H. S. Mosher, *J. Org. Chem.*, **32,** 1378 (1967).
67. Sadtler Standard Spectra, IR 34823.
68. H. H. Perkampus and A. Roders, *Z. Naturforsch.*, **15*b*,** 1 (1960).
69. Sadtler Standard Spectra, NMR 18420.
70. J. C. Cavagnol and G. Wilson, *J. Am. Chem. Soc.*, **72,** 3752 (1950).
71. J. S. Morley, *J. Chem. Soc.*, 4002 (1952).
72. R. Aguilera, J. C. Duplan, and C. Nofre, *Bull. Soc. Chim. Fr.*, 4491 (1968).
73. P. Vouros and K. Biemann, *Org. Mass. Spectrom.*, **2,** 375 (1969).
74. J. C. Cavagnol and F. Y. Wiselogle, *J. Am. Chem. Soc.*, **69,** 795 (1947).
75. H. Zellner and G. Zellner, *Helv. Chim. Acta*, **49,** 913 (1966).
76. A. S. Elina and I. S. Musatova, *Khim. Geterotsikl. Soedin., Akad. Nauk Latv. SSR*, 291 (1965); *Chem. Abstr.*, **63,** 7011 (1965).
77. P. Schuyler, F. D. Popp, and A. C. Noble, *J. Med. Chem.*, **9,** 774 (1966).
78. American Home Products Corp., Brit. Pat. 1,126,245; *Chem. Abstr.*, **70,** 115007*v* (1969).
79. B. Miller and T. H. Mladineo (to American Cyanamid Co.), U.S. Pat. 3,505,327; *Chem. Abstr.*, **72,** 132785*f* (1970).
80. F. H. Wilson (to Goodyear Tire and Rubber Co.), U.S. Pat. 3,580,884; *Chem. Abstr.*, **75,** 37633*h* (1971); U.S. Pat. 3,706,742; *Chem. Abstr.*, **78,** 98782*p* (1973).

81. B. Sturm and F. H. Wilson (to Goodyear Tire and Rubber Co.), Fr. Pat. 1,420,036; *Chem. Abstr.*, **65,** 18811 (1966).
82. Farbenfabriken Bayer A.-G., Fr. Demande 2,018,055; *Chem. Abstr.*, **75,** 7129*f* (1971).
83. G. Booth and B. N. Parsons (to I.C.I. Ltd.), Ger. Offen. 1,919,597; *Chem. Abstr.*, **72,** 122902*c* (1970).
84. E. Zeigler and T. Kappe (to J. R. Geigy A.-G.), Swiss Pat. 447,434; *Chem. Abstr.*, **69,** 37110*m* (1968).
85. A. I. Smith (to Monsanto), U.S. Pat. 3,163,675; *Chem. Abstr.*, **62,** 7656*g* (1965).
86. P. Beak and J. L. Miesel, *J. Am. Chem. Soc.*, **89,** 2375 (1967).
87. E. Brill and H. P. Schultz, *J. Org. Chem.*, **28,** 1135 (1963).
88. E. Brill and H. P. Schultz, *J. Org. Chem.*, **29,** 579 (1964).
89. K. M. Beck, K. E. Hamlin, and A. W. Weston, *J. Am. Chem. Soc.*, **74,** 605 (1952).
90. W. Christie, W. Rohde, and H. P. Schultz, *J. Org. Chem.*, **21,** 243 (1956).
91. H. Smith Broadbent, E. L. Allred, L. Pendleton, and C. W. Whittle, *J. Am. Chem. Soc.*, **82,** 189 (1960).
92. J. Adachi, *Nippon Kagaku Zasshi*, **76,** 311 (1955); *Chem. Abstr.*, **51,** 17936 (1957).
93. G. Henseke and R. Jacobi, *Justus Liebigs Ann. Chem.*, **684,** 146 (1965).
94. J. Klicnar and F. Kosek, *Collect. Czech. Chem. Commun.*, **30,** 3102 (1965).
95. E. L. Stogryn, *J. Med. Chem.*, **14,** 171 (1971).
96. J. Degutis and V. Bieksa, *Lietuvos TSR Mokslu Akad. Darbai, Ser. B*, 71 (1965); *Chem. Abstr.*, **64,** 4980 (1966).
97. B. A. Arbuzov and V. M. Zoroastrova, *Izv. Akad. Nauk SSSR, Otdel. Khim. Nauk*, 806 (1954); *Chem. Abstr.*, **49,** 13222 (1955).
98. A. S. Elina and L. G. Tsyrul'nikova, *J. Org. Chem. USSR* **1,** 1169 (1965).
99. R. G. Bass, D. D. Crichton, H. K. Meetz, and A. F. Johnson, *Tetrahedron Lett.*, 2073 (1975).
100. H. C. Cunningham and A. R. Day, *J. Org. Chem.*, **38,** 1225 (1973).
101. V. V. Titov and L. F. Kozhokina, *Khim. Geterotsikl. Soedin*, 1700 (1972); *Chem. Abstr.*, **78,** 4217*j* (1973).
102. M. D. Nair and S. R. Mehta, *Indian J. Chem.*, **6,** 490 (1968).
103. W. Schwaiger and J. P. Ward, *Rec. Trav. Chim.*, **91,** 1175 (1972).
104. S. C. Bell and S. J. Childress, *J. Org. Chem.*, **29,** 506 (1964).

CHAPTER XIX

Pyrrolo[1,2-*a*]pyrazines

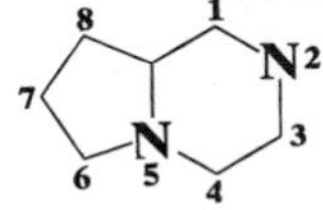

I. Pyrrolo[1,2-*a*]pyrazines (Not 1,4-Dioxo Compounds) 292
 1. Physical Properties 292
 2. Methods of Preparation 294
 3. Reactions 300
 4. Biological Activity 303
II. 1,4-Dioxopyrrolo[1,2-*a*]pyrazines 303
 1. Physical Properties 304
 2. Methods of Preparation 308
  A. Type A Synthesis 310
  B. Type B Synthesis 315
 3. Reactions 316
 4. Biological Activity 322
 5. Occurrence and Formation 323
 6. The Ergot Alkaloids 326
III. Tables 327
  Table 1. Pyrrolo[1,2-*a*]pyrazines (Not 1,4-Dioxo Compounds) 327
  Table 2. Perhydro-1,4-dioxopyrrolo[1,2-*a*]pyrazines 334
IV. References 342

Since the review by Mosby[1] there has been considerable interest in this ring system. In particular much work has been done on the perhydro-1,4-dioxo derivatives **1**, which are diketopiperazines derived from either proline or hydroxyproline and another α-amino acid. These compounds are of sufficient importance to be dealt with separately (see Section II).

O
N—R²
N
R¹
O

1

The ring system has also been referred to as 2,5-pyrrolopyrazine and as the 7-aza derivative of indolizine or pyrrocoline (**2**). Reduced derivatives are systematically named as 1,4-diazabicyclo[4.3.0]nonanes (**3**). Occasionally names such as 1,2-trimethylenepiperazine and pyrrolidino(1′:2′-1:2)piperazine (**4**) have been used. The 1,4-dioxo compounds (**1**) are usually described as derivatives of the constituent amino acids. Thus compound **5** may be referred to as cyclo(prolyl-glycine), glycyl-proline lactam, glycyl-proline anhydride, or glycyl-proline diketopiperazine. There is no specified way of ordering the constituent amino acids. The trioxo derivative **6** has been called 2,5-dioxopiperazine-3-propionic lactam.

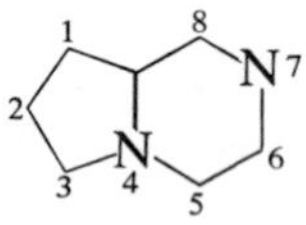

2 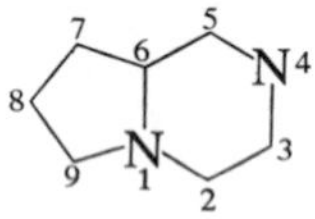 **3** 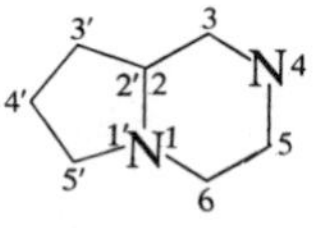 **4**

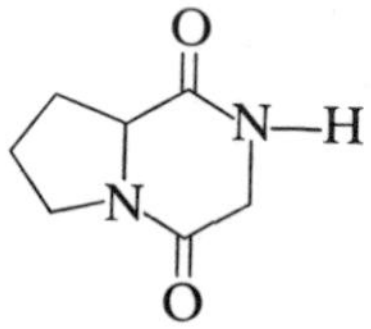

5 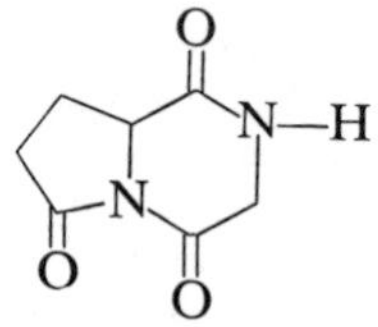 **6**

I. Pyrrolo[1,2-*a*]pyrazines

1. Physical Properties

Several sets of theoretical calculations have been performed on the parent ring system. HMO calculations of total π-electron densities and frontier electron densities successfully predicted that the nucleus would undergo electrophilic substitution at the 6- and 8-positions.[2] Two groups,[3,4] have compared the electronic structure of indolizine and various aza derivatives using the SCF or semiempirical antisymmetric configuration interaction method. The results allowed interpretations of the electronic spectrum to be made which were in good agreement with experiment.

The presence of a Bohlmann band at 2775 cm^{-1} in the infrared spectrum of compound **7** was taken to indicate that, as with the corresponding perhydropyrido[1,2-*a*]pyrazine, the lone pair of the bridgehead nitrogen and the hydrogen of the bridgehead carbon are in a trans orientation.[5] The infrared spectrum of compound **8** shows a strong band at 3500 cm^{-1} (ν O-H stretch), confirming the predominance of the hydroxy tautomer rather than the 2*H*-6-oxo form (**9**). Ultraviolet and PMR data are also provided for this and related compounds.[6] The parent heterocycle **10** is an unstable oil which darkens rapidly. In isooctane the ultraviolet spectrum of the parent has four absorption maxima, at 230, 280, 330, and 372 nm. This contrasts with that recorded in 95% ethanol for the pyrrolopyridine **11**, which has bands at 218, 263, and 296 nm.

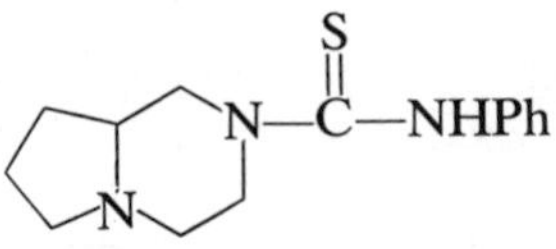

S
N—C—NHPh
N

7

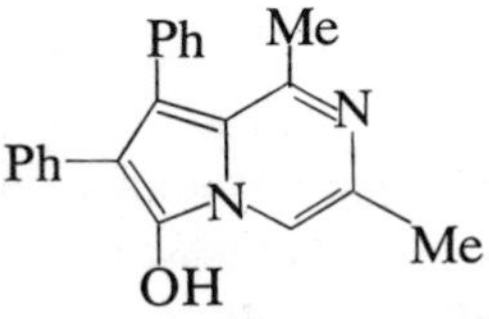

Ph Me
N
Ph
N
Me
OH

8

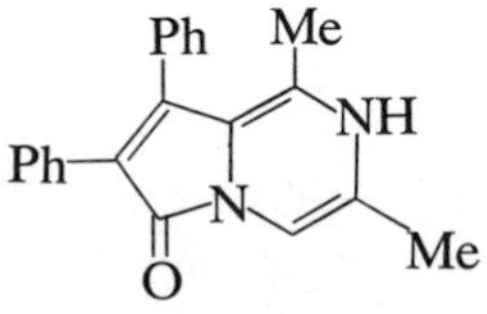

Ph Me
NH
Ph
N
Me
O

9

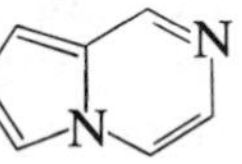

N
N

10

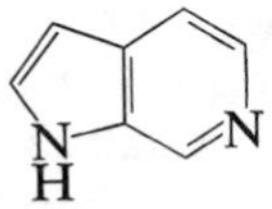

N
N
H

11

CO_2Me
N
MeO_2C
N
Me
CO_2Me

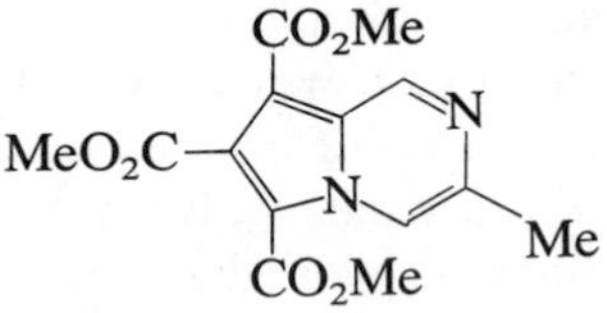

12

A detailed analysis of the PMR spectrum of the unsubstituted compound **10** has been performed (see Figure 1). This, together with a similar analysis of the 1,3-dibromo derivative, served to confirm the positions of electrophilic substitution in the molecule.[2] The structure **12** of the adduct from 2-methylpyrazine and dimethyl acetylenedicarboxylate was also established using PMR data.[8] PMR data are also available for related adducts[125] and for 1-phenylpyrrolo[1,2-*a*]pyrazine.[126]

FIGURE 1. PMR PARAMETERS OF PYRROLO-[1,2-*a*]PYRAZINE

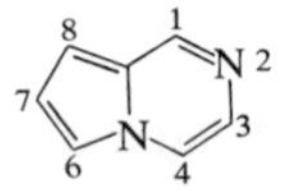

	$\delta(CDCl_3)$	Coupling constants, Hz
H-1	8.91	$J_{1,3}$ unresolved (broadening); $J_{1,4}=1.55$; $J_{1,6}=1.00$
H-3	7.58	$J_{3,4}=5.5$
H-4	7.89	$J_{4,8}=0.85$
H-6	7.46	$J_{6,7}=2.55$, $J_{6,8}=1.45$
H-7	6.97	$J_{7,8}=4.50$
H-8	6.85	

2. Methods of Preparation

2-Acylpyrroles (**13**) have been used in two ways to prepare examples of this ring system. Alkylation of the sodium salt of **13** in ethanol with the α-bromoacetal (**14**) followed by heating with ammonium acetate in acetic acid provides pyrrolo[1,2-*a*]pyrazines (**15**) in 25–50% yields.[9,10] Use of phenacyl bromide in place of the bromoacetal gave the corresponding 3-phenyl compound (**16**). Condensation of acylpyrroles with α-aminoacetaldehyde diethylacetal gave the imines **17**, which on treatment with phosphoryl chloride in polyphosphoric acid cyclize to the fused pyrazines **18** together with a small amount of the isomeric pyrido compounds **19** formed by attack at the 3-position of the pyrrole ring.[7]

Perhydro derivatives of the ring system may be obtained by treating aziridine with appropriate proline esters. Thus the ester **20**, (R = Et) gave the 1-oxo derivative **21** on reaction with aziridine in refluxing ethanol in the presence of a trace of hydrogen chloride.[5,11] The methyl ester **20**, (R = Me) also gives the expected oxo compound. When the cyclization was attempted using 1-*p*-nitrobenzoylaziridine (**22**), by analogy with work on the corresponding pyridopyrazines, the amino acid **23** was obtained, which was cyclized to the oxo derivative with sulfuric acid in ethanol.[12]

Reaction of the dichloro compound **24** with primary amines gave the perhydro compounds **25** together with, in certain cases, small amounts of the by-products **26**. The best solvent for reaction was suggested to be 50% aqueous acetone.[5] This approach has also been used to prepare various other substituted perhydro derivatives.[13,14]

R
C
O
NH
13
+ $BrCH_2CH(OBu)_2$ ⟶
14

R
C
O
N
$CH_2CH(OBu)_2$

$NH_2CH_2CH(OEt)_2$

R
C
N
NH
$CH_2CH(OEt)_2$
17

R
N
N
R^1

15 R = Me or Ph; R^1 = H
16 R = Ph; R^1 = Ph

R
N
N
+
N
H
N
R

18 R = H or Me **19** R = H or Me

CO_2R
NH
20
⟶
O
NH
N
21

NCO
NO_2
22

CO_2H
N
$CH_2CH_2NH_2$
23

CH_2Cl
N
CH_2CH_2Cl
24
+ RNH_2 ⟶
N—R
N
+
CH_2NHR
N
CH_2CH_2NHR

25 R = alkyl or aralkyl **26**

Recently the disubstituted pyrrolidine **27** has been reacted with chloracetylchloride, followed by treatment with sodamide in liquid ammonia, to give a pyrrolo[1,2-*a*]pyrazine.[127] The product was formulated as the 4-oxo compound **28**, but no attempt was made to exclude the 3-oxo isomer.

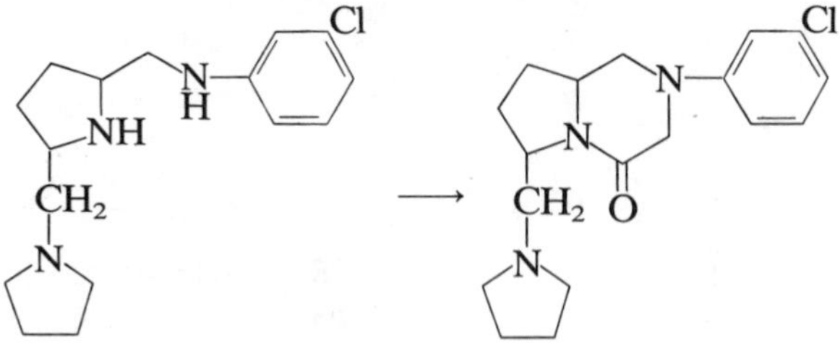

27 → **28**

Various complex derivatives of the ring system have been prepared starting from pyrazine intermediates. Frequently the yields from these reactions tend to be low and of little synthetic value. Thus reaction of dimethyl acetylenedicarboxylate with the ylid **29** gave a mixture of compounds **30a** and **30b** in 5 and 16% yields, respectively.[15] The corresponding reaction[16] with ylid **31** gave a 8.5% yield of a mixture containing two parts of **32** and one part of **33**.

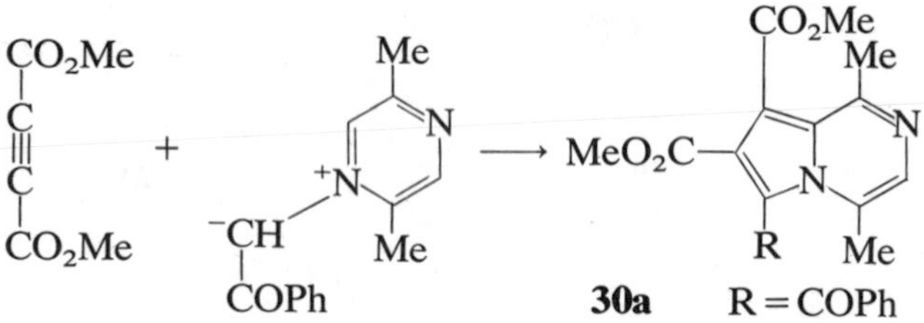

29 → **30a** R = COPh; **30b** R = H

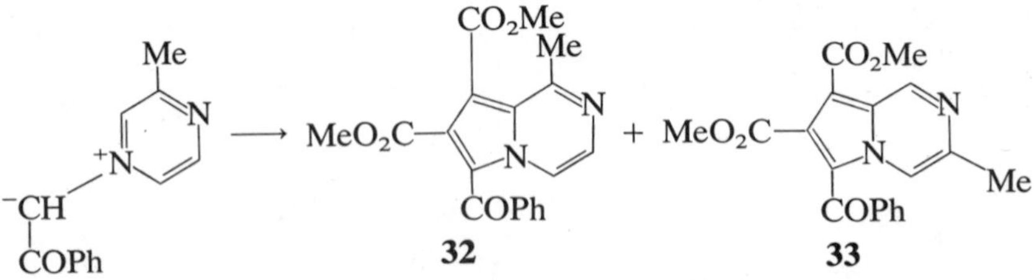

31 → **32** + **33**

Much higher yields (40–60%) were obtained from the reaction of the cyanoylids **34** with dimethyl acetylenedicarboxylate.[125] Apparently none of the 3-methoxy isomer of **35** (R = OMe) was detected. 2-Methylpyrazine and 2,6-dimethylpyrazine add two molecules of the acetylenic ester to give low yields of compounds **36a** and **36b**. No evidence was found of any isomer formation.[8]

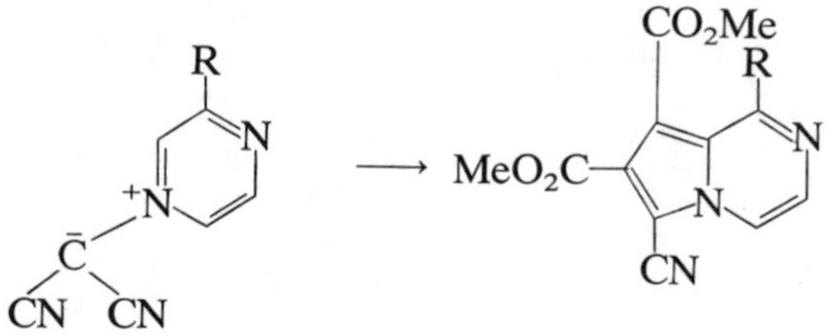

34 R = H or OMe **35** R = H or OMe

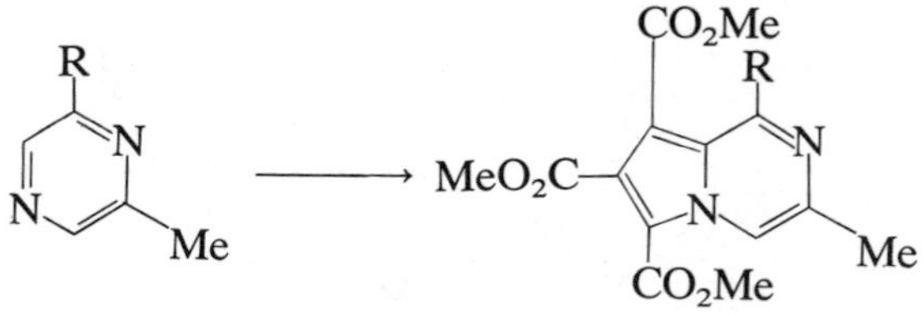

36a R = H
36b R = Me

Reaction of pyrazines with diphenylcyclopropenone provides a useful synthesis of 7,8-diphenyl-substituted compounds.[6] Thus pyrazine reacts with two molecules of the reagent to give the acryloyl ester **37** in 31% yield. 2,6-Dimethylpyrazine reacts with one molecular equivalent of the reagent to give an 80% yield of the hydroxy compound **8** on standing with the cyclopropenone in methanol for three days at room temperature. Similarly cycloheptenocyclopropenone reacts with 2,6-dimethylpyrazine to give the tricyclic system **38**. It has been assumed that 2-methylpyrazine reacts with both cyclopropenones to give the less hindered compounds **39** rather than the isomers **40.**

Reactions of pyrazines with diphenylcyclopropenethione (**41**) also provide good yields of 7,8-diphenylpyrrolo[1,2-*a*]pyrazines.[17] The reaction products, however, are somewhat different from those from diphenylcyclopropenone. Thus pyrazine itself gave a mixture of the thiol **42** and the disulfide **43** even under anaerobic conditions. In the presence of air the disulfide was the only product. The thiol **42** was quantitatively convertible into the corresponding disulfide simply by recrystallization from aerated chloroform. 2-Aminopyrazine gave a 70% yield of a disulfide formulated as **44**, although no evidence was presented to eliminate the 1-amino isomer of **44**. 2-Methylpyrazine reacted with diphenylcyclopropenethione

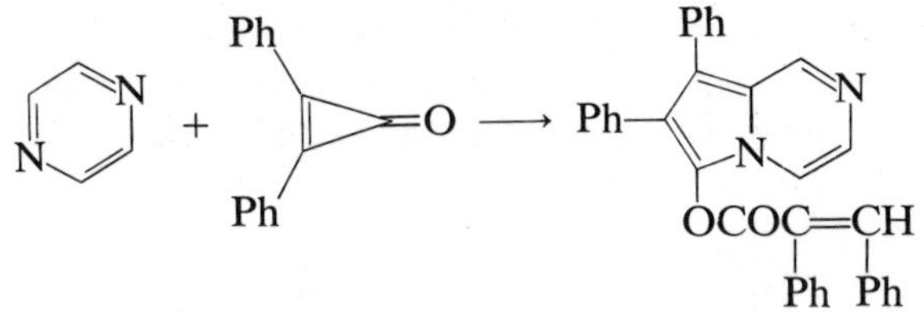

39 R, R = Ph or —$(CH_2)_5$—

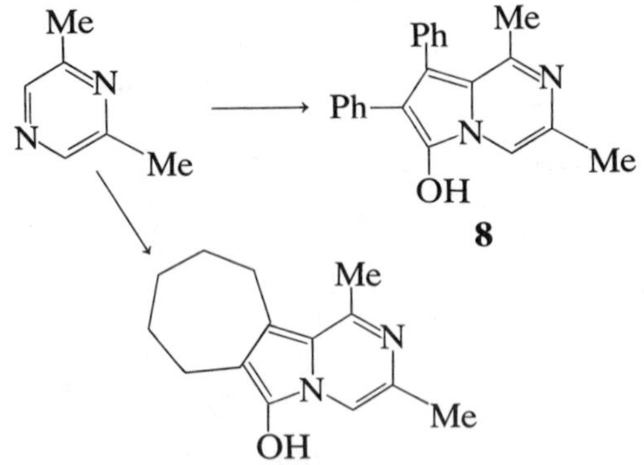

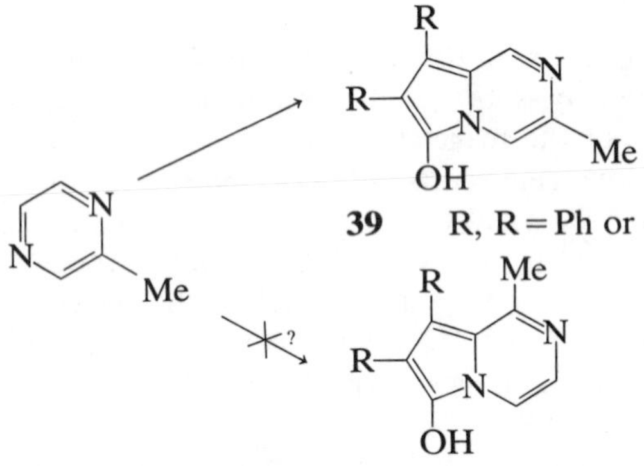

to give a single isomer, tentatively assigned the 3-methyl-6-thiol structure **45** on the basis that formation of the 1-methyl isomer **46** would be sterically hindered. No disulfide formation was reported in the reaction of 2-methylpyrazine or 2,6-dimethylpyrazine with the cyclopropenethione. The 2,6-dimethyl compound gave a 68% yield of the thiol **47**. The methylpyrazines thus behave like quinoxaline, which also forms a thiol on reaction with diphenylcyclopropenethione (see Chapter XXXV).

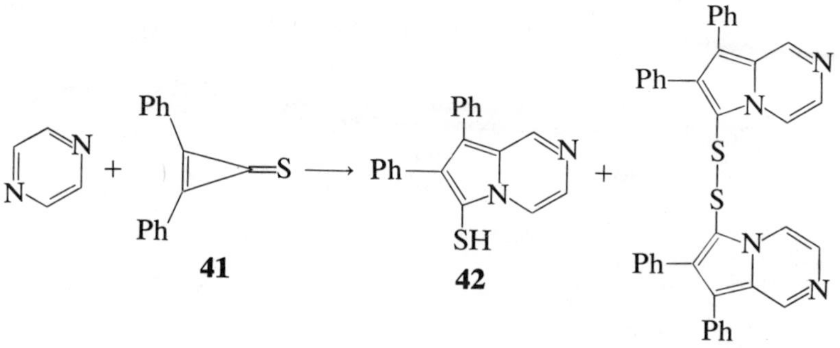

Ph, Ph, N, N, S, 41, 42, SH, 43, 44, NH$_2$, R, R^1

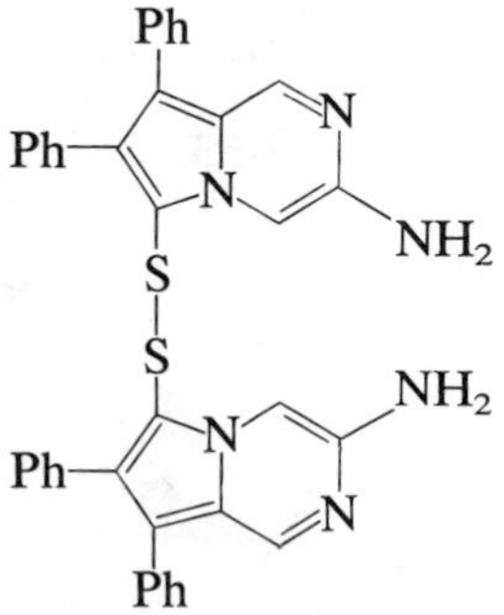

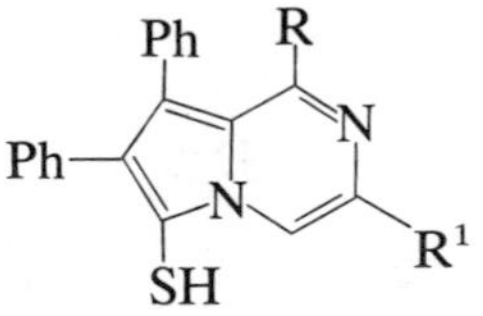

45 R = H, R^1 = Me
46 R = Me, R^1 = H
47 R = R^1 = Me

Several approaches, apart from those already considered, have provided partially or fully reduced derivatives of this ring system. Heating 2-acetylfuran (**48**) with ethylenediamine gave the 3,4-dihydro compound **49** in 45% yield.[18] *N*-Tetrahydrofurfurylethylenediamine (**50**) has been cyclized to the perhydro compound **51** at 300° over alumina.[19,20] 1-Oxoperhydro derivatives are obtained on reaction of ethylenediamine with α,δ-dihalocarboxylic acids.[21–24] This synthesis has been extended using suitable dihalo acids to a variety of diazabicyclic systems including pyrido[1,2-*a*]pyrazine (see Chapter XXVII).

Two examples have been reported of rearrangements that lead to pyrrolo[1,2-*a*]pyrazines. The Schmidt reaction on the ketone **52** gave only one product, the 3-oxo compound **53**, paralleling the behavior of the ketone **54**, which gave only the corresponding 3-oxopyrido[1,2-*a*]pyrazine on treatment with hydrazoic acid.[11] The 1,6-dioxo compound **56** was obtained on refluxing the fused imidazoline **55** in water.[25]

48 49 50 51 52 53 54 55 56

3. Reactions

There is only one report of the quaternization of an aromatic derivative of this ring system.[15] Compounds **57** reacted readily with phenacyl bromide to give the quaternary salts **58**. These were not isolated as such but were treated with sodium carbonate, resulting in formation of the dipyrrolopyrazines **59**. Although these cyclizations occurred in quite high yields, it did not prove possible to perform the same reaction on the related pyrazine **60** to give the corresponding pyrrolo[1,2-*a*]pyrazine. Monomethiodides have been obtained from the perhydro[5] compound **61a** and 1- and 3-oxoperhydropyrrolo[1,2-*a*]pyrazines.[11] Perhydro derivative **61b** formed a dimethiodide.[5] Although the structures of the monomethiodides have not been elucidated; comparison with work reported on perhydropyrido[1,2-*a*]pyrazines (see Chapter XXVII) suggests that the site of quaternization is probably the bridgehead nitrogen atom. This is confirmed by more recent studies.[155]

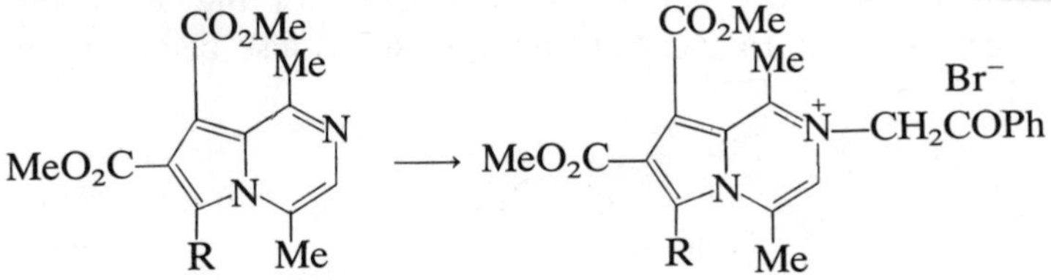

57 R = H or COPh

58

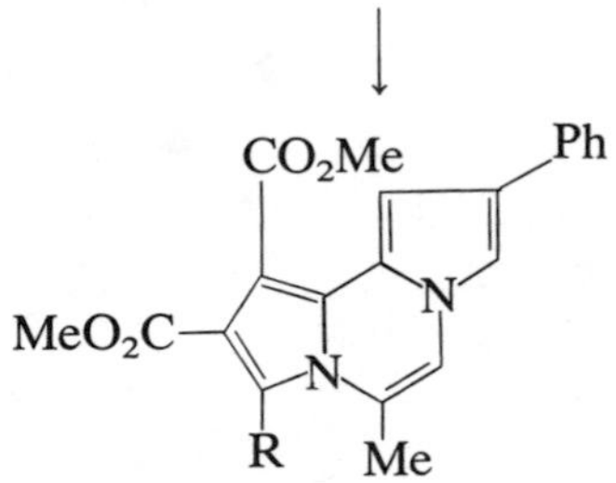

59

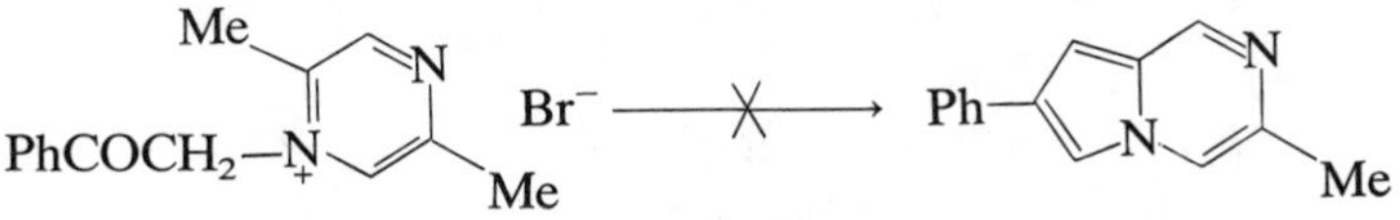

60

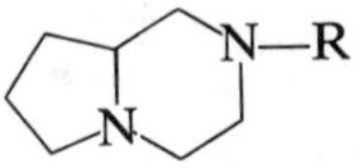

61a R = $CHMeCH_2Ph$
61b R = $CH_2CH_2CH_3$

Alkylation of the 6-hydroxy compounds **62** with triethyloxonium fluoroborate gave the corresponding 6-ethoxy derivatives.[6] The completely reduced compound **51** has been regarded as an analogue of piperazine and has therefore been alkylated to give a variety of 2-substituted derivatives with biological activity.[12,13,26] Compound **51** is readily acetylated by heating in acetic anhydride.[19] Acylations have also been performed with the azides of complex carboxylic acids.[27,28] For example, the cyclic amine **63**, obtained by lithium aluminum hydride reduction of the diketopiperazine from L-leucine and L-proline, reacts with dihydrolysergyl-D-alanylazide to give the acyl derivative **64**. However, when the reaction was repeated on the amine derived from L-leucine and D-proline, acylation was more difficult because of steric hindrance, and the acylated product was accompanied by some of the urea **65**. Presumably the rate of reaction had been slowed down sufficiently to allow the azide to rearrange to the corresponding isocyanate which then reacted with the cyclic amine.[28] The reaction of the amine **51** with phenylisothiocyanate to give the thiourea **66** has also been reported.[5]

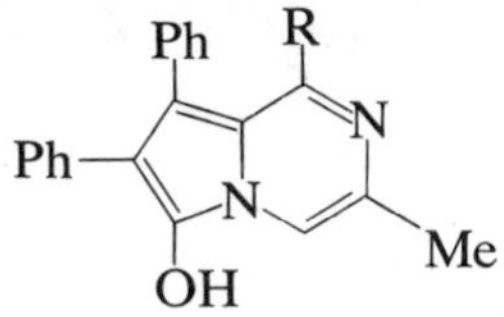

62 R = H or Me

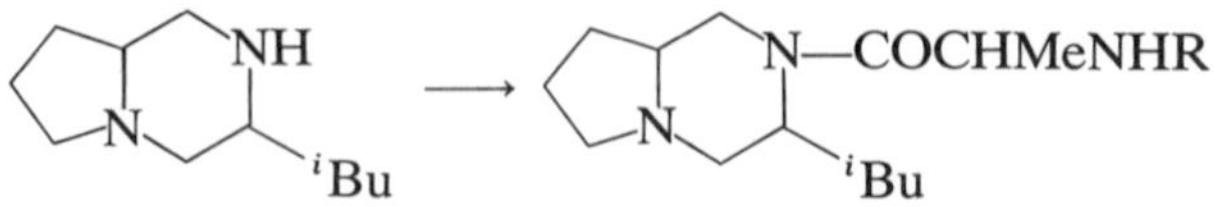

63

64 R = Dihydrolysergyl

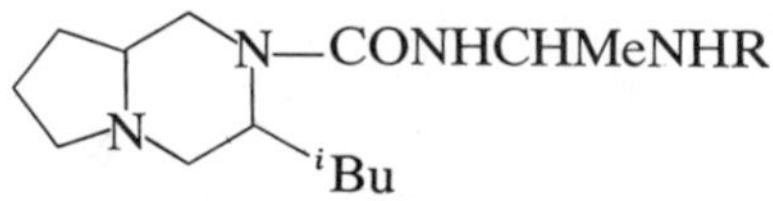

65 R = Dihydrolysergyl

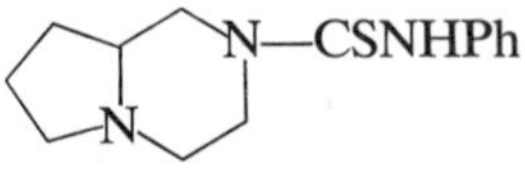

66

Bromination of the parent aromatic heterocycle occurs readily in the 6- and 8-positions to give the dibromo compound **67** as predicted by HMO calculations.[2] Even with only one equivalent of bromine it did not prove possible to isolate either of the monobromo derivatives. In contrast, treatment of pyrrolo[1,2-*a*]pyrazine with phenyllithium gives the 1-phenyl derivative of the heterocycle in 30% yield.[126] As in the related pyrido[1,2-*a*]pyrazine series, lithium aluminum hydride readily forms the perhydro compound **51** from the corresponding 3-oxo[11] or 1-oxo compounds.[5,11,12]

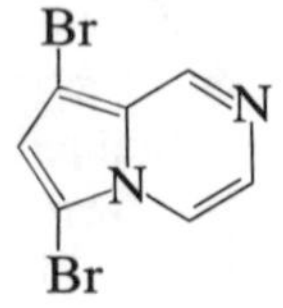

67

4. Biological Activity

Because of the relationship of the reduced ring system to piperazine, various derivatives have been studied and found to have activity on the central nervous system (CNS). Thus the compound **68** was found to be more active than the corresponding pyrido compound (**69**) in the mouse amphetamine-induced toxicity and hypermotility tests.[12] Certain phenothiazine derivatives are the subject of recent patents.[128,129] Various other 2-substituted perhydropyrrolo[1,2-*a*]pyrazines have been patented as sympatholytics,[21] antihistaminic, anticholinergic, and antitremorine compounds,[13] CNS depressants,[14,26] antiarrhythmics,[130] coronary dilators,[129] and anticonvulsants and analgesics.[14]

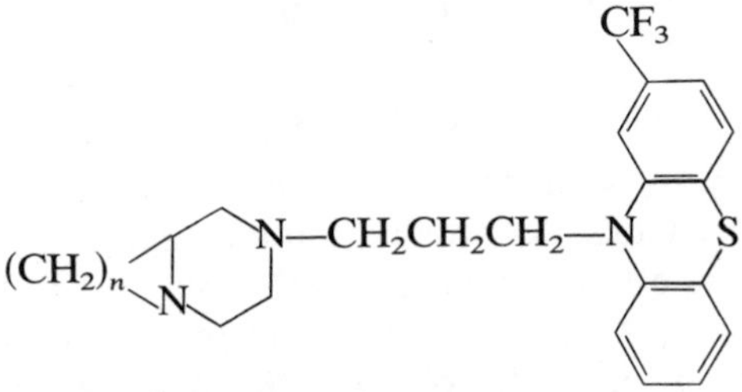

CF_3

$(CH_2)_n$ N—$CH_2CH_2CH_2$—N S

68 $n = 2$
69 $n = 3$

II. 1,4-Dioxopyrrolo[1,2-*a*]pyrazines

The major interest in pyrrolo[1,2-*a*]pyrazines has centered round the 1,4-dioxo compounds. These derivatives (**71**) are readily formed by the combination of the amino acid proline (**70**) and another α-amino acid. The reaction of two α-amino acids to give diketopiperazines is frequently encountered in peptide chemistry and in nature. When one of the components is proline, entropy considerations predict that the reaction should be facilitated, as the cyclic nature of the amino acid places two of the reacting groups in the optimum orientation. This is found to be so and explains the frequent occurrence of the bicyclic ring system in nature and hence the chemical and biochemical importance of these compounds. The derived ring system oxazolo[3,2-*a*]pyrrolo[2,1-*c*]pyrazine (**72**) is also of considerable significance as a constituent of the ergot alkaloids (see Section II6). Studies of these alkaloids touch in several places on the chemistry of 1,4-dioxopyrrolo[1,2-*a*]pyrazines.

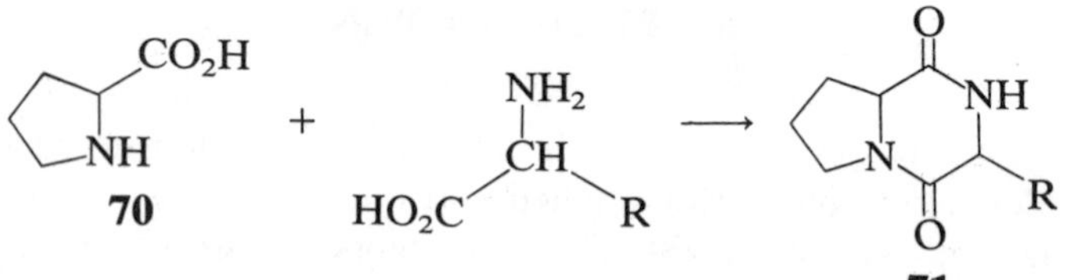

70 + → 71

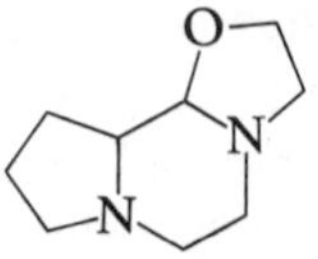

72

1. Physical Properties

X-Ray diffraction data on glycyl-L-proline anhydride (**73**)[30] and more recently on L-leucyl-L-proline anhydride (**74**),[133] have been obtained from the study of single crystals. Although 2,5-diketopiperazine itself possesses a planar ring, as successively bulkier substituents are introduced increasing deviation from planarity is observed. In compound **74** the six-membered ring is significantly folded with an angle between the two planar peptide units of 143°. The various angles and bond lengths in the diketopiperazine ring of **74** are almost identical to those observed in the tricyclic compound **75**. The five-membered ring of L-leucyl-L-proline

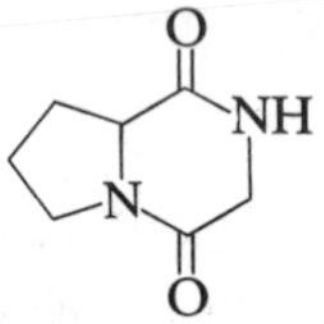

73 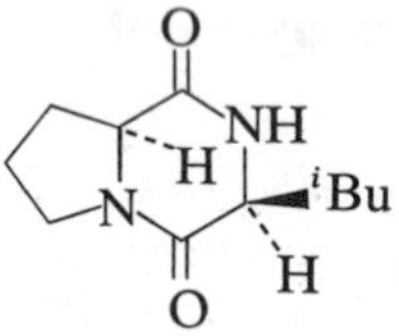 74 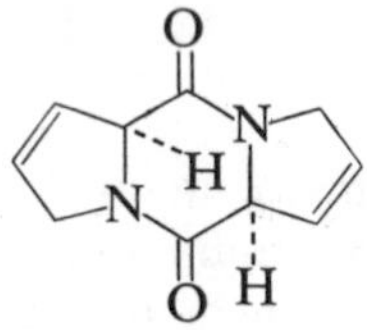 75

anhydride adopts the usual envelope conformation, with C-8 0.52 Å out of plane. PMR data on compound **73** have been interpreted as indicating that the diketopiperazine ring adopts a boat conformation in D_2O and DMSO-d_6, whereas in trifluoracetic acid the ring is planar. These conclusions were supported by comparison with the L-L and D-L forms of the dipyrrolo compound **76**, the former having the six-membered ring in a stable boat conformation and the latter containing a stable planar diketopiperazine ring.[31,134] In $CDCl_3$ the six-membered ring of compound **73** possesses a boat conformation as evidenced by its PMR spectrum. The depth of the boat conformation of the diketopiperazine ring varies with the size of the annelated ring so that the pyrido[1,2-*a*]pyrazine (**77**) is almost planar and the 4,6-fused system (**78**) is more folded than cyclo-L-prolylglycine (**73**). These changes have been related to differences in their PMR, IR, and CD spectra.[135] The solvent dependence of the chemical shifts of compound **73** has been used to clarify the conformations adopted by certain cyclic polypeptides in solution.[32]

76 **77** **78**

If the ring system has a substituent in the 3-position, then the molecule has two asymmetric centers, and two pairs of diastereoisomers are possible (see Fig. 2). The most commonly encountered system is that obtained from the naturally occurring L-α-amino acids. These acids have the *S* absolute configuration and thus provide the 3*S*, 8a*S* structure **79a**, which has the hydrogens at positions 3 and 8a cis-diaxially and the 3-substituent pseudoequatorially disposed. This isomer is frequently referred to as the cis isomer and is the mirror image of the DD isomer **79d**. Both the DL derivatives, **79b** and **79c**, are sometimes described as having trans geometry and have the 3-substituent in a pseudoaxial orientation.

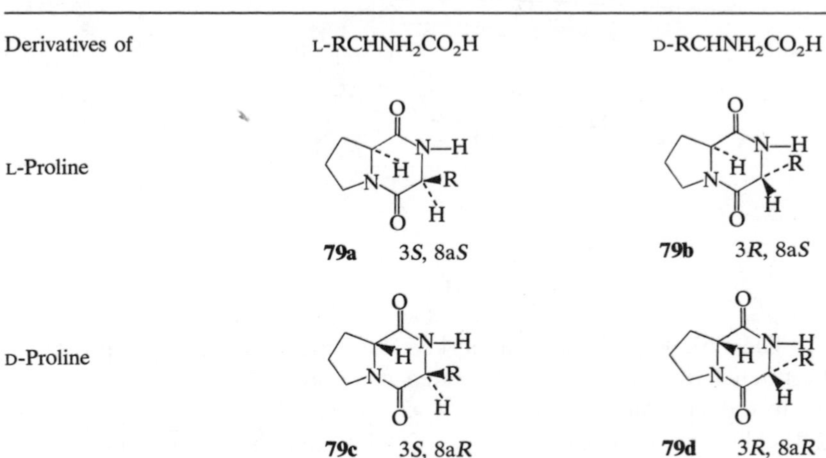

Fig. 2. Stereochemistry of perhydro-1,4-dioxopyrrolo[1,2-*a*]pyrazines

The stereochemistry at the 3-position of substituted derivatives of this ring system has been examined using PMR spectroscopy.[33,34,136] The 60 MHz spectrum of the 3-methyl derivative shows doublets in different positions due to the methyl group in the isomers **79a** and **79b** (R = Me). The LL (cis) form (**79a**) gives a signal at 97.5 Hz below TMS, whereas the DL isomer (**79b**), with the pseudoaxial methyl group, shows the corresponding signal at 100.5 Hz below TMS.[33] The same effect would obviously occur in the mirror images **79d** and **79c**, respectively.

6-Oxo compounds such as **80** and **81** have had their stereochemistry elucidated by means of PMR spectroscopy and optical rotations.[34] Steric interaction between the *N*-acetyl group and adjacent pseudoequatorial substituents favors formation of the pseudoaxial isomer when this is possible. This was demonstrated by the PMR spectrum of compound **81**

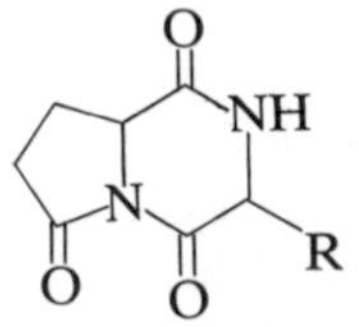

80 R = H or CH_2Ph 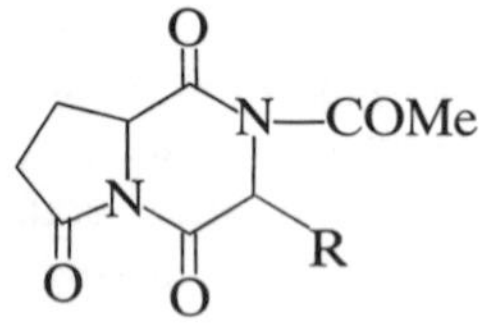 **81** R = H or CH_2Ph

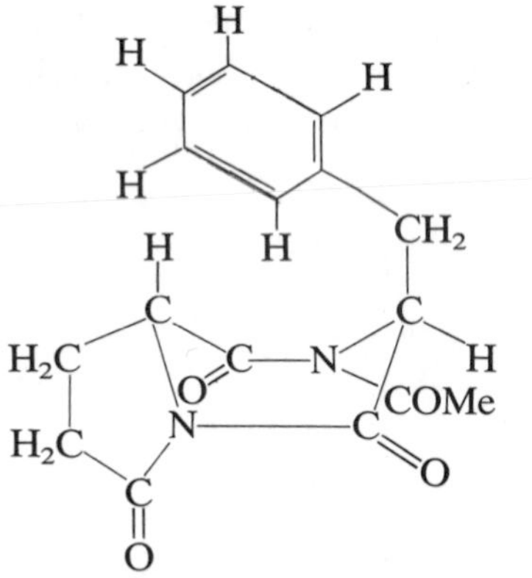

81a

(R = CH_2Ph), which showed a marked shielding effect of the 8a proton by the phenyl ring. In the unsubstituted compound (**81**, R = H) the triplet due to the 8a proton is found at δ 4.93, whereas in both of the DL isomers of the benzyl compound (**81**, R = CH_2Ph) the 8a proton resonates at δ 3.07. The explanation of this remarkable shielding effect is the proximity of the aromatic ring of the pseudoaxial 3-benzyl substituent to the 8a proton (see **81a**). The same effect is observed in the 3*R*, 8a*S* isomer of compound **80** ($R = CH_2Ph$) but not in the 3*S*,8a*S* isomer where the benzyl group is pseudoequatorially disposed. Recent infrared studies have supported and amplified the conformational conclusions drawn from PMR spectra.[137,138]

The different molecular shapes of diastereoisomeric diketopiperazines allows their ready separation on GLC columns. Whereas with most monocyclic diketopiperazines the DL compounds have shorter retention times than the LL isomers, the reverse is true for perhydro-1,4-dioxopyrrolo[1,2-*a*]pyrazines.[33] The high degree of diastereoisomeric resolution of these compounds has been studied on a variety of columns.[35] As diastereoisomeric diketopiperazines are more readily separated and studied as their *N*-methyl derivatives, GLC and PMR data have been collected for cyclo-L-prolylsarcosine.[139]

Earlier work on these compounds relied heavily on infrared spectra to determine structure. Glycyl-L-proline anhydride has had its infrared spectrum reported in some detail.[36] Particularly useful[37,38] for *N*-unsubstituted derivatives are the single band at 3260 cm^{-1} (ν N—H stretch), the two strong bands at 1670 and 1635 (or 1630) cm^{-1} (ν C=O stretch), and the absence of bands at ca. 1550 cm^{-1} and in the region 1250–1000 cm^{-1} (ν C—O stretch). The infrared spectra of the 6-oxo compounds **80** and **81** (R = H) have been used to correct an erroneous interpretation of infrared data.[39] The extra carbonyl groups incorporated into the molecules adjacent to the nitrogen atoms result in a raising of the carbonyl stretching frequencies from the range associated with lactams to frequencies (1785 or 1795 cm^{-1} and 1710 or 1730 cm^{-1}) corresponding to imides. The fine structure of the N—H and N—D absorption bands in crystals of cycloglycyl-L-proline (**73**) and a partially deuterated sample was used in a discussion of hydrogen bonding in the crystals of such compounds.[40]

Infrared spectra have been much used in the investigation of the tautomerism of hydroxyacyl compounds such as **83** to the oxazolo derivatives **84**. Studies of this tautomerism in relatively simple systems demonstrated the presence of the cyclol structure in the ergot alkaloids.[41–43] Hydrogenolysis of the benzyl ether **82** (IR: ν 1725, 1694 cm^{-1}) gave the hydroxyacyl derivative **83** (IR: ν 1730, 1693 cm^{-1}), which on keeping for ten days gave the cyclol **84** (IR: ν 1730, 1672 cm^{-1}). The infrared spectrum of the methyl ether of this cyclol has bands at 1720 and 1655 cm^{-1}. The carbonyl region in the infrared spectrum of the oxazolo derivative is similar to that of ergotamine (ν 1736, 1669 cm^{-1}).

On the basis of infrared spectral data[28] one of the thermal decomposition products of the ergot alkaloids was assigned the structure **85**. Later

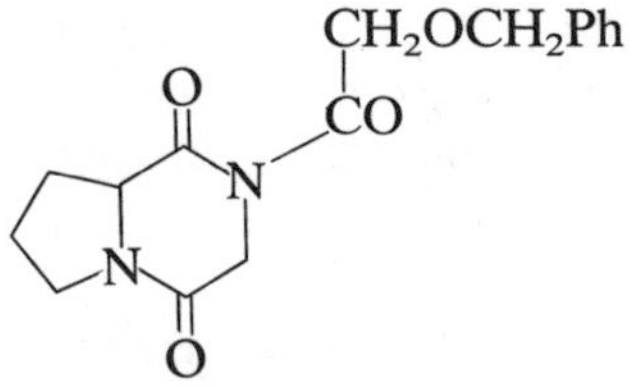

82 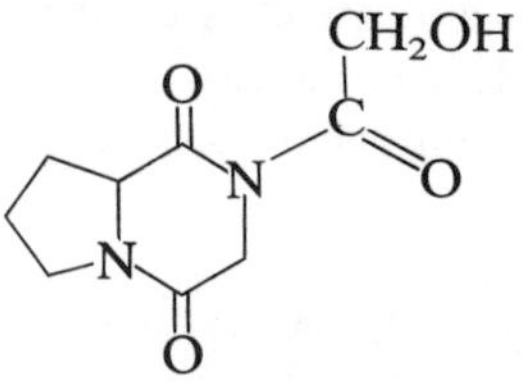 83 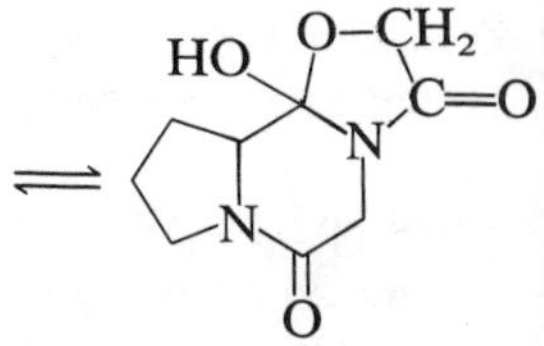 84

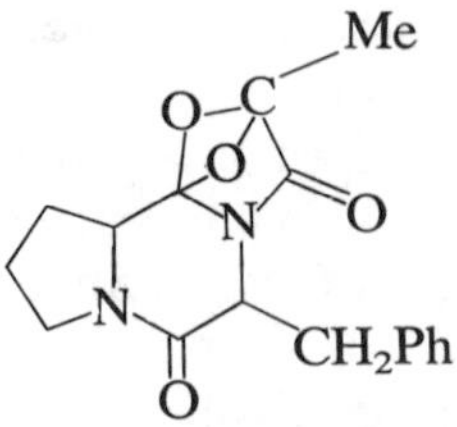

85 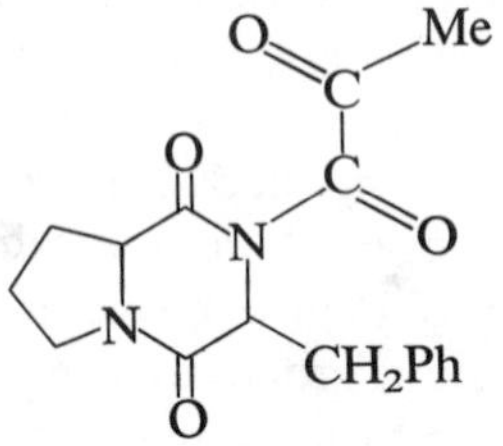 86

work[44] on the PMR spectrum of the product from ergotamine eliminated this structure in favor of the bicyclic system **86**. Presumably the related products described in the earlier paper should also be reformulated. They are listed in Table 2 as acyl derivatives of the ring system. Only one paper has dealt with the mass spectra of this group of compounds.[45] Perhydro-1,4-dioxopyrrolo[1,2-*a*]pyrazines were included in a study of the fragmentation of a series of LL-diketopiperazines, which concluded that the breakdown was dependent on the carbon substituents but almost independent of the substituents on nitrogen.

2. Methods of Preparation

As might be expected, all the routes to these compounds involve condensations of amino acids in one or more steps. Heating together a mixture of the appropriate amino acids results in the formation of mixtures,[46] and it seems that it is best first to prepare the intermediate dipeptide and then subsequently to form the diketopiperazine ring. The 6-oxo compounds **80** may, however, be obtained by formation of the five-membered ring either subsequent to, or simultaneously with, the cyclization of the six-membered ring.[34,47] Thus the piperazinepropionic acid **87** on treatment with refluxing trifluoroacetic anhydride (b.p. 39.5°) yields the pyrrolo compound **80**. Cyclization with acetic anhydride (b.p. 139.5°) under reflux results in the formation of the *N*-acetyl compound **81**, presumably as a result of the more vigorous reaction conditions.

Correspondingly, the use of other anhydrides of aliphatic acids provides the appropriate 2-acyl derivatives. Cyclization of glutamic acid dipeptides (**88**) with high-boiling anhydrides also provides good yields of the 2-acyl-1,4,6-trioxo compounds. The stereochemistry of the products indicated that a high degree of asymmetric induction was involved in the cyclization reaction.[34]

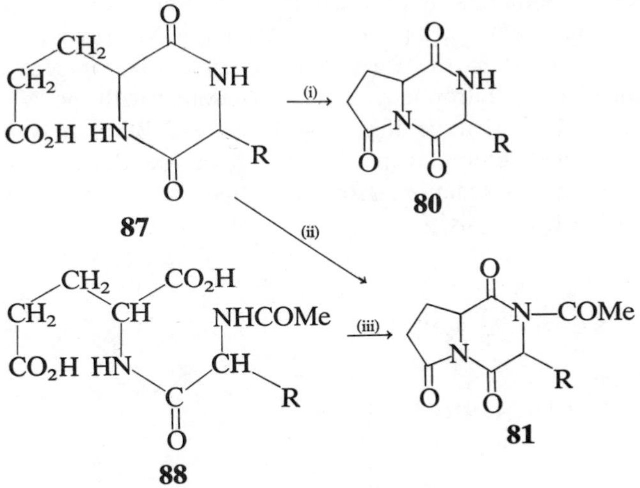

Reagents (i) $(CF_3CO)_2O$
(ii) $(CH_3CO)_2O$
(iii) $(CH_3CO)_2O$

The formation of the diketopiperazine ring from proline-containing peptides can occur in two ways (see type A and type B). Entropy considerations predict that type A syntheses should result in relatively easier cyclization than type B syntheses. In both approaches, the presence of a proline residue in the peptide allows ready cyclization to the diketopiperazine under conditions which would leave other nonproline-containing dipeptides unaffected.[48]

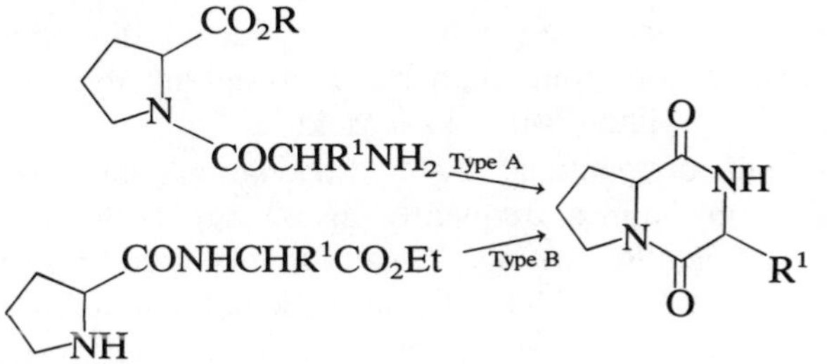

A. *Type A Syntheses*

In this type of synthesis four of the six atoms that form the diketopiperazine ring are already fixed in the correct positions for cyclization. It is not surprising therefore that perhydro-1,4-dioxopyrrolo[1,2-*a*]pyrazines are met as by-products in the synthesis of proline peptides. Even recrystallization of a peptide containing a *N*-terminal glycylproline residue will result in the formation of some diketopiperazine.[49] The treatment of the *p*-nitrophenyl ester of L-proline with the *N*-benzyloxycarbonyl (BOC) derivative of glycine gives the 2-BOC derivative (**90**) of glycyl-L-prolinediketopiperazine.[50] The dipeptide derivative (**89**), also obtained from this reaction, gives the same bicyclic derivative (**90**) on mild alkaline hydrolysis.

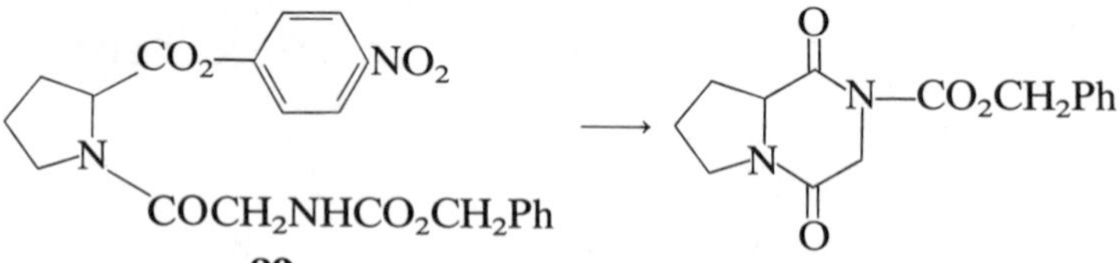

CO_2—C_6H_4—NO_2 ; $COCH_2NHCO_2CH_2Ph$ → N—CO_2CH_2Ph

89 **90**

Cyclization can be achieved by merely heating the appropriate dipeptide. Thus heating tyrosylproline at 105–110° gave the anhydride **91**.[51] If esters of dipeptides are used, then the condensation occurs under milder conditions. For example, glycyl-L-proline anhydride (**73**) is formed in 90% yield when the ester hydrochlorides **92** are basified with ammonia.[48,52] Similarly treatment of the bromoester **93** with methanolic ammonia provides a mixture of the diketopiperazine **94** and the corresponding dipeptide,[53] which could presumably be cyclized by gentle warming. Similarly the 2-phenyl compound **96** is obtained on treatment of the bromoester **95** with aniline. The same compound is also given when the anilide **97** is treated with sodamide in liquid ammonia.[140] Condensation of *N*-benzyloxycarbonyl-L-leucine azide (**99**) with L-proline benzyl ester (**98**) gave a product, presumably the protected dipeptide, which on repeated evaporation from methanol followed by hydrogenation gave cyclo-L-leucyl-L-proline (**74**) in 44% yield.[54]

Deprotection of peptides by hydrogenation of benzyl esters or benzyloxycarbonylamines frequently gives the diketopiperazine either directly or on gentle heating. Thus the protected dipeptide **100** on hydrogenation over 5% Pd–$BaSO_4$ in methanol containing a little acetic acid and water gave the corresponding pyrrolo[1,2-*a*]pyrazine (**101**) on

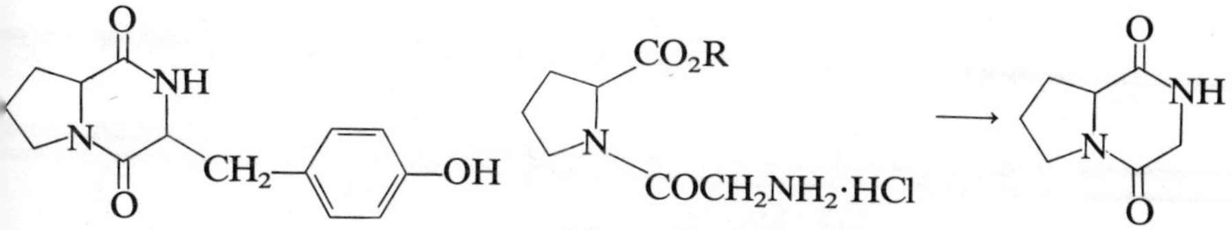

91

92 R = Me or Et

73

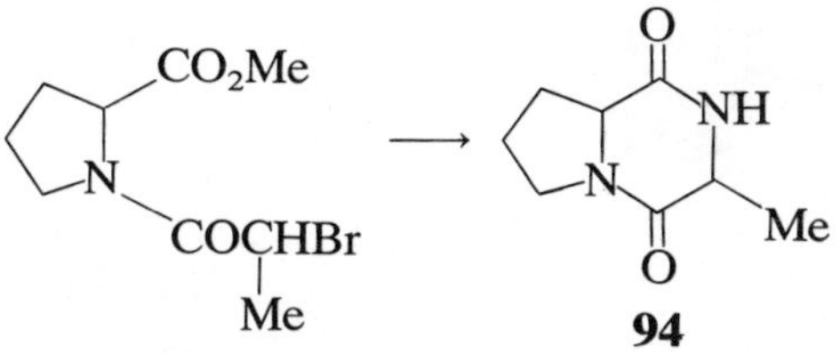

93

94

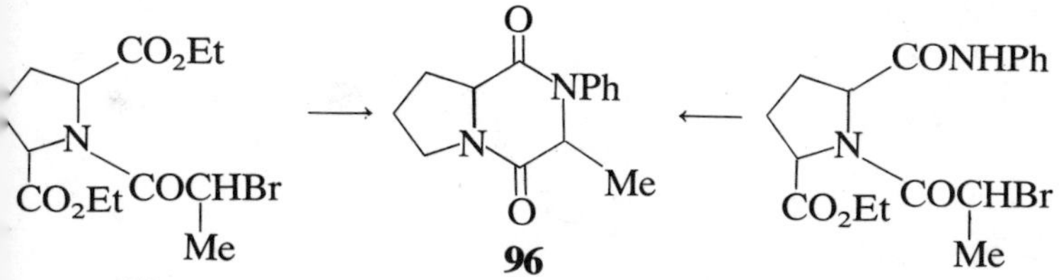

95

96

97

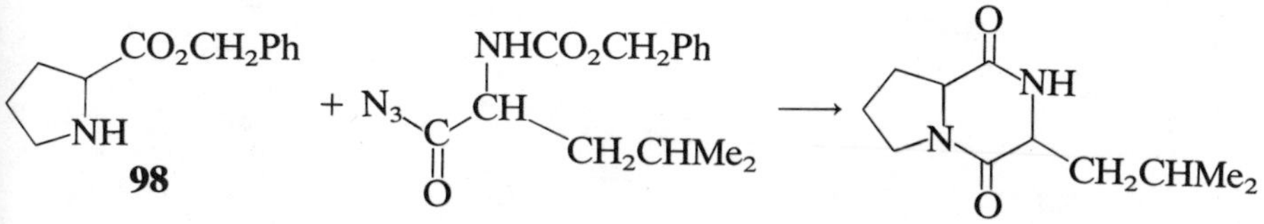

98

99

74

refluxing the filtered solution.[55] Similarly the 8a-methyl compound **103** was obtained when the benzyloxycarbonyl derivative **102** was hydrogenated in acetic acid followed by refluxing of the crude product in dioxan for 2 hr.[56] Catalytic hydrogenation of the protected tripeptide **104** gives D-valyl-L-prolylsarcosine (**105**), which spontaneously decomposes at room temperature to give sarcosine and the trans-diketopiperazine (**106**) in 50% yield.[57] In contrast, cyclization of the related tripeptide **107** with elimination of glycine to give the diketopiperazine **73** requires extended refluxing in ethanol and proceeds only in 40% yield.[48]

Although the diglycylproline ester **108** is somewhat resistant to cyclization, only a low yield of the anhydride **73** being obtained even on prolonged heating;[48] other *N*-acylaminoacetylproline derivatives have

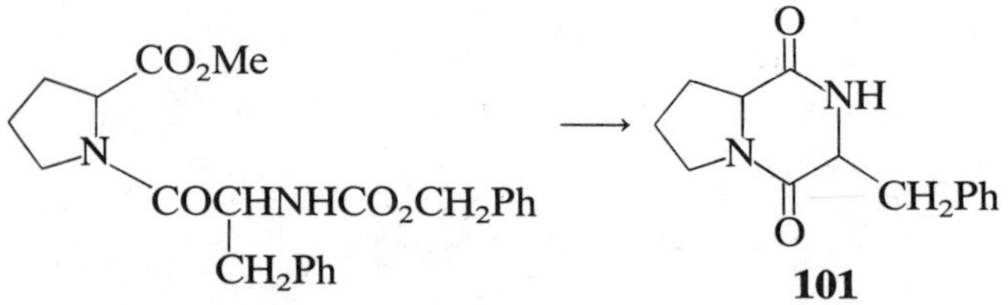

100 → **101**

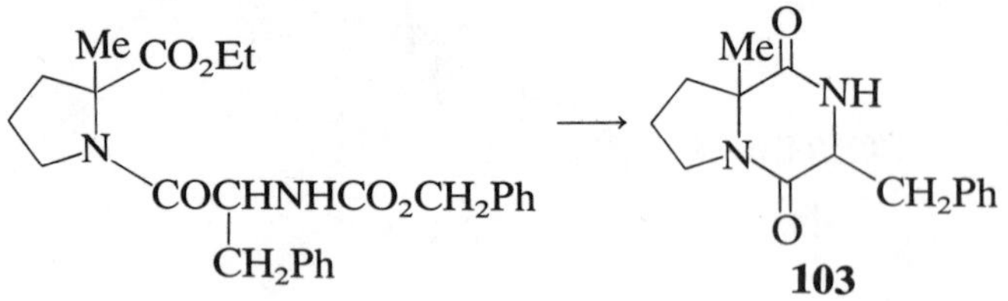

102 → **103**

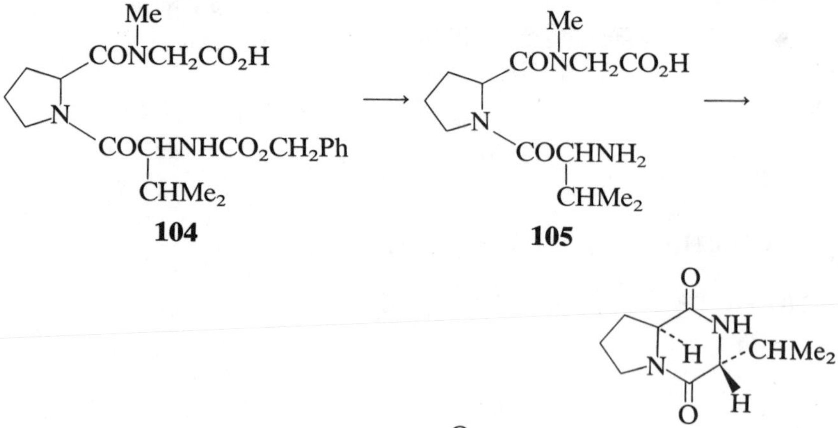

104 → **105** → **106**

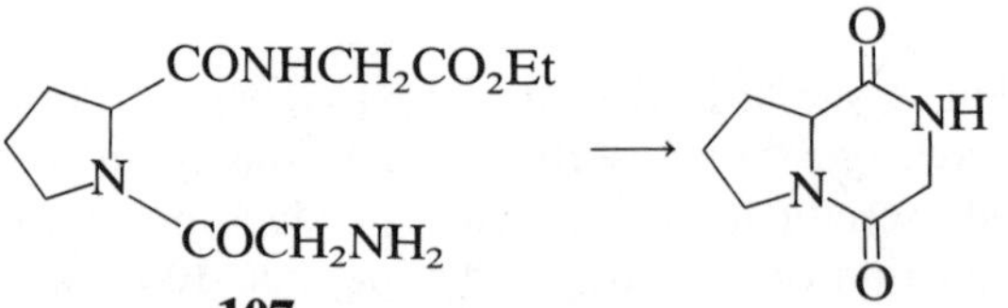

107 → **73**

been readily converted into 2-acyl derivatives of perhydro-1,4-dioxopyrrolo[1,2-*a*]pyrazine. Thus treatment of the pyruvoyl dipeptide **109** with sodium acetate in acetic anhydride at 100° gave the corresponding diketopiperazine (**110**) in good yield.[55] Similarly cyclization of the *N*-acetyl compound **111** proceeds in acetic anhydride to give compound **112** without loss of the acetyl group.[58,141] The aryl ester **113** readily cyclizes to the diketopiperazine **114** without loss of the *N*-benzyloxycarbonyl group. This reaction is performed under mild conditions at room temperature in the presence of sodium carbonate and sodium bicarbonate, or in the presence of pyridine and triethylamine. In contrast, treatment of the related tripeptide ester **115** with a mixture of sodium carbonate and sodium bicarbonate results in cyclization with loss of the *N*-terminal amino acid residue.[59]

CO_2Et / N / $COCH_2NHCOCH_2NH_2$ → O / NH / N / O

108 **73**

CO_2H / N / COCHNHCOCOMe / CH_2Ph → O / N—COCOMe / N / O / CH_2Ph

109 **110**

CO_2H / N / COCNHCOMe / CHPh → O / N—COMe / N / O / CHPh

111 **112**

CO_2—C_6H_4—NO_2 / N / $COCH_2NHCO_2CH_2Ph$ → O / N—CO_2CH_2Ph / N / O

113 **114**

CO_2—C_6H_4—NO_2 / N / $COCHNHCOCHNHCO_2CH_2Ph$ / CH_2Ph Me → O / NH / N / O / CH_2Ph

115 **101**

Although the dipeptide derivative **116** cyclized in pyridine to give the diketopiperazine **117** in good yield, the related proline derivative **118** was more stable in pyridine and underwent an alternative mode of cyclization in aqueous alkali.[142] The product (**119**) apparently is the only example of a 1,4-dioxoperhydropyrrolo[1,2-*a*]pyrazine containing a 2-α-aminoacetyl group which has been synthesized directly by ring closure of a tripeptide. Merely standing in methanol for 24 hr is sufficient to deacetylate (**119**) to give the 2-unsubstituted compound. During or subsequent to cyclization the 8a-position underwent epimerization, as the product from the L-L-L-tripeptide contained a D-prolyl residue and was identical with that from the corresponding D-pro-L-L-tripeptide. The preference for the "trans" product was further illustrated by cyclization of the corresponding L-pro-D-L-tripeptide, in which the product **120** contains an L-prolyl residue and no evidence of epimerization was obtained

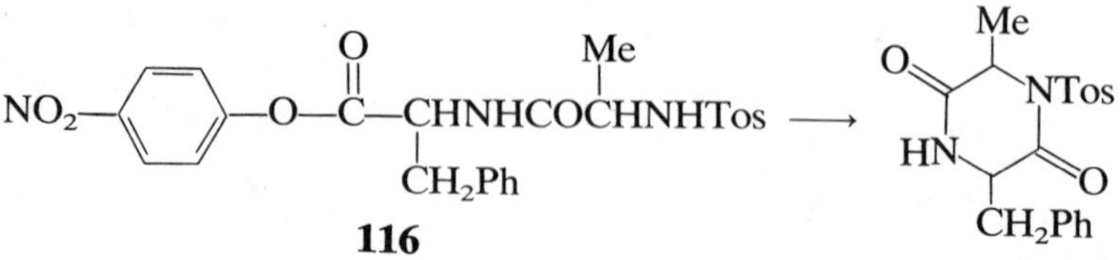

116 → **117**

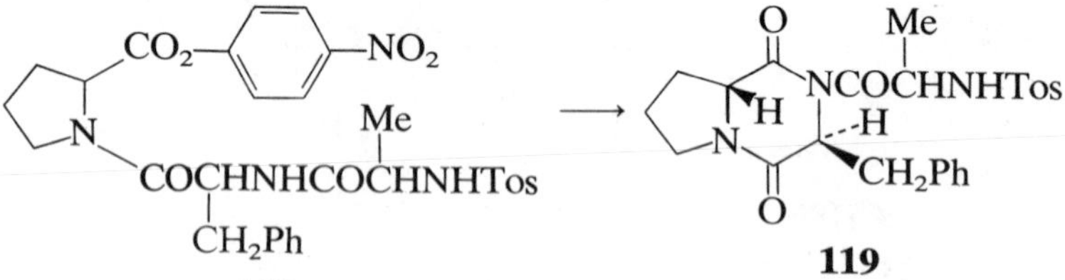

118 → **119**

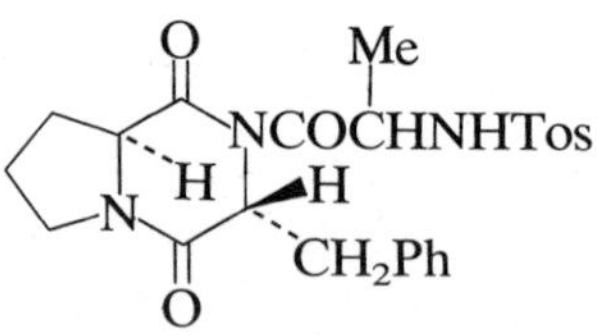

120

An interesting example of intramolecular catalysis of type A syntheses is provided by the ready condensation of certain histidylproline peptides.[60] Treatment of the ditrityl derivative of the tripeptide **121** with 50% acetic acid at room temperature results in rapid detritylation followed by cyclization to give the diketopiperazine **122** with elimination of phenylalanine. The corresponding cyclization with the nonhistidine-containing tripeptide **123** was much slower, and the difference is ascribed to neighboring group participation in the formation of the anhydride **122**.

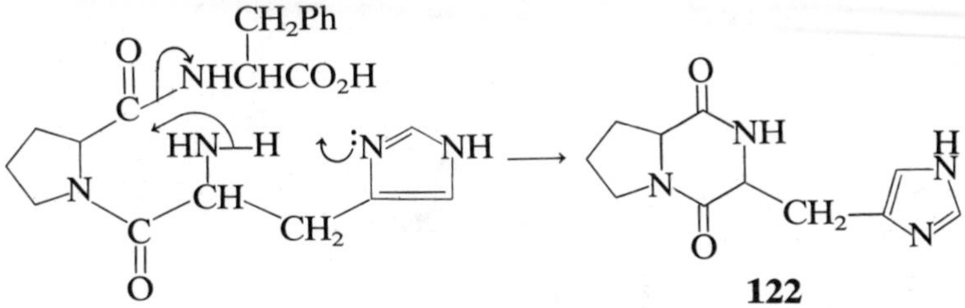

121

122

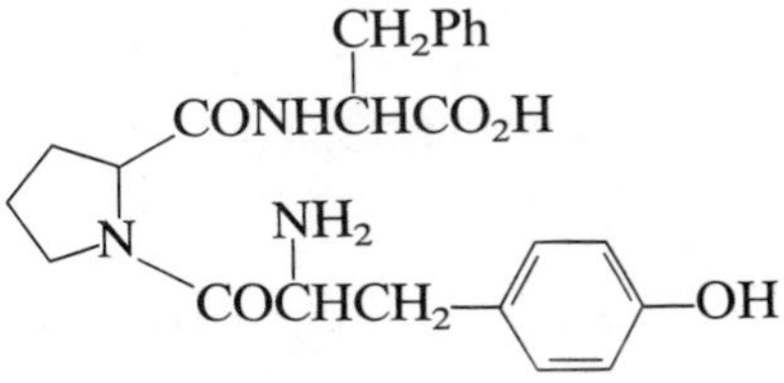

123

B. Type B Syntheses

Although syntheses by this approach are less favored by entropy considerations than type A syntheses, they nevertheless are capable of giving the required cyclic products in high yield. Prolylleucine (**124**) gave the diketopiperazine **74** when heated in vacuo at 180° over phosphorus pentoxide.[61] Similarly prolylalanine (**125**) was cyclized[62] to the anhydride by heating at 225°. As with type A syntheses, milder reaction conditions may be used if a dipeptide ester is used in place of the free acid. If the free base is allowed to stand for several weeks,[48,63,64] esters **126** gradually condense to give the corresponding diketopiperazines. Similar cyclizations have been performed more conveniently by heating the dipeptide ester to 105–120° in vacuo.[48,65,66] The formate salts of dipeptide methyl esters, which are readily prepared during the deblocking of the corresponding *t*-butyloxycarbonyl derivatives with formic acid, are easily converted into diketopiperazines by boiling in a mixture of toluene and *sec*-butyl alcohol for 2–3 hr.[67] This method is described as a simple route to sterically pure diketopiperazines as no racemization occurs during cyclization. It seems, however, that the yields are higher for the formation of monocyclic diketopiperazines than the 53% achieved for the bicyclic proline derivative **74**.

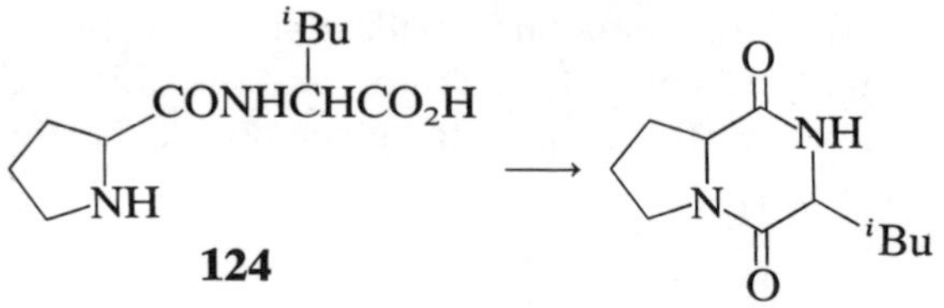

^{i}Bu
CONHCHCO$_2$H
NH
124
→
O
NH
N
^{i}Bu
O
74

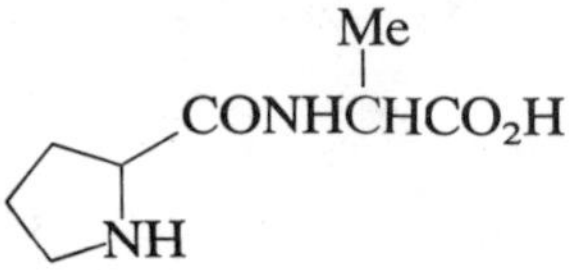

Me
CONHCHCO$_2$H
NH
125

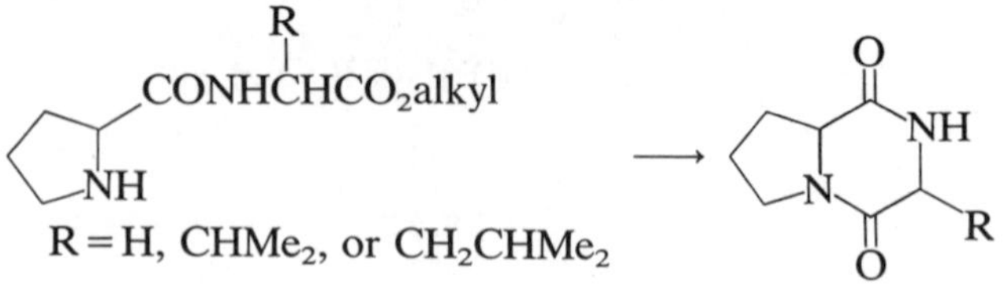

R
CONHCHCO$_2$alkyl
NH
126 R = H, CHMe$_2$, or CH$_2$CHMe$_2$
→
O
NH
N
R
O

Workers at Sandoz have prepared a series of perhydro-1,4-dioxopyrrolo[1,2-*a*]pyrazines in high yield using the methyl esters of *N*-prolylamino acids.[68–74] These esters (**128**) were obtained by hydrogenolysis of the *N*-benzyloxycarbonyl derivatives (**127**) over palladium in either acetic acid alone or admixed with ethylacetate. The cyclized products are formed by heating the filtered reaction mixture or by evaporation of the solution followed by heating the residue in vacuo.

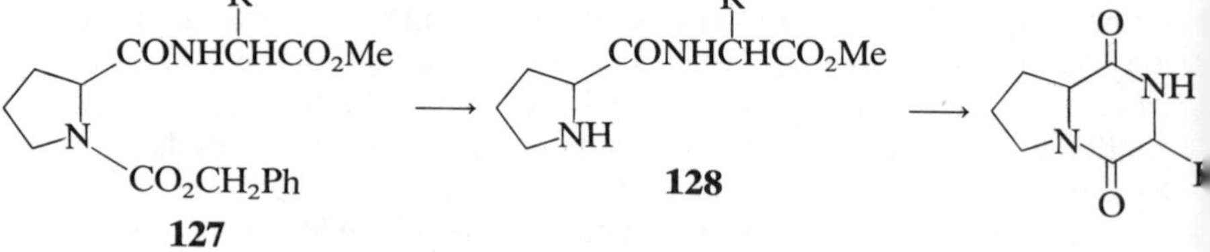

R
CONHCHCO$_2$Me
N
CO$_2$CH$_2$Ph
127
→
R
CONHCHCO$_2$Me
NH
128
→
O
NH
N
R
O

3. Reactions

Alkylation of perhydro-1,4-dioxopyrrolo[1,2-*a*]pyrazines with triethyloxonium fluoroborate gives 1-ethoxy-4-oxo compounds such as **129**. These have then been allowed to react with anthranilic acid to give fused quinazolines such as **130**.[132]

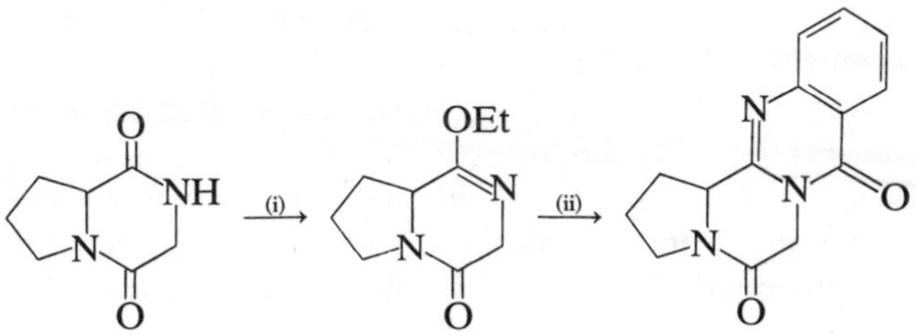

129 **130**

Reagents (i) $Et_3O^+BF_4^-$
(ii) Anthranilic acid

The 1,4,6-trioxo compound **131**, which is unaffected by refluxing trifluoroacetic anhydride, has been acylated by higher-boiling anhydrides of aliphatic acids.[34] A 37% yield of the 2-acetyl derivative is obtained using neat refluxing acetic anhydride, but the other anhydrides had to be diluted with dimethylformamide to dissolve the trioxo compound. This compound has been acylated to give the lauroyl derivative **132** by treatment with lauroyl chloride in refluxing xylene.[75]

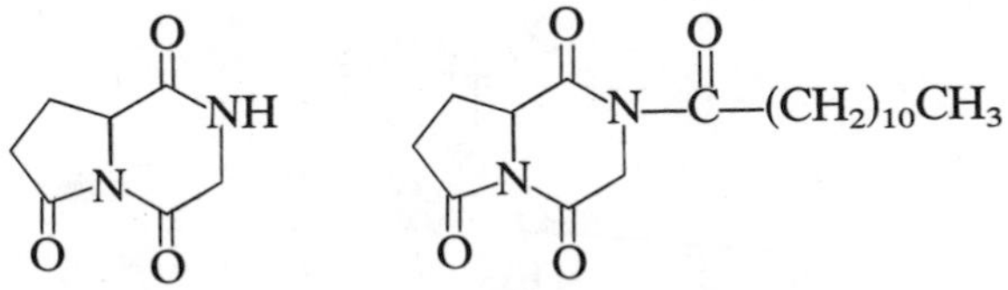

131 **132**

The principal interest in the acetylation of this ring system resides in the preparation of substituted protected α-hydroxy acyl derivatives **134** which on deprotection spontaneously cyclize to the fused oxazolo systems **135** which are present in certain ergot alkaloids and are of some pharmacological interest.[72] Frequently the oily benzyloxyacyl derivatives are not purified but are immediately hydrogenated to give the solid oxazolo compounds.

In addition to 2-methoxypropionyl chloride,[68] MeOCHMeCOCl, several benzyloxy acid chlorides based on acetic acid (**133**, $R^2 = H$) or malonic acid (**133**, $R^2 = CO_2Et$) have been used to acylate perhydro-1,4-dioxopyrrolo[1,2-*a*]pyrazines. Reactions with the derivatives of acetyl chloride were performed in refluxing toluene[42] or in a mixture of dioxan and pyridine[68] heated to 80°. Acylations with the half-ester acid chloride **133** ($R = Me$, $R^2 = CO_2Et$) have been carried out in pyridine[76–78] cooled to −20° or −30°, and in a mixture of dioxan and pyridine[56,66,69,70] at 70–80°. Refluxing toluene[79] has also been used with this reagent and with the acid chloride **133** (R = H, $R^2 = CO_2Et$). The related acyl chlorides **133** (R = Et or ^{i}Pr; $R^2 = CO_2Et$) react with perhydro-1,4-dioxopyrrolo[1,2-*a*]pyrazines in dioxan at 70–75° in the presence of pyridine,[80,81] *N*-methylmorpholine,[80] and ethyldiisopropylamine.[71,73,81] The benzyl group in the acyl compounds **134** is readily removed by hydrogenation at room

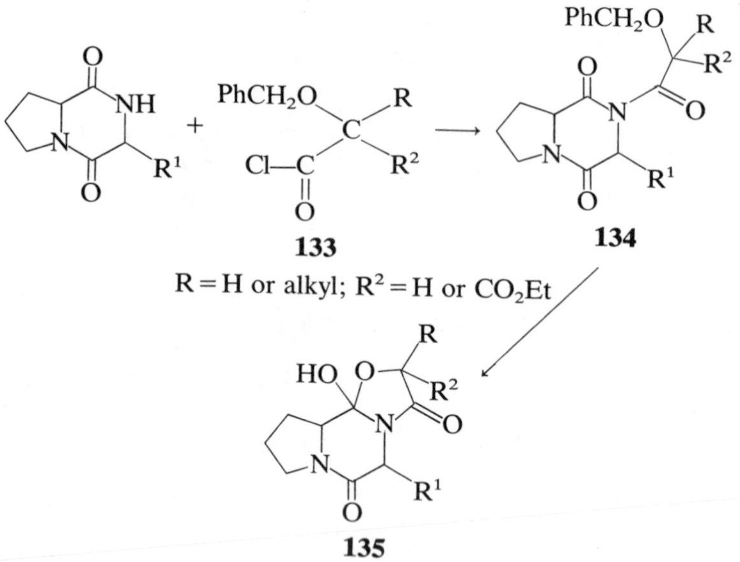

133

R = H or alkyl; R^2 = H or CO_2Et

134

135

temperature over palladium as catalyst. Cyclization of the resultant hydroxyacyl compound can occur spontaneously,[72] or it may be achieved by heating the compound in refluxing dioxan[56] or by keeping the compound in vacuo for several days.[42,79] On occasion cyclization is accompanied by racemization at the proline residue,[42,68] a phenomenon which also occurs on attempted purification[66] of compound **136** and on treatment of the diketopiperazine **137a** with alkali.[68,142].

The inversion of configuration at position 8a in compound **137a** occurs within 15 min in 0.5 *M* sodium hydroxide solution at 20°. An equilibrium mixture is attained in which the 3*S*,8a*S* compound (**137a**) is mixed with 90–95% of the 3*S*,8a*R* derivative (**137b**). It is suggested that an examination of Dreiding models demonstrates the greater thermodynamic stability of the trans form (**137b**) over the cis isomer.[68] Apparently the thermodynamically more stable 3*S*,8a*R* form (**137b**) is also obtained in the synthesis of the prolylphenylalanine diketopiperazine (**101**) from the

S,S,S tripeptide (**115**). Presumably the first-formed product is of structure **137a**, and this isomerizes even under the influence of a mixture of sodium carbonate and sodium bicarbonate.[59]

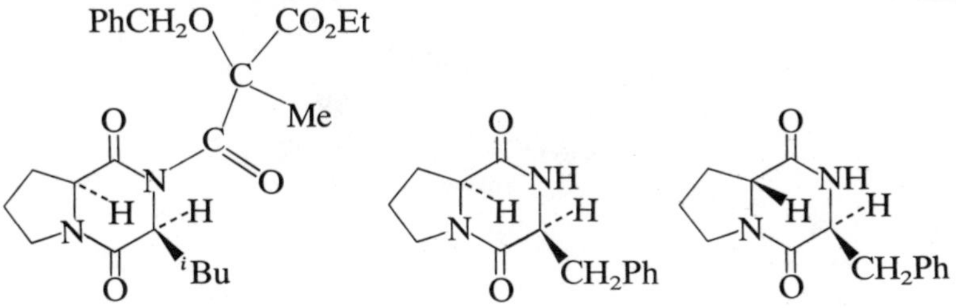

PhCH$_2$O CO$_2$Et C Me O C N O H H N ^{i}Bu O

136

O NH H H N CH$_2$Ph O

137a

O NH H H N CH$_2$Ph O

137b

The use of 0.1 *M* sodium hydroxide solution at 40° results in ring opening of the diketopiperazine **73** to give glycylproline (**138**). Ring opening in the opposite sense occurs with 0.1 *M* hydrochloric acid to give the isomeric dipeptide **139**.[65] Similarly refluxing 10% hydrochloric acid converts the anhydride **137a** into the prolyl dipeptide **140**.[58] Mild hydrolysis using 0.5 *M* barium hydroxide solution opens the diketopiperazine ring in compounds **141**, in the same direction as that noted for sodium hydroxide treatment, to give the α-aminoacetylproline derivatives.[51,61]

Hydrolysis of diketopiperazines has been used in the characterization of these compounds. Boiling with concentrated hydrochloric acid gives the two constituent amino acids, frequently without racemization.[63,82,83] Some racemization of L-prolyl-L-leucine diketopiperazine (**74**) was reported using 20% hydrochloric acid,[63] but this can be avoided with 33% sulfuric acid, which gave a mixture of the pure L-amino acids.[37]

2-Acyl derivatives of perhydro-1,4-dioxopyrrolo[1,2-*a*]pyrazines are rapidly hydrolyzed to the 2-unsubstituted compounds under very mild conditions. The 2-pyruvoyl compound **86** loses the *N*-acyl group on boiling in water or methanol, on treatment with ammonia in methylene chloride, on exposure to dilute sodium hydroxide solution, and on treatment with *p*-toluidine at 50–60° for 30 sec.[55] Similarly the methoxypropionyl compound **142** is extremely sensitive to alkali, immediately reacting with 0.1 *M* sodium hydroxide solution at 0°. This compound is also deacylated by ammonium hydroxide solution in a few minutes or by sodium bicarbonate solution after several hours. This behavior contrasts sharply with that of the isomeric oxazolo compound **143** and helps to justify the tricyclic formulation adopted for such compounds. Hydrolysis of the 3*S*,8a*S* isomer of compound **142** is accompanied by rapid inversion

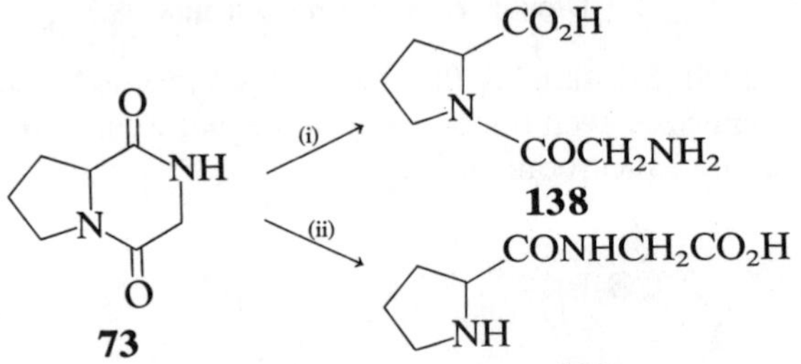

O
NH
N
O
73

(i) →
CO_2H
N
$COCH_2NH_2$
138

(ii) →
$CONHCH_2CO_2H$
NH
139

Reagents (i) NaOH
(ii) HCl

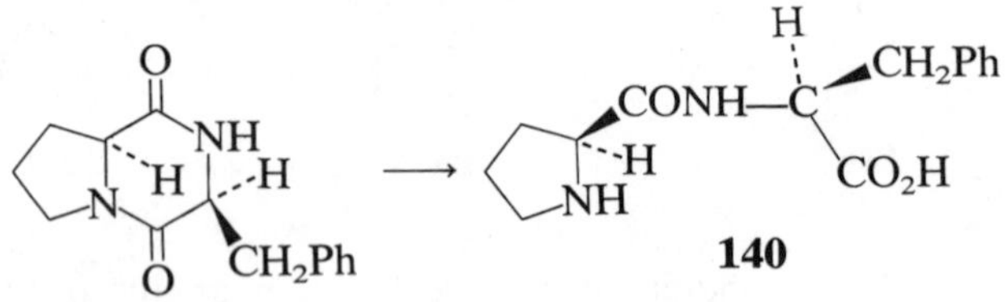

O
NH
H H
N
O CH_2Ph
137a

→

H
$CONH—C$ CH_2Ph
H
NH CO_2H
140

O
NH
N
O R

→

CO_2H
N
COCHNH$_2$
R

141 R = CH_2—C_6H_4—OH or CH_2CHMe_2

O
N—COCOMe
N
O CH_2Ph
86

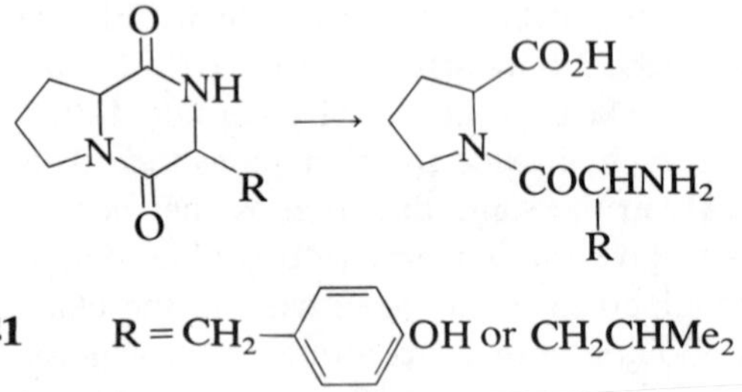

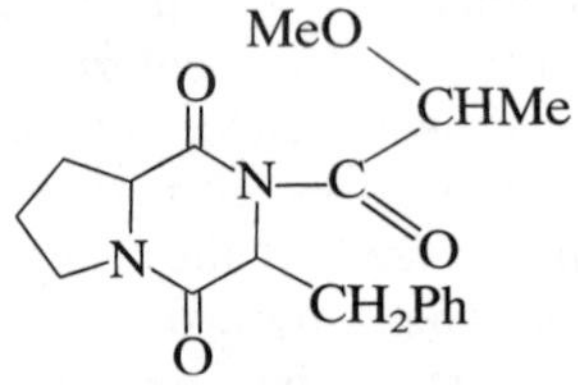

MeO
O CHMe
N—C
N =O
O CH_2Ph
142

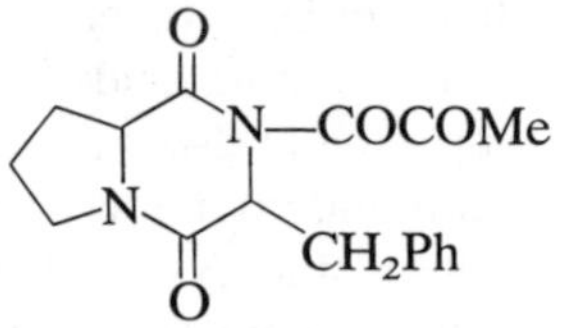

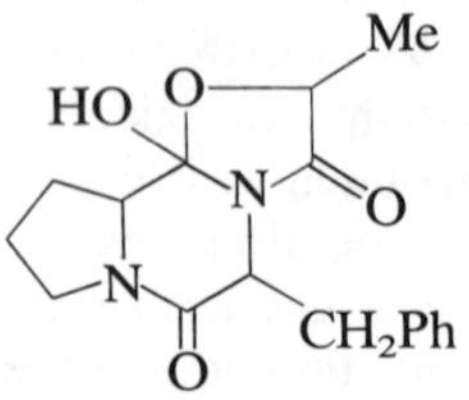

Me
HO O
N =O
N
O CH_2Ph
143

to the 3*S*,8a*R* form of the 2-unsubstituted compound **137b**. This isomerization occurs even more rapidly than the inversion of the 3*S*,8a*S* compound **137a**, and the increased rate is attributed to the enhanced electrophilicity of the carbonyl group in the 1-position of the acylated dioxopiperazine.[68]

Treatment of the benzylidene compound **144** with sodium hydroxide or sodium acetate provides the expected proline derivative **145** but without concomitant deacetylation. The use of ammonia provides the corresponding proline amide. When compound **144** is treated with sodium leucinate, the ring opens at the alternative site and deacetylation occurs, resulting in the formation of the prolyldipeptide **146** and *N*-acetylleucine. If sodium glycinate or ethyl glycinate is used instead, simple deacetylation takes place to give the diketopiperazine **147**.[58] Deacetylation may also be achieved with aromatic amines in chloroform at room temperature.[141] A

144 **145** **146** **147**

Reagents (i) NaOH, Na_3PO_4, NaOAc, or HCO_2Na
(ii) $NH_2CH_2CO_2Na$, $NH_2CH_2CO_2Et$, 1° or 2° amines, NH_2NH_2, or imidazole
(iii) $NH_2CH(CH_2CHMe_2)CO_2Na$

recent study has shown that the degree of deacetylation or ring opening depends on the nature of the nucleophile used as reagent.[143] Use of primary and secondary amines, hydrazine, and imidazole results in uncatalyzed deacetylation, whereas hydroxide, acetate, formate, and phosphate cause ring opening either directly, or with general base catalysis, to give the proline derivative **145**. Ammonia, semicarbazide, and *O*-methylhydroxylamine require general base or general acid catalysis and provide a mixture of the deacetylated and ring-opened products. The cleavage of the acetyl group in **144** by hydrazine has been studied kinetically in connection with studies on the α-effect.[144]

Reduction of perhydro-1,4-dioxopyrrolo[1,2-*a*]pyrazines with lithium aluminum hydride gives the expected completely reduced ring system.[27,28] Recently it has been shown that catalytic hydrogenation of the benzylidene compound **144** over palladium proceeds with high asymmetric induction to give the L-prolyl diketopiperazine of an L-amino acid.[141] This reaction, followed by acid hydrolysis of the product, provides a new synthesis of optically active amino acids such as L-dopa or L-phenylalanine.

4. Biological Activity

When diketopiperazines are administered to humans intravenously, they are not assimilated and are excreted unchanged in the urine. However, oral administration of compound **148** results in the breakdown of the compound, presumably in the digestive tract, and it therefore cannot be detected in the urine.[84] In contrast, people with phenylketonuria, who are fed on a low phenylalanine diet, excrete large amounts of the diketopiperazine **149**. Although this compound is found in small amounts in the urine of normal humans, the excessive excretion of this diketopiperazine has been shown not to be due to a metabolic defect. Instead the compound originates in the diet itself and is not metabolized in the human digestive tract.[85]

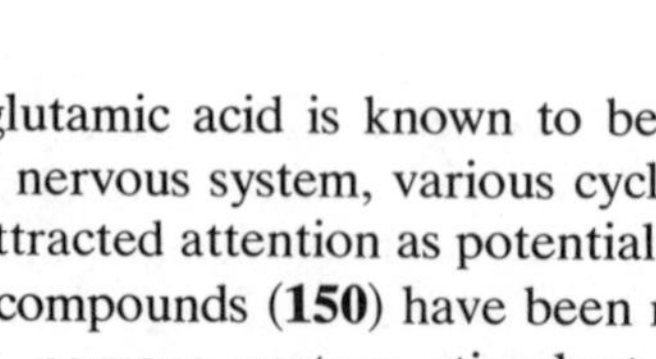

O, NH, N, O, $CH_2CH_2CO_2H$

148

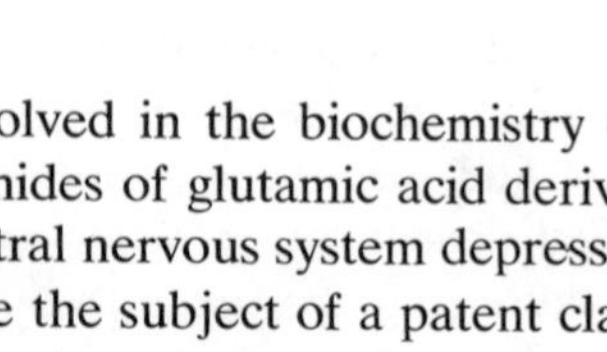

O, NH, H, N, CH_2, H, N, N, O, H

149

As glutamic acid is known to be involved in the biochemistry of the central nervous system, various cyclic imides of glutamic acid derivatives have attracted attention as potential central nervous system depressants.[34] These compounds (**150**) have been made the subject of a patent claiming central nervous system stimulant and depressant activity in mice.[47] Another 6-oxo compound (**132**) has been claimed to have antiviral activity.[75]

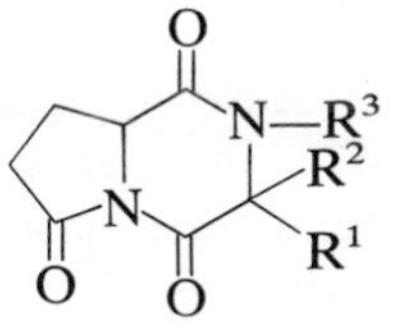

150 $R^1 = H$ or Me
$R^2 = H$, alkyl, or aralkyl
$R^3 = H$ or acyl

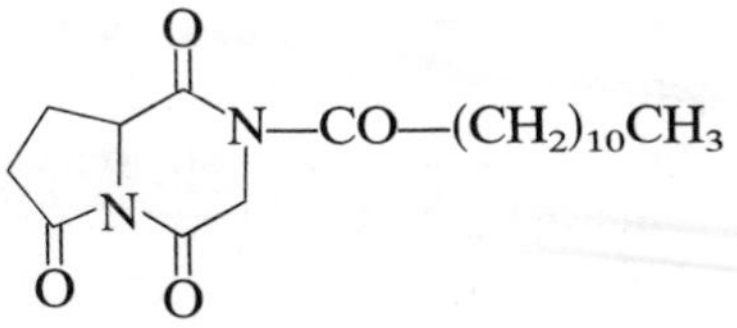

132

The diketopiperazine **151** is described as having pesticidal activity and the ability to inhibit germination.[86] In contrast, the anhydrides **152** have been shown to promote the germination of rice plant seeds.[87–89] However, compounds **152** and **137a** also inhibit the growth of the leaves and roots of rice seedlings.[90] The anhydrides **152** have also been found to be mildly toxic to silkworms.[91] A recent patent[140] has claimed a large series of 6-carboxylic acid derivatives of the ring system as central nervous system depressants and tranquilizers.

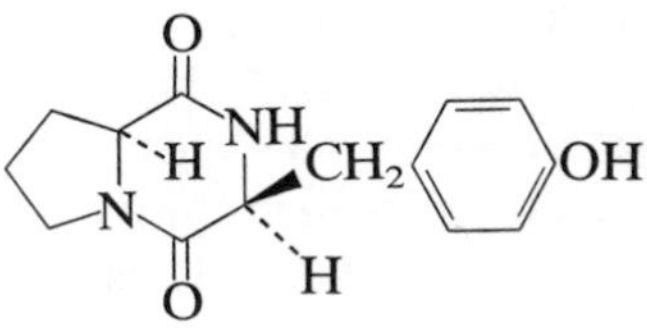

151

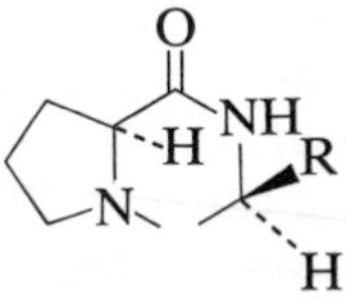

152 $R = CHMe_2$ or CH_2CHMe

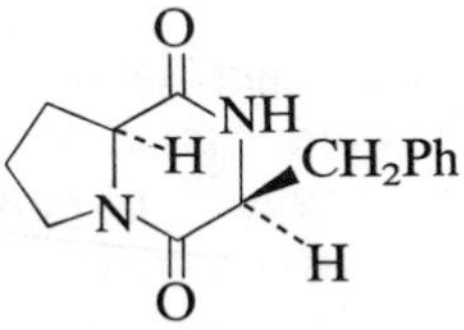

137a

5. Occurrence and Formation

The ubiquitous nature of proline-containing peptides results in the common appearance of perhydro-1,4-dioxopyrrolo[1,2-*a*]pyrazines as a result of metabolic processes, particularly of microorganisms. For example, the diketopiperazines **152** were found to be metabolic products of the algae species *Scenedesmus sp.*[92] These compounds have also been found in cultures of *Streptomyces.*[37,87,88,93,94] Glycyl-L-prolyl diketopiperazine has recently been isolated from the small sea animal *Luidia clathrata.*[147] The anhydrides **152** were not directly formed from the constituent amino acids but were obtained by hydrolysis of the casein, polypeptone, or gelatin in the culture medium by *Streptomyces* protease.[95] Peptone, a dietary essential for bacteria, has been shown by thin-layer and gas chromatography to contain a variety of prolyl diketopiperazines, and two of these (**152**) were isolated and identified.[96] Earlier work [38,97] had isolated L-leucine-L-proline anhydride (**152**, $R = CH_2CHMe_2$) from commercial peptone. This compound has also been obtained from cornsteep liquor[98] and from extracts of silkworm pupa.[99] The latter source also provided several other diketopiperazines.

The prevalence of perhydro-1,4-dioxopyrrolo[1,2-*a*]pyrazines in nutritional media complicates the study of metabolic products of microorganisms. Thus a recent report of the isolation of compound **151** from a culture of the fungus *Fusarium nivale*[86] could be explained either by assuming that the compound was a metabolic product of the fungus, or that it was merely derived from the peptone medium used. A variety of fungi have been claimed to give prolyl diketopiperazines as metabolic products.[90,91,100–105,146–148] More work is clearly needed to establish whether some of these metabolites are in fact artefacts from the culture medium. Of particular interest are the studies on the cultures of ergot fungi[103,104,149,150] from which alkaloids have been obtained containing the oxazolo[3,2-*a*]pyrrolo[2,1-*c*]pyrazine skeleton (see Section II6).

Thermal decomposition of the ergot alkaloids has provided several diketopiperazines of proline with other amino acids.[28,82,83,106] The ring system has also been obtained by cleavage of ergot alkaloids with hydrazine[107] or potassium hydroxide.[108] Degradation of proteins provided an early source of perhydro-1,4-dioxopyrrolo[1,2-*a*]pyrazines. Thus enzymic digestion of gelatin,[109] casein,[57,110] egg white and soybean proteins,[110] and edestin[111] all gave prolyl diketopiperazines. Although collagenase from *Clostridium histolyticum* has no effect on a wide variety of protein substrates,[112] it cleaves the peptide **153** to give the dipeptide **154** and the anhydride **73**. Diketopiperazine derivatives of proline have also been obtained from gelatin by pyrolysis[113] and acid hydrolysis.[114]

Degradative studies of peptide antibiotics have also provided examples of this ring system. The dehydro compound **155** was obtained by alkaline hydrolysis of the macrocyclic peptide lactone telomycin.[115,116] Treatment of gramicidin C with acid in an autoclave gave some proline phenylalanine anhydride,[117] and degradation of certain actinomycins with hydrazine at 140° gave proline valine anhydride.[118] The ready formation of diketopiperazines from neighboring pairs of amino acids has provided the basis for a method of sequence determination of small peptides. Pyrolysis–gas chromatography of actinomycins D and C_3 gave prolyl diketopiperazines which with the other diketopiperazines obtained

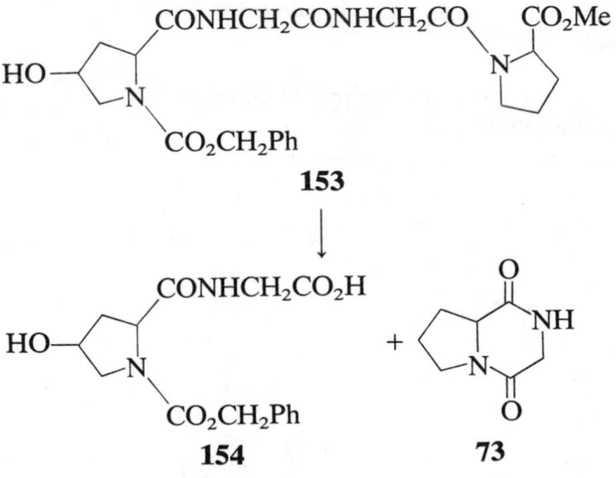

153

↓

154 **73**

indicated the structures **156** for these two antibiotics. The peaks in the gas chromatograms were identified by comparison with peaks from synthetic diketopiperazines and also by combined gas chromatography–mass spectometry.[119] Recently two reports have appeared describing failures in solid-phase peptide syntheses in which prolyl diketopiperazine formation was an important side reaction.[151,152]

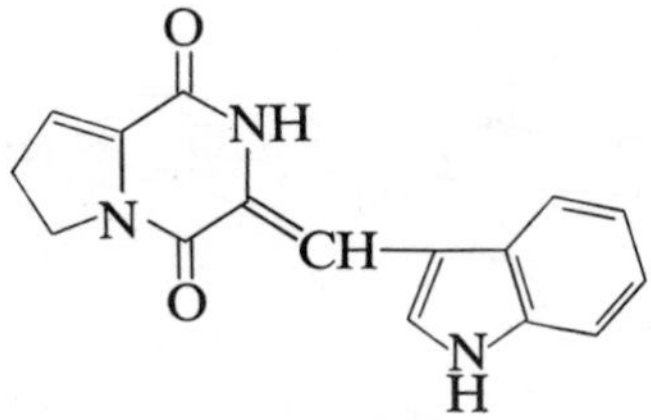

155

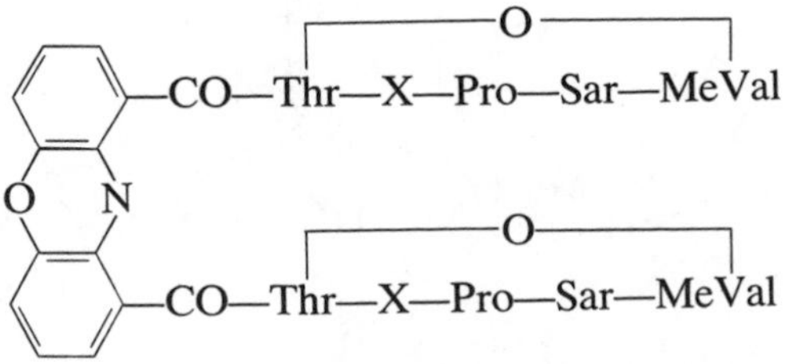

156 X = *D*-Val Actinomycin D
X = *D*-*allo*-Ileu Actinomycin C_3

6. The Ergot Alkaloids

Much of the interest in perhydro-1,4-dioxopyrrolo[1,2-*a*]pyrazines has centered around their use in the syntheses of peptide-containing ergot alkaloids and their analogues.[56,66,69–74,76–78,80,81,121,153] Typical is the stereospecific synthesis of ergotamine[78] from the 3*S*,8a*S*-diketopiperazine **137a**. Acylation and hydrogenation gave the cyclol **157**. Hydrolysis of the

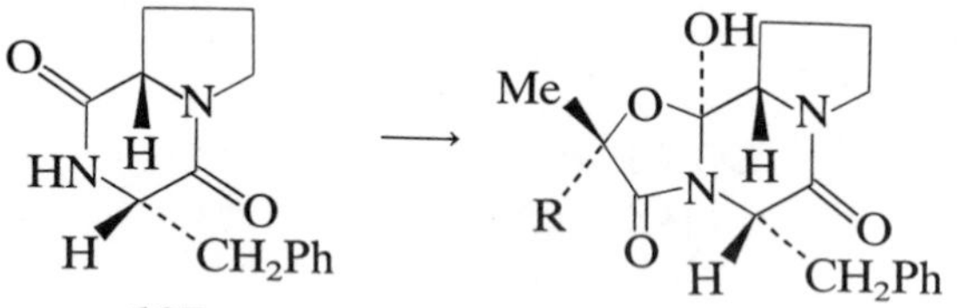

137a

157 R = CO_2Et
158 R = CON_3
159 R = $NHCO_2CH_2Ph$
160 R = NH_2

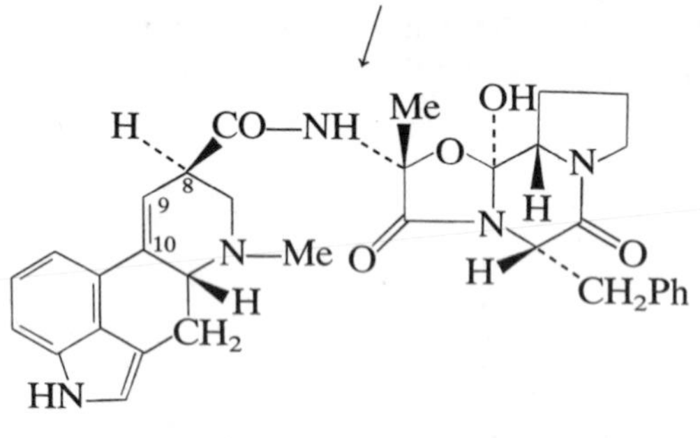

161 Ergotamine

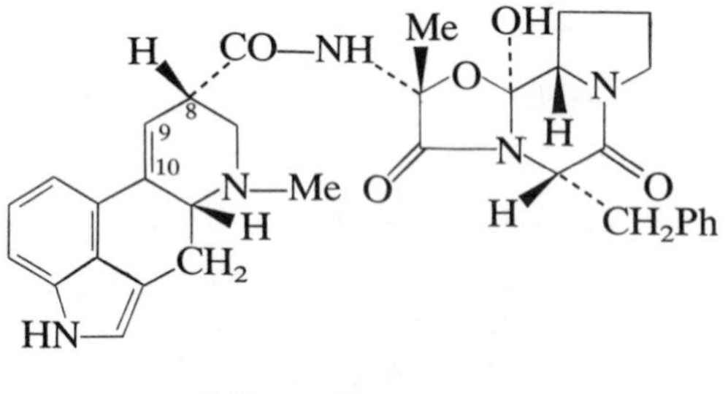

162 Ergotaminine

ester group and treatment with oxalyl chloride gave the acid chloride which was readily converted into the azide **158** by reaction with sodium azide. When the azide **158** was heated in benzyl alcohol it yielded the carbamate **159**, which was converted into the primary amine **160** by hydrogenation. Reaction of the amine with (+)−lysergic acid chloride gave a mixture of ergotamine (**161**) and ergotaminine (**162**).

The ergot alkaloids of the peptide type fall into two groups based on lysergic acid and isolysergic acid. The former group, analogues of ergotamine, are physiologically more potent than the isomers related to ergotaminine. The early work on these compounds resulted in some confusion over the names and identities of these alkaloids.[120] The chemistry of certain of these compounds has recently been reviewed,[123] and there has now been a study of their PMR and ^{13}C NMR spectra.[154] The pharmacology of the group has been much studied.[120] In particular, certain compounds are of use in the treatment of migraine[72,124] and to induce contraction of the puerperal uterus.[120] Toxicity problems limit the use of these alkaloids, but toxicity is considerably reduced in their 9,10-dihydro derivatives. These compounds, however, have almost no action on smooth muscle.

III. Tables

TABLE 1. PYRROLO[1,2-*a*]PYRAZINES (NOT 1,4-DIOXO COMPOUNDS)

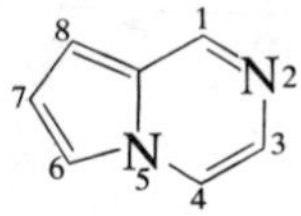

Molecular formula	Substituents or structure	m.p., °C	Ref.
$C_7H_4Br_2N_2$	6,8-Dibromo[a]	122–122.4	2
$C_7H_6N_2$	Unsubstituted[b]	b.p. 71/2 mm	7
	Picrate	212	
	Hydrochloride	160–161	9
$C_7H_{12}N_2O$	Perhydro-1-oxo	80–81	22, 24
		b.p. 110–120/0.4 mm	11
	Hydrochloride	242–243, 227–228 decomp.	5, 24, 155
	Methiodide	259–261	155
	Benzylbromide	226–227	
	Perhydro-3-oxo	120–121.5	11
	Methiodide	280–281.5 decomp.	

TABLE 1 (*contd.*)

Molecular formula	Substituents or structure	m.p., °C	Ref.
$C_7H_{14}N_2$	Perhydro[c]	b.p. 68–70/10 mm	5, 11, 19
	Dihydrochloride	204–205	
$C_7H_{14}N_2O$	Perhydro-7-hydroxy	190 decomp.	29
$C_8H_8N_2$	1-Methyl[d]	b.p. 82/1.5 mm	7
	Picrate	230	
$C_8H_{10}N_2$	3,4-Dihydro-1-methyl	b.p. 84–89/1.8–2.5 mm	18
$C_8H_{12}N_2O_3$	6-Carboxy-perhydro-1-oxo		
	Ethyl ester picrate	155–156 decomp.	21, 22
$C_8H_{14}N_2O$	Perhydro-6-methyl-1-oxo	93–96	22, 24
	Picrate	233–234 decomp.	
$C_8H_{16}N_2$	Perhydro-6-methyl[e]	b.p. 70–73/10 mm	19
$C_8H_{16}N_2O$	Perhydro-7-methoxy	195	29
$C_9H_{14}N_2O_2$	1-Ethoxy-3,4,6,7,8,8a-hexahydro-4-oxo[bb]	b.p. 138–140/3–4 mm	132
$C_9H_{16}N_2O$	2-Acetyl-perhydro[f]	b.p. 129–130/5.5 mm	19
$C_9H_{17}ClN_2$	2-(2-Chloroethyl)perhydro		
	Dihydrochloride[cc]		130
$C_9H_{18}N_2O$	Perhydro-2(2-hydroxyethyl)	b.p. 102–128/11.2–13 mm	130
$C_{10}H_5N_3O_4$	7,8-Dicarboxy-6-cyano		
	Dimethyl ester	162–163	125
$C_{10}H_{16}N_2O$	Perhydro-7,7,8a-trimethyl-1,6-dioxo	155.5–157	25
$C_{10}H_{18}N_2O$	2-Acetyl-perhydro-6-methyl[g]	b.p. 98–100/1 mm	19
$C_{10}H_{19}ClN_2$	2-(3-Chloropropyl)perhydro[h]	b.p. 87–90/1 mm	128
$C_{10}H_{20}N_2$	Perhydro-3-*n*-propyl	b.p. 130–135/25 mm	5
	Dihydrochloride hydrate	262–263	
	Dimethiodide	215–216	
$C_{11}H_7N_3O_5$	7,8-Dicarboxy-6-cyano-1-methoxy		
	Dimethyl ester	178–179	125
$C_{11}H_8N_2O_6$	6,7,8-Tricarboxy-3-methyl		
	Trimethyl ester	178–179	8
$C_{11}H_{10}N_2O_4$	7,8-Dicarboxy-1,4-dimethyl		
	Dimethyl ester	171–172	15
	Hydrobromide	219–221	
$C_{11}H_{12}N_2O_2$	7-Carboxy-1,6,8-trimethyl		
	Ethyl ester hydrochloride	180–182 decomp.	9
$C_{11}H_{22}N_2$	Perhydro-3-*i*-butyl	b.p. 110/15 mm	122
		b.p. 228	
	?Hydrochloride	181	29
	Chloroaurate	197	
	3*S*,8a*S* Form[i]	b.p. 120–125/12 mm	28
		b.p. 95–100/12 mm	27
	Dipicrate	185–186	

TABLE 1 (*contd.*)

Molecular formula	Substituents or structure	m.p., °C	Ref.
$C_{11}H_{22}N_2$ (*contd.*)	3*S*,8a*R* Form[j]		
	Dipicrate	222–224	28
$C_{12}H_{10}N_2O_6$	6,7,8-Tricarboxy-1,3-dimethyl		
	Trimethyl ester	129–130	8
$C_{12}H_{20}N_2O_2$	1-Ethoxy-3,4,6,7,8,8a-hexahydro-4-oxo-3-*i*-propyl		
	3*S*,8a*S* Form[dd]	b.p. 135–140/1 mm	132
	3*R*,8a*S* Form[ee]	b.p. 145–148/2 mm	
$C_{12}H_{24}N_2$	Perhydro-3-*n*-pentyl	b.p. 144–146/0.2 mm	5
	Dihydrochloride	220–221	
	Dipicrolonate	223–224	
$C_{12}H_{25}N_3$	3-(3-Dimethylaminopropyl)perhydro[k]	b.p. 138–140/25 mm	5
	Trihydrochloride hydrate	263–265	
$C_{13}H_{10}N_2$	1-Phenyl		
	Hydrochloride	185–187 decomp.	9
	Picrate	156.5	126
$C_{13}H_{22}N_2O_2$	3-*i*-Butyl-1-ethoxy-3,4,6,7,8,8a-hexahydro-4-oxo		
	3*S*,8a*S* Form[ff]	b.p. 150–154/1 mm	132
$C_{14}H_{18}N_2O$	2-Benzylperhydro-1-oxo		
	Hydrochloride	194–195	24
	8a-Benzylperhydro-1-oxo	144–145	155
$C_{14}H_{19}N_3O_2$	Perhydro-7-hydroxy-2-phenylaminocarbonyl	198 decomp.	29
$C_{14}H_{19}N_3S$	Perhydro-2-phenylaminothiocarbonyl	122–123	5, 11
$C_{14}H_{20}N_2$	3-Benzylperhydro	b.p. 108–115/0.05 mm	5
	Dihydrochloride	239–240	
	3*S*,8a*S* Form[l]	b.p. 185/11 mm	28
	Dipicrate	258	
	3*S*,8a*R* Form[m]	63–65	28
	Dipcrate	257–258	
	8a-Benzylperhydro	104–105	155
$C_{14}H_{27}N_3$	Perhydro-3-(2-piperidin-1-ylethyl)[n]	b.p. 120–125/0.5 mm	5
	Trihydrochloride	274–275	
$C_{15}H_{22}N_2$	Perhydro-3-(2-phenylethyl)[o]	b.p. 122–124/0.5 mm	5
	Dihydrochloride	225–226	
$C_{15}H_{24}N_2$	8a-Benzylperhydro-2-methyl	b.p. 115/1.0 mm	155
$C_{16}H_{22}N_2O$	2-Acetyl-8a-benzylperhydro	74–75	155
	Methiodide	246–249	
$C_{16}H_{24}N_2$	Perhydro-3-(1-methyl-2-phenylethyl)[p]	b.p. 117–120/0.2 mm	5
	Dihydrochloride	269–270	
	Monomethiodide	134–135	
$C_{16}H_{24}N_2O$	Perhydro-3-[2(4-methoxyphenyl)ethyl]	b.p. 155–158/0.25 mm	5
	Dihydrochloride	228–229	

TABLE 1 (*contd.*)

Molecular formula	Substituents or structure	m.p., °C	Ref.
	Dipicrate	214–215	
$C_{17}H_{12}N_2O_5$	6-Benzoyl-7,8-dicarboxy-1-methyl[q]		
	Dimethyl ester	141–142	16
	6-Benzoyl-7,8-dicarboxy-3-methyl[q]		
	Dimethyl ester	185–188	16
$C_{17}H_{23}FN_2O$	2-(3-(4-Fluorobenzoyl)propyl-perhydro[gg]	b.p. 155–165/0.2 mm	26, 131
	Dihydrochloride	180–181	
$C_{17}H_{26}N_2$	Perhydro-3-(1,1-dimethyl-2-phenylethyl)	b.p. 142–145/0.3 mm	5
	Dihydrochloride	231–232	
	Dipicrate	202–203	
$C_{17}H_{26}N_2O_2$	3-(2-(3,4-Dimethoxyphenyl)ethyl)-perhydro	b.p. 160–168/0.1 mm	5
	Dihydrochloride	235–236	
	Dipicrate	248–249	
$C_{18}H_{14}N_2O_5$	6-Benzoyl-7,8-dicarboxy-1,4-dimethyl		
	Dimethyl ester	128–129	15
$C_{18}H_{24}ClN_3O$	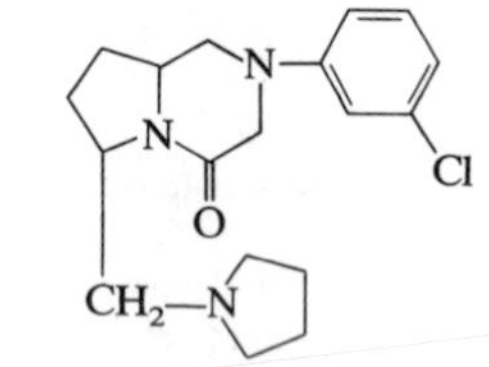	107–108	127
$C_{19}H_{14}N_2$	1,3-Diphenyl		
	Hydrochloride	115–116	9
$C_{19}H_{14}N_2O$	6-Hydroxy-7,8-diphenyl		
	cis-1,2-Diphenylacrylate	173	6
$C_{19}H_{14}N_2S$	6-Mercapto-7,8-diphenyl	148	17
$C_{20}H_{16}N_2O$	6-Hydroxy-3-methyl-7,8-diphenyl[r]	263	6
$C_{20}H_{16}N_2S$	6-Mercapto-3-methyl-7,8-diphenyl[r]	130	17
$C_{20}H_{23}ClN_2$	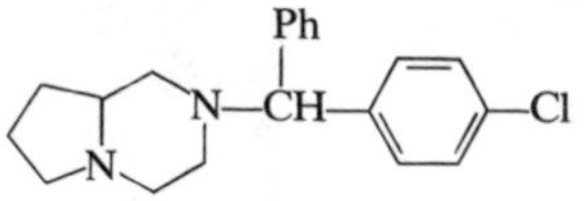		
	L(—) Form	b.p. 180–195/0.1 mm	13
	Monofumarate	222–223	
$C_{20}H_{24}N_2$	2-Benzhydrylperhydro		
	L(—) Form	b.p. 148–151/0.1 mm	13
	Monofumarate	226–227	
	2-Benzyl-perhydro-3-phenyl[s]		
	Dihydrochloride	224–225	14
$C_{21}H_{18}N_2O$	6-Hydroxy-1,3-dimethyl-7,8-diphenyl	216 decomp.	6
$C_{21}H_{18}N_2S$	6-Mercapto-1,3-dimethyl-7,8-diphenyl	144	17

TABLE 1 (*contd.*)

Molecular formula	Substituents or structure	m.p., °C	Ref.
$C_{21}H_{22}ClN_3OS$	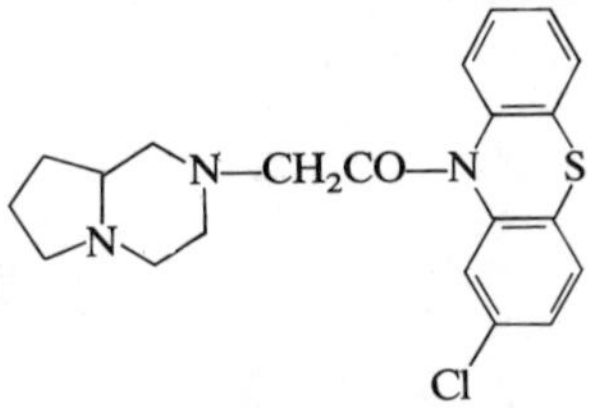 N—CH_2CO—N S N Cl	132.5–133.5	129
	Dihydrochloride	188–189.5	
$C_{21}H_{24}N_2O$	2-Benzoyl-8a-benzyl-perhydro	129–131	155
$C_{21}H_{26}N_2$	Perhydro-3-phenyl-2-(2-phenylethyl)[t]		
	Dihydrochloride	268–272 decomp.	14
$C_{22}H_{20}N_2O$	6-Ethoxy-3-methyl-7,8-diphenyl[r]	157	6
$C_{22}H_{24}ClN_3OS$	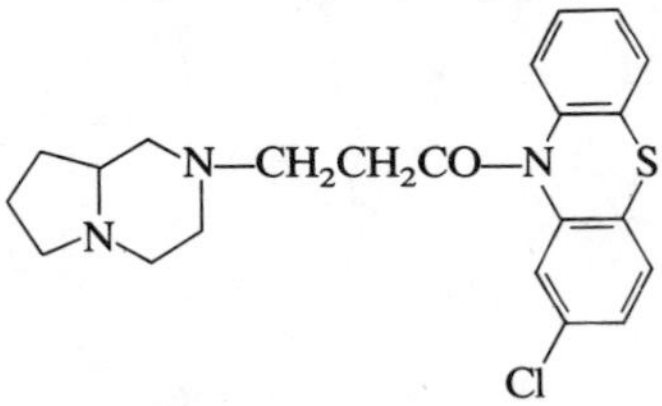 N—CH_2CH_2CO—N S N Cl		
	Dihydrochloride	215–216	129
	Difumarate	157–159	
	Dimaleate	169–170	
	Dimethiodide	146–148	
$C_{22}H_{25}N_3OS$	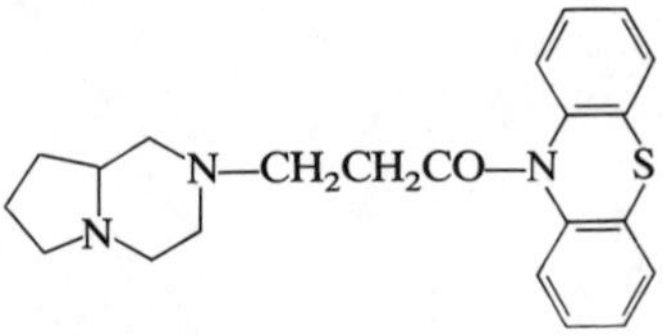 N—CH_2CH_2CO—N S N		
	Dihydrochloride	190.5–191	129
$C_{22}H_{28}N_4O$	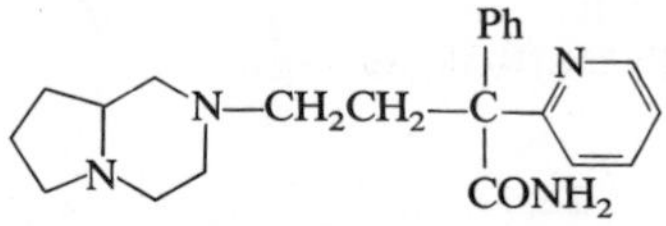 N—CH_2CH_2—C(Ph)(CONH$_2$)— N N	104–105	130
$C_{23}H_{22}N_2O$	6-Ethoxy-1,3-dimethyl-7,8-diphenyl	62	6
$C_{23}H_{24}F_3N_3OS$	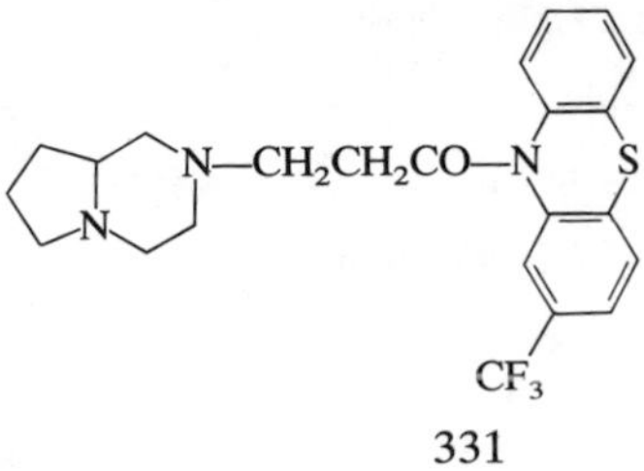 N—CH_2CH_2CO—N S N CF_3		

TABLE 1 (*contd.*)

Molecular formula	Substituents or structure	m.p., °C	Ref.
$C_{23}H_{24}F_3N_3OS$ (*contd.*)	Dihydrochloride	222–223	127
$C_{23}H_{26}ClN_3OS$	octahydropyrrolo[1,2-a]pyrazin-2-yl—N—$CH_2CH_2CH_2CO$—N (2-chlorophenothiazin-10-yl; S, Cl)		
	Dihydrochloride	224–226	129
$C_{23}H_{26}F_3N_3S$	octahydropyrrolo[1,2-a]pyrazin-2-yl—N—$CH_2CH_2CH_2$—N (2-trifluoromethylphenothiazin-10-yl; S, CF_3)		
	Maleate	147–150	12
	Dihydrochloride	220–222	128
$C_{23}H_{27}N_3$	N—$CH_2CH_2CPh_2$(CN)	b.p. 190–210/0.1 mm	130
$C_{23}H_{29}N_3O$	N—$CH_2CH_2CPh_2$($CONH_2$)	172–176	130
$C_{27}H_{37}N_3O$	N—$CH_2CH_2CPh_2$($CONEt_2$)	125–128	130
$C_{30}H_{43}N_5O_2$	N—COCH(Me)NH—dihydrolysergyl		
	3*S*,8a*S* Form[u] (a) 2′*S* isomer	133–135	28
	(b) 2′*R* isomer	168–170	
$C_{30}H_{44}N_6O_2$	N—CONHCH(Me)NH—dihydrolysergyl (ring substituent: ${}^{i}Bu$)		
	3*S*,8a*R* Form[v] 2′*S* isomer	201–202	28
$C_{30}H_{47}N_5$	N—$CH_2CH(Me)NHCH_2$—(6-methyl-8-ergolinyl) (ring substituent: ${}^{i}Bu$)		

TABLE 1 (*contd.*)

Molecular formula	Substituents or structure	m.p., °C	Ref.
$C_{30}H_{47}N_5$ (*contd.*)	3*S*,8a*S* Form[w](a) 2′*S* isomer	156–158	28
	(b) 2′*R* isomer	192–193	
	3*S*,8a*R* Form[w](a) 2′*S* Isomer	204–206	
	(b) 2′*R* isomer	205–206	
$C_{31}H_{49}N_5$	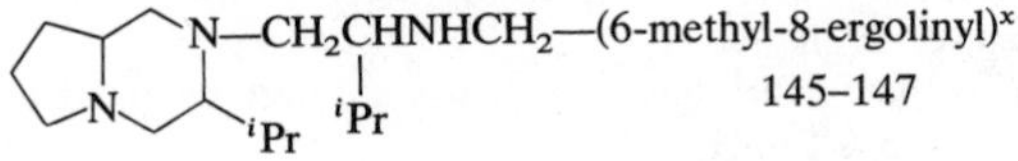 [structure: pyrrolo[1,2-a]pyrazine ring with 3-^{i}Pr, N—$CH_2CH(^iPr)NHCH_2$—(6-methyl-8-ergolinyl)[x]]	145–147	28
$C_{32}H_{47}N_5O_2$	[structure: pyrrolo[1,2-a]pyrazine ring with 3-^{i}Bu, N—$COCH(^iPr)NH$—dihydrolsergyl]		
	3*S*,8a*S* Form[y] 2′*R* isomer	203–204	28
$C_{32}H_{51}N_5$	[structure: pyrrolo[1,2-a]pyrazine ring with 3-^{i}Bu, N—$CH_2CH(^iPr)NHCH_2$—(6-methyl-8-ergolinyl)]		
	3*S*,8a*S* Form[z] (a) 2′*S* isomer	154–155	28
	(b) 2′*R* isomer	151–153	
$C_{35}H_{49}N_5$	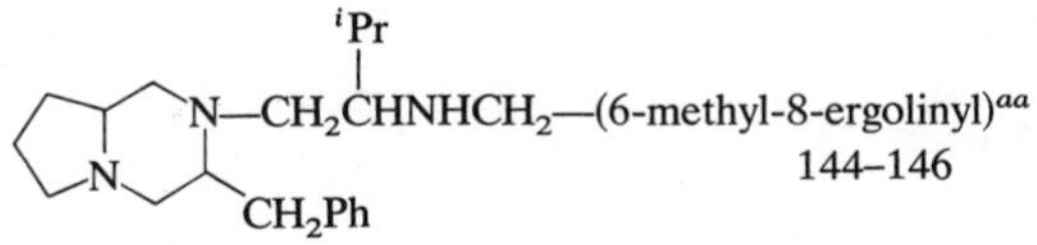 [structure: pyrrolo[1,2-a]pyrazine ring with 3-CH_2Ph, N—$CH_2CH(^iPr)NHCH_2$—(6-methyl-8-ergolinyl)[aa]]	144–146	28
$C_{38}H_{26}N_4S_2$	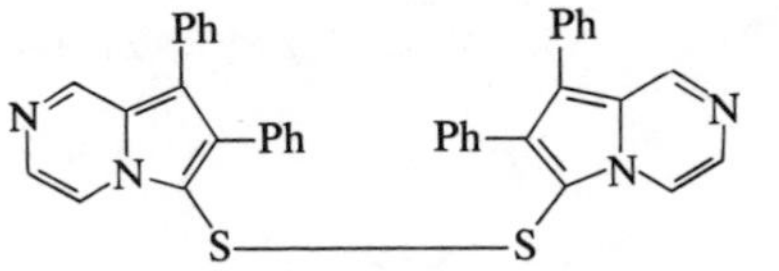 [structure: bis(7,8-diphenylpyrrolo[1,2-a]pyrazin-6-yl) disulfide, S—S]	230	17
$C_{38}H_{28}N_6S_2$	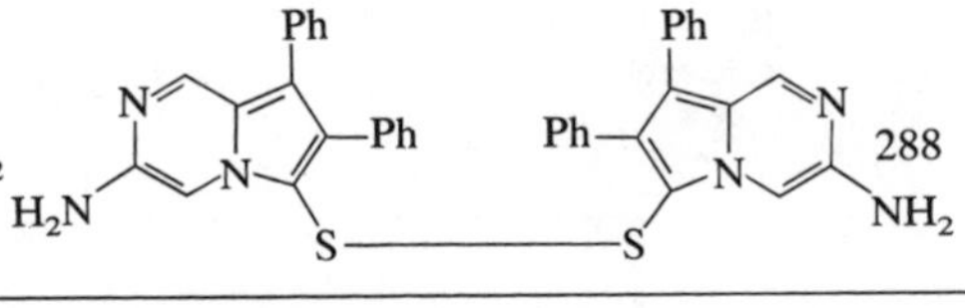 [structure: bis(3-amino-7,8-diphenylpyrrolo[1,2-a]pyrazin-6-yl) disulfide, H_2N, NH_2, S—S]	288	17

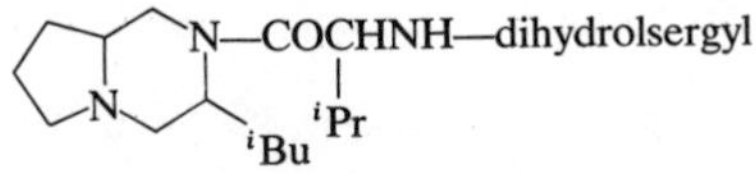

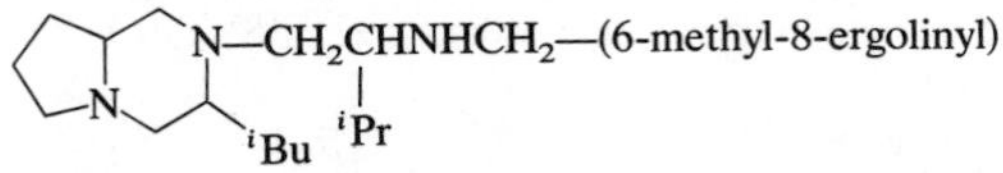

[a] A different numbering system was used in the original paper.
[b] Free base n_D^{20} 1.6176.
[c] Free base n_D^{20} 1.4940; n_D^{25} 1.4950; d_{20} 0.9679; 2 HCl$[\alpha]_D^{22} -2.84°$ (*c* 1, 95% EtOH).
[d] Free Base n_D^{20} 1.6910.
[e] n_D^{20} 1.4875; d_{20} 0.9452.
[f] n_D^{20} 1.5060.
[g] n_D^{20} 1.4998.
[h] n_D^{22} 1.4979.
[i] Free base $[\alpha]_D^{20}$ 15° (*c* 0.3, EtOH); dipicrate $[\alpha]_D^{20} -5°$.

Footnotes for Table 1

[j] Free base $[\alpha]_D^{20}$ 28° (*c* 0.3, EtOH); dipicrate $[\alpha]_D^{20}$ 13.5°.
[k] Free base n_D^{25} 1.4815; trihydrochloride $[\alpha]_D^{22}$ −9.16° (*c* 1, 95% EtOH).
[l] Free base $[\alpha]_D^{20}$ −17° (EtOH).
[m] Free base $[\alpha]_D^{20}$ 24°.
[n] Trihydrochloride $[\alpha]_D^{22}$ −12.44° (*c* 1, 95% EtOH).
[o] Free base n_D^{25} 1.5375.
[p] Dihydrochloride $[\alpha]_D^{22}$ −4.81° (*c* 1, 95% EtOH).
[q] There is some confusion over these melting points in the original. The most likely interpretation is given in the table.
[r] Assumed structure; the 1-substituted isomer has not been eliminated.
[s] $[\alpha]_D^{24}$ −45° (*c* 0.998, 95% EtOH).
[t] $[\alpha]_D^{24}$ −63° (*c* 1.002, 95% EtOH). Three other substituted phenylethyl compounds were also prepared.
[u] 2′*S* Isomer $[\alpha]_D^{20}$ −80° (*c* 0.2, pyridine); 2′*R* isomer $[\alpha]_D^{20}$ −28° (*c* 0.3, pyridine).
[v] $[\alpha]_D^{20}$ −86°.
[w] 3*S*,8a*S* Form, 2′*S* isomer, L isomer $[\alpha]_D^{20}$ −23°; 2′*R* isomer $[\alpha]_D^{20}$ −93°.
3*S*,8a*R* Form, 2′*S* isomer $[\alpha]_D^{20}$ 16°; 2′*R* isomer $[\alpha]_D^{20}$ 18° (*c* 0.2–0.4, pyridine).
[x] $[\alpha]_D^{20}$ −17° (pyridine).
[y] $[\alpha]_D^{20}$ −50° (*c* 0.3, pyridine).
[z] 2′*S* Isomer $[\alpha]_D^{20}$ −12°; 2′*R* isomer $[\alpha]_D^{20}$ −40° (*c* 0.2–0.4, pyridine).
[aa] $[\alpha]_D^{20}$ −23° (pyridine).
[bb] $[\alpha]_D$ −122° (EtOH).
[cc] Analysis but no melting point given.
[dd] $[\alpha]_D$ −97.2° (EtOH).
[ee] $[\alpha]_D$ −61.1° (EtOH).
[ff] $[\alpha]_D$ −56.7° (EtOH).
[gg] n_D^{20} 1.5284.

TABLE 2. PERHYDRO-1,4-DIOXOPYRROLO[1,2-*a*]PYRAZINES

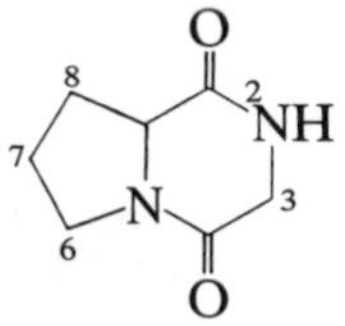

Molecular formula	Substituents or structure	m.p., °C	Ref.
$C_7H_8N_2O_3$	6-Oxo[a]	184–187	34, 47
$C_7H_{10}N_2O_2$	Unsubstituted		
	8a*S* Form[b]	213.5–215.5	48, 50, 65
		203–204	52
	8a*RS* Form	173	48
$C_8H_{10}N_2O_3$	3-Methyl-6-oxo		
	3*S*,8a*S* Form[c]	220–222	34, 47
	3*RS*,8a*S* Form	187–191	

TABLE 2 (*contd.*)

Molecular formula	Substituents or structure	m.p., °C	Ref.
$C_8H_{12}N_2O_2$	2-Methyl		
	8a*S* Form	b.p. 140/1 mm	139
	3-Methyl		
	3*S*,8a*S* Form[bbb]	153–156	136
	3*R*,8a*S* Form[ccc]	127–129	136
	3*RS*,8a*S* Form	114–115	53
$C_9H_{10}N_2O_4$	2-Acetyl-6-oxo[d]	148–150	34, 47
$C_9H_{12}N_2O_3$	3,3-Dimethyl-6-oxo[e]	258–261	34, 47
$C_9H_{12}N_2O_4$	6-Carboxy-3-methyl	238–242	140
	Ethyl ester	182–184	
	2-Hydroxyacetyl		
	8a*S* Form[f]	Oil	42, 43
	1-Naphthylurethane	216–220	43, 79
	Benzyl ether	150	43
$C_{10}H_{12}N_2O_4$	2-Acetyl-3-methyl-6-oxo		
	3*R*,8a*S* Form[g]	183–185	34, 47
	6-Oxo-2-propionyl[h]	158–160	34, 47
$C_{10}H_{14}N_2O_3$	3-Ethyl-3-methyl-6-oxo[i]	222–225	34, 47
	6-Oxo-3-*i*-propyl[j]	190–220	34
		190–192	47
$C_{10}H_{14}N_2O_4$	2-Methoxyacetyl	Oil	43
$C_{10}H_{16}N_2O_2$	3-*n*-Propyl		
	3*S*,8a*S* Form[k]	134	73
	3-*i*-Propyl		
	3*S*,8a*S* Form[l]	190–192	20, 66, 92
		169–172	136
		200	64
	3*S*,8a*R* Form[m]	147–149	83
	3*R*,8a*S* Form[n]	152–155	57, 136
$C_{11}H_{14}N_2O_4$	2-Acetyl-3,3-dimethyl-6-oxo[o]	163–166	34, 47
	2-*n*-Butyryl-6-oxo[p]	121–123	34, 47
	2-*i*-Butyryl-6-oxo[q]	148–150	34, 47
$C_{11}H_{14}N_4O_2$	3-Imidazol-4-ylmethyl[r]	168–170	60
$C_{11}H_{16}N_2O_2$	3-*i*-Butylidene		
	8a*S* Form[ddd]	165–166	141
$C_{11}H_{16}N_2O_3$	3-*i*-Butyl-6-oxo		
	3*S*,8a*S* Form[s]	215–218	34, 47
$C_{11}H_{18}N_2O_2$	3-*i*-Butyl		
	3*S*,8a*S* Form[t]	164–165	66, 81, 137
		157–158	67, 97
		170–171	28, 92
	3*S*,8a*R* Form[u]	148–150	82, 83
	3*RS*,8a*RS* Form	117–121	1
	3*R*,8a*S* Form[eee]	148–151	137
		119–122	136

TABLE 2 (*contd.*)

Molecular formula	Substituents or structure	m.p., °C	Ref.
$C_{12}H_{16}N_2O_4$	2-Acetyl-3-ethyl-3-methyl-6-oxo[v]	129–131	34, 47
	2-Acetyl-6-oxo-3-*i*-propyl[w]		
	3*R*,8a*S* Form	178–180	34, 47
$C_{12}H_{18}N_2O_3$	3-*i*-Butylidene-2-methyl		
	8a*S* Form	92–95	141
$C_{13}H_{18}N_2O_4$	2-Acetyl-3-*i*-butyl-6-oxo[x]		
	3*R*,8a*S* Form	158–160	34, 47,
$C_{13}H_{20}N_2O_4$	3-*i*-Butyl-2-hydroxyacetyl		
	3*S*,8a*S* Form benzyl ether	132	43
$C_{14}H_{14}N_2O_2$	3-Benzylidene		
	8a*S* Form[fff]	177–180	58, 141
$C_{14}H_{14}N_2O_3$	3-Benzyl-6-oxo		
	3*R*,8a*S* Form[z]	196–198	34, 47
	3*S*,8a*S* Form[aa]	192–194	
$C_{14}H_{16}N_2O_2$	3-Benzyl		
	3*R*,8a*S* Form[ggg]	147–149	135, 136, 142
	3*S*,8a*S* Form[bb]	135–136	58, 68
	3*S*,8a*R* Form[cc]	148–150	58, 59, 68
		130–132	58
	Hydrate	73–75	
	3*RS*,8a*RS* Form	148–150	55
	Unspecified stereochemistry	212–216.5	117
$C_{14}H_{16}N_2O_3$	3-(4-Hydroxybenzyl)[dd]	226–228	51
	Hydrate	187	
$C_{14}H_{20}N_2O_4$	3-*i*-Butyl-2-pyruvoyl[ee]	178	28
$C_{15}H_{16}N_2O_2$	3-Benzylidene-2-methyl		
	8a*S* Form[hhh]	165–170	141
$C_{15}H_{16}N_2O_4$	2-Benzyloxycarbonyl		
	8a*S* Form[ff]	110–110.5	50, 59
	6-Carboxy-2-(4-methylphenyl)		
	α-Form	230–232	140
	Ethyl ester	139	
	β-Form	127	
$C_{15}H_{18}N_2O_2$	3-Benzyl-2-methyl		
	3*S*,8a*R* Form	137	141, 152
	3*S*,8a*S* Form	127–130	141
	3-Benzyl-8a-methyl		
	3*S*,8a*S* Form[gg]	145–148	56
	3*S*,8a*R* Form[hh]	168–171	
$C_{15}H_{18}N_2O_3$	3-(4-Methoxybenzyl)	156–157	51
$C_{15}H_{22}N_2O_4$	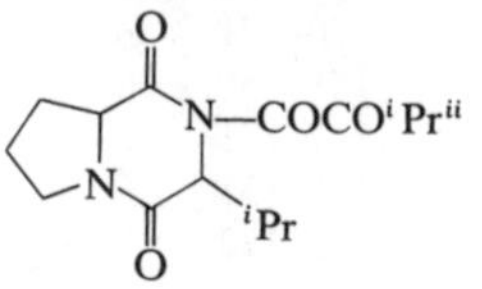 [structure: pyrrolo-pyrazinedione with N—COCO^{i}Pr[ii] and ^{i}Pr substituents]	111	28

TABLE 2 (*contd.*)

Molecular formula	Substituents or structure	m.p., °C	Ref.
$C_{16}H_{13}N_3O_2$	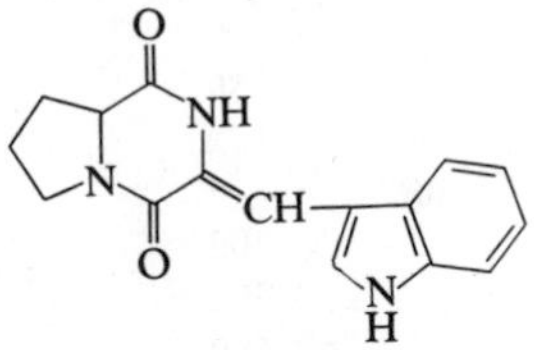 [structure: O, NH, N, CH, O, N H]	280–283 decomp.	115, 116
$C_{16}H_{16}N_2O_3$	2-Acetyl-3-benzylidene		
	8a*S* Form[jj]	172–173	58, 141
	8a*RS* Form	149–151	
$C_{16}H_{16}N_2O_4$	2-Acetyl-3-benzyl-6-oxo		
	3*S*,8a*S* Form[kk]	164–166	34, 47
	3*R*,8a*S* Form[ll]	164–166	
	Racemate	162–164	
$C_{16}H_{17}N_3O_2$	3-Indol-3-ylmethyl		
	3*S*,8a*S* Form[iii]	174	148
	3*S*,8a*R* Form[jjj]	204–206	
$C_{16}H_{18}N_2O_3$	2-Acetyl-3-benzyl		
	3S,8a*S* Form[mm]	123–125	58, 141
$C_{16}H_{18}N_2O_4$	2-Benzyloxyacetyl		
	8a*S* Form[nn]	150	42, 43, 79
	2-Benzyl-6-carboxy-3-methyl		
	Ethyl ester	b.p. 220–230/0.6 mm	140
	6-Carboxy-3-methyl-2-(4-methylphenyl)[kkk]		
	α-Form	153–156	140
	Ethyl ester	159–161	
	β Form	137–139	
	Ethyl ester	100–102	
	N,*N*-Diethylamide	125–129	
	p-Chloroanilide	188–189	
$C_{16}H_{20}N_2O_3$	6-Hydroxymethyl-3-methyl-2-(4-methylphenyl)	122–124	140
	Acetate	126–128	
	Carbamate	174–176	
$C_{16}H_{24}N_2O_4$	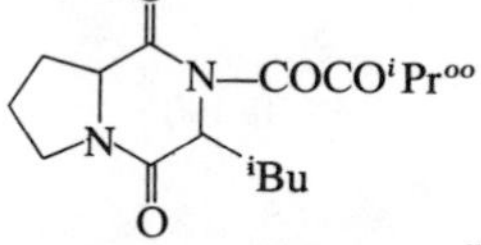 [structure: O, N—COCO^{i}Pr[oo], N, ^{i}Bu, O]	102	28
$C_{16}H_{26}N_2O_4$	2,3-Di-*n*-butyl-6-carboxy[lll]	163–166	140
$C_{17}H_{16}N_2O_5$	2-Acetyl-3-(3,4-methylenedioxybenzylidene)		
	8a*S* Form[mmm]	169–172	141
$C_{17}H_{18}N_2O_3$	2-Benzoyl-3-isopropylidene		
	Racemate	188–190	141

TABLE 2 (*contd.*)

Molecular formula	Substituents or structure	m.p., °C	Ref.
$C_{17}H_{18}N_2O_4$	3-Benzyl-2-pyruvoyl[pp]	183	28
	3*RS*,8a*RS* Form	157–158	55
	Hydrate	122–123	
$C_{17}H_{18}N_2O_5$	2-Acetyl-3-(4-methoxybenzyl)-6-oxo[qq]		
	3*S*,8a*S* Form[rr]	200–202	34
	3*R*,8a*S* Form[ss]	200–202	
	Racemate	168–169	
$C_{17}H_{20}N_2O_4$	2-Benzyl-6-carboxy-3-ethyl[nnn]		
	α Form	158–161	140
	Ethyl ester	b.p. 210/0.5 mm	
	β Form	180–182	
	Ethyl ester	135–139	
	3-Benzyl-2-(2-hydroxypropionyl)		
	Benzyl ether	135	43
$C_{17}H_{20}N_2O_5$	6-Carboxy-3-ethyl-2-(4-methoxy-phenyl)[ooo]	b.p. 240/0.6 mm	140
$C_{17}H_{22}N_2O_3$	2-Benzyl-3-ethyl-6-hydroxymethyl	93–95 b.p. 220–230/0.5 mm	140
	Acetate	b.p. 230/0.6 mm	
$C_{18}H_{20}N_2O_3$	2-Benzoyl-3-*i*-butylidene		
	8a*S* Form[ppp]	165–171	141
$C_{18}H_{20}N_2O_5$	2-Acetyl-3-(3,4-dimethoxy-benzylidene)		
	8a*S* Form[qqq]	194–196	141
$C_{18}H_{22}N_2O_4$	2-Benzyl-6-carboxymethyl-3-ethyl		
	Ethyl ester	b.p. 240–245/0.6 mm	140
	2-Benzyl-6-carboxy-3-*i*-propyl[rrr]	145–148	
	Ethyl ester	b.p. 210–215/0.5 mm	
	3-Benzyl-2-(2-methoxypropionyl)[tt]		
	Isomers	(i) 120–122 (ii) 153–155	68
$C_{19}H_{20}N_2O_6$	O; N—COMe; OMe; N; CH; OCOMe; O		
	8a*S* Form[sss]	170–190	141
$C_{19}H_{22}N_2O_4$	O; N—COCO^{i}Pr[uu]; N; CH$_2$Ph; O	126	28
$C_{19}H_{24}N_2O_4$	2-Benzyl-3-*n*-butyl-6-carboxy[ttt]		
	α Form	161–164	140

TABLE 2 (*contd.*)

Molecular formula	Substituents or structure	m.p., °C	Ref.
$C_{19}H_{24}N_2O_4$ (*contd.*)	Ethyl ester	b.p. 215–230/0.3 mm	
	β Form	144–146	
	2-Benzyl-3-*i*-butyl-6-carboxy[ttt]	159–160	140
	Ethyl ester	b.p. 230–235/0.5 mm	
$C_{19}H_{30}N_2O_4$	2-Lauroyl	95–96	75
$C_{20}H_{26}N_2O_4$	2-Benzyl-3-*n*-butyl-6-carboxy-methyl[uuu]		
	Ethyl ester	b.p. 220–230/0.6 mm	140
	2-Benzyloxyacetyl-3-*i*-butyl[vv]	131–132	43, 79
$C_{21}H_{23}N_3O_2$	O, NH, N, O, CH_2, N H, Me_2C, CH=CH_2		
	8a*S* Form[vvv]		148
$C_{21}H_{25}N_3O_2$	O, NH, N, O, CH_2, N H, Me_2C, CH=CH_2		
	3*S*,8a*S* Form[www]		148
$C_{21}H_{26}N_2O_6$	O, N—CO—C(Me)(OCH$_2$Ph)—CO$_2$H[ww], N, O, ^{i}Pr		
	Ethyl ester, 3*S*,8a*S* Form	110–111	66, 70
$C_{21}H_{27}N_3O_2$	3-[2-(2-Methylbut-2-yl)indol-3-ylmethyl][xxx]		148
$C_{22}H_{28}N_2O_6$	O, N—CO—C(Me)(OCH$_2$Ph)—CO$_2$H, N, O, ^{i}Bu		
	Ethyl ester[xx]	*ca.* 80	66

TABLE 2 (*contd.*)

Molecular formula	Substituents or structure	m.p., °C	Ref.
$C_{24}H_{26}N_2O_4$	3-Benzyl-2-(2-benzyloxypropionyl)[yy]		
	3*S*,8a*S* Form	137	43, 79
$C_{24}H_{27}N_3O_5S$	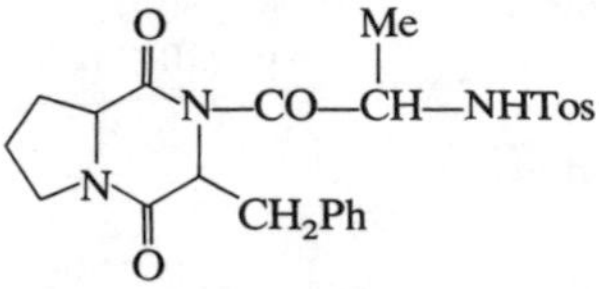 O, Me, N—CO—CH—NHTos, N, CH_2Ph, O		
	3*S*,8a*R* Form, 2′*S* Isomer[yyy]	191–192	142
	3*R*,8a*S* Form, 2′*S* Isomer[zzz]	178–179	
$C_{26}H_{28}N_2O_6$	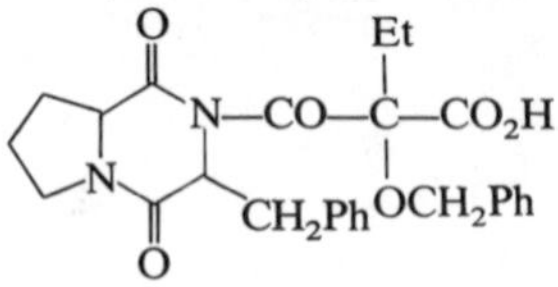 O, Et, N—CO—C—CO_2H, N, CH_2Ph OCH_2Ph, O		
	Ethyl ester[zz]		
	3*S*,8a*S* Form	124–126	121
$C_{26}H_{28}N_2O_6$	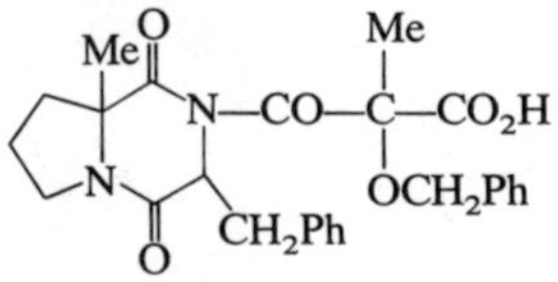 Me O, Me, N—CO—C—CO_2H, N, OCH_2Ph, CH_2Ph, O		
	Ethyl ester[aaa]		
	3*S*,8a*S* Form	104–105	56
$C_{28}H_{26}ClN_3O_3$	2,3-Dibenzyl-6-(4-chloroanilido)	198–200	140

[a] $[\alpha]_D^{25} -70.5°$ (DMF).
[b] $[\alpha]_D^{20} -217.4°$ (H_2O, $p=7.4$), $-202°$; $[\alpha]_D^{28} -185°$ (c 0.5, H_2O).
[c] $[\alpha]_D^{25} -84.0°$ (DMF).
[d] $[\alpha]_D^{25} -7.7°$ (DMF).
[e] $[\alpha]_D^{25} -9.2°$ (DMF).
[f] $[\alpha]_D^{20} -52°$ (c 0.5, THF).
[g] $[\alpha]_D^{25} +8.0°$ (DMF).
[h] $[\alpha]_D^{25} -3.9°$ (DMF).
[i] $[\alpha]_D^{25} -2.8°$ (DMF).
[j] $[\alpha]_D^{25} -13.7°$ (DMF).
[k] $[\alpha]_D^{20} -135°$ (EtOH).
[l] $[\alpha]_D^{25} -164.5°$ (dioxan); $[\alpha]_D^{20} -159°$ (c 0.7, EtOH); $[\alpha]_{5461}^{20} -180.5°$ (c 1, EtOH).
[m] $[\alpha]_D^{20} +88°$ (c 1, H_2O); $[\alpha]_{5461}^{20} +107°$ (c 1, H_2O).
[n] $[\alpha]_D^{20} -92.9°$ (MeOH); $[\alpha]_{5461}^{20} -121.2°$ (c 1, EtOH).
[o] $[\alpha]_D^{25} -25.4°$.
[p] $[\alpha]_D^{25} +0.5°$ (DMF).
[q] $[\alpha]_D^{25} +11.3°$ (DMF).
[r] $[\alpha]_D^{25} -66°$ (c 1, H_2O).
[s] $[\alpha]_D^{25} -10.5°$ (DMF).

Footnotes for Table 2 (*contd.*)

[t] $[\alpha]_D^{20} - 144°$ (c 0.4, H_2O); $-143.4°$ (EtOH, $p = 3$); $[\alpha]_D^{25} - 120°$ (dioxan); $[\alpha]_D^{23} - 146.7°$ (c 1, H_2O), $-143.7°$ (EtOH).

[u] $[\alpha]_D^{20} + 92°$ (c 1, H_2O); $[\alpha]_{5461}^{20} + 109°$ (c 1, H_2O).

[v] $[\alpha]_D^{25} - 3.2°$.

[w] $[\alpha]_D^{25} - 34.4°$.

[x] $[\alpha]_D^{25} + 0.2°$.

[y] $[\alpha]_D^{25.5} + 284.7°$ (c 5, pyridine).

[z] $[\alpha]_D^{25} + 25.0°$ (DMF).

[aa] $[\alpha]_D^{25} - 108.9°$ (DMF).

[bb] $[\alpha]_D^{20} - 100°$ (c 1, H_2O), $-83°$ (c 0.2, H_2O); $[\alpha]_D^{26} - 107.6°$; $[\alpha]_{5461}^{20} - 115.6°$ (c 1, EtOH).

[cc] $[\alpha]_D^{20} + 92°$; $[\alpha]_D^{25} + 83°$ (H_2O).

[dd] $[\alpha] + 67.1°$, hydrate $- 10.25°$.

[ee] Probably formulated incorrectly in ref. 28 (cf. ref. 44); $[\alpha]_D^{20} + 96°$, $[\alpha]_{5461}^{20} + 114°$ (c 0.25, EtOH).

[ff] $[\alpha]_D^{25} - 107°$, $-109.5°$ (c 1, EtOAc).

[gg] $[\alpha]_D^{20} - 173°$ (c 1, $CHCl_3$), $-87.8°$ (EtOH), $-85.1°$ (pyridine).

[hh] $[\alpha]_D^{20} + 35.2°$ (EtOH), $-14.3°$ (pyridine).

[ii] Probably formulated incorrectly in ref. 28 (cf. ref. 44); $[\alpha]_D^{20} + 109°$, $[\alpha]_{5461}^{20} + 132°$ (c 0.25, EtOH).

[jj] $[\alpha]_D^{25} + 49.3°$ (c 5, pyridine); $[\alpha]_D^{23} + 50.4°$ (c 2.5, $CHCl_3$).

[kk] $[\alpha]_D^{25} - 8.3°$ (DMF).

[ll] $[\alpha]_D^{25} + 8.2°$ (DMF).

[mm] $[\alpha]_D^{22} + 202.6°$ (c 5, pyridine).

[nn] $[\alpha]_D^{20} - 43.6°$ (c 1.0, THF).

[oo] Probably formulated incorrectly in ref. 28 (cf. ref. 44); $[\alpha]_D^{20} + 117°$, $[\alpha]_{5461}^{20} + 140°$ (c 0.25, EtOH).

[pp] Formulated incorrectly in ref. 28 (cf. ref. 44); $[\alpha]_D^{20} + 138°$, $[\alpha]_{5461}^{20} + 169°$ (c 0.25, EtOH).

[qq] The racemates of various di- and trimethoxybenzyl compounds have been prepared (refs. 34, 47).

[rr] $[\alpha]_D^{25} - 64.5°$ (DMF).

[ss] $[\alpha]_D^{25} + 64.3°$ (DMF).

[tt] Isomer (i) $[\alpha]_D^{20} + 138°$ (c 0.13, alcohol); isomer (ii) $[\alpha]_D^{20} + 101°$ (c 0.2, alcohol).

[uu] Probably formulated incorrectly in ref. 28 (cf. ref. 44); $[\alpha]_D^{20} + 145°$, $[\alpha]_{5461}^{20} + 175°$ (c 0.25, EtOH).

[vv] $[\alpha]_D^{20} + 103°$ (c 1, THF).

[ww] Prepared from the (+)-acid chloride; $[\alpha]_D^{20} + 138°$ (EtOH).

[xx] Prepared from the (+)-acid chloride.

[yy] $[\alpha]_D^{20} - 149°$ (c 0.2, alcohol), $-145°$ (c 1, THF).

[zz] Prepared from the (−)-acid chloride; $[\alpha]_D$ 97° (EtOH).

[aaa] Prepared from the (+)-acid chloride; $[\alpha]_D^{20} + 147°$ (c 0.5, EtOH).

[bbb] $[\alpha]_{5460}^{20} - 160°$ (c 1, EtOH).

[ccc] $[\alpha]_{5460}^{20} - 182.3°$ (c 1, EtOH).

[ddd] $[\alpha]_D^{23} + 21.6°$ (c 3, $CHCl_3$).

[eee] $[\alpha]_D^{25} - 93.9°$ (c 0.5, H_2O); $[\alpha]_{5461}^{20} - 78°$ (c 1, EtOH).

[fff] $[\alpha]_D^{25} + 284.3°$ (c 5, pyridine); $[\alpha]_D^{24} + 200.3°$ (c 2.8, $CHCl_3$).

[ggg] $[\alpha]_{5460}^{20} - 101.6°$ (c 1, EtOH); $[\alpha]_D^{25} - 94.2°$ (c 0.2, H_2O).

[hhh] $[\alpha]_D^{24} + 53.5°$ (c 3.15, $CHCl_3$).

[iii] $[\alpha]_D^{22} - 101°$ (c 1.1, AcOH).

[jjj] $[\alpha]_D^{20} - 101°$ (c 1.78, AcOH).

Footnotes for Table 2 (contd.).

[kkk] Reference 140 reports seven other substituted phenyl compounds.
[lll] Reference 140 reports three substituted amides of this acid.
[mmm] $[\alpha]_D^{22} + 113.5°$ (*c* 2.3, $CHCl_3$).
[nnn] Reference 140 reports four other substituted phenyl compounds and seven substituted amides of this acid.
[ooo] The 2-(4-chlorophenyl) compound is also reported.
[ppp] $[\alpha]_D^{24} + 116.5°$ (*c* 2, $CHCl_3$).
[qqq] $[\alpha]_D^{22} + 105°$.
[rrr] Reference 140 lists three other substituted benzyl derivatives.
[sss] $[\alpha]_D^{25} + 58.6°$ (*c* 3, $CHCl_3$).
[ttt] Reference 140 reports four other substituted benzyl derivatives.
[uuu] Reference 140 reports the 3,4-dimethylbenzyl compound also.
[vvv] Characterized by mass spectrometry, $[\alpha]_D^{22} - 59°$ (*c* 1.2, $CHCl_3$).
[www] Characterized by mass spectrometry, $[\alpha]_D^{22} - 38°$ (*c* 1.3, $CHCl_3$).
[xxx] Characterized by mass spectrometry, spectroscopic details only.
[yyy] $[\alpha]_D^{20} + 85°$ (*c* 0.6, EtOAc).
[zzz] $[\alpha]_D^{20} - 164°$ (*c* 1, EtOAc).

IV. References

1. W. L. Mosby, "Heterocyclic Systems with Bridgehead Nitrogen Atoms," Interscience, New York, 1961, p. 691.
2. W. W. Paudler and D. E. Dunham, *J. Heterocycl. Chem.*, **2,** 410 (1965).
3. V. Galasso, G. De Alti, and A. Bigotto, *Theor. Chim. Acta*, **9,** 222 (1968); *Chem. Abstr.*, **68,** 55108 (1968).
4. A. Gamba and G. Favini, *Gazz. Chim. Ital.*, **98,** 167 (1968); *Chem. Abstr.*, **68,** 118134 (1968).
5. M. E. Freed and A. R. Day, *J. Org. Chem.* **25,** 2108 (1960).
6. J. W. Lown and K. Matsumoto, *Can. J. Chem.*, **49,** 1165 (1971).
7. W. Herz and S. Tocker, *J. Am. Chem. Soc.*, **77,** 6355 (1955).
8. R. M. Acheson and M. W. Foxton, *J. Chem. Soc.* (*C*), 2218 (1966).
9. V. I. Shvedov, L. B. Altukhova, and A. N. Grinev, *Khim. Geterotsikl. Soedin.*, 1048 (1970); *Chem. Abstr.*, **74,** 125628 (1971).
10. V. I. Shvedov, L. B. Altukhova, A. V. Bocharnikova, and A. N. Grinev, USSR Pat. 237,153; *Chem. Abstr.*, **71,** 13142 (1969).
11. L. A. Paquette and M. K. Scott, *J. Org. Chem.*, **33,** 2379 (1968).
12. C. Casagrande, A. Galli, R. Ferrini, and G. Miragoli, *Arzneim.-Forsch.*, **21,** 808 (1971).
13. M. E. Freed (to American Home Products Corp.), U.S. Pat. 3,164,598; *Chem. Abstr.*, **62,** 9154 (1965).
14. M. E. Freed (to American Home Products Corp.), U.S. Pat. 3,531,485; *Chem. Abstr.*, **73,** 120672 (1970).
15. V. Boekelheide and K. Fahrenholtz, *J. Am. Chem. Soc.*, **83,** 458 (1961).
16. T. Sasaki, K. Kanematsu, Y. Yukimoto, and S. Ochiai, *J. Org. Chem.*, **36,** 813 (1971).
17. J. W. Lown and K. Matsumoto, *Can. J. Chem.*, **49,** 3119 (1971).
18. A. P. Dunlop and S. Swadesh (to Quaker Oats Co.), U.S. Pat. 2,655,512; *Chem. Abstr.*, **48,** 11495 (1954).

19. A. A. Ponomarev and I. M. Skvortsov, *Dokl. Akad. Nauk SSSR*, **148,** 860 (1963); *Chem. Abstr.*, **59,** 3918 (1963).
20. A. A. Ponomarev and I. M. Skvortsov, *Metody Poluch. Khim. Reactivov Prep.*, 5 (1967); *Chem. Abstr.*, **70,** 87778 (1969).
21. A. M. Likhosherstov, L. S. Nazarova, and A. P. Skoldinov, Fr. Pat. 1,510,781; *Chem. Abstr.*, **70,** 68427 (1969).
22. Scientific Research Institute of Pharmacology and Chemotherapy, Moscow, Brit. Pat. 1,144,749; *Chem. Abstr.*, **70,** 106558 (1969).
23. A. M. Likhosherstov, L. S. Nazarova, and A. P. Skoldinov, USSR Pat. 245,789; *Chem. Abstr.*, **71,** 124500 (1969).
24. A. M. Likhosherstov, L. S. Nazorova, and A. P. Skoldinov, *Zh. Org. Khim.*, **6,** 1729 (1970); *Chem. Abstr.*, **73,** 109724 (1970).
25. N. M. Bortnick and M. F. Fegley (to Rohm and Haas Co.), U.S. Pat. 2,984,666; *Chem. Abstr.*, **56,** 3483 (1962).
26. M. E. Freed (to American Home Products Corp.), U.S. Pat. 3,176,017; *Chem. Abstr.*, **62,** 16272 (1965).
27. A. Stoll and T. Petrzilka (to Sandoz Ltd.), U.S. Pat. 2,673,850; *Chem. Abstr.*, **49,** 4032 (1955).
28. A. Stoll, A. Hofmann, and T. Petrzilka, *Helv. Chim. Acta*, **34,** 1544 (1951).
29. E. Abderhalden and E. Schwab, *Z. Physiol. Chem.*, **148,** 254 (1925); *Chem. Abstr.*, **20,** 55 (1926).
30. N. S. Andreeva, V. I. Iveronova, T. D. Kozarenko, K. T. Poroshin, V. A. Shibnev, and N. E. Shutskever, *Izv. Akad. Nauk SSSR, Otd. Khim. Nauk*, 376 (1958); *Chem. Abstr.*, **52,** 12762 (1958).
31. I. Z. Siemion, *Justus Liebigs Ann. Chem.*, **748,** 88 (1971).
32. R. Schwyzer and U. Ludescher, *Helv. Chim. Acta*, **52,** 2033 (1969).
33. J. W. Westley, V. A. Close, D. N. Nitecki, and B. Halpern, *Anal. Chem.*, **40,** 1888 (1968).
34. M. R. Harnden, *J. Chem. Soc.* (*C*), 2341 (1967).
35. A. B. Mauger, *J. Chromatogr.*, **37,** 315 (1968).
36. P. W. G. Smith, *J. Chem. Soc.*, 3985 (1957).
37. J. L. Johnson, W. G. Jackson, and T. E. Eble, *J. Am. Chem. Soc.*, **73,** 2947 (1951).
38. T. Kosuge and H. Kamiya, *Nature*, **188,** 1112 (1960).
39. M. R. Harnden, *J. Heterocycl. Chem.*, **5,** 307 (1968).
40. Y. N. Chirgadze, *Optika i Spektroskopiya, Akad. Nauk SSSR, Otd. Fiz.-Mat. Nauk Sb. Statei*, **2,** 242 (1963); *Chem. Abstr.*, **60,** 117 (1964).
41. V. K. Antonov, A. M. Shkrob, and M. M. Shemyakin, "Peptides, Proc. European Symp., 5th, Oxford," 1962, p. 221; *Chem. Abstr.*, **62,** 4242 (1965).
42. M. M. Shemyakin, V. K. Antonov, A. M. Shkrob, Y. N. Sheinker, and L. B. Senyavina, *Tetrahedron Lett.*, 701 (1962).
43. M. M. Shemyakin, V. K. Antonov, A. M. Shkrob, V. I. Schelokov, and Z. E. Agadzhanyan, *Tetrahedron*, **21,** 3537 (1965).
44. M. Green and E. A. C. Lucken, *Helv. Chim. Acta*, **44,** 1417 (1961).
45. N. S. Vul'fson, V. A. Puchkov, Y. V. Denison, B. V. Rozynov, V. N. Bochkarev, M. M. Shemyakin, Y. A. Ovchinnikov, and V. K. Antonov, *Khim. Geterotsikl. Soedin*, 614 (1966); *Chem. Abstr.*, **66,** 28272 (1967).
46. N. A. Poddubnaya and G. I. Lavrenova, *Vestnik Moskov Univ., Ser. Mat. Mekh. Astron. Fiz. Khim.*, **13,** 165 (1958); *Chem. Abstr.*, **53,** 11396 (1959).
47. M. R. Harnden (to Abbott Laboratories), U.S. Pat. 3,563,992; *Chem. Abstr.*, **75,** 20443 (1971).

48. H. N. Rydon and P. W. G. Smith, *J. Chem. Soc.*, 3642 (1956).
49. E. Wuensch, "Peptides, Proc. European Symp., 5th, Oxford," 1962, p. 89; *Chem. Abstr.*, **62,** 5334 (1965).
50. M. Goodman and K. C. Stueben, *J. Am. Chem. Soc.*, **84,** 1279 (1962).
51. E. Abderhalden and H. Sickel, *Z. Physiol. Chem.*, **153,** 16 (1926); *Chem. Abstr.*, **20,** 3169 (1926).
52. V. A. Shibnev, T. D. Kozarenko, and K. T. Poroshin, *Izv. Akad. Nauk SSSR, Otd. Khim. Nauk*, 1500 (1960); *Chem. Abstr.*, **55,** 405 (1961).
53. E. Abderhalden and O. Zumstein, *Fermentforschung*, **12,** 1 (1930); *Chem. Abstr.*, **25,** 77 (1931).
54. R. E. Neumann and E. L. Smith, *J. Biol. Chem.*, **193,** 97 (1951).
55. C. A. Grob and W. Meier, *Helv. Chim. Acta*, **51,** 219 (1957).
56. Sandoz Ltd., Neth. Appl. 6,609,202; *Chem. Abstr.*, **67,** 82297 (1967).
57. J. Meienhofer, *J. Am. Chem. Soc.*, **92,** 3771 (1970).
58. M. Bergmann and J. E. Tietzman, *J. Biol. Chem.*, **155,** 535 (1944).
59. G. Lucente, G. Fiorentini, and D. Rossi, *Gazz. Chim. Ital.*, **101,** 109 (1971); *Chem. Abstr.*, **75,** 64266 (1971).
60. R. H. Mazur and J. M. Schlatter, *J. Org. Chem.*, **28,** 1025 (1963).
61. E. Abderhalden and H. Sickel, *Z. Physiol. Chem.*, **159,** 163 (1926); *Chem. Abstr.*, **21,** 390 (1927).
62. E. Fischer and U. Suzuki, *Ber.*, **37,** 2847 (1904).
63. E. Fischer and G. Reif, *Ann.*, **363,** 124 (1908).
64. R. A. Boissonnas, S. Guttmann, R. L. Huguenin, P. A. Jaquenoud, and E. Sandrin, *Helv. Chim. Acta*, **41,** 1867 (1958).
65. K. T. Poroshin, T. D. Kozarenko, and V. A. Shibnev, *Izv. Akad. Nauk SSSR, Otd. Khim. Nauk*, 1129 (1958); *Chem. Abstr.*, **53,** 3077 (1959).
66. P. A. Stadler, A. J. Frey, H. Ott, and A. Hoffmann, *Helv. Chim. Acta*, **47,** 1911 (1964).
67. D. E. Nitecki, B. Halpern, and J. W. Westley, *J. Org. Chem.*, **33,** 864 (1968).
68. H. Ott, A. J. Frey, and A. Hofmann, *Tetrahedron*, **19,** 1675 (1963).
69. A. Hofmann and P. Stadler (to Sandoz Ltd.), Fr. M. 6822; *Chem. Abstr.*, **74,** 88192 (1971).
70. Sandoz Ltd., Neth. Appl. 6,511,933; *Chem. Abstr.*, **65,** 7231 (1966).
71. S. Guttmann and R. Huguenin (to Sandoz Ltd.), Ger. Offen. 2,029,447; *Chem. Abstr.*, **74,** 88191 (1971).
72. P. Stadler, A. Hofmann, and F. Troxler (to Sandoz Ltd.), Fr. Pat. 1,583,797; *Chem. Abstr.*, **73,** 77459 (1970).
73. S. Guttmann and R. Huguenin (to Sandoz Ltd.), Ger. Offen. 1,931,081; *Chem. Abstr.*, **72,** 133049 (1970).
74. Sandoz Ltd., Neth. Appl. 6,513,688; *Chem. Abstr.*, **65,** 8980 (1966).
75. T. Ueda, S. Kato, and S. Toyoshima (to Dai-Nippon Drug Co.), Jap. Pat. 4823 (1959); *Chem. Abstr.*, **54,** 14279 (1960).
76. Sandoz Ltd., Belg. Pat. 617,499; *Chem. Abstr.*, **58,** 13970 (1963).
77. Sandoz Ltd., Belg. Pat. 617,451; *Chem. Abstr.*, **58,** 14029 (1963).
78. A. Hofmann, H. Ott, R. Griot, P. A. Stadler, and A. J. Frey, *Helv. Chim. Acta*, **46,** 2506 (1963).
79. V. K. Antonov, A. M. Shkrob, and M. M. Shemyakin, *Zh. Obshch. Khim.*, **37,** 2225 (1967); *Chem. Abstr.*, **69,** 10416 (1968).
80. P. Stadler, H. Hauth, G. Wersin, S. Guttmann, A. Hofmann, P. Stutz, and H. Willems (to Sandoz Ltd.), Fr. Pat. 1,588,541; *Chem. Abstr.*, **73,** 131033 (1970).

81. P. Stadler, H. Hauth, G. Wersin, S. Guttmann, A. Hofmann, P. Stutz, and H. Willems (to Sandoz Ltd.), Brit. Pat. 1,238,348; *Chem. Abstr.*, **75,** 88815 (1971).
82. S. Smith and G. M. Timmis, *J. Chem. Soc.*, 396 (1937).
83. A. Stoll, A. Hofmann, and B. Becker, *Helv. Chim. Acta*, **26,** 1602 (1943).
84. I. Chmielewska, B. Bulhak, and K. Toczko, *Bull. Acad. Pol. Sci., Ser. Sci. Biol.*, **15,** 719 (1967); *Chem. Abstr.*, **68,** 103216 (1968).
85. T. L. Perry, K. S. C. Richardson, S. Hansen, and A. J. D. Friesen, *J. Biol. Chem.*, **240,** 4540 (1965); *Chem. Abstr.*, **63,** 15350 (1965).
86. T. Tatsuno, M. Sato, Y. Kubota, and H. Tsunoda, *Chem. Pharm. Bull.* (Tokyo), **19,** 1498 (1971); *Chem. Abstr.*, **75,** 116967 (1971).
87. Y. Koaze, *Bull. Agr. Chem. Soc. Jap.* **21,** 197 (1957); *Chem. Abstr.*, **51,** 15718 (1957).
88. Y. Koaze, *Bull. Agr. Chem. Soc. Jap.*, **22,** 98 (1958); *Chem. Abstr.*, **52,** 20461 (1958).
89. Y. Koaze, *Nippon Nogeikaku Kaishi*, **34,** 371 (1960); *Chem. Abstr.*, **58,** 9569 (1963).
90. Y.-S. Chen, *Bull. Agr. Chem. Soc. Jap.*, **24,** 372 (1960); *Chem. Abstr.*, **55,** 680 (1961).
91. Y. Kodaira, *Agr. Biol. Chem.* (Tokyo), **25,** 261 (1961); *Chem. Abstr.*, **55,** 12678 (1961).
92. G. Luedemann, W. Charney, A. Woyciesjes, E. Pettersen, W. D. Peckam, M. J. Gentles, H. Marshall, and H. L. Herzog, *J. Org. Chem.*, **26,** 4128 (1961).
93. L. Ettlinger, E. Gäumann, R. Hütter, W. Keller-Schierlein, F. Kradolfer, L. Neip, V. Prelog, P. Reusser, and H. Zähner, *Monatsh. Chem.*, **88,** 989 (1957); *Chem. Abstr.*, **52,** 8282 (1958).
94. Y. Koaze, *Bull. Agr. Chem. Soc. Jap.*, **24,** 530 (1960); *Chem. Abstr.*, **55,** 3714 (1961).
95. Y. Koaze, *Bull. Agr. Chem. Soc. Jap.*, **24,** 449 (1960); *Chem. Abstr.*, **55,** 2792 (1961).
96. S. Tamura, A. Suzuki, Y. Aoki, and N. Otake, *Agr. Biol. Chem.* (Tokyo), **28,** 650 (1964); *Chem. Abstr.*, **62,** 5471 (1965).
97. T. Kosuge and H. Kamiya, *Chem. Pharm. Bull.* (Tokyo), **10,** 154 (1962); *Chem. Abstr.*, **58,** 2582 (1963).
98. L. A. Mitscher, M. P. Kunstmann, J. H. Martin, W. W. Andres, R. H. Evans, Jr., K. J. Sax, and E. L. Patterson, *Experientia*, **23,** 796 (1967); *Chem. Abstr.*, **67,** 115742 (1967).
99. A. Butenandt, P. Karlson, and W. Zillig, *Hoppe-Seyler's Z. Physiol. Chem.*, **288,** 279 (1951); *Chem. Abstr.*, **49,** 2628 (1955).
100. Y. Kodaira, *J. Fac. Tex. Sci. Technol., Shinshu Univ. Ser. E*, 1 (1961); *Chem. Abstr.*, **59,** 5522 (1963).
101. S. H. Eppstein, D. H. Peterson, H. M. Leigh, H. C. Murray, A. Weintraub, L. M. Reineke, and P. D. Meister, *J. Am. Chem. Soc.*, **75,** 421 (1953).
102. T. B. Lingappa, M. Prasad, Y. Lingappa, D. Hunt, and K. Biemann, *Science*, **163,** 192 (1969); *Chem. Abstr.*, **70,** 65415 (1969).
103. M. Abe, T. Fukuhara, S. Ohmomo, H. Masahiro, and T. Tabuchi, *Nippon Nogei Kagaku Kaishi*, **44,** 573 (1970); *Chem. Abstr.*, **74,** 136416 (1971).
104. M. Abe, T. Ohashi, S. Ohmomo, and T. Tabuchi, *Nippon Nogei Kagaku Kaishi*, **45,** 6 (1971); *Chem. Abstr.*, **75,** 60162 (1971).
105. P. S. Steyn, *Tetrahedron Lett.*, 3331 (1971).
106. A. Stoll, *Helv. Chim. Acta*, **28,** 1283 (1945).
107. A. Stoll, T. Petrzilka, and B. Becker, *Helv. Chim. Acta*, **33,** 57 (1950).
108. A. Stoll and A. Hofmann, *Helv. Chim. Acta*, **33,** 1705 (1950).
109. P. A. Levene and W. A. Beatty, *Ber.*, **39,** 2060 (1906).
110. I. Chmielewska, B. Bulhak, and K. Toczko, *Acta Biol. Pol.*, **14,** 409 (1967); *Chem. Abstr.*, **68,** 46264 (1968).

111. E. Abderhalden and E. Komm, *Z. Physiol. Chem.*, **145,** 308 (1925); *Chem. Abstr.*, **19,** 2810 (1925).
112. W. Grassmann, H. Hörmann, A. Nordwig, and E. Wünsch, *Z. Physiol. Chem.*, **316,** 287 (1959); *Chem. Abstr.*, **54,** 8958 (1960).
113. A. Fodor and S. Kuk, *Koll. Zeitschr.*, **74,** 66 (1936); *Chem. Abstr.*, **30,** 7966 (1936).
114. N. Gawrilow and K. Lawrowsky, *Biochem. Z.*, **190,** 278 (1927); *Chem. Abstr.*, **22,** 1593 (1928).
115. J. C. Sheehan, P. E. Drummond, J. N. Gardner, K. Maeda, D. Mania, S. Nakamura, A. K. Sen, and J. A. Stock, *J. Am. Chem. Soc.*, **85,** 2867 (1963).
116. J. C. Sheehan, D. Mania, S. Nakamura, J. A. Stock, and K. Maeda, *J. Am. Chem. Soc.*, **90,** 462 (1968).
117. N. A. Poddubnaya and M. I. Kiselev, *Zhur. Obshchei Khim.*, **26,** 1508 (1956); *Chem. Abstr.*, **50,** 14538 (1956).
118. H. Brockmann, G. Bohnsack, and C. H. Süling, *Angew. Chem.*, **68,** 66 (1956); *Chem. Abstr.*, **50,** 13025 (1956).
119. A. B. Mauger, *J. Chem. Soc.* (*D*), 39 (1971).
120. A. Stoll and A. Hofmann, in "The Alkaloids," Vol. 8, R. H. F. Manske, Ed., Academic Press, London, 1965, p. 725; T. A. Henry, "The Plant Alkaloids," 4th ed. Churchill, London, 1949 p. 517.
121. W. Schlientz, R. Brunner, P. A. Stadler, A. J. Frei, H. Ott, and A. Hofmann, *Helv. Chim. Acta*, **47,** 1921 (1964).
122. F. Werde, E. Bruch, and G. Feuerriegel. *Z. Physiol. Chem.*, **214,** 63 (1933); *Chem. Abstr.*, **27,** 983 (1933).
123. M. Semonsky, *Pharm. Ind.*, **32,** 899 (1970); *Chem. Abstr.*, **74,** 125876 (1971).
124. A. Cerletti and B. Berde, "Background Migraine, Migraine Symp., 2nd," 1967, p. 53; *Chem. Abstr.*, **75,** 97177 (1971).
125. Y. Kobayashi, T. Kutsuma, and K. Morinaga, *Chem. Pharm. Bull.* (Tokyo), **19,** 2106 (1971); *Chem. Abstr.*, **76,** 25222 (1972).
126. W. W. Paudler, C. I. P. Chao, and L. S. Helmick, *J. Heterocycl. Chem.*, **9,** 1157 (1972).
127. L. Fontanella, E. Occelli, and A. Perazzi, *Farmaco, Ed. Sci.*, **28,** 463 (1973); *Chem. Abstr.*, **79,** 78688 (1973).
128. V. V. Zakusov, A. P. Skoldinov, K. S. Raevskii, A. M. Likhosherstov, and L. M. Nazarova, USSR Pat. 367,102; *Chem. Abstr.*, **79,** 5354 (1973).
129. A. M. Likhosherstov, L. S. Nazarova, A. P. Skoldinov, G. A. Markova, and N. V. Kaverina, Ger. Offen. 2,210,382; *Chem. Abstr.*, **79,** 126514 (1973).
130. C. H. Lovell, Ger. Offen. 2,226,063; *Chem. Abstr.*, **78,** 58458 (1973).
131. V. V. Zakusov, A. P. Skoldinov, A. M. Likhosherstov, K. S. Raevskii, and L. S. Nazarova, USSR Pat. 352,877; *Chem. Abstr.*, **78,** 58053 (1973).
132. S. Rajappa and B. G. Advani, *Tetrahedron*, **29,** 1299 (1973).
133. I. L. Karle, *J. Am. Chem. Soc.*, **94,** 81 (1972).
134. I. Z. Siemion and A. Sucharda-Sobczyk, *Rocz. Chem.*, **46,** 1257 (1972); *Chem. Abstr.*, **77,** 140487 (1972).
135. K. Blaha, M. Budesinsky, I. Fric, J. Smolikova, and J. Vicar, *Tetrahedron Lett.*, 1437 (1972); J. Vicar, M. Budesinsky, and K. Blaha, *Collect. Czech. Chem. Commun.*, **38,** 1940 (1973); *Chem. Abstr.*, **79,** 126783 (1973).
136. I. Z. Siemion, *Org. Magn. Resonance*, **3,** 545 (1971).
137. J. Vicar, J. Smolikova, and K. Blaha, *Collect. Czech. Chem. Commun.*, **37,** 4060 (1972); *Chem. Abstr.*, **78,** 84810 (1973).
138. J. Vicar, J. Smolikova, and K. Blaha, *Collect. Czech. Chem. Commun.*, **38,** 1957 (1973); *Chem. Abstr.*, **79,** 126784 (1973).

139. A. B. Mauger, R. B. Desai, I. Rittner, and W. J. Rzeszotarski, *J. Chem. Soc., Perkin. I*, 2146 (1972).
140. L. Fontanella and L. Mariani (to Gruppo Lepetit S.p.A.), Ger. Offen. 2,345,056; *Chem. Abstr.*, **81,** 37576 (1974).
141. H. Poisel and U. Schmidt, *Chem. Ber.*, **106,** 3408 (1973).
142. G. Lucente and P. Frattesi, *Tetrahedron Lett.*, 4283 (1972).
143. T. C. Bruice and D. M. McMahon, *Biochemistry*, **11,** 1273 (1972).
144. J. E. Dixon and T. C. Bruice, *J. Am. Chem. Soc.*, **94,** 2052 (1972).
145. G. R. Pettit, R. B. Von Dreele, G. Bolliger, P. M. Traxler, and P. Brown, *Experientia*, **29,** 521 (1973).
146. E. P. White, *N. Z. J. Sci.*, **15,** 178 (1972); *Chem. Abstr.*, **77,** 98731 (1972).
147. A. J. Birch and R. A. Russell, *Tetrahedron*, **28,** 2999 (1972).
148. P. S. Steyn, *Tetrahedron*, **29,** 107 (1973).
149. T. Ohashi, H. Takahashi, and M. Abe, *Nippon Nogei Kagaku Kaishi*, **46,** 535 (1972); *Chem. Abstr.*, **78,** 39411 (1973).
150. P. Stuetz, R. Brunner, and P. A. Stadler, *Experientia*, **29,** 936 (1973).
151. B. F. Gisin and R. B. Merrifield, *J. Am. Chem. Soc.*, **94,** 3102 (1972).
152. M. C. Khosla, R. R. Smeby, and F. M. Bumpus, *J. Am. Chem. Soc.*, **94,** 4721 (1972).
153. P. Stadler, H. Hauth, G. Wersin, S. Guttmann, A. Hofmann, P. Stuetz, and H. Willems (to Sandoz Ltd.), Brit. Pat. 1,240,420; *Chem. Abstr.*, **76,** 14781 (1972).
154. N. J. Bach, H. E. Boaz, E. C. Kornfeld, C.-J. Chang, H. G. Floss, E. W. Hagaman, and E. Wenkert, *J. Org. Chem.*, **39,** 1272 (1974).
155. H. Kato, *Yakugaku Zasshi*, **95,** 830 (1975).

CHAPTER XX

Pyrrolo[*b*]pyrazines

I. Pyrrolo[2,3-*b*]pyrazines 348
II. Pyrrolo[3,4-*b*]pyrazines 352
III. Tables 356
Table 1. 5*H*-Pyrrolo[2,3-*b*]pyrazines 356
Table 2. 6*H*-Pyrrolo[3,4-*b*]pyrazines 357
IV. References 359

I. Pyrrolo[2,3-*b*]pyrazines

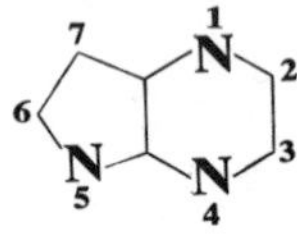

This ring system has also been referred to as 4,7-diazaindole and 1,4,7-triazaindene, and it is as indole analogues that derivatives of this heterocycle have attracted attention.

The parent unsubstituted compound has only recently been reported.[1–3] IR, UV, PMR, and mass spectral data have been tabulated for this compound and a series of methyl- and phenyl-substituted derivatives.[1] In ethanol the parent heterocycle shows two bands in its UV spectrum at 218 and 309 nm. The PMR spectrum[19] in $CDCl_3$, shows a broad singlet at δ 11.27 (NH) and doublets at δ 7.63 (H-2), 6.72 (H-3, $J_{2,3} = 3.8$ Hz), 8.47 (H-6), and 8.23 (H-7, $J_{6,7} = 2.6$ Hz). Fragmentation in the mass spectrometer takes place by initial loss of hydrogen followed by the successive loss of two molecules of HCN.

The preparation of this ring system involves the use of pyrazine intermediates. The parent heterocycle **3** has been prepared in two different ways from 2-amino-3-methylpyrazine (**1**). Formylation in a mixture of formic acid and acetic anhydride at 50° gives the formamido derivative **2**. This is cyclized by heating with sodium ethoxide at 325° to give pyrrolo[2,3-*b*]pyrazine in 38% yield after chromatographic separation of compound **1**, formed by hydrolysis of the intermediate **2**.[2] Reaction of the aminopyrazine **1** with triethyl orthoformate gives the formamidate **4**, which when immediately allowed to react with *N*-methylaniline, gives the amidine **5**. Treatment of this compound with the sodium salt of *N*-methylaniline at 180° gives the unsubstituted heterocycle **3** in 44% overall yield.[1,3,19]

Reagents (i) HCO_2H/Ac_2O
(ii) NaOEt
(iii) $(EtO)_3CH$
(iv) MeNHPh
(v) NaNMePh

Substituted derivatives of the ring system are prepared most conveniently by thermal indolization.[1,3,19] Reaction of hydrazinopyrazine (**6**) with carbonyl compounds gives the hydrazones **7**, which are cyclized in good yield to the pyrrolo compounds **8** by refluxing in diethylene glycol. Although this method has been shown to give 2,3-disubstituted derivatives and compounds substituted only in the 3-position (**8**, $R^1 = H$), attempts to prepare 2-monosubstituted derivatives (**8**, $R^2 = H$) were unsuccessful.[1] The hydrazones from cyclic ketones were also cyclized to give the tricyclic and tetracyclic systems **9** and **10**. No evidence was found in any of the thermal indolizations for cyclization onto the pyrazine nitrogen, which would have given compounds containing a bridgehead nitrogen atom.

7-Formyl derivatives **12** were obtained on Vilsmeier formylation of the 2-amino-3-methylpyrazines **11** with three molar quantities of phosphoryl chloride.[4] Although it seems reasonable to suggest that the reaction proceeds through intermediates such as **2** and **3**, attempts to formylate pyrrolo[2,3-*b*]pyrazine with the Vilsmeier reagent have proved unsuccessful.[3]

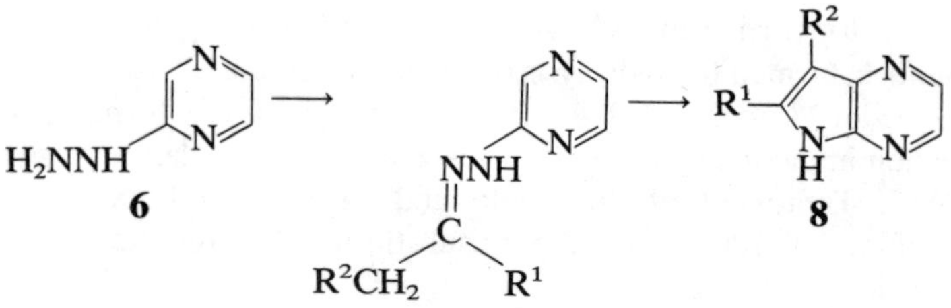

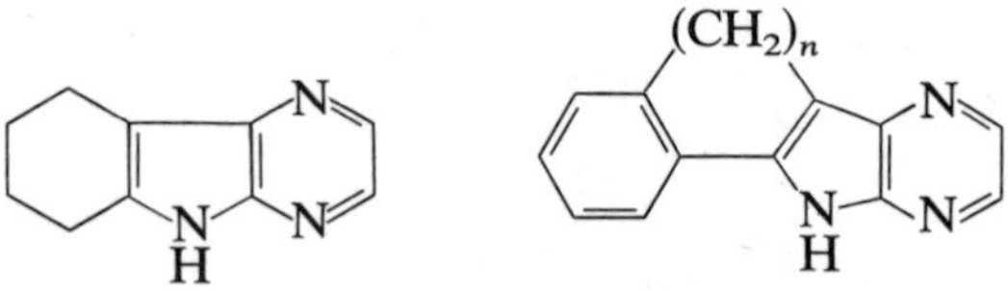

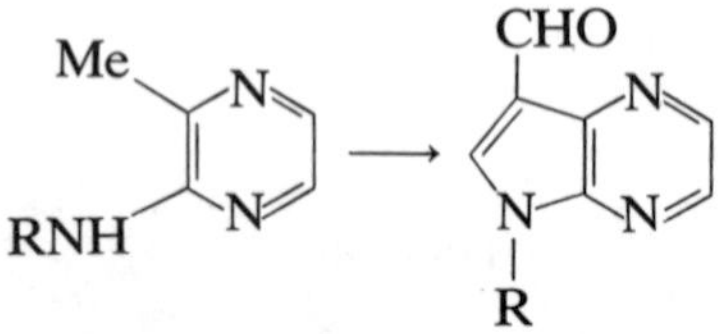

Both the 7-formyl compound and its 7-methyl analogue are readily methylated with methyl sulfate to give the 5-methyl derivatives.[4,5] Similarly the 7-methyl compound has been acetylated to form the 5-acetyl-7-formyl compound.[5] Although Vilsmeier formylation of the parent ring system was unsuccessful, the 7-formyl derivative was obtained using hexamethylenetetramine. Nitration and bromination both resulted in monosubstitution, and the products have been tentatively assigned as the 7-nitro and 7-bromo derivatives.[3]

Various reactions of the 7-substitutent have been carried out, all of which have their counterparts in indole chemistry.[4,5] Reduction of the imino compound **13**, obtained from DMF–phosphoryl chloride treatment of **11** (R = Me), with potassium borohydride gave the Mannich base **14** (R = Me). Reduction of the aldehyde **12** (R = H) with sodium borohydride in the presence of aqueous dimethylamine gave the amine **14** (R = H), whereas the analogous aldehyde **12** (R = Me) gave only the alcohol **15** (R = Me).[4] The related alcohol **15** (R = H) was obtained on reducing the aldehyde with sodium borohydride in the absence of dimethylamine.[5]

Treatment of the semicarbazone of the aldehyde **12** (R=H) with potassium hydroxide yielded the 7-methyl compound.[5] Condensation of the aldehyde with nitromethane gives the expected product **16**.

$CH{=}\overset{+}{N}Me_2$ (N-Me) **13** CH_2NMe_2 (N-R) **14** CH_2OH (N-R) **15** $CH{=}CHNO_2$ (N-H) **16**

Treatment of the amine **14** (R=H) with sodium ethylacetamidomalonate gave the tryptophan analogue **17** on acid hydrolysis. Decarboxylation of the amino acid **17** in acetophenone gave the Schiff base **18**, which was hydrolyzed by acid to the tryptamine analogue **19**. Methylation of the amine **14** (R=H) with methyl iodide gave the quaternary salt **20**,

$CH_2CH(NH_2)CO_2H$ **17** →(ii) $CH_2CH_2N{=}C(Ph)Me$ **18** →(iii) $CH_2CH_2NH_2$ **19**

CH_2NMe_2 →(i) **17**; →(iv) **20**

$CH_2\overset{+}{N}Me_3\ I^-$ **20** →(v) CH_2CN **21** →(vi) CH_2CO_2H **22**

Reagents (i) $NaC(CO_2Et)_2NHCOMe$, H^+
(ii) heat, MeCOPh
(iii) H^+
(iv) MeI
(v) KCN
(vi) OH^-

which readily reacted with potassium cyanide to give the acetonitrile **21**. Alkaline hydrolysis of the nitrile gave the acetic acid derivative **22**.[4,5]

II. Pyrrolo[3,4-b]pyrazines

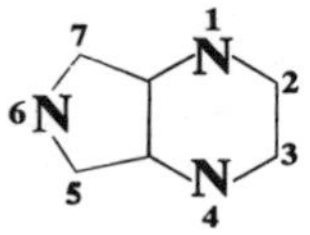

RRI 1262 (5H)
RRI 7988 (6H)

Most of the reported examples of this ring system are substituted imides derived from pyrazine-2,3-dicarboxylic acid. These imides are readily prepared by reaction of the corresponding anhydride **23** with appropriate amines. Thus 3,4-dimethoxyphenylethylamine in a mixture of acetic acid and acetic anhydride gave the pyrrolo compound **24** (R = 3,4-diMeO$C_6H_3CH_2CH_2$) in 45% yield.[6] Similarly methylamine has given the 6-methyl derivative **24** (R = Me); and acetamide, acting as a source of ammonia, provided the unsubstituted imide **24** (R = H).[7] Many amines give ring-opened amides (**25**) as intermediates, and these are isolated and cyclized by use of thionyl chloride[8] or sodium acetate in acetic acid.[9]

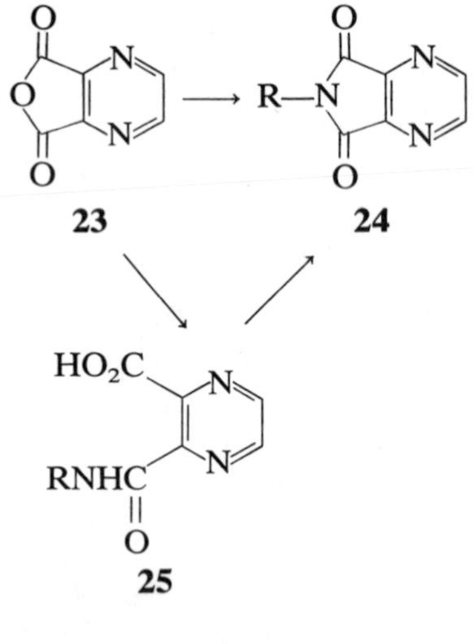

23 **24** **25**

The parent imide **24** (R = H) has also been prepared by ammonolysis of the diester **26**. The resultant diamide **27** gave the imide on pyrolysis.[10] Reaction of the imide **24** (R = H) with isopropylamine is reported to give the diamide **28**.[11] No pyrolysis experiment has, however, been reported on this compound. Treatment of the dinitrile **29** with ammonia in the presence of sodium methoxide yielded the diimine **30**, useful in the synthesis of azaphthalocyanines.[12]

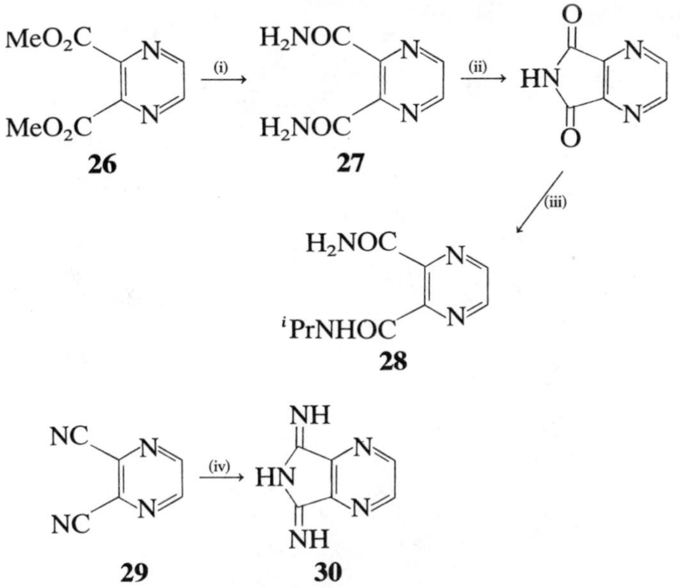

Reagents (i) NH_4OH
(ii) Heat
(iii) $^iPrNH_2$
(iv) NH_3–NaOMe

Recently the reaction between hydrazine and the dimethyl ester, or the anhydride, of pyrazine-2,3-dicarboxylic acid has been reexamined.[13] In glacial acetic acid in addition to products analogous to those formed with phthalic acid derivatives, anomalous products were identified resulting from reduction of the pyrazine ring. These products included **31** and **32**, the latter being obtained in 26% yield. When ethanol was used as a solvent, the main product was the bis-compound **33** which on heating at 135° for 16 hr was quantitatively converted into the pyrrolo compound **34**. The occurrence of decarboxylation during the ring-closure reaction prevented formation of the expected product **35**.

The only compound in this series to have been formed by synthesis of the pyrazine ring is the multisubstituted derivative **36**, formed from the appropriate diaminopyrrole and diacetyl.[14] Oxidation of this compound with chromium trioxide in acetic acid gave 2,3-dimethyl-5,6-dibenzoylpyrazine.

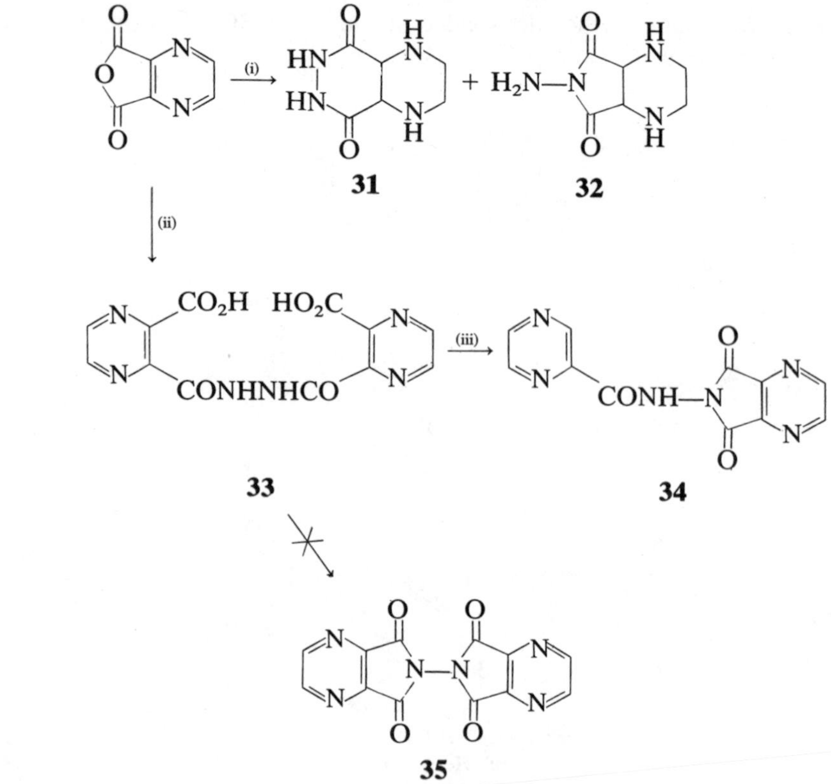

31 32

33 34

35

Reagents (i) N_2H_4, glacial HOAc
(ii) N_2H_4, EtOH
(iii) Heat

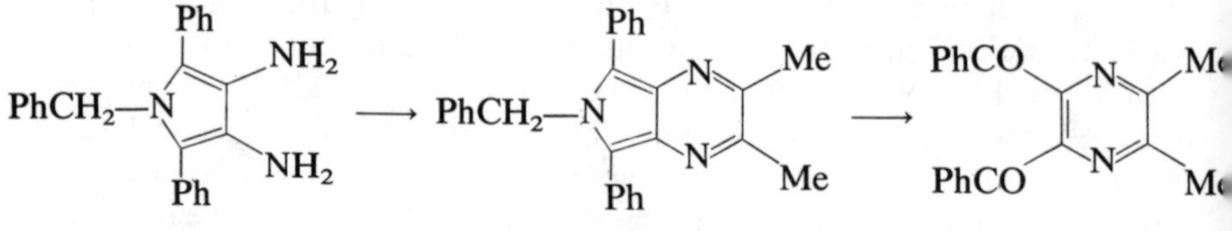

36

Catalytic hydrogenation of the imide **24** (R = H) over palladium on charcoal results in partial reduction of the pyrazine ring to give the tetrahydro compound **37**.[15] The anomalous products **31** and **32** from the reaction of the anhydride of pyrazine-2,3-dicarboxylic acid and hydrazine were presumed to have been formed by reduction of the pyrazine ring by diimide. This could have been formed in situ by oxidation of hydrazine with atmospheric oxygen.[13] In contrast, the pyrazine ring in compounds such as **38** is apparently unaffected by potassium borohydride. Reduction of imides such as **38** to the related hydroxy compounds such as **39** is effected by this reagent in aqueous dioxan at room temperature.[8]

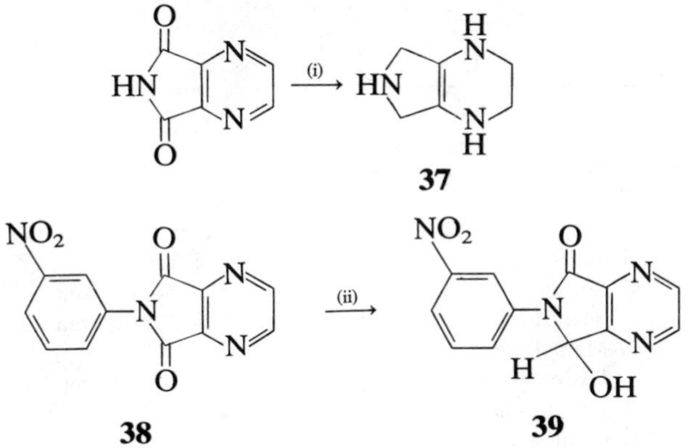

37

38 **39**

Reagents (i) H_2, Pd/C
(ii) KBH_4, aq. dioxan

Compounds exemplified by the piperazine derivative **40** have been patented as tranquilizers and anticonvulsants.[8] Certain 2,3-diamino-5,7-dioxo compounds are reported to be useful in forming thermally stable polymers.[16]

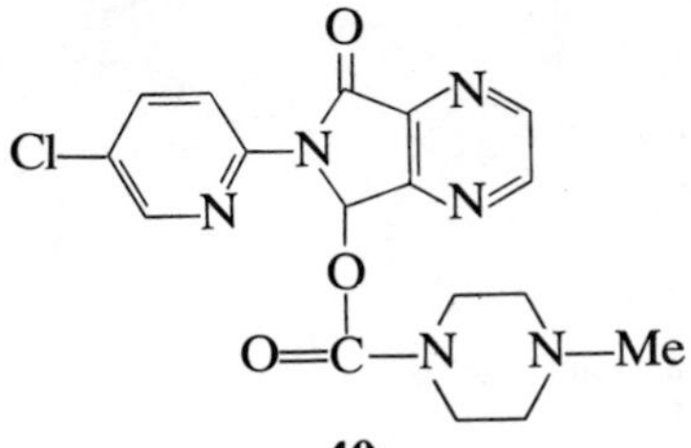

40

III. Tables

TABLE 1. 5*H*-PYRROLO[2,3-*b*]PYRAZINES

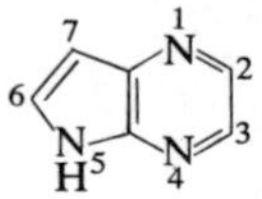

Molecular formula	Substituents or structure	m.p., °C	Ref.
$C_6H_4BrN_3$	7-Bromo	256–257	3, 19
$C_6H_4N_4O_2$	7-Nitro	>350	3, 19
$C_6H_5N_3$	Unsubstituted	155–156	1, 2, 3, 19
$C_7H_5N_3O$	7-Formyl	315–317 decomp.	4, 5
	Semicarbazone	315–325 decomp.	2
$C_7H_7N_3$	7-Methyl	221–222	1, 3, 5, 19
	Hydrochloride	231–235 decomp.	
$C_7H_7N_3O$	7-Hydroxymethyl	186–188	5
$C_8H_6N_4$	7-Cyanomethyl	221–223	5
$C_8H_6N_4O_2$	7-(2-Nitrovinyl)	245–247 decomp.	5
$C_8H_7N_3O$	7-Formyl-5-methyl	148–150	4
$C_8H_7N_3O_2$	7-Carboxymethyl	220–222	5
$C_8H_9N_3$	5,7-Dimethyl		
	Hydrochloride	217–219	5
	6,7-Dimethyl	191–193	1, 3, 19
$C_8H_9N_3O$	7-Hydroxymethyl-6-methyl	129–131	4
$C_8H_{10}N_4$	7-(2-Aminoethyl)		
	Sulfate	222–224	4, 5
$C_9H_9N_3O$	5-Acetyl-7-methyl	95–97	5
$C_9H_{10}N_4O_2$	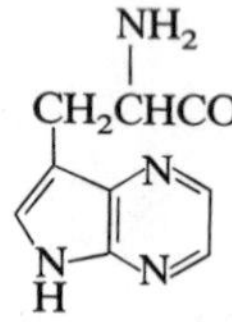 NH_2 / CH_2CHCO_2H (at C-7 of 5*H*-pyrrolo[2,3-*b*]pyrazine, N–H)	314–316 decomp.	4, 5
$C_9H_{12}N_4$	7-Dimethylaminomethyl	155–157	4, 5
$C_{10}H_{10}N_4O_4$	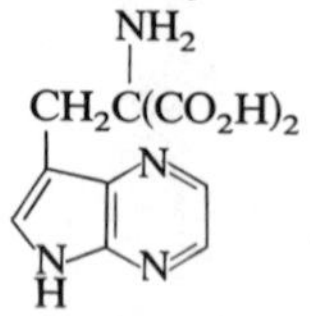 NH_2 / $CH_2C(CO_2H)_2$ (at C-7 of 5*H*-pyrrolo[2,3-*b*]pyrazine, N–H)		
	2′-Acetyl diethylester	165–167	4, 5
$C_{10}H_{11}N_3O$	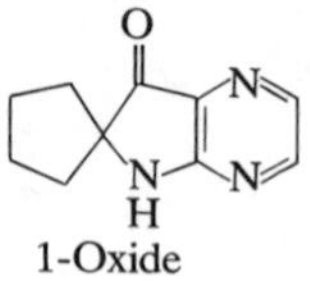 O, N, N, N–H (spiro structure) [a]		
	1-Oxide	217–218	1

TABLE 1 (*contd.*)

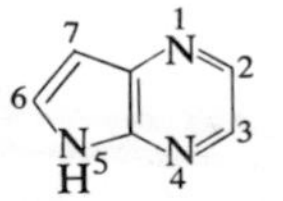

Molecular formula	Substituents or structure	m.p., °C	Ref.
$C_{10}H_{13}N_4$	$Me_2\overset{+}{N}{=}CH$–(7-position) 5-methyl-5H-pyrrolo[2,3-b]pyrazine (structure) Chloride	273–275	4
$C_{10}H_{14}N_4$	7-Dimethylaminomethyl-5-methyl Dihydrochloride	237–239	4
$C_{12}H_9N_3$	7-Phenyl	200.5–201.5	1, 3, 19
$C_{13}H_{11}N_3$	6-Methyl-7-phenyl	178.5–180	1, 3, 19
	7-Methyl-6-phenyl	193–194	1, 3, 19
$C_{16}H_{16}N_4$	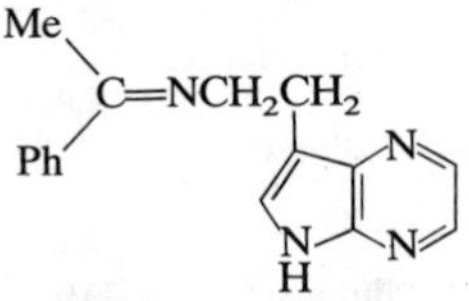 Me(Ph)C=NCH$_2$CH$_2$–(7-position) 5H-pyrrolo[2,3-b]pyrazine (structure)	213–215	4, 5
$C_{18}H_{13}N_3$	6,7-Diphenyl	240–241	1, 3, 19

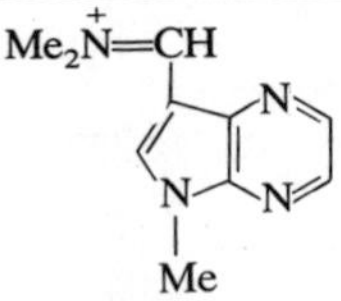

[a] Tentative structural assignment on the basis of spectral evidence. Obtained from *m*-chloroperoxybenzoic acid treatment of

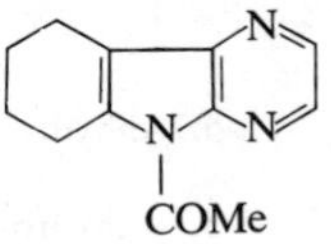

(structure: 5-acetyl-6,7,8,9-tetrahydro-5H-pyrazino[2,3-b]indole, N–COMe)

TABLE 2. 6*H*-PYRROLO[3,4-*b*]PYRAZINES

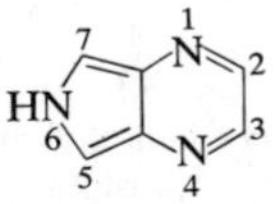

Molecular formula	Substituents or structure	m.p., °C	Ref.
$C_6H_3N_3O_2$	5,7-Dihydro-5,7-dioxo	245	7, 10
$C_6H_5N_5$	5,7-Dihydro-5,7-diimino	>320	12
$C_6H_7N_3O_2$	1,2,3,4,5,7-Hexahydro-5,7-dioxo	217–218 decomp.	15
$C_6H_{10}N_4O_2$	6-Aminoperhydro-5,7-dioxo	285 decomp.	13

TABLE 2 (*contd.*)

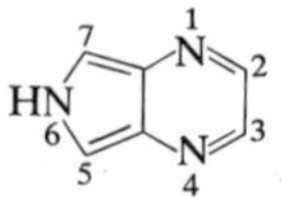

Molecular formula	Substituents or structure	m.p., °C	Ref.
$C_7H_5N_3O_2$	5,7-Dihydro-6-methyl-5,7-dioxo	183–184	7
$C_{11}H_5ClN_4O_2$	6-(5-Chloropyrid-2-yl)- 5,7-dihydro-5,7-dioxo[a]	237	8
$C_{11}H_6N_6O_3$	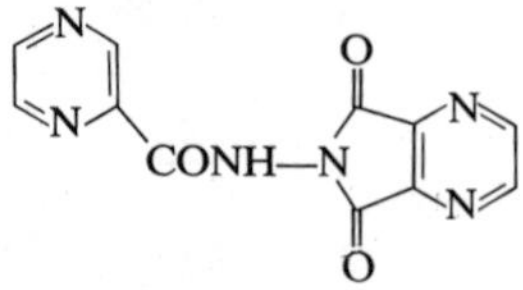	220	13
$C_{11}H_7ClN_4O_2$	6-(5-Chloropyrid-2-yl)-5,7-dihydro-5-hydroxy-7-oxo[b]	242	8
$C_{11}H_7N_5O_2$	5,7-Dihydro-6-(6-methylpyridazin-3-yl)-5,7-dioxo[a]	281	8
$C_{11}H_9N_5O_2$	5,7-Dihydro-5-hydroxy-6-(6-methylpyridazin-3-yl)-7-oxo[b]	268	8
$C_{12}H_6N_4O_4$	5,7-Dihydro-6-(3-nitrophenyl)-5,7-dioxo[a]	280	8
$C_{12}H_8N_4O_4$	5,7-Dihydro-5-hydroxy-6-(3-nitrophenyl)-7-oxo[b]	264	8
$C_{12}H_9N_5$	5,7-Dihydro-5-imino-7-(*N*-phenylimino)	219	18
$C_{14}H_9N_3O_4$	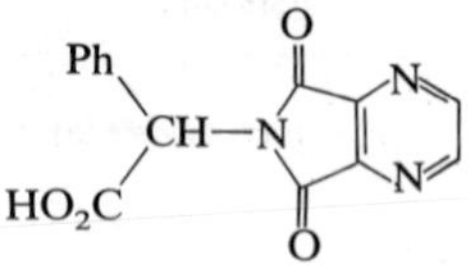 *dl* Form	215–218	9
$C_{15}H_7ClN_4O_2$	6-(7-Chloroquinolin-2-yl)-5,7-dihydro-5,7-dioxo[a]	253	8
$C_{15}H_9ClN_4O_2$	6-(7-Chloroquinolin-2-yl)-5,7-dihydro-5-hydroxy-7-oxo[b]	256–257	8
$C_{16}H_{15}N_3O_4$	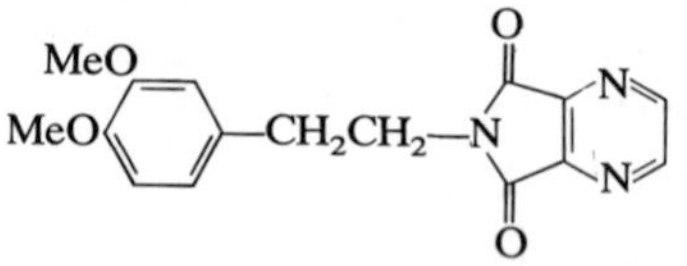	230	6
$C_{20}H_{17}N_5$	5,7-Dihydro-5,7-di[*N*-(4-methylphenylimino)]	178	18
$C_{27}H_{23}N_3$	6-Benzyl-2,3-dimethyl-5,7-diphenyl	163	14

[a] Reference 17 lists 12 other 6-aryl-5,7-dihydro-5,7-dioxo compounds.
[b] Phenyl carbonate and (4-methylpiperazin-1-yl)-carbonyloxy derivatives are also listed.

IV. References

1. B. A. J. Clark, Ph.D. Thesis, Brunel University, 1973.
2. V. A. Azimov and L. N. Yakhontov, *Khim. Geterotsikl. Soedin.*, 858 (1973); *Chem. Abstr.*, **79,** 105191 (1973). S. Ordzhonikidze (to Chem. Pharm. Res. Inst.), USSR Pat. 371228.
3. B. A. J. Clark, R. J. Dorgan, and J. Parrick, *Chem. Ind.* (London), 215 (1975).
4. S. Klutchko, H. V. Hansen, and R. I. Meltzer, *J. Org. Chem.*, **30,** 3454 (1965).
5. H. V. Hansen, S. Klutchko, and R. I. Meltzer, U.S. Pat. 3,382,245; *Chem. Abstr.*, **69,** 106744 (1968).
6. H. Von Dobeneck and B. Hansen, *Chem. Ber.*, **105,** 3611 (1972).
7. P. Hemmerich and S. Fallab, *Helv. Chim. Acta*, **41,** 498 (1958).
8. C. Cotrel, C. Jeanmart, and M. N. Messer (to Rhone-Poulenc S.A.), Ger. Offen. 2,300,491; *Chem. Abstr.*, **79,** 92284 (1973).
9. Beecham Group Ltd., Neth. Appl. 6,509,666; *Chem. Abstr.*, **65,** 3884 (1966).
10. S. Gabriel and A. Sonn, *Ber.*, **40,** 4856 (1907); S. Kushner, H. Dalalian, J. L. Sanjurjo, F. L. Bach, Jr., S. R. Safir, V. K. Smith, Jr., and J. H. Williams, *J. Am. Chem. Soc.*, **74,** 3617 (1952).
11. R. L. Jacobs (to Sherwin-Williams Co.), U.S. Pat, 3,732,224; *Chem. Abstr.*, **79,** 53369 (1973).
12. P. J. Brach, S. J. Grammatica, O. A. Ossana, and L. Weinberger, *J. Heterocycl. Chem.*, **7,** 1403 (1970).
13. D. B. Paul, *Austr. J. Chem.*, **27,** 1331 (1974).
14. V. Spiro, T. Aiello, and I. Fabra, *Ann. Chim.* (Rome), **56,** 866 (1966); *Chem. Abstr.*, **66,** 10902 (1967).
15. E. Felder, S. Maffei, S. Pietra, and D. Pitre, *Helv. Chim. Acta*, **43,** 888 (1960).
16. C. S. Marvel (to E. I. du Pont de Nemours and Co.), U.S. Pat. 3,736,299; *Chem. Abstr.*, **79,** 79500 (1973).
17. M. Augustin, M. Koehler, R. Harzer, G. Bernhard, and H. Brigsne, *Wiss. Z. Martin-Luther-Univ., Halle-Wittenberg, Math.-Naturwiss. Reihe*, **21,** 137 (1972); *Chem. Abstr.*, **77,** 75188 (1972).
18. M. Y. Kornilov and V. P. Makovetskii, *Ukr. Khim. Zh.*, **40,** 1328 (1974).
19. B. A. J. Clark, J. Parrick, and R. J. J. Dorgan, *J. Chem. Soc., Perkin I*, 1361 (1976).

CHAPTER XXI

Imidazopyrazines

I. Imidazo[1,2-*a*]pyrazines . . . 360
  1. Physical Properties . . . 360
  2. Methods of Preparation . . . 365
  3. Reactions . . . 370
II. Imidazo[1,5-*a*]pyrazines . . . 370
III. Imidazo[4,5-*b*]pyrazines . . . 375
IV. Tables . . . 382
  Table 1. Imidazo[1,2-*a*]pyrazines . . . 382
  Table 2. Imidazo[1,5-*a*]pyrazines . . . 385
  Table 3. 1*H*-2,3-Dihydro-2-oxoimidazo[4,5-*b*]pyrazines . . . 389
  Table 4. 1*H*-Imidazo[4,5-*b*]pyrazines . . . 394
V. References . . . 395

I. Imidazo[1,2-*a*]pyrazines

This ring system has also been called 1,3*a*,6-triazaindene and numbered as in formula **1**. A third alternative numbering system (**2**) has also been used on one occasion. Partly reduced derivatives of imidazo[1,2-*a*]pyrazine have been referred to as imidazo[1,2-*a*]piperazines and others, as imidazolino[1,2-*a*]pyrazines. The majority of papers on this heterocycle are of recent origin and concern the structure and oxidation products of *Cypridina* luciferin, a bioluminescent substance isolated from *Cypridina hilgendorfii*, a small crustacean.

1. Physical Properties

The electronic absorption spectra of indolizine and various azaderivatives, including imidazo[1,2-*a*]pyrazine, have been rationalized using SCF configuration interaction calculations. The results are in good agreement with experiment.[1] HMO calculations of the π-electron density at the 5-position in the heterocycle give a value close to that found for imidazo[1,2-*a*]pyridine.[2] Neither ring system undergoes a Dimroth-type rearrangement (**3** → **4**), which is presumed to occur by way of nucleophilic attack at the 5-position. The related imidazopyrimidines, however, readily rearrange at rates that reflect their much lower electron density at this position.

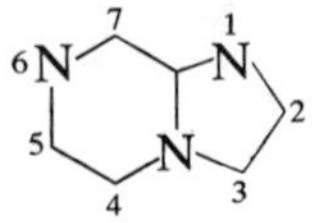

1 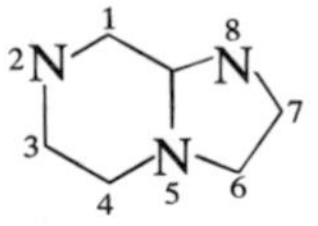 **2**

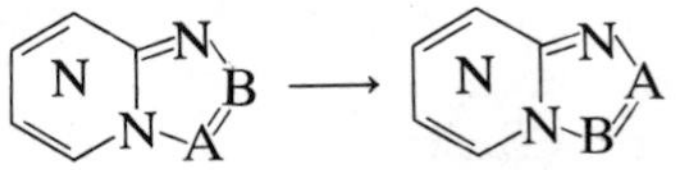

3 → **4**

The ultraviolet spectrum of the parent heterocycle in water shows three main peaks, at 221.5, 282, and 302 nm. The similarity of this spectrum to that of indolizine is apparently due to the fortuitous cancellation of the effects of the introduction of successive nitrogen atoms into indolizine.[3] UV data are available for several quaternary derivatives of this heterocycle and their 7-oxides.[4] 2,3-Dihydroimidazo[1,2-*a*]pyrazine (**5**) has only two peaks, at 258 and 394 nm, in its ultraviolet spectrum in water.[3] Both this compound and the unsubstituted aromatic heterocycle undergo large hypsochromic shifts on protonation. This contrasts with the bathochromic shift observed on protonation of aminopyrazines and is consistent with the hypsochromic effect obtained on protonation of the analogous imino compounds. These observations confirm the conclusion that protonation occurs at the 1-position in this heterocycle[3] to give the cation **6**.

The parent compound is a moderately weak base (pK_a 3.59), comparable with substituted aminopyrazines. The 2,3-dihydro compound **5**, however, is very much stronger (pK_a 9.32).[3] The base-weakening effect of the nitrogen in the 7-position is clearly demonstrated by comparison with the dihydroimidazopyridine (**7**). The latter compound has a pK_a of 12.51 units. Similar effects are found in the imidazopyridazine and imidazopyrimidine series. No evidence has been found for covalent hydration in any of the triazaindenes studied.[3]

The PMR spectrum of imidazo[1,2-*a*]pyrazine has recently been reported and discussed.[62] PMR data have been reported for the 2- and 3-methyl derivatives of the ring system.[2] The differences are sufficient to indicate that the two compounds do not interconvert by a Dimroth-type

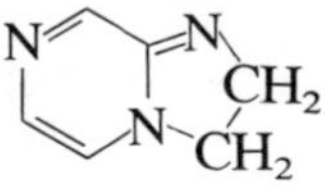

5

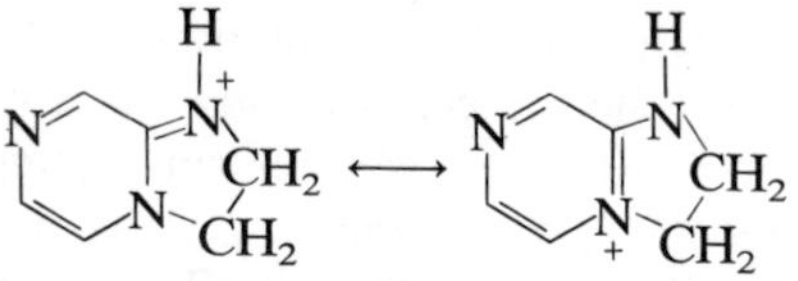

6

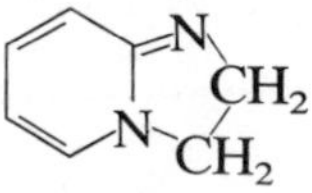

7

rearrangement in the presence of base. The similarities in the PMR, UV, and mass spectra of 3-hydroxyimidazo[1,2-*a*]pyrazine (**8**) and the 7-methyl compound **10** indicate that in neutral solution the oxo form **9** predominates instead of the hydroxy tautomer. In contrast, the 3-amino derivative is formulated as such.[5]

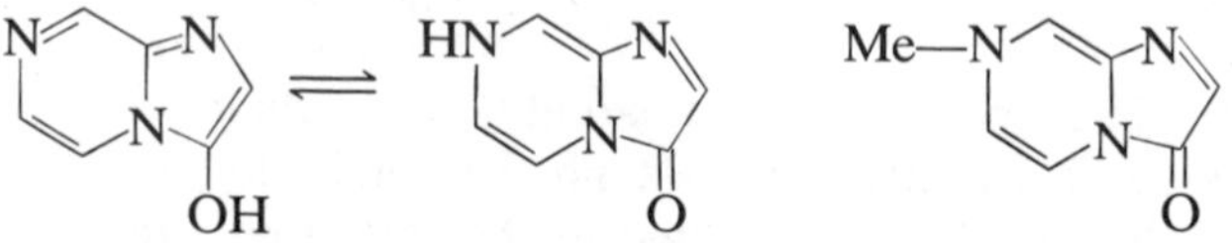

8

9

10

Recent PMR and UV studies[59] have identified the positions of protonation of compounds **11** and **15**. In methanol, protonation of **11** occurs initially at the oxo group to give the monocation **12**. In water, however, monoprotonation gives the 1-protonated cation **13**. In stronger acid the same dication (**14**) is formed in both solvents. In both methanol and water the 3-amino compound **15** gives the monocation **16**, and then the diprotonated species **17** is formed at greater acidities.

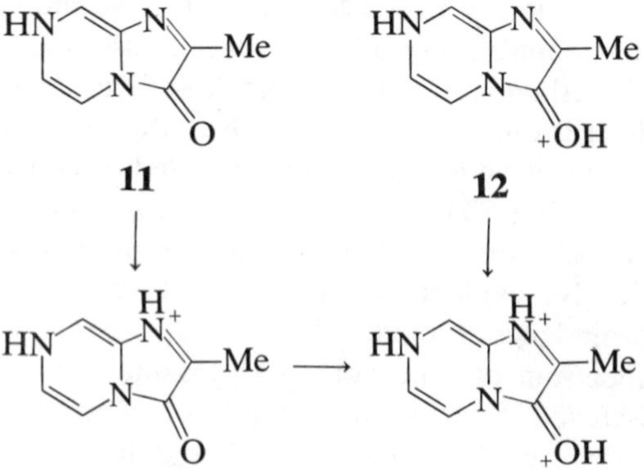

11

12

13

14

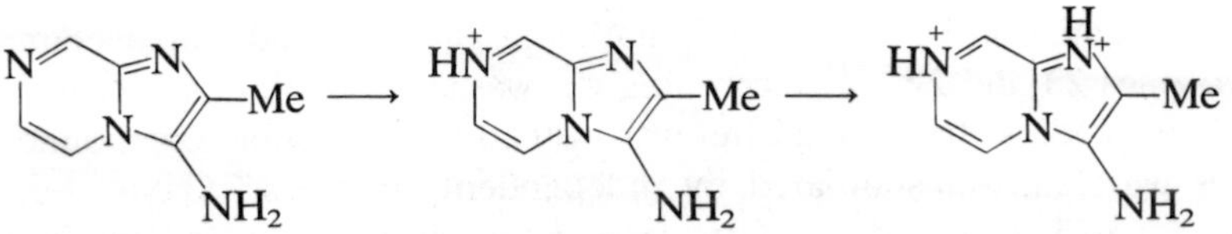

15 **16** **17**

Interestingly, crystalline salts of *Cypridina* luciferin (**18**) have been described as possessing two forms with different colors.[6–8] The color differences have been interpreted as indicating that the brilliant orange dihydrochloride has structure **19** and the faint brown dihydrobromide has structure **20**. In solution both salts are orange yellow, corresponding to the protonated forms **19** and **12**.

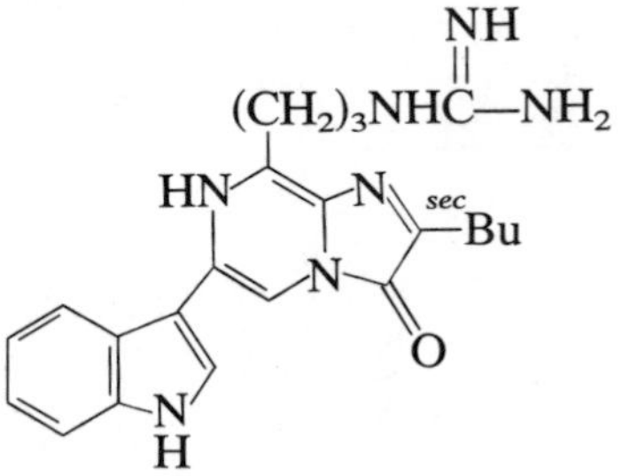

18

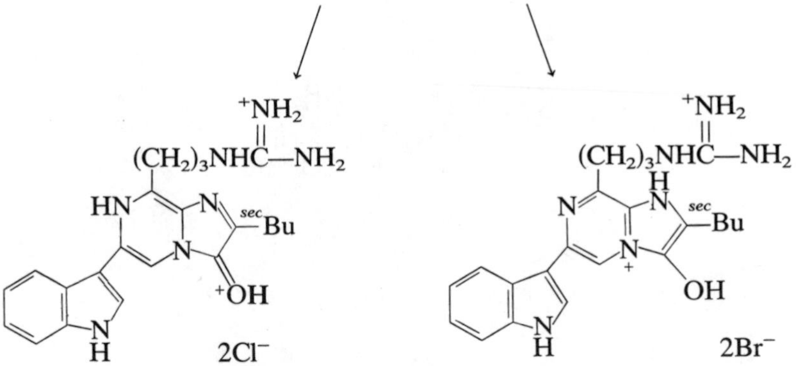

19 Orange

20 Brown

Cypridina luciferin, in the presence of oxygen and the enzyme *Cypridina* luciferase, is oxidized with emission of light to give products called oxyluciferin and etioluciferin.[6] On the basis of PMR and mass spectra the structure **21** was assigned to oxyluciferin.[6,7] This is now known to be incorrect.[9] Work on simple analogues of luciferin in aprotic solvents showed that oxidation of compounds such as **22** and **11** was accompanied by the emission of light (λ 455 nm) and the formation of acetamidopyrazines **23** and **24**.[10,11] Accordingly it was suggested[12] that oxyluciferin did not possess structure **21** but was instead the acylamino compound **25**. This was later substantiated by independent synthesis.[9] Oxyluciferin is first formed in an excited state from which light is emitted by fluorescence.

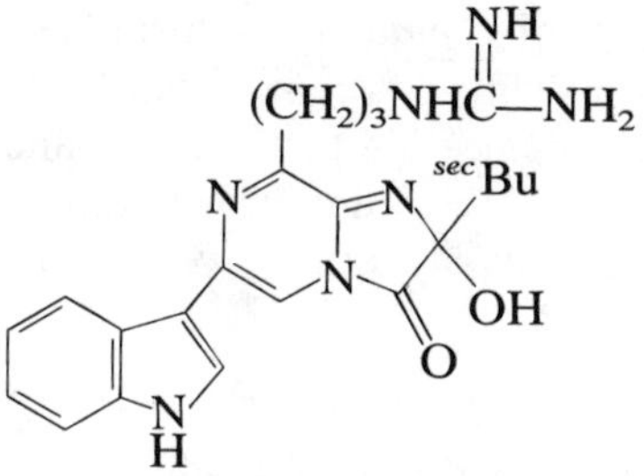

21

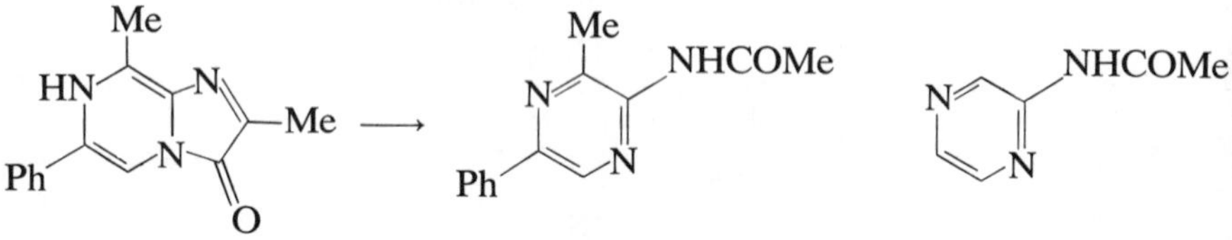

22 → **23** **24**

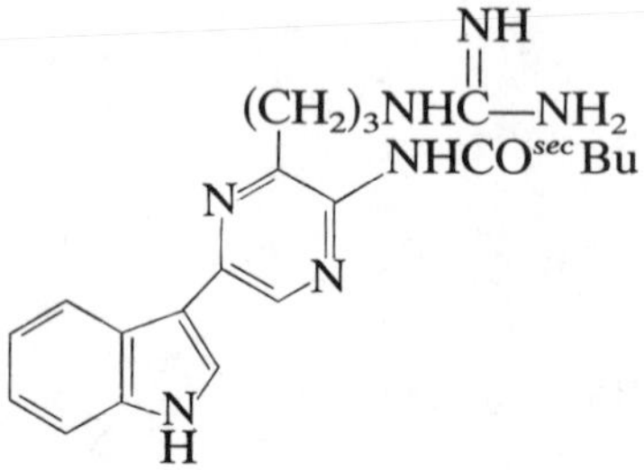

25

The structural requirements for chemiluminescence of luciferin analogues in aprotic solvents have been examined,[13] and the structure of the emitting species has been discussed.[12] It has also been demonstrated that the chemiluminescence of luciferin is markedly increased in micelle solutions.[14] The structural unit **26** is necessary for light production. The corresponding imidazo[1,2-*a*]pyridin-3-one was less luminescent than the fused pyrazine, and 3-aminoimidazo[1,2-*a*]pyrazines produced hardly any light. The final ambiguity in the structure of luciferin (**18**) has been recently resolved by an unambiguous synthesis which confirmed that the 3-indolyl group is in the 6-position, not in the 5-position, which could also have been possible.[15] The mechanism of luminescence of *Cypridina* luciferin has been revised.[16]

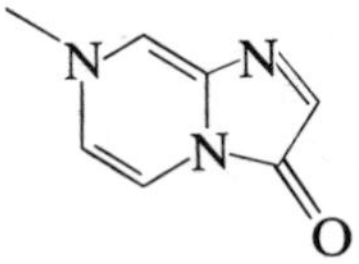

26

2. Methods of Preparation

The major methods for the preparation of this ring system involve formation of the imidazo ring, either by cyclization of appropriately substituted aminopyrazines or by the reaction of aminopyrazines with bifunctional reagents.

Although 2-aminoquinoxaline does not react with phenacyl bromide (**28**), treatment of the aminopyrazine **27** with this reagent in refluxing alcohol gave a 17% yield of the 2-phenyl compound **29**.[17] The orientation of the phenyl substituent was assigned by analogy with the product from the corresponding reaction of 2-aminopyridine. The reaction of 2-aminopyrazine with the appropriate α-halocarbonyl compounds similarly affords the methyl-substituted products **30** (from $ClCH_2COMe$) and **31** (from BrCHMeCHO) in only moderate yields.[2] The parent heterocycle has been obtained in low yield from cyclization of the aminoacetal **32** with concentrated sulfuric acid.[3] The aminoacetal **32** is itself formed in low yield from 2-chloropyrazine and aminoacetaldehyde diethylacetal. Recently imidazo[1,2-*a*]pyrazine has been obtained directly in 39% yield by reaction of 2-aminopyrazine with $ClCH_2CH(OEt)_2$ in dioxan–hydrochloric acid.[62]

Cyclization of the pyrazinylaminoacetic acids **33** affords 3-hydroxyimidazo[1,2-*a*]pyrazines, more properly regarded as the corresponding 3,7-dihydro-3-oxo derivatives. Thus treatment of the acids **33** (R = Me or Ph) with sodium alkoxide solutions gave the expected products **34** (R = Me or Ph).[18] Basic conditions are superior to treatment with acid[10] for this cyclization, and the corresponding esters and amides also provide the oxoimidazo compounds.[11,18] The acid **33** (R = H) has been cyclized to the corresponding oxo compound by treatment with

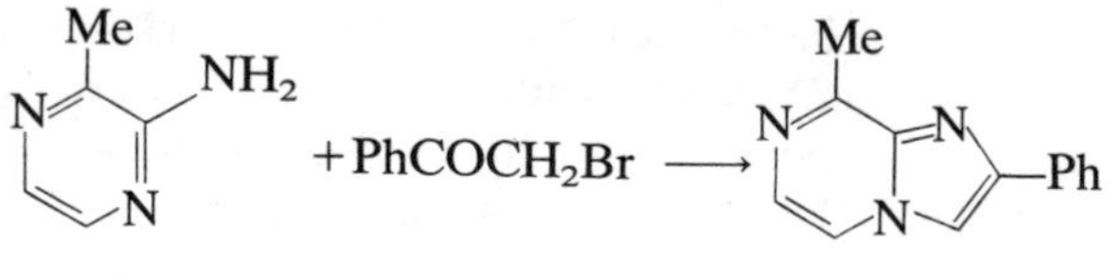

27 **28** **29**

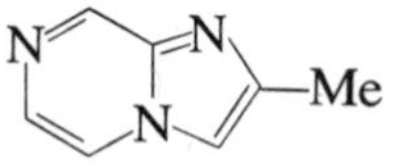

30

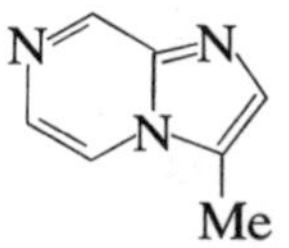

31

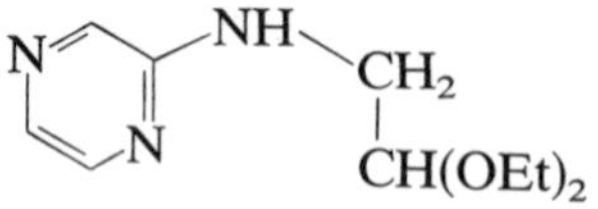

32

dicyclohexylcarbodiimide.[11] The related 3-amino derivative **35** was obtained by reaction of 2-aminopyrazine with a mixture of sodium cyanide and formaldehyde as its bisulfite compound.[11]

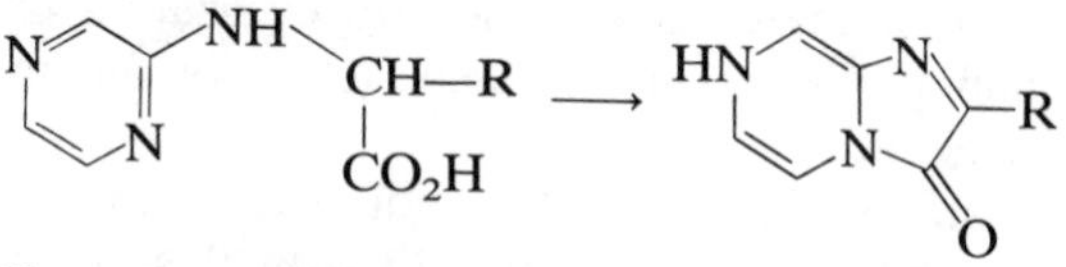

33 R = H, Me, or Ph **34**

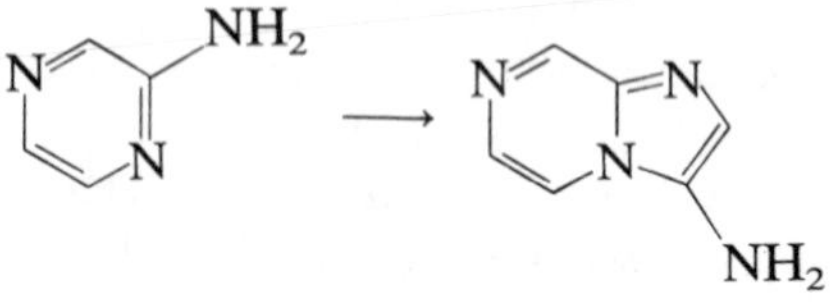

35

The initial syntheses of *Cypridina* luciferin and its analogues were performed in low yields by reaction of appropriate 2-aminopyrazines with α-keto acids, followed by reduction with aluminum amalgam or catalytic hydrogenation, and treatment of the product with dicyclohexylcarbodiimide.[19–21] For example, 2-amino-5-phenylpyrazine (**36**) on reaction with pyruvic acid gave the product **37**, which was reduced to give an intermediate formulated as **38**, which was then cyclized to give the oxo compound **39** in 7% yield.[21] It was later discovered that these products could be obtained in high yield in one step by reaction of aminopyrazines such as **36** with α-keto aldehydes such as pyruvaldehyde (MeCOCHO).[21,22] Condensation of the appropriate aminopyrazine with

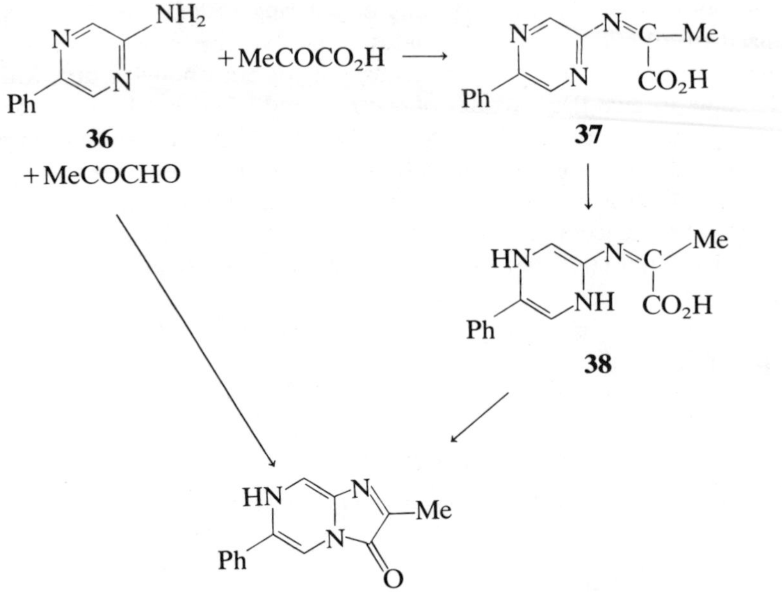

EtMeCHCOCHO in aqueous acid gave the luciferin from *Cypridina* in 80% yield,[21] which contrasts with the yield of <1% from the multistage synthesis.

Treatment of 2-(2-hydroxyethylamino)pyrazine (**40**) with thionyl chloride, followed by cyclization of the resultant chloro compound in boiling ethanol, gave a considerable amount of tar together with 6.5% of the 2,3-dihydroimidazo compound (**41**).[3] The 5,6-diphenyl analogue of **41** has been similarly prepared in undisclosed yield.[23] The tetrahydro derivative **44** was obtained when the tetrahydropyrazine **42** was heated in ethanolic hydrochloric acid with α-amino-α-cyanoacetamide (**43**).[24]

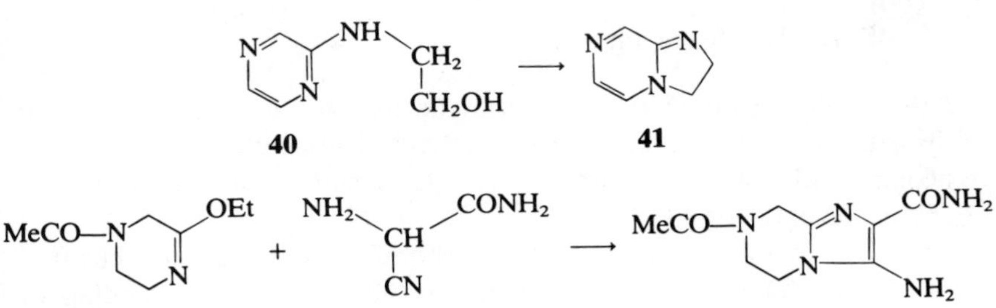

Various 1-alkyl quaternary salts of the ring system have been synthesized from imidazole intermediates.[4] The pyrazine ring was formed by quaternization, followed by condensation of a carbonyl group with an aldoxime to give the 7-oxides of the required products. Two approaches were used. Quaternization of the oximes **45** with α-bromocarbonyl compounds gave the salts **46**, which cyclized in strong acid to give the 7-oxides **47**. The second procedure involved quaternization of the acetals **48** with bromoacetaldehyde oxime. The resultant quaternary salts **49** on treatment with concentrated sulfuric acid gave the required 7-oxides **47** ($R^1 = H$).

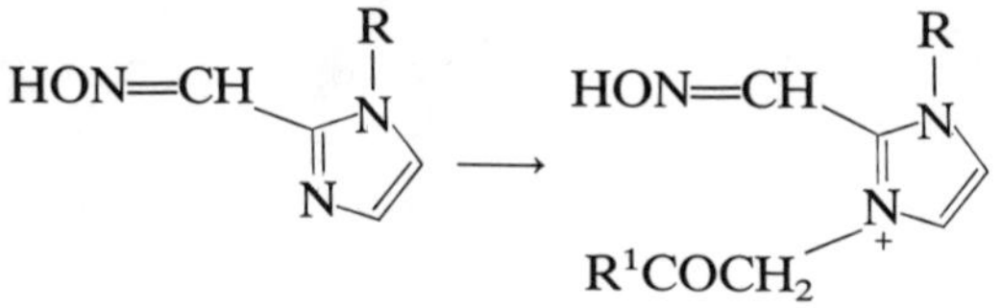

45 $R = Me$ or $PhCH_2$ **46** $R^1 = H$, Me or Ph

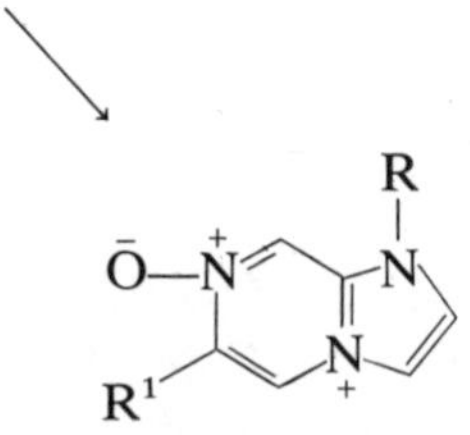

47

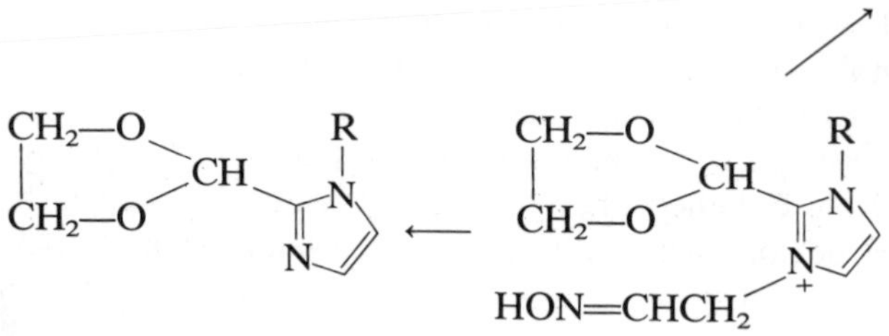

48 $R = H$, Me, or Ph **49**

2-Benzoylimidazole (**50**) has been made to react as its sodium salt in DMF with the bromoacetal **51**. The resultant disubstituted product **52** on refluxing with ammonium acetate in acetic acid afforded 8-phenylimidazo[1,2-*a*]pyrazine (**53**).[25]

Reductive removal of the benzyloxycarbonyl group on the side chain of the oxazolone **54** gives the unstable amino compound **55**. On standing, or more rapidly by treatment with acetic anhydride, the oxazolone gave a compound originally formulated as the imidazopyrazine **56**,[26] which was

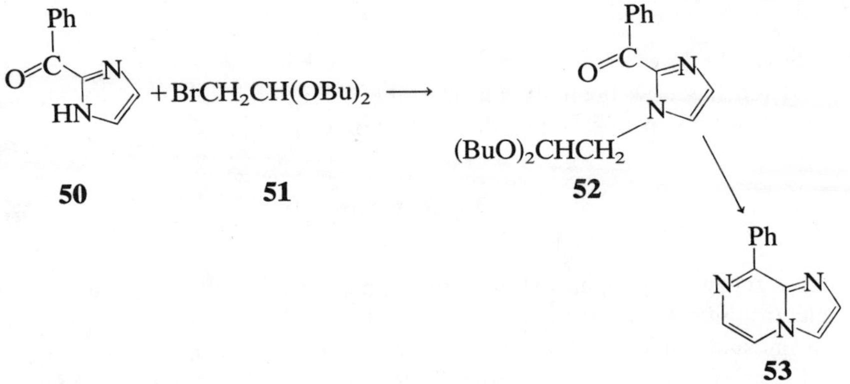

50 51 52 53

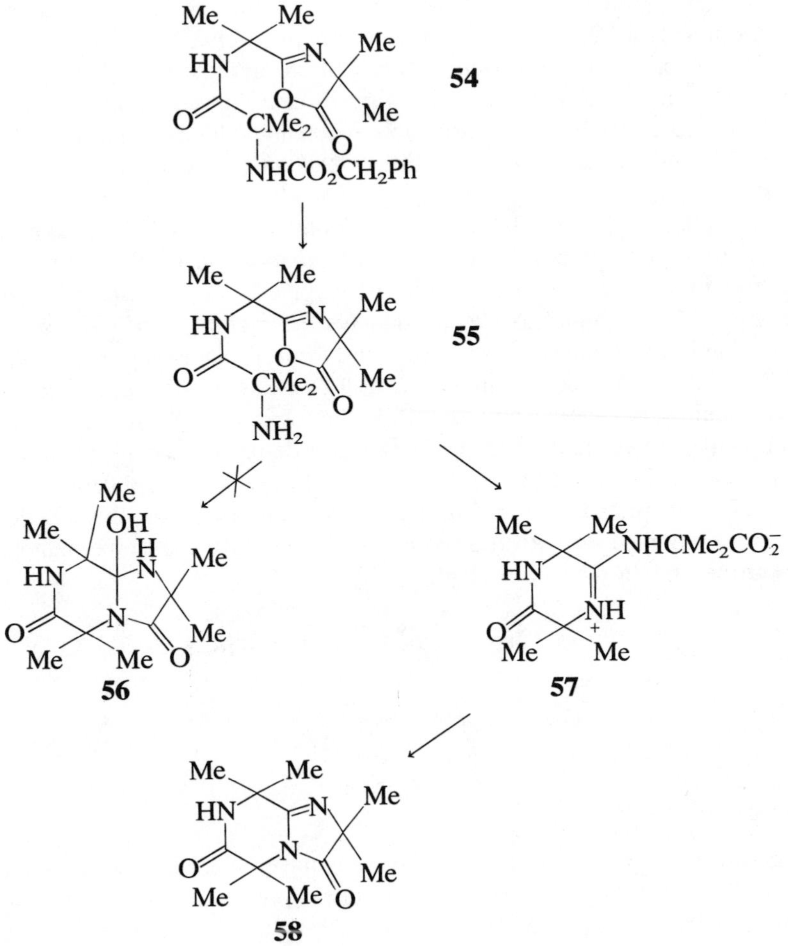

54 55 56 57 58

thought to lose water to give the product **58**. The mechanism of this reaction has now been studied in some detail, and the structure **56** has been withdrawn in favor of the zwitterionic structure **57**.[27]

3. Reactions

A recent paper has shown that imidazo[1,2-*a*]pyrazine undergoes electrophilic substitution, as expected, at the 3-position. Thus *N*-bromosuccinimide afforded a 72% yield of 3-bromoimidazo[1,2-*a*]pyrazine from the parent heterocycle. Base-catalyzed proton–deuterium exchange was shown to occur at the 3- and 5-positions (PMR evidence). On heating with phosphorus pentachloride at 265° in a sealed tube, imidazo[1,2-*a*]pyrazine yielded its perchloro derivative in 55% yield. Reaction of this product with sodium methoxide resulted in nucleophilic substitution at the 5- and 8-positions. The remaining chlorines in the molecule proved to be removable by catalytic hydrogenation.[62] Apart from this study there has been little work on the reactions of this ring system.

No reports of *N*-oxidation are available; all the oxides known have been prepared directly by cyclization.[4] Acetylation of the 3-oxo compound **11** occurred at the 7-position[5] to give the compound **59**. Treatment of **11** with phosphoryl chloride gave the 3-chloro derivative **60**. The chlorine in this compound was apparently not displaceable.[5] Oxidation of 3,7-dihydro-3-oxo compounds such as **11** results in the formation of the ring-opened derivative and the emission of light (see above). Various substituted quaternary 7-oxides have been deoxygenated to give quaternary derivatives of the parent heterocycle. Refluxing phosphorus trichloride or phosphorus tribromide has been used for this purpose.[4] Condensation of **44** with triethyl orthoformate gave the "piperazino[1,2-*e*]purine" **61** in moderate yield.[24]

II. Imidazo[1,5-*a*]pyrazines

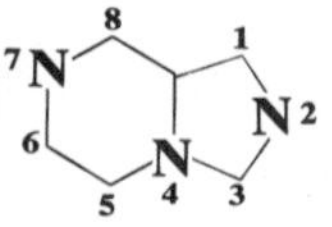

RRI 1184

Saturated derivatives of this heterocycle have been referred to as 1,4,8-triazabicyclo[4.3.0]nonanes (**62**). The ultraviolet spectrum of the parent aromatic ring system **63** has three bands, at 269, 280, and 331 nm. The

PMR spectrum (in $CDCl_3$) of imidazo[1,5-*a*]pyrazine and its 7-oxide are give in Figure 1. Ultraviolet and PMR details have also been reported for the 3-methyl,[28] the 8-amino,[29] and the 3-(2-pyrazinyl)[30] compounds. More recently PMR data have been published for a variety of 3-substituted compounds.[60] Several substituted imidazo[1,5-*a*]pyrazines have been studied using ^{13}C NMR spectroscopy.[64]

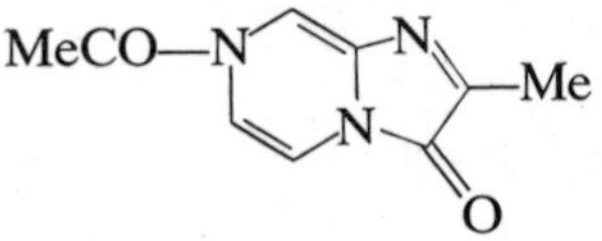

MeCO—N ... Me, O

59

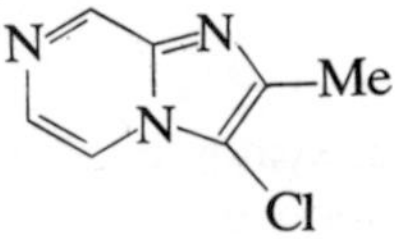

N ... Me, Cl

60

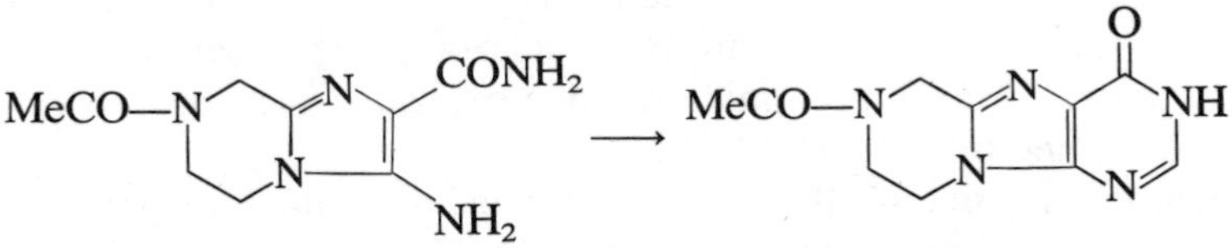

MeCO—N ... $CONH_2$, NH_2 → MeCO—N ... O, NH, N

44 **61**

4 N, 5, 6, 7, N 8, 3, N 1, 2, 9

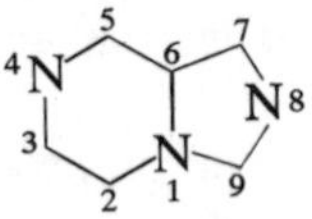

62

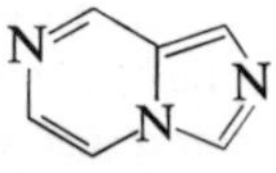

63

The 8-amino-3-mercapto compound **64** is probably more correctly formulated as the thioxo tautomer **65**. The compound behaves as a weak acid (pK_a 8.74).[29] The desulfurized compound **66** is a moderately strong base (pK_a 6.55), apparently protonated on the ring nitrogen atom in the 2-position rather than on the 8-amino group.

Figure 1. PMR SPECTRA OF IMIDAZO[1,5-*a*]PYRAZINE AND ITS 7-OXIDE.[28]

	Chemical shift (δ)					Coupling constants, Hz	
	H-1	H-3	H-5	H-6	H-8	$J_{5,6}$	$J_{6,8}$
Parent	7.83*s*	8.28*s*	7.58*d*	7.91*d*	9.03*s*	5	
7-Oxide	7.63*s*	8.25*s*	7.97*m*	7.38*q*	8.56*m*	5	1

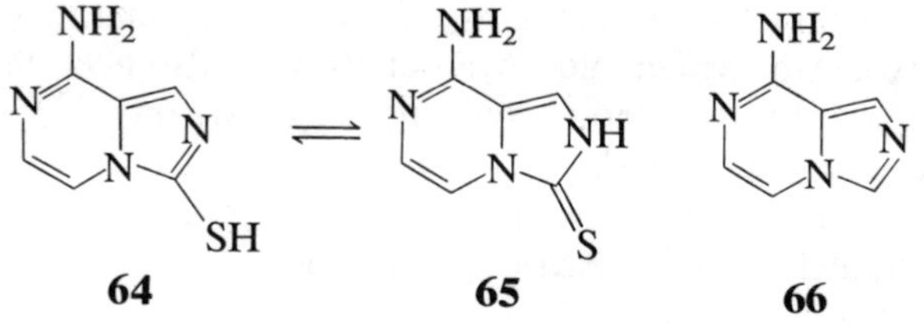

64 65 66

The most useful synthesis of aromatic derivatives of this ring system involves condensation of aminomethylpyrazines **67** with anhydrides or carboxylic acids. Thus the amine **67** ($R^1 = H$) was made to react with formic acid or acetic anhydride to give amides which were cyclized to the 3-substituted compounds **68**.[28] Clearly this synthesis could be extended to give a wide variety of substituted imidazo[1,5-*a*]pyrazines. Recently several more 3-substituted imidazo[1,5-*a*]pyrazines have been prepared from the amine **67** ($R^1 = H$).[60,61,66]

When the diamine **67** ($R^1 = NH_2$) is treated with, for example, triethyl orthoacetate, the products are usually 3,4-dihydropteridines. However, when the diamine is heated with carbon disulfide in pyridine, none of the expected product is formed. Instead the 8-aminoimidazo[1,5-*a*]pyrazine **65** is formed in 90% yield.[29] 2,3-Dihydro-3-thioxoimidazo[1,5-*a*]pyrazine may be prepared in excellent yield from the aminomethylpyrazine **67** ($R^1 = H$) using this method.[66]

The pyrazinyl compound **69** was obtained as a by-product (10%)

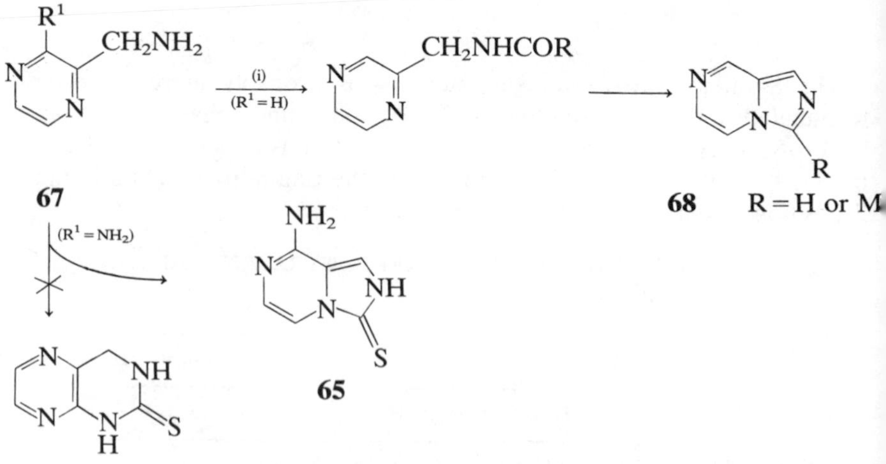

67 68 R = H or Me

65

Reagents (i) HCO_2H or $(MeCO)_2O$
(ii) CS_2

during the ammonium chloride-catalyzed preparation of pyrazinecarboxaldehyde dimethylacetal from pyrazinecarboxaldehyde and methyl orthoformate.[30]

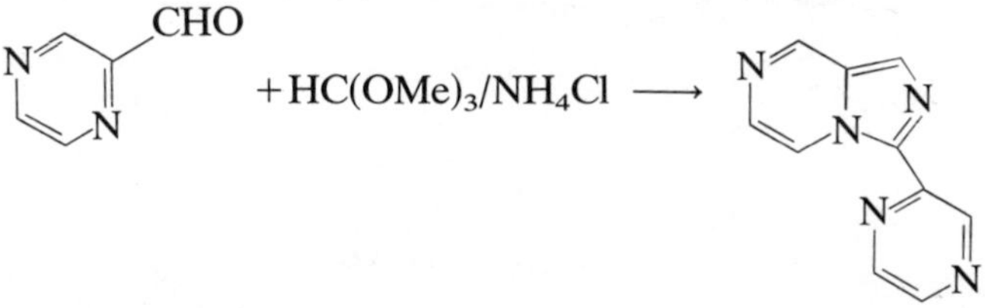

CHO N N + $HC(OMe)_3/NH_4Cl$ ⟶ N N N N N

69

Prolonged heating of 7-theophyllineacetic acid (**70**) with barium hydroxide,[31] or treatment of the acid with aqueous sodium hydroxide solution at room temperature,[32] results in fission of the pyrimidine ring and recyclization to give the imidazo[1,5-*a*]pyrazine **71**.

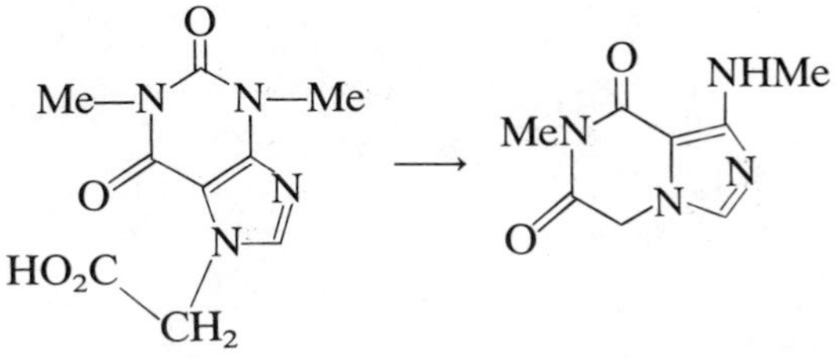

O Me—N N—Me O N N HO_2C CH_2 ⟶ O NHMe MeN N N O

71

Treatment of the piperazine ester **72** with phenyl isocyanate, followed by refluxing with hydrochloric acid, gave the perhydrodioxo derivative **73**.[33] Other perhydro derivatives (**74**) have been obtained by cyclization of anilinomethylpiperazines with aldehydes.[34,35]

The 1-methylamino derivative **71** behaves like a typical secondary amine, forming a benzoyl derivative with benzoyl chloride and a *N*-nitroso compound with nitrous acid.[32] Alkaline hydrolysis results in ring opening to give caffeidine (**75**). The parent heterocycle and its 3-methyl derivative form 7-oxides on treatment with *m*-chloroperbenzoic acid.[28] The 3-thioxo compound **65** is readily desulfurized by Raney nickel in ammonium hydroxide solution.[29]

Recent work on this ring system has been concerned with reactions with electrophilic reagents.[64-66] Chlorination of imidazo[1,5-*a*]pyrazine with *N*-chlorosuccinimide in carbon tetrachloride produces a complex mixture containing the 1-chloro, 3-chloro, 1,3-dichloro, and (probably) 1,3,5-trichloro derivatives. Under these conditions the 3-methyl compound yields a mixture of the 1-chloro and 1,5-dichloro-3-methylimidazo[1,5-*a*]pyrazines. At 0° bromine in carbon tetrachloride

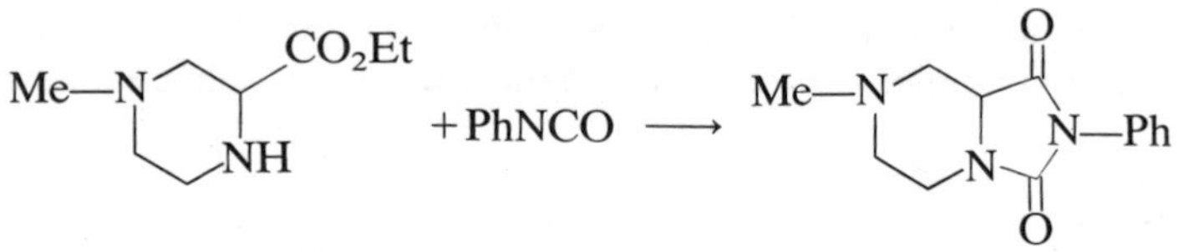

72 + PhNCO ⟶ 73

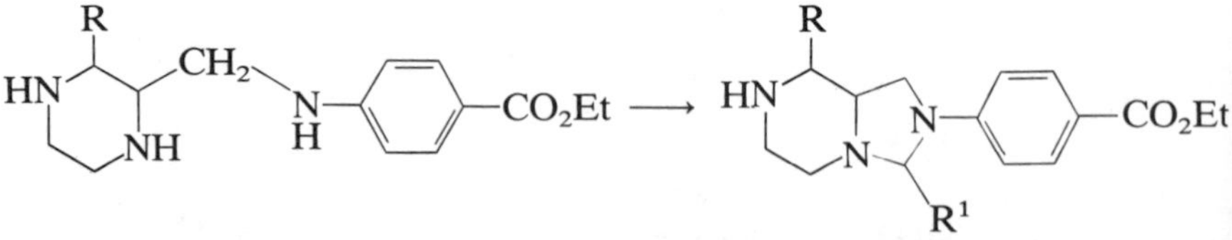

R = H or Me ⟶ 74 R^1 = H, Ph, or

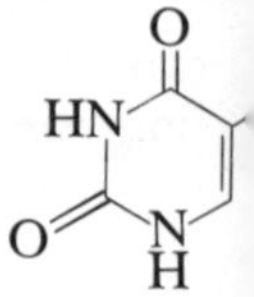

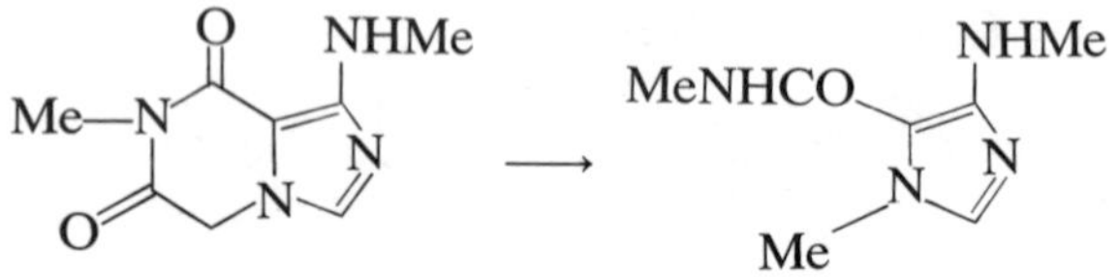

71 ⟶ 75

gives 1-bromo-3-methylimidazo[1,5-*a*]pyrazine (**76b**), but at higher temperatures an unseparable mixture of polybrominated products was obtained. The Mannich reaction on the 3-methyl compound gives the expected product **76a** in 38% yield. The halogen in compounds **76b** and **76c** could not be displaced by a variety of nucleophiles, and attempts to convert the monobromo compound **76b** into a dibromo derivative failed when using bromine water followed by an excess of dilute alkali. The only product isolated was the imidazole **77**, presumably formed by addition of water across the 7,8-double bond and alkaline decomposition of the product.[64]

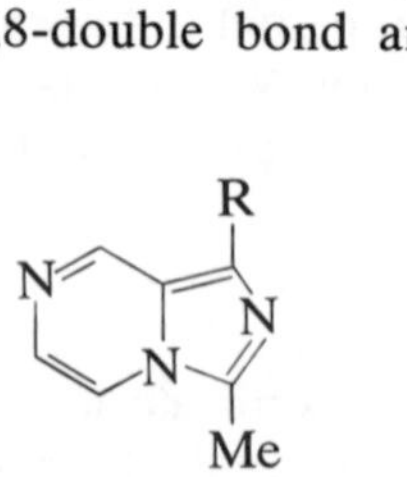

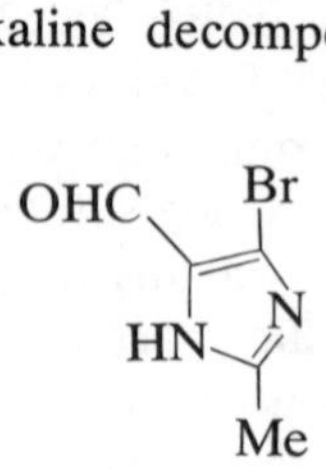

76a $R = CH_2NMe_2$
76b R = Br
76c R = Cl

77

Addition across the 7,8-double bond proved a convenient route to a series of adducts such as **78**. Treatment of imidazo[1,5-*a*]pyrazine and its 3-methyl derivative with acid chlorides and acid anhydrides resulted in addition at the 7,8-double bond in a manner analogous to the Reissert reaction. Variation of the quenching agent as well as the acylating agent resulted in a series of 7,8-dihydro-7,8-disubstituted compounds.[65]

Alkylation of 2,3-dihydro-3-thioxoimidazo[1,5-*a*]pyrazine with a variety of alkylating agents provided the expected 3-alkylthio compounds. The 3-methylthio and 3-ethylthio derivatives were easily oxidized by standard methods to the corresponding sulfoxides and sulfones. Both the thioether and sulfone groups could not be displaced by nucleophiles. 3-Methylthioimidazo[1,5-*a*]pyrazine was treated with a variety of electrophilic reagents. Nitrosonium fluoroborate provided the 1-nitro-3-methylthio derivative, whereas aqueous nitric acid resulted in pyrazine ring cleavage. Reaction with ethyl chloroformate followed by quenching with ethanol gave the 3-methylthio analogue of **78**. In contrast, the 2,3-dihydro compound **79** was formed from reaction with *p*-nitrobenzoyl chloride followed by quenching with water.[66]

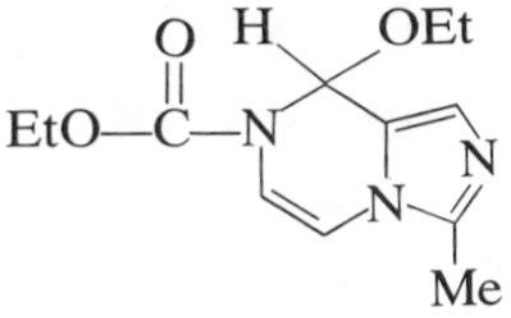

O H OEt
EtO—C—N N
N
Me

78

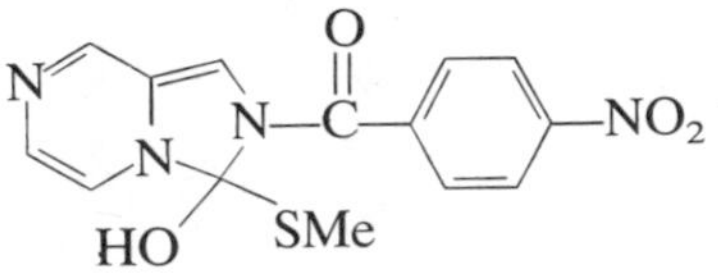

O
N N—C—(C₆H₄)—NO_2
N
HO SMe

79

Several perhydro derivatives (**74**) of the ring system were synthesized as analogues of tetrahydrofolic acid.[34,35] They were not as effective as aminopterin as inhibitors of either thymidylate synthetase or dihydrofolate reductase. They are therefore presumably of little interest as anticancer agents[34] and were found to have no inhibitory effect on the growth of various bacteria.[35]

III. Imidazo[4,5-*b*]pyrazines

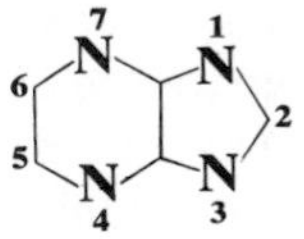

7 1
6 N N
2
5 N N
4 3

1*H* RRI 1176

The UV spectrum of the 1-methyl compound **80** in methanol shows a maximum at 290 nm with a shoulder at 255 nm, and the 1-methyl-4-oxide **81** has two maxima, at 227 and 304 nm. Comparison of these spectra with those of more complex derivatives has been used to substantiate both the positions of substitution and the site of *N*-oxidation in the ribonucleosides of imidazo[4,5-*b*]pyrazines.[36] Ultraviolet data have also been published for several 5,6-diphenyl compounds together with details of their x-ray powder diffraction patterns.[23] X-Ray crystallography has also been used to determine the crystal structure of the inosine analogue 1-(β-D-ribofuranosyl)imidazo[4,5-*b*]pyrazine 4-oxide (**82**).[36] PMR details have been reported for compounds **80** and **81**[36] as well as for more complex derivatives of the ring system.[36,37] The IR spectrum in chloroform of the "2-hydroxy" derivative **83** exhibits bands at 3441 cm^{-1} (ν N—H stretch) and 1752 cm^{-1} (ν C=O stretch) which confirm that the compound is predominantly in the amide form (**84**) analogous to benzimidazolone.[38]

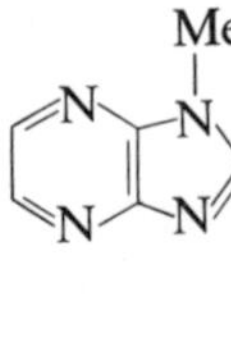

80 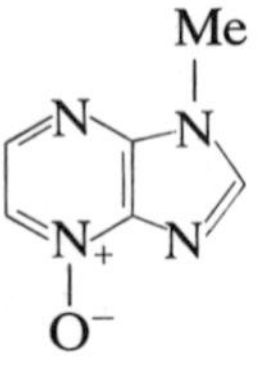 **81** 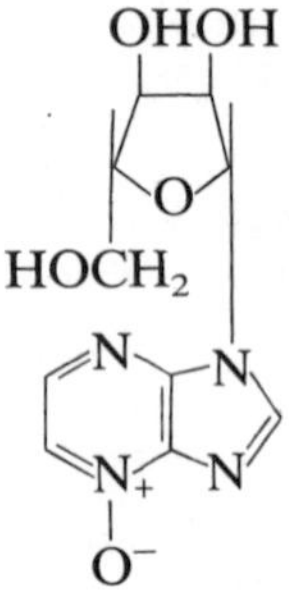 **82**

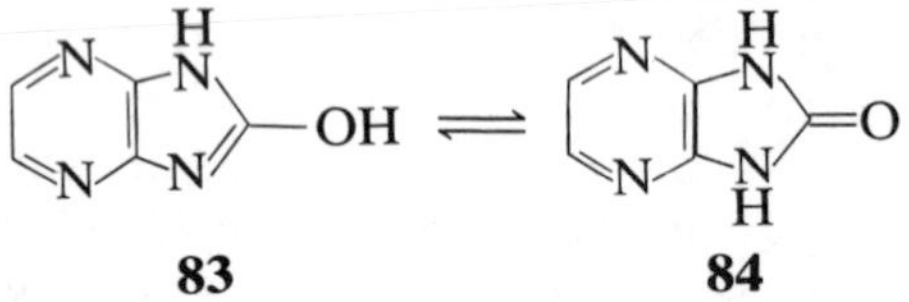

83 **84**

In general, derivatives of this ring system have been synthesized by formation of the imidazole ring. Only two examples have been reported of syntheses starting from imidazole derivatives.[37,54] The earliest syntheses utilized 2,3-diaminopyrazines as intermediates,[39,40] but more recently alternative approaches have been examined, particularly for "2-hydroxy" derivatives, as the preparation of diaminopyrazines is tedious.[41]

The reaction of 2,3-diaminopyrazine (**85**, R = H)[40] or its 5,6-diphenyl[23] or dichloro[42] derivatives with acetic anhydride gives the corresponding 2-methylimidazo[4,5-*b*]pyrazine (**86**, R^1 = Me). Similarly, acid chlorides[39] or carboxylic acids[40] afford 2-substituted compounds. The 2-unsubstituted derivatives are formed when the diaminopyrazines are reacted with triethyl orthoformate.[23,39,40] This reagent with the substituted *N*-oxide **87** provides the 1-substituted 4-oxide **81**.[36] Other pyrazines possessing substituted amino groups have been used as intermediates to 1-substituted[36] or 1,2-disubstituted imidazo[4,5-*b*]-pyrazines.[43,44] Thus reaction of the phenethylamino compound **88** with propionic anhydride at 130–140° resulted in the formation of the disubstituted derivative **89**.[43]

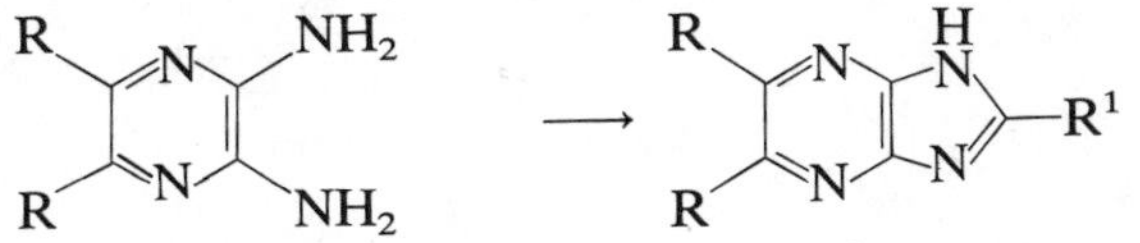

85 R = H, Me, Ph, or Cl → **86** R^1 = H, alkyl, or phenyl

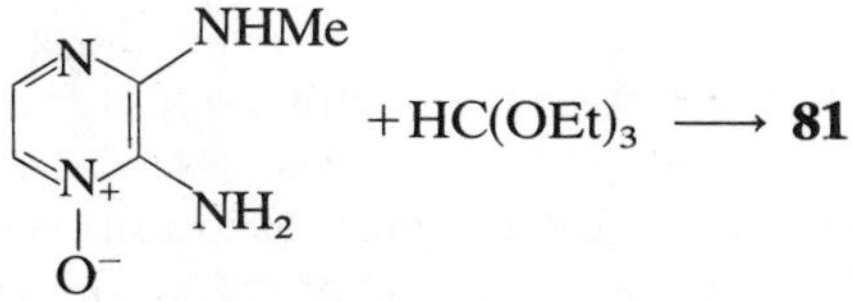

87 + $HC(OEt)_3$ ⟶ **81**

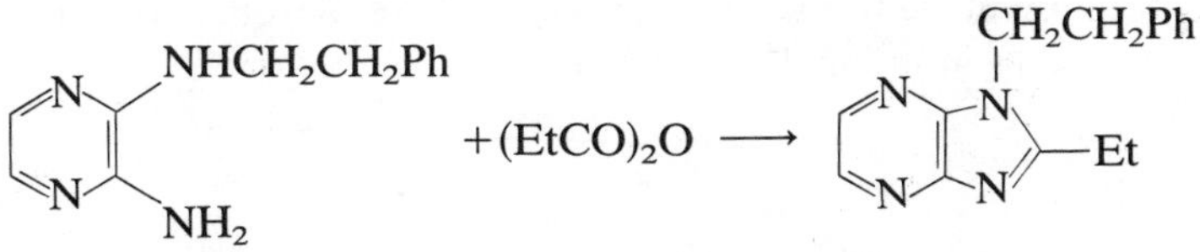

88 + $(EtCO)_2O$ ⟶ **89**

The reaction of diaminopyrazines **85** with urea gives the 2-oxo derivatives such as the parent **84**.[40,23,39] Because of the great interest in 2-oxoimidazo[4,5-*b*]pyrazines as antihypertensives (see below), alternative routes to these compounds have been developed. Treatment of 2,3-diaminopyrazines with a wide variety of reagents has been shown to give the 2-oxo compounds.[45] Typical reagents that may be used for this purpose are phosgene, diethyl carbonate, potassium cyanate, carbamoyl chlorides, and methyl chloroformate. The observation[41] that the hydroxamic acid **90**, on treatment with benzenesulfonyl chloride, rearranges to the 2-oxo compound **84** has been extended to the synthesis of a large number of such compounds.[46] The rearrangement is carried out under basic conditions, and it is possible, although not necessary, to isolate the intermediate benzenesulfonates **91**. Various other sulfonyl chlorides and carboxylic acid chlorides have also been used in this rearrangement. Simple fusion of the hydroxamic acid **92** results in the formation of the corresponding 2-oxo compound **93**.[46]

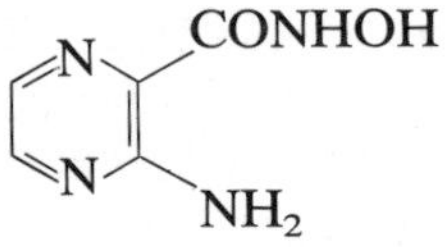

90 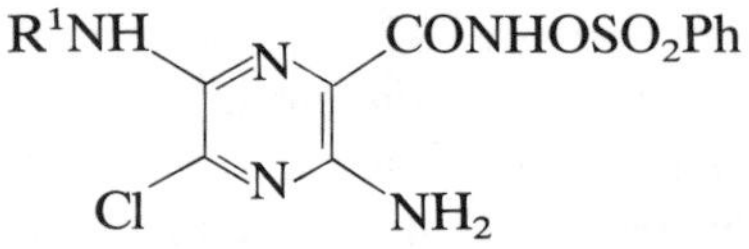 **91**

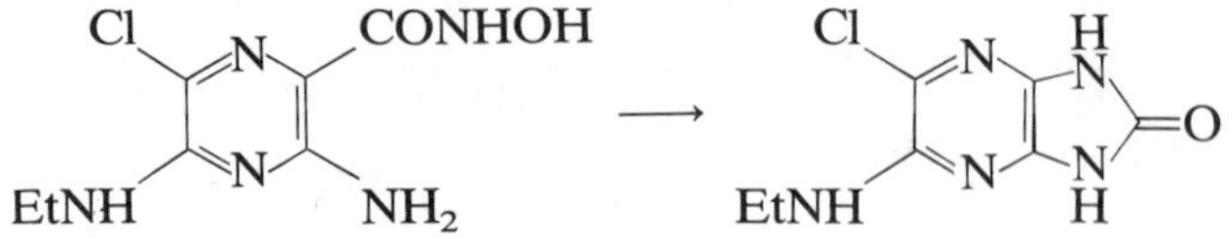

92 → **93**

The Curtius rearrangement of the acyl azides **95**, obtained by diazotization of the hydrazides **94**, also gives rise to 2-oxoimidazo[4,5-*b*]pyrazines **96** in good yields.[47] The rearrangements are preferably carried out by heating the azide in 2-methoxyethanol. Basic side chains in the more complex derivatives are protected during the diazotization stage by protonation or acylation. Various modifications of this approach are covered by a group of patents.[48–51] Such variations include using isoamyl nitrite[50] or sodium nitrite in hydrochloric acid[48,49] for diazotization, and pyrolysis[51] or refluxing in ethanol[49] to effect rearrangement.

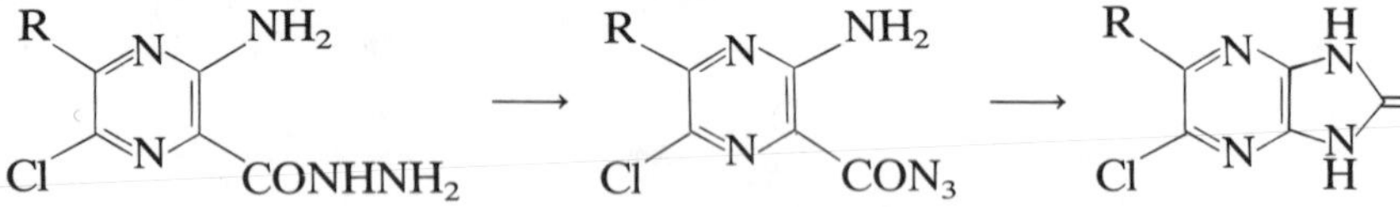

94 → **95** → **96**

R = H, SH, NH_2, substituted amino, or substituted thio

Other routes to 2-oxo compounds have also been patented. The action of heat on the derivatives **97**[52] and **98**[53] provides the required imidazo[4,5-*b*]pyrazin-2-ones.

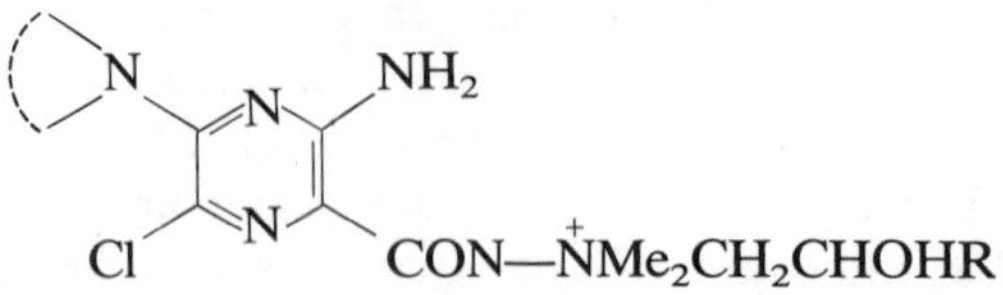

97 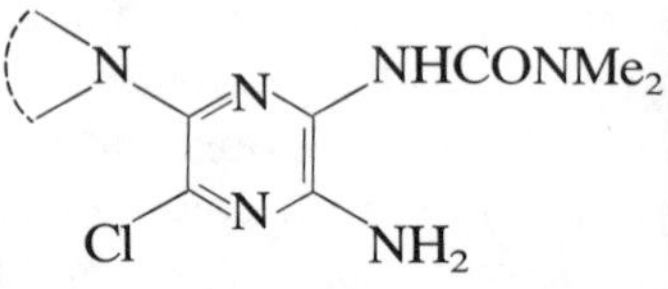 **98**

An example of the formation of the pyrazine ring as the final step in a synthesis of imidazo[4,5-*b*]pyrazines involves the reaction of diacetyl with the diamine **99**, which was generated in situ. The condensation gave the nucleoside analogue **100**.[37] The only other synthesis using an imidazole starting material results from the condensation of ethylenediamine with **101** to give the perhydro derivative **102** in 95% yield.[54]

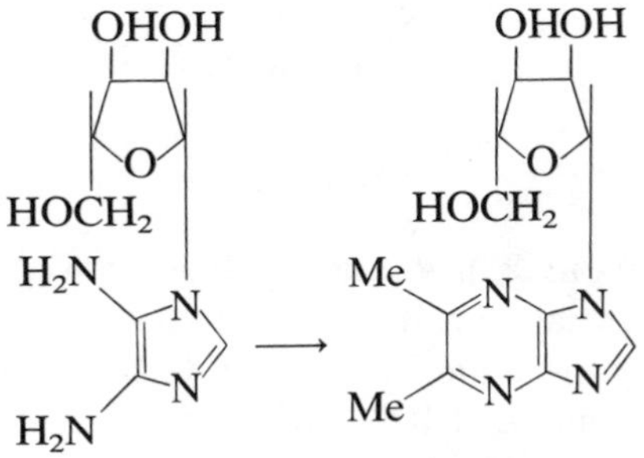

99 → **100**

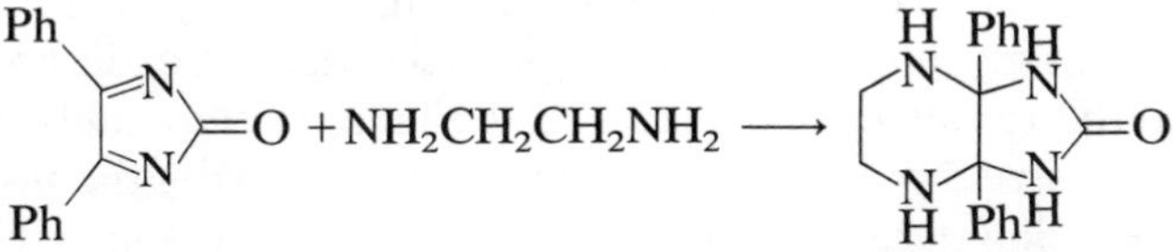

101 + $NH_2CH_2CH_2NH_2$ → **102**

Treatment of the unsubstituted compound **103** with hydrogen peroxide in glacial acetic acid resulted in the isolation of two *N*-oxides. The minor product (8%) was the 4,7-dioxide **105**, and the major product (50%) was suggested to be the 4-oxide **104**. This assignment was supported by the similarity of the compound's UV spectrum to that of the 1-methyl-4-oxide **81**, which had been obtained from the pyrazine **87** by an unambiguous route.[36] A 4-oxide was also obtained when the acetylated nucleoside analogue **106** was oxidized with *meta*-chloroperbenzoic acid.[36] There has been a recent Russian communication on the chemistry of imidazo[4,5-*b*]pyrazine *N*-oxides.[68]

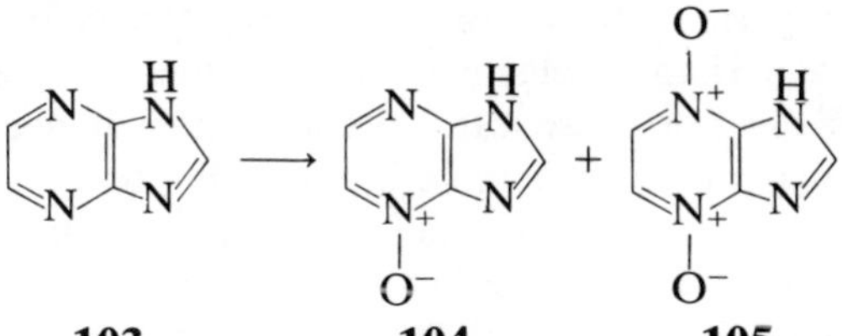

103 → **104** + **105**

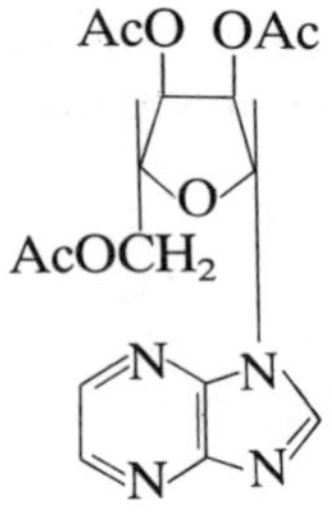

AcO OAc

AcOCH$_2$

106

Various 2-oxo compounds have been subjected to a variety of reactions. Alkylation and acylation result in the formation of 1,3-disubstituted products.[48,49] Similarly the Mannich reaction with morpholine gives products such as the bis compound **107**. Halogenation of the oxo compound **108** with bromine in acetic acid,[49] or chlorine in acetic acid,[50] or with sulfuryl chloride[50] results in formation of the monohalo derivatives **109**. The chlorine atom in **109** (R = Cl) can be removed by hydrogenation over palladium on charcoal.[49] Nitration of the chloro compound **110** with a mixture of nitric and sulfuric acids provides the nitro derivative **111**, which may be catalytically reduced to the amino compound **112**. If the reduction is carried out in the presence of an aldehyde, the product is a substituted amino derivative (**113**).[51] Alternatively the amine **112** can be condensed with an aldehyde and the resultant Schiff base reduced to give the product **113**.

The most interesting derivatives of this ring system from a biological point of view are the 5-substituted amino-6-chloro-2-oxo compounds, which have been covered by an extensive series of patents[45,46,48–53] and discussed in a recent paper.[47] They are generally highly effective antihypertensives in dogs and primates and are active orally, possessing only low toxicity. They show a moderate degree of diuretic and saluretic activity. Two representative compounds, **93** and **114**, were selected for further study.[55] The group generally are relatively potent inhibitors of cyclic AMP phosphodiesterase. Compound **93** is two to five times more potent than theophylline as a phosphodiesterase inhibitor, whereas **114** is less active than theophylline. The compounds are also active as bronchodilators and peripheral vasodilators. However, compound **114** was found to be remarkably selective in that, unlike compound **93** or theophylline, it did not cause cardiac stimulation at doses that caused marked hypotension or considerable bronchodilation.[47,55]

Compounds of type **115** are claimed to be muscle relaxants, tranquilizers, and sedatives,[43] and the related compounds **116** are analgesics, sedatives, and muscle relaxants.[44] The effect of the unsubstituted parent

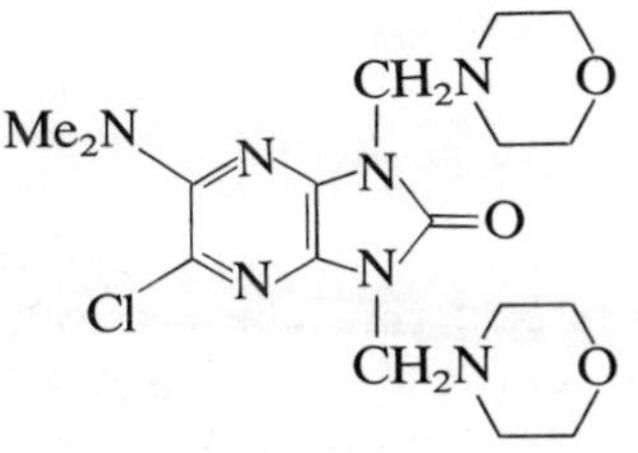

107

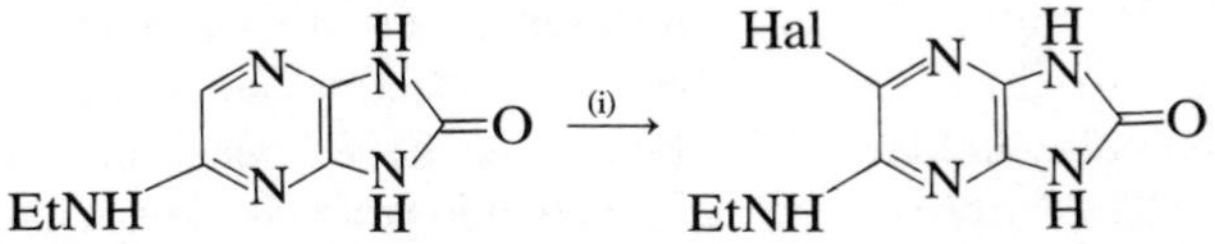

Hal = Cl or Br

108 **109**

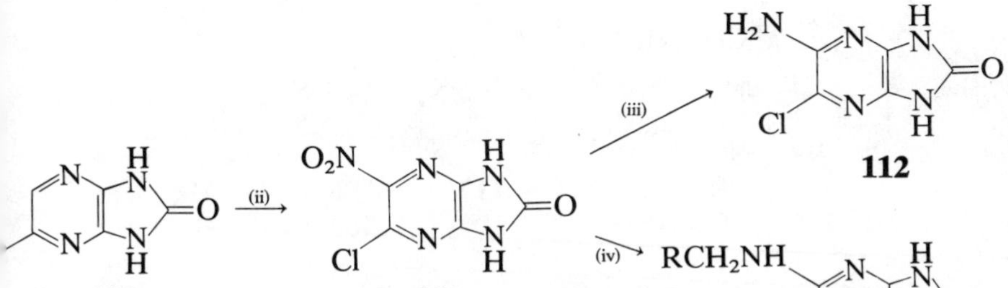

110 **111** **112** **113**

Reagents (i) Br_2 or Cl_2–HOAc or SO_2Cl_2
(ii) HNO_3–H_2SO_4
(iii) Catalytic reduction
(iv) Catalytic reduction/RCHO.

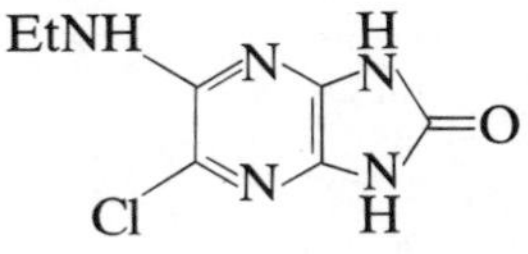

93 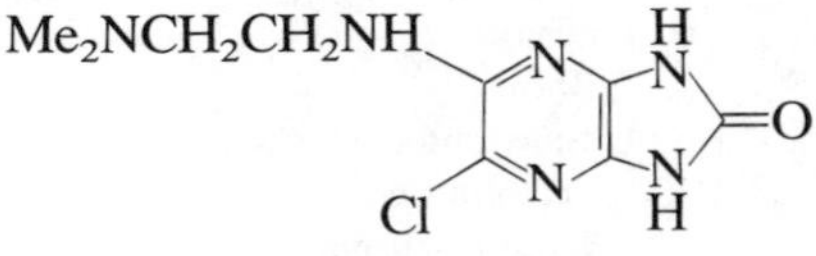 **114**

heterocycle on *Chlorella* respiration has been studied, and it seems capable of replacing purines in the respiratory stimulation of nitrogen-starved cells.[56] Similarly the effect of 2,5,6-trimethylimidazo[4,5-*b*]pyrazine on the growth of bacteria and cancer cells has been studied.[57] The nebularine analogue **117** and its 4-oxide **82**, together with the simpler compounds **80** and **103** and their *N*-oxides, were also examined for their effect as growth inhibitors on microbial and cancer cells.[36] Only imidazo[4,5-*b*]pyrazine 4-oxide and compound **82** were active against *Escherichia coli*, and none of the compounds affected the growth of cancer cells in spite of the considerable activity of nebularine. The antibacterial *N*-oxides were shown to be acting as antimetabolites.[36] Certain 2-haloalkylimidazo[4,5-*b*]pyrazines have measurable antitumor activity.[69] Some derivatives of 1*H*-imidazo[4,5-*b*]pyrazin-5-ol are claimed to be useful in photographic emulsions.[58]

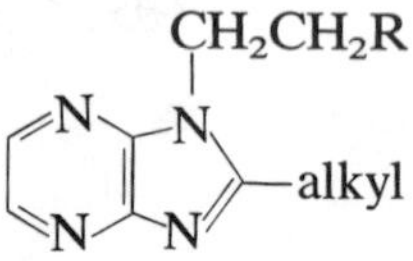

115 R = Ph
116 R = NEt_2
117
80 R = Me 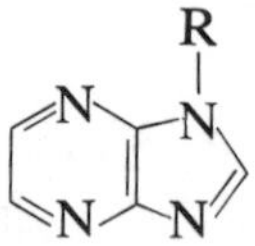
103 R = H

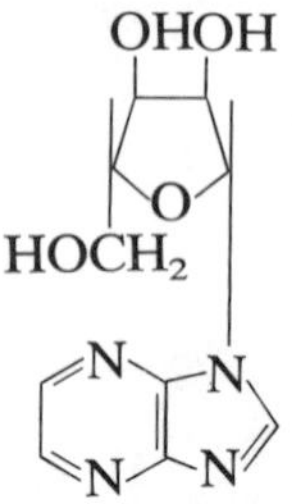

IV. Tables

TABLE 1. IMIDAZO[1,2-*a*]PYRAZINES

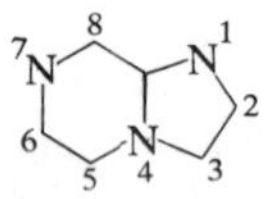

Molecular formula	Substituents or structure	m.p., °C	Ref.
$C_6Cl_5N_3$	Perchloro	134–136	62
$C_6H_4BrN_3$	3-Bromo	194–196	62
$C_6H_5N_3$	Unsubstituted	83–85	62
	Perchlorate	157–158	3
	1-Methyl bromide	>330	4
	1-Benzyl perchlorate	118–121	

TABLE 1 (*contd.*)

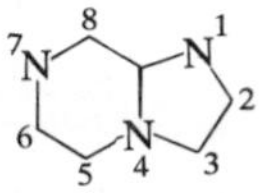

Molecular formula	Substituents or structure	m.p., °C	Ref.
$C_6H_5N_3$ (*contd.*)	7-Oxide hydrobromide	168–172[a]	
	7-Oxide perchlorate	204–206	
	1-Methyl-7-oxide bromide	>330	
	1-Methyl-7-oxide picrate	218–221	
	1-Benzyl-7-oxide perchlorate	214–217	
$C_6H_5N_3O$	3,7-Dihydro-3-oxo	>300	11, 18
$C_6H_6N_4$	3-Amino	208–209 decomp., 217 decomp.	11, 18
$C_6H_7N_3$	2,3-Dihydro		
	Hydrochloride monohydrate	207–208 decomp.	3
$C_7H_6ClN_3$	3-Chloro-2-methyl	135–136	5
$C_7H_7N_3$	2-Methyl	89	2
	3-Methyl	159	2
	6-Methyl		
	1-Methyl bromide	>330	4
	1-Methyl perchlorate	201–204	
	1-Methyl picrate	162–164	
	1-Methyl-7-oxide bromide monohydrate	264–267 decomp.	
	1-Methyl-7-oxide perchlorate	252–255 decomp.	
$C_7H_7N_3O$	3,7-Dihydro-2-methyl-3-oxo	239–240	11, 18, 22
$C_7H_8N_4$	3-Amino-1-methyl	204–206 decomp.	18
$C_8H_6Cl_3N_3O_2$	2,3,6-Trichloro-5,8-dimethoxy	185–187	62
$C_8H_9N_3O$	3,7-Dihydro-2,6-dimethyl-3-oxo	260 decomp.	22
	3,7-Dihydro-2,7-dimethyl-3-oxo		
	Hemihydrate	182 decomp.	5
$C_8H_9N_3O_2$	5,8-Dimethoxy	126–128	62
$C_9H_9N_3O_2$	7-Acetyl-3,7-dihydro-2-methyl-3-oxo	208–209 decomp.	5
$C_9H_{13}N_5O_2$	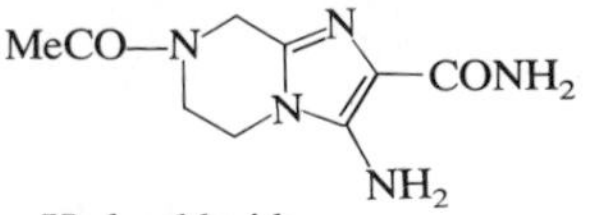 [structure: MeCO–N (tetrahydroimidazo ring) with CONH₂ and NH₂ substituents]	243–244	24
	Hydrochloride	229–231	
$C_{12}H_9N_3$	6-Phenyl		
	1-Methyl perchlorate	209–211	4
	1-Methyl picrate	214–217	
	1-Methyl-7-oxide perchlorate	225–228 decomp.	
	1-Methyl-7-oxide picrate	153–155	
	8-Phenyl[b]		
	Hydrochloride	188–189	25
$C_{12}H_9N_3O$	3,7-Dihydro-3-oxo-2-phenyl	258 decomp.	18, 22
$C_{12}H_{10}N_4$	3-Amino-2-phenyl	169–170 decomp.	18

TABLE 1 (*contd.*)

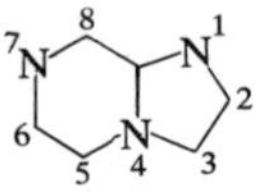

[Structure: imidazo[1,2-a]pyrazine ring numbering 1–8]

Molecular formula	Substituents or structure	m.p., °C	Ref.
$C_{12}H_{19}N_3O_2$	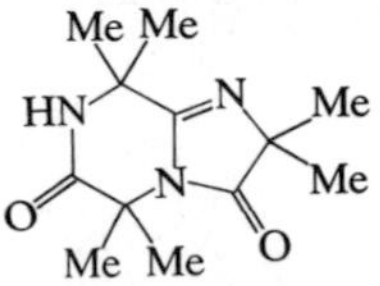 [Structure: HN, Me Me, =N, Me, Me, N, O, O, Me Me]	255	26, 27
$C_{13}H_{11}N_3$	8-Methyl-2-phenyl		
	Hydrobromide	315–325 decomp.	17
$C_{13}H_{11}N_3O$	3,7-Dihydro-2-methyl-3-oxo-6-phenyl	259 decomp.	18, 21, 22
$C_{13}H_{21}N_3O_2$	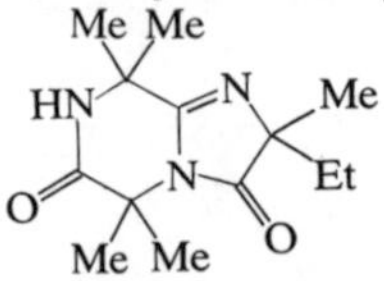 [Structure: HN, Me Me, =N, Me, Et, N, O, O, Me Me]		27
$C_{14}H_{12}N_4$	3-Benzylideneamino-2-methyl	117–119	5
$C_{14}H_{13}N_3O$	3,7-Dihydro-2,8-dimethyl-3-oxo-6-phenyl	[c]	10
$C_{14}H_{14}N_4$	3-Benzylamino-2-methyl	116–117	5
$C_{15}H_{12}N_4O$	3,7-Dihydro-6-(3-indolyl)-2-methyl-3-oxo		
	Hydrochloride	296	21
$C_{18}H_{13}N_3O$	3,7-Dihydro-3-oxo-2,6-diphenyl	277 decomp.	22
$C_{18}H_{15}N_3$	2,3-Dihydro-5,6-diphenyl	183.5–184.5	23
	Picrate	187–188 decomp.	
$C_{22}H_{29}N_7O_2$	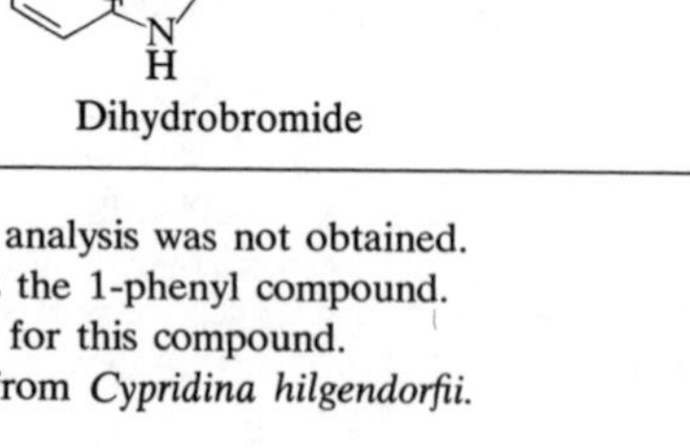 [Structure: $CH_2CH_2CH_2NHC(=NH)NH_2$ at 8; HN, N, CH(Me)Et at 2, O, 3-indolyl (N H) at 6] [d]		
	Dihydrobromide	252–254	6, 21

[a] A satisfactory analysis was not obtained.

[b] Referred to as the 1-phenyl compound.

[c] No data given for this compound.

[d] The luciferin from *Cypridina hilgendorfii.*

TABLE 2. IMIDAZO[1,5-*a*]PYRAZINES

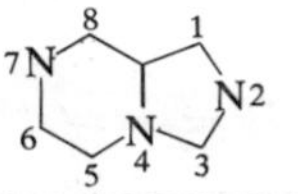

Molecular formula	Substituents or structure	m.p., °C	Ref.
$C_6H_2Cl_3N_3$	1,3,5-Trichloro	89	64
$C_6H_3Cl_2N_3$	1,3-Dichloro	109	64
$C_6H_4ClN_3$	1-Chloro	128–129	64
	3-Chloro	136–137	64
	8-Chloro	105–106	65
$C_6H_5N_3$	Unsubstituted	103–104	28
$C_6H_5N_3O$	7,8-Dihydro-8-oxo	270 decomp.	65
$C_6H_5N_3S$	2,3-Dihydro-3-thioxo	209 decomp.	66
	7,8-Dihydro-8-thioxo		
	Hydrochloride	>310	65
$C_6H_6N_4$	8-Amino	159	29
$C_6H_6N_4S$	8-Amino-2,3-dihydro-3-thioxo	decomp. >220	29
$C_7H_5Cl_2N_5$	1,5-Dichloro-3-methyl	90–91	64
$C_7H_5N_3O_3$	7-Carboxy-7,8-dihydro-8-oxo		
	Ethyl ester	171–172	65
$C_7H_6BrN_3$	1-Bromo-3-methyl	119–120	64
$C_7H_6ClN_3$	1-Chloro-3-methyl	102	64
	8-Chloro-3-methyl	126–127	65
$C_7H_6N_4O_2S$	3-Methylthio-1-nitro	180–181	66
$C_7H_6N_6$	3-Azidomethyl	80–81	60
$C_7H_7N_3$	3-Methyl	142–143	28
	7-Oxide	165–166	
$C_7H_7N_3O$	7,8-Dihydro-3-methyl-8-oxo	280 decomp.	65
	3-Hydroxymethyl	101	60
$C_7H_7N_3OS$	3-Methylsulfinyl	76–77	66
$C_7H_7N_3O_2S$	3-Methylsulfonyl	138–139	66
$C_7H_7N_3O_3$	7-Carboxy-7,8-dihydro-8-hydrox		
	Ethyl ester	163–165	65
$C_7H_7N_3S$	7,8-Dihydro-3-methyl-8-thioxo		
	Hydrochloride	305	65
	3-Methylthio	34–35 b.p. 102–105/ 0.05 mm	66
$C_7H_8N_4$	3-Methylamino	115	60
$C_7H_{13}N_3O$	Perhydro-7-methyl-3-oxo	99–103	63
$C_8H_7N_3OS$	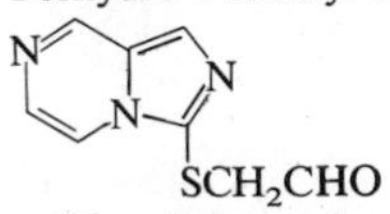 SCH_2CHO		
	Dimethyl acetal	b.p. 130/0.02 mm	66
$C_8H_7N_3O_2S$	3-Carboxymethylthio		
	Methyl ester	81–82	66

TABLE 2 (*contd.*)

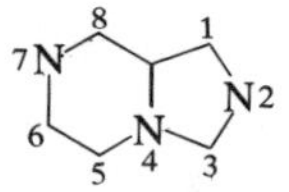

Molecular formula	Substituents or structure	m.p., °C	Ref.
$C_8H_9N_3O$	3-Methoxymethyl	78–79	60
$C_8H_9N_3OS$	3-Ethylsulfinyl	81–82	66
	3-(2-Hydroxyethyl)thio	99–100	66
$C_8H_9N_3O_2S$	3-Ethylsulfonyl	63–64	66
$C_8H_9N_3O_3$	7-Carboxy-7,8-dihydro-8-hydroxy-3-methyl		
	Ethyl ester	208 decomp	65
$C_8H_9N_3S$	7-Ethylthio	35–36	66
		b.p. 97–98/0.01 mm	
$C_8H_{10}N_4O_2$	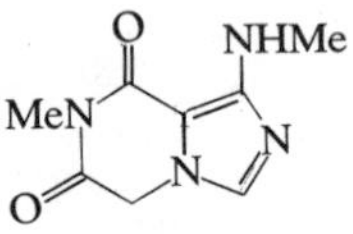 (structure: MeN, O, NHMe, N, N)	208	31
	Monohydrate	220	32
	Hydrochloride	270–280 sublimes	
	N-Benzoyl	210–212	
	N-Nitroso	200 decomp.	
$C_9H_9N_3O_2$	7-Acetyl-7,8-dihydro-3-methyl-8-oxo	174–176	65
$C_9H_{11}N_3O_2$	7-Acetyl-7,8-dihydro-8-hydroxy-3-methyl	168–169	65
$C_9H_{17}N_3O$	Perhydro-2-ethyl-7-methyl-3-oxo	b.p. 65/0.015 mm	63
$C_{10}H_7N_3O$	3-Furan-2-yl	79–81	60
$C_{10}H_7N_5$	3-Pyrazin-2-yl	202–203	30
$C_{10}H_{13}N_3O_2$	7-Acetyl-7,8-dihydro-8-methoxy-3-methyl	129–131	65
$C_{10}H_{13}N_3O_3$	7-Carboxy-8-ethoxy-7,8-dihydro-3-methyl		
	Ethyl ester	97–98	65
$C_{10}H_{13}N_3O_3S$	7-Carboxy-8-ethoxy-7,8-dihydro-3-methylthio		
	Ethyl ester	Oil	66
$C_{10}H_{17}N_3O$	2-Allyl-perhydro-7-methyl-3-oxo[a]	b.p. 80/0.04 mm	63
$C_{11}H_{14}N_4$	1-Dimethylaminomethyl-3-methyl		
	Methiodide	214–215	64
$C_{11}H_{15}N_3O_2$	7-Acetyl-8-ethoxy-7,8-dihydro-3-methyl	96–97	65
$C_{11}H_{21}N_3O$	2-*n*-Butyl-perhydro-7-methyl-3-oxo	b.p. 110/0.04 mm	63
$C_{12}H_7Cl_2N_3$	3-(3,4-Dichlorophenyl)	141	60
$C_{12}H_8ClN_3$	3-(2-Chlorophenyl)	84–85	60
$C_{12}H_8N_4O_2$	3-(4-Nitrophenyl)	252	60
$C_{12}H_9N_3$	3-Phenyl	75–76	60
$C_{12}H_{10}N_4$	3-Anilino	155–156	60
$C_{12}H_{15}N_3O_5S$	β-D-Glucopyranos-1-ylthio	112–116	66
	Tetraacetate	153–154	66
$C_{13}H_{10}BrN_3S$	8-Bromo-3-phenylthiomethyl	101	61

TABLE 2 (*contd.*)

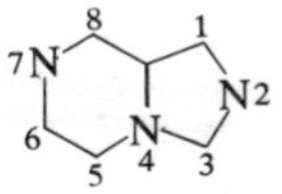

Molecular formula	Substituents or structure	m.p., °C	Ref.
$C_{13}H_{11}N_3S$	3-Phenylthiomethyl[b]	80	61
	Sulfoxide	118	
	Sulfone	204–206	
$C_{13}H_{15}N_3O_2$	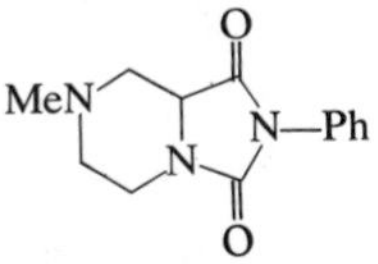 [structure: MeN, N—Ph, O, O]	102.5–104 b.p. 165–175/ 0.03 mm	33
$C_{13}H_{16}ClN_3O$	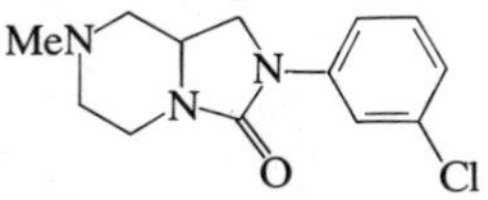 [structure: MeN, N, O, Cl]	136–138	63
$C_{13}H_{17}N_3O$	Perhydro-7-methyl-3-oxo-2-phenyl	80–81	63
$C_{13}H_{17}N_3O_2$	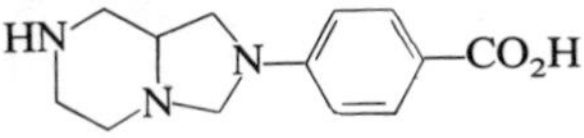 [structure: HN, N, CO_2H]		
	Ethyl ester	174–175.5	34
$C_{14}H_{12}N_4O_4$	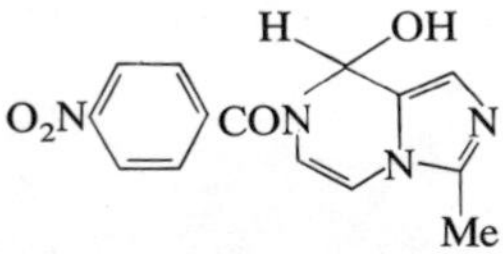 [structure: H, OH, O_2N, CON, N, Me]	166–167	65
	O-*p*-Nitrobenzoyl	153–155	65
$C_{14}H_{12}N_4O_4S$	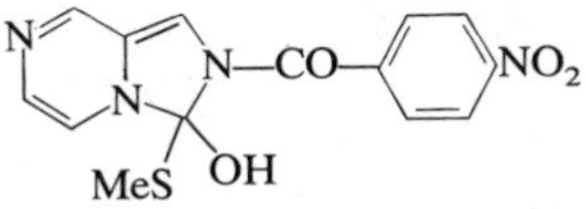 [structure: N, N—CO, NO_2, MeS, OH]	163	66
$C_{14}H_{12}N_6Se$	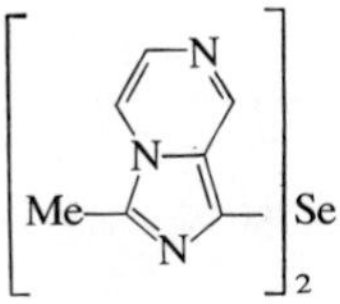 [structure: N, N, Me, N, Se, 2]	239–240	64
$C_{14}H_{19}N_3O$	2-Benzylperhydro-7-methyl-3-oxo	b.p. 135/0.04 mm	63
	Perhydro-7-methyl-3-oxo-2-*o*-tolyl	69–70	63
	Perhydro-7-methyl-3-oxo-*m*-tolyl	94–95	
	Perhydro-7-methyl-3-oxo-*p*-tolyl	110–111	
$C_{14}H_{19}N_3O_2$	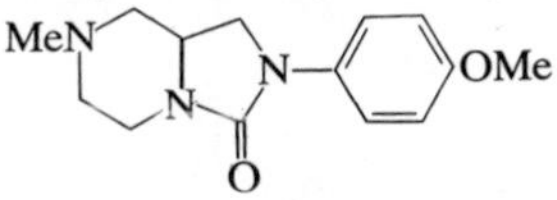 [structure: MeN, N, OMe, O]	118–120	63

TABLE 2 (*contd.*)

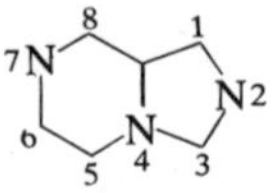

Molecular formula	Substituents or structure	m.p., °C	Ref.
$C_{16}H_{16}N_4O_4$	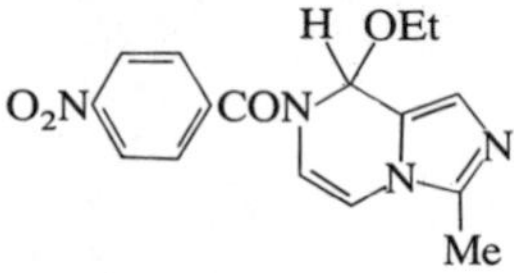 [structure]	145–147	65
$C_{17}H_{19}N_5O_5$	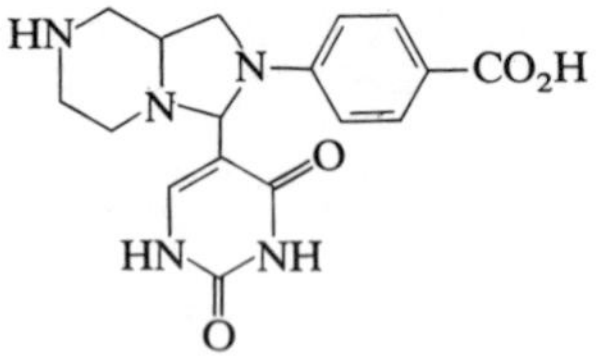 [structure] Ethyl ester ethanolate	219–220	34
$C_{18}H_{21}N_5O_5$	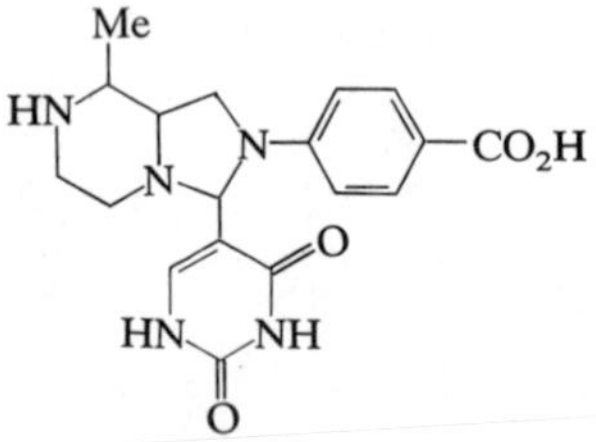 [structure] Ethyl ester	240 decomp.	35
$C_{19}H_{21}N_3O_3$	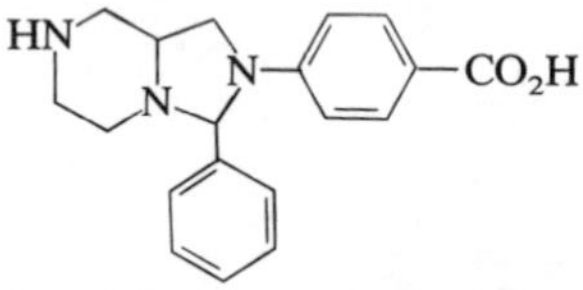 [structure] Ethyl ester	165–166.5	34

[a] Four substituted allyl derivatives were also reported.
[b] The 3-phenylthio(monodeuterio)methyl compound m.p. 80° was also reported.

TABLE 3. 1*H*-2,3-DIHYDRO-2-OXOIMIDAZO[4,5-*b*]PYRAZINES

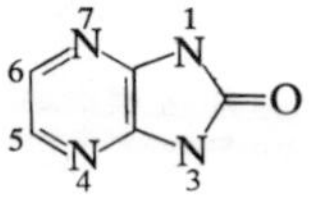

Molecular formula	Substituents or structure	m.p., °C	Ref.
$C_5H_2ClN_7O$	5-Azido-6-chloro	188 decomp.	49
$C_5H_3ClN_4O$	5-Chloro	280 decomp.	47
$C_5H_3ClN_4OS$	5-Chloro-6,7-dihydro-6-thio[a]	>290	48
	Hemihydrate	300	47
$C_5H_3ClN_4O_2$	5-Chloro-6,7-dihydro-6-oxo[b]	>300	47
$C_5H_4ClN_5O$	5-Amino-6-chloro	>300	46, 47, 49, 52
$C_5H_4N_4O$	Unsubstituted	334–336	40, 41
$C_6H_5ClN_4OS$	5-Chloro-6-methylthio	308 decomp.	47
$C_6H_5ClN_4O_2$	5-Chloro-6-methoxy	315 decomp.	47
$C_6H_5ClN_5O$	5-Chloro-6-methylamino	>280	45, 46, 47, 49, 50, 51, 52, 53
$C_6H_6ClN_7O$	5-Chloro-6-guanidino		
	Hydrochloride	>300	47, 49, 52
$C_7H_4ClN_5OS$	5-Chloro-6-cyanomethylthio	250–252	47
$C_7H_5ClN_4O_3S$	5-Carboxymethylthio-6-chloro		
	Ethyl ester	206–208	47
$C_7H_7ClN_4OS$	5-Chloro-6-ethylthio	256–258	47
$C_7H_7ClN_4O_2S$	5-Chloro-6-(2-hydroxyethyl)thio	236–238	47
$C_7H_8ClN_5O$	5-Chloro-6-dimethylamino	216–217, 222–224	45, 46, 47, 49, 50, 52, 53
	5-Chloro-6-ethylamino	248–249	45, 46, 47, 49, 50, 51, 53
	N-Acetyl	286–288 decomp.	47
$C_7H_8ClN_5O_2$	5-Chloro-6-(2-hydroxyethyl)amino	250–251	47, 49, 52
$C_7H_8N_4O$	5,6-Dimethyl	418	39
$C_7H_9ClN_6O$	5-(2-Aminoethyl)amino-6-chloro		
	Hydrochloride	>300	47, 49
	N-Acetyl[c]	235–238 decomp.	46, 47, 49, 50, 51, 52
$C_7H_9N_5O$	5-Ethylamino	250–252	49
$C_8H_8ClN_5O$	5-Allylamino-6-chloro	224–225, 284–285	46, 47, 49, 50, 52
$C_8H_9ClN_4OS$	5-Chloro-6-*n*-propylthio	244–246	47
$C_8H_{10}ClN_5O$	5-Chloro-6-ethylamino-1-methyl	248–249	49
	5-Chloro-6-(*N*-ethyl-*N*-methyl)amino	203–205	46, 47, 49, 50, 52
	5-Chloro-6-*n*-propylamino	215–216	45, 46, 47, 49, 50, 51, 52, 53

TABLE 3 (*contd.*)

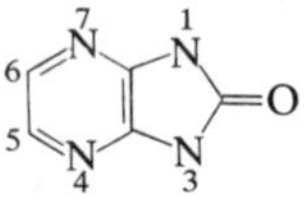

Molecular formula	Substituents or structure	m.p., °C	Ref.
$C_8H_{10}ClN_5O$ (*contd.*)	5-Chloro-6-isopropylamino	247–249	46, 47, 49, 50, 52
$C_8H_{10}ClN_5O_2$	5-Chloro-6-[*N*-(2-hydroxyethyl)-*N*-methyl]amino	188–190 decomp. 250–251	47, 49, 52
	5-Chloro-6-(2-methoxyethyl)amino	229–231	45, 46, 47, 49, 50, 51, 52, 53
$C_8H_{10}N_4O_2$	5-Hydroxymethyl-1,6-dimethyl		
	4-Oxide	247–248	68
$C_8H_{11}ClN_6O$	5-(*N*-(2-Aminoethyl)-*N*-methyl)-amino-6-chloro		
	Hydrochloride	304–306 decomp.	47
	5-(3-Aminopropyl)amino-6-chloro		
	N-Acetyl[d]	222–223, 226–227	46, 47, 50, 51, 52
	5-Chloro-6-(2-methylamino-ethyl)amino		
	Hydrochloride	236–237, 247–248	47, 49
	N-Acetyl[e]	227–228	46,47,49,52
$C_9H_8ClN_5O_5$	1,3-Dicarboxy-5-chloro-6-ethylamino		
	Diethyl ester	202–204	49
$C_9H_9ClN_6OS$	5-Chloro-6-imidazolin-2-ylmethylthio		
	Hydrochloride hemihydrate	292–293	47
$C_9H_{10}ClN_5O$	5-Chloro-6-pyrrolidin-1-yl	229 decomp.	47, 49
$C_9H_{10}ClN_5O_2$	5-Chloro-6-morpholin-1-yl	272–273	46, 47, 49, 52
$C_9H_{12}ClN_5O$	5-*n*-Butylamino-6-chloro	195–198, 215–216	45, 46, 47, 49, 50, 51, 52,
	5-*t*-Butylamino-6-chloro	235	46, 47, 49, 52
	5-Chloro-6-diethylamino	207–208	45, 46, 47, 50, 52, 53
	5-Chloro-6-dimethylamino-1-ethyl	175–178	49
	5-Chloro-6-dimethylamino-1,3-dimethyl	158–160	49
	5-Chloro-6-(*N*-methyl-*N*-*n*-propyl) amino	177–178, 215–216	47, 49, 52
$C_9H_{12}ClN_5O_2$	5-Chloro-6-dimethylamino-1-(2-hydroxyethyl)	190	49
	5-Chloro-6-ethylamino-1-(2-hydroxyethyl)	213	49

TABLE 3 (*contd.*)

Molecular formula	Substituents or structure	m.p., °C	Ref.
$C_9H_{13}ClN_6O$	5-[*N*-(2-aminoethyl)-*N*-ethyl]amino-6-chloro		
	Hydrochloride	281–282	47
	5-Chloro-6-(2-dimethylaminoethyl)-amino	219–220	45, 47, 49, 50, 51, 52, 53
	5-Chloro-6-(2-ethylaminoethyl)amino		
	Hydrochloride	247–248	47
$C_{10}H_{12}ClN_5O$	1-Allyl-5-chloro-6-ethylamino	181–182	49
	5-Chloro-6-cyclopentylamino	256–257	46, 47, 49, 52
	5-Chloro-6-piperidin-1-yl	229	46, 52
$C_{10}H_{12}N_4O_3$	3-Acetyl-5-hydroxymethyl-1,6-dimethyl		
	O-Acetyl	144.5–145	68
$C_{10}H_{13}ClN_4OS$	5-Chloro-6-pentylthio[f]	232–233	47
$C_{10}H_{13}ClN_6O$	5-Chloro-6-(4-methylpiperazin-1-yl)	275 decomp.	46, 47, 49, 52
$C_{10}H_{14}ClN_5O_2$	5-Chloro-6-ethylamino-1-(2-methoxyethyl)	205–207	49
$C_{10}H_{14}ClN_7O$			
	Hydrochloride	285	49, 52
$C_{10}H_{15}ClN_6O$	5-[*N*-(2-Aminoethyl)-*N*-isopropyl]-amino-6-chloro		
	Hydrochloride	274–276	47
	5-Chloro-6-(3-dimethylaminopropyl)-amino	229–230	46
	Maleate hemihydrate	188–190	49, 50, 52
$C_{11}H_8ClN_5O$	5-Anilino-6-chloro	257 decomp.	46, 47, 49, 52
$C_{11}H_9ClN_6O$	5-Chloro-6-pyrid-2-ylmethylamino		
	Hydrochloride	280	45, 46, 47, 49, 50, 51, 52, 53
	5-Chloro-6-pyrid-3-ylmethylamino		
	Hydrochloride hydrate	271–272	45, 46, 47, 49, 50, 51, 52, 53
	5-Chloro-6-pyrid-4-ylmethylamino		
	Hydrochloride	291–293 decomp.	46, 47, 49, 50, 51, 52, 53

TABLE 3 (*contd.*)

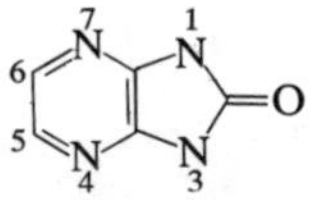

Molecular formula	Substituents or structure	m.p., °C	Ref.
$C_{11}H_{12}ClN_5O_3$	1,3-Diacetyl-5-chloro-6-ethylamino	198–200	49
$C_{11}H_{12}ClN_5O_5$	1,3-Dicarboxymethyl-5-chloro-6-ethylamino		
	Diethyl ester	150–152	49
	Dihydrazide	250–253 decomp.	
$C_{11}H_{15}ClN_6O$	5-Chloro-6-(4-ethylpiperazin-1-yl)		
	Hydrochloride	307 decomp.	46, 47, 49, 52
	5-Chloro-6-(2-pyrrolidin-1-ylethyl)amino		
	Maleate hydrate	148–149	46, 47, 49, 52
$C_{11}H_{17}ClN_6O$	5-Chloro-6-(2-diethylaminoethyl)-amino	244	46
	Hydrochloride	276–278	45, 47, 49, 50, 51, 52, 53
	5-Chloro-6-(*N*-(3-dimethylamino-propyl)-*N*-methyl)amino		
	Hydrochloride hemihydrate	229–230	47, 49, 52
	5-Chloro-6-(4-dimethylaminobutyl)-amino		
	Hydrate	278–279	47
	Hydrochloride	278–279	45, 46, 49, 50, 51, 53
	5-Chloro-6-(3-isopropylaminopropyl)-amino	190–192	46, 49, 50, 52
$C_{12}H_9Cl_2N_5O$	5-Chloro-6-*p*-chlorobenzylamino	270–272	45, 46, 47, 49, 50, 51, 52, 53
$C_{12}H_{11}ClN_6O$	5-Chloro-6-(*N*-methyl-*N*-pyrid-2-ylmethyl)amino		
	Hydrochloride	255–256	47
	5-Chloro-6-(2-pyrid-2-ylethyl)amino	235–236	47
	5-Chloro-6-(2-pyrid-4-ylethyl)amino		
	Hydrochloride	256–257	47
$C_{12}H_{17}ClN_6O_2$	5-Chloro-6-(3-morpholin-1-ylpropyl)-amino		
	Maleate hemihydrate	154–156, 158–160	45, 46, 47, 49, 50, 51, 52,
$C_{12}H_{19}ClN_6O$	5-Chloro-6-[*N*-(2-diethylaminoethyl)-*N*-methyl]-amino		
	Hydrochloride	244 decomp.	47, 52

TABLE 3 (*contd.*)

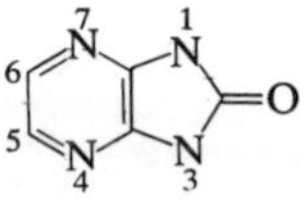

Molecular formula	Substituents or structure	m.p., °C	Ref.
$C_{12}H_{19}ClN_6O$ (*contd.*)	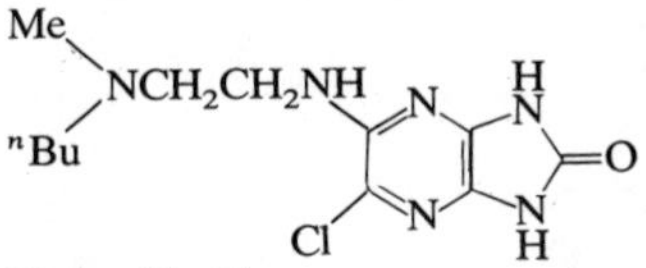 [structure: 5-chloro-6-[2-(N-methyl-N-n-butylamino)ethyl]amino derivative]		
	Hydrochloride	254–255, 257–258	46, 47, 49, 52
$C_{12}H_{19}ClN_6O_2$	5-Chloro-6-(3-diethylamino-2-hydroxypropyl)amino		
	Hydrochloride	267–268	49, 52
$C_{13}H_9ClN_4O_2S$	5-Chloro-6-phenacylthio	254–257	47
$C_{13}H_{13}ClN_6O$	5-Chloro-6-[*N*-methyl-*N*-(2-pyrid-2-ylethyl)amino		
	Hydrochloride	251–252	47
$C_{13}H_{13}N_5O$	5-Dimethylamino-6-phenyl	281–282	49
$C_{13}H_{20}ClN_7O$	5-Chloro-6-[3-(4-methylpiperazin-1-yl)propyl]amino	167–169 decomp., 244–245	45, 46, 47, 49, 50, 52, 53
$C_{17}H_{12}N_4O$	5,6-Diphenyl	311–318	23
$C_{17}H_{20}ClN_7O_3$	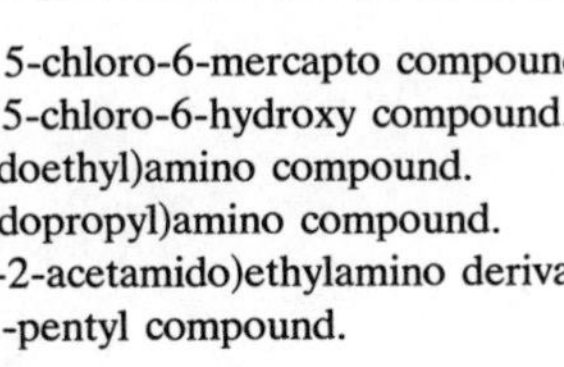 [structure: 5-chloro-6-dimethylamino-1,3-bis(morpholinomethyl) derivative]	126–128	49

[a] Described as the 5-chloro-6-mercapto compound.
[b] Described as the 5-chloro-6-hydroxy compound.
[c] The 5-(2-acetamidoethyl)amino compound.
[d] The 5-(3-acetamidopropyl)amino compound.
[e] The 6-(*N*-methyl-2-acetamido)ethylamino derivative.
[f] Presumably the *n*-pentyl compound.

TABLE 4. 1*H*-IMIDAZO[4,5-*b*]PYRAZINES

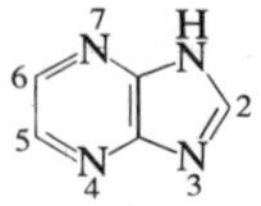

Molecular formula	Substituents	m.p., °C	Ref.
$C_5H_4N_4$	Unsubstituted	257	40
	4-Oxide	>300	36
	4,7-Dioxide	>300	
$C_6HBr_2F_3N_4$	5,6-Dibromo-2-trifluoromethyl[a]	191–192	69
$C_6HCl_2F_3N_4$	5,6-Dichloro-2-trifluoromethyl[a]	168–170	69
$C_6HF_5N_4$	5,6-Difluoro-2-trifluoromethyl[a]	150–152	69
$C_6H_3Cl_3N_4$	5,6-Dichloro-2-chloromethyl	204	69
$C_6H_3F_3N_4$	2-Trifluoromethyl	220–222	69
$C_6H_4Cl_2N_4$	5,6-Dichloro-2-methyl	295–297	42
$C_6H_4Cl_2N_4O$	5,6-Dichloro-2-hydroxymethyl	280–281	69
$C_6H_5BrN_4$	6-Bromo-1-methyl	154.5–155	67
	4-Oxide	265–266	68
$C_6H_5ClN_4$	2-Chloromethyl	178–180	67
$C_6H_6N_4$	1-Methyl	192–194	36
	4-Oxide	290–291	
	2-Methyl	370 decomp.	40, 69
$C_6H_6N_4O$	2-Hydroxymethyl	309 decomp.	40
$C_7H_6Cl_2N_4$	5,6-Dichloro-2-ethyl[b]	286–290	69
$C_7H_8N_4$	5,6-Dimethyl	209	39
$C_7H_8N_4O$	5-Hydroxymethyl-6-methyl		
	7-Oxide	276–277	68
	7-Oxide, *O*-acetyl	265–266	68
$C_8H_5N_5S$	2-Thiazol-4-yl	333	69
$C_8H_6Cl_2N_4$	5,6-Dichloro-2-cyclopropyl	276–280	69
$C_8H_9ClN_4$	2-Chloro-1,5,6-trimethyl	163–163.5	67
$C_8H_{10}N_4$	2,5,6-Trimethyl	259	39
$C_8H_{10}N_4O$	5-Hydroxymethyl-1,6-dimethyl		
	Acetate	119–120	68
$C_9H_{12}N_4$	2-Ethyl-5,6-dimethyl	256	39
$C_9H_{12}N_4O$	2-Methoxy-1,5,6-trimethyl	116.5–117	67
$C_9H_{13}N_7$	2-Guanidino-1,5,6-trimethyl		
	Hydrochloride	291–292	67
$C_{10}H_{12}N_4O_4$	1-(β-D-Ribofuranosyl)[c]	205–206	36
	4-Oxide[d]	228–229	
$C_{10}H_{13}N_5O$	1,5,6-Trimethyl-2-morpholin-4-yl		
	Hydrochloride	232–233	67
$C_{10}H_{14}N_4$	5,6-Dimethyl-2-*n*-propyl	227	39
$C_{11}H_5Cl_2FN_4$	5,6-Dichloro-2-(4-fluorophenyl)	377	69
$C_{11}H_{16}N_4$	2-Butyl-5,6-dimethyl	186	39
$C_{11}H_{16}N_6$	2-(4-Methylpiperazin-1-yl)methyl		
	Hydrochloride	233.5–234	67

TABLE 4 (*contd.*)

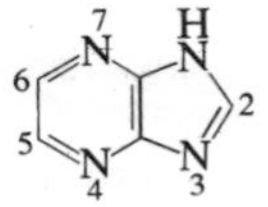

Molecular formula	Substituents	m.p., °C	Ref.
$C_{11}H_{17}N_5$	1,5,6-Trimethyl-2-isopropylamino	180–181	67
	Hydrochloride	255.5–256.5	
$C_{12}H_5Cl_2F_3N_4$	5,6-Dichloro-2-(3-trifluoromethylphenyl)	173–174	69
$C_{12}H_{16}N_4O_4$	5,6-Dimethyl-1-(β-D-ribofuranosyl)[e]	234–235	37
$C_{12}H_{19}N_5$	2-Diethylamino-1,5,6-trimethyl	74.5-75.5	67
$C_{13}H_{12}N_4$	5,6-Dimethyl-2-phenyl	310	39
$C_{13}H_{20}N_6$	1,5,6-Trimethyl-2-(4-methylpiperazin-1-yl)		
	Dihydrochloride	261.5–262.5	67
$C_{13}H_{21}N_5$	1-(2-Diethylaminoethyl)-2-ethyl		
	Hydrochloride	180–185	44
$C_{14}H_{14}N_4$	2-Methyl-1-(2-phenylethyl)	111–113	43
$C_{15}H_{16}Cl_2N_4$	2-Adamantyl-5,6-dichloro	315–317	69
$C_{15}H_{16}N_4$	2-Ethyl-1-(2-phenylethyl)	106–108	43
$C_{17}H_{12}N_4$	5,6-Diphenyl	282–284 decomp.	23
$C_{17}H_{18}N_4O$	2*H*-1,3,3a,4,5,6,7,7a-Octahydro-2-oxo-		
	3a,7a-diphenyl	213–213.5	54
$C_{18}H_{14}N_4$	2-Methyl-5,6-diphenyl	345–352 decomp.	23
$C_{20}H_{19}N_4$	2-Benzyl-1-(2-phenylethyl)	104–106	43
$C_{20}H_{27}N_4$	2-Benzyl-1-(2-diethylaminoethyl)		
	Hydrochloride	172–174	44

[a] Seven other 2-polyhaloalkyl derivatives listed.
[b] Five other 2-alkyl substituents reported.
[c] $[\alpha]_D^{26} -49.6$ (*c* 0.2, DMSO).
[d] $[\alpha]_D^{26} -56.3$ (*c* 0.2 DMSO).
[e] $[\alpha]_D^{27} -69.76$.

V. References

1. V. Galasso, G. De Alti, and A. Bigotto, *Theor. Chim. Acta*, **9,** 222 (1968); *Chem. Abstr.*, **68,** 55108 (1968); E. M. Evleth, *Theor. Chim. Acta*, **16,** 22 (1970); *Chem. Abstr.*, **72,** 60988 (1970).
2. P. Guerret, R. Jacquier, and G. Maury, *J. Heterocycl. Chem.*, **8,** 643 (1971).
3. W. L. F. Armarego, *J. Chem. Soc.*, 2778 (1965).
4. J. Adamson and E. E. Glover, *J. Chem. Soc.* (*C*), 2748 (1971).
5. S. Sugiura, S. Inoue, and T. Goto, *Yakugaku Zasshi*, **90,** 431 (1970); *Chem. Abstr.*, **73,** 45460 (1970).
6. Y. Kishi, T. Goto, Y. Hirata, O. Shimomura, and F. H. Johnson, *Tetrahedron Lett.*, 3427 (1966); *Chem. Abstr.*, **65,** 13701 (1966).

7. Y. Kishi, T. Goto, Y. Hirata, O. Shimomura, and F. H. Johnson, *Biolumin. Progr., Proc., Kanagawa-Ken, Jap.*, 89 (1965); *Chem. Abstr.*, **67,** 32674 (1967).
8. Y. Hirata and T. Goto, *Kagaku* (Tokyo), **37,** 640 (1967); *Chem. Abstr.*, **68,** 4625 (1968).
9. S. Sugiura, S. Inoue, and T. Goto, *Yakugaku Zasshi*, **90,** 711 (1970); *Chem. Abstr.*, **73,** 98906 (1970).
10. F. McCapra and Y. C. Chang, *Chem. Commun.*, 1011 (1967).
11. T. Goto, S. Inoue, and S. Sugiura, *Tetrahedron Lett.*, 3873 (1968); *Chem. Abstr.*, **69,** 86951 (1968).
12. T. Goto, S. Inoue, S. Sugiura, K. Nishikawa, M. Isobe, and Y. Abe, *Tetrahedron Lett.*, 4035 (1968); *Chem. Abstr.*, **69,** 86948 (1968).
13. S. Sugiura, S. Inoue, H. Fakatsu, and T. Goto, *Yakugaku Zasshi*, **90,** 1475 (1970); *Chem. Abstr.*, **74,** 141691 (1971).
14. T. Goto and H. Fukatsu, *Tetrahedron Lett.*, 4299 (1969); *Chem. Abstr.*, **72,** 62753 (1970).
15. T. P. Karpetsky and E. H. White, *Tetrahedron*, **29,** 3761 (1973).
16. T. Goto, *Pure Appl. Chem.*, **17,** 421 (1968); *Chem. Abstr.*, **73,** 32730 (1970); T. Goto, *Kagaku To Seibutsu*, **7,** 445 (1969); *Chem. Abstr.*, **72,** 62631 (1970).
17. L. M. Werbel and M. L. Zamora, *J. Heterocycl. Chem.*, **2,** 287 (1965).
18. S. Sugiura, S. Inoue, and T. Goto, *Yakugaku Zasshi*, **90,** 423 (1970); *Chem. Abstr.*, **73,** 45459 (1970).
19. Y. Kishi, T. Goto, S. Inoue, S. Sugiura, and H. Kishimoto, *Tetrahedron Lett.*, 3445 (1966); *Chem. Abstr.*, **65,** 13703 (1966).
20. Y. Kishi, S. Sugiura, S. Inoue, and T. Goto, *Yakugaku Zasshi*, **89,** 1657 (1969); *Chem. Abstr.*, **72,** 90406 (1970).
21. S. Sugiura, S. Inoue, and T. Goto, *Yakugaku Zasshi*, **90,** 707 (1970); *Chem. Abstr.*, **73,** 98904 (1970).
22. S. Sugiura, H. Kakoi, S. Inoue, and T. Goto, *Yakugaku Zasshi*, **90,** 441 (1970); *Chem. Abstr.*, **73,** 45462 (1970).
23. R. H. Martin and Z. Tarasiejska, *Bull. Soc. Chim. Belg.*, **66,** 136 (1957).
24. V. G. Granik and R. G. Glushkov, *Khim.-Farm. Zh.*, **1,** 16 (1967); *Chem. Abstr.*, **68,** 68955 (1968); USSR Pat. 196,876; *Chem. Abstr.*, **68,** 95859 (1968).
25. V. I. Schvedov, L. B. Altukhova, L. A. Chernyshkova, and A. N. Grinev, *Khim.-Farm. Zh.*, **3,** 15 (1969); *Chem. Abstr.*, **72,** 66899 (1970).
26. D. S. Jones, G. W. Kenner, and R. C. Sheppard, *Experientia*, **19,** 126 (1963); *Chem. Abstr.*, **59,** 1751 (1963).
27. D. S. Jones, G. W. Kenner, J. Preston, and R. C. Sheppard, *Tetrahedron*, **21,** 3209 (1965); *Chem. Abstr.*, **64,** 5205 (1966).
28. E. Abushanab, A. P. Bindra, L. Goodman, and H. Peterson, Jr., *J. Org. Chem.*, **38,** 2049 (1973).
29. A. Albert and K. Ohta, *J. Chem. Soc.* (*C*), 1540 (1970).
30. E. Abushanab, *Tetrahedron Lett.*, 1441 (1971).
31. G. B. Crippa and A. Crippa, *Il Farmaco* (Pavia) *Ed. Sci.*, **10,** 616 (1955); *Chem. Abstr.*, **50,** 1840 (1956).
32. A. M. Bellini, *Ann. Chim.* (Rome), **51,** 1409 (1961); *Chem. Abstr.*, **56,** 15509 (1962).
33. E. Jucker and E. Rissi, *Helv. Chim. Acta*, **45,** 2383 (1962).
34. M. P. Mertes and N. R. Patel, *J. Med. Chem.*, **9,** 868 (1966).
35. M. P. Mertes and A. J. Lin, *J. Med. Chem.*, **13,** 77 (1970).
36. R. A. Sharma, M. Bobek, F. E. Cole, and A. Bloch, *J. Med. Chem.*, **16,** 643 (1973).

37. R. P. Panzica and L. B. Townsend, *Tetrahedron Lett.*, 1013 (1970).
38. S. F. Mason, *J. Chem. Soc.*, 4874 (1957).
39. E. Schipper and A. R. Day, *J. Am. Chem. Soc.*, **74,** 350 (1952).
40. F. L. Muehlmann and A. R. Day, *J. Am. Chem. Soc.*, **78,** 242 (1956).
41. F. M. Hershenson, L. Bauer, and K. F. King, *J. Org. Chem.*, **33,** 2543 (1968).
42. G. Palamidessi and F. Luini, *Farmaco, Ed. Sci.*, **21,** 811 (1966); *Chem. Abstr.*, **66,** 37886 (1967).
43. S. Kawano, S. Zoga, H. Watanabe, and T. Sato (to Eisai Co. Ltd.), Jap. Pat. 71 33,957; *Chem. Abstr.*, **75,** 151835 (1971).
44. S. Kawano, S. Zoga, H. Watanabe, and T. Sato (to Eisai Co. Ltd.), Jap. Pat. 71 33,956; *Chem. Abstr.*, **76,** 3902 (1972).
45. R. J. Tull and P. I. Pollak (to Merck & Co., Inc.), U.S. Pat. 3,541,093; *Chem. Abstr.*, **74,** 141878 (1971).
46. E. J. J. Grabowski, E. W. Tristram, and R. J. Tull (to Merck & Co., Inc.), U.S. Pat. 3,549,633; *Chem. Abstr.*, **74,** 53847 (1971).
47. J. H. Jones, W. J. Holtz, and E. J. Cragoe, Jr., *J. Med. Chem.*, **16,** 537 (1973).
48. J. H. Jones and E. J. Cragoe, Jr. (to Merck & Co., Inc.), U.S. Pat. 3,461,123; *Chem. Abstr.*, **71,** 101883 (1969).
49. E. J. Cragoe, Jr., and J. H. Jones (to Merck & Co., Inc.), Fr. Pat. 1,578,366; *Chem. Abstr.*, **72,** 111514 (1970).
50. R. J. Tull and P. I. Pollak (to Merck & Co., Inc.), Brit. Pat. 1,248,146; *Chem. Abstr.*, **75,** 140887 (1971).
51. R. J. Tull and P. I. Pollak (to Merck & Co., Inc.), Brit. Pat. 1,238,105; *Chem. Abstr.*, **75,** 88639 (1971).
52. E. J. J. Grabowski, E. W. Tristram, and R. J. Tull (to Merck & Co., Inc.), Ger. Offen. 1,957,711; *Chem. Abstr.*, **73,** 66621 (1970).
53. R. J. Tull and P. I. Pollak (to Merck & Co., Inc.), Brit. Pat. 1,232,758; *Chem. Abstr.*, **75,** 49134 (1971).
54. G. Tuchtenhagen and K. Ruehlmann, *Justus Liebigs Ann. Chem.*, **711,** 174 (1968).
55. L. R. Mandel, *Biochem. Pharmacol.*, **20,** 3413 (1971).
56. M. K. Bach and J. Fellig, *Plant Physiol.*, **35,** 36 (1960); *ibid.*, **36,** 85 (1961); *Chem. Abstr.*, **54,** 16566 (1960); *ibid.*, **55,** 9590 (1961).
57. D. B. McNair Scott, M. L. Rogers Batcheler, E. Chu Lesher, and A. M. Pakoskey, *Cancer Res., Suppl.*, **23,** 1235 (1963).
58. Ilford Ltd., Neth. Appl. 6,407,905; *Chem. Abstr.*, **62,** 15631 (1965).
59. T. Goto, M. Isobe, Y. Kishi, S. Inoue, and S. Sugiura, *Tetrahedron*, **31,** 939 (1975).
60. E. Abushanab, A. P. Bindra, D.-Y. Lee, and L. Goodman, *J. Heterocycl. Chem.*, **12,** 211 (1975).
61. E. Abushanab, A. P. Bindra, and L. Goodman, *J. Heterocycl. Chem.*, **12,** 207 (1975).
62. M. F. Depompei and W. W. Paudler, *J. Heterocycl. Chem.*, **12,** 861 (1975).
63. A. Omodei-Sale and E. Toja, *Farmaco, Ed. Sci.*, **30,** 650 (1975).
64. E. Abushanab, A. P. Bindra, and D.-Y. Lee, *J. Org. Chem.*, **40,** 3373 (1975).
65. E. Abushanab, D.-Y. Lee, and L. Goodman, *J. Org. Chem.*, **40,** 3376 (1975).
66. E. Abushanab, A. P. Bindra, and L. Goodman, *J. Org. Chem.*, **40,** 3379 (1975).
67. A. S. Elina, I. S. Musatova, M. A. Muratov, and M. D. Mashkovskii, *Khim. Farm. Zh.*, **9**(3), 10 (1975); *Chem. Abstr.*, **82,** 170839 (1975).
68. A. S. Elina and I. S. Musatova, *Khim. Geterotsikl. Soedin.*, 690 (1975); *Chem. Abstr.*, **83,** 114336 (1975).
69. Y. C. Tong, *J. Heterocycl. Chem.*, **12,** 1127 (1975).

CHAPTER XXII

Pyrazolopyrazines

I. Pyrazolo[1,5-*a*]pyrazines 398
II. Pyrazolo[3,4-*b*]pyrazines 400
III. Tables 406
Table 1. Pyrazolo[1,5-*a*]pyrazines 406
Table 2. Pyrazolo[3,4-*b*]pyrazines 407
IV. References 408

I. Pyrazolo[1,5-*a*]pyrazines

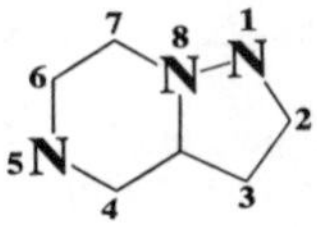

Only two reports have been found dealing with the chemistry of this ring system. In the first of these the amination of several simple pyrazine derivatives such as the dimethyl compound (**1**) with hydroxylamine *O*-sulfonic acid to give the corresponding *N*-amino compounds is described. Treatment of the latter with derivatives of acetylenecarboxylic acid provides pyrazolo[1,5-*a*]pyrazines such as **2** in 2–10% yields. In contrast, the *N*-oxide of **1** gave a 41% yield of the 5-oxide **3**. Condensation of the

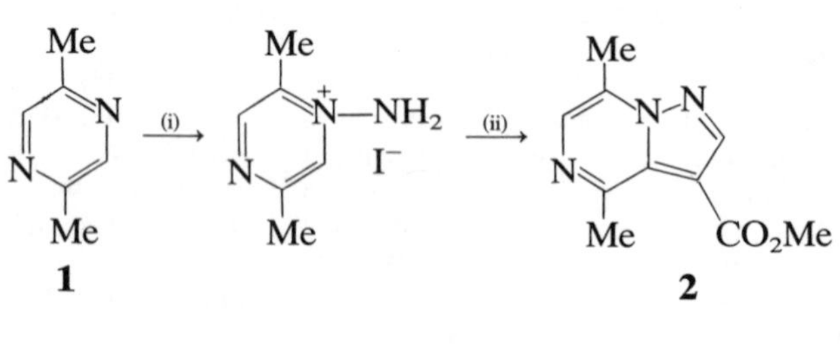

Reagents (i) NH_2OSO_3H, HI
(ii) $HC{\equiv}CCO_2Me$

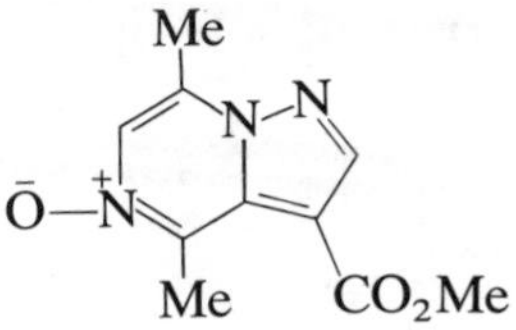

3

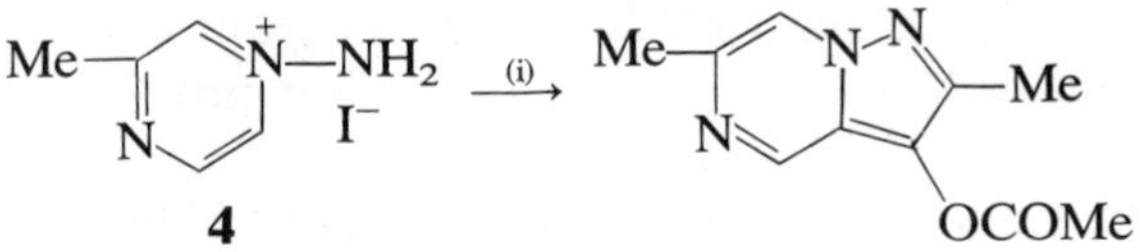

4 → 5

Reagents (i) Ac_2O, NaOAc

N-aminopyrazinium compound **4** with acetic anhydride and sodium acetate afforded the ester **5** in 5% yield.[19]

The only other report of a synthesis of this ring system concerns the formation of compound **9** by treatment of the pyrazole-5-carboxylic acid **8** with thionyl chloride. The ethyl ester of **8** is formed together with **7** on alkylation of the pyrazole **6** with β-dimethylaminoethyl chloride.[1]

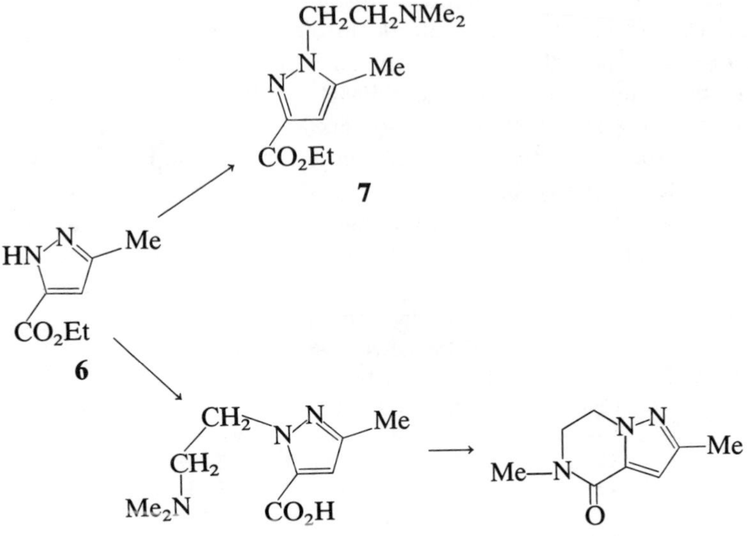

6 → 7

6 → 8 → 9

II. Pyrazolo[3,4-*b*]pyrazines

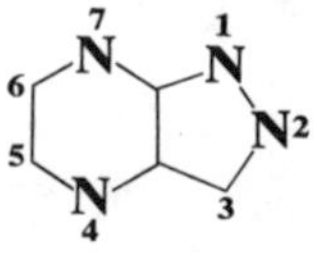

1*H* RRI 1171
3*H* RRI 7965

Pyrazolo[3,4-*b*]pyrazines have been the subject of a variety of reports dealing mainly with the synthesis and physical properties of the ring system.

The UV spectrum of the unsubstituted compound **10** at pH 7 has three maxima, at 208, 290, and 300 nm, with an inflection at 320 nm. At $H_0 = -2.8$, the spectrum consists of two bands at 212 and 302 nm. Measurements of the change in optical density at 275 nm indicated that the parent heterocycle is a weak base of $pK_a = -0.64$. In *M*-deuterohydrochloric acid the PMR spectrum exhibits the expected pattern with a singlet at δ 8.72 (H-3) and an AB quartet (H-5, H-6) centered at δ 9.00 and 9.16 ($J_{5,6} = 3$ Hz).[2] As the acidic solvent used was not strong enough to effect complete protonation of the molecule, these PMR measurements must have been made on a mixture of the free base and cation of compound **10**. Brief details have been reported of the PMR spectrum of the free base in methanol.[3] More recently the free base and its 4-oxide have been compared using UV and PMR spectroscopy.[20]

The PMR spectra of various methyl-substituted compounds have also been recorded.[2,3] The proton at the 1-position in compound **11** is evident as an exchangeable singlet in deuterochloroform at δ 8.80. The 3-proton at δ 8.29 is also apparent.[3] Methylation of 1*H*-pyrazolo[3,4-*b*]pyrazine gave two monomethyl derivatives whose PMR spectra are given in Figure 1. The structural assignments of the isomers as the 1-methyl derivative **12**, (m.p. 102–103°) and the 2-methyl compound **13** (m.p. 152–154°) are based on alternative syntheses. However, at approximately the same time another paper[2] appeared describing a monomethyl pyrazolo[3,4-*b*]pyrazine (m.p. 158–159°) as the 1-methyl derivative. As neither report

Figure 1. PMR SPECTRA IN $CDCl_3$ OF COMPOUNDS **12** AND **13**

Compound	Methyl	H-3	H-5[a]	H-6[a]	$J_{5,6}$
12	δ 4.20	8.29	8.49	8.58	2.3 Hz
13	δ 4.37	8.27	8.55	8.64	2.0 Hz

[a] These assignments could be reversed.

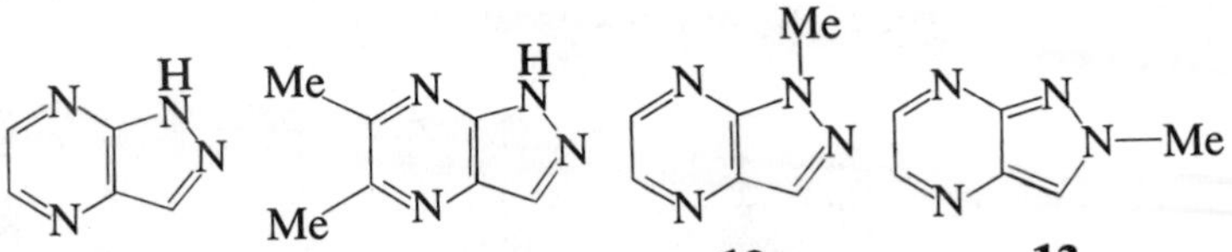

10 **11** **12** **13**

described a completely unambiguous synthesis of either isomer, the inconsistency remains to be resolved.

Although UV data are available for the 3-hydroxy derivatives of this ring system,[4] no IR details have been published. Accordingly it is unclear which of the possible tautomeric structures (**14–16**) best represents these compounds. Possibly, as with 5-hydroxypyrazoles, the hydroxy form **14** is less stable than oxo tautomers such as **16**. Ultraviolet and infrared data are, however, available for the 6-hydroxy derivatives (**17**).[5] In the solid state there is surprisingly no evidence for a carbonyl group in the infrared. The presence of a broad band from 2300 to 3100 cm^{-1} (ν C=$\overset{+}{N}$H) is taken as indicating the importance of the dipolar structure **18** at the expense of the neutral forms **17** and **19**.

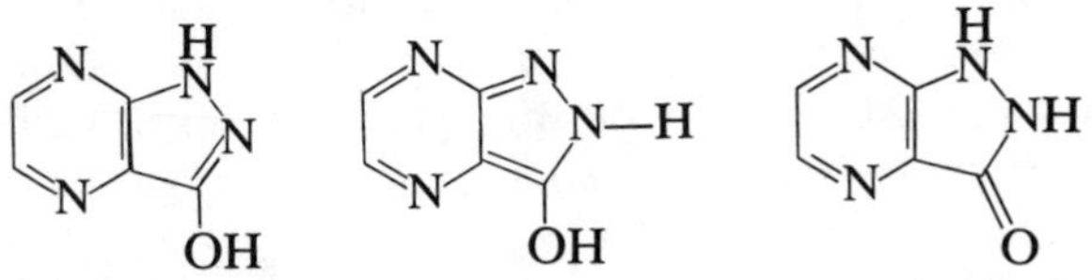

14 **15** **16**

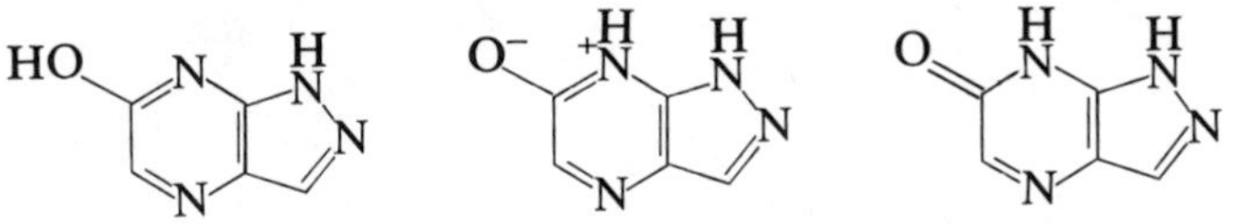

17 **18** **19**

The position of protonation of pyrazolo[3,4-*b*]pyrazine has not been elucidated. The parent heterocycle (pK_a −0.64) is a weaker base than either pyrazine (pK_a 0.4) or pyrazole (pK_a 2.5). The progressive base-strengthening effect of the successive introduction of methyl groups into the molecule is evident from Figure 2. The small change in pK_a on introduction of a 1-methyl group (0.19, 0.21 units) contrasts with the much larger effect caused by the introduction of methyl groups at the 5- and 6-positions (1.37 units). This suggests that protonation occurs in the pyrazine ring, possibly at the 7-position.

Figure 2. BASICITIES OF METHYLPYRAZOLO[3,4-*b*]PYRAZINES

Compound	pK_a	Ref.
Unsubstituted	−0.64	6
1-Methyl	−0.45	2
5,6-Dimethyl	0.73	6
1,5,6-Trimethyl	0.94	2

Compound **20** is included in a study of 300 pyrazoles whose UV fluorescence colors have been reported.[7]

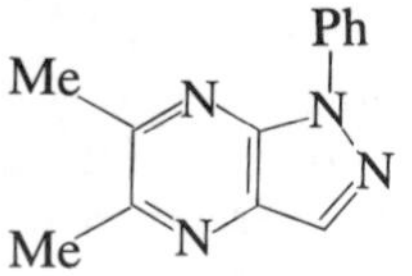

20

The most frequently adopted approach to the synthesis of this ring system involves condensation of appropriately substituted 4,5-diaminopyrazoles **21** with α-dicarbonyl compounds such as benzil, diacetyl, and glyoxal. Such condensations have given pyrazolo[3,4-*b*]pyrazines unsubstituted in the pyrazole ring.[2,3,6] Similarly, 1-methyl compounds,[2,3] 1-phenyl-3-methyl derivatives,[8,9] and 3-hydroxy compounds, both unsubstituted in the 1-position and substituted by methyl or phenyl groups,[4,10] have been obtained by this approach.

21 + R^2COCOR^2 **22** $R^2 = H$, Me, or Ph → **23**

The reaction of 4,5-diamino-3-hydroxypyrazole (**24**) with pyruvaldehyde (MeCOCHO) afforded a mixture of compounds **25** and **26** in a ratio of 3:1. This is the only example of a synthesis of this ring system starting from an unsymmetrical dicarbonyl compound.[4] Condensation of the 3,4-diaminopyrazole **27** with glyoxal or diacetyl gave the 2-substituted derivatives **28**.[4]

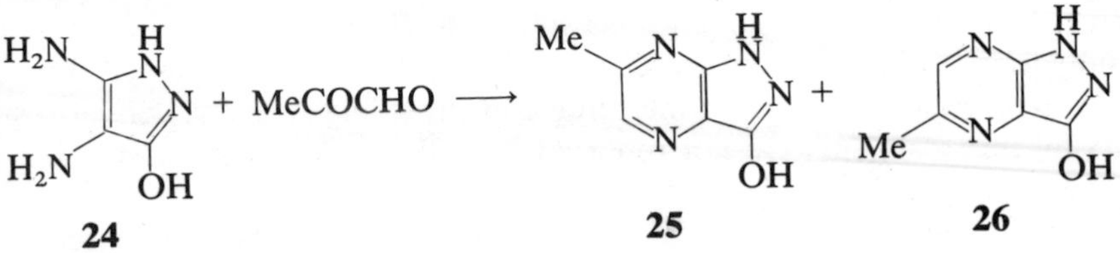

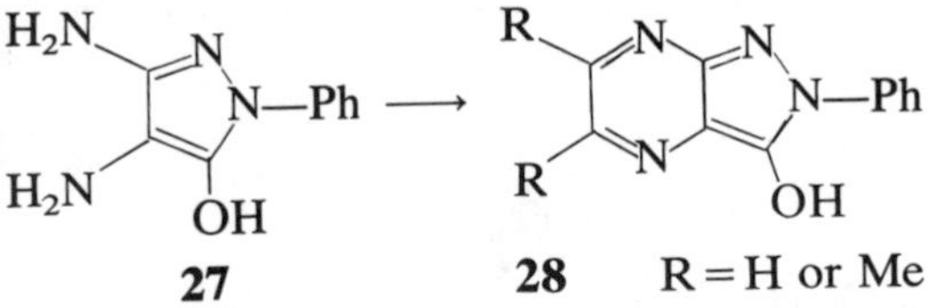

H₂N, H, N, N + MeCOCHO ⟶ Me, N, H, N, N + N, H, N, N, Me, N, OH

24 **25** **26**

27 ⟶ **28** R = H or Me

Treatment of the diaminopyrazole **29** with bromoacetic acid and zinc results in cyclization to the oxo compound **30** identical with that obtained unambiguously from compound **32**. Apparently none of the 5-oxo isomer **31** was isolated.[5] Reaction of the nitrosopyrazole **32** with acetic anhydride provided the 6-acetoxy derivative **34**.[5] Various substituted analogues were also made by this approach, alkaline hydrolysis providing the

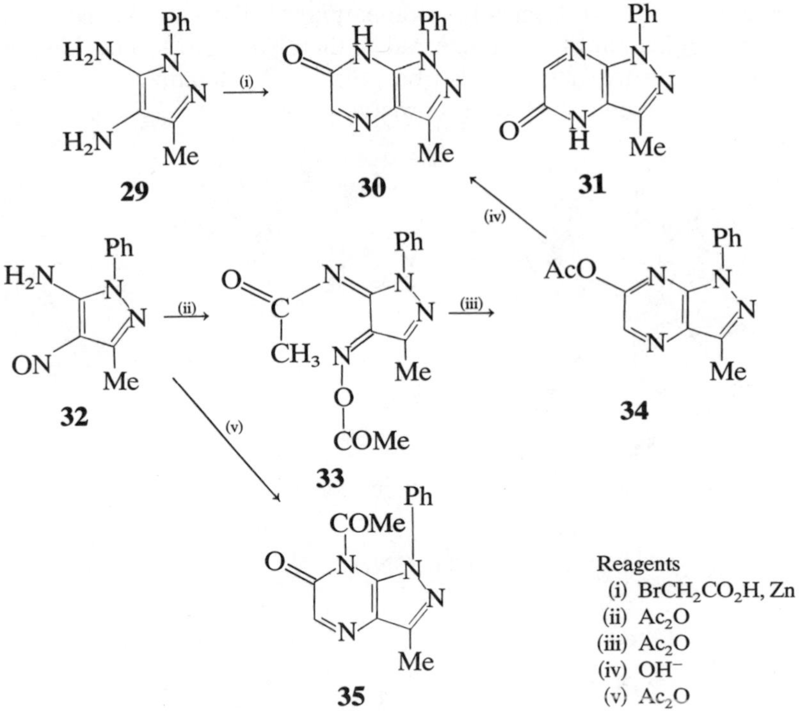

29 →(i) **30** **31**

32 →(ii) **33** →(iii) **34** →(iv) **30**

32 →(v) **35**

Reagents
(i) $BrCH_2CO_2H$, Zn
(ii) Ac_2O
(iii) Ac_2O
(iv) OH^-
(v) Ac_2O

corresponding oxo compounds such as **30**. The reaction with acetic anhydride proceeds through a diacetyl derivative formulated as **33**. Earlier experiments which were interpreted as providing *N*-acetyl derivatives **35** by this reaction were not reported in sufficient detail to substantiate this claim.[11]

Unsymmetrical derivatives of this ring system have also been prepared in unambiguous fashion by condensation of the arylalkyl ketones **37** with the azo derivatives **36**.[12,13]

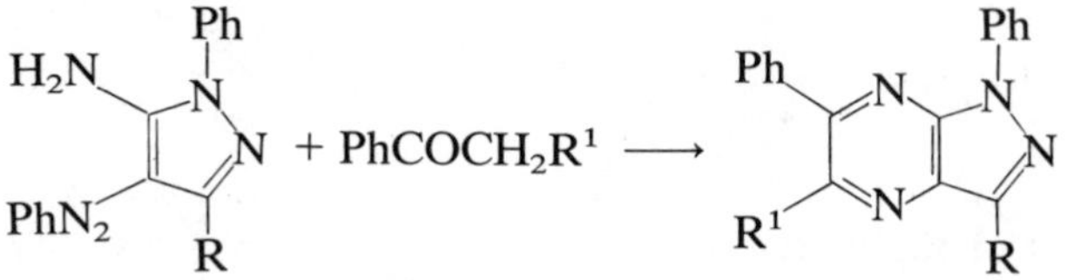

36 + $PhCOCH_2R^1$ ⟶

37 R^1 = H or Me

When the thiomorpholide **38** is heated, hydrogen sulfide is eliminated and the morpholino compound **39** is formed.[14] However acid treatment results in an alternative mode of decomposition to give the "mercapto" compound **40**.[15] Presumably protonation of the morpholine residue favors elimination of the amine rather than hydrogen sulfide. The morpholino derivative **39** may also be prepared by treatment of the dichloroacetyl compound **41** with morpholine and sulfur.[16]

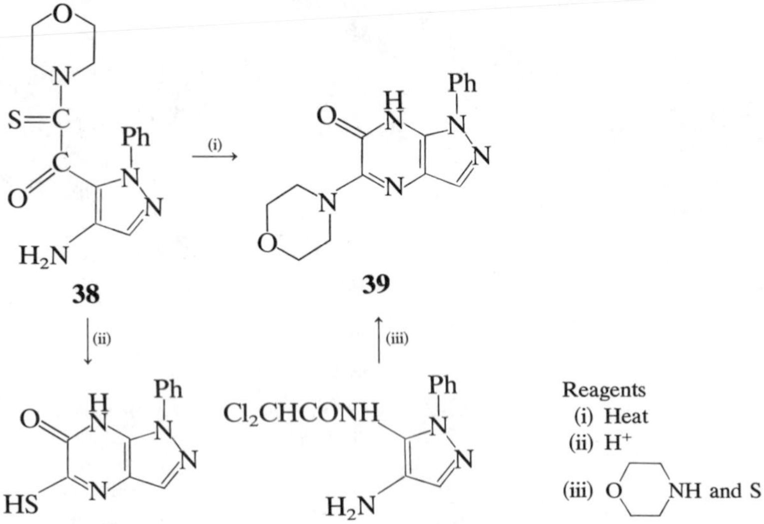

38 —(i)→ **39**

38 —(ii)→ **40**

41 —(iii)→ **39**

38 **39** **40** **41**

Reagents
(i) Heat
(ii) H^+
(iii) Morpholine (O⟨ ⟩NH) and S

Although no details have been given, it seems that treatment of the cyanopyrazine **42** with hydrazine in the presence of potassium iodide affords the 3-aminopyrazolo[3,4-*b*]pyrazine **43**.[17] Oxidation of the pyrazoloquinoxaline **44** with aqueous potassium permanganate gave the dicarboxylic acid **45**.[18]

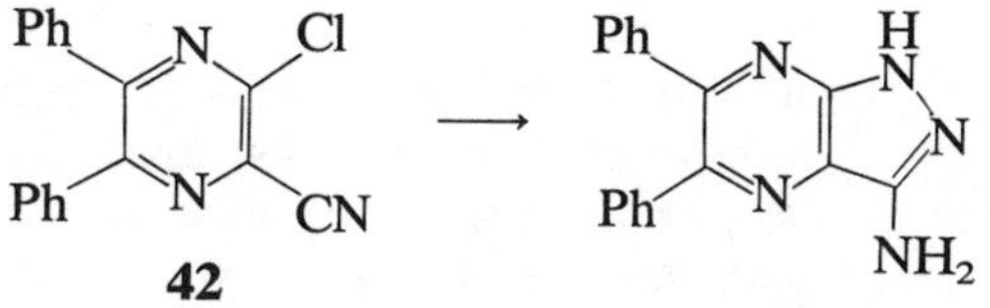

Whereas acetylation of the unsubstituted heterocycle **10** gave the 1-acetyl derivative,[2] methylation using alkaline methyl sulfate gave a mixture from which were isolated compounds formulated as the 1-methyl derivative (26%) and the 2-methyl compound (34%). Similar results were obtained with the 5,6-dimethyl compound.[3] Methylation of the 3-hydroxy derivative **14** with alkaline methyl iodide gave the 1-methyl compound. Similarly benzyl chloride yielded the 1-benzyl derivative **46**.[4]

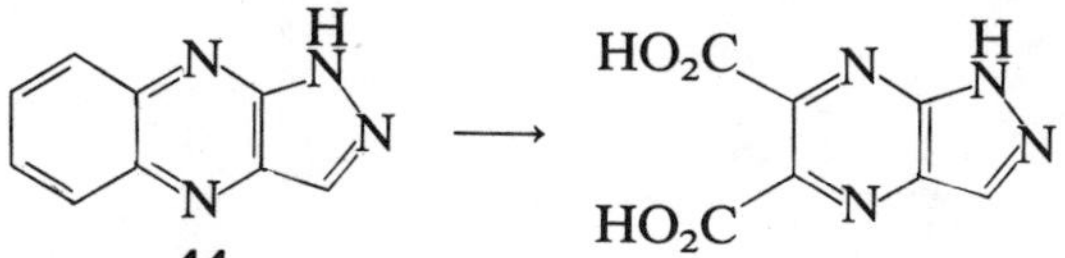

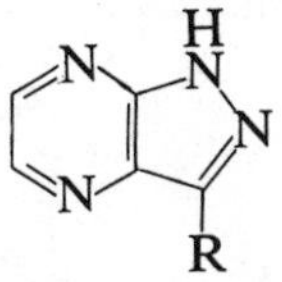

10 R=H
14 R=OH

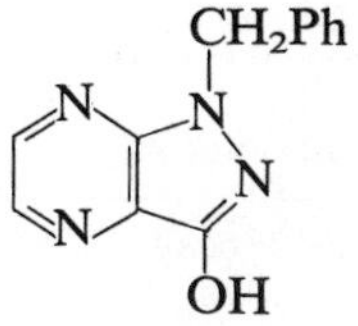

46

Acetylation of a 6-"hydroxy" pyrazolo[3,4-*b*]pyrazine with acetic anhydride gave the *O*-acetyl compound **34**, which, as mentioned earlier, can be hydrolyzed with sodium hydroxide to the parent oxo compound **30**.[5] An earlier claim that acetylation of (**30**) gave the *N*-acetyl derivative was not substantiated.[11] Treatment of compound **30** with phosphoryl chloride afforded the 6-chloro derivative **47**.[5]

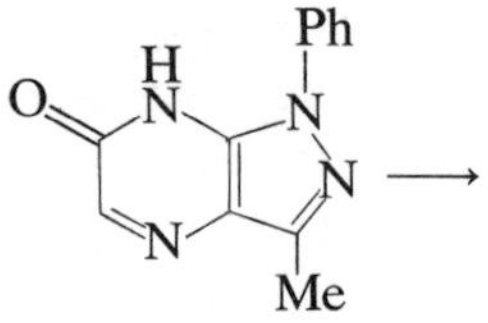

Pyrazolo[3,4-*b*]pyrazines are useful intermediates for the preparation of 2-aminopyrazine-3-carboxamides such as **49**, much used in pteridine syntheses.[4,10] For example, reductive cleavage of the 1-substituted 3-hydroxy compound **48** with Raney nickel results in smooth breakdown of the pyrazole ring to give the product **49**.

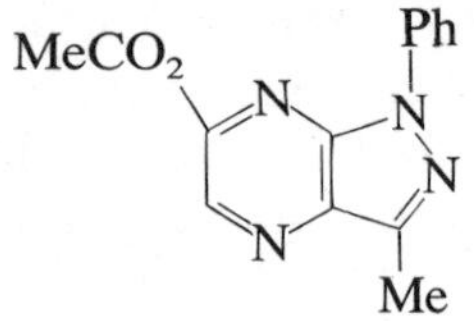

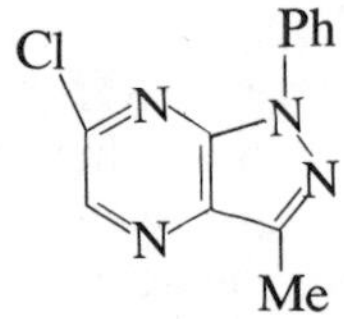

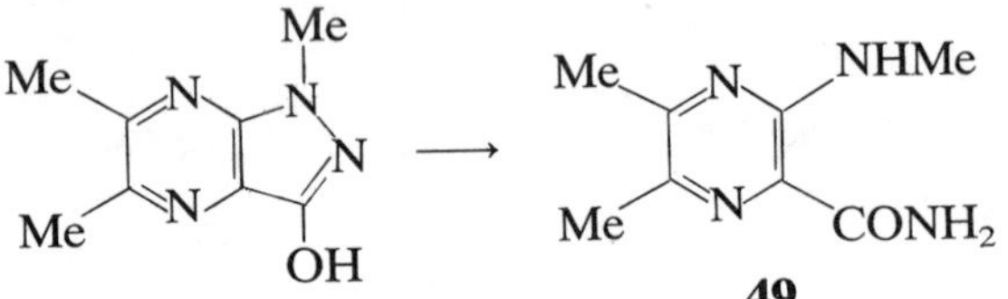

The 6-oxo compounds **39** and **40** have been patented as diuretics.[14–16]

III. Tables

TABLE 1. PYRAZOLO[1,5-*a*]PYRAZINES

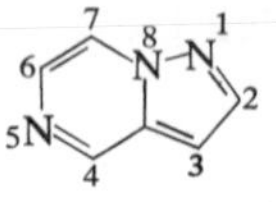

Molecular formula	Substituents	m.p., °C	Ref.
$C_7H_5N_3O_2$	3-Carboxy		
	Methyl ester	137–138	19
$C_8H_5N_3O_4$	2,3-Dicarboxy		
	Dimethyl ester	139–140	19
$C_8H_7N_3O_2$	3-Carboxy-4-methyl		
	Methyl ester	137–139	19
$C_8H_9N_3O$	3-Hydroxy-2,6-dimethyl	204–205	19
	Acetate	137–138	
	Benzoate	148–149	
$C_8H_{11}N_3O$	4,5,6,7-Tetrahydro-5-methyl-4-oxo	106	1
$C_9H_9N_3O_2$	3-Carboxy-4,7-dimethyl		
	Methyl ester	130–131	19
	5-Oxide methyl ester	238–240	

TABLE 2. PYRAZOLO[3,4-*b*]PYRAZINES

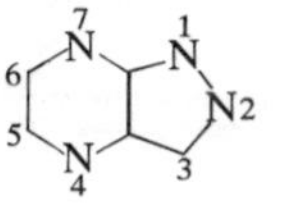

Molecular formula	Substituents	m.p., °C	Ref.
$C_5H_4N_4$	Unsubstituted	198–200	2, 3
	4-Oxide	248–250	20
$C_5H_4N_4O$	3-Hydroxy	314–315 decomp.	4, 10
$C_5H_4N_4O_2$	4,5,6,7-Tetrahydro-5,6-dioxo[a]	>360	3
$C_6H_6N_4$	1-Methyl[b]	158–159, 102–103[c]	2, 3
	2-Methyl[b]	152–154	3
$C_6H_6N_4O$	3-Hydroxy-1-methyl	242–243	4, 10
	3-Hydroxy-5-methyl	234–235	4
	3-Hydroxy-6-methyl	312–321	4
$C_7H_4N_4O_4$	5,6-Dicarboxy		18
$C_7H_6N_4O$	1-Acetyl	154–155	2
$C_7H_8N_4$	5,6-Dimethyl	163–165, 169-171	2, 3
$C_7H_8N_4O$	3-Hydroxy-5,6-dimethyl	325 decomp.	4, 10
$C_8H_{10}N_4$	1,5,6-Trimethyl	106–108	2, 3
	2,5,6-Trimethyl	96–98	3
$C_8H_{10}N_4O$	3-Hydroxy-1,5,6-trimethyl	268–269	4, 10
$C_{11}H_8N_4OS$	4,5,6,7-Tetrahydro-6-oxo-5-thio[d]	249–251	15
$C_{11}H_{10}N_4O$	2,3-Dihydro-3-oxo-2-phenyl	232–233.5	4
	3-Hydroxy-1-phenyl	227–229	4
$C_{12}H_9ClN_4$	6-Chloro-3-methyl-1-phenyl	77	5
$C_{12}H_{10}N_4$	3-Methyl-1-phenyl	68–69	9
$C_{12}H_{10}N_4O$	1-Benzyl-3-hydroxy	175–176	4
	6,7-Dihydro-1-methyl-6-oxo-3-phenyl	291	5
	6,7-Dihydro-3-methyl-6-oxo-1-phenyl	298	5
$C_{13}H_{12}N_4O$	6,7-Dihydro-1,5-dimethyl-6-oxo-3-phenyl	310	5
	6,7-Dihydro-3,5-dimethyl-6-oxo-1-phenyl	231	5
	2,3-Dihydro-5,6-dimethyl-3-oxo-2-phenyl	239–240[e]	4
$C_{14}H_{12}N_4O_2$	6-Acetoxy-1-methyl-3-phenyl	143–144	5
	6-Acetoxy-3-methyl-1-phenyl	98	5
$C_{14}H_{14}N_4$	3,5,6-Trimethyl-1-phenyl	123.5–124	9
$C_{14}H_{14}N_4O$	5-Ethyl-6,7-dihydro-1-methyl-6-oxo-3-phenyl	284–285	5
	5-Ethyl-6,7-dihydro-3-methyl-6-oxo-1-phenyl	175	5
$C_{15}H_{15}N_4O_2$	6,7-Dihydro-5(morpholin-1-yl)-6-oxo-1-phenyl	238–240	14, 16
$C_{16}H_{16}N_4O_2$	1,5-Dimethyl-3-phenyl-6-propionoxy	136	5
	3,5-Dimethyl-1-phenyl-6-propionoxy	129	5
$C_{17}H_{12}N_4$	1,6-Diphenyl	130–132	13
$C_{17}H_{12}N_4O$	6,7-Dihydro-6-oxo-1,3-diphenyl	216	5
	3-Hydroxy-5,6-diphenyl	269 decomp.	4, 10
$C_{17}H_{13}N_5$	3-Amino-5,6-diphenyl	[f]	17
$C_{18}H_{14}N_4$	3-Methyl-1,6-diphenyl	186	12
$C_{18}H_{14}N_4O$	6,7-Dihydro-5-methyl-6-oxo-1,3-diphenyl	305–306	5
	3-Hydroxy-1-methyl-5,6-diphenyl	258–260	4

TABLE 2 (*contd.*)

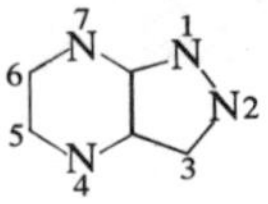

Molecular formula	Substituents	m.p., °C	Ref.
$C_{18}H_{20}N_4O_2$	6-*n*-Butyryloxy-5-ethyl-1-methyl-3-phenyl	81	5
	6-*n*-Butyryloxy-5-ethyl-3-methyl-1-phenyl	60	5
$C_{19}H_{14}N_4O_2$	6-Acetoxy-1,3-diphenyl	145	5
$C_{19}H_{16}N_4$	3,5-Dimethyl-1,6-diphenyl	186	12
	3-Methyl-1-phenyl-6-*p*-tolyl	192	12
$C_{19}H_{16}N_4O$	5-Ethyl-6,7-dihydro-6-oxo-1,3-diphenyl	257–258	5
$C_{21}H_{18}N_4O_2$	5-Methyl-1,3-diphenyl-6-propionoxy	130	5
$C_{23}H_{22}N_4O_2$	6-*n*-Butyryloxy-5-ethyl-1,3-diphenyl	115	5
$C_{24}H_{18}N_4$	3-Methyl-1,5,6-triphenyl	183, 190	8, 9

[a] Described as the 5,6-dihydroxy compound.
[b] Assignments unclear (see text).
[c] Boiling point 216–217°.
[d] Referred to as the 6-hydroxy-5-mercapto derivative.
[e] A polymorphic form of this compound has m.p. 193–195°.
[f] No data have been reported for this compound.

IV. References

1. M. Mühlstädt and J. Zschiedrich, *J. Prakt. Chem.*, **311,** 363 (1969).
2. M. E. C. Biffin, D. J. Brown, and Q. N. Porter, *J. Chem. Soc.* (*C*), 2159 (1968).
3. H. Dorn and H. Dilcher, *Justus Liebigs Ann. Chem.*, **717,** 118 (1968).
4. E. C. Taylor, J. W. Barton, and T. S. Osdene, *J. Am. Chem. Soc.*, **80,** 421 (1958).
5. M. Guarneri and P. Giori, *Gazz. Chim. Ital.*, **99,** 463 (1969); *Chem. Abstr.*, **71,** 81306 (1969).
6. M. E. C. Biffin, D. J. Brown, and Q. N. Porter, *Tetrahedron Lett.*, 2029 (1967).
7. I. I. Grandberg, S. V. Tabak, and A. N. Kost, *Zh. Obshch. Khim.*, **33,** 525 (1963); *Chem. Abstr.*, **59,** 1616 (1963).
8. Mohr, J. pr. [2], **79,** 43 (1909); Beilstein, **26,** 381^9.
9. I. I. Grandberg and G. V. Klyuchko, *Zh. Obshch. Khim.*, **32,** 1898 (1962); *Chem. Abstr.*, **58,** 4537 (1963).
10. T. S. Osdene and E. C. Taylor, *J. Am. Chem. Soc.*, **78,** 5451 (1956).
11. M. Guarneri and P. Giori, *Boll. Chim. Farm.*, **103,** 242 (1964); *Chem. Abstr.*, **61,** 8308 (1964).
12. G. B. Crippa and G. Perroncito, *Gazz. Chim. Ital.*, **66,** 649 (1936); *Chem. Abstr.*, **31,** 3488 (1937).
13. M. Guarneri and L. Duda, *Ann. Chim.* (Rome), **51,** 446 (1961); *Chem. Abstr.*, **55,** 27275 (1961).
14. Ciba Ltd., Fr. Pat. 1,350,002; *Chem. Abstr.*, **62,** 9153 (1965).

15. Ciba Ltd., Fr. Pat. 1,350,003; *Chem. Abstr.*, **62,** 7780 (1965).
16. Ciba Ltd., Brit. Pat. 1,021,195; *Chem. Abstr.*, **65,** 13738 (1966).
17. E. C. Taylor, Jr., and W. W. Paudler, *Chem. Ind.* (London), 1061 (1955).
18. W. Sauer and G. Henseke, *Z. Chem.*, **10,** 381 (1970); *Chem. Abstr.*, **74,** 13096 (1971).
19. K. Kasuga, M. Hirobe, and T. Okamoto, *Chem. Pharm. Bull.* (Tokyo), **22,** 1814 (1974).
20. I. A. Korbukh, N. G. Yakunina, H. Dorn, and M. N. Preobrazhenskaya, *Zh. Org. Khim.*, **12,** 900 (1976).

CHAPTER XXIII

Triazolo- and Tetrazolopyrazines

I. s-Triazolo[1,5-a]pyrazines . . . 410
II. s-Triazolo[4,3-a]pyrazines . . . 415
III. v-Triazolo[4,5-b]pyrazines . . . 423
IV. Tetrazolo[1,5-a]pyrazines . . . 423
V. Tables . . . 425
Table 1. s-Triazolo[1,5-a]pyrazines . . . 425
Table 2. s-Triazolo[4,3-a]pyrazines . . . 426
Table 3. v-Triazolo[4,5-b]pyrazines . . . 428
Table 4. Tetrazolo[1,5-a]pyrazines . . . 429
VI. References . . . 429

I. *s*-Triazolo[1,5-*a*]pyrazines

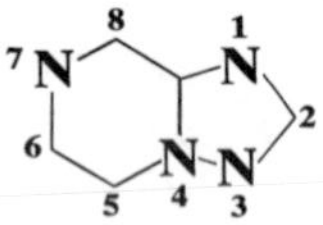

This ring system has been referred to on several occasions as *s*-triazolo[2,3-*a*]pyrazine and numbered as shown in formula **1** below. Ultraviolet spectra of several simple derivatives of the heterocycle have been recorded in cyclohexane and are useful in differentiating between derivatives of this ring system and the isomeric *s*-triazolo[4,3-*a*]pyrazines.[1] In cyclohexane, 5,6-dimethyl-2-phenyl-*s*-triazolo[1,5-*a*]pyrazine has maxima at λ (log ε) 249 (4.65), 258 (4.62), 281 *sh* (3.91), 294 (3.82), 301 *sh* (3.78), and 315 (3.55) nm. However, differences in UV spectra must be interpreted with caution. At least one supposed synthesis of *s*-triazolo[1,5-*a*]pyrazines, which was apparently supported by UV spectral data,[2] has now been shown to give 1*H*-imidazo[2,1-*c*]-*s*-triazoles[3] (see Section II).

The PMR spectrum[25] of *s*-triazolo[1,5-*a*]pyrazine in $CDCl_3$ shows single-proton signals at δ 8.50 (*s*, H-2), 8.58 (*dd*, H-5), 8.19 (*d*, H-6), and 9.37 (*d*, H-8), with $J_{5,6} = 4.5$ Hz, $J_{5,8} = 1.5$ Hz. The mass spectra of several *s*-triazolo[1,5-*a*]pyrazines have been recorded and compared with the fragmentation of the triazolo[4,3-*a*]system.[4] The unsubstituted compound **2** fragments by successive loss of two molecules of HCN from the molecular ion. The trimethyl compound **3** behaves analogously to derivatives of the [4,3-*a*] system, as it loses MeCN followed by a hydrogen atom and rearrangement to a diazatropylium ion.

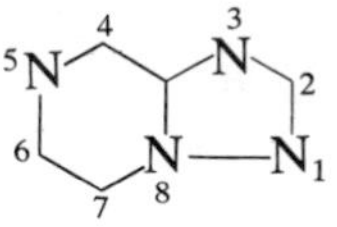

1 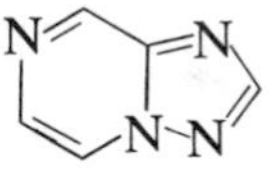 **2** 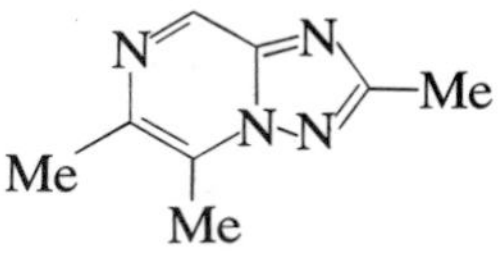 **3**

The only generally useful method of preparation of members of this ring system involves reaction of aminopyrazines **4** with nitriles in the presence of aluminum chloride to give the intermediates **5**. Treatment of these compounds with lead tetraacetate gives 2-substituted derivatives of the heterocycle **6**. The method has been successfully applied using phenyl cyanide;[1] and in spite of an earlier report to the contrary,[1] methyl cyanide has also been used.[5,6]

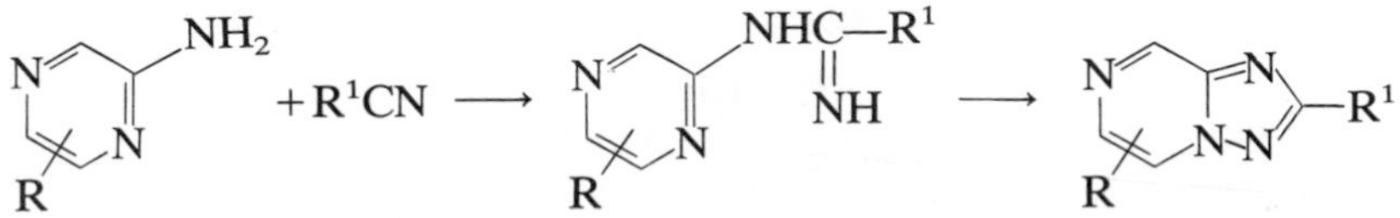

$+R^1CN \longrightarrow$ $\longrightarrow$

4 **5** **6**

The parent heterocycle and its 5- and 6-monochloro derivatives have been recently prepared from the corresponding aminopyrazines. For example, reaction of 2-amino-6-chloropyrazine with dimethylformamide dimethylacetal gave the intermediate **7**, which on reaction with hydroxylamine gave the oxime **8**. Cyclization of the oximes with polyphosphoric acid afforded the *s*-triazolo[1,5-*a*]pyrazines in 61–76% yields.[25,27]

Attempts to methylate 2,5,6-triphenyl-*s*-triazolo[1,5-*a*]pyrazine with methyl iodide were unsuccessful.[1] The 2-methyl derivative of the ring system, on treatment with hydrogen peroxide in acetic acid, gave the 7-oxide **9**.[5] Reaction of this *N*-oxide with phosphorus oxychloride gave a mixture of the chloro compounds **10** and **11**, which were separated and reacted with benzylamine, benzyl alcohol, and benzyl mercaptan to give the expected products such as **12** from nucleophilic displacement of the chlorine atom.

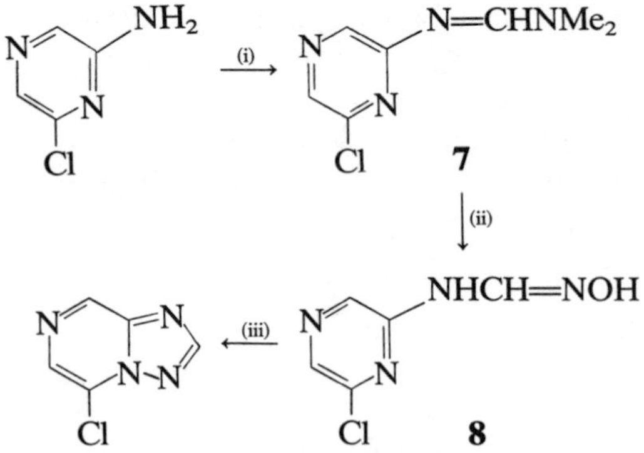

Reagents (i) $(MeO)_2CHNMe_2$
(ii) H_2NOH
(iii) Polyphosphoric acid

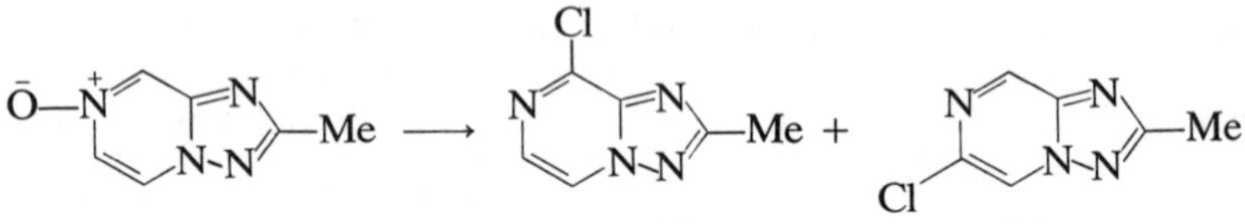

Bromination of *s*-triazolo[1,5-*a*]pyrazine with bromine in methylene chloride gives the 5-bromo compound **13**. The bromine atom is readily displaced by different nucleophiles to give various 5-substituted derivatives. Potassium hydrosulfide in ethanol at room temperature afforded the 5-mercapto compound **14**. Other substituents such as ethoxy, hydrazino, and methylthio could be introduced at moderately elevated temperatures.[26] The 5-amino compound **17**, not obtainable from the bromo compound **13** and ammonia, was prepared by treatment of the hydrazino derivative **15** with nitrous acid to give the azide **16**, which was then reduced with hydrogen sulfide.

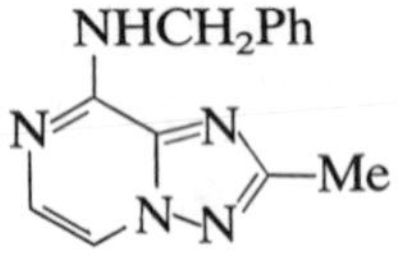

If 5-bromo-*s*-triazolo[1,5-*a*]pyrazine is treated with hydroxylamine or hydrazine at room temperature overnight, the abnormal substitution products **18** and **19** are formed by substitution at the 8-position. Similarly

Br

13

SH

14

(iii)

NHNH$_2$

15

N$_3$

16

NH$_2$

17

Reagents (i) Br_2–CH_2Cl_2
(ii) KHS–EtOH
(iii) NH_2NH_2
(iv) HNO_2
(v) H_2S

NHOH

18

NHNH$_2$

19

20

NHN=CHPh

21

Ph

Reagents (i) $(EtO)_3CH$
(ii) PhCHO
(iii) $Pb(OAc)_4$

8-amino-*s*-triazolo[1,5-*a*]pyrazine is obtained from the 5-bromo compound in liquid ammonia.[26] The hydrazino derivative **19** gave the unsubstituted heterocycle on treatment with hot aqueous cupric sulfate solution. Triethyl orthoformate converted **19** into the tricyclic system **20**. The 3-phenyl derivative of this heterocycle was obtained from lead tetraacetate oxidation of the benzylidene compound **21**.

Ring system **24**, an isomer of **20**, was prepared from 8-amino-*s*-triazolo[1,5-*a*]pyrazine by way of the intermediates **22** and **23**. The heterocycle **24** is most easily distinguished from the isomer **20** by its PMR spectrum.[26] The protons at positions 2 and 9 are magnetically equivalent giving a 2-proton singlet at δ 8.69. In contrast, the nonequivalent protons at positions 3 and 9 in **20** give singlets at δ 9.21 and 8.44.

NH_2 → $N{=}CHNMe_2$ **22** → $NHCH{=}NOH$ **23** → **24**

The susceptibility of *s*-triazolo[1,5-*a*]pyrazines to nucleophilic attack at the 5-position explains the decomposition products obtained when the parent heterocycle is heated with 50% aqueous sodium hydroxide solution.[26]

HO^- → H_2N—CH_2— (N-H triazole) + $HOCH_2CO_2H$

Substitution at the 5-position by a methyl group presumably inhibits this reaction to some extent as the 5,8-dimethyl compound (**26**) has been obtained in 20% yield by heating the isomeric *s*-triazolo[4,3-*a*]pyrazine **25** in 10% sodium hydroxide solution for 72 hr.[7]

8-Benzylamino-2-methyl-*s*-triazolo[1,5-*a*]pyrazine (**12**) has been found to possess marked cytokinin activity.[5,8] The only patent on this ring system claims various silver complexes as heat-developable, light-sensitive materials.[9]

Me Me 8 8a N N N 5 N Me → Me N N N N Me

25 **26**

II. s-Triazolo[4,3-a]pyrazines

7 N 8 1 N N 2 6 N 5 4 3

RRI 11829

Hückel molecular orbital calculations have shown that the 5-position in this ring system has associated with it a greater π-electron density than is present in the corresponding positions in *s*-triazolo[4,3-*a*]pyridine, *s*-triazolo[4,3-*a*]pyrimidine, and *s*-triazolo[4,3-*c*]pyrimidine.[7] This fact has been used to rationalize the difficulty in rearranging this heterocycle to its *s*-triazolo[1,5-*a*] analogue, in contrast to the ease with which the other *s*-triazolo systems rearrange.

The presence of bands in the infrared spectrum of hydroxy compound **27** in the region 1710–1760 cm^{-1} has been taken to mean that the compound exists as the oxo compound **28**.[2] However, the isomeric compound **29** is reported to have bands in the infrared characteristic of hydroxyl groups and amide carbonyl functions.[7] Accordingly it is suggested that this is evidence for the presence of the equilibrium **29** ⇌ **30**. Similarly the mercapto derivative **31** shows evidence of the equilibrium **31** ⇌ **32**. In contrast, the quinoxaline analogue **33** is completely in the thioamide form.[7]

The infrared spectrum of the acetyl derivative **34** of 3-amino-6-methyl-8-propyl-*s*-triazolo[4,3-*a*]pyrazine shows a band at 1708 cm^{-1} assigned to the acetamido group.[10] However the trifluoroacetyl derivative absorbs instead at 1620 cm^{-1}, which is interpreted as evidence for the isomeric formulation **35**.

Ultraviolet absorption data in either ethanol or cyclohexane have been tabulated for ten derivatives of the ring system.[1] In ethanol, *s*-triazolo[4,3-*a*]pyrazine has three main bands, at 206, 253, and 292 nm, and the spectrum is very similar to that of *s*-triazolo[4,3-*a*]pyrimidine and the corresponding fused pyridine. Replacement of CH by N in the pyridine ring of *s*-triazolo[4,3-*a*]pyridine results in only small shifts in the

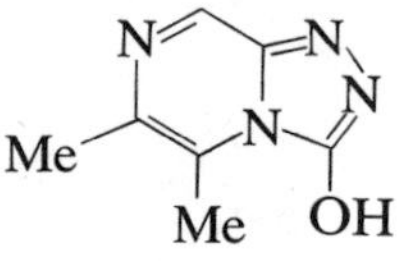

27

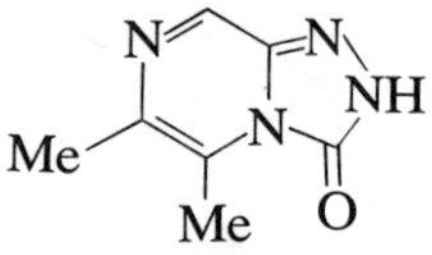

28

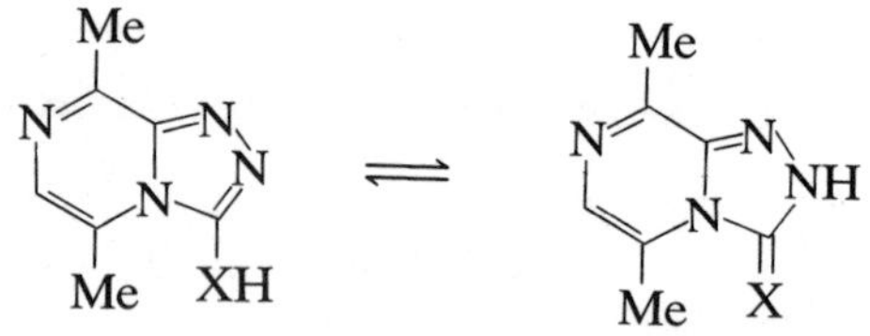

29 X=O
31 X=S

30 X=O
32 X=S

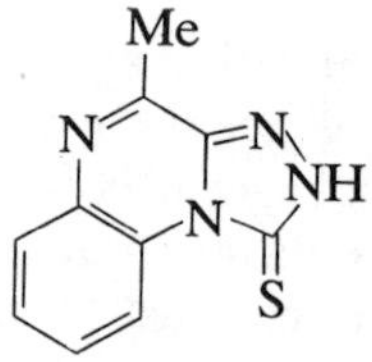

33

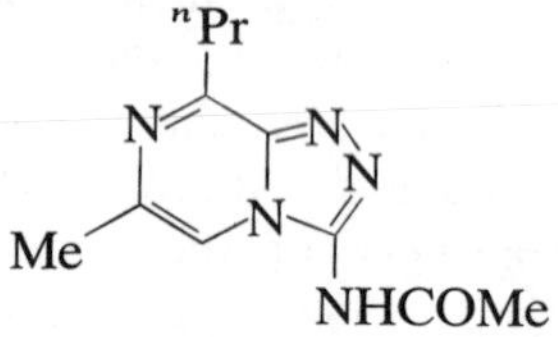

34

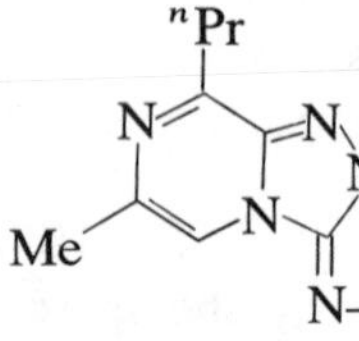

35

spectra (pyrazine and pyrimidine systems) but much larger shifts when a pyridazine ring is formed. The differences in the ultraviolet spectra of *s*-triazolo[4,3-*a*]pyrazines and *s*-triazolo[1,5-*a*]pyrazines are stated to be extremely useful in establishing the structures of unknown products in these ring systems.[1] However, reliance on changes in ultraviolet spectra can be misleading.[2,3]

Insertion of a phenyl group in the *s*-triazole ring of *s*-triazolo[4,3-*a*]quinoxaline to give compound **36** produces a marked change in its ultraviolet spectrum, due to conjugation with the coplanar phenyl ring. However the corresponding change is not evident in the spectrum of the dimethylphenylpyrazine **37**. Steric interaction prevents the rings from becoming coplanar. This interaction is demonstrated by the PMR spectrum of compound **37**, in which the 5-methyl signal has moved upfield from δ 2.65 in the 3-unsubstituted compound to δ 2.16 in the phenyl compound.[7]

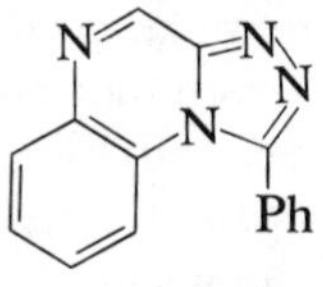

36 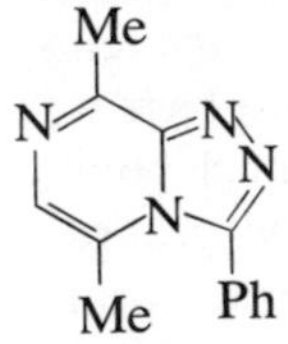 **37**

The mass spectra of various derivatives of this ring system have been examined.[4] Generally the breakdown patterns are very similar to those of analogous s-triazolo[4,3-a]pyridines. The unsubstituted compound breaks down by loss of N_2 to give the ion **38**, which then loses HCN to give the fragment **39**. 3-Alkyl derivatives lose alkyl cyanide to give the ion **40**, which again loses HCN. The breakdown products of alkyl derivatives of the ring system are more prominent in the mass spectrum than the fragments from the unsubstituted compound. Methyl groups in the pyrazine ring result in ring expansion after loss of the 3-substituent. 3-Amino compounds behave similarly after losing NH_2CN. 2,3-Dihydro-3-oxo compounds lose the triazole ring altogether by hydrogen transfer followed by loss of CH_2N_2O. 5,6-Diphenyl substitution in the pyrazine ring permits a 1,2-phenyl migration after loss of all the nitrogen atoms, resulting in formation of ion **41**.

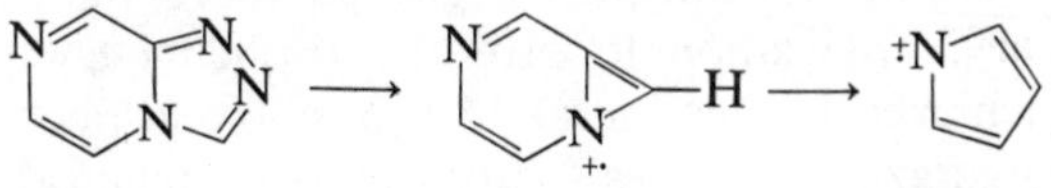

38 **39**

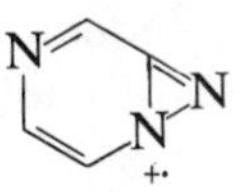

40

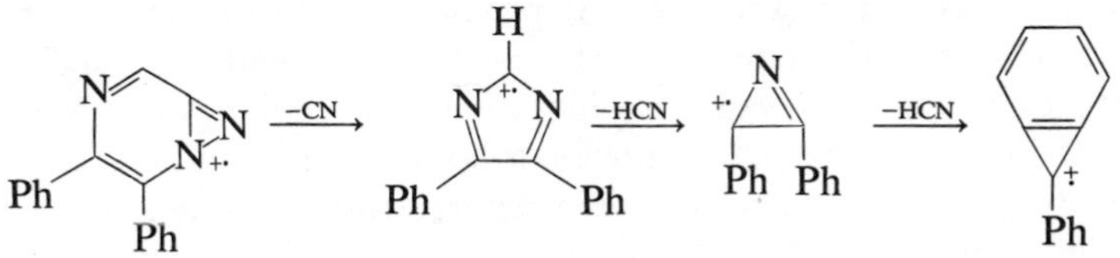

−CN −HCN −HCN

41

Many of the methods used to prepare *s*-triazolopyridines and pyrimidines have been applied with little success to the synthesis of *s*-triazolo[4,3-*a*]pyrazines. In general the methods rely on the cyclization of α-hydrazino azines with one carbon insertion from suitable electrophilic reagents. The substituents on the pyrazine ring exert marked effects. Phenyl groups are base weakening and tend to make cyclization onto the already weakly nucleophilic pyrazine nitrogen even more difficult. On the other hand, alkyl groups seem sufficiently base strengthening to allow ready cyclization to form *s*-triazolo[4,3-*a*]pyrazines. The nucleophilicity of the pyrazine nitrogen is not the sole controlling force, inasmuch as the corresponding *s*-triazolo[4,3-*a*]quinoxalines are more readily formed, even though hydrazinoquinoxalines are just as weakly nucleophilic as the corresponding pyrazines. Cyclizations in the pyrazine series must be more sensitive to steric influences and the loss of aromaticity in the diazine ring.

The most general synthesis of *s*-triazolo[4,3-*a*]pyrazines involves reaction of suitably substituted hydrazinopyrazines with triethyl orthoesters. The products are the corresponding 3-substituted compounds **42**.[7,10,11] In contrast to the corresponding quinoxaline series, attempts to react hydrazine compounds with carboxylic acids, esters, or anhydrides were largely unsuccessful or gave the products in less than 10% yield.[7,11] Similarly, attempts at cyclization of the acylhydrazines **43** with phosphorus oxychloride proved much more difficult than in the quinoxaline series.[7] The presence of alkyl groups in the pyrazine ring allowed cyclization of the benzoylhydrazine **44** (R = alkyl) with phosphorus oxychloride to give the triazolo compounds **45** (R = alkyl).[7] However, when the pyrazine ring was substituted with phenyl groups, more vigorous conditions were necessary. Cyclization to give the triphenyl compound **45** (R = phenyl) was achieved by heating at 150° in polyphosphoric acid.[1]

Treatment of hydrazinopyrazines with cyanogen chloride[2,10,12] or cyanogen bromide[7] provides 3-amino-*s*-triazolo[4,3-*a*]pyrazines (**46**). The reactions readily take place in buffered aqueous acetic acid at room temperature, although occasionally sodium carbonate solution is preferred.[10] Thus reaction of 2-hydrazino-5-methyl-3-*n*-propylpyrazine with cyanogen chloride in acetate buffer at room temperature gave the 3-acetamidotriazolopyrazine **34**. Since this compound can only be prepared from the parent amine by heating with acetic anhydride, it is suggested that the formation of the acylated product is due to the intermediate formation of acetyl isocyanate. The latter compound is thought to be formed by reaction of cyanogen chloride with acetate ion and rearrangement of the first-formed acetyl cyanate to acetyl isocyanate.[10]

The 3-"hydroxy" compounds are prepared from hydrazinopyrazines by reaction with urea[7] at 150° or by treatment with phosgene at room

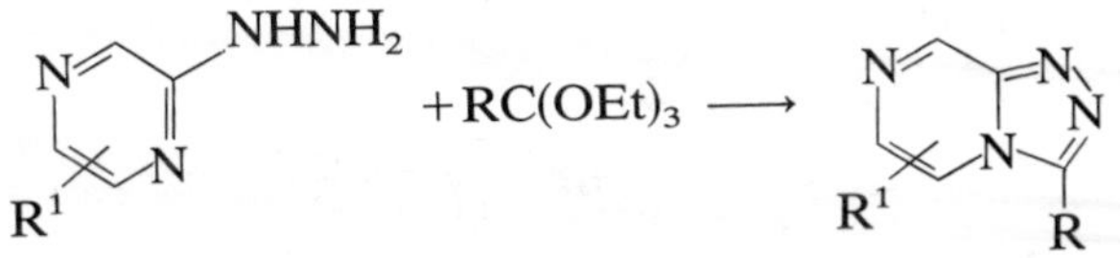

R^1 = H, alkyl, or phenyl + $RC(OEt)_3$ ⟶ **42** R = H or alkyl

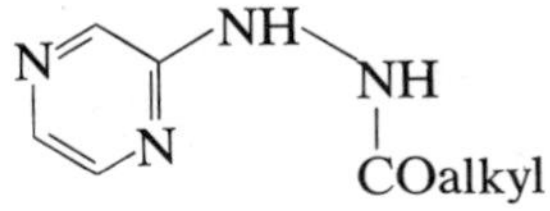

43

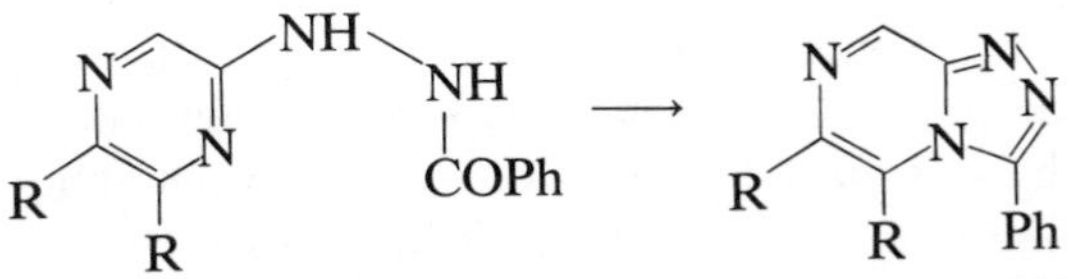

44 ⟶ **45** R = alkyl or phenyl

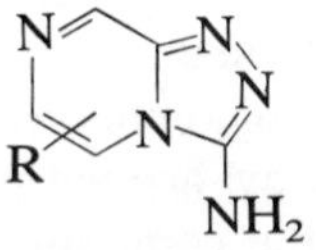

46 R = H, alkyl, or phenyl

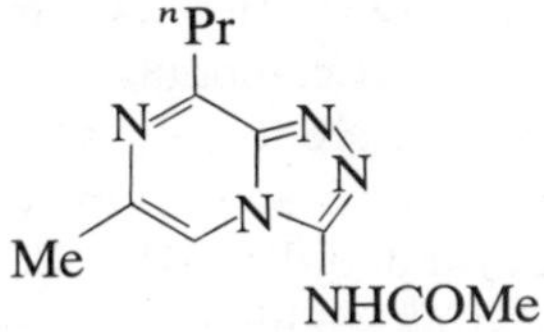

34

temperature[10] or below.[2,12] Alternatively the hydrazine may be reacted with ethyl chloroformate and the resultant urethan cyclized in sodium hydroxide solution[12] or by refluxing in *o*-dichlorobenzene.[2] Synthesis of 3-mercapto derivatives of this ring system is particularly sensitive to the nature of the substituents in the pyrazine ring. Thus hydrazinopyrazines (**47**) containing alkyl substituents react with carbon disulfide in refluxing chloroform after several days,[7] or by heating in a sealed tube in *t*-butanol at 140° for 10 hr,[10] to give the mercapto derivatives **48**. Attempts to prepare phenyl-substituted analogues either in butanol at 140° in a sealed tube[2] or in refluxing pyridine[11] have been unsuccessful. Although the carbon disulfide reacted with the hydrazino group, the pyrazine ring was insufficiently activated for cyclization, and the only products isolated were the pyrazines **49** and **50**.

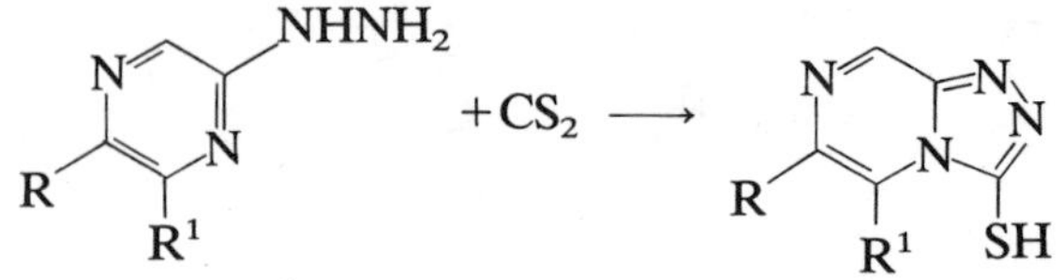

47 R, R^1 = Alkyl

48

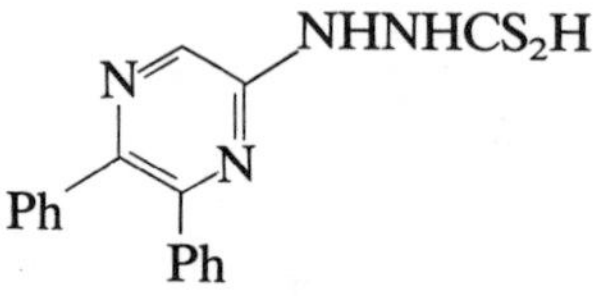

49

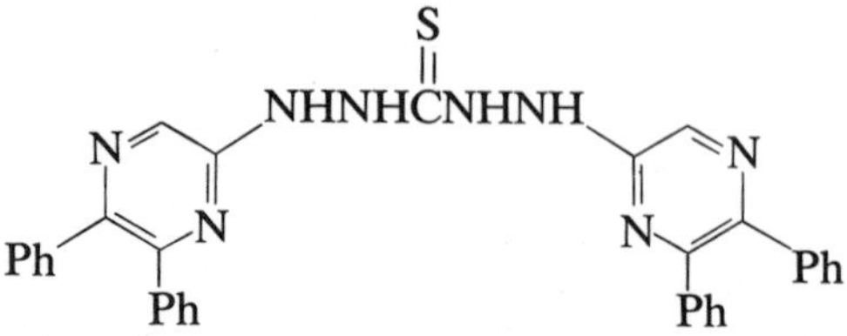

50

Methylation of 3-mercapto-5,8-dimethyl-*s*-triazolo[4,3-*a*]pyrazine with methyl iodide gave the expected methylthio compound.[7] Bromination of 5,8-dimethyl-*s*-triazolo[4,3-*a*]pyrazine with bromine in acetic acid resulted in substitution in the 3-position.[7] The 3-amino compound **51** has been acylated with a wide variety of agents to give 3-substituted amino compounds.[10,12] Acetic anhydride when warmed on a steam bath gave the acetamido derivative **34**. Formic acid and dicyclohexylcarbodiimide gave the *N*-formyl compound **52**.[12] The presence of a 5-methyl group has a marked steric effect on acylation.[10] Thus although hot acetic anhydride converts amine **53** into the expected acetamido compound **54**, the use of acetic acid and dicyclohexylcarbodiimide in ethyl acetate results in acetylation at the unhindered 2-position to give the compound **55** (R = ^{n}Pr). Similarly the dimethyl compound **56** gives **55** (R = Me) on acetylation with the sterically demanding acetic acid–dicyclohexylcarbodiimide complex. The action of acetic anhydride on **56** gives the diacetyl derivative **57**, which is hydrolyzed in dilute sodium hydroxide solution to the 3-acetamido compound.

The *s*-triazolo[4,3-*a*]pyrazine ring system has been described as being relatively stable to hot potassium permanganate solution. Attempts to diazotize 3-amino derivatives were unsuccessful, and the 3-hydroxy group does not react with phosphorus oxychloride.[7] Treatment of 5,8-dimethyl-*s*-triazolo[4,3-*a*]pyrazine (**58**) with refluxing 10% sodium hydroxide solution results in isomerization to the *s*-triazolo[1,5-*a*] compound **59**.[7] This reaction is analogous to similar rearrangements in other *s*-triazolo-fused systems. An early report that 3-amino-*s*-triazolo[4,3-*a*]pyrazines **60** could be rearranged in hot acid with concomitant hydrolysis to the oxo-*s*-triazolo[1,5-*a*] compounds **61** has now been corrected.[3] Treatment of the

^{n}Pr

N N N N

Me NHCOMe

34

(i) ← **51** → (ii)

^{n}Pr

Me NH_2

51

^{n}Pr

Me NHCHO

52

^{n}Pr

Me NH_2

53

(iii) →

^{n}Pr

Me NHCOMe

54

(iv) ↓

R

N—COMe

Me NH

55 R = ^{n}Pr or Me

← (v)

Me

Me NH_2

56

(vi) ↙

Me

Me $N(COMe)_2$

57

Reagents (i) Ac_2O
(ii) HCO_2H–DCC
(iii) Ac_2O
(iv) HOAc–DCC
(v) HOAc–DCC
(vi) Ac_2O

amino compound **51** with cold dilute hydrochloric acid gave the triazole **62**, which cyclized on heating to give the imidazo[2,1-*c*]-*s*-triazole **63**. The kinetics and mechanism of the ring fission of *s*-triazolo[4,3-*a*]pyrazines have been discussed more recently.[28]

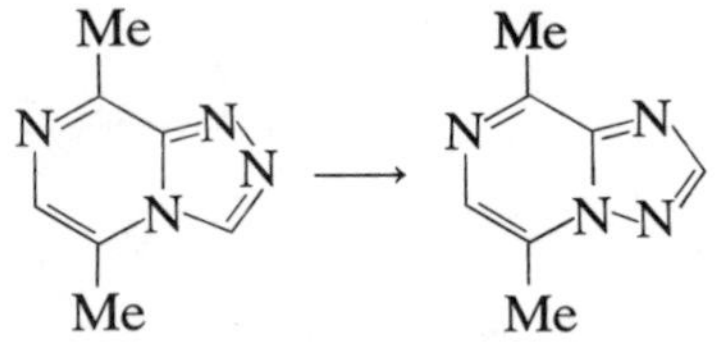

Me Me Me Me
58 **59**

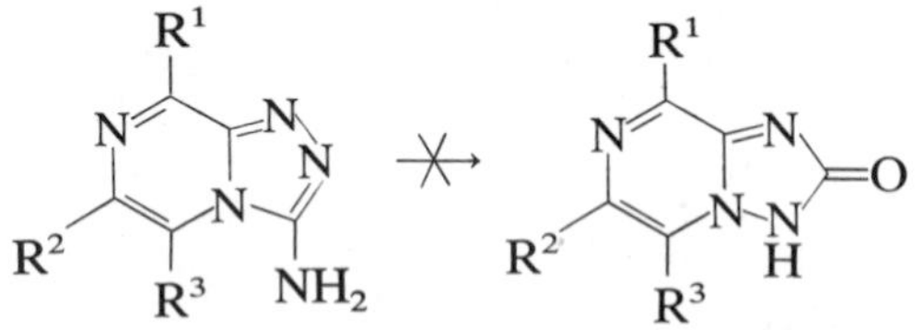

R^1 R^2 R^3 NH_2 O H
60 **61**

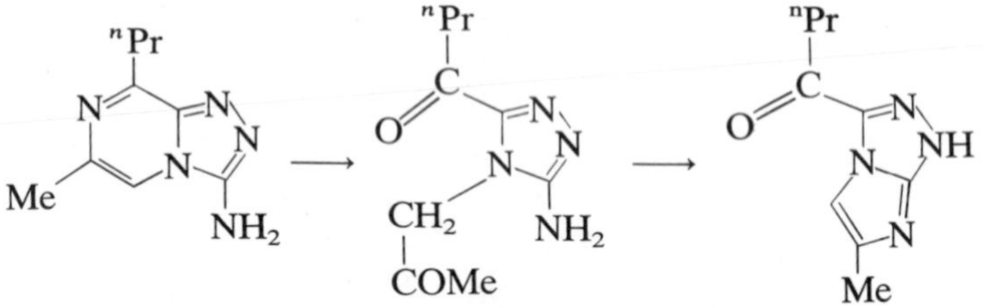

ⁿPr Me NH_2 CH_2 COMe NH
51 **62** **63**

Various derivatives of the ring system have been synthesized as purine analogues, particularly as relatives of theophylline, a known bronchodilator. Structure–activity relationships have been discussed for a series of 3-amino compounds.[10] It seems that the most active compounds are substituted in the 6- and 8-positions by small alkyl groups. The presence of a substituent in the 5-position completely destroys activity. Patents claim a series of compounds, in particular **34**, for the treatment of asthma and bronchospasm.[12,14] As the inhibition of cyclic AMP phosphodiesterase is thought to be responsible for the bronchodilator action of theophylline, compound **34** was tested and found to be a potent inhibitor of this enzyme.[15,16] The amino compound **51** has been found to be active against asthma in the clinic, but it caused nausea and vomiting.[16] The acetamido derivative **34** was selected as a possible alternative[17] and was found not to have bronchodilator activity, although it may be useful as a preventative of bronchoconstriction.[18]

III. *v*-Triazolo[4,5-*b*]pyrazines

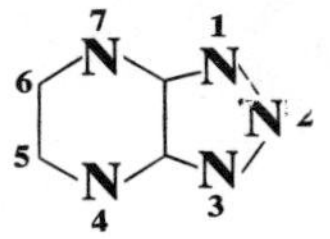

2*H* RRI 1082

The triphenyl compound **64**, m.p. 217°, is the only reported member of this ring system. Referred to as a derivative of 2,1,3,4,7-benzpentazole, it was prepared by the reaction of benzil with the appropriate diaminotriazole.[19]

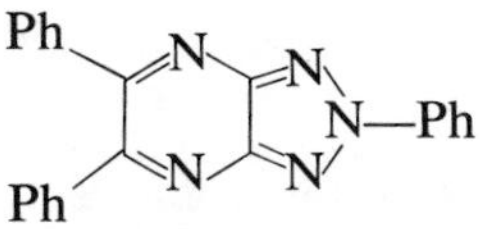

64

IV. Tetrazolo[1,5-*a*]pyrazines

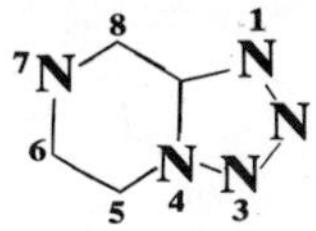

These compounds have been referred to as tetrazolo[1,5-*c*]pyrazines, and reduced derivatives have been named tetrazolo[1,5-*a*]piperazines. This ring system has attracted attention mainly in the study of tetrazolo azide tautomerism and has been used to generate nitrenopyrazine. Although tetrazolo derivatives of pyridazine and pyridine are stable in solution, tetrazolo[1,5-*a*]pyrazine isomerizes readily in chloroform or trifluoracetic acid. Although an earlier report[20] indicated that no azide (**66**) was present in chloroform solutions of the unsubstituted compounds (**65**), this has now been contradicted.[21] Both compound **65** and the diphenyl derivative **67** show bands in the infrared spectrum (at ca. 2140 cm^{-1}) in chloroform, which indicate the presence of a significant amount of azido tautomer.[22] These observations are confirmed by PMR spectra, which indicate the presence of the azide **66** to a small extent in chloroform and to the extent of 55% in trifluoroacetic acid.[21] The diphenyl derivative **67** is present entirely as azide in trifluoroacetic acid and exists 100% in the tetrazolo form in dimethyl sulfoxide.[22] These conclusions, which are based on PMR data, are confirmed by infrared spectra determined in the respective solvents. Recently the synthesis and infrared spectrum of 1(3)-[^{15}N]-tetrazolo[1,5-*a*]pyrazine has been reported.[28]

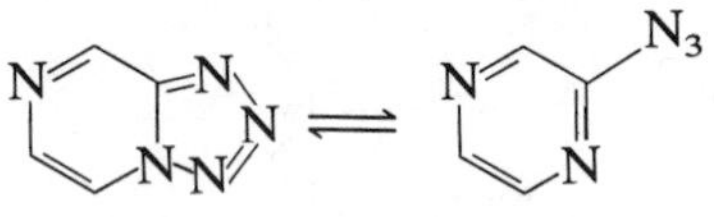

65 ⇌ **66**

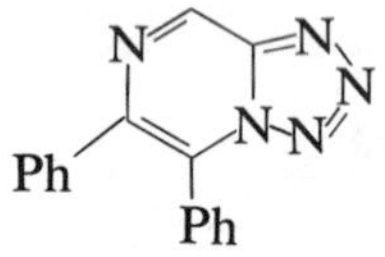

67

The importance of the azido tautomer is reflected in the mass spectrum of the unsubstituted compound.[21] As with the analogous tetrazolopyridine, there is an abundant peak at *m/e* M-28 due to loss of nitrogen and another peak at M-26, which is interpreted as due to loss of nitrogen, followed by the addition of two hydrogen atoms to the nascent nitrene, to give 2-aminopyrazine (**68**).

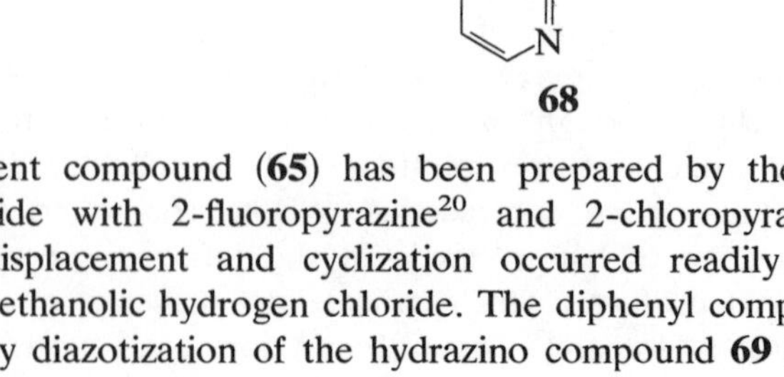

68

The parent compound (**65**) has been prepared by the reaction of sodium azide with 2-fluoropyrazine[20] and 2-chloropyrazine.[21] Nucleophilic displacement and cyclization occurred readily in refluxing ethanol or ethanolic hydrogen chloride. The diphenyl compound **67** was prepared by diazotization of the hydrazino compound **69** in acetic acid with sodium nitrite. Similarly the reduced derivative **71** was obtained by reaction of the ethoxy compound **70** with hydrazine and then sodium nitrite in sulfuric acid.[23]

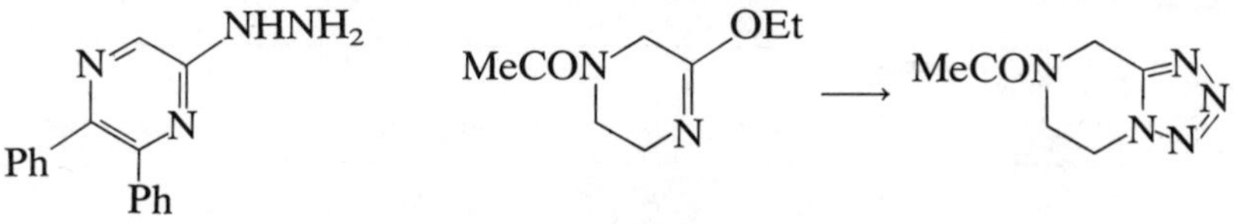

69 **70** **71**

The acetyl group in compound **71** is readily removed by acid hydrolysis, and the resulting 7-unsubstituted compound has been benzoylated with benzoyl chloride.[23] Refluxing the diphenyl compound **67** with acetic acid or acetic anhydride results in decomposition via the azide and nitrene to give the imidazoles **72**.[22] Thermal decomposition of tetrazolo compounds has been used as a method of nitrene generation, and the pyrolysis of tetrazolo[1,5-*a*]pyrazine is similar to that of other tetrazolo compounds. Pyrolysis gave a mixture of 1-cyanoimidazole (**73**) and two unidentified products.[24] The diphenyltetrazolo compound **67** reacts as the azide in refluxing chloroform to give the pyrazinyltriazole **74** with dimethylacetylene dicarboxylate.[22]

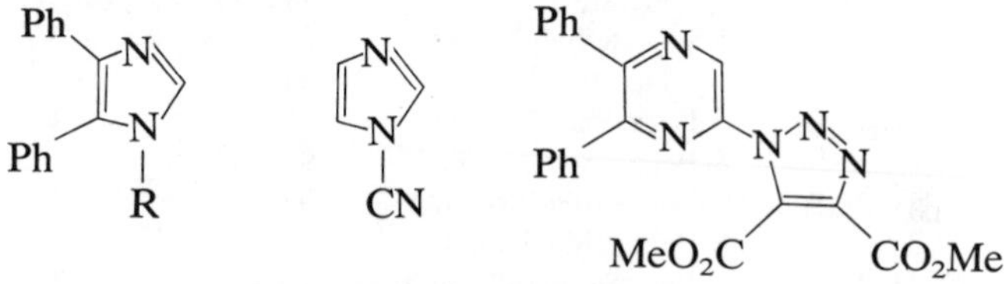

72 R = H or COMe **73** **74**

V. Tables

TABLE 1. *s*-TRIAZOLO[1,5-*a*]PYRAZINES

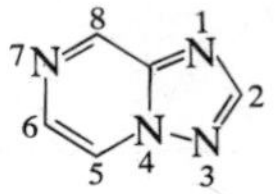

Molecular formula	Substituents	m.p., °C	Ref.
$C_5H_3BrN_4$	5-Bromo	143–145	26
$C_5H_3ClN_4$	5-Chloro	108–110	25, 27
	6-Chloro	140–141	27

TABLE 1 (*contd.*)

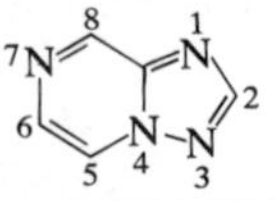

Molecular formula	Substituents	m.p., °C	Ref.
$C_5H_3N_7$	5-Azido	95–96	26
$C_5H_4N_4$	Unsubstituted	127	25, 27
$C_5H_4N_4S$	5-Mercapto	>250 decomp.	26
$C_5H_5N_5$	5-Amino	>250	26
	8-Amino	218–219	26
$C_5H_5N_5O$	8-Hydroxyamino	242–244	26
$C_5H_6N_6$	5-Hydrazino	217 decomp.	26
$C_6H_5ClN_4$	6-Chloro-2-methyl	112	5
	8-Chloro-2-methyl	111.5–112	5
$C_6H_6N_4$	2-Methyl	132	5, 6
	7-Oxide	228 decomp.	
$C_6H_6N_4S$	5-Methylthio	147–148	26
$C_7H_8N_4$	5,8-Dimethyl	77–78	7
$C_7H_8N_4O$	5-Ethoxy	90–91	26
$C_{11}H_8N_4S$	5-Phenylthio	95	26
$C_{13}H_{12}N_4$	5,6-Dimethyl-2-phenyl	133–134	1
	Picrate	192–193	
	5,8-Dimethyl-2-phenyl	103–104	1
	Picrate	216–217	
$C_{13}H_{12}N_4O$	6-Benzyloxy-2-methyl	182	5
	8-Benzyloxy-2-methyl	93–94	5
$C_{13}H_{12}N_4S$	6-Benzylthio-2-methyl	97–98	5
	8-Benzylthio-2-methyl	76–77	5
$C_{13}H_{13}N_5$	6-Benzylamino-2-methyl	131	5
	8-Benzylamino-2-methyl	136–137	5
$C_{23}H_{16}N_4$	2,5,6-Triphenyl	238–239	1

TABLE 2. *s*-TRIAZOLO[4,3-*a*]PYRAZINES

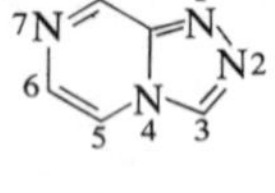

Molecular formula	Substituents or structure	m.p., °C	Ref.
$C_5H_4N_4$	Unsubstituted		
	Hemihydrate	194–195	11
	Picrate	177 decomp.	
$C_5H_5N_5$	3-Amino	>320	2

TABLE 2. (*contd.*)

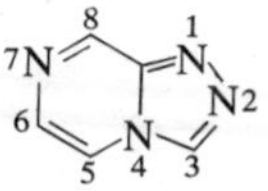

Molecular formula	Substituents or structure	m.p., °C	Ref.
$C_6H_6N_4$	3-Methyl	239	11
	Picrate	156–157	
	8-Methyl	128–129	10
	Picrate	133–135	7
$C_6H_7N_5$	3-Amino-8-methyl	258–259	10
	N-Acetyl	271–272	
$C_7H_7BrN_4$	3-Bromo-5,8-dimethyl		
	Hydrobromide	275–276 decomp.	7
$C_7H_8N_4$	3-Ethyl	158	11
	Picrate	100–101	
	3,8-Dimethyl	204–206	7
	5,6-Dimethyl	190	11
	Picrate	136–137	
	5,8-Dimethyl	171	7
$C_7H_8N_4O$	2,3-Dihydro-5,6-dimethyl-3-oxo	249–250	2
	2,3-Dihydro-5,8-dimethyl-3-oxo	252	7
$C_7H_8N_4S$	2,3-Dihydro-5,8-dimethyl-3-thioxo	278	7
$C_7H_9N_5$	3-Amino-5,6-dimethyl	261–262	2, 10
	N-Acetyl	227–228	
	N,N-Diacetyl	162–163	
	3-Amino-5,8-dimethyl	290–292	7
	Hydrobromide	268 decomp.	
	3-Amino-6,8-dimethyl	230–232	10, 12
$C_8H_{10}N_4$	3-Ethyl-8-methyl	159–161	7
	3,5,6-Trimethyl		
	Hydrate	126–127	11
	Picrate	134–135	
	3,5,8-Trimethyl	153–154 decomp.	7
$C_8H_{10}N_4O$	2,3-Dihydro-5,6,8-trimethyl-3-oxo	272–274	10
	2,3-Dihydro-3-oxo-8-*n*-propyl	210–212	10
$C_8H_{10}N_4S$	5,8-Dimethyl-3-methylthio	140	7
$C_8H_{11}N_5$	3-Amino-5,6,8-trimethyl	245–247 decomp.	10
$C_9H_{11}N_5O$	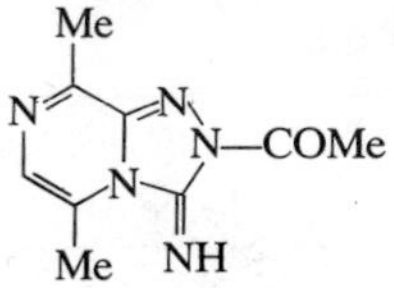 [structure: 5,8-dimethyl, 2-COMe, 3-imino (=NH) 2,3-dihydro derivative]	131–132	10
$C_9H_{12}N_4$	3-Ethyl-5,6-dimethyl	93–94	11
	Picrate	127–128	
	3-Ethyl-5,8-dimethyl	128–130	7
	6-Methyl-8-*n*-propyl	104–106	10

TABLE 2 (*contd.*)

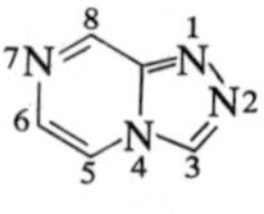

Molecular formula	Substituents or structure	m.p., °C	Ref.
$C_9H_{12}N_4O$	2,3-Dihydro-6-methyl-3-oxo-8-*n*-propyl	183–184	10, 12
$C_9H_{12}N_4S$	2,3-Dihydro-6-methyl-8-*n*-propyl-3-thioxo	173–174	10
$C_9H_{13}N_5$	3-Amino-5-methyl-8-*n*-propyl	184–185	10
	N-Acetyl	141–142	
	3-Amino-6-methyl-8-*n*-propyl	199–200	12
		195–196	10
	N-Acetyl[a]	264–265	
	3-Amino-8-methyl-6-*n*-propyl	216–217	10, 12
$C_{10}H_{15}N_5$	3-Amino-5,6-dimethyl-8-*n*-propyl	229–231	10
$C_{11}H_{15}N_5O$	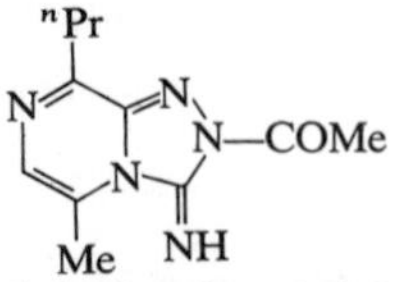 ${}^nPr$; N—COMe; Me; NH	76–77	10
$C_{12}H_{11}N_5$	3-Amino-6-methyl-8-phenyl	255–257 decomp.	10
$C_{13}H_{12}N_4$	5,8-Dimethyl-3-phenyl	188–190	7
$C_{17}H_{12}N_4$	5,6-Diphenyl	187–188	11
	Picrate	145–146	
$C_{17}H_{12}N_4O$	2,3-Dihydro-3-oxo-5,6-diphenyl	224–226	2
$C_{17}H_{13}N_5$	3-Amino-5,6-diphenyl	248–250	2
$C_{18}H_{14}N_4$	3-Methyl-5,6-diphenyl	200–201	11
	Picrate	158–159	
$C_{19}H_{16}N_4$	3-Ethyl-5,6-diphenyl	234–235	11
	Picrate	132–133	
$C_{23}H_{16}N_4$	3,4,6-Triphenyl	240–241	1

[a] Reference 10 lists eight other *N*-substituted derivatives.

TABLE 3. *v*-TRIAZOLO[4,5-*b*]PYRAZINES

Molecular formula	Structure	m.p., °C	Ref.
$C_{22}H_{15}N_5$	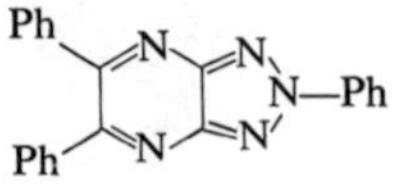 Ph, Ph; N—Ph	217	19

TABLE 4. TETRAZOLO[1,5-*a*]PYRAZINES

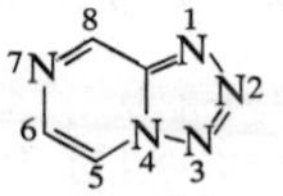

Molecular formula	Substituents	m.p., °C	Ref.
$C_4H_3N_5$	Unsubstituted	90–91	20, 21
$C_4H_7N_5$	5,6,7,8-Tetrahydro	61–63	23
	Hydrochloride	225.5–226	
$C_6H_9N_5O$	7-Acetyl-5,6,7,8-tetrahydro	134–138	23
$C_{11}H_{11}N_5O$	7-Benzoyl-5,6,7,8-tetrahydro	190.5–191	23
$C_{16}H_{11}N_5$	5,6-Diphenyl[a]	171–173	22

[a] Described as the 7,8-diphenyl compound.

VI. References

1. G. M. Badger, P. J. Nelson, and K. T. Potts, *J. Org. Chem.*, **29,** 2542 (1964).
2. S. E. Mallett and F. L. Rose, *J. Chem. Soc.* (*C*), 2038 (1966).
3. F. L. Rose, G. J. Stacey, P. J. Taylor, and T. W. Thompson, *J. Chem. Soc.* (*D*), 1524 (1970).
4. K. T. Potts and E. Brugel, *Org. Mass Spectrom.*, **5,** 663 (1971).
5. T. Okamato, Y. Torigoe, M. Sato, and Y. Isogai, *Chem. Pharm. Bull.* (Tokyo), **16,** 1154 (1968); *Chem. Abstr.*, **70,** 4058 (1969).
6. T. Okamato and Y. Torikoshi, Jap. Pat. 69 12,149; *Chem. Abstr.*, **71,** 81408 (1969).
7. K. T. Potts and S. W. Schneller, *J. Heterocycl. Chem.*, **5,** 485 (1968).
8. K. Syono and T. Furuya, *Plant Cell Physiol.*, **12,** 61 (1971); *Chem. Abstr.*, **74,** 123896 (1971).
9. Fuji Photo Film Co Ltd., Fr. Pat. 1,553,229; *Chem. Abstr.*, **72,** 73174 (1970).
10. J. Maguire, D. Paton, and F. L. Rose, *J. Chem. Soc.* (*C*), 1593 (1969).
11. P. J. Nelson and K. T. Potts, *J. Org. Chem.*, **27,** 3243 (1962).
12. J. A. Maguire and F. L. Rose (to I.C.I. Ltd.), Brit. Pat. 1,146,770; *Chem. Abstr.*, **71,** 39005 (1969).
13. S. Nicholson, G. J. Stacey and P. J. Taylor, *J. Chem. Soc.*, *Perkin II*, 4 (1972).
14. G. E. Davies and T. P. Johnston (to I.C.I. Ltd.), Ger. Offen. 1,961,967; *Chem. Abstr.*, **73,** 48545 (1970).
15. A. R. Somerville, M.-L. Rabouhans, and A. A. Smith, *Biochem. J.*, **120,** 11P (1970).
16. G. E. Davies, F. L. Rose, and A. R. Somerville, *Nature New Biol.*, **234,** 50 (1971).
17. G. E. Davies and F. L. Rose, "Progr. Antimicrob. Anticancer Chemother., Proc. Int. Congr. Chemother., 6th," Vol. 1, 1969, p. 244; *Chem. Abstr.*, **74,** 97819 (1971).
18. A. J. Richards, S. R. Walker, and J. W. Paterson, *Brit. J. Dis. Chest.*, **65,** 247 (1971); E. Middleton, Jr., and R. G. Coffey, *Ann. Rep. Med. Chem.*, **8,** 276 (1973).
19. Thiele and Schleussner, *Ann.* **295,** 145 (1897).
20. H. Rutner and P. E. Spoerri, *J. Heterocycl. Chem.*, **3,** 435 (1966).
21. C. Wentrup, *Tetrahedron,* **26,** 4969 (1970).

22. T. Sasaki, K. Kanematsu, and M. Murata, *J. Org. Chem.*, **36,** 446 (1971).
23. V. G. Granik and R. G. Glushkov, *Khim.-Farm. Zh.*, **1**(4), 16 (1967); *Chem. Abstr.*, **68,** 68955 (1968).
24. C. Wentrup and W. D. Crow, *Tetrahedron*, **26,** 4915 (1970).
25. S. Polanc, B. Vercek, B. Stanovnik, and M. Tisler, *Tetrahedron Lett.*, 1677 (1973).
26. B. Vercek, B. Stanovnik, and M. Tisler, *Tetrahedron Lett.*, 4539 (1974).
27. S. Polanc, B. Vercek, B. Sek, B. Stanovnik, and M. Tisler, *J. Org. Chem.*, **39,** 2143 (1974).
28. C. Thetaz, F. W. Wehrli, and C. Wentrup, *Helv. Chim. Acta*, **59,** 259 (1976).

CHAPTER XXIV

Furo-, Oxazolo-, Isoxazolo-, and Oxadiazolopyrazines

I. Furo[3,4-*b*]pyrazines 431
II. Oxazolo[3,2-*a*]pyrazines 433
III. Oxazolo[3,4-*a*]pyrazines 435
IV. Isoxazolo[4,5-*b*]pyrazines 435
V. [1,2,3]Oxadiazolo[3,4-*a*]pyrazines 435
VI. [1,2,5]Oxadiazolo[3,4-*b*]pyrazines 436
VII. Tables 437
Table 1. Spirofuro[3,4-*b*]pyrazines 437
Table 2. Furo[3,4-*b*]pyrazines 437
Table 3. Oxazolo[3,2-*a*]pyrazines 438
Table 4. Oxazolo[3,4-*a*]pyrazines 438
Table 5. Isoxazolo[4,5-*b*]pyrazines 439
Table 6. [1,2,3]Oxadiazolo[3,4-*a*]pyrazines 439
Table 7. [1,2,5]Oxadiazolo[3,4-*b*]pyrazines 439
VIII. References 439

I. Furo[3,4-*b*]pyrazines

This ring system is most frequently encountered as derivatives of 2,3-pyrazinedicarboxylic anhydride (**1**). Moreover the properties, preparation, and reactions of the 5,7-dioxo derivatives are strictly analogous to those of phthalic anhydride. Reduced derivatives of type **2** are also known, and their ultraviolet spectra have been recorded.[1] The mass spectrum of compound **1** corresponds to that of phthalic anhydride in that the molecular ion loses both carbon dioxide and then carbon monoxide to give an ion corresponding to 2,3-dehydropyrazine (**3**).[2]

The anhydride **1** is readily prepared from pyrazine-2,3-dicarboxylic acid by treatment with thionyl chloride[3] or acetic anhydride.[3–5] The 2,3-dimethyl-substituted analogue of **1** has been similarly prepared using

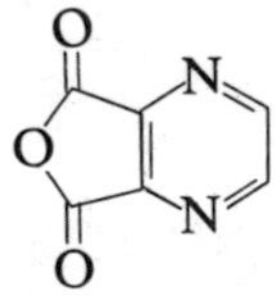

1

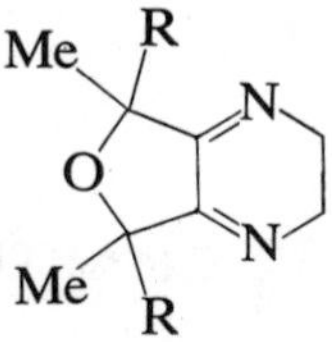

2 R = Me or Et

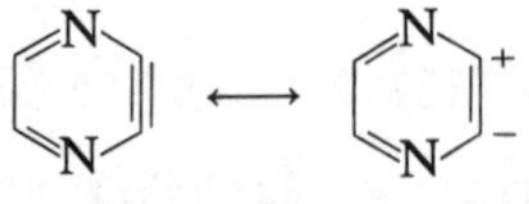

3

acetic anhydride.[6] Pyrazine-2,3-dicarboxylic acid on reaction with various phenols and aromatic amines has provided a series of spiro derivatives of furo[3,4-*b*]pyrazine.[7] For example, reaction with resorcinol gave compound **4**, analogous to the product from phthalic acid.

The only other derivatives of this ring system to be reported are the tetraalkyl compounds **2** which were obtained by reaction of the dioxo furans **5** with ethylenediamine.[8]

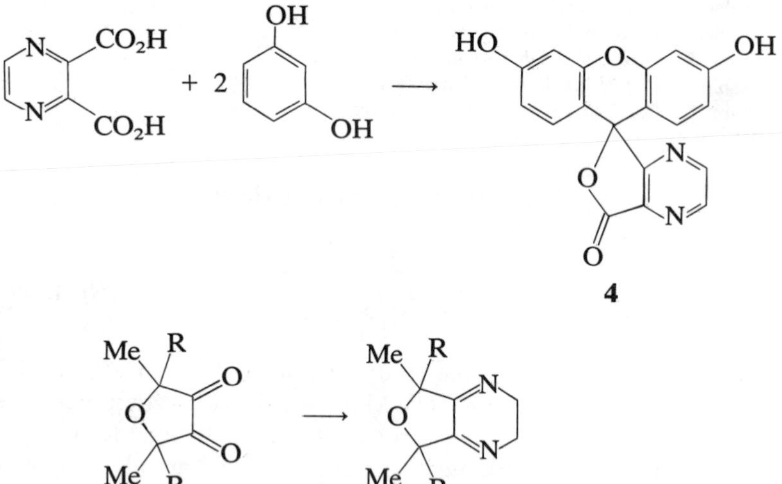

4

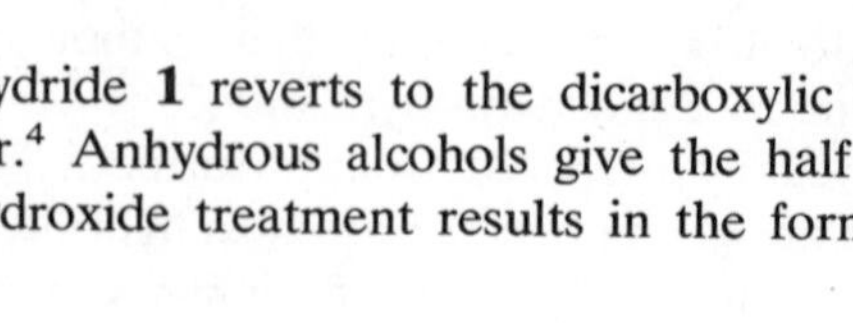

5 R = Me or Et 2

The anhydride **1** reverts to the dicarboxylic acid on treatment with warm water.[4] Anhydrous alcohols give the half acid-esters **6**, and ammonium hydroxide treatment results in the formation of the half acid-amide **7**.

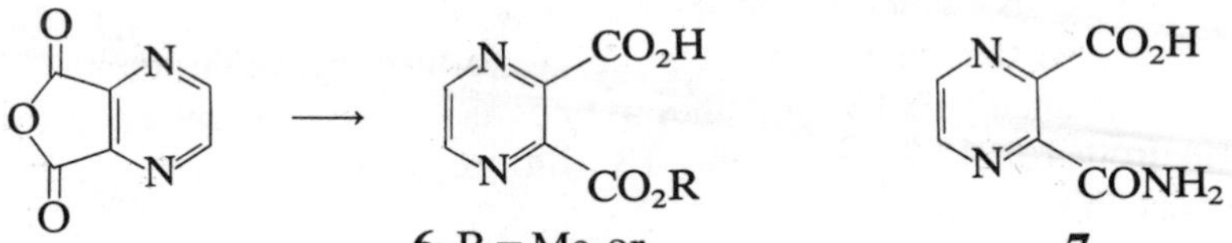

1 **6** R = Me or $CH_2CH_2Nalkyl_2$ **7**

The results of pyrolysis experiments[2,9] indicate that compound **1** decomposes with the formation of the pyrazyne **3**. Pyrolysis in the presence of benzene results in a small but definite yield of 2-phenylpyrazine (**8**) from an insertion reaction with pyrazyne.[9] This result is similar to results from pyrolysis experiments with various aromatic and heterocyclic anhydrides. Pyrazyne is also invoked to explain the formation of maleonitrile (**9**) and fumaronitrile (**10**) by pyrolysis of anhydride **1** over silica.[2]

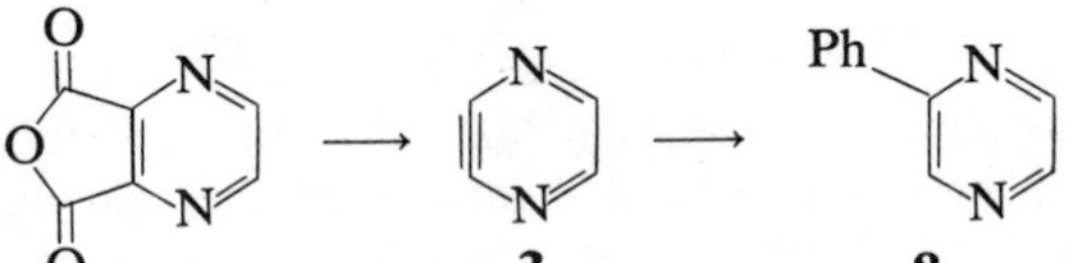

1 **3** **8**

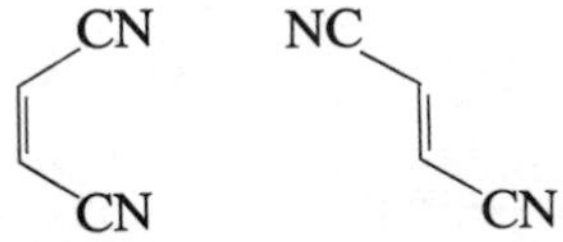

9 **10**

Complex derivatives of the ring system have been used instead of phthalic anhydrides to form polyimides by reaction with suitable amines. The resultant polymers are claimed to have improved chemical and abrasion resistance and improved mechanical and electrical properties.[10]

II. Oxazolo[3,2-*a*]pyrazines

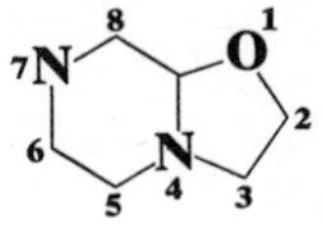

RRI 11837

The work on this ring system has been confined to perhydro derivatives related to constituents of certain ergot alkaloids. The saturated heterocycle has been described as 1-oxa-6,9-diazabicyclo[4.3.0]nonane (**11**). Attention has been directed towards the study of the tautomerism of these compounds. In studies[12,13] of the behavior of N-(α-hydroxyacyl) amides it was shown that they slowly equilibrate to form cyclol structures. Thus the piperazinedione **12** on standing undergoes partial conversion to the perhydrooxazolo[3,2-*a*]pyrazine **13**. Infrared and ultraviolet evidence support this conclusion.

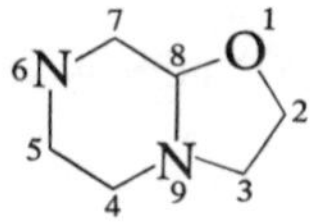

11

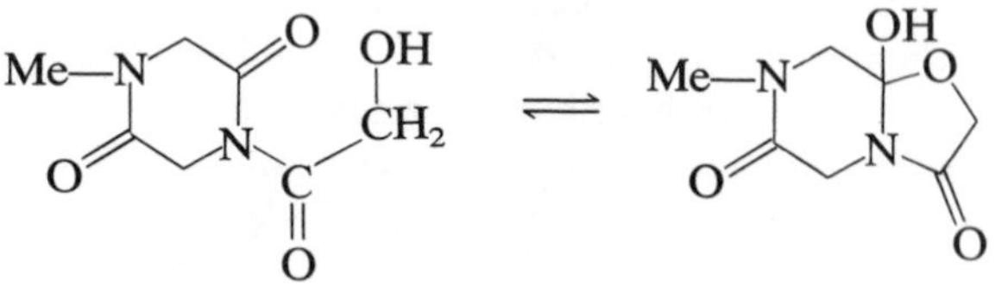

12 ⇌ **13**

The only syntheses of the ring system to be reported involve hydrogenolysis of benzyl ethers of hydroxyacylpiperazinediones **14** over palladium on charcoal. Using these conditions, compounds **15**[14] and **16**[15] were obtained.

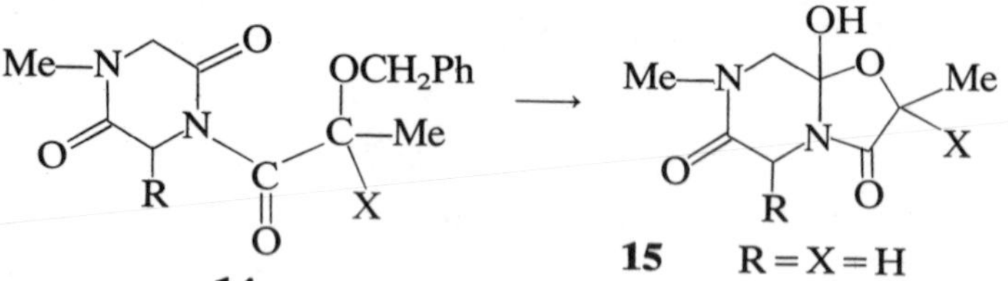

14 → **15** R = X = H; **16** R = CH_2Ph, X = CO_2Et

The ester **16** has been converted in a series of steps to the amine **17**, which on reaction with *d*-lysergic acid chloride (or its 9,10-dihydro derivative) gave the compound **18** (or its 9′,10′-dihydro analogue), which is claimed as a uterotomic and sympatholytic agent.[15]

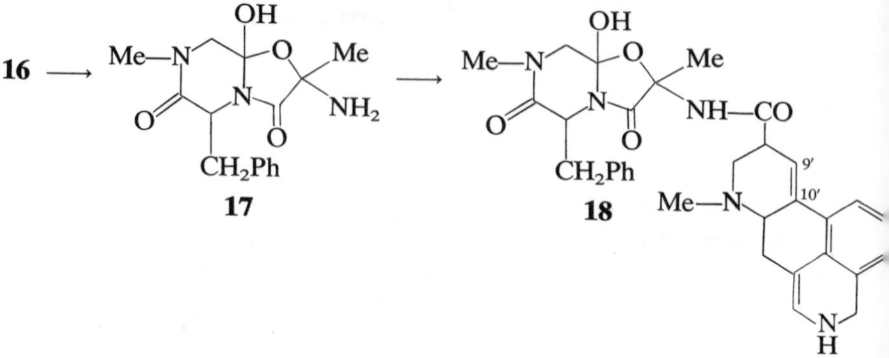

16 → **17** → **18**

III. Oxazolo[3,4-*a*]pyrazines

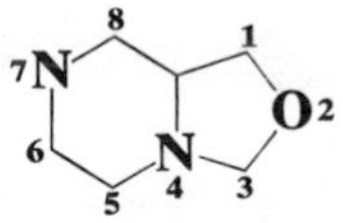

RRI 11838

Only one member of this ring system is known.[16] Referred to as 1,4-diaza-4-methyl-8-oxa-9-oxobicyclo[4.3.0]nonane, compound **20** was prepared from the hydroxymethylpiperazine **19** by treatment with sodium ethoxide and ethyl chloroformate. The product has m.p. 151–151.5° and shows bands in the infrared at 1750 and 1040 cm.$^{-1}$

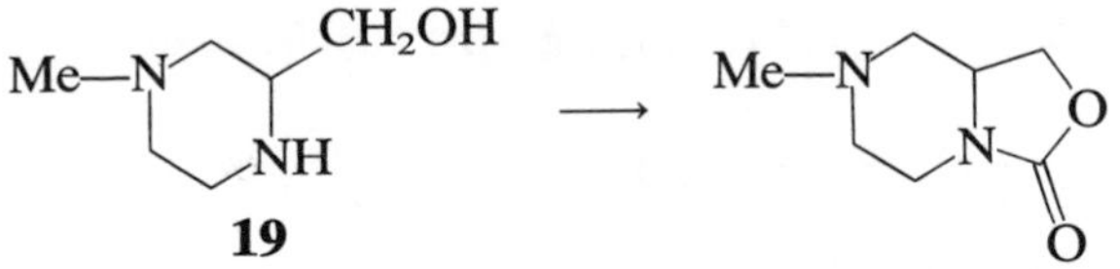

19 → **20**

IV. Isoxazolo[4,5-*b*]pyrazines

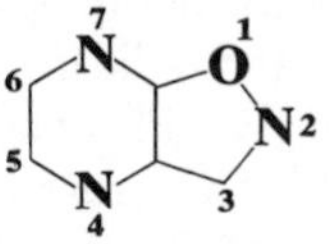

Four 3-phenylisoxazolo[4,5-*b*]pyrazines (**22**) have been prepared by reacting the diamine **21** with appropriate α-dicarbonyl compounds.[17] Compound **22** (R$=CH_2Br$) was also obtained by bromination of the corresponding dimethyl compound with *N*-bromosuccinimide.

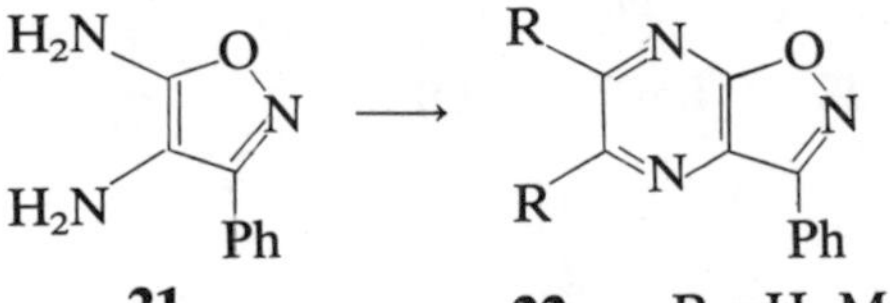

21 → **22** R=H, Me, $BrCH_2$, or Ph

V. [1,2,3]Oxadiazolo[3,4-*a*]pyrazines

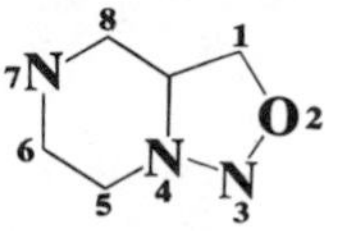

The sydnone **23** on treatment with aqueous sodium hydroxide solution provides the condensed mesoionic pyrazine **24**, which is claimed to be an antiinflammatory and analgesic agent.[18]

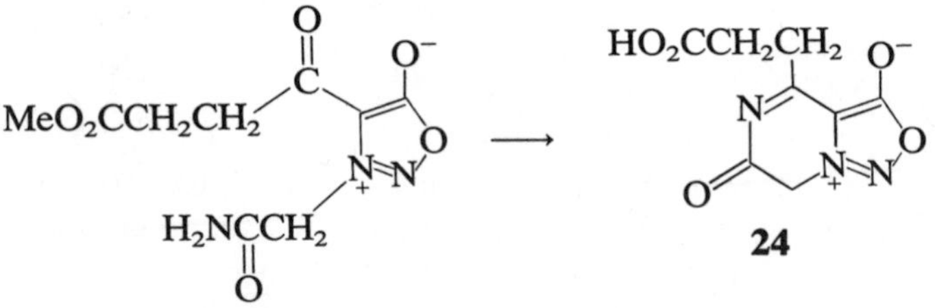

23 → 24

VI. [1,2,5]Oxadiazolo[3,4-*b*]pyrazines

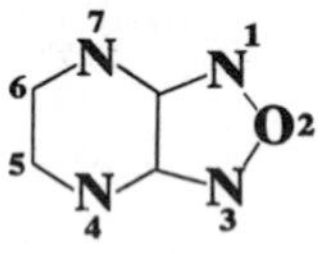

The only reference to this ring system describes the heterocycle as a furazono[3,4-*b*]pyrazine.[19] The compounds are prepared by condensation of 3,4-diaminofurazan (**25**) with α-dicarbonyl compounds. Benzil gave the 5,6-diphenyl derivative **26** (R = Ph), and oxalic acid gave the "dihydroxy" compound **26** (R = OH). The presence of a band at 1695 cm^{-1} in the solution infrared spectrum of **26** (R = OH) is taken to indicate the presence of the oxo tautomer (**27**).

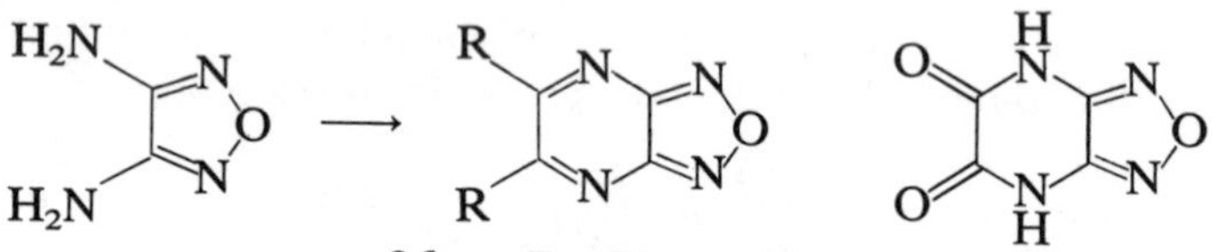

25 → **26** R = Ph or OH **27**

VII. Tables

TABLE 1. SPIROFURO[3,4-*b*]PYRAZINES

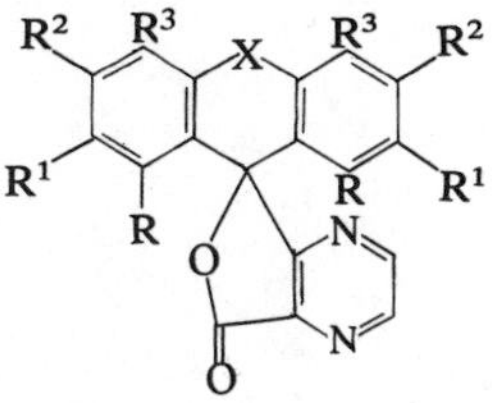

Molecular formula	Substituents X	R	R^1	R^2	R^3	m.p., °C	Ref.
$C_{18}H_6Br_4N_2O_5$	O	H	Br	OH	Br	>280	7
$C_{18}H_{10}N_2O_5$	O	H	H	OH	H	>295	7
$C_{18}H_{10}N_2O_7$	O	OH	H	OH	H	235	7
$C_{18}H_{11}N_3O_4$	NH	H	H	OH	H	>295	7
$C_{18}H_{13}N_5O_2$	NH	H	H	NH_2	H	>295	7
$C_{20}H_{14}N_2O_5$	O	Me	H	OH	H	190	7
$C_{22}H_{20}N_4O_3$	O	H	H	NMe_2	H	178–180	7
$C_{24}H_{24}N_4O_3$	O	H	H	NEt_2	H	278	7

TABLE 2. FURO[3,4-*b*]PYRAZINES

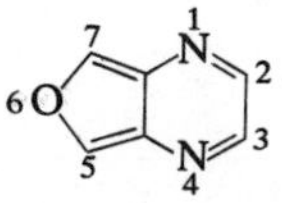

Molecular formula	Substituents	m.p., °C	Ref.
$C_6H_2N_2O_3$	5,7-Dihydro-5,7-dioxo	207 decomp.[a]	3, 4, 5, 11
$C_8H_6N_2O_3$	5,7-Dihydro-2,3-dimethyl-5,7-dioxo	171–171.5	6
$C_{10}H_{16}N_2O$	2,3,5,7-Tetrahydro-5,5,7,7-tetramethyl	b.p. 75.5–75.7/9 mm	8
$C_{12}H_{20}N_2O$	5,7-Diethyl-2,3,5,7-tetrahydro-5,7-dimethyl	b.p. 98–98.2/9 mm	8

[a] Sublimes at 160° at 0.01 mm.

TABLE 3. OXAZOLO[3,2-*a*]PYRAZINES

Molecular formula	Structure	m.p., °C	Ref.
$C_8H_{12}N_2O_4$	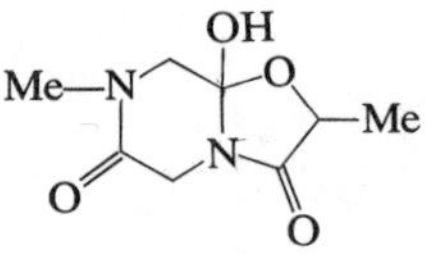	149–150	13, 14
$C_{15}H_{19}N_3O_4$	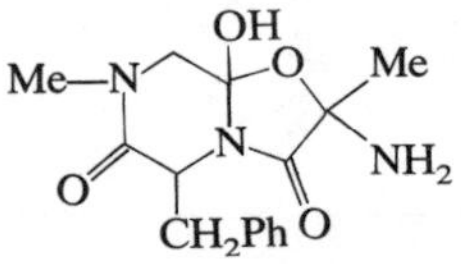	75 decomp.	15
	Benzylcarbamate[a,b]	145–147	
$C_{16}H_{18}N_2O_6$	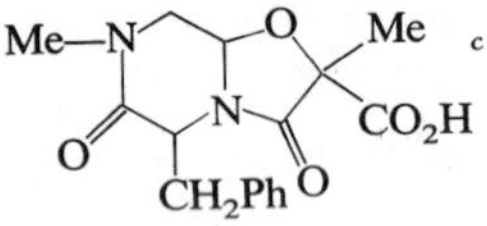 [c]	102–103	15
	Ethyl ester[d]	136–138	
	Acid chloride	134–135	
	Acid azide	105	

[a] $[\alpha]_D^{20}$ 13.8 (*c* 1, EtOH), 12.2 (*c* 1, $CHCl_3$).
[b] Several amides with lysergic acid are reported in Reference 15.
[c] $[\alpha]_D^{20}$ 2.2 (*c* 1, EtOH), 15.2 (*c* 1, $CHCl_3$).
[d] $[\alpha]_D^{20}$ −13.4 (*c* 1, EtOH), −8.8 (*c* 2, $CHCl_3$).

TABLE 4. OXAZOLO[3,4-*a*]PYRAZINES

Molecular formula	Structure	m.p., °C	Ref.
$C_7H_{12}N_2O_2$	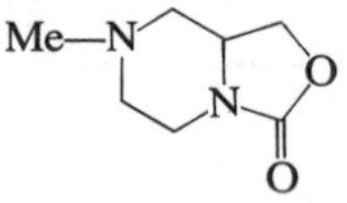	151–151.5	16

TABLE 5. ISOXAZOLO[4,5-*b*]PYRAZINES

Molecular formula	Substituents	m.p., °C	Ref.
$C_{11}H_7N_3O$	3-Phenyl	143.5–144.5	17
$C_{13}H_9Br_2N_3O$	5,6-Bis(bromomethyl)-3-phenyl	166–167	17
$C_{13}H_{11}N_3O$	5,6-Dimethyl-3-phenyl	134–135	17
$C_{23}H_{15}N_3O$	3,5,6-Triphenyl	216–217	17

TABLE 6. [1,2,3]OXADIAZOLO[3,4-*a*]PYRAZINES

Molecular formula	Structure	m.p., °C	Ref.
$C_8H_7N_3O_5$	$HO_2CCH_2CH_2$ O^- N O O $\overset{+}{N}$=N	182–183	18

TABLE 7. [1,2,5]OXADIAZOLO[3,4-*b*]PYRAZINES

Molecular formula	Substituents	m.p., °C	Ref.
$C_4H_2N_4O_3$	4,5,6,7-Tetrahydro-5,6-dioxo	306–310 decomp.	19
$C_{16}H_{10}N_4O$	5,6-Diphenyl	195–196	19

VIII. References

1. I. K. Korobitsyna, L. A. Kazitsyna, and Yu. K. Yur'ev, *Zhur. Obshchei Khim.*, **25,** 1394 (1955); *Chem. Abstr.*, **49,** 15481 (1955).
2. R. F. C. Brown, W. D. Crow, and R. K. Solly, *Chem. Ind.* (London), 343 (1966).
3. S. Gabriel and A. Sonn, *Ber.*, **40,** 4850 (1907).
4. J. A. Solomons and P. E. Spoerri, *J. Am. Chem. Soc.*, **75,** 679 (1953).
5. N. Sugimoto and T. Tanaka, *J. Pharm. Soc.* Jap., **76,** 470 (1956); *Chem. Abstr.*, **50,** 14777 (1956).

6. R. A. Pages and P. E. Spoerri, *J. Org. Chem.*, **28,** 1702 (1963).
7. S. C. De and P. C. Datta, *Ber.*, **64B,** 2606 (1931).
8. I. K. Korobitsyna, Yu. K. Yur'ev, and O. I. Nefedova, *J. Gen. Chem. USSR*, **24,** 191 (1954); *Chem. Abstr.*, **49,** 3197, 4618 (1955).
9. E. K. Fields and S. Meyerson, *J. Org. Chem.*, **31,** 3307 (1966).
10. Toyo Rayon Co. Ltd., Fr. Pat. 1,527,947; *Chem. Abstr.*, **71,** 31059 (1969).
11. P. Hemmerich and S. Fallab, *Helv. Chim. Acta*, **41,** 498 (1958).
12. M. M. Shemyakin, V. K. Antonov, A. M. Shkrob, Yu. N. Sheinker, and L. B. Senzavina, *Tetrahedron Lett.*, 701 (1962).
13. M. M. Shemyakin, A. M. Shkrob, V. I. Schelokov, and Z. E. Agadzhanyan, *Tetrahedron*, **21,** 3537 (1965).
14. V. K. Antonov, A. M. Shkrob, and M. M. Shemyakin, *Zh. Obshch. Khim.*, **37,** 2225 (1967); *Chem. Abstr.*, **69,** 10416 (1968).
15. Sandoz Ltd., Neth. Appl. 6,412,779; *Chem. Abstr.*, **63,** 13269 (1965).
16. E. Jucker and E. Rissi, *Helv. Chim. Acta*, **45,** 2383 (1962).
17. G. Desimoni and G. Minoli, *Tetrahedron*, **24,** 4907 (1968).
18. Fujisawa Pharmaceutical Co. Ltd., Jap. Pat. 73 29,239.
19. A. Gasco, G. Rua, E. Menziani, G. M. Nano, and G. Tappi, *J. Heterocycl. Chem.*, **6,** 769 (1969).

CHAPTER XXV

Thieno-, Thiazolo-, Isothiazolo-, and Thiadiazolopyrazines

I. Thieno[2,3-*b*]pyrazines 441
II. Thieno[3,4-*b*]pyrazines 443
III. Thiazolo[3,2-*a*]pyrazines 444
IV. Thiazolo[3,4-*a*]pyrazines 447
V. Thiazolo[4,5-*b*]pyrazines 448
VI. Isothiazolo[4,5-*b*]pyrazines 449
VII. 1,2,5-Thiadiazolo[3,4-*b*]pyrazines 450
VIII. Tables 451
Table 1. Thieno[2,3-*b*]pyrazines 451
Table 2. Thieno[3,4-*b*]pyrazines 452
Table 3. Thiazolo[3,2-*a*]pyrazin-4-ium Salts 452
Table 4. Perhydro-5,8-dioxothiazolo[3,4-*a*]pyrazines 453
Table 5. Thiazolo[4,5-*b*]pyrazines 453
Table 6. Isothiazolo[4,5-*b*]pyrazines 454
Table 7. 1,2,5-Thiadiazolo[3,4-*b*]pyrazines 454
IX. References 454

I. Thieno[2,3-*b*]pyrazines

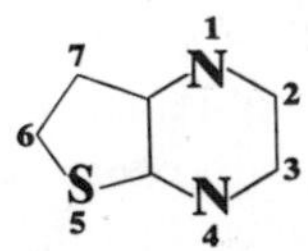

Although the correct numbering above has been used, the only group to investigate the chemistry of thieno[2,3-*b*]pyrazines has preferred to use the numbering shown in formula **1**. Molecular orbital calculations on the parent ring system and the isomeric [3,4-*b*]system suggest some bond fixation in the pyrazine ring as represented in structures **1** and **2**, the fixation being greater in the latter heterocycle.[15] These calculations suggest that thieno[2,3-*b*]pyrazine should be most susceptible to nucleophilic substitution in position 6 and to attack by free radicals and electrophiles in position 7. All these results parallel the conclusions derived for the corresponding thienopyridine.

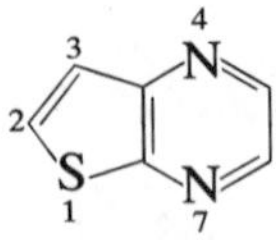

1 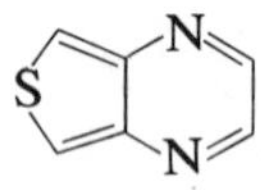 **2**

The parent heterocycle[16] shows a band in its infrared spectrum at 1536 cm^{-1} and its PMR spectrum (d_6-DMSO) comprises the expected four doublets at δ 7.59 ($J=6$ Hz, H-6); 8.30 ($J=6$ Hz, H-7); 8.59 ($J=2$ Hz, H-2 or H-3); and 8.74 ($J=2$ Hz, H-2 or H-3).

Preparation of thieno[2,3-*b*]pyrazines involves appropriate substituted pyrazines as intermediates. Possible alternative routes starting from thiophenes were not attempted as 2,3-diaminothiophene is only accessible with difficulty.[17] Reaction of 2-chloro-3-cyanopyrazine (**3**) with α-mercaptocarbonyl compounds in the presence of sodium carbonate gave high yields of the 7-amino derivatives **4** and **5**.

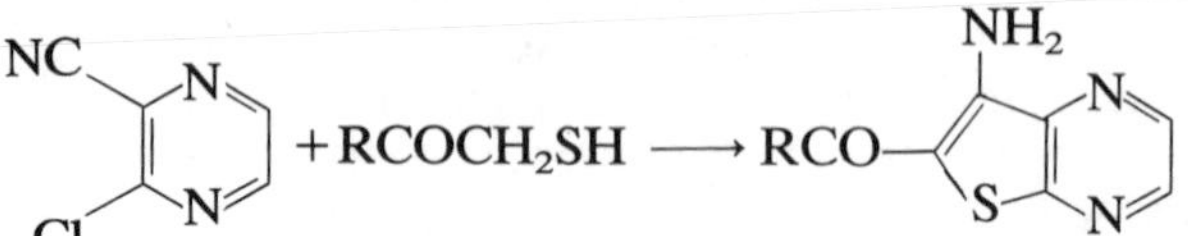

3 + $RCOCH_2SH$ ⟶ RCO–(7-aminothieno[2,3-*b*]pyrazine)

4 R = OEt
5 R = NH_2

The ester **4** gave the tricyclic system **6** on treatment with formamide. The related heterocycle **8** was obtained by saponification of the ester **4** and reaction of the resulting carboxylic acid **7** with acetic anhydride.[17] Decarboxylation of the acid **7** by heating with copper powder gave 7-aminothieno[2,3-*b*]pyrazine (**9**), which on diazotization gave only high-melting intractable solids. In contrast diazotization of the amino ester **4** gave the ester **10** after treatment of the diazonium salt with hypophosphorous acid.[16] Subsequent hydrolysis and decarboxylation afforded the unsubstituted heterocycle **1** in 32% overall yield from the pyrazine starting material **3**.

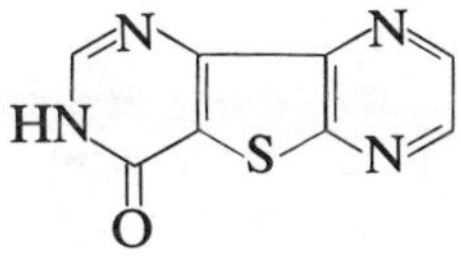

6 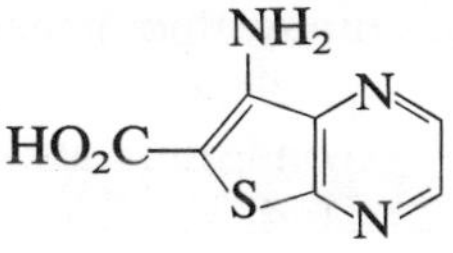 **7**

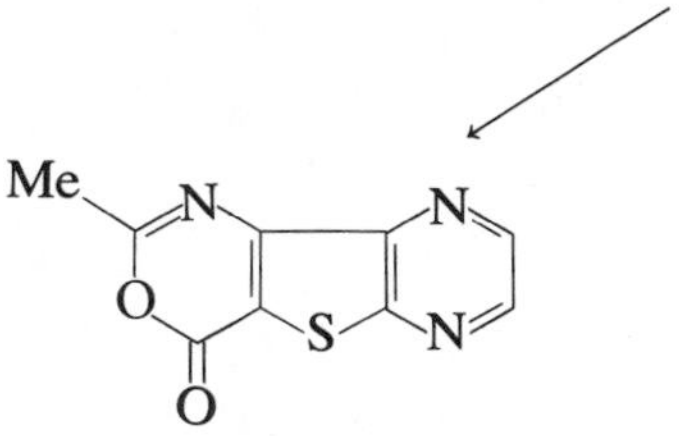

8 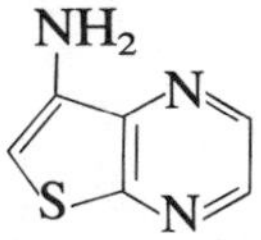 **9** 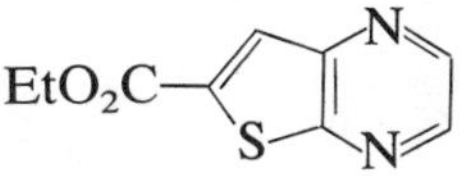 **10**

II. Thieno[3,4-*b*]pyrazines

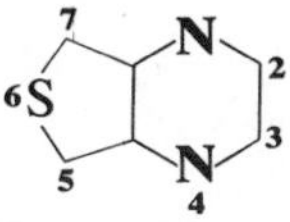

RRI 9966

Recently, molecular orbital calculations have been performed on the parent heterocycle.[15] The results of the calculations indicate C-5 and C-7 should be the most reactive sites to electrophilic, nucleophilic, and free-radical attack. The degree of bond fixation evident from bond length calculations suggested that these positions should also act as the termini of a *s*-butadiene fragment rendering the molecule liable to Diels-Alder-type reactions.

Several derivatives of this ring system have been prepared by the reaction of α-dicarbonyl compounds with various 3,4-diaminothiophenes. Thus the diamine **11** has given the diphenyl compound **12** and the "dihydroxy" compound **13**,[1] presumably more correctly formulated as the dioxo compound **14**.

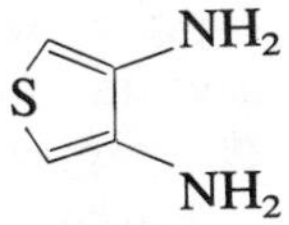

11 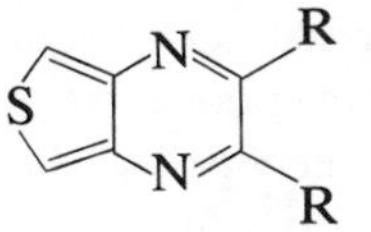 **12** R = Ph **13** R = OH **14**

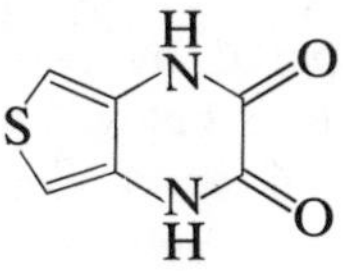

Reduction of the 3,4-dinitrothiophenes **15** with tin and hydrochloric acid in the presence of benzil provides the 2,3-diphenyl compounds **16** and **17**.[2,3] The reducing conditions also serve to remove the bromine atom in the starting materials, thus forming a useful route to 7-unsubstituted compounds.

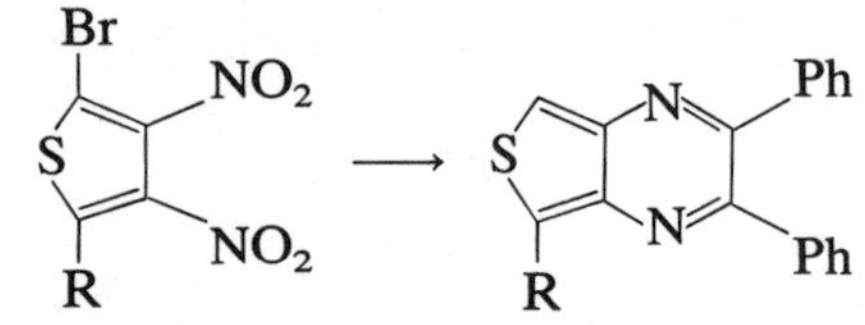

15 R = Me or $(CH_2)_3CO_2H$

16 R = Me
17 R = $(CH_2)_3CO_2H$

III. Thiazolo[3,2-*a*]pyrazines

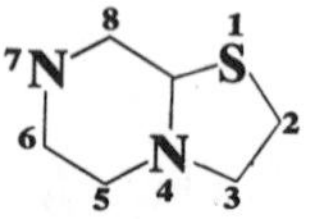

The only report[4] of the synthesis and properties of this bicyclic heterocycle adopts a different numbering system to that used by *Chemical Abstracts* (shown above). The alternative system (**18**) corresponds to that used for the related pyrido[1,2-*a*]pyrazine system (**19**).

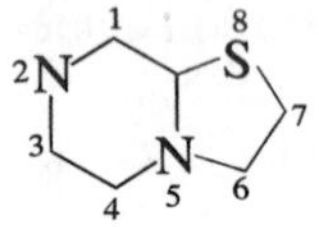

18 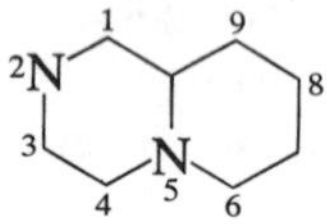 **19**

The ultraviolet spectra of thiazolo[3,2-*a*]pyrazin-4-ium bromide, its 7-oxide, and several 6- or 8-substituted derivatives have been recorded. The parent quaternary salt has absorption maxima at 233, 315, and 326 nm in aqueous solution.

The parent heterocycle and its 8-methyl and 8-phenyl derivatives have been prepared by two methods. Deoxygenation of the corresponding 7-oxides (see below) with phosphorus tribromide gave the quaternary bromides in good yield. Alternatively the anils **20** were cyclized in 22–36% yield by treatment with a 50% solution of hydrogen bromide in acetic acid at −78°. The anils were readily formed by heating together 2-acylthiazoles and aminoacetaldehyde dimethylacetal.

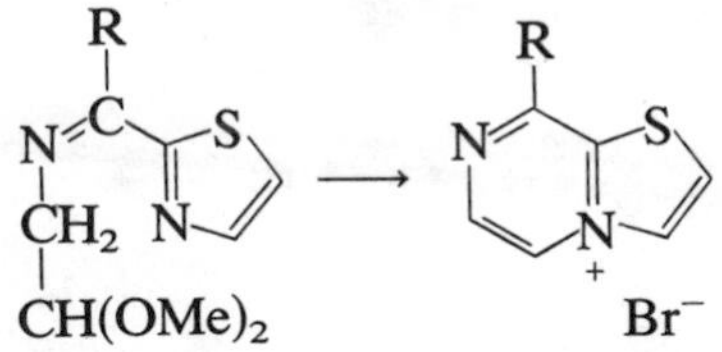

20 R = H, Me, or Ph **21**

The 7-oxides of thiazolo[3,2-*a*]pyrazin-4-ium salts are available by analogous methods to those used for pyrido[1,2-*a*]pyrazines. The *N*-oxide (**25**) of the parent was obtained by quaternization of the ethylene acetal **22** (R = H) with bromoacetaldehyde oxime (**23**) followed by the cyclization of the quaternary compound **24** with concentrated sulfuric acid. Attempts to quaternize the acetal **22** (R = Me) with bromoacetaldehyde oxime were unsuccessful.

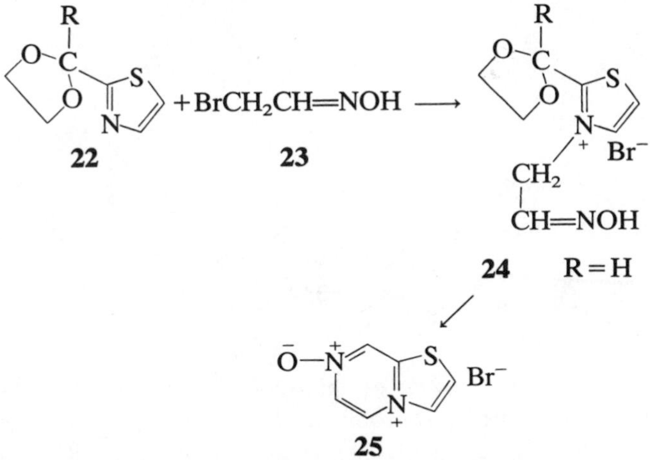

22 + $BrCH_2CH{=}NOH$ **23** → **24** R = H → **25**

The *N*-oxides **27** were obtained directly by boiling solutions of the oximes **26** with either bromoacetaldehyde or bromoacetone in acetonitrile containing 48% hydrobromic acid. Phenacyl bromide in acetonitrile provided the quaternized intermediates **28**, which were cyclized to the corresponding 6-phenyl-7-oxides (**29**) by treatment with concentrated sulfuric acid.

When the oxime **30** was heated with bromoacetaldehyde in acetonitrile containing no hydrobromic acid, a mixture of the aromatic compound **32** and the intermediate hydroxydihydro derivative **31** was obtained. Compound **31** could be dehydrated to **32** by heating under reflux in methanolic hydrogen bromide.

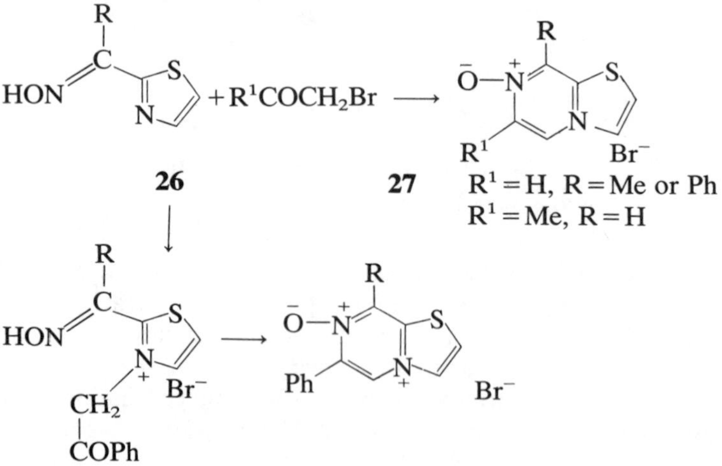

$+ R^1COCH_2Br \longrightarrow$

26 **27** $R^1 = H$, $R = Me$ or Ph
$R^1 = Me$, $R = H$

28 **29** $R = H$ or Me

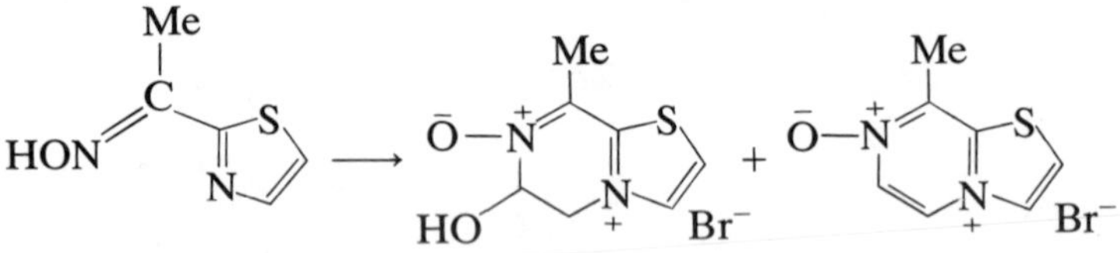

30 **31** **32**

Bromination of the 7-oxide of thiazolo[3,2-*a*]pyrazin-4-ium bromide by bromine in aqueous hydrobromic acid gives a monobromo derivative. By analogy with the pyrido[1,2-*a*]pyrazin-5-ium series, it is suggested that substitution has taken place in the 8-position to give the bromo-*N*-oxide **33**. All the *N*-oxides prepared have been deoxygenated in good yield by heating with phosphorus tribromide. The 8-bromo compound (**33**) and the 8-methyl compound (**32**) were the most resistant to deoxygenation, requiring heating in a sealed tube to effect reaction.

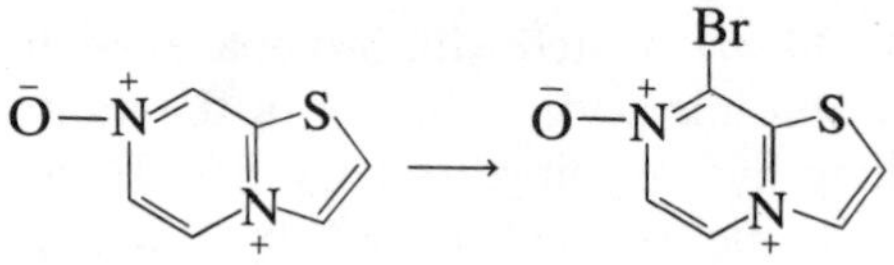

33

IV. Thiazolo[3,4-*a*]pyrazines

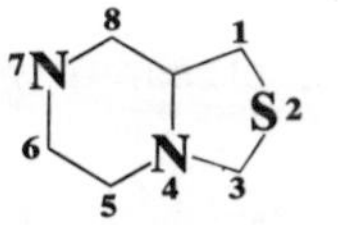

3*H* RRI 1140

Several perhydro derivatives of this ring system have been obtained as by-products in the study of thiazolidine chemistry related to the penicillin nucleus (**34**). Thus treatment of the chloroacetyl thiazolidine **35** with ammonium hydroxide in the cold yielded the diketopiperazine **36**.[5,6] This compound has also been obtained by boiling the glycylthiazolidine **37** in ethanol.[7] The 1-unsubstituted analogue of **36** was similarly obtained.[8] Treatment of the *d* form of the thiazolidine **38** with *N*-acetylglycyl chloride (**39**) gave the hoped for acetylglycylthiazolidine **40** together with the *d*-form of the diketopiperazine **41**. Corresponding products were obtained from the *l* form of **38**, but no data were provided for the *l*-diketopiperazine.[9]

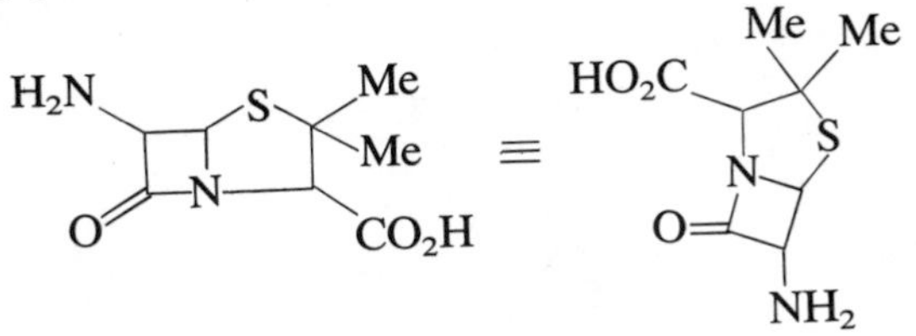

34

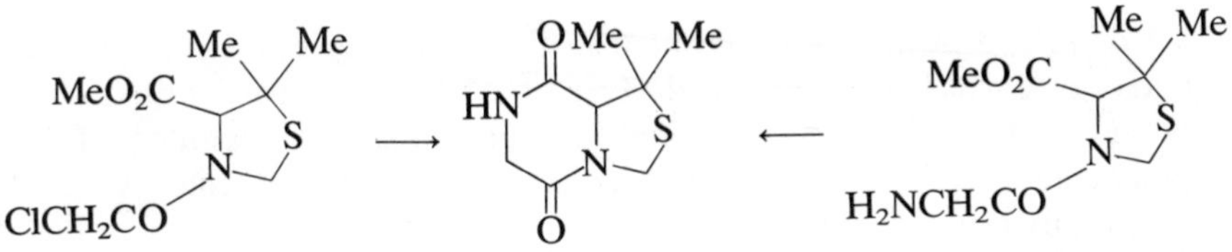

35 **36** **37**

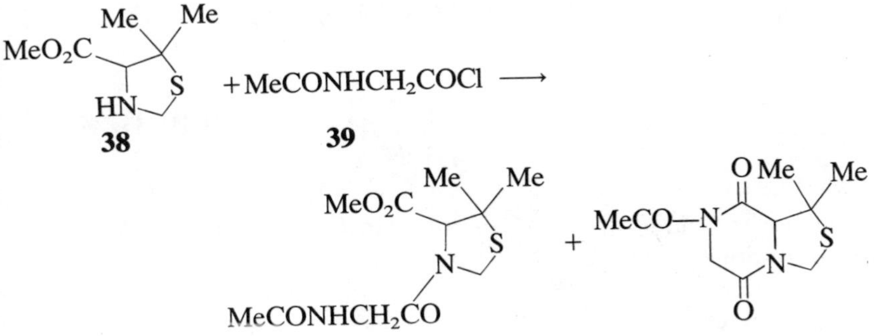

38 + $MeCONHCH_2COCl$ → **39**

40 + **41**

Treatment of the anhydride **42** with benzylamine gave a mixture of two products, one of which was tentatively assigned the structure **43**.[10] The diketopiperazine **45** has been prepared by thermal cyclization of the acylglycine **44**.[11]

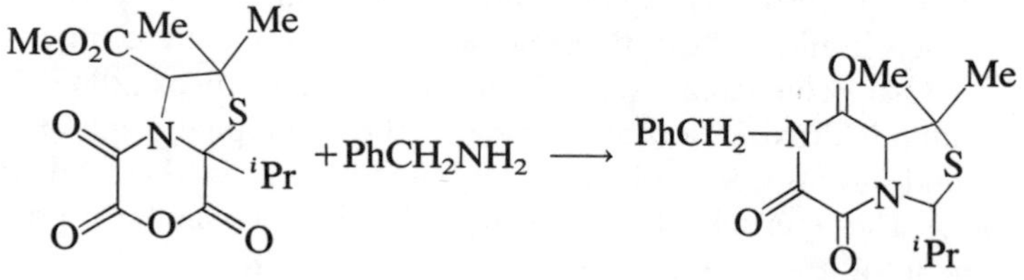

Compound **36** has been acylated in the cold to give a 7-phenylacetyl derivative, which is readily hydrolyzed back to the unsubstituted compound.[5] More vigorous acylation conditions result in opening of the piperazine ring.[6] Ring opening also occurs when **36** is warmed with *M* sodium hydroxide solution.[7]

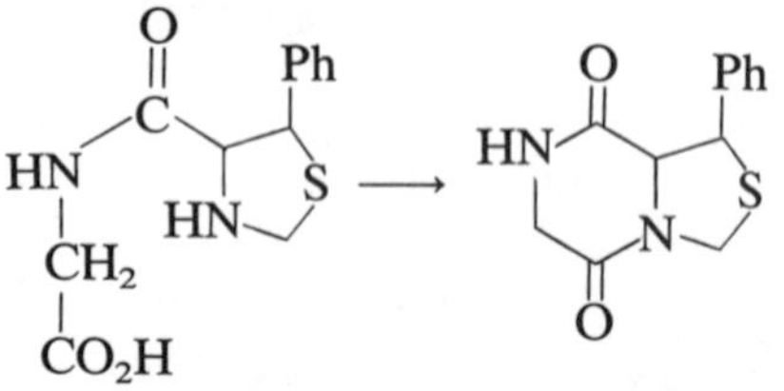

V. Thiazolo[4,5-*b*]pyrazines

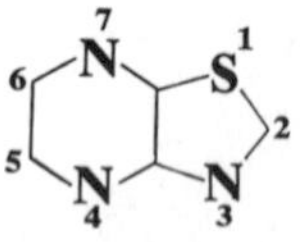

RRI 11849

Treatment of the pyrazine **46** with ethyl orthoformate provides the 2-unsubstituted compound **47**.[12] The use of acetic anhydride results in the 2-methylthiazolo[4,5-*b*]pyrazine **48** being obtained.[12,13] Reaction of this compound with sodium methoxide provides the 6-methoxy derivative **49**.[12,13]

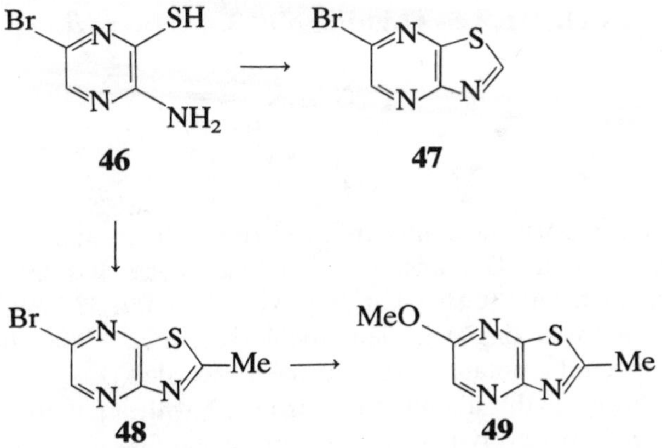

46 → 47

↓

48 → 49

VI. Isothiazolo[4,5-*b*]pyrazines

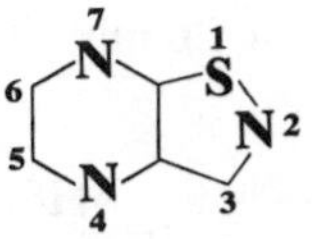

Condensation of the diaminoisothiazole **50** with *n*-butyl glyoxylate is reported to give the fused pyrazine **51**. No comments were made about possible formation of the 5-hydroxy isomer.[14] Presumably the compound is more accurately described as the oxo tautomer **52**.

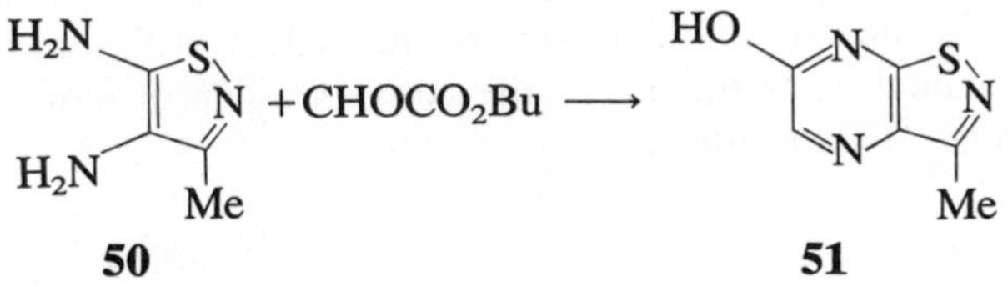

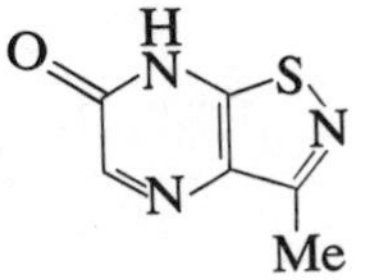

50 + $CHOCO_2Bu$ → 51

52

VII. 1,2,5-Thiadiazolo[3,4-*b*]pyrazines

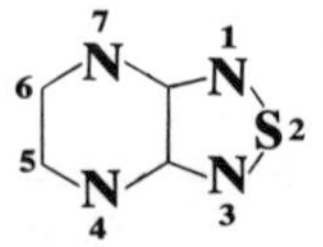

Two recent reports have appeared describing different syntheses of the parent ring system. The unsubstituted heterocycle has maxima in its ultraviolet spectrum[19] ($CHCl_3$) at 304, 311, 317, and 324 nm. The PMR spectrum[18] shows a singlet in deuterochloroform at δ 9.06. Infrared[18,19] and mass-spectral[18] details have also been recorded.

Condensation of the diaminothiadiazole **53** with aqueous glyoxal and with diacetyl respectively gave the parent system **56** and its 5,6-dimethyl derivative.[18] The condensation with aqueous glyoxal takes place in two stages, probably involving a covalent hydrate **54** or **55** of the unsubstituted heterocycle. The intermediate precipitates from the hot reaction mixture on the addition of glyoxal. Dehydration is accomplished by refluxing the aqueous mixture for 20 min.

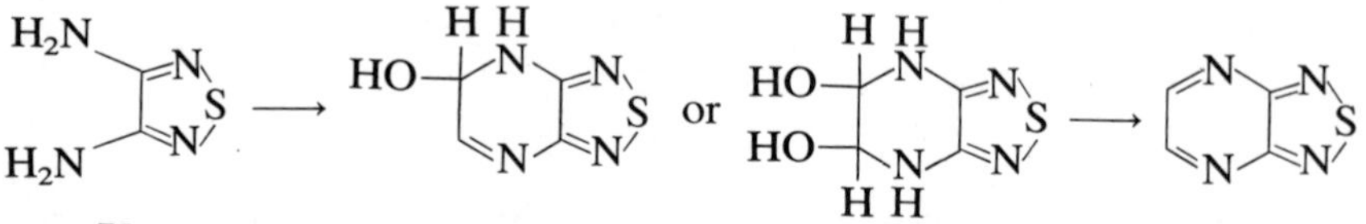

2,3-Diaminopyrazine (**57**) and its 5,6-dichloro derivative react with thionyl chloride to give thiadiazolopyrazines.[19] Attempts to react 2,3-diamino-5,6-dibromopyrazine with thionyl chloride have resulted in partial or complete chlorine exchange. Thionyl bromide gave only tars on reaction with the dibromodiaminopyrazine. The 5,6-dibromo derivative **58** was obtained by treatment of the corresponding dichloro compound with bromine in chloroform.

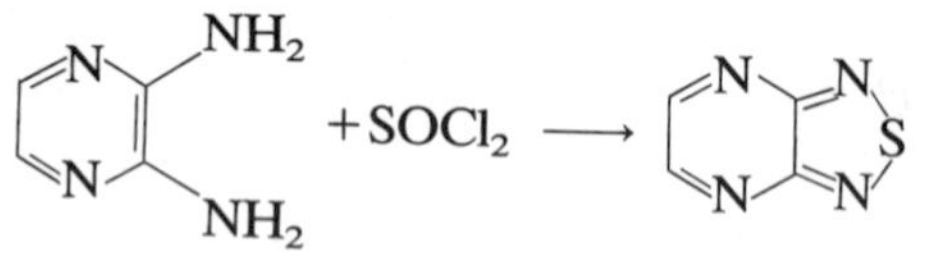

$+SOCl_2 \longrightarrow$

57

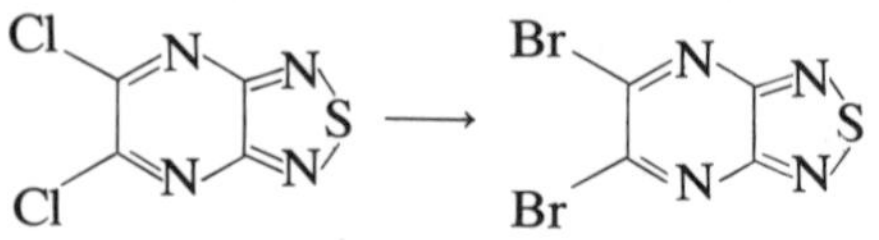

58

Reaction of 5,6-dichloro-1,2,5-thiadiazolo[3,4-*b*]pyrazine with di-*n*-propylamine gave good yields of the mono and disubstituted compounds **59** and **60**, depending on the ratio of chloro compound to amine. In contrast, reaction with *n*-propanethiol gave only the disubstituted product **61** irrespective of the ratio of nucleophile to dichloro compound.[19]

59 ($^nPr_2N$, Cl)   **60** ($^nPr_2N$, $^nPr_2N$)   **61** (^nPrS, ^nPrS)

VIII. Tables

TABLE 1. THIENO[2,3-*b*]PYRAZINES

Molecular formula	Substituents	m.p., °C	Ref.
$C_6H_3ClN_2S$	7-Chloro[a]	70–72	16
$C_6H_4N_2S$	Unsubstituted[b]	43–44.5	16
$C_6H_5N_3S$	7-Amino[c]	81–82	16
$C_7H_3ClN_2O_2S$	6-Carboxy-7-chloro[d]	248–250	16
	Ethyl ester monohydrate	104–105	
$C_7H_4N_2O_2S$	6-Carboxy		
	Monohydrate	261–262	16
	Ethyl ester	73–74.5	
$C_7H_4N_4S$	7-Amino-6-cyano	204–206	17
$C_7H_5N_3O_2S$	7-Amino-6-carboxy	218–220	17
	Ethyl ester	114–116	
$C_7H_6N_4OS$	7-Amino-6-carboxamido	284–286	17

[a] Sublimes at 50°/2 mm.
[b] Sublimes at 35–40°/3 mm.
[c] Sublimes at 78–80°/3 mm.
[d] No elemental analysis reported.

TABLE 2. THIENO[3,4-*b*]PYRAZINES

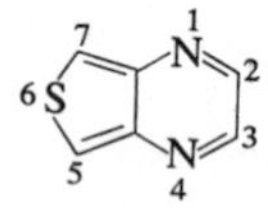

Molecular formula	Substituents	m.p., °C	Ref.
$C_6H_4N_2O_2S$	1,2,3,4-Tetrahydro-2,3-dioxo	Sublimes 300	1
$C_{18}H_{12}N_2S$	2,3-Diphenyl	190–191	1
$C_{19}H_{14}N_2S$	5-Methyl-2,3-diphenyl	171–172	2
$C_{22}H_{18}N_2O_2S$	5-(3-Carboxypropyl)-2,3-diphenyl	205–208 decomp.	3

TABLE 3. THIAZOLO[3,2-*a*]PYRAZIN-4-IUM SALTS[a]

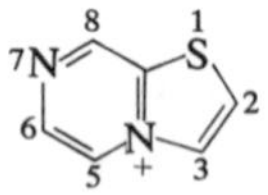

Molecular formula[b]	Substituents	m.p., °C	Ref.
$C_6H_4BrN_2S$	8-Bromo		
	Bromide	Chars < 330	4
	7-Oxide bromide	Chars < 330	
$C_6H_5N_2S$	Unsubstituted		
	Bromide	Chars < 330	4
	perchlorate	134–135 decomp.	
	7-Oxide bromide	Chars < 350	
	Perchlorate	195–196 decomp.	
$C_7H_7N_2S$	6-Methyl		
	7-Oxide bromide	293–294 decomp.	4
	picrate	256–257 decomp.	
	8-Methyl		
	Bromide	292–293 decomp.	4
	Picrate	184 decomp.	
	7-Oxide bromide	307–308 decomp.	
$C_7H_9N_2OS$	5,6-Dihydro-6-hydroxy-8-methyl		
	7-Oxide bromide	284–285	4
$C_{12}H_9N_2S$	6-Phenyl		
	Bromide hemihydrate	276–278 decomp.	4
	Picrate	205–207 decomp.	
	7-Oxide bromide	254–255 decomp.	
	8-Phenyl		
	Bromide	266–268 decomp.	4
	7-Oxide bromide	234–235 decomp.	
	perchlorate	270–272 decomp.	

TABLE 3 (*contd.*)

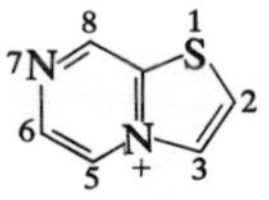

Molecular formula[b]	Substituents	m.p., °C	Ref.
$C_{13}H_{11}N_2S$	8-Methyl-6-phenyl		
	7-Oxide bromide	230–231 decomp.	4

[a] Reference 4 uses a different numbering system (see text).
[b] Anions are not included in the molecular formulae.

TABLE 4. PERHYDRO-5,8-DIOXOTHIAZOLO[3,4-*a*]PYRAZINES

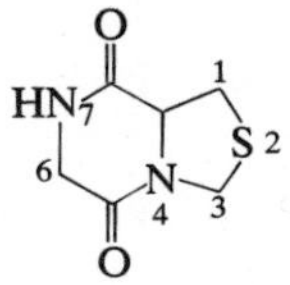

Molecular formula	Substituents	m.p., °C	Ref.
$C_6H_8N_2O_2S$	Unsubstituted	190–192	8
$C_8H_{12}N_2O_2S$	1,1-Dimethyl	167	5, 6
		161–163	7
$C_{10}H_{14}N_2O_3S$	7-Acetyl-1,1-dimethyl		
	d Form[a]	125–128	9
$C_{12}H_{12}N_2O_2S$	1-Phenyl	203–204	11
$C_{15}H_{18}N_2O_3S$	1,1-Dimethyl-7-phenylacetyl	143	5
$C_{18}H_{22}N_2O_3S$	7-Benzyl-1,1-dimethyl-6-oxo-3-isopropyl	189 decomp.	10

[a] $[\alpha]_D^{22} + 140°$.

TABLE 5. THIAZOLO[4,5-*b*]PYRAZINES

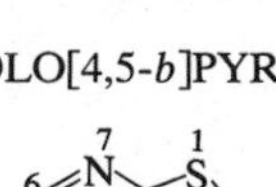

Molecular formula	Substituents	m.p., °C	Ref.
$C_5H_2BrN_3S$	6-Bromo	151	12, 13
$C_6H_4BrN_3S$	6-Bromo-2-methyl	158	12, 13
$C_7H_7N_3OS$	6-Methoxy-2-methyl	71	12, 13

TABLE 6. ISOTHIAZOLO[4,5-*b*]PYRAZINES

Molecular formula	Structure	m.p., °C	Ref.
$C_6H_5N_3OS$	3-Methylisothiazolo[4,5-*b*]pyrazin-5(4H)-one (O=, N–H, S, N, N, Me)	292–293 decomp.	14

TABLE 7. 1,2,5-THIADIAZOLO[3,4-*b*]PYRAZINES

(Ring numbering: N1, S2, N3, N4, 5, 6, N7)

Molecular formula	Substituents	m.p., °C	Ref.
$C_4Br_2N_4S$	5,6-Dibromo	150–151	19
$C_4Cl_2N_4S$	5,6-Dichloro	180–182	19
$C_4H_2N_4S$	Unsubstituted	162–165	18, 19
$C_6H_6N_4S$	5,6-Dimethyl	124–125	18
$C_{10}H_{14}N_4S_3$	5,6-Di-*n*-propylthio	69–70	19
$C_{10}H_{14}ClN_5S$	5-Chloro-6-di-*n*-propylamino	63–64	19
$C_{16}H_{28}N_6S$	5,6-Bis(di-*n*-propylamino)	54–56	19

IX. References

1. R. Motoyama and E. Imoto, *Nippon Kagaku Zasshi*, **78,** 793 (1957); *Chem. Abstr.*, **54,** 22560 (1960).
2. S. Nishimura and E. Imoto, *Nippon Kagaku Zasshi*, **82,** 1540 (1961); *Chem. Abstr.*, **57,** 15051 (1962).
3. S. Nishimura and E. Imoto, *Nippon Kagaku Zasshi*, **82,** 1680 (1962); *Chem. Abstr.*, **59,** 1680 (1963).
4. J. Adamson, E. C. Campbell, and E. E. Glover, *J. Chem. Soc.* (*C*), 2270 (1969).
5. O. Süs, *Justus Liebigs Ann. Chem.*, **561,** 31 (1948); *Chem. Abstr.*, **43,** 4668 (1949).
6. O. Süs and S. Rosenberger, *Justus Liebigs Ann. Chem.*, **564,** 54 (1949); *Chem. Abstr.*, **44,** 1971 (1950).
7. R. Neher, M. Spillmann, L. H. Werner, A. Wettstein, and K. Miescher, *Helv. Chim. Acta*, **29,** 1874 (1946).
8. R. Neher, A. Wettstein, and K. Miescher, *Helv. Chim. Acta*, **29,** 1815 (1946).
9. L. H. Werner, A. Wettstein, and K. Miescher, *Helv. Chim. Acta*, **30,** 432 (1947).
10. R. Bentley, A. H. Cook, and J. A. Elvidge, *J. Chem. Soc.*, 2357 (1949).
11. D. O. Holland and P. Mamalis, *J. Chem. Soc.*, 4584 (1958).
12. L. Bernardi, G. Palamidessi, A. Leone, and G. Larini, *Gazz. Chim. Ital.*, **91,** 1431 (1961); *Chem. Abstr.*, **57,** 2224 (1962).

13. Società Farmaceutici Italia, Brit. Pat. 958,626; *Chem. Abstr.*, **61,** 5668 (1964).
14. A. Holland, R. Slack, and T. F. Warren, *J. Chem. Soc.*, 7277 (1965).
15. S. W. Schneller and F. W. Clough, *J. Heterocycl. Chem.*, **13,** 581 (1976).
16. S. W. Schneller, F. W. Clough, and L. E. Hardee, *J. Heterocycl. Chem.*, **13,** 273 (1976).
17. S. W. Schneller and F. W. Clough, *J. Heterocycl. Chem.*, **12,** 513 (1975).
18. A. P. Komin and M. Carmack, *J. Heterocycl. Chem.*, **13,** 13 (1976).
19. Y. C. Tong, *J. Heterocycl. Chem.*, **12,** 451 (1975).

CHAPTER XXVI

Miscellaneous 5,6-Systems

I.	Cyclopentapyrazines	456
II.	1,3-Dithiolo[4,5-*b*]pyrazines	459
III.	Tables	460
	Table 1. Cyclopentapyrazines	460
	Table 2. 1,3-Dithiolo[4,5-*b*]pyrazines	461
IV.	References	461

I. Cyclopentapyrazines

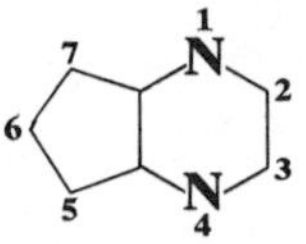

2*H* RRI 1364

Apart from two recent papers[1,8] very little data have been published on this ring system. The fully unsaturated heterocycle (**1**) has not been reported, and all known cyclopentapyrazines are dihydro or tetrahydro compounds. Various 5*H*-6,7-dihydrocyclopentapyrazines have attracted attention as components of the flavor and aroma of roasted peanuts.[1–3,8] It has been pointed out, however, that a great deal of this work is unsubstantiated. Confusion has arisen because certain examples of this ring system were mistakenly identified as isopropenylpyrazines.[1]

Details of the PMR and mass spectra of alkylated compounds such as **2** and **3** have been published,[1,8] together with a discussion of their infrared spectra.[1] The hydrogen atoms attached to the five-membered ring in **2** appear in the PMR ($CDCl_3$/CCl_4) as an eight-proton multiplet extended over the range δ 1.30–3.00. In compounds of formula **3** the PMR signals

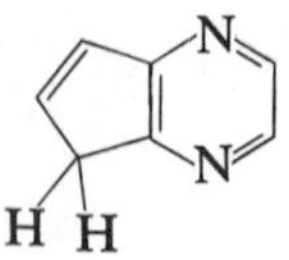

1 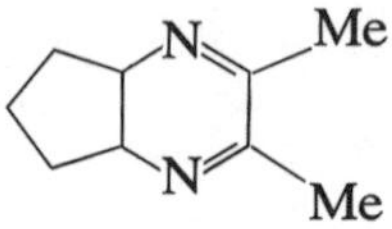 **2** 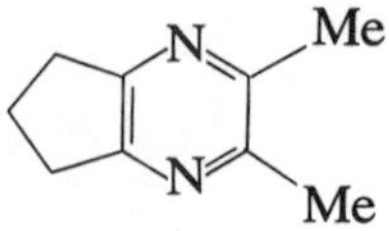 **3**

are resolved into a four-proton triplet at δ 2.95 (H-5, H-7) and a two-proton multiplet at δ 2.15 (H-6).

The unsubstituted tetrahydro compound (**4**) was too unstable to be adequately purified and characterized. However the 5-methyl derivative (**5**) could be kept for a few days at 0°C in the absence of air and light.[1] The PMR of this compound (d_6-DMSO) confirmed the presence of the N—H group (broad singlet, δ 5.06) and the 5-methyl substituent (singlet δ 1.70). The methylated isomers **6** and **7** could be differentiated and identified by PMR. Compound **6** in CCl_4 shows two three-proton singlets from the methyl groups at δ 2.81 (N—Me) and at δ 1.96 (C—Me). In contrast the *C*-methyl group in compound **7** exhibits a doublet in the PMR spectrum at δ 1.13 as well as the expected *N*-methyl singlet at δ 2.58.

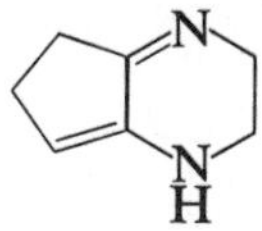

N H **4**

N N Me H **5**

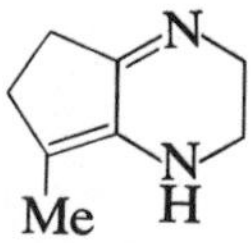

N N Me Me **6**

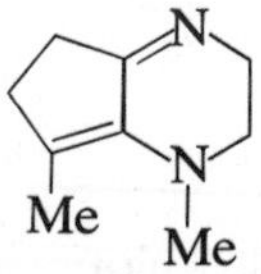

Me N N Me **7**

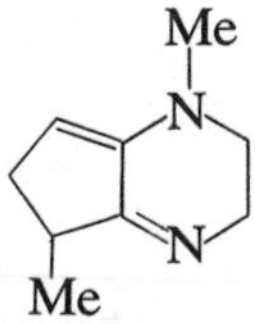

The main synthetic route to cyclopentapyrazines involves formation of the six-membered ring by condensation of appropriate 1,2-diamines and α-dicarbonyl compounds. For example, reaction of diacetyl with cyclopentane-1,2-diamine yielded the tetrahydro compound **2**, which was dehydrogenated to the dihydroderivative **3**. Suitable dehydrogenation methods include heating under reflux in xylene with palladium on charcoal or heating to 300° over copper chromite. Condensation of ethylenediamine or 1,2-propanediamine with cyclopentanediones such as **8** afforded tetrahydro compounds of formula **9**, which on dehydrogenation with copper chromite at 300° afforded the dihydro derivatives **10**. As expected, mixtures of isomers are formed when appropriate unsymmetrical starting materials are used.

The cyclopentane derivative **11** on reaction with ethylenediamine gave a tetrahydro derivative formulated as **12**, which was hydrolyzed to give the diketo compound **13**.[4] The alternative tetrahydro formulation **14** may

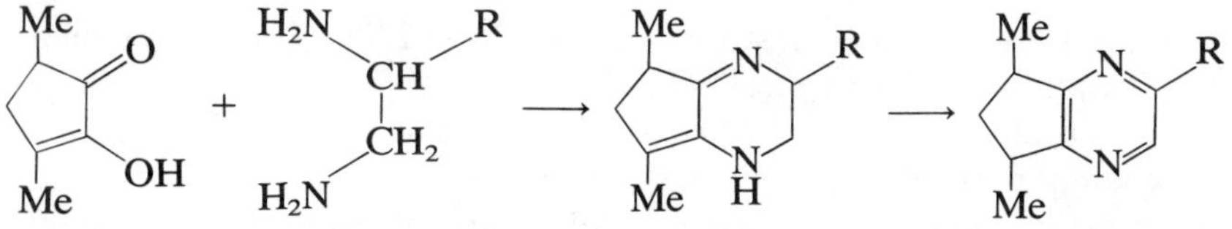

8 R = H or Me 9 10

be a more accurate representation of the product **12** from the ethylenediamine reaction, particularly as analogous structures such as **5** and **9** are supported by PMR and infrared evidence.

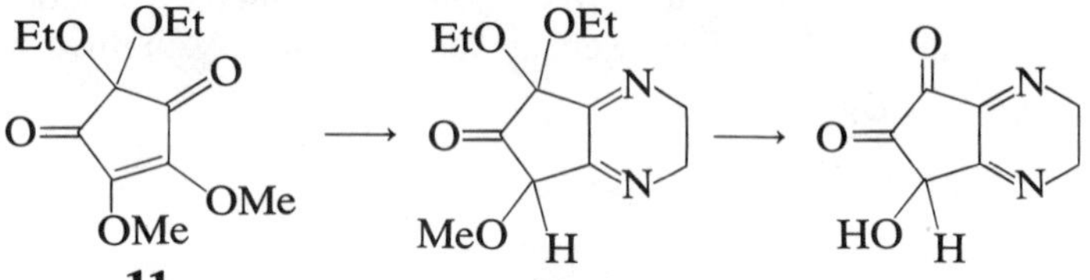

11 12 13

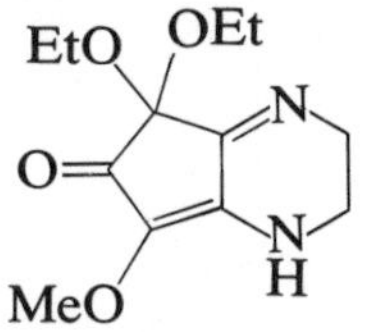

14

Treatment of the diester **15** with hydrochloric acid is reported to give the α-oxo acid **16** although no melting point is given for the product.[5] This condensation is the only reported synthesis of this ring system which involves formation of the cyclopentane ring. Again a formulation analogous to **5** or **9** may be a more exact representation of the structure of compound **16**.

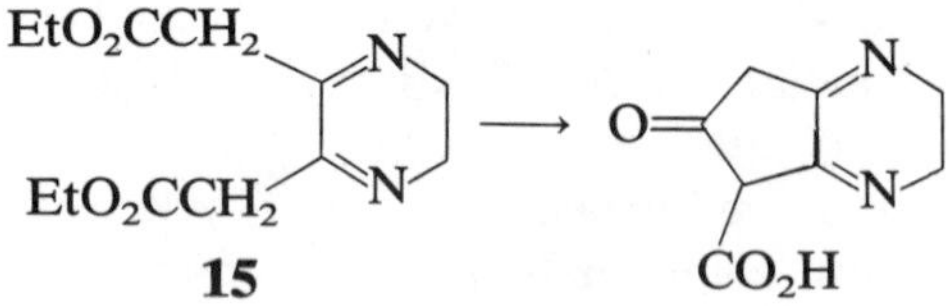

15 16

Quaternization of the tetrahydro compound **5** with ethereal methyl iodide gave a quaternary salt which on treatment with base afforded the product **7**, evidenced by PMR and mass spectra alone. Treatment of **5** with sodamide in liquid ammonia, to form the sodium salt, followed by methylation with methyl iodide gave a mixture of products which were separated by GLC. On the basis of PMR and mass-spectral evidence, the products were formulated as compounds **6**, **17**, and **18**.[1]

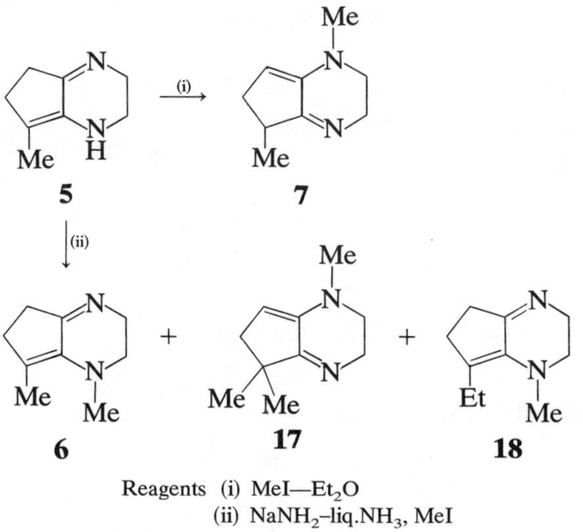

5 → (i) → **7**

5 → (ii) → **6** + **17** + **18**

Reagents (i) MeI—Et_2O
(ii) $NaNH_2$–liq.NH_3, MeI

Patents on this ring system concentrate on the utility of 5*H*-6,7-dihydrocyclopentapyrazines as flavoring agents for coffee and cocoa[3] or beef, tea, and tobacco.[7]

II. 1,3-Dithiolo[4,5-*b*]pyrazines

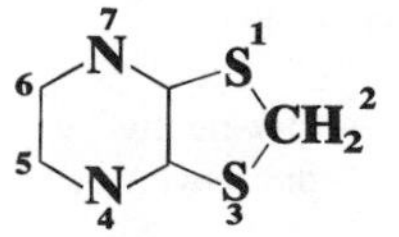

RRI 11874

The only derivatives of this ring system to be reported are the 2-oxo compounds **20** and the corresponding 2-thiones **21**.[6] They are described by *Chemical Abstracts* as cyclic esters of dithiocarbonic acid and trithiocarbonic acid, respectively. The compounds are prepared by treatment of the dithioxo compounds **19** with phosgene or thiophosgene in the presence of aqueous sodium hydroxide solution. Alternatively the disodium salt of **19** may be reacted in benzene with the appropriate acylating agent. The products are useful as acaricides[6] and thus have similar biological properties to the corresponding quinoxaline derivatives.

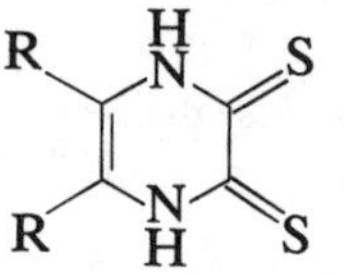

19 R = H or Me

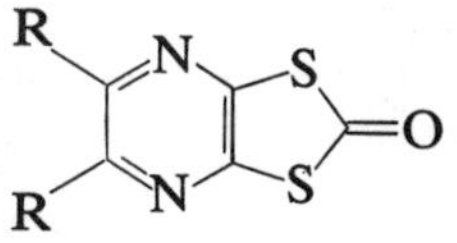

20 R = H or Me

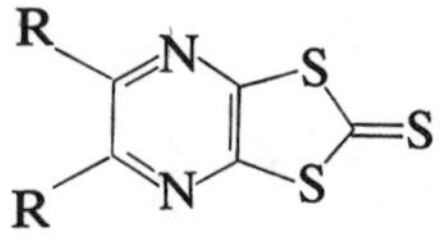

21 R = H or Me

III. Tables

TABLE 1. CYCLOPENTAPYRAZINES

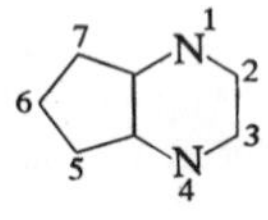

Molecular formula	Substituents	m.p., °C	Ref.
$C_7H_6N_2O_3$	5*H*-2,3,6,7-Tetrahydro-5-hydroxy-6,7-dioxo[a]	300 decomp.	4
$C_7H_8N_2$	5*H*-6,7-Dihydro	b.p. 115–120/10 mm	1
		b.p. 100/12 mm	3
$C_8H_8N_2O_3$	5*H*-5-Carboxy-2,3,6,7-tetrahydro-6-oxo[a]		5
$C_8H_{10}N_2$	5*H*-6,7-Dihydro-2-methyl	b.p. 125–130/10 mm	1, 3
	5*H*-6,7-Dihydro-5-methyl	b.p. 78–80/10 mm	1
		b.p. 66/2 mm	3
$C_8H_{12}N_2$	1*H*-2,3,5,6-Tetrahydro-7-methyl[b]	115–117 decomp.	1
$C_9H_{12}N_2$	5*H*-6,7-Dihydro-2,3-dimethyl	25–27	1
		b.p. 102–104/10 mm	
	5*H*-6,7-Dihydro-3,5(or 7)-dimethyl[c]	b.p. 90–91/12 mm	1
	5*H*-6,7-Dihydro-5,7-dimethyl	b.p. 82–86/10 mm	1
$C_9H_{14}N_2$	1*H*-2,3,5,6-Tetrahydro-3,5(or 7)-dimethyl[b,c]	72–85 decomp.	1
	1*H*-2,3,5,6-Tetrahydro-5,7-dimethyl[b]	70–74 decomp.	1
	5*H*-4a,6,7,7a-Tetrahydro-2,3-dimethyl	57–59	1
$C_{10}H_{14}N_2$	2-Ethyl-5*H*-6,7-dihydro-3-methyl	b.p. 114–116/10 mm	1
	5*H*-6,7-Dihydro-2,3,5-trimethyl	b.p. 105–110/10mm	1
	5*H*-6,7-Dihydro-2,5,7-trimethyl	b.p. 94–98/10 mm	1
$C_{10}H_{16}N_2$	2-Ethyl-5*H*-4a,6,7,7a-tetrahydro-3-methyl	b.p. 98–103/10 mm	1
	1*H*-2,3,5,6-Tetrahydro-3,5,7-trimethyl[b]	50–55 decomp.	1
	5*H*-4a,6,7,7a-Tetrahydro-2,3,5-trimethyl	b.p. 97–100/10 mm	1

TABLE 1 (*contd.*)

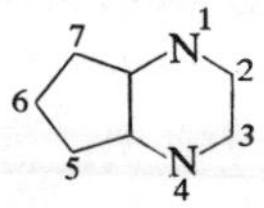

Molecular formula	Substituents	m.p., °C	Ref.
$C_{11}H_{16}N_2$	2,3-Diethyl-5*H*-6,7-dihydro	b.p. 121–123/10 mm	1
	5*H*-6,7-Dihydro-2,3,5,7-tetramethyl	b.p. 113–118/15 mm	1
	5*H*-6,7-Dihydro-2-methyl-3-*n*-propyl	b.p. 124–127/10 mm	1
	5*H*-6,7-Dihydro-2-methyl-3-*i*-propyl	b.p. 115–120/10 mm	1
$C_{11}H_{18}N_2$	2,3-Diethyl-5*H*-4a,6,7,7a-tetrahydro	b.p. 108–110/10 mm	1
	5*H*-4a,6,7,7a-Tetrahydro-2,3,5,7-tetramethyl	b.p. 105–108/12 mm	1
	5*H*-4a,6,7,7a-Tetrahydro-2-methyl-3-*n*-propyl	[d]	1
	5*H*-4a,6,7,7a-Tetrahydro-2-methyl-3-*i*-propyl	b.p. 100–105/10 mm	1
$C_{12}H_{18}N_2$	2,3-Diethyl-5*H*-6,7-dihydro-5-methyl	b.p. 125–130/10 mm	1
$C_{12}H_{18}N_2O_4$	5,5-Diethoxy-5*H*-2,3,6,7-tetrahydro-7-methoxy-6-oxo[a]	124–125	4
$C_{12}H_{20}N_2$	2,3-Diethyl-5*H*-4a,6,7,7a-tetrahydro-5-methyl	b.p. 115–120/10 mm	1

[a] An alternative tetrahydro formulation is possible (see text).
[b] Described as 4*H*-2,3,6,7-tetrahydro compound.
[c] A mixture of the 5- and 7-methyl isomers.
[d] Boiling point given as 109–113° but no pressure given (presumably 10 mm).

TABLE 2. 1,3-DITHIOLO[4,5-*b*]PYRAZINES

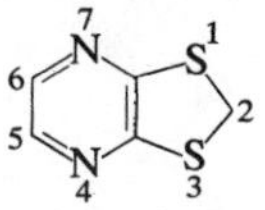

Molecular formula	Substituents	m.p., °C	Ref.
$C_5H_2N_2OS_2$	2-Oxo	81	6
$C_5H_2N_2S_3$	2-Thioxo	169–171	6
$C_7H_6N_2OS_2$	5,6-Dimethyl-2-oxo	78	6
$C_7H_6N_2S_3$	5,6-Dimethyl-2-thioxo	136	6

IV. References

1. I. Flament, P. Sonnay, and G. Ohloff, *Helv. Chim. Acta*, **56,** 610 (1973).
2. J. P. Walradt, A. O. Pittet, T. E. Kinlin, R. Muralidhara, and A. Sanderson, *J. Agr. Food Chem.*, **19,** 972 (1971); *Chem. Abstr.*, **75,** 108666 (1971).

3. Polaks Frutal Works N.V. and Douwe Egberts Koninklijke Tabaksfabrick-Koffiebranderijon Theehandel N.V., Neth. App. 68 12,899; *Chem. Abstr.*, **71,** 61421 (1969).
4. R. Malachowski and S. Prebendowski, *Ber.*, **71B,** 2241 (1938).
5. R. Thomas-Mamert and S. Weil, *Bull. Soc. Chim. Fr.* [3], **23,** 439 (1900).
6. K. Dickore, K. Sassi, and R. Wegler, Belg. Pat. 610,601; *Chem. Abstr.*, **57,** 13774 (1962).
7. I. Flament (to Firmenich S.A.), Ger. Offen. 2,353,467; *Chem. Abstr.*, **81,** 37578 (1974).
8. O. G. Vitzthum and P. Werkhoff, *J. Agr. Food Chem.*, **23,** 510 (1975).

CHAPTER XXVII

Pyrido[1,2-*a*]pyrazines

I. Physical Properties . . . 464
II. Methods of Preparation . . . 465
1. Type A Syntheses . . . 466
2. Type B Syntheses . . . 467
3. Type C Syntheses . . . 469
4. Type D Syntheses . . . 470
5. Type E Syntheses . . . 471
6. Miscellaneous Syntheses . . . 471
III. Reactions . . . 475
IV. Biological Properties . . . 480
V. Tables . . . 482
Table 1. Perhydropyrido[1,2-*a*]pyrazines . . . 482
Table 2. Pyrido[1,2-*a*]pyrazines . . . 487
VI. References . . . 493

2*H* RRI 1612
4*H* RRI 8103
8*H* RRI 12014

The preferred numbering of the pyrido[1,2-*a*]pyrazine ring is given above. Derivatives of the aromatic system have been referred to as azaquinolizinium ions. The completely reduced compounds are the subject of a variety of names and numbering systems. They have been called 1,4-diazabicyclo[4.4.0]decanes (**1**), 2-azaquinolizidines (**2**), 8-azaquinolizidines (**3**), and, surprisingly, 7,10-diazaperhydronaphthalenes (**4**).

1 **2**

3 **4**

Partially unsaturated systems have been described as pyridopiperazines, piperazino[9:10]pyridines also numbered as (**4**), and lactams of *N*-β-aminoethylchelidamic acid.

I. Physical Properties

The pyrido[1,2-*a*]pyrazinium cation has been compared with the quinolizinium ion and the other monoazaquinolizinium cations, using the Pariser-Parr-Pople method to calculate electronic structures and absorption spectra.[1] The ultraviolet spectrum of pyrido[1,2-*a*]pyrazinium bromide in water[2] shows maxima at 232, 276, 287, 322, and 336 nm. Ultraviolet spectra have been recorded for various substituted compounds and their *N*-oxides.[2,3] The bands at 1685 and 1655 cm^{-1} in the infrared spectrum (KBr disc) of the oxo compound **5** (R = H) confirm its formulation as the oxo form rather than as the corresponding hydroxy tautomer.[2] Similarly the dioxo derivative **6** exists as such, in preference to the hydroxy form **5** (R = OH). The presence of a two-proton singlet at δ 6.1 in the PMR spectrum (in trifluoroacetic acid) of compound **6** is strong evidence in favor of this formulation. Ultraviolet spectral data are consistent with structures **6** and **5** (R = H).[2]

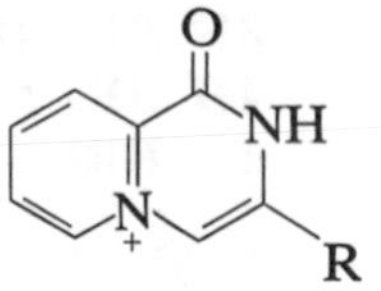

O NH N+ R
5

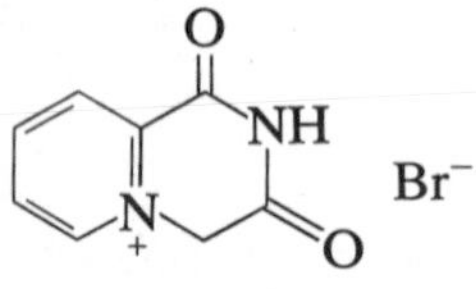

O NH N+ O Br⁻
6

Perhydro derivatives of this ring system have attracted interest as analogues of piperazine, and the conformational analysis of several such compounds has been discussed.[4–6] The conclusions drawn rest entirely on analogy with quinolizidines. The presence of bands ("Bohlmann bands") in the range of 2700–2800 cm^{-1} in the infrared spectra of such compounds is taken to indicate a trans orientation of the nitrogen lone pair and the bridgehead carbon-hydrogen bound. Compounds of structures **7** and **8** show bands in this region and have been assigned trans configurations. Similarly the pyrrolo compound **9** has a trans orientation, although its isomer **10** has a cis configuration.[4] The *d*(+) isomer of **8** (R = H) has been assigned the *S*-configuration by unambiguous synthesis. Optical rotatory dispersion studies are consistent with this assignment.[5]

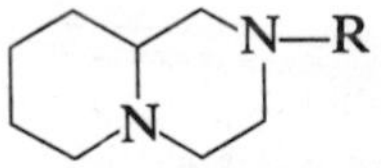

7 R=H, CH_2Ph, or CSNHPh

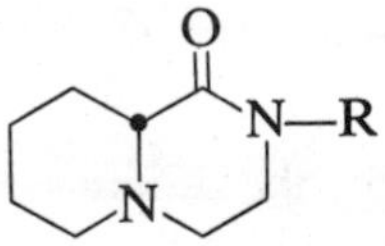

8 R=H or $PhCH_2$

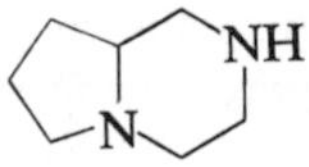

9

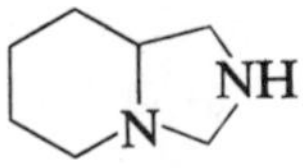

10

II. Methods of Preparation

All the synthetically useful methods for the preparation of pyrido[1,2-*a*]pyrazines start from pyridine intermediates. The closure of the pyrazine ring is completed by carbon–nitrogen bond formation. The commonest syntheses involve, as the final step, formation of the 2,3 bond either alone (type A synthesis) or in combination with formation of the 1,2 or the 4,5 bonds. These approaches are described here as syntheses of type B and type C, respectively. Less frequently preparations are completed by formation of the bonds between positions 1 and 2 (type D synthesis), and between positions 4 and 5 (type E synthesis).

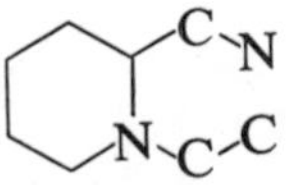

A

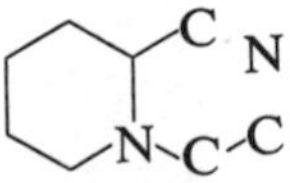

B

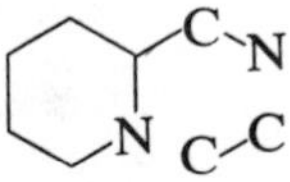

C

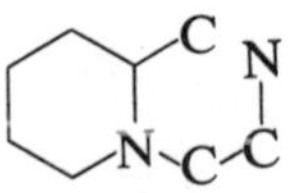

D

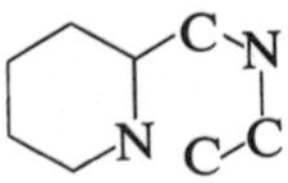

E

1. Type A Syntheses

Treatment of the amides **11** with phenacyl bromide in refluxing acetonitrile[3,7] provides the 1-oxo compounds **13** without isolation of the quaternized intermediates **12**. Use of 2-cyanopyridine in wet acetonitrile also yields the oxo compound **13** (R = H).[7] Treatment of the amide **11** (R = H) with ethyl bromoacetate provides the dioxo compound **15**. This was also formed when attempts were made to recrystallize the quaternary salt **14**.[2] Dehydration of the piperidine **16** gives the reduced compound **17**. A variety of 3-aryl derivatives were prepared using this approach.[8]

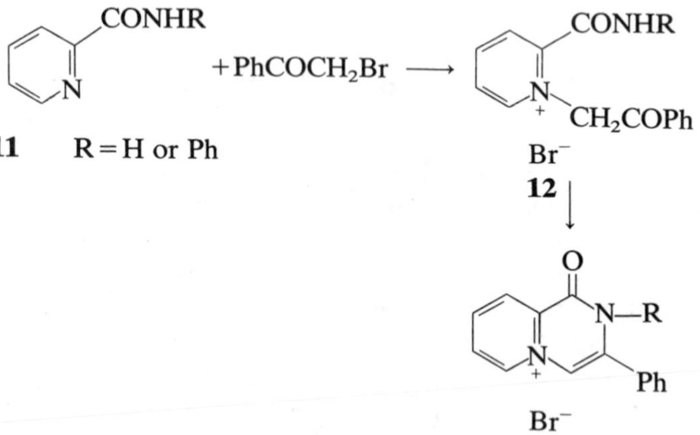

CONHR + $PhCOCH_2Br$ ⟶ CONHR, CH_2COPh, Br^-

11 R = H or Ph **12**

↓

O, N—R, Ph, Br^-

13

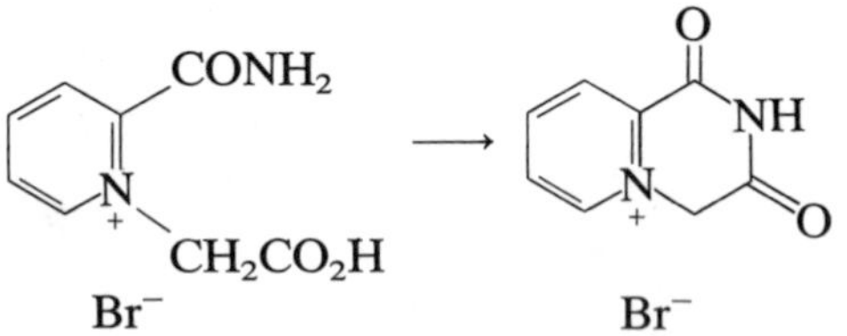

$CONH_2$, CH_2CO_2H, Br^- ⟶ O, NH, O, Br^-

14 **15**

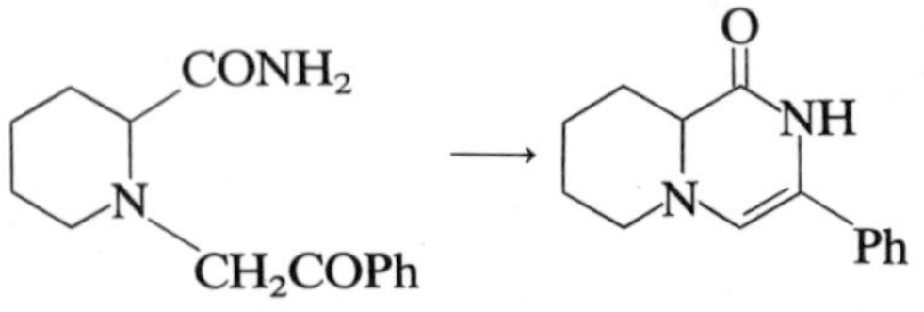

$CONH_2$, CH_2COPh ⟶ O, NH, Ph

16 **17**

Compound **18** apparently undergoes a Beckmann rearrangement in concentrated sulfuric acid to give the amide **19**. Oxidation of bromide ions in the solution results in bromination of the cyclized derivative to give the 4-bromo compound **20** as the final product.[3] If the oxime **18** is treated with boiling hydrobromic acid, no rearrangement occurs, and rapid ring closure provides the dihydro compound **21**, which may be isolated after several minutes. Prolonged acid treatment causes dehydration to the aromatic compound **22**.[3]

NOH Ph CH_2COPh Br^- **18** → CONHPh CH_2COPh Br^- **19** → O N—Ph Ph Br Br^- **20**

18 → Ph N—O OH Ph **21** → Ph N—O Ph **22**

Quaternization of the oximes **23** with α-bromoketones **24** provides a useful route to the 2-oxides **25** of the ring system.[3,9] Treatment of the aldoxime **23** (R=H) with phenacyl bromide yields the intermediate quaternary salt **26**, which may be cyclized with hydrobromic acid.[9] Similarly 3-unsubstituted compounds are available from haloacetaldehyde derivatives. Thus bromoacetaldehyde in refluxing acetonitrile converts the oxime **23** (R=Et) into the dihydro compound **27**, which on boiling in hydrobromic acid provides the aromatic compound **28**.[3] Sequential treatment of the aldoxime **23** (R=H) with chloroacetaldehyde oxime and concentrated hydrobromic acid yields the *N*-oxide **29** of pyrido[1,2-*a*]pyrazinium bromide without isolation of the intermediate quaternary salt.[9]

2. Type B Syntheses

Various 1,5-diketones of type **30** have been cyclized using ammonium acetate in acetic acid to give 1,3-disubstituted pyrido[1,2-*a*]pyrazinium salts (**31**).[7] Similar treatment of the ester **32** gave the oxo compound **33**.

+ $BrCH_2COR^1$ ⟶

23 R = H or alkyl **24** R^1 = Me or Ph **25**

CH=NOH

CH_2COPh

26

27 **28** **29** Br^-

Alternatively the action of hydrogen bromide in acetic acid–acetic anhydride converted the ester into the azoniaisocoumarin **34**, which again provided the pyrido[1,2-*a*]pyrazine on treatment with ammonium acetate.[7]

COR

$CH_2COAryl$

30 R = Alkyl or phenyl **31**

CO_2Et

CH_2COPh

32 **33**

34

The dioxo compound **36** has been prepared by reaction of the δ-chloroester **35** with methylamine.[10] Ammonia treatment of a related ester gave the 3,3-diphenyl compound **37**.[11] There have been many examples of the reaction of 1,5-dihalo compounds **38** with primary amines to give 2-substituted perhydro derivatives **39** of the ring system.[6,12–14] Similarly the pyridone **40** provided the 8-oxo derivatives **41** with primary amines. The use of secondary amines results in cyclization to the quaternary salts **42**.[15]

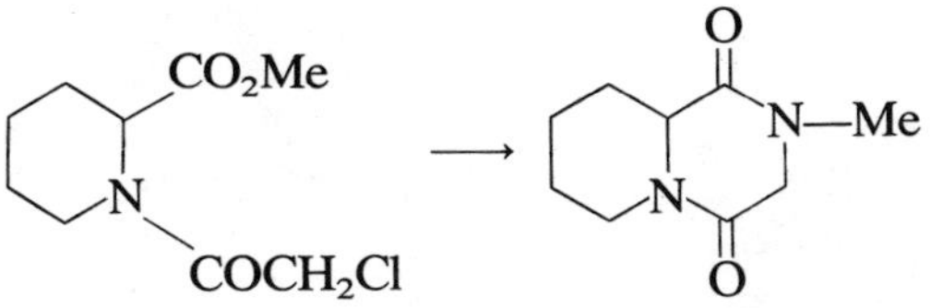

35 **36** 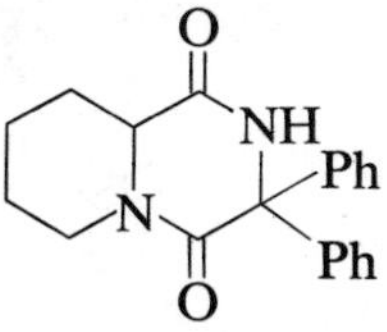 **37**

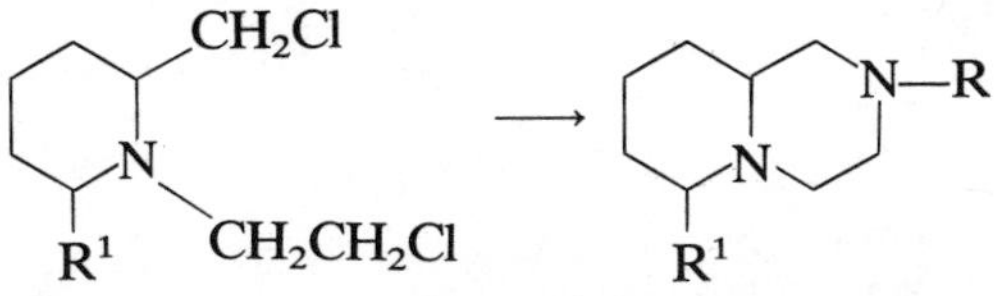

38 $R^1 = H$ or Me **39** R = Alkyl or phenyl

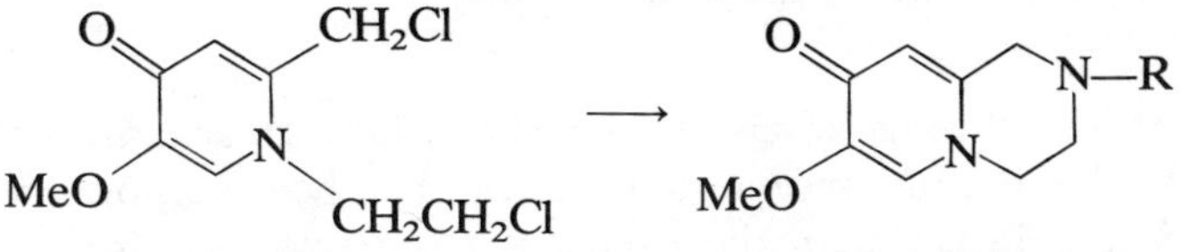

40 **41** R = Alkyl

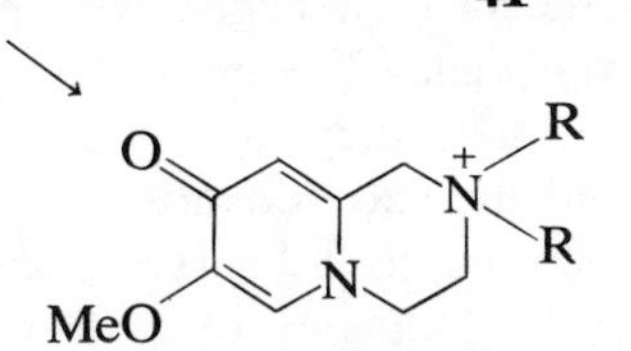

42 R = Et or Pr

3. Type C Syntheses

The tetrahydro compounds **44** and **46** have been obtained in good yields by treating the amines **43** and **45** with ethylene dibromide in refluxing alcoholic solvents.[16,17] Diethyl oxalate has been reacted with the piperidylamine **47** to give the dioxo compound **48**.[18]

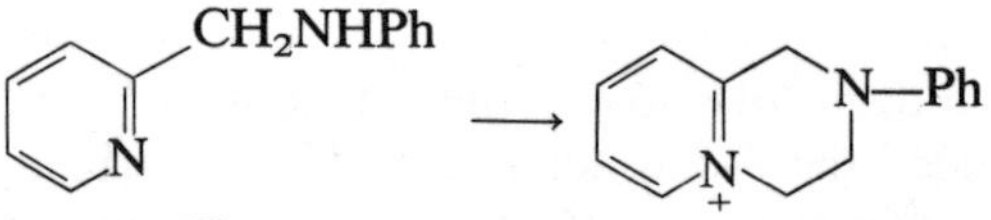

CH_2NHPh N—Ph N + **43** → **44**

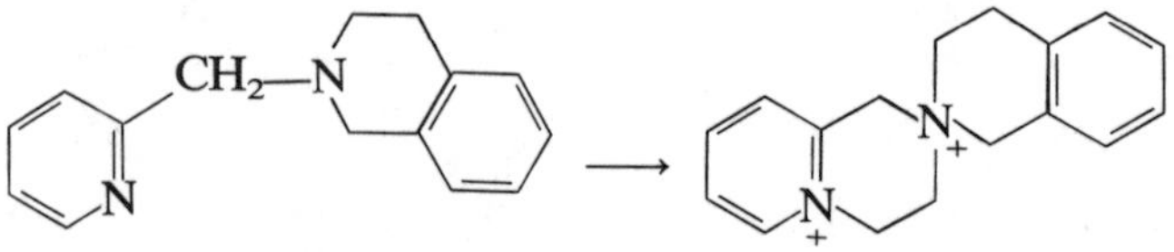

CH_2—N N N N + + **45** → **46**

CH_2NH_2 NH NH N O O **47** → **48**

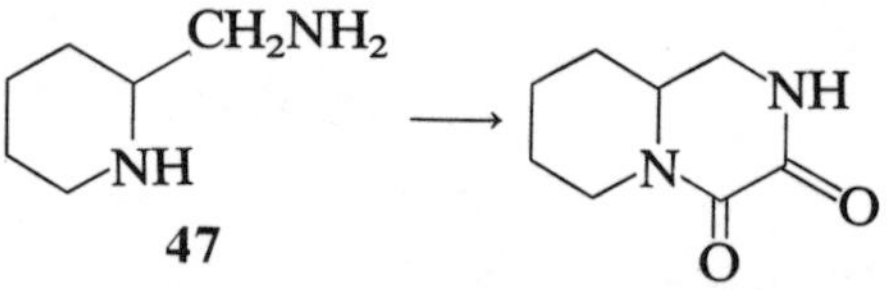

4. Type D Syntheses

Treatment of the ethylene ketals **49** (R = H or Me) with acid gives the *N*-oxides **50**.[2,9] The use of concentrated sulfuric acid, however, produces the quinolizine **51**, rather than the expected fused pyrazine, from the ketal **49** (R = Et).[3] The nitrile **52** on reaction with bromoacetaldehyde oxime (**53**) in dry acetonitrile yields the imino compound **54** without isolation of the intermediate quaternary salt.[2] The imino compound is hydrolyzed by aqueous acid to the oxo derivative **55**.

Thionyl chloride has been used to convert the alcohol **56** into the cyclized quaternary salt **57**.[15,19] Reduction of the nitrile **58** with Raney nickel and hydrogen gave the 1-oxo compound **60** (R = H),[20] presumably by way of the intermediate **59**. This intermediate is also involved in the preparation of compound **60** (R = H) by treatment of the ester **61** with ethyleneimine.[4,5] The optically active ester **61** has been converted by this route into perhydropyrido[1,2-*a*]pyrazines of known configuration.[5] When this reaction was repeated with the methyl-substituted piperidine

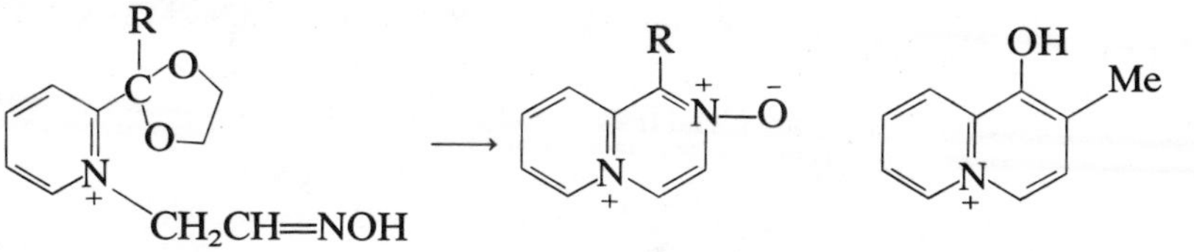

62, only a poor yield of the required pyrido[1,2-*a*]pyrazine was obtained.[13] The use of *p*-nitrobenzoylethyleneimine (**63**), followed by acid hydrolysis of the intermediate amide **64**, gave a good yield of the required product.[13,14]

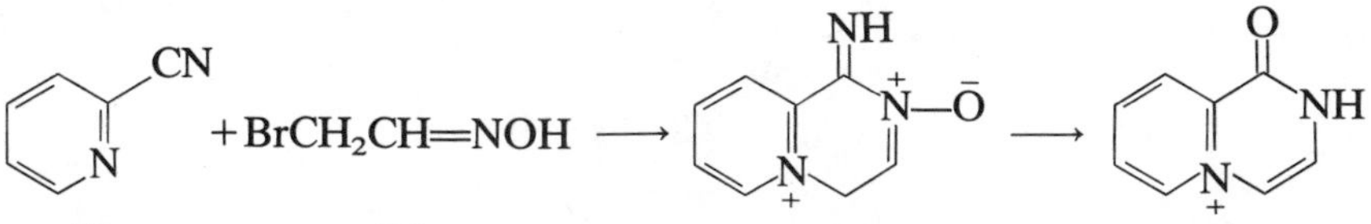

5. Type E Syntheses

Treatment of the acetal **65** with hydrogen bromide in acetic acid at −78° results in cyclization to form the 1-phenyl compound **66**. Attempts to prepare other derivatives of this ring system from such intermediates have been unsuccessful.[21]

Heating the chloroacetyl compound **67** under reflux in ethanol yields the 3-oxo compound **68**. This was not isolated as such but was hydrogenated over platinum to give the perhydro compound **69**.[20] The 1-oxo compound **71** has similarly been prepared by treatment of the alcohol **70** with phosphorus tribromide[2] or thionyl chloride and base.[22].

Reductive cyclization of the esters **72** and **74** provides the 3,4-dioxo compound **73**[18] and the 4-oxo derivatives **75**.[23] The reductions are performed using hydrogen over platinum in acetic acid. If milder conditions are used, the intermediates **76** to the 4-oxo compounds can be isolated.[24] Distillation then affords the cyclized products **75**.

6. Miscellaneous Syntheses

The earliest synthesis of the pyrido[1,2-*a*]pyrazine ring system involved the reaction of chelidonic acid (**77**) and ethylenediamine[25,26] to give the

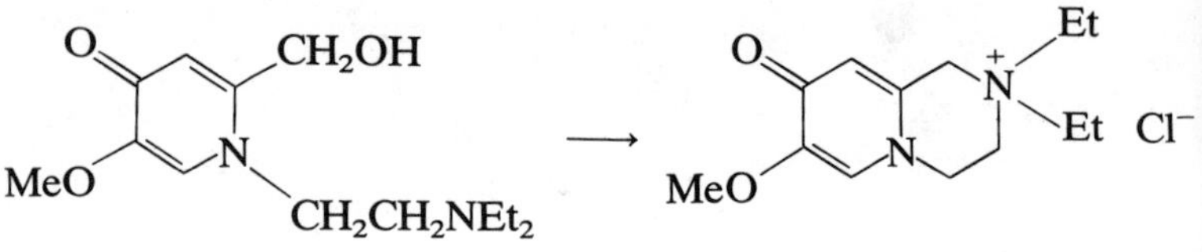

56 57

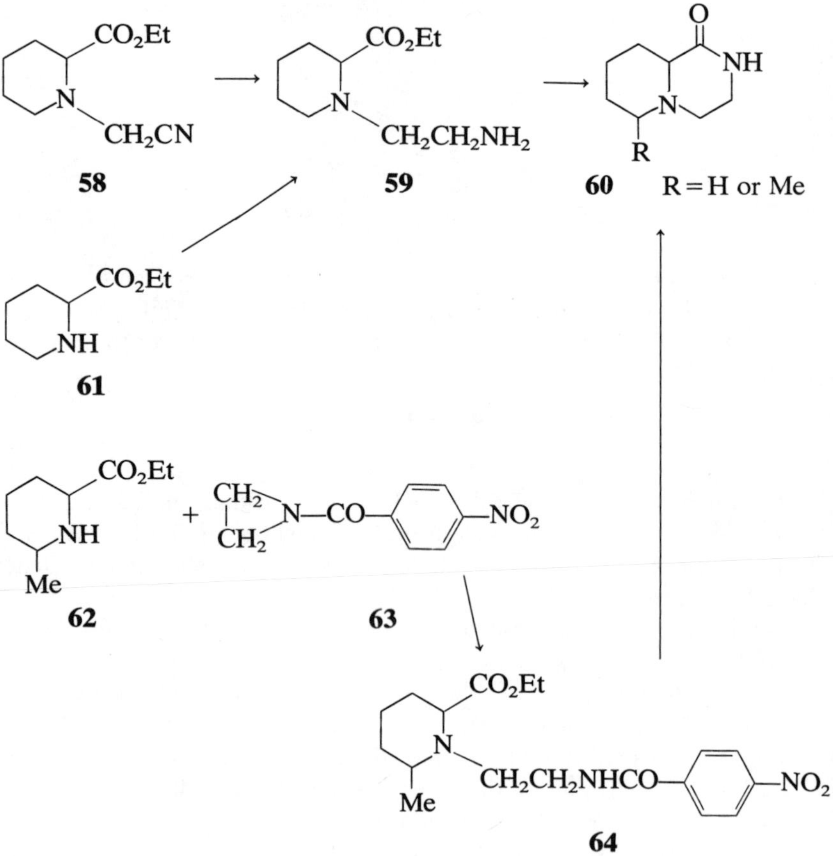

58 59 60 R = H or Me

61

62 63

64

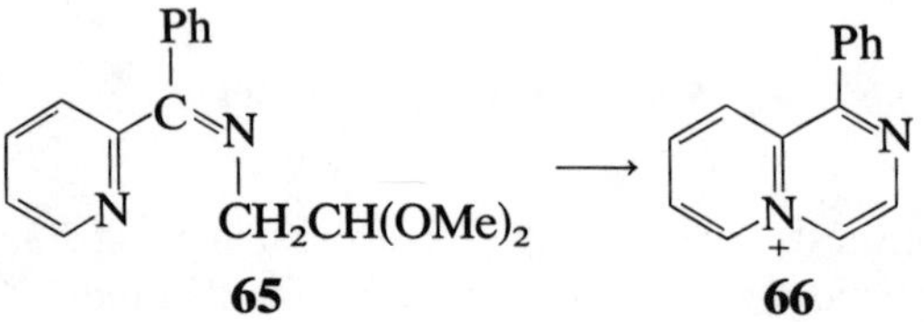

65 66

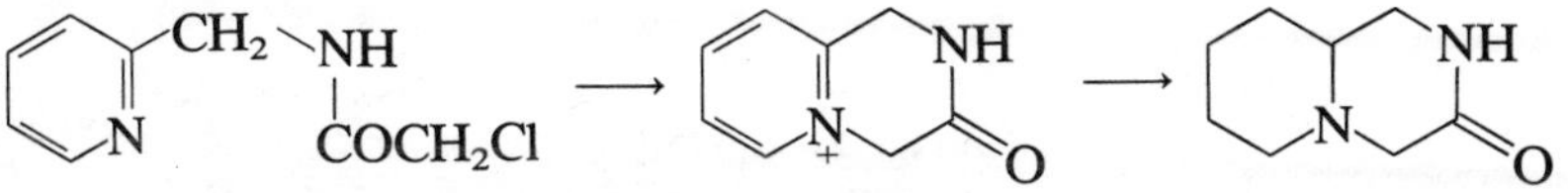

67 → 68 → 69

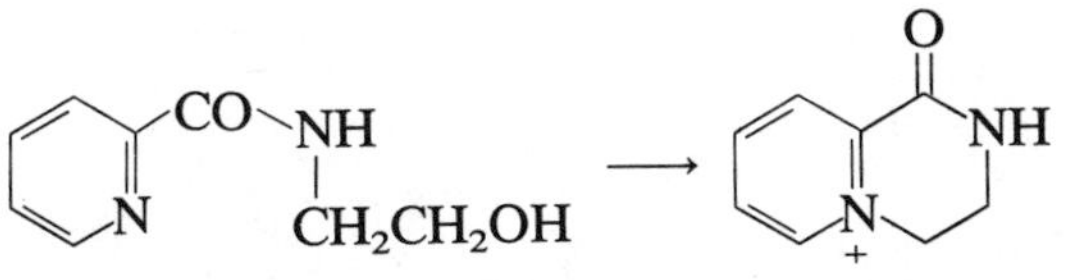

70 → 71

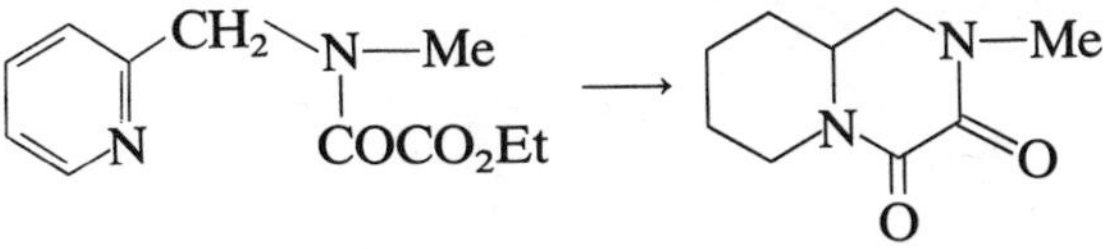

72 → 73

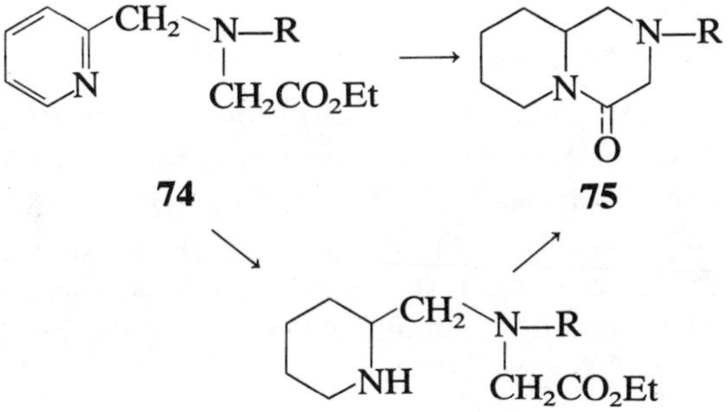

74 → 75

74 → 76 → 75

compound **78**. The reaction has been repeated and applied to various complex derivatives.[27–29] Optimum yields are attained by performing the reaction in aqueous acetic acid.[27] Ethylenediamine reacts with a variety of α,ω-dihalocarboxylic acids or esters to provide various perhydro fused pyrazines. When α,ε-dihalo compounds **79** are used, the products are pyrido[1,2-*a*]pyrazines **80**.[30–32] The Schmidt reaction on ketone **81** has been shown to give the 3-oxo compound **82** exclusively.[33]

Apart from the synthetically useful methods of synthesis described above, derivatives of this ring system have been obtained in low yields by a variety of methods. Aminolysis of sucrose with ethylenediamine,[34] pyrolysis[35] of actinomycin Pip-2, and photochemical degradation of diquat[22] have all produced pyrido[1,2-*a*]pyrazines. Hydrogenation of

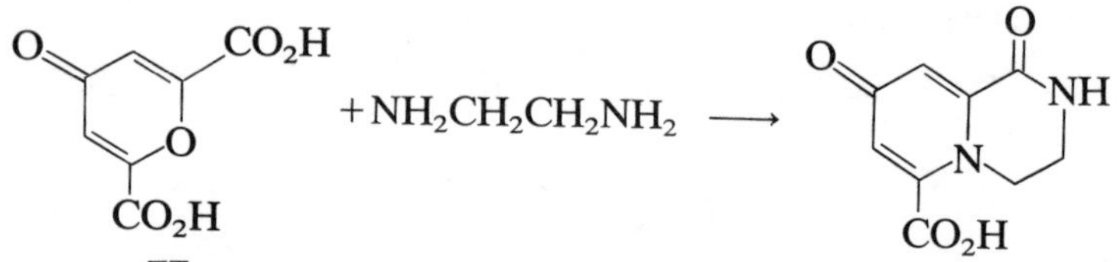

77 + $NH_2CH_2CH_2NH_2$ ⟶ 78

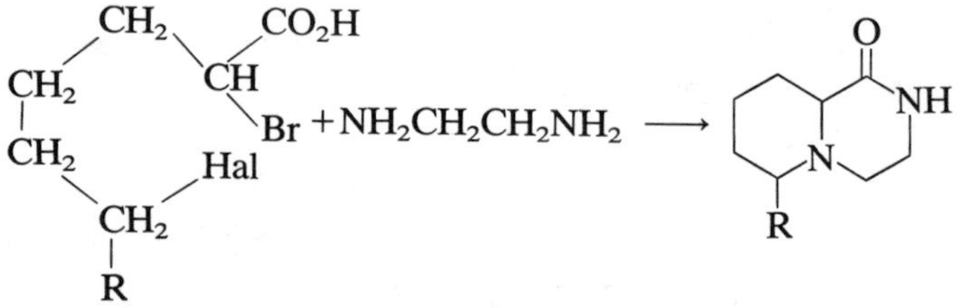

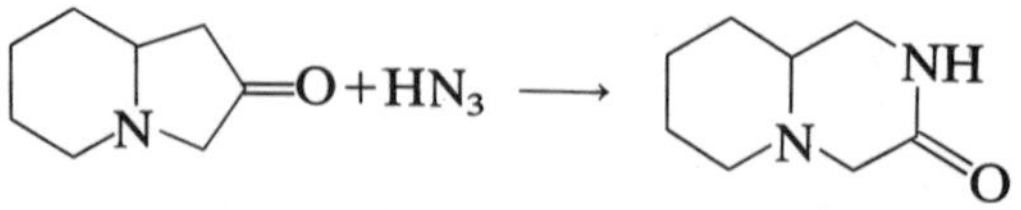

79 + $NH_2CH_2CH_2NH_2$ ⟶ 80

81 + HN_3 ⟶ 82

complex fused pyrazinium salts has also provided derivatives of the ring system.[36] The betaine **83** is reported to react in toluene with the isocyanates **84** to give the pyrido[1,2-*a*]pyrazinium betaines **85** in 15–18% yield.[37] The only example of the use of a piperazine intermediate in the synthesis of this heterocycle involves acid-catalyzed cyclization of compound **86**. The perhydro compound **87** is among the products isolated from this reaction.[38,39]

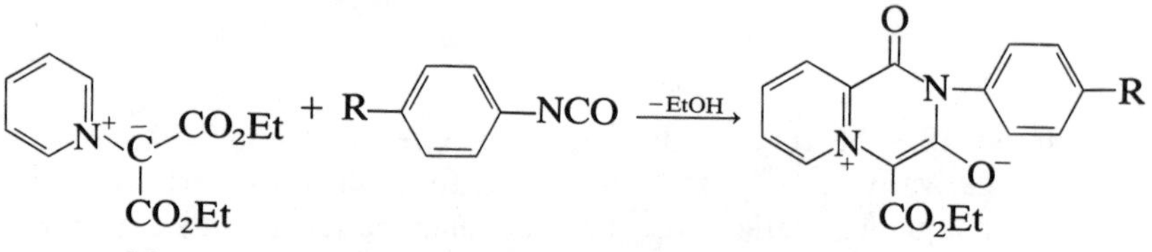

83 + 84 (R = H or Me) —(−EtOH)→ 85

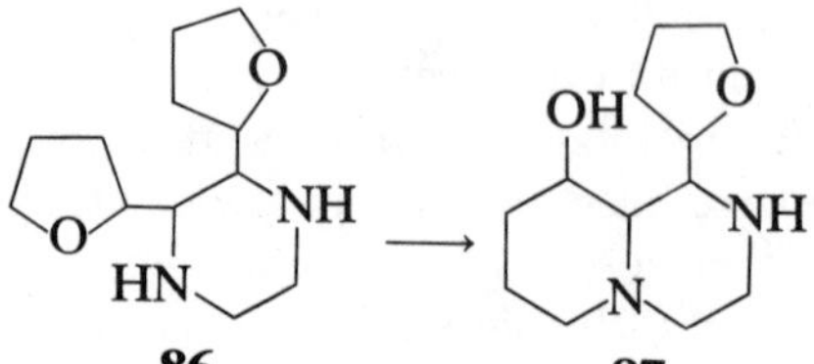

86 ⟶ 87

Thermal rearrangement of the addition product **88** from 2-ethyl-1-methylimidazole and two molecules of dimethyl acetylenedicarboxylate gives a mixture of the pyrido[1,2-*a*]pyrazines **89** and **90**. Both of these products rearrange to the dioxo compound **91** in hot dilute hydrochloric acid.[40] In the rearrangement of **89**, intermediates such as **88** and **90** are presumably involved.

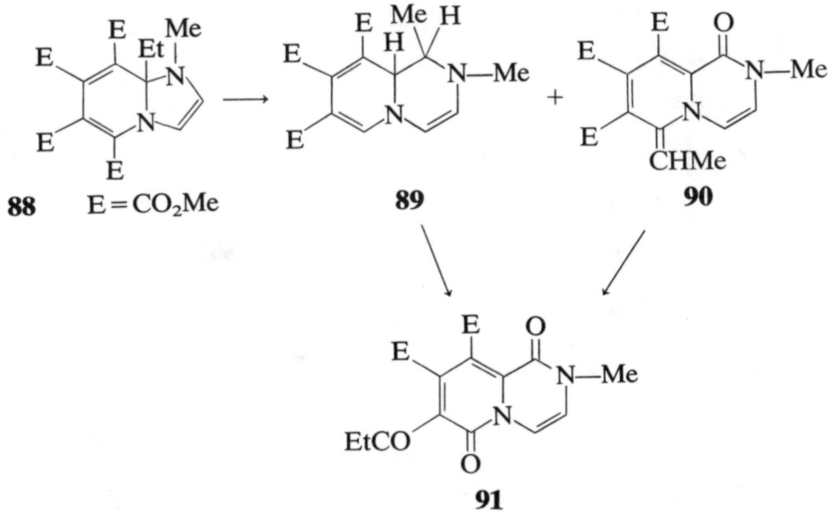

88 E = CO_2Me **89** **90**

91

III. Reactions

Very little is known of the chemistry of the pyrido[1,2-*a*]pyrazinium cation. More emphasis has been placed on the reactions of *N*-oxides and the partly or wholly reduced derivatives of the ring system. There are no reports of methylation of the aromatic system **92** ↔ **93**, presumably because of the deactivating influence of the canonical form **93**, which would ensure a greatly reduced electron density at the 2-position. Treatment of the oxo compounds **94** and **95** with methyl iodide in methanol has recently been reported to give 5-methiodides.[6] The product **96** from the perhydro compound **94** was separable into cis and trans isomers. The cis compound was converted to the trans methiodide by heating at 245°. Alkylation of the sodium salt of the amide **94** with benzyl chloride gave the 2-substituted product **97**.[6] Similarly treatment of the sodium salt of the amide **98** with methyl sulfate gave the 2-methyl compound **99** (X = H). Reaction of the amide **98** with formaldehyde gave the hydroxymethyl compound **99** (X = OH).[8]

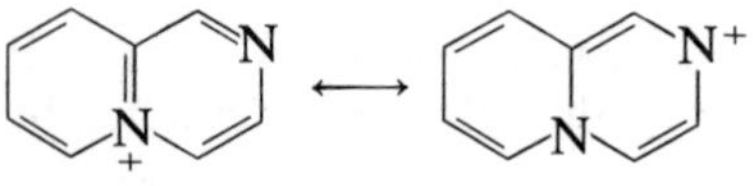

92 93

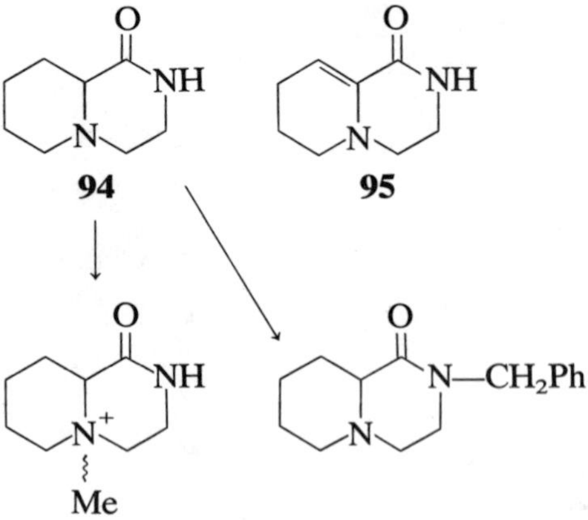

94 95

96 97

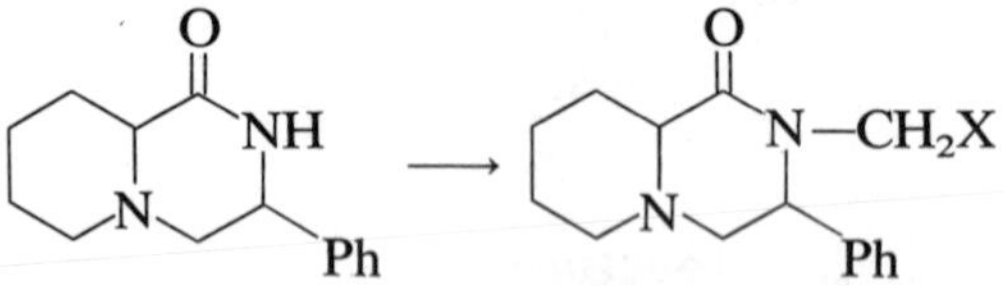

98 99 X = H or OH 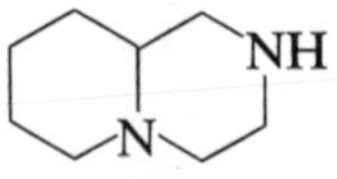 100

The reactions of the completely reduced compounds have attracted much attention because of their close relationship to piperazine. Thus it has been shown, as might be expected, that the 2-position in compounds such as the parent **100** can be readily acetylated and benzoylated[41] and ureas are formed with phenylisocyanate[23] and phenylisothiocyanate.[4] Compound **100** also reacts at the 2-position with *p*-acetamidobenzene-sulfonyl chloride to give a sulfonamide.[23] There are several examples of the alkylation of the amine in the 2-position. Various alkyl halides[5,13,42] have been used, as well as ethylene oxide,[4] *p*-nitrobenzoyl-ethyleneimine,[14] and 1-guanyl-3,5-dimethylpyrazole nitrate.[5] Compound **100** also undergoes the Eschweiler-Clarke modification of the Leuckart reaction to give the 2-methyl compound.[41]

Bromination of the 2-oxide of the pyrido[1,2-*a*]pyrazinium cation is reported to give the 1-bromo compound **101**.[2] The 4-bromo derivative

102 is obtained on bromination of the 1,2-dihydro-1-oxo-2,3-diphenylpyrido[1,2-*a*]pyrazinium cation.[3]

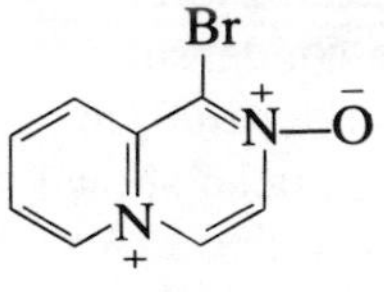

101 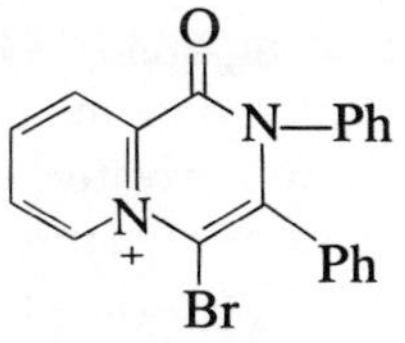 **102**

There are no reports of oxidation of the aromatic system. The various *N*-oxides that are known have been synthesized directly by ring closure rather than by way of the parent compound. Treatment of the perhydro compounds **103** and **105** with mercuric acetate provides the oxo enamines **104**.[6]

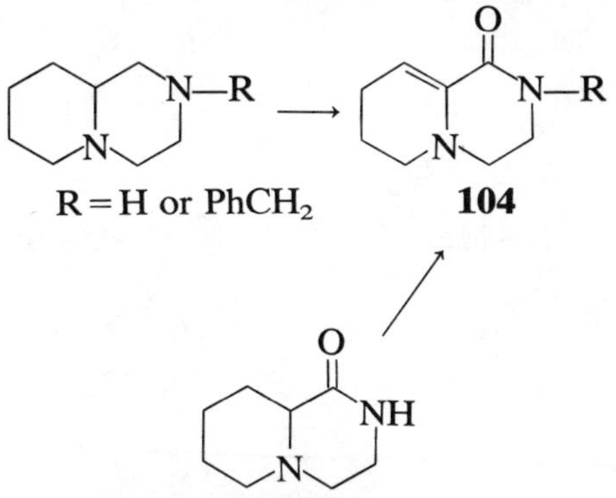

103 R = H or $PhCH_2$ **104**

105

Treatment of the oxo compounds **106** and **108** with base results in formation of the betaines **107**[7] and **109**.[2]

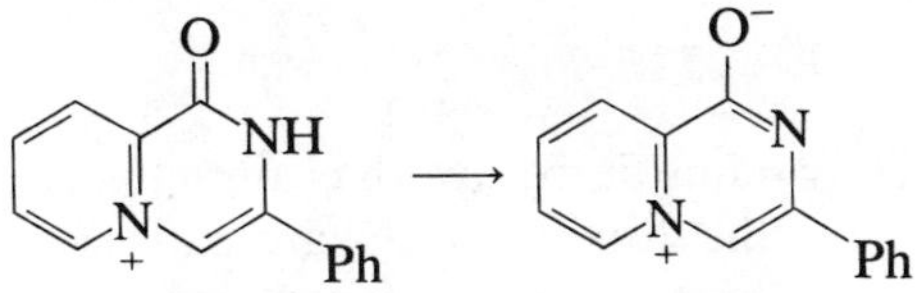

106 **107**

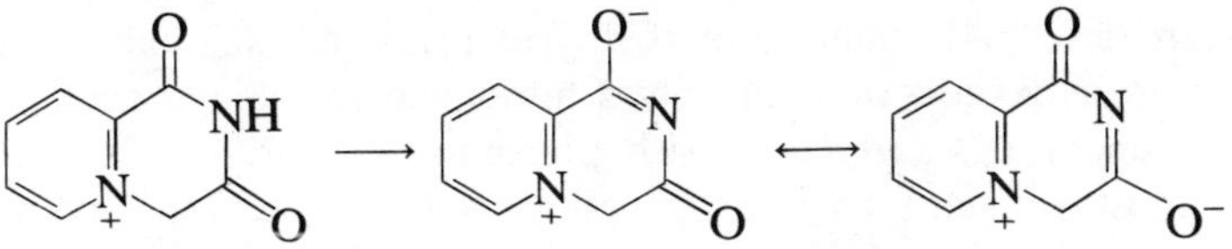

108 **109**

The 1,8-dioxo compound **110**, obtained by decarboxylation of the 6-carboxylic acid,[27] reacts with activated methylene groups in compounds such as malononitrile or 1,3-dimethylbarbituric acid to give condensed compounds, such as compound **111**. The condensations are performed in acetic acid–acetic anhydride mixtures and result in acetylation at the 2-position.[29] The dioxo derivative **112** has been shown to condense with *o*-phenylenediamine to give the fused quinoxaline **113**.[18] Protonation of the enamine **104** (R = H) yields the cation **114**, which reacts with potassium cyanide to give the compound **115**.[6]

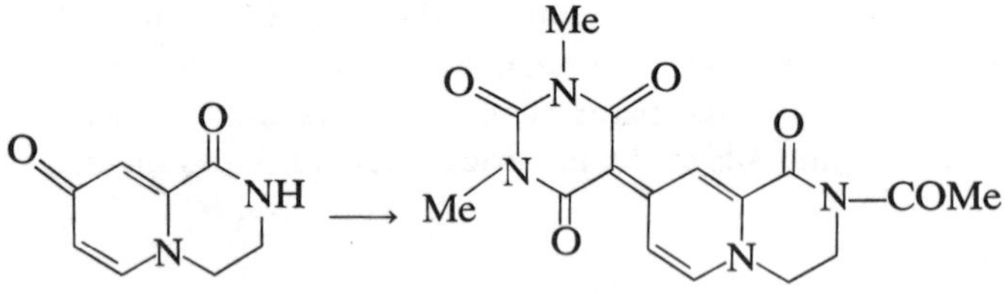

Me O N O O O NH → Me N N—COMe O N N

110 **111**

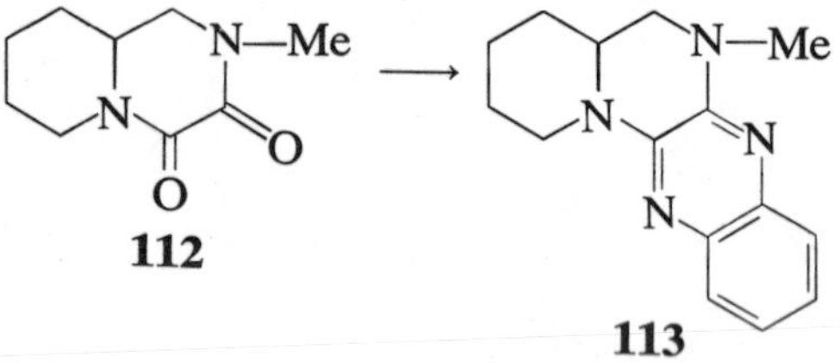

N—Me → N—Me N N N O O N

112 **113**

O CN O NH → NH N + N

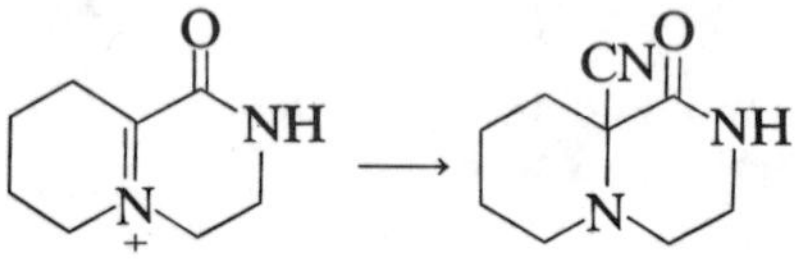

114 **115**

Reduction of the ring system and its derivatives has attracted considerable attention. Deoxygenation of the 2-oxides of the aromatic system has been investigated.[2,3,9] In general these *N*-oxides may be deoxygenated to the parent heterocycles by heating with phosphorus tribromide or phosphorus trichloride. Large substituents in the 1- or 3- positions facilitate deoxygenation.[3] Although 3-methylpyrido[1,2-*a*]pyrazinium 2-oxide is resistant to deoxygenation with phosphorus trichloride and acetic anhydride,[9] it undergoes reduction with phosphorus tribromide.[3] The compound most resistant to deoxygenation is the 1-methyl 2-oxide **116**, requiring treatment with phosphorus tribromide at 190° to effect reaction.

The resistance of this compound to deoxygenation is ascribed to the stability of the intermediate **117**. Larger substituents in the 1-position are sterically prevented from adopting the stable, planar configuration.[3] Carefully controlled hydrogenation has been reported to give the parent pyrido[1,2-*a*]pyrazinium cation **120** from its 2-oxide (**119**). It is also possible to remove the bromine selectively from the 1-bromo 2-oxide (**118**) under mild conditions.[2] Further hydrogenation results in complete reduction of the ring to the perhydro compound **121**.[2] Similarly hydrogenation of the oxo compounds **122** provides the reduced compound **123**.[3]

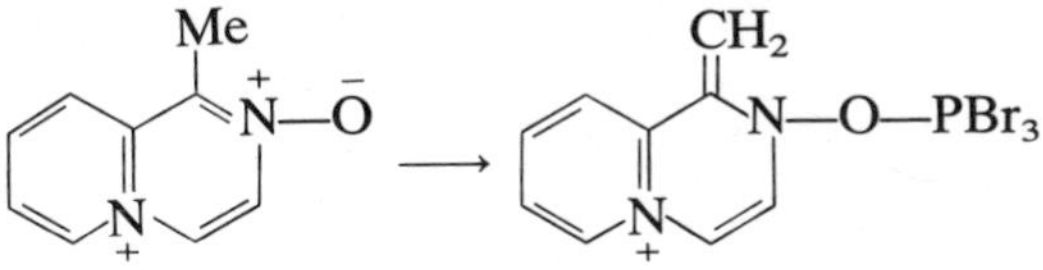

116 → **117**

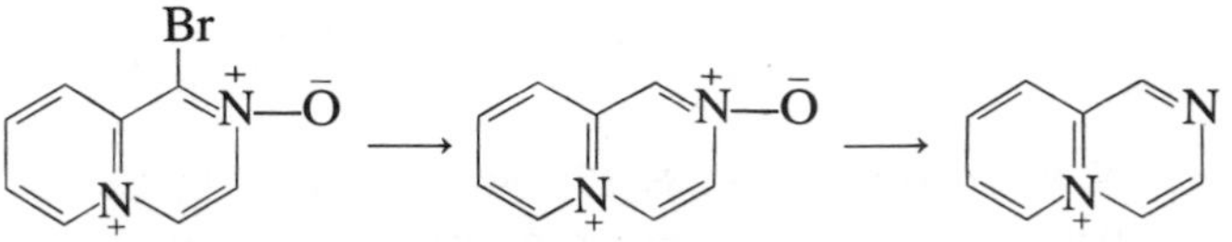

118 → **119** → **120** 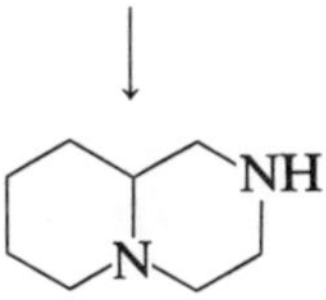 → **121**

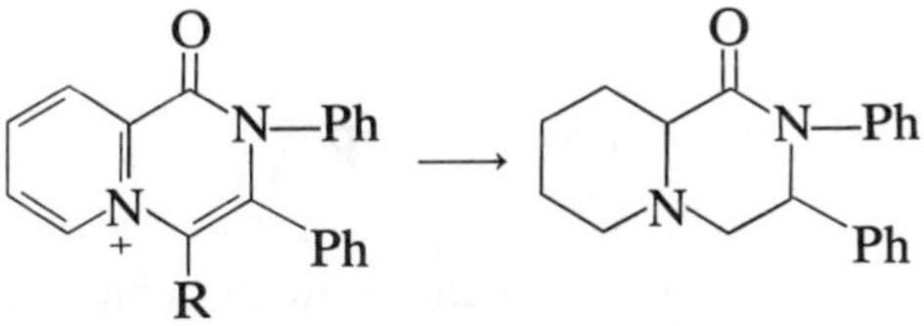

122 R = H or Br → **123**

The ease of reduction of the aromatic systems such as **118**–**120** is demonstrated by the fact that the product **121** can be obtained from them by hydrogenation over platinum.[2,9] These conditions also serve to reduce the tetrahydro compound **124** to the perhydro oxo derivative **125**.[2] Partially unsaturated compounds such as **126** and **127** have been reduced to similar perhydro oxo derivatives using hydrogen over palladium[8] and sodium borohydride,[6] respectively. Lithium aluminum hydride has been much used to reduce the oxo group in perhydro oxo compounds to produce the piperazine analogues such as **121**.[2,4,13,20,23,24] This reagent similarly reduces the dioxo compound **128** directly to **121**.[2]

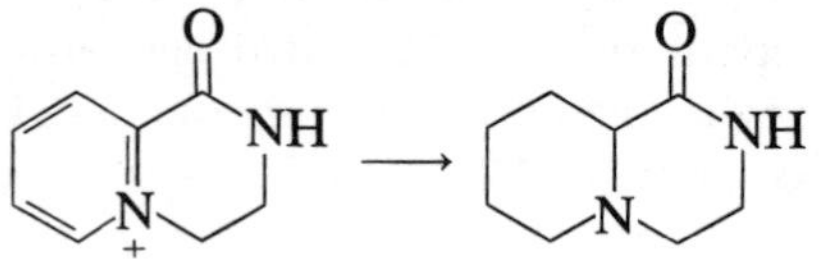

124 → **125**

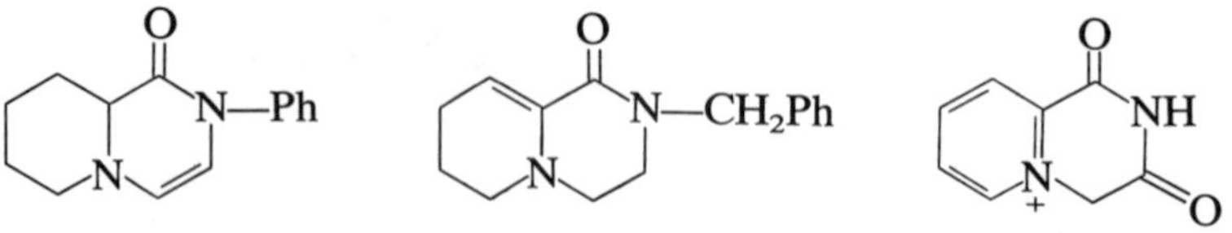

126 **127** **128**

Reduction of the oxo compounds **129** and **130** with lithium in liquid ammonia has been shown to give the ten-membered ring compound **131**.[6]

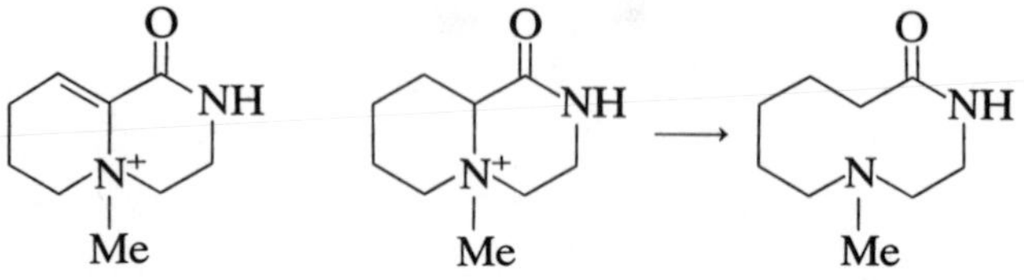

129 **130** → **131**

IV. Biological Properties

There are no reports of biological activity in the completely aromatic derivatives of this ring system. The perhydro compounds have attracted attention as piperazine analogues.[30,42,43] Various perhydro derivatives substituted in the 2-position by arylalkyl and diarylalkyl groups are claimed to be useful in the treatment of intestinal spasms and migraine.[42] Other derivatives such as **132** are described as neuroleptics.[43] Structure–activity relationships in this series are discussed, and **132** is shown to be the most active derivative, with a pharmacological profile similar to that of perphenazine (**133**). Compound **134** is claimed to be useful in human and veterinary medicine.[38]

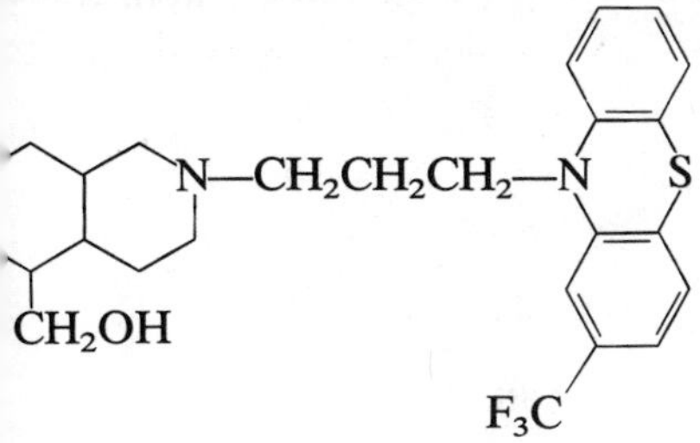

132

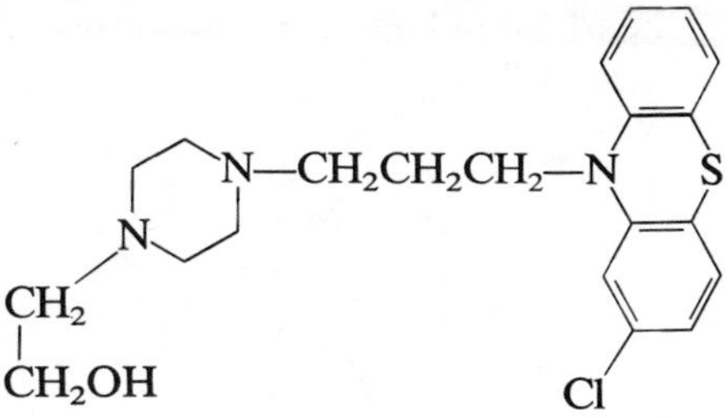

133

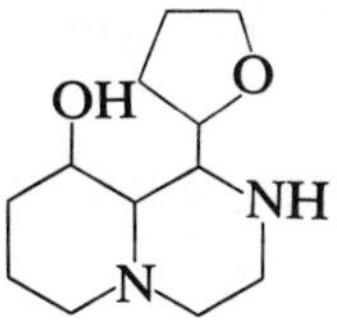

134

Derivatives such as **135** and its 3,4-dihydro analogue have been patented as having sedative, antiepileptic, and muscle-relaxing properties.[8] The related compound **136** is claimed as having anticonvulsant, depressant, and analgesic activity.[11] Various 2,6-disubstituted perhydropyrido[1,2-*a*]pyridines are described as having marked hypotensive activity[13] and are patented as anticholinergics and antihistamines as well.[14] The most active hypotensive compounds were **137** and **138**.[13] A study has been started to see if the optical isomers of this ring system possess different hypotensive activity. The pharmacology has yet to be reported.[5]

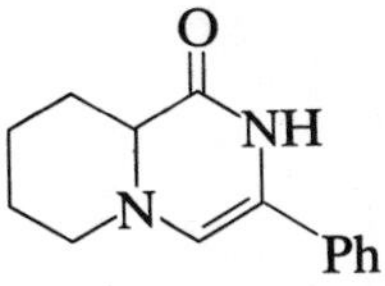

135

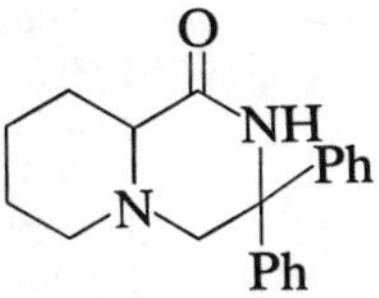

136

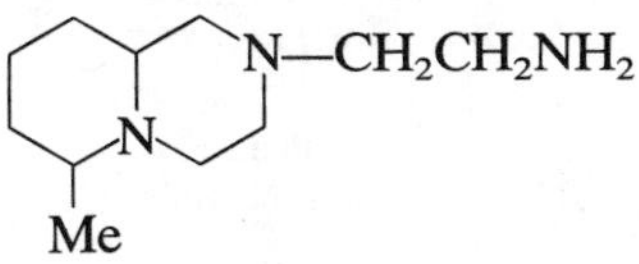

137

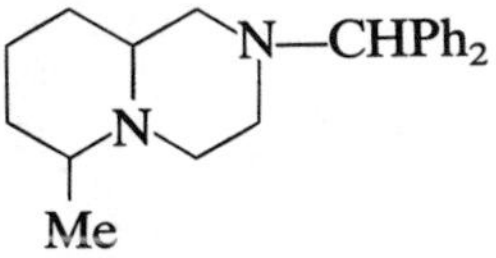

138

The partially reduced compound **139** is mentioned in a paper on piperazines as a tranquilizer.[17] The compound showed moderate activity in several animal tests. The oxo compound **140** is claimed in a patent for heart stimulants, hyaluronidase inhibitors, CNS depressants, and fungicides.[19]

139

140

V. Tables

TABLE 1. PERHYDROPYRIDO[1,2-*a*]PYRAZINES

Molecular formula	Substituents or structure	m.p., °C	Ref.
$C_8H_{12}N_2O_2$	3,4-Dioxo	168.5–170	18
$C_8H_{14}N_2O$	1-Oxo	136–137	2, 4, 20, 32
	d and *l* forms[a]	149–151	5
	Hydrobromide	>350	2
	Hydrochloride	271–272 decomp.	31
	cis-Methiodide	177–178	6
	trans-Methiodide	221–223	6
	3-Oxo	114–115	20, 33
	Picrate	213 decomp.	20
	Methiodide	273–275	33
	4-Oxo	b.p. 111–113/ 0.05 mm	23, 24
	Picrate	187–190	23
$C_8H_{16}N_2$	Unsubstituted	b.p. 98–99/25 mm	4
		b.p. 60/1 mm	2, 6, 9, 12, 23
	Dihydrochloride	213–214	4
	Dipicrate	275–278	2
		253–254 decomp.	20, 6
$C_9H_{13}N_3O$	9a-Cyano-1-oxo	137–139	6
$C_9H_{14}N_2O_2$	2-Methyl-3,4-dioxo	134–135	18
$C_9H_{14}N_2O_3$	6-Carboxy-1-oxo[b]		
	Hydrochloride	272–274	13, 14

TABLE 1 (*contd.*)

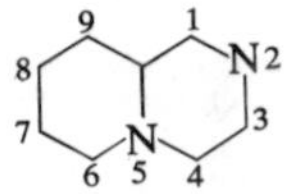

Molecular formula	Substituents or structure	m.p., °C	Ref.
	Methyl ester hydrochloride	155.5–156.5	13
$C_9H_{16}N_2O$	3-Methyl-4-oxo	b.p. 117/0.2 mm	23
	6-Methyl-1-oxo[b]	124–127	13, 14
	Hydrochloride	266 decomp.	13
$C_9H_{18}N_2$	2-Methyl	b.p. 82–83/11 mm	12
		b.p. 120–122/65 mm	17
	Dihydrochloride	245–246 decomp.	12
	Dipicrate	275–276 decomp.	12
	3-Methyl	b.p. 90/1.5 mm	9
		b.p. 89–90/13 mm	23
	Dipicrate	232–245	9
	6-Methyl[b]	b.p. 66/0.63 mm	13
	Dihydrochloride	255–258 decomp.	13, 14
$C_9H_{18}N_2O$	3-Hydromethyl	110.5	23
		b.p. 111/0.2 mm	
	6-Hydroxymethyl[b]	b.p. 158–166/2 mm	13, 14
$C_9H_{18}N_4$	2-Amidino		
	Sulfate	314–317	5
$C_{10}H_{16}N_2O_3$	2-Carboxymethyl-4-oxo		
	Ethyl ester	b.p. 169–170/0.4 mm	23
		b.p. 194/3 mm	24
	Ethyl ester perchlorate	204–207	23
$C_{10}H_{17}N_3$	2-Cyanomethyl	60.5–62	5
$C_{10}H_{18}N_2O$	2-Acetyl	b.p. 110/0.8 mm	41
$C_{10}H_{18}N_2O_2$	2-Carboxymethyl	180 decomp.	12
	Dipicrate	215–216 decomp.	
	Ethyl ester	b.p. 90–91/0.01 mm	
	Ethyl ester dipicrate	195 decomp.	
$C_{10}H_{20}N_2$	2-Ethyl	b.p. 95–97/11 mm	12
	Dihydrochloride	270–273 decomp.	
	Dipicrate	264–267 decomp.	
$C_{10}H_{20}N_2O$	2-(2-Hydroxyethyl)	b.p. 105–106/0.6 mm	4, 23
	Dipicrate	250.5 decomp.	23
	Dimethiodide	195–196	4
$C_{10}H_{21}N_3$	2-(2-Aminoethyl)[c]	b.p. 92/0.23 mm	5
$C_{11}H_{20}N_2O_2$	3-Ethoxymethyl-4-oxo	b.p. 134–135/0.3 mm	23
$C_{11}H_{22}N_2O$	3-Ethoxymethyl	b.p. 72/0.05 mm	23
	Dipicrate	203–206	
	2-(2-Hydroxyethyl)-6-methyl	b.p. 128/0.85 mm	13
	Dihydrochloride	255–257	

TABLE 1 (*contd.*)

[Structure: ring system numbered 1–9, N at positions 2 and 5]

Molecular formula	Substituents or structure	m.p., °C	Ref.
$C_{11}H_{23}N_3$	2-(2-Aminoethyl)-6-methyl	b.p. 105–110/0.43 mm	13
	Trihydrochloride	265–268	
$C_{11}H_{23}N_5$	2-(2-Guanidinoethyl)		
	Mononitrate	143–144.5	5
$C_{12}H_{22}N_2O_2$	9-Hydroxy-1-(2-tetrahydrofuryl)[d]	b.p. 143–144/0.7 mm	38, 39
	Dipicrate	240	38, 39
	Bisdiphenylacetyl hydrochloride	138–140	39
$C_{13}H_{26}N_2$	2-*n*-Butyl-6-methyl	b.p. 158/0.85 mm	13
	Dipicrate	240 decomp.	
$C_{14}H_{18}N_2O$	1-Oxo-3-phenyl[e]	146–148	8
$C_{14}H_{19}N_3O$	2-(4-Nitrosophenyl)		
	Dipicrate	124 decomp.	12
$C_{14}H_{20}N_2$	2-Phenyl	b.p. 113–114/0.02 mm	12
	Dihydrochloride	209–210	
	Dipicrate	179–181	
$C_{14}H_{21}N_3O_2S$	2-(4-Aminophenylsulfonyl)	154.5	23
$C_{14}H_{25}N_3O$	1-Oxo-2-(2-piperidylmethyl)	b.p. 135/0.1 mm	36
	Dipicrate	120–122	
$C_{14}H_{27}N_3$	2-(2-Piperidylmethyl)	b.p. 115–120/0.15 mm	36
	Tripicrate	223–225 decomp.	
$C_{14}H_{29}N_3$	2-(2-Diethylaminoethyl)[f]		
	Trihydrochloride	253.5–255	5
$C_{15}H_{19}N_3O_2$	6-Carboxamido-1-oxo-3-phenyl	283–284	8
$C_{15}H_{19}N_3O_3$	2-Methyl-3-(4-nitrophenyl)-1-oxo	205–207	8
$C_{15}H_{20}N_2O$	2-Benzoyl	b.p. 153–154/0.35 mm	6, 41
	Picrate	203–204	6
	Methiodide	213–214	6
	2-Benzyl-1-oxo	111	6
	2-Methyl-1-oxo-3-phenyl	84–86	8
$C_{15}H_{20}N_2O_2$	2-Hydroxymethyl-1-oxo-3-phenyl	131–132	8
$C_{15}H_{21}N_3S$	2-Phenylaminothiocarbonyl	118–119	4
$C_{15}H_{22}N_2$	2-Benzyl	b.p. 110–120/0.2 mm	6
	Dihydrobromide	266–267	
$C_{15}H_{22}N_2O$	2-(2-Methoxyphenyl)	b.p. 116–118/0.04 mm	12
	Dihydrochloride	189–191	
	Diperchlorate	210–212	
	Dipicrate	184–185	
$C_{15}H_{24}N_3$	2-(2-Pyridiniumethyl)		
	p-Toluenesulfonate	134–135	4
$C_{15}H_{31}N_3$	2-(2-Diethylaminoethyl)-6-methyl[b]		
	Trihydrochloride	236–242	13, 14

TABLE 1 (*contd.*)

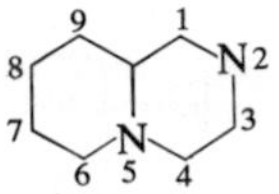

Molecular formula	Substituents or structure	m.p., °C	Ref.
$C_{15}H_{31}N_3O$	2-(2-Diethylaminoethyl)-6-hydroxymethyl[b]		
	Trihydrochloride	247–249	13, 14
$C_{16}H_{22}N_2O_2$	2-(3,4-Methylenedioxybenzyl)[g]		
	Hydrochloride	239–241	42
$C_{16}H_{23}N_3O$	3-Methyl-2-phenylaminocarbonyl	174–176	23
$C_{16}H_{23}N_3S$	6-Methyl-2-phenylaminothiocarbonyl	168.5–170 decomp.	13
$C_{16}H_{24}N_2$	2-(2-Phenylethyl)[h]		
	Dihydrochloride	210–212	42
	2-Benzyl-6-methyl[b]	b.p. 158–162/3 mm	13, 14
	Dipicrate	240–260 decomp.	13, 14
$C_{16}H_{31}N_3$	6-Methyl-2-(2-piperid-1-ylethyl)[b]		
	Trihydrochloride	[i]	13, 14
$C_{16}H_{31}N_3O$	6-Hydroxymethyl-2-(2-piperid-1-ylethyl)[b]		
	Trihydrochloride	277.5–279 decomp.	13, 14
$C_{17}H_{24}N_2O_2$	3-Methyl-2-(3,4-methylenedioxybenzyl)		
	Dihydrochloride hemihydrate	187–205 decomp.	42
$C_{17}H_{26}N_2O$	6-Hydroxymethyl-2-(2-phenylethyl)		
	Dihydrochloride	274–276	13
$C_{18}H_{26}N_4O_3$	2-[2-(4-Nitrobenzoyl)aminoethyl]-6-methyl[b]	128.5–129.5	13, 14
$C_{18}H_{28}N_4O$	2-[2-(4-Aminobenzoyl)aminoethyl]-6-methyl[b]		
	Trihydrochloride	250–253	13
$C_{20}H_{20}N_2O_2$	1,4-Dioxo-3,3-diphenyl	221–223	11
$C_{20}H_{22}N_2O$	1-Oxo-2,3-diphenyl	173–176	3
	Hydrobromide	251–254 decomp.	
	1-Oxo-3,3-diphenyl	135	11
$C_{21}H_{26}N_2$	2-Diphenylmethyl[j]		
	Dihydrochloride	245–246, 195–210 decomp.	6, 42
$C_{21}H_{32}N_2$	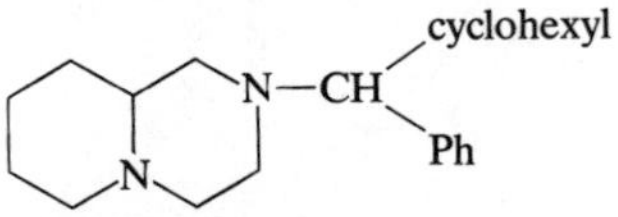 N—CH(cyclohexyl)(Ph)		
	Dihydrochloride	239–240	42
$C_{22}H_{28}N_2$	2-Diphenylmethyl-3-methyl		
	Fumarate	236–245 decomp.	42
	2-Diphenylmethyl-6-methyl[b]		
	Dihydrochloride	228–229	13, 14

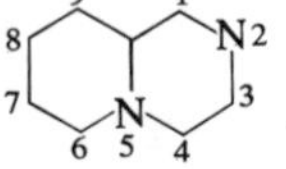

TABLE 1 (*contd.*)

Molecular formula	Substituents or structure	m.p., °C	Ref.
$C_{22}H_{28}N_2O$	2-(2-Hydroxy-2,2-diphenylethyl)		
	Dihydrochloride	210	42
	6-Hydroxymethyl-2-diphenylmethyl[b]	153–154	13, 14
	Dihydrochloride	213–214 decomp.	
$C_{23}H_{28}ClN_3S$	2-[3-(4-Chlorophenothiazin-6-yl)-propyl		
	Dimaleate	165–166	43
$C_{23}H_{28}N_4O_3$	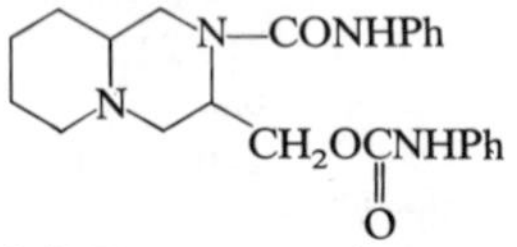 N—CONHPh; $CH_2OCNHPh$ (C=O)	179	23
$C_{23}H_{29}N_3S$	2-(3-Phenothiazin-6-ylpropyl)		
	Dimaleate	157–160	43
$C_{23}H_{30}N_2$	2-(3,3-Diphenylpropyl)		
	Dihydrochloride	200–207 decomp.	42
$C_{23}H_{30}N_2O$	2-(3-Hydroxy-3,3-diphenylpropyl)		
	Difumarate	189–198 decomp.	42
$C_{24}H_{28}F_3N_3S$	2-[3-(4-Trifluoromethylpheno-thiazin-6-yl)propyl		
	Dimaleate	173–175	43
$C_{25}H_{30}F_3N_3OS$	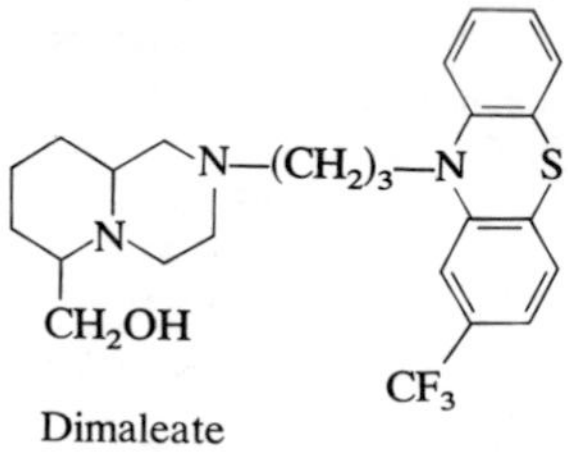 N—$(CH_2)_3$—N (phenothiazine, S, CF_3); CH_2OH		
	Dimaleate	152–154	43
$C_{26}H_{32}N_2$	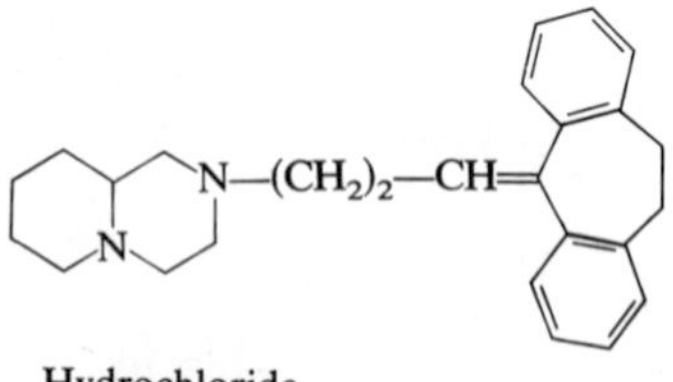 N—$(CH_2)_2$—CH= (dibenzocycloheptene)		
	Hydrochloride	263–265	43
$C_{29}H_{34}N_2$	2-(3,3,3-Triphenylpropyl)		
	Difumarate	165–170	42

[a] *d* Form $[\alpha]_D^{25}+118.7$; *l* form $[\alpha]_D^{25}-93.8$ (both *c* 3.8, HOH).[5]

TABLE 1 (*contd.*)

[b] Incorrectly numbered in reference 14 (cf. ref. 13).
[c] $n_D^{26.5}$ 1.5021.
[d] n_D^{20} 1.524.
[e] Seven other 3-aryl derivatives are also described.
[f] *d* Form $[\alpha]_D^{25} + 8.4$ (*c* 5.3, HOH); *l* form $[\alpha]_D^{25} - 12.0$ (*c* 4.0, HOH).
[g] 3,4-Dimethoxy- and 2,3,4-trimethoxybenzyl derivatives also described.
[h] 3,4-Methylenedioxyphenylethyl compound also described.
[i] Conflicting melting points are given in a paper (ref. 13, m.p. 168.5–170°) and a patent (ref. 14, m.p. 301° dec.) by the same authors. Possibly the lower m.p. refers to the free base. Five other α-arylbenzyl derivatives are also described.[42]

TABLE 2. PYRIDO[1,2-*a*]PYRAZINES[a]

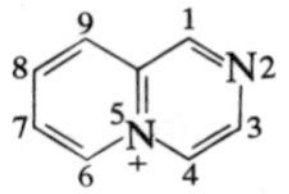

Molecular formula	Substituents or structure	m.p., °C	Ref.
$C_8H_6BrN_2$	1-Bromo		
	Bromide	Chars < 320	2
	Picrate	142	
	2-Oxide bromide	Chars < 320	
	picrate	155–156	
$C_8H_6N_2O$	3,4-Dihydro-1-hydroxy-3-oxo betaine	>320	2
$C_8H_7N_2$	Unsubstituted		
	Bromide hemihydrate	272–274	2
	Picrate	192–194	
	2-Oxide chloride	>240 decomp.	9
	bromide	280 decomp.	
	picrate	199	2
$C_8H_7N_2O$	1,2-Dihydro-1-oxo		
	Bromide	Chars < 320	2
	Picrate	250–254	
$C_8H_8N_2O_2$	1,2,3,4,5,8-Hexahydro-1,8-dioxo	307–310 decomp.	27
$C_8H_8N_3$	1,4-Dihydro-1-imino		
	2-Oxide bromide	Chars <320	2
	picrate	183–185	
$C_8H_9N_2O$	1,2,3,4-Tetrahydro-1-oxo		
	Chloride	>300	22
	Bromide	>350	2
	Picrate	184–186	
$C_8H_{12}N_2O$	1,2,3,4,5,6,7,8-Octahydro-1-oxo	146–148	6
	Perchlorate	186–188	
	Methiodide	209–211	
$C_9H_8N_2O_4$	6-Carboxy-1,2,3,4,5,8-hexahydro-1,8-dioxo	335–338 decomp.	25, 26, 27
	p-Bromophenacyl deriv. dihydrate	172–174	25

TABLE 2 (*contd.*)

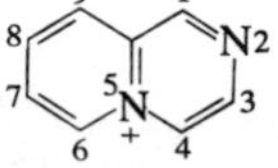

Molecular formula	Substituents or structure	m.p., °C	Ref.
$C_9H_9N_2$	1-Methyl		
	Perchlorate	192–194 decomp.	3
	Picrate	174–175 decomp.	
	2-Oxide bromide	254–256 decomp.	2
	picrate	185–186	
	3-Methyl		
	Bromide	>330 decomp.	3
	Picrate	173–174	
	2-Oxide bromide	260 decomp.	9
	picrate	207–210 decomp.	
$C_{10}H_{11}N_2$	1-Ethyl		
	Hexafluorophosphate	181–184	3
	2-Oxide bromide hemihydrate	191–194 decomp.	
	perchlorate	184–186	
	1,3-Dimethyl		
	2-Oxide bromide monohydrate	207 decomp.	9
$C_{10}H_{13}N_2O$	1-Ethyl-3,4-dihydro-3-hydroxy		
	2-Oxide bromide	153–155 decomp.	3
	picrate	172–174 decomp.	
$C_{11}H_{13}N_2$	1-Isopropyl		
	Perchlorate	138–140	3
	2-Oxide bromide monohydrate	162–164	
	picrate	199–201 decomp.	
$C_{11}H_{14}N_4O_3$	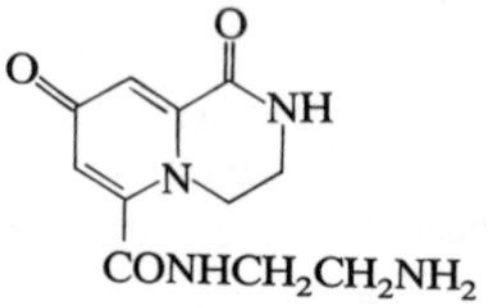 [structure] CONHCH$_2$CH$_2$NH$_2$	>320	28
$C_{11}H_{16}N_2O_2$	2-Ethyl-1,2,3,4,5,8-hexahydro-7-methoxy-8-oxo		
	Dihydrochloride	222–227 decomp.	15
	Ethochloride	258–261 decomp.	15, 19
	Hydrochloride	222–225	
$C_{11}H_{16}N_2O_3$	1,2,3,4,5,8-Hexahydro-2-(2-hydroxyethyl)-7-methoxy-8-oxo		
	Dihydrochloride	194–196	15
$C_{12}H_{15}N_2$	1-*t*-Butyl		
	Perchlorate	172–175	3
	Picrate	161–162	

TABLE 2 (*contd.*)

Molecular formula	Substituents or structure	m.p., °C	Ref.
$C_{12}H_{15}N_2$ (*contd.*)	2-Oxide bromide	193–196 decomp.	
	perchlorate	176–179 decomp.	
	picrate	168–170 decomp.	
$C_{12}H_{18}N_2O_2$	1,2,3,4,5,8-Hexahydro-7-methoxy-8-oxo-2-propyl		
	Propochloride	229–232 decomp.	15
$C_{12}H_{18}N_2O_3$	O; N—$CH_2CH_2CH_2OH$; N; MeO	158–160	15
	Dihydrochloride	205–209	
	Diphenylacetylester dihydrochloride	185–190	
$C_{13}H_{10}N_4O_2$	CN; C; NC; O; N—COMe; N	300 decomp.	29
$C_{13}H_{11}N_3O_4$	CO_2H; C; NC; O; N—COMe; N		
	Ethyl ester	225	29
$C_{14}H_{10}N_2O$	1-Hydroxy-3-phenyl betaine	252–255	7
$C_{14}H_{11}N_2$	1-Phenyl		
	Bromide	259–260	2, 21
	Picrate	162–164	
	2-Oxide bromide	263–264	2
	picrate	212–213	
	3-Phenyl		
	Bromide	298–300	3
	2-Oxide picrate	167–168	9
$C_{14}H_{11}N_2O$	1,2-Dihydro-1-oxo-3-phenyl		
	Bromide monohydrate	309	7
$C_{14}H_{12}N_2O_7$	CO_2H; O; HO_2C; N—Me; N; EtCO; O		
	Dimethyl ester	173	40
	Dimethyl ester oxime	255–257	

TABLE 2 (*contd.*)

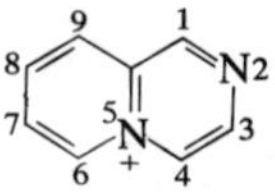

Molecular formula	Substituents or structure	m.p., °C	Ref.
$C_{14}H_{12}N_2O_7$ (*contd.*)	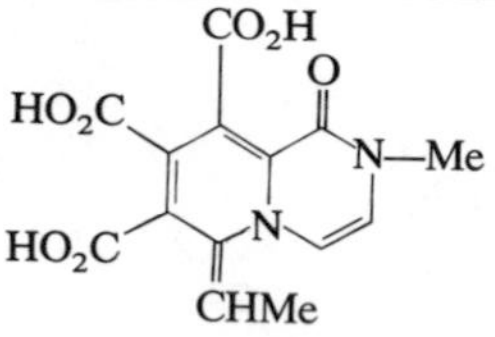		
	Trimethyl ester	212–214	40
$C_{14}H_{14}N_2O_4$	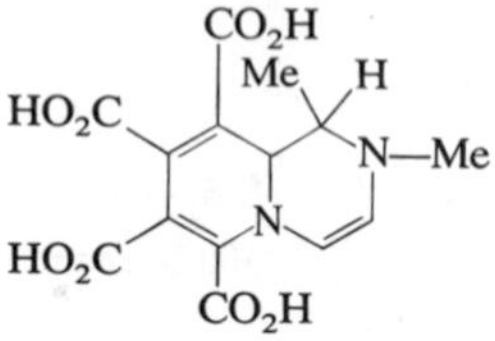		
	Tetramethyl ester	206	40
$C_{14}H_{14}N_4O_4$	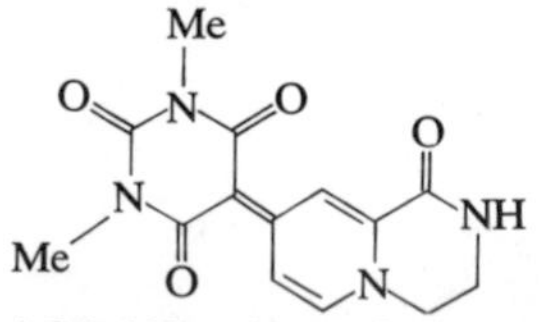	>340	29
$C_{14}H_{15}N_2$	1,2,3,4-Tetrahydro-2-phenyl		
	Chloride hydrochloride	184–185	17
$C_{14}H_{16}N_2O$	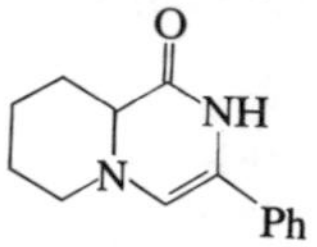 [b]	176–180	8
$C_{14}H_{20}N_2O_3$	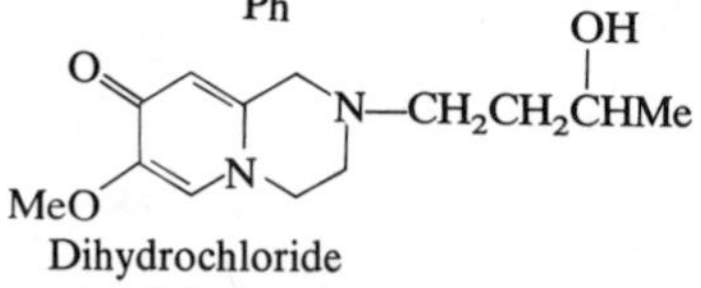		
	Dihydrochloride	175–179	15
$C_{14}H_{22}N_2O_2$	1,2,3,4,5,8-Hexahydro-7-methoxy-8-oxo-2-pentyl	206–212 decomp.	15
$C_{15}H_{10}N_2O_4$	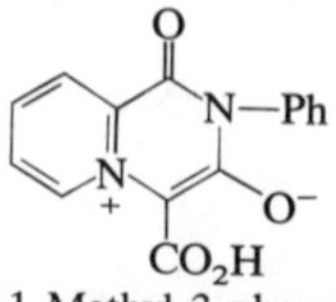		
$C_{15}H_{13}N_2$	1-Methyl-3-phenyl		
	Perchlorate	252–254	7

TABLE 2 (*contd.*)

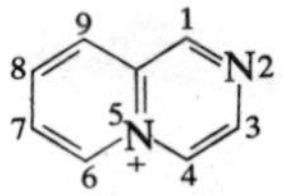

Molecular formula	Substituents or structure	m.p., °C	Ref.
$C_{15}H_{17}N_3O_2$	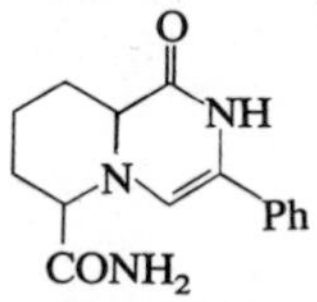 [structure]	290–292	8
$C_{15}H_{18}N_2O$	2-Benzyl-1,2,3,4,5,6,7,8-octahydro-1-oxo	b.p. 170–180/0.2 mm	6
$C_{16}H_{12}N_2O_4$	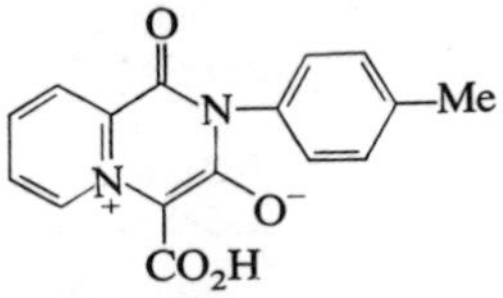 [structure]		
	Ethyl ester	229–231 decomp.	37
$C_{16}H_{15}N_2$	1-Ethyl-3-phenyl		
	Picrate	185–187 decomp.	7
$C_{16}H_{15}N_2O$	3-(4-Methoxyphenyl)-1-methyl		
	Perchlorate	179–181	7
$C_{17}H_{17}N_2$	3-Phenyl-1-propyl		
	Perchlorate	154–156	7
$C_{17}H_{20}N_2$	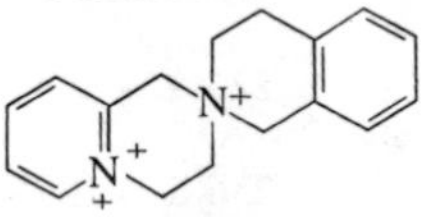 [structure]		
	Dibromide	275	16
$C_{17}H_{20}N_6O_5$	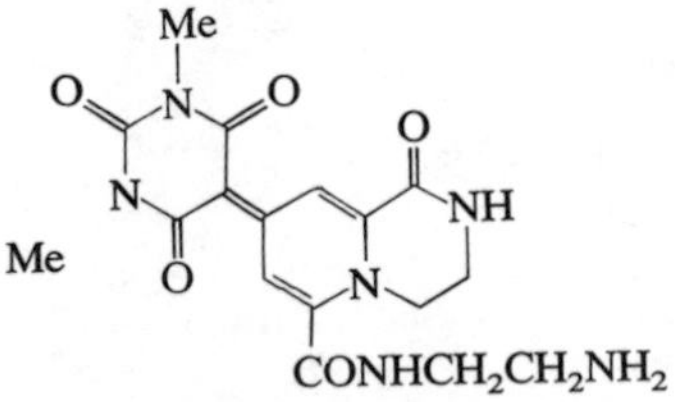 [structure]	>320	28
$C_{17}H_{21}N_2$	1,2,3,4-Tetrahydro-2-(1-methyl-2-phenylethyl)		
	Bromide hydrobromide	226–230	17
$C_{20}H_{13}Br_2N_2O$	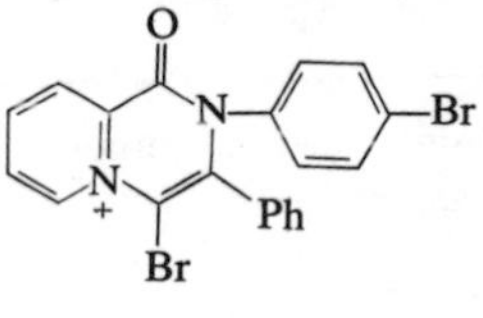 [structure]		

TABLE 2 (*contd.*)

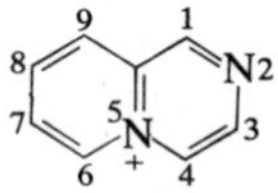

Molecular formula	Substituents or structure	m.p., °C	Ref.
$C_{20}H_{13}Br_2N_2O$ (*contd.*)	Bromide	decomp. <330	3
	Perchlorate	181–184	
	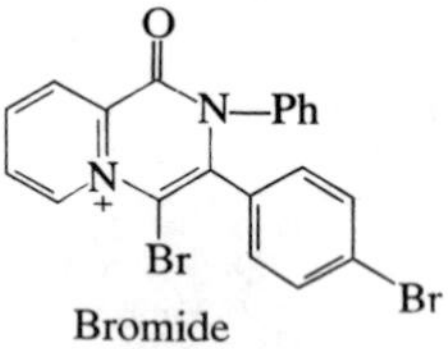		
	Bromide	decomp. <330	3
	Perchlorate	290–292	
$C_{20}H_{14}BrN_2$	3-(4-Bromophenyl)-1-phenyl		
	Bromide monohydrate	363–366	3
	Perchlorate	355–357 decomp.	7
$C_{20}H_{14}BrN_2O$	4-Bromo-1,2-dihydro-1-oxo-2,3-diphenyl		
	Bromide	decomp. <330	3
	Perchlorate	317–320 decomp.	
$C_{20}H_{15}N_2$	1,3-diphenyl		
	Bromide monohydrate	266–268 decomp.	3
	Perchlorate	264–265	
	Picrate	237–238 decomp.	
$C_{20}H_{15}N_2O$	1,2-Dihydro-1-oxo-2,3-diphenyl		
	Perchlorate	273–275	3
$C_{20}H_{17}N_2O$	3,4-Dihydro-3-hydroxy-1,3-diphenyl		
	Bromide	256–259 decomp.	7
$C_{25}H_{20}N_4O_4$	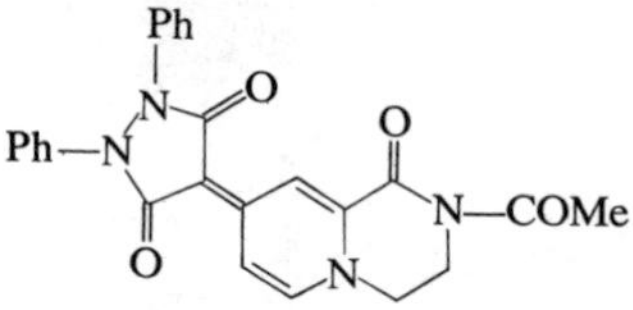	333	29
$C_{26}H_{20}N_4O_5$	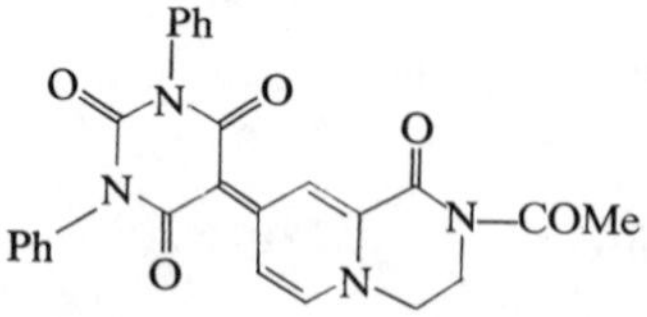	>340	29

[a] Anions of quaternary salts are not included in the molecular formulae.
[b] Seven other 3-aryl derivatives are described.

VI. References

1. V. Galosso, *Theor. Chim. Acta,* **11,** 417 (1968); *Chem. Abstr.,* **69,** 101284 (1968).
2. E. E. Glover and M. J. R. Loadman, *J. Chem. Soc.* (*C*), 2391 (1967).
3. E. E. Glover and J. Adamson, *J. Chem. Soc.* (*C*), 861 (1971).
4. M. E. Freed and A. R. Day, *J. Org. Chem.,* **25,** 2108 (1960).
5. R. L. Peck and A. R. Day, *J. Heterocycl. Chem.,* **6,** 181 (1969).
6. Y. Arata and Y. Nakagawa, *Chem. Pharm. Bull.* (Tokyo), **21,** 1248 (1973).
7. F. Kröhnke, H. Schnegelberger, and W. Weis, *Chem. Ber.,* **97,** 3566 (1964).
8. Ciba Ltd., Neth. Appl. 6,613,937; *Chem. Abstr.,* **68,** 49650 (1968)
9. C. K. Bradsher and S. A. Telang, *J. Org. Chem.,* **31,** 941 (1966).
10. A. B. Mauger, R. B. Desai, I. Rittner, and W. J. Rzeszotarski, *J. Chem. Soc., Perkin I,* 2146 (1972).
11. M. E. Freed (to American Home Products), U.S. Pat. 3,531,485; *Chem. Abstr.,* **73,** 120672 (1970).
12. K. Winterfeld, K. Küllmar, and W. Göbel, *Chem. Ber.,* **92,** 1510 (1959).
13. A. D. Lourie and A. R. Day, *J. Med. Chem.,* **9,** 311 (1966).
14. A. R. Day and A. D. Lourie, U.S. Pat. 3,388,128; *Chem. Abstr.,* **69,** 67419 (1968).
15. C. P. Krimmel (to G. D. Searle), U.S. Pat. 2,999,860; *Chem. Abstr.,* **56,** 4780 (1962).
16. G. Buchmann and L. Krahnert, *J. Prakt. Chem.,* **30,** 241 (1965); *Chem. Abstr.,* **64,** 11172 (1966).
17. K. Lanyi, I. Erdelyi, E. Institoris, L. Tardos, and Z. Szabo, *Arzneim.-Forsch.,* **18,** 1431 (1968).
18. K. Winterfeld, H. Lampke, and H. Franzke, *Justus Liebigs Ann. Chem.,* **685,** 181 (1965); *Chem. Abstr.,* **63,** 8361 (1965).
19. C. P. Krimmel (to G. D. Searle), U.S. Pat. 2,965,641; *Chem. Abstr.,* **55,** 13457 (1961).
20. T. Yamazaki, M. Nagata, K. Ogawa, and F. Nohara, *Yakugaku Zasshi,* **87,** 668 (1967); *Chem. Abstr.,* **67,** 90770 (1967).
21. J. Adamson, E. C. Campbell, and E. E. Glover, *J. Chem. Soc.* (*C*), 2270 (1969).
22. P. Slade and A. E. Smith, *Nature,* **213,** 919 (1967).
23. K. Winterfeld and G. Gierenz, *Chem. Ber.,* **92,** 240 (1959).
24. K. Winterfeld and G. Gierenz, *Naturwissenschaften,* **44,** 62 (1957).
25. A. W. Schwab, *J. Am. Chem. Soc.,* **76,** 1189 (1954).
26. A. W. Schwab (to U.S. Sec. of Agriculture), U.S. Pat. 2,740,786; *Chem. Abstr.,* **50,** 10796 (1956).
27. G. A. Garkusha, *Zh. Org. Khim.,* **1,** 2222 (1965); *Chem. Abstr.,* **64,** 11209 (1966).
28. F. Eiden and P. Peter, *Arch. Pharm.,* (Weinheim), **299,** 139 (1966); *Chem. Abstr.,* **64,** 15878 (1966).
29. F. Eiden and M. Beuttenmueller, *Arch. Pharm.* (Weinheim), **304,** 342 (1971).
30. A. M. Likhosherstov, L. S. Nazarova, and A. P. Skoldinov, *Zh. Org. Khim.,* **6,** 1729 (1970); *Chem. Abstr.,* **73,** 109724 (1970).
31. A. M. Likhosherstov, L. S. Nazarova, and A. P. Skoldinov, Fr. Pat. 1,510,781; *Chem. Abstr.,* **70,** 68427 (1969); A. M. Likhosherstov, L. S. Nazarova, and A. P. Skoldinov, USSR Pat. 245,789; *Chem. Abstr.,* **71,** 124500 (1969).
32. Scientific Research Institute of Pharmacology and Chemotherapy., Moscow, Brit. Pat. 1,144,749; *Chem. Abstr.,* **70,** 106558 (1969).
33. L. A. Paquette and M. K. Scott, *J. Org. Chem.,* **33,** 2379 (1968).
34. I. Jezo and I. Luzak, *Chem. Zvesti,* **18,** 186 (1964); *Chem. Abstr.,* **62,** 5322 (1965).
35. A. B. Mauger, *J. Chem. Soc.* (*D*), 39 (1971).

36. E. C. Campbell, E. E. Glover, and G. Trenholm, *J. Chem. Soc.* 1987 (1969).
37. H. Wittmann, J. Kuhn-Kuhnenfelf, and E. Ziegler, *Monatsh. Chem.* **102,** 1120 (1971).
38. R. E. Paul and P. J. C. Buisson (to Rhône-Poulenc), Brit. Pat. 763,990; *Chem. Abstr.,* **52,** 2099 (1958).
39. R. Paul and P. Buisson, *Bull. Soc. Chim. Fr.,* 2135 (1961); *Chem. Abstr.,* **56,** 14285 (1962).
40. A. Crabtree and A. W. Johnson, *J. Chem. Soc.,* 1510 (1962).
41. E. E. Mikhlina, A. D. Yanina, and M. V. Rubstov, *Khim. Geterotsikl. Soedin.,* 547 (1969); *Chem. Abstr.,* **71,** 124366 (1969).
42. G. Regnier, R. Canevari, and J. C. LeDouarec, Brit. Pat. 1,125,112; *Chem. Abstr.,* **70,** 28940 (1969).
43. C. Casagrande, A. Galli, R. Ferrini, and G. Miragoli, *Arzneim.-Forsch.,* **21,** 808 (1971).

CHAPTER XXVIII

Pyrido[2,3-*b*]pyrazines

I. Physical Properties 496
II. Methods of Preparation 497
1. From Symmetrical α-Dicarbonyl Compounds 498
2. From Unsymmetrical α-Dicarbonyl Compounds 500
A. α-Keto Aldehydes 502
B. α-Keto Esters and Acids 503
C. Other Dicarbonyl Compounds 507
3. From 2-Amino-3-nitrosopyridines 508
4. From 3-Nitropyridine Derivatives 509
5. From Pyrazine Precursors 510
III. Reactions 512
1. Reactions with Methylating Reagents 512
2. Reactions of Active Methyl Groups 513
3. Reactions with Oxidizing Agents 514
4. Reactions at Electron-Deficient Carbon Atoms 516
5. Decarboxylation 518
6. Reactions with Reducing Agents 518
7. Intramolecular Cyclizations 521
8. Miscellaneous Reactions 523
IV. Biological Properties 524
V. Tables 525
Table 1. 6-Aminopyrido[2,3-*b*]pyrazines 525
Table 2. Pyrido[2,3-*b*]pyrazines (not 6-Amino Compounds) 530
VI. References 541

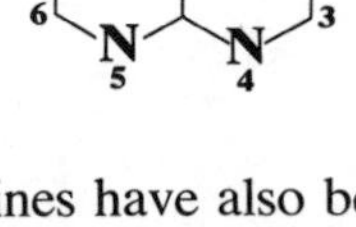

RRI 1606

Pyrido[2,3-*b*]pyrazines have also been named 1,4,5-triazanaphthalenes and 5-azaquinoxalines. Occasionally they have been referred to as pyridino[2′.3′:2.3]pyrazines and numbered as shown in formula **1**. The preferred numbering is given above.

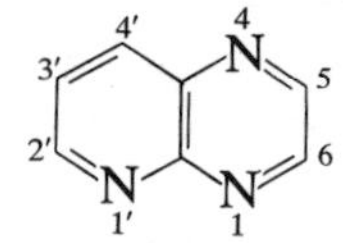

1

I. Physical Properties

The Pariser-Parr-Pople method,[1] self-consistent molecular orbital calculations,[2] and the Hückel approach[3] have been applied to the theoretical calculation of transition energies and intensities in the ultraviolet spectra of pyrido[2,3-*b*]pyrazine and other azanaphthalenes. Good agreement with experimental data was obtained. The infrared spectra of the parent heterocycle[4] and its 8-hydroxy derivative[5] have been discussed; the oxo form is preferred to the hydroxy form both in chloroform solution and in the solid state. As with other α-hydroxyazines, the oxo form is also favored in the 2-, 3-, and 6-hydroxy compounds.[6] Ultraviolet spectra have been measured for the parent heterocycle[7,8] and various substituted compounds.[7,9] The spectrum of the parent is substantially similar to other azanaphthalenes. In ethanol the principal bands are at *ca.* 260, 300, and 350 nm. Both the infrared and ultraviolet spectra of about 60 derivatives of the ring system have been assembled.[10]

Ultraviolet spectral and pK_a data have been used to show that the cation and the neutral species of 2- and 3-"hydroxy"pyrido[2,3-*b*]pyrazine undergo covalent hydration in the pyrazine ring in aqueous solution. Although extensive, the hydration is less than that found for pyrido[3,4-*b*]pyrazines.[6] Covalent hydration of the parent heterocycle, or other "hydroxy" derivatives, could not be demonstrated by this technique. However PMR studies[11] have shown that in aqueous solution the cation of pyrido[2,3-*b*]pyrazine contains 10% of the dihydrated species **2**.

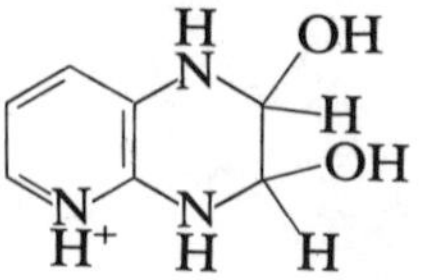

2

The ring system is weakly basic,[12] the parent heterocycle having a pK_a of only 1.20. Acidity constants are also available for various simple derivatives.[6,12,13] The fact that quinoxaline (pK_a 0.56) is a weaker base

Figure 1. BASIC STRENGTHS OF AZA-QUINOLINES

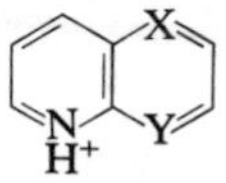

3

Compound	X	Y	pK_a
Quinoline	CH	CH	4.93
1,5-Naphthyridine	N	CH	2.91
1,8-Naphthyridine	CH	N	3.39
Pyrido[2,3-*b*]pyrazine	N	N	1.20

than pyrido[2,3-*b*]pyrazine indicates that protonation occurs at the 5-position in the pyrido compound. The base-weakening effect of the successive introduction of =N— groups into the 5- and 8-positions of quinoline is demonstrated in Figure 1. The pK_a of pyrido[2,3-*b*]pyrazine is very close to that which would be predicted from the effects on the pK_a of quinoline (**3**, X = Y = CH) of the introduction of aza groups at appropriate positions. This confirms the site of protonation of pyrido[2,3-*b*]pyrazine. The influence of the small amount of covalent hydration is not noticeable. This contrasts sharply with the [3,4-*b*]system discussed later (see Chapter XXIX).

Polarographic reduction of the parent gave a value for the half-wave potential ($E_{1/2} = -0.85$ v) which is intermediate between the values for quinoxaline ($E_{1/2} = -1.09$ v) and pteridine ($E_{1/2} = -0.52$ v).[14] The electrochemical reduction of substituted pyrido[2,3-*b*]pyrazines has been the subject of a recent study.[61] Along with other fused pyrazine systems, the 2,3-bis(2-pyridyl) and the 2,3-bis(6-methyl-2-pyridyl) compounds have been studied as potential copper(I) chelators.[15] The 1:1 adduct of the parent heterocycle and copper(II) nitrate has been characterized using magnetic susceptibility measurements with electronic and infrared spectroscopy.[62]

II. Methods of Preparation

The major route to these compounds involves condensation of 2,3-diaminopyridines with α-dicarbonyl compounds. Other methods of forming the pyrazine ring have occasionally been employed. These are particularly useful when the products from the standard condensation are of uncertain orientation. Such approaches have made use of the condensation of 2-amino-3-nitrosopyridines with compounds containing active methylene groups, or the reduction, in situ cyclization, and mild oxidation of 2-substituted amino-3-nitropyridines. Only recently have reports appeared of the synthesis of this ring system from pyrazine rather than pyridine intermediates.

Usually any necessary substitutions are performed on the pyridine precursor before the eventual synthesis of the pyrido[2,3-*b*]pyrazine. The ring system is now easily accessible, as an improved route is available to 2,3-diaminopyridine. Thus bromination of 2-aminopyridine followed by nitration, reduction, and hydrogenolysis of the product gives 2,3-diaminopyridine in 26–43% yield.[16] A more direct synthesis is not possible since nitration of 2-aminopyridine provides mainly the unwanted 5-nitro compound, which must be laboriously separated before reduction of the desired 2-amino-3-nitropyridine.

1. From Symmetrical α-Dicarbonyl Compounds

2,3-Diaminopyridine has been condensed with glyoxal,[7,17] benzil,[18,19] diacetyl,[17,19] and the pyridils **4**.[20] Ethyl oxalate yields the 2,3-dioxo compound.[19] The conditions employed for these condensations can be critical. Thus glyoxal in neutral solution yields an unidentified, insoluble, infusible solid, whereas in the presence of acetic acid the reaction readily provides the parent heterocycle.[10] The preferred method makes use of glyoxal sodium bisulfite.[7] Many symmetrical dicarbonyl compounds have been condensed with a variety of 2,3-diaminopyridines substituted in the pyridine ring by bromo,[10,17] chloro,[7,21,22] methyl,[23,24] or combinations of these substituents.[21,25]

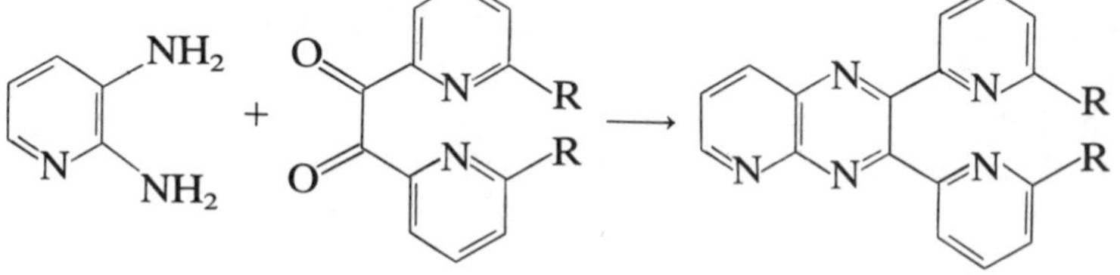

4 R = H or Me

Many examples have been reported of condensations resulting in pyrido[2,3-*b*]pyrazines containing substituted amino groups in the 6- and/or 8-positions.[17,19,26–30]

Condensations with 2,3,4-triaminopyridine[31] and 2,3,4,6-tetraaminopyridine[30] are complicated because of the possible formation of pyrido[3,4-*b*]pyrazine derivatives. The direction of cyclization is pH dependent. Thus polyglyoxal reacts with 2,3,4-triaminopyridine (**6**) at pH

2 to give the [2,3-*b*] system (**7**) exclusively. At pH 7 only the [3,4-*b*] system (**5**) is isolated.[31]

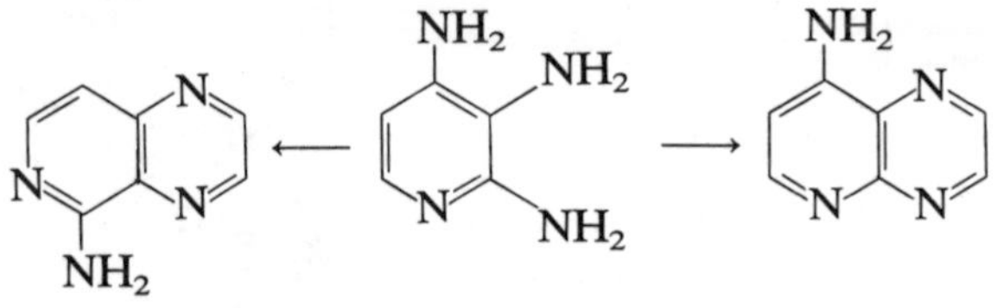

Similarly the condensation of the tetraamino compound **8** with pyruvic aldehyde in 0.1 *M* hydrochloric acid results in a mixture from which a 49% yield of the [2,3-*b*] compound (**9**), and only 6% of the [3,4-*b*] isomer (**10**) was obtained. The orientation of the methyl group is discussed in Section II2A. The use of neutral conditions with the amine **8** results in the formation of complex mixtures, possibly because the free amine is unstable.[30]

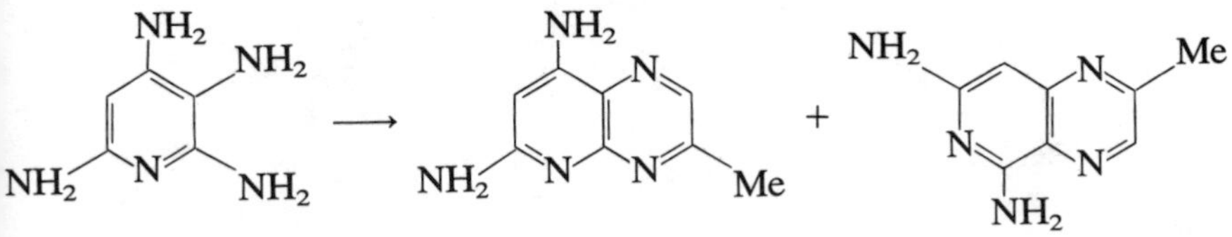

In the reactions of polyamines **6** and **8** with dicarbonyl compounds it is likely that initial condensation occurs at the 3-amino group. The amino groups in the 2- and 4-positions will be deactivated, as their lone pair electrons suffer conjugation into the pyridine ring. The results described above can then be rationalized by assuming that the reaction is kinetically controlled. The orientation of the products will then be determined by reaction with the most reactive amino group. In neutral solution the 4-amino group in the intermediate **11** will be more nucleophilic than the 2-amino group, and the product will be the [3,4-*b*] system **12**. In dilute acid the intermediate would be monoprotonated as in formula **13**, and the 4-amino group would be largely prevented from taking part in the cyclization. The lone pair on the 2-amino group would be less affected than the lone pair on the 4-amino group, as 2-aminopyridine (pK_a 6.86) is a weaker base than 4-aminopyridine (pK_a 9.17). As a result, cyclization would provide the [2,3-*b*] system **14**. Even stronger acid would diprotonate the pyridine as in formula **15**, permitting initial reaction at only the 2-position. Cyclization of this must also provide only the [2,3-*b*] system.

$$\textbf{11} \xrightarrow{\text{pH 7}} \textbf{12}$$

$$\textbf{13} \xrightarrow{\text{pH 2}} \textbf{14}$$

$$\textbf{15} \xrightarrow{\text{pH} < 1}$$

2. From Unsymmetrical α-Dicarbonyl Compounds

A second type of ambiguity is introduced when 2,3-diamino pyridines are condensed with unsymmetrical α-dicarbonyl compounds such as α-keto aldehydes or α-keto esters. Originally the orientation of the products was suggested by analogy with pteridine chemistry. More recently unambiguous syntheses have served to confirm the parallel with the pteridine series. Often mixtures are obtained although, depending on the reactions conditions, one isomer usually predominates. It is now possible to suggest the likely orientation of several compounds prepared by these routes, although further work is necessary to confirm these assignments.

If we again assume that the reaction products, and their orientation, are determined by kinetic rather than thermodynamic considerations, then the two important features of the condensation are the pH of the reaction mixture and the difference in reactivity of the two carbonyl groups in the α-dicarbonyl compound. In general aldehydes are more reactive than ketones, which are in turn more reactive than carboxylate groups. Thus applying the arguments used in Section II1, at neutral pH the products **16** and **17**, respectively, should be obtained from α-keto aldehydes (RCOCHO) and α-keto esters ($RCOCO_2R^1$), in each case by initial reaction at the 3-amino group. Similarly at pH values sufficient only to monoprotonate the diaminopyridine, products **16** and **17** should still be obtained.

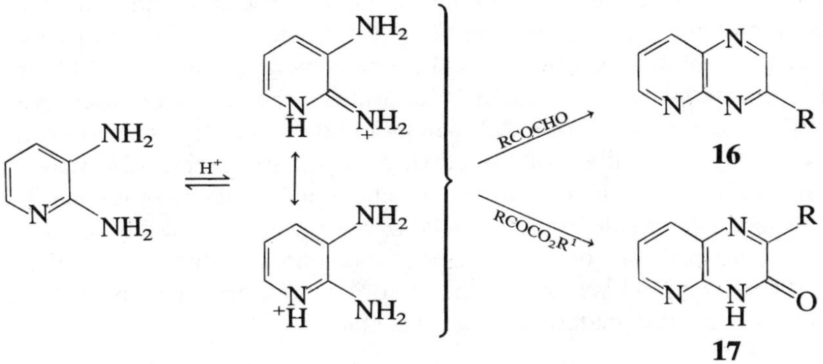

Under strongly acidic conditions the diaminopyridine will be diprotonated as in formula **18**, thus preventing initial reaction of the more reactive carbonyl group with the 3-amino substituent. In consequence the isomers **19** and **20** should be obtained. It has been suggested that the situation is a good deal more complicated than simple explanations such as the above will allow.[10] Certainly more work is required to elucidate the mechanisms of these reactions, in particular to explain the discrepancies mentioned below. Nevertheless, the simple concepts outlined above serve to rationalize much of the chemistry to be described.

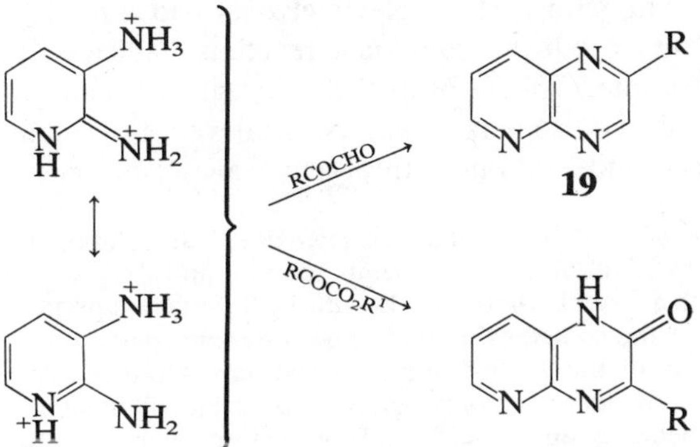

A. α-Keto Aldehydes

Diaminopyridines containing strong electron-withdrawing groups such as NO_2 or SO_3H fail to condense at all with pyruvaldehyde or glyoxal.[21] In contrast unambiguous synthesis of 6,8-diamino-2-methylpyrido[2,3-*b*]pyrazine[32] (see Section II4) indicated, as expected, that the major product of the reaction between 2,3,4,6-tetraminopyridine and pyruvaldehyde in dilute hydrochloric acid is the 3-methyl compound **21**. Similarly reactions of pyruvaldehyde in water or dilute acetic acid yield 3-methyl compounds from 2,3-diaminopyridine and its 5-bromo derivative.[10] The instability[10] of 3-methylpyrido[2,3-*b*]pyrazine (**22**) may explain an early comment that the reaction of 2,3-diaminopyridine with pyruvaldehyde yielded only intractable red pigments, possibly related to methylquinoxaline orange.[7] The successful preparation of 3-methylpyrido[2,3-*b*]pyrazine depends on rapid extraction and purification, as the crude material is very unstable.*

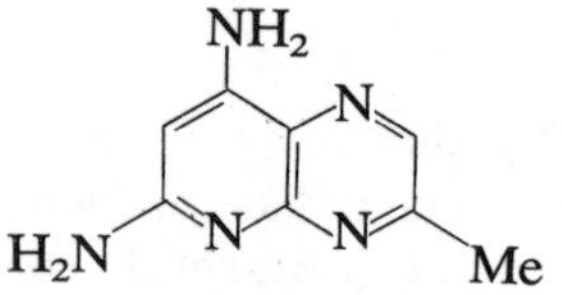

NH2 H2N N N N Me **21** N N N Me **22**

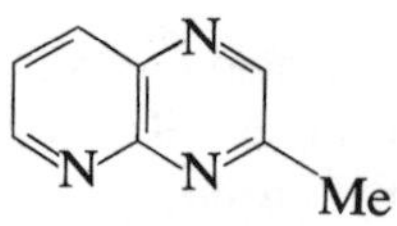

By analogy with the preparation of compound **21** it was suggested that the reaction of several complex 2,3-diaminopyridines with a variety of substituted phenylglyoxals resulted in 3-aryl derivatives.[26,33,34] More recently this has been confirmed by the unambiguous synthesis of the 2-aryl compound **23**.[27] This compound did not melt below 300°, whereas the product from the condensation of the appropriate 2,3-diaminopyridine with *p*-chlorophenylglyoxal had a m.p. 215–217°. PMR analysis of the products from such reactions indicated that the 3-aryl isomers constituted 75–100% of the recrystallized material. If necessary, separation of the isomers could be achieved by column chromatography.[18] The condensations with arylglyoxals were generally carried out

* *Preparation of* 3-*Methylpyrido*[2,3-*b*]*pyrazine.*[10] A solution of pyruvaldehyde bisulfite (from 15 ml of a 10% aqueous solution and 2.5 g sodium metabisulfite) was heated on a water bath for 0.75 hr with 1 g 2,3-diaminopyridine in 10 ml water. Basification of the cooled solution, followed by immediate ether extraction, gave on evaporation of the washed and dried extracts, a red crystalline residue. The residue was immediately chromatographed on a Florisil column, eluted with 1:1 chloroform–benzene, and crystallized from benzene–light petroluem (b.p. 60–80°) as pale-yellow cubes m.p. 106.5–107.5°.

in ethanol with the amine or its hydrochloride. It seems reasonable to suggest that the products from the reaction of 2,3-diaminopyridine in ethanol with the aldehydes **24**[35] and **25**[36] are 3-substituted compounds rather than the 2-aryl isomers proposed.

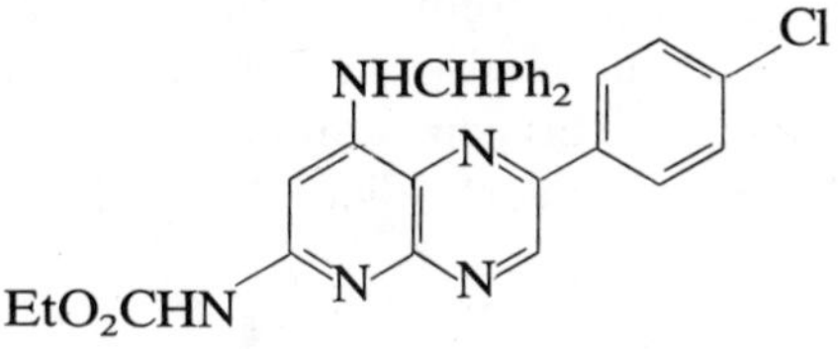

23

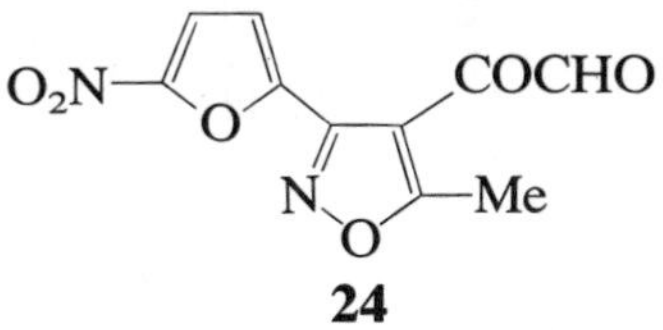

24 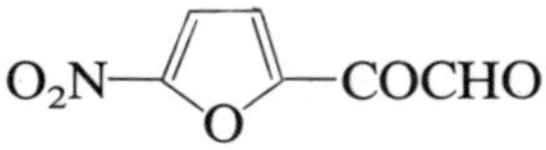 **25**

B. α-Keto Esters and Acids

By analogy with the synthesis of hydroxypteridines it was suggested that the ester **27** gave the 3-hydroxy compound **28** when treated with the diaminopyridine **26** in acetic acid. When the reaction was performed in *M* sulfuric acid, the isomer was obtained and assigned the structure **29**. In common with most hydroxyazines, such hydroxy compounds are more properly regarded as dihydro oxo derivatives, such as **30**, and are listed as such in the tables in this book.

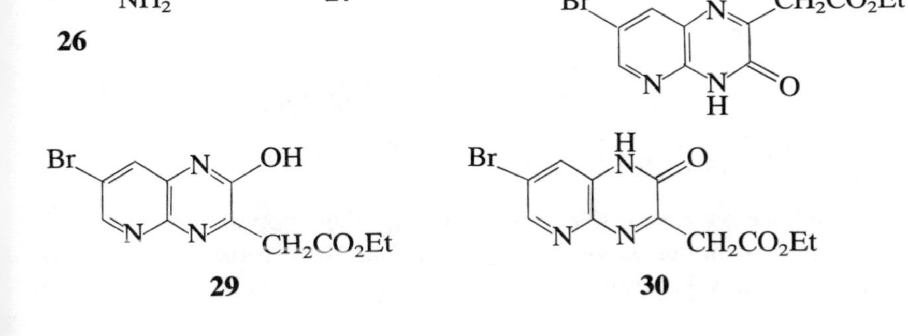

26 + $EtO_2CCOCH_2CO_2Et$ **27** $\xrightarrow{HOAc}$ **28**

29 **30**

Although the orientation of the products **28** and **29** has not been confirmed, they are consistent with the observation that ethyl mesoxalate (**31**) reacts under neutral conditions with 2,3-diaminopyridine to give the 3-oxo compound **32**, mixed with a small proportion of its 2-oxo isomer.[37] The orientation of the principal product was established by sequential hydrolysis, decarboxylation, and methylation of compound **32** to give the 4-methyl-3-oxo derivative **34**. This compound was synthesized unambiguously from 3-amino-2-methylaminopyridine (**33**) as described at the end of this section.

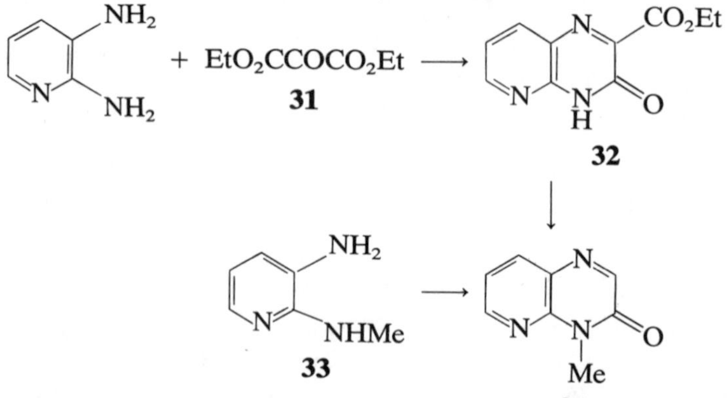

+ EtO_2CCOCO_2Et ⟶
31
32
33
34

The ethyl ester of pyruvic acid under neutral conditions reacts with 2,3-diaminopyridine to give the 2-methyl-3-oxo compound **35** (R=H), which decomposes at 240°. The isomer **36** (R=H), decomp. 270°, is obtained when the reaction is performed in dilute mineral acid.[7,10] It seems likely therefore that the product (decomp. 205°) obtained from 2,3-diaminopyridine using aqueous pyruvic acid[29] is not the 2-oxo compound as suggested but is probably a mixture. Analogous products (R=Br) are obtained from 2,3-diamino-5-bromopyridine.[10]

35 R=H or Br

36 R=H or Br

Ethyl acetopyruvate (**37**) reacts with 2,3-diaminopyridine in a variety of acidic conditions to yield mainly the 2-acetonyl-3-oxo compound **38**. This is readily separated from the much more soluble 2-oxo isomer, which is formed in small amounts.[10] The orientation of the major product was confirmed by its ready alkaline hydrolysis to the 2-methyl-3-oxo compound **35**. The route by way of the acetonyl derivative is recommended as the most satisfactory approach to the synthesis of compound **35**.[10] Moreover it is suggested that a product previously described[38] as 3-acetonyl-2-hydroxypyrido[2,3-*b*]pyrazine is in fact the 3-oxo isomer.[10] A study of the effect of pH on the isomer ratios in the product of the reaction of 5-bromo-2,3-diaminopyridine (**39**) with ethyl acetopyruvate[10] demonstrated that at pH 1 the only product was the 3-oxo compound **40**, whereas at pH 5 the isomer ratio was 70:30 in favor of the 2-oxo form **41**. The acidity constants of **39** are such that at pH 1 the diamine will be mainly monoprotonated, and at pH 5 about 50% will be in the neutral form.[10] Although the monoprotonated species evidently reacts as expected, the neutral molecule does not give the same product, which it should do if the simple rationalization applied.

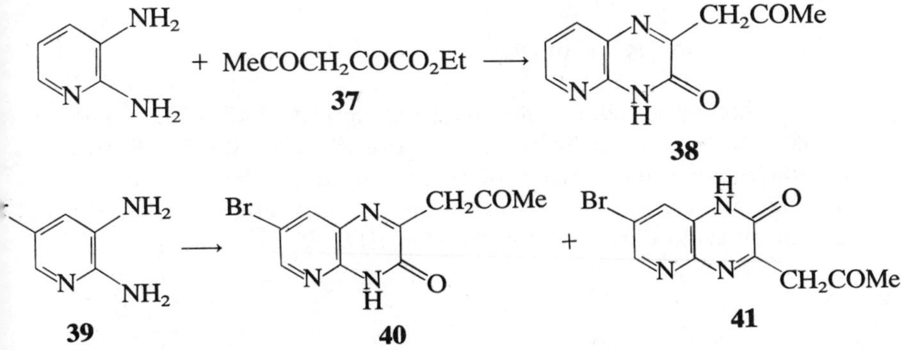

NH_2 NH_2 + $MeCOCH_2COCO_2Et$ **37** ⟶ CH_2COMe N N H O **38**

NH_2 NH_2 **39** ⟶ Br N N H O CH_2COMe **40** + Br H N O N N CH_2COMe **41**

Another example of the failure of the simple rationalization of the products from 2,3-diaminopyridine and α-keto acids is evidenced by the reactions of glyoxylic acid and its ethyl ester ($CHOCO_2Et$) with 2,3-diaminopyridine and its 5-bromo derivative.[10] Under a wide variety of experimental conditions the only product obtained is the 2-oxo compound **42**, identical with a sample prepared unambiguously (see Section II4). Only when the reaction mixture is sufficiently acidic to diprotonate the diaminopyridine is any of the 3-oxo compound **43** obtained. These anomalies may be explained, it is suggested,[10] by the involvement of intermediates such as **44**. From the foregoing discussion it seems reasonable tentatively to assign the 3-oxo structure to the products **45** and **46** from the condensation in ethanolic acetic acid of 2,3-diaminopyridine with cinnamoylformic acids ($ArCH{=}CHCOCO_2H$) and benzoylpyruvic acid ($PhCOCH_2COCO_2H$).[39]

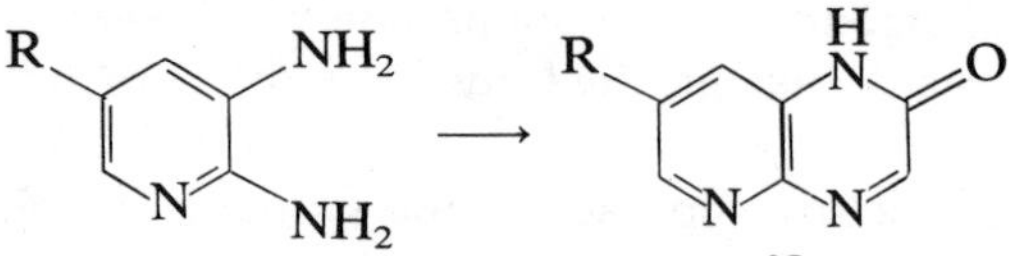

R = H or Br

42

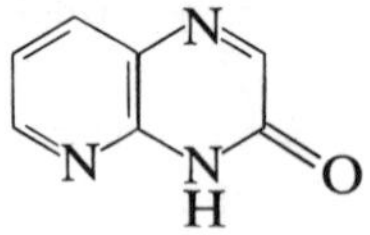

43

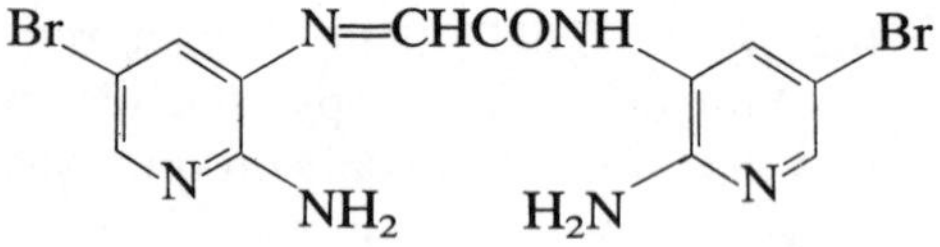

44

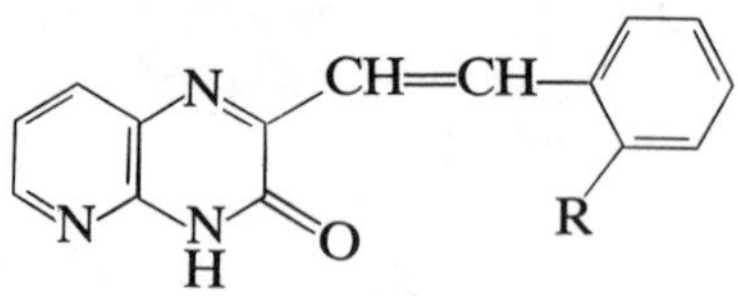

45 R = H or NO_2

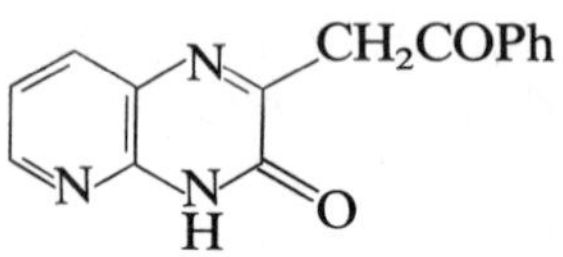

46

Surprisingly it is claimed that the 2-oxo derivative **48** is the product of the condensation of dehydro-L-ascorbic acid (**47**) with 2,3-diamino-5-bromopyridine in the presence of phenylhydrazine.[40] No proof is given of the orientation of **48**, which must be in doubt together with that of the derived pyrazolo derivative (see Section III7).

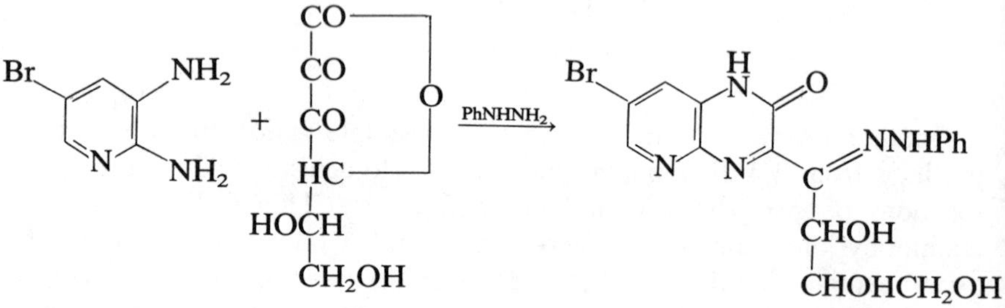

47

48

Doubts over the orientation of compounds such as those above can frequently be resolved by oxidation of the side chain, decarboxylation of the resulting carboxylic acid, *N*-methylation, and comparison with the *N*-methyl oxo compounds **50** and **51**. These compounds have been prepared unambiguously from methylamino aminopyridines by reaction with diethyl mesoxalate (EtO_2CCOCO_2Et),[37,41] followed by hydrolysis and decarboxylation. Thus 2-amino-3-methylaminopyridine (**49**) gave the

1-methyl-2-oxo compound **50** m.p. 223–224°. Similarly 3-amino-2-methylaminopyridine yielded the isomer **51**, m.p. 115–117°.

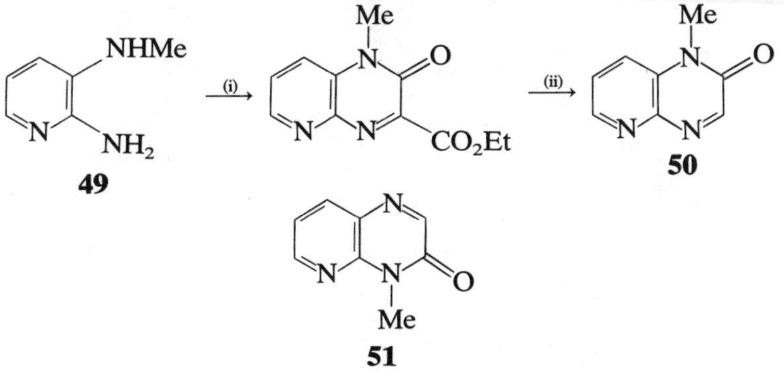

Reagents (i) EtO_2CCOCO_2Et
(ii) OH^-, heat

C. *Other Dicarbonyl Compounds*

3-Amino-2-anilinopyridine (**52**) reacts with α-diketones containing an acetyl group to give the methylene compounds **53**.[42]

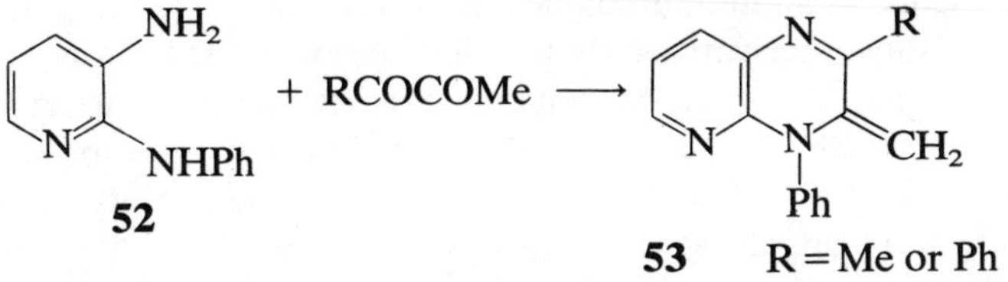

53 R = Me or Ph

Early work[43,44] on the reaction of 2-substituted amino-3-amino-pyridines (**54**) with alloxan (**55**) in neutral solution has been corrected.[37] The initial yellow product from the condensation is a ureide of formula **56**, which readily rearranges to the colorless spirohydantoin **57** on treatment with acid or base. The isomeric pyridine **49** yields a second series of ureides and hydantoins. When 2,3-diaminopyridine itself is used, a mixture of products is obtained in which the 3-oxo compound **56** (R = H) predominates over its 2-oxo isomer. Alkaline hydrolysis followed by decarboxylation and methylation served to confirm the orientation[37] (see Section II2B). If the reaction with alloxan is carried out in alkaline solution, the only product obtained is the 3-oxo-2-carboxylic acid.[45]

3. From 2-Amino-3-nitrosopyridines

The reaction of 2-amino-3-nitrosopyridines with compounds containing an activated methylene group permits unambiguous synthesis of various derivatives of pyrido[2,3-*b*]pyrazine. For example, the pyridine **58** reacts in the presence of sodium ethoxide with a variety of arylacetonitriles and cyanoacetic acid derivatives to provide various 2-substituted 3-amino compounds (**59**).[9,46,47] Diethyl malonate reacts similarly to give the 2-carboxylic acid **60**, its ester being presumably hydrolyzed in the alkaline reaction conditions.[7,9] Ethyl acetoacetate[7] yields the 2-acetyl-3-oxo compound **61**, and acetylacetone[10] provides the 2-acetyl-3-methyl compound **62**. The latter condensation proceeds poorly in ethanolic sodium ethoxide, but heating the nitroso compound with acetylacetone under reflux in pyridine gives a 59% yield of the product **62**.[10]

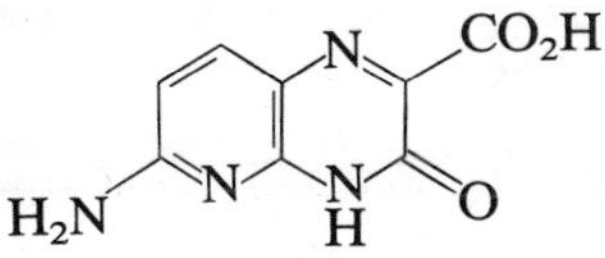

60

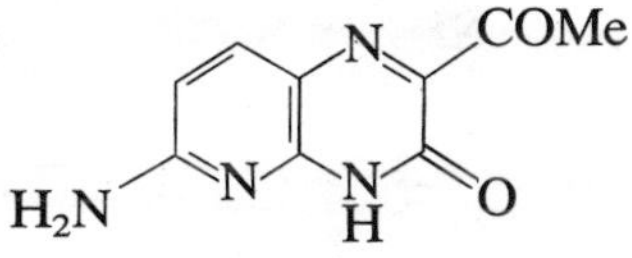

61

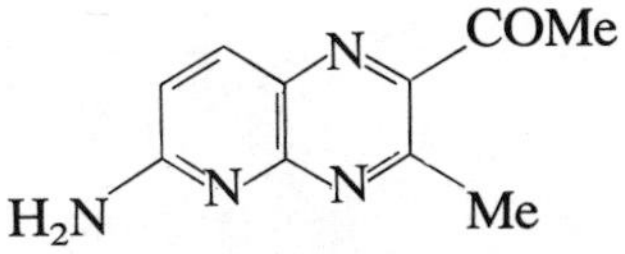

62

4. From 3-Nitropyridine Derivatives

A useful route for the unequivocal synthesis of complex derivatives of this ring system involves cyclization of compounds of general formula **63**. These are readily prepared from the appropriate 2-chloro-3-nitropyridine and an aminoketone. If the requisite amino ketone is too unstable, the corresponding amino alcohol may be used and oxidized to the ketone after reaction with the chloropyridine.

Compounds of formula **63** are readily cyclized by reduction of the nitro group with Raney nickel and hydrogen. The amine thus formed condenses intramolecularly, with the carbonyl function providing the 3,4-dihydro system **64**. This may be smoothly oxidized with iodine[6] or an acetone solution of potassium permanganate.[27,32,48,49] This route has been used for the unambiguous synthesis of compounds **65**,[6] **66**,[32] **67**,[27] and **68**.[48,49] Of particular interest are compounds **66** and **67**, which have been used to confirm the orientation of products from the reaction of polyaminopyridines with pyruvaldehyde[30,32] and *p*-chlorophenylglyoxal.[27]

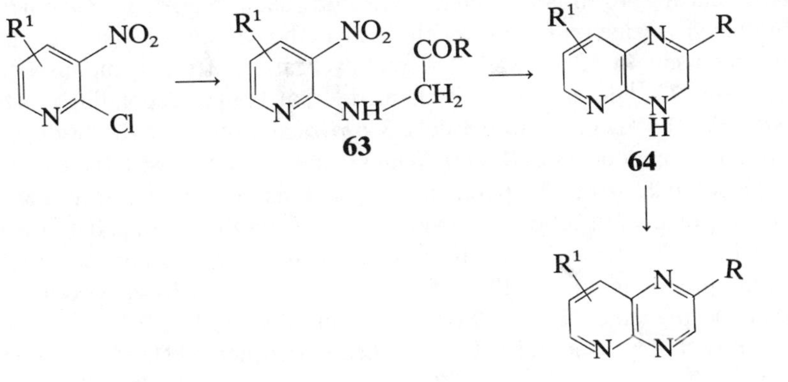

63

64

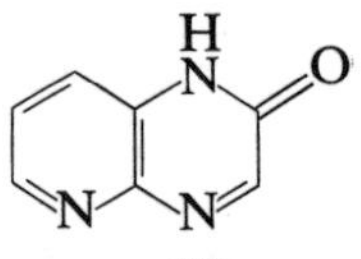

65

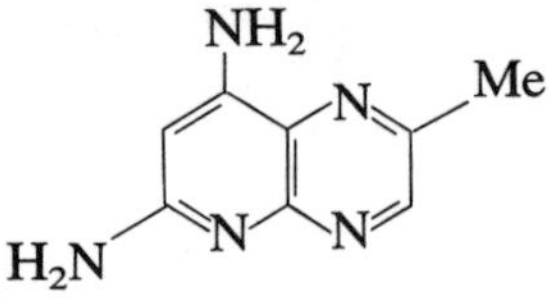

66

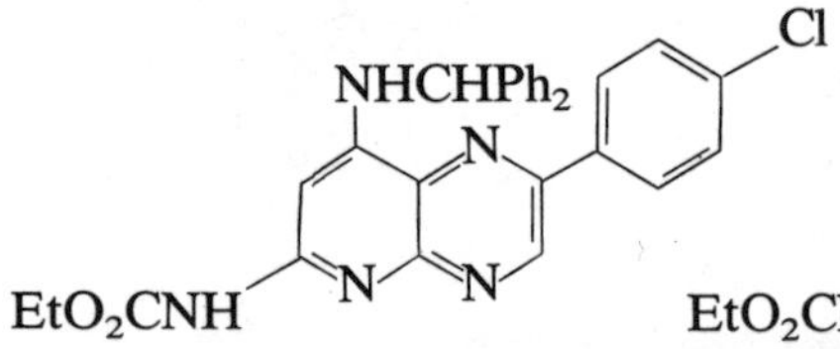

67

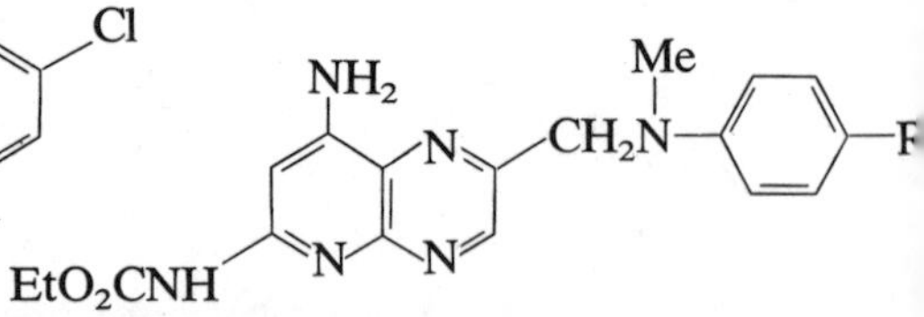

68 R = H or CO_2Me

Partial reduction of the nitro group in compounds **69** with either sodium borohydride or hydrogen over a catalyst of palladium on charcoal results in the 1-hydroxy-2-oxo compounds **70**.[50]

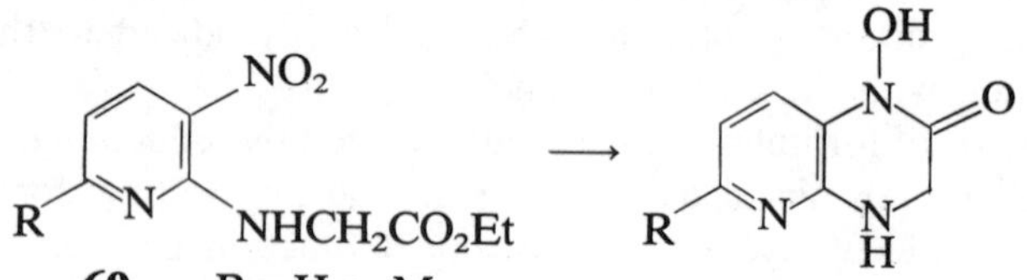

69 R = H or Me

70

5. From Pyrazine Precursors

Only recently has the synthesis of the pyrido[2,3-*b*]pyrazine ring system starting from substituted pyrazines been reported. 2-Amino-6-methoxypyrazine on reaction with diethyl ethoxymethylenemalonate gave the pyrazine **71** (R = OMe). Thermal cyclization of this compound gave the ester **72**.[51] In contrast it is claimed[10] that the corresponding reaction with **71** (R = H) gives no pyrido[2,3-*b*]pyrazine but only the bridgehead nitrogen compound **74** (R = H). While this is consistent with the electron-deficient nature of the pyrazine ring and its consequent resistance to electrophilic substitution, the evidence on which the claim is based is not conclusive, particularly as it is not clear if the hydrogen-bonded form **72** would predominate over the γ-pyridone form **73**. PMR was used to show that the pyrazine **71** (R = OMe) gave only the product **72** or **73**. More recently this cyclization has been studied with other substituted pyrazines, and both pyrido[2,3-*b*]pyrazines and pyrazino[1,2-*a*]pyrimidines have been obtained.[63]

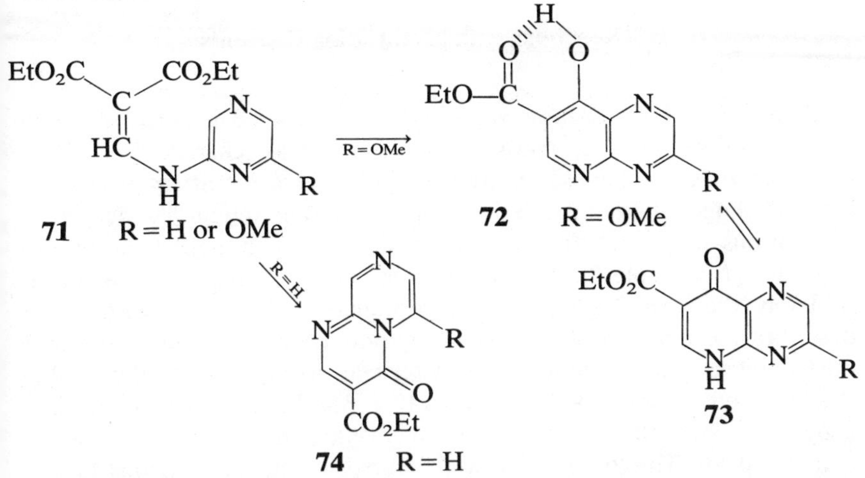

71 R = H or OMe **72** R = OMe **73** **74** R = H

Use of the aldehyde **75** avoids condensation directly onto the pyrazine ring. Reaction of this compound with the activated acetonitriles **76** at room temperature in the presence of *N*-methylpiperidine gave the condensation products **77**, which on treatment with alkali gave the 6-amino compound **78**. It was also found that pteridine (**79**) gave pyrido[2,3-*b*]pyrazines directly on treatment with the acetonitriles.[52]

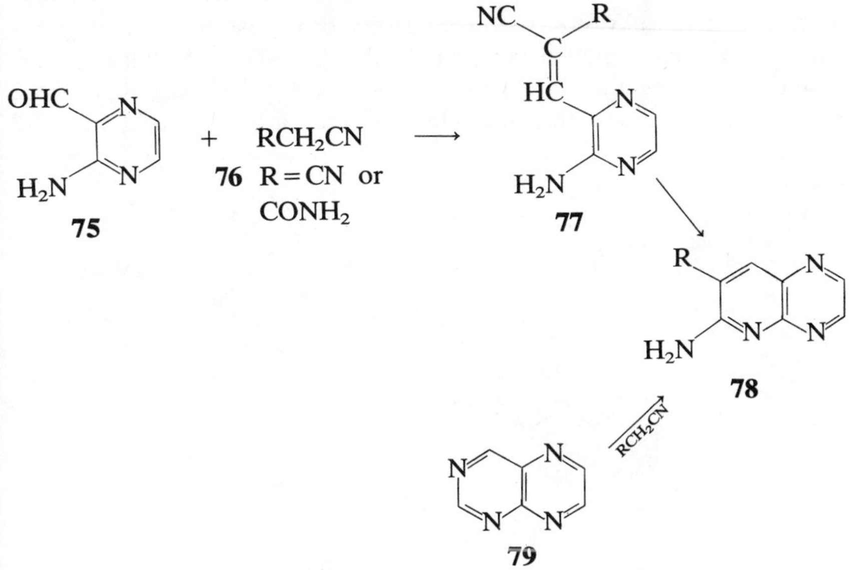

75 **76** **77** **78** **79**

III. Reactions

1. Reactions with Methylating Reagents

Methylation of 7-bromo-2,3-diphenylpyrido[2,3-*b*]pyrazine with methyl sulfate gave a methosulfate which was characterized as a methotriiodide.[17] The use of methyl iodide in acetonitrile gives only monoalkylated quaternary derivatives of the ring system.[10] The parent and various simple substituted compounds have been methylated by this procedure (see Table 2). The site of monomethylation is not clear but is probably the same as the site of protonation, *N*-5. A recent paper describes the reaction of 5-alkylated pyrido[2,3-*b*]pyrazinium salts with nucleophiles. The products are described as 1,2,3,4-tetrahydro-2,3-disubstituted compounds.[64] Treatment of 3,4-dihydro-3-oxopyrido[2,3-*b*]pyrazine with ethereal diazomethane gave a 15% yield of the 4-methyl compound **80**. The low yield was attributed to the simultaneous formation of the .*O*-methylated product **81,** which was not isolated.[37] The *N*-methylated compound has a surprisingly low melting point (115–117°) when compared with the 1-methyl isomer (**82**), which melts at 223–224°. The structures of both *N*-methyl oxo compounds have been confirmed by unambiguous synthesis (see Section II2B). It would be interesting to compare their properties with the corresponding *O*-methyl isomers which should be readily available from the corresponding chloro compounds. Methyl sulfate alkylates 6-amino-3,4-dihydro-3-oxopyrido[2,3-*b*]pyrazine to give the 4-methyl compound **83**. The related 6-hydroxy compound (**84**) is resistant to further methylation by this reagent. A more recent paper[60] shows that alkylation of the carboxylic acid from **72**

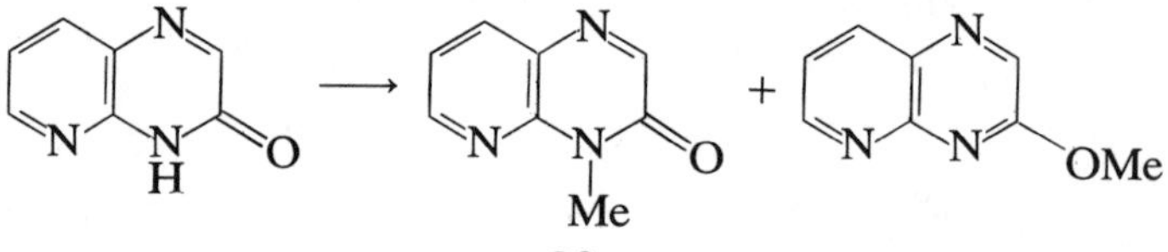

80 **81**

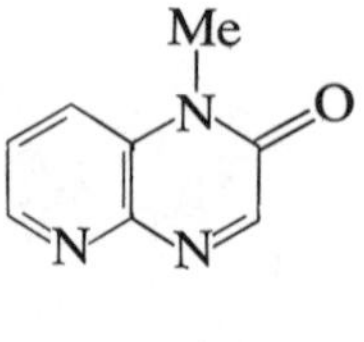

82 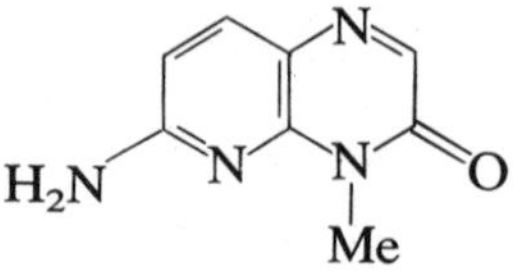 **83**

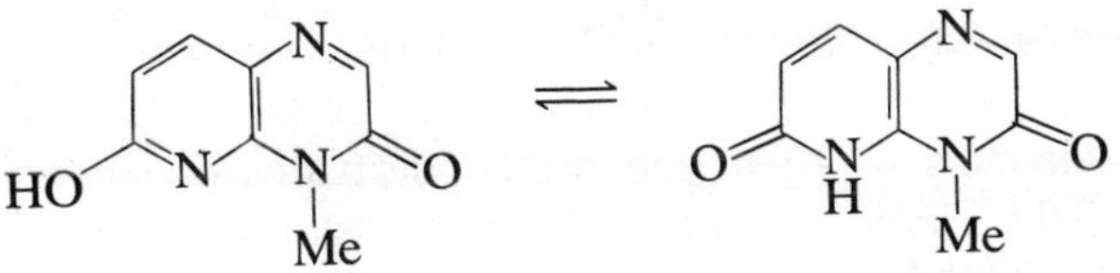

84

(R = OMe) with methyl iodide occurs at the 5-position. If ethyl iodide is used the 3-methoxy group is converted to ethoxy during *N*-ethylation.

2. Reactions of Active Methyl Groups

As in the quinoxaline series, compounds with a methyl group in the 2- or 3-position can undergo electrophilic substitution on the methyl group.[10] The "active" methyl groups in compounds such as **85** and **87** undergo Claisen reactions with ethyl oxalate in refluxing ethanolic sodium ethoxide to give the expected ethoxalyl condensation products **86** and **88**. The dimethyl compound **89** reacts at only one methyl with simultaneous reaction at the 6-acetamido group. The product is assumed to be the 3-ethoxalylmethyl derivative (**90**) rather than the 2-ethoxalylmethyl isomer. It seems reasonable on electronic grounds that the 3-methyl position will be more activated than the 2-methyl to electrophilic substitution.[10] More work is needed to confirm this point.

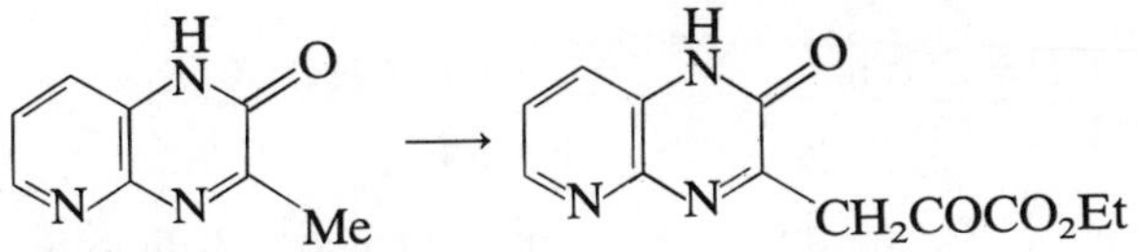

85 → **86** (CH_2COCO_2Et)

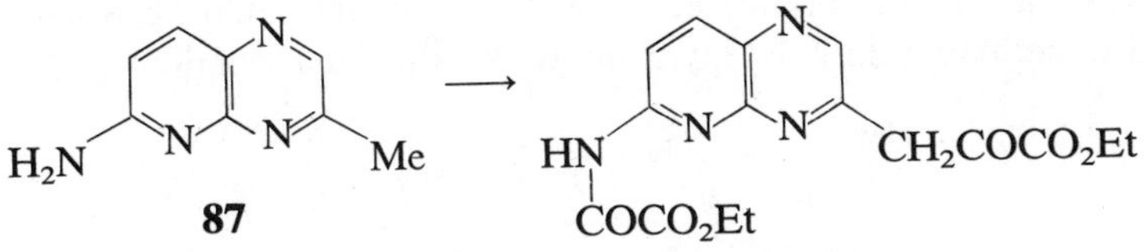

87 → **88** (CH_2COCO_2Et; $HN–COCO_2Et$)

89 → **90** (CH_2COCO_2Et; $HN–COCO_2Et$)

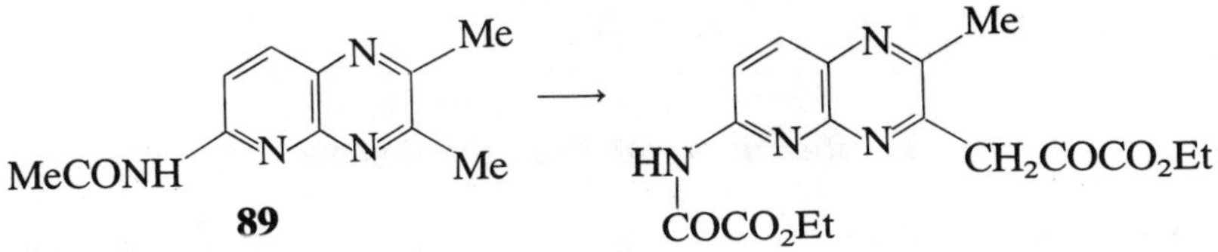

The 2-or 3-methyl derivatives also undergo aldol condensation in the presence of acidic or basic catalysts.[10] The products obtained are either secondary alcohols or their dehydrated derivatives, depending on the structure of the aldehyde used in the condensation. In acetic anhydride only nitrobenzaldehydes gave any products. 2,3-Dimethylpyrido[2,3-*b*]pyrazine and its 7-bromo derivative gave styryl derivatives assumed to be of type **91**. *p*-Nitrobenzaldehyde gave the alcohols **92**. Again substitution at the 3-position is assumed. In ethanolic sodium ethoxide solution, 2,3-dimethylpyrido[2,3-*b*]pyrazine reacted with benzaldehyde to give the bis-styryl derivative **93**. Chloral in pyridine gave the diol **94**.

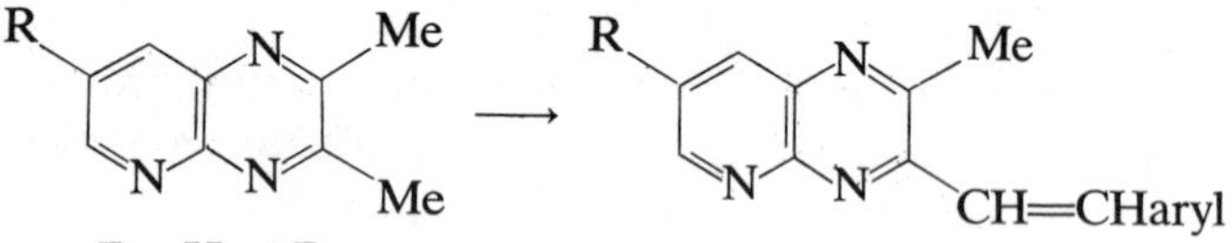

R = H or Br

91 (Me; CH=CHaryl)

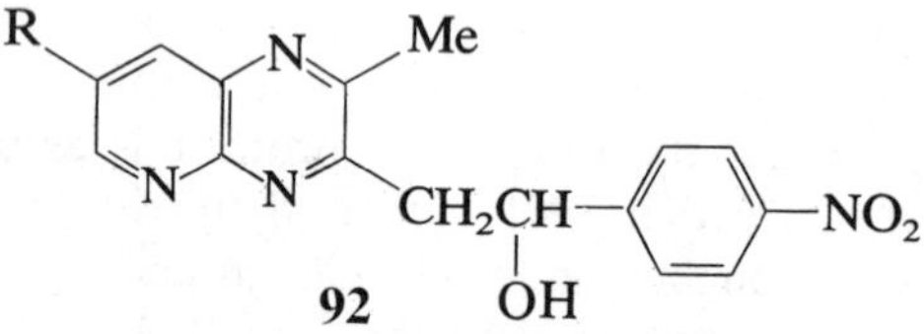

92 (Me; $CH_2CH(OH)-C_6H_4-NO_2$)

93 (CH=CHPh; CH=CHPh)

94 ($CH_2CHOHCCl_3$; $CH_2CHOHCCl_3$)

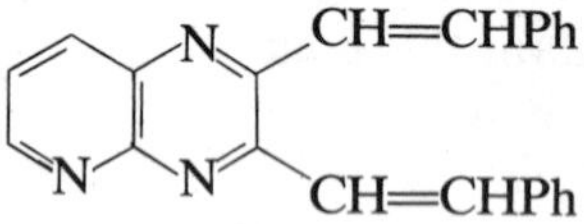

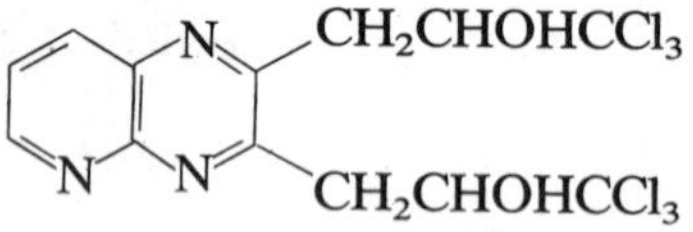

One but not both of the methyl groups in 7-bromo-2,3-dimethylpyrido[2,3-*b*]pyrazine reacts with amyl nitrite in the presence of a trace of acid to yield an oxime formulated as **95**. No product could be isolated from 2,3-dimethylpyrido[2,3-*b*]pyrazine under the same conditions.[10]

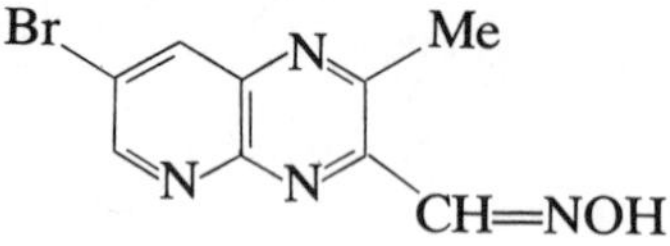

95 (Br; Me; CH=NOH)

3. Reactions with Oxidizing Agents

Although *N*-oxides are less easily formed in this ring system than they are in the quinoxaline series,[21] 7-bromopyrido[2,3-*b*]pyrazine does form a mono-*N*-oxide using hydrogen peroxide in acetic acid at 70°. It is not certain, however, which position is oxidized. These reaction conditions have no effect on 2,3-dimethylpyrido[2,3-*b*]pyrazine.[17] Treatment of the parent molecule with hydrogen peroxide in 2.5 *M* sulfuric acid at 20° results in the formation of the 6-oxo compound rather than an *N*-oxide.[6] As azine *N*-oxides readily rearrange to α-oxo compounds, this observation suggests that *N*-5 is the most likely site of oxidation, which fits well with our conclusions on protonation and methylation.

Attempted oxidation of the 3-oxo compound **96** with potassium permanganate resulted only in ring breakdown.[7] Nevertheless a solution of this reagent in acetone is commonly used to oxidize compounds such as the dihydro derivative **97** to the aromatic system without further oxidation occurring.[48,49] Oxidation of the tetrahydro oxo compound **98** with potassium permanganate in 2 *M* sodium hydroxide solution at room temperature gives the 2-oxo compound **99**.[10] This compound is sensitive to oxidation, and the use of iodine has been recommended for this conversion.[6] If alkaline potassium ferricyanide is used, the product formed is the dioxo derivative **100**. The dioxo compound is also obtained from 1,2-dihydro-2-oxo- or 3,4-dihydro-3-oxopyrido[2,3-*b*]pyrazine under the same conditions.[6] Recently ferric chloride has been used to aromatize a series of dihydropyrido[2,3-*b*]pyrazines.[65]

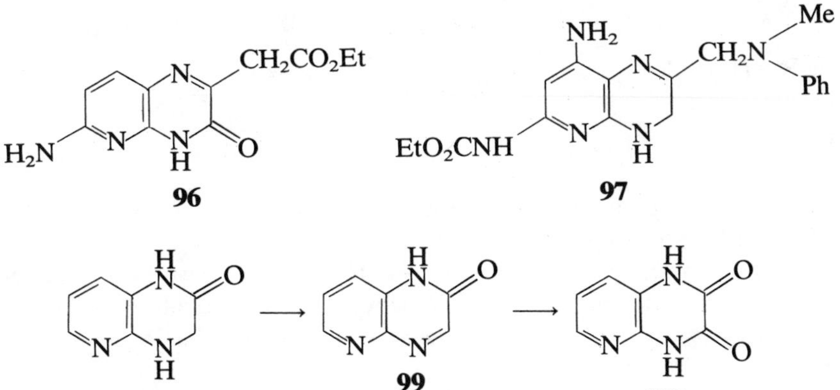

96 **97**

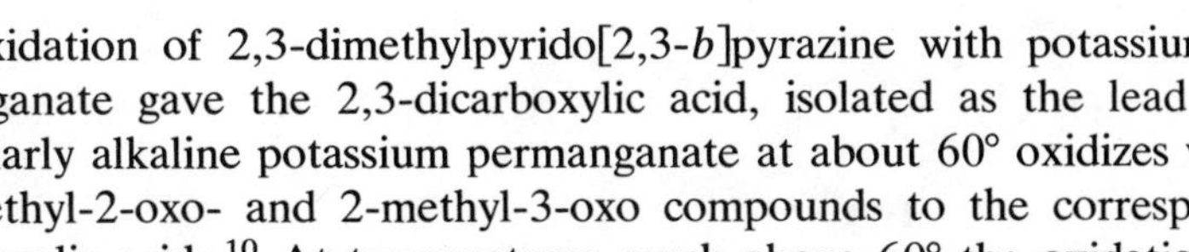

98 **99** **100**

Oxidation of 2,3-dimethylpyrido[2,3-*b*]pyrazine with potassium permanganate gave the 2,3-dicarboxylic acid, isolated as the lead salt.[19] Similarly alkaline potassium permanganate at about 60° oxidizes various 3-methyl-2-oxo- and 2-methyl-3-oxo compounds to the corresponding carboxylic acids.[10] At temperatures much above 60° the oxidation continues, resulting in the dioxo compound being obtained.

4. Reactions at Electron-Deficient Carbon Atoms

Only limited work has been done on the displacement of halogen by nucleophiles in this ring system, partly because the requisite displacements have usually been carried out on the pyridine precursors of the ring system. Acid hydrolysis of 6-chloropyrido[2,3-*b*]pyrazine provides the 6-oxo compound,[22] and the 2,3-dichloro compound is reported to give 84% of the 3-ethoxy-2-oxo derivative on treatment with ethanolic potassium hydroxide. In contrast the 2,3,7-trichloro compound gave 7-chloro-2,3-diethoxypyrido[2,3-*b*]pyrazine, also in 84% yield.[53] No evidence is given to show that the product from the 2,3-dichloro compound is not the 2-ethoxy-3-oxo isomer. Both 2,3-dichloro- and 2,3,7-trichloropyrido[2,3-*b*]pyrazine have been reacted with the piperidine **101** to give the tetracyclic compounds **102** or their isomers **102a**.[54]

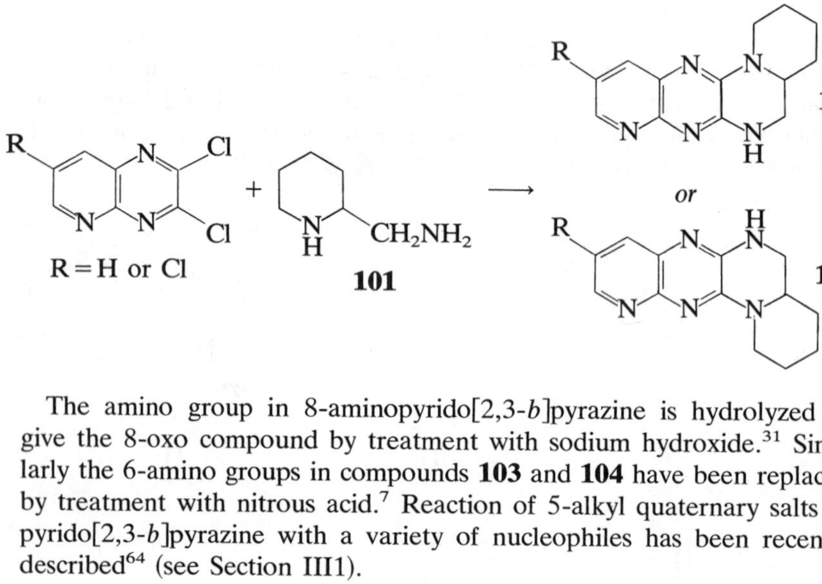

R = H or Cl

101

102

or

102a

The amino group in 8-aminopyrido[2,3-*b*]pyrazine is hydrolyzed to give the 8-oxo compound by treatment with sodium hydroxide.[31] Similarly the 6-amino groups in compounds **103** and **104** have been replaced by treatment with nitrous acid.[7] Reaction of 5-alkyl quaternary salts of pyrido[2,3-*b*]pyrazine with a variety of nucleophiles has been recently described[64] (see Section III1).

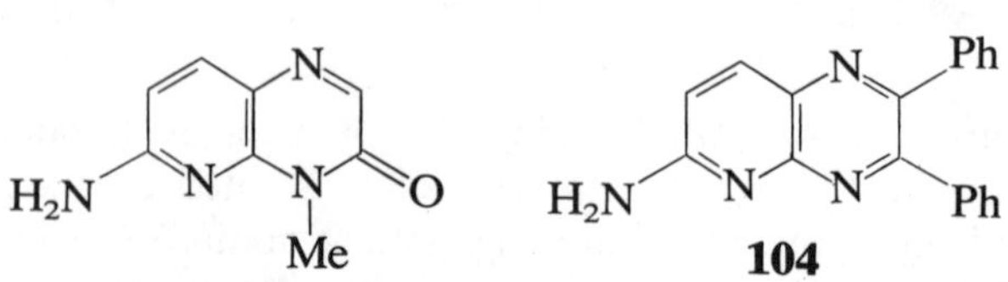

103 **104**

A recent study of the reaction of pyrido[2,3-*b*]pyrazines with organomagnesium compounds showed that the products of reaction depend on the substitution pattern in the original heterocycle.[65] Pyrido[2,3-*b*]pyrazine and its 7-bromo derivative gave 1,2,3,4-tetrahydro-*trans*-2,3-disubstituted products, whereas 2,3-diphenylpyrido[2,3-*b*]pyrazine and its 7-bromo compound afforded 5,6-dihydro-6-substituted derivatives. Yields varied from 30–70%. 3-Phenylpyrido[2,3-*b*]pyrazine gave only 7–10% yields of 1,2-dihydro-2-substituted products and the 7-bromo-3-phenyl compound with methylmagnesium iodide gave a derivative tentatively identified as 7-bromo-5,8-dihydro-8-methyl-3-phenylpyrido[2,3-*b*]pyrazine (**105**). This compound was obtained in 55% yield.

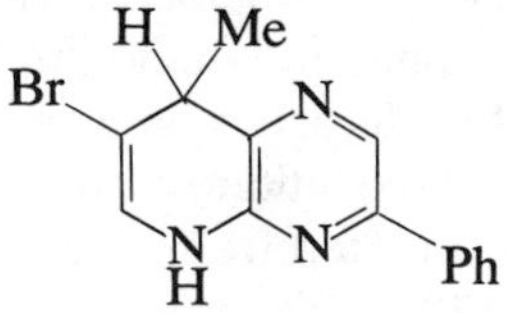

105

Very little work has been done on the preparation of chloro compounds from the corresponding oxo derivatives. In the pyridine ring this would be largely superfluous as the requisite chloropyridines themselves are readily available. However in the pyrazine ring the oxo compounds are more readily synthesized and could be used as intermediates to 2- and 3-chloro derivatives. Neither monochloro compound has yet been prepared. Dioxo compounds (**106**, R = H, Cl) have however been converted in high yield to the corresponding dichloro compounds (**107**, R = H, Cl) by treatment with phosphorus oxychloride.[53,54] A mixture of this reagent with phosphorus pentachloride converted **106** (R = Br) into its dichloro compound.[10] However the use of this mixture with **106** (R = H) resulted in the introduction of three chlorine atoms with the formation of **107** (R = Cl).[10] The 2- and 3-oxo compounds **108** and **110** are readily converted into the corresponding thioxo derivatives **109** and **111** by treatment with phosphorus pentasulfide in pyridine.[10]

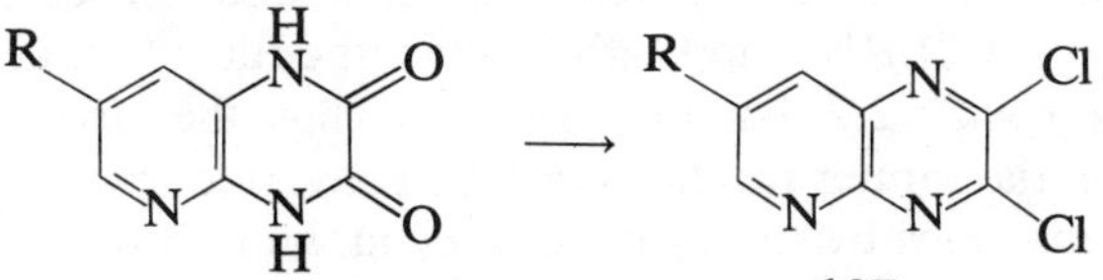

106 R = H, Cl, or Br → **107**

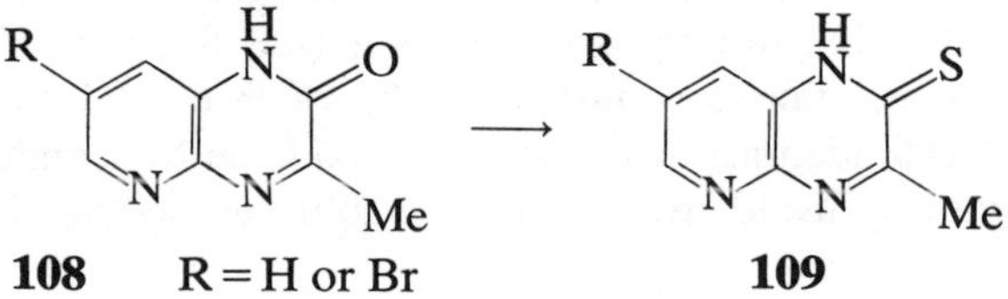

108 R = H or Br → **109**

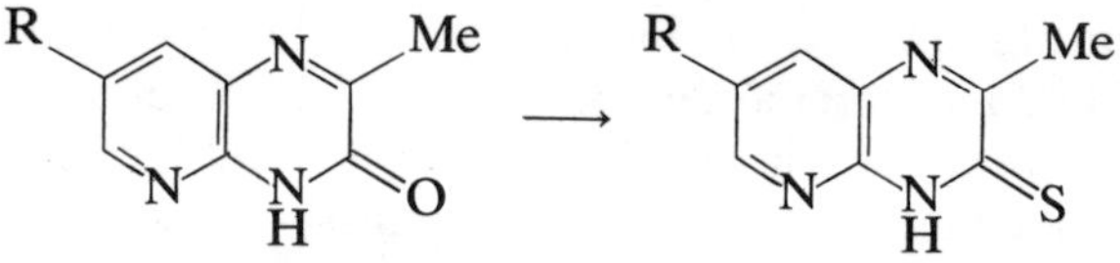

110 R=H or Br **111**

5. Decarboxylation

Decarboxylation of the 3,6-diamino-2-carboxylic acid is readily achieved by heating with copper and quinoline.[9] Simple heating above the melting point serves to decarboxylate various 3-oxo-2-carboxylic acids.[7,37] During the decarboxylation of the acid **112** a red by-product was formed. This was also isolated from acid or alkaline hydrolysis of the ethyl ester of the acid **112**. This product was formulated as **113** by analogy with indigo. A similar product was obtained from the 2-oxo-3-carboxylic acid.[37]

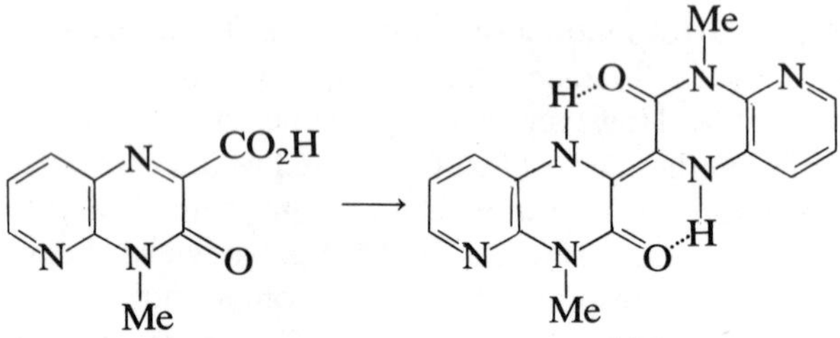

112 **113**

6. Reactions with Reducing Agents

The thioxo compounds **109** and **111** are readily desulfurized by shaking with Raney nickel.[10] This reaction provides a link for the establishment of orientations of compounds obtained by condensation reactions of α-keto esters ($RCOCO_2R^1$) and the corresponding α-keto aldehydes (RCOCHO) with 2,3-diaminopyridines. Thus the oxo compounds obtained from the former reactions may be converted into the products from the latter reactions by successive treatment with phosphorus pentasulfide and Raney nickel. For example, the compound **114**, obtained as the only product from the condensation of 2,3-diamino-5-bromopyridine with ethyl pyruvate in strongly acid solution (see Section II2B), has been converted to the thioxo compound **115**, which on desulfurization yields compound **116**, identical with that obtained from the condensation of the pyridine with pyruvaldehyde in neutral solution (see Section II2A).

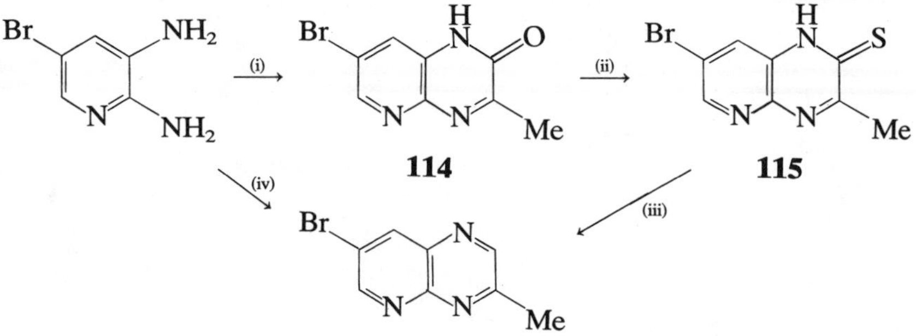

Br, NH_2, NH_2 →(i) **114** →(ii) **115** →(iii) **116**; (iv) → **116**

Reagents (i) $MeCOCO_2Et$
ii) P_4S_{10}
(iii) Raney Ni
(iv) MeCOCHO

Hydrogenolysis of 7-bromopyrido[2,3-*b*]pyrazine over palladized strontium carbonate gave the parent heterocycle. Similar treatment provided the 2-methyl-3-oxo compound from its 7-bromo derivative, without reduction of the pyrazine ring.[7] A variety of other 7-bromo compounds have also been dehalogenated by this technique.[10] On the other hand, catalytic hydrogenation has also been used to reduce 2- and 3-oxo compounds to the tetrahydro derivatives **117** and **118**. Reduction of the

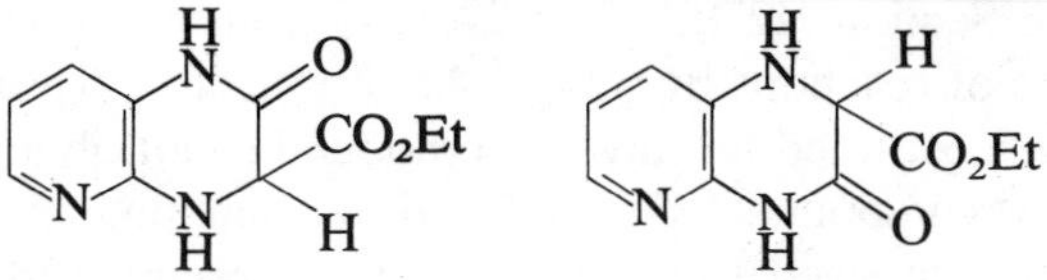

117 **118**

ring generally requires a more active catalyst than that used for debromination. Compounds **117** and **118** were obtained using palladium on charcoal.[41] However, the debromination conditions (5% palladium on strontium carbonate in *M* sodium hydroxide under hydrogen at room temperature and pressure) are sufficient to reduce the 2-oxo compound **119** to the tetrahydro derivative **120**. The 3-oxo compound **121** remains unaffected under these conditions.[10]

Sodium amalgam and aluminum amalgam have been investigated as reagents for the reductions of 2- or 3-oxopyrido[2,3-*b*]pyrazines.[10] Oxo compounds **119** and **121** are both reduced to tetrahydro compounds by aluminum amalgam. Sodium amalgam reduces the dioxo compounds **122** and **123** to the tetrahydro derivative **120**, possibly by way of the 2-oxo

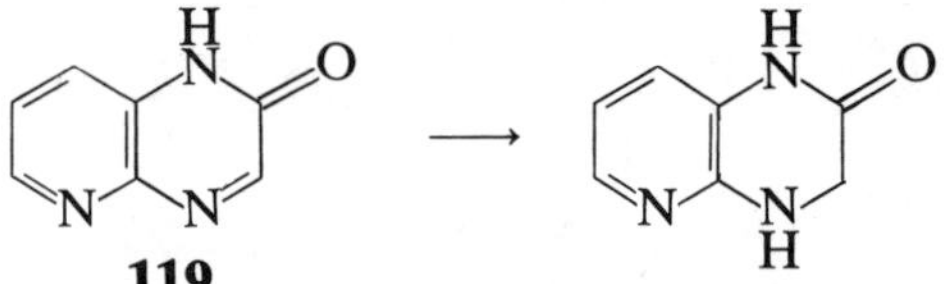

119 → **120**

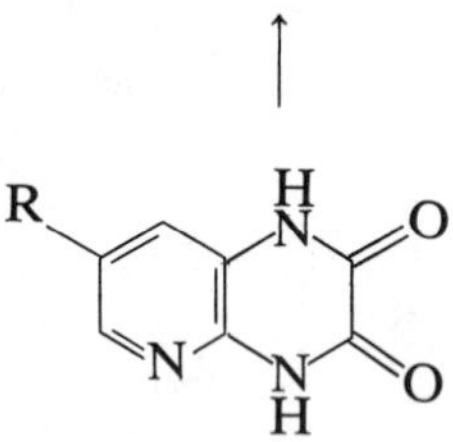

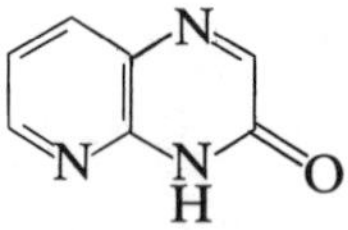

121

122 R=H
123 R=Br

compound **119**, which is also reduced by this reagent. The greater ease of reduction of 1,2-dihydro-2-oxopyrido[2,3-*b*]pyrazine (**119**) compared to the 3-oxo isomer (**121**) has been explained by the greater polarity of the 3,4 double bond in **119** compared to the 1,2 double bond in **121**.[10]

Reduction of the acetonyl compound **124** with aluminum amalgam in cold alkali gave the tetrahydro compound **125**, which on boiling in alkali yielded the 3-oxo compound **121**. This route is recommended as the best available to this compound, whose preparation by condensation methods is difficult (see Section II2B).[10] The 2-oxo-3-acetonyl isomer undergoes the same series of reactions by way of the tetrahydro derivative **126**. The corresponding substituted tetrahydro compound is not obtained when the ethoxalylmethyl compound **127** is reduced by aluminum amalgam. In this case the reaction proceeds further, yielding compound **128**.

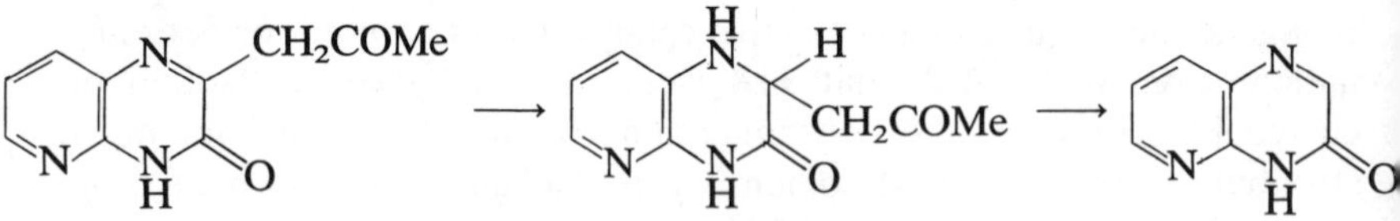

124 → **125** → **121**

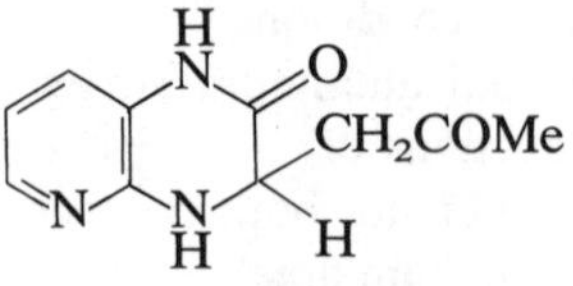

126

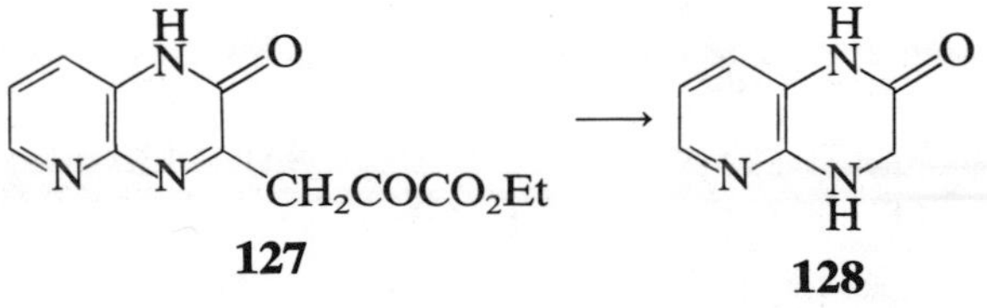

Reduction of 2,3-dimethylpyrido[2,3-*b*]pyrazine with lithium aluminum hydride gave the (±) *cis*-dimethyl-1,2,3,4-tetrahydro compound,[55] whose stereochemistry and conformation were established by PMR.[56]

Recently electrochemical reduction of pyrido[2,3-*b*]pyrazines has been shown to give 1,2-(or possibly 3,4-)dihydro derivatives.[61] These partially reduced compounds are isolated in good yield under controlled conditions. It is suggested that 1,4-dihydro compounds are initially formed, and these rapidly isomerize before they can be isolated.

7. Intramolecular Cyclizations

Although the side chain reactions in this series are similar to those in quinoxaline chemistry,[37] the cyclization of ureides such as compound **129** to the hydantoin **130** occurs more readily in the pyrido[2,3-*b*]pyrazine series. Boiling dilute hydrochloric acid converts **129** into acid-stable **130**, whereas it hydrolyzes the quinoxaline analogue of **129** to the corresponding carboxylic acid. Compounds of both series are readily hydrolyzed to the acid by aqueous alkali.[37]

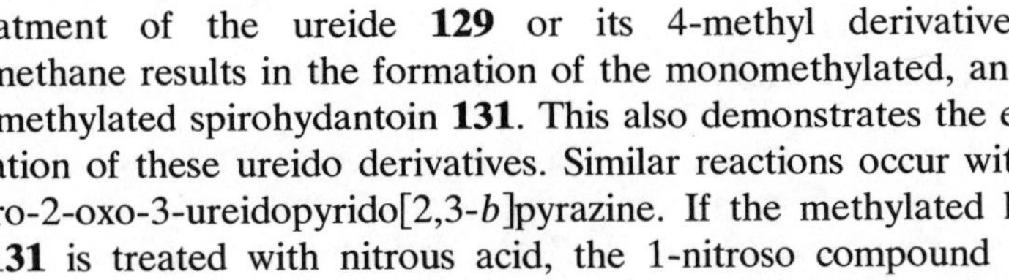

Treatment of the ureide **129** or its 4-methyl derivative with diazomethane results in the formation of the monomethylated, and then the dimethylated spirohydantoin **131**. This also demonstrates the ease of cyclization of these ureido derivatives. Similar reactions occur with 1,2-dihydro-2-oxo-3-ureidopyrido[2,3-*b*]pyrazine. If the methylated hydantoin **131** is treated with nitrous acid, the 1-nitroso compound **132** is formed. On alkaline hydrolysis this yields the 2-methylamino compound **133**. The corresponding reactions with the 2-oxo isomer of **131** failed as

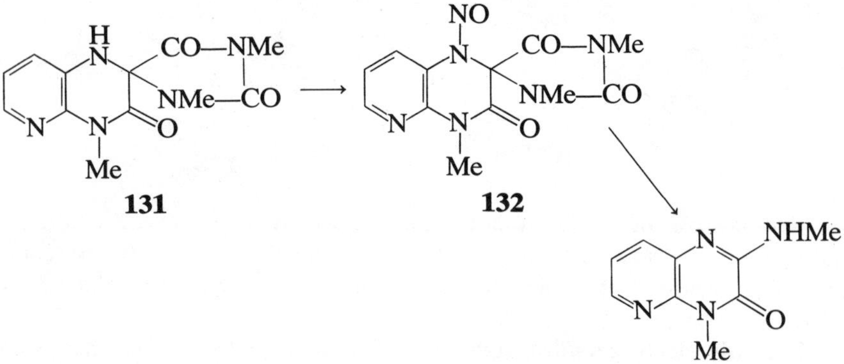

131 **132** **133**

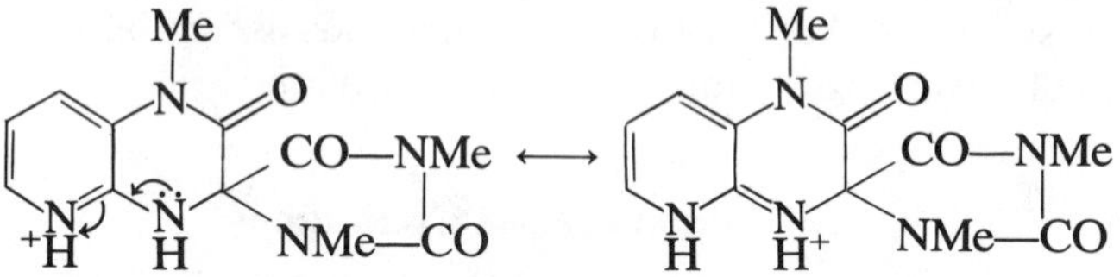

134

the NH group was resistant to nitrosation. This difference in reactivity was ascribed to amidinium-type stablization of the conjugate acid **134** of the 2-oxo isomer of hydantoin **131**.[37]

Treatment of the ureides described earlier with boric acid results in cyclization with loss of water to form azaalloxazines (**136**).[45,57] It has been shown[45] that the compound originally thought to be the ureide **135** (R = Cl)[57] is in fact the hydrate of the cyclized compound **136** (R = Cl). The only simple way to differentiate between the various isomeric structures is by use of ultraviolet spectra in acid solution.[45]

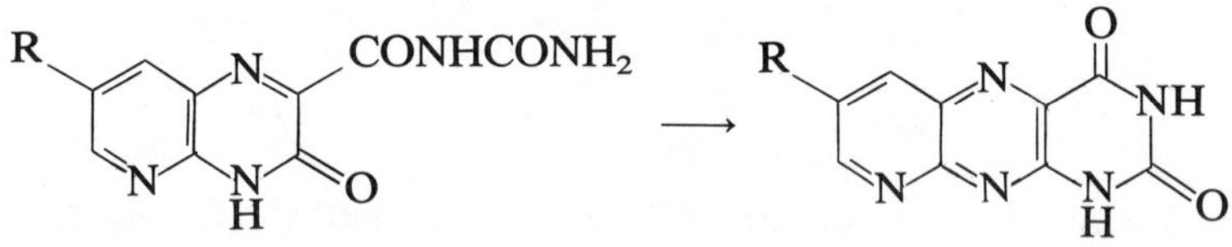

135 **136** R = H or Cl

Further work is required to establish the orientation of two other fused systems derived from pyrido[2,3-*b*]pyrazines. As the assignment of the phenacyl compound **137** as the 3-oxo derivative is uncertain, the structure of the furo compound **138**, obtained on treatment of compound **137** with phosphorus pentoxide in concentrated sulfuric acid at 100° must also

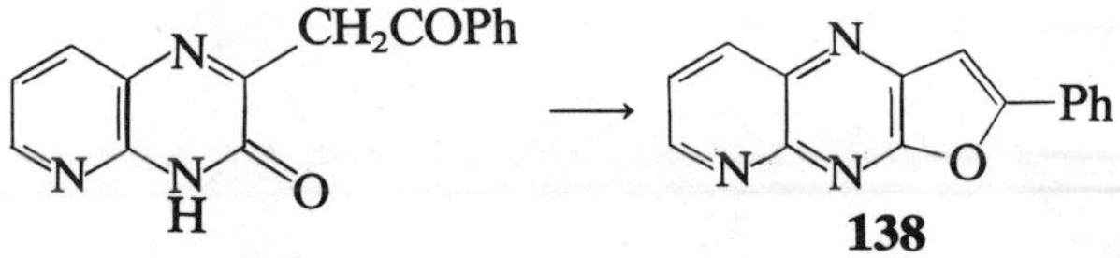

137 → 138

be in doubt. These conditions result in some sulfonation of the furo compound, although the position of substitution has not been ascertained.[39]

Cyclization in dilute alkali of the phenylhydrazone **139**, which may in fact be the 3-oxo isomer (see Section II2B), provides the pyrazolo system **140**.[40] If the authors' assignment of the 2-oxo structure to **139** is incorrect, then the pyrazolo compound will require reformulation as well. It seems that the oxo function is not necessary for the formation of pyrazolo compounds, as it is reported that compound **141** can be cyclized in dilute acid to give a 50% yield of the pyrazolo compound **142**.[40] No details are given for the preparation of **141**, whose orientation may also be in doubt.

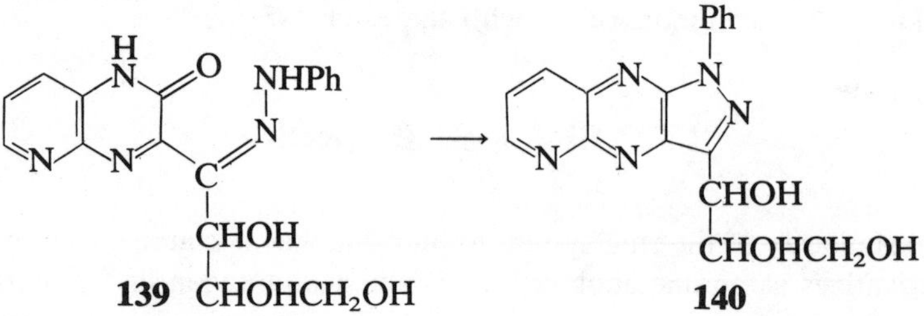

139 → 140

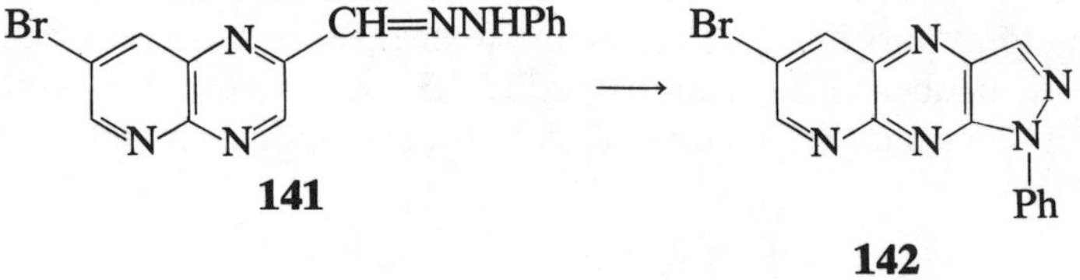

141 → 142

8. Miscellaneous Reactions

The 2-oxo compound **119** undergoes an addition reaction across its 3,4 double bond with acetoacetic acid in aqueous solution. Carbon dioxide is eliminated during the reaction, and the tetrahydro compound **144** is formed. The ester **145** provides the dihydro oxo compound **146** by a mechanism that is not clear.[10] Both reactions occur in solutions of sufficient acidity to ensure that the active methylene groups in **143** and **145** are un-ionized, thus apparently precluding nucleophilic attack on the ring system. The compound **146** is identical with that obtained from the reaction of 2,3-diaminopyridine with the ester **145**.

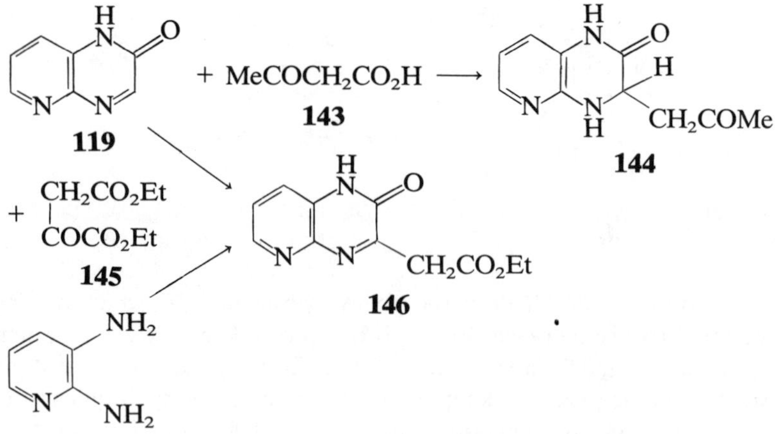

IV. Biological Properties

The similarity of the ring system to pteridine has encouraged syntheses of derivatives as antimetabolites[21] and folic acid antagonists.[9,49] Various 8-substituted amino derivatives of pyrido[2,3-*b*]pyrazine show antibacterial action.[58–60] However it seems that the successive introduction of nitrogen atoms into 8-hydroxyquinoline (oxine) reduces the antibacterial effect.[58] A number of 3,6-diaminopyrido[2,3-*b*]pyrazines have been the subject of patents claiming antiviral, diuretic, and antiinflammatory properties.[46,47]

V. Tables

TABLE 1. 6-AMINOPYRIDO[2,3-*b*]PYRAZINES[a]

[Structure: pyrido[2,3-b]pyrazine ring numbered 1–8, with RNH at position 6]

Molecular formula	Substituents or structure	m.p., °C	Ref.
$C_7H_6N_4$	6-Amino	268–270	6, 19, 28
$C_7H_6N_4O$	6-Amino-3,4-dihydro-3-oxo	>360	7
	6-Acetyl	>300	9
$C_7H_6N_4O_2$	6-Amino-1,2,3,4-tetrahydro-2,3-dioxo	>300	19, 28
	6-Acetyl	>300	19
$C_7H_7N_5$	3,6-Diamino	238	9
	Picrate	260 decomp.	
	3,6-Diacetyl	300	
$C_8H_5N_5$	6-Amino-7-cyano	>330	52
$C_8H_6N_4O_3$	6-Amino-2-carboxy-3,4-dihydro-3-oxo	>360	7, 9
$C_8H_7N_5O$	6-Amino-7-carboxamido	>350	52
$C_8H_7N_5O_2$	3,6-Diamino-2-carboxy	284 decomp.	9
$C_8H_8N_4$	6-Amino-2(or 3)-methyl	254–256	10
$C_8H_8N_4O$	6-Amino-3,4-dihydro-2-methyl-3-oxo	>360	7
	6-Amino-3,4-dihydro-4-methyl-3-oxo	278–288	7
$C_8H_8N_6O$	3,6-Diamino-2-carboxamido	>300	9
$C_8H_9N_5$	6,8-Diamino-2-methyl		
	Hydrochloride	>260	32
	6-Carbamate	>260	
	6,8-Diamino-3-methyl		
	Hydrochloride	>330	30
$C_8H_{11}N_5$	6,8-Diamino-3,4-dihydro-2-methyl		
	6-Carbamate	>260	32
$C_9H_6N_4O_4$	6-Amino-2,3-dicarboxy		
	Pb^{2+} Salt	>300	19
$C_9H_8N_4O_2$	2-Acetyl-6-amino-3,4-dihydro-3-oxo	>360	7
$C_9H_8N_4O_3$	6-Amino-2-carboxymethyl-3,4-dihydro-3-oxo		
	Ethyl ester	>360	7
$C_9H_{10}N_4$	6-Amino-2,3-dimethyl	231–231.5	19, 28
	6-Acetyl	200, 278–279	10, 19
$C_9H_{10}N_4O$	6-Amino-3,4-dihydro-2,4-dimethyl-3-oxo	296–297	7
$C_{10}H_8N_4O_3$	[Structure: H_2N-pyrido[2,3-b]pyrazine-CH_2COCO_2H]		
	Diethylester of oxalylamido deriv.	178–180	10

TABLE 1 (*contd.*)

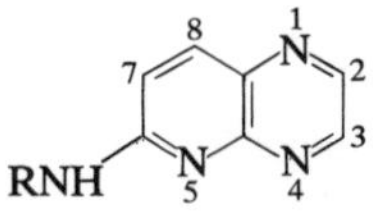

RNH, N 5, N 4, 3, 2, N 1, 8, 7

Molecular formula	Substituents or structure	m.p., °C	Ref.
$C_{10}H_{10}N_4O$	2-Acetyl-6-amino-3-methyl	284–286	7
$C_{11}H_{10}N_4O_3$	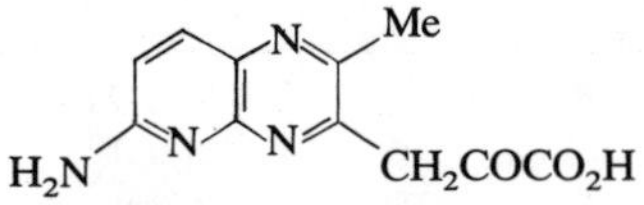 Me; H_2N; N; N; N; CH_2COCO_2H		
	Diethylester of oxalylamido deriv.	234–235	10
$C_{11}H_{14}N_6O$	3,6-Diamino-2-*n*-propylcarboxamido	283–284	46
$C_{12}H_{16}N_6O_2$	3,6-Diamino-2-(2-ethoxyethyl)-carboxamido	216	46
$C_{13}H_8Cl_2N_4$	6-Amino-8-chloro-3-(*p*-chlorophenyl)	>300 decomp.	27
	6-Carbamate	259–261	
$C_{13}H_{10}ClN_5$	6,8-Diamino-3-(*p*-chlorophenyl)[b]		
	Hydrochloride	>360	27
	6-Carbamate	>340	
$C_{13}H_{11}N_5$	3,6-Diamino-2-phenyl[c]	245	9
$C_{13}H_{19}N_7O$	3,6-Diamino-2-(3-dimethylaminopropyl)-carboxamido	209	46
$C_{14}H_{20}N_6O_3$	3,6-Diamino-2-(2-diethoxyethyl)-carboxamido	230–231	47
$C_{14}H_{21}N_7O$	3,6-Diamino-2-(2-diethylaminoethyl)-carboxamido	216–217	46
$C_{15}H_{16}N_6$	6,8-Diamino-2-(*N*-methylanilino)methyl	79–86	48
	6-Carbamate	>350	
$C_{15}H_{18}N_6$	6,8-Diamino-3,4-dihydro-2-(*N*-methyl-anilino)methyl		
	6-Carbamate	ca. 115	48
$C_{16}H_{16}N_6O_2$	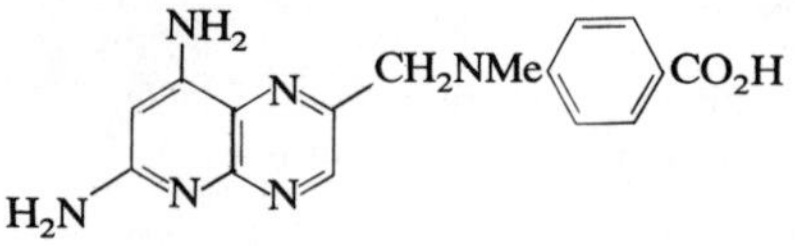 NH_2; N; CH_2NMe; CO_2H; H_2N; N; N		
	Hydrochloride	265 decomp.	49
	6,8-Diacetyl[d]	275–283 decomp.	
	6-Carbamate methyl ester	>360	
$C_{16}H_{18}N_6O_2$	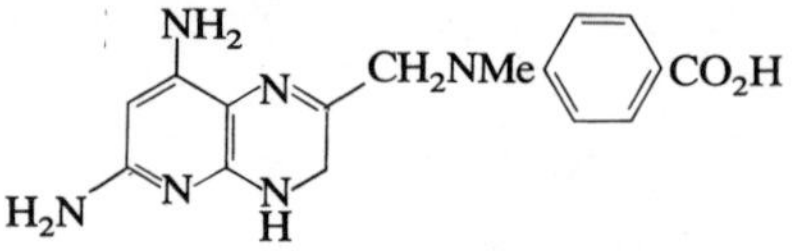 NH_2; N; CH_2NMe; CO_2H; H_2N; N; N H		

TABLE 1 (*contd.*)

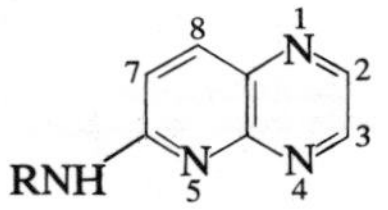

RNH (6-substituted pyrido[2,3-b]pyrazine; ring positions 1–8)

Molecular formula	Substituents or structure	m.p., °C	Ref.
$C_{16}H_{18}N_6O_2$ (*contd.*)	6-Carbamate methyl ester	ca. 250 decomp.	49
$C_{16}H_{26}N_6$	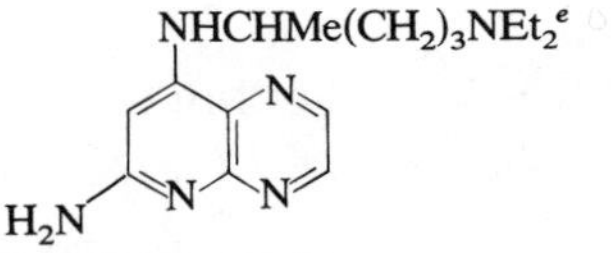 6-Amino-8-[NHCHMe$(CH_2)_3NEt_2$][e] (structure)		
	Dihydrochloride 6-Carbamate	Indefinite glass	26
$C_{17}H_{28}N_6$	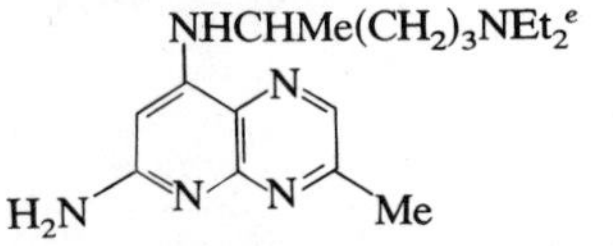 6-Amino-8-[NHCHMe$(CH_2)_3NEt_2$][e]-3-Me (structure)		
	Dihydrochloride 6-Carbamate	Indefinite glass	26
$C_{19}H_{14}N_4$	6-Amino-2,3-diphenyl	271	7, 17
	Hydrochloride	293.5–294.5	19
	6-Acetyl	268–269	17, 19
$C_{19}H_{14}N_4O$	6-Amino-5,8-dihydro-8-oxo-2,3-diphenyl	310–315 decomp.	59
$C_{19}H_{15}N_5$	6,8-Diamino-2,3-diphenyl	ca. 148–153	30
	6-Carbamate	>280	
	6-Carbamate hydrobromide	245–248 decomp.	
$C_{20}H_{17}N_5O$	6-Amino-8-NHCHPh$_2$-1H-2-oxo (structure)		
	6-Carbamate	275–277 decomp.	27
	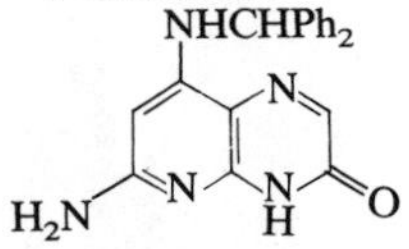 6-Amino-8-NHCHPh$_2$-4H-3-oxo (structure)		
	6-Carbamate	210–212	27
$C_{21}H_{18}N_4O$	6-Amino-8-ethoxy-2,3-diphenyl	268	30
	6-Carbamate	111–114	
$C_{21}H_{21}N_5$	 6-Amino-8-NHCHPh$_2$-2-Me-3,4-dihydro (structure)		

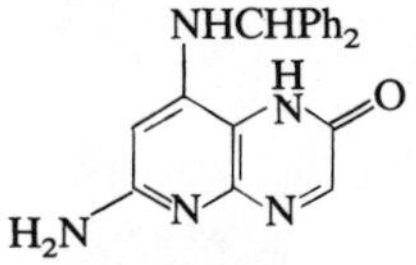

TABLE 1 (*contd.*)

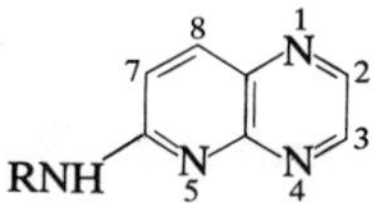

Molecular formula	Substituents or structure	m.p., °C	Ref.
$C_{21}H_{21}N_5$ (*contd.*)	6-Carbamate	197	32
$C_{21}H_{23}N_7O_5$	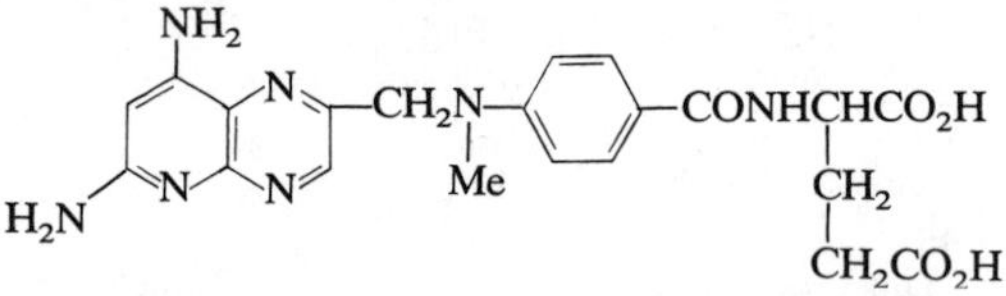	ca. 228[f]	49
	6,8-Diacetyl diethylester	ca. 239 decomp.	
$C_{22}H_{29}ClN_6$	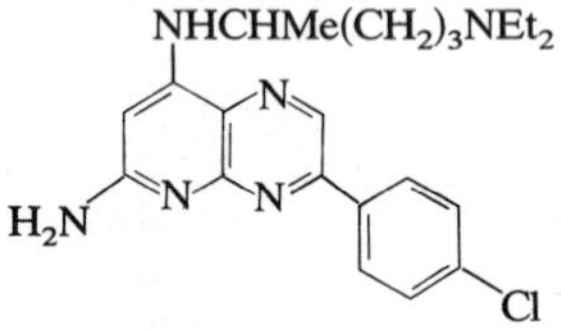 [g]		
	Dihydrochloride	116–118	26
	6-Carbamate hydrochloride	136–139	
	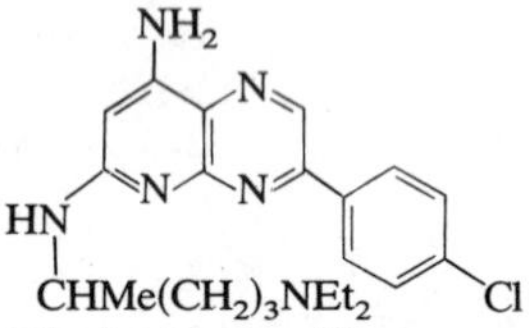		
	Dihydrochloride dihydrate	250–252 decomp.	34
	8-Carbamate dihydrochloride	decomp. >145	
$C_{23}H_{29}F_3N_6$	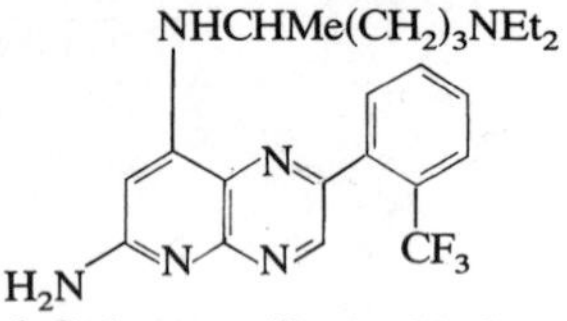		
	6-Carbamate dihydrochloride	122–124	27
$C_{23}H_{32}N_6O$	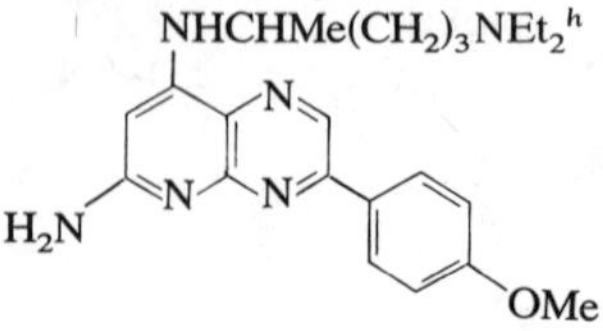 [h]	241–242 decomp.	27
	6-Carbamate	210–211 decomp.	

TABLE 1 (*contd.*)

RNH, N1, 2, 3, N4, N5, 7, 8

Molecular formula	Substituents or structure	m.p., °C	Ref.
$C_{25}H_{25}ClN_4O_6$	Cl; OMe; OMe; OMe; OMe; OMe; OMe; H_2N; N; N; N [i]	210–212	27
	6-Carbamate	206–208 decomp.	
$C_{26}H_{20}ClN_5$	$NHCHPh_2$; Cl; H_2N; N; N; N [b]		
	6-Carbamate	>300	27
	$NHCHPh_2$; H_2N; N; N; N; Cl [b]		
	6-Carbamate	215–217	27
$C_{28}H_{32}Cl_2N_6$	$NHCHMe(CH_2)_3NEt_2$; Cl; Cl; H_2N; N; N; N [j]		
	6-Carbamate	80–82	26
$C_{28}H_{34}N_6$	$NHCHMe(CH_2)_3NEt_2$; Ph; Ph; H_2N; N; N; N [e]	183–184	26
	6-Carbamate hydrochloride	211–213	

TABLE 1 (*contd.*)

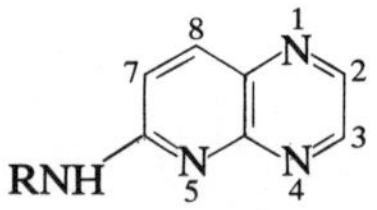

Molecular formula	Substituents or structure	m.p., °C	Ref.
$C_{32}H_{25}N_5$	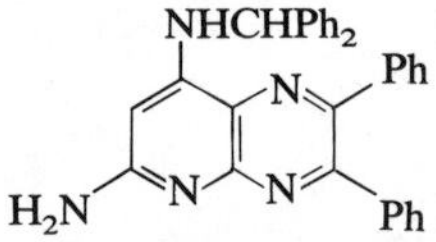 6-Amino-8-(diphenylmethylamino)-2,3-diphenyl[b] (structure) 6-Carbamate	ca. 225	30

[a] Carbamates and acetyl derivatives of amines are listed with the amines. Carbamate refers to ethyl carbamate.
[b] The 3,4,5-trimethoxyphenyl analogue was also prepared.[18]
[c] Seventeen other 2-aryl derivatives were prepared.
[d] Analysed as 0.1 HCl 1.6 H_2O.
[e] Also 8-NHC_6H_4-Cl-*p*, 8-$NHCH_2C_6H_4SO_2NH_2$-*p*, 8-$NHCH_2CH_2C_6H_4SO_2NH_2$-*p*, and 8-$NHCH_2CH_2C_6H_4SO_2NEt_2$-*p* analogues were prepared.
[f] Analysed as 0.1 HCl 1.0 H_2O.
[g] Also 8-NHC_6H_4Cl-*p* and 8-$NHCH_2CH_2C_6H_4SO_2NEt_2$-*p* analogues were prepared.
[h] Various other aryl analogues were prepared, many as mixtures with the 2-aryl compound.
[i] 8-Amino compound hydrochloride melting point indefinite.
[j] Also 8-NHC_6H_4Cl-*p* analogue was prepared.

TABLE 2. PYRIDO[2,3-*b*]PYRAZINES (NOT 6-AMINO COMPOUNDS)

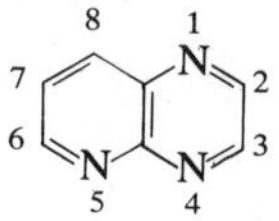

Molecular formula	Substituents or structure	m.p., °C	Ref.
$C_7H_2BrCl_2N_3$	7-Bromo-2,3-dichloro	190–192	10
$C_7H_2Cl_3N_3$	2,3,7-Trichloro	198	53
$C_7H_3Cl_2N_3$	2,3-Dichloro	144–145	53, 54
	6,7-Dichloro	162–164 decomp.	21
$C_7H_4BrN_3$	7-Bromo	167	17
	Mono-*N*-oxide	286	
	Methiodide	186–187	10
	Picrate	103.5–104 decomp.	

TABLE 2 (*contd.*)

Molecular formula	Substituents or structure	m.p., °C	Ref.
$C_7H_4BrN_3O$	7-Bromo-1,2-dihydro-2-oxo	315–318	10
	7-Bromo-3,4-dihydro-3-oxo	274–275.5	10
$C_7H_4BrN_3O_2$	7-Bromo-1,2,3,4-tetrahydro-2,3-dioxo	368–370 decomp.	10
$C_7H_4ClN_3$	6-Chloro	156	22
	7-Chloro	161	7, 21
$C_7H_4ClN_3O_2$	7-Chloro-1,2,3,4-tetrahydro-2,3-dioxo	360 decomp.	53
$C_7H_5N_3$	Unsubstituted	147–148	7, 17
	Methiodide	165–166	10
	Adduct: Cu(II) nitrate	285 decomp.	62
$C_7H_5N_3O$	1,2-Dihydro-2-oxo	>300	6
		318–320	10
	3,4-Dihydro-3-oxo	239–240	37
	5,6-Dihydro-6-oxo	246–248	6, 22
	5,8-Dihydro-8-oxo	290–310 decomp.	31
$C_7H_5N_3O_2$	1,2,3,4-Tetrahydro-2,3-dioxo	>310	6, 19, 53
		413–415 decomp.	
$C_7H_6N_4$	8-Amino		
	Monohydrate	254–255 decomp.	31
$C_7H_7N_3O$	1,2,3,4-Tetrahydro-2-oxo	287–289	6
	1,2,3,4-Tetrahydro-3-oxo	276–278	10
$C_7H_7N_3O_2$	1,2,3,4-Tetrahydro-1-hydroxy-2-oxo	192	50
$C_8H_4BrN_3O_3$	7-Bromo-2-carboxy-3,4-dihydro-3-oxo	242–245	45
	7-Bromo-3-carboxy-1,2-dihydro-2-oxo	310–320 decomp.	10
$C_8H_4ClN_3O_3$	3-Chloro-7-carboxy-8-hydroxy		
	Ethyl ester	257	63
	7-Chloro-2-carboxy-3,4-dihydro-3-oxo	238–239	45
$C_8H_5N_3O_3$	2-Carboxy-3,4-dihydro-3-oxo	229–232	37, 43, 45
	Ethyl ester	213–214	37
	3-Carboxy-1,2-dihydro-2-oxo		
	Ethyl ester	247–254 decomp.	49
$C_8H_6BrN_3$	7-Bromo-2-methyl	133–135	10
	7-Bromo-3-methyl	167–168	10
	7-Bromo-6-methyl	165–166 decomp.	21
	7-Bromo-8-methyl	204–206	21

TABLE 2 (*contd.*)

Molecular formula	Substituents or structure	m.p., °C	Ref.
$C_8H_6BrN_3O$	7-Bromo-1,2-dihydro-3-methyl 2-oxo	>319	10
	7-Bromo-3,4-dihydro-2-methyl-3-oxo	240 decomp., 252–253	7, 10
$C_8H_6BrN_3S$	7-Bromo-1,2-dihydro-3-methyl-2-thioxo	Chars 280–290	10
	7-Bromo-3,4-dihydro-2-methyl-3-thioxo	253–256 decomp.	10
$C_8H_6N_4O_2$	2-Carboxamido-3,4-dihydro-3-oxo	>300 decomp.	37
$C_8H_6N_4O_3$	3-Amino-2-carboxy-5,6-dihydro-6-oxo		9
$C_8H_7N_3$	2-Methyl	106.5–108.5	10
	3-Methyl	106.5–107.5	10
	6-Methyl	95–96.5	10
	Monohydrate	56–57	
$C_8H_7N_3O$	1,2-Dihydro-1-methyl-2-oxo	223–224	37
	1,2-Dihydro-3-methyl-2-oxo	270 decomp.	7
		306–308	10
	3,4-Dihydro-2-methyl-3-oxo	240 decomp.	7
		252–253	10
	3,4-Dihydro-4-methyl-3-oxo	115–117	37
$C_8H_7N_3O_2$	3,4,5,6-Tetrahydro-4-methyl-3,6-dioxo	297	7
$C_8H_7N_3O_3$	1,2,3,4-Tetrahydro-1-hydroxy-6-methyl-2,3-dioxo	252–253	50
$C_8H_7N_3S$	1,2-Dihydro-3-methyl-2-thioxo	291–296 decomp.	10
	3,4-Dihydro-2-methyl-3-thioxo	292–294 decomp.	10
$C_8H_9N_3$	3,4-Dihydro-2-methyl	108–111	10
$C_9H_6BrN_3O$	7-Bromo-3-formyl-2-methyl		
	Oxime	267–268	10
$C_9H_6BrN_3O_3$	7-Bromo-2-carboxy-3,4-dihydro-6-methyl-3-oxo	180–183	45
	7-Bromo-2-carboxy-3,4-dihydro-8-methyl-3-oxo	221–223	45
	7-Bromo-2-carboxymethyl-3,4-dihydro-3-oxo		
	Ethyl ester	260 decomp.	7
$C_9H_7Cl_2N_3$	6,7-Dichloro-2,3-dimethyl	197–199 decomp.	21
$C_9H_7N_3O_3$	2-Carboxy-3,4-dihydro-4-methyl-3-oxo	186 decomp.	37

TABLE 2 (*contd.*)

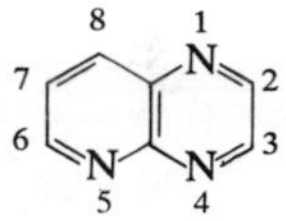

Molecular formula	Substituents or structure	m.p., °C	Ref.
$C_9H_7N_3O_3$ (*contd.*)			
	Ethyl ester	119.5–120	
	3-Carboxy-1,2-dihydro-1-methyl-2-oxo		
	1 H_2O	163 decomp.	37
	1.5 H_2O	146–148 decomp.	
	Ethyl ester	161–162	
	7-Carboxy-8-hydroxy-3-methyl		
	Ethyl ester	274	63
	2-Carboxymethyl-3,4-dihydro-3-oxo		
	Ethyl ester	204.5–207	10
	3-Carboxymethyl-1,2-dihydro-2-oxo		
	Ethyl ester	223–224.5	10
$C_9H_7N_3O_4$	7-Carboxy-8-hydroxy-3-methoxy		
	Ethyl ester	275	63
$C_9H_7N_5O_3$	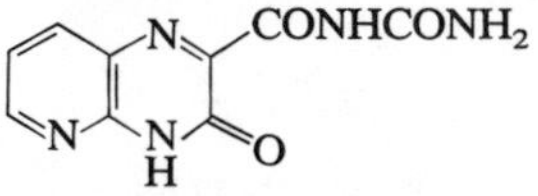 [structure: 1,2-dihydro-2-oxo-pyrido[2,3-b]pyrazine with 3-CONHCONH$_2$]	283–285	37, 43, 45, 56
	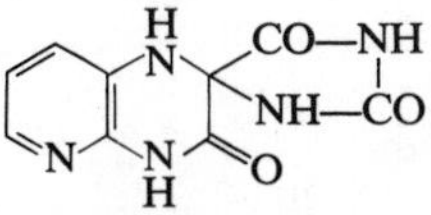 [structure: 3,3-spiro-hydantoin of 1,2,3,4-tetrahydro-2-oxopyrido[2,3-b]pyrazine, with CO—NH / NH—CO]	277–278, 306	37, 43
$C_9H_8BrN_3$	7-Bromo-2,3-dimethyl	150 decomp.	17
	Methiodide	153–154	10
	Perbromide	138–140 decomp.	
	Adducts with:		
	1,3,5-trinitrobenzene	125–126	
	2,4-dinitrotoluene	95–96	
	7-Bromo-6,8-dimethyl	174–174.5 decomp.	21
$C_9H_8ClN_3$	7-Chloro-2,3-dimethyl	155–156	7, 21
$C_9H_8N_4O_2$	2-Carboxamido-3,4-dihydro-4-methyl-3-oxo	284–285	37
	3-Carboxamido-1,2-dihydro-1-methyl-2-oxo	264–265	37
$C_9H_9N_3$	2,3-Dimethyl	143.5–144	19
		148–149	17
	Methiodide	196–198 decomp.	10
$C_9H_9N_3O_2$	3-Ethoxy-1,2-dihydro-2-oxo[a]	265	53

TABLE 2 (*contd.*)

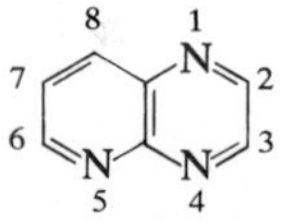

Molecular formula	Substituents or structure	m.p., °C	Ref.
$C_9H_9N_3O_2$ (*contd.*)	3,4,5,6-Tetrahydro-2,4-dimethyl-3,6-dioxo	266–268	7
$C_9H_{10}BrN_3$	7-Bromo-1,2(or 3,4)-dihydro-2,3-dimethyl	160	61
$C_9H_{10}N_4O$	3,4-Dihydro-4-methyl-2-methylamino-3-oxo	211–212	37
$C_9H_{11}N_3$	1,2(or 3,4)-Dihydro-2,3-dimethyl	106	61
$C_9H_{12}BrN_3$	7-Bromo-1,2,3,4-tetrahydro-2,3-dimethyl[b]	162–163	65
$C_9H_{13}N_3$	1,2,3,4-Tetrahydro-2,3-dimethyl[b]	126–128	65
	(±)*cis*-1,2,3,4-Tetrahydro-2,3-dimethyl	111.5–112.5	54, 55
$C_{10}H_6BrN_3O_3$	7-Bromo-3-oxalylmethyl		
	Ethyl ester	154–154.5	10
$C_{10}H_6BrN_3O_4$	7-Bromo-1,2-dihydro-3-oxalylmethyl-2-oxo		
	Ethyl ester	235–238	10
$C_{10}H_7N_3O_3$	6-Oxalylmethyl		
	Ethyl ester	171–172 decomp.	10
$C_{10}H_7N_3O_4$	1,2-Dihydro-3-oxalylmethyl-2-oxo		
	Ethyl ester	245–247	10
	3,4-Dihydro-2-oxalylmethyl-3-oxo		
	Ethyl ester	290–292	10
$C_{10}H_8BrN_3O_3$	7-Bromo-2-carboxy-3,4-dihydro-6,8-dimethyl-3-oxo	208–210	45
$C_{10}H_8ClN_3O_3$	7-Carboxy-3-chloro-5-ethyl-5,8-dihydro-8-oxo	245	63
	Ethyl ester	188–189	63
$C_{10}H_9N_3O$	2-Acetyl-3-methyl	116–116.5	10
$C_{10}H_9N_3OS$	2-Acetyl-5,6-dihydro-3-methyl-6-thioxo	216–217 decomp.	10
$C_{10}H_9N_3O_2$	2-Acetonyl-3,4-dihydro-3-oxo	299.5–301.5	10
	3-Acetonyl-1,2-dihydro-2-oxo	280–283	10
	2-Acetyl-5,6-dihydro-3-methyl-6-oxo	275–290	10
$C_{10}H_9N_3O_3$	7-Carboxy-8-hydroxy-2,3-dimethyl		
Ethyl ester	Ethyl ester	280	63
$C_{10}H_9N_3O_4$	7-Carboxy-5,8-dihydro-3-methoxy-5-methyl-8-oxo	>275	60

TABLE 2 (*contd.*)

Molecular formula	Substituents or structure	m.p., °C	Ref.
$C_{10}H_9N_5O_3$	N, CONHCONH$_2$, N, N, O, Me	232 decomp.	37, 43
	Me, N, O, N, N, CONHCONH$_2$	284	37
	H, N, CO—NH, NH—CO, N, N, O, Me	240	37, 43
	1-Acetyl	263–264	
	Me, N, O, NH—CO, 1′, N, N, H, CO—NH	284–287	37
	4,1′-Diacetyl	204	
$C_{10}H_{10}BrN_3$	7-Bromo-2,3,6-trimethyl	161–162	21
	7-Bromo-2,3,8-trimethyl	141.5–143	21
$C_{10}H_{11}N_3O_2$	2-Acetonyl-1,2,3,4-tetrahydro-3-oxo	225–227	10
	3-Acetonyl-1,2,3,4-tetrahydro-2-oxo	233–234.5	10
$C_{11}H_6N_4O_3$	3(or 2)-(5-Nitro-2-furyl)	264–265	36
$C_{11}H_8BrN_3O_3$	7-Bromo-2-methyl-3-oxalylmethyl		
	Ethyl ester	193–194	10
$C_{11}H_9N_3O_3$	2-Methyl-3-oxalylmethyl		
	Ethyl ester	104.5–105	10
$C_{11}H_{10}N_4O_3$	N, CONHCOMe, N, N, O, Me	204–205	37
	Me, N, O, N, N, CONHCOMe	195–196	37
$C_{11}H_{11}N_3O_3$	7-Carboxy-5-ethyl-5,8-dihydro-3-methyl-8-oxo	282	63

TABLE 2 (*contd.*)

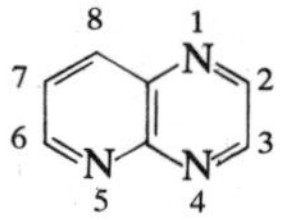

Molecular formula	Substituents or structure	m.p., °C	Ref.
	Ethyl ester	149–150	63
$C_{11}H_{11}N_3O_4$	7-Carboxy-5-ethyl-5,8-dihydro-3-methoxy-8-oxo	241	63
	Ethyl ester	146–147	63
$C_{11}H_{11}N_5O_3$	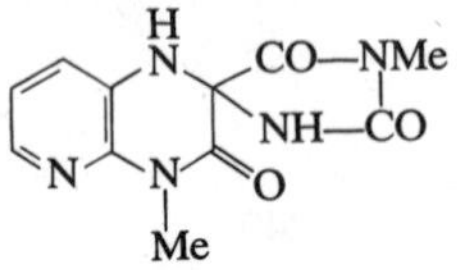 H N CO—NMe NH—CO N N O Me	254–255	37
	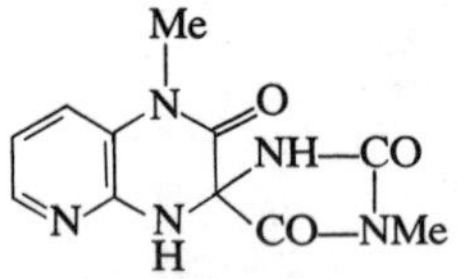 Me N O NH—CO N N H CO—NMe	287	37
$C_{11}H_{12}BrN_3$	7-Bromo-1,2,3,4-tetrahydro-2,3-divinyl[b]	118–120	65
	Picrate	192 decomp.	
	7-Bromo-2,3,6,8-tetramethyl	166–167	21
$C_{11}H_{12}ClN_3O_2$	7-Chloro-2,3-diethoxy	125	53
$C_{11}H_{13}N_3$	1,2,3,4-Tetrahydro-2,3-divinyl[b]	75	65
	Picrate	172–173	
$C_{11}H_{16}BrN_3$	7-Bromo-2,3-diethyl-1,2,3,4-tetrahydro[b]		
	Picrate	225 decomp.	65
$C_{11}H_{17}N_3$	2,3-Diethyl-1,2,3,4-tetrahydro[b]	94–95	65
$C_{12}H_{12}N_4O_3$	7-Carboxy-8-hydroxy-3-pyrrolidin-1-yl		
	Ethyl ester	305	63
$C_{12}H_{12}N_4O_4$	7-Carboxy-8-hydroxy-3-morpholin-1-yl		
	Ethyl ester	308	63
$C_{12}H_{13}N_3O_3$	7-Carboxy-5-ethyl-5,8-dihydro-2,3-dimethyl-8-oxo	269	63
	Ethyl ester	145–147	
$C_{12}H_{13}N_3O_4$	7-Carboxy-3-ethoxy-5-ethyl-5,8-dihydro-8-oxo	261	60
$C_{12}H_{13}N_5O_3$	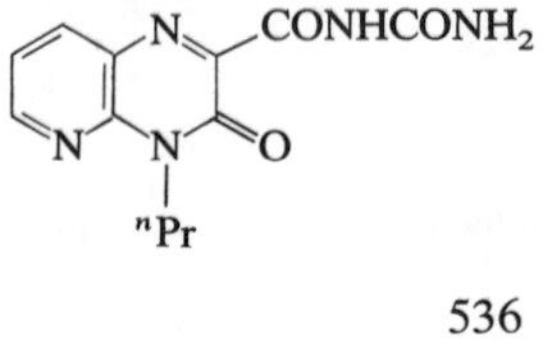 N $CONHCONH_2$ N N O ^{n}Pr	243 decomp.	37, 43

TABLE 2 (*contd.*)

Molecular formula	Substituents or structure	m.p., °C	Ref.
$C_{12}H_{13}N_5O_3$ (*contd.*)	H N CO—NH NH—CO N N O ^{n}Pr	243 decomp.	37, 43
	H N CO—NMe NMe—CO N N O Me	174–175	37
	1-Nitroso	154–155	
	Me N O NMe—CO N N H CO—NMe	218–219	37
$C_{12}H_{14}N_4O_3$	7-Carboxy-3-dimethylamino-5-ethyl-5,8-dihydro-8-oxo	325	63
	Ethyl ester	190–191	63
$C_{13}H_8BrN_3$	7-Bromo-2-phenyl	152–153	10
	7-Bromo-3-phenyl	167–168	10
	Methiodide	218	
	Picrate	149–150	
$C_{13}H_9N_3$	2-Phenyl	125.5–126.5	10
	3-Phenyl	140–141	10
$C_{13}H_9N_3O$	3,4-Dihydro-3-oxo-4-phenyl	245	44
$C_{13}H_{11}Cl_6N_3O_2$	N $CH_2CHOHCCl_3$ N N $CH_2CHOHCCl_3$	203–205 decomp.	10
$C_{13}H_{11}N_5O_3$	7-Carboxy-5-ethyl-5,8-dihydro-3-imidazol-1-yl-8-oxo	308	63
	Ethyl ester	241–242	
$C_{13}H_{14}N_4O_3$	7-Carboxy-8-hydroxy-3-piperidin-1-yl		
	Ethyl ester	290	63
$C_{13}H_{16}N_4O_4$	O HO_2C N N N Me N Et CH_2CH_2OH	286	63
	Ethyl ester	215–216	

TABLE 2 (*contd.*)

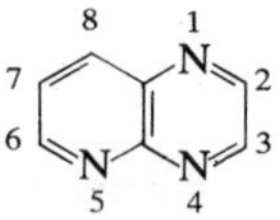

8 1 7 N 2 6 N N 3 5 4

Molecular formula	Substituents or structure	m.p., °C	Ref.
$C_{14}H_{10}BrN_3$	7-Bromo-6-methyl-3-phenyl	142	65
	7-Bromo-8-methyl-3-phenyl	208	65
$C_{14}H_{10}BrN_3O$	7-Bromo-2(or 3)-(*p*-methoxy-phenyl)	185–186	10
	Methiodide	252–255	
$C_{14}H_{11}N_3O$	2(or 3)-(*p*-Methoxyphenyl)	150–151	10
$C_{14}H_{12}BrN_3$	7-Bromo-5,8-dihydro-8-methyl-3-phenyl	178–180 decomp.	65
$C_{14}H_{13}N_3$	1,2-Dihydro-2-methyl-3-phenyl	176	65
$C_{14}H_{15}N_3O$	5,6,7,8-Tetrahydro-6-methoxy-2-phenyl	183	61
$C_{14}H_{16}N_4O_3$	7-Carboxy-5-ethyl-5,8-dihydro-8-oxo-3-pyrrolidin-1-yl	315	63
	Ethyl ester	239–240	
$C_{14}H_{16}N_4O_4$	7-Carboxy-5-ethyl-5,8-dihydro-3-morpholin-1-yl-8-oxo	315	63
	Ethyl ester	245–246	
$C_{15}H_9N_5O_4$	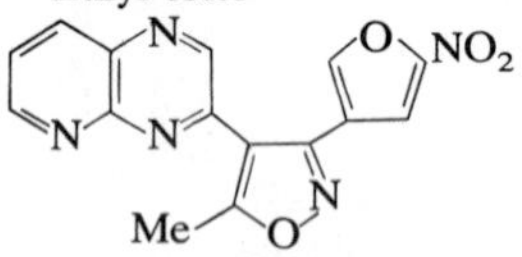 [structure: N, N, N, O, NO_2, N, Me, O] (or 2-isomer)	217–218	35
$C_{15}H_{10}N_4O_3$	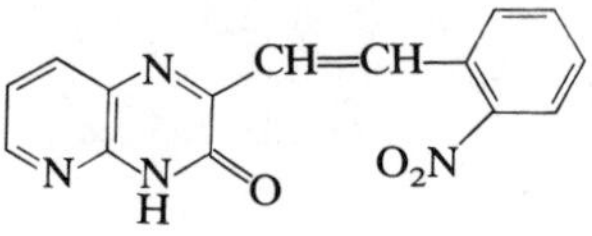 [structure: N, CH=CH, N, H, O, O_2N] (or 2 -oxo-isomer)	275 decomp.	39
$C_{15}H_{11}N_3O$	3,4-Dihydro-3-oxo-2-styryl (or 2-oxo isomer)	251	39
$C_{15}H_{11}N_3O_2$	3,4-Dihydro-3-oxo-2-phenacyl (or 2-oxo isomer)	ca. 283	39
$C_{15}H_{11}N_5O_3$	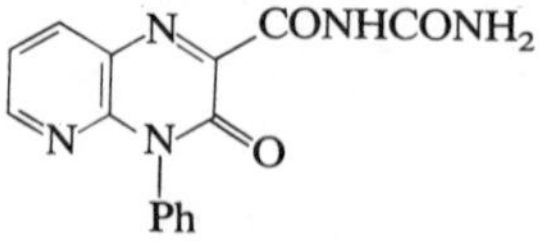 [structure: N, $CONHCONH_2$, N, N, O, Ph]	225	37, 44
	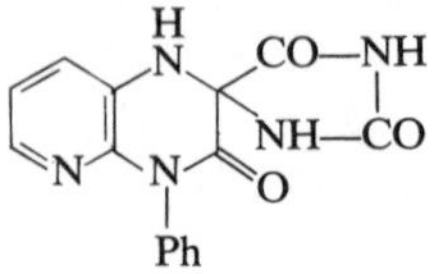 [structure: H, N, CO—NH, NH—CO, N, N, O, Ph]	252 decomp.	37, 44

TABLE 2 (*contd.*)

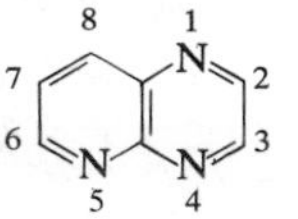

Molecular formula	Substituents or structure	m.p., °C	Ref.
$C_{15}H_{13}N_3$	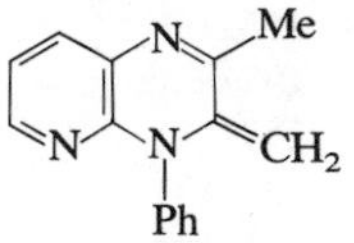 N, N, Me, CH_2, Ph	152	42
$C_{15}H_{17}N_3O$	6-Ethoxy-5,6,7,8-tetrahydro-2-phenyl	165	61
$C_{15}H_{18}N_4O_3$	7-Carboxy-5-ethyl-5,8-dihydro-8-oxo-3-piperidin-1-yl	308	63
	Ethyl ester	202–203	
$C_{16}H_{10}BrN_5O_4$	7-Bromo-2-methyl-3-(2,4-dinitrostyryl)	Darkens > 235	10
$C_{16}H_{11}N_5O_4$	2-Methyl-3-(2,4-dinitrostyryl)	260–265 decomp.	10
$C_{16}H_{12}N_4O_2$	2-Methyl-3-(*o*-nitrostyryl)	204–206.5	10
	2-Methyl-3-(*m*-nitrostyryl)	211–213	10
$C_{16}H_{14}N_4O_3$	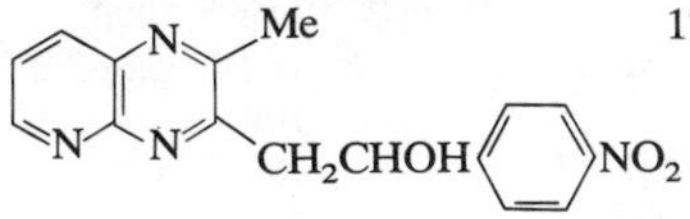 Me, CH_2CHOH, NO_2	170–171	10
$C_{16}H_{14}N_6O_2$	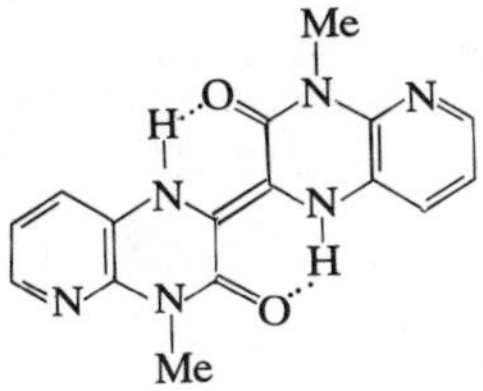 Me, H, O, N, Me	303–304 decomp.	37
	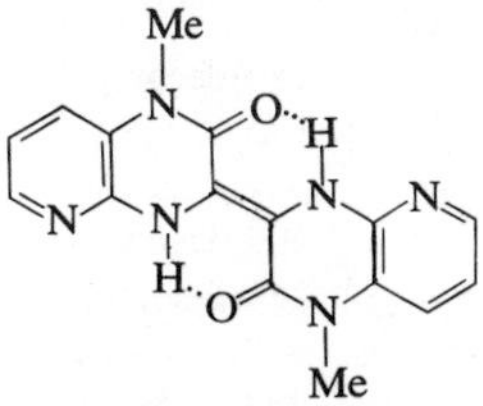 Me, O, H, N, Me	>360 decomp.	37
$C_{16}H_{16}BrN_3O$	1-Benzoyl-7-bromo-1,2,3,4-tetrahydro-2,3-dimethyl[b]	227–229	65
$C_{17}H_{11}N_5$	2,3-Di(2-pyridyl)	146	20
$C_{17}H_{16}BrN_5O_4$	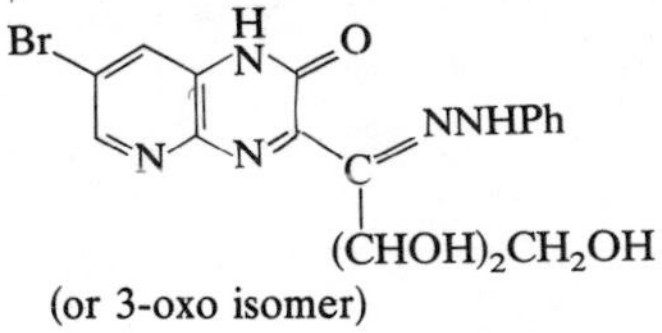 Br, H, O, NNHPh, $(CHOH)_2CH_2OH$ (or 3-oxo isomer)	281–283	40

TABLE 2 (*contd.*)

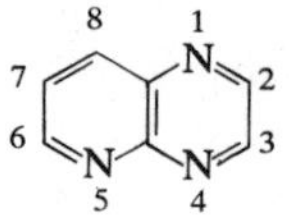

Molecular formula	Substituents or structure	m.p., °C	Ref.
$C_{19}H_{12}BrN_3$	7-Bromo-2,3-diphenyl	156–158	17
	Methotriiodide	192 decomp.	
$C_{19}H_{12}ClN_3$	7-Chloro-2,3-diphenyl	159–161	21
		136–137	7
$C_{19}H_{13}N_3$	2,3-Diphenyl	141–142	18, 19, 39
	Methiodide	230–232	10
	3,7-Diphenyl	218	65
$C_{19}H_{13}N_3O$	5,6-Dihydro-6-oxo-2,3-diphenyl	273–274	7
$C_{19}H_{13}N_3O_2$	2,3-Bis(*p*-hydroxyphenyl)	315 decomp.	39
$C_{19}H_{15}N_3$	1,2-Dihydro-2,3-diphenyl	230	65
	1,2(or 3,4)-Dihydro-2,3-diphenyl	208	61
$C_{19}H_{15}N_5$	2,3-Bis(6-methyl-2-pyridyl)	155	20
$C_{19}H_{16}BrN_3$	7-Bromo-1,2,3,4-tetrahydro-2,3-diphenyl[b]	220–222	65
$C_{19}H_{17}N_3$	1,2,3,4-Tetrahydro-2,3-diphenyl[b]	186 decomp.	65
$C_{20}H_{14}BrN_3$	7-Bromo-6-methyl-2,3-diphenyl	160 decomp.	21, 65
		165	
	7-Bromo-8-methyl-2,3-diphenyl	199.5–201	21
$C_{20}H_{15}N_3$	6-Methyl-2,3-diphenyl	169–170	23, 65
	7-Methyl-2,3-diphenyl	160–161	23
	8-Methyl-2,3-diphenyl	143–144	23
	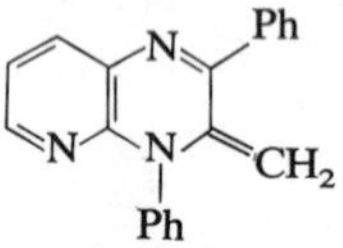 [structure: 3-Methylene-2,4-diphenyl-3,4-dihydropyrido[2,3-b]pyrazine; labels: N, Ph, N, N, CH$_2$, Ph]	325–327	42
$C_{20}H_{16}BrN_3$	7-Bromo-5,6-dihydro-6-methyl-2,3-diphenyl	215 decomp.	65
$C_{20}H_{17}N_3$	5,6-Dihydro-6-methyl-2,3-diphenyl	175	65
$C_{20}H_{19}N_3O$	5,6,7,8-Tetrahydro-6-methoxy-2,3-diphenyl	200 decomp.	61
$C_{21}H_{16}BrN_3$	7-Bromo-6-ethyl-2,3-diphenyl	240 decomp.	65
	7-Bromo-6,8-dimethyl-2,3-diphenyl	222	25
$C_{21}H_{17}N_3$	6,8-Dimethyl-2,3-diphenyl	193	24
$C_{21}H_{18}BrN_3$	7-Bromo-6-ethyl-5,6-dihydro-2,3-diphenyl	192	65
$C_{22}H_{28}ClN_5$	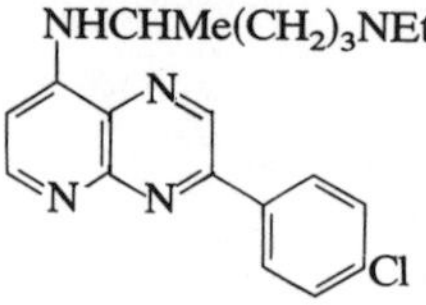 [structure: 8-NHCHMe(CH$_2$)$_3$NEt$_2$-3-(p-chlorophenyl)pyrido[2,3-b]pyrazine; labels: NHCHMe(CH$_2$)$_3$NEt$_2$, N, N, N, Cl]		

TABLE 2 (*contd.*)

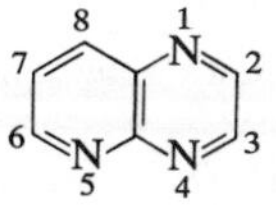

Molecular formula	Substituents or structure	m.p., °C	Ref.
$C_{22}H_{28}ClN_5$(*contd.*)	Hydrochloride dihydrate	220–230 decomp.	33
$C_{23}H_{17}N_3$	2,3-Distyryl	198–199	10
$C_{25}H_{17}N_3$	2,3,7-Triphenyl	282–284	65
$C_{26}H_{21}N_3O$	1-Benzoyl-1,2,3,4-tetrahydro-2,3-diphenyl[b]	245	65

[a] Orientation may be in doubt.
[b] 2,3-*trans*-Substituted.

VI. References

1. G. Favini, I. Vandoni, and M. Simonetta, *Theor. Chim. Acta*, **3,** 45, 418 (1965); *Chem. Abstr.*, **62,** 8525 (1965); *ibid.*, **63,** 17326 (1965).
2. B. Tinland, *Theor. Chim. Acta*, **8,** 361 (1967); *Chem. Abstr.*, **68,** 100246 (1968).
3. S. C. Wait, Jr., and J. W. Wesley, *J. Mol. Spectrom.*, **19,** 25 (1966); *Chem. Abstr.*, **64,** 9073 (1966).
4. W. L. F. Armarego, G. B. Barlin, and E. Spinner, *Spectrochim. Acta*, **22,** 117 (1966); *Chem. Abstr.*, **64,** 7536 (1966).
5. S. F. Mason, *J. Chem. Soc.*, 4874 (1957).
6. A. Albert and G. B. Barlin, *J. Chem. Soc.*, 5737 (1963).
7. C. L. Leese and H. N. Rydon, *J. Chem. Soc.*, 303 (1955).
8. S. F. Mason, *J. Chem. Soc.*, 493 (1962).
9. T. S. Osdene and G. M. Timmis, *J. Chem. Soc.*, 2033 (1955).
10. T. L. Threlfall, Ph.D. Thesis, University of London, 1971.
11. T. J. Batterham, *J. Chem. Soc.* (*C*), 999 (1966).
12. A. Albert and J. N. Phillips, *J. Chem. Soc.*, 1294 (1956).
13. A. Albert and A. Hampton, *J. Chem. Soc.*, 505 (1954).
14. K. Wiberg and T. P. Lewis, *J. Am. Chem. Soc.*, **92,** 7154 (1970).
15. W. I. Stephen and P. C. Uden, *Anal. Chim. Acta*, **39,** 357 (1967); *Chem. Abstr.*, **67,** 96513 (1967).
16. B. A. Fox and T. L. Threlfall, *Org. Synth.*, **44,** 34 (1964).
17. V. Petrow and J. Saper, *J. Chem. Soc.*, 1389 (1948).
18. A. E. Chichibabin and A. V. Kirsamov, *Ber.*, **60B,** 766 (1927).
19. I. A. Kaye, *J. Med. Chem.*, **7,** 240 (1964).
20. N. P. Buu-Hoi and G. Saint-Ruf, *J. Chem. Soc.*, 2258 (1961).
21. M. Israel and A. R. Day, *J. Org. Chem.*, **24,** 1455 (1959).
22. A. Albert and F. Reich, *J. Chem. Soc.*, 1370 (1960).
23. G. R. Lappin and F. B. Slezak, *J. Am. Chem. Soc.*, **72,** 2806 (1950).
24. T. Batkowski, *Roczniki Chem.*, **37,** 385 (1963); *Chem. Abstr.*, **59,** 11487 (1963).

25. T. Batkowski and M. Tuszynska, *Roczniki Chem.*, **38,** 585 (1964); *Chem. Abstr.*, **61,** 10654 (1964).
26. C. Temple, Jr., J. D. Rose, R. D. Elliott, and J. A. Montgomery, *J. Med. Chem.*, **11,** 1216 (1968).
27. C. Temple, Jr., J. D. Rose, R. D. Elliott, and J. A. Montgomery, *J. Med. Chem.*, **13,** 853 (1970).
28. J. Bernstein, B. Stearns, E. Shaw, and W. A. Lott, *J. Am. Chem. Soc.*, **69,** 1151 (1947).
29. F. Korte, *Chem. Ber.*, **85,** 1012 (1952).
30. R. D. Elliott, C. Temple, Jr., and J. A. Montgomery, *J. Org. Chem.*, **33,** 2393 (1968).
31. A. Albert and A. Hampton, *J. Chem. Soc.*, 4985 (1952).
32. R. D. Elliott, C. Temple, Jr., and J. A. Montgomery, *J. Org. Chem.*, **31,** 1890 (1966).
33. C. Temple, Jr., J. D. Rose, and J. A. Montgomery, *J. Med. Chem.*, **13,** 1234 (1970).
34. C. Temple, Jr., R. D. Elliott, J. D. Rose, and J. A. Montgomery, *J. Heterocycl. Chem.*, **7,** 451 (1970).
35. M. Giannella, F. Gualtieri, and M. Pigini, *Farmaco, Ed. Sci.*, **22,** 333 (1967); *Chem. Abstr.*, **67,** 108581 (1967).
36. C. Caradonna, F. Gualtieri, and F. M. Riccieri, *Gazz. Chim. Ital.*, **92,** 1459 (1962); *Chem. Abstr.*, **59,** 623 (1963).
37. J. W. Clark-Lewis and M. J. Thompson, *J. Chem. Soc.*, 430 (1957).
38. F. Korte and R. A. Wallace, in "Pteridine Chemistry," W. Pfleiderer and E. C. Taylor, Eds., Pergamon, Oxford, 1964, p. 75.
39. S. Bodforss, *Justus Liebigs Ann. Chem.*, **676,** 136 (1964); *Chem. Abstr.*, **62,** 560 (1965).
40. G. Henseke and D. Lehmann, *Angew. Chem., Int. Ed. Engl.*, **3,** 802 (1964).
41. E. C. Taylor, M. J. Thompson, and W. Pfleiderer, "Pteridine Chem., Proc. Int. Symp., 3rd, Stuttgart," 1962, p. 181; *Chem. Abstr.*, **62,** 16245 (1965).
42. W. Ried and J. Grabosch, *Chem. Ber.*, **89,** 2684 (1956).
43. H. Rudy and O. Majer, *Ber.*, **71B,** 1323 (1938).
44. H. Rudy and O. Majer, *Ber.*, **72B,** 940 (1939).
45. M. Israel, L. C. Jones, and E. J. Modest, *J. Med. Chem.*, **15,** 559 (1972).
46. T. S. Osdene and A. A. Santilli (to American Home Products), U.S. Pat. 3,180,868; *Chem. Abstr.*, **63,** 7022 (1965).
47. A. A. Santilli and T. S. Osdene (to American Home Products), U.S. Pat. 3,209,004; *Chem. Abstr.*, **63,** 18110 (1965).
48. R. D. Elliott, C. Temple, Jr., and J. A. Montgomery, *J. Org. Chem.*, **33,** 533 (1968).
49. C. Temple, Jr., R. D. Elliott, J. L. Frye, and J. A. Montgomery, *J. Org. Chem.*, **36,** 2818 (1971).
50. R. T. Coutts, D. Noble, and D. G. Wibberley, *J. Pharm. Pharmacol.*, **16,** 773 (1964).
51. R. Albrecht and G. A. Hoyer, *Chem. Ber.*, **105,** 3118 (1972).
52. A. Albert and H. Mizuno, *J. Chem. Soc., Perkin I,* 1615 (1973).
53. K. Winterfeld and M. Wildersohn, *Arch. Pharm.* (Weinheim), **303,** 44 (1970); *Chem. Abstr.*, **72,** 78970 (1970).
54. K. Winterfeld and M. Wildersohn, *Arch. Pharm.* (Weinheim), **303,** 1002 (1970); *Chem. Abstr.*, **74,** 42336 (1971).
55. R. C. De Selms and H. S. Mosher, *J. Am. Chem. Soc.*, **82,** 3762 (1960).
56. R. A. Archer and H. S. Mosher, *J. Org. Chem.*, **32,** 1378 (1967).
57. J. B. Ziegler, *J. Am. Chem. Soc.*, **71,** 1891 (1949).
58. A. Albert and A. Hampton, 2nd Congr. Int. Biochim., Résumés Communs, Paris, 1952, p. 444; *Chem. Abstr.*, **49,** 4648 (1955).

59. B. S. Gordon, J. M. Ravel, and W. Shrive, *J. Biol. Chem.*, **231,** 331 (1958).
60. R. Albrecht and G. A. Hoyer, *Chim. Ther.*, **8,** 346 (1973).
61. J. Armand, K. Chekir, J. Pinson, and N. Vinot, *C.R. Acad. Sci. Paris*, **281,** 547 (1975).
62. H. J. Stoklosa, J. R. Wasson, E. V. Brown, H. W. Richardson, and W. E. Hatfield, *Inorg. Chem.*, **14,** 2378 (1975).
63. T. Tanaka and S. Narita, *Yakugaku Zasshi*, **95,** 1092 (1975); *Chem. Abstr.*, **84,** 31002 (1976).
64. V. N. Charushin, O. N. Chupakhin, and I. Y. Postovskii, *Khim. Geterotsikl. Soedin.*, 1146 (1976); *Chem. Abstr.*, **86,** 5408 (1977).
65. N. Vinot and P. Maitte, *Bull. Chem. Soc. Fr.*, 251 (1976).

CHAPTER XXIX

Pyrido[3,4-*b*]pyrazines

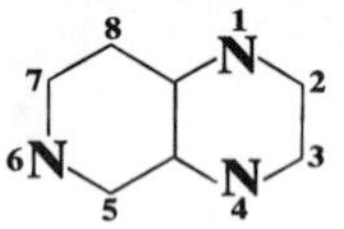

RRI 1607

I. Physical Properties . . . 545
II. Methods of Preparation . . . 547
1. From Symmetrical α-Dicarbonyl Compounds . . . 547
2. From Unsymmetrical α-Dicarbonyl Compounds . . . 549
A. α-Keto Aldehydes . . . 550
B. α-Keto Esters . . . 552
C. Alloxan . . . 553
3. From Substituted Alkylaminopyridines . . . 554
III. General Reactions . . . 556
1. Reaction with Electrophilic Reagents . . . 557
2. Oxidation . . . 557
3. Reaction with Nucleophilic Reagents . . . 558
4. Reduction . . . 559
IV. Biological Properties . . . 560
V. Tables . . . 561
Table 1. Aminopyrido[3,4-*b*]pyrazines . . . 561
Table 2. Pyrido[3,4-*b*]pyrazines (not amino compounds) . . . 564
VI. References . . . 566

This ring system has also been referred to as 1,4,6-triazanaphthalene and pyridino[3′,4′-2,3]pyrazine. In the latter formulation the heterocycle was numbered as in formula **1**. The standard numeration is depicted above.

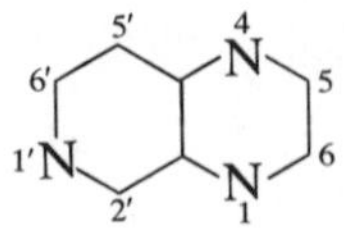

I. Physical Properties

Molecular orbital calculations have been performed on pyrido[3,4-*b*]pyrazine using several approaches.[1] Along with pyrido[2,3-*b*]pyrazine and other azanaphthalenes, these calculations provide good estimates of the transition energies and intensities of the ultraviolet spectrum of the parent heterocycle. The infrared spectra of the parent and other azanaphthalenes have been determined and discussed.[2] Infrared data for several α- and γ-hydroxy compounds confirm their existence, in solution and in the solid state, as oxo compounds.[3,4] The visible and ultraviolet spectrum of the parent has been analyzed.[5] In 95% ethanol the principal peaks are at 232 and 309 nm. The spectrum is substantially the same as that of quinoxaline, and analogous derivatives of the two heterocycles also possess similar spectra.[6] Comparative ultraviolet data have been used to differentiate between amino derivatives of pyrido[3,4-*b*]pyrazine and pyrido[2,3-*b*]pyrazine[7] and between isomeric dihydro and tetrahydro derivatives of the [3,4-*b*]system.[6] PMR data are available for certain 5,8-diamino compounds.[7,8] The ^{13}C NMR spectrum of the cation of pyrido[3,4-*b*]pyrazine has now been published.[36]

Covalent hydration is much more evident in this ring system than it is in pyrido[2,3-*b*]pyrazines.[4,9,10] The neutral parent heterocycle is stable as the anhydrous species. However the anhydrous cation is unstable and is rapidly hydrated. Originally it was thought, on the basis of chemical reactivity, that the cation was monohydrated across the 1,2 double bond.[4] Now PMR examination has shown that the cation is dihydrated as in formula **2**. There is no evidence for the presence of any of the monohydrated species in acid solution.[11] These conclusions have recently been confirmed using ^{13}C NMR spectroscopy.[36]

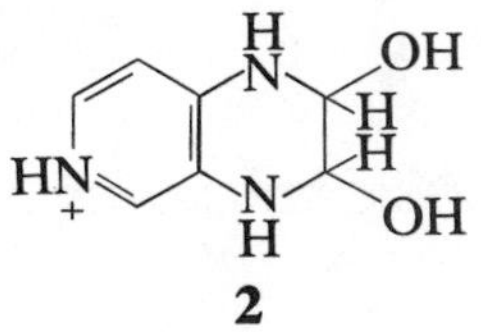

H OH
H
H
HN N OH
\+ H

2

Several substituted derivatives of pyrido[3,4-*b*]pyrazine have been shown to undergo covalent hydration.[4] Probably many more such compounds, as yet unexamined, also exhibit this behavior. The 2,3-dimethyl compound is not hydrated, whereas a monomethyl compound, now known to be the 2-methyl compound (see Section II2A), is 90% hydrated as its cation.[4] This compound, originally formulated as the 3-methyl isomer,[4,11] was shown by PMR evidence to be partially hydrated across what was thought to be the 1,2 bond. The supposed 3,4 bond was shown to be either completely hydrated or completely anhydrous. These conclusions must now be revised. Accordingly it seems that the 3,4 bond is only partially hydrated in the cation of 2-methylpyrido[3,4-*b*]pyrazine. The 3-oxo compound **3** (R=H) is strongly hydrated,[12] although its 2-oxo isomer gave no evidence of hydration.[4] The blocking effect of a 2-methyl group is demonstrated by the observation that compound **3** (R=H) is more hydrated than its 2-methyl derivative (**3**, R=Me).[4]

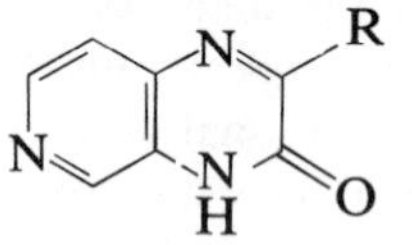

3

Interpretation of pK_a values of pyrido[3,4-*b*]pyrazines is complicated by the presence of variable amounts of covalent hydration. Indeed the study of acidity constants is one of the major methods for determining the extent of hydration. The base-strengthening effect of covalent hydration can be eliminated by determining pK_a values using rapid reaction methods. The use of this approach[13] with the parent heterocycle provides a pK_a value of 2.47. This value is more accurate than the previously accepted value[4] of 2.62 and contrasts sharply with the equilibrium value (affected by covalent hydration) of 4.60. As in the pyrido[2,3-*b*]pyrazines, the position of protonation must be on the pyridine nitrogen. The base-weakening effect of the successive introduction of nitrogen atoms into the other ring is indicated in Figure 1. As with isoquinoline, these compounds are all stronger bases than their quinoline analogues. The acidity constants of various substituted derivatives of the ring system are available.[4,14,15]

Figure 1. BASIC STRENGTHS OF AZAISOQUINOLINES

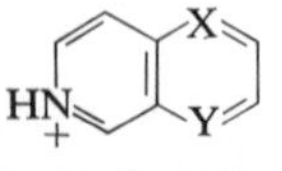

Compound	X	Y	pK_a
Isoquinoline	CH	CH	5.46
1,6-Naphthyridine	N	CH	3.78
1,7-Naphthyridine	CH	N	3.63
Pyrido[3,4-*b*]pyrazine	N	N	2.47

II. Methods of Preparation

Compared with the isomeric pyrido [2,3-*b*]pyrazines, relatively little work has been done on this ring system. All the syntheses of the heterocycle so far reported start from pyridine intermediates. A key intermediate for the preparation of pyrido[3,4-*b*]pyrazines is 3,4-diaminopyridine, which is prepared by nitration of 4-aminopyridine, followed by reduction of the 4-amino-3-nitro derivative using either sodium sulfide or hydrogen over palladium on charcoal.[4,16] Reduction of 4-amino-3-nitropyridine with stannous chloride and hydrochloric acid gives 3,4-diamino-2-chloropyridine[6] rather than the 6-chloro compound originally claimed.[16] The pyrido[3,4-*b*]pyrazine derived from the chlorodiamine must therefore also be reformulated. The earliest report of the synthesis of this ring system[17] could not be repeated when authentic 3,4-diaminopyridine was used.[16]

1. From Symmetrical α-Dicarbonyl Compounds

The condensation of 3,4-diaminopyridine with glyoxal hydrate in alcohol[18] gives a 50% yield of the parent heterocycle m.p. 97°. This method is more convenient than the use of the bisulfite compound in aqueous acetic acid, which provides only a low yield of the required compound.[16] Benzil and diacetyl[6] and 1,4-dibromobutane-2,3-dione[19] provide the expected 2,3-disubstituted products **4** in good yields. Dimethyl oxalate reacts with 3,4-diaminopyridine in refluxing *M* hydrochloric acid to give the open-chain compound **5**.[4] This may be cyclized to the dioxo compound **6** by heating at 230°.

3,4-Diaminopyridine $+ RCOCOR \longrightarrow$ **4**

4 R = Me, Ph, or CH_2Br

3,4-Diaminopyridine $+ MeO_2CCO_2Me \longrightarrow$ **5** (3-$NHCOCO_2H$-4-NH_2-pyridine) $\longrightarrow$ **6**

Pyrido[3,4-*b*]pyrazines substituted in the pyridine ring are available from the reactions of suitably substituted 3,4-diaminopyridines and the appropriate dicarbonyl compounds. No problems are encountered with diamines substituted with halogen[6,20] or hydroxy groups.[21] Similarly the pyridones **7** provide the oxo compounds **8** and **9**.[22] In contrast, 3,4-diamino-5-nitropyridine did not react with glyoxal, and only a low yield of 2,3-dimethyl-8-nitropyrido[3,4-*b*]pyrazine could be obtained using diacetyl.[23]

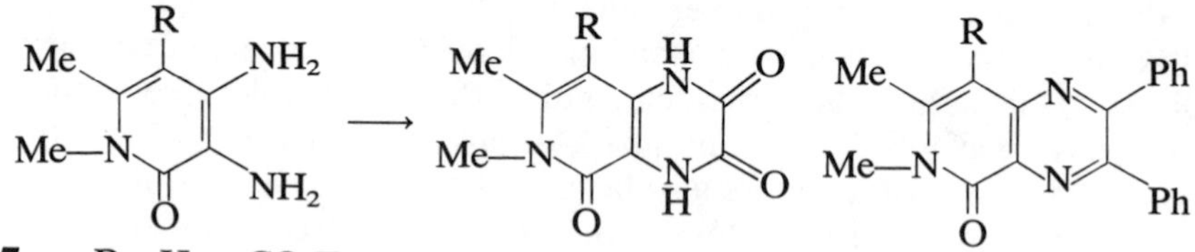

7 R = H or CO_2Et **8** **9**

The reaction of dicarbonyl compounds with 2,3,4-triaminopyridines[21] has been discussed as a route to aminosubstituted pyrido[2,3-*b*]pyrazines (see Chapter XXVIII). The direction of cyclization, and hence the proportions of the [2,3-*b*] and [3,4-*b*] systems formed, is dependent on the pH of the reaction medium. At pH 7 only the 5-aminopyrido[3,4-*b*]pyrazine (**10**) is isolated from the reaction with polyglyoxal, albeit in only 15% yield. The reaction of the symmetrical triamino compound **11** with glyoxal or diacetyl[23] or various benzils[24] readily yields the corresponding 8-amino compounds (**12**).

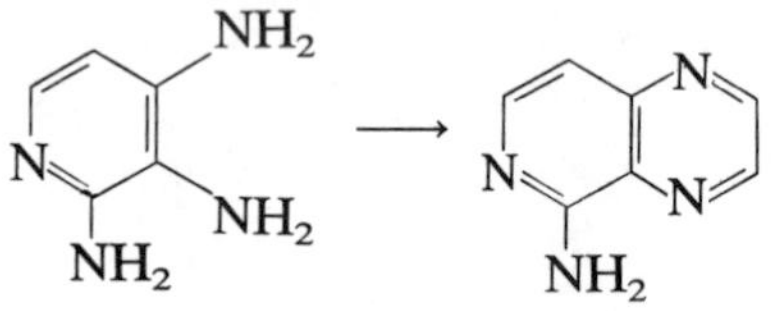

10

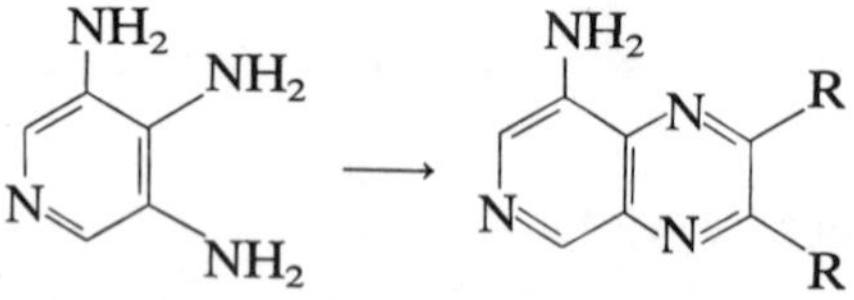

11 **12**

Diethyl oxalate reacts with 3,4-diaminopyridines having an alkyl or aryl substituent on the 4-amino group to provide 1-substituted 2,3-dioxo compounds. Thus the pyridines **13** yield the compounds **14**,[25] and the pyridones **15** give the corresponding trioxo products **16**.[22]

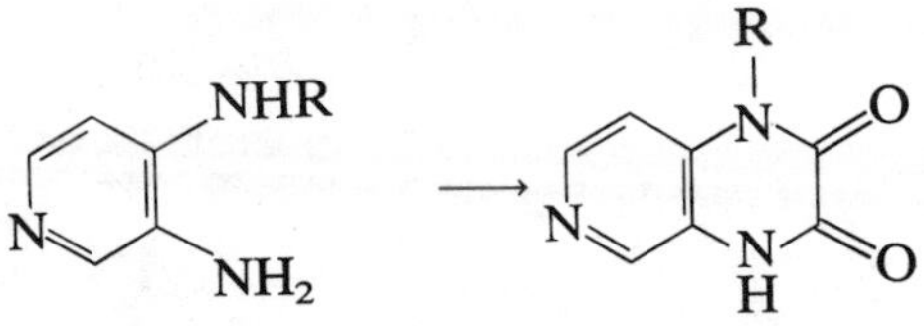

13 R = Bu or Ph → **14**

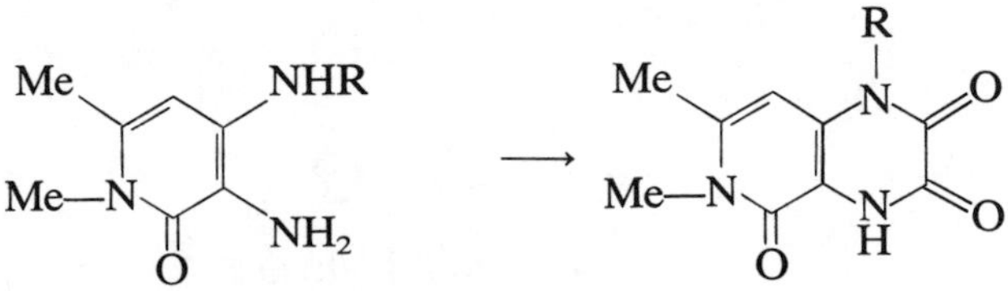

15 R = Bu or CH_2CH_2OH → **16**

2. From Unsymmetrical α-Dicarbonyl Compounds

The arguments used to suggest likely orientations in the pyrido[2,3-*b*]pyrazine series may also be applied to pyrido[3,4-*b*]pyrazines. Thus in neutral conditions one would expect initial reaction at the 3-amino group of 3,4-diaminopyridine with the aldehyde group of an α-keto aldehyde or the keto group of an α-keto ester. Such reaction would produce products of orientation **17** and **18**, respectively. The same products should be

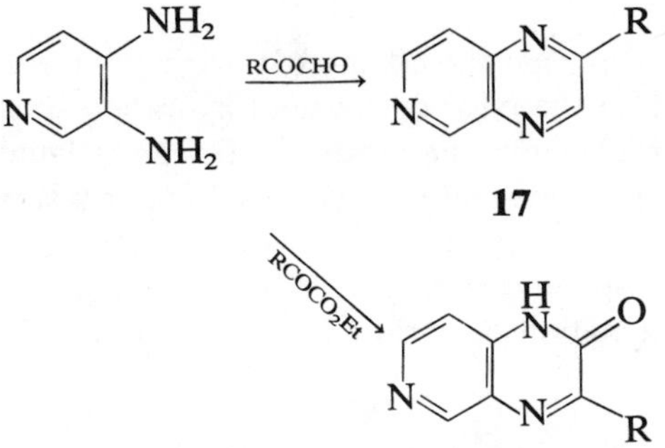

RCOCHO → **17**; $RCOCO_2Et$ → **18**; **19**

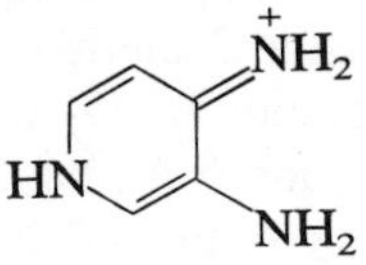

obtained in weakly acidic solutions, as the monocation **19** of 3,4-diaminopyridine can react in the same manner. In strongly acidic solutions ($pH \leqslant 1$) appreciable quantities of the diprotonated amine **20** will be present, favoring the formation of the isomers **21** and **22**.

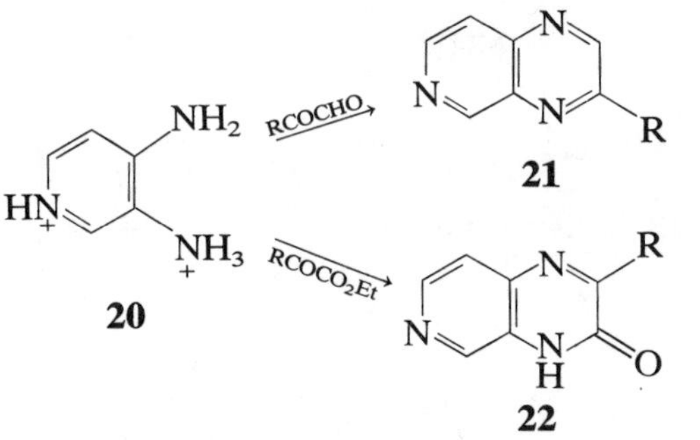

NH$_2$ RCOCHO → 21
HN$^+$ NH$_3^+$ RCOCO$_2$Et → 22
20

A. α-Keto Aldehydes

Reaction of 3,4-diaminopyridine with pyruvaldehyde (MeCOCHO) under neutral conditions gives the expected 2-methyl compound **23**, identical with a sample obtained from the 2-methyl-3-thioxo derivative **24**.[26] The infrared spectrum[26] of the compound is different to that of 3-methylpyrido[3,4-*b*]pyrazine prepared by an unambiguous route[27] (see

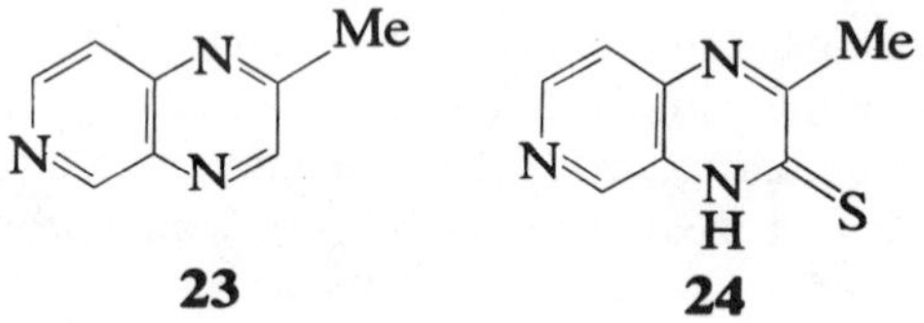

23 **24**

Section II3). It seems therefore that the compound originally formulated as the 3-methyl compound on the basis of covalent hydration studies[4] is in fact the 2-methyl isomer **23**. Only one product has been isolated from the reaction of 3,4-diaminopyridine and phenylglyoxal under a variety of conditions.[26] As the isomer was obtained under weakly acidic conditions it was formulated as the 2-phenyl derivative **25** by analogy with the arguments used for the orientation of pyrido[2,3-*b*]pyrazines.

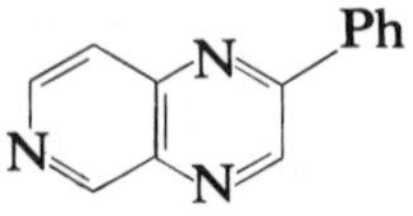

25

As mentioned in Chapter XXVIII condensation of the tetra-aminopyridine **26** (R = H) with pyruvaldehyde in 0.1 *M* hydrochloric acid resulted in a mixture from which the minor product **27** was obtained in only 6% yield.[7] The same product was also obtained from **26** (R = CO_2Et) and pyruvaldehyde followed by hydrolysis. The identity of the 2-methyl compound **27** was established by comparison with a sample of its 3-methyl isomer (**28**) prepared by an unambiguous route[8] (see Section II3).

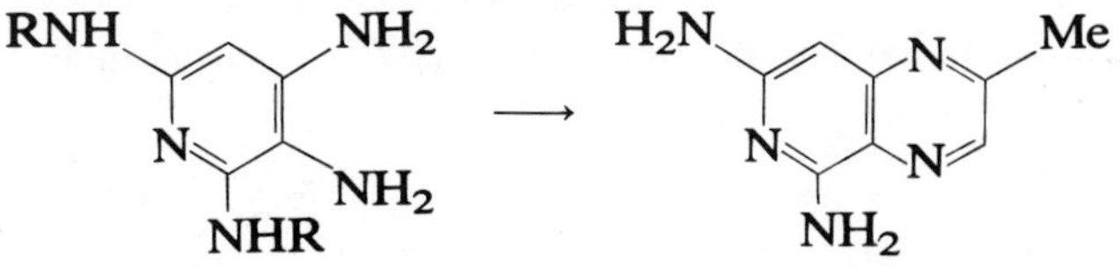

RNH NH$_2$ NH$_2$ NHR → H_2N Me NH$_2$

26 R = H or CO_2Et **27**

H_2N NH$_2$

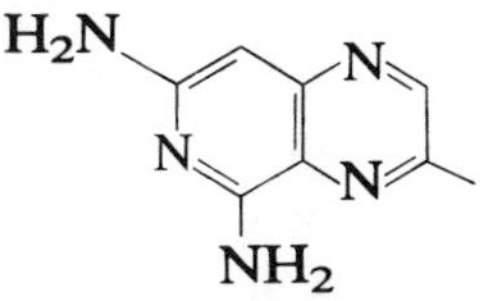

28

By analogy with the reaction of the tetraamino compound **26** and pyruvaldehyde, it has been suggested that the 2-aryl compounds **30** and **32** are obtained on reacting the triaminopyridines **29** and **31** with arylglyoxals.[24] In contrast, it is suggested that the reaction of 3,4,5-triaminopyridine (**31**) with pyruvaldehyde gives a mixture of the 2- and 3-methyl compounds, although neither compound was separated.[23]

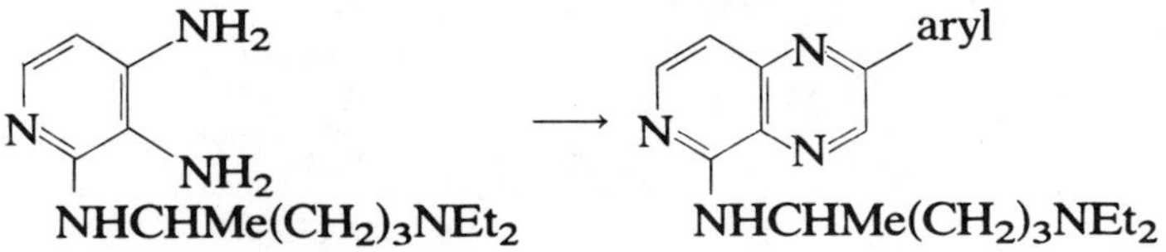

NH$_2$ NH$_2$ $NHCHMe(CH_2)_3NEt_2$ → aryl $NHCHMe(CH_2)_3NEt_2$

29 **30**

NH$_2$ NH$_2$ NH$_2$ → NH$_2$ aryl

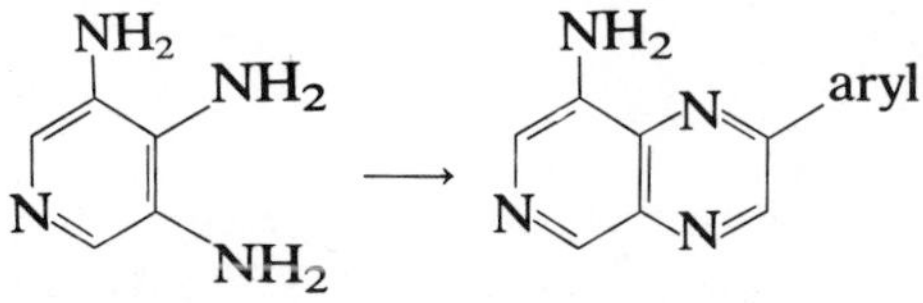

31 **32**

B. α-Keto Esters

The original report[6] of the condensation of ethyl pyruvate ($MeCOCO_2Et$) with 3,4-diaminopyridine contained several errors. The confusion resulting from this has now been clarified,[4] and a correction has appeared.[28] The reaction in ethanol of ethyl pyruvate and 3,4-diaminopyridine gives a mixture of the 2-methyl-3-oxo compound **33** m.p. 280° and the 3-methyl-2-oxo isomer **34** m.p. 265°. Surprisingly the

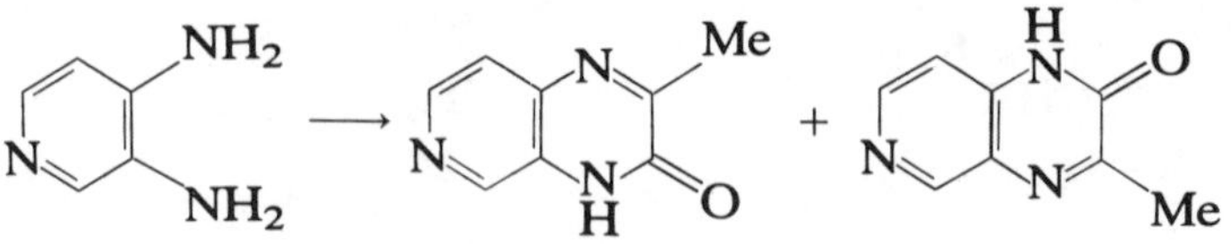

33 **34**

major product is claimed to be the 3-oxo compound,[4,6] although neutral reaction conditions should favor formation of the 2-oxo isomer. The proportions of the isomers obtained from this reaction have been disputed.[26] Although the *N*-methyl compounds **36** and **38** cannot be obtained by methylation of **33** and **34** using diazomethane,[4] they are available from the corresponding methylaminopyridines **35** and **37**. Thus reaction with ethyl pyruvate gives compound **36** m.p. 114–115° and compound **38** m.p. 141–142°.

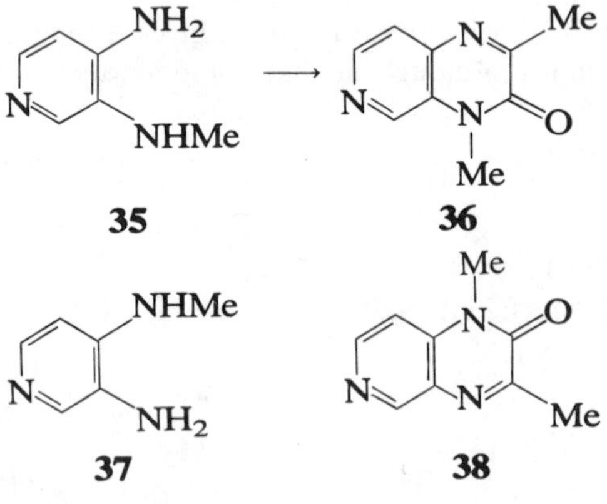

35 **36**

37 **38**

Ethyl mesoxalate (EtO_2CCOCO_2Et) reacts with 3,4-diaminopyridine in ethanol to give a mixture in which the expected 2-oxo compound **39** predominates. Separation of the isomers gave 65% of the 2-oxo compound and only 0.6% of the 3-oxo isomer.[6] Use of *M* hydrochloric acid as a reaction medium gave an 80% yield of the 2-oxo compound, which was hydrolyzed in situ to the carboxylic acid. Reaction of the appropriate methylaminopyridines gave the *N*-methyl derivatives **40** and **41** in alcohol and the corresponding carboxylic acids in acid solution.[6]

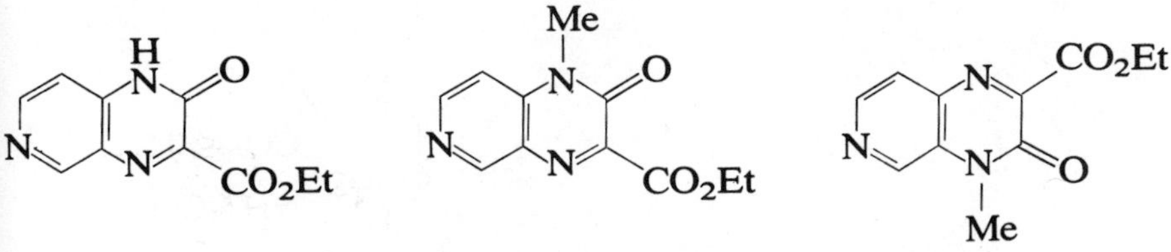

40 **41**

The reaction of ethyl glyoxylate with 3,4-diaminopyridine in benzene–dioxan gave only the 2-oxo compound **42** on work-up.[26] In water a mixture of the 2-oxo compound (56%) and its 3-oxo isomer **43** (35%)

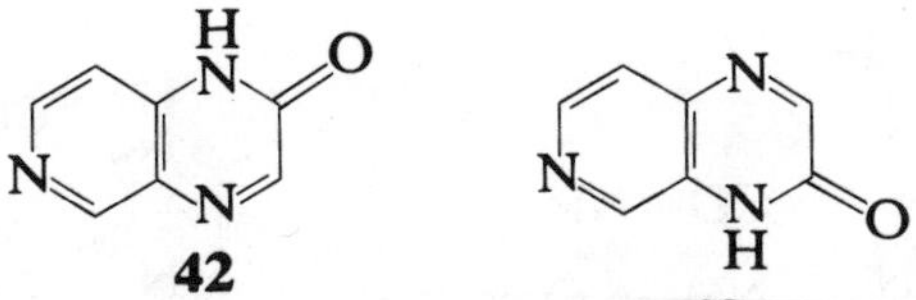

42 **43**

was obtained using ethyl glyoxylate hemiacetal.[4] In *M* sulfuric acid the expected 3-oxo compound was formed. In spite of the simple rationalization which may be applied to the formation of these products, it has been suggested that in reality the condensations are much more complex.[26] Further work will be needed to clarify this point.

C. *Alloxan*

As in the pyrido[2,3-*b*]pyrazine series, a group of ureides and spirohydantoins are available from the condensation of alloxan (**44**) with 3,4-diaminopyridines. If the reaction is performed at room temperature in ethanol, the ureide **45** is obtained from 3,4-diaminopyridine. Treatment of the ureide with acid or alkali causes cyclization to the hydantoin **46**.

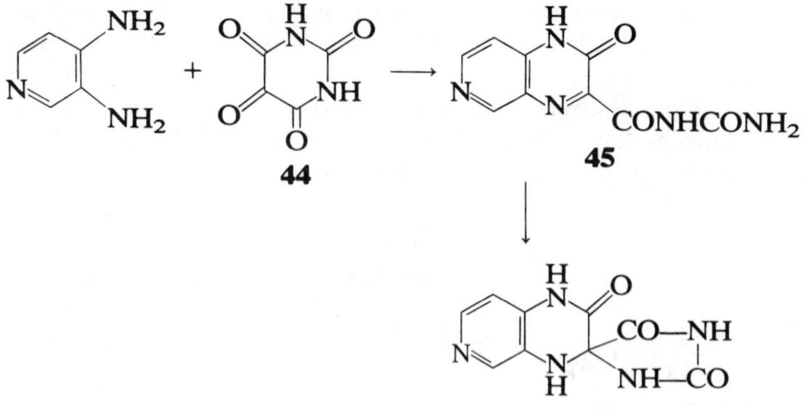

44

45

46

The hydantoins **47** and **48** are formed directly from the appropriate methylaminopyridine.[6] Methylation, acetylation, or hydrolysis of these compounds gives products analogous to those in the [2,3-*b*] series.[6]

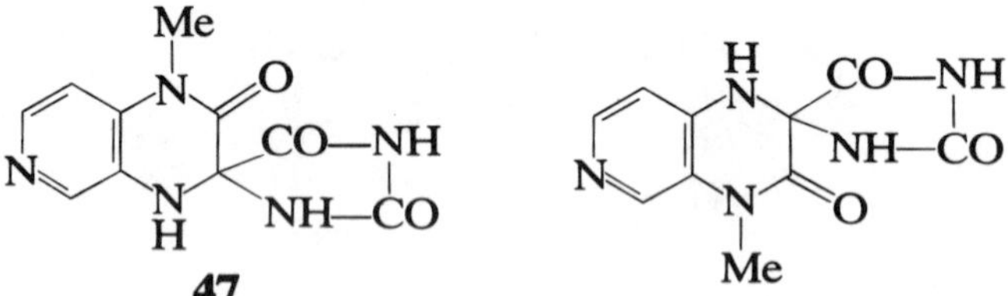

47

48

3. From Substituted Alkylaminopyridines

The reactivity of 4-chloro-3-nitropyridines allows ready replacement of the chlorine with suitably substituted alkylamines. Reduction of the 3-nitro group then permits ready cyclization in situ between the newly formed 3-amino groups and an appropriate center (e.g., a carbonyl group) on the alkylamino function in the 4-position. By this means various pyrido[3,4-*b*]pyrazines can be synthesized unambiguously. Thus hydrogenation of the nitro compound **49** over Raney nickel gave the dihydropyrido[3,4-*b*]pyrazine **50**, which was readily oxidized to the aromatic compound **51** using potassium permanganate in acetone.[8] This unambiguous synthesis confirmed the orientation of the product from the condensation of the tetraaminopyridine **26** (R = CO_2Et) and pyruvaldehyde (see Section II2A). The same approach has been used for the synthesis of more complex analogues of **51**.[29,30]

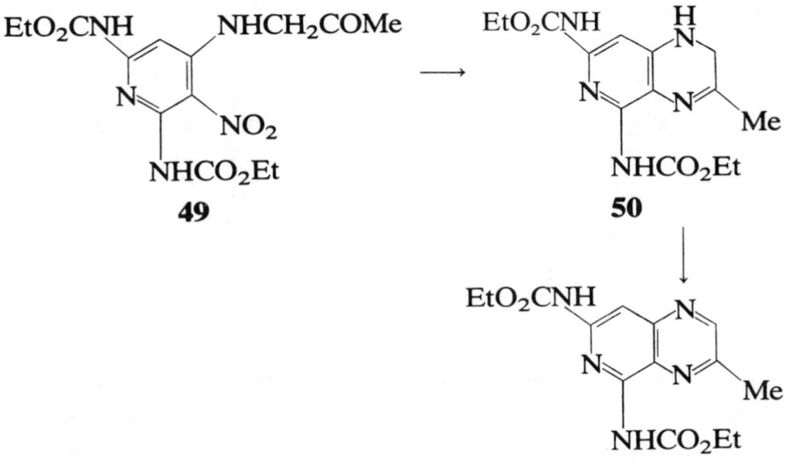

Similarly the orientation of 3-oxo derivatives of the ring system was established by the unambiguous synthesis of the tetrahydro derivatives **52**.[4] Compound **53** was obtained by the same approach.[22]

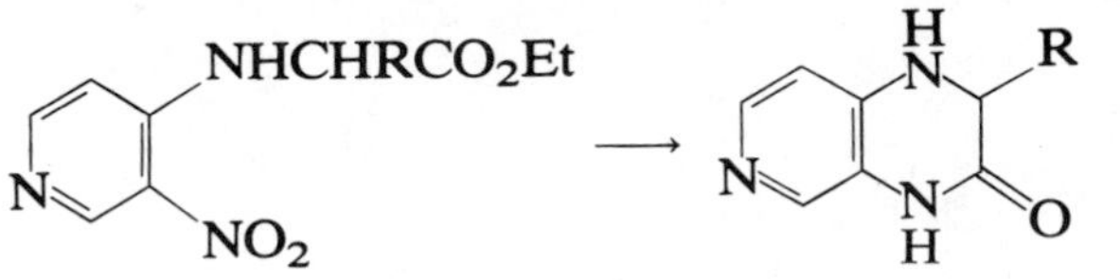

Cyclization of the alcohol **54** using hydrogen bromide was used in the unambiguous synthesis of 3-methylpyrido[3,4-*b*]pyrazine (**56**), obtained from the tetrahydro compound **55** by dehydrogenation over palladium on charcoal.[27] Reduction of the nitro compound **57** resulted in cyclization to form the reduced derivative **58** of the ring system.[22]

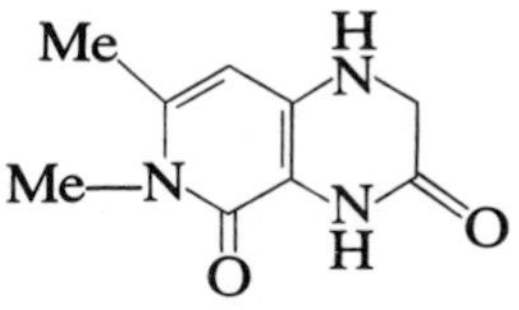

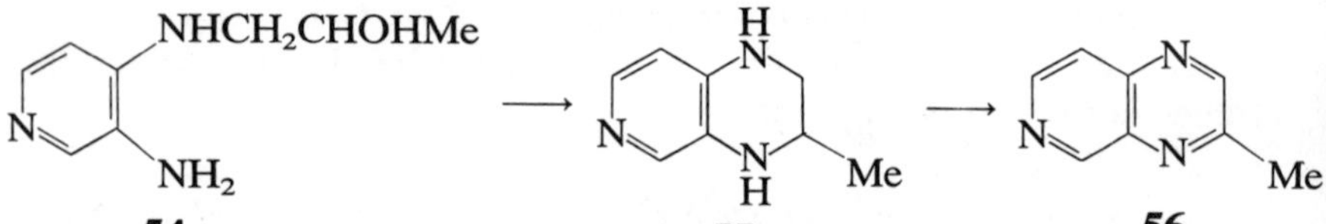

The reaction of 3,4-diaminopyridine and chloroacetic acid is reported[31] to give the 2-oxo compound **59** m.p. 284° and thus parallels the corresponding reaction of *o*-phenylenediamine with chloroacetic acid, which yields 1,2,3,4-tetrahydro-2-oxoquinoxaline.[32] No imidazopyridine (**60**) was isolated from the reaction. The possibility of the isomer **52** (R = H) being formed is not discussed, although the melting point of the latter (>250° decomp.[4]) makes it seem that this was not the compound isolated.

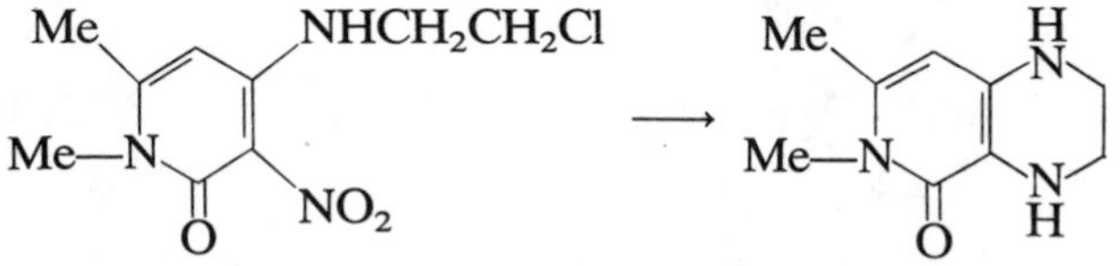

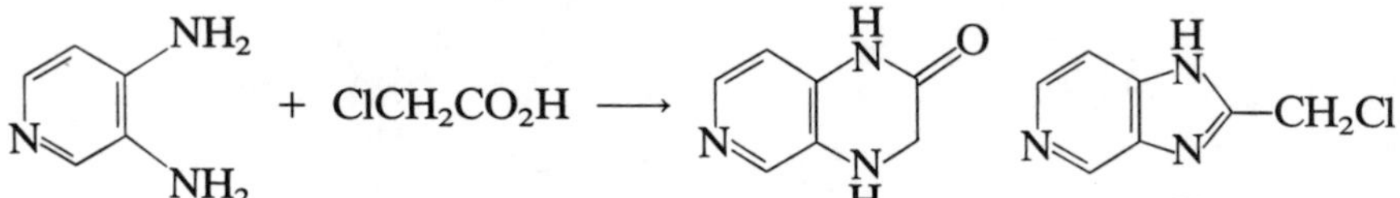

III. General Reactions

Although much less work has been done on this ring system, as might be expected the chemical properties are generally similar to those of the pyrido[2,3-*b*]pyrazines.

1. Reaction with Electrophilic Reagents

In spite of an earlier report to the contrary,[6] 1,2-dihydro-3-methyl-2-oxopyrido[3,4-*b*]pyrazine (**34**) and its 3-oxo isomer (**33**) cannot be methylated with diazomethane. Methyl iodide and methyl sulfate are similarly ineffective.[4] This lack of reactivity contrasts with the easy methylation of the [2,3-*b*] series. A solution of diazomethane in ether is sufficient however to *N*-methylate the 2-oxo compounds **61**.[6] Surprisingly the carboxylic acid group in **61** (R = H) is unaffected by this reagent.

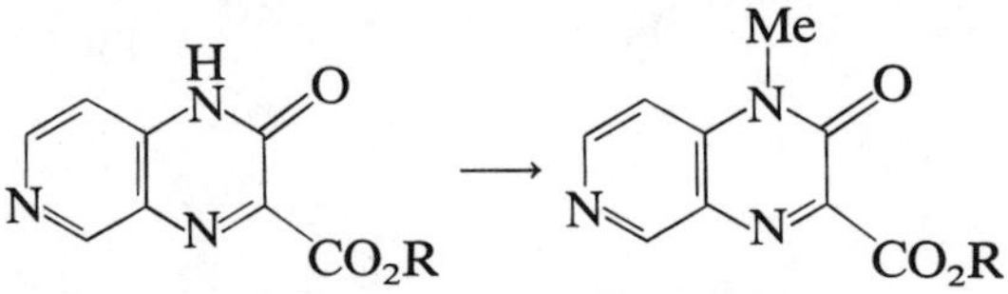

61 R = H or Et → **62** R = H or Et

Bromination of the compounds **63** using bromine in acetic acid provides the 8-bromo compounds **64**.[24] Although a solution of *N*-bromosuccinimide in carbon tetrachloride does not brominate the amine **65** (R = H), it does brominate, in an unspecified yield, the carbamate **65** (R = CO_2Et). The position of substitution was identified as the 8-position on the basis of PMR evidence.[8] There was no evidence of bromination of the methyl group. Bromination of a methyl phenyl compound, described without proof as **66**, using bromine in carbon tetrachloride gave the dibromo compound **67**. Treatment of **66** with iodine in pyridine gave the pyridinium salt **68**, presumably by way of the iodomethyl compound.[19] The orientation of these products remains to be established.

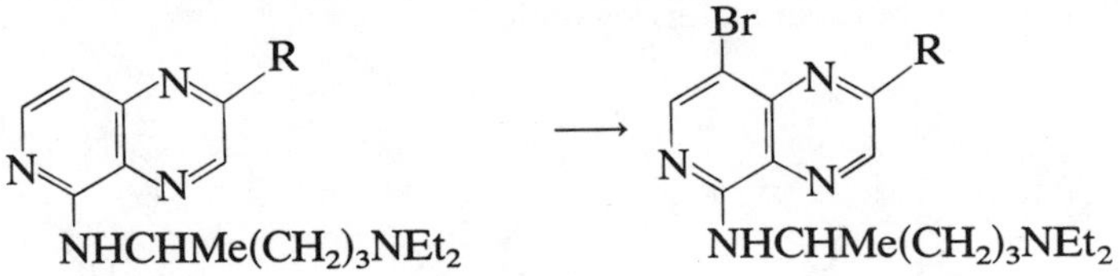

$NHCHMe(CH_2)_3NEt_2$ → $NHCHMe(CH_2)_3NEt_2$

63 R = Cl or 2-chlorophenyl → **64**

2. Oxidation

Treatment of the parent heterocycle with cold potassium permanganate solution gave the 1,2-dihydro-2-oxo compound. Similar treatment of 2-methylpyrido[3,4-*b*]pyrazine (thought at the time to be the 3-methyl compound) resulted only in extensive decomposition. Hydrogen peroxide

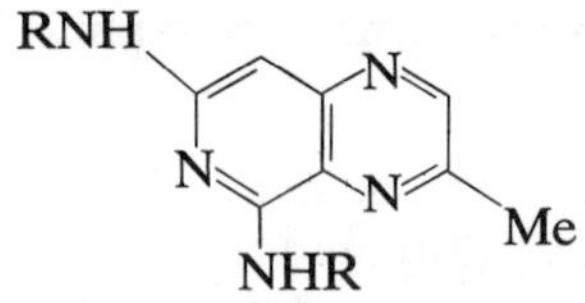

65 R = H or CO_2Et

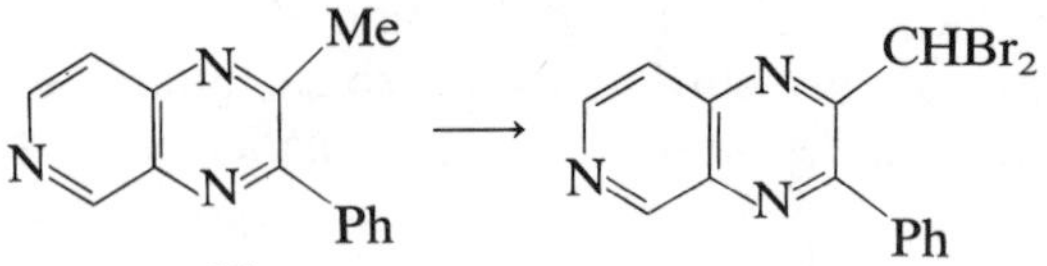

66 **67**

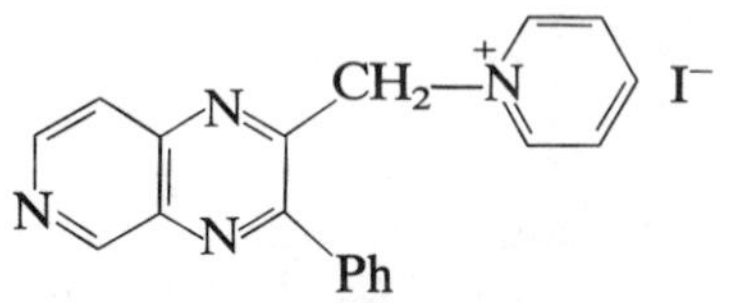

68

also caused the methyl compound to decompose.[4] Mild oxidizing agents such as iodine[4] and potassium permanganate in acetone[8,29,30] have been used to oxidize 1,2-dihydro compounds to the aromatic heterocycle, and palladium on charcoal treatment dehydrogenated 1,2,3,4-tetrahydro-3-methylpyrido[3,4-*b*]pyrazine to the aromatic system.[27]

Although mild oxidation with iodine of compound **69** gave the mono-oxo compound **70**, more vigorous oxidation with the same reagent provided the dioxo compound **71**. This could also be obtained from **70** by oxidation with potassium ferricyanide.[4]

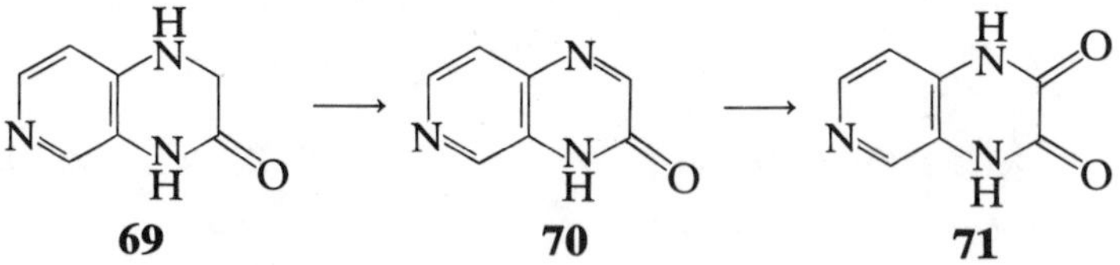

69 **70** **71**

3. Reaction with Nucleophilic Reagents

Although 3,4-dihydro-3-oxopyrido[2,3-*b*]pyrazine does not react with phosphoryl chloride or phosphorus pentasulfide to give the 3-chloro and 3-thioxo compounds,[4] its 2-methyl derivative **72** reacts with phosphorus pentasulfide in pyridine to give the 3-thioxo compound **73**.[26] Similarly the

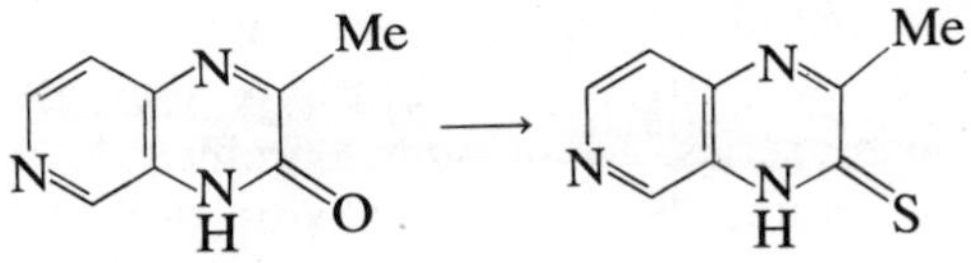

72 → **73**

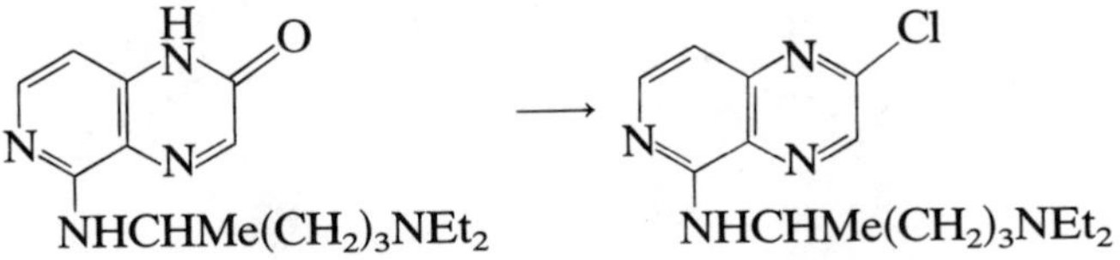

$NHCHMe(CH_2)_3NEt_2$ **74** → $NHCHMe(CH_2)_3NEt_2$ **75**

2-oxo compound **74** reacts with phosphoryl chloride to give the 2-chloro derivative **75**.[24]

No other nucleophilic reactions at the 2- or 3-position have been reported. The 7-chloro compound **76** has been treated with liquid ammonia to give the 7-amino compound **77**. The chloro compound does not, however, react with aqueous alkali or acid, neither is it displaced by methoxide ion in methanol.[4]

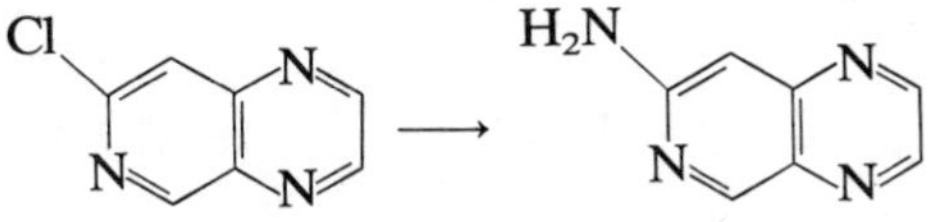

76 → **77**

4. Reduction

Treatment of the thioxo compound **73** with Raney nickel readily results in desulfurization to give 2-methylpyrido[3,4-*b*]pyrazine.[26] Hydrogenation of the chloro compound **76** over palladized calcium carbonate gives the parent heterocycle.[6] In contrast to the ease of reduction of 1,2,3,4-tetrahydro-2,3-dioxopyrido[2,3-*b*]pyrazine, its pyrido[3,4-*b*] analogue (**78**), on treatment with sodium amalgam, extensively decomposes.[26]

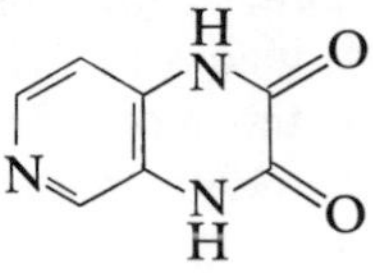

78

Lithium aluminum hydride reduces 2,3-dimethylpyrido[3,4-*b*]pyrazine to the *cis*-tetrahydro derivative, analogous to that described in the pyrido[2,3-*b*]pyrazine series.[33] Potassium borohydride has been used to reduce the oxo compound **79** to the tetrahydro derivative **80**.[4]

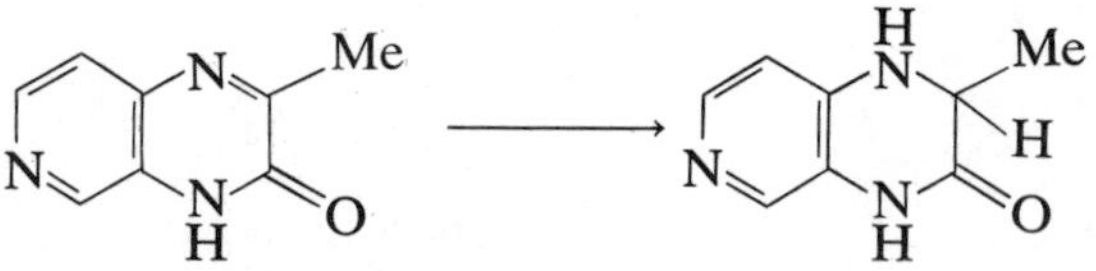

79 **80**

IV. Biological Properties

This ring system has attracted little interest as a source of compounds of possible pharmacological interest. The synthesis of the potential folic acid antagonist **81** has been described.[30] Analogues of chloroquine such

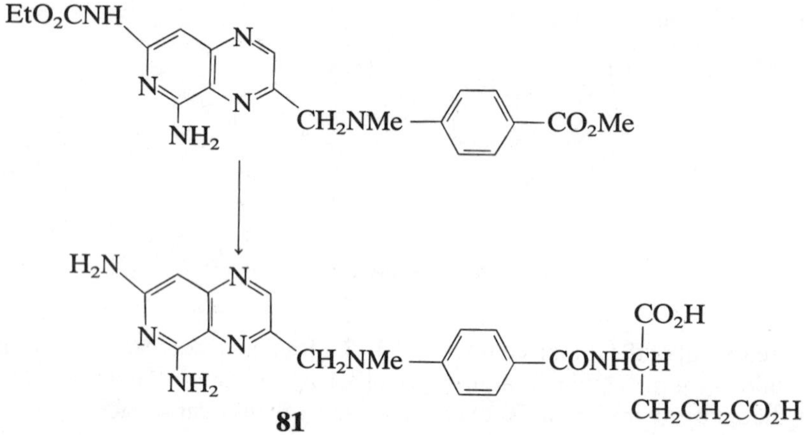

81

as **82** are inactive as antimalarials.[24] As the corresponding pyrido[2,3-*b*]pyrazines do show some activity, it seems that activity requires the chloroquine side chain to be para to the ring nitrogen. As in the [2,3-*b*] series the oxine analogue **83** is inactive as an antibacterial; increasing the number of nitrogen atoms in the molecule decreases the activity.[34] A recent patent[37] claims, without data, a series of amino derivatives of pyrido[3,4-*b*]pyrazinecarboxylic acids and esters as having antiinflammatory and tranquilizing properties.

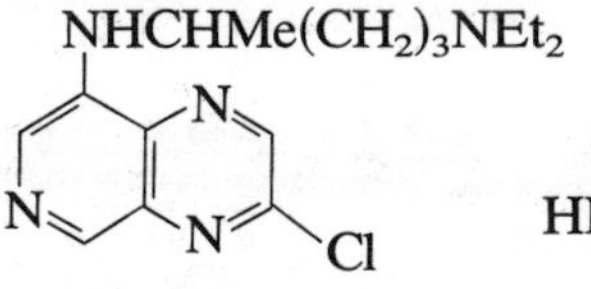

$NHCHMe(CH_2)_3NEt_2$ … Cl

82

HN … O

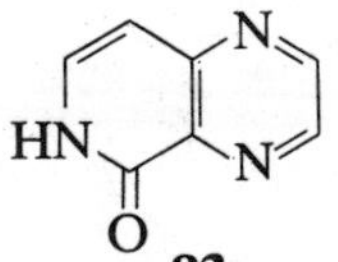

83

V. Tables

TABLE 1. AMINOPYRIDO[3,4-*b*]PYRAZINES

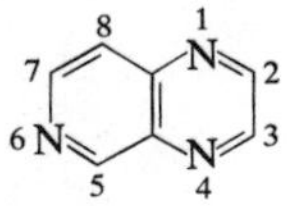

Molecular formula	Substituents or structure	m.p., °C	Ref.
$C_7H_6N_4$	5-Amino	187–188	21
	7-Amino	210–212	4
	8-Amino	149–150	23
$C_8H_8BrN_5$	5,7-Diamino-8-bromo-3-methyl		
	Dicarbamate	222 decomp.	8
$C_8H_8N_4$	8-Amino-2-(and 3)-methyl (mixture)	137–162	23
$C_8H_9N_5$	5,7-Diamino-2-methyl	237 decomp.	7
	Dicarbamate	199	
	5,7-Diamino-3-methyl	226	8
	Monocarbamate	184	
	Dicarbamate	206	
$C_8H_{11}N_5$	5,7-Diamino-1,2-dihydro-3-methyl		
	Dicarbamate	188–190 decomp.	8
$C_9H_{10}N_4$	8-Amino-2,3-dimethyl	187–189	23
$C_{13}H_8Cl_2N_4$	8-Amino-2-(3,4-dichlorophenyl)	236–238	24
$C_{13}H_9ClN_4$	8-Amino-2-(4-chlorophenyl)	200–202	24
$C_{15}H_{14}N_6O_2$	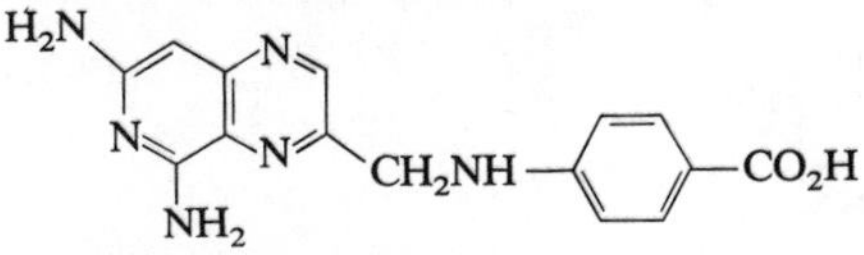 7-H_2N, 5-NH_2, 3-CH_2NH-C_6H_4-CO_2H		
	7-Carbamate ethyl ester	ca. 201 decomp.	30
$C_{15}H_{16}N_6$	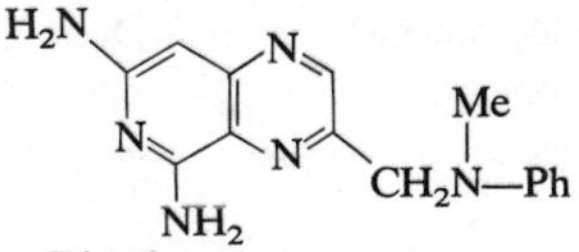 7-H_2N, 5-NH_2, 3-$CH_2N(Me)$—Ph	195 decomp.	29
	Dicarbamate	173	

TABLE 1 (*contd.*)

8 1 7 N 2 6 N 3 N 5 4

Molecular formula	Substituents or structure	m.p., °C	Ref.
$C_{15}H_{16}N_6O_2$	H_2N H N N N CH_2NH CO_2H NH_2		
	Carbamate ethyl ester	ca. 168	30
$C_{16}H_{16}N_6O_2$	H_2N N N N Me CH_2N CO_2H NH_2		
	5,7-Diacetyl	ca. 370	30
	7-Carbamate methyl ester	217	
$C_{16}H_{18}N_6O_2$	H_2N H N N N Me CH_2N CO_2H NH_2		
	7-Carbamate methyl ester	172 decomp.	30
$C_{16}H_{23}BrClN_5$	Br N Cl N N $NHCHMe(CH_2)_3NEt_2$		
	Hydrochloride	ca. 223 decomp.	24
$C_{16}H_{24}ClN_5$	N Cl N N $NHCHMe(CH_2)_3NEt_2$	Oil	24
$C_{16}H_{25}N_5O$	H N O N N $NHCHMe(CH_2)_3NEt_2$	118–120	24
$C_{19}H_{10}Cl_4N_4$	8-Amino-2,3-bis(3,4-dichlorophenyl)	Decomp. >116	24
$C_{19}H_{12}Cl_2N_4$	8-Amino-2,3-bis(4-chlorophenyl)	180–181	24
$C_{19}H_{15}N_5$	5,6-Diamino-2,3-diphenyl	ca. 131 decomp.	7
	Dicarbamate	170	
$C_{21}H_{18}N_4$	8-Amino-2,3-bis(2-methylphenyl)	179–181	24

TABLE 1 (*contd.*)

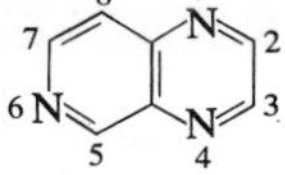

Molecular formula	Substituents or structure	m.p., °C	Ref.
$C_{21}H_{18}N_4O$	7-Amino-5-ethoxy-2,3-diphenyl Carbamate	205	7
$C_{21}H_{23}N_7O_5$	H_2N, NH_2, Me, CH_2N, $CONHCHCO_2H$, $CH_2CH_2CO_2H$ 5,7-Diacetyl diethyl ester	ca. 91	30
$C_{22}H_{27}BrClN_5$	Cl, Br, $NHCHMe(CH_2)_3NEt_2$ Hydrochloride	223–225 decomp.	24
$C_{22}H_{28}ClN_5$	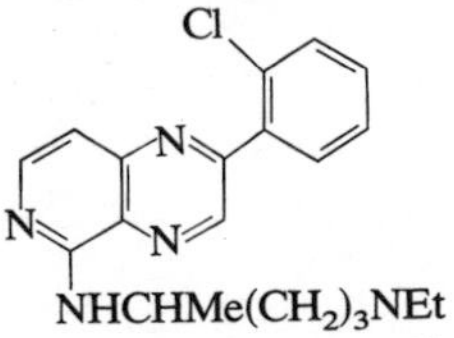 Cl, $NHCHMe(CH_2)_3NEt$	Oil	24
	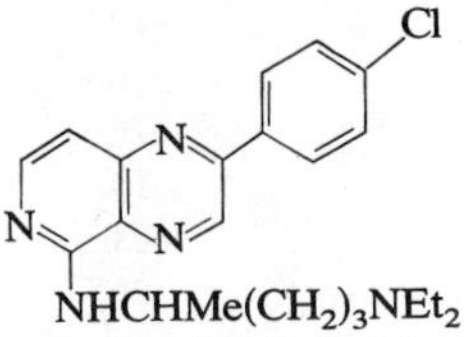 Cl, $NHCHMe(CH_2)_3NEt_2$	Oil	24
$C_{25}H_{26}N_4O_6$	8-Amino-2,3-bis(3,4,5-trimethoxyphenyl)	217–219	24

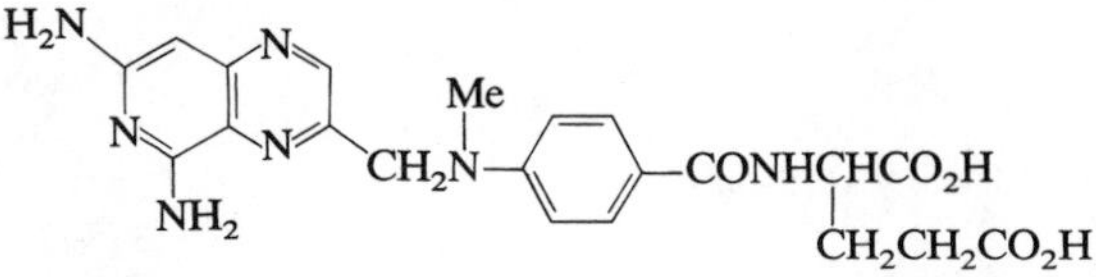

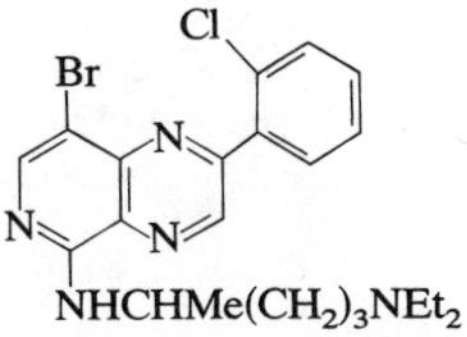

TABLE 2. PYRIDO[3,4-*b*]PYRAZINES (NOT AMINO COMPOUNDS)

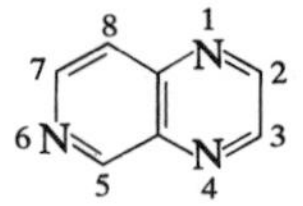

Molecular formula	Substituents or structure	m.p., °C	Ref.
$C_7H_4ClN_3$	5-Chloro[a]	138–139	6, 16
	Hydrochloride	>250	16
	7-Chloro	114–115	4
$C_7H_5N_3$	Unsubstituted[b]	97	18
		100–101	16
	Picrate	185 decomp.	16
	Methiodide	198–204	26
$C_7H_5N_3O$	1,2-Dihydro-2-oxo	284	4, 31
		315–316	26
	3,4-Dihydro-3-oxo	Color change 280	4
	5,6-Dihydro-5-oxo		
	Hemihydrate	270 decomp.	3, 21
$C_7H_5N_3O_2$	1,2,3,4-Tetrahydro-2,3-dioxo	>320	4
$C_7H_7N_3O$	1,2,3,4-Tetrahydro-3-oxo	>250	4
$C_8H_5N_3O_3$	2-Carboxy-3,4-dihydro-3-oxo		
	Ethyl ester	153–154	6
	3-Carboxy-1,2-dihydro-2-oxo		
	Dihydrate	>380	6
	Ethyl ester	227–228	
$C_8H_6N_4O_2$	2-Carboxamido-3,4-dihydro-3-oxo	355 decomp.	6
	3-Carboxamido-1,2-dihydro-2-oxo	>315 decomp.	6
$C_8H_7N_3$	2-Methyl	76–77	26
		78–79	4[c]
	3-Methyl	84.5–85	27
$C_8H_7N_3O$	1,2-Dihydro-1-methyl-2-oxo	177	6
	1,2-Dihydro-3-methyl-2-oxo[d]	265	4, 28
	3,4-Dihydro-2-methyl-3-oxo[e]	280	4, 28
	3,4-Dihydro-4-methyl-3-oxo	145	6
$C_8H_7N_3S$	3,4-Dihydro-2-methyl-3-thioxo	Chars 249–252	26
$C_8H_9N_3O$	1,2,3,4-Tetrahydro-2-methyl-3-oxo	265	4
$C_8H_{11}N_3$	1,2,3,4-Tetrahydro-3-methyl		
	Hydrobromide	185–186	27
	Picrate	168–169	
$C_9H_7Br_2N_3$	2,3-Bisbromomethyl	[f]	19
$C_9H_7N_3O_3$	2-Carboxy-3,4-dihydro-4-methyl-3-oxo	286–288 decomp.	6
	Ethyl ester	127	
	3-Carboxy-1,2-dihydro-1-methyl-2-oxo	214–215	6
	Ethyl ester	176	
$C_9H_7N_5O_3$	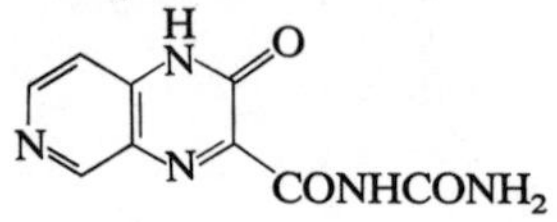 $CONHCONH_2$	238–240	6

TABLE 2 (*contd.*)

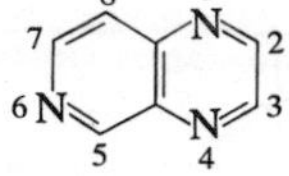

Molecular formula	Substituents or structure	m.p., °C	Ref.
$C_9H_7N_5O_3$ (*contd.*)	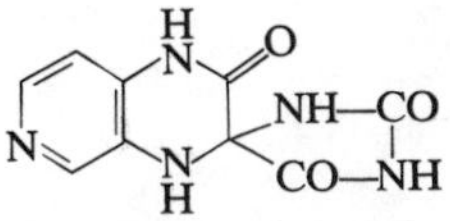	257–258	6
$C_9H_8N_4O_2$	2-Carboxamido-3,4-dihydro-4-methyl-3-oxo	262–263	6
	3-Carboxamido-1,2-dihydro-1-methyl-2-oxo	253–254 decomp.	6
	2,3-Dimethyl-8-nitro	109.5–110	23
$C_9H_9N_3$	2,3-Dimethyl	126	6
$C_9H_9N_3O$	1,2-Dihydro-1,3-dimethyl-2-oxo[g]	141–142	4, 28
	3,4-Dihydro-2,4-dimethyl-3-oxo[h]	114–115	4, 28
$C_9H_9N_3O_3$	1,2,3,4,5,6-Hexahydro-6,7-dimethyl-2,3,5-trioxo	chars > 350	22
$C_9H_{11}N_3O_2$	1,2,3,4,5,6-Hexahydro-6,7-dimethyl-3,5-dioxo	313–314 decomp.	22
$C_9H_{13}N_3$	1,2,3,4-Tetrahydro-2,3-dimethyl, cis isomer	149–150	33
$C_9H_{13}N_3O$	1,2,3,4,5,6-Hexahydro-6,7-dimethyl-5-oxo	198–199	22
$C_{10}H_9N_3O_2$	3-Acetonyl-1,2-dihydro-2-oxo	267–269	26
$C_{10}H_9N_3O_5$	8-Carboxy-1,2,3,4,5,6-Hexahydro-6,7-dimethyl-2,3,5-trioxo		
	Ethyl ester	311 decomp.	22
$C_{10}H_9N_5O_3$	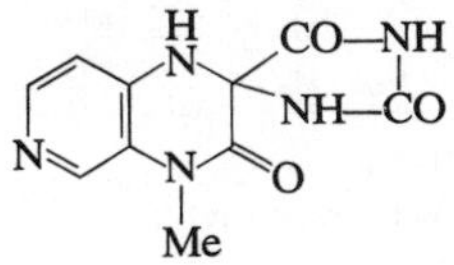	167	6
	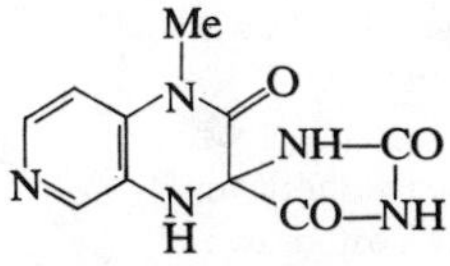	225–226 decomp.	6
$C_{11}H_9N_3O_3$	3-Methyl-2-oxalylmethyl		
	Ethyl ester	175–176.5	26
$C_{11}H_{13}N_3O_2$	1-Butyl-1,2,3,4-tetrahydro-2,3-dioxo	256	25
$C_{11}H_{13}N_3O_4$	1,2,3,4,5,6-Hexahydro-1-(2-hydroxyethyl)-6,7-dimethyl-2,3,5-trioxo	291–292	22
$C_{12}H_{13}N_5O_3$	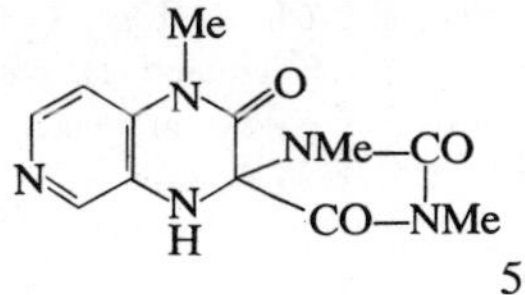	185	6

TABLE 2 (*contd.*)

Molecular formula	Substituents or structure	m.p., °C	Ref.
$C_{13}H_9N_3$	2(or 3)-Phenyl	125.5–126	26
$C_{13}H_9N_3O_2$	1,2,3,4-Tetrahydro-2,3-dioxo-1-phenyl	>325	25
$C_{13}H_{17}N_3O_3$	1-Butyl-1,2,3,4,5,6-hexahydro-6,7-dimethyl-2,3,5-trioxo	270	22
$C_{14}H_9Br_2N_3$	2-Dibromomethyl-3-phenyl[i]	70 decomp	19
$C_{14}H_{13}N_5O_5$	Me; N; O; COMe; N—CO; N; N; CO—NH; COMe	272	6
$C_{19}H_{12}BrN_3$	8-Bromo-2,3-diphenyl	198–200	20
$C_{19}H_{13}N_3$	2,3-Diphenyl	177	6
$C_{19}H_{15}N_3$	1,2(or 3,4)-Dihydro-2,3-diphenyl	247	61
$C_{19}H_{15}N_4$ Iodide	3-Phenyl-2-pyridiniummethyl[i] Iodide	174	19
$C_{21}H_{17}N_3O$	5,6-Dihydro-6,7-dimethyl-5-oxo-2,3-diphenyl	246–247	22
$C_{22}H_{19}N_5O$	O^-; N; CH=$\overset{+}{N}$—C_6H_4—NMe_2[i]; N; N; Ph	100	19

[a] Incorrectly formulated as 7-chloro in reference 16.
[b] Reference 17 reports m.p. 153° but this could not be repeated.
[c] Incorrectly formulated as the 3-methyl compound.
[d] Reference 6 gave m.p. 276–278° but this was later withdrawn.[28]
[e] Reference 6 gave m.p. 262° but this was later withdrawn.[28]
[f] Decomposes on warming.
[g] Reference 6 gave m.p. 276–277° but this was later withdrawn.[28]
[h] Reference 6 gave m.p. 228–230° but this was later withdrawn.[28]
[i] Possibly the 2-phenyl isomer.

VI. References

1. G. Favini, I. Vandoni, and M. Siminetta, *Theoret. Chim. Acta*, **3,** 45, 418 (1965); *Chem. Abstr.*, **62,** 8525 (1965); *ibid.*, **63,** 17326 (1965); B. Tinland, *Theoret. Chim. Acta*, **8,** 361 (1967); *Chem. Abstr.*, **68,** 100246 (1968); S. C. Wait, Jr., and J. W. Wesley, *J. Mol. Spectrom.*, **19,** 25 (1966); *Chem. Abstr.*, **64,** 9073 (1966).

2. W. L. F. Armarego, G. B. Barlin, and E. Spinner, *Spectrochim. Acta*, **22,** 117 (1966); *Chem. Abstr.*, **64,** 7536 (1966).
3. S. F. Mason, *J. Chem. Soc.*, 4874 (1957).
4. A. Albert and G. B. Barlin, *J. Chem. Soc.*, 5156 (1963)
5. S. F. Mason, *J. Chem. Soc.*, 493 (1962).
6. J. W. Clark-Lewis and R. P. Singh, *J. Chem. Soc.*, 3162 (1962).
7. R. D. Elliott, C. Temple, Jr., and J. A. Montgomery, *J. Org. Chem.*, **33,** 2393 (1968).
8. J. A. Montgomery and N. F. Wood, *J. Org. Chem.*, **29,** 734 (1964).
9. D. D. Perrin and Y. Inoue, *Proc. Chem. Soc.*, 342 (1960).
10. J. W. Bunting and D. D. Perrin, *J. Chem. Soc.* (*B*), 950 (1967).
11. T. J. Batterham, *J. Chem. Soc.* (*C*), 999 (1966).
12. Y. Inoue, *Tetrahedron*, **20,** 243 (1964).
13. J. W. Bunting and D. D. Perrin, *J. Chem. Soc.* (*B*), 436 (1966).
14. A. Albert and A. Hampton, *J. Chem. Soc.*, 505 (1954).
15. A. Albert and J. N. Phillips, *J. Chem. Soc.*, 1294 (1956).
16. E. Koenigs, H. Bueren, and G. Jung, *Ber.*, **69B.** 2690 (1936).
17. E. Koenigs, G. Kinne, and W. Weiss, *Ber.*, **57B,** 1174 (1924).
18. A. Albert and C. Pedersen, *J. Chem. Soc.*, 4683 (1956).
19. E. Lippmann and J. Spindler, *Z. Chem.*, **11,** 256 (1971).
20. J. S. Wieczorek and T. Talik, *Roczniki Chem.*, **36,** 967 (1962); *Chem. Abstr.*, **58,** 5676 (1963).
21. A. Albert and A. Hampton, *J. Chem. Soc.*, 4985 (1952).
22. A. Dornow and H. V. Plessen, *Chem. Ber.*, **99,** 244 (1966).
23. M. Israel and A. R. Day, *J. Org. Chem.*, **24,** 1455 (1959).
24. C. Temple, Jr., A. G. Laseter, J. D. Rose, and J. A. Montgomery, *J. Heterocycl. Chem.*, **7,** 1195 (1970).
25. O. Bremer, *Justus Liebigs Ann. Chem.*, **529,** 290 (1937); *Chem. Abstr.*, **31,** 5360 (1937).
26. T. L. Threlfall, Ph.D. Thesis, University of London, 1971.
27. J. D. Hepworth and E. Tittensor, *J. Chem. Soc.*, 1558 (1965).
28. J. W. Clark-Lewis and R. P. Singh, *J. Chem. Soc.*, 2825 (1964).
29. R. D. Elliott, C. Temple, Jr., and J. A. Montgomery, *J. Org. Chem.*, **33,** 533 (1968).
30. C. Temple, Jr., R. D. Elliott, J. L. Frye, and J. A. Montgomery, *J. Org. Chem.*, **36,** 2818 (1971).
31. W. Knobloch and H. Kuehne, *J. Prakt. Chem.*, **17,** 199 (1962).
32. F. Cuiban, M. Ionesco, H. Bala, and M. Steresco, *Bull. Soc. Chim. Fr.*, 356 (1963).
33. R. C. De Selms and H. S. Mosher, *J. Am. Chem. Soc.*, **82,** 3762 (1960); R. A. Archer and H. S. Mosher, *J. Org. Chem.*, **32,** 1379 (1967).
34. A. Albert and A. Hampton, "2nd Congr. Int. Biochim., Résumés Communs, (Paris," 1952), p. 444; *Chem. Abstr.*, **49,** 4648 (1955).
35. J. Armand, K. Chekir, J. Pinson, and N. Vinot, *C.R. Acad. Sci. Paris*, **281,** 547 (1975).
36. U. Ewers, H. Günther, and L. Jaenicke, *Angew. Chem., Int. Ed.*, **14,** 354 (1975).
37. T. Denzel and H. Hoehn (to von Heyden) Ger. Offen. 2,530,681; *Chem. Abstr.*, **84,** 180292 (1976).

CHAPTER XXX

Pyrazinopyrimidines*

I. Pyrazino[1,2-*a*]pyrimidines 568
II. Pyrazino[1,2-*c*]pyrimidines 570
III. Tables 572
Table 1. Pyrazino[1,2-*a*]pyrimidines 572
Table 2. Perhydropyrazino[1,2-*c*]pyrimidines 573
IV. References 573

I. Pyrazino[1,2-*a*]pyrimidines

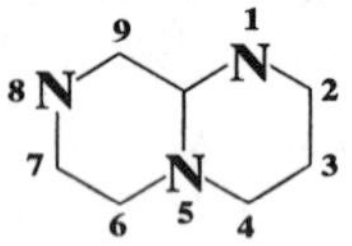

Little work has been reported on either of the two possible pyrazinopyrimidines containing a bridgehead nitrogen atom. This ring system has on one occasion been numbered as shown in formula **1**.

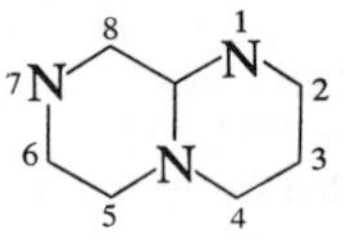

1

Reaction of 2-aminopyrazine (**2**) with diethylphenyl malonate (**3**) provides the 3-phenyl-4-oxo compound **4** in unspecified yield.[2,3] The betaine formulation for the product is supported by its infrared spectrum,[2] which does not show a band in the OH stretching region but exhibits a broad absorption from 2300 to 3200 cm^{-1}, with a maximum at 2650 cm^{-1}. Condensation of 2-aminopyrazine with the cyano ester **5** gave a product

* Pyrazino[2,3-*d*]pyrimidines (pteridines) are to be the subject of a separate monograph[1] and are not discussed in this work.

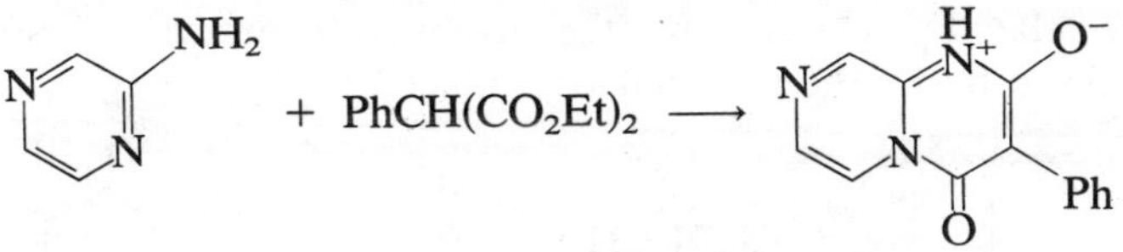

2 + $PhCH(CO_2Et)_2$ (**3**) ⟶ **4**

whose structure is less clear.[2] Tentatively structure **6** was assigned to the substance, but other protomeric forms, or the isomer **7**, could not be

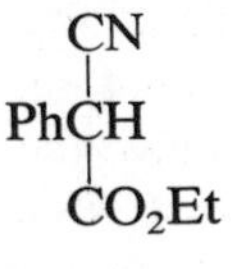

5 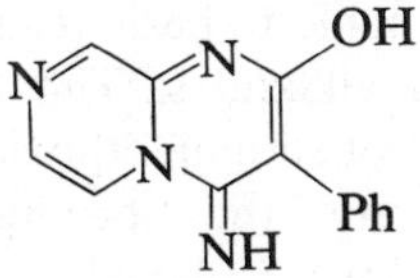 **6** 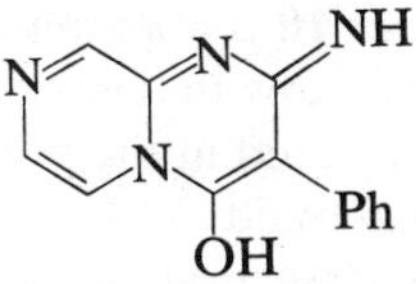 **7**

excluded by the spectral data considered. Heating the compound under reflux in acetic anhydride gave a diacetyl derivative, possibly of structure **8**.

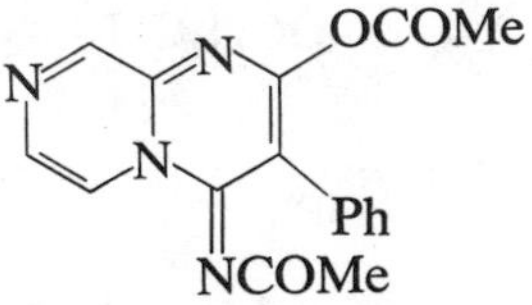

8

Treatment of 2-aminopyrazine with esters of ethoxymethylenemalonic acid gives the substituted amino compounds **9**, which may be cyclized to 4-oxopyrazino[1,2-*a*]pyrimidines by heating in diphenyl ether at 250°. When the diethyl ester **9** ($R^1 = R^2 = Et$) is cyclized, the 3-ethoxycarbonyl derivative **10** is obtained.[2] However when the cyclic isopropylidene ester **9** (R^1 plus $R^2 = CMe_2$) is cyclized,[4] the product is the decarboxylated compound **11**.

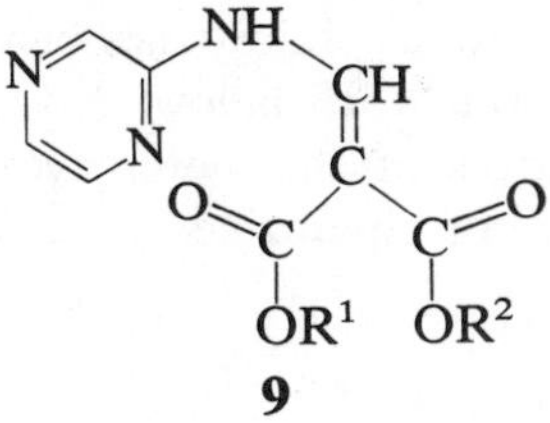

9 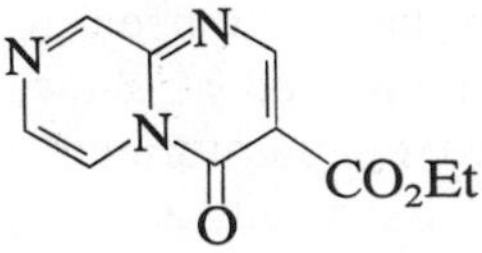 **10** 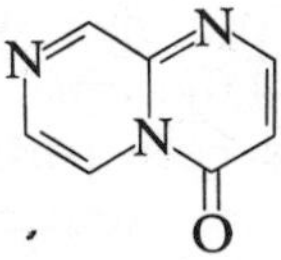 **11**

Quaternization of 2-aminopyrazine with ethyl 3-chloropropionimidate (**12**) in ethanol gives the 1,2,3,4-tetrahydro-2-oxo derivative **13** in line with the related products from other α-amino heterocycles.[5] No attempt was made to basify the protonated product **13**.

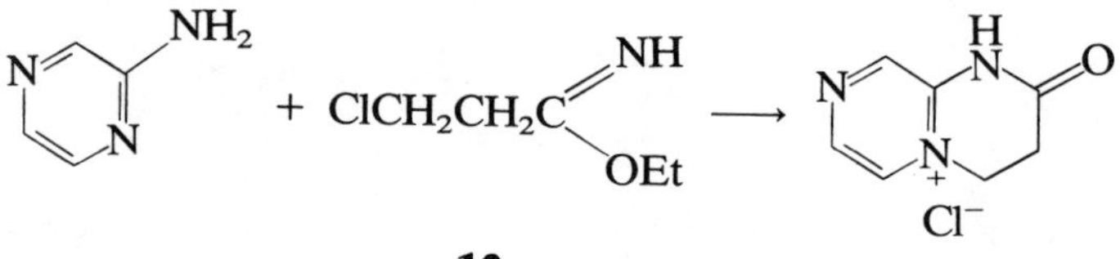

12 **13**

The ester **10** is apparently unstable to basic reagents and gives only tars on treatment with amines and alkoxides. The betaine **4**, however, is readily alkylated in the presence of sodium ethoxide to give the 2-alkoxy-4-oxo compounds.[2,3] Examination of these products for CNS activity has shown that several 2-alkoxy-4-oxo-3-phenylpyrazino[1,2-*a*]pyrimidines have antidepressant profiles similar to those of imipramine and amitriptyline in rats and mice.[2,3] The most active compound appears to be the morpholinoethyl compound **14**.

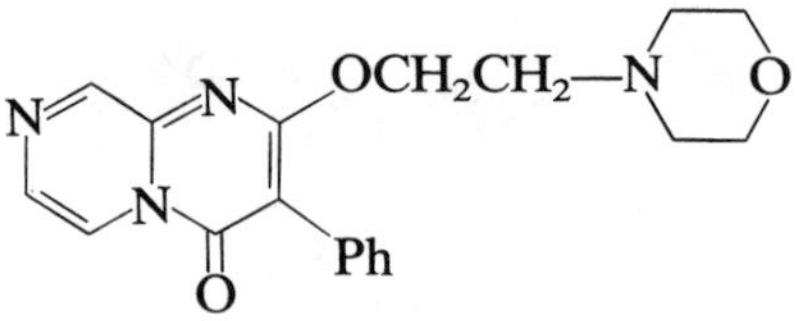

14

II. Pyrazino[1,2-*c*]pyrimidines

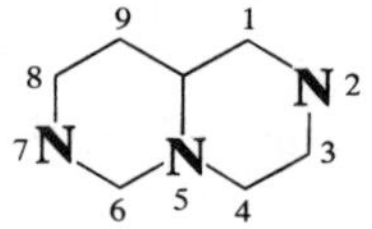

This ring system has also been referred to as 1,3,8-triazabicyclo-[4.4.0]decane and numbered as shown in formula **15** below. No work has been performed on unsaturated derivatives of this heterocycle, it having attracted attention only as a derivative of piperazine.

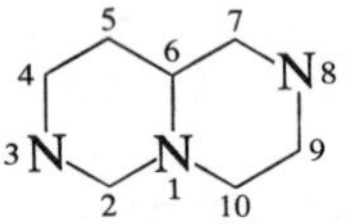

15

Fusion of the piperazinyl esters **16** with urea at 180–190° resulted in cyclization to the trioxobicyclic compounds **17**. Reaction of the ester **16** (R = H) with phenyl isocyanate gave the ureide **18**, which on refluxing in hydrochloric acid condensed to give the 7-phenyl derivative **19** in 56% yield.[6]

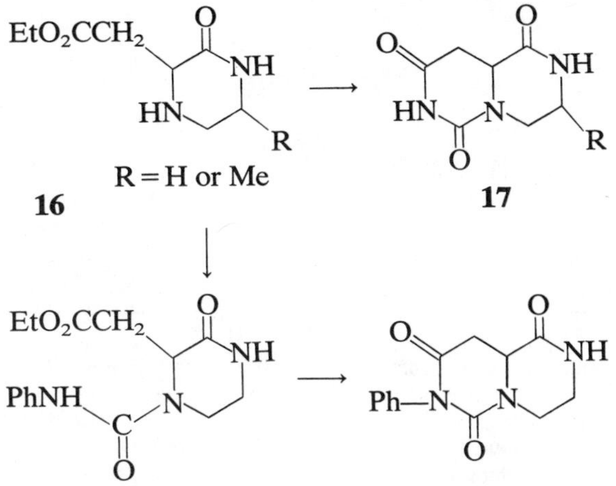

R = H or Me

16 **17**

18 **19**

The only other derivative of this ring system to be prepared is the monooxo compound **21**. This was obtained by treatment of the ethylaminoethylpiperazine **20** with ethyl chloroformate to give a monocarbamate which cyclized to the pyrazino[1,2-*c*]pyrimidine on refluxing with sodium ethoxide.[7–9]

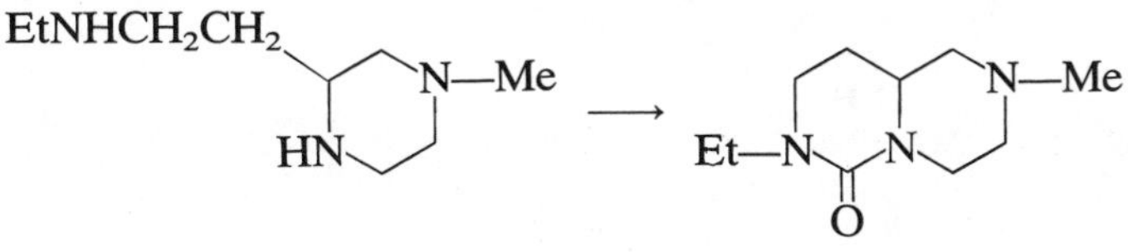

20 **21**

Compound **21** is a ring-closed analogue of the filaricide diethylcarbamazine (**22**); and, unlike other cyclic analogues of this drug, it was found to have marked activity in cotton rats infected with *Litomosoides carinii.* Moreover the bicyclic compound is longer acting and has a higher safety margin than diethylcarbamazine.[7–10] Antiinflammatory activity in the carrageenin rat paw volume test is also shown by compound **21**, which also inhibits passive cutaneous anaphylaxis.[7]

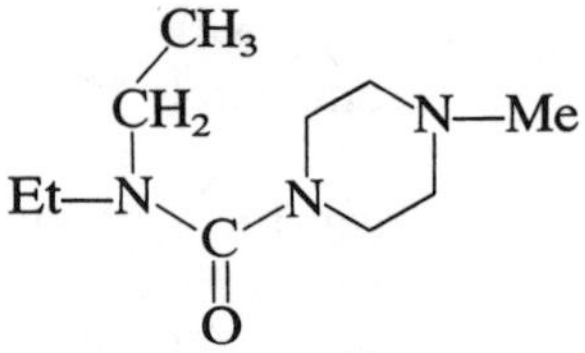

Et—N(CH$_2$CH$_3$)—C(=O)—N(piperazine)N—Me

22

III. Tables

TABLE 1. PYRAZINO[1,2-*a*]PYRIMIDINES

2*H* (ring numbering 1–9) 4*H*

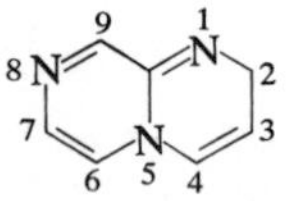

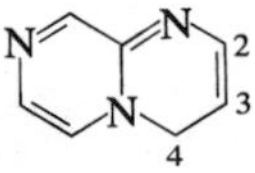

Molecular formula	Substituents or structure	m.p., °C	Ref.
$C_7H_5N_3O$	4*H*-4-Oxo	171.8–173.8	4
$C_7H_7N_3O$	2*H*-3,4-Dihydro-2-oxo		
	Hydrochloride	>270	5
$C_8H_5N_3O_3$	4*H*-3-Carboxy-4-oxo		
	Ethyl ester	164–165	2, 11
$C_9H_7N_3O_3$	4*H*-3-Carboxy-6-methyl-4-oxo		
	Ethyl ester	92–94	11
$C_{10}H_9N_3O_3$	4*H*-3-Carboxy-6,7-dimethyl-4-oxo		
	Ethyl ester	124–125	11
$C_{13}H_8ClN_3O_2$	4*H*-3-(4-Chlorophenyl)-2-hydroxy-4-oxo[a]	281–282 decomp.	2,3
$C_{13}H_9N_3O_2$	4*H*-2-Hydroxy-4-oxo-3-phenyl[a]	278–279 decomp.	2, 3
$C_{13}H_{10}N_4O$	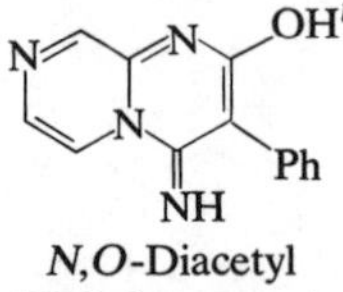 (structure: 2-OH[b], 3-Ph, 4-NH)	202–207	2
	N,O-Diacetyl	250–252	
$C_{15}H_{10}N_4O_2$	4*H*-2-Cyanomethoxy-4-oxo-3-phenyl	164–165	2, 3
$C_{15}H_{11}N_3O_4$	4*H*-2-Carboxymethoxy-4-oxo-3-phenyl[c]		
	Ethyl ester	102–103	2
$C_{16}H_{13}N_3O_4$	4*H*-2-(2-Carboxyethoxy)-4-oxo-3-phenyl[c]		
	Ethyl ester	124.5–125.5	2, 3
$C_{17}H_{18}N_4O_2$	4*H*-2-(2-Dimethylaminoethoxy)-4-oxo-3-phenyl[d]		
	Hydrobromide	200–201	2, 3
$C_{18}H_{20}N_4O_2$	4*H*-2-(3-Dimethylaminopropoxy)-4-oxo-3-phenyl[d]		
	Hydrobromide	213–214 decomp.	2, 3
$C_{19}H_{20}N_4O_3$	4*H*-2-(2-Morpholinoethoxy)-4-oxo-3-phenyl[d,e]		
	Hydrobromide	240–241 decomp.	2, 3

TABLE 1. (*contd.*)

2*H* (ring structure with positions 1–9) 4*H* (ring structure with positions 2, 3, 4)

Molecular formula	Substituents or structure	m.p., °C	Ref.
$C_{19}H_{22}N_4O_2$	4*H*-2-(2-Diethylaminoethoxy)-4-oxo-3-phenyl[d]		
	Hydrobromide	210–211	2, 3
$C_{20}H_{24}N_4O_2$	4*H*-2-(3-Diethylaminopropoxy)-4-oxo-3-3-phenyl[d,e]		
	Hydrobromide	197–198	2, 3

[a] The betaine formulation (see text) is more accurate.
[b] Structure not proved (see text).
[c] Two other carboxy alkoxy derivatives are described but not tabulated here.
[d] Seven other dialkylamino alkoxy derivatives are described but not tabulated here.
[e] The 4-chlorophenyl analogue is also described but not tabulated here.

TABLE 2. PERHYDROPYRAZINO[1,2-*c*]PYRIMIDINES

Molecular formula	Substituents	m.p., °C	Ref.
$C_7H_9N_3O_3$	1,6,8-Trioxo	300–305 decomp.	6
$C_8H_{11}N_3O_3$	3-Methyl-1,6,8-trioxo	270 decomp.	6
$C_{10}H_{19}N_3O$	7-Ethyl-2-methyl-6-oxo	b.p. 130–132/ 0.05 mm	7
	Hydrochloride hemihydrate	142	8, 9
$C_{13}H_{13}N_3O$	1,6,8-Trioxo-7-phenyl	255–256	6

IV. References

1. W. Pfleiderer, "Pteridines," Wiley-Interscience, New York, in preparation.
2. D. L. Trepanier, L. W. Rampy, K. L. Shriver, J. N. Eble, and P. J. Shea, *J. Med. Chem.*, **11,** 1045 (1968).
3. D. L. Trepanier (to Dow Chemical Co.), U.S. Pat. 3,558,618; *Chem. Abstr.*, **74,** 141870 (1971).
4. Sterling Drug Inc., Brit. Pat. 1,147,760; *Chem. Abstr.*, **71,** 49967 (1969).
5. Y. Okamoto, A. Takada, and T. Ueda, *Chem. Pharm. Bull.* (Tokyo), **19,** 764 (1971).

6. K. Masuzawa, M. Masaki, and M. Ohta, *Chem. Pharm. Bull.* (Tokyo), **14,** 194 (1966); *Chem. Abstr.*, **64,** 14188 (1966).
7. R. Saxena, S. Sharma, R. N. Iyer, and N. Anand, *J. Med. Chem.*, **14,** 929 (1971).
8. N. Anand, R. K. Chatterjee, R. N. Iyer, R. Sakena, and A. B. Sen, U.S. Pat. 3,560,503; *Chem. Abstr.*, **74,** 141845 (1971).
9. N. Anand, R. Sakena, R. K. Chatterjee, R. Narayana, and A. B. Sen, Brit. Pat. 1,284,582; *Chem. Abstr.*, **77,** 152226 (1972).
10. R. Saxena, R. N. Iyer, N. Anand, R. K. Chatterjee, and A. B. Sen, *J. Pharm. Pharmacol.*, **22,** 306 (1970).
11. T. Tanaka and S. Narita, *Yakugaku Zasshi*, **95,** 1092 (1975); *Chem. Abstr.*, **84,** 31002 (1976).

CHAPTER XXXI

Pyrazinopyrazines

I. Pyrazino[1,2-*a*]pyrazines 575
II. Pyrazino[2,3-*b*]pyrazines 576
III. Tables 580
Table 1. Pyrazino[1,2-*a*]pyrazines 580
Table 2. Pyrazino[2,3-*b*]pyrazines 581
IV. References 582

I. Pyrazino[1,2-*a*]pyrazines

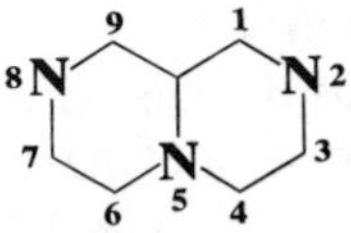

2*H* RRI 12010

4*H* RRI 12011

The only reference to this ring system names the perhydro heterocycle as piperazino[1,2-*a*]piperazine.[1] Reduction of the 2,4-dinitrophenylhydrazone **1** with Raney nickel gave the dioxo derivative **2** in 26% yield. This compound was reduced with lithium aluminum hydride to give the product **3**.

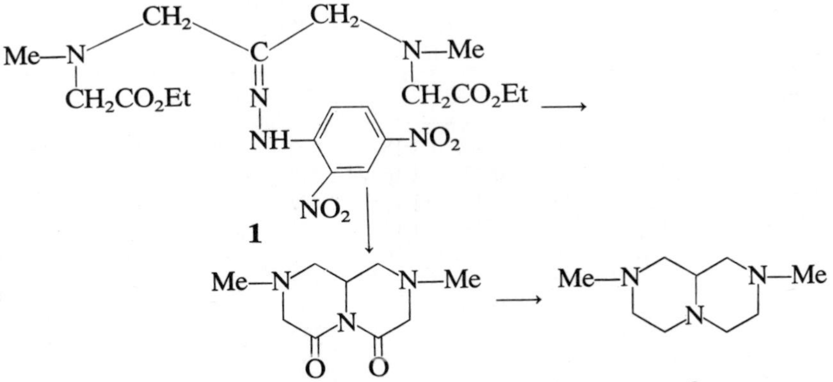

II. Pyrazino[2,3-*b*]pyrazines

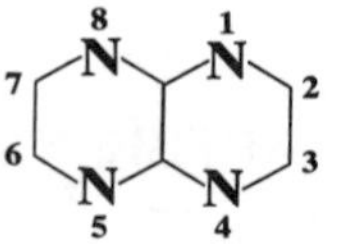

RRI 1590

This symmetrical molecule is the 5,8-diaza derivative of quinoxaline. An alternative name frequently used for this heterocycle is 1,4,5,8-tetraazanaphthalene. Completely reduced derivatives of the ring system have been called piperazinopiperazines and naphthopiperazines.

As a tetraaza derivative of naphthalene the parent heterocycle has been subjected to several theoretical treatments. Simple Hückel molecular orbital calculations of charge densities, energy levels, and wave functions gave results in qualitative agreement with experimental data for the lowest singlet–singlet transition in the electronic spectrum.[2] This agreement was not as good as that obtained by other workers[3] using the Pariser-Parr and Pople method with limited configuration interaction. Even better results were obtained using SCF MO calculations with limited configuration interaction.[4]

Figure 1 lists electron density data extracted from a theoretical study of all 74 azanaphthalenes that do not have nitrogen atoms at the central bond.[2] It was suggested that the Hückel charge densities listed would be of some value in predicting positions of electrophilic and nucleophilic substitution. Clearly accurate predictions cannot be made without considering the attacking species and the solvent.

Figure 1. TOTAL ELECTRON DENSITIES OF AZAQUINOXALINES[2]

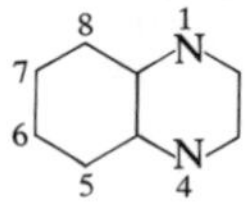

Compound	1	2	3	4	5	6	7	8
Quinoxaline	1.151	0.902	0.902	1.151	1.002	0.987	0.987	1.002
5-Aza	1.140	0.906	0.888	1.163	1.217	0.884	0.996	0.934
6-Aza	1.154	0.886	0.902	1.136	0.897	1.187	0.933	1.011
5,6-Diaza	1.142	0.887	0.883	1.149	1.121	1.088	0.938	0.940
5,7-Diaza	1.126	0.905	0.872	1.166	1.224	0.829	1.194	0.831
5,8-Diaza	1.153	0.889	0.889	1.153	1.153	0.889	0.889	1.153
6,7-Diaza	1.139	0.886	0.886	1.139	0.902	1.137	1.137	0.902
5,6,7-Triaza	1.129	0.887	0.869	1.151	1.122	1.030	1.139	0.841
5,6,8-Triaza	1.155	0.869	0.885	1.138	1.052	1.096	0.838	1.160
5,6,7,8-Tetraaza	1.141	0.868	0.868	1.141	1.060	1.039	1.039	1.060

The symmetry of the parent aromatic heterocycle results in a very simple infrared spectrum being obtained. Bond assignments have been discussed and compared with those for naphthalene.[5] Strong infrared absorption bands at 1720 and 1680 cm^{-1} confirm that the 2,3-dihydroxy derivative is more properly regarded as the dioxo tautomer. Similarly the 2-hydroxy compound (ν C=O 1700 cm^{-1}) exists as the 1,2-dihydro-2-oxo derivative.[6] The ultraviolet spectrum of the unsubstituted heterocycle has been determined in cyclohexane, and the intensity of the long wavelength (402 nm, $n \rightarrow \pi$) transition has been discussed.[7] In water at pH 7 the compound exhibits two main bands, each containing a series of maxima, and an additional flat maximum at long wavelength.[6] The maxima appear at 237, 245, 290, 296, 302, 308, 315, 322, and 375 nm.

In aqueous solution the cations of the parent heterocycle and its 2-methyl and 2,3-dimethyl derivatives are covalently hydrated. The neutral molecules are anhydrous. The tri- and tetramethyl derivatives are not hydrated either as neutral species or as cations. These conclusions were reached on the basis of acidity constant measurements and ultraviolet studies[6] and confirmed by PMR spectroscopy.[8] The PMR spectra show that hydration occurs by the addition of two molecules of water to give cations of formula **4**. Substitution by a methyl group at either of the sites of hydration renders the cation completely anhydrous. Moreover the introduction of methyl groups to give compound **4** (R = Me) reduces the extent of hydration. Thus the cations of pyrazino[2,3-*b*]pyrazine and its 2-methyl derivative are completely hydrated in aqueous solution, whereas the protonated 2,3-dimethyl compound exists as a 3 : 1 mixture of the dihydrate and the anhydrous species.

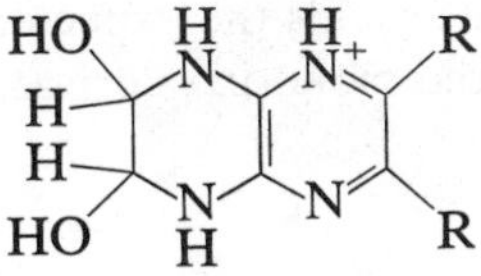

4 R = H or Me

The acidity constants of pyrazino[2,3-*b*]pyrazine and its methyl derivatives demonstrate the apparent base-weakening effect of methyl groups in molecules undergoing covalent hydration.[6] The usual base-strengthening effect of methyl groups is overshadowed by their inhibition of covalent hydration. Unhindered covalent hydration would produce very much stronger bases, reminiscent of aminopyrazines, and it is the progressive loss of this contribution that causes the trend shown in Figure 2.

The ESR spectra of the radical anions of pyrazino[2,3-*b*]pyrazine obtained by electrolytic and by chemical reduction have been examined,

Figure 2. ACIDITY CONSTANTS OF PYRAZINO[2,3-*b*]PYRAZINE AND ITS METHYL DERIVATIVES

Compound	pK_a	Covalent hydration
Pyrazino[2,3-*b*]pyrazine	2.51	Complete
2-Methyl	1.14	Almost complete
2,3-Dimethyl	−0.25	75%
2,3,6-Trimethyl	−0.53	Anhydrous
2,3,6,7-Tetramethyl	−0.02	Anhydrous

and differences in the hyperfine structure have been discussed in terms of ion pairing.[9] More recently a comparison has been made of the ESR spectra of the radical anions of pyrazino[2,3-*b*]pyrazine (**5**), pyrazino[2,3-*d*]pyridazine (**6**), and pyridazino[4,5-*d*]pyridazine (**7**). Whereas the symmetry of compounds **5** and **7** allowed straightforward interpretation of their ESR spectra, the pyrazinopyridazine **6** demonstrated an inadequacy in the assignments made by MO calculations.[10]

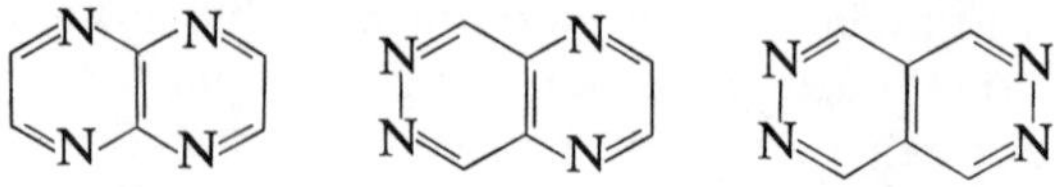

5 **6** **7**

The aromatic ring system is conveniently obtained by reaction of α-dicarbonyl compounds with 2,3-diaminopyrazines. The parent heterocycle and simple methyl derivatives have been obtained by this approach.[6,11] Diethyl oxalate yields the expected dioxo compound **8** from 2,3-diaminopyrazine, and the monooxo derivative **10** was obtained using the acetal **9**.[6]

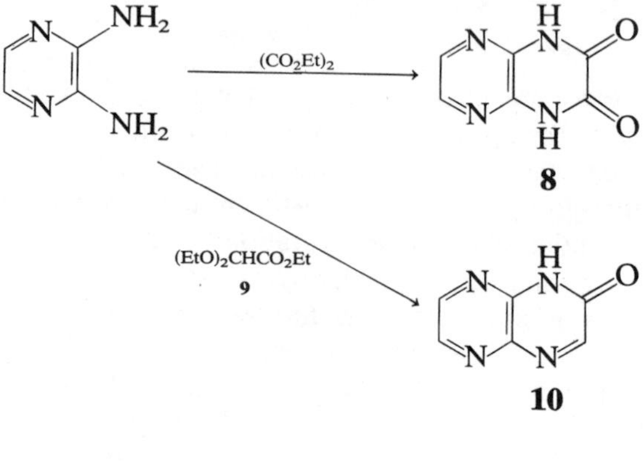

$(CO_2Et)_2$

$(EtO)_2CHCO_2Et$ **9**

8

10

Halo-substituted diaminopyrazines have been successfully condensed with diethyl oxalate[12] and with diacetyl[13] to give pyrazino[2,3-*b*]pyrazines. 2,3-Diamino-5,6-dimethylpyrazines have been reacted with a large number of diaryl α-diketones and aryl α-keto aldehydes to give aryl-substituted derivatives of the heterocycle.[14]

Perhydropyrazino[2,3-*b*]pyrazine (**13**) has been obtained by reacting ethylenediamine with the α-chloro ethers **11**[15] and **12**.[16] These ethers act effectively as a source of glyoxal. Analogously, diacetyl reacts with ethylenediamine to give the dimethyl derivative **14**.[17]

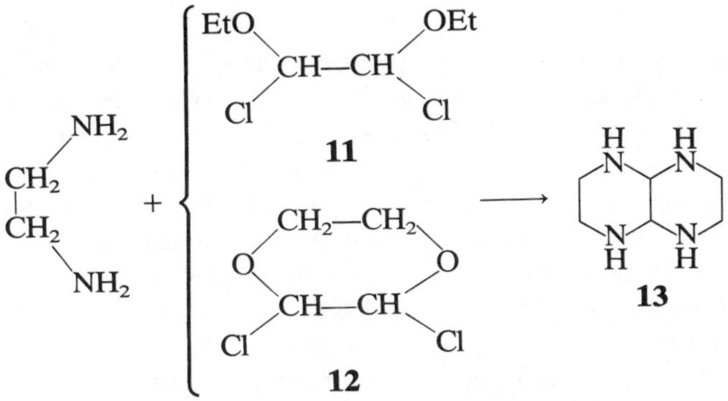

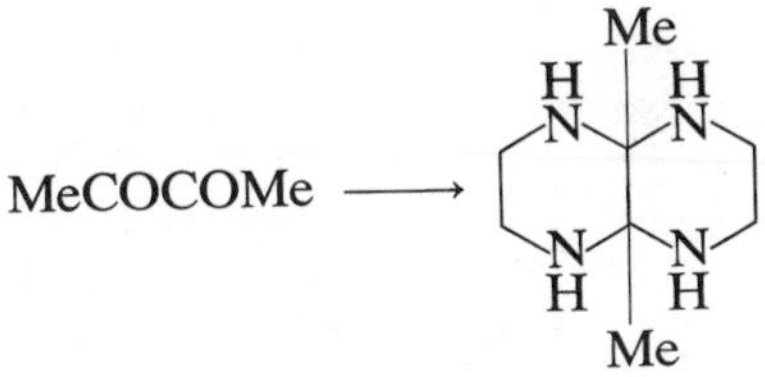

$NH_2CH_2CH_2NH_2$ + { EtO(Cl)CH–CH(Cl)OEt **11** ; cyclic CH_2–CH_2–O–CH(Cl)–CH(Cl)–O **12** } ⟶ **13**

MeCOCOMe ⟶ **14**

Although the product from the ethers **11** and **12** is the same, nothing is known of the stereochemistry of the compound. Moreover the product has a variable melting point depending on the temperature of insertion.[16] The dimethyl derivative **14** was only obtained as a dihydrate; attempts at removing the water by heating resulted in decomposition and loss of ethylenediamine.[17] Very small yields of the perhydro compound **13** have been obtained by oxidation of ethylenediamine with hydrogen peroxide or with lanthanum trichloride and oxygen.[18]

Oxidation of the parent aromatic heterocycle with hydrogen peroxide at room temperature[6] yields the dioxo compound **8**, in line with the susceptibility of the ring system to covalent dihydration. The bromine atom in compound **15** is readily displaced by amines in refluxing ethanol to give the expected amino derivatives such as the morpholinopyrazino[2,3-*b*]pyrazine **16**.[13]

15 → **16**

Catalytic hydrogenation of perhydropyrazino[2,3-*b*]pyrazines provides monocyclic piperazines.[19] In other ways the compound behaves as a tetramine, forming a tetrahydrochloride and reacting with four moles of acetic anhydride, phenyl isocyanate, and phenyl isothiocyanate.[15]

Amino derivatives of the ring system have been patented as pharmaceuticals,[13] and 2,3-dioxo compounds are claimed to have antiviral activity.[12] A large series of pyrazino[2,3-*b*]pyrazines have been described as provoking ataxia by lowering CNS activity.[14] More complex derivatives have found use in the formation of polymers.[20] Perhydro derivatives are useful in the preparation of dyes and pharmaceuticals[21] and in accelerating vulcanization.[22]

III. Tables

TABLE 1. PYRAZINO[1,2-*a*]PYRAZINES

Molecular formula	Substituents	m.p., °C	Ref.
$C_9H_{15}N_3O_2$	Perhydro-2,8-dimethyl-4,6-dioxo	83–85	1
$C_9H_{19}N_3$	Perhydro-2,8-dimethyl	b.p. 145–147	1
	Tripicrate	260 decomp.	1
	Triperchlorate	222 decomp.	1

TABLE 2. PYRAZINO[2,3-*b*]PYRAZINES

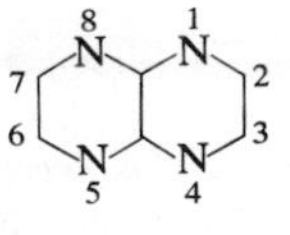

Molecular formula	Substituents	m.p., °C	Ref.
$C_6H_2Cl_2N_4O_2$	6,7-Dichloro-1,2,3,4-tetrahydro-2,3-dioxo	360 decomp.	12
$C_6H_3BrN_4O_2$	6-Bromo-1,2,3,4-tetrahydro-2,3-dioxo	350–360 decomp.	12
$C_6H_4N_4$	Unsubstituted	>270 decomp.	6
$C_6H_4N_4O$	1,2-Dihydro-2-oxo		
	Monohydrate	>200 decomp.	6
$C_6H_4N_4O_2$	1,2,3,4-Tetrahydro-2,3-dioxo	>360	6
$C_6H_{14}N_4$	Perhydro	195[a] decomp.	15, 16
		232–234	21
	Tetraacetyl	201–202	15
	Tetra(PhNHCO)	195	15
	Tetra(PhNHCS)	170	15
$C_7H_6N_4$	2-Methyl	196–197	6
$C_8H_8N_4$	2,3-Dimethyl	215–216	6, 11
		170–173	14
$C_8H_{18}N_4$	Perhydro-4a,8a-dimethyl		
	Dihydrate	56	17
$C_9H_{10}N_4$	2,3,6-Trimethyl	152–153	6
$C_{10}H_{12}N_4$	2,3,6,7-Tetramethyl	254–256	6, 11
$C_{12}H_{15}N_5O$	2,3-Dimethyl-6-morpholin-4-yl	207–208	13
$C_{12}H_{18}N_6$	6-(2-Dimethylaminoethylamino)-2,3-dimethyl		
	Hydrobromide	220 decomp.	13
$C_{13}H_{17}N_5$	2,3-Dimethyl-6-piperidin-1-yl	113–114	13
$C_{13}H_{18}N_6$	2,3-Dimethyl-6-(1-methylpiperazin-4-yl)		
	Hydrobromide	302 decomp.	13
$C_{14}H_{12}N_4$	2,3-Dimethyl-6-phenyl[b]	209–210	14
$C_{15}H_{14}N_4$	2,3,6-Trimethyl-7-phenyl	125–126	14
$C_{19}H_{14}N_4$	6-Methyl-2,3-diphenyl	192–193	14
$C_{20}H_{16}N_4$	2,3-Dimethyl-6,7-diphenyl[c]	228–230	14
	2,6-Dimethyl-3,7-diphenyl[d]	233–234	14
$C_{22}H_{20}N_4$	2,3-Diethyl-6,7-diphenyl	217–218	14
$C_{25}H_{18}N_4$	7-Methyl-2,3,6-triphenyl	220–221	14

[a] This compound has a variable melting point dependent on the temperature of insertion of the sample,[16] for example, 162–187° (140°), 200–205° (190°), 200–210° (200°).

[b] Four other 6-(substituted phenyl) derivatives were reported.

[c] Twenty-two other 6,7-diaryl compounds were described.

[d] Possibly the 2,7-dimethyl-3,6-diphenyl compound.

IV. References

1. M. Rink and K. Feiden, *Arch. Pharm.* (Weinheim), **295,** 121 (1962); *Chem. Abstr.*, **57,** 11192 (1962).
2. S. C. Wait, Jr., and J. W. Wesley, *J. Mol. Spectrosc.*, **19,** 25 (1966).
3. G. Favini, I. Vandoni, and M. Simonetta, *Theoret. Chim. Acta*, **3,** 45, 418 (1965); *Chem. Abstr.*, **62,** 8525, **63,** 17326 (1965).
4. A. Hinchcliffe, M. A. Ali, and E. Farmer, *Spectrochim. Acta*, A**23,** 501 (1967).
5. W. L. F. Armarego, G. B. Barlin, and E. Spinner, *Spectrochim. Acta*, **22,** 117 (1966).
6. W. L. F. Armarego, *J. Chem. Soc.*, 4304 (1963).
7. S. F. Mason, *J. Chem. Soc.*, 493 (1962).
8. T. J. Batterham, *J. Chem. Soc. (C)*, 999 (1966).
9. F. Gerson and W. L. F. Armarego, *Helv. Chim. Acta*, **48,** 112 (1965).
10. R. Danieli, L. Lunazzi, and G. Placucci, *J. Am. Chem. Soc.*, **93,** 5850 (1971).
11. R. C. Ellingson and R. L. Henry, *J. Am. Chem. Soc.*, **70,** 1257 (1948).
12. Società Farmaceutici Italia, Brit. Pat. 1,145,730; *Chem. Abstr.*, **71,** 30505 (1969).
13. G. Palamidessi, M. A. Verini, and A. Fioretti (to Società Farmaceutici Italia), Brit. Pat. 1,159,412; *Chem. Abstr.*, **71,** 91535 (1969).
14. C. K. Cain, Fr. M. 6,196; *Chem. Abstr.*, **72,** 12767 (1970).
15. H. Baganz, L. Domaschke, and G. Kirchner, *Chem. Ber.*, **94,** 2676 (1961).
16. L. A. Cort and N. R. Francis, *J. Chem. Soc.*, 2799 (1964).
17. H. Stetter, *Chem. Ber.*, **86,** 69 (1953).
18. C. Castellani-Bisi, S. Facchetti, H. Knoeppel, and A. Copet, *Chim. Ind.* (Milan), **52,** 1110 (1970); *Chem. Abstr.*, **74,** 87919 (1971).
19. H. C. Chitwood, U.S. Pat. 2,345,236; *Chem. Abstr.*, **38,** 4274 (1944).
20. C. S. Marvel (to Research Corp.), U.S. Pat. 3,563,917; *Chem. Abstr.*, **74,** 126521 (1971); F. E. Arnold and R. L. Van Deusen, *J. Appl. Polym. Sci.*, **15,** 2035 (1971); *Chem. Abstr.*, **75,** 118842 (1971).
21. H. C. Chitwood and R. W. McNamee, U.S. Pat. 2,345,237; *Chem. Abstr.*, **38,** 4247 (1944).
22. Rhone–Poulenc S.A., Fr. Pat. 1,310,524; *Chem. Abstr.*, **58,** 7013 (1963).

CHAPTER XXXII

Pyrazinotriazines and Pyrazinotetrazines

I. Pyrazino[1,2-*a*]-*s*-triazines 583
II. Pyrazino[2,3-*e*]-*as*-triazines 583
III. Pyrazino[2,3-*d*]-*v*-triazines 584
IV. Pyrazino[2,3-*e*]-*v*-tetrazines 584
V. References 584

I. Pyrazino[1,2-*a*]-*s*-triazines

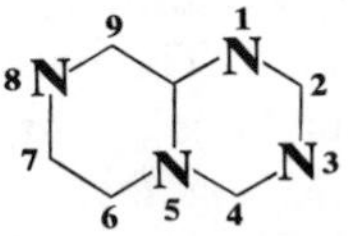

A recent report[1] confirms, by x-ray analysis of its acetyl derivative, that compound **2** is a product of reaction of molecular oxygen and the tetrahydropterin **1**.

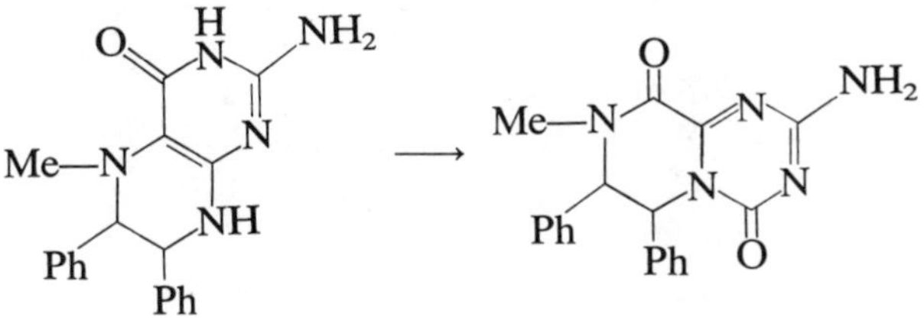

1 → **2**

II. Pyrazino[2,3-*e*]-*as*-triazines

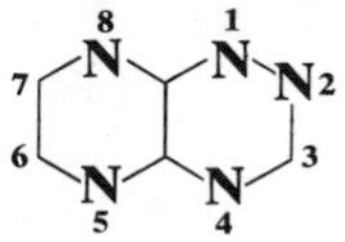

This ring system is also referred to as 1,4,5,7,8-pentaazanaphthalene (see Section IV).

III. Pyrazino[2,3-*d*]-*v*-triazines

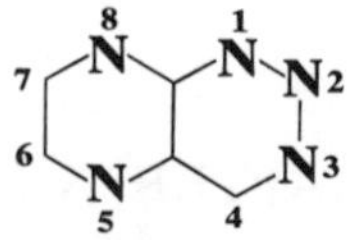

This ring system is also referred to as 1,4,6,7,8-pentaazanaphthalene (see Section IV).

IV. Pyrazino[2,3-*e*]-*v*-tetrazines

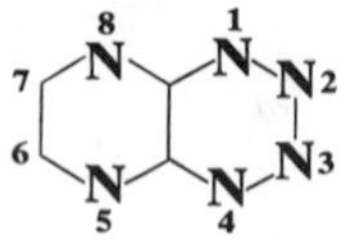

This ring system has been described as 1,4,5,6,7,8-hexaazanaphthalene. Only one report concerning this ring system and the related pyrazino triazines has appeared in the literature. No examples of any of these heterocycles have been prepared. The parent nuclei have however been the subject of HMO calculations giving energy levels and charge densities for each heterocycle together with over 70 other azanaphthalenes[2] (see Chapter XXXI, Fig 1.).

V. References

1. H. van Koningsveld *Tetrahedron,* **31,** 541 (1975).
2. S. C. Wait, Jr., and J. W. Wesley, *J. Mol. Spectrosc.,* **19,** 25 (1966).

CHAPTER XXXIII

Pyrazinooxazines

I. Pyrazino[2,1-*b*][1,3]oxazines 585
II. Pyrazino[1,2-*c*][1,3]oxazines 586
III. Pyrazino[2,3-*d*][1,3]oxazines 587
IV. Pyrazino[2,1-*c*][1,4]oxazines 588
V. Tables 589
Table 1. Perhydropyrazino[2,1-*b*][1,3]oxazines 589
Table 2. Pyrazino[2,3-*d*][1,3]oxazin-4-ones 589
Table 3. Pyrazino[2,1-*c*][1,4]oxazines 590
VI. References 590

I. Pyrazino[2,1-*b*][1,3]oxazines

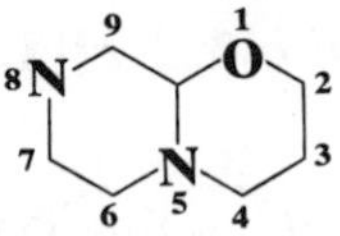

Only saturated derivatives of this heterocycle are known, and they have been named as 1-oxa-7,10-diazabicyclo[4.4.0]decanes, numbered as in formula **1**.

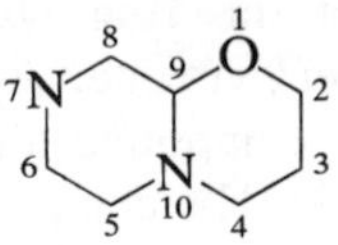

1

In contrast to the ready cyclization of the hydroxyacetyl compound **2** (see Chapter XIX), the hydroxypropionyl derivative **3** did not give the cyclol **4** on standing in solution at room temperature. However treatment of compound **3** with ethanol in the presence of dry hydrogen chloride gave the ethyl ether **5**.[1,2] The corresponding methyl derivative **6** was obtained by methylation of **3** with a methyl iodide–silver oxide mixture.[1] In both reactions small amounts of the corresponding esters (**7a** and **7b**) were obtained.

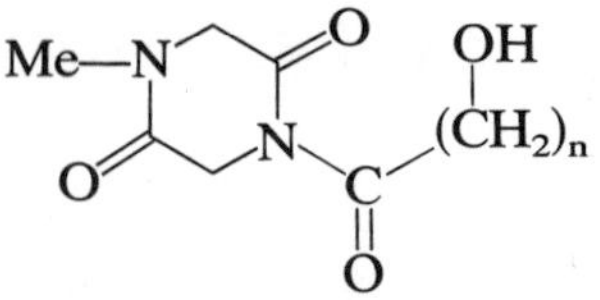

2 $n = 1$
3 $n = 2$

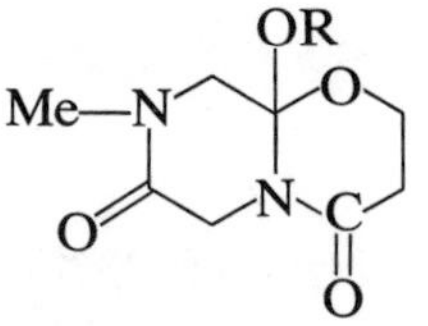

4 R = H
5 R = Et
6 R = Me

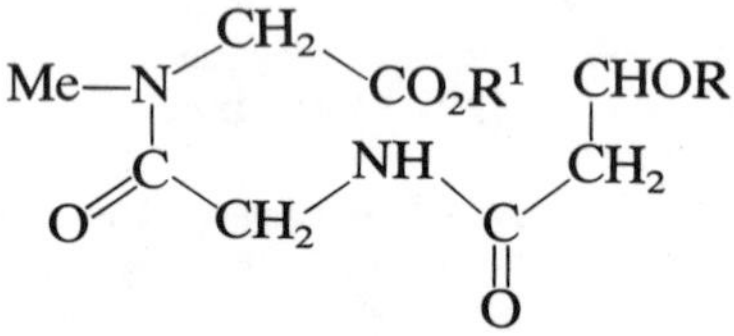

7a R = H, R^1 = Et
7b R = R^1 = Me

II. Pyrazino[1,2-*c*][1,3]oxazines

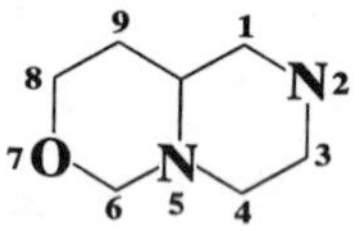

RRI 11991

Only one derivative (**8**) of this ring system is known, and this was described as an oxazino[3,4-*a*]pyrazine. Apparently no preparative details have been given for this compound although it has attracted attention as a pharmacological tool.[3] The compound, frequently referred to as COP, is a spinal convulsant producing clonic and tonic extensor seizures.[4,5] Originally thought to be strychnine-like in its action,[4,6] differences have now been demonstrated.[7]

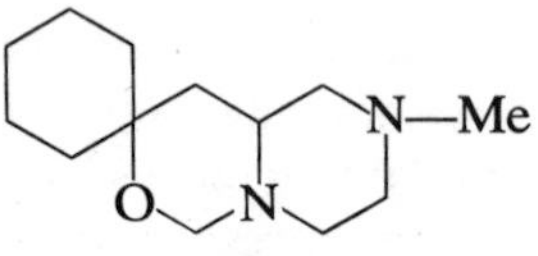

8

III. Pyrazino[2,3-*d*][1,3]oxazines

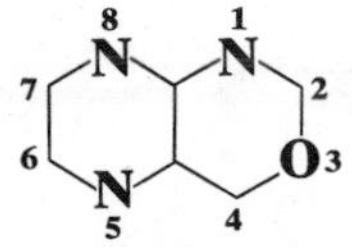

Only 4*H*-4-oxo derivatives of this heterocycle have been prepared. The infrared spectra[8] of these compounds show a characteristic carbonyl absorption at 1780–1790 cm^{-1}. A variety of 2-methyl compounds have been obtained by reaction of the appropriately substituted aminopyrazinecarboxylic acids (**9**) with acetic anhydride,[8–10] an approach which has been described in many patents.[11–20] Other anhydrides such as propionic[8] and butyric[11–13] anhydrides react to give the expected products. No reports of 2-unsubstituted compounds have appeared.

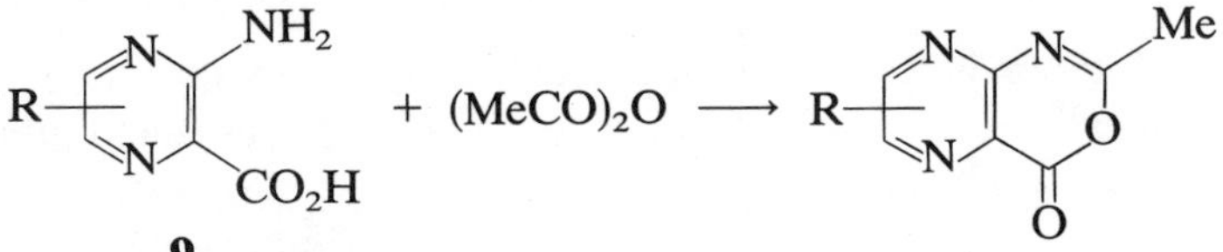

$R = H, Cl, MeO, MeSO_2, Me_2N$ etc

Compounds of structure **10** react readily with nucleophiles. Thus moisture causes rapid hydrolysis to the pyrazine amino acid, and aliphatic amines give the amides (**11**), which may be cyclized to pteridinones (**12**) by dissolving in phosphoryl chloride. Ammonia, hydroxylamine, and aniline give the pteridinones (**13**) directly.[8]

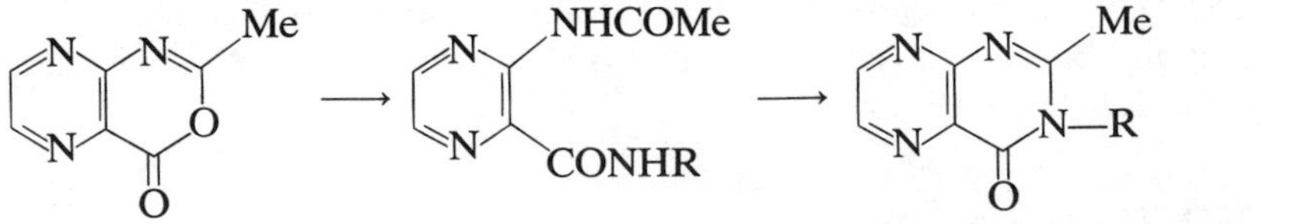

11 R = Me or CH_2Ph **12** R = Me or CH_2Ph
13 R = H, OH, or Ph

Benzamidine reacts with the pyrazinooxazinone **14** at room temperature to give the acylamidine **15**. Guanidines react similarly to give the acylguanidines **16**. This approach to acylguanidines is particularly useful when the guanidine component is insufficiently nucleophilic to react directly with esters of pyrazine amino acids.[9] The reaction has been extended to other substituted pyrazine amino acids.[10]

The principal use of 4-oxopyrazino[2,3-*d*][1,3]oxazines is as reactive intermediates to diuretic pyrazinoylguanidines.[9–11,14] They are also used

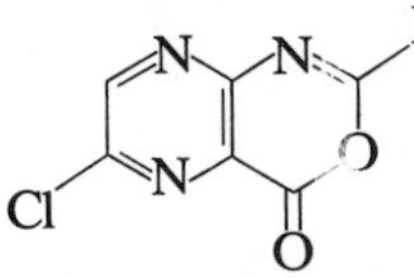

14 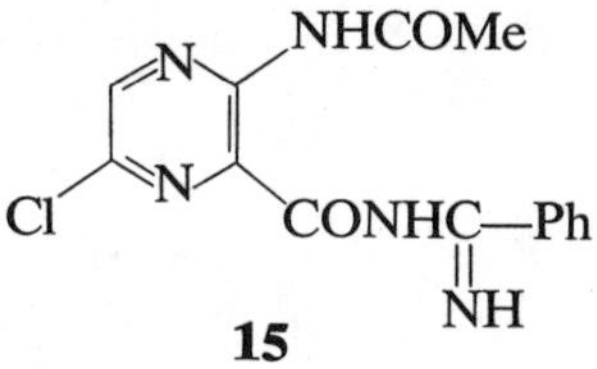 **15**

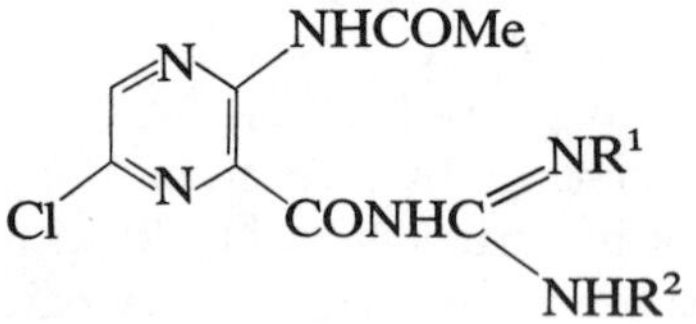

16

$R^1 = H$ or Ph
$R^2 = Ph$, PhCO, MeCO, PhCH=N or $Me_2C{=}N$

as intermediates to certain acetamidopyrazines not accessible by other routes[10] as well as intermediates to 2,3-disubstituted pteridinones as described above.[8]

IV. Pyrazino[2,1-*c*][1,4]oxazines

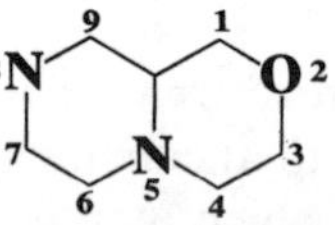

Treatment of the piperazine acid **17** with thionyl chloride resulted in cyclization to give the chloroethylperhydropyrazinooxazin-1-one **18**, which was inactive against HeLa cells in vitro.[21]

The only other reference[22] to this heterocycle suggested that the 6-oxo compound **20** may have been formed from heating morpholine with the glycine derivative **19**.

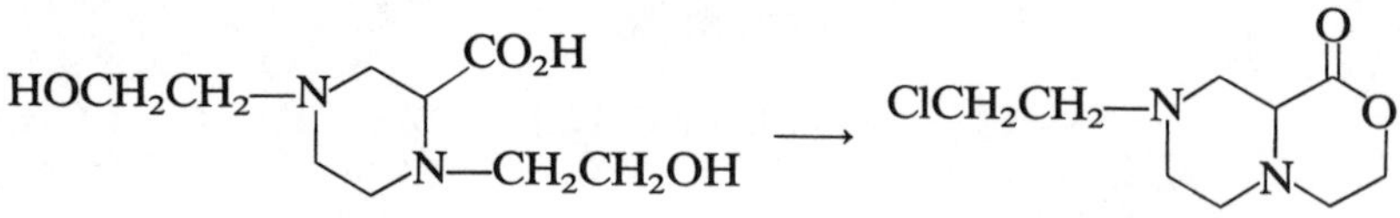

17 → **18**

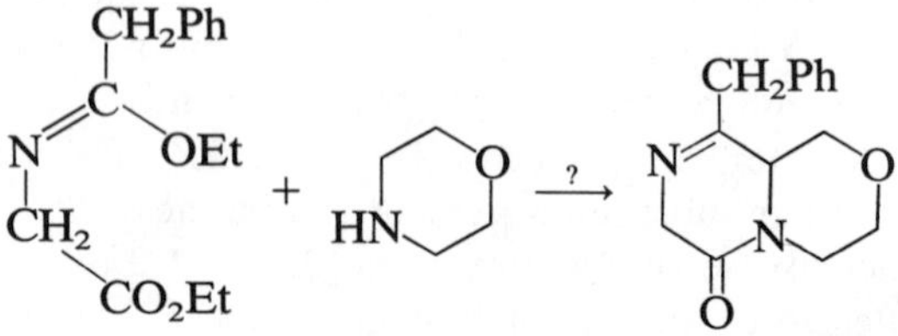

19 + morpholine →(?) **20**

V. Tables

TABLE 1. PERHYDROPYRAZINO[2,1-*b*][1,3]OXAZINES

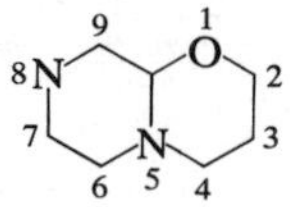

Molecular formula	Substituents	m.p., °C	Ref.
$C_9H_{14}N_2O_4$	9a-Methoxy-8-methyl-4,7-dioxo	118	1
$C_{10}H_{16}N_2O_4$	9a-Ethoxy-8-methyl-4,7-dioxo	147–149	1, 2

TABLE 2. PYRAZINO[2,3-*d*][1,3]OXAZIN-4-ONES

Molecular formula	Substituents	m.p., °C	Ref.
$C_7H_4ClN_3O_2$	6-Chloro-2-methyl	158–160 decomp.	9, 11, 12, 13, 16, 17, 19
$C_7H_5ClN_4O_2$	7-Amino-6-chloro-2-methyl	280 decomp.	11, 13, 19
$C_7H_5N_3O_2$	2-Methyl	150	8
$C_8H_7N_3O_2$	2-Ethyl	142	8
$C_8H_7N_3O_2S$	2-Methyl-6-methylthio	189–191	10, 11, 13, 14, 15, 18, 19, 20
$C_8H_7N_3O_3$	6-Methoxy-2-methyl	190–192	10, 11, 13, 14
$C_8H_7N_3O_4S$	2-Methyl-6-methylsulfonyl	214–216	10, 11, 13, 14 18, 19, 20
$C_8H_8N_4O_3$	6-Methoxyamino-2-methyl	190–192	14
$C_9H_7ClN_4O_3$	7-Acetamido-6-chloro-2-methyl	200–204	11, 13, 19
$C_9H_{10}N_4O_2$	6-Dimethylamino-2-methyl	212 decomp.	10, 11, 13, 14, 19
$C_{10}H_{12}N_4O_2$	2-Methyl-6-isopropylamino	212–214	10
	Nitrate	203–205	14
$C_{12}H_{14}N_4O_2$	2-Methyl-6-piperidin-1-yl	172–174	10, 14
$C_{13}H_{15}ClN_4O_3$	7-Butyramido-6-chloro-2-propyl	169–170	11, 13
$C_{14}H_{11}N_3O_2S$	6-Benzylthio-2-methyl	116.5–118.5	10, 11, 13, 14, 15, 18, 19, 20
$C_{14}H_{12}N_4O_2$	6-Benzylamino-2-methyl	168-170	10, 14

TABLE 3. PYRAZINO[2,1-*c*][1,4]OXAZINES

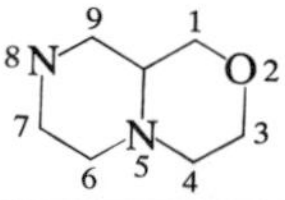

Molecular formula	Substituents	m.p., °C	Ref.
$C_9H_{15}ClN_2O_2$	8-(2-Chloroethyl)perhydro-1-oxo		
	Dipicrate ethanolate	183–185	21
$C_{14}H_{16}N_2O_2$	9-Benzyl-1,3,4,6,7,9a-hexahydro-6-oxo[a]	219	22

[a] Structure in doubt.

VI. References

1. M. M. Shemyakin, V. K. Antonov, A. M. Shkrob, V. I. Shchelokov, and Z. E. Agadzhanyan, *Tetrahedron* **21,** 3537 (1965).
2. V. K. Antonov, A. M. Shkrob, and M. M. Shemyakin, *Zh. Obshch. Khim.*, **37,** 2225 (1967); *Chem. Abstr.*, **69,** 10416 (1968).
3. D. R. Curtis and G. A. R. Johnston, *Nature,* **215,** 1502 (1967).
4. G. Chen and F. P. Hauck, *Fed. Proc.*, **20,** 323 (1961).
5. G. Chen, *Proc. Soc. Exp. Biol. Med.*, **112,** 611 (1963).
6. V. G. Longo and S. Chiavarelli, "Proc. Int. Pharmacol. Meeting, 1st, Stockholm," 1961, Vol. 8, p. 189; *Chem. Abstr.*, **60,** 11242 (1964).
7. J. E. Desmedt and V. La Grutta, *Arch. Int. Pharmacodyn.*, **149,** 263 (1964).
8. W. J. Irwin and D. G. Wibberley, *Tetrahedron Lett.*, 3359 (1972).
9. J. B. Bicking, J. W. Mason, O. W. Woltersdorf, Jr., J. H. Jones, S. F. Kwong, C. M. Robb, and E. Cragoe Jr., *J. Med. Chem.*, **8,** 638 (1965).
10. J. H. Jones, J. B. Bicking, and E. Cragoe, Jr., *J. Med. Chem.*, **10,** 899 (1967).
11. E. J. Cragoe and J. B. Bicking, U.S. Pat. 3,305,552; *Chem. Abstr.*, **67,** 108670 (1967).
12. Merck & Co., Inc., Neth. Appl. 6,413,272; *Chem. Abstr.*, **63,** 14885 (1965).
13. E. J. Cragoe and J. B. Bicking (to Merck & Co., Inc.), U.S. Pat. 3,361,748; *Chem. Abstr.*, **69,** 19220 (1968).
14. E. J. Cragoe, Jr., (to Merck & Co., Inc.), U.S. Pat. 3,360,517; *Chem Abstr.*, **69,** 27459 (1968).
15. E. J. Cragoe, Jr., (to Merck & Co., Inc.), U.S. Pat. 3,313,813; *Chem. Abstr.*, **69,** 36172 (1968).
16. E. J. Cargoe, Jr. (to Merck & Co., Inc.), Fr. Pat. 1,447,967; *Chem. Abstr.*, **69,** 59279 (1968).
17. E. J. Cragoe, Jr., (to Merck & Co., Inc.), Belg. Pat. 639,393; *Chem. Abstr.*, **62,** 7778 (1965).
18. E. J. Cragoe, Jr., (to Merck & Co., Inc.), Belg. Pat. 639,386; *Chem. Abstr.*, **62,** 14700 (1965).
19. Merck & Co., Inc., Neth. Appl. 6,409,717; *Chem. Abstr.*, **64,** 8209 (1966).
20. Merck & Co., Inc., Neth. Appl. 6,409,714; *Chem. Abstr.*, **64,** 12699 (1966).
21. Chun-Chao Tsai and I-Sheng Kao, *Yao Hsueh Hsueh Pao,* **12,** 11 (1965); *Chem. Abstr.*, **63,** 8361 (1965).
22. W. L. Mosby, "Heterocyclic Systems with Bridgehead Nitrogen Atoms," Interscience, New York, 1961, p. 1231.

CHAPTER XXXIV

Pyrazinothiazines

I. Pyrazino[2,1-*c*][1,4]thiazines 591
II. Pyrazino[2,3-*b*][1,4]thiazines 592
III. Tables 595
Table 1. Pyrazino[2,1-*c*][1,4]thiazines 595
Table 2. 4*H*-Pyrazino[2,3-*b*][1,4]thiazines 595
IV. References 596

I. Pyrazino[2,1-*c*][1,4]thiazines

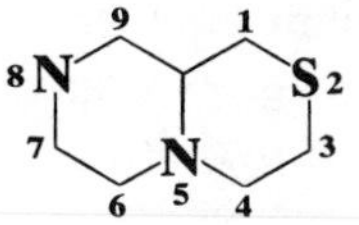

Only one report of this ring system has appeared.[1] Treatment of phenacylthiazole quaternary salts such as **1** with sodium methoxide gave the tricyclic compound **2** by dimerization of a betaine intermediate. Treatment of **2** with concentrated hydrobromic acid results in ring opening at *c* and reaction with the aryl ketone moiety following rotation around bond *d*. The acidic conditions cleave the remaining thiazole ring at *a* and *b* resulting in the formation of the pyrazino[2,1-*c*][1,4]thiazine **3**. The PMR spectra of the hydrobromides of compounds such as **3** were analyzed using double-resonance techniques.

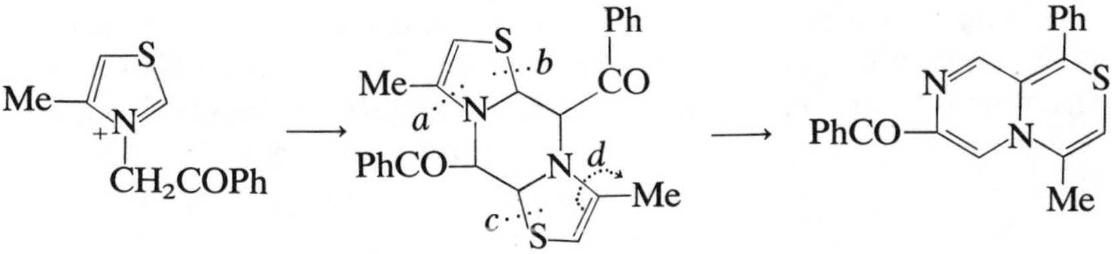

II. Pyrazino[2,3-*b*][1,4]thiazines

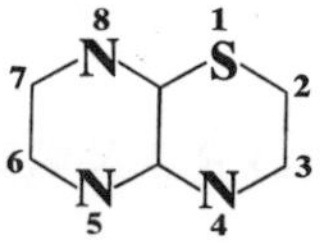

Although the systematic numbering system is given above, the alternative shown in formula **4** has frequently been employed, resulting in some confusion in abstracts of the patent literature.[2]

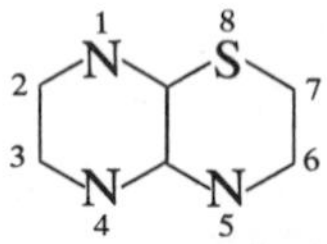

4

The 4*H*-tautomers **5** are favored more than the 2*H*-forms **6** in simple 3-aryl derivatives of this heterocycle. This conclusion is supported by infrared and PMR evidence.[3,4] An earlier formulation of such compounds as 2*H*-derivatives[5] should therefore be amended. More recently the 3-amino derivatives **7** have been described as being in the 2*H*-form, presumably stabilized by an amidinium type of resonance.[6] The CH_2 group in compound **7** ($R = Me$, $R^1 = H$) gives a singlet, apparently of two protons, in the PMR at δ 3.73 (deuteropyridine).

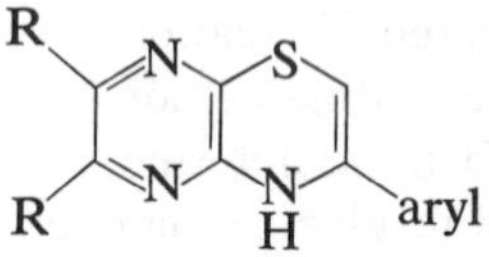

5 R = H, Me, or Ph

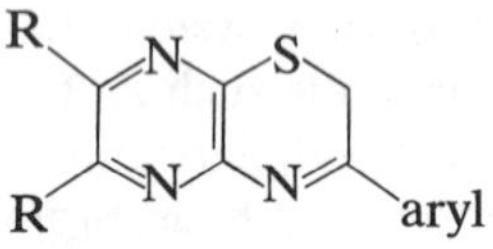

6 R = Ph

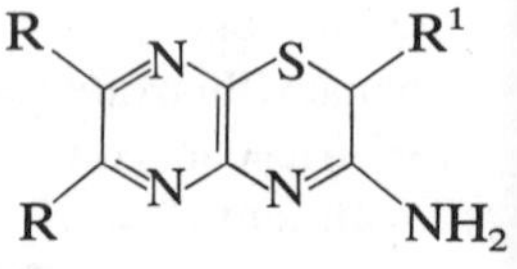

7 R = H, Me, or Ph
R^1 = H or Ph

Treatment of 2-amino-3-"mercapto"-pyrazines **8** with α-halo ketones has been used on several occasions to prepare derivatives of this ring system.[3–5] When the reaction is performed in the presence of alcoholic potassium hydroxide at <0°C, the dihydro compounds **9** are obtained. At room temperature with substituted phenacyl bromides, the ring-opened derivatives **10** are the main products.[3,4] The latter compounds may be cyclized in acetic anhydride[4] to give the substituted pyrazinothiazines **11**. These may also be obtained directly from the pyrazine **8** by performing the reaction in a chloroform–aqueous potassium hydroxide mixture at room temperature.[3]

8 R = H, Me, or Ph **9** R^1 = Alkyl

10 R^1 = Aryl **11**

Reagents (i) R^1COCH_2Br, KOH, EtOH < 0°
(ii) $ArylCOCH_2Br$, room temperature, KOH, EtOH
(iii) R^1COCH_2Br, KOH, H_2O, $CHCl_3$, room temperature
(iv) $(MeCO)_2O$

Reaction of compound **8** with α-halo nitriles ($R^1CHHalCN$) gave the 3-amino compounds **7** after cyclization of the intermediate *S*-alkylated products **12** with aqueous potassium hydroxide.[6] Similarly reaction of 2-amino-3-chloropyrazine (**13**) with the mercapto esters **14** in the presence of sodium ethoxide gave the 3-oxo derivatives **16** on refluxing in dioxan.[7] The intermediates **15** are involved in the reaction. The most convenient synthesis of these compounds involves reaction of 2-amino-3-chloropyrazines with thioglycollic acid in alkali and acid-catalyzed cyclization of the carboxymethylthio derivative.[13]

The 2-unsubstituted 3-oxo compound **17** is readily halogenated in the 2-position.[8,9] Treatment of the pyrazinothiazine with an equimolar amount of bromine provides the 2-monobromo compound **18** (R = Br), and an equimolar quantity of sulfuryl chloride gives the 2-chloro derivative **18** (R = Cl). The use of two moles of sulfuryl chloride results in the formation of the dichloro compound **19**.

Compound **17** and its 2-methyl derivative have been oxidized to the corresponding 1,1-dioxides with potassium permanganate in aqueous acetic acid.[10] Hydrogen peroxide results in ring opening and the formation of pyrazinesulfonic acids.[13] The halogens in compounds **18** and **19** are readily displaced by nucleophiles, and this approach has provided a variety of 2-(substituted amino) derivatives.[2,9] 3-Hydrazinopyrazino[2,3-*b*][1,4]thiazine has been prepared by reaction of the corresponding

3-amino compound with hydrazine hydrate.[11] The oxo group in **17** is readily reduced to —CH_2— with lithium aluminum hydride.[13]

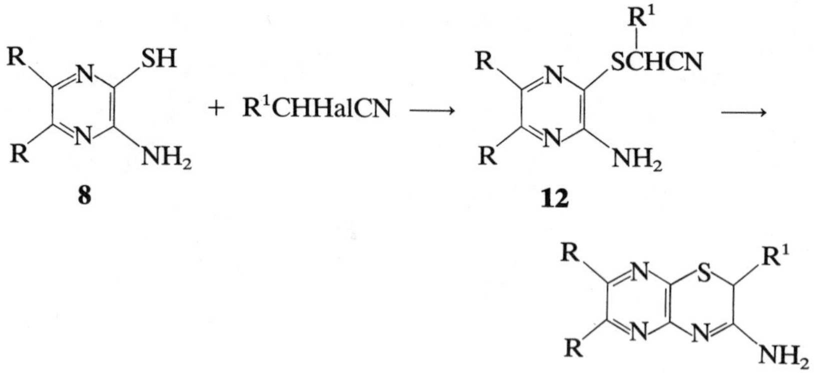

8 + $R^1CHHalCN$ ⟶ **12** ⟶ **7**

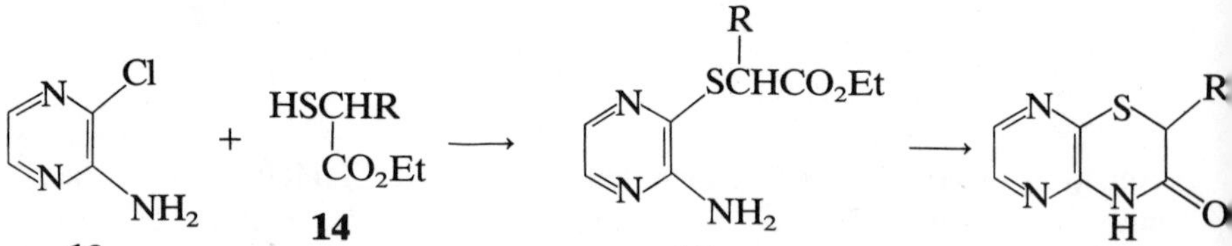

13 + **14** ⟶ **15** ⟶ **16**

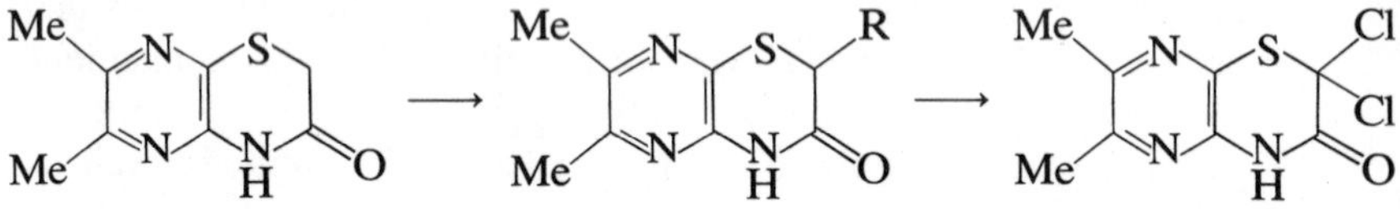

17 ⟶ **18** $R = Br$ or Cl ⟶ **19**

The 3-oxo compounds **20** and their 1,1-dioxides are claimed to possess stronger diuretic activity than the thiazide diuretics.[7,10] Various other 3-oxo compounds have attracted interest as potential oncostatic agents.[9] Recently a paper has appeared describing analgesic, antiinflammatory, and antipyretic effects produced by 4-alkyl-3-oxo derivatives of this heterocycle.[12]

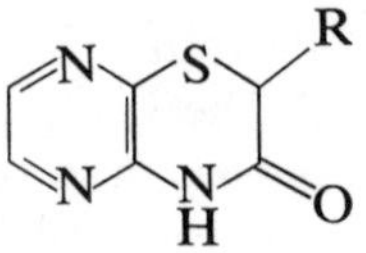

20 $R = H$ or Me

III. Tables

TABLE 1. PYRAZINO[2,1-*c*][1,4]-THIAZINES

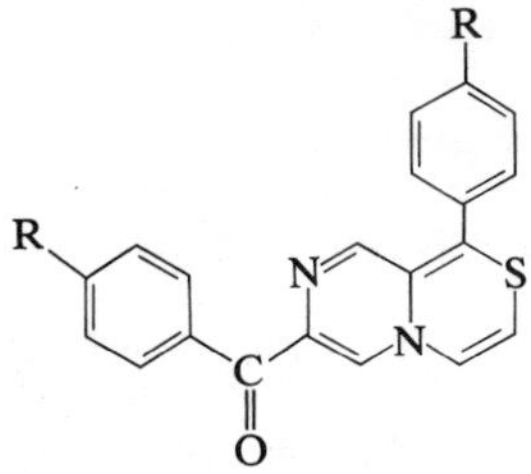

R		Ref.
H	High-melting hydrobromides	1
Br		
Cl		
MeO		

TABLE 2. 4*H*-PYRAZINO[2,3-*b*][1,4]THIAZINES

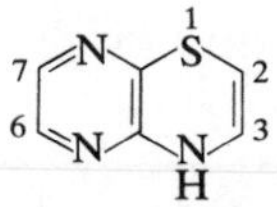

Molecular formula	Substituents or structure	m.p., °C	Ref.
$C_6H_5N_3OS$	2,3-Dihydro-3-oxo	217–218	7
	1,1-Dioxide	238–240	10
$C_7H_7N_3OS$	2,3-Dihydro-2-methyl-3-oxo	206–207	7
	1,1-Dioxide	221–222	10
$C_8H_7Cl_2N_3OS$	2,2-Dichloro-2,3-dihydro-6,7-dimethyl-3-oxo	185–186	8, 9
$C_8H_8BrN_3OS$	2-Bromo-2,3-dihydro-6,7-dimethyl-3-oxo	180–181	8, 9
$C_8H_8ClN_3OS$	2-Chloro-2,3-dihydro-6,7-dimethyl-3-oxo	175–176	8, 9
$C_8H_9N_3OS$	2,3-Dihydro-6,7-dimethyl-3-oxo	186–187	13
$C_8H_{10}N_4S$	3-Amino-6,7-dimethyl[a]	200–201	6
$C_8H_{11}N_3OS$	2,3-Dihydro-3-hydroxy-2,3-dimethyl	78.5–80.5	3
$C_8H_{11}N_3S$	2,3-Dihydro-6,7-dimethyl	145–146	13
$C_8H_{11}N_5S$	3-Hydrazino-6,7-dimethyl[a]	238–239	11
$C_9H_{13}N_3OS$	2,3-Dihydro-3-hydroxy-3,6,7-trimethyl	88–90	3
$C_{10}H_{12}N_4OS$	2-Aziridin-1-yl-2,3-dihydro-6,7-dimethyl-3-oxo	195–196	2, 9
$C_{10}H_{13}N_3O_2S$	2-Ethoxy-2,3-dihydro-6,7-dimethyl-3-oxo	144–145	9
$C_{12}H_8N_4O_2S$	3-(4-Nitrophenyl)	224	3
$C_{12}H_{15}N_5OS$	2,2-Diaziridin-1-yl-2,3-dihydro-6,7-dimethyl-3-oxo	220–221	2, 9

TABLE 2 (*contd.*)

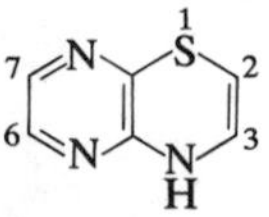

Molecular formula	Substituents or structure	m.p., °C	Ref.
$C_{12}H_{16}N_4OS$	2,3-Dihydro-6,7-dimethyl-2-morpholin-4-yl-3-oxo	246–247	2, 9
$C_{12}H_{18}N_4OS$	2-Butylamino-2,3-dihydro-6,7-dimethyl-3-oxo	169–170	2, 9
$C_{13}H_{18}N_4OS$	2,3-Dihydro-6,7-dimethyl-3-oxo-2-piperidin-1-yl	223–225	2, 9
$C_{14}H_{12}BrN_3S$	3-(4-Bromophenyl)-6,7-dimethyl	199–201	3
$C_{14}H_{12}N_4O_2S$	6,7-Dimethyl-3-(4-nitrophenyl)	222–223	3
$C_{14}H_{13}N_3S$	6,7-Dimethyl-3-phenyl	129–131	3
$C_{14}H_{14}N_4OS$	2-Anilino-2,3-dihydro-6,7-dimethyl-3-oxo	217–218	2, 9
$C_{14}H_{14}N_4S$	3-Amino-6,7-dimethyl-2-phenyl[a]	227–228	6
$C_{14}H_{22}N_4OS$	2-Dipropylamino-2,3-dihydro-6,7-dimethyl-3-oxo	110–112	2, 9
$C_{15}H_{14}N_4O_3S$	2-(4-Carboxyanilino)-2,3-dihydro-6,7-dimethyl-3-oxo		
	Ethyl ester	222–223	9
$C_{15}H_{16}N_4OS$	2-Benzylamino-2,3-dihydro-6,7-dimethyl-3-oxo	191–192	2, 9
$C_{16}H_{23}N_5O_3S$	2,3-Dihydro-6,7-dimethyl-2,2-bis(morpholin-4-yl)-3-oxo	181–183	2, 9
$C_{17}H_{18}N_4OS$	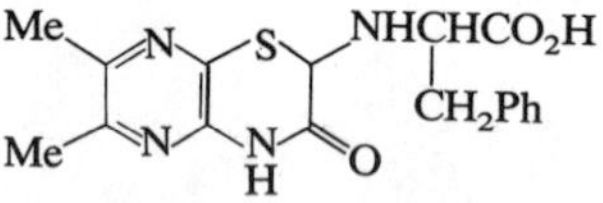 6,7-Dimethyl-2-[NHCH(CH$_2$Ph)CO$_2$H]-3-oxo-2,3-dihydro (structure: Me, Me, NHCHCO$_2$H, CH$_2$Ph, O, N, S, NH)		
	Ethyl ester	138–140	9
$C_{18}H_{13}N_3OS$	2,3-Dihydro-3-oxo-6,7-diphenyl	253–255	13
$C_{18}H_{14}N_4S$	3-Amino-6,7-diphenyl[a]	249–250	6
$C_{19}H_{16}ClN_3OS$	3-Chloromethyl-2,3-dihydro-3-hydroxy-6,7-diphenyl	101–102	4
$C_{19}H_{17}N_3OS$	2,3-Dihydro-3-hydroxy-3-methyl-6,7-diphenyl	120–121	4
$C_{24}H_{15}Cl_2N_3S$	3-(2,5-Dichlorophenyl)-6,7-diphenyl	295–298	4, 5
$C_{24}H_{16}BrN_3S$	3-(4-Bromophenyl)-6,7-diphenyl	246–247	4, 5
$C_{24}H_{16}N_4O_2S$	3-(3-Nitrophenyl)-6,7-diphenyl	222–224	4, 5
	3-(4-Nitrophenyl)-6,7-diphenyl	301–303	4, 5
$C_{24}H_{17}N_3S$	3,6,7-Triphenyl	266	4, 5
$C_{25}H_{19}N_3OS$	3-(4-Methoxyphenyl)-6,7-diphenyl	265–266	5

[a] Formulated as the 2*H* compound.

IV. References

1. J. Fröhlich, U. Habermalz, and F. Kröhnke, *Tetrahedron Lett.*, 271 (1970).
2. L. A. Myshkina and T. S. Safonova, USSR Pat. 283,224; *Chem. Abstr.*, **77**, 34580 (1972).

3. T. S. Safonova and L. A. Myshkina, *Khim. Geterotsikl. Soedin.*, 1092 (1970); *Chem. Abstr.*, **74,** 53744 (1971).
4. L. A. Myshkina and T. S. Safonova, *Khim. Geterotsikl. Soedin.*, 1101 (1970); *Chem. Abstr.*, **74,** 53742 (1971).
5. T. S. Safonova and L. A. Myshkina, USSR Pat. 240,709; *Chem. Abstr.*, **71,** 49955 (1969).
6. T. S. Safonova, M. P. Nemeryuk, L. A. Myshkina, and N. I. Traven, *Khim. Geterotsikl. Soedin.*, 944 (1972); *Chem. Abstr.*, **77,** 126570 (1972).
7. S. Kawano, H. Watanabe, and T. Sato, Jap. Pat. 71 32,670; *Chem. Abstr.*, **75,** 140888 (1971).
8. T. S. Safonova and L. A. Myshkina, USSR Pat. 280,482; *Chem. Abstr.*, **76,** 72556 (1972).
9. T. S. Safonova, L. A. Myshkina, V. A. Chernov, and A. S. Sokolova, *Khim. Geterotsikl. Soedin.*, 1498 (1971); *Chem. Abstr.*, **77,** 34453 (1972).
10. S. Kawano, S. Zoga, H. Watanabe, and T. Sato, Jap. Pat. 71 32,671; *Chem. Abstr.*, **75,** 140890 (1971).
11. T. S. Safonova, M. P. Nemeryuk, and L. A. Myshkina, USSR Pat. 265,110; *Chem. Abstr.*, **73,** 25492 (1970).
12. K. Kubota, Y. Sato, and T. Urono, *Jap. J. Pharmacol.*, Suppl. **23,** 59 (1973).
13. L. A. Myshkina and T. S. Safonova, *Khim. Geterotsikl. Soedin.*, 66 (1975); *Chem. Abstr.*, **83,** 9982 (1975).

CHAPTER XXXV

Pyrroloquinoxalines

I. Pyrrolo[1,2-*a*]quinoxalines . . . 598
  1. Physical Properties . . . 598
  2. Methods of Preparation . . . 601
    A. Syntheses from *o*-Aminophenylpyrroles and Related Compounds . . . 601
    B. Syntheses from *o*-Phenylenediamine . . . 604
    C. The Cyclization of β-Quinoxalinylpropionic Acids . . . 605
    D. Syntheses Involving Quaternization of Quinoxalines . . . 610
    E. Miscellaneous Syntheses . . . 611
  3. Reactions . . . 615
    A. Alkylation . . . 615
    B. Electrophilic Substitution Reactions . . . 616
    C. Oxidation . . . 618
    D. Reactions with Nucleophilic Reagents . . . 618
    E. Reduction . . . 621
  4. Uses . . . 622
II. Pyrrolo[2,3-*b*]quinoxalines . . . 622
III. Pyrrolo[3,4-*b*]quinoxalines . . . 625
IV. Pyrrolo[1,2,3-*de*]quinoxalines . . . 630
V. Tables . . . 633
  Table 1. Pyrrolo[1,2-*a*]quinoxalines . . . 633
  Table 2. 2-Oxopyrrolo[2,3-*b*]quinoxalines . . . 643
  Table 3. 1*H*-Pyrrolo[2,3-*b*]quinoxalines . . . 644
  Table 4. Pyrrolo[3,4-*b*]quinoxalines . . . 646
  Table 5. Pyrrolo[1,2,3-*de*]quinoxalines . . . 649
VI. References . . . 650

I. Pyrrolo[1,2-*a*]quinoxalines

1 2 9 N 3 8 7 4 N 6 5

1. Physical Properties

The parent aromatic heterocycle has been subjected to theoretical calculations in order to compare the predicted and the found sites of electrophilic substitution.[1] HMO calculations suggest that the most susceptible position is the 2-position, whereas a consideration of localization energies indicates the susceptibility to electrophilic substitution decreases in order 1>3>6>2. The latter prediction is regarded as the more valid and agrees with the experimental results described below.

The ultraviolet spectrum of pyrrolo[1,2-*a*]quinoxaline in aqueous buffer at pH 6.9 shows three maxima at 224, 247, and 334 nm. The long-wavelength band moves to 352 nm on protonation (pH 1.0).[2,3] This bathochromic shift is also observed on protonation of various substituted derivatives of this heterocycle.[4] The infrared spectra of several salts of pyrrolo[1,2-*a*]quinoxalines suggest that protonation occurs at the 5-position.[4]

Several 1-oxo-1,5-dihydro compounds (**1**) have had their ultraviolet spectra determined in 96% ethanol.[4] They show three characteristic bands, at ca. 235, 300, and 420–430 nm. Apparently certain 2,3-diphenyl derivatives exist in the tautomeric hydroxy form **2**, as evidenced by a prominent OH stretching frequency (no details given) in their infrared spectra.[5] It is suggested that the compounds undergo a keto–enol type of

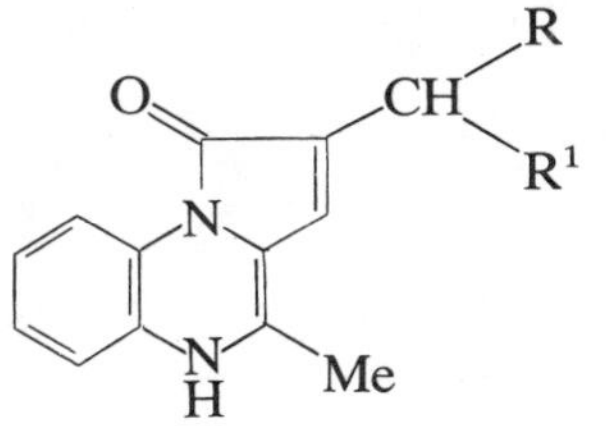

1 R = H, CO_2H or CO_2Me; R^1 = Me or Ph

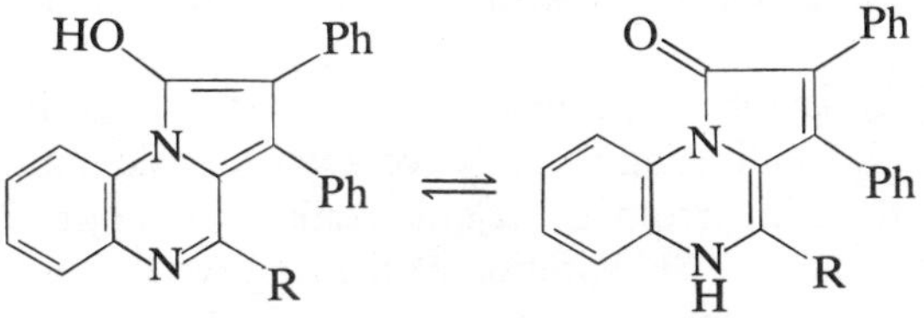

2 R = H or Me **2a**

tautomerism (**2** ⇌ **2a**) which is influenced by changes in pH. In acetonitrile, however, there seem to be marked differences between the ultraviolet spectra of compounds **2** (R = H) and **2** (R = Me) and the *O*-ethyl derivative **3**.[5]

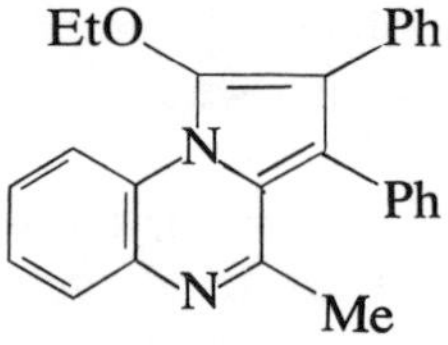

3

The PMR spectrum (CCl_4) of the parent heterocycle[2] shows signals at δ 6.95 and 6.99 (H-2, H-3), multiplets in the ranges δ 7.51–7.69 (H-7, H-8) and δ 7.87–8.30 (H-1, H-6, H-9), and a singlet at δ 8.95 (H-4). The full analysis of this complex 8-proton PMR spectrum has recently been reported.[72] The PMR spectra of several 4-methyl compounds have been discussed,[4] and the marked deshielding (0.8–1.0 ppm) of the proton in position 9 by chlorine in the 1-position has been reported.[6] The 1-bromo derivative shows an even greater deshielding of the 9-proton.[10] The structure of the product obtained from the reaction of the benzimidazolium salt **4** with dimethyl acetylenedicarboxylate was shown by its ^{13}C NMR spectrum to be the pyrrolo[1,2-*a*]quinoxaline **5** (R = CO_2Me). The structure was established by comparison with the ^{13}C NMR spectrum of authentic **5** (R = H).[73]

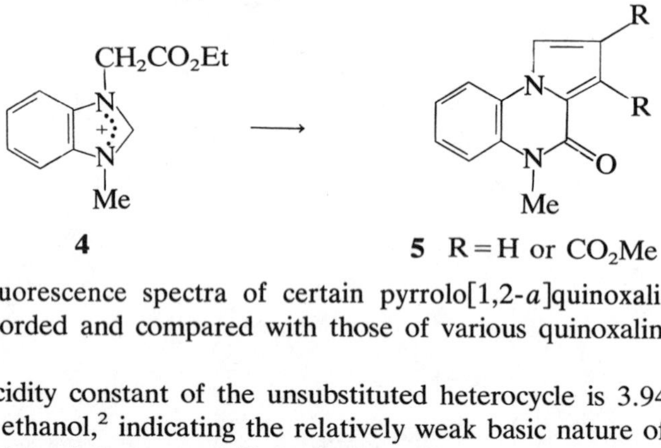

4 → **5** R = H or CO_2Me

The fluorescence spectra of certain pyrrolo[1,2-*a*]quinoxalines have been recorded and compared with those of various quinoxaline derivatives.[7]

The acidity constant of the unsubstituted heterocycle is 3.94 in 50% aqueous ethanol,[2] indicating the relatively weak basic nature of the ring system. Nevertheless pyrrolo[1,2-*a*]quinoxalines are appreciably stronger bases than quinoxalines. Substitution of the ring system by methyl groups results in a base-strengthening effect dependent on the site of substitution. The 4-methyl compound has a pK_a of 4.58 in 50% alcohol,[4] which is consistent with protonation at the 5-position, as the increase in pK_a (0.64) is characteristic of the effect of methyl substitution in six-membered rings α to the site of protonation.[8] The much smaller increase in pK_a between the 4-methyl compound (4.58) and 2,4-dimethyl compound (4.89) is also consistent with protonation at the 5-position.

2. Methods of Preparation

Several different approaches have been adopted for the synthesis of pyrrolo[1,2-*a*]quinoxalines. Particularly useful are approaches involving cyclization reactions of *o*-aminophenylpyrroles. Alternative syntheses have involved the use of *o*-phenylenediamine, the cyclization of β-quinoxalinylpropionic acids, quaternization of quinoxaline derivatives, as well as several other routes less amenable to classification.

A. *Syntheses from o-Aminophenylpyrroles and Related Compounds*

Routes via *o*-aminophenylpyrroles present the most convenient syntheses of a wide variety of pyrrolo[1,2-*a*]quinoxalines. Thus reaction of the amino compound **6** with acetic anhydride in acetic acid gave the acetamido derivative which was cyclized with phosphoryl chloride to give the 4-methyl compound **7** (R = Me) in 56% yield.[2,3] The 4-phenyl compound **7** (R = Ph) has been prepared similarly.[2,3] An even more convenient synthesis of 4-aryl compounds is achieved by reaction of compound **6** with aromatic aldehydes to give the 4,5-dihydro derivatives **8**.[9,74] These are readily oxidized to 4-arylpyrrolo[1,2-*a*]quinoxalines **9** with manganese dioxide.[9] This approach may be carried out in one step by reaction of compound **6** with aromatic aldehydes (e.g., benzaldehyde) in the presence of cupric acetate.[10] Reaction of the aminophenylpyrrole **6** with 90% formic acid gave pyrrolo[1,2-*a*]quinoxaline (**7**, R = H) directly in 98% yield.[2,3] Pyrrolo[1,2-*a*]quinoxalines substituted in the 1-position[11] and the 7-position[12] have also been prepared from appropriately substituted *o*-aminophenylpyrroles.

4,5-Dihydro-4-oxopyrrolo[1,2-*a*]quinoxaline (**11**) is obtained on reaction of the amino compound **6** with phosgene.[10] Ethyl chloroformate gives the carbamate **10**, which could not be cyclized with phosphoryl chloride but reacted with zinc chloride in refluxing *o*-dichlorobenzene to give the 4-oxo compound **11**.[3] Treatment of compound **6** with carbon disulfide provides the 4-thioxo compound **12**, and cyanogen bromide gives 4-aminopyrrolo[1,2-*a*]quinoxaline directly, albeit in low yield. Thiophosgene on reaction with the aminopyrrole **6** did not give the expected thioxo derivative **12** but the sulfide **13**.[10]

The adduct **14**, from the reaction of the amino compound **6** with dimethyl acetylenedicarboxylate, was cyclized to the pyrrolo[1,2-*a*]quinoxaline **15** using polyphosphoric acid. The anilino compound **17**

Reagents (i) $(RCO)_2O$
(ii) $POCl_3$
(iii) ArCHO
(iv) MnO_2

Reagents (i) $ClCO_2Et$
(ii) $ZnCl_2$, *o*-$Cl_2C_6H_4$, Δ
(iii) CS_2
(iv) $CSCl_2$

was obtained by treatment of the urea **16** with phosphoryl chloride.[10] An attempt to formylate the pyrrole **18** with phosphoryl chloride in DMF gave the 4-phenyl compound **19**, presumably by way of Beckmann rearrangement of the oxime, followed by cyclization of the resultant amide.[12] Catalytic reduction of the nitrocyanide **20** gave the 4-amino-5-oxide **21**.[9]

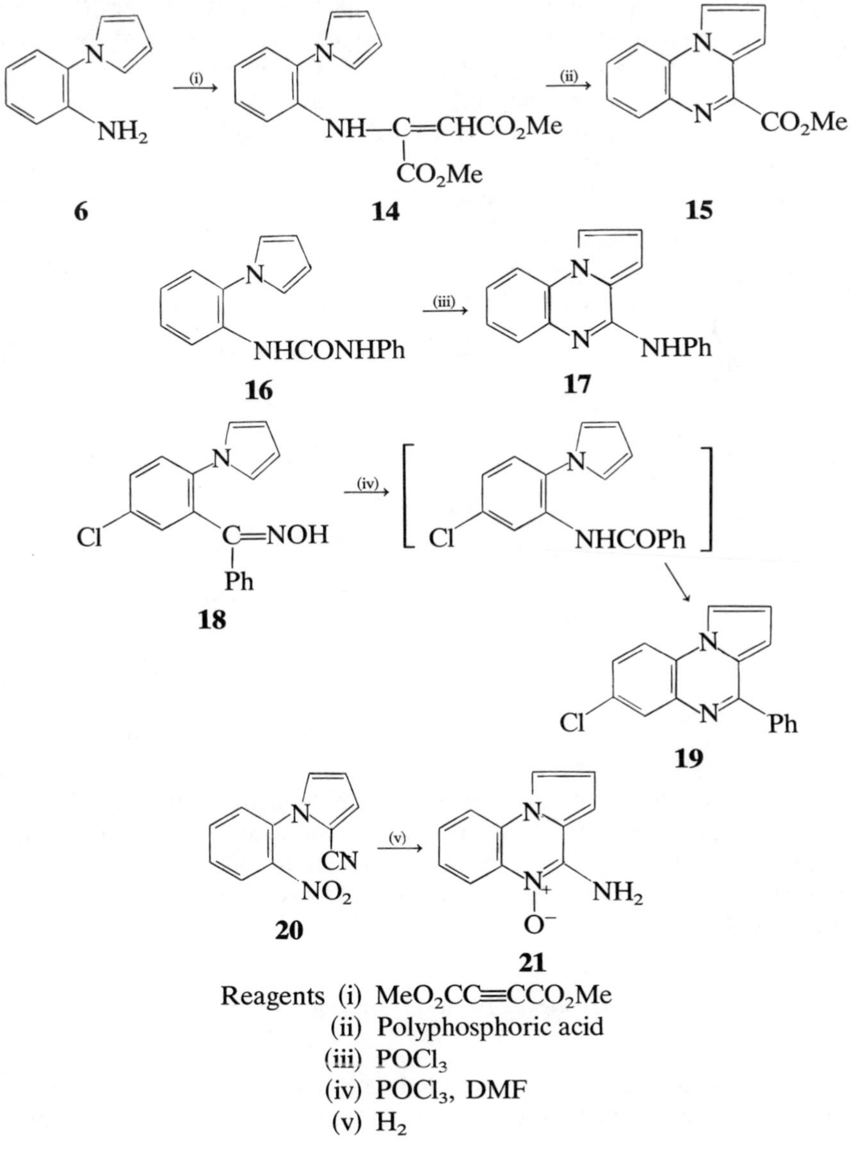

6 → (i) → **14** → (ii) → **15**

16 → (iii) → **17**

18 → (iv) → [intermediate] → **19**

20 → (v) → **21**

Reagents (i) $MeO_2CC{\equiv}CCO_2Me$
(ii) Polyphosphoric acid
(iii) $POCl_3$
(iv) $POCl_3$, DMF
(v) H_2

Several partially reduced pyrrolo[1,2-*a*]quinoxalines have been obtained as derivatives of proline in studies on the stepwise degradation of peptides. These compounds (**25**) are obtained when a peptide (**22**) with an *N*-terminal proline residue is condensed with an appropriately activated aryl halide, and the resultant *N*-aryl derivative **23** or **24** is cyclized with expulsion of the peptide residue **26**.

The *N*-aryl derivatives **23** and **24** spontaneously cyclize to the pyrrolo[1,2-*a*]quinoxalines when the neighboring substituent on the phenyl ring is converted into an NH_2 function. This may be achieved by deblocking of a *t*-butyloxycarbonylamino group (**23**) with trifluoroacetic acid[13] or by catalytic reduction of a nitro group (**24**).[14,15]

22 R = Peptide

23

24 X = CO_2Me or SO_2Me

25a + RNH_2 **26**

25b + RNH_2 **26**

Reaction of the pyrrolidinoaniline **27** with *N*-methylisatin (**28**) gives a variety of products (**29–31**) dependent on the acidity of the alcoholic solution used for the reaction.[16] The products formed in weakly acidic solution (**30** and **31**) rearrange in excess acid to give the spiropyrrolo[1,2-*a*]quinoxaline **29**.

B. *Syntheses from o-Phenylenediamine*

The dihaloester **33** has been reacted with *o*-phenylenediamine to give the hexahydrooxo compound **34**. Ethylenediamine gave the analogous pyrrolo[1,2-*a*]pyrazines.[17] The disubstituted phenylenediamine **35** has been condensed with methyl acrylate[18] and methyl crotonate[19] to give the cyclized Michael adducts **36**. The product **36** (R = H) was also prepared by reaction of the phenylenediamine **35** with ethyl 3-bromopropionate to give the triester **37**, followed by cyclization with sodium ethoxide. Acid cyclization of the intermediate **37** results in hydrolysis and decarboxylation of the 3*a*-ester group to give compound **38**.[18]

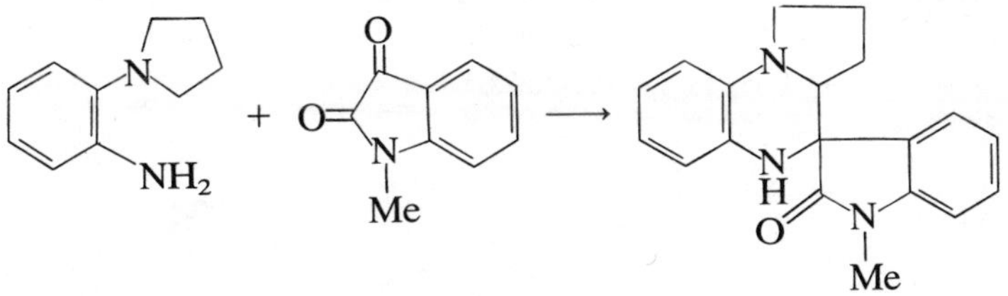

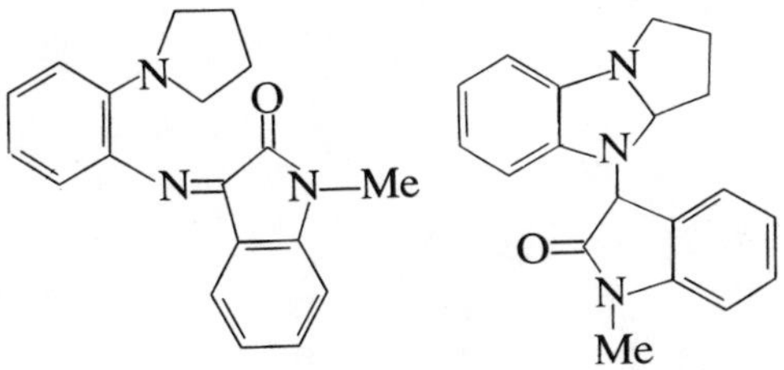

Two moles of *o*-phenylenediamine condense with the pyrone derivative **39** to give the anilide **40** in 75% yield.[20] This reaction is most easily understood by regarding the pyrone **39** as an internal ester of the enolized form of the dicarbonyl compound **39a**.

C. *The Cyclization of β-Quinoxalinylpropionic Acids*

1,5-Dihydro-1-oxopyrrolo[1,2-*a*]quinoxaline (**42**) is obtained in 85% yield when the quinoxalinylpropionic acid **41** is heated with polyphosphoric acid.[21] Various 2- and 4-substituted analogues of **42** have been prepared by cyclization of appropriately substituted quinoxalinylpropionic acids using either acetic anhydride in sulfuric acid[22–24] or polyphosphoric acid[24] as cyclization reagent. Reaction of the propionic acid **41** with phosphoryl chloride gives the 1-chloro compound **43**.[11] 1-Chloro-4-methylpyrrolo[1,2-*a*]quinoxaline has been prepared similarly.[24] Treatment of the ethyl ester **44** with phosphoryl chloride gives the 1-ethoxy

$$\text{32} + Cl(CH_2)_3CHBrCO_2Et \ (\textbf{33}) \longrightarrow \textbf{34}$$

32 **33** **34**

$$\textbf{35} + RCH{=}CHCO_2Et \longrightarrow \textbf{36}$$

35

36 R = H or Me

37 **38**

Reagents (i) $BrCH_2CH_2CO_2Et$
(ii) NaOEt
(iii) H^+

39 **39a**

40 R = $NHC_6H_4NH_2$-*o*

compound **45** as a minor product.[25] In contrast to the reaction with phosphoryl chloride, phosphoryl bromide reacts with the acid **41** to give, not the 1-bromo analogue of **43**, but a mixture of the 2,3-dibromo compound and 1,2,3-tribromopyrrolo[1,2-*a*]quinoxaline.[26] This mixture of polybromo compounds is also obtained on treatment of the parent heterocycle with bromine in refluxing hydrobromic acid.

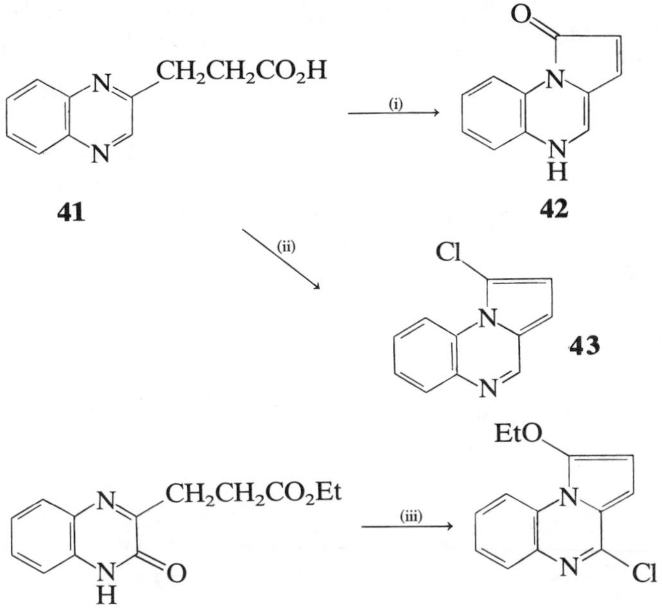

$CH_2CH_2CO_2H$ **41** —(i)→ **42**

41 —(ii)→ **43**

$CH_2CH_2CO_2Et$ **44** —(iii)→ **45**

Reagents (i) Polyphosphoric acid
(ii) $POCl_3$
(iii) $POCl_3$

A 1,5-dihydro-1-oxo derivative (**47**) of pyrrolo[1,2-*a*]quinoxaline has been shown to be the product of a supposed Diels-Alder reaction between 2,3-dimethylquinoxaline (**46**) and maleic anhydride.[23] 2-Methylquinoxaline and 2-methyl-3-phenylquinoxaline give the corresponding derivatives **48**.[22,24] The reaction between 2,3-dimethyl quinoxaline and maleic anhydride proceeds in 80% yield when the reactants are fused together.[24]

Reaction of 2,3-dimethylquinoxaline with monosubstituted maleic anhydrides (**49**) gives only one of the two possible isomers. The products formed are of formula **51**. This implies that the reaction proceeds through an intermediate **50** rather than the isomeric **52**. With disubstituted maleic anhydrides, where reaction would necessitate a product corresponding to **52**, no reaction occurs.[4]

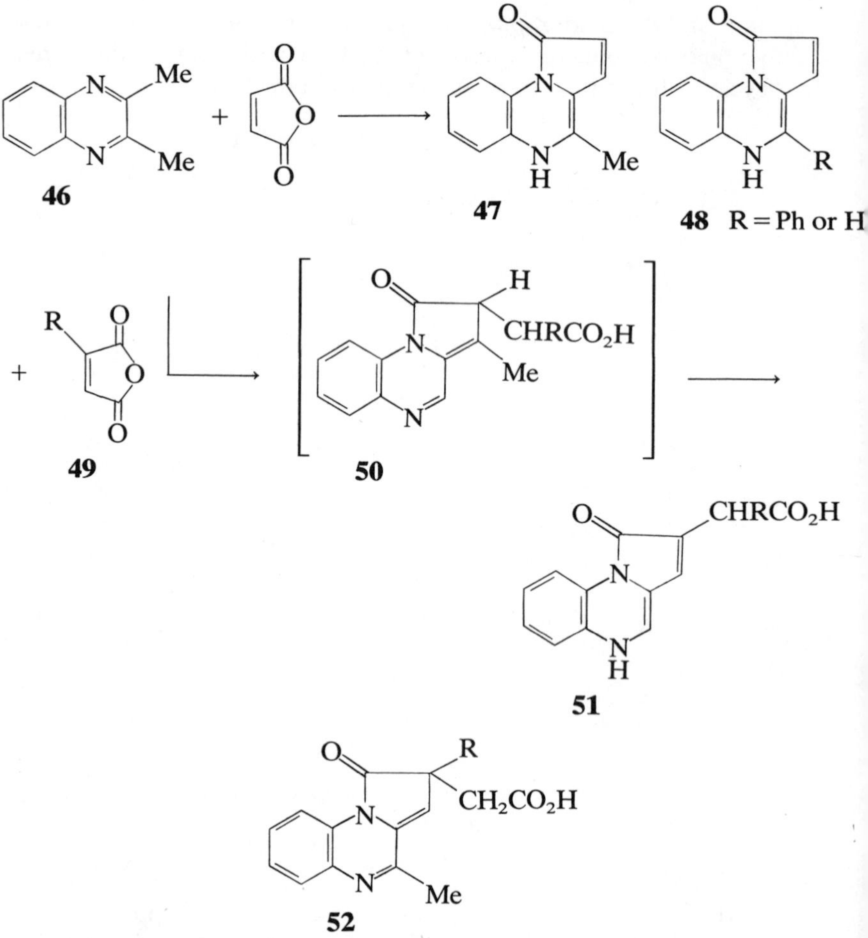

Me Me N N **46** + O O O → O N N H Me **47** O N N H R **48** R = Ph or H

+ R O O O **49** → [O H CHRCO$_2$H N Me N **50**] →

O CHRCO$_2$H N N H **51**

O R CH$_2$CO$_2$H N N Me **52**

Although 1,5-dihydro-1-oxo derivatives of this ring system are quite stable to reduction, reductive cyclization of quinoxalinylpropionic acids yields hexahydro-1-oxo compounds. Thus reduction with hydrogen and Raney nickel of the acids **53** yields the hexahydro compounds **54**.[27] The earliest preparation of the ring system involved hydrogenation of the keto ester **55** at high temperature and pressure over copper chromite.[28] The perhydro compound **56** was obtained by this approach. The use of hydrogen and Raney nickel resulted in the synthesis of compound **57**.[29]

The most carefully studied reductive approach concerned the preparation of dioxo compounds **59** from the propionic acids **58**. The best reducing agent for this synthesis was found to be alkaline sodium borohydride.[30]

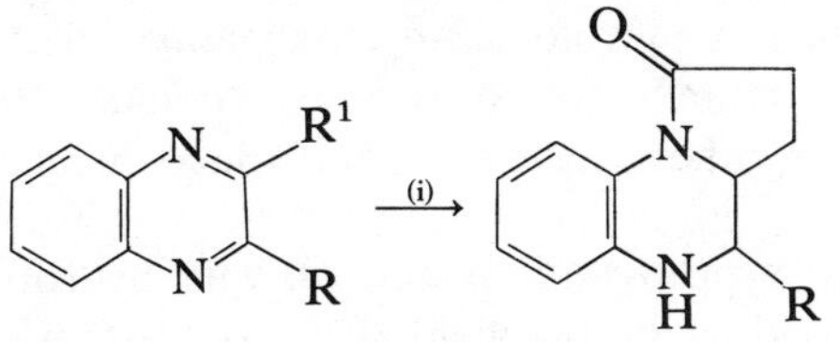

53 $R^1 = CH_2CH_2CO_2H$ or $CH{=}CHCO_2H$
$R = H$, Me, or Ph

54

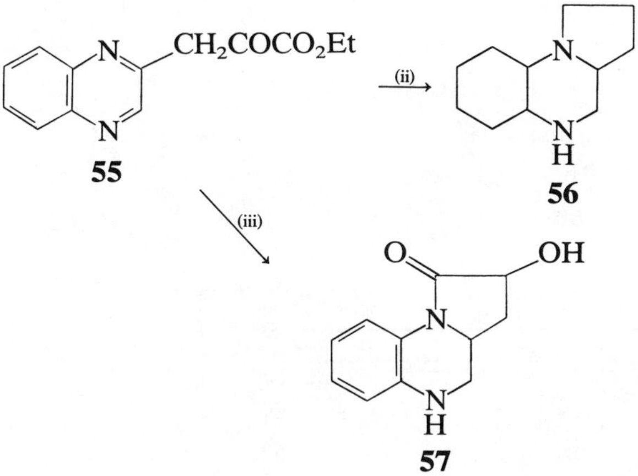

55 CH_2COCO_2Et

56

57

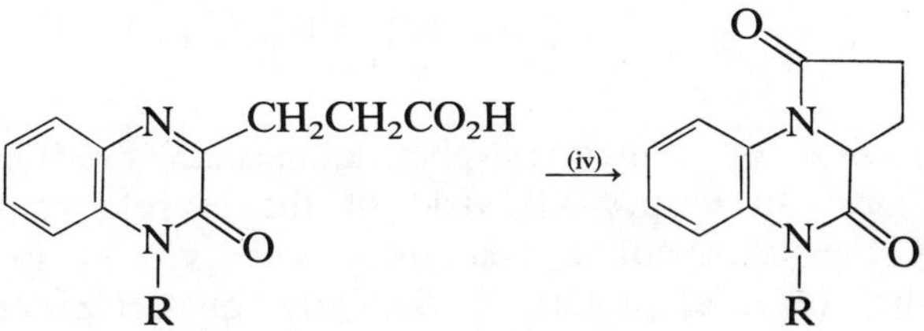

58 $R = H$ or Me

59

Reagents (i) H_2, Raney Ni
(ii) H_2, Copper chromite
(iii) H_2, Raney Ni
(iv) $NaBH_4$, OH^-

D. *Syntheses Involving Quaternization of Quinoxalines*

In general the quinoxaline ring system is quaternized only with difficulty, so that approaches to pyrrolo[1,2-*a*]quinoxalines that involve a quaternization step are unlikely to be very convenient. Nevertheless several different types of synthesis have been performed using quinoxaline quaternary salts.

Treatment of the hydroxypropylquinoxaline **60** with hydrobromic acid gave the pyrrolo compound **61** in low yield. This contrasts with the high yields obtained from this approach in the more readily quaternized quinoline series.[4] 2,3-Diphenylquinoxaline (**62**) forms a dianion on treatment with sodium, and this has been alkylated with 1,3-diiodopropane to give the tetrahydro compound **63**.[31]

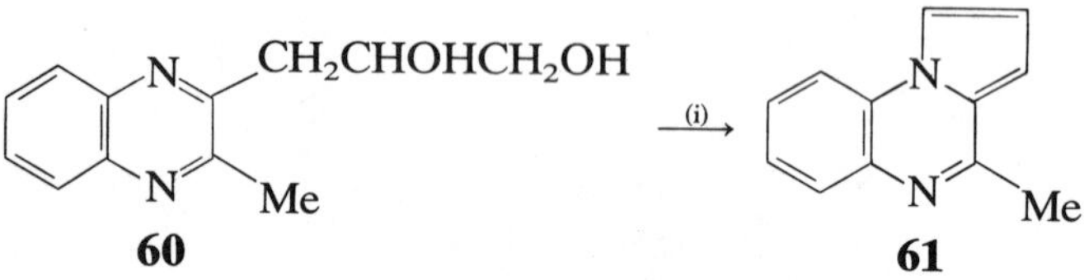

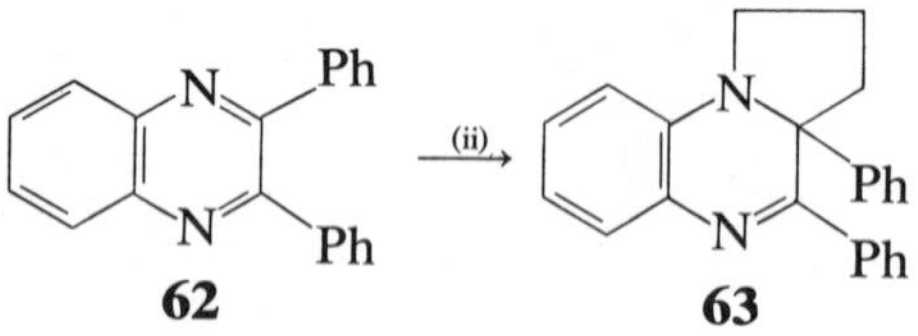

$CH_2CHOHCH_2OH$ Me **60** —(i)→ Me **61**

Ph Ph **62** —(ii)→ Ph Ph **63**

Reagents (i) HBr
(ii) Na, $ICH_2CH_2CH_2I$

Quaternization of 2-methyl-3-phenylquinoxaline (**64**) with chloroacetone gave an unspecified yield of the pyrrolo compound **65**. Apparently the intermediate quaternary salt cyclized spontaneously to give the tricyclic product.[24] Similarly quaternization of 2,3-dimethylquinoxaline with phenacyl bromide gave a low yield of the pyrrolo[1,2-*a*]quinoxaline **66**.[4]

Phenacylquinoxalinium bromide (**67**) reacts with nitromethane in the presence of sodium carbonate to give the 3-nitro compound **68** in 41% yield.[32] The quaternary salt **67** has also been reacted with dimethyl acetylenedicarboxylate to give the 1:1 adduct **69**.[33]

64 + $ClCH_2COMe$ ⟶ **65**

66

67 $\xrightarrow{(i)}$ **68**

67 $\xrightarrow{(ii)}$ **69**

Reagents (i) CH_3NO_2, Na_2CO_3
(ii) $MeO_2CC{\equiv}CCO_2Me$

E. *Miscellaneous Syntheses*

Quaternization of quinoxaline is neatly avoided by the synthesis of the azirino[1,2-*a*]quinoxalines **71** from *o*-phenylenediamine and the dibromo ketones **70**. The azirino compound apparently acts as a 1,3 dipole and undergoes addition with a variety of dipolarophiles to give fused quinoxalines containing a bridgehead nitrogen atom.[34] Thus the acetylene derivatives **72** give the pyrrolo derivatives **73**. The yields are much improved by the use of chloranil in situ to oxidize the presumed dihydro intermediates **74**.

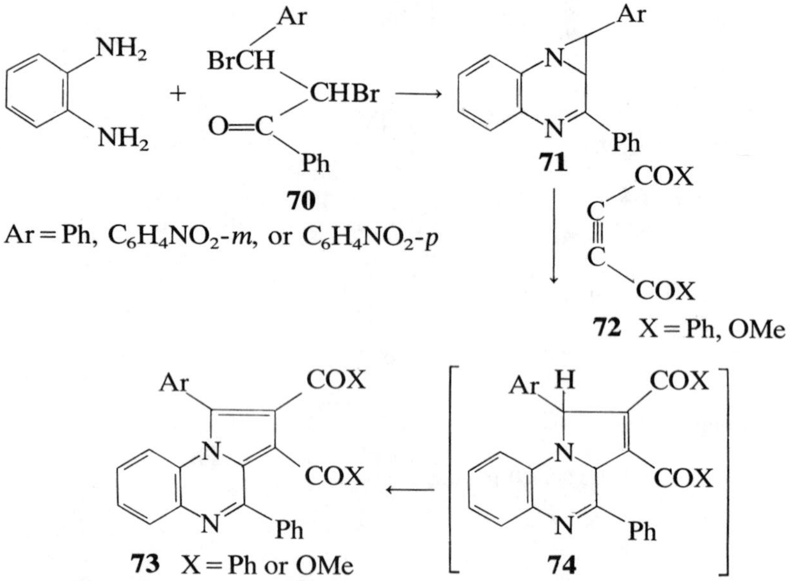

70

$Ar = Ph, C_6H_4NO_2$-*m*, or $C_6H_4NO_2$-*p*

71

72 X = Ph, OMe

73 X = Ph or OMe

74

Quinoxaline and 2-methylquinoxaline react with diphenylcyclopropenone (**75**) to give the products **76**, analogous to that obtained from the reaction of **75** with pyrazine (see Chapter XIX).[5] The formulation of the products as 1-hydroxy rather than 1,5-dihydro-1-oxo compounds is discussed in Section I1. The thioxo analogue of **75** gives the 1-mercapto derivative **77** on reaction with quinoxaline.[35] No spectral details are given to permit discussion of possible tautomerism of this compound to the 1,5-dihydro-1-thioxo derivative.

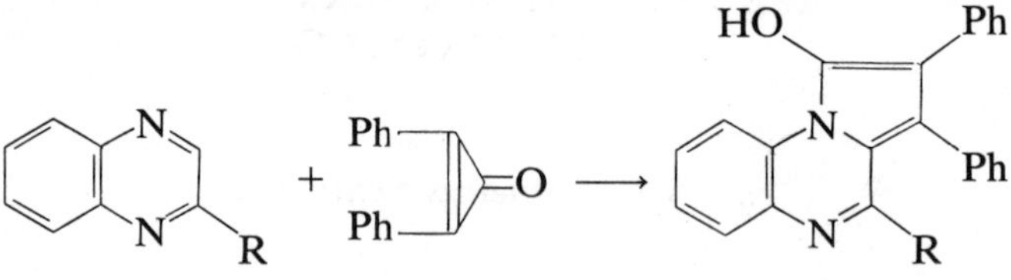

R = H or Me

75

76

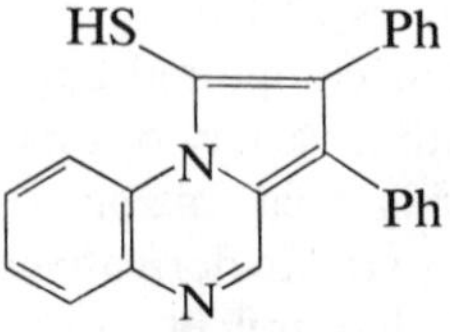

77

Reaction of the β-diketone **78** with ethyl orthoformate did not give the hoped for pyrano[2,3-*b*]quinoxaline **79** but gave a mixture of the pyrrolo[1,2-*a*]quinoxalines **80**.[36]

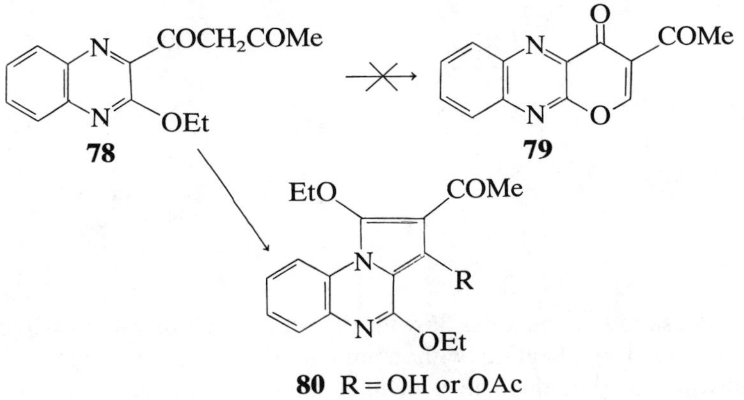

78 **79**

80 R = OH or OAc

Treatment of 1-methylbenzotriazole (**81**) with dimethyl acetylenedicarboxylate gave the 2:1 adduct **82**, which on reduction with stannous chloride in hydrochloric acid gave the pyrrolo compound **83**.[37] It has recently been demonstrated that the product from the reaction of the acetylenic ester with the benzimidazole **4** is the pyrroloquinoxalinone **5**.[73]

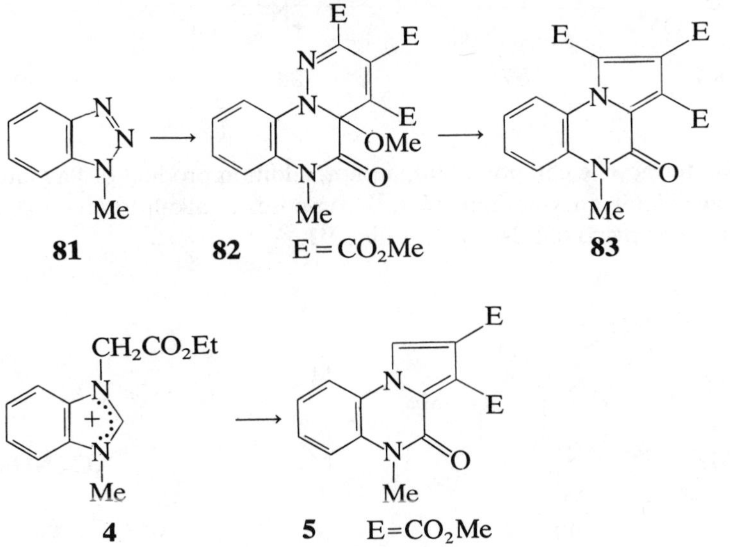

81 **82** E = CO_2Me **83**

4 **5** E = CO_2Me

The pyrrolobenzoxazine **84**, prepared in two steps from *o*-aminophenol, gave the corresponding pyrroloquinoxaline **85** on reaction with ammonium acetate in acetic acid.[38]

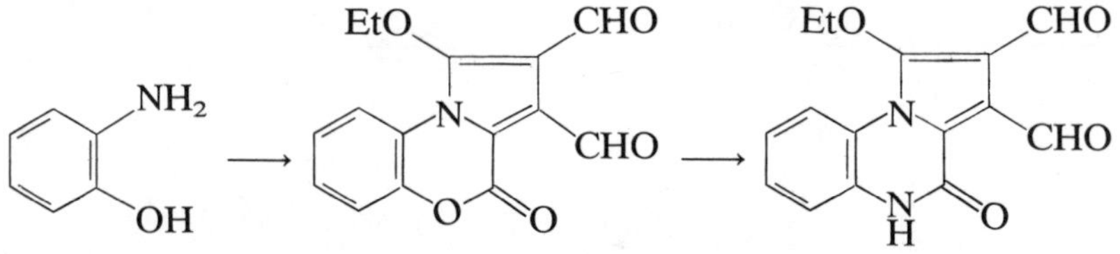

84 **85**

Condensation of the ketal **86** with the sodium salt of 2-benzoylpyrrole (**87**) followed by heating with ammonium acetate gave the 6,7,8,9-tetrahydro compound **88** in a manner analogous to that described for pyrrolo[1,2-*a*]pyrazines (see Chapter XIX). The aromatic compound **89** was obtained in 96% yield from the tetrahydro derivative by refluxing with nickel in xylene.[39]

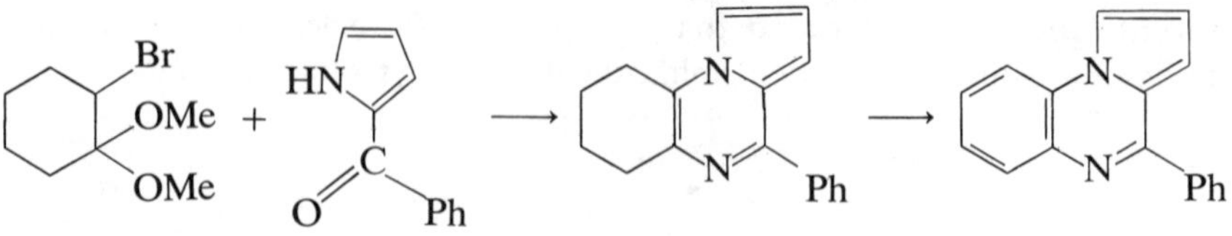

86 **87** **88** **89**

The tetracyclic compound **90**, a photoaddition product of flavoquinone and acetylenic enzyme inhibitors, decomposes in alkali to give the highly fluorescent pyrrolo[1,2-*a*]quinoxaline **91**.[40]

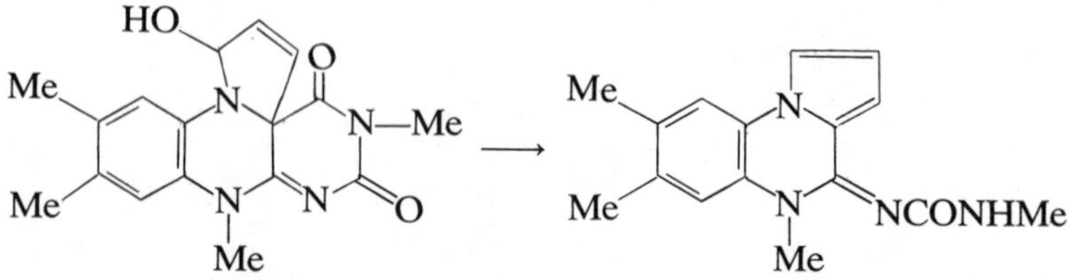

90 **91**

3. Reactions

A. *Alkylation*

The parent aromatic heterocycle undergoes ready quaternization with methyl iodide to give the 5-methyl quaternary salt. Quaternization of the 4-methyl compound to give the 4,5-dimethyl derivative proceeds more slowly, presumably because of steric hindrance to the reaction.[3] The 1,5-dihydro-1-oxo derivative **92** reacts with methyl iodide in methanolic sodium methoxide to give a monomethyl derivative formulated as the 5-methyl derivative **93** on the basis of UV spectral data.[24] In contrast treatment of the presumed 1-hydroxy compound **94** with triethyloxonium fluoroborate gave a product described as the 1-ethoxy derivative **3**.[5]

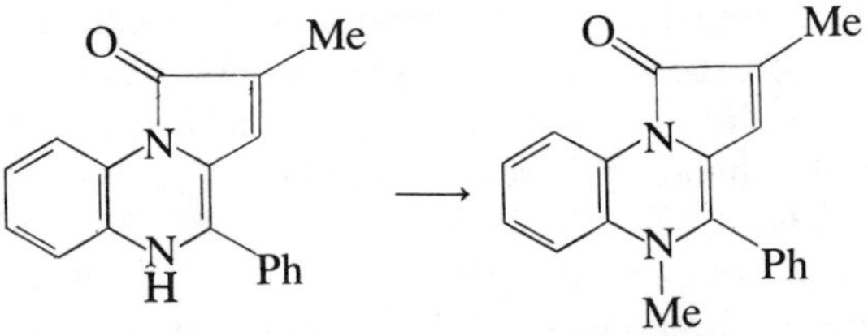

92 → **93**

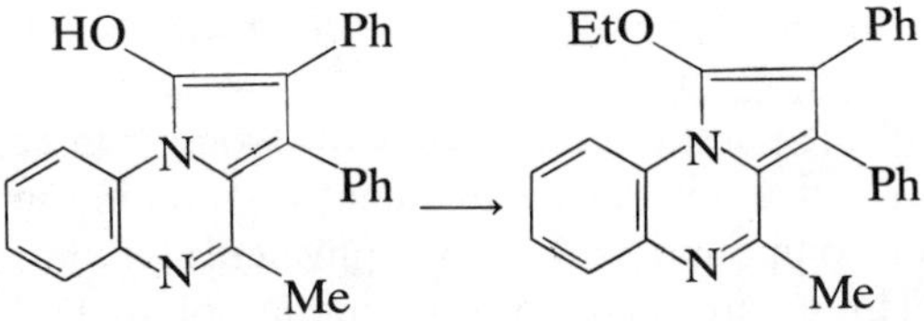

94 → **3**

The 4-oxo derivative **95** has been methylated with methyl sulfate to give the 5-methyl compound **96**. The corresponding 4-thioxo compound on methylation with methyl iodide gave the 4-methylthio derivative **97**, as expected.[3] Both these reactions have been extended by the use of more complex alkylating agents to give a series of dialkylaminoalkyl compounds for biological evaluation.[10]

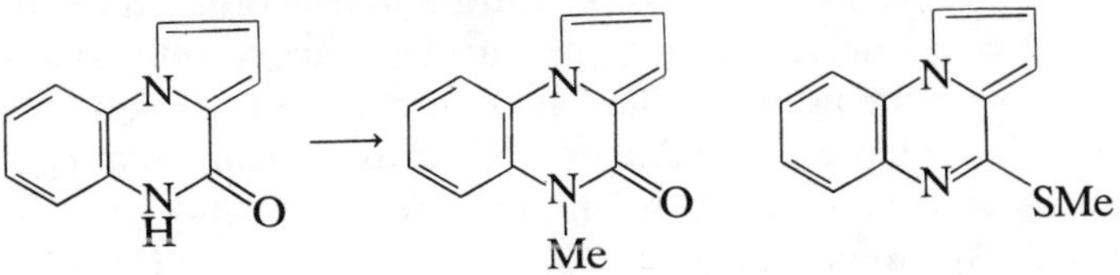

95 → **96** **97**

B. *Electrophilic Substitution Reactions*

Pyrrolo[1,2-*a*]quinoxaline undergoes smooth electrophilic substitution. The distribution of products obtained depends on the size of the incoming electrophile, as with bulky reagents steric interactions tend to inhibit reaction at the most electron-rich carbon-1 atom.[11] Theoretical calculations predict that the most favored positions for electrophilic substitution, in decreasing order of susceptibility, are $1 > 3 > 6 > 2$.[1] Reports have appeared describing electrophilic substitution at the 1,3- and 2-positions, but no reaction at the 6-position has yet been observed.

Halogenation of pyrrolo[1,2-*a*]quinoxaline provides mixtures of products which increasingly disfavor substitution at the 1-position, as the electrophiles increase in size in the order $Cl^+ < Br^+ < I^+$. All three halogenations, with one equivalent of halogenating agent, provide mixtures of the 1-halo, 3-halo, and 1,3-dihalo compounds. Reaction at room temperature with *N*-chlorosuccinimide in 50% sulfuric acid[11] gives 75% 1-chloro, 2.5% 3-chloro, and 0.5% 1,3-dichloro compounds. *N*-Bromosuccinimide under the same conditions[26] gives 42% 1-bromo, 2.7% 3-bromo, and 0.7% 1,3-dibromo compounds. Bispyridineiodonium nitrate[26] at room temperature yields 18% 1-iodo, 26% 3-iodo, and 0.11% 1,3-diiodo compounds. 2,4-Dimethylpyrrolo[1,2-*a*]quinoxaline gives similar mixtures on bromination and chlorination. All the monohalo compounds can be dihalogenated by use of excess of the halogenating agent.

Bromination of the parent heterocycle with bromine in acetic acid gives the 1-bromo derivative with one equivalent of bromine in the cold and the 1,3-dibromo compound with two equivalents of bromine at room temperature.[10] Three equivalents of bromine in boiling hydrobromic acid give a mixture of the 2,3-dibromo and the 1,2,3-tribromo compounds. High-temperature chlorination with a mixture of phosphoryl chloride and phosphorus pentachloride gives a mixture of the 1-chloro and 1,2,3-trichloro derivatives.[26]

Nitration of pyrrolo[1,2-*a*]quinoxaline by adding a mixture of the compound and potassium nitrate to concentrated sulfuric acid yields a mixture of 25% 1-nitro and 48% 3-nitro derivatives. No significant reaction was observed with potassium nitrate in trifluoroacetic acid, or in fuming nitric acid alone or fuming nitric acid and acetic anhydride.[11] The attempts to avoid the use of sulfuric acid were made because of the ease with which the heterocycle undergoes sulfonation. Thus treatment with concentrated sulfuric acid at room temperature readily gave the 3-sulfonic acid. Apparently the electrophile is too large to allow formation of detectable amounts of the isomeric 1-sulfonic acid. The 1-methyl and

2,4-dimethyl derivatives of pyrrolo[1,2-*a*]quinoxaline were also readily sulfonated at room temperature.[11] Sulfonation in concentrated sulfuric acid at 130° gives the 2-sulfonic acid, and it has been shown that the 3-sulfonic acid rearranges to give the 2-substituted product on heating. Although 3-aminopyrrolo[1,2-*a*]quinoxaline gives the 3-amino-2-sulfonic acid on sulfonation, the 3-nitro compound does not react.[21]

The 3-sulfonic acid nitrates readily in the 1-position, and the 1-chloro-[11] and 1-bromo-3-sulfonic acids[26] may be obtained by halogenation of the 3-sulfo compound. Sulfonation of 1-chloro- and 1-bromo-pyrrolo[1,2-*a*]-quinoxaline takes place in the 3-position to give the 1-halo-3-sulfonic acids, identical with those obtained by halogenation of the 3-sulfo derivative.

The 3-sulfo compounds **98** (R = H, NH_2, NO_2, or Cl) give the corresponding 3-unsubstituted derivatives **99** on refluxing in aqueous sulfuric acid.[21,26] However the 1-bromo-3-sulfonic acid **98** (R = Br) gives instead 3-bromopyrrolo[1,2-*a*]quinoxaline (**100**). The difference in behavior between the 1-chloro- and the 1-bromo-3-sulfonic acids is attributed to steric interaction between the 1-halo substitutent and the 9-hydrogen. It has also been shown that although 1-chloropyrrolo[1,2-*a*]quinoxaline is stable under the reaction conditions, the 1-bromo compound **101** rearranges to give the 3-substituted product **100**. Similarly 1-bromo-2,4-dimethylpyrrolo[1,2-*a*]quinoxaline rearranges to the 3-bromo compound. These rearrangements may also be carried out in boiling hydrobromic acid.[26]

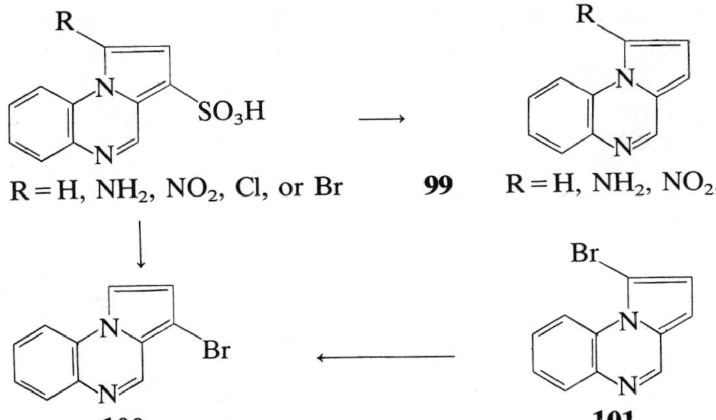

R, N, N, SO_3H

98 R = H, NH_2, NO_2, Cl, or Br

R, N, N

99 R = H, NH_2, NO_2, or Cl

N, N, Br

100

Br, N, N

101

The above transformations seem to occur by an intermolecular rearrangement. Thus refluxing in hydrobromic acid a mixture of 1-bromopyrrolo[1,2-*a*]quinoxaline and an excess of the 2,4-dimethyl compound

102 gives the crossover product **103**. Also if potassium iodide is added to a solution of the 1-bromo compound **101** in boiling hydrobromic acid, the parent heterocycle is obtained together with a sublimate of iodine, presumably formed by the reaction of iodide ions with the released bromonium ions. Direct evidence for the release of Br^+ is provided by the conversion in acid solution of the 1,2,3-tribromo compound into 2,3-dibromopyrrolo[1,2-*a*]quinoxaline. The 1- and 3-iodo derivatives of this ring system are also converted into the parent heterocycle in refluxing hydrobromic acid.[26]

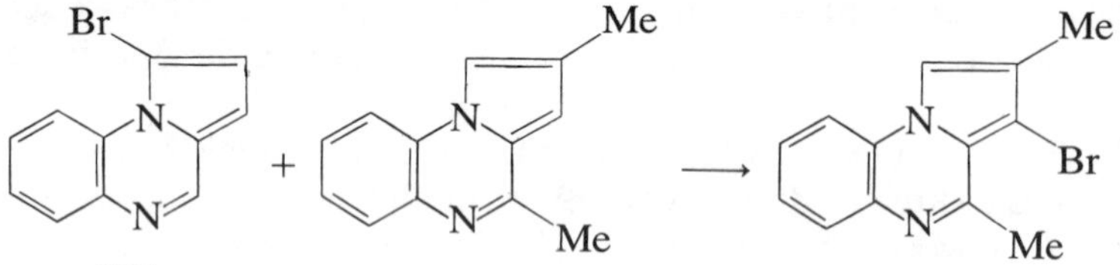

101 + **102** ⟶ **103**

C. *Oxidation*

Little work has been done on the oxidation of this ring system. There are no reports of attempts at *N*-oxidation; the only *N*-oxide known was prepared by direct synthesis.[9] 4,5-Dihydro-4-phenylpyrrolo[1,2-*a*]quinoxaline is readily oxidized to the 4-phenyl derivative of the aromatic ring system by manganese dioxide.[9] Palladium on charcoal at 270° dehydrogenates 1,2,3,3a,4,5-hexahydro compounds to give the aromatic ring system,[18,19] and nickel in refluxing xylene has been used to aromatize a 6,7,8,9-tetrahydro compound.[39] Treatment of the 1-oxo-1,5-dihydro compound **47** with alkaline potassium ferricyanide gave the ring-opened product **104**.[22]

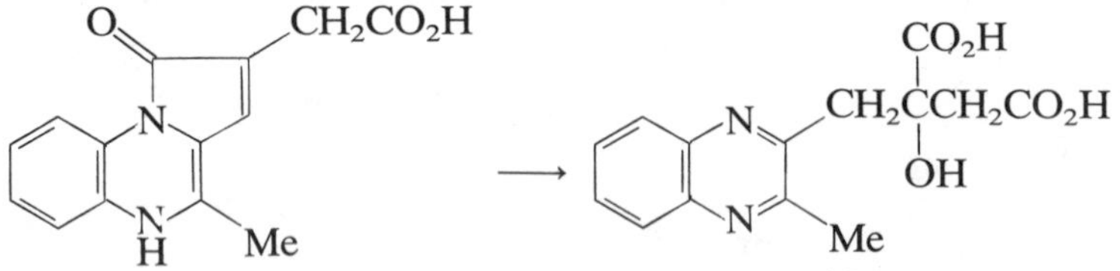

47 ⟶ **104**

D. *Reactions with Nucleophilic Reagents*

Reaction of pyrrolo[1,2-*a*]quinoxaline with potassium amide in liquid ammonia affords the 4-amino derivative in 56% yield. Treatment of the methiodide of the parent heterocycle with sodium carbonate gives the 5-methyl-4-oxo compound, thus establishing the site of quaternization.[3]

Phosphoryl chloride reacts with 1,5-dihydro-1-oxo compounds to give the 1-chloro derivatives, which are apparently very stable. 1-Chloro compounds are however hydrolyzed to 1,5-dihydro-1-oxo derivatives by refluxing with potassium hydroxide in ethylene glycol[24] or dimethyl sulfoxide.[21] In contrast, 4-chloro derivatives of pyrrolo[1,2-*a*]quinoxalines, obtained from 4,5-dihydro-4-oxo compounds with phosphoryl chloride, are readily converted into other 4-substituted derivatives by nucleophilic substitution. A wide variety of nucleophiles have been used, including MeO^-, HS^-, N_3^-, NH_3,[3] NH_2NH_2,[25] EtO^-,[20,25] as well as several more complex reagents.[10] Sodium azide reacts with 4-chloropyrrolo[1,2-*a*]quinoxaline to give the tetrazolo compound **105**.[3] The triazolo compound **106** was obtained by reaction of the corresponding 4-hydrazinopyrrolo[1,2-*a*]quinoxaline with formic acid.[25] Reaction of the chloro compound **107** with hydrazine gave the fused pyridazine **108**.[20]

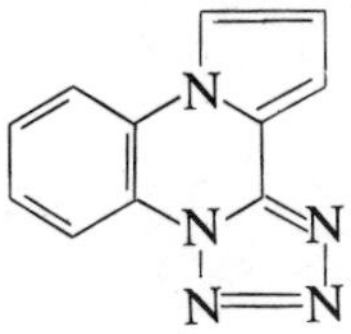

105 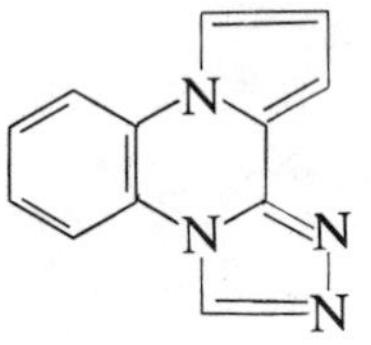 **106**

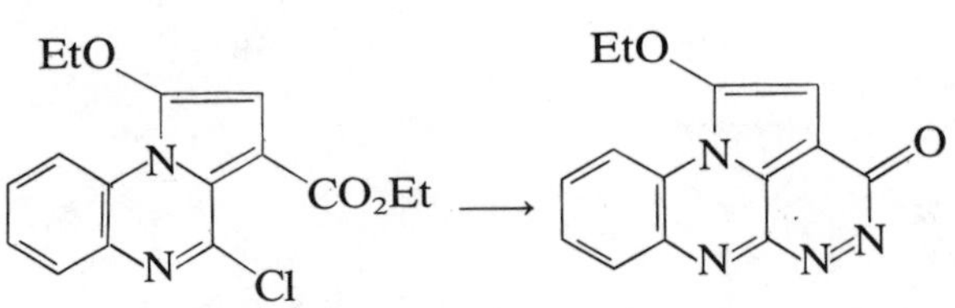

107 → **108**

Although refluxing the nitrosulfonic acid **109** in aqueous sulfuric acid gives simply the 1-nitro compound **110**, in concentrated hydrochloric acid the sulfonic group is stable, but the nitro group undergoes nucleophilic displacement by chloride to give the chlorosulfonic acid **111**. Lithium chloride in boiling DMF also effects this transformation in excellent yield. The enhancement of reactivity by the 3-sulfonic acid function is important for this reaction, as 1-nitropyrrolo[1,2-*a*]quinoxaline is only slowly converted into the 1-chloro compound in refluxing hydrochloric acid, and with lithium chloride there is no reaction.[21]

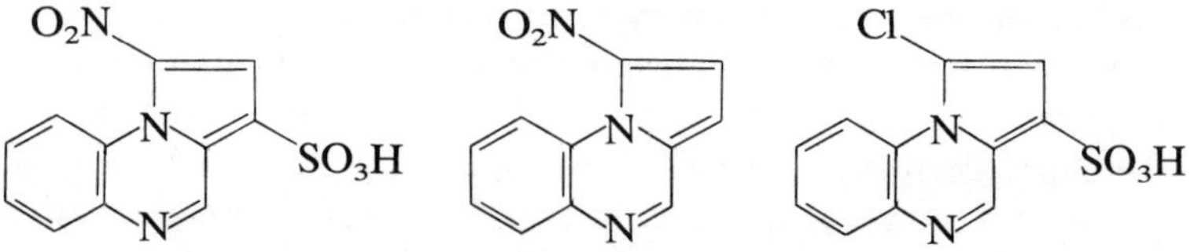

109 110 111

The lability of the bromine in 1-bromopyrrolo[1,2-*a*]quinoxaline (**101**) makes the reaction of the nitrosulfonic acid **109** more complex in both concentrated hydrobromic acid and in DMF containing lithium bromide. In the reaction of **109** with lithium bromide the 1-bromo-3-sulfonic acid **112** may be isolated after 10 min. However after 1 hr only the 3-bromo compound **100** is obtained. It was also shown that the intermediate **112**, as well as the 1-bromo compound **101** and 1,3-dibromopyrrolo[1,2-*a*]quinoxaline, form the 3-bromo compound **100** on reaction with lithium bromide in DMF.[21]

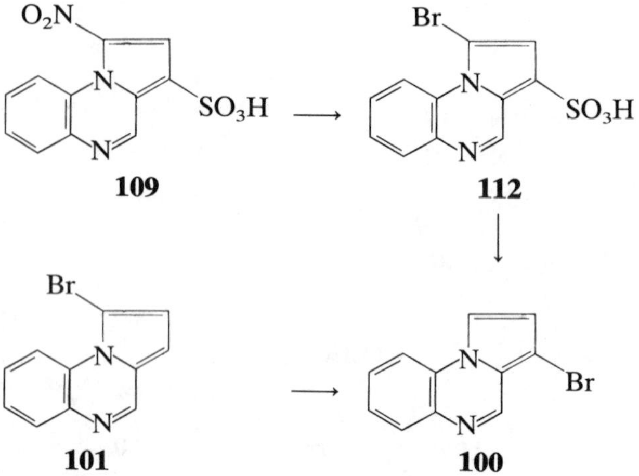

109 → 112 → 100

101 → 100

Concentrated hydrobromic acid may be used as well as sulfuric acid to convert the 3-sulfonic acid **113** into the parent heterocycle **114**. However with the former reagent a small amount of the 3-bromo derivative **100** is also obtained. When the reaction is repeated with the 1-nitro-3-sulfonic acid **109**, a mixture of the di- and tribrominated products **115** and **116** is obtained. The same two products in different proportions are also obtained by refluxing the 1-nitro compound **110** in concentrated hydrobromic acid. It seems that a brominating species such as bromine is formed in situ by reaction between hydrobromic acid and the displaced nitrite ions. The displacement of the nitro group apparently requires protonation of the attacked species, as no reaction occurs when 1-nitropyrrolo[1,2-*a*]quinoxaline is refluxed with lithium bromide in DMF.[21]

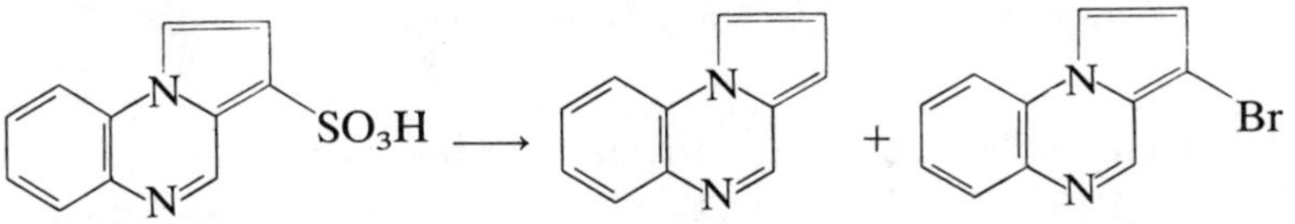

113 → **114** + **100**

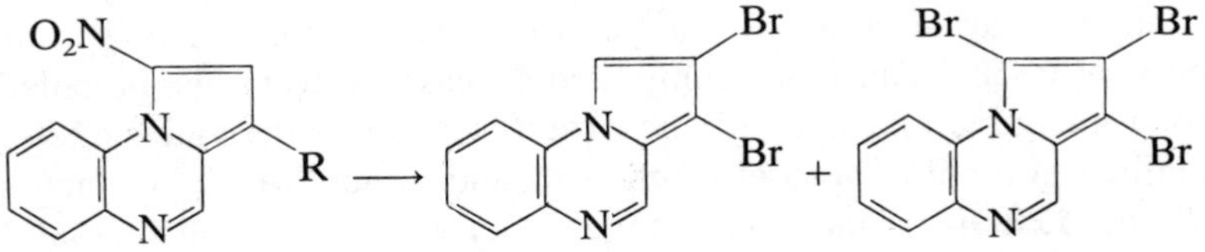

109 $R = SO_3H$
110 $R = H$
→ **115** + **116**

E. Reduction

Catalytic hydrogenation of 1-chloropyrrolo[1,2-*a*]quinoxalines results in removal of the halogen atom. Apparently the dechlorinated products are themselves reducible, and care must be taken in dechlorinations to allow the absorption of only one mole of hydrogen.[4,24] A recent paper describes reduction of the 4,5 bond in pyrrolo[1,2-*a*]quinoxalines.[75] 1,5-Dihydro-1-oxo compounds are resistant to hydrogenation at atmospheric pressure. Reduction of the 1-oxo compound **117** over palladium on charcoal at 2000 psi gave an unidentified tetrahydro derivative.[24] The dioxo compounds **118** have been reduced with lithium aluminum hydride to give the hexahydro products **119**.[18,19] Heating 4,5-dihydro-4-thioxo derivatives with Raney nickel results in formation of the aromatic, desulfurized compound.[25] Stannous chloride has been used to reduce 1-nitro- and 3-nitropyrrolo[1,2-*a*]quinoxalines to the corresponding

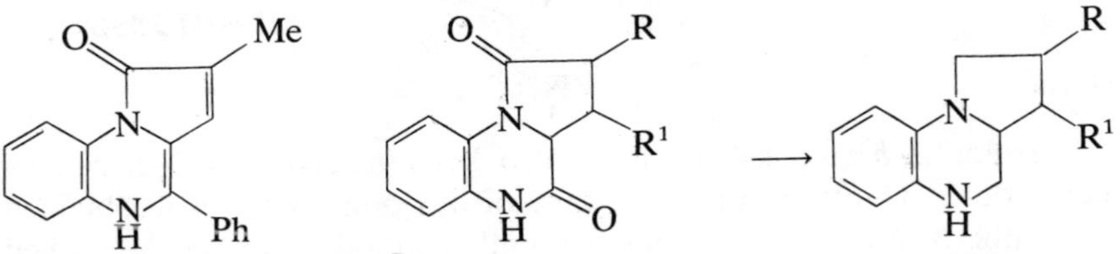

117

118 R = H or Me
R^1 = H, Me, or Ph
→ **119**

amino compounds,[11] and 4-aminopyrrolo[1,2-*a*]quinoxaline was obtained from its 5-oxide by reduction with sodium dithionite.[9]

4. Uses

Although a large number of derivatives of this ring system are known, surprisingly few uses of pyrrolo[1,2-*a*]quinoxalines have been described. An extended series of derivatives of the heterocycle have been described as uninteresting in antileukemia[75] and in a wide variety of (unspecified) biological tests.[10] The 1-oxo compound **57** has however been patented as having analgesic and hypnotic activity.[29] The spiro derivatives **120** are described in a patent as useful relaxants and tranquilizers.[76] Compounds **121** and **122** are readily characterized derivatives of proline, useful in peptide sequencing.[13–15]

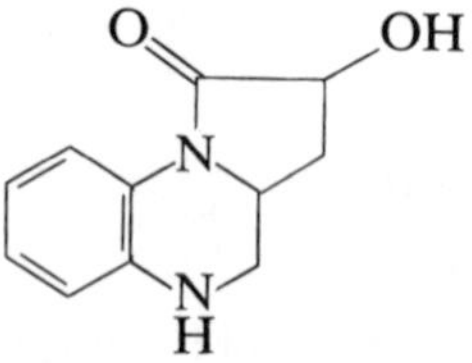

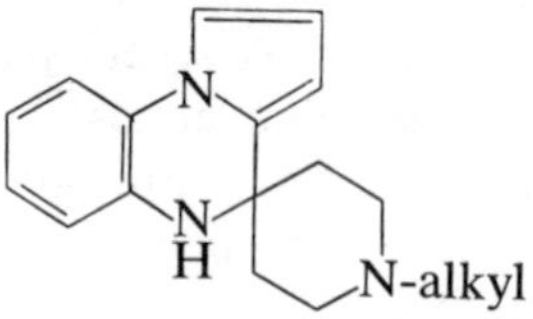

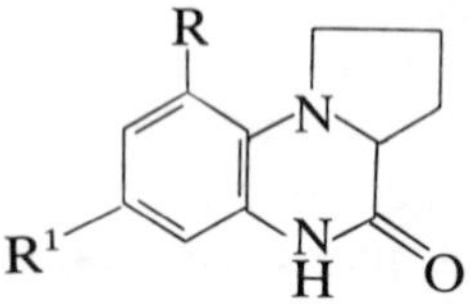

121 $R = H$; $R^1 = CO_2Me$ or SO_2Me
122 $R = R^1 = NO_2$

II. Pyrrolo[2,3-*b*]quinoxalines

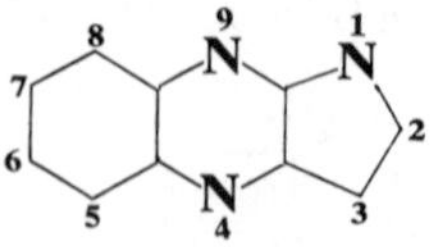

1*H* RRI 2856

Pyrrolo[2,3-*b*]quinoxalines have also been named as 4,7-diaza-5,6-benzindoles. Until recently derivatives of this ring system were obtained by synthesis of the pyrazine ring. Recently methods have been described starting from suitably substituted quinoxalines. The first reported synthesis of this heterocycle gave the diphenyl compound **124** and its 2-*p*-tolyl analogue.[41] These were obtained from *o*-phenylenediamine by condensation with dioxopyrroles such as **123**. The same approach has given a series of 1*H*-2,3-dihydro derivatives from dioxopyrrolidines. Examples of such compounds are the 1,2,3-triaryl derivatives **125**,[42,43] the 1,2-diaryl-3-ethoxycarbonyl compounds **126**,[44–46] and the 1-butyl-3-cyano derivative **127**.[47] The trioxopyrrolidines **128** gave the 2-oxo heterocycles **129**.[46]

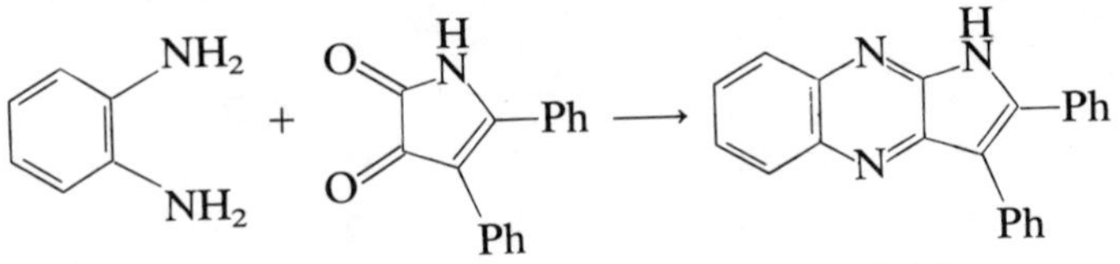

123 **124**

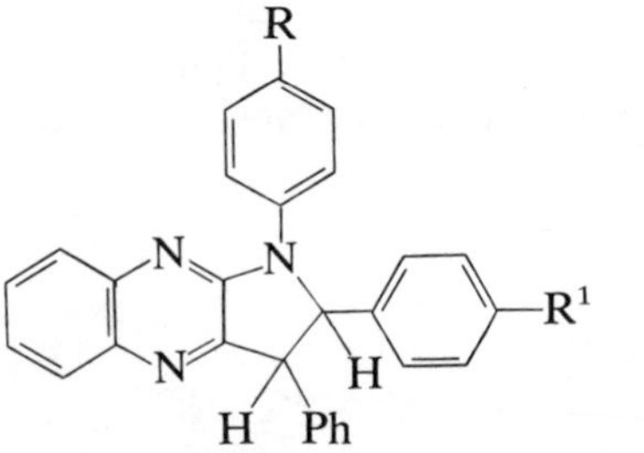

125 R or R^1 = Cl, OMe, or Me

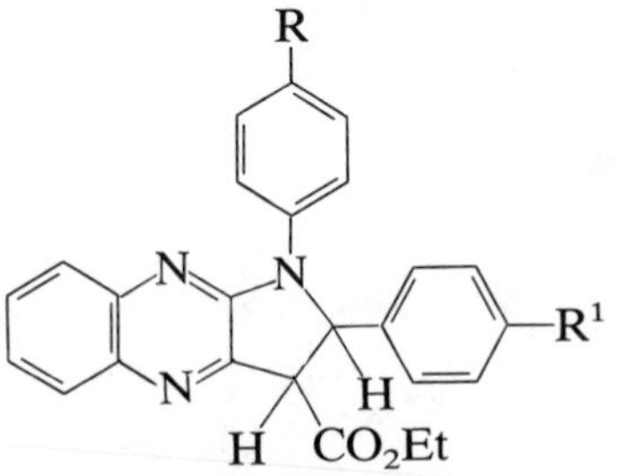

126

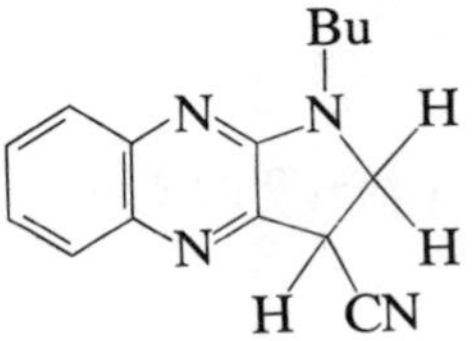

127

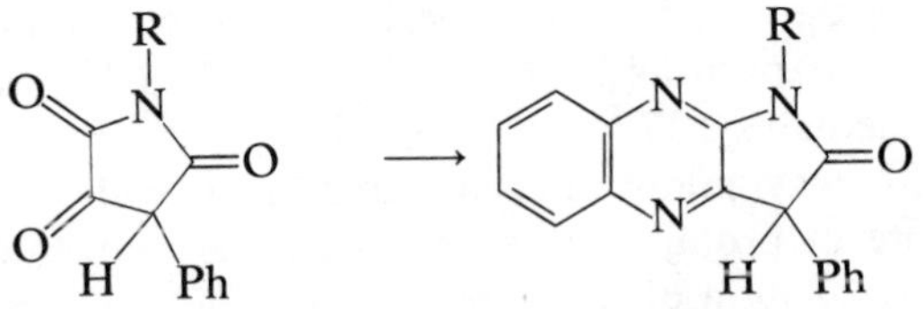

128 R = H or Ph

129

The second method of formation of the pyrazine ring involves the reaction of *o*-phenylenediamine with a chloromaleimide of formula **130**.[48–50] Many of the resultant 2-oxo compounds **131** are the subject of patents claiming their use as fungicides[49] and as intermediates to pigments.[51]

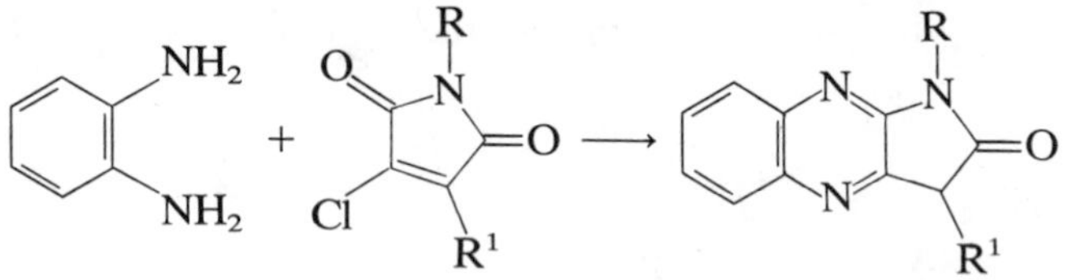

130 **131** R = H, Ph, or alkyl
R^1 = Cl or CN

The pyrrolo ring has been synthesized by reaction of the quinoxalinylacetonitriles **132** with ammonia or primary amines to give the 2-amino derivatives **133** of the heterocycle.[52] Similarly the quinoxalinylacetic acid ester **134** (R = Ph) reacts with aniline to give the oxo diphenyl compound **135**. The reaction does not proceed with the unsubstituted compound **134** (R = H).[53]

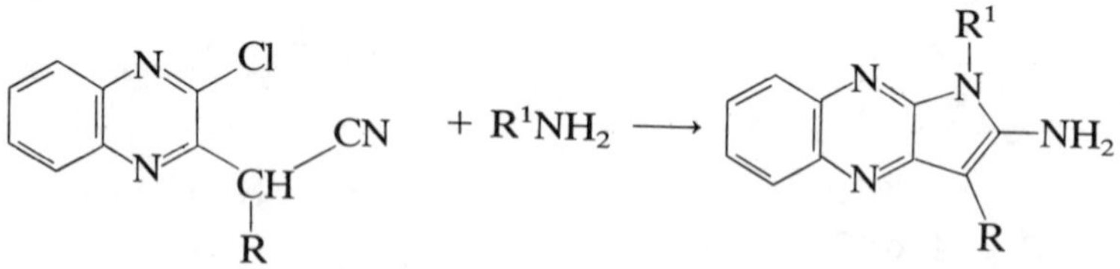

132 R = CN or CO_2Et + R^1NH_2 → **133**

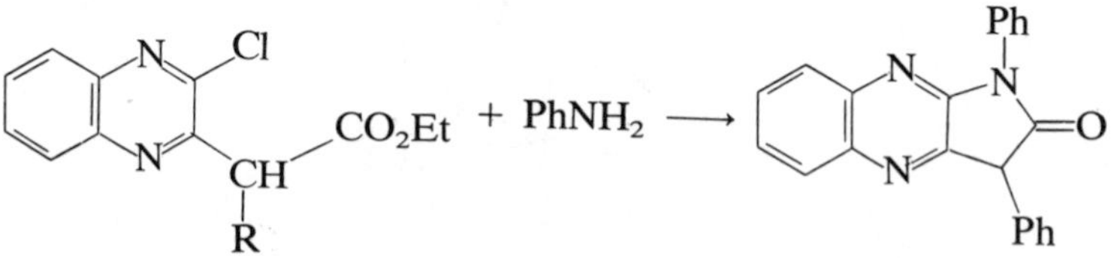

134 R = Ph + $PhNH_2$ → **135**

Reduction of the benzimidazole derivatives **136** with lithium aluminum hydride gave the pyrrolo[2,3-*b*]quinoxalines **138** by way of the ring-opened intermediates **137**.[54]

2-Amino-3-benzylquinoxaline on treatment with diethyl carbonate gave a mixture of products including the three pyrrolo compounds **139** and **140a,b**. Alkylation of the pyrrolo compound **139** results in substitution at the 1- and the 3-positions.[55]

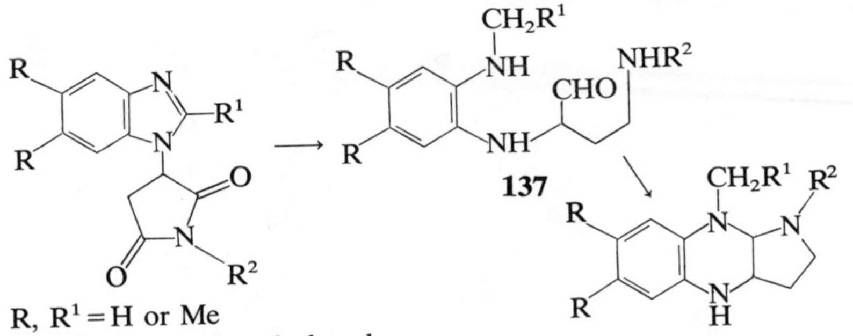

36 R, R^1 = H or Me
R^2 = Ph or substituted phenyl

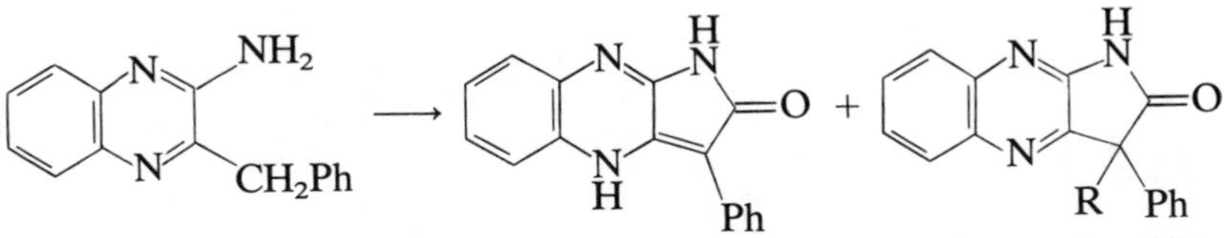

137

138

139

140a R = OH
140b R = Et

Various complex derivatives of the ring system have been patented as photographic sensitizers.[56] Others are claimed as isoindigo pigments.[51]

III. Pyrrolo[3,4-*b*]quinoxalines

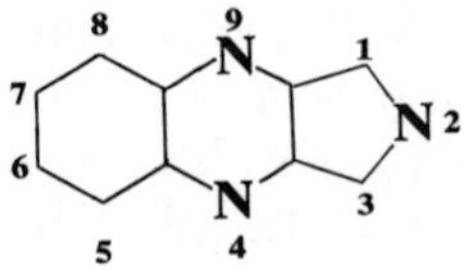

2*H* RRI 2857

This ring system has also been called quinoxalino[2,3-*c*]pyrrole. Reaction of 3,4-dioxopyrrolidines with *o*-phenylenediamine has given 2*H*-1,3-dihydro compounds such as **141a,b**.[58] The 1,3-dioxo derivatives of the ring system are imides of quinoxaline-2,3-dicarboxylic acid. Accordingly, heating the amide **142** by itself gives the compound **143** (R = H), whereas heating the amide in acetic anhydride yields the *N*-acetyl derivative **143** (R = COMe).[59] Treatment of quinoxaline-2,3-dicarboxylic acid[60] or its anhydride[61] with primary amines gives substituted half-acid amides, which may be cyclized in acetic anhydride[60] or sulfuryl chloride[61] to give 2-substituted 1,3-dioxo compounds such as **144**.

Reaction of the amide **145** with primary amines in boiling ethanol gave the imino derivatives **146** in 31–86% yield. On acid hydrolysis the

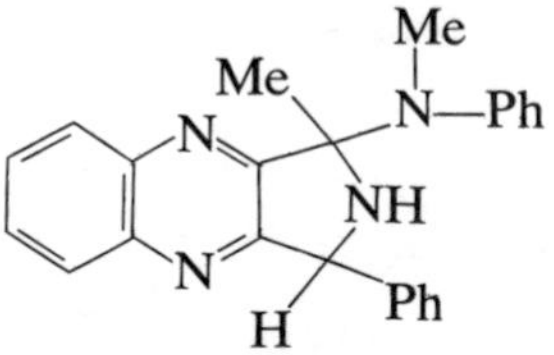

141a

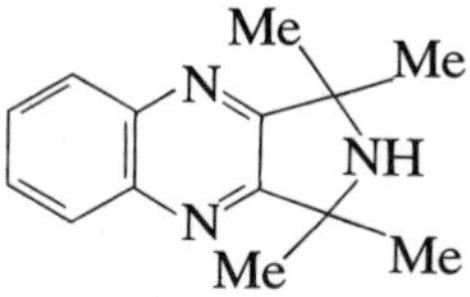

141b

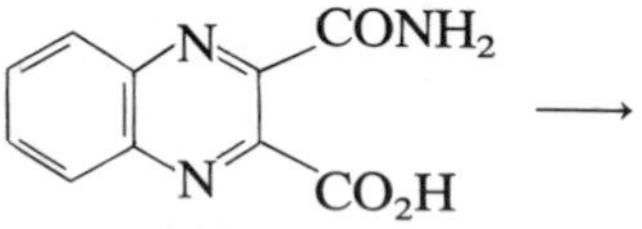

142

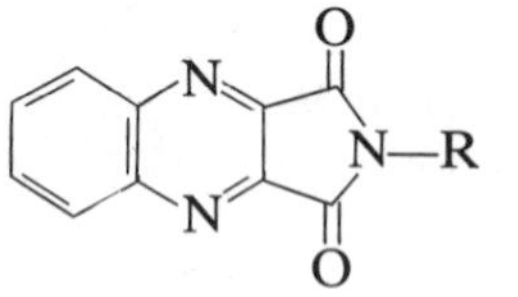

143 R=H or COMe

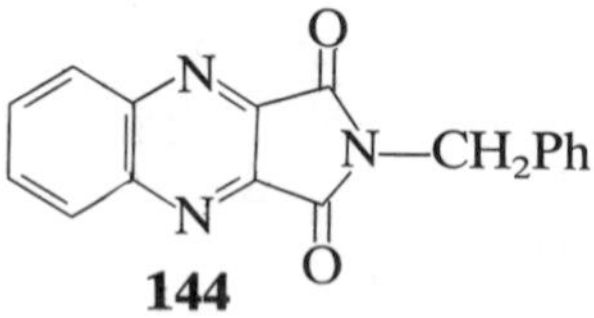

144

corresponding dioxo derivatives were obtained. Condensation of the nitrile **147** with cyclic amines (e.g., morpholine) gave amino compounds such as **148** in 60–86% yield.[78]

Cyclodehydration of the nitrosoamines **149** in acetic anhydride at 100° did not give the expected sydnones but afforded 1,3-dioxopyrrolo[3,4-*b*]quinoxalines in 40–60% yield.[62] No attempt was made to suggest a mechanism for this novel synthesis of pyrrolo[3,4-*b*]quinoxalines.

Benzofuroxan (**150**) on reaction with ethyl acetoacetate provides the methylquinoxaline **151**, which on bromination gives the bromomethyl derivative **152**. Treatment of this compound with amines gives the monooxopyrrolo compound **154**. The 2-methyl derivative **154** (R=Me) was also prepared directly from benzofuroxan by reaction with the *N*-methylpyrrolinone **153**. The products **154** are claimed to be antibacterials and animal growth promotants.[63]

The bis bromomethyl compound **155** reacts with primary amines to give 2-substituted 1,3-dihydro compounds such as **156**.[64] Secondary amines provide cyclic quaternary salts.[64–66,79] Piperazine[64,65] and morpholine[64] yield the spiro quaternary derivatives **157**. The mono- and di-*N*-oxides of the bromomethylquinoxaline **155** react with liquid methylamine to give the 1,3-dihydro-2-methyl compounds in which one of the *N*-oxide functions has been reduced.[67] Thus the mono-*N*-oxide of **155** gives the *N*-methyl derivative **158**. This compound is also obtained

145 → (i) → 146 → (ii)

147 → (iii) → 148

Reagents (i) R^1NH_2, EtOH, Δ
(ii) H^+
(iii) morpholine

149 R = Substituted phenyl
R^1 = H, MeO, or Me

by treatment of the mono-*N*-oxide product derived from the di-*N*-oxide of **155** with phosphorus trichloride or Raney nickel. The bispyridinium salt **159** reacts with substituted *p*-nitrosoanilines in the presence of sodium cyanide to give the aromatic derivatives **160**. Presumably the bisnitrones **161** are intermediates, as treatment of the compound **161** (R = Me) with sodium cyanide produces the heterocycle **160** (R = Me).[68]

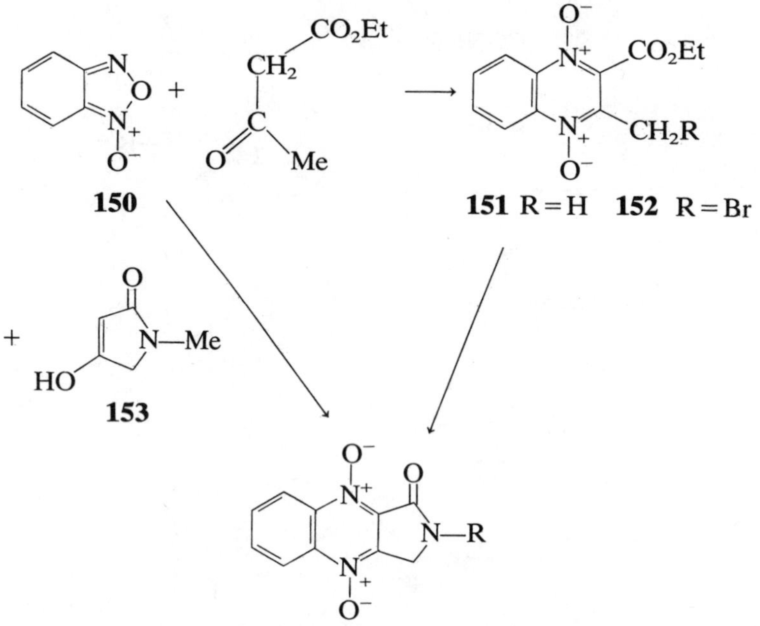

150 + $CH_2(CO_2Et)C(=O)Me$ ⟶ **151** R = H **152** R = Br

+ **153** (HO, N—Me, O)

154 R = H, Me, or Et

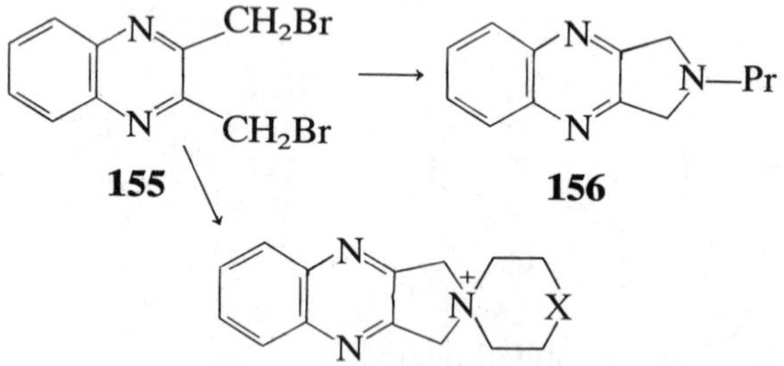

155 (CH_2Br, CH_2Br) ⟶ **156** (N—Pr)

157 X = CH_2 or O

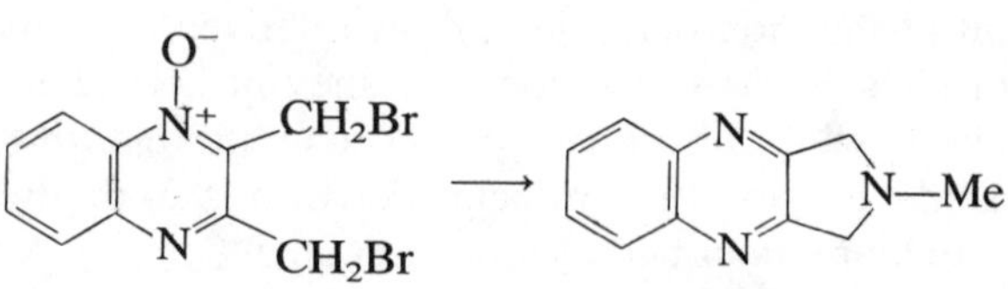

(CH_2Br, CH_2Br, O^-, N^+) ⟶ **158** (N—Me)

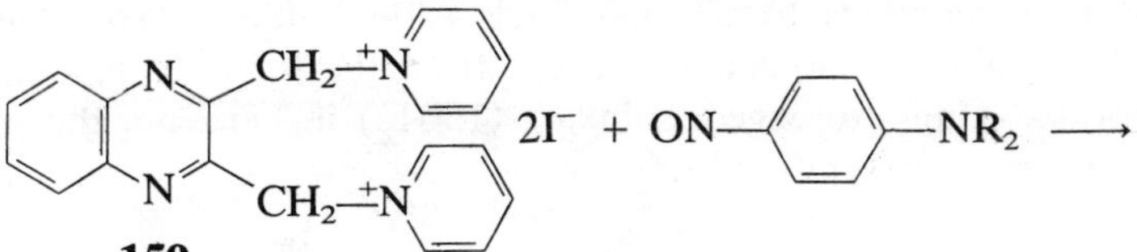

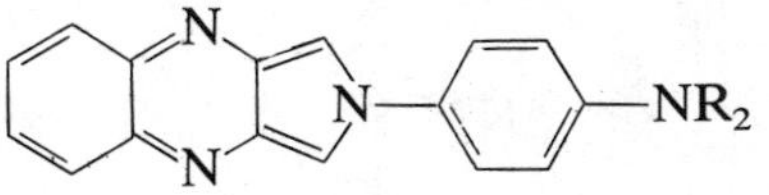

160 R = Me, CH_2CH_2OH, or CH_2CH_2Cl

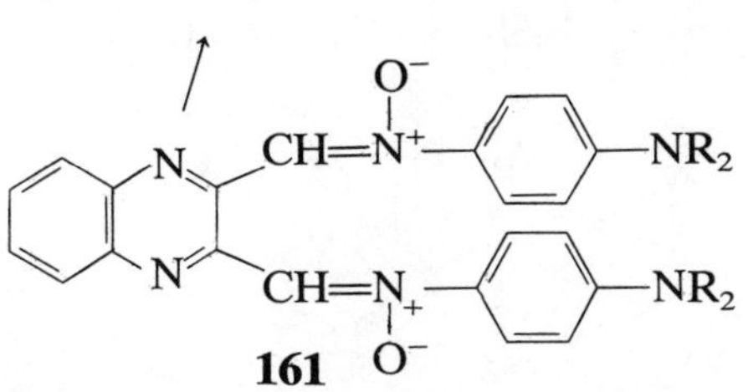

A study has been made of the catalytic reduction of the 1,3-dioxo compound **162**.[69] An unidentified blue product was obtained using palladium on charcoal in THF. The use of acetic acid as solvent, followed by acetylation, provided the mono-, di-, and triacetyl derivatives of the partially reduced compounds **163** and **164**.

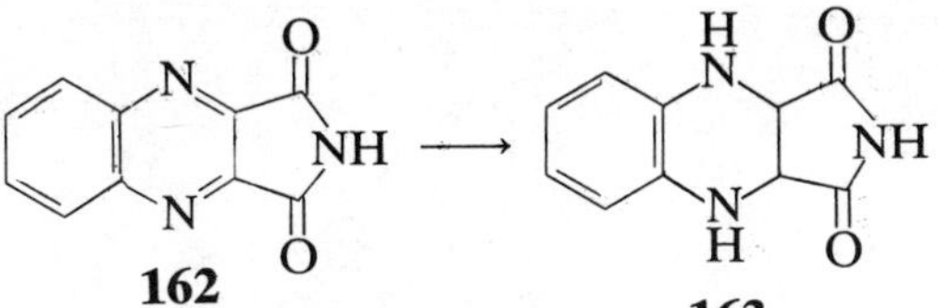

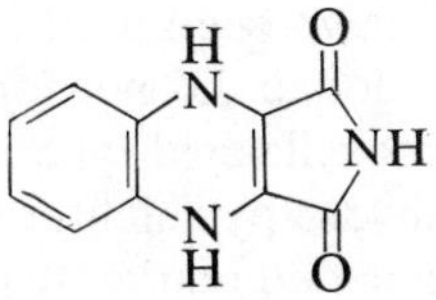

Treatment of a hot methanolic solution of 2,3-dibenzoylquinoxaline and methylamine with sodium borohydride or sodium dithionite results in the speedy formation of a blue color. This is followed by precipitation of blue **165a** in 90% yield. The corresponding thieno compound (**165b**) is red. Both compounds form endo Diels-Alder adducts (**166**) with *N*-phenylmaleimide. Although the adduct from the thieno compound was only formed after prolonged refluxing (150 hr) in benzene, the pyrrolo compound reacted within 5 hr.[80]

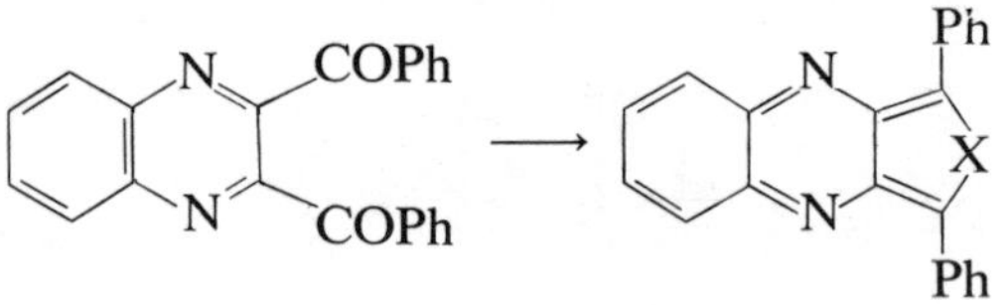

165a X = NMe (blue)
165b X = S (red)

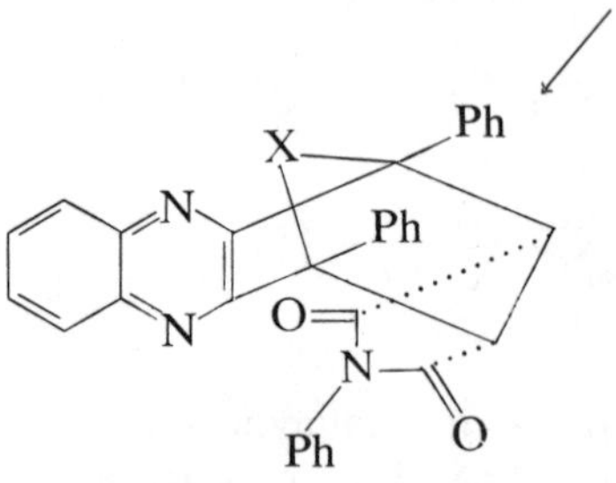

166 X = N—Me or S

IV. Pyrrolo[1,2,3-*de*]quinoxalines

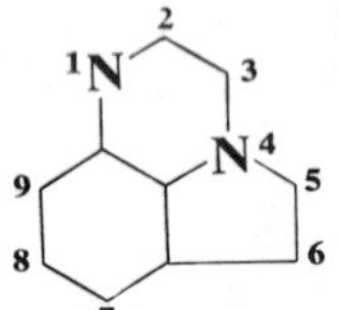

Two separate approaches have been used for the synthesis of this ring system. Although 7-aminoindoline did not form fused 5,5,6-systems such as **167** by reaction with carboxylic acid-type reagents, condensation with benzoins **168** gave the tetrahydro pyrrolo[1,2,3-*de*]quinoxalines **169**. The yield from this condensation varied markedly with the benzoin used. Only the benzoins **168** (Ar = Ph or C_6H_4OMe-*p*) gave reasonable yields of product. Diethyl oxalate gave the dioxo compound **170** in 19% yield. Reaction of 7-aminoindoline with benzil gave the dihydro derivative **171** in low yield.[70] Reaction of dimethyl acetylenedicarboxylate with 7-amino-2-methylindoline and 7-amino-2-methylindole gave the pyrroloquinoxalines **172** and **174** which on hydrolysis gave **173** and **175**.[81]

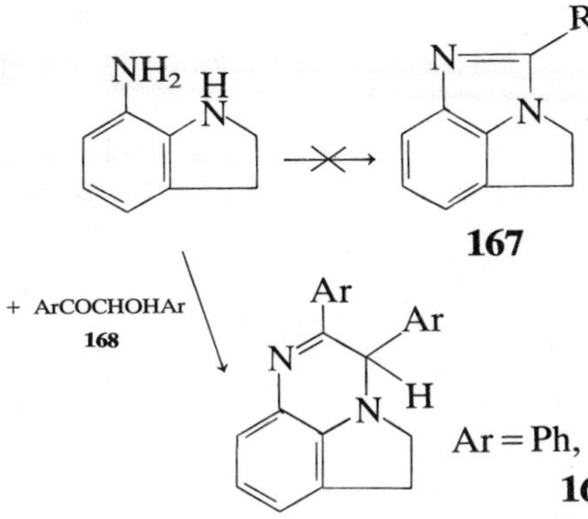

167

Ar = Ph, C_6H_4OMe-*p*, C_6H_4Et-*p*, or 2-furyl

169

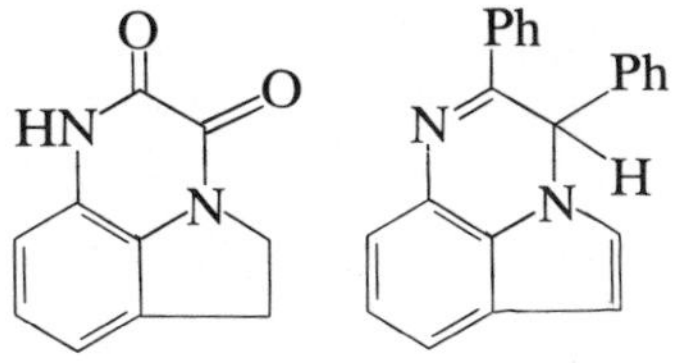

170 **171**

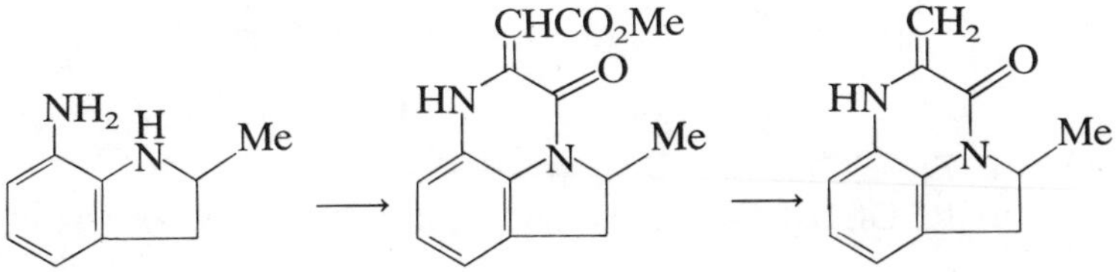

172 **173**

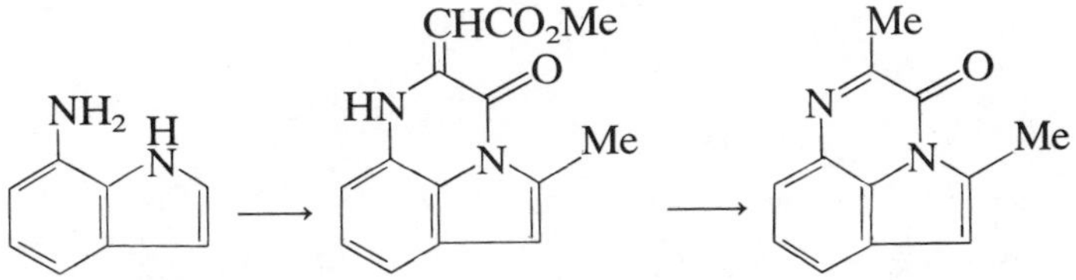

174 **175**

The formulation of compounds **169** and **171** as 3*H*- and not 1*H*-derivatives is supported by their PMR spectra. In $CDCl_3$, compound **171** shows a one-proton singlet at δ 6.59 which is not deuterium exchangeable in sodium deuteroxide–deuterium oxide. This signal is assigned to the benzylic proton in the 3-position. The tetrahydro analogue **169** (Ar = Ph) shows a similar signal at δ 5.54. The ultraviolet spectra of compounds **169** and **171** have also been discussed. In methanol the dihydro derivative **171** has maxima at 262 and 377 nm, whereas the tetrahydro analogue **169** (Ar = Ph) possesses maxima at 276 and 428 nm, with inflections at 258 and 318 nm.

A second approach to this heterocycle was disclosed in patents claiming the products **178** as CNS depressants. It seems that certain of the

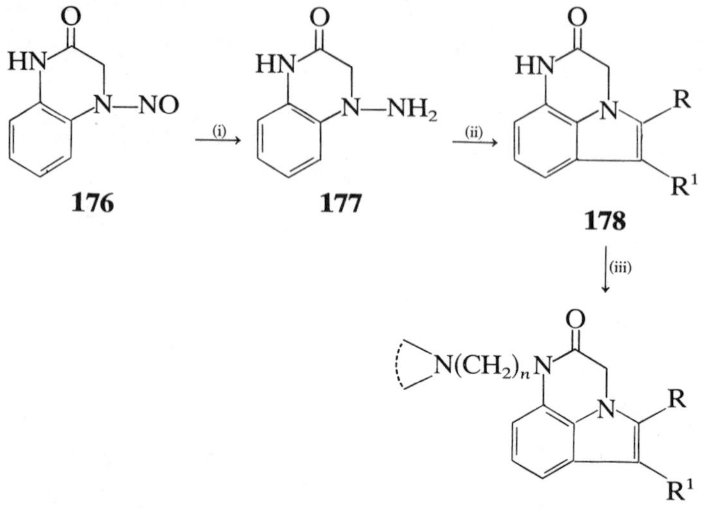

176 → (i) → **177** → (ii) → **178** → (iii) → **179** $n = 2$ *or* 3

Reagents (i) Zn, HOAc
(ii) $RCOCH_2R^1$ (R, R^1 = alkyl, aryl, CO_2Et, CN, or CH_2CO_2Et)
(iii) NaH, DMF, $N(CH_2)_nCl$ ($n = 2$ or 3)

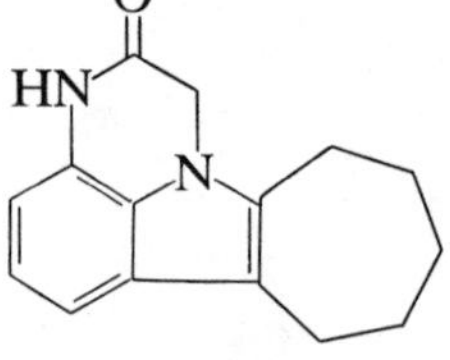

180

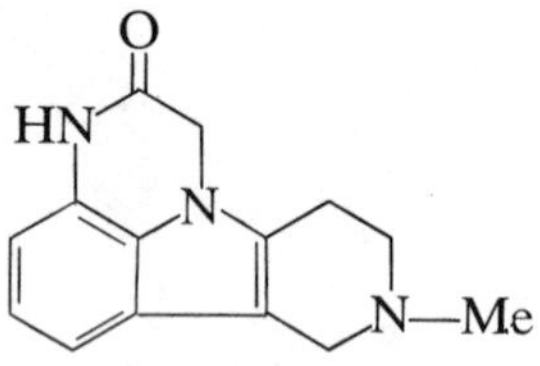

181

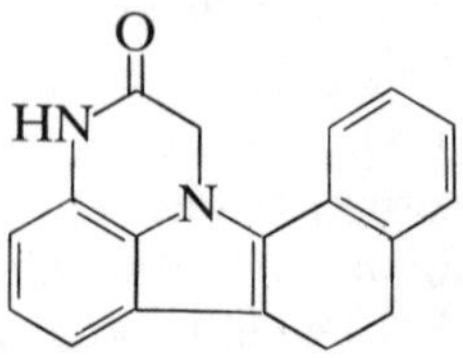

182

1-substituted derivatives **179** also possess carcinostatic and immunosuppressant activity in rats.[71] The ring system was synthesized by reduction of the nitroso compound **176** with zinc dust in acetic acid. The reduced product **177** was not isolated but was condensed with a ketone in a Fischer indolization reaction to give the 1,2,3,4-tetrahydro-2-oxopyrrolo[1,2,3-*de*]quinoxalines **178**. Alkylation of these compounds was achieved using sodium hydride in DMF followed by a suitably substituted aminoalkyl chloride. The condensation with the ketone was also carried out with various cyclic ketones to give a variety of tetracyclic and pentacyclic systems such as **180–182**.

V. Tables

TABLE 1. PYRROLO[1,2-*a*]QUINOXALINES

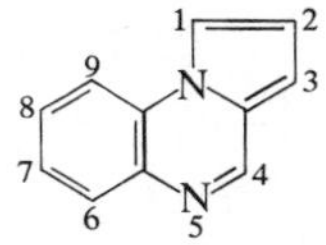

Molecular formula	Substituents or structure	m.p., °C	Ref.
$C_{11}H_5Br_3N_2$	1,2,3-Tribromo	169–170	26
$C_{11}H_5Cl_3N_2$	1,2,3-Trichloro	155–156	26
$C_{11}H_6Br_2N_2$	1,3-Dibromo	162–163	10, 26
	2,3-Dibromo	235–236	26
$C_{11}H_6Cl_2N_2$	1,3-Dichloro	132–133	11
	1,4-Dichloro	110–111	10
$C_{11}H_6I_2N_2$	1,3-Diiodo	229–230	26
$C_{11}H_7BrN_2$	1-Bromo	107–108	26
	Hydrobromide	187–188	10
	3-Bromo	182–183	26
$C_{11}H_7BrN_2O_3S$	1-Bromo-3-sulfo		
	Monohydrate	Decomp. >255	26
$C_{11}H_7ClN_2$	1-Chloro	97–98	11, 26
	3-Chloro	169–172	11
	4-Chloro	171–172	3, 10
$C_{11}H_7ClN_2O$	1-Chloro-4,5-dihydro-4-oxo	304–305	10
$C_{11}H_7ClN_2O_3S$	1-Chloro-2-sulfo	>350	21
	1-Chloro-3-sulfo		
	Monohydrate	>300	11
$C_{11}H_7IN_2$	1-Iodo	164–165	26
	3-Iodo	180–181	26
$C_{11}H_7N_3O_2$	1-Nitro	139–145 decomp.	11
	3-Nitro	234–237	10, 11

TABLE 1 (*contd.*)

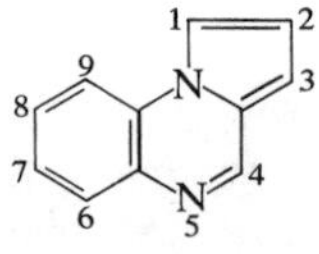

Molecular formula	Substituents or structure	m.p., °C	Ref.
$C_{11}H_7N_3O_5S$	1-Nitro-3-sulfo		
	Monohydrate	Decomp. >250	11
$C_{11}H_8N_2$	Unsubstituted	129–130, 133–134	2, 3, 18
	Methiodide	244–245 decomp	
$C_{11}H_8N_2O$	1,5-Dihydro-1-oxo		
	Monohydrate	Decomp. >230	21
	4,5-Dihydro-4-oxo	268–269, 276–278	3, 10
$C_{11}H_8N_2O_3S$	2-Sulfo	>360	21
	3-Sulfo		
	Monohydrate	>300	11
$C_{11}H_8N_2S$	4,5-Dihydro-4-thioxo	265 decomp., 278	3, 10
$C_{11}H_9ClN_4$	1-Chloro-4-hydrazino[a]		
	Hydrochloride hydrate	214–216	10
$C_{11}H_9N_3$	1-Amino	140–145 decomp.	11
	3-Amino	177–179	11
	4-Amino	223–224	3, 9, 10
	5-Oxide	262–263 decomp.	
$C_{11}H_9N_3O_2S$	3-Sulfonamido	275	11
$C_{11}H_9N_3O_3S$	3-Amino-2-sulfo		
	Monohydrate	Decomp. >350	21
$C_{11}H_{10}N_2$	4,5-Dihydro	115–117	74, 75
$C_{11}H_{10}N_2O_2$	1,2,3,3a,4,5-Hexahydro-1,4-dioxo	237–240, 253–257	18, 30
$C_{11}H_{10}N_4$	4-Hydrazino	169–171	10
$C_{11}H_{10}N_4O_5$	1,2,3,3a,4,5-Hexahydro-7,9-dinitro-4-oxo	205–209 decomp.	13
$C_{11}H_{12}N_2O$	1,2,3,3a,4,5-Hexahydro-1-oxo	164–166	27
	1,2,3,3a,4,5-Hexahydro-4-oxo	174.5–176	17
	Hydrochloride	161–162.5 decomp.	
$C_{11}H_{12}N_2O_2$	1,2,3,3a,4,5-Hexahydro-2-hydroxy-1-oxo	212–214	29
$C_{11}H_{14}N_2$	1,2,3,3a,4,5-Hexahydro	97–98	18
$C_{11}H_{20}N_2$	Perhydro		
	Picrolonate	213–232	28
$C_{12}H_8N_2O_2$	4-Carboxy		
	Methyl ester	160–161	10
	Ethyl ester	98–100	10
$C_{12}H_9ClN_2$	1-Chloro-4-methyl	96–97.5	24
	4-Chloromethyl	143–144	10

TABLE 1 (*contd.*)

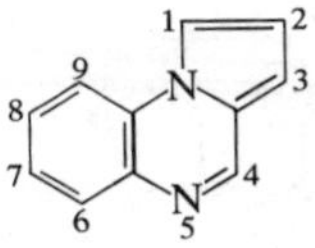

Molecular formula	Substituents or structure	m.p., °C	Ref.
$C_{12}H_{10}N_2$	1-Methyl	70–72	11
	2-Methyl	100–101	19
	3-Methyl	123–134	19
	4-Methyl	136–138	2, 3, 4
	Methiodide	258–260 decomp.	
$C_{12}H_{10}N_2O$	1-Hydroxymethyl	186–187	75
	Acetate	106–107	
	1,5-Dihydro-2-methyl-1-oxo	Chars 250	22
	1,5-Dihydro-4-methyl-1-oxo	>340	22, 23
	4,5-Dihydro-5-methyl-4-oxo	132–134	3
	4-Methoxy	87–89	3
$C_{12}H_{10}N_2O_3S$	1-Methyl-3-sulfo		
	Monohydrate	>300	11
$C_{12}H_{10}N_2S$	4-Methylthio	62–64	3
$C_{12}H_{10}N_2O_4$	3a-Carboxy-1,2,3,3a,4,5-hexahydro-1,4-dioxo		
	Ethyl ester	160–162	18
$C_{12}H_{11}NO_3$	1-Carboxamido-4,5-dihydro	220–221	75
$C_{12}H_{12}N_2$	4,5-Dihydro-4-methyl[h]	b.p. 106–107/ 0.1 mm	74
	4,5-Dihydro-5-methyl	65–66	74
$C_{12}H_{12}N_2O_2$	1,2,3,3a,4,5-Hexahydro-2-methyl-1,4-dioxo	222–223	19
	1,2,3,3a,4,5-Hexahydro-3-methyl-1,4-dioxo	216–217	19
	1,2,3,3a,4,5-Hexahydro-5-methyl-1,4-dioxo	167–169.5	30
$C_{12}H_{12}N_2O_3$	7-Carboxy-1,2,3,3a,4,5-hexahydro-4-oxo[b]		
	Methyl ester	182–192 decomp.	14
$C_{12}H_{12}N_2O_4$	7-Carboxy-1,2,3,3a,4,5-hexahydro-2-hydroxy-4-oxo[c]		
	Methyl ester	227–235	14
$C_{12}H_{14}N_2O$	1,2,3,3a,4,5-Hexahydro-4-methyl-1-oxo	151–154	27
$C_{12}H_{14}N_2O_3S$	1,2,3,3a,4,5-Hexahydro-7-methylsulfonyl-4-oxo	286–288	15
$C_{12}H_{16}N_2$	1,2,3,3a,4,5-Hexahydro-2-methyl	89–91	19
	1,2,3,3a,4,5-Hexahydro-3-methyl	125–126	19
$C_{12}H_{22}N_2$	Perhydro-4-methyl[d]	b.p. 75/0.5 mm	28
$C_{13}H_{10}Br_2N_2$	1,3-Dibromo-2,4-dimethyl	142–143	26
$C_{13}H_{10}Cl_2N_2O$	5-Dichloroacetyl-4,5-dihydro	113–114	75

TABLE 1 (*contd.*)

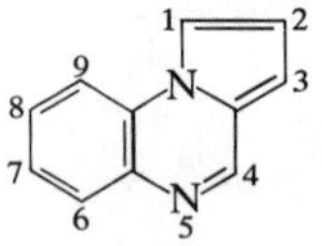

Molecular formula	Substituents or structure	m.p., °C	Ref.
$C_{13}H_{10}N_2O_3$	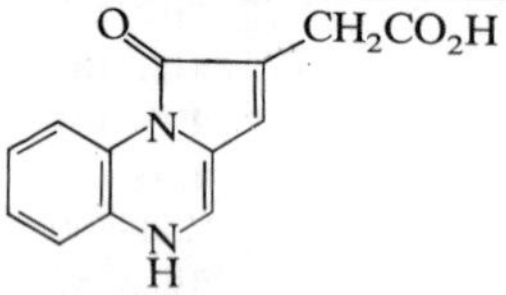	229 decomp.	22
	Methyl ester	Chars 240	
$C_{13}H_{11}ClN_2$	1-Chloro-2,4-dimethyl	108–110	11, 24
	3-Chloro-2,4-dimethyl	142.5–143.5	11
$C_{13}H_{11}ClN_2O$	4-Chloro-1-ethoxy	138.5–139.5	25
$C_{13}H_{12}N_2$	2,4-Dimethyl	86–87	4
$C_{13}H_{12}N_2O$	5-Acetyl-4,5-dihydro	99–100	75
	1-Ethoxy	88	25
	1,5-Dihydro-2,4-dimethyl-1-oxo	>300	22
$C_{13}H_{12}N_2OS$	1-Ethoxy-4,5-dihydro-4-thioxo	250 decomp.	25
$C_{13}H_{12}N_2O_2$	4-Carboxy-4,5-dihydro-4-methyl		
	Ethyl ester	142–145	74
	7,8-Dimethoxy	124–125	75
$C_{13}H_{12}N_2O_3S$	2,4-Dimethyl-3-sulfo	267–269	11
$C_{13}H_{12}N_2O_4$	3a-Carboxy-1,2,3,3a,4,5-hexahydro-2-methyl-1,4-dioxo		
	Ethyl ester	128	19
	3a-Carboxy-1,2,3,3a,4,5-hexahydro-3-methyl-1,4-dioxo		
	Ethyl ester	155–157	19
$C_{13}H_{14}N_2O_2$	5-Acetyl-1,2,3,3a,4,5-hexahydro-1-oxo	131–139	27
	4,5-Dihydro-7,8-dimethoxy	129-130	75
$C_{13}H_{14}N_4O$	1-Ethoxy-4-hydrazino	157–159	25
$C_{13}H_{24}N_2$	5-Ethyl-perhydro[d]	b.p. 84–87/0.5 mm	28
	Dipicrate	260–263 decomp.	
$C_{14}H_{10}N_2O_5$	2,3-Dicarboxy-4,5-dihydro-5-methyl-4-oxo		
	Dimethyl ester	255–257	73
$C_{14}H_{11}N_3O$	5-Acetyl-1-cyano-4,5-dihydro	138–139	75
$C_{14}H_{12}N_2O_2$	5-Acetyl-1-formyl-4,5-dihydro	120–121	75
	Semicarbazone	156–158	
	Thiosemicarbazone	216–218	
$C_{14}H_{12}N_2O_3$	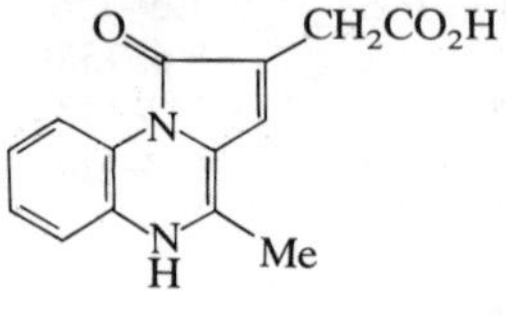	>300	22, 23

TABLE 1 (*contd.*)

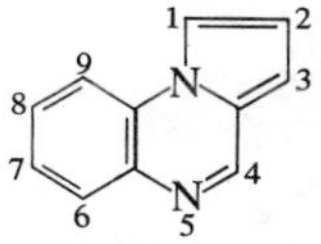

Molecular formula	Substituents or structure	m.p., °C	Ref.
$C_{14}H_{12}N_2O_3$ (*Contd.*)	Methyl ester	Chars 250	
$C_{14}H_{13}ClN_2$	1-Chloro-2-ethyl-4-methyl	85.5–86	4
$C_{14}H_{13}N_3O_2$	5-Acetyl-1-carboxamido-4,5-dihydro	242–243	75
$C_{14}H_{14}N_2$	2-Ethyl-4-methyl		
	Hydrochloride	[e]	4
$C_{14}H_{14}N_2O$	2-Ethyl-1,5-dihydro-4-methyl-1-oxo	315	4
$C_{15}H_{10}N_2O_7$	1,2,3-Tricarboxy-4,5-dihydro-4-methyl-5-oxo		
	Trimethyl ester	169–170	37
$C_{15}H_{12}N_2O_4$	1-Ethoxy-2,3-diformyl-4,5-dihydro-4-oxo	256–257	38
$C_{15}H_{14}ClN_3O$	1-Chloro-4-morpholin-4-yl		
	Hydrochloride	249–250	10
$C_{15}H_{14}Cl_2N_2O_3$	5-Dichloroacetyl-4,5-dihydro-7,8-dimethoxy	116–117	75
$C_{15}H_{14}N_2O_3$	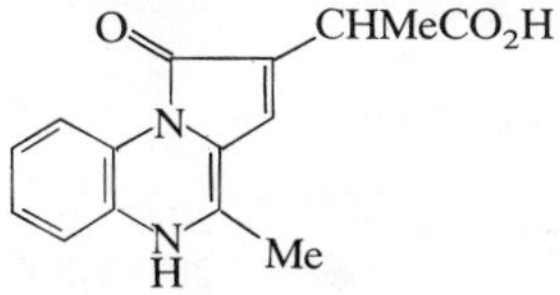 O; $CHMeCO_2H$; N; NH; Me	Decomp. >200	4
$C_{15}H_{14}N_2O_4$	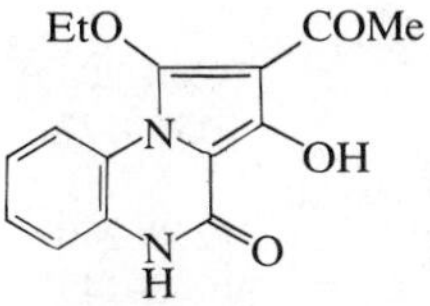 EtO; COMe; N; OH; NH; O	228–230 decomp.	36
	Oxime	255–260 decomp.	
$C_{15}H_{15}ClN_2O_3$	5-Chloroacetyl-4,5-dihydro-7,8-dimethoxy	125–126	75
$C_{15}H_{15}N_3O$	4-Morpholin-4-yl		
	Hydrochloride	278–280	10
$C_{15}H_{16}ClN_3O$	1-Chloro-5-(2-dimethylaminoethyl)-4,5-dihydro-4-oxo		
	Hydrochloride	301–302	10
$C_{15}H_{16}N_2O_2$	1,4-Diethoxy	74–75	25
$C_{15}H_{16}N_2O_3$	5-Acetyl-4,5-dihydro-7,8-dimethoxy	129–130	75
$C_{15}H_{17}N_3S$	4-(2-Dimethylaminoethylthio)[f]		
	Hydrochloride	226–228	10
$C_{16}H_{16}N_2O_4$	5-Acetyl-1-formyl-4,5-dihydro-7,8-dimethoxy	143–146	75

TABLE 1 (*contd.*)

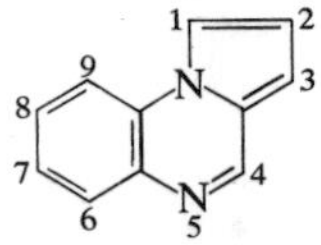

Molecular formula	Substituents or structure	m.p., °C	Ref.
$C_{16}H_{17}N_3O$	4-Morpholin-4-ylmethyl		
	Dihydrochloride	232–234	10
$C_{16}H_{18}N_2$	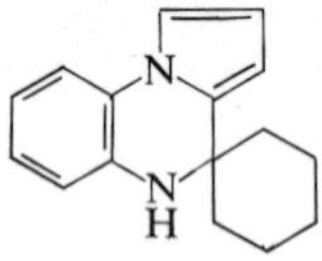 [structure]	99–100	74
$C_{16}H_{18}N_4$	4-(1-Methylpiperazin-4-yl)		
	Dihydrochloride hydrate	281–283	10
	Dipicrate	216–217	
$C_{16}H_{18}N_4O$	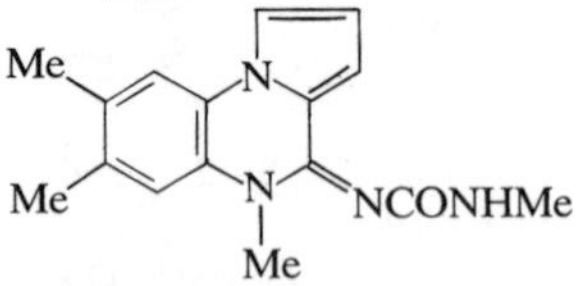 [structure] Me, Me, N, NCONHMe, Me	g	40
$C_{16}H_{19}N_3O$	4-(3-Dimethylaminopropyloxy)		
	Picrate	210–211	10
	5-(3-Dimethylaminopropyl)-4,5-dihydro-4-oxo		
	Hydrochloride monohydrate	209–210	10
	Picrate	204–205	
$C_{16}H_{19}N_3S$	4-(3-Dimethylaminopropylthio)		
	Hydrochloride	200–204	10
	Dipicrate	176–177	
$C_{16}H_{20}N_4$	4-(3-Dimethylaminopropylamino)		
	Dihydrochloride monohydrate	238–240	10
$C_{17}H_{11}ClN_2$	1-Chloro-4-phenyl	143–145	24
	7-Chloro-4-phenyl	154–156	12
$C_{17}H_{11}N_3O_2$	3-Nitro-2-phenyl	231	32
$C_{17}H_{12}N_2$	3-Phenyl	177–178	19
	4-Phenyl	98–100	2, 10, 39
$C_{17}H_{13}N_3$	4-Anilino	114–116	10
$C_{17}H_{13}N_3O_3$	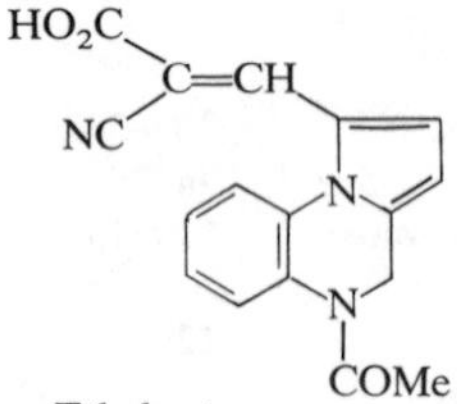 [structure] HO_2C, C=CH, NC, N, N, COMe		
	Ethyl ester	172–173	75

TABLE 1 (*contd.*)

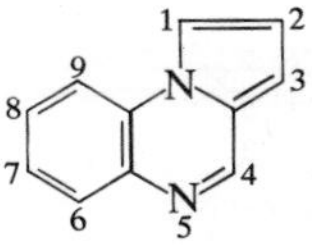

Molecular formula	Substituents or structure	m.p., °C	Ref.
$C_{17}H_{14}N_2$	4,5-Dihydro-4-phenyl	97–98	9, 74
$C_{17}H_{14}N_2O$	4,5-Dihydro-4-(2-hydroxyphenyl)	169–171	74
$C_{17}H_{14}N_2O_2$	1,2,3,3a,4,5-Hexahydro-1,4-dioxo-3-phenyl	250–251	19
$C_{17}H_{14}N_4O_2$	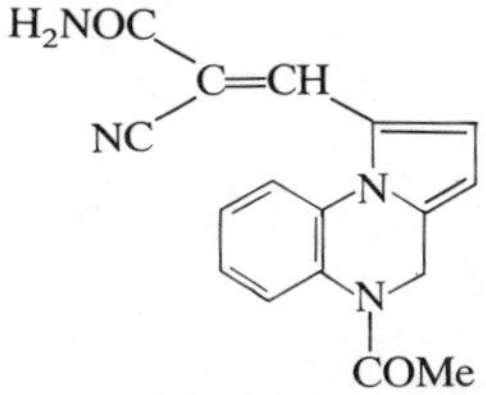 H_2NOC, C=CH, NC, N, N, COMe	254–255	75
$C_{17}H_{16}N_2$	6,7,8,9-Tetrahydro-4-phenyl		
	Hydrochloride	202–204 decomp.	39
$C_{17}H_{16}N_2O$	1,2,3,3a,4,5-Hexahydro-1-oxo-4-phenyl	202–203	24
$C_{17}H_{18}ClN_3$	1-Chloro-4-(1-methylpiperazin-4-yl)	44–46	10
$C_{17}H_{18}ClN_3O_2$	1-Chloro-4,5-dihydro-5-(2-morpholin-4-ylethyl)-4-oxo		
	Hydrochloride	309–310 decomp.	10
$C_{17}H_{18}N_2$	1,2,3,3a,4,5-Hexahydro-3-phenyl	134–135	19
$C_{17}H_{18}N_2O_4$	2-Acetyl-1,4-diethoxy-3-hydroxy	143–144	36
$C_{17}H_{19}N_3O$	4,5-Dihydro-4-oxo-5-(2-pyrrolidin-1-ylethyl)		
	Hydrochloride	235–238	10
$C_{17}H_{19}N_3O_2$	4,5-Dihydro-5-(2-morpholin-4-ylethyl)-4-oxo		
	Hydrochloride monohydrate	269–271 decomp.	10
$C_{18}H_{12}Cl_2N_4$	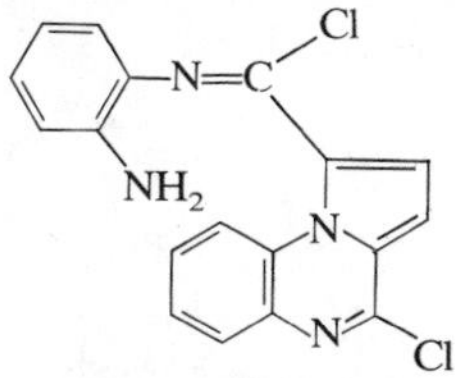 Cl, N=C, NH_2, N, N, Cl	218.5–219.5	20
$C_{18}H_{13}ClN_2$	1-Chloro-2-methyl-3-phenyl	157–158	24
$C_{18}H_{14}N_2$	2-Methyl-4-phenyl		
	Picrate	248 decomp.	24
	4-Methyl-2-phenyl		
	Hydrobromide	100–102	4
$C_{18}H_{14}N_2O$	4-(4-Methoxyphenyl)	118–119	9
	1,5-Dihydro-2-methyl-1-oxo-4-phenyl	Decomp. >250	24

TABLE 1 (*contd.*)

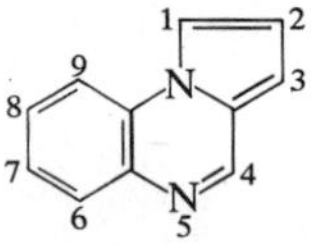

Molecular formula	Substituents or structure	m.p., °C	Ref.
$C_{18}H_{14}N_2O_4$	3a-Carboxy-1,2,3,3a,4,5-hexahydro-1,4-dioxo-3-phenyl	235	19
$C_{18}H_{14}N_4O_2$	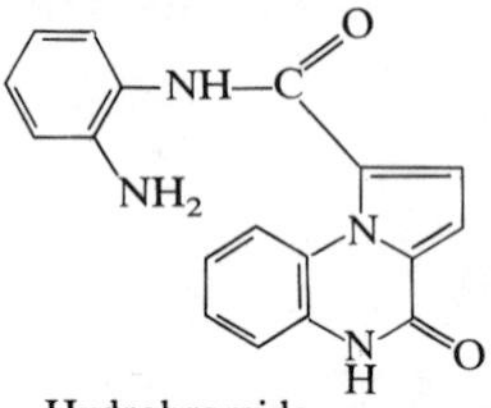	>290	20
	Hydrobromide	290	
$C_{18}H_{16}N_2$	4,5-Dihydro-7-methyl-4-phenyl	123–125	74
$C_{18}H_{16}N_2O$	4,5-Dihydro-4-(3-methoxyphenyl)	132–134	74
	4,5-Dihydro-4-(4-methoxyphenyl)	130–132	9
	4,5-Dihydro-7-methoxy-4-phenyl	138–141	74
$C_{18}H_{18}N_2O$	Tetrahydro derivative from 1,5-dihydro-2-methyl-1-oxo-4-phenyl	170–176	24
$C_{19}H_{14}N_2O_3$	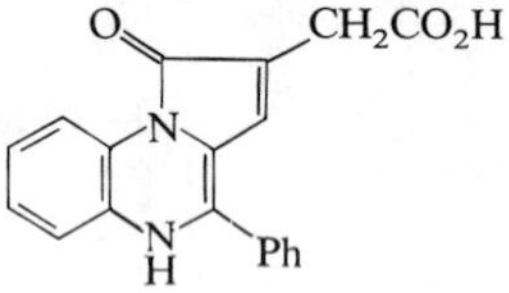	Decomp. >240	24
	Methyl ester	Decomp. >220	
$C_{19}H_{14}N_4O_4$	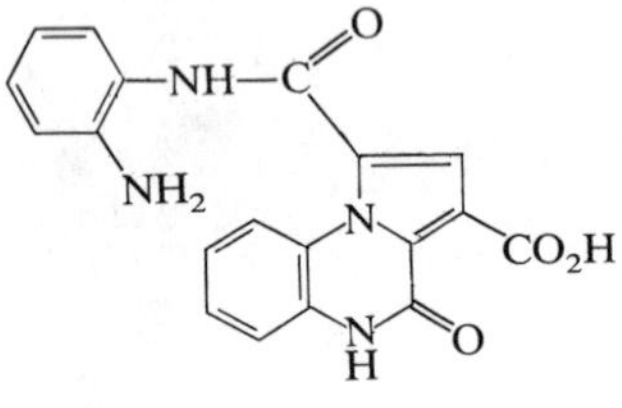		
	Ethyl ester hydrate	>290	20
$C_{19}H_{15}ClN_2$	2-Benzyl-1-chloro-4-methyl	111–112.5	4
	Hydrochloride	248–249 decomp.	
$C_{19}H_{16}N_2$	2-Benzyl-4-methyl	85–86	4
	Picrate	196–198 decomp.	
$C_{19}H_{16}N_2O$	2-Benzyl-1,5-dihydro-4-methyl-1-oxo	>250	4
	1,5-Dihydro-2,5-dimethyl-1-oxo-4-phenyl	235–237	24
$C_{19}H_{16}N_2O_2$	4-(3,4-Dimethoxyphenyl)	149–150	9

TABLE 1 (*contd.*)

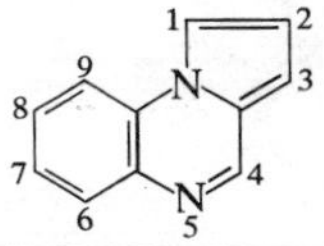

Molecular formula	Substituents or structure	m.p., °C	Ref.
$C_{19}H_{17}N_3O_5$	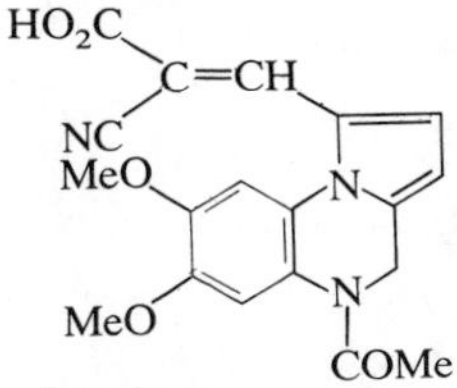 [structure]		
	Ethyl ester	198–199	75
$C_{19}H_{18}N_2O_2$	4-(3,4-Dimethoxyphenyl)-4,5-dihydro	198–199	9
$C_{19}H_{18}N_4O_4$	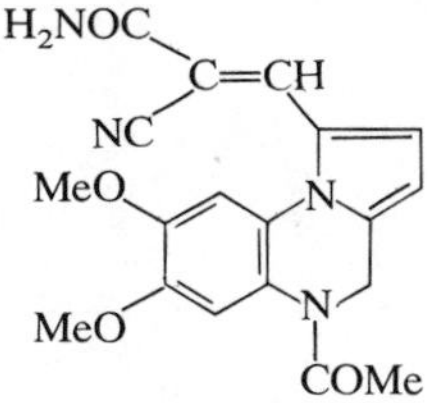 [structure]	210–212	75
$C_{19}H_{19}N_3O$	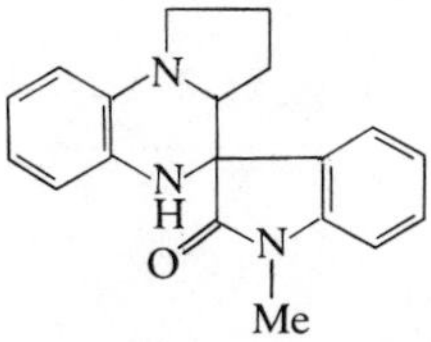 [structure]	215	16
	Hydrochloride	218 decomp.	
$C_{19}H_{20}N_2O_5$	3-Acetoxy-2-acetyl-1,4-diethoxy	124–125	36
$C_{20}H_{14}N_2O_5$	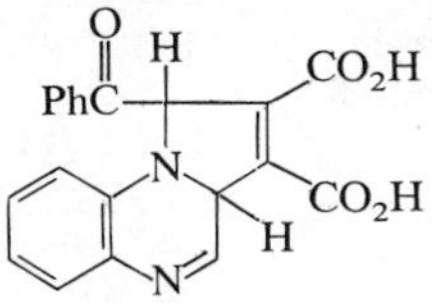 [structure]		
	Dimethyl ester	183	33
$C_{20}H_{16}N_2O_3$	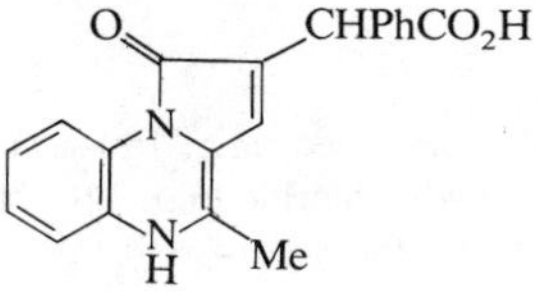 [structure]	Decomp. >300	4
	Methyl ester	>250	

TABLE 1 (*contd.*)

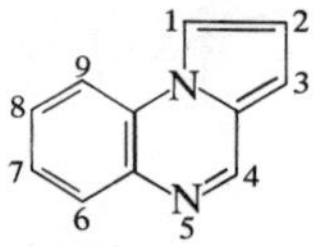

Molecular formula	Substituents or structure	m.p., °C	Ref.
$C_{20}H_{16}N_4O$	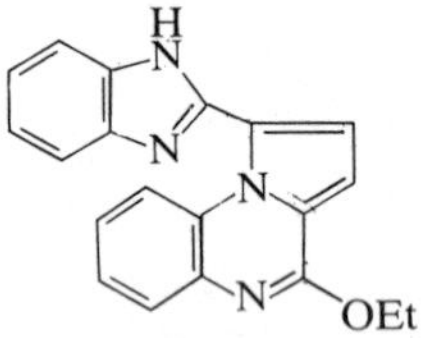	168–169	20
$C_{20}H_{20}N_2O_2$	Tetrahydro derivative from 5-acetyl-1,5-dihydro-1-oxo-4-phenyl	136–138	24
$C_{20}H_{20}N_2O_3$	4,5-Dihydro-4-(3,4,5-trimethoxyphenyl)	133–134	74
$C_{22}H_{14}N_4S$	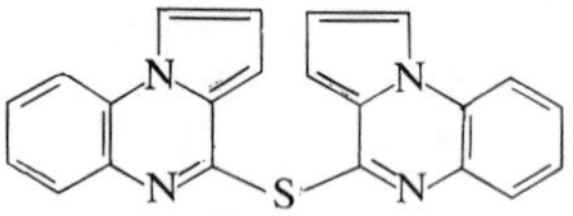	314–317	10
$C_{22}H_{22}N_4$	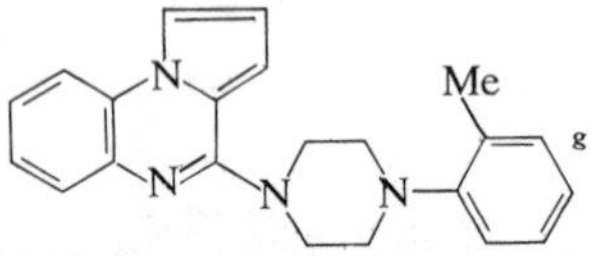 [g]		
	Dihydrochloride hydrate	275–278 decomp.	10
$C_{23}H_{15}N_3O_2$	1-(4-Nitrophenyl)-4-phenyl	222–224	34
$C_{23}H_{16}N_2O$	1-Hydroxy-2,3-diphenyl	285	5
$C_{23}H_{16}N_2S$	1-Mercapto-2,3-diphenyl	255	35
$C_{23}H_{20}N_2$	1,2,3,3a-Tetrahydro-3a,4-diphenyl		31
$C_{24}H_{18}N_2O$	1-Hydroxy-4-methyl-2,3-diphenyl	240	5
$C_{25}H_{15}N_3O_6$	2,3-Dicarboxy-1-(4-nitrophenyl)		
	Dimethyl ester	234–237	34
$C_{26}H_{22}N_2O$	1-Ethoxy-4-methyl-2,3-diphenyl	152	5
$C_{37}H_{23}N_3O_4$	2,3-Dibenzoyl-1-(4-nitrophenyl)-4-phenyl	269.5–271.5	34

[a] *p*-Anisylidene derivative m.p. 179–180°.

[b] $[\alpha]_D^{23} -270°$ (*c* 0.5, 1.0, MeOH).

[c] $[\alpha]_D^{23} -270°$ (*c* 0.5, MeOH).

[d] n_D^{20} 1.5033, *d* 0.981.

[e] No melting point given; good analysis.

[f] The 4-(2-pyrrolidin-1-ylthio) compound also described, m.p. 190–193°; no analysis.

[g] The 1-chloro analogue was also described (hydrochloride m.p. 110–112°).

[h] Twenty other 4-alkyl derivatives are listed in reference 74.

TABLE 2. 2-OXOPYRROLO[2,3-*b*]QUINOXALINES[a]

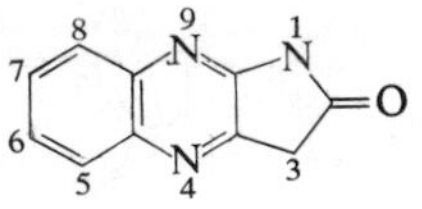

Molecular formula	Substituents or structure	m.p., °C	Ref.
$C_{10}H_6ClN_3O$	3-Chloro-1,3-dihydro-2-oxo	184–192	49
$C_{11}H_6N_4O$	3-Cyano-1,3-dihydro-2-oxo	>360	48
$C_{11}H_8ClN_3O$	3-Chloro-1,3-dihydro-1-methyl-2-oxo	196–208	49
$C_{12}H_{10}ClN_3O_2$	3-Chloro-1,3-dihydro-1-(2-hydroxyethyl)-2-oxo	162–174	49
$C_{14}H_{14}ClN_3O$	1-Butyl-3-chloro-1,3-dihydro-2-oxo	92–94	49
$C_{14}H_{15}N_3O$	1-Butyl-1,3-dihydro-2-oxo	192–198	49
$C_{16}H_9ClN_4O_3$	3-Chloro-1,3-dihydro-1-(4-nitrophenyl)-2-oxo	>250	49
$C_{16}H_9Cl_2N_3O$	3-Chloro-1-(3-chlorophenyl)-1,3-dihydro-2-oxo	>250	49
$C_{16}H_{10}ClN_3O$	3-Chloro-1,3-dihydro-2-oxo-1-phenyl	238–246	49, 50
	1-(4-Chlorophenyl)-1,3-dihydro-2-oxo	>250	49
$C_{16}H_{10}ClN_3O_2$	3-Chloro-1,3-dihydro-1-(4-hydroxyphenyl)-2-oxo	>250	49
$C_{16}H_{11}N_3O$	1,3-Dihydro-2-oxo-1-phenyl	>250	49
	1,3-Dihydro-2-oxo-3-phenyl	269	46
	1,4-Dihydro-2-oxo-3-phenyl	296–300	55
$C_{16}H_{11}N_3O_2$	1,3-Dihydro-1-(3-hydroxyphenyl)-2-oxo	>250	49
	1,3-Dihydro-3-hydroxy-2-oxo-3-phenyl	297–298	55
$C_{16}H_{16}ClN_3O$	3-Chloro-1-cyclohexyl-1,3-dihydro-2-oxo	210–216	49
$C_{17}H_{12}ClN_3O$	2-Benzyl-3-chloro-1,3-dihydro-2-oxo	212–218	49
$C_{17}H_{12}ClN_3O_2$	3-Chloro-1,3-dihydro-1-(4-methoxyphenyl)-2-oxo	>250	49
$C_{17}H_{13}N_3O_2$	1,3-Dihydro-1-(4-methoxyphenyl)-2-oxo	>250	49
$C_{18}H_{15}N_3O$	3-Ethyl-1,3-dihydro-2-oxo-3-phenyl	217–218	55
	1,3-Dihydro-1,3-dimethyl-2-oxo-3-phenyl	129–130	55
$C_{20}H_{12}ClN_3O$	3-Chloro-1,3-dihydro-1-naphthyl-2-oxo	>250	49
$C_{22}H_{15}ClN_4O$	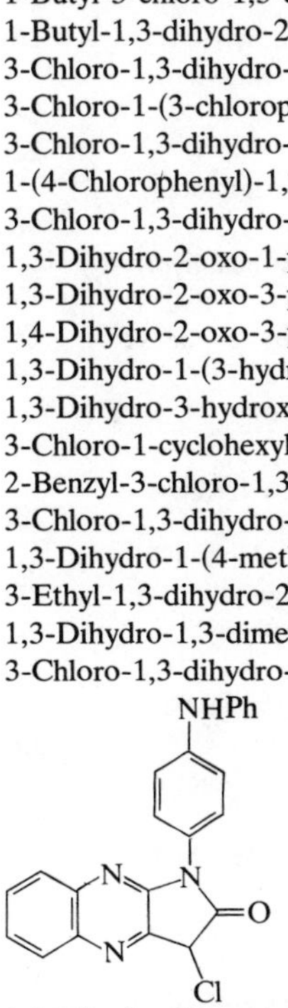 NHPh [structure] N Cl O	>250	49
$C_{22}H_{15}N_3O$	1,3-Dihydro-2-oxo-1,3-diphenyl	218, 287–289	46, 53
$C_{23}H_{16}N_4O$	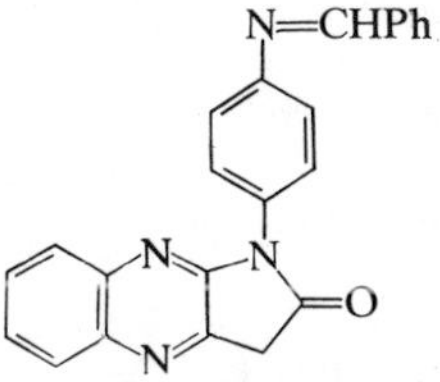 N=CHPh [structure] N O	>250	49

[a] Several compounds have also been prepared with substituents in the 6(or 7)-position.[49] Their orientation is uncertain, and they are not listed in the table.

TABLE 3. 1*H*-PYRROLO[2,3-*b*]QUINOXALINES

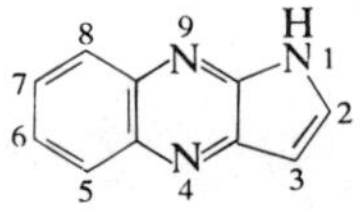

Molecular formula	Substituents or structure	m.p., °C	Ref.
$C_{11}H_8N_4O_2$	2-Amino-3-carboxy		
	Ethyl ester	360	52
$C_{12}H_9N_5$	2-Amino-3-cyano-1-methyl	236–237	52
$C_{12}H_{10}N_4O_2$	2-Amino-3-carboxy-1-methyl		
	Ethyl ester	237–238	52
$C_{13}H_{11}N_5O$	2-Amino-3-cyano-1-(2-hydroxyethyl)	268	52
$C_{13}H_{12}N_4O_3$	2-Amino-3-carboxy-1-(2-hydroxyethyl)		
	Ethyl ester	239–240	52
$C_{15}H_{16}N_4$	1-Butyl-3-cyano-2,3-dihydro	230	47
$C_{16}H_{19}N_5O_2$	2-Amino-3-carboxy-1-(3-dimethylaminopropyl)		
	Ethyl ester	164	52
$C_{17}H_{18}FN_3$	1-(4-Fluorophenyl)-2,3,3a,4,9,9a-hexahydro-9-methyl	150–151	54
$C_{17}H_{19}N_3$	2,3,3a,4,9,9a-Hexahydro-9-methyl-1-phenyl	140–142	54
$C_{18}H_{13}N_5$	2-Amino-1-benzyl-3-cyano	328	52
$C_{18}H_{14}N_4O_2$	2-Amino-1-benzyl-3-carboxy		
	Ethyl ester	248	52
$C_{18}H_{20}FN_3$	9-Ethyl-1-(4-fluorophenyl)-2,3,3a,4,9,9a-hexahydro	128–130	54
$C_{18}H_{21}N_3$	9-Ethyl-2,3,3a,4,9,9a-hexahydro-1-phenyl	129–131	54
$C_{19}H_{15}N_5O$	2-Amino-3-cyano-1-(4-ethoxyphenyl)	278	52
$C_{19}H_{16}N_4O_3$	2-Amino-3-carboxy-1-(4-ethoxyphenyl)		
	Ethyl ester	242	52
$C_{19}H_{20}FN_3O$	4-Acetyl-1-(4-fluorophenyl)-2,3,3a,4,9,9a-hexahydro-9-methyl	173–175	54
$C_{19}H_{21}N_3O$	4-Acetyl-2,3,3a,4,9,9a-hexahydro-9-methyl-1-phenyl	180–182	54
$C_{19}H_{22}ClN_3$	1-(4-Chlorophenyl)-2,3,3a,4,9,9a-hexahydro-6,7,9-trimethyl	160	54
$C_{19}H_{22}FN_3$	1-(4-Fluorophenyl)-2,3,3a,4,9,9a-hexahydro-6,7,9-trimethyl	142–143	54
$C_{19}H_{23}N_3$	2,3,3a,4,9,9a-Hexahydro-6,7,9-trimethyl-1-phenyl	142–143	54
$C_{20}H_{22}FN_3O$	4-Acetyl-9-ethyl-1-(4-fluorophenyl)-2,3,3a,4,9,9a-hexahydro	165–166	54
$C_{20}H_{23}N_3O$	4-Acetyl-9-ethyl-2,3,3a,4,9,9a-hexahydro-1-phenyl	173–175	54
$C_{20}H_{23}N_3O_2$	4-Carboxy-2,3,3a,4,9,9a-hexahydro-9-methyl-1-phenyl		
	Ethyl ester	180–181	54
$C_{20}H_{25}N_3$	2,3,3a,4,9,9a-Hexahydro-6,7,9-trimethyl-1-(4-methylphenyl)	123–125	54

TABLE 3 (*contd.*)

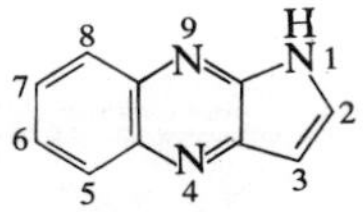

Molecular formula	Substituents or structure	m.p., °C	Ref.
$C_{21}H_{24}ClN_3O$	4-Acetyl-1-(4-chlorophenyl)-2,3,3a,4,9,9a-hexahydro-6,7,9-trimethyl	158–159	54
$C_{21}H_{24}FN_3O$	4-Acetyl-1-(4-fluorophenyl)-2,3,3a,4,9,9a-hexahydro-6,7,9-trimethyl	181–183	54
$C_{21}H_{25}N_3O$	4-Acetyl-2,3,3a,4,9,9a-hexahydro-6,7,9-trimethyl-1-phenyl	177	54
$C_{22}H_{15}N_3$	2,3-Diphenyl	240–241	41
$C_{22}H_{27}N_3O$	4-Acetyl-2,3,3a,4,9,9a-hexahydro-6,7,9-trimethyl-1-(4-methylphenyl)	190	54
$C_{23}H_{15}ClN_4O_4$	3-Carboxy-1-(3-chlorophenyl)-2,3-dihydro-2-(3-nitrophenyl)		
	Ethyl ester		45
$C_{23}H_{15}Cl_2N_3O_2$	3-Carboxy-2-(2,4-dichlorophenyl)-2,3-dihydro-1-phenyl		
	Ethyl ester	295	45
$C_{23}H_{15}N_5O_6$	3-Carboxy-2,3-dihydro-2-(3-nitrophenyl)-1-(4-nitrophenyl)		
	Ethyl ester	234–235	45
$C_{23}H_{16}ClN_3O_2$	3-Carboxy-2-(4-chlorophenyl)-2,3-dihydro-1-phenyl		
	Ethyl ester	278	46
$C_{23}H_{17}N_3$	2-(4-Methylphenyl)-3-phenyl	258	41
$C_{23}H_{17}N_3O_2$	3-Carboxy-2,3-dihydro-1,2-diphenyl		
	Ethyl ester	215–216	44
		283–284	45
$C_{24}H_{18}ClN_3O_3$	3-Carboxy-2-(4-chlorophenyl)-2,3-dihydro-1-(4-methoxyphenyl)		
	Ethyl ester	270	46
$C_{24}H_{18}N_4O_4$	3-Carboxy-2,3-dihydro-1-(4-methylphenyl)-2-(3-nitrophenyl)		
	Ethyl ester	260–261	45
	3-Carboxy-2,3-dihydro-1-(3-methylphenyl)-2-(3-nitrophenyl)		
	Ethyl ester	260	45
$C_{24}H_{18}N_4O_5$	3-Carboxy-2,3-dihydro-2-(4-methoxyphenyl)-1-(3-nitrophenyl)		
	Ethyl ester	219	45
$C_{24}H_{20}N_4$	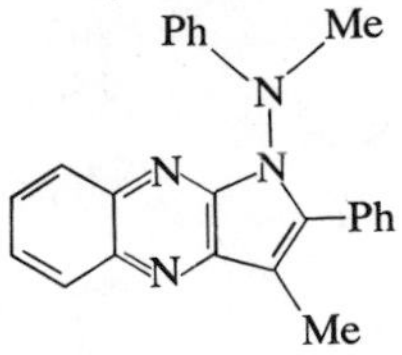	189–190	77

TABLE 3 (*contd.*)

Molecular formula	Substituents or structure	m.p., °C	Ref.
$C_{24}H_{21}N_3$	2,3-Dihydro-3-methyl-1-(4-methylphenyl)-2-phenyl	195	43
$C_{24}H_{21}N_3O$	2,3-Dihydro-1-(4-methoxyphenyl)-3-methyl-2-phenyl	225	43
$C_{28}H_{19}Cl_2N_3$	2,3-Dihydro-1,2-di(4-chlorophenyl)-3-phenyl	238	43
$C_{28}H_{21}N_3$	2,3-Dihydro-1,2,3-triphenyl	252	42
$C_{29}H_{22}ClN_3$	2-(4-Chlorophenyl)-2,3-dihydro-1-(4-methylphenyl)-3-phenyl	244	43
$C_{29}H_{22}N_4$	1-Diphenylamino-3-methyl-2-phenyl	162–163	77
$C_{29}H_{22}N_4O_3$	2,3-Dihydro-2-(4-methoxyphenyl)-1-(3-nitrophenyl)-3-phenyl	315	42
$C_{30}H_{23}N_3O_2$	2,3-Dihydro-2-(3,4-methylenedioxyphenyl)-1-(3-methylphenyl)-3-phenyl	219	42
$C_{30}H_{25}N_3O$	2,3-Dihydro-2-(4-methoxyphenyl)-1-(4-methylphenyl)-3-phenyl	235	42
$C_{31}H_{24}N_4O_6$	3-(3,4-Dimethoxyphenyl)-2,3-dihydro-2-(3,4-methylenedioxyphenyl)-1-(4-nitrophenyl)	270 decomp.	42
$C_{31}H_{27}N_3O_2$	2-(3,4-Dimethoxyphenyl)-2,3-dihydro-1-(4-methylphenyl)-3-phenyl	157	42
$C_{32}H_{29}N_3O_3$	3-(3,4-Dimethoxyphenyl)-2,3-dihydro-1-(4-methylphenyl)-2-(4-methoxyphenyl)	193	42
$C_{35}H_{26}N_4$	3-Benzyl-1-diphenylamino-2-phenyl	158–159	77

TABLE 4. PYRROLO[3,4-*b*]QUINOXALINES

1*H*

Molecular formula	Substituents or structure	m.p., °C	Ref.
$C_{10}H_5N_3O_2$	1*H*-2,3-Dihydro-1,3-dioxo	Ca. 206 decomp.	59
$C_{10}H_7N_3O$	1*H*-2,3-Dihydro-1-oxo	183–184	63
$C_{10}H_7N_3O_2$	1*H*-2,3,4,9-Tetrahydro-1,3-dioxo		
	4-Acetyl	239	69
	2,4-Diacetyl	202–206 decomp.	
	2,4,9-Triacetyl	180 decomp.	
$C_{10}H_9N_3O_2$	1*H*-2,3,3a,4,9,9a-Hexahydro-1,3-dioxo		
	Monoacetyl	290–291	69
	2,4-Diacetyl		
	2,4,9-Triacetyl	220–223 decomp.	
$C_{11}H_9N_3O$	1*H*-2,3-Dihydro-2-methyl-1-oxo	199–200	63

TABLE 4 (*contd.*)

1*H* [structure: pyrrolo[3,4-b]quinoxaline ring numbered 1–9]

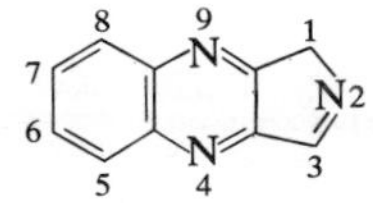

Molecular formula	Substituents or structure	m.p., °C	Ref.
$C_{11}H_{11}N_3$	1*H*-2,3-Dihydro-2-methyl	Decomp. >170	67
	Mono-*N*-oxide	174 decomp.	
$C_{12}H_7N_3O_3$	1*H*-2-Acetyl-2,3-dihydro-1,3-dioxo	ca. 220 decomp.	59
$C_{12}H_{11}N_3O$	1*H*-2-Ethyl-2,3-dihydro-1-oxo	193–194	63
$C_{13}H_{11}N_3O_2$	1*H*-2,3-Dihydro-1,3-dioxo-2-*n*-propyl	246	78
$C_{13}H_{13}N_3$	1*H*-2-Allyl-2,3-dihydro	172	64
	Picrate	177	
$C_{13}H_{15}N_3$	1*H*-2,3-Dihydro-2-propyl	160–161	64
$C_{14}H_{12}N_4O$	1*H*-1-Oxo-3-pyrrolidin-1-yl	293	78
$C_{14}H_{12}N_4O_2$	1*H*-3-(Morpholin-1-yl)-1-oxo	237	78
$C_{14}H_{16}N_3O$	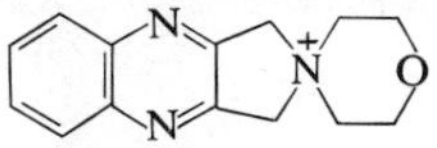 [structure: spiro morpholinium quinoxaline]		
	Bromide	265–266	65
	Bromide hemihydrate	258	64
	Chloride	257–258	64
	Picrate	254–255	64
$C_{14}H_{17}N_3$	1*H*-2,3-Dihydro-2-butyl	118–119	64
	1*H*-2,3-Dihydro-1,1,3,3-tetramethyl	118–120	58
$C_{15}H_{14}N_4O$	1*H*-1-Oxo-3-piperidin-1-yl	218	78
$C_{15}H_{18}N_3$	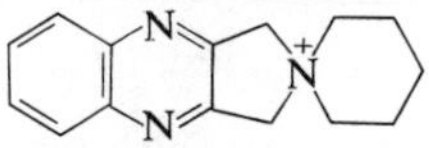 [structure: spiro piperidinium quinoxaline]		
	Bromide	266–268	64
	Picrate	189–190	
$C_{16}H_9N_3O_2$	1*H*-2,3-Dihydro-1,2-dioxo-2-phenyl[a]	>300	62
$C_{16}H_{10}N_4O_2$	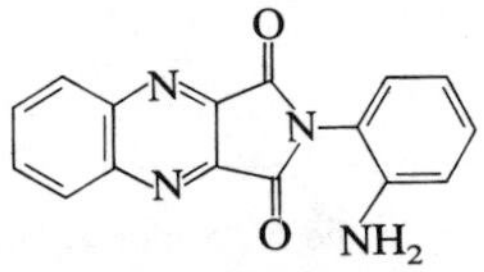 [structure: 1,3-dioxo, N-(2-aminophenyl)]		
	Acetyl derivative	310–315	60
$C_{16}H_{19}N_3$	1*H*-2-Cyclohexyl-2,3-dihydro	233–235	64
$C_{16}H_{19}N_3O$	1*H*-2-Acetyl-2,3-dihydro-1,1,3,3-tetramethyl	212–214	58
$C_{17}H_{11}N_3O_2$	1*H*-2-Benzyl-2,3-dihydro-1,3-dioxo	270–272	61
		255–258	62
$C_{17}H_{15}N_3$	1*H*-2-Benzyl-2,3-dihydro	212.5–214.5	64
$C_{17}H_{18}N_4O$	1*H*-3-Cyclohexylimino-2,3-dihydro-2-methyl-1-oxo[b]	168	78
$C_{18}H_{12}N_4O_3$	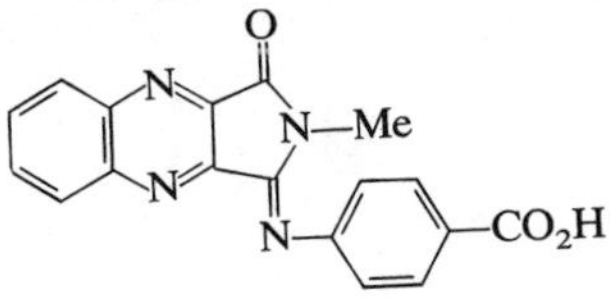 [structure: 1-oxo, 2-Me, 3-(4-carboxyphenylimino)]	305	78

TABLE 4 (*contd.*)

1*H* [ring structure with numbering 1–9, N2, N4, N9]

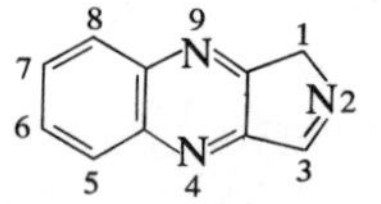

Molecular formula	Substituents or structure	m.p., °C	Ref.
$C_{18}H_{20}N_4O$	1*H*-3-Ethylimino-2-cyclohexyl-2,3-dihydro-1-oxo[b]	246	78
$C_{18}H_{24}N_3$	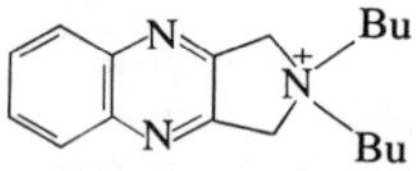 [structure: 2,2-dibutyl-pyrrolo[3,4-b]quinoxalinium] Bromide hemihydrate	205–206	64
$C_{21}H_{17}Cl_2N_5$	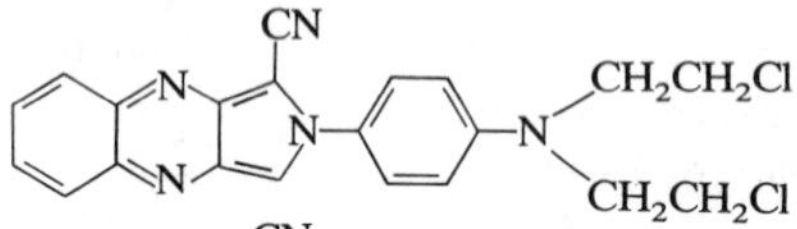 [structure: 1-CN, 2-(4-N(CH₂CH₂Cl)₂-phenyl)] CN; CH_2CH_2Cl; CH_2CH_2Cl	217–218	68
$C_{21}H_{19}N_5O_2$	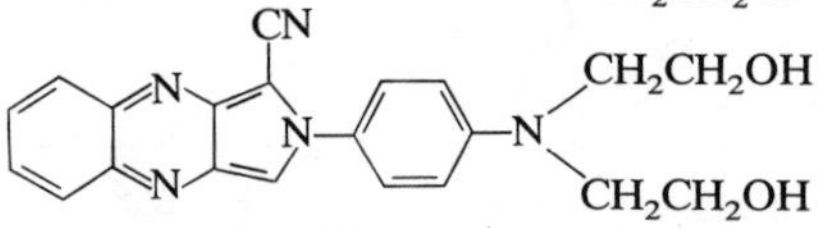 [structure: 1-CN, 2-(4-N(CH₂CH₂OH)₂-phenyl)] CN; CH_2CH_2OH; CH_2CH_2OH	238–239	68
$C_{22}H_{19}ClN_4O$	1*H*-2-(4-Chlorophenyl)-3-cyclohexylimino-2,3-dihydro-1-oxo	140	78
$C_{23}H_{17}N_3$	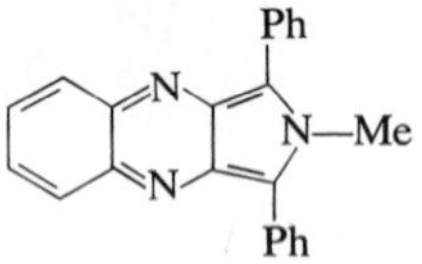 [structure: 1,3-Ph, N—Me]	224–225	80
$C_{24}H_{22}N_4$	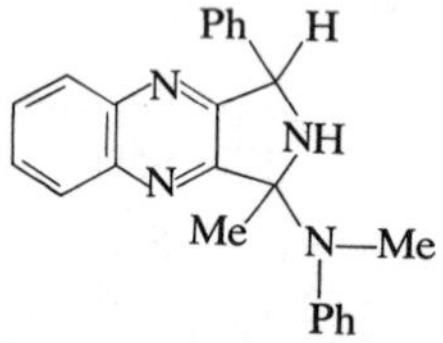 [structure: 1-Ph, 1-H, NH, 3-Me, 3-N(Me)Ph]	182	57

[a] Eight other 2-aryl compounds, several substituted in the 6-position by methyl, methoxy, or ethoxy, are described; all have melting points >275°.
[b] Six other 2-alkyl-3-alkylimino compounds are described.

TABLE 5 PYRROLO[1,2,3-*de*]QUINOXALINES[a]

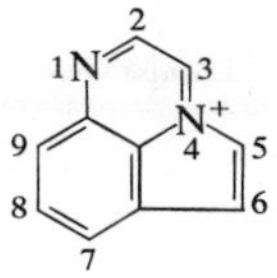

Molecular formula	Substituents	m.p., °C	Ref.
$C_{10}H_8N_2O_2$	1,2,3,4,5,6-Hexahydro-2,3-dioxo	303–305	70
$C_{12}H_{10}N_2O$	3,4-Dihydro-2,5-dimethyl-3-oxo	151–152	81
$C_{12}H_{10}N_2O_3$	6-Carboxy-1,2,3,4-tetrahydro-5-methyl-2-oxo		
	Ethyl ester	266.5–267.5	71
$C_{12}H_{12}N_2O$	1,2,3,4,5,6-Hexahydro-5-methyl-2-methylene-3-oxo	171–173	81
$C_{13}H_{10}N_2O_3$	2-Carboxymethylene-1,2,3,4-tetrahydro-5-methyl-3-oxo		
	Methyl ester	163–164 decomp.	81
$C_{13}H_{12}N_2O_3$	6-Carboxymethyl-1,2,3,4-tetrahydro-5-methyl-2-oxo	286–287	71
	2-Carboxymethylene-1,2,3,4,5,6-hexahydro-5-methyl-2-oxo	216–218	81
$C_{13}H_{14}N_2O$	6-Ethyl-1,2,3,4-tetrahydro-5-methyl-2-oxo	217–218	71
$C_{17}H_{11}N_3O$	6-Cyano-1,2,3,4-tetrahydro-2-oxo-5-phenyl	312–313	71
$C_{17}H_{13}ClN_2O$	5-(2-Chlorophenyl)-1,2,3,4-tetrahydro-6-methyl-2-oxo	219–222	71
	5-(3-Chlorophenyl)-1,2,3,4-tetrahydro-6-methyl-2-oxo	273–276	71
	5-(4-Chlorophenyl)-1,2,3,4-tetrahydro-6-methyl-2-oxo	333–336	71
$C_{17}H_{14}N_2O$	1,2,3,4-Tetrahydro-5-methyl-2-oxo-6-phenyl	282–284	71
	1,2,3,4-Tetrahydro-6-methyl-2-oxo-5-phenyl	288–290	71
$C_{18}H_{14}N_2O_2$	2,3-Difur-2-yl-3,3a,4,5-tetrahydro	107–109	70
$C_{18}H_{14}N_2O_3$	6-Carboxymethyl-1,2,3,4-tetrahydro-2-oxo 5-phenyl	301–303	71
$C_{18}H_{16}N_2O$	6-Ethyl-1,2,3,4-tetrahydro-2-oxo-5-phenyl	245–246	71
$C_{18}H_{16}N_2O_2$	1,2,3,4-Tetrahydro-5-(4-methoxyphenyl)-6-methyl-2-oxo	278–281.5	71
$C_{18}H_{25}N_3O$	1-(3-Dimethylaminopropyl)-6-ethyl-1,2,3,4-tetrahydro-5-methyl-2-oxo	88–90.5	71
$C_{19}H_{18}N_2O$	1,2,3,4-Tetrahydro-2-oxo-5-phenyl-6-propyl	230–230.5	71
$C_{19}H_{18}N_2O_3$	1,2,3,4-Tetrahydro-6-(2-hydroxyethyl)-5-(4-methoxyphenyl)-2-oxo	231–233	71
	Acetate	199–201	
$C_{21}H_{23}N_3O$	1-(2-Dimethylaminoethyl)-1,2,3,4-tetrahydro-6-methyl-2-oxo-5-phenyl	150–152	71
$C_{21}H_{23}N_3O$	1-(3-Dimethylaminopropyl)-1,2,3,4-tetrahydro-2-oxo-5-phenyl	131–132	71
$C_{22}H_{16}N_2$	3,4-Dihydro-2,3-diphenyl	151.5–153	70

TABLE 5 (*contd.*)

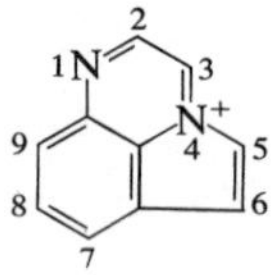

Molecular formula	Substituents	m.p., °C	Ref.
$C_{22}H_{18}N_2$	3,4,5,6-Tetrahydro-2,3-diphenyl	150–152	70
$C_{22}H_{24}ClN_3O$	5-(2-Chlorophenyl)-1-(3-dimethylaminopropyl)-1,2,3,4-tetrahydro-6-methyl-2-oxo	190.5–192.5	71
	5-(3-Chlorophenyl)-1-(3-dimethylaminopropyl)-1,2,3,4-tetrahydro-6-methyl-2-oxo	107–109	71
	5-(4-Chlorophenyl)-1-(3-dimethylaminopropyl)-1,2,3,4-tetrahydro-6-methyl-2-oxo	186–188	71
$C_{22}H_{25}N_3O$	1-(3-Dimethylaminopropyl)-1,2,3,4-tetrahydro-6-methyl-2-oxo-5-phenyl	127.5–129.5	71
$C_{22}H_{25}N_3O_2$	1-(3-Dimethylaminopropyl)-1,2,3,4-tetrahydro-5-(4-hydroxyphenyl)-6-methyl-2-oxo	172.5–175.5	71
$C_{23}H_{18}N_2O$	5-Benzyl-1,2,3,4-tetrahydro-2-oxo-6-phenyl	240–242	71
$C_{23}H_{27}N_3O$	1-(3-Dimethylaminopropyl)-6-ethyl-1,2,3,4-tetrahydro-2-oxo-5-phenyl[b]	114–116	71
$C_{23}H_{27}N_3O_2$	1-(3-Dimethylaminopropyl)-1,2,3,4-tetrahydro-5-(4-methoxyphenyl)-6-methyl-2-oxo	169.5–171	71
$C_{24}H_{20}N_2O_2$	5-(4-Benzyloxyphenyl)-1,2,3,4-tetrahydro-6-methyl-2-oxo	270–273	71
$C_{24}H_{22}N_2O_2$	3,4-Dihydro-2,3-bis(4-methoxyphenyl)	158.5–160	70
$C_{24}H_{29}N_3O$	1-(3-Dimethylaminopropyl)-1,2,3,4-tetrahydro-2-oxo-5-phenyl-6-propyl	119–121.5	71
$C_{27}H_{27}N_3O$	5-Benzyl-1-(2-dimethylaminoethyl)-1,2,3,4-tetrahydro-2-oxo-6-phenyl	152–153	71
$C_{29}H_{31}N_3O_2$	5-(4-Benzyloxyphenyl)-1-(3-dimethylaminopropyl)-1,2,3,4-tetrahydro-6-methyl-2-oxo	138–140	71

[a] Compounds are named as reduced derivatives of the pyrrolo[1,2,3-*de*]quinoxalinium cation.

[b] Ten other 1-(aminoalkyl) substituents are described.

VI. References

1. G. W. H. Cheeseman, M. Rafiq, P. D. Roy, C. J. Turner, and G. V. Boyd, *J. Chem. Soc.* (*C*), 2018 (1971).
2. G. W. H. Cheeseman and B. Tuck, *Chem. Ind.* (London), 1382 (1965).
3. G. W. H. Cheeseman and B. Tuck, *J. Chem. Soc.* (*C*), 852 (1966).
4. G. W. H. Cheeseman and B. Tuck, *J. Chem. Soc.*, 3678 (1965).
5. J. W. Lown and K. Matsumoto, *Can. J. Chem.*, **49,** 1165 (1971).

6. R. C. Fort, Jr., G. W. H. Cheeseman, and E. C. Taylor, *J. Org. Chem.*, **29,** 2440 (1964).
7. I. Kumashiro, *Nippon Kagaku Zasshi,* **82,** 1224 (1961); *Chem. Abstr.*, **58,** 4048 (1963).
8. R. F. Cookson and G. W. H. Cheeseman, *J. Chem. Soc., Perkin II,* 392 (1972).
9. G. W. H. Cheeseman and M. Rafiq, *J. Chem. Soc.* (*C*), 2732 (1971).
10. K. Nagarajan, V. R. Rao, and A. Venkateswarlu, *Indian J. Chem.*, **10,** 344 (1972).
11. G. W. H. Cheeseman and B. Tuck, *J. Chem. Soc.* (*C*), 1164 (1967).
12. E. E. Garcia, J. G. Riley, and R. I. Fryer, *J. Org. Chem.*, **33,** 1359 (1968).
13. K. L. Kirk and L. A. Cohen, *J. Org. Chem.*, **34,** 395 (1969).
14. R. W. Holley and A. D. Holley, *J. Am. Chem. Soc.*, **74,** 5445 (1952).
15. P. de la Llosa, M. Jutisz, and E. Scoffone, *Bull. Soc. Chim. Fr.*, 1621 (1960); *Chem. Abstr.*, **55,** 6393 (1961).
16. R. K. Grantham and O. Meth-Cohn, *Chem. Commun.*, 500 (1968).
17. A. M. Likhoshertov, L. S. Nazarova, and A. P. Skoldinov, Fr. Pat. 1,510,781; *Chem. Abstr.*, **70,** 68427 (1969); Scientific Research Institute of Pharmacology and Chemotherapy, Moscow, Brit. Pat. 1,144,749; *Chem. Abstr.*, **70,** 106558 (1969).
18. M. Artico, G. De Martino, and V. Nacci, *Ann. Chim.* (Rome), **57,** 1431 (1967); *Chem. Abstr.*, **68,** 105157 (1969).
19. M. Artico, V. Nacci, and G. De Martino, *Ann. Chim.* (Rome), **58,** 136 (1968); *Chem. Abstr.*, **69,** 36072 (1968).
20. I. Kumashiro, *Nippon Kagaku Zasshi,* **82,** 934 (1961); *Chem. Abstr.*, **57,** 12489 (1962).
21. G. W. H. Cheeseman and P. D. Roy, *J. Chem. Soc.* (*C*), 856 (1969)
22. E. C. Taylor and E. S. Hand, *J. Am. Chem. Soc.*, **85,** 771 (1963).
23. E. C. Taylor and E. S. Hand, *Tetrahedron Lett.*, 1225 (1962).
24. E. C. Taylor and G. W. H. Cheeseman, *J. Am. Chem. Soc.*, **86,** 1830 (1964).
25. I. Kumashiro, *Nippon Kagaku Zasshi,* **82,** 1068 (1961); *Chem. Abstr.*, **59,** 621 (1963).
26. G. W. H. Cheeseman and P. D. Roy, *J. Chem. Soc.* (*C*), 2848 (1968).
27. E. C. Taylor and A. McKillop, *J. Am. Chem. Soc.*, **87,** 1984 (1965).
28. N. J. Leonard and J. H. Boyer, *J. Am. Chem. Soc.*, **72,** 2980 (1950).
29. K. Okumura and K. Shigemitsu (to Tanabe Seiyaku Co., Ltd.), Jap. Pat. 67 20,069; *Chem. Abstr.*, **69,** 19217 (1968).
30. E. C. Taylor, A. McKillop, and R. E. Ross, *J. Am. Chem. Soc.*, **87,** 1990 (1965).
31. J. G. Smith and E. M. Levi, *J. Organometal. Chem.*, **36,** 215 (1972); *Chem. Abstr.*, **76,** 140723 (1972).
32. W. Kieland and F. Kröhnke, *Chem. Ber.*, **105,** 3709 (1972).
33. M. Ungureanu, I. Druta, M. Petrovanu, and I. Zugravescu, *An. Stiint. Univ. "Al. I. Cuza" Iasi, Sect. Ic*, **18,** 49 (1972); *Chem. Abstr.*, **78,** 29712 (1973).
34. H. W. Heine and R. P. Henzel, *J. Org. Chem.*, **34,** 171 (1969).
35. J. W. Lown and K. Matsumoto, *Can. J. Chem.*, **49,** 3119 (1971).
36. F. Eiden and G. Bachmann, *Arch. Pharm.* (Weinheim), **306,** 876 (1973).
37. P. J. Abbott, R. M. Acheson, M. W. Foxton, N. R. Raulius, and G. E. Robinson, *J. Chem. Soc., Perkin I,* 2182 (1972).
38. I. Kumashiro, *Nippon Kagaku Zasshi,* **82,** 1072 (1961); *Chem. Abstr.*, **59,** 607 (1963).
39. V. I. Shvedov, L. B. Altukhova, and A. N. Grinev, *Khim. Geterotsikl. Soedin.*, 1048 (1970); *Chem. Abstr.*, **74,** 125628 (1971).
40. E. A. Zeller, B. Gaertner, and P. Hemmerich, *Z. Naturforsch. B*, **27,** 1050 (1972); *Chem. Abstr.*, **78,** 29714 (1973).
41. S. Ruhemann, *J. Chem. Soc.*, **97,** 1438 (1910).

42. R. J. Shah and J. R. Merchant, *Curr. Sci.* (India), **27,** 441 (1958); *Chem. Abstr.*, **53,** 13134 (1959).
43. J. R. Merchant, R. J. Shah, and R. M. Bhandarkar, *Rec. Trav. Chim.*, **81,** 131 (1962); *Chem. Abstr.*, **57,** 4626 (1962).
44. W. R. Vaughan and R. C. Tripp, *J. Am. Chem. Soc.*, **82,** 4370 (1960).
45. J. R. Merchant and V. Srinivasan, *Rec. Trav. Chim.*, **81,** 144 (1962); *Chem. Abstr.*, **57,** 4627 (1962).
46. J. R. Merchant and R. M. Bhandarkar, *J. Indian Chem. Soc.*, **40,** 353 (1963); *Chem. Abstr.*, **59,** 9952 (1963).
47. M. D. Nair and P. A. Malik, *Indian J. Chem.*, **5,** 603 (1967); *Chem. Abstr.*, **69,** 59022 (1968).
48. R. H. Wiley and S. C. Slaymaker, *J. Am. Chem. Soc.*, **80,** 1385 (1958).
49. P. Dimroth and K. Dury, Ger. Pat. 1,196,205; *Chem. Abstr.*, **63,** 14884 (1965).
50. P. Dimroth and F. Reicheneder, *Angew. Chem., Int. Ed. Engl.*, **8,** 751 (1969).
51. P. Dimroth and K. Dury, Fr. Pat. 1,410,733; *Chem. Abstr.*, **65,** 9065 (1966).
52. H. Otomasu, S. Omiya, T. Sekiguchi, and H. Takahashi, *Chem. Pharm. Bull.* (Tokyo), **18,** 2065 (1970); *Chem. Abstr.*, **74,** 22799 (1971).
53. J. Klicnar, M. Hajek, J. Hoffmann, and M. Vecera, *Collect. Czech. Chem. Commun.*, **36,** 262 (1971); *Chem. Abstr.*, **74,** 76393 (1971).
54. Q. Ahmed, T. Wagner-Jauregg, E. Pretsch, and J. Seibl, *Helv. Chim. Acta*, **56,** 1646 (1973).
55. D. D. Chapman, *J. Org. Chem.*, **37,** 2498 (1972).
56. J. D. Mee, D. W. Heseltine, and W. S. Gaugh, Ger. Offen. 1,907,618; *Chem. Abstr.*, **73,** 30660 (1970).
57. O. Diels, O. Buddenberg, and S. Wang, *Ann.*, **451,** 223 (1927).
58. C. Sandris and G. Ourisson, *Bull. Soc. Chim. Fr.*, 345 (1958).
59. F. D. Chattaway and W. D. Humphrey, *J. Chem. Soc.*, 645 (1929).
60. B. Crippa and A. Aguzzi, *Gazz. Chim. Ital.*, **67,** 352 (1937); *Chem. Abstr.*, **32,** 1265 (1938).
61. A. Cesari, *Ann. Ist. Super. Sanità*, **1,** 555 (1965); *Chem. Abstr.*, **65,** 10588 (1966).
62. S. I. Burmistrov, N. E. Kul'chitskaya, and V. D. Romanenko, *Zh. Org. Khim.*, **8,** 1095 (1972); *Chem. Abstr.*, **77,** 61947 (1972).
63. Pfizer Inc., Brit. Pat. 1,303,372; *Chem. Abstr.*, **78,** 124629 (1973).
64. W. Ried and J. Grabosch, *Chem. Ber.*, **91,** 2485 (1958).
65. J. K. Landquist and J. A. Silk, *J. Chem. Soc.*, 2052 (1956).
66. W. E. Hahn and J. Z. Lesiak, *Soc. Sci. Lodz., Acta Chim.*, **17,** 201 (1972); *Chem. Abstr.*, **78,** 58357 (1973).
67. R. C. Anderson and R. H. Fleming, *Tetrahedron Lett.*, 1581 (1969).
68. W. Schulze, *J. Prakt. Chem.*, **19,** 212 (1963); *Chem. Abstr.*, **59,** 3925 (1963).
69. G. F. Bettinetti and E. Tisselli, *Ann. Chim.* (Rome), **51,** 1102 (1961); *Chem. Abstr.*, **57,** 11194 (1962).
70. A. Richardson, Jr., *J. Org. Chem.*, **30,** 2589 (1965).
71. J. H. Sellstedt and M. Wolf, (to American Home Products), U.S. Pat. 3,813,392; U.S. Pat. 3,891,643; U.S. Pat. 3,892,746; *Chem. Abstr.*, **81,** 49700 (1974); **83,** 147502, 147503 (1975).
72. M. L. Heffernan and G. M. Irvine, *Austr. J. Chem.*, **29,** 837 (1976).
73. O. Meth-Cohn, *Tetrahedron Lett.*, 413 (1975).
74. S. Raines, S. Y. Chai, and F. P. Palopoli, *J. Heterocycl. Chem.*, **13,** 711 (1976).
75. G. De Martino, S. Massa, and M. Scalzo, *Farmaco, Ed. Sci.*, **30,** 581 (1975); *Chem. Abstr.*, **83,** 131549 (1975).

76. H. Dornauer and V. B. Anderson (to American Hoechst Corp.), U.S. Pat. 3,939,159; *Chem. Abstr.*, **84,** 180293 (1976).
77. G. Kollenz, *Justus Liebigs Ann. Chem.*, **762,** 13 (1972); *Chem. Abstr.*, **78,** 29719 (1973).
78. U. Eholzer, H. Heitzer, F. Seng, and K. Ley, *Synthesis*, 296 (1974).
79. W. Hahn and J. Lesiak, Pol. Pat. 71,049; *Chem. Abstr.*, **83,** 28279 (1975); *ibid.*, **85,** 177483 (1976).
80. M. J. Haddadin, N. C. Chelhot, and M. Pieridou, *J. Org. Chem.*, **39,** 3278 (1974).
81. H. Suschitzky, B. J. Wakefield, and R. A. Whittaker, *J. Chem. Soc.*, *Perkin I*, 401 (1975).

CHAPTER XXXVI

Imidazoquinoxalines

I. Imidazo[1,2-*a*]quinoxalines . . . 654
II. Imidazo[1,5-*a*]quinoxalines . . . 658
III. Imidazo[4,5-*b*]quinoxalines . . . 663
IV. Imidazo[1,5,4-*de*]quinoxalines . . . 672
V. Tables . . . 674
Table 1. Imidazo[1,2-*a*]quinoxalines . . . 674
Table 2. Imidazo[1,5-*a*]quinoxalines . . . 675
Table 3. Cyanine Dyes based on Imidazo[4,5-*b*]quinoxalines . . . 677
Table 4. Imidazo[4,5-*b*]quinoxalines (Not Cyanine Dyes) . . . 684
Table 5. Imidazo[1,5,4-*de*]quinoxalines . . . 687
VI. References . . . 689

I. Imidazo[1,2-*a*]quinoxalines

1 2 9 N N3 7 N 4 6 5

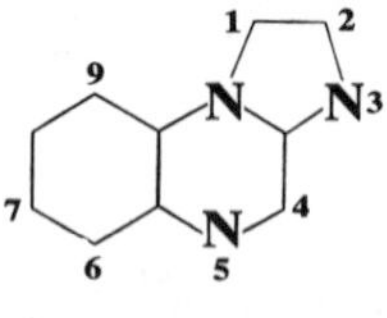

LCAO calculations have been performed on this heterocycle, and the resulting electron density map suggests that the 4-position is the most susceptible to nucleophilic substitution.[1] The ultraviolet spectrum in methanol[2] of imidazo[1,2-*a*]quinoxaline has a maximum at 315 nm and a minimum at 266 nm. The compound is a moderately weak base with a pK_a (50% ethanol) of 4.6. No evidence for the site of protonation has been published, although the 3-position seems favored both from the theoretical calculations and by comparison with imidazo[1,2-*a*]pyrazines (see Chapter XXI).

The aromatic ring system has been prepared both from quinoxaline intermediates and from 1-phenylimidazoles. Reaction of the 2-chloroquinoxalines **1** with aminoacetaldehyde dimethylacetal followed by acid treatment of the resultant aminoacetals gave imidazo[1,2-*a*]quinoxaline (**2**) and its 4-phenyl derivative (**3**). Compound **3** was the major product from the reaction of the quinoxalinone **4** with ethylenediamine monotosylate (EDMT) at 200°. The reaction also gave a small amount of the somewhat more strongly basic dihydro compound **5**. The products were readily separated by partitioning between chloroform and aqueous buffer (pH 6). Apparently the aromatic product is formed by oxidation of the dihydro derivative during the reaction. When the reaction of EDMT at 200° was attempted with quinoxalone itself, the only products isolated were benzimidazole and the pyrazinoquinoxaline **6**.[3]

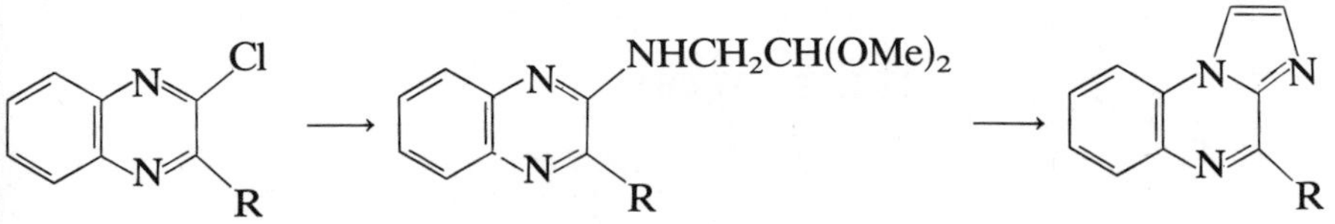

1 R = H or Ph

2 R = H
3 R = Ph

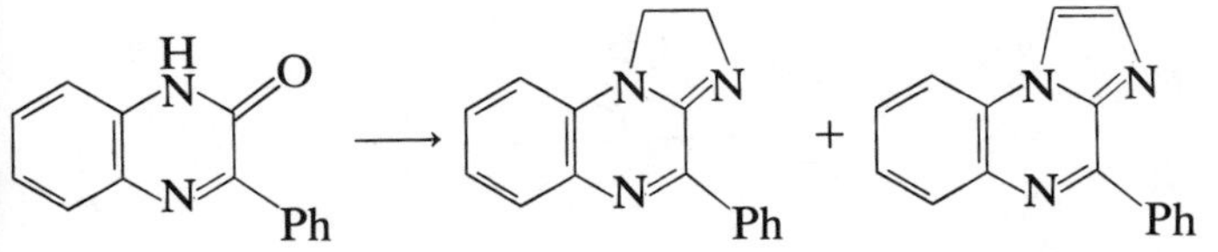

4 **5** **3**

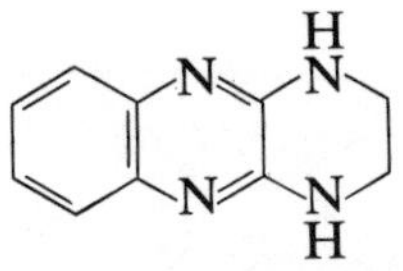

6

Reduction of the nitrophenylimidazole **7** with sodium dithionite in 5% aqueous ammonia solution gave the parent heterocycle **2**.[2] The yield is improved by using 22% ammonia solution, but a small amount of the 4-amino compound **8** is also formed under these conditions.[1] The reaction has been extended to provide several 7-substituted derivatives of imidazo[1,2-*a*]quinoxaline.[5] 2-Methyl-1-(*o*-nitrophenyl)imidazole gives **2** in low yield when heated with ferrous oxalate.[57]

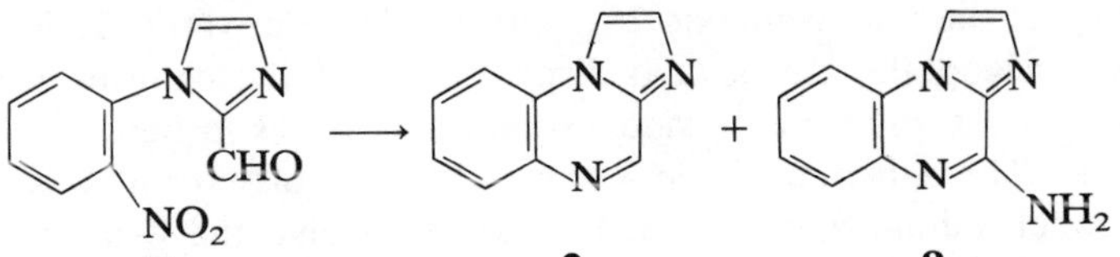

7 **2** **8**

1,2-Dihydro derivatives of the ring system have been prepared by several of the standard routes to such systems. Treatment of the hydroxyethylamino compounds **9** with phosphoryl chloride and then base provides the cyclized compounds **10**.[6,7] Similarly reaction of the bromoethyl derivative **11** with sodium methoxide gives the dihydroimidazo compound **12**, the chlorine in the quinoxaline ring being displaced under the reaction conditions.[8] The aziridinylquinoxalines **13** are rearranged to the dihydroimidazo derivatives **14** merely by warming in acetone containing sodium iodide.[6,8]

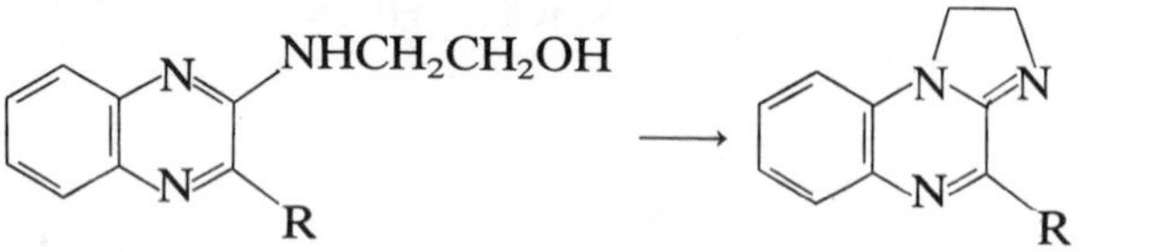

$NHCH_2CH_2OH$ R → R

9 **10** R = H, Me, or Ph

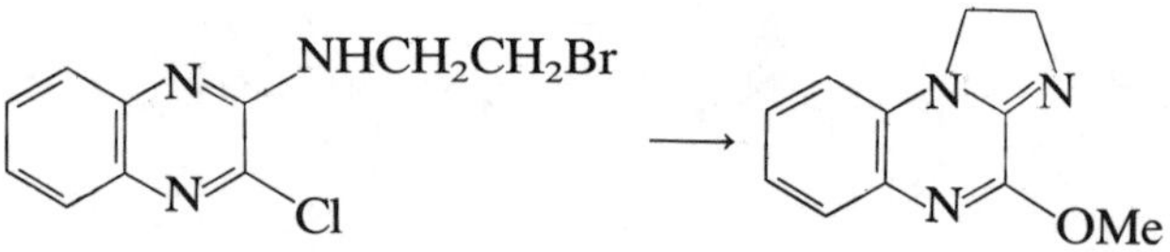

$NHCH_2CH_2Br$ Cl → OMe

11 **12**

R → R

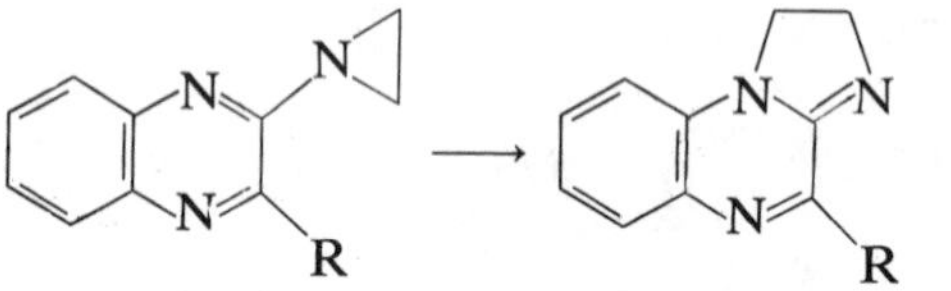

13 **14** R = Cl, Me, or OMe

Two other syntheses of a less general nature have also been reported. Reaction of phenacylquinoxalinium bromide (**15**) with phenyl isocyanate in the presence of triethylamine gave the tetrahydro compound **16**.[9] Condensation of 2-aminoquinoxaline with the perfluoro compound **17** gave the bisimino derivative **18**.[10]

Treatment of the dihydroimidazo[1,2-*a*]quinoxaline **19** with sodium hydroxide solution results in opening of the five-membered ring and formation of the quinoxalinone **20**.[6] The 4-chloro derivative **21** is susceptible to nucleophilic attack, and displacement of the chlorine has been achieved with sodium methoxide,[8] sodium alkoxides in benzene,[6,11] and hydrazine. The 4-hydrazino derivative was not isolated but was treated with copper sulfate and sodium hydroxide to give the 4-unsubstituted compound.[6]

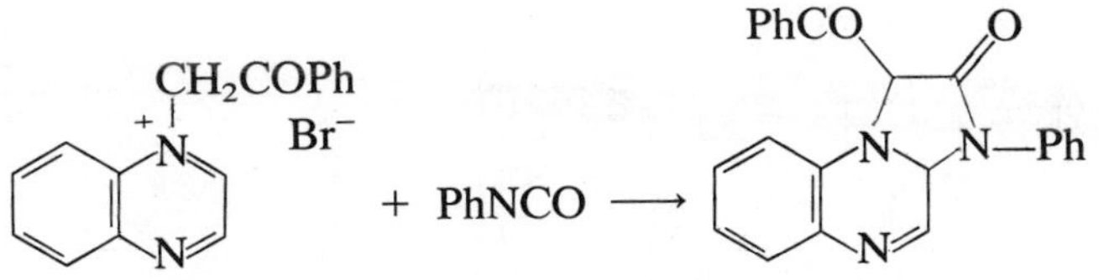

The parent aromatic heterocycle reacts with sodamide in dimethylaniline at 180° to give the 4-amino compound **8** in 44% yield. When the reaction is performed in xylene at 150°, the yield of 4-amino product drops to 11% and a 46% yield of bis compound **22** is obtained.[1]

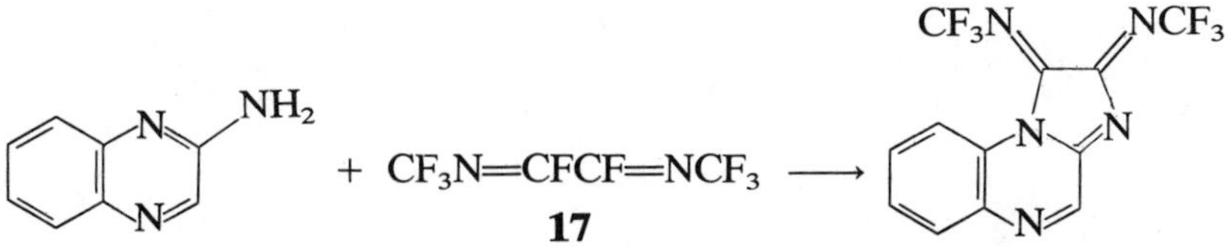

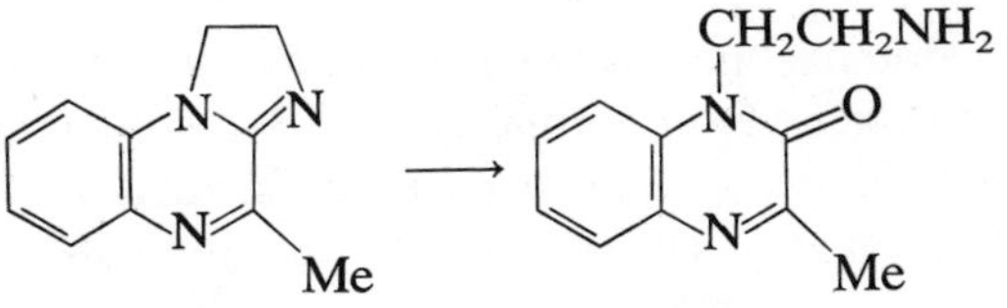

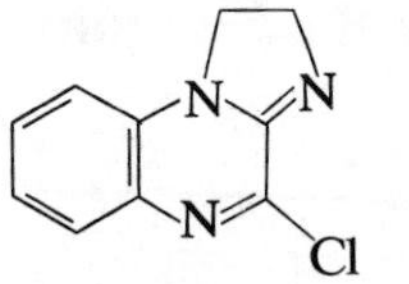

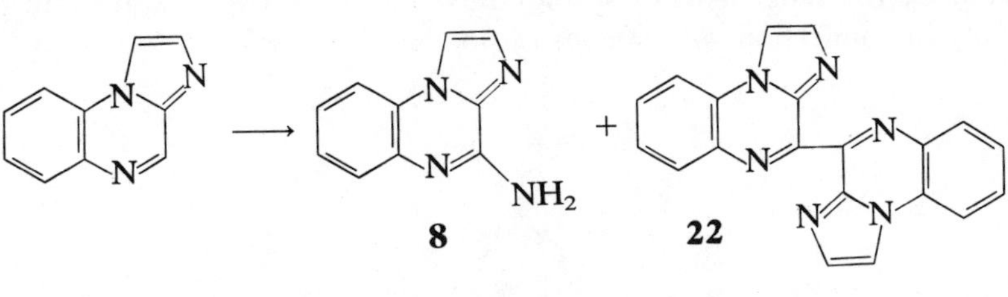

Reduction of imidazo[1,2-*a*]quinoxaline with phosphorus and hydrogen iodide gives the 4,5-dihydro derivative **23**. Under these conditions the 7-methoxy compound **24** is demethylated as well as reduced to give the hydroxy derivative **25**.[5]

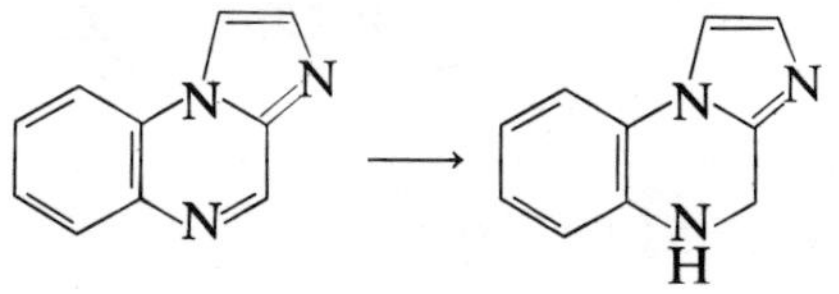

23

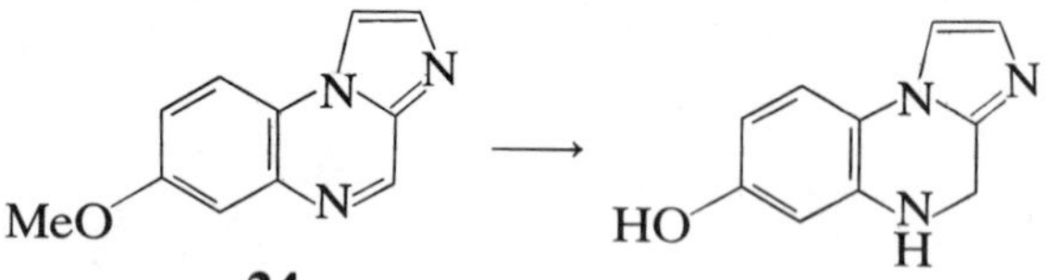

MeO 24 HO 25

Imidazo[1,2-*a*]quinoxaline has been shown by two independent groups to have hypotensive activity in normal rats.[4,12] 1,2-Dihydro derivatives of the ring system have been claimed as antiinflammatory agents[7,11] and autonomic nerve blockers.[7] The polyfluoro compound **18** is patented as a fungicide.[10]

II. Imidazo[1,5-*a*]quinoxalines

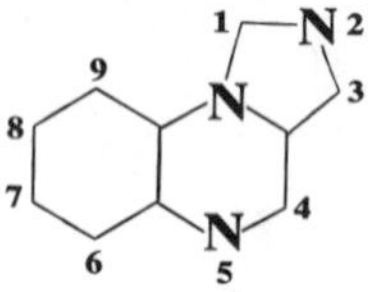

RRI 2746

This ring system has also been called glyoxalino(1′:5′-1:2)quinoxaline and numbered as shown in formula **26**. No systematic study of this heterocycle has appeared although several compounds have been prepared by rearrangement of spiroquinoxaline derivatives[13] and more recently in connection with studies on models for tetrahydrofolic acid.

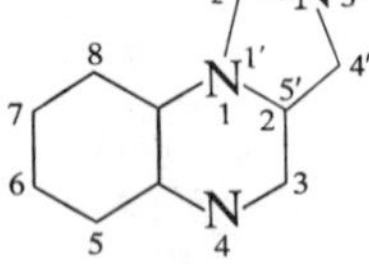

26

Spectral measurements have been made only to confirm structural assignments. Interestingly the protons in the 1-position of the hexahydro compound **27** appear in the PMR spectrum as an AB quartet with a geminal coupling constant of 4 Hz. This rather small J value seems characteristic of five-membered rings of similar structure.[14] The hexahydro dioxo compounds **28** and **29** have had their PMR spectra measured, and parameters for the aliphatic protons have been refined by computation.[15] Compound **28** showed an ABX pattern from the 4-methylene protons (AB) and the 3a-hydrogen (X) (δ_A 3.77, δ_B 3.57, δ_X 4.37; J_{AB} 11.8, J_{AX} 2.0, J_{BX} 11.5 Hz). The methyl derivative **29** gave a spectrum which was analyzed as an ABX_3 spin system and which showed that the compound possessed a 3a,4-trans configuration (J_{AB} 9.0 Hz).

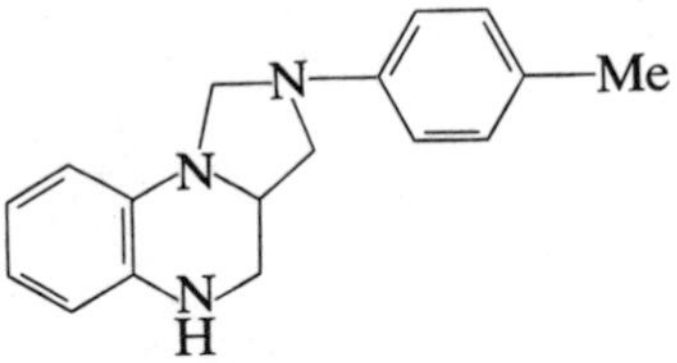

27

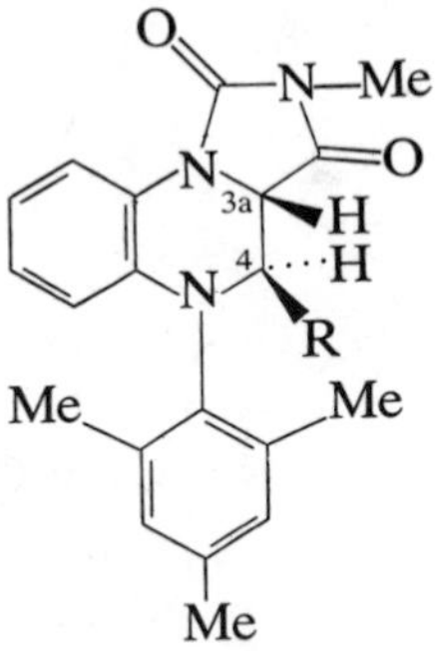

28 R = H
29 R = Me

The mass spectrum of compound **30** favored the structure shown rather than the isomeric pyrazoloquinoxaline **31**.[16] The observed loss of PhCN could occur from either structure, but the concomitant loss of MeCN from the molecular ion is unlikely from isomer **31**. Moreover the pyrazolo derivative would also be expected to lose PhC≡CMe, but no such process was apparent in the fragmentation pattern.

Compound **30** was prepared by the Beckmann rearrangement in polyphosphoric acid of the diphenylhydrazone **32** (R = Me). The reaction also worked for the benzyl analogue **32** (R = CH_2Ph), but the 3-unsubstituted analogue could not be prepared from **32** (R = H), as this compound on acid treatment hydrolyzed back to the ketone **33**. Phosphoryl chloride converted **30** into the corresponding 4-chloroimidazo[1,5-*a*]quinoxaline.

The hydantoin **34** on treatment with ethyl formate in the presence of sodium gave the tricyclic compound **35**, which could be methylated with diazomethane to the 2-methyl derivative **37a**. This compound could also be obtained from the hydantoin **36** using ethyl formate. Ethyl acetate gave the corresponding 4-methyl derivative **37b**. No reaction occurred during attempts to prepare **37** (R = Ph) using ethyl benzoate.[17]

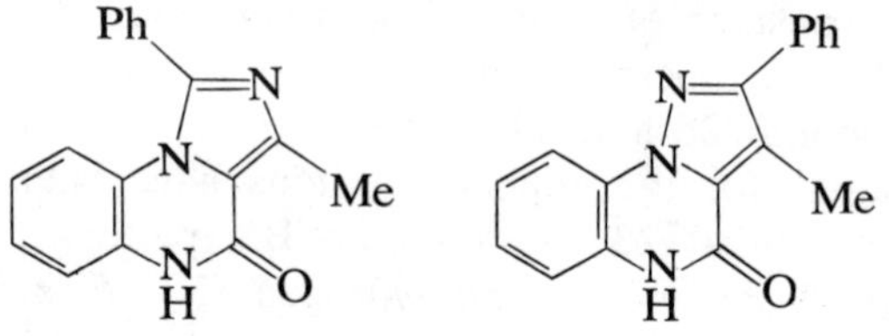

30 **31**

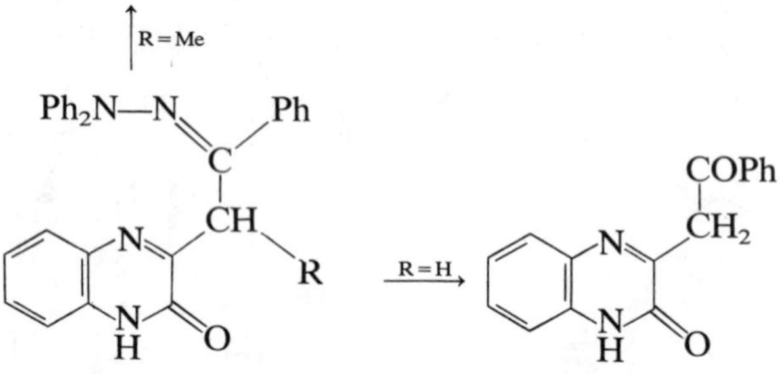

32 R = Me or CH_2Ph **33**

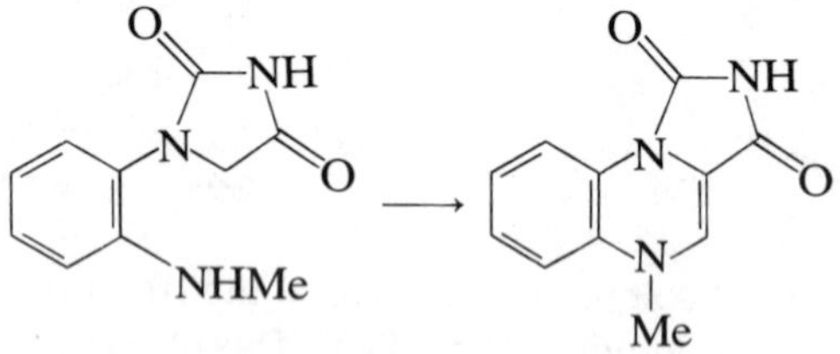

34 **35**

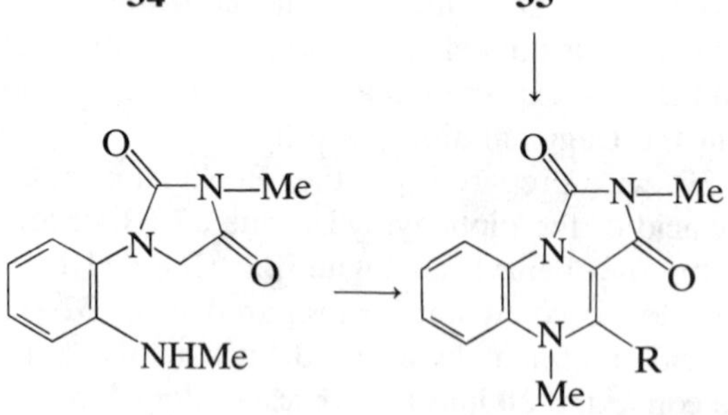

36 **37a** R = H

37b R = Me

Reduction of compounds **35** and **37a** with tin and hydrochloric acid gave the products **38**.[17] Oxidation of **38** (R = Me) with oxygen in refluxing acetic acid gave back the unsaturated compound **37a**.[18] The synthesis of compounds **35**, **37**, and **38** permitted the reformulation of the product from the condensation of alloxan **39** with *o*-dimethylaminoaniline and its alkaline hydrolysis product.[17] It was shown that the alloxan condensation product was the spiroquinoxaline **40**, not the originally suggested anil **41**, and this gave the imidazo[1,5-*a*]quinoxaline **38** (R = H) rather than the previously claimed compound **42** on boiling in 30% aqueous sodium hydroxide solution.

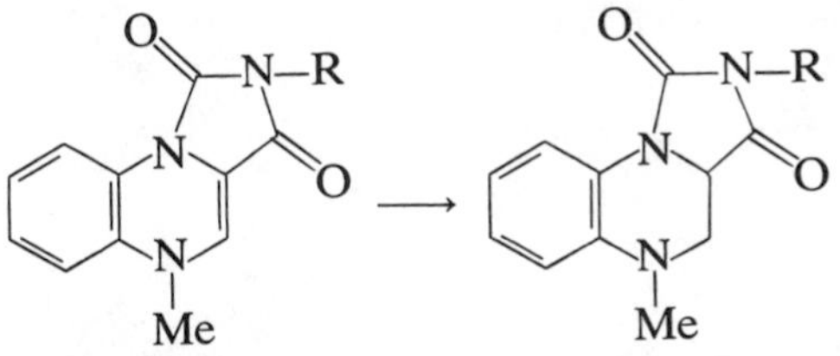

35 R = H
37a R = Me

38 R = H or Me

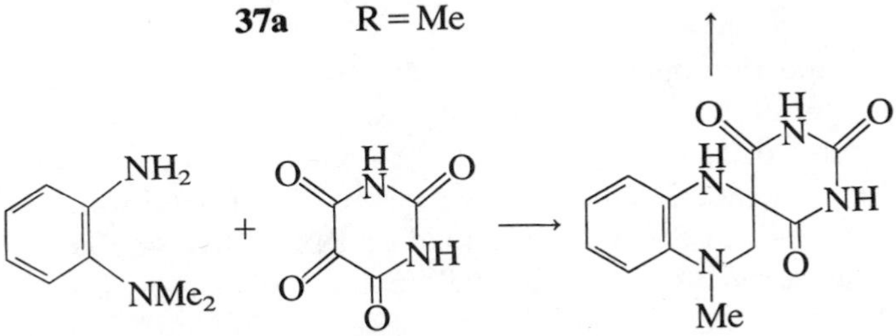

39

40

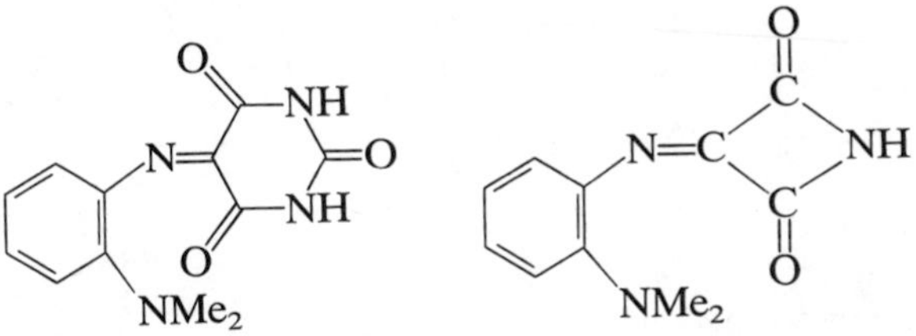

41

42

The spiroquinoxalines **43** rearranged in ethanolic potassium hydroxide to give the imidazoquinoxalines **28** and **29** in high yield.[15] The PMR spectra of the products were discussed earlier in this chapter.

Reaction of the tetrahydroquinoxalines **44** with triethyl orthoformate gave the bright-yellow formamidinium salts **45**. Reduction of **45** (R = CO_2Et) with sodium borohydride in THF gave the colorless derivative **46**.[19] The latter compounds were also prepared from **44** by reaction with formaldehyde.[14] The use of 5-formyluracil gave the substituted compound **47**.[20] These compounds were of interest as model systems for N^5,N^{10}-methylenetetrahydrofolic acid (**48**). During these investigations the hydrolysis of the formamidinium salts **45** as a function of pH was

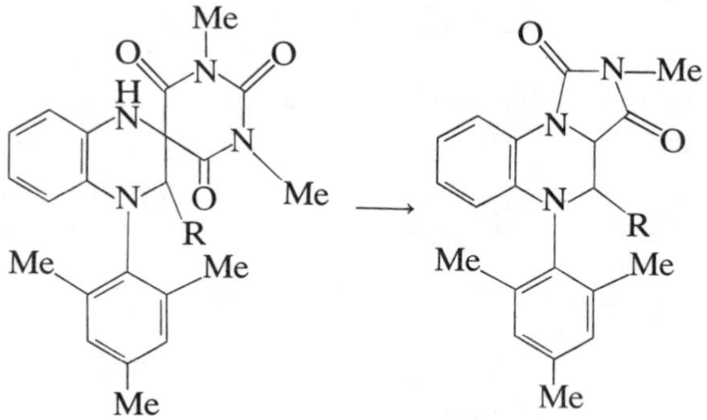

43

28 $R = H$

29 $R = Me$

examined,[19] the ring-opened products **49** being formed in differing ratios at different pH values. The kinetics and mechanism of the formation of compounds **46** from the quinoxaline **44** and formaldehyde were also examined in some detail. This work, because of the structural similarity of compounds **45–47** to the tetrahydrofolic acid derivative **48**, has resulted in some clarification of the essential features of tetrahydrofolic acid biochemistry.[14,19] Compound **47** was assayed for inhibitory activity against thymidylate synthetase and dihydrofolate reductase and was found to have slight but not marked activity.[20]

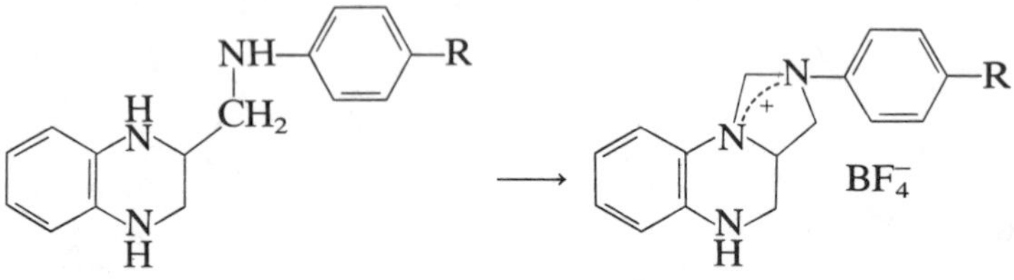

44

45 $R = Me$ or CO_2Et

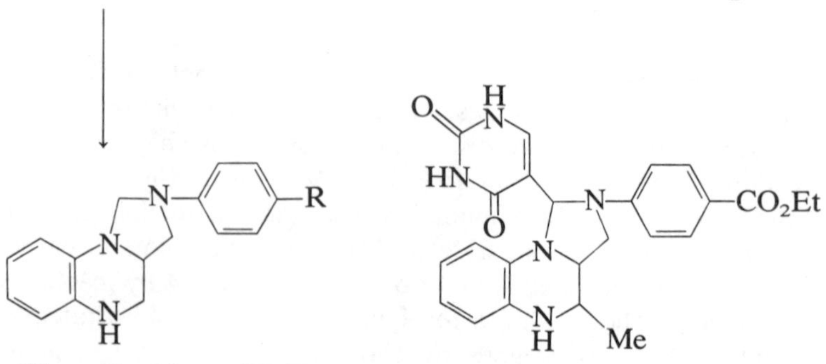

46 $R = Me$ or CO_2Et

47

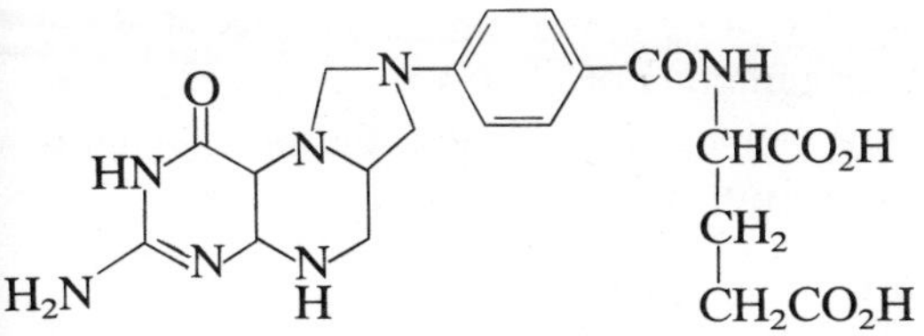

48

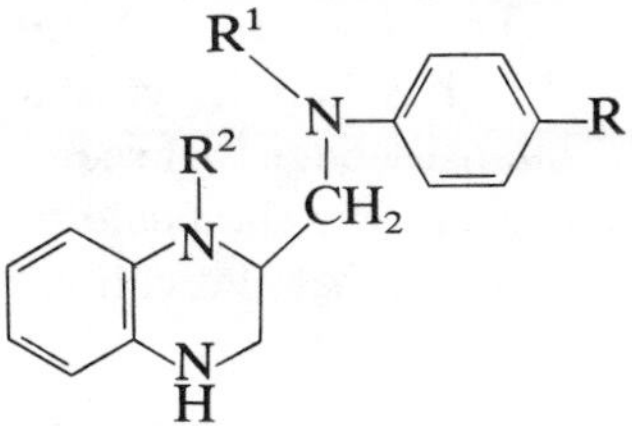

49 $R^1 = H, R^2 = CHO$
$R^1 = CHO, R^2 = H$

III. Imidazo[4,5-*b*]quinoxalines

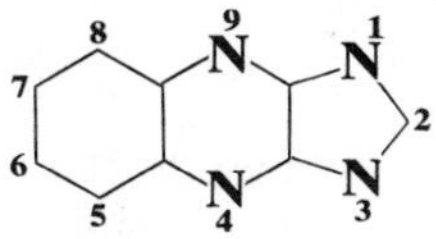

RRI 2736

This nucleus has also been referred to as 4,5-quinoxalinoimidazole and imidazolo-5′,4′:2,3-quinoxaline. Most of the reported derivatives of this ring system are cyanine dyestuffs and may be named accordingly. For example, *Chemical Abstracts* indexes compound **50** as an imidazo[4,5-*b*]quinoxalinocarbocyanine, as well as by using more systematic nomenclature. Similarly **51** may also be called an imidazo[4,5-*b*]quinoxalino-indocarbocyanine, and compound **52** is referred to as an imidazo[4,5-*b*]quinoxalinothiacarbocyanine.

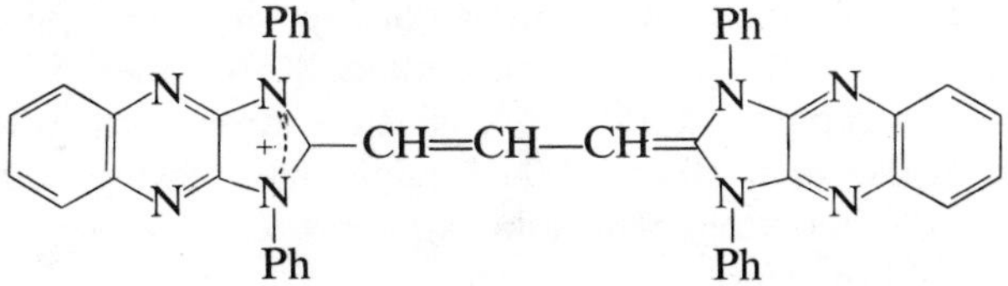

50

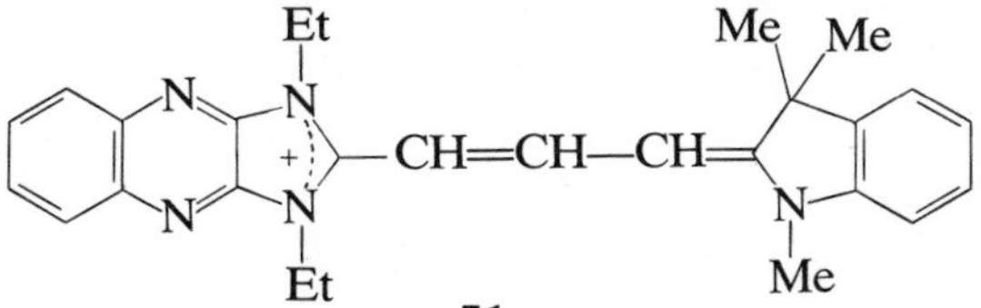

51

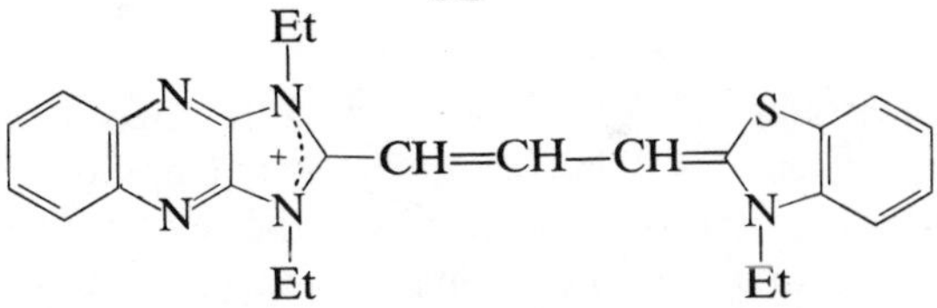

52

The dye 1,1′,3,3′-tetraethylimidazo[4,5-*b*]quinoxalinocyanine iodide (**53**) has had its crystal structure determined from single-crystal x-ray intensity data.[21] It seems that steric interactions between the ethyl groups prevent the molecule from being planar. Nevertheless the bond lengths indicate that the cation is highly conjugated.

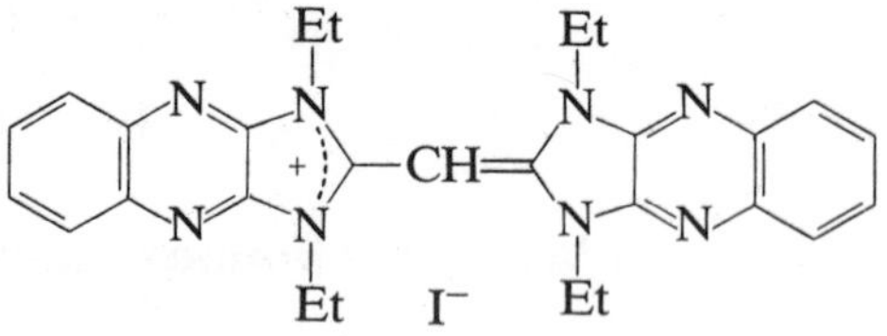

53

The ultraviolet spectra of simple imidazo[4,5-*b*]quinoxalines have been described as resembling those of quinoxalines. Apparently fusion of the imidazole ring onto the quinoxaline nucleus results in only a small contribution to the ultraviolet spectrum.[22] The parent 1*H*-imidazo[4,5-*b*]quinoxaline (**54**) is amphoteric, forming stable sodium salts, hydrochlorides, and picrates. It is a moderately weak acid with a pK_a (50% ethanol) of 9.5. Substitution by alkyl groups in the 2-position produces only a small acid-weakening effect,[22] but the trifluoromethyl derivative **55** is a much stronger acid,[23] with a pK_a (50% ethanol) of 4.5. Chlorine substitution in the benzo ring of **55** produces an acid-strengthening effect which is small for substituents in the 6- or 7-position but quite marked in the 5-position. The pK_a (50% ethanol) of the 5,7-dichloro-2-trifluoromethyl compound **56** is 3.3, which is stronger than that of most carboxylic acids. In a study of NH-acidic π-excessive heteroaromatic compounds, the imidazo[4,5-*b*]quinoxaline **56** was shown to be one of the strongest inhibitors studied of oxygen evolution from chloroplast photosynthesis. This activity was correlated with pK_a values and partition coefficients.[23,24]

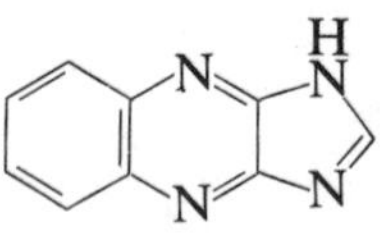

54 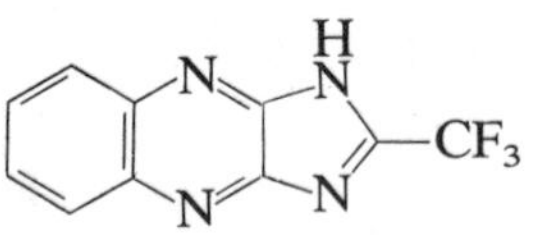 **55** 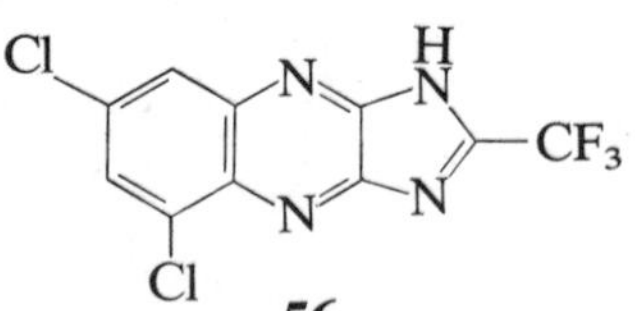 **56**

Imidazo[4,5-*b*]quinoxalines have invariably been prepared from quinoxaline intermediates. Almost twenty years elapsed between the first reported synthesis of the ring system[25] and the next publication containing an extensive study of the utility of reagents for converting 2,3-diaminoquinoxalines into 2-substituted imidazo[4,5-*b*]quinoxalines.[22]

Although the parent compound **54** may be prepared from 2,3-diaminoquinoxaline* (**57**) and formic acid, trimethyl orthoformate gives a much better yield of the unsubstituted heterocycle. Carboxylic acids other than formic acid could not be reacted with **57** to give 2-substituted compounds. 2-Methylimidazo[4,5-*b*]quinoxaline was obtained using triethyl orthoacetate, but better yields were obtained using acetyl chloride or acetic anhydride. Reaction of diamine **57** with acid chlorides or anhydrides in pyridine or xylene provides a general method for the synthesis of 2-alkyl derivatives of this ring system.[22] The trifluoromethyl compound **55** was obtained from 2,3-diaminoquinoxaline by reaction with trifluoroacetic acid and trifluoroacetic anhydride.[23] 2-Alkyl-substituted compounds have been prepared by heating 2,3-dichloroquinoxalines with alkyl cyanides and ammonia in the presence of cuprous chloride.[58]

2-Aryl substituents are best introduced by reaction of **57** with aromatic aldehydes in pyridine. Heating the two reactants without solvent provided 1,2-diaryl-substituted compounds.[25] In general, reaction of the diamine **57** with aroyl chlorides gave none of the required 2-aryl derivatives. However a small yield of the furyl compound **58** was obtained by this approach. Most aliphatic aldehydes possess active α hydrogens and polymerize rather than react with **57**, as the reactions are usually carried out in the presence of sodium hydroxide. Reaction of 2,3-diaminoquinoxaline with phosgene gives the 2-hydroxy compound, which is more correctly formulated as the imidazolone **59**. Better yields of this compound were obtained by reaction of **57** with urea although, in contrast to the corresponding syntheses of benzimidazoles, thiourea did not afford the corresponding thioxo derivative.[22] Compound **59** has also been prepared by Curtius degradation of the acid azide **60**.[26]

Other 1*H*-2,3-dihydro 2-oxo compounds have been prepared by more specialized routes. Reaction of quinoxaline *N*-oxide (**61**) with phenyl isocyanate gives a mixture of anilinoquinoxaline **62** and the (presumably) derived diphenyl oxo compound **63**.[27] Recently it has been shown that the alloxazine derivative (**64**) on treatment with di-*n*-butylamine undergoes a Lossen rearrangement to give the 2-oxo compound **65**.[28] Alkali treatment of compound **65** gives the 1,3-unsubstituted product **59**.

4-Substituted derivatives of imidazo[4,5-*b*]quinoxaline have also been obtained. The 1-methyl-2-iminoquinoxaline **66** reacts with triethyl orthoformate to give 4-methylimidazo[4,5-*b*]quinoxaline (**67**), and the 4-methyl-2-oxo derivative **68** is produced by reaction of **66** with urea.[22]

The bis(secondary amine) **69** on reaction with acetyl chloride in dimethylacetamide affords the imidazo[4,5-*b*]quinoxalinium salt **70**,[29,30]

* The chemistry of 2,3-diaminoquinoxaline is discussed in Chapter XI.

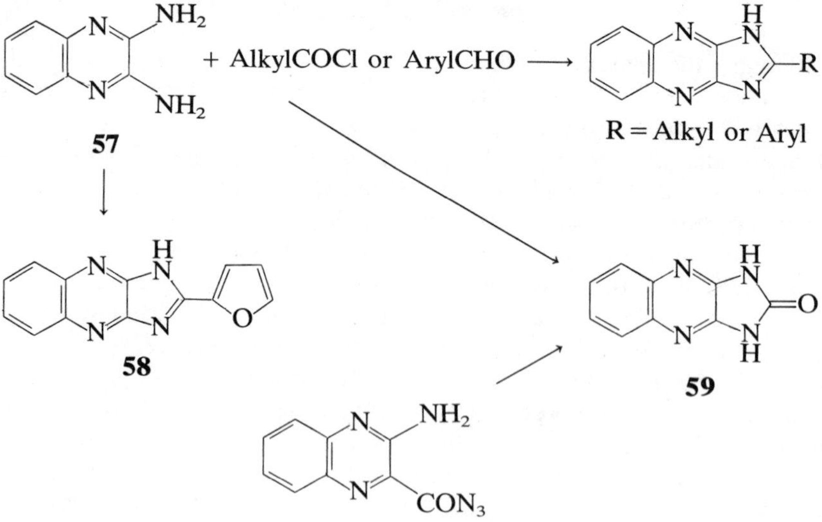

+ AlkylCOCl or ArylCHO ⟶

R = Alkyl or Aryl

57

58

59

60

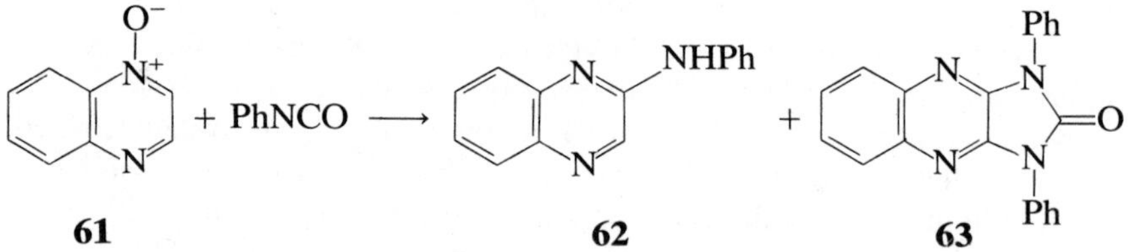

+ PhNCO ⟶

+

61

62

63

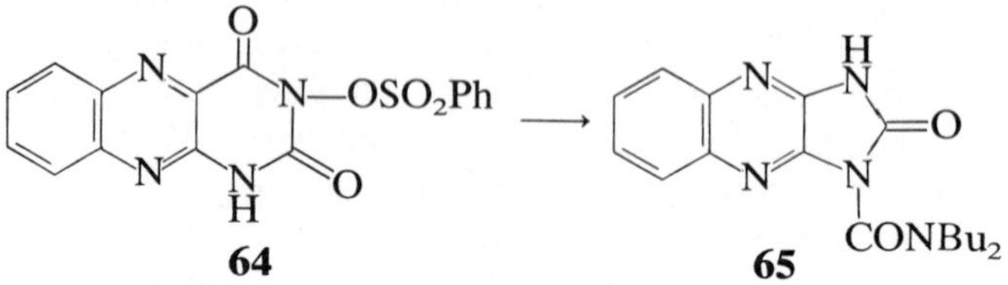

64

65

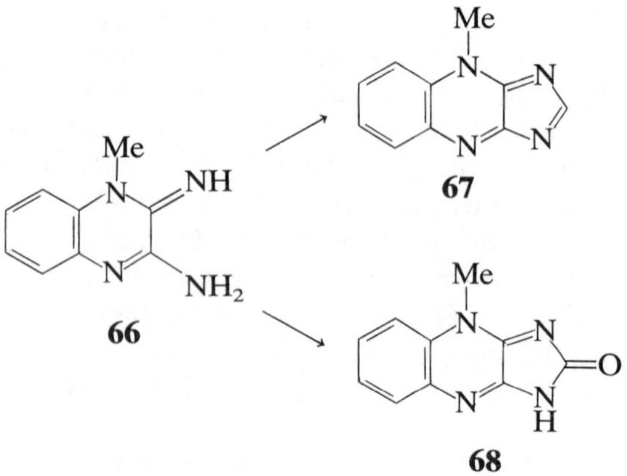

66

67

68

which is a key intermediate in the synthesis of cyanine dyes derived from this ring system (see below). The reaction has been extended to provide a variety of 2-methylimidazo[4,5-*b*]quinoxalinium salts.[31]

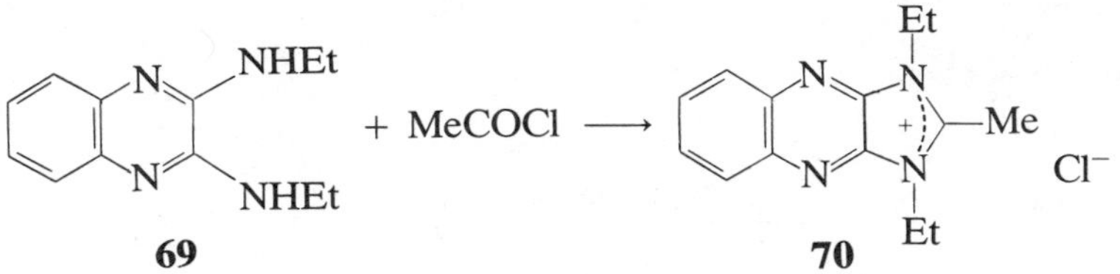

69 + MeCOCl ⟶ 70

Acetylation of 2-methylimidazo[4,5-*b*]quinoxaline occurs at the 1-position to give the product **71**, which may also be prepared directly from 2,3-diaminoquinoxaline and an excess of acetic anhydride.[22] Oxidation of compound **71** with hydrogen peroxide in acetic acid at room temperature gives the deacetylated mono-*N*-oxide **72**. The use of peracetic acid at 65° gives the 4,9-dioxide **73**. 2-Styrylimidazo[4,5-*b*]quinoxaline gives similar products.[32] Both the 2-styryl compound[22,32] and its 4-oxide[32] on treatment with potassium permanganate at room temperature give the corresponding 2-unsubstituted imidazo[4,5-*b*]quinoxalines. The heterocyclic ring is not affected by this treatment, although the first-formed 2-carboxylic acid must spontaneously decarboxylate under the reaction conditions. Selenium dioxide has also been used to convert 2-methyl derivatives into the corresponding 2-unsubstituted compounds.[32]

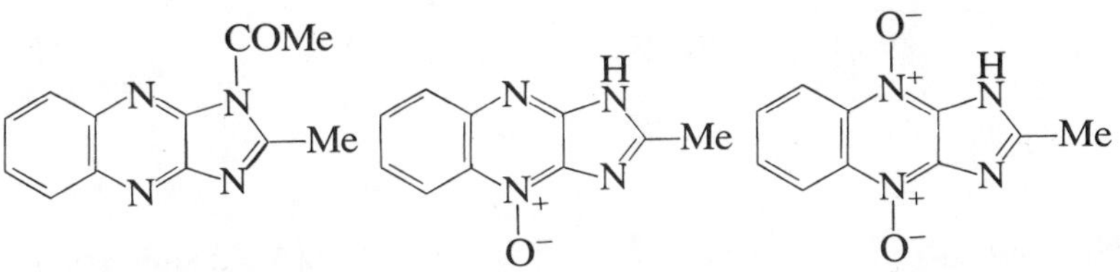

71 **72** **73**

Although the 1,3-diphenyl 2-oxo compound **63** is stable to acid hydrolysis,[27] imidazo[4,5-*b*]quinoxaline itself undergoes very ready hydrolytic ring opening.[22] Hot dilute hydrochloric acid gives 3-aminoquinoxalin-2-one (**74**), and hot alkali affords 2,3-diaminoquinoxaline (**57**). Similarly acid treatment of the 2-methyl *N*-oxides **72** and **73** provides the quinoxaline *N*-oxides **75** and **76**, respectively.[32]

Condensation of benzaldehyde with the 2-methyl compound **77** affords the 2-styryl derivative **78**.[22] The reactivity of the 2-methyl group is even greater in compounds such as the quaternary salt **70**, and this and related

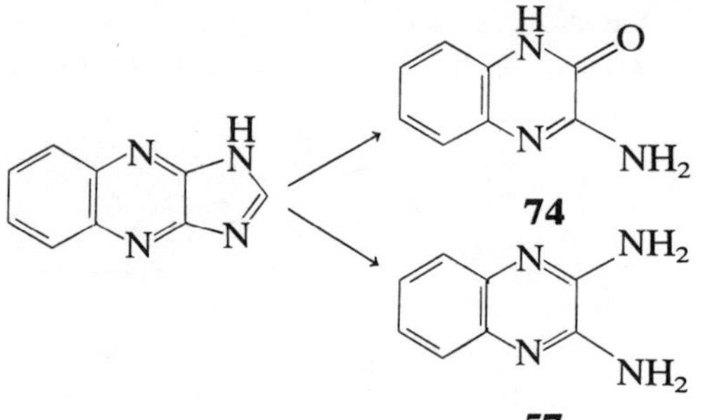

74

57

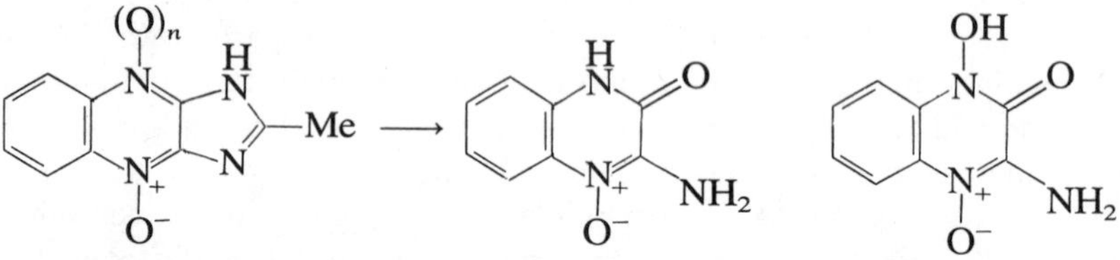

72 $n = 0$
73 $n = 1$

75

76

compounds have been condensed in a variety of standard reactions to provide a series of imidazo[4,5-*b*]quinoxaline-based cyanine dyes.[29–31]

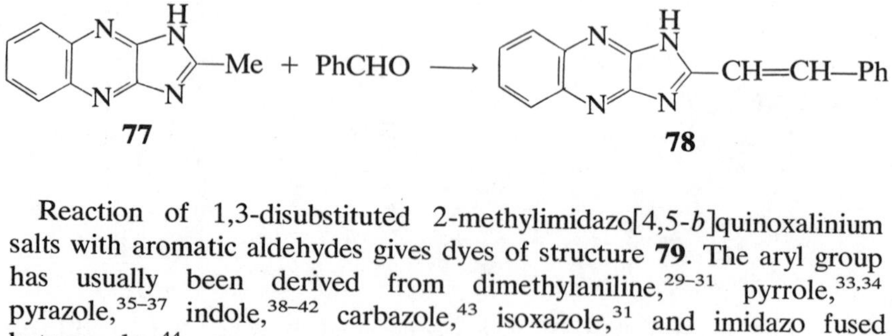

77

78

Reaction of 1,3-disubstituted 2-methylimidazo[4,5-*b*]quinoxalinium salts with aromatic aldehydes gives dyes of structure **79**. The aryl group has usually been derived from dimethylaniline,[29–31] pyrrole,[33,34] pyrazole,[35–37] indole,[38–42] carbazole,[43] isoxazole,[31] and imidazo fused heterocycles.[44]

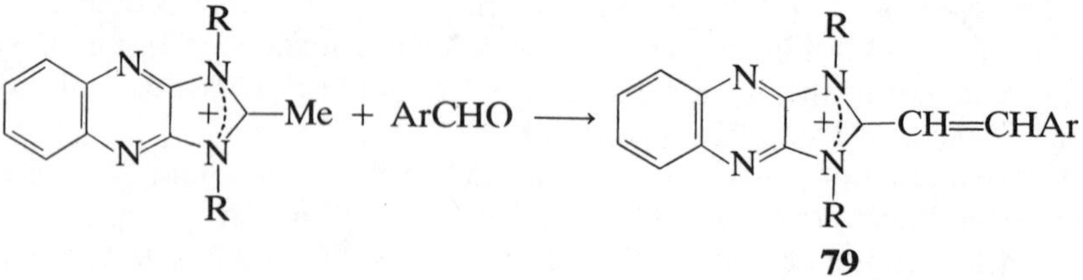

79

If condensation is performed using a heterocyclic aldehyde possessing an α hydrogen, a proton is lost from the product and neutral merocyanine dyes such as **80**[45] and **81**[46,47] are formed.

Merocyanine dyes are also obtained by reaction of compounds such as **70** with the acetanilidomethylene heterocycles **82**, **83**, and **84**.[29–31]

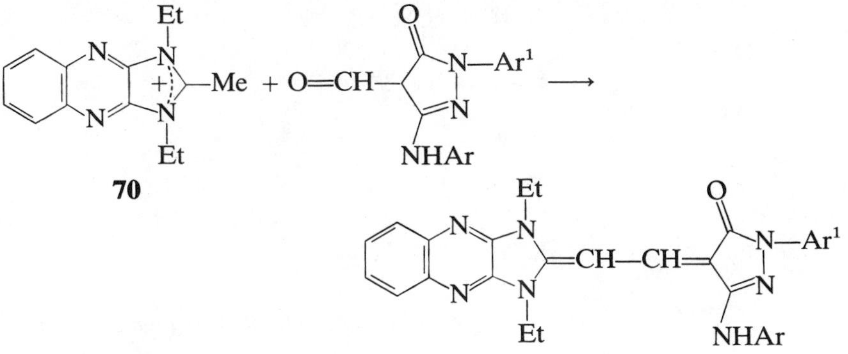

70

80

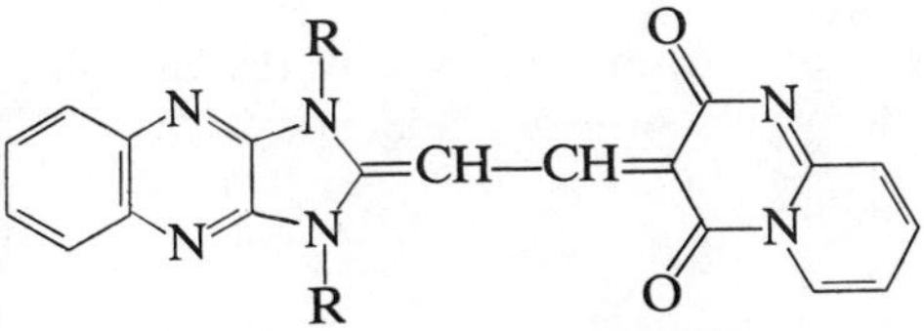

81 R = Me, Et, or Ph

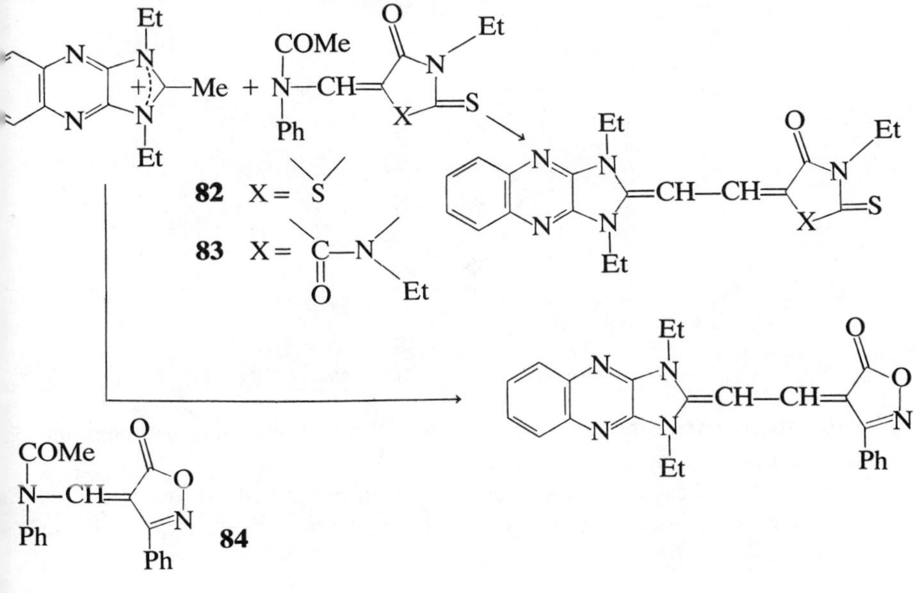

82 X = S

83 X = C(=O)—N(Et)

84

Acetanilidovinyl heterocyclic quaternary salts **85** have been used to give carbocyanine dyes of formula **86**, in which various heterocycles have been incorporated.[29–31]

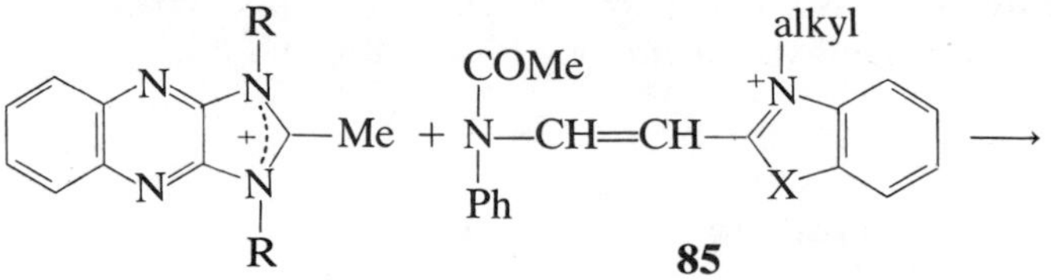

85

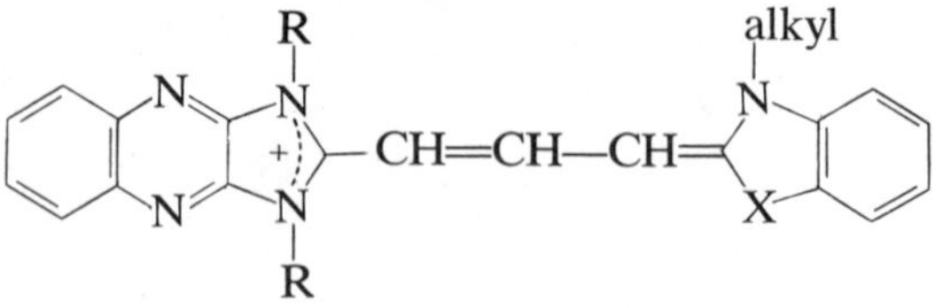

86 R = Et, Ph, or $CH_2CH{=}CH_2$
X = O, S, CMe_2, or CH=CH

Related dyes of formula **88** have also been prepared by reacting the appropriate quaternized methylheterocycle with the aldehyde **87**.[48]

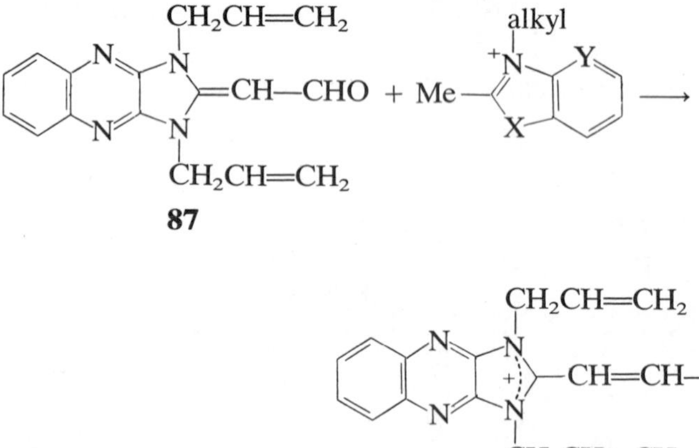

87

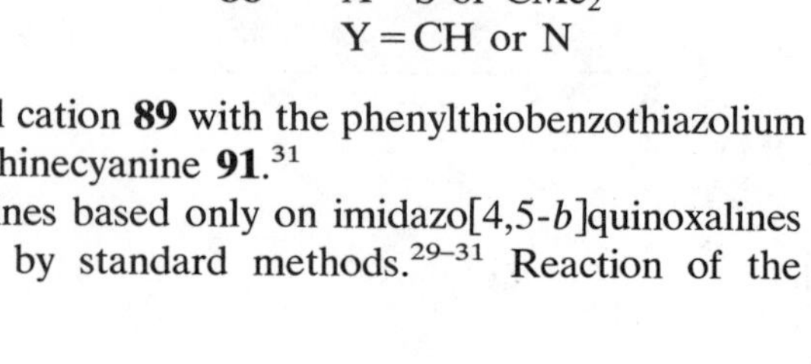

88 X = S or CMe_2
Y = CH or N

Reaction of the diphenyl cation **89** with the phenylthiobenzothiazolium salt **90** gave the monomethinecyanine **91**.[31]

Mono- and dicarbocyanines based only on imidazo[4,5-*b*]quinoxalines have also been obtained by standard methods.[29–31] Reaction of the

89 Tos$^-$ **90** I^-

91

quaternized 2-methyl compounds **92** with $MeCO_2CH(OEt)_2$ in pyridine gave the monocarbocyanines **93**, and β-methoxyacrolein dimethylacetal **94** gave the dicarbocyanines **95**.

92 X = H, Cl, or NO_2
R = Et, CH_2CH_2OH, $CH_2CH{=}CH_2$, or Ph

$MeCO_2CH(OEt)_2 \rightarrow$ **93** $n = 1$
$MeOCH_2CH{=}C(OMe)_2 \rightarrow$ **95** $n = 2$
94

As will be apparent, the principal interest in this ring system is as a source of carbocyanine dyes. In the last few years over 40 patents have been listed in *Chemical Abstracts* describing their use as sensitizers and desensitizers of direct positive photographic emulsions. A further dozen or so patents claim their use as sensitizers for electrophotographic recording materials. Other uses for the ring system are as fungicides[58,59] and as initiators of photochemical polymerization of vinyl monomers.[49]

IV. Imidazo[1,5,4-*de*]quinoxalines

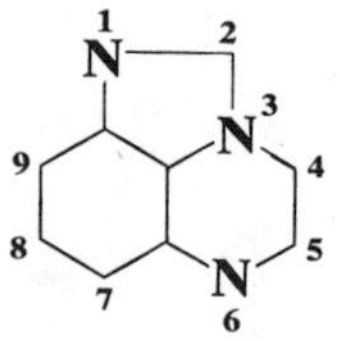

This heterocycle has also been called 1,3,6-triazaacenaphthylene and piperazino[*hi*]benzimidazole. Only partially reduced derivatives of this ring system are known, and their main interest lies in their preparation, as the only reactions reported are those typical of secondary amines.

Dehydration of the aminobenzimidazoles **96** with polyphosphoric acid gave the cyclized products **97**.[50] Compound **98** has also been prepared by the same approach.[51] Chloroethylbenzimidazoles such as **99** have been cyclized to imidazo[1,5,4-*de*]quinoxalines,[52,60] and the bis(chloroethyl)-amine **101** was similarly cyclized to the quaternized heterocycle **102**.[53]

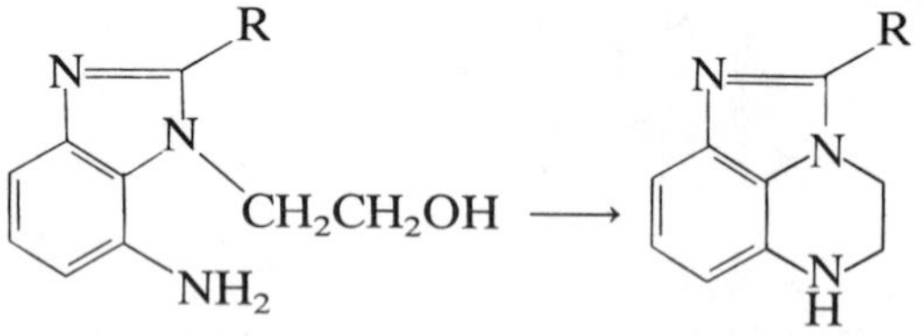

96 R = H or Me

97

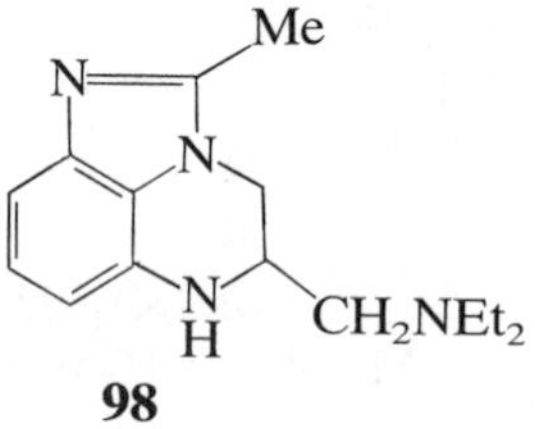

98

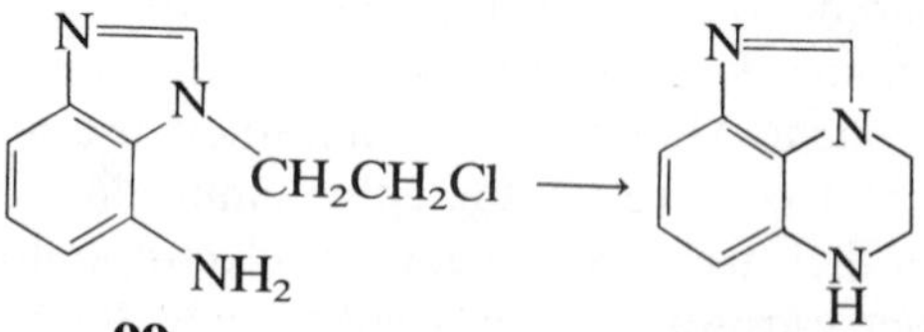

99

100

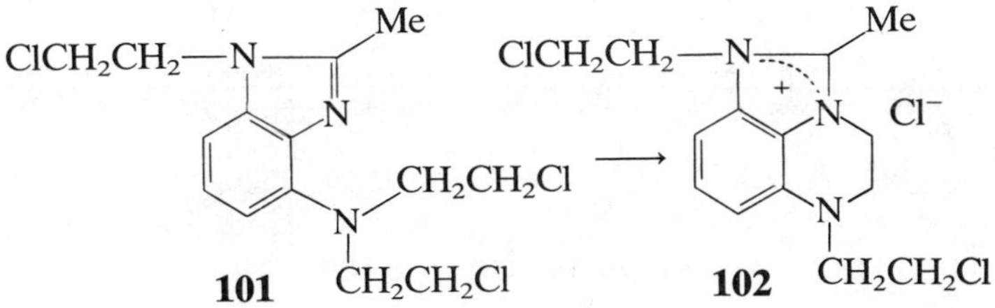

101 **102**

Treatment of compound **103** with ammonium acetate in acetic acid caused cyclization to the imidazoquinoxaline **104**.[54]

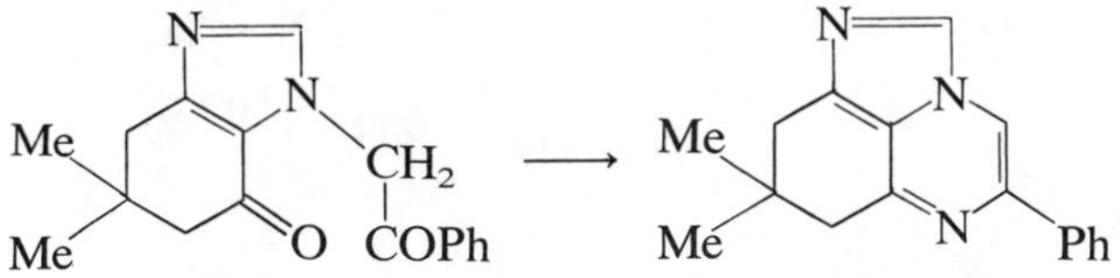

103 **104**

Condensation of carboxylic acids with the tetrahydroaminoquinoxalines **105** gave directly the tricyclic products **106**.[55] The use of urea afforded the 2-oxo derivatives **107**. Catalytic hydrogenation of the 5-acetamidoquinoxalines **108** gave the partially reduced imidazo[1,5,4-*de*]quinoxalines **109** in one step.[56] The formamido compound **110** gave the 6-formyl derivative **111**, which gave the 6-unsubstituted heterocycle on acid hydrolysis.

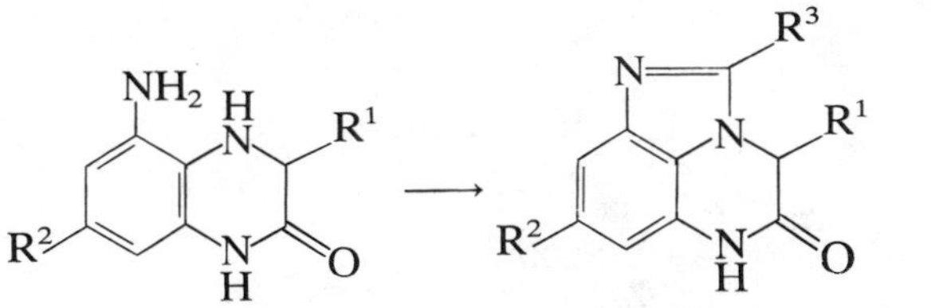

105 $R^1 = H$ or Me
$R^2 = H$ or Cl

106 $R^3 = H$, Me, or Ph

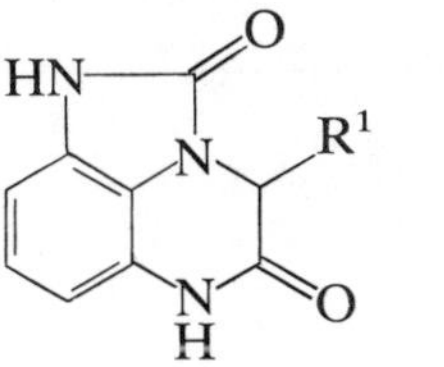

107 $R^1 = H$ or Me

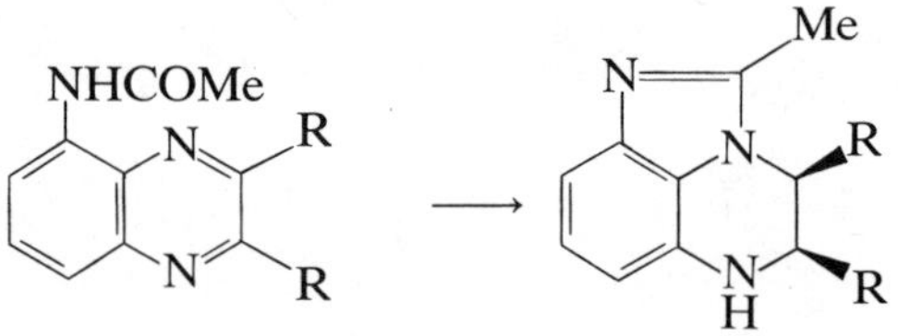

108 R = Me or Ph **109**

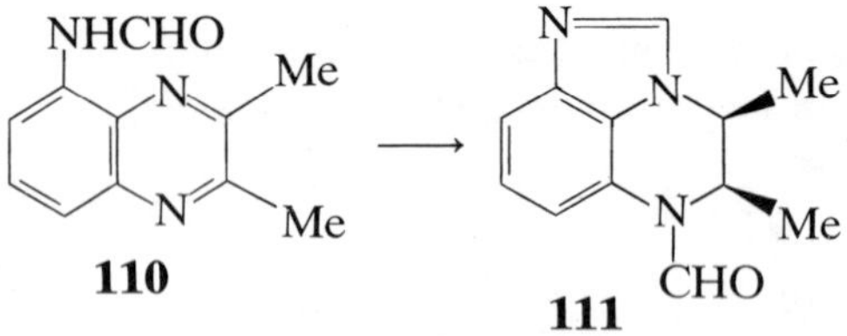

110 **111**

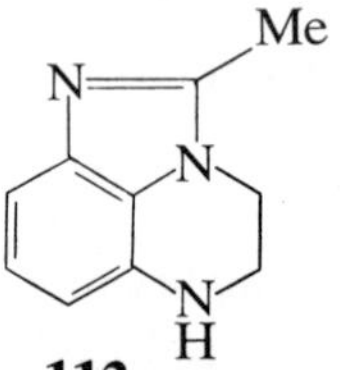

112

Compound **112** undergoes a variety of reactions at the 6-position to give 6-nitroso, 6-acetyl, and 6-allylthiourea derivatives. Methyl iodide gave a quaternary methiodide, and heating under pressure in methanolic hydrogen bromide gave the 2,3,6-trimethyl quaternary bromide.[50]

V. Tables

TABLE 1. IMIDAZO[1,2-*a*]QUINOXALINES

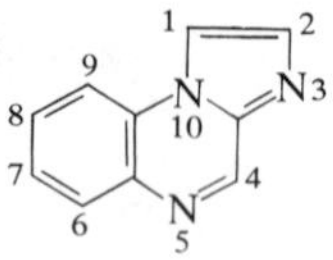

Molecular formula	Substituents or structure	m.p., °C	Ref.
$C_{10}H_6BrN_3$	7-Bromo	229	5
$C_{10}H_7N_3$	Unsubstituted	124	2, 3
	Hydrochloride	267–269	
	Picrate	250	

TABLE 1 (*contd.*)

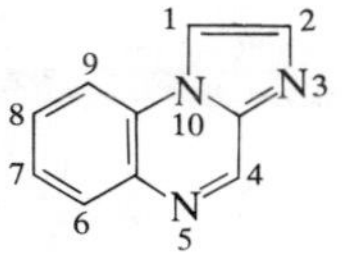

Molecular formula	Substituents or structure	m.p., °C	Ref.
$C_{10}H_8ClN_3$	4-Chloro-1,2-dihydro	151–153	8
$C_{10}H_8N_4$	4-Amino	209–210	1
$C_{10}H_9N_3$	1,2-Dihydro	101–102	6
	4,5-Dihydro	114	5
$C_{10}H_9N_3O$	4,5-Dihydro-7-hydroxy	285	5
$C_{11}H_9N_3$	7-Methyl	184	5
	Methiodide	268	
$C_{11}H_9N_3O$	7-Methoxy	167	5
	Methiodide	264	
$C_{11}H_{11}N_3$	1,2-Dihydro-4-methyl	82	6
$C_{11}H_{11}N_3O$	1,2-Dihydro-4-methoxy	139–141	8
$C_{14}H_{18}N_4O$	4-(2-Dimethylaminoethoxy)-1,2-dihydro		
	Dihydrochloride	200–202	6
$C_{15}H_{21}N_5$	4-(3-Dimethylaminopropylamino)-1,2-dihydro		
	Dihydrochloride	275–277	6, 11
$C_{16}H_{11}N_3$	4-Phenyl	154–157	3
$C_{16}H_{13}N_3$	1,2-Dihydro-4-phenyl	115	3, 6, 7
	Hydrochloride	268–270	
$C_{20}H_{12}N_6$	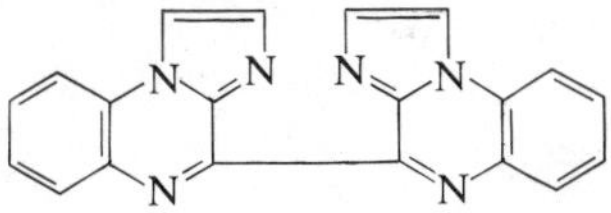 [structure: 4,4'-bi(imidazo[1,2-a]quinoxaline)]	360	1
$C_{23}H_{17}N_3O_2$	1-Benzoyl-1,2,3,3a-tetrahydro-2-oxo-3-phenyl	160	9

TABLE 2. IMIDAZO[1,5-*a*]QUINOXALINES

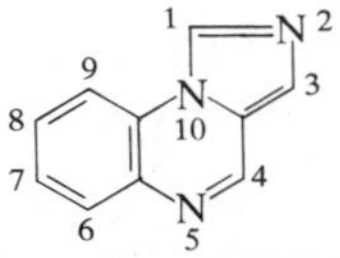

Molecular formula	Substituents or structure	m.p., °C	Ref.
$C_{11}H_9N_3O_2$	1,2,3,5-Tetrahydro-5-methyl-1,3-dioxo	270	17
$C_{11}H_{11}N_3O_2$	1,2,3,3a,4,5-Hexahydro-5-methyl-1,3-dioxo	240	17
$C_{12}H_{11}N_3O_2$	1,2,3,5-Tetrahydro-2,5-dimethyl-1,3-dioxo	260	17
$C_{12}H_{13}N_3O_2$	1,2,3,3a,4,5-Hexahydro-2,5-dimethyl-1,3-dioxo	154	17
	Picrate	133	
$C_{13}H_{13}N_3O_2$	1,2,3,5-Tetrahydro-2,4,5-trimethyl-1,3-dioxo	190–191	17

TABLE 2 (*contd.*)

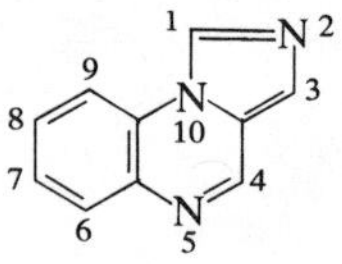

Molecular formula	Substituents or structures	m.p., °C	Ref.
$C_{16}H_5ClN_3$	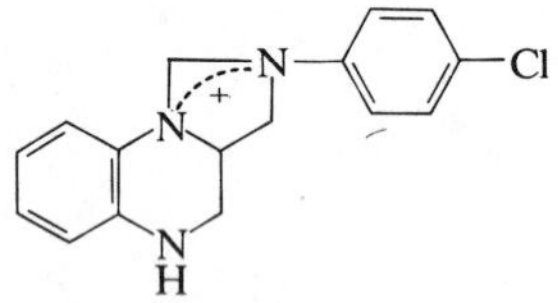		
	Fluoroborate	290–294	19
$C_{17}H_{12}ClN_3$	4-Chloro-3-methyl-1-phenyl	147–148	16
$C_{17}H_{13}N_3O$	4,5-Dihydro-3-methyl-4-oxo-1-phenyl	278–280	16
$C_{17}H_{16}N_3O_2$	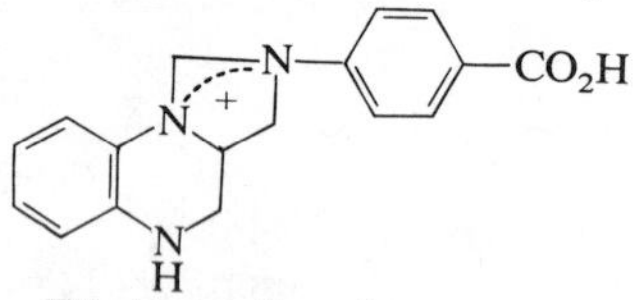		
	Ethyl ester fluoroborate	260–265	19
$C_{17}H_{17}N_3O_2$	2-(4-Carboxyphenyl)-1,2,3,3a,4,5-hexahydro		
	Ethyl ester	168–169	14
$C_{17}H_{18}N_3$	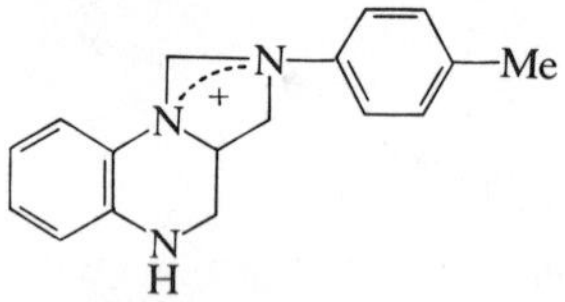		
	Fluoroborate	274–277	19
$C_{17}H_{19}N_3$	1,2,3,3a,4,5-Hexahydro-2-(4-methylphenyl)	195–196	14
$C_{20}H_{21}N_3O_2$	1,2,3,3a,4,5-Hexahydro-2-methyl-1,3-dioxo-5-(2,4,6-trimethylphenyl)	188–189	15
$C_{21}H_{23}N_3O_2$	1,2,3,3a,4,5-Hexahydro-2,4-dimethyl-1,3-dioxo-5-(2,4,6-trimethylphenyl)[a]	211–212	15
$C_{22}H_{21}N_5O_4$	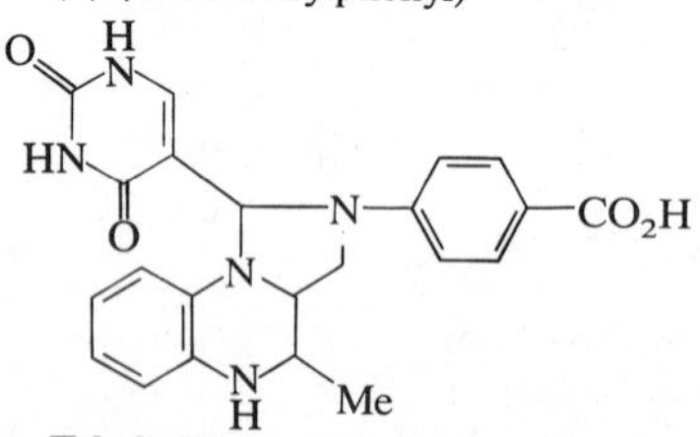		
	Ethyl ester	237–238	20
$C_{23}H_{17}N_3O$	3-Benzyl-4,5-dihydro-4-oxo-1-phenyl	290	16

[a] Described on the basis of PMR evidence as the 3a,4-trans compound.

TABLE 3. CYANINE DYES BASED ON IMIDAZO[4,5-*b*]QUINOXALINES[a]

Type A   Type B

Molecular formula	Type	R^1	R^2	m.p., °C	Ref.
$C_{20}H_{21}N_5OS_2$	B	Et	=CHCH= (3-ethyl-4-oxo-2-thioxothiazolidin-5-ylidene; O, N—Et, S, =S)	285–286 decomp.	29, 30, 31
$C_{20}H_{24}N_4$	A	Et	—CH=CHCH= (3-methylthiazolidin-2-ylidene; S, N, Me)		
		Iodide		281–282 decomp.	29, 30, 31
$C_{22}H_{21}N_5OS_2$	B	$CH_2CH{=}CH_2$	=CHCH= (3-ethyl-4-oxo-2-thioxothiazolidin-5-ylidene; O, N—Et, S, =S)	227–228	31
$C_{23}H_{25}N_6$	A	$CH_2CH{=}CH_2$	—CH=CH— (1,3,5-trimethylpyrazol-4-yl; Me, N—Me, N, Me)	[b]	
$C_{23}H_{26}N_5$	A	Et	—CH=CH—C_6H_4—NMe_2		
		Iodide		262–263	29, 30, 31
$C_{23}H_{26}N_5O_2$	A	CH_2CH_2OH	—CH=CH—C_6H_4—NMe_2		
		Tosylate		280–281	31
$C_{25}H_{23}N_6O_2$	A	Et	—CH=CH—(pyrrol-2-yl; N—$C_6H_4NO_2$-*p*)		
		Iodide		236–237	33
$C_{25}H_{24}N_5$	A	Et	—CH=CH—(pyrrol-2-yl; N—Ph)		
		Perchlorate		280–281	33
$C_{25}H_{26}N_5$	A	$CH_2CH{=}CH_2$	—CH=CH—C_6H_4—NMe_2		
		Tosylate		238–239	31

TABLE 3 (*contd.*)

Type A: 2-R²-1,3-di-R¹-imidazo[4,5-b]quinoxalinium (positions 5, 6, 7, 8 numbered on the benzo ring)

Type B: 2(=R²)-1,3-di-R¹-imidazo[4,5-b]quinoxaline (positions 5, 6, 7, 8 numbered on the benzo ring)

Molecular formula	Type	R^1	R^2	m.p., °C	Ref.
$C_{25}H_{26}N_5O$	A	Et	—CH=CHCH= (benzoxazole ring: O, N—Et)		
		Iodide		282–283	29, 3[illegible], 31
$C_{25}H_{26}N_5S$	A	Et	—CH=CHCH= (benzothiazole ring: S, N—Et)		
		Iodide		284–285	29, 3[illegible], 31
$C_{26}H_{23}N_6$	A	$CH_2CH{=}CH_2$	—CH=CH— (pyrazole ring: N—Ph, N)		
		Tosylate		232–233 decomp.	35, 3[illegible]
$C_{26}H_{27}N_6$	A	Et	—CH=CH— (pyrazole ring: Me, N—Ph, N, Me)		
		Iodide		237–238 decomp.	35, 3[illegible]
$C_{27}H_{25}N_5$	A	Et	—CH=CHCH= (quinoline ring: N—Et)		
		Iodide		291–292 decomp.	29, 3[illegible], 31
$C_{27}H_{26}N_5O$	A	$CH_2CH{=}CH_2$	—CH=CHCH= (benzoxazole ring: O, N—Et)		
		Iodide		253–254	31
$C_{27}H_{26}N_5S$	A	$CH_2CH{=}CH_2$	—CH=CHCH= (benzothiazole ring: S, N—Et)		
		Iodide		250–251	31

TABLE 3 (*contd.*)

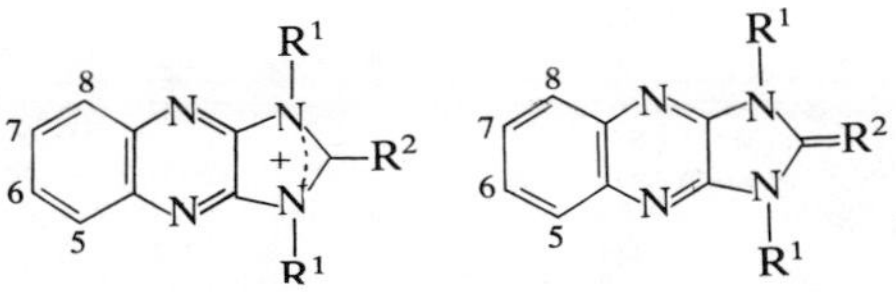

Molecular formula	Type	R^1	R^2	m.p., °C	Ref.
$C_{27}H_{30}N_5$	A	Et	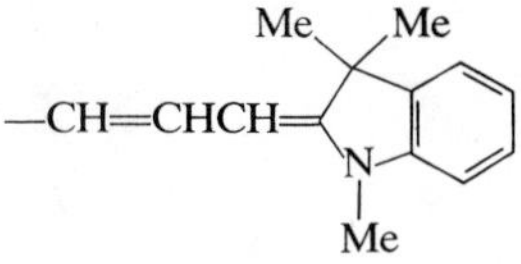 —CH=CHCH= (3,3-Me, Me; N-Me indoline)		
		Iodide		273–274	29, 30, 31
$C_{27}H_{30}N_5O_2$	A	CH_2CH_2OH	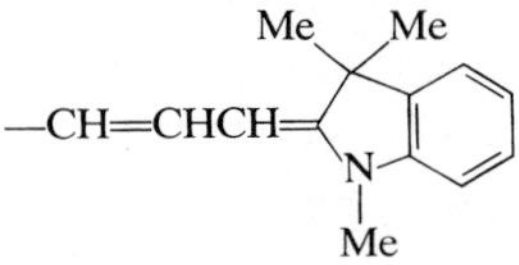 —CH=CHCH= (3,3-Me, Me; N-Me indoline)		
		Iodide		273–274	31
$C_{28}H_{21}N_5OS_2$	B	Ph	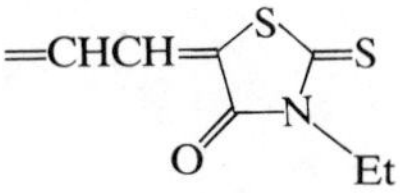 =CHCH= (S, =S, O, N-Et thiazolidinone)	>320	31
$C_{28}H_{26}N_5$	A	Et	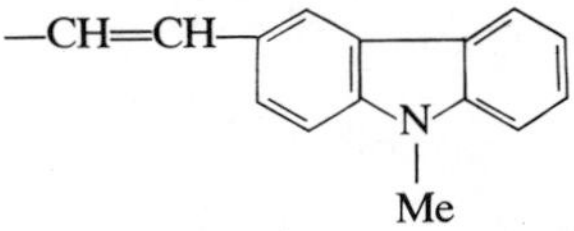 —CH=CH— (N-Me carbazole)		
		Iodide		270–273	43
$C_{28}H_{27}N_6$	A	$CH_2CH=CH_2$	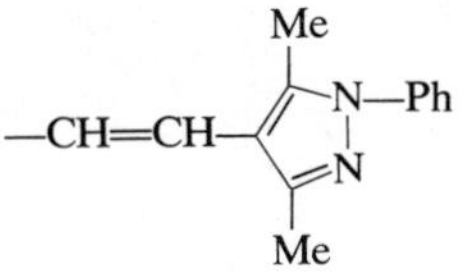 —CH=CH— (Me, N—Ph, N, Me pyrazole) [c]		
		Iodide		202–203 decomp.	35, 36
$C_{29}H_{25}N_5$	A	$CH_2CH=CH_2$	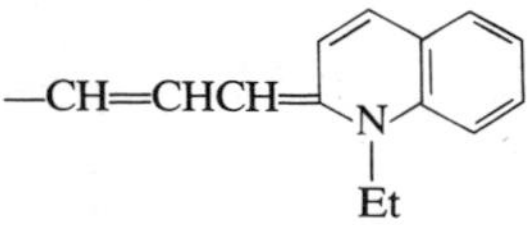 —CH=CHCH= (N-Et quinoline)		
		Iodide		261–262	31
$C_{29}H_{26}N_7S$	A	$CH_2CH=CH_2$	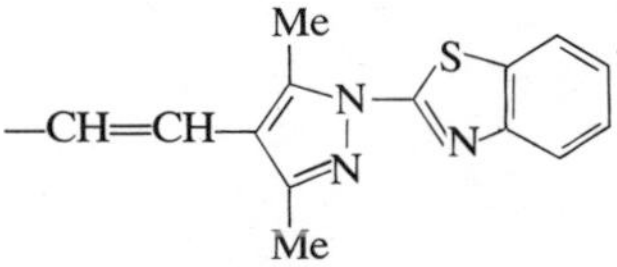 —CH=CH— (Me, N—benzothiazolyl, N, Me pyrazole) [d]		
		Tosylate		219–220	35, 37

TABLE 3 (*contd.*)

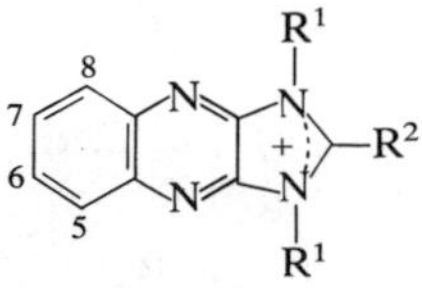

Type A

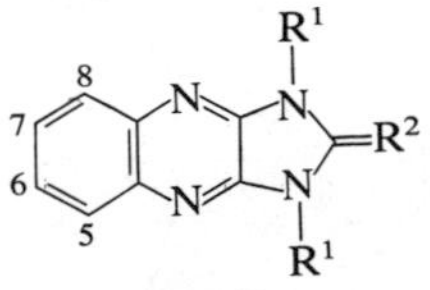

Type B

Molecular formula	Type	R^1	R^2	m.p., °C	Ref.
$C_{29}H_{27}N_6$	A	Et	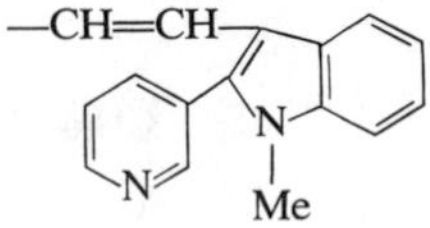		
		Iodide		223 decomp.	38
$C_{29}H_{30}N_5$	A	$CH_2CH{=}CH_2$	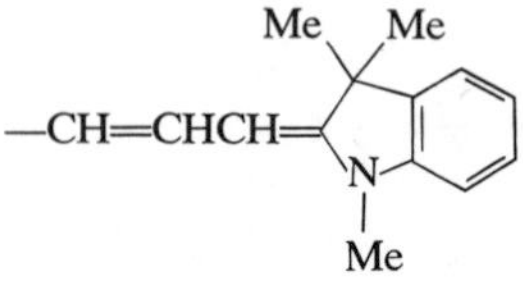		
		Iodide		246–247	31
$C_{29}H_{31}N_8$	A	Et	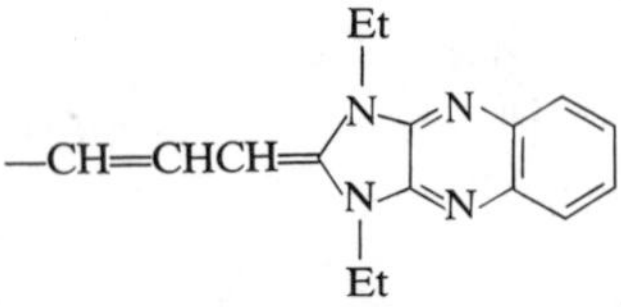		
		Chloride		250–252 decomp.	29, 3[illegible], 31
$C_{29}H_{31}N_8O_4$	A	CH_2CH_2OH	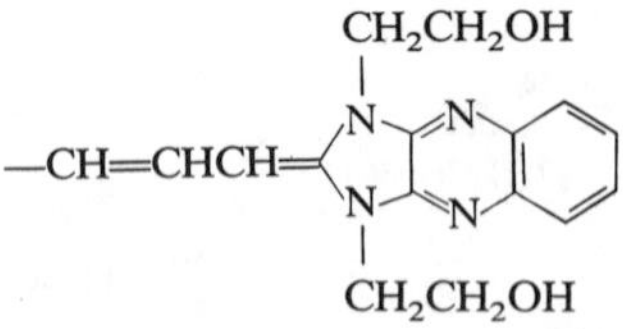		
		Iodide		254–255 decomp.	31
$C_{31}H_{24}N_5S$	A	Ph	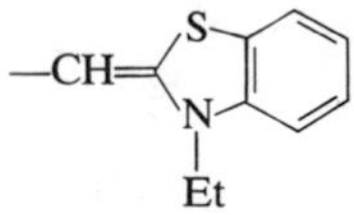		
		Iodide		288–289	31
$C_{31}H_{26}N_5$	A	Ph	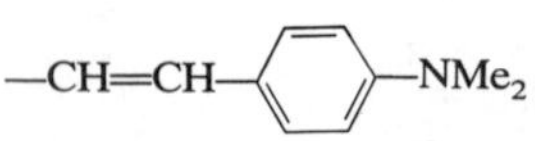		
		Tosylate		250–251	31
$C_{31}H_{28}N_5$	A	Et	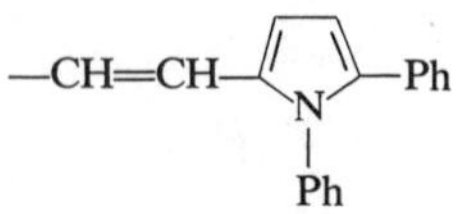		
		Perchlorate		293–294	33

TABLE 3 (*contd.*)

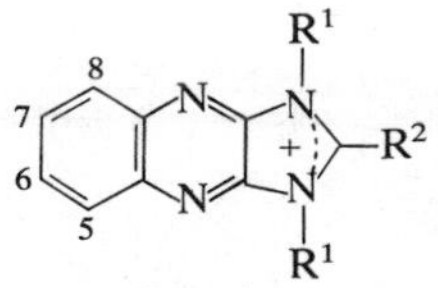

Type A

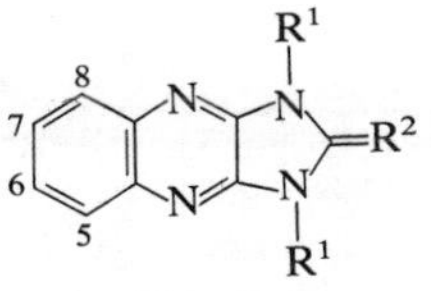

Type B

Molecular formula	Type	R^1	R^2	m.p., °C	Ref.
$C_{31}H_{28}N_5S$	A	$CH_2CH{=}CH_2$	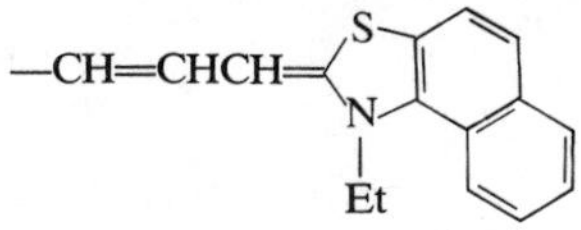		
		Tosylate		243–244	31
$C_{31}H_{33}N_8$	A	Et	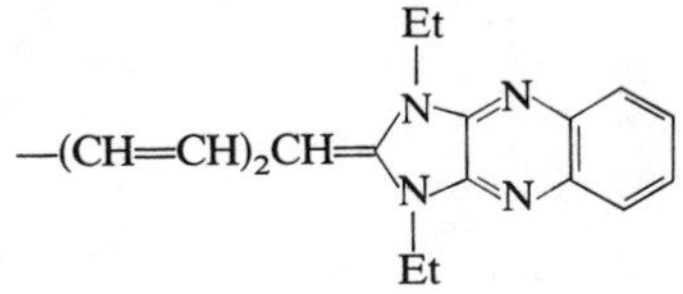		
		Chloride		231–232	29, 30, 31
$C_{32}H_{21}N_5O_2$	B	Ph	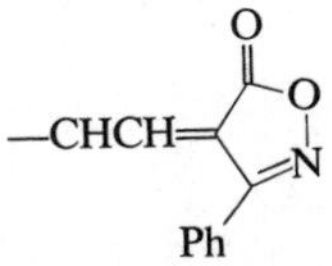	[f]	
$C_{32}H_{23}N_6$	A	Ph	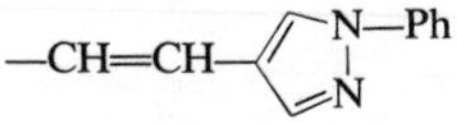	[g]	
$C_{32}H_{28}N_5$	A	$CH_2CH{=}CH_2$	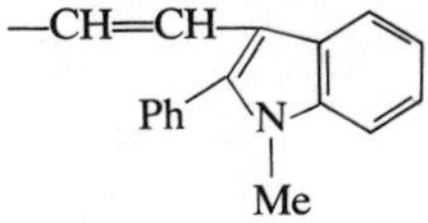	[h]	
$C_{32}H_{30}N_5$	A	Et	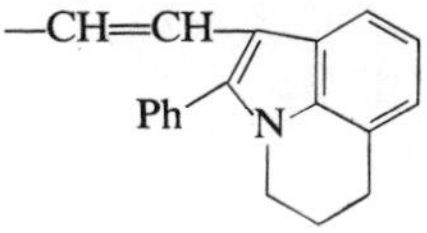		
		Iodide		230–231	38
$C_{33}H_{24}N_5O_2$	A	Ph	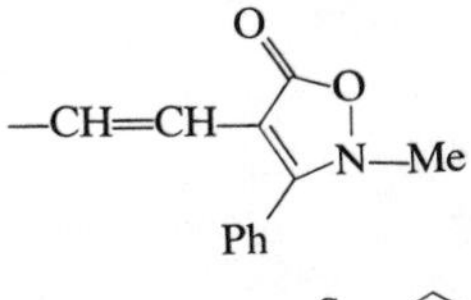	[i]	
$C_{33}H_{25}N_6O_2S$	A	Ph	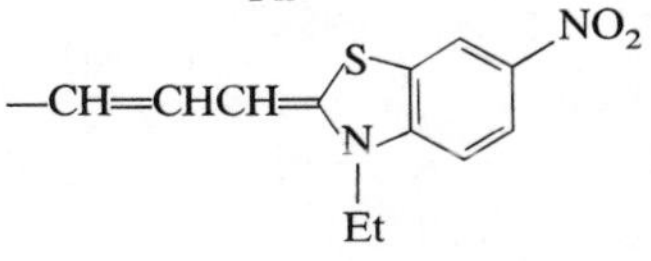		

TABLE 3 (*contd.*)

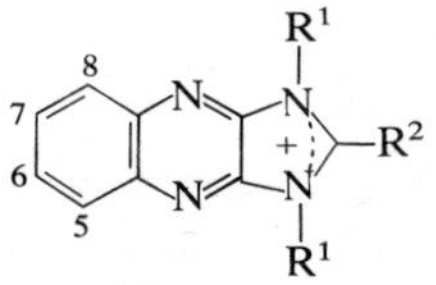

Type A 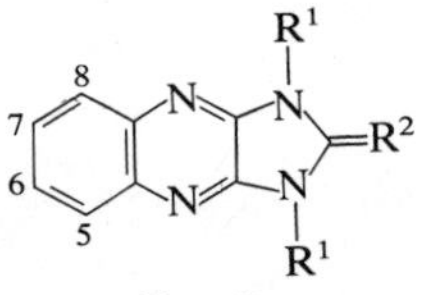   Type B

Molecular formula	Type	R^1	R^2	m.p., °C	Ref.
$C_{33}H_{26}N_5O$	A	Ph	—CH=CHCH= (O, N, Et)		
		Iodide		289–290	31
$C_{33}H_{26}N_5S$	A	Ph	—CH=CHCH= (S, N, Et)		
		Iodide		288–289	31
$C_{33}H_{31}N_8$	A	$CH_2CH{=}CH_2$	—CH=CHCH= (N, N, N, N; $CH_2CH{=}CH_2$, $CH_2CH{=}CH_2$) [k]		
		Tosylate		245–246	31
$C_{34}H_{27}N_6$	A	Ph	—CH=CH— (Me, N—Ph, N, N, Me) [l]		
		Iodide		319–320 decomp.	35, 36
$C_{34}H_{30}N_5$	A	Et	—CH=CH— (Ph, N, Me)		
		Iodide		293–294	38
$C_{35}H_{25}N_5$	A	Ph	—CH=CHCH= (N, Et) [m]		
		Iodide		284–285	31
$C_{35}H_{27}N_5SO_5$	A	Ph	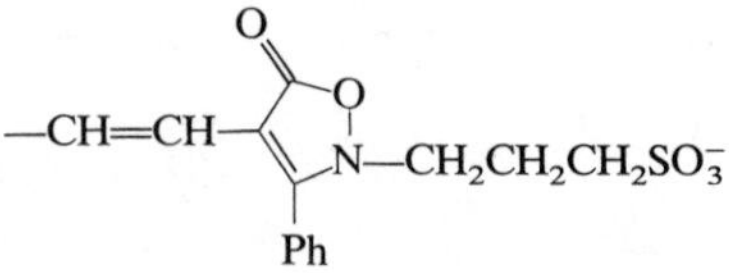 —CH=CH— (O, O, N—$CH_2CH_2CH_2SO_3^-$, Ph)		

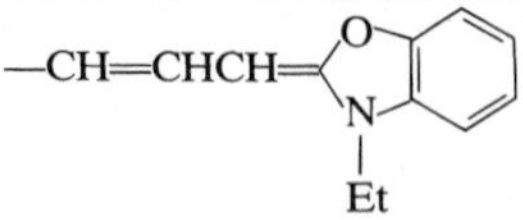

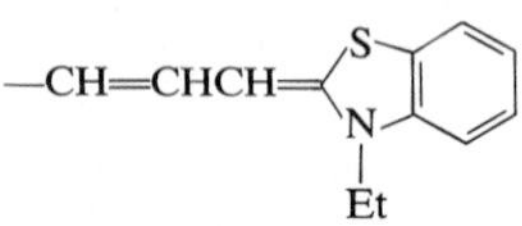

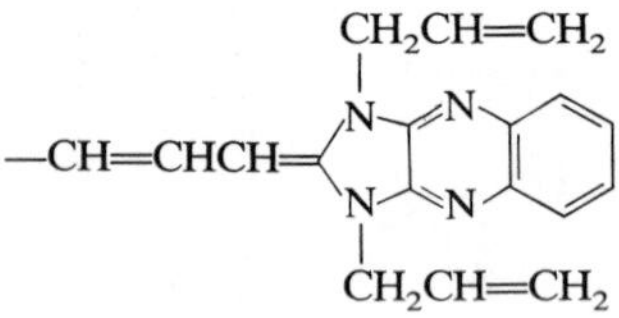

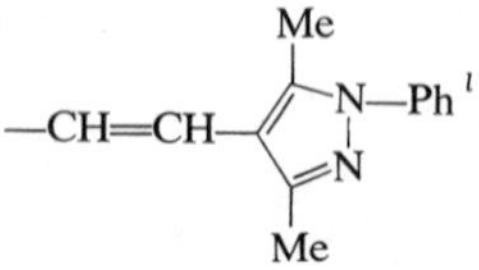

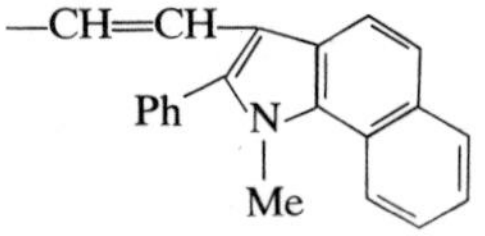

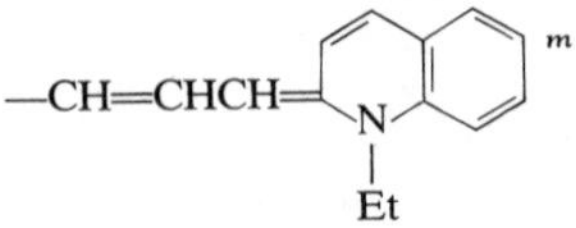

TABLE 3 (*contd.*)

[Structures: Type A (2-R²-1,3-di-R¹-imidazo[4,5-b]quinoxalinium, positions 5–8 numbered) and Type B (2-R²-ylidene-1,3-di-R¹-imidazo[4,5-b]quinoxaline)]

Type A Type B

Molecular formula	Type	R¹	R²	m.p., °C	Ref.
$C_{35}H_{30}N_5$	A	Ph	—CH=CHCH=(3,3-dimethyl-1-methylindolin-2-ylidene) [o]		
		Iodide		299–300	31
$C_{36}H_{26}N_5$	A	Ph	—CH=CH—(9-methylcarbazol-3-yl) [p]		
$C_{37}H_{31}N_8$	A	Ph	—CH=CHCH=(1,3-diphenylimidazo[4,5-b]quinoxalin-2-ylidene) [q]		
		Tosylate		286–288	31
$C_{38}H_{28}N_5$	A	Ph	—CH=CH—(1-methyl-2-phenylindol-3-yl)		
$C_{39}H_{30}N_9O_2$	A	$CH_2CH{=}CH_2$	—CH=CHCH=(1,3-diphenylimidazo[4,5-b]quinoxalin-2-ylidene)		
		Tosylate		180–183	31

[a] Only a representative selection of compounds is listed.
[b] References 35 and 36 only list the 6-chloro derivative, tosylate m.p. 122–124° decomp.
[c] References 35 and 36 also list the 6-chloro derivative, tosylate m.p. 189–190° decomp.
[d] References 35 and 37 also list the 6-chloro derivative, tosylate m.p. 230–231° decomp.
[e] Reference 31 also lists the 6-nitro derivative, tosylate m.p. 257–258°; the 6-chloro derivative, tosylate m.p. 280–281°; and the 6,7-dichloro derivative, tosylate, m.p. 293–294°.
[f] Reference 31 only lists the 6,7-dichloro derivative, tosylate m.p. >320°.
[g] References 35 and 36 only list the 6-chloro derivative, tosylate m.p. 291–292° decomp.
[h] Reference 31 only lists the 6-chloro derivative, tosylate m.p. 240–241°.
[i] Reference 31 only lists the 6,7-dichloro derivative, tosylate m.p. 274–275°.

TABLE 3 (*contd.*)

[j] Reference 31 only lists the 6-nitro derivative, tosylate m.p. 250–252°.

[k] Reference 31 also lists the 6,6′-dichloro compound, tosylate m.p. 251–252°.

[l] References 35 and 36 also list the 6-chloro derivative, tosylate m.p. 283–284° decomp.

[m] Reference 31 also lists the 6-chloro derivative, tosylate m.p. 185–187°.

[n] Reference 31 only lists the anhydro salt of the 6,7-dichloro derivative, m.p. 247–250°.

[o] Reference 31 also lists the iodides of the 6-nitro derivative, m.p. 285–286°; the 6-chlo derivative, m.p. 283–284°; and the 6,7-dichloro derivative, m.p. 310–311°.

[p] Reference 31 only lists the tosylates of the 6-nitro derivative, m.p. 291–292°; the 6-chlo derivative, m.p. 287–288°; and the 6,7-dichloro derivative, m.p. 312–313°.

[q] Reference 31 also lists the tosylates of the 6,6′-dinitro compound, m.p. 206–207°; the 6, dichloro compound, m.p. 293–294°; and the 6,7,6′7′-tetrachloro compound, m.p. 312–313°.

[r] Reference 31 only lists the tosylates of the 6-nitro derivative, m.p. 303–304°; the 6-chlo derivative, m.p. 288–289°; and the 6,7-dichloro derivative, m.p. 300–310°.

TABLE 4. IMIDAZO[4,5-*b*]QUINOXALINES (NOT CYANINE DYES[a])

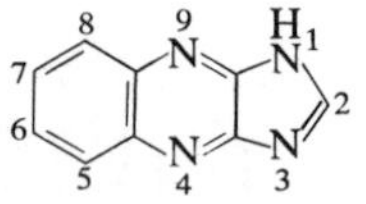

Molecular formula	Substituents or structure	m.p., °C	Ref.
$C_9H_5ClN_4O$	6-Chloro-2,3-dihydro-2-oxo	362	22
$C_9H_6N_4$	Unsubstituted	286	22, 32
	4-Oxide	260.5	
$C_9H_6N_4O$	2,3-Dihydro-2-oxo	450	22, 28
		380	26
$C_{10}H_3Cl_2F_3N_4$	5,7-Dichloro-2-trifluoromethyl	360	23
	6,7-Dichloro-2-trifluoromethyl	272	23
$C_{10}H_4ClF_3N_4$	6-Chloro-2-trifluoromethyl	237	23
$C_{10}H_5F_3N_4$	2-Trifluoromethyl	265	23
$C_{10}H_7ClN_4$	6-Chloro-2-methyl	344	22
$C_{10}H_7ClN_4S$	6-Chloro-2-methylthio	269–270	23
$C_{10}H_8N_4$	2-Methyl	322	22, 32
	4-Oxide	273	
	4,9-Dioxide	259	
	4-Methyl	210–213 decomp.	22
	6-Methyl	246	22
$C_{10}H_8N_4O$	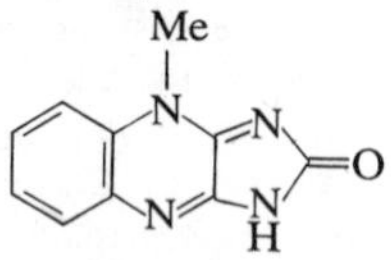 [structure]	355	22
	1,2-Dihydro-4-methyl-2-oxo	400	22
$C_{11}H_{10}N_4$	2-Ethyl	313	22
	2,6-Dimethyl	305	22
$C_{11}H_{10}N_4O$	2-Methoxymethyl	233	22

TABLE 4 (*contd.*)

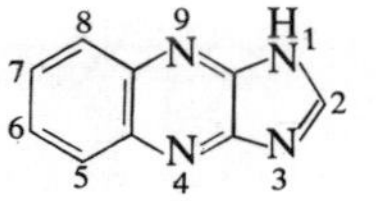

Molecular formula	Substituents or structure	m.p., °C	Ref.
$C_{12}H_{10}N_4O_2$	2-(2-Carboxyethyl)	284	22
$C_{12}H_{12}N_4$	1-Ethyl-2-methyl		
	3-Ethochloride	198–200 decomp.	29, 30, 31
	2-Ethyl-6-methyl	273	22
	2-*n*-Propyl	286	22
$C_{13}H_8N_4O$	2-(2-Furyl)	>300	25
$C_{13}H_{12}N_4O_2$	2-(2-Carboxyethyl)-6-methyl	283	22
$C_{13}H_{14}N_4$	2-*n*-Butyl	250	22
	6-Methyl-2-*n*-propyl	271	22
	6-Methyl-2-*i*-propyl	250	22
$C_{14}H_{16}N_4$	2-(1-Ethyl-*n*-propyl)	246	22
	2-*n*-Pentyl	242	22
	2-*i*-Pentyl	223	22
$C_{14}H_{17}N_4$	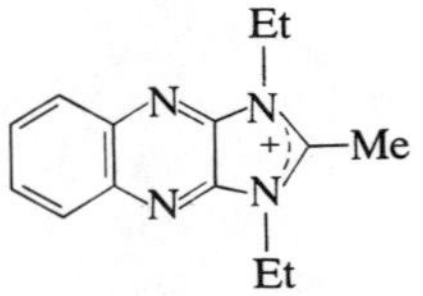 1,3-Diethyl-2-methyl cation (Et, Et, Me)		
	Chloride	198–200 decomp.	29, 30, 31
$C_{14}H_{17}N_4O_2$	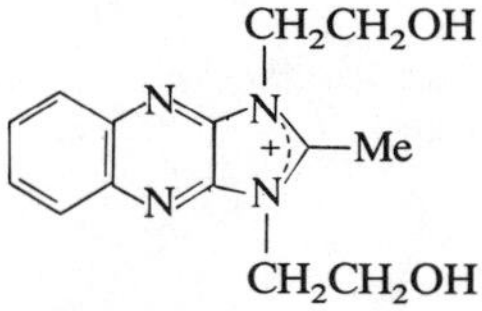 1,3-bis(CH_2CH_2OH)-2-methyl cation (Me)		
	Tosylate	[b]	31
$C_{15}H_{10}N_4$	2-Phenyl[c]	324	22
		290	25
$C_{15}H_{18}N_4$	2-(1-Ethyl-*n*-propyl)-6-methyl	247	22
	6-Methyl-2-*n*-pentyl	224	22
$C_{16}H_{12}N_4$	2-Benzyl	276	22
$C_{16}H_{16}ClN_4$	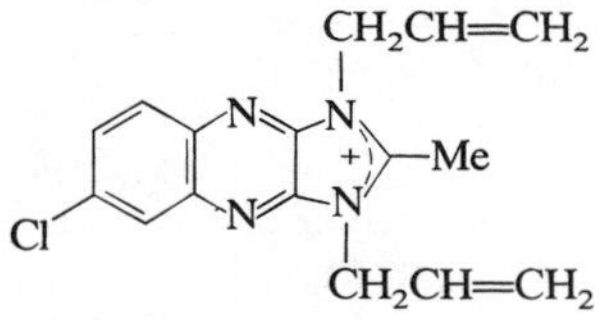 Cl; 1,3-bis($CH_2CH=CH_2$)-2-methyl cation (Me)		
	Tosylate	173–175	31

TABLE 4 (*contd.*)

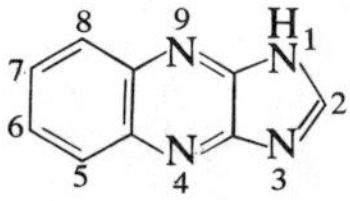

[Structure: 1H-imidazo[4,5-b]quinoxaline, positions numbered 1–9]

Molecular formula	Substituents or structure	m.p., °C	Ref.
$C_{16}H_{17}N_4$	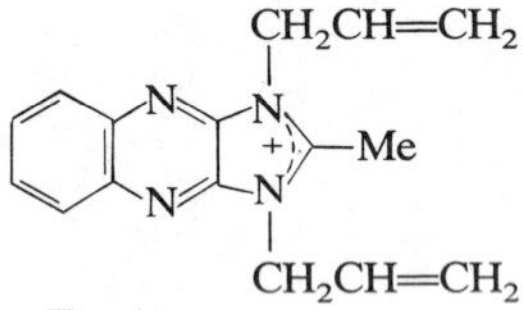 [Structure: 1,3-di($CH_2CH{=}CH_2$)-2-Me-imidazo[4,5-b]quinoxalinium]		
	Tosylate	157–159	31
$C_{16}H_{18}N_4$	2-Cyclohexylmethyl	324	22
$C_{16}H_{20}N_4$	2-(1-Ethyl-*n*-pentyl)	221	22
$C_{17}H_{12}N_4$	2-Styryl	311	22, 32
	4-Oxide	258	
	4,9-Dioxide	253	
$C_{17}H_{12}N_4O$	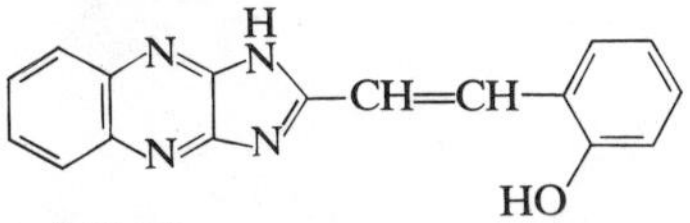 [Structure: 2-(CH=CH-(2-HO-phenyl))-1H-imidazo[4,5-b]quinoxaline]	253.5	32
	4-Oxide	343	
$C_{18}H_{23}N_5O_2$	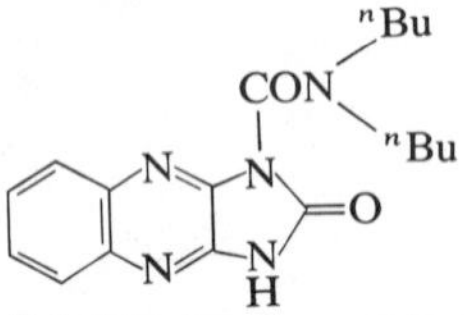 [Structure: 1-($CON(^nBu)_2$)-2,3-dihydro-2-oxo-3H-imidazo[4,5-b]quinoxaline]	177–178	28
$C_{21}H_{14}N_4O$	2,3-Dihydro-2-oxo-1,3-diphenyl	235–236	27
$C_{22}H_{15}Cl_2N_4$	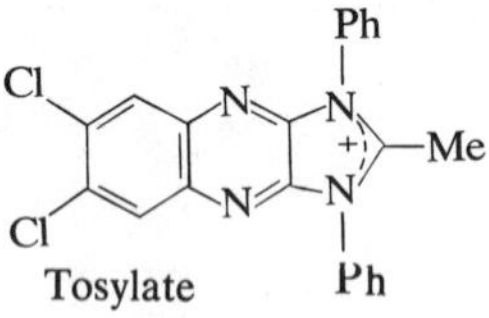 [Structure: 6,7-dichloro-1,3-diphenyl-2-Me-imidazo[4,5-b]quinoxalinium]		
	Tosylate	210–245	31
$C_{22}H_{16}ClN_4$	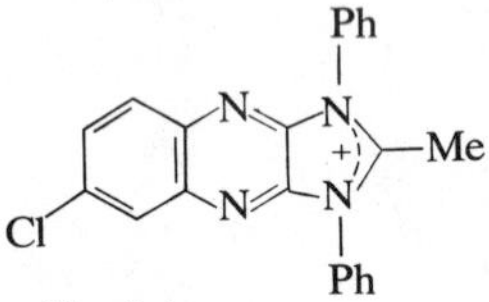 [Structure: 6-chloro-1,3-diphenyl-2-Me-imidazo[4,5-b]quinoxalinium]		
	Tosylate	278–280	31

TABLE 4 (*contd.*)

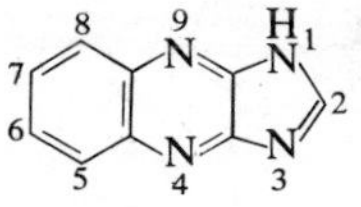

Molecular formula	Substituents or structure	m.p., °C	Ref.
$C_{22}H_{16}N_5O_2$	[structure: 2-methyl-1,3-diphenyl-6(7)-nitroimidazo[4,5-b]quinoxalinium] Tosylate	284–285	31
$C_{22}H_{17}N_4$	[structure: 2-methyl-1,3-diphenylimidazo[4,5-b]quinoxalinium] Tosylate	275–285	31
$C_{22}H_{18}N_4$	1-Benzyl-2-phenyl	>300	25
$C_{22}H_{18}N_4O$	1-(2-Hydroxybenzyl)-2-phenyl	>300	25
$C_{24}H_{22}N_4O_2$	1-(4-Methoxybenzyl)-2-(4-methoxyphenyl)	243–245	25

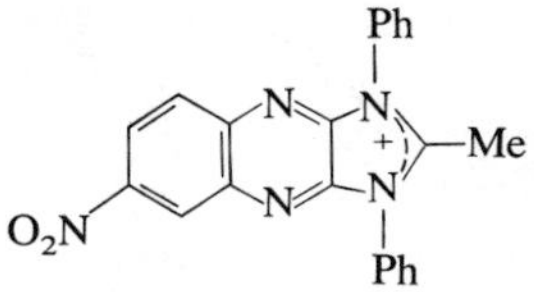

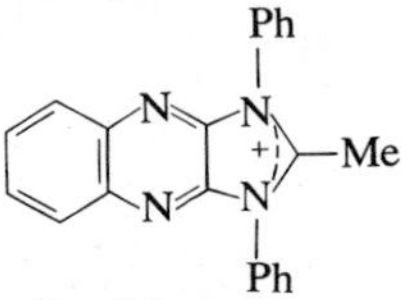

[a] A representative selection of cyanine dyes are listed in Table 3.

[b] No melting point given.

[c] Seven other 2-(substituted phenyl) derivatives listed in reference 25, all m.p.s >300.

TABLE 5. IMIDAZO[1,5,4-*de*]QUINOXALINES

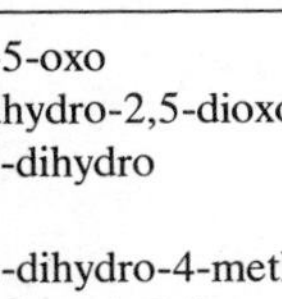

(4*H*)

Molecular formula	Substituents or structure	m.p., °C	Ref.
$C_9H_7N_3O$	4*H*-5,6-Dihydro-5-oxo	288	55
$C_9H_7N_3O_2$	4*H*-1,2,5,6-Tetrahydro-2,5-dioxo	250	55
$C_9H_8ClN_3$	8-Chloro-4*H*-5,6-dihydro	208–210	50
$C_9H_9N_3$	4*H*-5,6-Dihydro	158	50
$C_{10}H_8ClN_3O$	8-Chloro-4*H*-5,6-dihydro-4-methyl-5-oxo	323	55
$C_{10}H_9N_3O$	4*H*-5,6-Dihydro-2-methyl-5-oxo	255	55
	4*H*-5,6-Dihydro-4-methyl-5-oxo	300	55
$C_{10}H_9N_3O_2$	4*H*-1,2,5,6-Tetrahydro-4-methyl-2,5-dioxo	270–272	55
$C_{10}H_{10}ClN_3$	6-Chloro-4*H*-5,6-dihydro-2-methyl	199–200	50
$C_{10}H_{10}N_4O$	4*H*-5,6-Dihydro-2-methyl-6-nitroso	79–80	50

TABLE 5 (*contd.*)

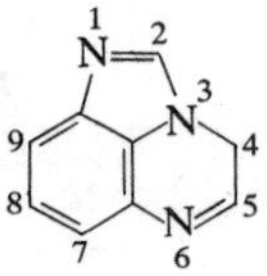

1
N
2
3
N
4
5
N
6
7
8
9

(4*H*)

Molecular formula	Substituents or structure	m.p., °C	Ref.
$C_{10}H_{11}N_3$	4*H*-5,6-Dihydro-2-methyl	185	50
	Methiodide	257–259	
$C_{11}H_{11}N_3O$	4*H*-5,6-Dihydro-2,4-dimethyl-5-oxo	228	55
$C_{11}H_{13}N_3$	4*H*-5,6-Dihydro-2,6-dimethyl		
	3-Methobromide	250	50
	4*H*-5,6-Dihydro-*cis*-4,5-dimethyl	150	56
$C_{12}H_{13}N_3O$	6-Acetyl-4*H*-5,6-dihydro-2-methyl	175–176	50
	6-Formyl-4*H*-5,6-dihydro-*cis*-4,5-dimethyl	175	56
$C_{12}H_{15}N_3$	4*H*-5,6-Dihydro-2,(*cis*)4,5-trimethyl	178	56
$C_{13}H_{16}N_4S$	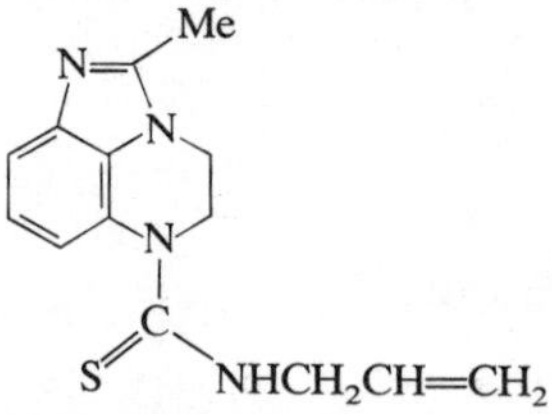Me N N N C S NHCH2CH=CH2	175–176.5	50
$C_{13}H_{17}N_3O$	4*H*-5,6-Dihydro-7-methoxy-2,(*cis*)4,5-trimethyl	165	56
	4*H*-5,6-Dihydro-9-methoxy-2,(*cis*)4,5-trimethyl	138	56
$C_{15}H_{11}N_3O$	4*H*-5,6-Dihydro-5-oxo-2-phenyl	277	55
$C_{15}H_{21}N_5O$	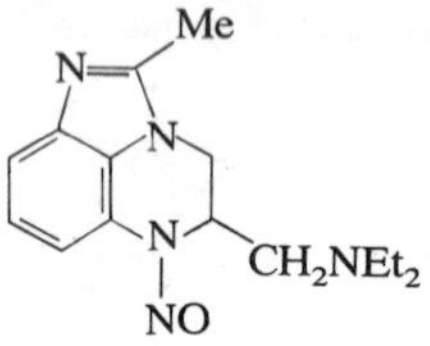Me N N N CH2NEt2 NO	95–98	51
$C_{15}H_{22}N_4$	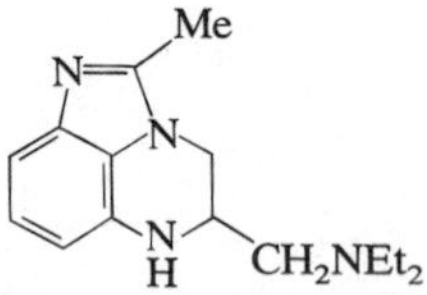Me N N N H CH2NEt2	89–91	51
$C_{16}H_{13}N_3O$	4*H*-5,6-Dihydro-4-methyl-5-oxo-2-phenyl	216	55
$C_{16}H_{19}N_3$	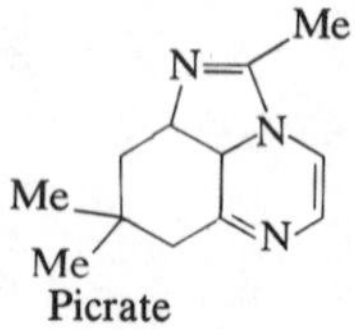Me N N Me Me N		
	Picrate	260–265	54
$C_{22}H_{19}N_3$	4*H*-5,6-Dihydro-2-methyl-*cis*-4,5-diphenyl	195	56

VI. References

1. A. M. Simonov, I. G. Uryukina, and A. F. Pozharsky, *Khim. Geterotsikl. Soedin.*, 416 (1972); *Chem. Abstr.*, **77,** 88433 (1972).
2. A. M. Simonov and I. G. Uryukina, *Khim. Geterotsikl. Soedin.*, 570 (1971); *Chem. Abstr.*, **76,** 25242 (1972).
3. R. F. Cookson and R. E. Rodway, *J. Chem. Soc., Perkin I,* 1854 (1975).
4. R. E. Rodway and R. F. Cookson, Brit. Pat. 1,347,493; *Chem. Abstr.*, **81,** 13559 (1974).
5. I. G. Uryukina and A. M. Simonov, *Khim. Geterotsikl. Soedin.*, 1558 (1972); *Chem. Abstr.*, **78,** 58345 (1973).
6. H. Otomasu, K. Yoshida, and H. Takahashi, *Yakugaku Zasshi,* **90,** 1391 (1970); *Chem. Abstr.*, **74,** 53730 (1971).
7. Chugai Pharmaceutical Co., Jap. Pat. 74 010,677; *Chem. Abstr.*, **81,** 91574 (1974).
8. H. W. Heine and A. C. Brooker, *J. Org. Chem.*, **27,** 2943 (1962).
9. M. Ungureanu, I. Druta, M. Petrovanu, and I. Zugravescu, *An. Stiint. Univ. "Al. I. Cuza" Iasi, Sect.* 1c, **18,** 49 (1972); *Chem. Abstr.*, **78,** 29712 (1973).
10. H. J. Scholl and E. Klauke (to Farbenfabriken Bayer A.-G.), Ger. Offen. 2,062,347; *Chem. Abstr.*, **77,** 152194 (1972).
11. Chugai Pharmaceutical Co., Jap. Pat. 74 010,678; *Chem. Abstr.*, **81,** 77967 (1974).
12. G. V. Kovalev, S. M. Gofman, S. V. Ivanovskaya, M. V. Pan'shina, V. I. Petrov, A. M. Simonov, and I. N. Tyurenkov, *Farmakol. Toksikol.* (Moscow), **36,** 232 (1973); *Chem. Abstr.*, **78,** 154693 (1973).
13. W. L. Mosby, "Heterocyclic Systems with Bridgehead Nitrogen Atoms," Interscience, New York, 1961, p. 836.
14. S. J. Benkovic, P. A. Benkovic, and D. R. Comfort, *J. Am. Chem. Soc.*, **91,** 5270 (1969).
15. J. W. Clark-Lewis, K. Moody, and M. J. Thompson, *Austr. J. Chem.*, **23,** 1249 (1970).
16. G. Kollenz, *Justus Liebigs Ann. Chem.*, **762,** 23 (1972).
17. F. E. King, J. W. Clark-Lewis, and C. R. P. Morgan, *J. Chem. Soc.*, 3080 (1951).
18. F. E. King and J. W. Clark-Lewis, *J. Chem. Soc.*, 172 (1953).
19. S. J. Benkovic, W. P. Bullard, and P. A. Benkovic, *J. Am. Chem. Soc.*, **94,** 7542 (1972).
20. M. P. Mertes and A. J. Lin, *J. Med. Chem.*, **13,** 77 (1970).
21. D. L. Smith and E. K. Barrett, *Acta Cyrstallogr., Sect. B,* **27,** 969 (1971); *Chem. Abstr.*, **75,** 54843 (1971).
22. E. Schipper and A. R. Day, *J. Am. Chem. Soc.*, **73,** 5672 (1951)
23. H. Roechling and K. H. Buechel, *Chem. Ber.*, **104,** 344 (1971).
24. K. H. Buechel and W. Draber, "Progr. Photosyn, Res., Proc. Int. Congr., Vol. 3," 1968, p. 1777; *Chem. Abstr.*, **74,** 30954 (1971).
25. A. C. Sircar and I. B. Pal, *J. Indian Chem. Soc.*, **9,** 527 (1932).
26. A. Dornow and E. Hinz, *Chem. Ber.*, **91,** 1834 (1958).
27. C. Iijima, *Yakugaku Zasshi,* **87,** 164 (1967); *Chem. Abstr.*, **67,** 3067 (1967)
28. K.-Y. Tserng and L. Bauer, *J. Heterocycl. Chem.*, **11,** 637 (1974).
29. L. G. S. Brooker and E. J. Van Lare (to Eastman Kodak), Fr. Pat. 1,401, 594; *Chem. Abstr.*, **64,** 19853 (1966).
30. Kodak, Belg. Pat. 660,253; *Chem. Abstr.*, **64,** 3745 (1966).
31. L. G. S. Brooker and E. J. Van Lare (to Eastman Kodak), U.S. Pat. 3,431,111; *Chem. Abstr.*, **72,** 68222 (1970).
32. A. S. Elina, I. S. Musatova, and L. G. Tysrul'nikova, *Khim. Geterotsikl. Soedin.*, 1266 (1972); *Chem. Abstr.*, **78,** 4216 (1973).

33. A. J. Fumia, Jr., and D. W. Heseltine (to Eastman Kodak), Ger. Offen. 1,800,421; *Chem. Abstr.*, **71,** 126006 (1969).
34. D. D. Chapman (to Eastman Kodak), Fr. Pat. 2,030,875; *Chem. Abstr.*, **75,** 65308 (1971).
35. B. D. Illingsworth and E. J. Van Lare (to Eastman Kodak), Fr. Pat. 1,519,916; *Chem. Abstr.*, **71,** 92669 (1969).
36. B. D. Illingsworth and E. J. Van Lare (to Eastman Kodak), Fr. Pat. 1,513,841; *Chem. Abstr.*, **71,** 49984 (1969).
37. B. D. Illingsworth and E. J. Van Lare (to Eastman Kodak), Fr. Pat. 1,522,355; *Chem. Abstr.*, **71,** 40235 (1969).
38. R. A. Litzerman, J. D. Mee, and D. W. Heseltine (to Eastman Kodak), Fr. Pat. 1,520,819; *Chem. Abstr.*, **71,** 62295 (1969).
39. J. D. Mee, P. W. Jenkins, and D. W. Heseltine (to Eastman Kodak), Ger. Offen. 1,950,746; *Chem. Abstr.*, **73,** 100079 (1970).
40. D. D. Chapman (to Eastman Kodak), Ger. Offen. 1,937,251; *Chem. Abstr.*, **73,** 89168 (1970).
41. J. D. Mee (to Eastman Kodak), Fr. Pat. 1,599,796; *Chem. Abstr.*, **74,** 100619 (1971).
42. W. S. Gaugh (to Eastman Kodak), U.S. Pat. 3,687,675; *Chem. Abstr.*, **78,** 73667 (1973).
43. B. D. Illingsworth, H. E. Spencer, J. D. Mee, and D. W. Heseltine (to Eastman Kodak), Fr. Pat. 1,520,823; *Chem. Abstr.*, **71,** 112988 (1969).
44. J. W. Carpenter, J. D. Mee, and D. W. Heseltine (to Eastman Kodak), Ger. Offen. 1,804,465; *Chem. Abstr.*, **74,** 100620 (1971).
45. F. G. Webster (to Eastman Kodak), Fr. Pat. 1,597,509; *Chem. Abstr.*, **74,** 113271 (1971).
46. C. H. Eldredge and J. D. Mee (to Eastman Kodak), Fr. Pat. 1,565,255; *Chem. Abstr.*, **73,** 89163 (1970).
47. C. H. Eldredge and J. D. Mee (to Eastman Kodak), U.S. Pat. 3,674,782; *Chem. Abstr.*, **77,** 128093 (1972).
48. J. D. Mee and D. W. Heseltine (to Eastman Kodak), Fr. Pat. 1,565,912; *Chem. Abstr.*, **73,** 110914 (1970).
49. J. D. Mee, P. W. Jenkins, and D. W. Heseltine (to Eastman Kodak), Ger. Offen. 1,950,749; *Chem. Abstr.*, **73,** 4359 (1970).
50. I. Molnar, *Chimia*, **14,** 364 (1960); *Chem. Abstr.*, **55,** 9400 (1961).
51. I. Molnar, *Pharm. Acta Helv.*, **39,** 288 (1964); *Chem. Abstr.*, **62,** 1662 (1965).
52. W. Knobloch and G. Lietz, *J. Prakt. Chem.*, **36,** 113 (1967).
53. A. Dikciuviene, V. Bieksa, and J. Degutis, *Liet. TSR Mokslu Akad. Darb. Ser. B*, 105 (1973); *Chem. Abstr.*, **79,** 115497 (1973).
54. V. I. Shvedov, L. B. Altuklova, and A. N. Grinev, *Khim. Geterotsikl. Soedin.*, 131 (1972); *Chem. Abstr.*, **76,** 153710 (1972).
55. H. Otomasu, S. Ohmiya, H. Takahashi, K. Yoshida, and S. Sato, *Chem. Pharm. Bull.* (Tokyo), **21,** 353 (1973); *Chem. Abstr.*, **78,** 136218 (1973).
56. H. Otomasu, H. Takahashi, and K. Yoshida, *Chem. Pharm. Bull.* (Tokyo), **21,** 492 (1973); *Chem. Abstr.*, **78,** 159555 (1973).
57. R. G. R. Bacon and S. D. Hamilton, *J. Chem. Soc., Perkin I,* 1970 (1974).
58. W. Lunkenheimer and K. E. Buechel (to Bayer A.-G.), Ger. Offen. 2,342,724; *Chem. Abstr.*, **82,** 156379 (1975).
59. D. J. Sam and M. A. Wuonola (to E. I. du Pont de Nemours), Ger. Offen. 2,339,012; *Chem. Abstr.*, **80,** 108578 (1974).
60. A. Dikciuviene, V. Bieksa, and J. Degutis, *Liet. TSR Mokslu Akad. Darb. Ser. B*, 81 (1974); *Chem. Abstr.*, **83,** 43243 (1975).

CHAPTER XXXVII

Pyrazoloquinoxalines

I. Pyrazolo[1,5-*a*]quinoxalines . . . 691
II. Pyrazolo[3,4-*b*]quinoxalines . . . 693
III. Tables . . . 704
Table 1. Pyrazolo[1,5-*a*]quinoxalines . . . 704
Table 2. Pyrazolo[3,4-*b*]quinoxalines . . . 704
IV. References . . . 710

I. Pyrazolo[1,5-*a*]quinoxalines

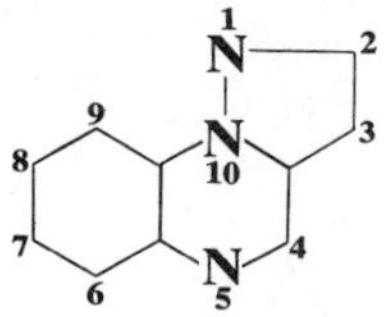

This ring system has also been called pyrazolo[2,3-*a*]quinoxaline and pyrazolo[*a*]quinoxaline. Little work has however been done on this heterocycle. Catalytic reduction of the *o*-nitrophenylpyrazole **1** results in ring closure to give the pyrazolo[1,5-*a*]quinoxaline **2**. Sequential decarboxylation, treatment with phosphoryl chloride, and catalytic hydrogenation give the parent heterocycle **3**. The unsubstituted compound is reported to have a broad band at 240–245 nm in its ultraviolet spectrum in 95% ethanol.[1] A recent patent describes several 4-substituted amino derivatives of pyrazolo[1,5-*a*]quinoxaline-3-carboxylic acid as possessing antiinflammatory properties.[55]

When 4-oxo-5-methyl-1*H*,4,5-dihydro-1,2,5-benzotriazepine is heated with hydrochloric acid, isomerization occurs with the formation of the *N*-iminoquinoxaline **4**. The structure of this product was demonstrated by reaction with diethyl acetylenedicarboxylate which on hydrolysis and methylation gave the pyrazolo[1,5-*a*]quinoxaline **5**, identical with authentic material prepared from the appropriately substituted phenylpyrazole.[2]

CO_2H N N CO_2H NO_2 **1** → CO_2H N N N H O **2**

↓

N N N **3** ← N N N Cl ← N N N H O

NH—N CH N—C Me O $\xrightarrow{(i)}$ ^{-}NH ^{+}N N Me O **4**

↓ (ii)

CO_2Et N N CO_2Et CO_2H NH_2 $\xrightarrow{(iii)}$ CO_2Me N N CO_2Me N Me O **5**

Reagents (i) Δ, HCl
(ii) $EtO_2CC{\equiv}CCO_2Et$, OH^-, CH_2N_2
(iii) OH^-, H^+, CH_2N_2

II. Pyrazolo[3,4-*b*]quinoxalines

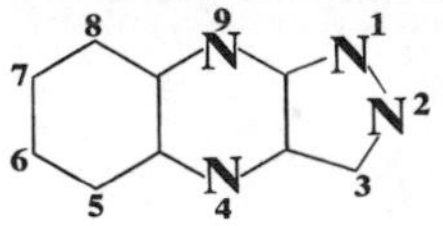

RRI 2735

The main interest in this ring system stems from the use of 1-phenyl compounds as derivatives of reducing sugars. The yellow color of these derivatives has led to the heterocycle being referred to frequently as flavazole. Other names that have been used on isolated occasions are pyrazolo-4′5′,2,3-quinoxalines, pyrazo[3,4-*b*]quinoxalines, and, surprisingly, pyrazolo[2,3-*b*]quinoxaline. Since the original survey of this nucleus by Simpson,[3] a brief review has appeared in German describing the synthesis of flavazoles from sugars and their derivatives.[4]

Although no details have been given, it seems that LCAO MO calculations of π-electron density on 1-phenylpyrazolo[3,4-*b*]quinoxaline do not predict the observed positions of electrophilic substitution.[5,6]

Ultraviolet spectral data have been recorded on the parent molecule[7] and on sugar flavazoles.[8] The unsubstituted compound has a three-banded spectrum with maxima at 233, 331, and 376 nm and an inflection at 393 nm. Three-banded spectra are also shown by the 1-phenylflavazoles of glucose (λ_{max} 267, 335, and 410 nm), maltose, and lactose. The PMR spectrum of the parent heterocycle ($CDCl_3$) has peaks at δ 14.06 (N-H), 8.8 (H-3), and a multiplet in the range of δ 7.6–8.4, which has been partially analyzed as an ABCD pattern.[7] Mass fragmentation details have been reported for 3-methylpyrazolo[3,4-*b*]quinoxaline[7] and 1-phenylflavazoles derived from monosaccharides.[9] The 1-phenylflavazoles of oligosaccharide peracetates have sufficiently stable molecular ions to permit oligosaccharide sequencing by mass spectrometry.[10] Flavazole and 1-phenylflavazole derivatives of sugars exhibit Cotton effects at 390 and 410 nm, respectively,[11] and the sign of this effect may be used to assign the configuration at the C-4 atom of the sugar.

Many 1-phenylflavazoles derived from sugars have been characterized by chromatographic R_f values[12–14] and by x-ray diffraction patterns.[15–17] Most syntheses of pyrazolo[3,4-*b*]quinoxalines involve polyhydroxyalkylquinoxalines as intermediates and, by reaction with phenylhydrazine, result in the formation of 1-phenyl substituted derivatives of the ring system.* Thus the D-arabotetrahydroxybutyl compound **6**, obtained from reaction of D-glucose with *o*-phenylenediamine, reacts with three moles

* The chemistry of polyhydroxyalkylquinoxalines is reviewed in Chapter XVI.

of phenylhydrazine in acetic acid at 100° to give the D-erythrotrihydroxypropyl compound **7** in 97% yield.

N N HOCH HCOH HCOH CH_2OH + $3PhNHNH_2$ ⟶

6

Ph N N N N N + $2PhNH_2 + 2NH_3 + H_2O$
HCOH HCOH CH_2OH

7

As in osazone formation, the phenylhydrazine functions as an oxidizing agent as well as a reactant.[18] The corresponding reaction with hydrazine gives a much lower yield of the 1-unsubstituted compound, and the analogous formation of 1-methylflavazoles using methylhydrazine is even poorer.[19] No pyrazolo compound is formed with hydrazines substituted with benzoyl, tosyl, or carbamyl groups. The L-threo isomer **9** of the flavazole **7** is obtained in good yield from the L-xylotetrahydroxybutyl compound **8**.[20]

N N HOCH HCOH HOCH CH_2OH ⟶ Ph N N N N N HCOH HOCH CH_2OH

8 **9**

1-Phenylflavazole derivatives of sugars may be conveniently prepared without isolation of the intermediate hydroxyalkylquinoxalines[21] and as such provide a convenient means for the characterization of sugars and sugar derivatives,[22] particularly on the semimicro scale.[23] Fully characterized derivatives have been described for many monosaccharides,[21,24] disaccharides,[13,17,24,25] and trisaccharides.[13,16,17,24–27] 1-Phenylflavazoles have also been prepared from various higher order oligosaccharides.[13,14,24] In general these derivatives are prepared by heating one mole of *o*-phenylenediamine and five moles of phenylhydrazine with one mole of reducing sugar. The reaction is carried out in glacial acetic acid, by heating on a boiling water bath in the absence of air.[14]

The phenylhydrazone **10** of quinoxaline-2-aldehyde has been cyclized to 1-phenylpyrazolo[3,4-*b*]quinoxaline (**11**) by heating with phenylhydrazine in acetic acid[28,29] or merely by heating with a mixture of acetic acid and hydrochloric acid in *n*-propanol.[30] In spite of the general difficulty in performing intranuclear cyclizations on the quinoxaline nucleus, **11** is obtained in 60–70% yield.

CH=NNHPh

10 → **11**

Quinoxalinones of general formula **12** have been cyclized to 1-phenylflavazoles by treatment with dilute sodium hydroxide solution. The reaction works well for **12** (R = CH=NNHPh or CH<(NPh—CH_2)(NPh—CH_2))[31] and for **12** (R = $CHOHCHOHCH_2OH$),[29] but **12** (R = CHO) does not undergo ring closure but loses the CHO group instead.[4]

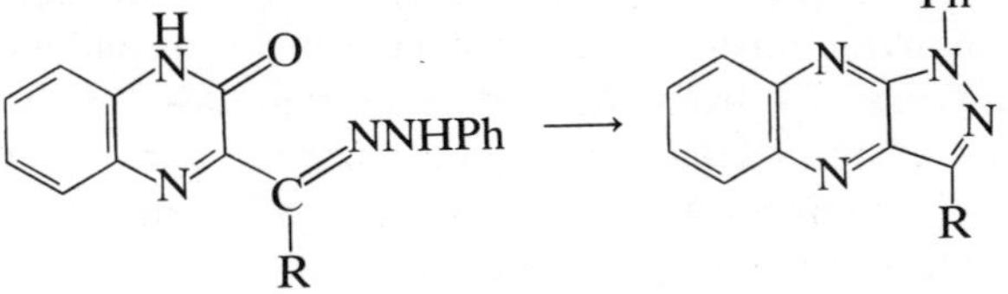

NNHPh →

12 R = CH=NNHPh, CH<(N(Ph)—CH_2—CH_2—N(Ph)) or $CHOHCHOHCH_2OH$

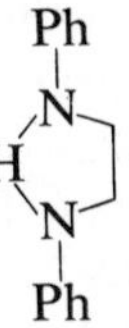

Analogously the furoquinoxaline **13**, obtained from the reaction of dehydro-L-ascorbic acid with *o*-phenylenediamine and phenylhydrazine, can be converted into the pyrazolo derivative **14** by a ring opening–ring closure sequence either by heating in dilute alkali or by heating the sodium salt of the compound in water.[32]

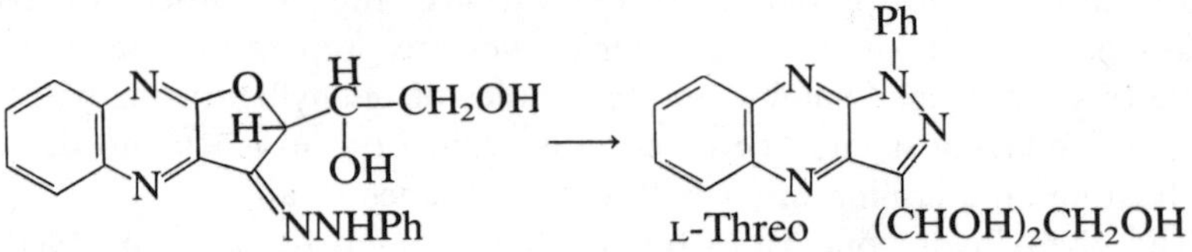

13 → **14** (L-Threo)

Although the reaction of 2-benzoylquinoxaline (**15**) with excess phenylhydrazine gives 1,3-diphenylpyrazolo[3,4-*b*]quinoxaline (**18**) in 89% yield, reaction with only one equivalent of phenylhydrazine yields only 38% of the diphenylpyrazolo compound. Excess phenylhydrazine also converts the substituted compounds **16** into diphenyl compound **18** in greater than 90% yield.[33] The benzyl and benzoyl analogues **17** give the same product but in lower yield.[34]

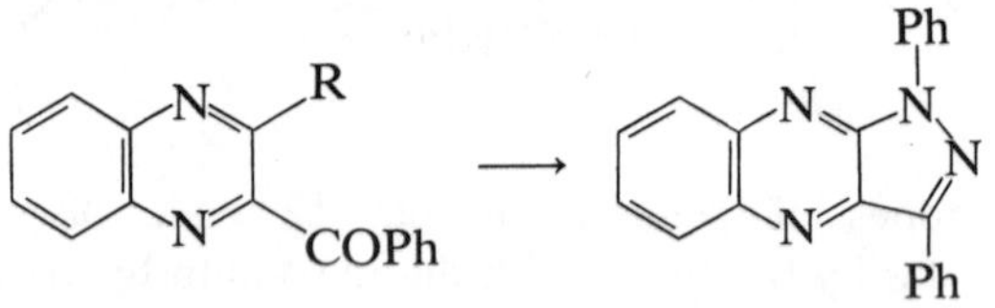

15 R = H
16 R = CO_2H or $CONH_2$
17 R = CH_2Ph or COPh

→ **18**

A study of the scope of this reaction has been performed which shows that excess phenylhydrazine will convert the disubstituted quinoxalines **19** into the pyrazolo[3,4-*b*]quinoxaline **20**. Analogously hydroxylamine gave the isoxazolo compound **21** from these starting materials.[34] The compounds **19** (R = Me or Ph) did not undergo cyclization with phenylhydrazine. A reaction mechanism has been suggested (see Scheme 1) which is analogous to osazone formation in sugar chemistry.

Reaction of 2-chloro 3-substituted quinoxalines **22** with hydrazines provides another simple preparation of pyrazolo[3,4-*b*]quinoxalines. The parent heterocycle **23** has been obtained by treating the aldehyde **22** (R = CHO)[35] or the dibromomethyl derivative **22** (R = $CHBr_2$)[36] with hydrazine. Similarly the nitrile **22** (R = CN) has given the 3-amino compound **24** in 72% yield,[37] and methylhydrazine with the ester **22** (R = CO_2Et) gave the 3-hydroxy derivative **25**.[36] In spite of the ready cyclization of the aldehyde **22** (R = CHO) with hydrazine, the corresponding

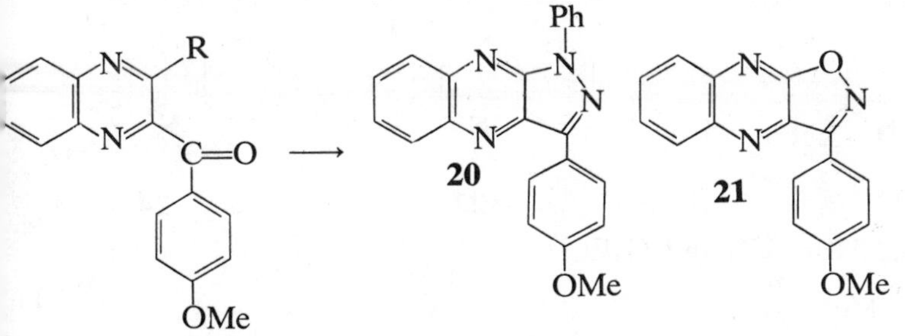

20 **21**

19 R = H, OH, Cl, CN, $CONH_2$, CO_2H

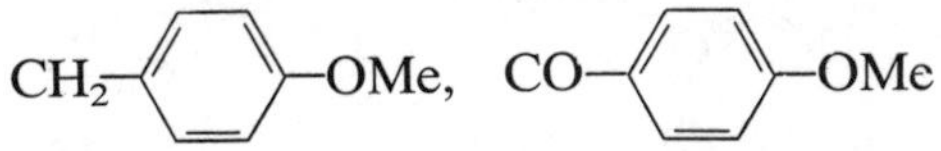

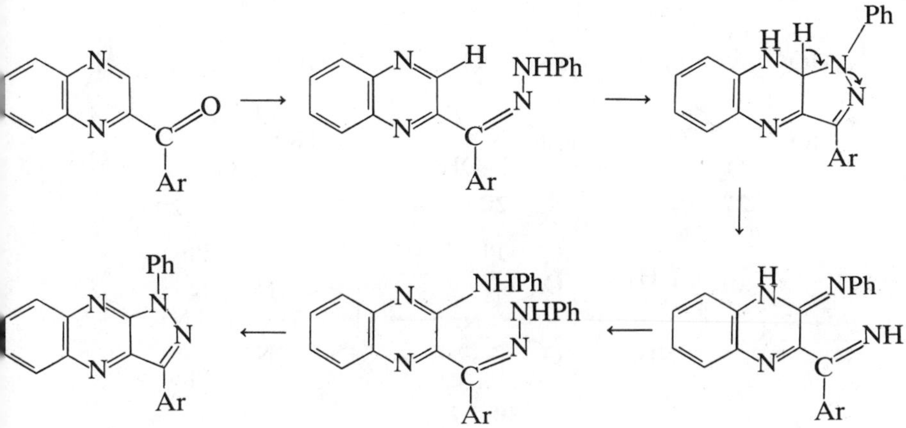

CH_2–C_6H_4–OMe, CO–C_6H_4–OMe

Scheme 1

reaction with phenylhydrazine was much more difficult even at 180°. On the other hand, the phenylhydrazone **26** readily cyclized to the diphenyl compound **18** on refluxing in acetic acid.[35]

Several syntheses of pyrazolo[3,4-*b*]quinoxalines have been reported which do not involve quinoxaline intermediates. Reaction of the diaminopyrazole **28** with rhodizonic acid (**27**) gives the trihydroxydione **29**.[38] Condensation of *o*-phenylenediamine with the pyrazoledione **30** in alcohol yields an anil which on heating in acetic acid gives the 1,3-diphenyl compound **18**.[39] 1-Phenyl-3-methylpyrazolo[3,4-*b*]quinoxaline[40] and 6- and 7-substituted derivatives[41] have been similarly prepared. The anils **31** on heating in acetic acid gave low yields of the pyrazoloquinoxalines **32**.[42]

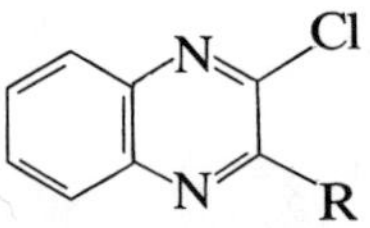

22 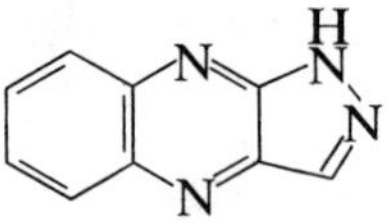 23 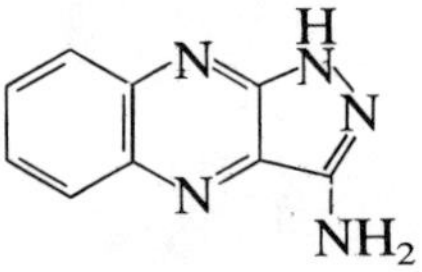 24

R = CHO, $CHBr_2$, CN, or CO_2Et

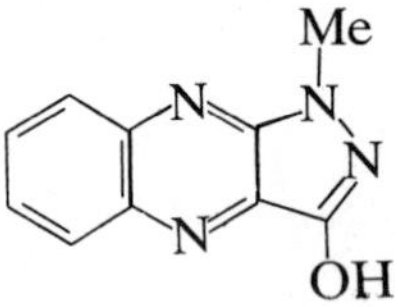

25 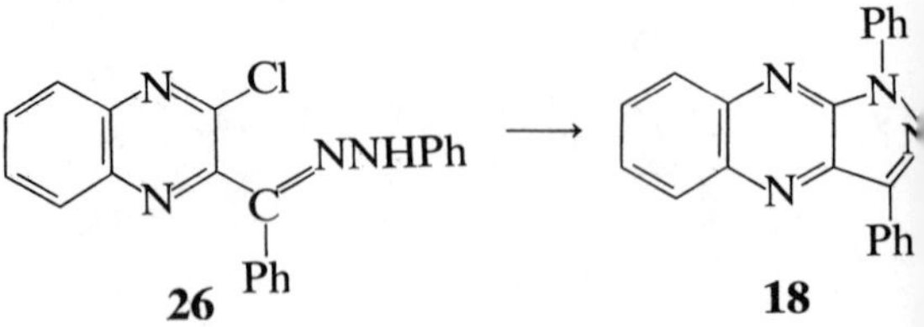 26 18

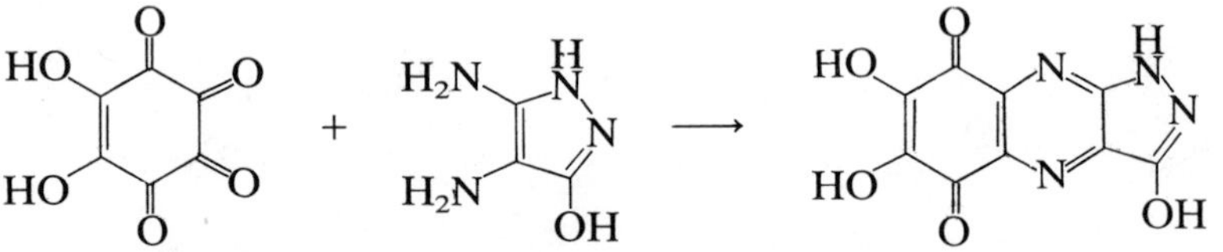

27 28 29

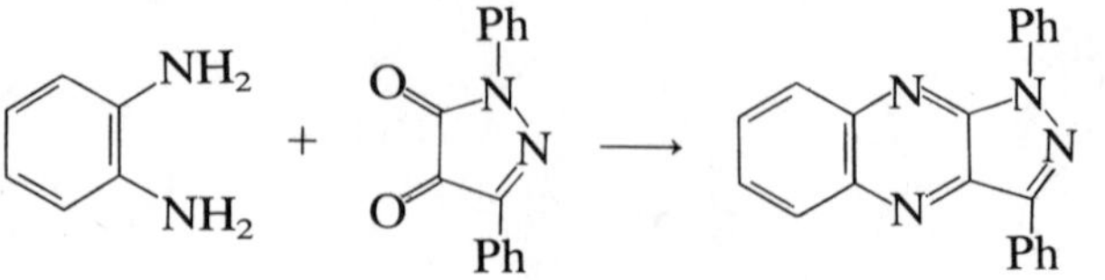

30 18

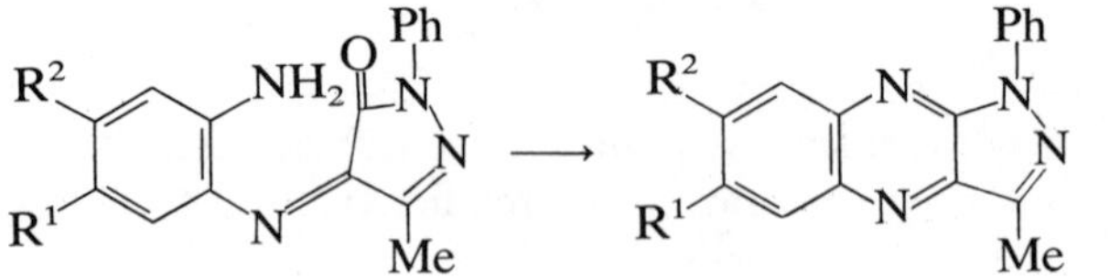

31 32 R^1 = H or Me
R^2 = NH_2 or NMe_2

The anils **33** on heating or on treatment with acid cyclize to give the fused quinoxalines **34**.[43] The compound **34** (R = Me) has also been prepared[44] by reaction of the pyrazoline **36** with the aniline **35** in the presence of $K_2S_2O_8$ or AgBr.

Although the unsubstituted heterocycle is readily acylated by acid chlorides and anhydrides,[36] the position of substitution has not been

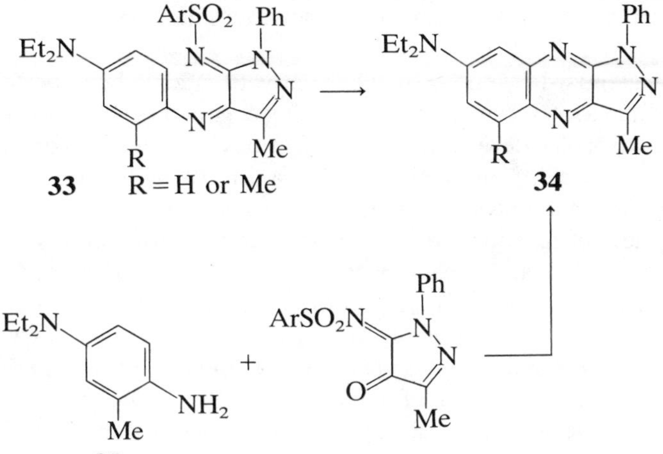

33 R = H or Me → **34**

35 + **36** → **34**

specified. The parent is unaffected by methyl iodide at 100° and by diazomethane at room temperature.[19] Methyl sulfate in sodium hydroxide solution gives a monomethyl derivative shown by independent synthesis to be the 1-methyl compound. 1-Alkyl derivatives are readily prepared by reaction with alkyl halides in DMF in the presence of potassium carbonate. The 1-unsubstituted compound is sufficiently acidic to undergo the Mannich reaction with formaldehyde and secondary amines.[36] In the preliminary disclosure of these results no data have been given on any of the products of these reactions. Although no quantitative data are available on the acidity of the parent ring system, 1-unsubstituted compounds have been shown to form sodium salts in aqueous solution from which the free acid may be precipitated by carbon dioxide.

It seems that treatment of the parent compound with NBS or NCS in DMF, or reaction with aqueous sodium hypobromite, results in halogenation at the 3-position.[36] The ring system has been described as being as inert to electrophilic substitution as indazole but more reactive than quinoxaline. Nitration under unspecified conditions of the heterocycle results in a mixture of mononitro compounds, the major product being either the 6- or 7-nitro derivative and the minor product having a nitro group in either the 5- or the 8-position.[6] Both nitro compounds were reduced to the corresponding primary amines with stannous chloride.

Nitration of 1-phenylpyrazolo[3,4-*b*]quinoxalines occurs in the para-position of the pendent phenyl ring.[5] Reduction of the nitro group followed by diazotization and coupling with, for example, β-naphthol has given several azo dyes. Sulfonation of the parent heterocycle with the pyridine–sulfur trioxide adduct again provides 3-substituted products.

1-Phenylpyrazolo[3,4-*b*]quinoxalines however sulfonate at the para position of the phenyl ring.[45]

Many standard reactions of sugar derivatives have been performed on 1-phenylflavazoles derived from reducing sugars. For example, glycosidic links in flavazoles derived from disaccharides are readily hydrolyzed with 2 *M* sulfuric acid,[25] although the mechanism of hydrolysis seems to be different to that encountered with other disaccharide derivatives.[46] A wide variety of reactions such as the formation of trityl ethers, benzoates, and acetone derivatives have been described for the 1-phenylflavazole **7** derived from glucose.[18] Treatment of compound **7** with lead tetraacetate gives the 3-formyl compound **38**.[18] This oxidation does not work well with the corresponding 1-unsubstituted compound **37**, but periodic acid gives a quantitative yield of the aldehyde **39**.[19] The oxidation of the 1-phenyl flavazole may also be performed using sodium periodate in methanol.[16]

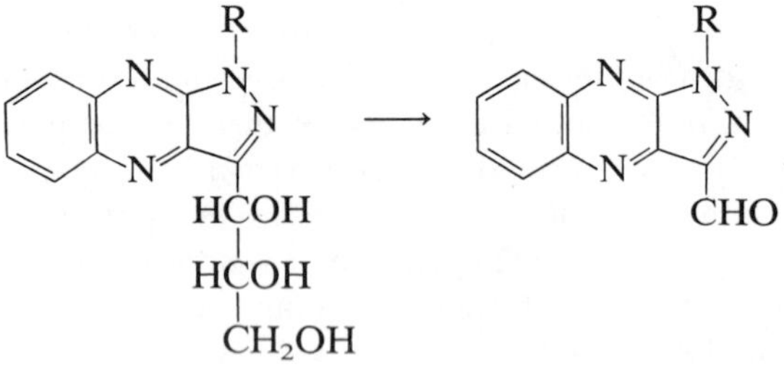

R
N N N
N
HCOH
HCOH
CH_2OH
→
R
N N N
N
CHO

7 R = Ph
37 R = H

38 R = Ph
39 R = H
40 R = Me

Oxidation of the trihydroxypropyl compound **7** or the aldehydes **38** and **40** with chromium trioxide in acetic acid or 50% sulfuric acid furnishes the 3-carboxylic acids, which readily decarboxylate to the 3-unsubstituted compounds at their melting points.[19,40] In contrast, the 3-methyl-1-phenyl compound is not appreciably attacked by chromium trioxide, potassium permanganate, or selenium dioxide. Chromyl chloride oxidation of **7** does not give the aldehyde **38** but apparently provides the tetraflavazolyl ether **41**. The inertness of the 3-methyl group is emphasized by its stability toward bromine and its lack of reaction with benzaldehyde.[40]

Pyrazolo[3,4-*b*]quinoxalin-3-ylcarboxylic acid (**42**) is only sparingly soluble in alcohols and therefore esterifies with difficulty. Moreover treatment of the methyl ester with methanolic ammonia did not provide the amide **44**. This amide was obtained, however, by acetylation of the

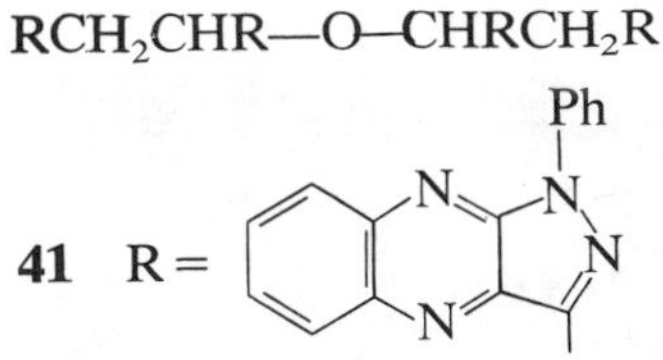

RCH$_2$CHR—O—CHRCH$_2$R

41 R =

acid with acetic anhydride to give the 1-acetyl derivative, which with thionyl chloride formed the acid chloride **43**. Reaction of compound **43** with methanolic ammonia smoothly gave the required amide with simultaneous deacetylation under the reaction conditions.[19]

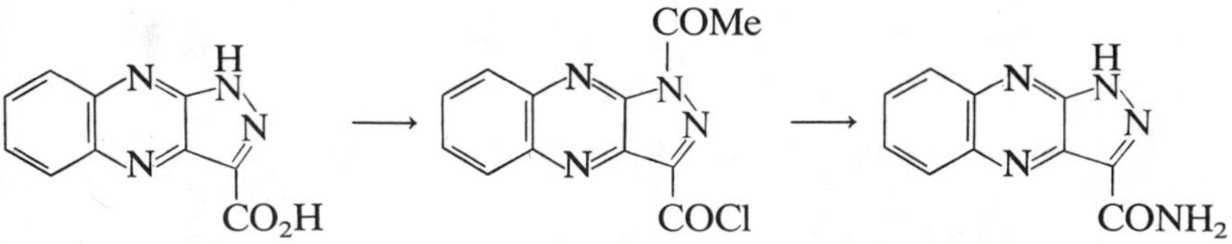

42 → **43** → **44**

The aldehydes **38** and **39** behave similarly to other aromatic aldehydes under a variety of conditions. Treatment of **39** with boiling concentrated alkali results in the Cannizzaro reaction providing the corresponding carboxylic acid and the 3-hydroxymethyl derivative. Boiling hydrazine converts the aldehyde into 3-methylpyrazolo[3,4-*b*]quinoxaline in a Wolff-Kishner-type reduction.[9] Compound **39** readily forms hydrazones[7,47] and gives olefins on reaction with malonic acid and *p*-nitrobenzylcyanide.[47] 1-Phenylpyrazolo[3,4-*b*]quinoxalin-3-yl aldehydes have been reacted with potassium cyanide to provide the corresponding so-called flavazoins **45**, which on aerial oxidation gave the analogous flavazils.[31,48] Reactions of the aldehyde **38** with the diamine **46** gave the imidazolidine **47**.[31]

As already indicated, the principal use of 1-phenylflavazoles is in the characterization of oligosaccharides containing reducing sugars.[21–23] The method has been used to identify the products of enzymic digestion of polysaccharides[14,15,17,27] as well as in the structural elucidation of stachyose[16] and an exotoxin from a bacillus.[49] The advantages of flavazoles as derivatives of sugars have been listed.[14] Apart from their ease of preparation, 1-phenylflavazoles may be chromatographed in the same solvent as the parent sugars, and their yellow color and ultraviolet fluorescence make for easy detection. Not only are 1-phenylflavazoles useful in characterization, they can provide evidence on the configuration of the sugars from which they are derived.[11,21] Flavazole chemistry has

45 R = H or Cl

38 + PhNHCH$_2$CH$_2$NHPh (**46**) ⟶ **47**

been used to measure the extent of ^{14}C labeling in the individual carbon atoms of glucose.[51] Carbon atoms in positions 1 to 3 of glucose become incorporated into positions 9a, 3a, and 3 of the tricyclic system, and the carbon atoms in positions 4 to 6 of glucose form the side chain at position 3 in the flavazole **48**.

Sodium periodate oxidation converts **48** into the corresponding aldehyde **38** and releases one mole of formaldehyde (from C-6) and one mole of formic acid (from C-5), both of which can be estimated independently. Chromium trioxide oxidation of **38** gives the corresponding acid which decarboxylates readily to the 3-unsubstituted compound **11** and

48 ⟶ (CHO) + HCO_2H[C-5] + CH_2O[C-6] ⟶ **11** + CO_2[C-4]

carbon dioxide (from C-4), which may then be measured. Comparison of the residual flavazole with unsubstituted quinoxaline, obtained by oxidation of the flavazole followed by decarboxylation, gives an indication of the extent of labeling at C-3 of glucose. Independent estimations of the label at C-1 then allow calculation by difference of the remaining label on C-2.

1-Phenylflavazoles have found little application in chemotherapeutics,[4] however certain 1-unsubstituted compounds show tuberculostatic activity.[7,47] Compounds **49** and **50** show analgesic activity at 100 mg/kg orally in the mouse acetic acid antiwrithing test but are inactive in the rat paw volume test.[7] The quinone **29** has been patented as having antiinflammatory and diuretic activity.[38]

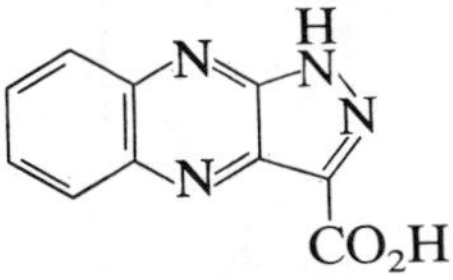

49 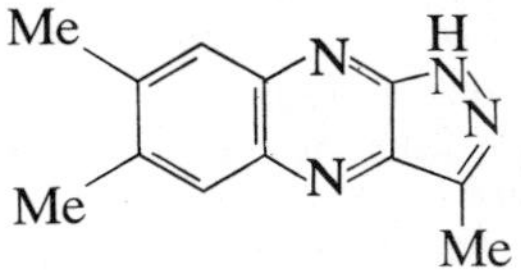 **50** 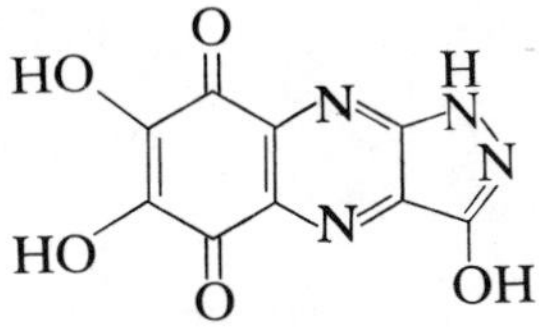 **29**

Certain meta-substituted phenylflavazoles of mono- and polysaccharides have found application as immunogens.[12,52] Azo dyes have been obtained from *para*-aminophenylflavazoles,[5] and the pyrazoloquinoxalines **51** have been patented as fluorescent dyes.[44]

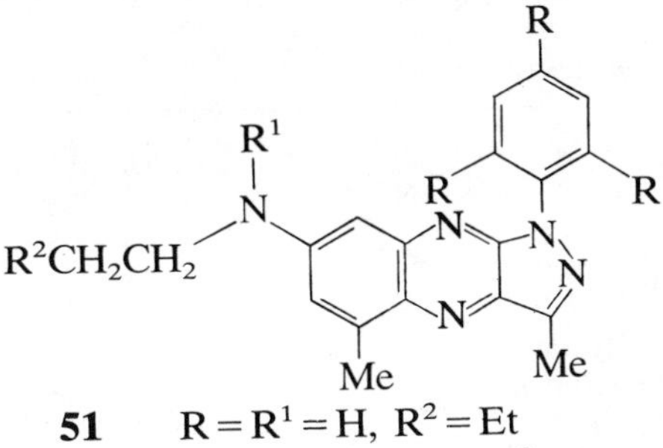

51 $R = R^1 = H$, $R^2 = Et$
$R = Cl$, $R^1 = OH$, $R^2 = H$

III. Tables

TABLE 1. PYRAZOLO[1,5-*a*]QUINOXALINES

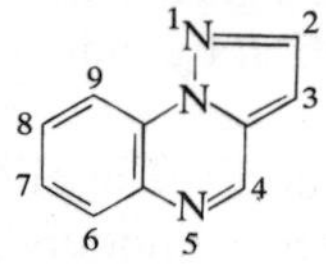

Molecular formula	Substituents	m.p., °C	Ref.
$C_{10}H_6ClN_3$	4-Chloro	120–122	1
$C_{10}H_7N_3$	Unsubstituted	88–90	1
	Picrate	175–178	1
$C_{10}H_7N_3O$	4,5-Dihydro-4-oxo	266–268	1
$C_{11}H_7N_3O_3$	2-Carboxy-4,5-dihydro-4-oxo	>315	1
$C_{12}H_7N_3O_5$	2,3-Dicarboxy-4,5-dihydro-4-oxo		
	Dimethyl ester	141–142	2
$C_{13}H_9N_3O_5$	2,3-Dicarboxy-4,5-dihydro-5-methyl-4-oxo	284–285	2
	Dimethyl ester	165–166	

TABLE 2. PYRAZOLO[3,4-*b*]QUINOXALINES

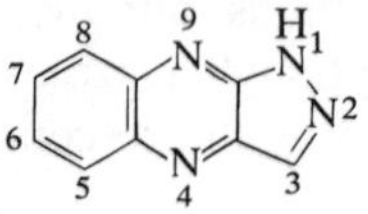

Molecular formula	Substituents or structure	m.p., °C	Ref.
$C_9H_4N_4O_5$	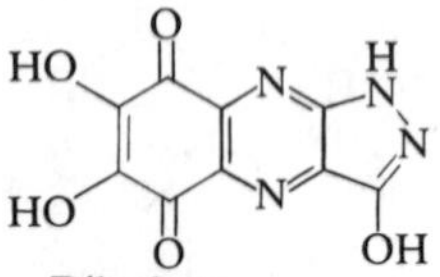		
	Dihydrate	>300	38
$C_9H_6N_4$	Unsubstituted	283.5–284.5	7, 19, 35, 36
$C_{10}H_5N_5$	3-Cyano	273	7
$C_{10}H_6N_4O$	3-Formyl	285	7
		256–258	19
	Oxime	>360	
	Phenylhydrazone[a]	199	47
$C_{10}H_6N_4O_2$	3-Carboxy	272–273	19
	Methyl ester	257–258	
$C_{10}H_7N_5O$	3-Carboxamido		19
$C_{10}H_8N_4$	1-Methyl	165	19
	3-Methyl	226	7 19

TABLE 2 (*contd.*)

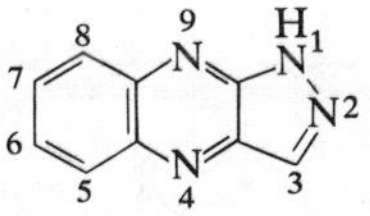

Molecular formula	Substituents or structure	m.p., °C	Ref.
$C_{10}H_8N_4O$	3-Hydroxymethyl	244–246	19
$C_{11}H_8N_4O$	1-Acetyl	160–162	3
$C_{11}H_8N_4O_2$	3-Carboxy-1-methyl	249	19
$C_{11}H_{10}N_4$	6,7-Dimethyl	265	7
$C_{12}H_8N_4O_2$	3-(2-Carboxyvinyl)	266 decomp.	47
$C_{12}H_8N_4O_3$	1-Acetyl-3-carboxy	213–214	19
	Acid chloride	162–163	
$C_{12}H_9N_5$	3-Cyano-6,7-dimethyl	306	7
$C_{12}H_{10}N_4O$	3-Formyl-6,7-dimethyl	283	7
	Oxime	>360	
	Phenylhydrazone[a]	313	47
	2,4-Dinitrophenyldrazone	>362	
$C_{12}H_{10}N_4O_2$	3-Carboxy-6,7-dimethyl	362	7
$C_{12}H_{11}N_5O$	3-Carboxamido-6,7-dimethyl	>360	47
$C_{12}H_{12}N_4$	3,6,7-Trimethyl	252	7
$C_{12}H_{12}N_4O_3$	3-(D-Erythro)trihydroxypropyl[b]	225–226 decomp.	19
	Monoacetate[c]	195	
	Tetracetate[d]	99–100	
	Monobenzoate[e]	215–216	
	Tetrabenzoate[f]	96–98	
$C_{13}H_{14}N_4O_3$	1-Methyl-3-(D-erythro)trihydroxypropyl	151–152	19
$C_{14}H_{12}N_4O_3$	3-Acetoxymethyl-1-acetyl	185–187	19
$C_{14}H_{16}N_4O_3$	3-(D-Erythro)trihydroxypropyl-6,7-dimethyl	272	7
$C_{15}H_{10}N_4$	1-Phenyl	152.5–153.5	28, 30, 40
$C_{16}H_9ClN_4O$	7-Chloro-3-formyl-1-phenyl	242	48
$C_{16}H_{10}N_4O$	3-Formyl-1-phenyl	144	16, 18
	Phenylhydrazone	196–197[g], 225[h]	
	2,4-Dinitrophenylhydrazone	271–272	
$C_{16}H_{10}N_4O_2$	3-Carboxy-1-phenyl	244 decomp.	31, 40
	Ethyl ester	168	
$C_{16}H_{11}N_5O_2$	3-Methyl-7-nitro-1-phenyl	216–218	41
$C_{16}H_{12}N_4$	3-Methyl-1-phenyl	133.5–134[i]	40
$C_{16}H_{13}N_5$	7-Amino-3-methyl-1-phenyl	234–235	40, 41
$C_{17}H_{12}N_4O$	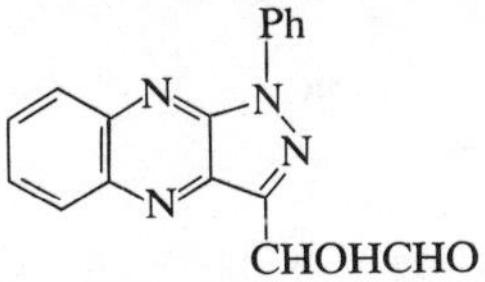 (1-Ph, 3-CHOHCHO structure)		
	1-Benzoate[j]	147	18

TABLE 2 (*contd.*)

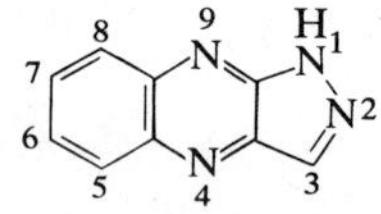

Molecular formula	Substituents or structure	m.p., °C	Ref.
$C_{17}H_{14}N_4O_2$	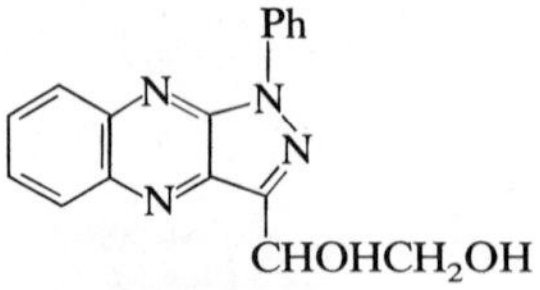 1-Ph, 3-$CHOHCH_2OH$		
	D-Erythro[k]	212	21
	L-Erythro[l]	215	21, 24
$C_{17}H_{15}N_5$	7-Amino-3,6-dimethyl-1-phenyl	312–314	42
$C_{18}H_{16}N_4O_2$	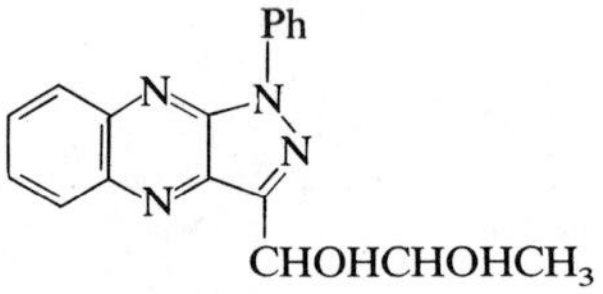 1-Ph, 3-$CHOHCHOHCH_3$		
	L-Erythro[m]	215–217	21, 24, 50
	[n]	199–200.5	50
	[o]	213–215	50
$C_{18}H_{16}N_4O_3$	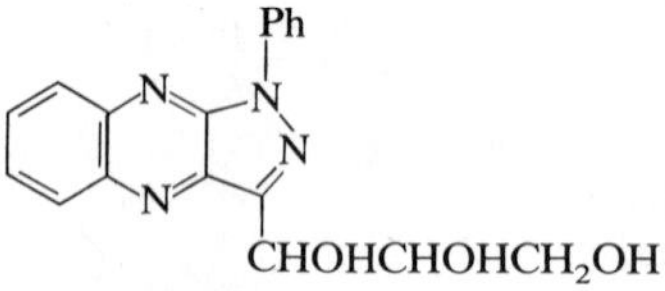 1-Ph, 3-$CHOHCHOHCH_2OH$		
	D-Erythro[p,q]	218	24, 25, 54
	Triacetyl[r]	123–124	54
	D-Threo[s]	194–195	21, 22, 24
	Acetyl deriv.	97	21
	L-Threo[t]	194	
	Acetyl deriv.	97	
$C_{18}H_{17}N_5$	7-Dimethylamino-3-methyl-1-phenyl	203–204	42
$C_{19}H_{18}N_4O_3$	6(7)-Methyl-1-phenyl-3-(D-erythro)-trihydroxypropyl[u]	228 decomp.	53
$C_{20}H_{14}N_6O_2$	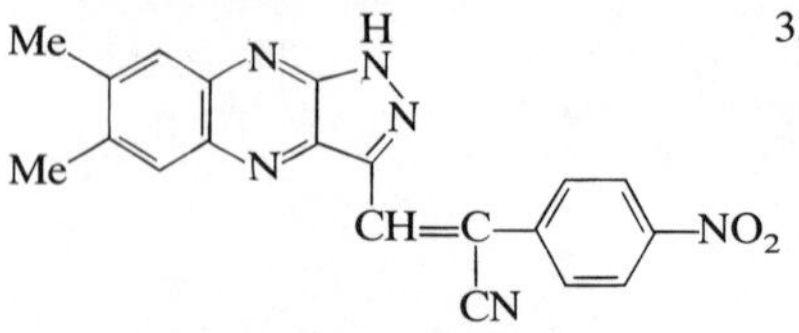 6,7-Me₂, 1-H, 3-$CH{=}C(CN)$-C_6H_4-NO_2	334	47
$C_{20}H_{21}N_5$	7-Diethylamino-3-methyl-1-phenyl	203	43
$C_{21}H_{14}N_4$	1,3-Diphenyl	233–234	33, 39
$C_{21}H_{23}N_5$	7-Diethylamino-3,5-dimethyl-1-phenyl	160	43

TABLE 2 (*contd.*)

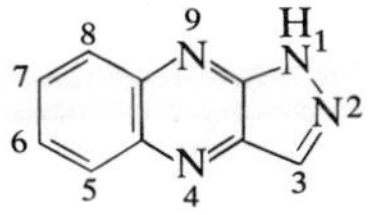

Molecular formula	Substituents or structure	m.p., °C	Ref.
$C_{24}H_{26}N_4O_8$	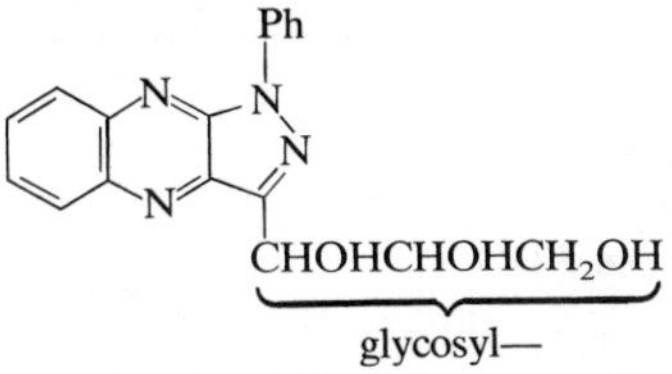 1-Ph, 3-CHOHCHOHCH$_2$OH (glycosyl—)		
	Derivatives of maltose[v]	265	24, 25
	gentiobiose[w]	245–247	24
	melibiose[x]	218–221	24
	lactose[y]	272	24, 25
	vicianose[z]	216–220	24
	cellobiose[aa]	280–282	25
$C_{30}H_{24}N_6$	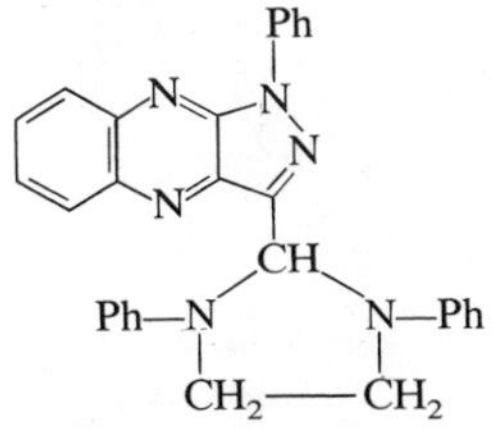 1-Ph, 3-(1,3-diphenylimidazolidin-2-yl) [CH(N(Ph)CH$_2$CH$_2$N(Ph))]	241	31
$C_{30}H_{36}N_4O_{13}$	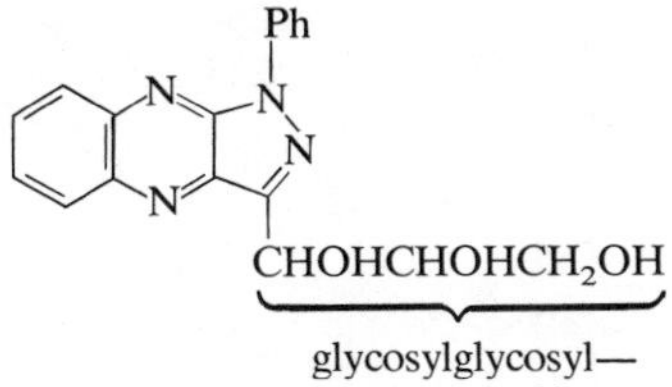 1-Ph, 3-CHOHCHOHCH$_2$OH (glycosylglycosyl—)		
	Derivatives of manninotriose[bb]	236–238	24
	maltotriose[cc]	180–192	26
	a trisaccharide containing isomaltose[dd]	206–209	27
$C_{32}H_{16}Cl_2N_8O_2$	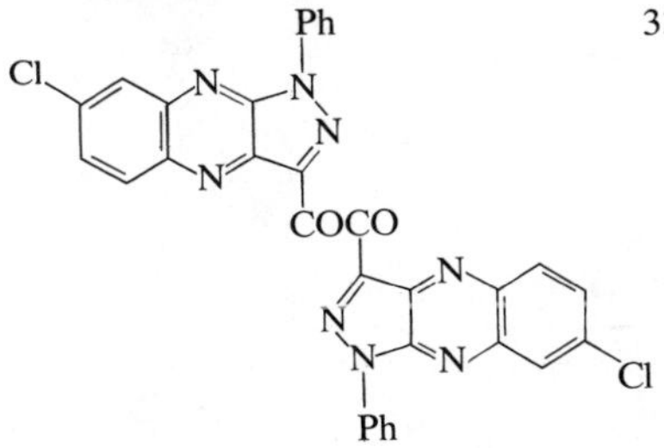 bis(7-Cl-1-Ph-pyrazolo[3,4-b]quinoxalin-3-yl) ketone (COCO bridge)	334	48

TABLE 2 (*contd.*)

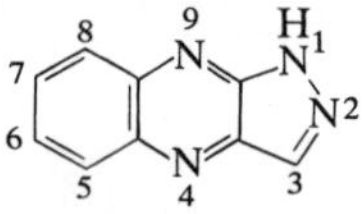

Molecular formula	Substituents or structure	m.p., °C	Ref.
$C_{32}H_{18}Cl_2N_8O_2$	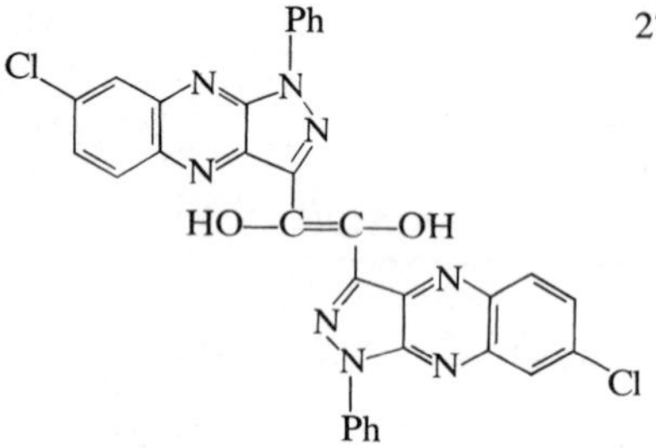 HO—C=C—OH (bis(7-chloro-1-phenylpyrazolo[3,4-b]quinoxalin-3-yl))	275	48
$C_{32}H_{18}N_8O_2$	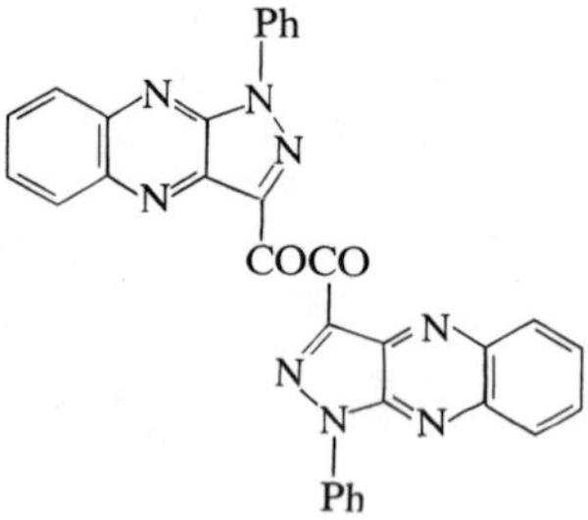 COCO (bis(1-phenylpyrazolo[3,4-b]quinoxalin-3-yl))	300	48
	Monophenylhydrazone	195–198	
$C_{32}H_{20}N_8O_2$	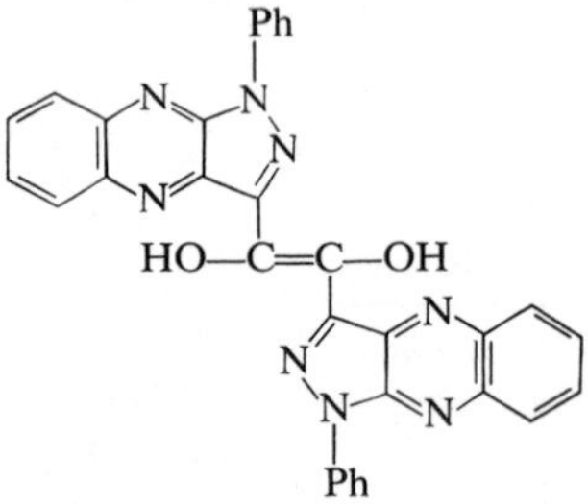 HO—C=C—OH (bis(1-phenylpyrazolo[3,4-b]quinoxalin-3-yl))	264	31
		288	48
	Diacetate	335–340	48
	Dibenzoate	355–360	48
	Bisphenylhydrazone	160	48
$C_{36}H_{46}N_4O_{18}$	1-Phenylpyrazolo[3,4-b]quinoxalin-3-yl CHOHCHOHCH$_2$OH; glycosylglycosylglycosyl—		
	Derivative of trigalactosidoglucose[ee]	257–262	24

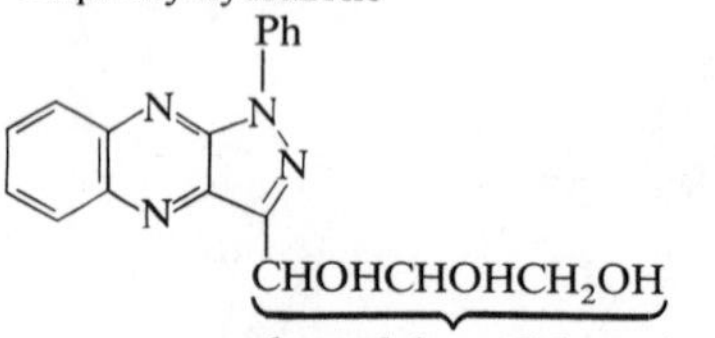

TABLE 2 (*contd.*)

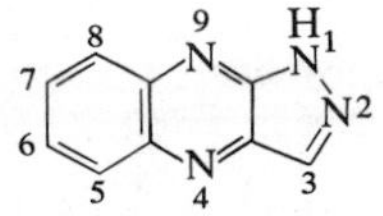

Molecular formula	Substituents or structure	m.p., °C	Ref.
$C_{38}H_{22}N_{10}$	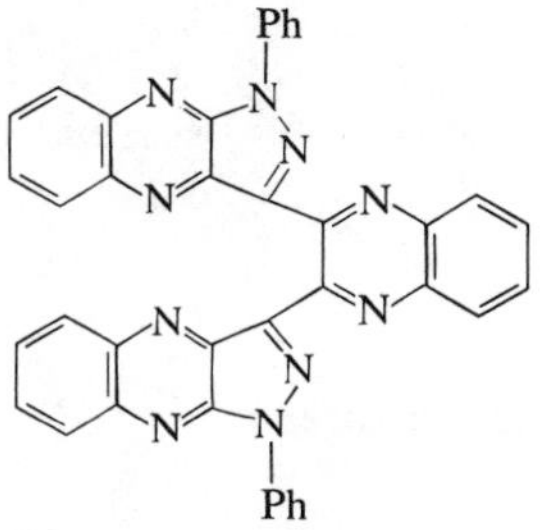 Ph, N, N, N, N, N, N, N, N, N, N, Ph	238	48
$C_{64}H_{42}N_{16}O$	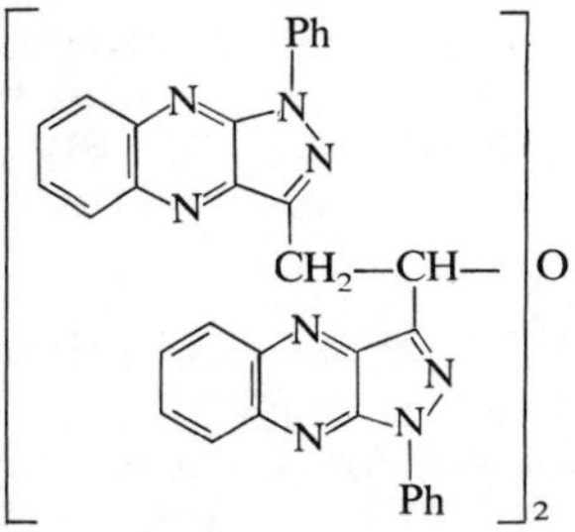 [Ph, N, N, N, N, CH_2—CH—, N, N, N, N, Ph]$_2$ O	356–358	40

[a] Various other hydrazones are reported in reference 47.
[b] $[\alpha]_D^{24.5} -12.8°$ (c 0.703, pyridine).
[c] $[\alpha]_D^{20} -11°$ (c 1.364, AcOH).
[d] $[\alpha]_D^{23} +46.0°$ (c 1.304, $CHCl_3$).
[e] $[\alpha]_D^{20} -58.4°$ (c 0.372, pyridine).
[f] $[\alpha]_D^{25} -106.4°$ (c 2.208, $CHCl_3$).
[g] Light-red isomer.
[h] Violet-red isomer.
[i] Not 225°, as claimed in *Ber.*, **35,** 1437 (1902).
[j] $[\alpha]_D^{20} +101.1°$ (c 1.068, $CHCl_3$).
[k] From D-xylose; $[\alpha]_D +8°$ (pyridine).
[l] From L-arabinose; $[\alpha]_D^{20} -8°$ (pyridine).
[m] From L-rhamnose; $[\alpha]_D^{23} +40°$ (c 0.8, pyridine).
[n] From L-fucose; $[\alpha]_D^{23} +36°$ (c 1.4, pyridine).
[o] $[\alpha]_D^{22} -42.5°$ (c 0.8, pyridine).
[p] From D-glucose; $[\alpha]_D^{20} -20°$ (c 1, pyridine).
[q] Many esters and other derivatives reported with melting points and $[\alpha]_D$ values in references 18 and 22.
[r] $[\alpha]_D$ 64.4° ($CHCl_3$).
[s] From D-galactose; $[\alpha] -53°$ (c 1, pyridine).

TABLE 2 (*contd.*)

[t] From L-sorbose; $[\alpha]+53.5°$ (*c* 1, pyridine).
[u] Unclear if a mixture or a single isomer.
[v] $[\alpha]_D^{20}+53.5°$ (*c* 1, pyridine).
[w] $[\alpha]_D^{20}-43°$ (*c* 1, pyridine).
[x] $[\alpha]_D^{20}+45°$ (*c* 1, pyridine).
[y] $[\alpha]_D^{20}-88°$ (*c* 1, pyridine).
[z] $[\alpha]_D^{20}$ not reported.
[aa] $[\alpha]_D^{20}-73°$ (*c* 1, pyridine).
[bb] $[\alpha]_D^{20}+60°$ (*c* 1, pyridine).
[cc] $[\alpha]_D^{25}+87.0$ (*c* 0.5, pyridine).
[dd] $[\alpha]_D+98°$ (pyridine).
[ee] $[\alpha]_D^{20}+80°$ (*c* 1, pyridine).

IV. References

1. V. Evodokimoff, *Rend. Ist. Super. Sanità.*, **23,** 542 (1960); *Chem. Abstr.*, **55,** 5517 and 9414 (1961); *Gazz. Chim. Ital.*, **90,** 1133 (1960); *Chem. Abstr.*, **55,** 22332 (1961).
2. S. Rossi, O. Pirola, and F. Selva, *Tetrahedron*, **24,** 6395 (1968).
3. J. C. E. Simpson, "Condensed Pyridazine and Pyrazine Rings," Interscience, New York, 1953, p. 348.
4. G. Henseke, *Z. Chem.*, **6,** 329 (1966).
5. B. Kohlstock, G. Henseke, and R. Starke, *Z. Chem.*, **13,** 11 (1973).
6. B. Kohlstock and G. Henseke, *Z. Chem.*, **12,** 385 (1972).
7. N. P. Buu-Hoi, J. N. Vallat, G. Saint-Ruf, and G. Lambelin, *Chim. Ther.*, **6,** 245 (1971).
8. P. Nordin and M. Doty, *Science*, **134,** 112 (1961).
9. L. Dolejs, Z. Veisova, and J. Farkas, *Org. Mass Spectrom.*, **3,** 1535 (1970).
10. G. S. Johnson, W. S. Ruliffson, and R. G. Cooks, *J. Chem. Soc.* (*D*), 587 (1970).
11. W. S. Chilton and R. C. Krahn, *J. Am. Chem. Soc.*, **89,** 4129 (1967).
12. B. Teichmann, K. Himmelspach, and O. Westphal, *J. Prakt. Chem.*, **313,** 940 (1971).
13. T. Kobayshi and M. Saito, *Nippon Nogeikagaku Kaishi*, **36,** 189 (1962); *Chem. Abstr.*, **61,** 10759 (1964).
14. P. Nordin and D. French, *J. Am. Chem. Soc.*, **80,** 1445 (1958).
15. J. H. Pazur, *J. Am. Chem. Soc.*, **77,** 1015 (1955).
16. D. French, G. M. Wild, and W. J. James, *J. Am. Chem. Soc.*, **75,** 3664 (1953).
17. J. H. Pazur and D. French, *J. Biol. Chem.*, **196,** 265 (1952).
18. H. Ohle and G. A. Melkonian, *Ber.*, **74B,** 279 (1941).
19. H. Ohle and A. Iltgen, *Ber.*, **76B,** 1 (1943).
20. G. Henseke and C. Bauer, *Chem. Ber.*, **92,** 501 (1959).
21. H. Ohle and R. Liebig, *Ber.*, **75B,** 1536 (1942).
22. H. Ohle and J. J. Kruyff, *Ber.*, **77B,** 507 (1944).
23. P. Nordin, *Meth. Carbohyd. Chem.*, **2,** 136 (1963).
24. J. E. Courtois and U. Ariyoshi, *Ann. Pharm. Fr.*, **16,** 385 (1958); *Chem. Abstr.*, **53,** 8003 (1959).
25. G. Neumüller, *Arkiv. Kemi, Mineral. Geol.*, **23A**(No. 19), (1946); *Chem. Abstr.*, **41,** 1210 (1947).

26. R. L. Whistler and J. L. Hickson, *J. Am. Chem. Soc.*, **76,** 1671 (1954).
27. U. Rosenquist, G. Neumüller, and K. Myrbäck, *Arkiv. Kemi, Mineral. Geol.*, **24A**(No. 14), (1946); *Chem. Abstr.*, **42,** 5425 (1948).
28. G. Henseke and W. Lemke, *Chem. Ber.*, **91,** 113 (1958).
29. G. Henseke, W. Dose, and K. Dittrich, *Angew. Chem.*, **69,** 479 (1957).
30. H. Dahn and J. P. Fumeaux, *Bull. Soc. Vaudoise Sci. Natur.*, **70,** 313 (1970); *Chem. Abstr.*, **75,** 140791 (1971).
31. G. Henseke, D. Lehmann, and K. Dittrich, *Chem. Ber.*, **94,** 1743 (1961).
32. G. Henseke and K. Dittrich, *Chem. Ber.*, **92,** 1550 (1959).
33. H. Dahn and H. Moll, *Helv. Chim. Acta*, **49,** 2426 (1966).
34. H. Dahn and J. Nussbaum, *Helv. Chim. Acta*, **52,** 1661 (1969).
35. V. D. Romanenko and S. I. Burmistrov, *Khim. Geterotsikl. Soedin.*, 852 (1973); *Chem. Abstr.*, **79,** 92158 (1973).
36. W. Sauer and G. Henseke, *Z. Chem.*, **10,** 381 (1970).
37. G. Henseke and W. Sauer, Ger. (East) Pat. 83,361; *Chem. Abstr.*, **78,** 58459 (1973).
38. G. R. Wendt and K. W. Ledig, U.S. Pat. 3,431,262; *Chem. Abstr.*, **70,** 106512 (1969).
39. F. Sachs and P. Becherescu, *Ber.*, **36,** 1132 (1903).
40. H. Ohle and G. A. Melkonian, *Ber.*, **74B,** 398 (1941).
41. J. Klicnar, *Collect. Czech. Chem. Commun.*, **30,** 3087 (1965); *Chem. Abstr.*, **63,** 16349 (1965).
42. V. Vanicek, *Collect. Czech. Chem. Commun.*, **27,** 2699 (1962); *Chem. Abstr.*, **58,** 3526 (1963).
43. H. H. Credner, *Chem. Ber.*, **104,** 2640 (1971).
44. H. H. Credner and W. Pueschel, Ger. Offen. 2,109,455; *Chem. Abstr.*, **78,** 5406 (1973).
45. B. Kohlstock and G. Henseke, *Z. Chem.*, **13,** 100 (1973).
46. J. N. BeMiller and R. K. Mann, *Carbohyd. Res.*, **2,** 70 (1966).
47. N. P. Buu-Hoi, J. N. Vallat, G. Saint-Ruf, and G. Lambelin, *Chim. Ther.*, **7,** 210 (1972).
48. H. J. Binte, G. Henseke, W. Bauer, and K. Koehnke, *Z. Chem.*, **8,** 104 (1968).
49. J. Farkas, K. Sebesta, K. Horska, Z. Samek, L. Dolejs, and F. Sorm, *Collect. Czech. Chem. Commun.*, **34,** 1118 (1969); *Chem. Abstr.*, **70,** 97091 (1969).
50. M. Von Saltza, J. D. Dutcher, J. Reid, and O. Wintersteiner, *J. Org. Chem.*, **28,** 999 (1963).
51. F. Weygand, K. Fehr, and J. F. Klebe, *Z. Naturforsch.*, **14b,** 217 (1959); *Chem. Abstr.*, **54,** 1327 (1960).
52. B. Teichmann, K. Himmelspach, and O. Westphal, *J. Prakt. Chem.*, **314,** 877 (1972); *Chem. Abstr.*, **78,** 58712 (1973); V. Harisdangkul and E. A. Kabat, *J. Immunol.*, **108,** 1232 (1972); *Chem. Abstr.*, **77,** 3661 (1972); B. Teichmann, K. Himmelspach, and O. Westphal, *Z. Chem.*, **11,** 380 (1971); K. Himmelspach, O. Westphal, and B. Teichmann, *Eur. J. Immunol.*, **1,** 106 (1971); *Chem. Abstr.*, **75,** 61385 (1971).
53. G. Henseke and K. J. Bähner, *Chem. Ber.*, **91,** 1605 (1958).
54. H. Ohle and M. Hielscher, *Ber.*, **74B,** 13 (1941).
55. U. D. Treuner (to E. R. Squibb and Sons, Inc.), U.S. Pat. 3,994,893; *Chem. Abstr.*, **86,** 106651 (1977).

CHAPTER XXXVIII

Benzo[*f*]quinoxalines

RRI 3393

I. Physical Properties 712
II. Methods of Preparation 713
  1. From 1,2-Diaminonaphthalene 713
  2. From *o*-Aminonitroso- and *o*-Aminophenylazo Compounds 715
  3. Miscellaneous Syntheses 718
III. Reactions 719
IV. Uses 723
V. Tables 723
  Table 1. Benzo[*f*]quinoxalines 723
VI. References 727

Benzo[*f*]quinoxalines were formerly named 5,6-benzoquinoxalines, and thus 3-substituted benzo[*f*]quinoxalines were formulated as 2-substituted 5,6-benzoquinoxalines (**1**). A potential source of confusion is the alternative designation of 2-substituted benzo[*f*]quinoxalines as 2-substituted 7,8-benzoquinoxalines (**2**).

1 **2**

I. Physical Properties

The ultraviolet spectrum of benzo[*f*]quinoxaline has been measured in ethanol and shows the expected three-banded structure; these bands are at about 227, 279, and 357 nm.[1] Ultraviolet spectra for benzo[*f*]quinoxaline in methanol and cyclohexane have also been published.[42] The phosphorescence[1,2] and luminescence[43] of this compound have been studied. The mass spectrum of benzo[*f*]quinoxaline shows a prominent molecular ion and fragment ions corresponding to the loss of atomic hydrogen and one and two molecules of hydrogen cyanide.[44] The base has been titrated with Lewis acids ($FeCl_3$ and $SbCl_3$) in nitrobenzene,[3] and its polarographic reduction in methyl cyanide has been studied. The measured half-wave potentials are −1.744, −2.128, and −2.673 volts.[4]

II. Methods of Preparation

The main substrates for the synthesis of benzo[*f*]quinoxalines are 1,2-diaminonaphthalene, 2-amino-1-nitrosonaphthalene, the isomeric 1-amino-2-nitroso derivative, and the corresponding aminophenylazo compounds. These are ring closed with a suitable two-carbon component.

1. From 1,2-Diaminonaphthalene

The condensation of 1,2-diaminonaphthalene with an α-dicarbonyl compound to give a benzo[*f*]quinoxaline represents a simple extension of the classical quinoxaline synthesis. As a result of the unsymmetrical structure of the diamine, isomeric mixtures of benzo[*f*]quinoxalines can in principle be formed when the dicarbonyl compound is also unsymmetrical (Scheme 1). In practice only one isomer may be formed in appreciable quantities and be thus readily isolated. The orientation of a number of the products obtained from the condensation reactions of 1,2-diaminonaphthalene is uncertain, though, as discussed in the following sections, unambiguous syntheses of benzo[*f*]quinoxalines may be achieved using *o*-aminonitroso- and *o*-aminophenylazonaphthalenes.

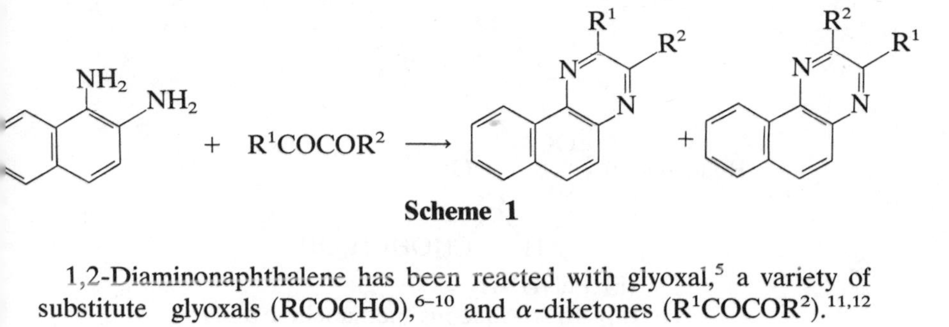

NH$_2$ NH$_2$ + R^1COCOR^2 ⟶ (benzo[*f*]quinoxaline, 2-R^1, 3-R^2) + (benzo[*f*]quinoxaline, 2-R^2, 3-R^1)

Scheme 1

1,2-Diaminonaphthalene has been reacted with glyoxal,[5] a variety of substitute glyoxals (RCOCHO),[6–10] and α-diketones (R^1COCOR^2).[11,12]

When sodium 2-nitro-3-oxosuccinaldehydate [$NaO_2CCH(NO_2)COCHO$] is used,[9] a nitromethyl derivative is obtained as a result of condensation and decarboxylation. The diamine has been converted into 2-D-arabotetrahydroxybutylbenzo[*f*]quinoxaline in 67% yield by reaction with D-fructosone methylphenylhydrazone (**3**); but with D-fructosone (**4**) itself the yield is in the range of 30–40%.[10]

$RCOCH{=}NMePh$ **3** $RCOCHO$ **4** $RCOCH_2NHC_6H_4Me(p)$ **5**

R = —CH(OH)—CH(OH)—CH(OH)—CH_2OH

p-Tolyl-D-isoglucosamine (**5**) when treated with hydrazine in dilute acetic acid gives a mixture of the osone **4**, the mono- and dihydrazone of **4**, and *p*-toluidine. After removal of the amine from this mixture, reaction with 1,2-diaminonaphthalene gives a 87% yield of a mixture of 2- and 3-(D-arabotetrahydroxybutyl)benzo[*f*]quinoxalines. This has been separated into its components by fractional crystallization and found to contain mainly the 2-isomer.[10] The benzo[*f*]quinoxaline **7** is obtained by condensation of the diamine with dehydroascorbic acid (**6**) and treatment of the product with phenylhydrazine.[13] Dehydroascorbic acid is generated in situ by oxidation of L-ascorbic acid with a molar proportion of benzoquinone. Treatment of compound **7** with aqueous methanolic sodium hydroxide gives the phenylhydrazone of the 2-formyl 3-oxo derivative **8**.

1,2-Diaminonaphthalene (NH_2, NH_2) —(i), (ii)→ **7** (PhNHN=, $CHOHCH_2OH$, O, N) —(iii)→ **8** ($CH{=}NNHPh$, O, NH)

Reagents (i) **6** (OC, OC, C=O, O, C, H, $CHOHCH_2OH$)

(ii) $PhNHNH_2$

(iii) 10% Aqueous methanolic NaOH

Condensation of the diamine with α-keto acids ($RCOCO_2H$)[14] and α-oximino esters [$RC(=NOH)CO_2R^1$][15] leads to mixtures of isomeric 2- and 3-oxo compounds. Reaction of the diamine with diethyl acetylenedicarboxylate gives the 2-oxo ester **9** or the isomeric 3-oxo ester[16]; and with cyanogen di-N-oxide ($\bar{O}-\overset{+}{N}\equiv C-C\equiv\overset{+}{N}-\bar{O}$), the dioximino derivative **10** is obtained.[17]

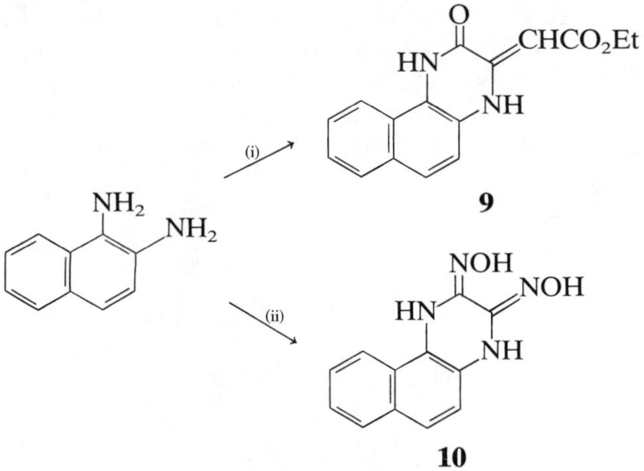

Reagents (i) $EtO_2CC\equiv CCO_2Et$
(ii) $\bar{O}-\overset{+}{N}\equiv C-C\equiv\overset{+}{N}-\bar{O}$

2. From o-Aminonitroso- and o-Aminophenylazo Compounds

The use of these substrates leads to unambiguous syntheses of benzo[*f*]quinoxalines as illustrated in Schemes 2 and 3. Products of

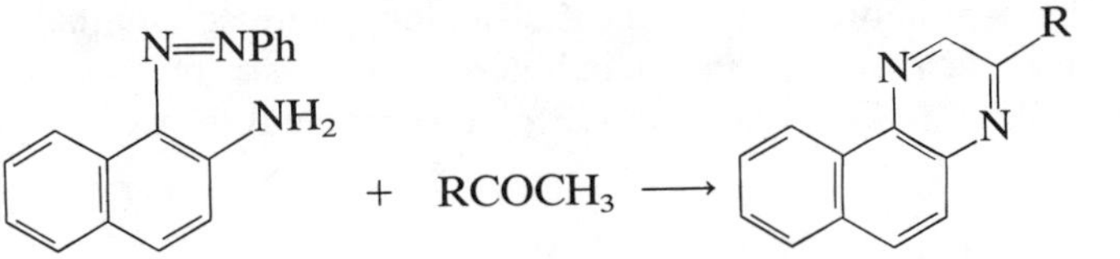

Scheme 2

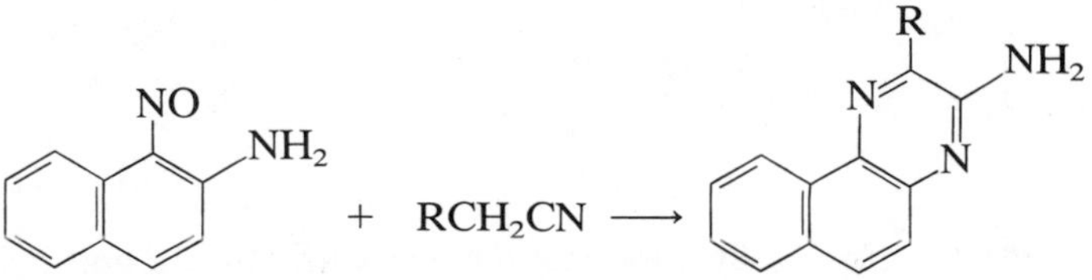

Scheme 3

reversed orientation can be obtained from the corresponding 1-amino-2-phenylazo- and 1-amino-2-nitrosonaphthalenes. 3-Methylbenzo-[f]quinoxaline, m.p. 81°, is formed by the condensation of 2-amino-1-phenylazonaphthalene with acetone, indicating that the methylbenzoquinoxaline of m.p. 95°, isolated in 80% yield from reaction of 1,2-diaminonaphthalene and isonitrosoacetone ($CH_3COCH{=}NOH$), is the isomeric 2-methyl derivative.[18] Reaction of 2-amino-1-phenylazonaphthalene with ethyl pyruvate gives the 3-ethoxycarbonyl derivative **11**,[19] but the corresponding reaction with pyruvic acid gives the condensed imidazole **12**. The ester **11** has been converted by alkaline hydrolysis, followed by decarboxylation of the resulting acid, into the parent heterocycle of m.p. 62°.

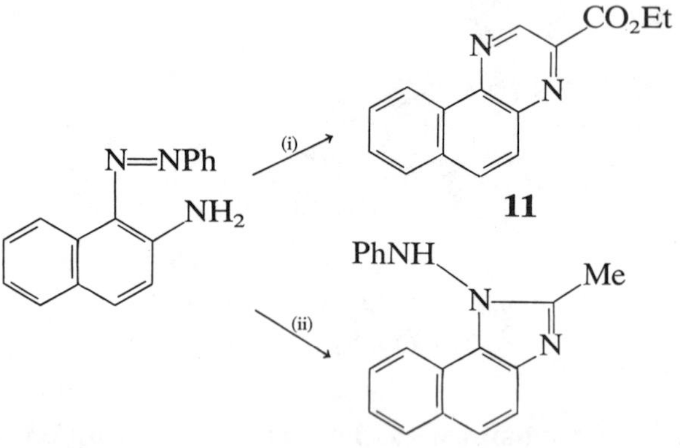

N=NPh NH$_2$ (i) → **11** (CO_2Et); (ii) → **12** (PhNH, Me)

Reagents (i) $MeCOCO_2Et$
(ii) $MeCOCO_2H$

The base-promoted reaction of 2-amino-1-nitrosonaphthalene with cyanoacetic acid gives an unambiguous synthesis of 3-aminobenzo-[f]quinoxaline-2-carboxylic acid (**13**),[20] a compound previously prepared

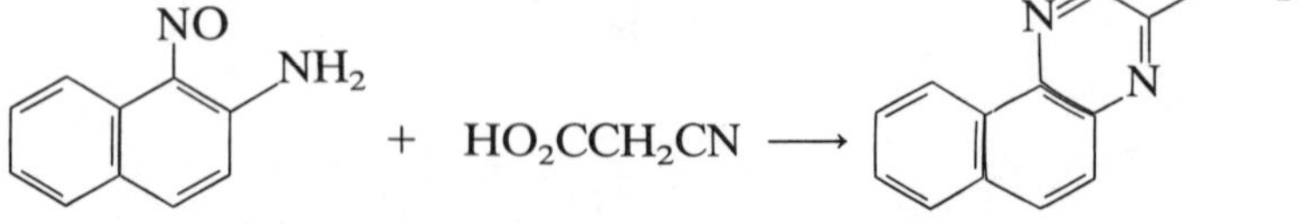

NO NH$_2$ + HO_2CCH_2CN ⟶ **13** (CO_2H, NH_2)

by condensation of 1,2-diaminonaphthalene with alloxan and hydrolysis of the resulting mixture of condensation products.[21] 1-Amino-2-nitrosonaphthalene undergoes analogous condensation reactions with derivatives of cyanoacetic acid and condenses with benzyl cyanide to give 2-amino-3-phenylbenzo[*f*]quinoxaline (**14**).[22] 3-Amino-2-phenylbenzo[*f*]-quinoxaline is similarly prepared from 2-amino-1-nitrosonaphthalene.[22] Both aminonitroso naphthalenes give *N*-oxides on treatment with the pyridinium ylid **15**, as illustrated in Scheme 4.[23]

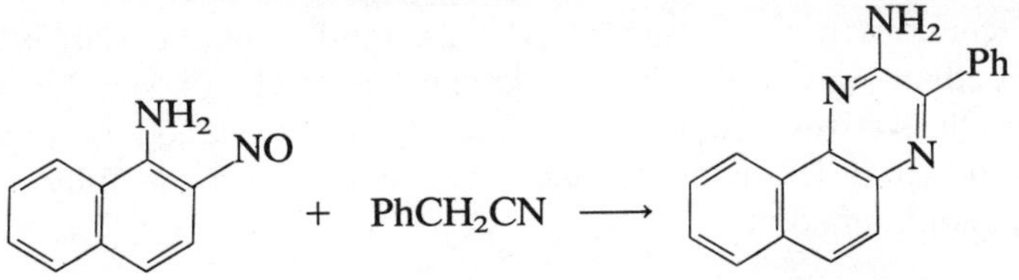

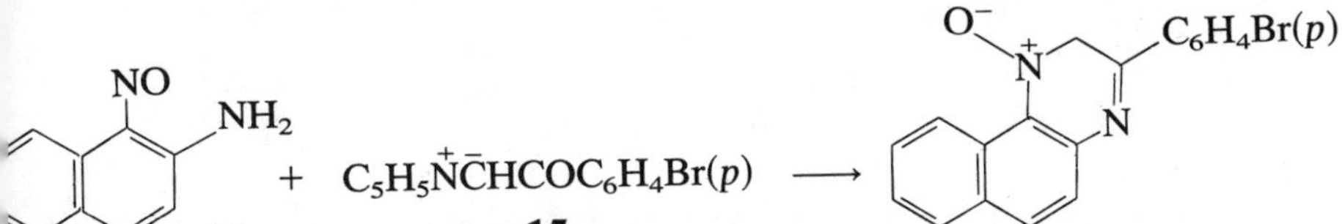

Scheme 4

Reaction of 1-nitroso-2-aminonaphthalene with the aziridine **16** gives a mixture of 3-phenylbenzo[*f*]quinoxaline (**18**) and the naphthoimidazole **19**. The reaction is initiated by thermal cleavage of the aziridine into the corresponding azomethine **17** and cycloaddition of the azomethine to the nitroso group. Other examples of this type of reaction have been reported.[24]

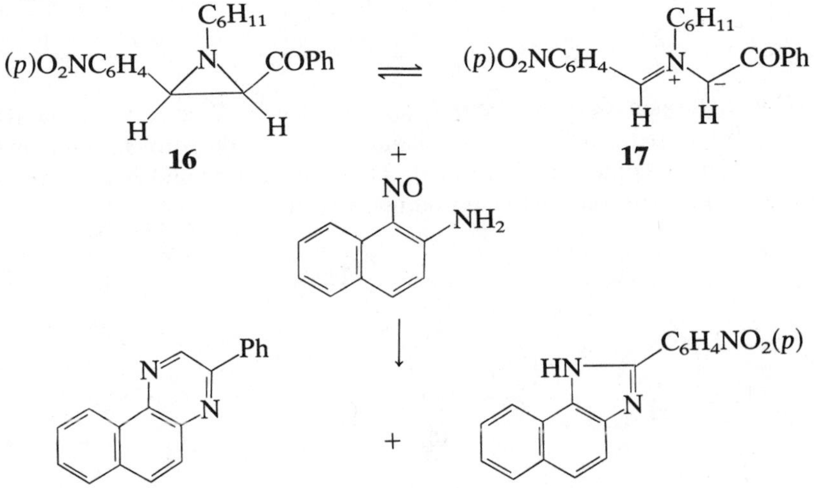

3. Miscellaneous Syntheses

The condensation of 4-aziridinyl-1,2-naphthoquinone (**20**) with 1,2-diaminoethane gives the benzo[f]quinoxaline **21**. In simpler systems dihydro derivatives are isolated from the condensation of dicarbonyl compounds and 1,2-diaminoethane, but in this case the fully aromatic product is obtained.[25]

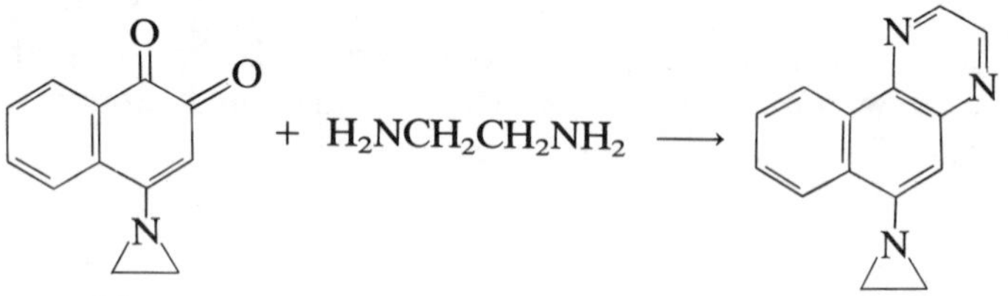

+ $H_2NCH_2CH_2NH_2$ ⟶

20 **21**

There have been two reports of the photochemical synthesis of benzo[f]quinoxalines. On irradiation of *trans*-2-styrylpyrazine (**22**), photoisomerization occurs, and the resulting cis isomer **23** then undergoes

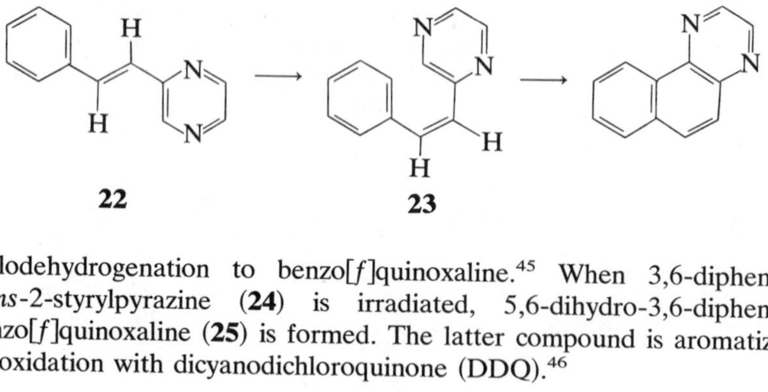

22 **23**

cyclodehydrogenation to benzo[f]quinoxaline.[45] When 3,6-diphenyl-*trans*-2-styrylpyrazine (**24**) is irradiated, 5,6-dihydro-3,6-diphenylbenzo[f]quinoxaline (**25**) is formed. The latter compound is aromatized by oxidation with dicyanodichloroquinone (DDQ).[46]

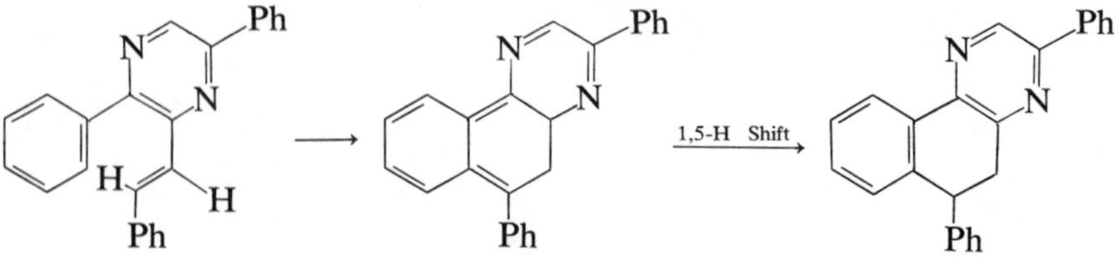

1,5-H Shift

24 **25**

An interesting reaction sequence based on 1-tetralone (**26**) leads to an unambiguous synthesis of 2-amino-3-cyano-5,6-dihydro-benzo[*f*]-quinoxaline (**27**). The isomeric 3-amino-2-cyano compound is similarly obtained from 2-tetralone.[47]

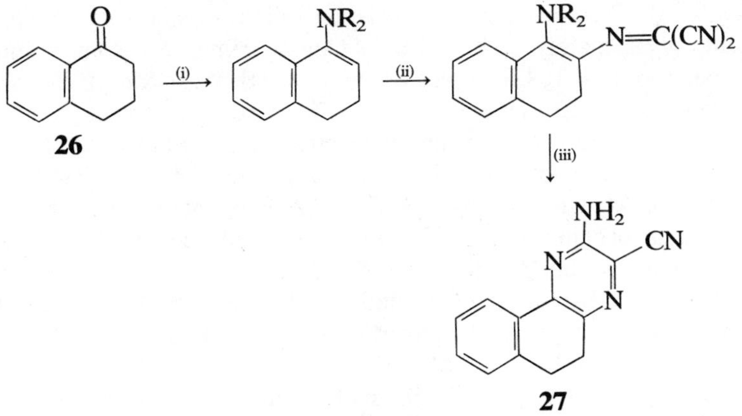

Reagents (i) Morpholine
(ii) $TosN{=}C(CN)_2$
(iii) NH_3

III. Reactions

There is little systematic knowledge of the chemistry of this ring system. Oxidation of 3-phenylbenzo[*f*]quinoxaline with chromic anhydride in boiling acetic acid gives 3-phenylbenzo[*f*]quinoxaline-5,6-quinone (**28**).[26] The quinone reacts with aromatic aldehydes to form addition products **29** on exposure to light. Reduction of **28** with phenylhydrazine gives the corresponding hydroquinone, which is characterized as a diacetate.[26]

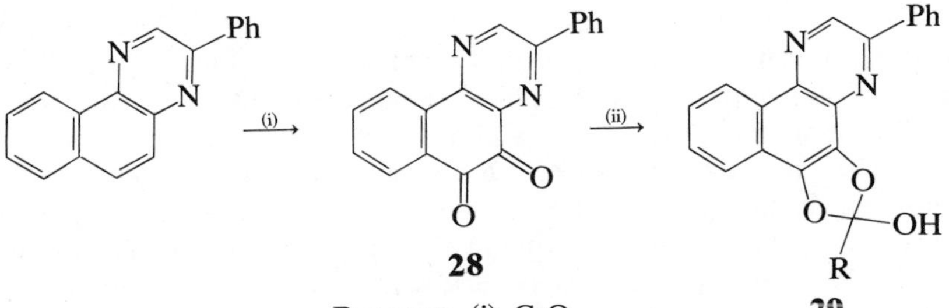

Reagents (i) CrO_3
(ii) RCHO

Oxidation of the parent base with hydrogen peroxide in formic acid gives a high yield of mono-*N*-oxide and a little di-*N*-oxide.[27] *N*-Oxidation is assumed to proceed preferentially at the less hindered 4-nitrogen. A number of substituted benzo[*f*]quinoxaline 4-oxides have been prepared, the majority by oxidation of the appropriate benzo[*f*]quinoxaline with peracids.[38,41] Treatment of benzo[*f*]quinoxaline 4-oxide with phosphoryl chloride gives mainly 3-chlorobenzo[*f*]quinoxaline. An isomeric monochloride, m.p. 104–104.5°, is also isolated from this reaction which is shown by independent synthesis not to be the 2-chloro derivative.[27] A dichloro derivative, m.p. 187–188°, of unknown structure is obtained by treatment of the parent base with chlorine in glacial acetic acid.[27] A disubstitution product of unknown orientation and m.p. 288° is obtained by nitration of 2-thienylbenzo[*f*]quinoxaline with a mixture of nitric and sulfuric acids at 10–15°. This is thought to be a 6,*x*-dinitro derivative.[7] Chlorine at positions 2 and 3 shows the expected lability to nucleophilic substitution, and both isomers have been converted into the corresponding piperidino derivatives of m.p. 101.5–102.5° and 124–125°, respectively.[27]

A comparative study has been made of the reactivity of the methyl group of a number of 2-methylbenzodiazines. This has shown that the methyl group in a methylquinoxalinone such as 3-methylbenzo[*f*]-quinoxalin-2-one (**30**) is readily condensed with benzaldehydes, furfural, and chloral to give compounds of type **31**.[28,29] The condensation of

O HN Me N —(i)→ O HN CH═CHR N

30 **31**

Reagents (i) RCHO, Ac_2O

quaternary salts of the quinoxalinone **30** and quinaldine with triethyl orthoformate in pyridine yields both a symmetrical (**32**) and unsymmetrical (**33**) trimethine; the latter is a photographic sensitizer.[30]

Several transformations of 2- and 3-D-arabotetrahydroxybutyl benzo[*f*]quinoxalines have been reported. Thus oxidation of the 2-tetrahydroxybutyl derivative with hydrogen peroxide in dilute sodium hydroxide gives the 2-carboxylic acid, m.p. 185°, and oxidation with potassium periodate in dilute acetic acid gives the 2-aldehyde,

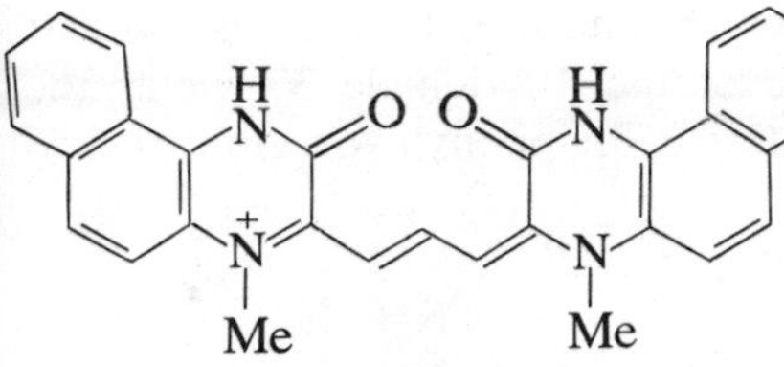

ClO$_4^-$
32

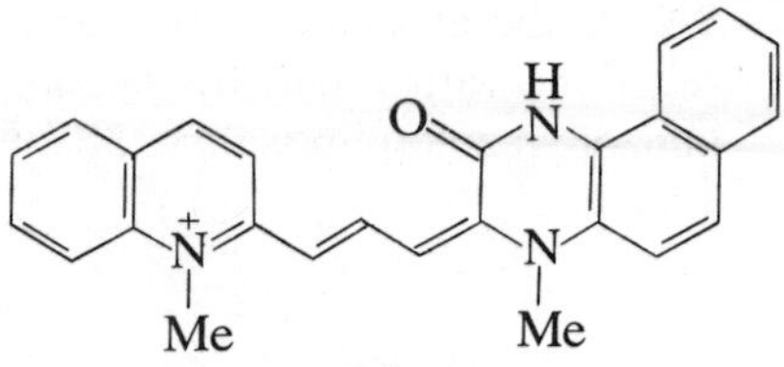

ClO$_4^-$
33

m.p. 132°.[10] The aldehyde is characterized by the preparation of a phenylhydrazone, m.p. 217°, and gives the expected α-hydroxy ketone on treatment with alcoholic potassium cyanide. The latter compound is oxidized to the corresponding α-diketone in boiling dimethylformamide. The 3-tetrahydroxybutyl derivative is similarly converted into the 3-carboxylic acid, m.p. 234°, and the 3-aldehyde, m.p. 162°. The aldehyde gives a phenylhydrazone of m.p. 237°.[10] It has been found that the melting points of the 3-substituted derivatives are higher than those of the corresponding 2-substituted compounds. The melting points of the isomeric 2- and 3-chloro- and methylbenzo[*f*]quinoxalines do however not conform to this generalization.

The phenylhydrazone **35** of the 2-aldehyde is also obtained directly from 2-D-arabotetrahydroxybutylbenzo[*f*]quinoxaline (**34**) by treatment with phenylhydrazine in neutral solution; however in acid solution phenylhydrazine treatment gives the benzopyrazoloquinoxaline **36**.[31]

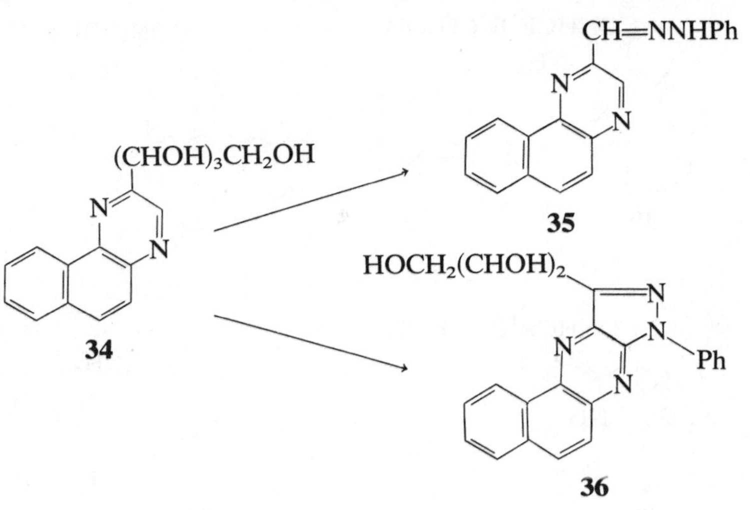

CH=NNHPh
35

$(CHOH)_3CH_2OH$
34

$HOCH_2(CHOH)_2$
Ph
36

3-D-Arabotetrahydroxybutylbenzo[f]quinoxaline fails to undergo similar reactions with phenylhydrazine. However the 3-aldehyde **37** yields the benzopyrazoloquinoxaline **38** on reaction with phenylhydrazine in acid solution.[31] 2- and 3-Aminobenzo[f]quinoxalines were originally obtained by condensation of 1,2-diaminonaphthalene with alloxan, followed by hydrolysis of the product to a mixture of aminocarboxylic acids and by decarboxylation.[21] The isomer of m.p. 150–152° was subsequently identified as the 2-amino compound, and the isomer of m.p. 215–217° as the 3-amino compound. This assignment of structure was made since the higher-melting isomer was also obtained by decarboxylation of the 3-amino-2-carboxylic acid prepared by unambiguous synthesis.[20] The amino amides of type **39** are converted into the corresponding chlorides **40** on treatment with thionyl chloride, and these on treatment with sodium carbonate give 3-amino-2-(2-oxazolin-2-yl)benzo[f]quinoxalines (**41**).

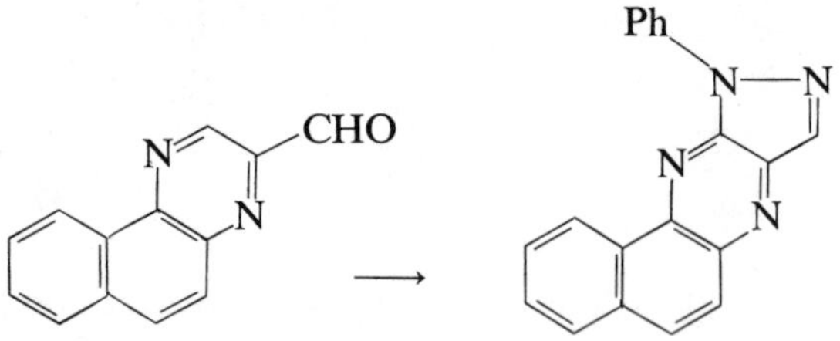

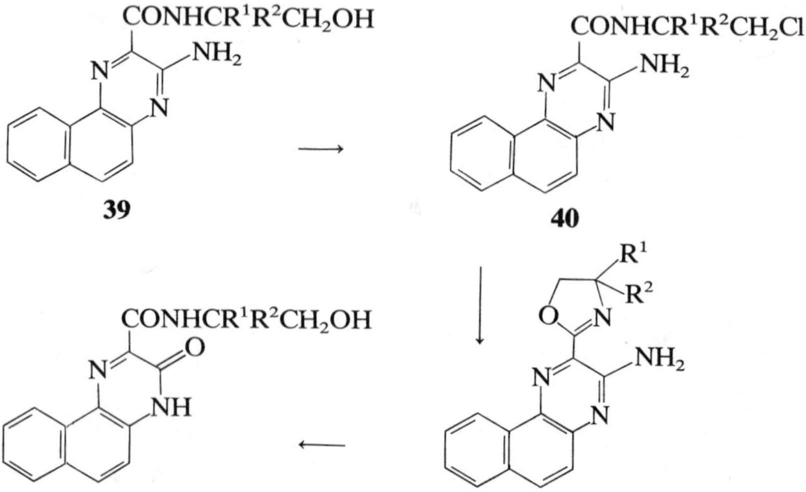

The oxazolines were originally incorrectly formulated as diazepinones. Nitrous acid treatment of both the oxazoline **41** ($R^1 = R^2 = H$) and its precursor **39** ($R^1 = R^2 = H$) gives a common product, the 3-oxo compound **42**, thus supporting the revised formulation. The aminooxazolines are reported to have anticonvulsant and hypotensive properties.[32–35]

IV. Uses

Benzo[*f*]quinoxalines have been patented for use as catalysts for the silver dye bleaching process[36] and as dyestuff intermediates.[37] The latter compounds have been derived by nitration, sulfonation, or chlorosulfonation of either 2,3-dioxo- or 2,3-dichlorobenzo[*f*]quinoxaline.

V. Tables

TABLE 1. BENZO[*f*]QUINOXALINES

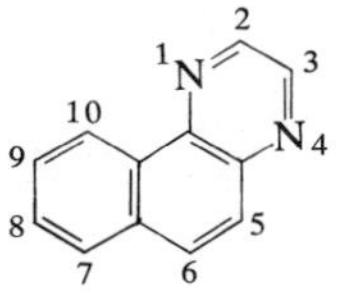

Molecular formula	Substituents or structure	m.p., °C	Ref.
$C_{12}H_6Cl_2N_2$	2,3-Dichloro		37
$C_{12}H_7ClN_2$	2-Chloro	128–129	27
	3-Chloro	120.5, 121.5–122	27, 38
$C_{12}H_8N_2$	Unsubstituted	56.5–57.5, 62	5, 19, 38, 45
	4-Oxide	158–159	27, 38
	1,4-Dioxide	215–216	27
$C_{12}H_8N_2O$	1,2-Dihydro-2-oxo	273–274	38
	Hydrate	275–275.5	27
	3,4-Dihydro-3-oxo	311–313	38
$C_{12}H_8N_2O_2$	1,2,3,4-Tetrahydro-2,3-dioxo	>320	39
$C_{12}H_9N_3$	2-Amino	150–152	21
	3-Amino	215–217, 219–220	20, 21
	Hydrochloride	277–278 decomp.	20
$C_{12}H_{10}N_2O$	1,2,3,4-Tetrahydro-2-oxo	197–198	27
$C_{12}H_{10}N_4O_2$	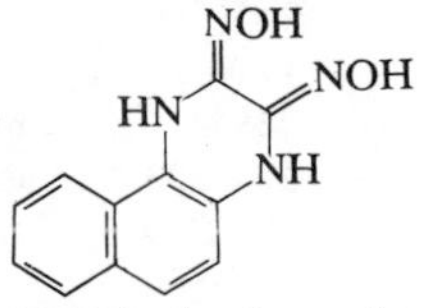 NOH, NOH, HN, NH	212–214 decomp.	17
$C_{13}H_7F_3N_2O$	1,2-Dihydro-2-oxo-3-trifluoromethyl		53

TABLE 1 (*contd.*)

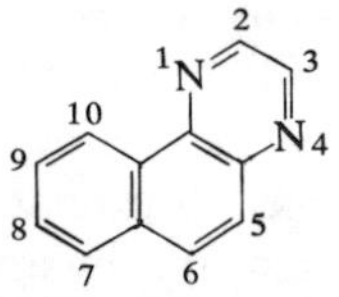

Molecular formula	Substituents or structure	m.p., °C	Ref.
$C_{13}H_7N_3$	3-Cyano	171–172.5	38
$C_{13}H_8N_2O$	2-Formyl	132	10
	Phenylhydrazone	217	10
	3-Formyl	162	10
	Phenylhydrazone	237	10
$C_{13}H_8N_2O_2$	2-Carboxy	185	10
	3-Carboxy	195, 234	10, 19
	Ethyl ester	126	19
	2-Formyl-3,4-dihydro-3-oxo		
	Phenylhydrazone	299 decomp.	13
$C_{13}H_8N_2O_3$	2-Carboxy-3,4-dihydro-3-oxo (or 3-Carboxy-1,2-dihydro-2-oxo)		
	Methyl ester	249–250	15
	Ethyl ester	227–228	16
$C_{13}H_9ClN_2$	3-Chloro-2-methyl	110–111	38
$C_{13}H_9N_3O_2$	2(or 3)-Nitromethyl	139–140	9
	2-Amino-3-carboxy		
	Amide	255–256	20
	3-Amino-2-carboxy	232 decomp.	20
	Methyl ester	225–226	20
	Amide	282–283	20
	3-Acetyl deriv. of the amide	264–265	20
$C_{13}H_{10}N_2$	2-Methyl	95	6, 18, 38
	Picrate	180–182	38
	4-Oxide	189–189.5	38
	3-Methyl	81	18
	4-Oxide	156–158	38
$C_{13}H_{10}N_2O$	1,2-Dihydro-3-methyl-2-oxo[a]	288, 295–296	15, 16
	Chloral deriv.	237–238	28
	p-Dimethylaminobenzylidene deriv.	176	29
	o-Hydroxybenzylidene deriv.	295	28
	p-Methoxybenzylidene deriv.	297	28
	3,4-Dihydro-2-methyl-3-oxo	242–244, 245–247	38, 51
$C_{13}H_{10}N_4$	2-Amino-3-cyano-5,6-dihydro	247	47
	3-Amino-2-cyano-5,6-dihydro	241	47
$C_{13}H_{12}N_2O$	2,3,4-Trihydro-5-methyl-6-oxo	174–175	52
$C_{14}H_8N_2O_4$	2,3-Dicarboxy	228–230 decomp.	50
	Anhydride	220–223 decomp.	50
	N-Methylimide	306–309	50
	N-Phenylimide	335–340	50
	N-1-Naphthylimide	318–321	50
	N-2-Naphthylimide	355–360	50

TABLE 1 (*contd.*)

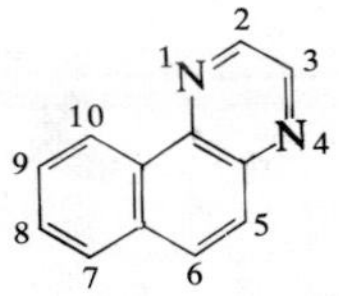

Molecular formula	Substituents or structure	m.p., °C	Ref.
$C_{14}H_{10}N_2O_3$[b]	3-Carboxymethyl-1,2-dihydro-2-oxo (or 2-Carboxymethyl-3,4-dihydro-3-oxo)		
	Methyl ester	236	15
$C_{14}H_{11}N_3$	6-(1-Aziridinyl)	151–153	25
$C_{14}H_{12}N_2$	2,3-Dimethyl	101	48
	4-Oxide	165–167	48
	4-Methyl perchlorate	270–272 decomp.	55
	4-Ethyl tetrafluoroborate	235–239 decomp.	55
$C_{14}H_{12}N_2O$	3-Ethoxy	88–89.5	38
$C_{15}H_{12}N_4O$[c]	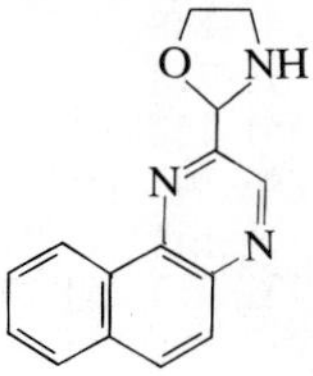 [structure: 2-(2-oxazolidinyl) with O, NH]	231–232.5	32
$C_{15}H_{13}N_3O_3$	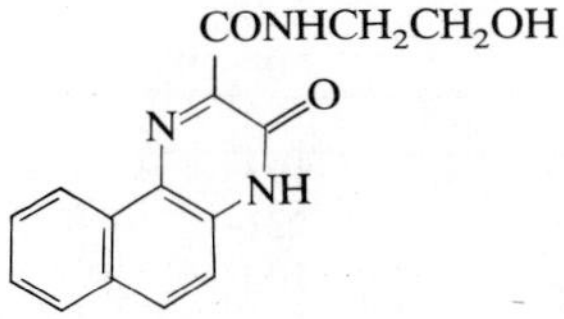 [structure: $CONHCH_2CH_2OH$, O, NH]	284–286	33
$C_{15}H_{14}N_4O_2$	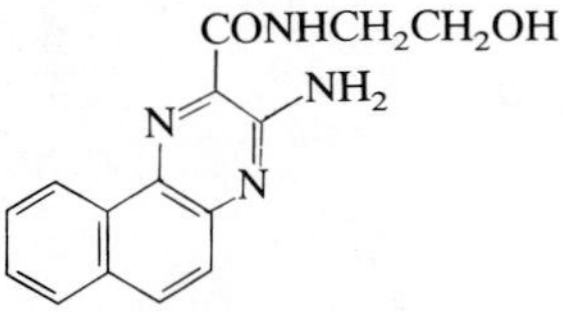 [structure: $CONHCH_2CH_2OH$, NH_2]	181–182	32
$C_{16}H_{10}N_2O$	3-(2-Furanyl)		49
$C_{16}H_{10}N_2S$	2(or 3)-(2-Thienyl)	213–215	7
	Dinitro deriv.	288	7
$C_{16}H_{14}N_2O_5$	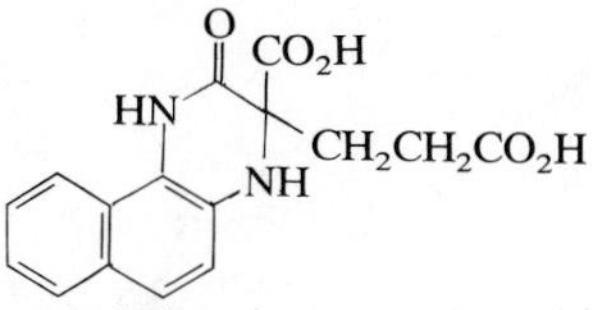 [structure: O, CO_2H, HN, NH, $CH_2CH_2CO_2H$]		
	Diethyl ester		54
$C_{16}H_{16}N_2O_4$	2-(D-Arabotetrahydroxybutyl)	183 decomp.	10
	Tetraacetate	149	10
	3-(D-Arabotetrahydroxybutyl)	231 decomp.	10

TABLE 1 (*contd.*)

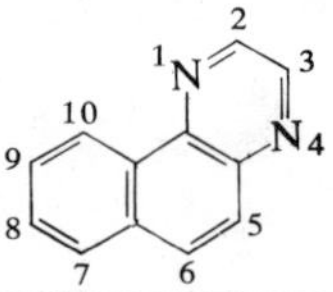

Molecular formula	Substituents or structure	m.p., °C	Ref.
$C_{17}H_{17}N_3$	2-Piperidino	101.5–102.5	27
	3-Piperidino	124–125	27
$C_{18}H_{10}N_2O_2$	3-Phenyl-5,6-dioxo	250	26
$C_{18}H_{11}BrN_2$	2-*p*-Bromophenyl		
	4-Oxide	245	23
	3-*p*-Bromophenyl	200–201.5	24
	1-Oxide	301	23
$C_{18}H_{11}N_3O_2$	2-*m*-Nitrophenyl[d]	196	40
	2-*p*-Nitrophenyl[d]	237	40
	3-*m*-Nitrophenyl	247–248	24
	3-*p*-Nitrophenyl	216–218	24
$C_{18}H_{12}N_2$	2-Phenyl	151–152	38
	4-Oxide	203–204	38
	3-Phenyl	163–164	38
	4-Oxide	231–233, 234–235	24, 38
$C_{18}H_{12}N_2O_2$	5,6-Dihydroxy-3-phenyl	217	26
	Diacetyl	210	26
$C_{18}H_{18}N_2$	2(or 3)-Cyclohexyl	104–108	45
$C_{18}H_{18}N_3$	2-Amino-3-phenyl	180	22
	3-Amino-2-phenyl	193	22
$C_{19}H_{14}N_2$	2-Methyl-3-phenyl[e]	121–122	38, 41
	3-Methyl-2-phenyl	102–103	38
	4-Oxide	169–170	38
$C_{20}H_{16}N_2O_2$	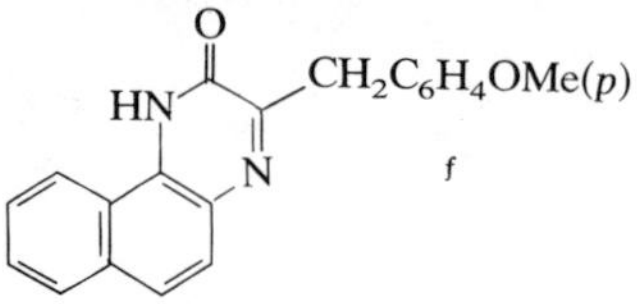 [structure: HN, O, $CH_2C_6H_4OMe(p)$, N] [f]		14
	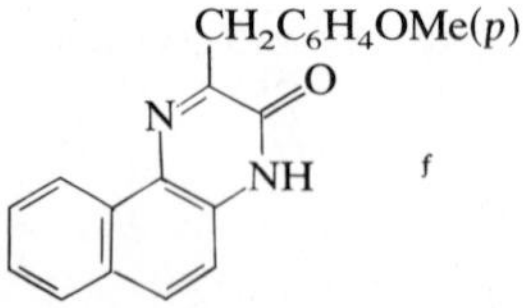 [structure: $CH_2C_6H_4OMe(p)$, N, O, NH] [f]		14
$C_{21}H_{11}N_3$	2(or 3)-(3-Quinolyl)	181–183	8
$C_{22}H_{14}N_4$	2,3-Bis(2-pyridyl)	146	12
$C_{24}H_{16}N_2$	2,3-Diphenyl	148–149	11
	3,6-Diphenyl	190–191	46
$C_{24}H_{18}N_2$	5,6-Dihydro-3,6-diphenyl	148–149	46

TABLE 1 (*contd.*)

Molecular formula	Substituents or structure	m.p., °C	Ref.
$C_{26}H_{14}N_2O_2$	COCO	255 decomp.	10
$C_{26}H_{16}N_2O_2$	OH, HO	310 decomp.	10

[a] A series of homologous compounds have been prepared.[15]
[b] A series of ω-carboxyalkyl methyl esters have been prepared.[15]
[c] A series of substituted oxazolinyl derivatives have been prepared.
[d] These compounds were prepared from 1,2-diaminonaphthalene and the corresponding nitrophenacyl bromide ($ArCOCH_2Br$). They appear to be 2-substituted derivatives, since their melting points are different from those subsequently published for the corresponding 3-substituted compounds in reference 24.
[e] Two mono-*N*-oxides are obtained on oxidation, one of m.p. 217–218°, the other of m.p. 160–162°.
[f] Two isomers of m.p. 243–245° decomp. and 298–299° decomp. have been isolated. A number of analogous compounds have been prepared.

VI. References

1. F. Doerr, H. Gropper, and N. Mika, *Ber. Bunsenges. Phys. Chem.*, **67,** 202 (1962).
2. F. Doerr and H. Gropper, *Angew. Chem.*, **74,** 354 (1962).
3. C. B. Riolo, T. F. Soldi, and F. Martinotti, *Ann. Chim.* (Rome), **53,** 1531 (1963); *Chem. Abstr.*, **60,** 15204 (1964).
4. S. Millefiore, *J. Heterocycl. Chem.*, **7,** 145 (1970).
5. O. Hinsberg, *Chem. Ber.*, **23,** 1394 (1890).
6. C. Neuberg and M. Schuer, *Chem. Ber.*, **63,** 3068 (1930).
7. C. Musante and V. Parrini, *Sperim. Sez. Chim. Biol.*, **3,** 140 (1952); *Chem. Abstr.*, **48,** 4553 (1954).
8. A. Stener, *Gazz. Chim. Ital.*, **90,** 1365 (1960); *Chem. Abstr.*, **56,** 7309 (1962).
9. P. E. Fanta and C.-S. Wang, *J. Heterocycl. Chem.*, **3,** 525 (1966).

10. G. Henseke and W. Lemke, *Chem. Ber.*, **91,** 101 (1958).
11. N. Whittaker and T. S. G. Jones, *J. Chem. Soc.*, 1565 (1951).
12. N. P. Buu-Hoi and G. Saint-Ruf, *J. Chem. Soc.*, 2258 (1961).
13. G. Henseke, D. Lehmann, and K. Dittrich, *Chem. Ber.*, **94,** 1743 (1961).
14. M. Pailer and E. Kesler, *Monatsh. Chem.*, **95,** 1193 (1964).
15. H. Reinheckel, *Monatsh. Chem.*, **99,** 2203 (1968).
16. Y. Iwanami, Y. Kenjo, K. Nishibe, M. Kajiura, and S. Isoyama, *Bull. Chem. Soc. Jap.*, **37,** 1740 (1964).
17. N. E. Alexandrou and D. N. Nicolaides, *J. Chem. Soc.* (*C*), 2319 (1969).
18. S. Maffei, *Boll. Sci. Fac. Chim. Ind. Bologna*, **8,** 93 (1950); *Chem. Abstr.*, **45,** 8017 (1951).
19. G. B. Crippa and S. Sacerdoti, *Ann. Chim.* (Rome), **46,** 348 (1956); *Chem. Abstr.*, **51,** 3613 (1957).
20. D. G. I. Felton, T. S. Osdene, and G. M. Timmis, *J. Chem. Soc.*, 2895 (1954).
21. F. J. Wolf, R. H. Beutel, and J. R. Stevens, *J. Am. Chem. Soc.*, **70,** 2572 (1948).
22. T. S. Osdene and G. M. Timmis, *J. Chem. Soc.*, 4349 (1955).
23. K. Gerlach and F. Kröhnke, *Chem. Ber.*, **95,** 1124 (1962).
24. J. W. Lown and M. H. Akhtar, *Can. J. Chem.*, **49,** 1610 (1971).
25. W. Gauss and S. Petersen, *Angew. Chem.*, **69,** 252 (1957).
26. A. Mustafa, A. H. E. Harhash, A. K. E. Mansour, and S. M. A. E. Omran, *J. Am. Chem. Soc.*, **78,** 4306 (1956).
27. J. K. Landquist, *J. Chem. Soc.*, 2816 (1953).
28. L. Mandasescu and L. Stoicescu-Crivat, *Acad. Rep. Pop. Romine, Filiala Iasi, Studii Cercetari Stiint. Chim.*, **11,** 75 (1960); *Chem. Abstr.*, **56,** 4764 (1962).
29. L. Mandasescu, L. Stoicescu-Crivat, and I. Gabe, *Acad. Rep. Pop. Romine, Filiala Iasi, Studii Cercetari Stiint. Chim.*, **11,** 311 (1960); *Chem. Abstr.*, **56,** 11751 (1962).
30. L. Mandasescu, L. Stoicescu-Crivat, I. Gabe, S. Lica, and M. Stefanescu, *Acad. Rep. Pop. Romine, Filiala Iasi, Studii Cercetari Stiint. Chim.*, **13,** 115 (1962); *Chem. Abstr.*, **59,** 4070 (1963).
31. G. Henseke and W. Lemke, *Chem. Ber.*, **91,** 113 (1958).
32. A. A. Santelli and T. S. Osdene, *J. Org. Chem.*, **29,** 2066 (1964).
33. A. A. Santelli and T. S. Osdene, *J. Org. Chem.*, **30,** 2100 (1965).
34. T. S. Osdene (to American Home Products), U.S. Pat. 3,145,205; *Chem. Abstr.*, **61,** 16079 (1964).
35. A. A. Santelli and T. S. Osdene (to American Home Products), U.S. Pat. 3,334,097; *Chem. Abstr.*, **67,** 90840 (1967).
36. Ciba, Neth. Appl. 6,515,065; *Chem. Abstr.*, **65,** 11596 (1966).
37. See, for example, W. Yamatani, K. Nei, and S. Yamamoto (to Mitsubishi Chem. Ind. Co. Ltd.), Jap. Pat. 70/25,506; *Chem. Abstr.*, **73,** 131028 (1970).
38. E. Hayashi and H. Saito, *Yakugaku Zasshi*, **86,** 687 (1966); *Chem. Abstr.*, **65,** 16970 (1966).
39. L. G. S. Brooker and E. J. Van Lare (to Eastman Kodak), U.S. Pat. 3,431,111; *Chem. Abstr.*, **72,** 68222 (1970).
40. N. P. Buu-Hoi and N. H. Khoi, *Bull. Soc. Chim. Fr.*, 753 (1950).
41. S. Maffei, *Gazz. Chim. Ital.*, **76,** 239 (1946); *Chem. Abstr.*, **42,** 911 (1948).
42. H. H. Perkampus, Th. Bluhm and J. V. Knop, *Spectrochim. Acta, Part A*, **28,** 2179 (1972).
43. Th. Bluhm, H. H. Perkampus, and J. V. Knop, *Ber. Bunsenges. Phys. Chem.*, **76,** 1251 (1972).
44. G. Schmidtberg and Th. Bluhm, *Org. Mass Spectrom.*, **9,** 449 (1974).

45. H. H. Perkampus and Th. Bluhm, *Tetrahedron*, **28,** 2099 (1972).
46. A. Padwa and L. Gehrlein, *J. Am. Chem. Soc.*, **94,** 4933 (1972).
47. M. Lang and J. P. Fleury, *Tetrahedron Lett.*, 3967 (1974).
48. J. K. Landquist and G. J. Stacey, *J. Chem. Soc.*, 2822 (1953).
49. N. P. Buu-Hoi, J. N. Vallat, G. Saint-Ruf, and G. Lambelin, *Chim. Ther.*, **7,** 210 (1972); *Chem. Abstr.*, **77,** 152114 (1972).
50. V. D. Romanenko, N. E. Kul'chitskaya, and S. I. Burmistrov, *Chem. Heterocycl. Comp.*, **9,** 241 (1973).
51. K. Zaitsu and Y. Ohkura, *Chem. Pharm. Bull.* (Tokyo), **23,** 1057 (1975).
52. H. J. Kallmayer, *Arch. Pharm.* (Weinheim), **307,** 806 (1974).
53. J. E. Nottke (to E. I. du Pont de Nemours), U.S. Pat. 3,928,950; *Chem. Abstr.*, **84,** 105652 (1976).
54. V. Nacci, G. Stefancich, G. Filacchioni, R. Giuliano, and M. Artico, *Farmaco, Ed. Sci.*, **28,** 49 (1973); *Chem. Abstr.*, **78,** 84634 (1973).
55. D. Schelz, *Helv. Chim. Acta*, **57,** 1075 (1974).

CHAPTER XXXIX

Benzo[g]quinoxalines

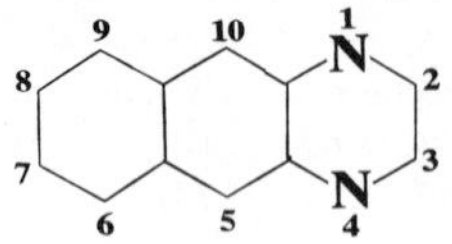

RRI 3388

Contents

I. Physical Properties 730
II. Methods of Preparation 731
III. Reactions 734
IV. Tables 736
Table 1. Benzo[g]quinoxaline-5,10-diones 736
Table 2. Benzo[g]quinoxalines 739
V. References 741

Benzo[g]quinoxalines were formerly named 6,7-benzoquinoxalines.

I. Physical Properties

The ultraviolet spectrum of benzo[g]quinoxaline has been measured in ethanol and in heptane. In ethanol the principal absorption maxima are at 227, 264, and 364 nm.[1] Proton chemical shifts and coupling constants are tabulated in Figure 1.[2] Values calculated by HMO theory are found to be close to the experimental ones.[39]

Figure 1. PMR DATA FOR BENZO[g]QUINOXALINE

Chemical shifts (δ)[a]		Coupling constants[a]	
H2(3)	8.86	$J_{6,7}$	8.63
H5(10)	8.68	$J_{7,8}$	0.48
H6(9)	8.11	$J_{6,8}$	1.20
		$J_{6,9}$	0.60
H7(8)	7.58	$J_{5,10}$	0.60
		$J_{9,10}$	0.50

[a] Measured in $CDCl_3$.

The ultraviolet and PMR spectra of some 1,2-dihydro-2-oxo-benzo[*g*]-quinoxalines have been recorded in trifluoroacetic acid and dimethyl sulfoxide; the 3-acetonyl derivative **1** is predominantly in the enamine form **2** in both solvents. Singlets due to H-5 and H-10 are observed in the region of δ 7.5–8.0.[3]

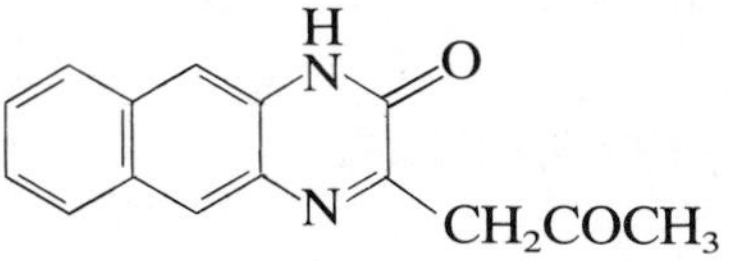

In an earlier paper the ultraviolet spectrum of the 2,3-diphenyl derivative is reported in ethanol, in ethanolic HCl, and in 95% H_2SO_4.[4] 2,3-Bis(2-pyridyl)benzo[*g*]quinoxaline and the corresponding 2,3-bis[2-(6-methylpyridyl)] derivative have been suggested as possible spectrophotometric reagents for copper(I) because of their structural similarity to cuproin and ferroin.[5] The polarographic reduction of a number of benzo[*g*]quinoxaline-5,10-diones has been investigated. The quinonoid ring is reduced first and then the heterocyclic ring. The fusion of a pyrazine ring onto a 1,4-naphthoquinone has hardly any effect on the half-wave potential of the quinone, as shown by the data for benzo[*g*]-quinoxaline-5,10-dione (**3**) and 1,4-naphthoquinone (**4**). As expected, electron-donor substituents lower the half-wave potential of benzo[*g*]quinoxaline-5,10-diones and electron-attracting substituents raise it.[6]

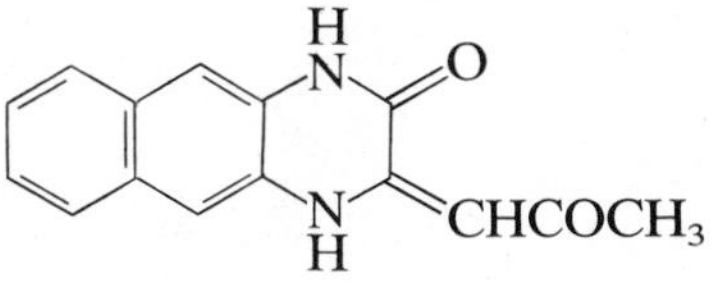

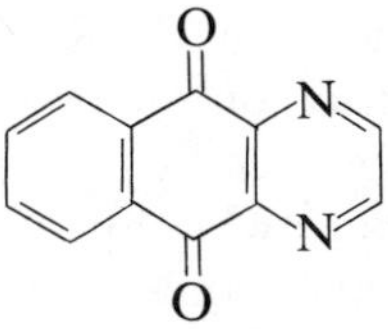

3

$-E_{1/2}$I = 0.73 V
II = 1.37 V
III = 1.90 V

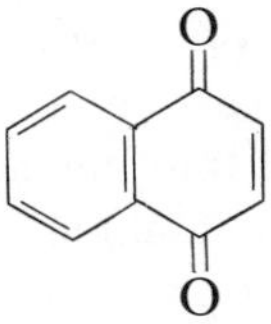

4

$-E_{1/2}$I = 0.72 V
II = 1.63 V

II. Methods of Preparation

The major substrate for the preparation of benzo[*g*]quinoxalines is 2,3-diaminonaphthalene. This has been condensed with a range of α-dicarbonyl compounds including glyoxal,[7] diketones,[8–10,40] keto aldehydes,[11] keto acids,[12,13] keto esters,[14] and osones.[15,16] Reaction of the diamine with diethyl acetylenedicarboxylate gives compound **5**.[17] Benzo[*g*]-quinoxaline-5,10-diones **6** have been prepared from 2,3-diamino-1,4-naphthoquinone by reaction with α-dicarbonyl compounds.[18]

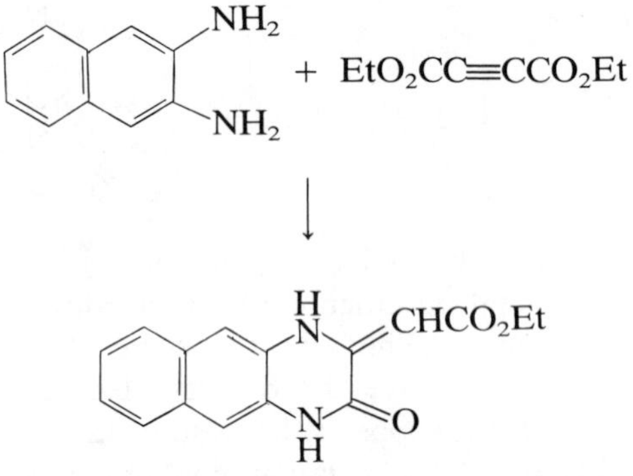

5

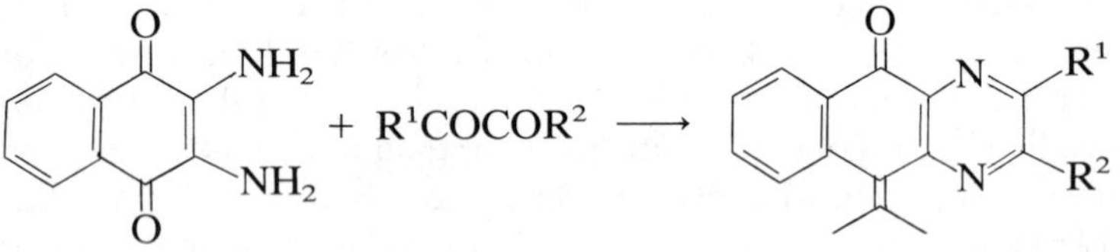

6

Treatment of substituted diaminoquinones (e.g., **7**, $R^1 = R^2 = H$) with alkali results in ring closure to reduced benzoquinoxaline diones (e.g., **8**, $R^1 = R^2 = H$). Intermediates of type **7** are obtained by ring cleavage of the corresponding aminoaziridinonaphthoquinone with hydriodic acid, followed by neutralization of the first-formed hydriodide. In the presence of excess of alkali, the substituted diaminoquinone reverts to the aminoaziridinonaphthoquinone (Scheme 1).[19,41] A detailed study has been made of the stereochemical aspects of these reactions.[41]

The aziridine **9**, prepared from 2,3-dichloro-1,4-naphthoquinone by displacement of the first chlorine with alanine, acetylation, and replacement of the remaining chlorine with aziridine, is converted into the benzoquinoxaline dione **10** on treatment with acid.[20]

2-Amino-3-chloro-1,4-naphthoquinone is converted by successive treatment with chloroacetyl chloride and diethylamine into the benzoquinoxaline trione **12**. The intermediate **11** is not isolated, but with other

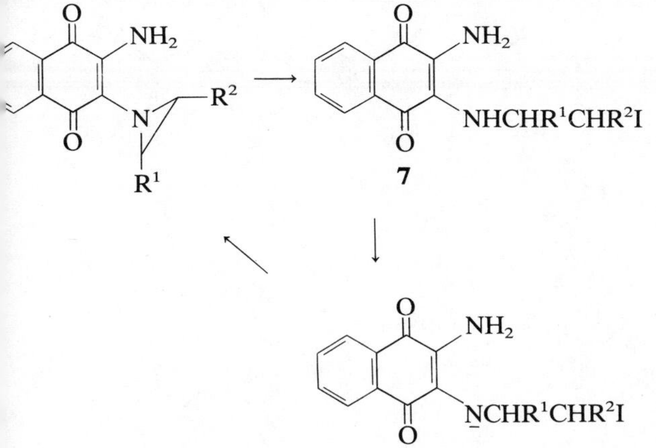

Scheme 1

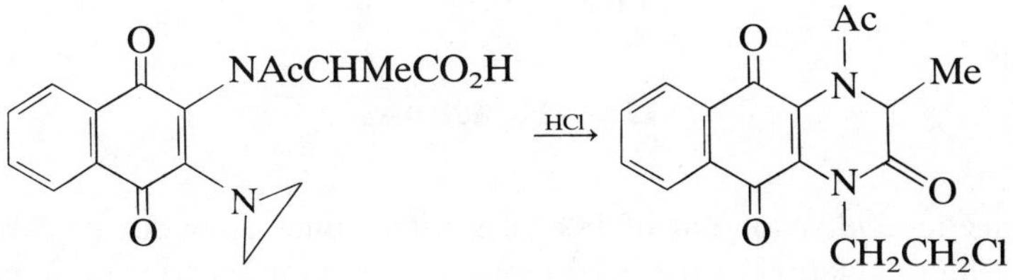

secondary amines it is possible to isolate the intermediate dialkylamino compound.[21] Benzo[*g*]quinoxaline diones have also been prepared by Diels-Alder addition of dienes to 2,3-dimethylquinoxaline-5,8-dione (**13**)[22] and the corresponding 2,3-diphenyl compound.[42] The initial adducts are tautomerized by acid and finally oxidized to the desired quinones (e.g., **14**).

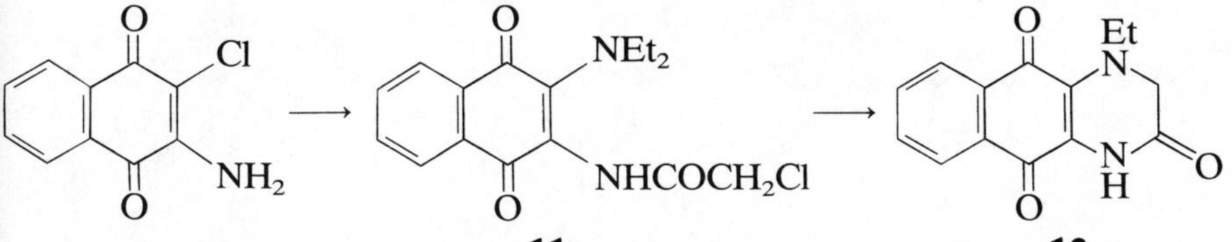

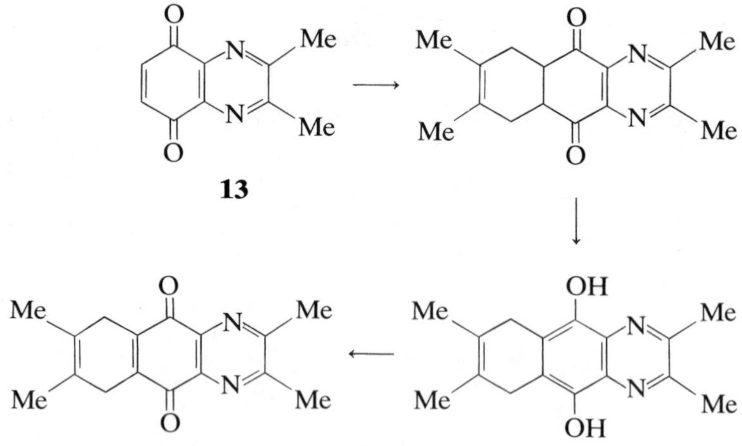

13

14

Reagents (i) $CH_2{=}CMe{-}CMe{=}CH_2$
(ii) 40% Ethanolic HCl
(iii) Ag_2O

III. Reactions

Peracetic acid oxidation of benzo[*g*]quinoxaline does not give an *N*-oxide,[7] and oxidation of the 5,10-dione with chromium trioxide in glacial acetic acid, or 40% sulfuric acid, yields the 2,3-dioxo derivative **15**.[23] The latter compound gives 2,3,5,10-tetrachlorobenzo[*g*]quinoxaline (**16**)* on treatment with phosphorus pentachloride.[18,31]

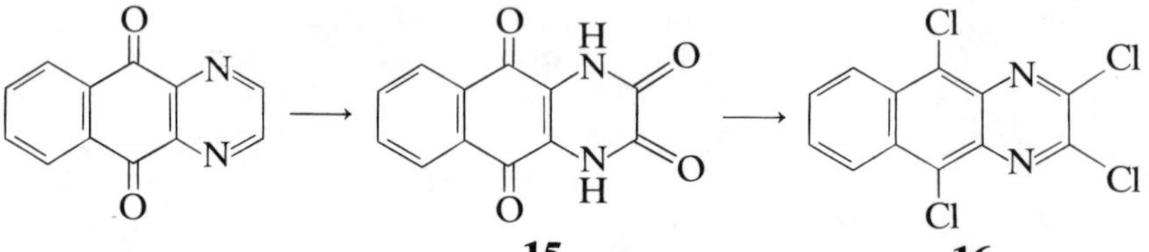

15

16

α-Methyl derivatives of benzo[*g*]quinoxaline-5,10-diones show enhanced reactivity and give condensation products with aromatic aldehydes.[24] Compound **17** reacts with aryldiazonium compounds to give the corresponding arylhydrazones **18**.[24]

* This compound has been indexed incorrectly as 2,3,5,6-tetrachlorobenzo[*g*]-quinoxaline.

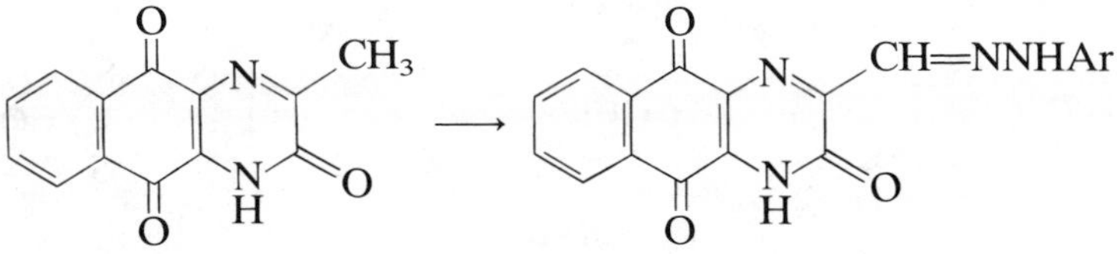

17 18

2-D-Arabotetrahydroxybutylbenzo[*g*]quinoxaline (**19**) has been oxidized to the 2-carboxylic acid with potassium permanganate and to the 2-aldehyde with potassium metaperiodate.[15] It is converted into the phenylhydrazone of the 2-aldehyde **20** on treatment with phenylhydrazine under neutral conditions and into the pyrazoloquinoxaline **21** on treatment with phenylhydrazine under acid conditions.[25]

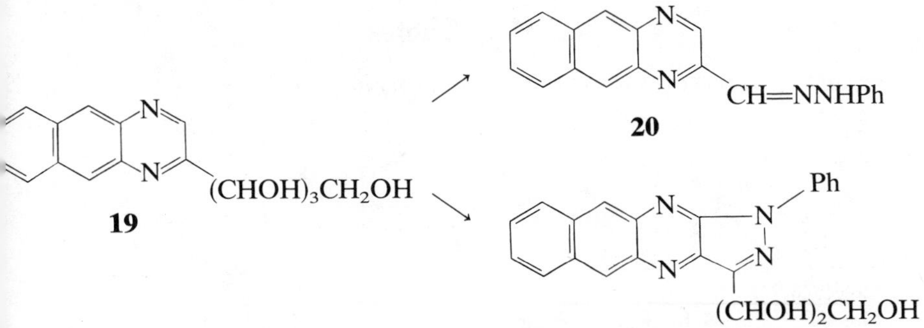

19 20 21

2,3-Diaminobenzo[*g*]quinoxaline (**22**), prepared by the reaction of the corresponding dichloro compound with liquid ammonia, is converted into the dibenzofluorubin **23** when boiled with *p*-toluenesulfonic acid in

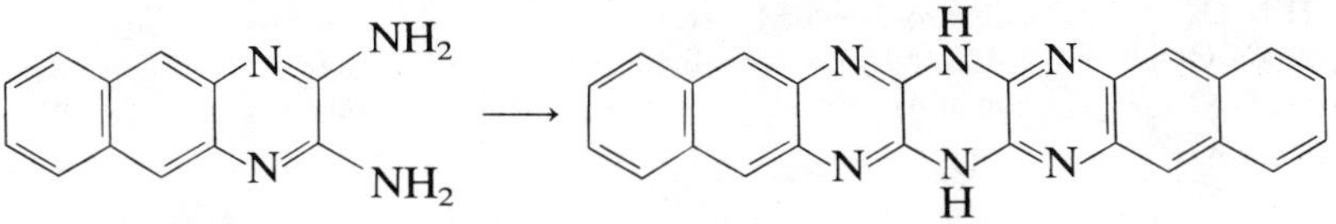

22 23

α-methylnaphthalene.[26] The cyclization of benzo[*g*]quinoxaline-2,3-dithiones with $COCl_2$ (or $CSCl_2$) yields products which are useful as acaricides (Scheme 2).[27–30] 2,3,5,10-Tetrachlorobenzo[*g*]quinoxaline (**16**) has been condensed with 2-pipecolylamine (**24**) to give the polycyclic compound **25**.[31]

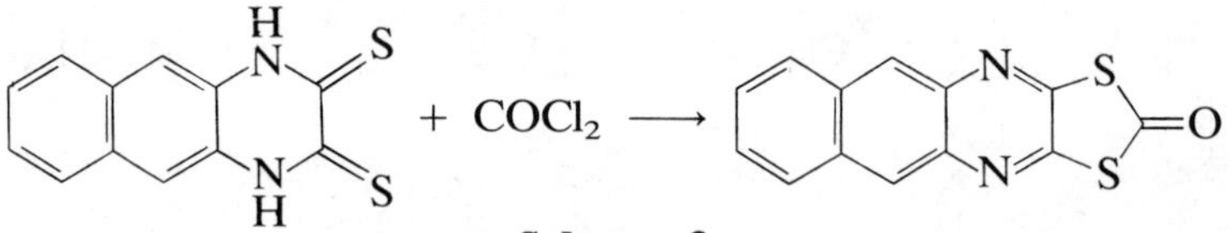

Scheme 2

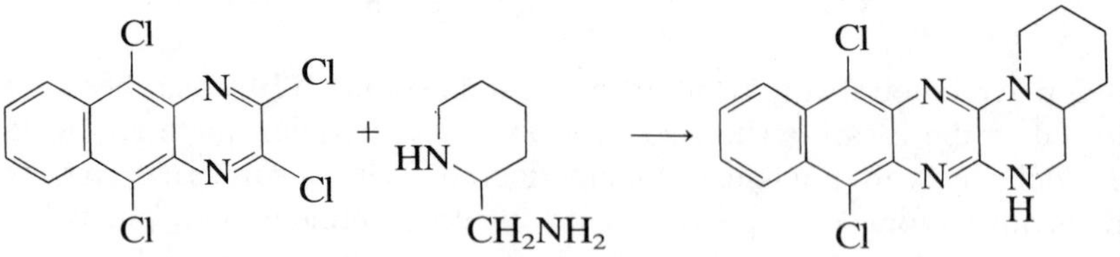

16 **24** **25**

IV. Tables

TABLE 1. BENZO[*g*]QUINOXALINE-5,10-DIONES

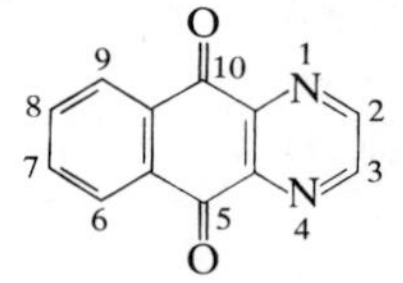

Molecular formula	Substituents or structure	m.p., °C	Ref.
$C_{12}H_6N_2O_2$	Unsubstituted	317–318 decomp.	18
$C_{12}H_6N_2O_4$	1,2,3,4-Tetrahydro-2,3-dioxo	>360	18, 23
$C_{12}H_{10}N_2O_2$	1,2,3,4-Tetrahydro	220–223	19
	1-Acetyl	240–243	19
	1-β-Chloroethyl	>260	37
$C_{13}H_8N_2O_2$	2-Methyl	250–251 decomp.	18
$C_{13}H_8N_2O_3$	1,2-Dihydro-3-methyl-2-oxo	266–268 decomp.	24
$C_{13}H_{12}N_2O_2$	1,2,3,4-Tetrahydro-2-methyl	167–169	41
$C_{14}H_{10}N_2O$	2,3-Dimethyl	240–241	18
$C_{14}H_{12}N_2O_3$	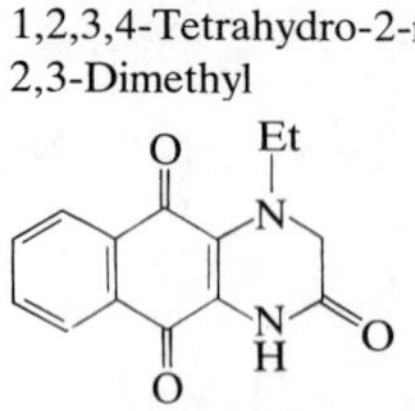	207–208	21
	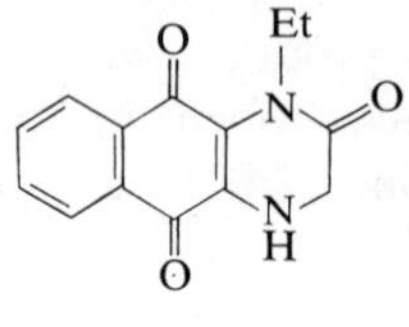		38

TABLE 1 (*contd.*)

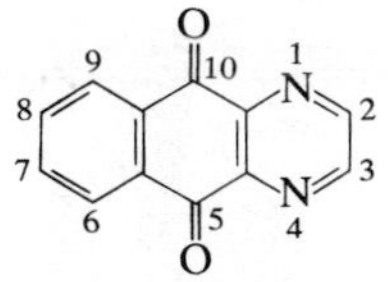

Molecular formula	Substituents or structure	m.p., °C	Ref.
$C_{14}H_{14}N_2O_2$	2-Ethyl-1,2,3,4-tetrahydro	142–144	41
	trans(?)-1,2,3,4-Tetrahydro-2,3-dimethyl	210–211	41
$C_{15}H_{14}N_2O_3$	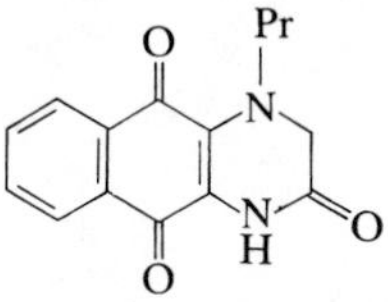	201–202	21
$C_{16}H_{15}ClN_2O_4$	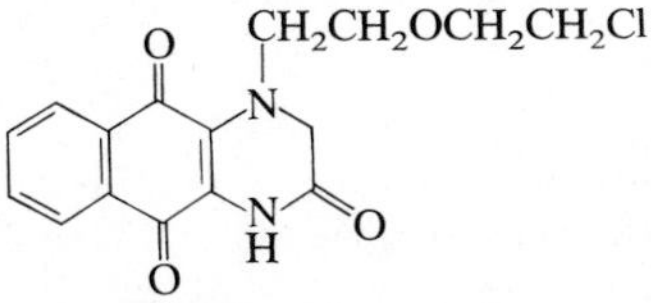	155.5–156	21
$C_{16}H_{16}N_2O_2$	6,9-Dihydro-2,3,7,8-tetramethyl	310–312	22
$C_{16}H_{16}N_2O_3$	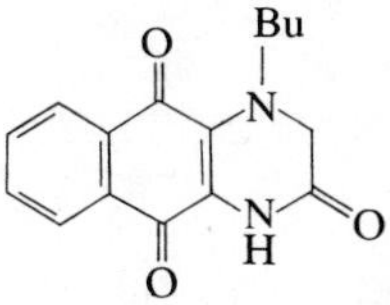	197–198	21
$C_{17}H_{15}ClN_2O_4$	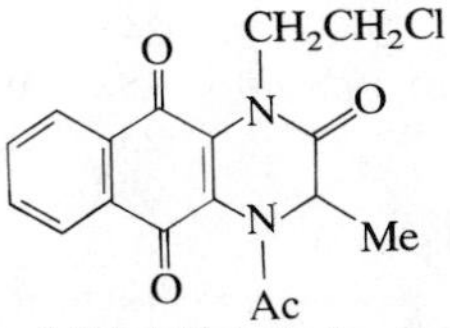		20
$C_{18}H_{14}N_2O_2$	1,2,3,4-Tetrahydro-2-phenyl	203–205	41
$C_{18}H_{22}N_2O_2$	*trans*(?)-1,2,3,4-Tetrahydro-2,3-dipropyl	124–125	41
$C_{19}H_{11}BrN_4O_3$	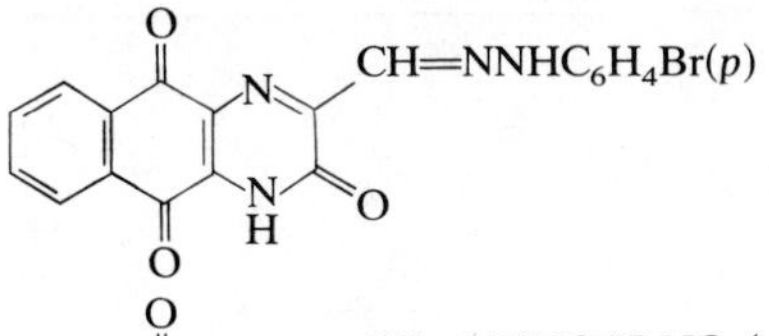	300 decomp.	24
$C_{19}H_{11}N_5O_5$	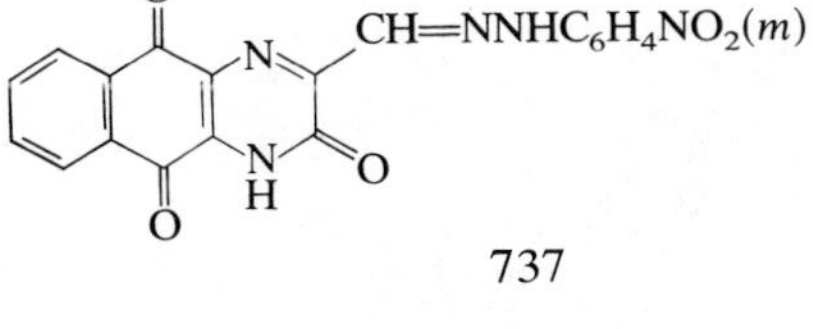	305 decomp.	24

TABLE 1 (*contd.*)

[Structure: 5,10-dioxo-5,10-dihydrobenzo[g]quinoxaline ring with positions numbered 1–10; O, N labels]

Molecular formula	Substituents or structure	m.p., °C	Ref.
$C_{19}H_{11}N_5O_5$ (*contd.*)	[Structure: 1,2-dihydro-2-oxo-benzo[g]quinoxaline-5,10-dione bearing CH=NNHC$_6$H$_4$(NO$_2$)*p* at C-3; labels O, N, N, H, O, O]	325–326 decomp.	24
$C_{19}H_{12}N_2O_3$	2-Methyl-3-phenyl	231–232 decomp.	24
$C_{20}H_{11}ClN_2O_2$	*p*-Chlorostyryl	256–257 decomp.	24
$C_{20}H_{11}N_3O_4$	*m*-Nitrostyryl	319–320 decomp.	24
	p-Nitrostyryl	325–326 decomp.	24
$C_{20}H_{12}N_2O_2$	2-Styryl	254–255 decomp.	24
$C_{20}H_{12}N_2O_3$	1,2-Dihydro-2-oxo-3-styryl	278–280 decomp.	24
$C_{20}H_{18}N_2O_2$	*trans*-2-Ethyl-1,2,3,4-tetrahydro-3-phenyl	158–160	41
$C_{22}H_{17}N_3O_2$	*p*-Dimethylaminostyryl	259–260 decomp.	24
$C_{23}H_{19}ClN_2O_4$	[Structure: labels O, CH$_2$CH$_2$Cl, N, O, N, Ph, O, Ac]		20
$C_{24}H_{14}N_2O_2$	2,3-Diphenyl	268–269	18
	Hemihydrate	242	42
$C_{25}H_{20}N_2O_2$	*trans*-2-Benzyl-1,2,3,4-tetrahydro-3-phenyl	188–190	41
$C_{26}H_{16}N_2O_2$	2-Phenyl-3-styryl	234–235 decomp.	24
$C_{26}H_{18}N_2O_2$	7,8-Dimethyl-2,3-diphenyl	290	42
$C_{26}H_{20}N_2O_2$	6,9-Dihydro-7,8-dimethyl-2,3-diphenyl		
	Hemihydrate	Thermally labile	42
$C_{28}H_{18}N_2O_2$	2,3-Distyryl	276–277	18
$C_{28}H_{21}N_3O_2$	2-*p*-Dimethylaminostyryl-3-phenyl	292–293 decomp.	24

TABLE 2. BENZO[g]QUINOXALINES

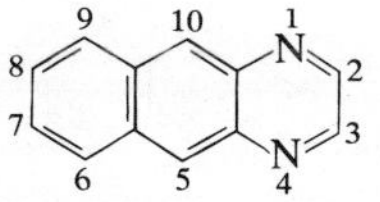

Molecular formula	Substituents or structure	m.p., °C	Ref.
$C_{12}H_4Cl_4N_2$	2,3,5,10-Tetrachloro	207–208, 215	18, 31
$C_{12}H_6Cl_2N_2$	2,3-Dichloro		26
$C_{12}H_7ClN_2S_2$	5-Chloro-1,2,3,4-tetrahydro-2,3-dithioxo		26
$C_{12}H_8N_2$	Unsubstituted	117–118, 125–126	7, 15
$C_{12}H_8N_2O_2$	1,2,3,4-Tetrahydro-2,3-dioxo	415 corr.	9
$C_{12}H_{10}N_4$	2,3-Diamino		26
$C_{12}H_{15}ClN_2$	(+)-(5a*S*,9a*S*)-2-Chloro-*trans*-5,5a,6,7,8,9,9a,10-octahydro[a]	71–73	40
$C_{12}H_{16}N_2$	(+)–5a*S*,9a*S*)-*trans*-5,5a,6,7,8,9,9a,10-octahydro[b]	86–87	40
$C_{12}H_{16}N_2O$	(±)-*trans*-1,2,5,5a,6,7,8,9,9a,10-decahydro-2-oxo	265–267	40
	(+)-*trans*-1,2,5,5a,6,7,8,9,9a,10-decahydro-2-oxo[c]	268–270	40
$C_{12}H_{16}N_2S$	(+)-(5a*S*,9a*S*)-1,2,5,5a,6,7,8,9,9a,10-decahydro-2-thioxo[d]	188–190	40
$C_{12}H_{18}N_4$	(+)-(5a*S*,9a*S*)-2-Hydrazino-*trans*-5,5a,6,7,8,9,9a,10-octahydro[e]	155–157	40
$C_{13}H_8N_2O$	2-Formyl	174	15
	Phenylhydrazone	276 decomp.	15
	Methylphenylhydrazone	225 decomp.	15
$C_{13}H_8N_2O_2$	2-Carboxy	194 decomp.	15
$C_{13}H_9BrN_2O$	3-Bromomethyl-1,2-dihydro-2-oxo	230 decomp.	14
$C_{13}H_9N_3O_2$	2-Nitromethyl	216–218	33
$C_{13}H_{10}N_2O$	1,2-Dihydro-3-methyl-2-oxo	290 corr., 295 decomp.	9, 14, 17
$C_{13}H_{18}N_2S$	(+)-(5a*S*,9a*S*)-2-Methylthio-*trans*-5,5a,6,7,8,9,9a,10-octahydro[f]	65–66	40
$C_{14}H_8N_2O_4$	2,3-Dicarboxy		
	Dimethyl ester	170–172	34, 43
	Diamide	360–362 decomp.	43
	Dinitrile	250 decomp.	43
$C_{14}H_{10}N_2O_3$	3-Carboxymethyl-1,2-dihydro-2-oxo		
	Ethyl ester	232–233 corr.	3, 17
$C_{14}H_{12}N_2$	2,3-Dimethyl	211 corr.	9
	1-Methyl perchlorate	227–230	45
$C_{14}H_{20}N_2O$	(+)-(5a*S*,9a*S*)-2-Ethoxy-*trans*-5,5a,6,7,8,9,9a,10-octahydro[g]	68–69	40
$C_{15}H_{12}N_2O_2$	2-(2-Carboxyethyl)	198–200 decomp.	11
$C_{15}H_{12}N_2O_3$	3-Carboxymethyl-1,2-dihydro-1-methyl-2-oxo		

TABLE 2 (*contd.*)

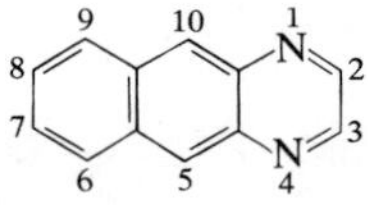

Molecular formula	Substituents or structure	m.p., °C	Ref.
$C_{15}H_{12}N_2O_3$ (*contd.*)	Ethyl ester	165–167	3
$C_{15}H_{13}N_2O_2$	3-Acetonyl-1,2-dihydro-2-oxo	315 decomp.	3
$C_{15}H_{14}N_2O_3$	2-(D-Threotrihydroxypropyl)[h]	159–160 decomp.	16
	2-(L-Erythrotrihydroxypropyl)		
	Hydrate	unstable	16
$C_{16}H_{12}N_2O_4$	5,10-Diacetoxy	238–240	19
$C_{16}H_{16}N_2O_4$	2-(D-Arabotetrahydroxybutyl)[i]	183 decomp.	15
	Tetraacetyl	148	15
	2-(D-Lyxotetrahydroxybutyl)[j]	170 decomp.	16
	2-(L-Xylotetrahydroxybutyl)[k]	163 decomp.	16
$C_{16}H_{18}N_2O_5$	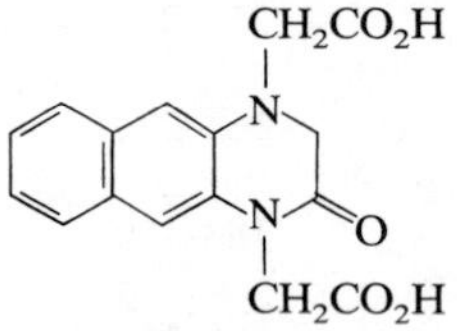 CH_2CO_2H, N, N, O, CH_2CO_2H	205–217	32
$C_{17}H_{25}N_3$	(+)-(5a*S*,9a*S*)-2-(*N*-Piperidino)-*trans*-5,5a,6,7,8,9,9a,10-octahydro[l]	121–122	40
$C_{19}H_{14}N_2O$	3-Benzyl-1,2-dihydro-2-oxo		44
$C_{20}H_{13}N_5$	2-Amino-3-(2-quinoxalinyl)	300	36
$C_{20}H_{14}N_2O_2$	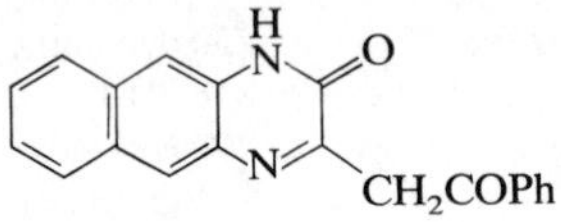 H, N, O, N, CH_2COPh	343	35
$C_{20}H_{16}N_2O_2$	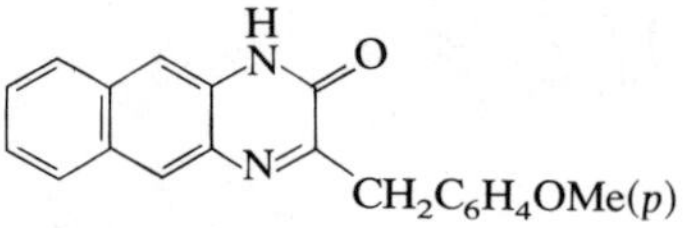 H, N, O, N, $CH_2C_6H_4OMe(p)$	264	12
$C_{20}H_{20}N_2O_8$	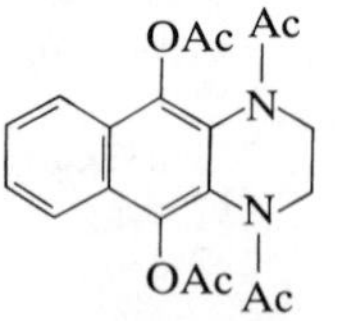 OAc, Ac, N, N, OAc, Ac	170–172	19
$C_{22}H_{14}N_4$	2,3-Bis(2-pyridyl)	172	5, 10
$C_{24}H_{16}N_2$	2,3-Diphenyl	187–189, 189.5–190, 192 corr.	4, 8, 9
$C_{24}H_{18}N_2O_2$	6,9-Dihydro-5,10-dihydroxy-2,3-diphenyl	240	42
$C_{24}H_{18}N_4$	2,3-Bis[2-(6-methylpyridyl)]	177	5

TABLE 2 (*contd.*)

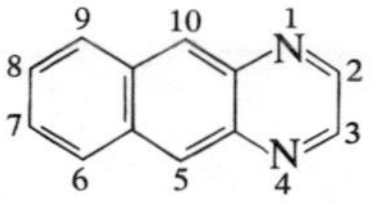

Molecular formula	Substituents or structure	m.p., °C	Ref.
$C_{24}H_{30}N_4S$	(+)-Bis(5a*S*,9a*S*)-5,5a,6,7,8,9,9a,10-octahydrobenzo[g]quinoxalin-2-yl disulfide[m]	141–143	40
$C_{26}H_{22}N_2O_2$	6,9-Dihydro-5,10-dihydroxy-7,8-dimethyl-2,3-diphenyl	218	42

[a] $[\alpha]_D^{23}$ +158° (*c* 0.3, EtOH).
[b] $[\alpha]_D^{23}$ +168° (*c* 0.5, EtOH).
[c] $[\alpha]_D^{23}$ +140° (*c* 0.8, $MeCO_2H$).
[d] $[\alpha]_D^{23}$ +167° (*c* 1.7, EtOH).
[e] $[\alpha]_D^{23}$ +138° (*c* 0.3, EtOH).
[f] $[\alpha]_D^{23}$ +130° (*c* 0.6, EtOH).
[g] $[\alpha]_D^{23}$ +130° (*c* 0.4, EtOH).
[h] $[\alpha]_D^{20}$ −177° (*c* 0.5, C_5H_5N).
[i] $[\alpha]_D^{18}$ −84° (*c* 1.0, C_5H_5N).
[j] $[\alpha]_D^{20}$ −51° (*c* 0.5, C_5H_5N).
[k] $[\alpha]_D^{20}$ −182° (*c* 0.5, C_5H_5N).
[l] $[\alpha]_D^{23}$ +96° (*c* 0.5, EtOH).
[m] $[\alpha]_D^{23}$ +130° (*c* 3.5, EtOH).

V. References

1. F. Kummer and H. Zimmermann, *Ber. Bunsenges. Phys. Chem.*, **71,** 1119 (1967).
2. H. H. Limbach, W. Seiffert, E. Ohmes, and H. Zimmermann, *Ber. Bunsenges. Phys. Chem.*, **74,** 966 (1970).
3. R. Mondelli and L. Merlini, *Tetrahedron*, **22,** 3253 (1966).
4. E. Sawicki, B. Chastain, H. Bryant, and A. Carr, *J. Org. Chem.*, **22,** 625 (1957).
5. W. I. Stephen and P. C. Uden, *Anal. Chim. Acta*, **39,** 357 (1967); *Chem. Abstr.*, **67,** 96531 (1967).
6. G. A. Efimova and L. S. Efros, *J. Org. Chem. USSR*, **3,** 2001 (1967).
7. J. K. Landquist, *J. Chem. Soc.*, 2816 (1953).
8. R. Thomas-Mamert and S. Weil, *Bull. Soc. Chim. Fr.* [3], **23,** 454 (1900); *Beilstein*, **23,** 338.
9. H. Goldstein and M. Streuli, *Helv. Chim. Acta*, **20,** 650 (1937).
10. R. E. Jensen and R. T. Pflaum, *J. Heterocycl. Chem.*, **1,** 295 (1964).
11. H. J. Kissel and L. Heilmeyer, *Biochem. Biophys. Acta*, **177,** 78 (1969).
12. M. Pailer, G. Pruckmayr, H. Zellner, and G. Zellner, *Monatsh. Chem.*, **93,** 1005 (1962).
13. H. Reinheckel, *Monatsber. Deut. Akad. Wiss. Berlin*, **1,** 698 (1959); *Chem. Abstr.*, **54,** 21119 (1960).

14. R. Kuhn and K. Dury, *Justus Liebigs Ann. Chem.*, **571,** 44 (1951).
15. G. Henseke and W. Lemke, *Chem. Ber.*, **91,** 101 (1958).
16. G. Henseke and C. Bauer, *Chem. Ber.*, **92,** 501 (1959).
17. Y. Iwanami, *Nippon Kagaku Zasshi*, **83,** 597 (1962); *Chem. Abstr.*, **59,** 3920 (1963).
18. G. A. Efimova and L. S. Efros, *J. Org. Chem. USSR*, **2,** 533 (1966).
19. G. Casini, F. Claudi, M. Felici, M. Ferappi, and M. Grifantini, *Farmaco, Ed. Sci.*, **24,** 732 (1969); *Chem. Abstr.*, **71,** 61343 (1969).
20. T. Kolesnikov, B. G. Boldyrev, and M. P. Romashkin, *Biol. Aktiv. Soedin.*, 271 (1968); *Chem. Abstr.*, **72,** 132356 (1970).
21. J. R. E. Hoover and A. R. Day, *J. Am. Chem. Soc.*, **77,** 35 (1955).
22. W. F. Gum, Jr., and M. M. Joullie, *J. Org. Chem.*, **30,** 2583 (1965).
23. G. A. Efimova and L. S. Efros, *J. Org. Chem. USSR*, **2,** 1867 (1966).
24. G. A. Efimova and L. S. Efros, *J. Org. Chem. USSR*, **3,** 372 (1967).
25. G. Henseke and W. Lemke, *Chem. Ber.*, **91,** 113 (1958).
26. W. Deuschel, W. Vilsmeier, and G. Reidel (to B.A.S.F. A.-G.), Belg. Pat. 612092; *Chem. Abstr.*, **57,** 16635 (1962).
27. K. Sasse, *Hoefchen-Briefe*, **13,** 197 (1960); *Chem. Abstr.*, **56,** 1802 (1962).
28. K. Sasse, R. Wegler, G. Unterstenhoefer, and F. Grewe, *Angew. Chem.*, **72,** 973 (1960).
29. K. Sasse R. Wegler, and G. Unterstenhoefer (to Farbenfabriken Bayer A.-G.), Ger. Pat. 1,088,965; *Chem. Abstr.*, **57,** 12507 (1962).
30. K. Sasse, R. Wegler, and G. Unterstenhoefer (to Farbenfabriken Bayer A.-G.), U.S. Pat. 3,141,886; *Chem. Abstr.*, **61,** 9509 (1964).
31. K. Winterfeld and M. Wildersohn, *Arch. Pharm.* (Weinheim), **303,** 1002 (1970).
32. T. Yano, H. Kobayashi, and K. Ueno, *Bull. Chem. Soc. Jap.*, **43,** 3167 (1970).
33. P. E. Fanta and Chun-Shan Wang, *J. Heterocycl. Chem.*, **3,** 525 (1966).
34. M. G. Gal'pern and E. A. Luk'yanets, *Chem. Heterocycl. Comp.*, **7,** 257 (1971).
35. Y. Iwanami, T. Seki, and T. Inagaki, *Bull. Chem. Soc. Jap.*, **44,** 1316 (1971).
36. P. E. Fanta and Chun-Shan Wang, *J. Heterocycl. Chem.*, **3,** 367 (1966).
37. V. T. Kolesnikov, B. G. Boldyrev, and M. E. Yarish, *J. Org. Chem. USSR*, **7,** 1051 (1971).
38. A. Wagner, W. Beck, and A. Diskus (to Oesterreichische Stickstoffwerke A.-G.), Austrian Pat. 231,220; *Chem. Abstr.*, **60,** 12027 (1964).
39. G. Ege and H. Vogler, *Z. Naturforsch.*, **27,** 1164 (1972).
40. G. Snatzke and G. Hajos, *Heterocycles*, **5,** 299 (1976).
41. G. Casini, F. Claudi, M. Grifantini, and S. Martelli, *J. Heterocycl. Chem.*, **11,** 377 (1974).
42. G. Kumar and A. P. Bhaduri, *Indian J. Chem.*, **13,** 1009 (1975).
43. M. G. Gal'pern and E. A. Luk'yanets, *J. Gen. Chem. USSR*, **41,** 2579 (1971).
44. T. Hayashi, T. Sugiura, H. Terada, S. Kawai, and T. Ohno, *J. Chromatogr.*, **118,** 403 (1976); *Chem. Abstr.*, **84,** 147095 (1976).
45. D. Schelz, *Helv. Chim. Acta*, **57,** 1075 (1974).

CHAPTER XL

Pyridoquinoxalines

I. Pyrido[1,2-*a*]quinoxalines 743
II. Pyrido[2,3-*b*]quinoxalines 746
III. Pyrido[3,4-*b*]quinoxalines 749
IV. Pyrido[1,2,3-*de*]quinoxalines 752
V. Tables 753
Table 1. Pyrido[1,2-*a*]quinoxalines 753
Table 2. Pyrido[2,3-*b*]quinoxalines 755
Table 3. Pyrido[3,4-*b*]quinoxalines 757
Table 4. Pyrido[1,2,3-*de*]quinoxalines 758
VI. References 758

I. Pyrido[1,2-*a*]quinoxalines

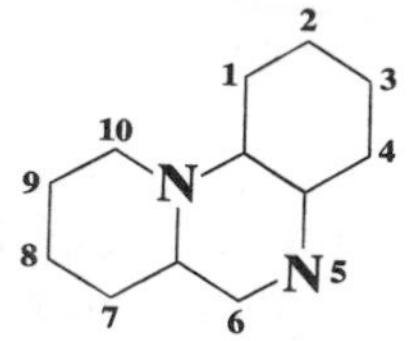

As mentioned by Mosby,[1] the earliest claims to the synthesis of this ring system have been shown to be incorrect. Compounds described as glucazidones and formulated as **1** are now known to be furylquinoxalines such as **2**.[2] More recently several authentic representatives of this heterocycle have been synthesized. The nucleus has been referred to as pyridino[1,2-*a*]quinoxaline (**3**). A third numbering system (**4**) has also been used.

Treatment of the quinoxalinylalkanoic acids **5** with alkaline sodium borohydride results in cyclization to give the pyrrolo or pyrido fused systems **6**. Although both these products are obtained in high yields, corresponding cyclizations with higher homologues (**5**, $n = 4$–6) could not be effected.[3]

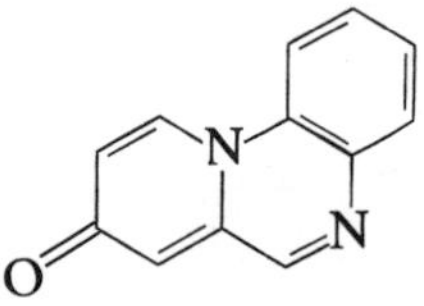

1 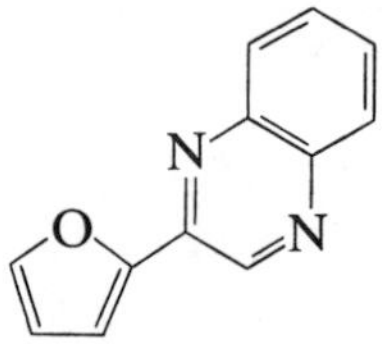 **2** 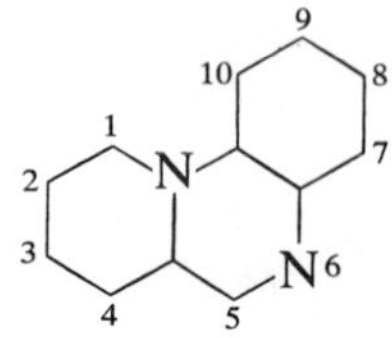 **3**

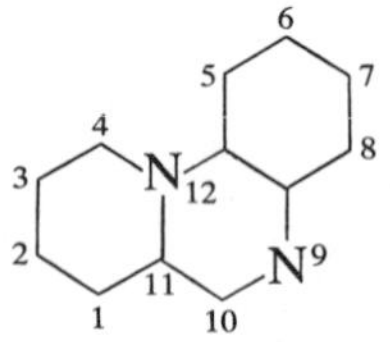

4

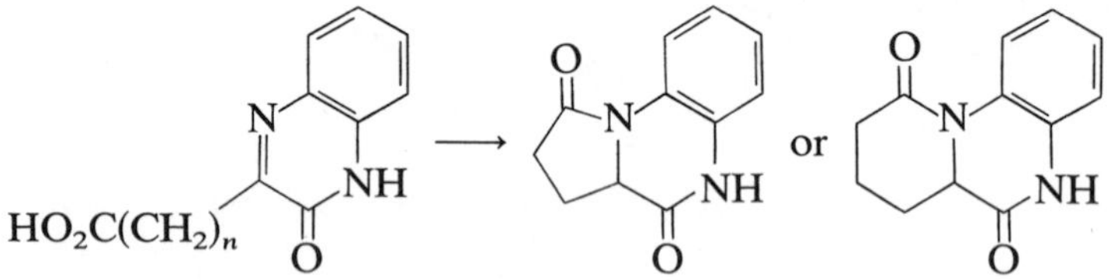

5 $n=2$ or 3 **6**

Both D-threo-2,5-hexodiulose (**7**) and kojic acid (**10**) afford the pyrido[1,2-*a*]quinoxaline **11** on treatment with *o*-phenylenediamine. The pathway via **8** and **9** explains the formation of the tricyclic compound from the hexodiulose. The product **11** forms the *N*-benzenesulfonate **12** on reaction with benzenesulfonyl chloride. Treatment with acetic anhydride in pyridine gives the diacetyl derivative **13**. Quinoxaline-2-carboxylic acid is produced when **11** is oxidized with potassium permanganate.[21] The condensation of *o*-phenylenediamine with the pyrylium salt **14** affords the pyridoquinoxalinium salt **16** by way of the intermediate *N*-phenylpyridinium salt **15**.[22]

A series of complex derivatives (**18**) of the ring system have been prepared by reaction of the pyranylidene compounds **17** with *o*-phenylenediamine.[4] The only other pyrido[1,2-*a*]quinoxaline to be reported is compound **19**, obtained from the reaction of quinoxaline with two molecular equivalents of dimethyl acetylenedicarboxylate.[5]

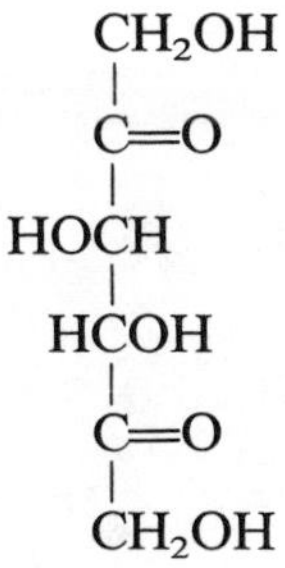

7

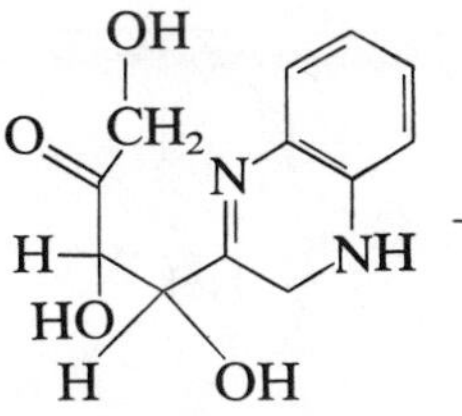

8

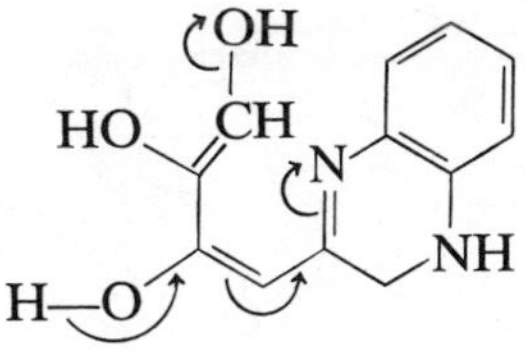

9

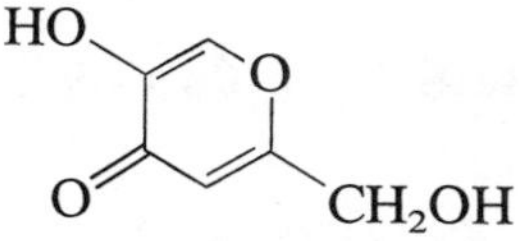

10

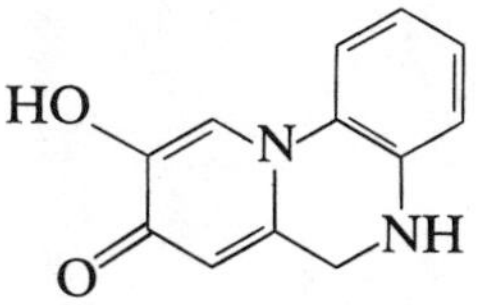

11

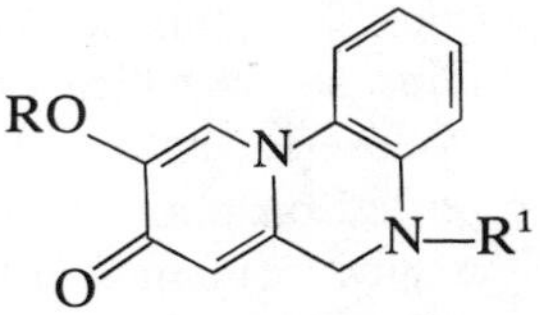

12 R = H, $R^1 = SO_2Ph$

13 $R = R^1 = Ac$

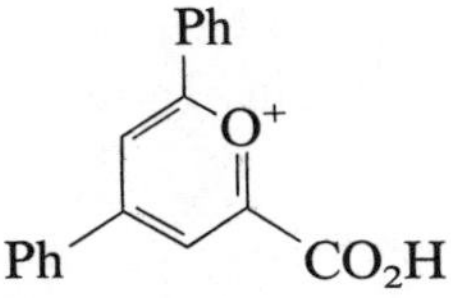

14

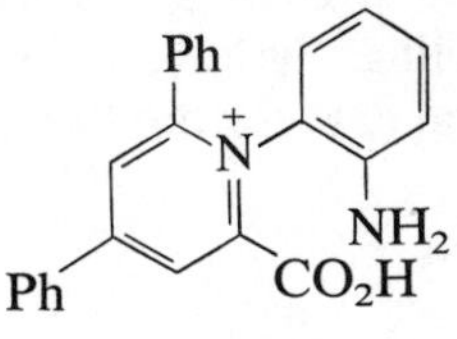

15

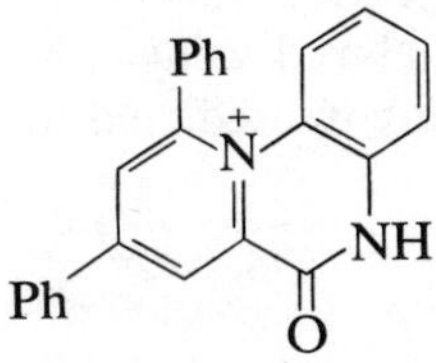

16

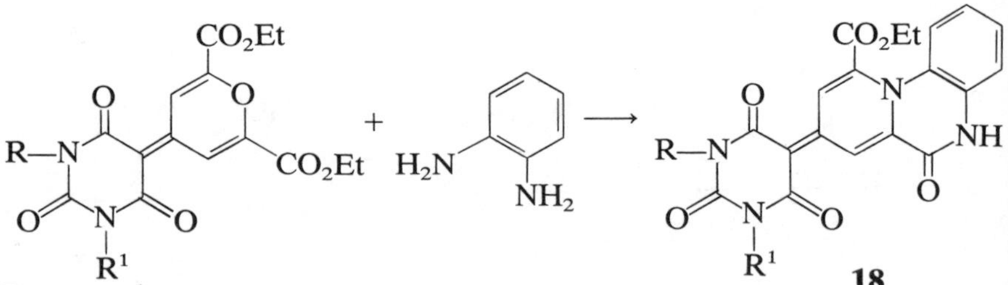

17 R, R^1 = H, Me, or Ph

18

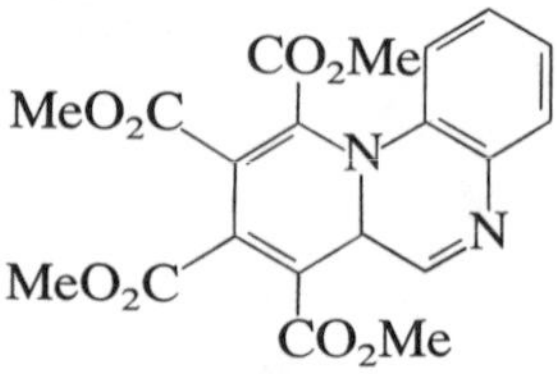

19

II. Pyrido[2,3-*b*]quinoxalines

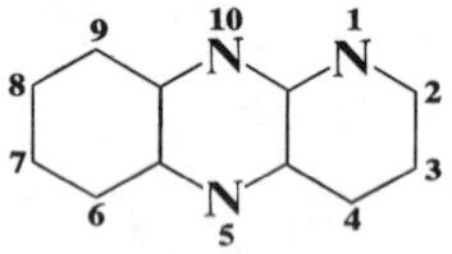

RRI 3360

This ring system has also been described as pyridino-2′.3′:2.3-quinoxaline. Both pyrido[2,3-*b*]quinoxaline (**20**) and its [3,4-*b*] isomer (**21**) are yellow crystalline solids which darken slowly on storage. The melting points are very close to that of phenazine. Brief spectral details have been recently reported on both of the pyrido[*b*]quinoxalines.[6] The parent pyrido[2,3-*b*]pyrazine has bands in the infrared spectrum at 1505, 1410, and 740 cm^{-1}. The mass spectrum of the compound, as with phenazine, shows peaks due to the molecular ion, and peaks due to the loss of HCN and the loss of HCN and C_2H_2. The ultraviolet spectrum possesses absorption maxima in methanol at 208, 245, and 358 nm. Pyrido[2,3-*b*]quinoxaline has a PMR spectrum which has been partially analyzed using spin decoupling. Pyrido[3,4-*b*]quinoxaline has been similarly treated, and the spectral data for both isomers are given in Figure 1.

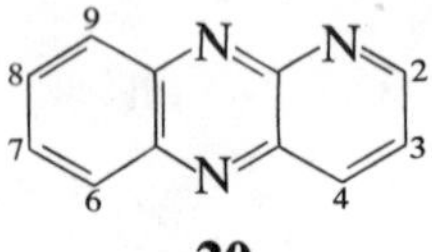

20 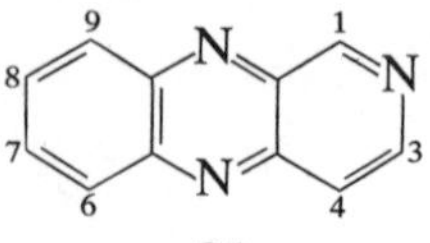 **21**

Figure 1. PMR DATA FOR LINEAR PYRIDOQUINOXALINES **20** AND **21**

Assignment	Pyrido[2,3-*b*]quinoxaline (**20**)	Pyrido[3,4-*b*]quinoxaline (**21**)
	δ ($CDCl_3$)	δ ($CDCl_3$)
H-1		9.72*d* $J_{1,4}$ 0.8 Hz
H-2	9.39*q* $J_{2,3}$ 3.9, $J_{2,4}$ 1.95 Hz	
H-3	7.79*q* $J_{3,4}$ 8.7 Hz	8.78*d* $J_{3,4}$ 6.0 Hz
H-4	8.61*q*	7.99*q*
H-6 and H-9	8.2–8.5 m	8.15–8.35 m
H-7 and H-8	7.8–8.0 m	7.8–7.95 m

Spin decoupling results on various methyl-substituted pyridoquinoxalines suggested that there is a marked degree of bond fixation in both the [2,3-*b*] and [3,4-*b*] isomers.[6] It seems that considerable double-bond character is possessed by the 6,7 and 8,9 bonds in both ring systems **20** and **21**.

The parent heterocycle and methyl- and methoxy-substituted derivatives have been obtained in low yields by reactions of anilinonitropyridines with ferrous oxalate at 260–280°. This extension of a conventional phenazine synthesis is the only reported approach to simple examples of this ring system.[6]

Although no cyclized product was obtained from the nitroanilino compound **22**, the isomeric nitropyridine (**23**) cyclized to give pyrido[2,3-*b*]quinoxaline in 20% yield. Cyclization onto the 2-position of the 3-anilinopyridine **24** gave 27% of the [2,3-*b*] compound, but this also resulted in a 10% yield of the [3,4-*b*] isomer from competing cyclization at the 4-position of the pyridine ring. The route **23**→**20** was also used to obtain pyrido[2,3-*b*]quinoxalines substituted in the 6-, 7-, 8-, or 9-positions by methyl or methoxy. Depending on the symmetry of the starting aniline, either mixtures of isomers or pure products were obtained.

o-Phenylenediamine has been made to react with the azaquinone **25** to give the oxo compound **26**.[7] 3-Hydroxy-2-pyridone condenses with benzofurazan 1-oxide to give the di-*N*-oxide **27**; apparently monohydroxypyridines are insufficiently nucleophilic to undergo reaction with benzofurazan 1-oxide.[8] The trioxo compound **28** affords a pyrido[2,3-*b*]quinoxaline on reaction with *o*-phenylenediamine.[9]

Base-catalyzed cyclization of the quinoxalinylaminopropionic esters **29** has recently been shown to give the pyrido[2,3-*b*]quinoxalines **30**. These may be dehydrogenated by sequential treatment with bromine and triethylamine to give the compounds **30a**. The carboxylic acids derived

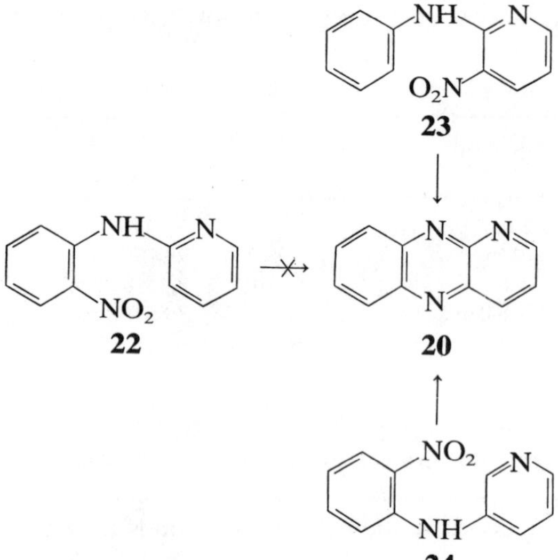

23

22 20

24

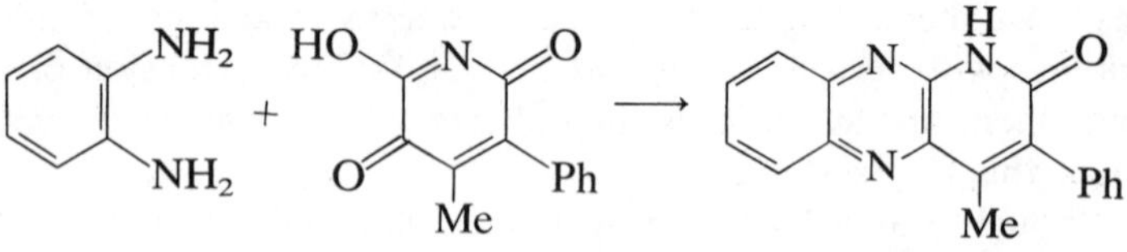

25 26

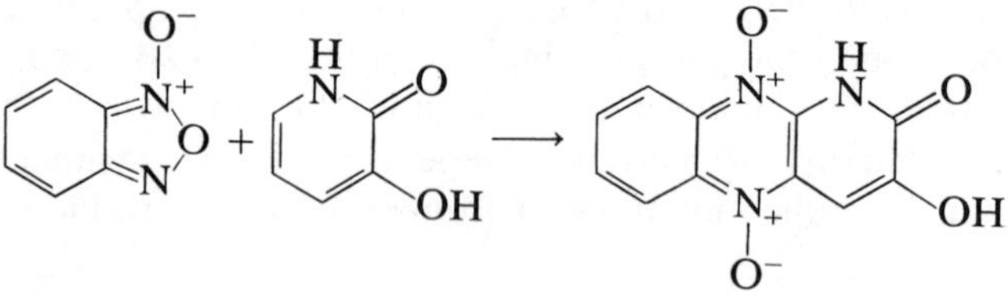

27

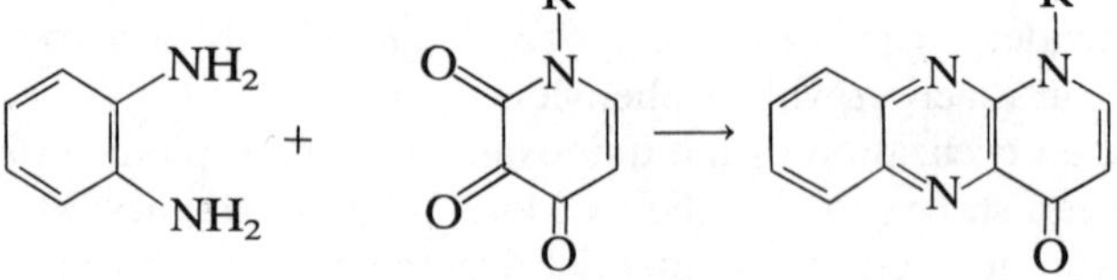

28 R = H or NHPh

from these esters possess antibacterial properties.[23] A more complex derivative of this ring system has been described as having antitumor properties.[10]

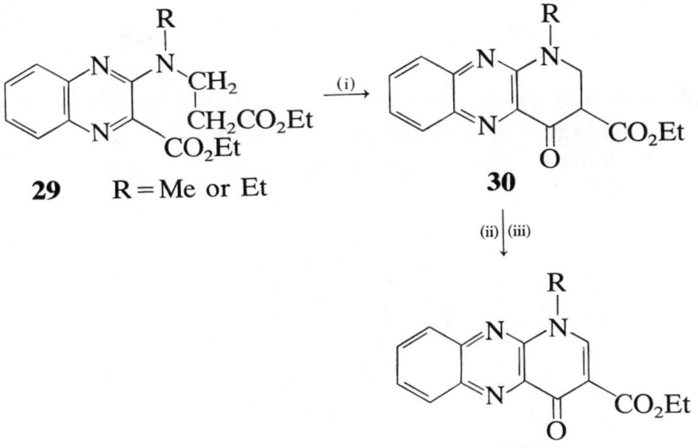

Reagents (i) KO^{t}Bu
(ii) Br$_2$
(iii) Et$_3$N

III. Pyrido[3,4-*b*]quinoxalines

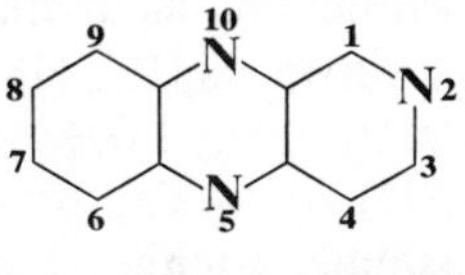

This heterocycle has also been described as 3-azaphenazine (**31a**) and 2,9,10-triazaanthracene (**31b**). As mentioned in Section II, a recent paper[6] has given spectral details on the parent heterocycle. The infrared spectrum of the nucleus has bands at 1590, 820, and 745 cm^{-1} and the mass spectrum shows similar fragmentation to that observed for pyrido[2,3-*b*]quinoxaline. The ultraviolet spectrum in methanol also resembles that of the [2,3-*b*] isomer, with maxima at 209, 248, and 353 nm. The PMR spectrum of the unsubstituted compound is given in Figure 1. As in the [2,3-*b*] series there is evidence for bond fixation in the benzo ring of this heterocycle (see Section II).

Heating the 4-anilinopyridine **32** with ferrous oxalate affords a 21% yield of the parent heterocycle **21**.[6] The corresponding *p*-toluidino derivative gave only 10% of 8-methylpyrido[3,4-*b*]quinoxaline. The yield of

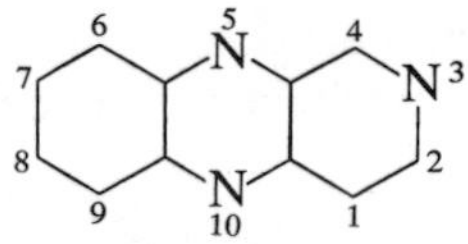

31a

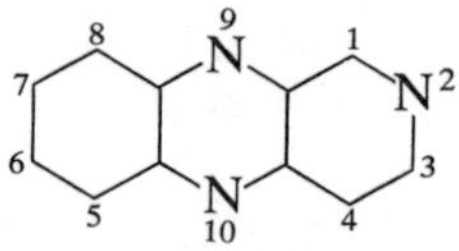

31b

the unsubstituted heterocycle from the nitroaniline **33** was only 6%. As mentioned earlier, although the isomeric nitroaniline **24** gave 10% of pyrido[3,4-*b*]quinoxaline, the major product (27%) was the [2,3-*b*] fused system.

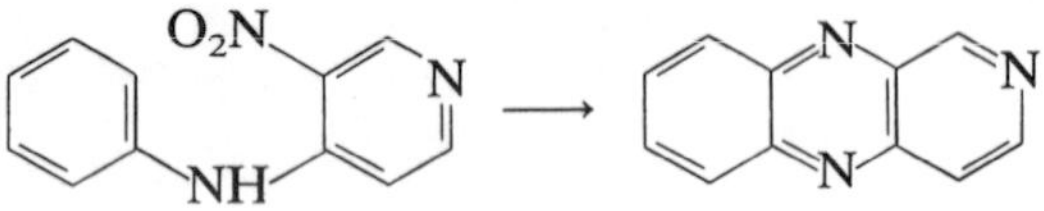

32 → 21

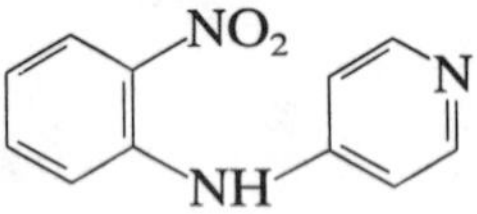

33

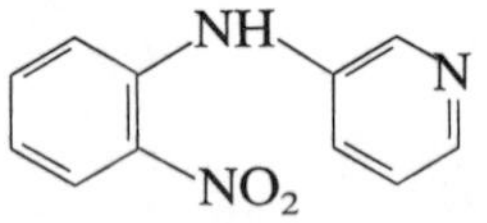

24

The diamine **34** (R = H) on heating as its dihydrochloride gave a 58% yield of the 5,10-dihydro derivative **35** (R = H), which was oxidized to the aromatic parent heterocycle using hydrogen peroxide.[11] Although the methyl compound **34** (R = Me) could also be converted into the corresponding pyrido[3,4-*b*]quinoxaline, attempted cyclizations of the methoxy analogue **34** (R = OMe) were unsuccessful.

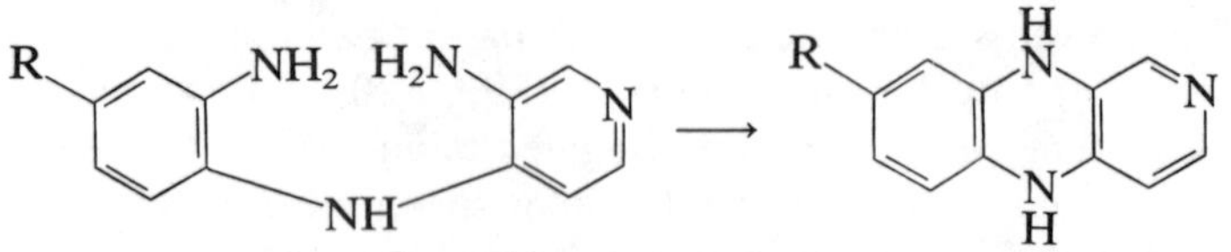

34 R = H or Me → **35**

Alkaline treatment of **36** gave the 5,10-dihydro compound **37**. This was also obtained directly by refluxing the diamine **38** and the chlorodinitropyridine **39** in quinoline.[12] Reaction of **37** with iron and ferric chloride gave the charged species **40**.

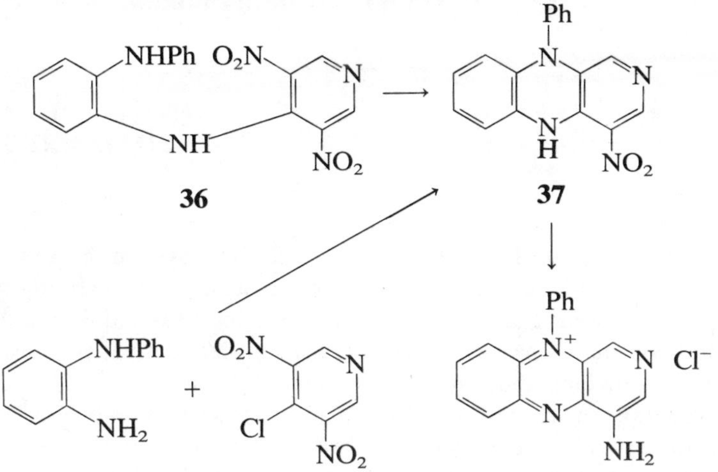

36 **37**

38 **39** **40**

The partially reduced compound **42** was prepared by reaction of the bisoxime **41** with *o*-phenylenediamine.[13] Reaction of the quinoxalinyl esters **43** with primary amines has given a variety of 2-substituted-1,2-dihydro-1-oxo derivatives (**44**).[14,15] It is not clear if the reaction with hydrazine gave the 2-amino compound **44** ($R = NH_2$) or the diazepino compound **45**.[15]

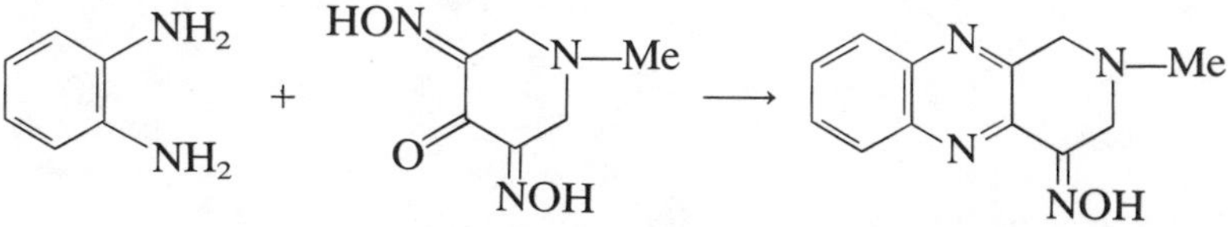

41 **42**

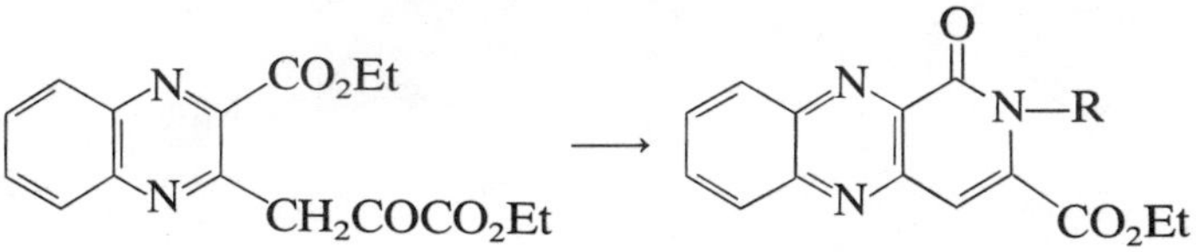

43 **44**

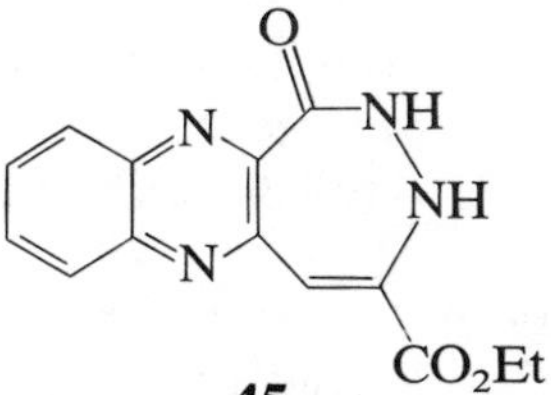

45

IV. Pyrido[1,2,3-*de*]quinoxalines

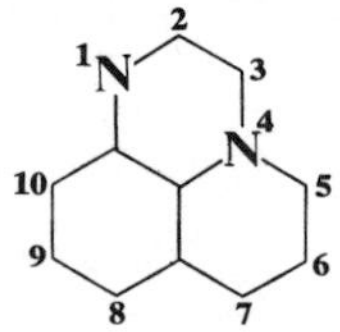

1*H*-4-ium RRI 3380
3*H*,5*H* RRI 3379

The only reported syntheses of this ring system have utilized 8-aminoquinoline derivatives as intermediates. Two such intermediates have been used: 8-amino-6-methoxyquinoline (**46**) and 8-amino-1,2,3,4-tetrahydroquinoline (**51**). The latter furnished derivatives described as 1,8-trimethylenequinoxalines or quinolo-1,4-diazines.

Reaction of **46** with ethylene oxide followed by phosphoryl chloride treatment of the resultant bis(hydroxyethyl)amine afforded the chloroethyl compound **47**.[16] Heating the chloroacetyl derivative **48** of the amine results in cyclization to the oxo derivative **49**, and refluxing the amine in ethanol with phenacyl bromide provides the phenyl compound **50**.[17]

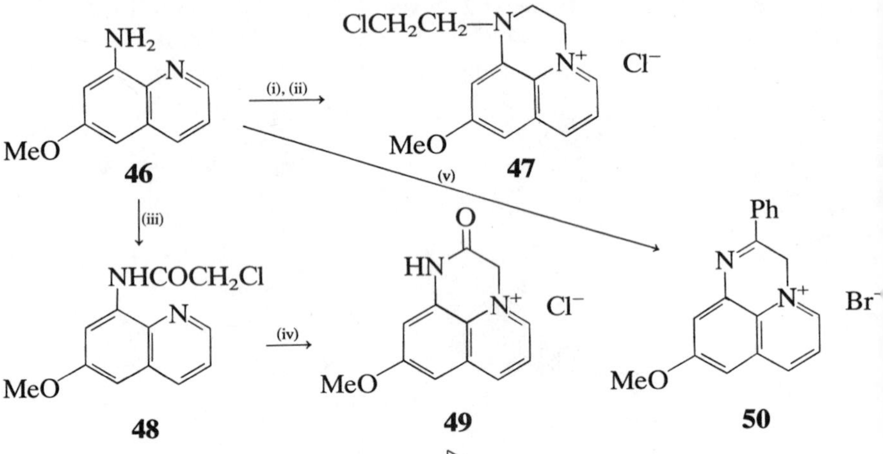

Reagents (i) ethylene oxide;
(ii) $POCl_3$
(iii) $ClCOCH_2Cl$
(iv) heat
(v) $PhCOCH_2Br$

Uncharged derivatives of the ring system have been obtained from the tetrahydroquinoline **51**. Pyruvic acid converts the aminoquinoline into the pyridoquinoxaline **52**. Similarly benzil affords the diphenyl compound **53**.[18] This diphenyl compound has also been obtained by reacting **51** with benzoin. Various substituted diphenyl compounds have been prepared by this approach.[19,20] Reaction of the aminoquinoline **51** with oxalyl chloride[19] or ethyloxalyl chloride[20] affords a compound described as the hydroxypyridoquinoxalinone **54**.

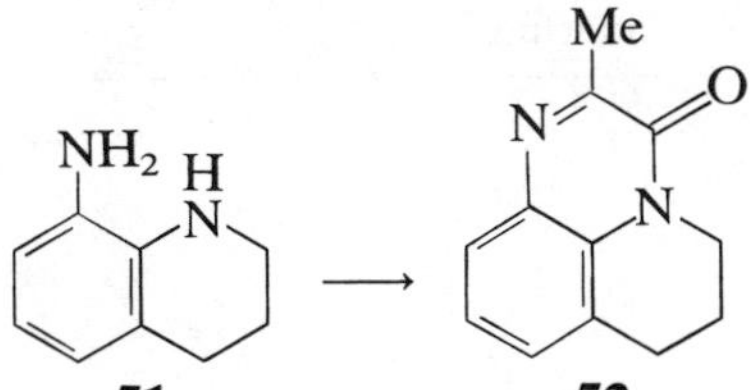

51 **52**

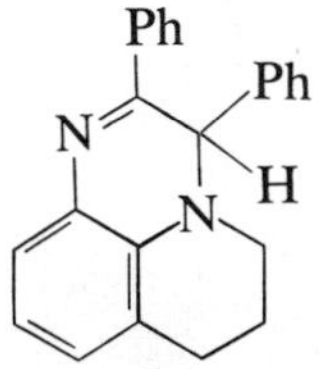

53 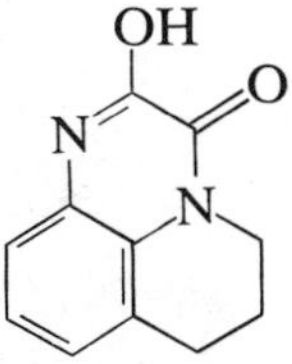 **54**

V. Tables

TABLE 1. PYRIDO[1,2-*a*]QUINOXALINES

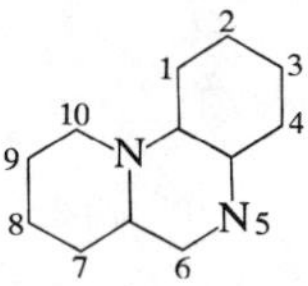

Molecular formula	Structure	m.p., °C	Ref.
$C_{12}H_{10}N_2O_2$	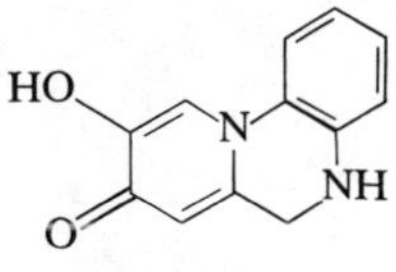		
	Hydrate	>300°	21
	Hydrochloride	>300°	

TABLE 1 (*contd.*)

Molecular formula	Structure	m.p., °C	Ref.
$C_{12}H_{12}N_2O_2$		235–237	3
$C_{16}H_{10}N_2O_8$			
	Tetramethyl ester	172	5
$C_{16}H_{14}N_2O_2$		169.5 decomp.	21
$C_{17}H_{10}N_4O_6$			
	Ethyl ester[a]	320	4
$C_{18}H_{14}N_2O_4S$		193 decomp.	21
$C_{24}H_{17}N_2O$			
	Perchlorate	319–320	22

TABLE 1 (*contd.*)

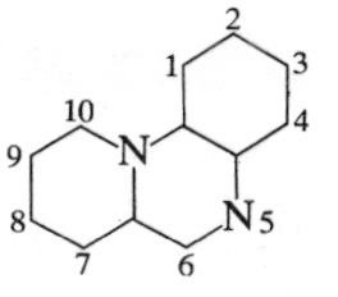

Molecular formula	Structure	m.p., °C	Ref.
$C_{29}H_{18}N_4O_5S$	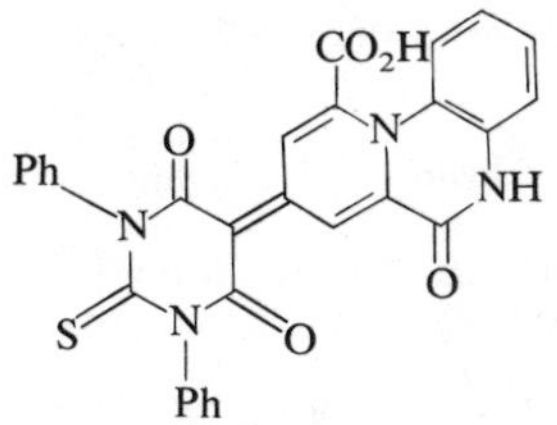		
	Ethyl ester	300–305 decomp.	4

[a] Reference 4 lists the ethyl esters (all m.p. >320°) of derivatives substituted at the 1′ and/or 3′-position by methyl or phenyl.

TABLE 2. PYRIDO[2,3-*b*]QUINOXALINES

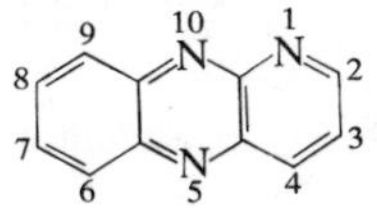

Molecular formula	Substituents	m.p., °C	Ref.
$C_{11}H_7N_3$	Unsubstituted	176	6
$C_{11}H_7N_3O_2$	1,2-Dihydro-3-hydroxy-2-oxo		
	5,10-Dioxide	>300	8
$C_{12}H_9N_3$	7-Methyl	140–141	6
	9-Methyl	135	6
$C_{12}H_9N_3O$	7-Methoxy	200	6
$C_{13}H_9N_3O_3$	3-Carboxy-1,4-dihydro-1-methyl-4-oxo	328–330 decomp.	23
	Ethyl ester	234	
$C_{13}H_{11}N_3O_3$	3-Carboxy-1,2,3,4-tetrahydro-1-methyl-4-oxo		
	Ethyl ester	213	23
$C_{14}H_{10}ClN_3O_3$	3-Carboxy-7-chloro-1-ethyl-1,4-dihydro-4-oxo	300 decomp.	23
	Methyl ester	257	
$C_{14}H_{11}N_3O_3$	3-Carboxy-1-ethyl-1,4-dihydro-4-oxo	318–320 decomp.	23
	Ethyl ester	209	
$C_{14}H_{12}ClN_3O_3$	3-Carboxy-7-chloro-1-ethyl-1,2,3,4-tetrahydro-4-oxo		
	Methyl ester	194	23

TABLE 2 (*contd.*)

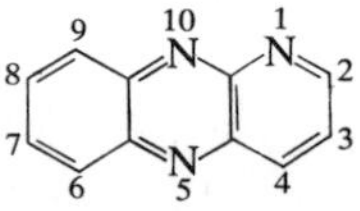

Molecular formula	Substituents	m.p., °C	Ref.
$C_{14}H_{13}N_3O_3$	3-Carboxy-1-ethyl-1,2,3,4-tetrahydro-4-oxo		
	Ethyl ester	171	23
$C_{15}H_{13}N_3O_3$	3-Carboxy-1-ethyl-1,4-dihydro-7-methyl-4-oxo	300 decomp.	23
	Methyl ester	219	
	3-Carboxy-1,4-dihydro-1,7,8-trimethyl-4-oxo	340 decomp.	23
	Ethyl ester	259	
$C_{15}H_{15}N_3O_3$	3-Carboxy-1-ethyl-1,2,3,4-tetrahydro-7-methyl-4-oxo		
	Methyl ester	189	23
	3-Carboxy-1,2,3,4-tetrahydro-1,7,8-trimethyl-4-oxo		
	Ethyl ester	185	23
$C_{16}H_{13}N_3O_4$	7-Acetyl-3-carboxy-1-ethyl-1,4-dihydro-4-oxo	320	23
	Methyl ester	239–240	
$C_{16}H_{15}N_3O_3$	3-Carboxy-1-ethyl-1,4-dihydro-7,8-dimethyl-4-oxo	280 decomp.	23
	Ethyl ester	200	
$C_{16}H_{15}N_3O_4$	7-Acetyl-3-carboxy-1-ethyl-1,2,3,4-tetrahydro-4-oxo		
	Methyl ester	182	23
$C_{16}H_{17}N_3O_3$	3-Carboxy-1-ethyl-1,2,3,4-tetrahydro-7,8-dimethyl-4-oxo		
	Ethyl ester	183	23
$C_{18}H_{13}N_3O$	1,2-Dihydro-4-methyl-2-oxo-3-phenyl	275	7
$C_{18}H_{15}N_3O_5$	7,8-Diacetyl-3-carboxy-1-ethyl-1,4-dihydro-4-oxo	354–356 decomp.	23
	Ethyl ester	326	
$C_{18}H_{17}N_3O_5$	7,8-Diacetyl-3-carboxy-1-ethyl-1,2,3,4-tetrahydro-4-oxo		
	Ethyl ester	210	23

TABLE 3. PYRIDO[3,4-*b*]QUINOXALINES

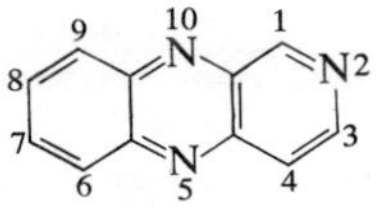

Molecular formula	Substituents or structure	m.p., °C	Ref.
$C_{11}H_7N_3$	Unsubstituted	172	6
		181–182	11
$C_{11}H_9N_3$	5,10-Dihydro		
	Hydrochloride hydrate	290	11
$C_{12}H_8N_4O_2$	3-Carboxamido-1,2-dihydro-1-oxo	360 decomp.	15
$C_{12}H_8N_4O_3$	2-Amino-3-carboxy-1,2-dihydro-1-oxo[a]		
	Ethyl ester	263–265	15
$C_{12}H_9N_3$	8-Methyl	152–153	6, 11
$C_{12}H_{12}N_4O$	1,2,3,4-Tetrahydro-4-hydroxyimino-2-methyl		
	Hydrochloride	[b]	13
$C_{13}H_9N_3O_3$	3-Carboxy-1,2-dihydro-2-methyl-1-oxo		
	Ethyl ester	158	14
		177–178	15
$C_{14}H_{11}N_3O_3$	3-Carboxy-2-ethyl-1,2-dihydro-1-oxo		
	Ethyl ester	158	15
$C_{14}H_{12}N_4O_2$	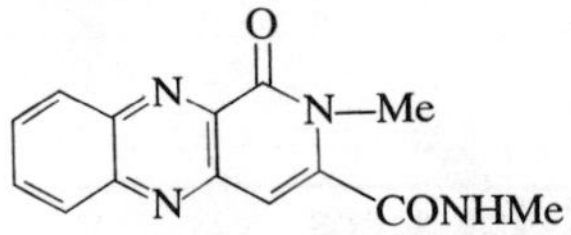	305 decomp.	15
$C_{17}H_{12}N_4O_2$	5,10-Dihydro-4-nitro-10-phenyl	247–247.5	12
$C_{17}H_{13}N_4$	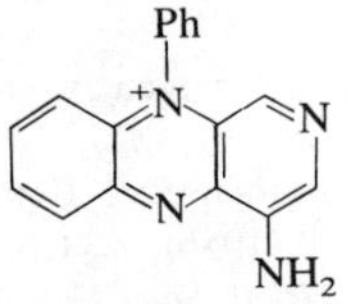		
	Chloride	>310	12
	Acetyl derivative	275 decomp.	
$C_{18}H_{12}N_4O_3$	2-(2-Aminophenyl)-3-carboxy-1,2-dihydro-1-oxo		
	Ethyl ester	>350	15
$C_{19}H_{13}N_3O_3$	3-Carboxy-1,2-dihydro-2-(4-methylphenyl)-1-oxo		
	Ethyl ester	207	15

[a] Possibly the isomeric diazepinoquinoxaline.

[b] Darkens at 190°, does not melt.

TABLE 4. PYRIDO[1,2,3-*de*]QUINOXALINES

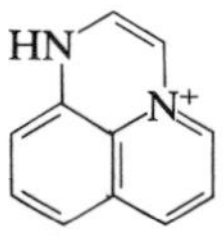

1H-4-ium

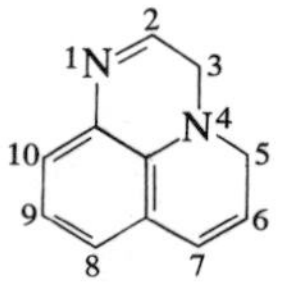

3H,5H

Molecular formula	Substituents	m.p., °C	Ref.
$C_{11}H_{10}N_2O_2$	3*H*,5*H*-6,7-Dihydro-2-hydroxy-3-oxo	258–261	20
$C_{12}H_{11}N_2O_2$	1*H*-2,3-Dihydro-9-methoxy-2-oxo		
	Chloride	>300	17
$C_{12}H_{12}N_2O$	3*H*,5*H*-6,7-Dihydro-2-methyl-3-oxo	113	18
$C_{13}H_{14}ClN_2O$	1*H*-1-(2-Chloroethyl)-2,3-dihydro		
	Nitrate	177–179	16
$C_{18}H_{15}N_2O$	3*H*-9-methoxy-2-phenyl		
	Bromide·$\frac{1}{4}$ H_2O	233–236	17
$C_{23}H_{20}N_2$	3*H*,5*H*-6,7-Dihydro-2,3-diphenyl	148–148.5	18, 19
	Picrate	150–151 decomp.	
$C_{25}H_{24}N_2$	3*H*,5*H*-6,7-Dihydro-2,3-bis(4-methylphenyl)	130–131	20
$C_{25}H_{24}N_2O_2$	3*H*,5*H*-6,7-Dihydro-2,3-bis(4-methoxyphenyl)	148–149	19
	Picrate	166	
$C_{27}H_{28}N_2$	3*H*,5*H*-2,3-Bis(4-ethylphenyl)-6,7-dihydro	136–138	20

VI. References

1. W. L. Mosby, "Heterocyclic Systems with Bridgehead Nitrogen Atoms," Interscience, New York, 1961, p. 1191.
2. A. Gómez-Sánchez and M. Yruela Antiñolo, *An. Real Soc. Españ. fis. y Quím.* (Madrid), **51B,** 423 (1955); *Chem. Abstr.*, **50,** 10108 (1956); A. Gómez-Sánchez, M. Yruela Antiñolo, and F. Garciá González, *An. Real Soc. Españ. Fis. y Quím.* (Madrid), **50B,** 431 (1954); *Chem. Abstr.*, **52,** 11078 (1958).
3. E. C. Taylor and A. McKillop, *J. Am. Chem. Soc.*, **87,** 1984 (1965).
4. F. Eiden and P. Peter, *Arch. Pharm.* (Weinheim), **299,** 139 (1966); *Chem. Abstr.*, **64,** 15878 (1966).
5. M. Ungureanu, I. Druta, M. Petrovanu, and I. Zugravescu, *An. Stiint. Univ.* "*Al. I. Cuza*" *Iasi, Sect.* 1*c*, **18,** 49 (1972); *Chem. Abstr.*, **78,** 29712 (1973).
6. R. G. R. Bacon and S. D. Hamilton, *J. Chem. Soc., Perkin I,* 1965 (1974).
7. J. A. Moore and F. J. Marascia, *J. Am. Chem. Soc.*, **81,** 6049 (1959).
8. M. J. Abu El-Haj, B. W. Dominy, J. D. Johnston, M. J. Haddadin, and C. H. Issidorides, *J. Org. Chem.*, **37,** 589 (1972).
9. E. Peratoner, *Gazz. Chim. Ital.*, **41,** 629 (1911).
10. W. Parri, *Giorn. Ital. Chemioterap.*, **3,** 588 (1956); *Chem. Abstr.*, **52,** 11289 (1958).
11. V. Petrow, J. Saper, and B. Sturgeon, *J. Chem. Soc.*, 2540 (1949).
12. V. A. Petrow and J. Saper, *J. Chem. Soc.*, 588 (1946).

13. G. H. Cookson, *J. Chem. Soc.*, 1328 (1953).
14. F. Eiden, H. Mueller, and G. Bachmann, *Tetrahedron Lett.*, 2903 (1968).
15. F. Eiden, H. Mueller, and G. Bachmann, *Arch. Pharm.* (Weinheim), **305,** 2 (1972); *Chem. Abstr.*, **76,** 126920 (1972).
16. R. M. Peck, R. K. Preston, and H. J. Creech, *J. Am. Chem. Soc.*, **81,** 3984 (1959).
17. L. M. Werbel, J. Battaglia, and M. L. Zamora, *J. Heterocycl. Chem.*, **5,** 371 (1968).
18. S. J. Hazlewood, G. K. Hughes, and F. Lions, *J. Proc. Roy. Soc. N.S. Wales*, **71,** 462 (1938); *Chem. Abstr.*, **33,** 610 (1939).
19. A. Richardson, Jr., and E. D. Amstutz, *J. Org. Chem.*, **25,** 1138 (1960).
20. A. Richardson, Jr., *J. Org. Chem.*, **30,** 2589 (1965).
21. K. Imada, *Carbohydr. Res.*, **39,** 379 (1975).
22. P. Yu. Andreichikov, N. V. Kholodova, and G. N. Dorofeenko, *Khim. Geterotsikl. Soedin.*, 1578 (1975); *Chem. Abstr.*, **84,** 43980 (1976).
23. M. Pesson, P. De Lajudie, M. Antoine, S. Chabassier, and P. Girard, *C. R. Acad. Sci., Ser. C*, **282,** 861 (1976); *Chem. Abstr.*, **85,** 63035 (1976).

Author Index

Numbers in parentheses are reference numbers and indicate that the author's work is referred to although his name is not mentioned in the text. Numbers in italics show the pages on which the complete references are listed.

Abbott, P. J., 613 (37), 637 (37), *651*

Abderhalden, E., 310 (51, 53), 315 (61), 319 (51, 61), 324 (111), 328 (29), 329 (29), 335 (53), 336 (51), *343, 344, 346*

Abdulla, R. F., 91 (63a), 98 (63a), *107*, 173 (93), *178*

Abe, M., 324 (103, 104, 149), *345, 347*

Abe, Y., 364 (12), *396*

Abezgauz, F. I., 268 (51), 278 (51), *289*

Abramovitch, R. A., 49 (85), 56 (85), *75*

Abuel-Haj, 64 (134), 65 (134), *76*

Abu El-Haj, M. J., 748 (8), 755 (8), *758*

Abushanab, E., 37 (37), 38 (42), 39 (37, 42), 42 (54), 51 (87), 56 (42), 58 (42, 54, 87), 63 (42, 54, 87, 121, 125), 64 (54, 87, 121, 125), 65 (42, 87, 121), 66 (87), 67 (54, 125), 69 (125), 70 (87), *73, 74, 75, 76*, 224 (84), *231*, 371 (28, 30), 372 (28, 60, 61, 64, 66), 373 (28, 30, 64), 374 (64), 375 (65, 66), 385 (28, 60, 64, 65, 66), 386 (28, 60, 61, 64, 65, 66), 387 (61, 65, 66), 388 (65), *396, 397*

Acheson, R. M., 80 (8), 97 (8), *106*, 152 (52), *160*, 165 (27), 174 (27), *176*, 180 (8), 189 (8), *191*, 219 (56), *230*, 268 (54), 281 (54), *289*, 293 (8), 297 (8), 328 (8), 329 (8), *342*, 613 (37), 637 (37), *651*

Adachi, J., 21 (25), 22 (25), 24 (25), 25 (25), *26*, 65 (133), *76*, 149 (40), 158 (40), *160*, 277 (92), *290*

Adachi, K., 202 (18), *204*, 207 (15), 228 (15), *229*

Adamek, M., 153 (57), *160*

Adams, R., 227 (118), *232*

Adamson, J., 361 (4), 368 (4), 370 (4), 382 (4), 383 (4), *395*, 444 (4), 452 (4), 453 (4), *454*, 464 (3), 466 (3), 467 (3), 471 (21), 477 (3), 478 (3), 479 (3), 488 (3), 489 (21), 492 (3), *493*

Adger, B., 88 (50), *107*

Advani, B. G., 316 (132), 328 (132), 329 (132), *346*

Agadzhanyan, Z. E., 434 (13), 438 (13), *440*

Agai, B., 249 (13), 254 (13), *259*

Aguggini, G., 184 (33), *192*

Aguilera, R., 271 (72), 279 (72), 282 (72), 283 (72), *289*

Aguzzi, A., 625 (60), 647 (60), *652*

Ahmad, Y., 18 (6), 22 (6), *25*, 29 (8), 33 (21, 27, 28), 44 (66), 45 (8), 46 (66), 47 (66), 51 (88), 52 (90, 91), 54 (95), 56 (8, 27), 57 (8, 27), 58 (27), 59 (8, 27), 60 (21, 28, 66, 90), 61 (28, 66, 90), 62 (28), 70 (66), *73, 74, 75*, 97 (108, 110), 101 (148, 156, 171a), 102 (163, 171a), 103 (163, 171a), 105 (148), *109, 110*, 138 (2), 152 (2), 153 (75, 77), 154 (2, 77), 155 (75, 77), 156 (77), *158, 160*, 163 (8), 173 (8), *175*, 224 (89), *231*

Ahmed, Q., 624 (54), 644 (54), 645 (54), *652*

Aidoo, A. S. K., 220 (57), *230*

Aiello, T., 353 (14), 358 (14), *359*

Akhtar, M. H., 717 (24), 726 (24), 727 (24), *728*

Aksel'rod, Zh. I., 39 (44), 56 (44), 57 (44), 58 (44), 60 (44), *74*, 153 (68), *160*

Al-Azawe, S., 61 (117), *76*, 154 (85), *161*, 233 (1), 237 (1), 238 (1), *238*

Albert, A., 8 (4, 7), 9 (19), *15*, 19 (9), 21 (9), 22 (48), *26, 27*, 114 (11), 119 (11), *121*, 255 (34), *260*, 371 (29), 372 (29), 373 (29), 385 (29), *396*, 496 (6, 12, 13), 498 (22, 31), 499 (31), 509 (6), 511

(52), 515 (6), 516 (31), 524 (58), 525 (6, 52), 531 (6, 31), *541, 542,* 545 (4), 546 (4, 14, 15), 547 (4, 18), 548 (21), 552 (4), 553 (4), 556 (4), 557 (4), 558 (4), 559 (4), 560 (4, 34), 561 (4, 21), 564 (4, 18, 21), *567*
Albrecht, R., 510 (51), 512 (60), 524 (60), 534 (60), 536 (60), *542, 543*
Aldous, A., 166 (32), *176*, 207 (16), *229*
Alexandrou, N. E., 715 (17), 723 (17), *728*
Ali, M. A., 576 (4), *582*
Alkaysi, H. N., 267 (32), *288*
Allen, G., 13 (63), *16*
Allison, C. G., 22 (39), *26*, 95 (89), 96 (89), 104 (89), *108*, 168 (48), 170 (48), 172 (48), 174 (48), *177*, 203 (19), *204*
Allred, E. L., 11 (40), *16*, 276 (91), 287 (91), *290*
Alteri, N. A., 37 (37), 39 (37), *73*
Altermatt, H., 168 (47), 170 (47), 171 (47), *177*
Altukhova, L. B., 294 (9, 10), 327 (9), 328 (9), 329 (9), 330 (9), *342*, 368 (25), 383 (25), *396*, 614 (39), 618 (39), 638 (39), *651,* 673 (54), 688 (54), *690*
Alwair, K., 19 (15), *26*
Alwani, D. W., 267 (41), 279 (41), 280 (41), *289*
Amarego, W. L. F., 361 (3), 365 (3), 367 (3), 382 (3), 383 (3), *395*
Ambroz, M., 207 (13), *229*
American Cyanamid Co., 183 (30), *192*
American Home Products Corp., 274 (78), *289*
Ames, D. E., 62 (120), 72 (120), *76*, 170 (60), *177*, 203 (20), 204 (20), *204*
Amstutz, E. D., 222 (80), *231*, 753 (19), 758 (19), *759*
Anand, N., 571 (7, 8, 9, 10), 573 (7, 8, 9), *574*
Anderson, J. D., 269 (63), 283 (63), *289*
Anderson, R. C., 12 (46), *16*, 626 (67), *652*
Anderson, R. K., 165 (30), *176*, 200 (6), 201 (6), 203 (6), 204 (6), *204*
Anderson, V. B., 622 (76), *653*
Andreeva, N. S., 304 (30), *343*
Andreichikov, P. Yu, 744 (22), 754 (22), *759*
Andreichikov, S. Yu, 92 (67), *107*
Andres, W. W., 324 (98), *345*
Antinola, M. Y., 101 (151), *110*
Antoine, M., 749 (23), 755 (23), 756 (23), *759*
Antonov, V. K., 307 (41, 42), 308 (45), 317 (42), 318 (42, 79), 335 (42, 79), 337 (42, 79), 339 (79), 340 (79), *343, 344*, 434 (12, 14), 438 (14), *440*
Aoki, Y., 324 (96), *345*
Arata, Y., 464 (6), 469 (6), 475 (6), 477 (6), 478 (6), 480 (6), 482 (6), 484 (6), 485 (6), 491 (6), *493*
Arbuzov, B. A., 59 (150), *77*, 224 (87, 93), *231,* 285 (97), *290*
Archer, R. A., 270 (66), 271 (66), 282 (66), 286 (66), *289*, 521 (56), 533 (56), *542*
Ariyoshi, U., 695 (24), 706 (24), 707 (24), 708 (24), *710*
Armand, J., 262 (3), 276 (3), 277 (3), 278 (3), 283 (3), 284 (3), 286 (3), *288*, 497 (61), 521 (61), 534 (61), 538 (61), 539 (61), 540 (61), *543, 567* (35)
Armarego, W. L. F., 496 (4), *541*, 545 (2), *567*, 577 (5, 6), 578 (6, 9), 579 (6), 581 (6), *582*
Arnold, F. E., 580 (20), *582*
Artico, M., 605 (18, 19), 618 (18, 19), 621 (18, 19), 634 (18), 635 (18, 19), 636 (19), 638 (19), 639 (19), 640 (19), *651,* 725 (54), *729*
Asai, M., 12 (43), *16*, 29 (2), *72*, 174 (101), *178*
Asai, S., 95 (90, 91a), 97 (112), 98 (112, 119), *108, 109*, 113 (7, 8), 114 (2), 116 (7, 21), 117 (27), 118 (27), 119 (7, 8, 12, 37), 120 (21, 37, 40), 121 (21), *121*, *122*, 153 (65), *160*, 163 (17), 171 (73), 172 (17), *176, 177,* 194 (1, 5), 195 (1), 196 (1, 5), *197*, 202 (16), *204*
Asano, K., 95 (90), 97 (112), 98 (112, 119), *108, 109,* 113 (7, 8), 114 (2), 116 (7, 19, 20, 21), 117 (27, 37), 118 (27), 119 (7, 8, 12, 37), 120 (19, 21, 37, 40), 121 (21), *121, 122*, 126 (18), *129*, 153 (65), *160*, 163 (17), 171 (73), 172 (17), *176*, *177*, 194 (1, 5, 6), 195 (1), 196 (1, 5, 6), 197 (6), *197*, 202 (16), *204*

Asinger, F., 225 (106), *231*
Atkinson, C. M., 12 (51), *16*, 18 (5), 25 (5, 72), *25, 27,* 79 (2), 97 (2), 98 (2), 100 (2), *106*, 171 (72), 172 (72), *177*, 180 (13), 181 (13), 183 (13), 187 (13), *191*, 199 (4), 200 (4), 203 (4), *204*, 247 (4), 255 (4), *259*
Augustin, M., *359* (17)
Austrup, R., 101 (157a), *110*
Azimov, V. A., 349 (2), 356 (2), *359*

Bach, F. L., 128 (29), *129*, 172 (84), *178*
Bach, F. L., Jr., 352 (10), 357 (10), *359*
Bach, M. K., 382 (56), *397*
Bach, N. J., 327 (154), *347*
Bachmann, G., 89 (54), *107*, 131 (15), 134 (20), 135 (15), *136*, 152 (54), 156 (54), *159*, 225 (130), *232*, 613 (36), 637 (36), 639 (36), 641 (36), *651*, 751 (14, 15), 757 (14, 15), *759*
Backeberg, O. G., 234 (2), *238*
Bacon, R. G. R., 655 (57), *690*, 746 (6), 747 (6), 749 (6), 755 (6), 757 (6), *758*
Badger, G. M., 9 (17), *15*, 410 (1), 411 (1), 415 (1), 416 (1), 417 (1), 426 (1), 428 (1), *429*
Baganz, H., 82 (20), *106*, 579 (15), 580 (15), 581 (15), *582*
Bahner, C. T., 247 (2, 3), 255 (2, 3), 256 (2), 257 (2, 3), 258 (2), *259*
Bähner, K. H., 245 (19), *246*
Bähner, K.-J., 124 (4), 127 (4), 128 (4), *128*, 131 (4), 134 (4), *136*, 154 (86), *161*, 214 (37), 215 (37), 221 (67), *230* 706 (53), *711*
Baizer, M. M., 269 (63), 283 (63), *289*
Bajwa, G. S., 237 (10), *239*
Bakhtiari, B., 44 (66), 46 (66), 47 (66), 52 (91), 60 (66), 61 (66), 70 (66), *74, 75,* 102 (163), 103 (163), *110*, 163 (8), 173 (8), *175,* 224 (89), *231*
Bala, H., 81 (16), 97 (16), 98 (16), *106*, 114 (13), 119 (13), *121*, 166 (31), *176,* 556 (32), *567*
Baldwin, J. J., 171 (65), *177*
Bambury, R. E., 38 (41), 55 (97), 58 (41), 59 (41), 65 (41, 97), 66 (97), 67 (41, 140), 68 (97), 69 (41), 70 (41), 71 (41), *73, 75*
Banks, C. K., 99 (129), 100 (129), *109*, 172 (86), *178*, 180 (12), 189 (12), *191*
Bannore, S. N., 10 (30), *15*, 41 (53), 60 (53), 61 (53), *74*, 234 (3), 238 (38), *238, 239*
Baranov, S. N., 47 (79), 62 (79), *75,* 98 (122), *109*, 102 (167), 103 (167), 104 (167), *110*, 126 (17), 127 (23), *129*, 214 (32), *229*
Barath, M., 56 (101), *75*
Barclay, L. R., 247 (3), 255 (3), 257 (3), *259*
Barlin, G. B., 100 (145), *110*, 114 (11), 119 (11), *121*, 181 (17), 187 (17), *191,* 255 (37), *260*, 496 (4, 6), 509 (6), 515 (6), 525 (6), 531 (6), *541*, 545 (2, 4), 546 (4), 547 (4), 552 (4), 553 (4), 556 (4), 557 (4), 558 (4), 559 (4), 560 (4), 561 (4), *567*, 577 (5), *582*
Barlow, R. B., 141 (11), 153 (11), *159*
Barltrop, J. A., 99 (131), 101 (131), *109*, 131 (2), 132 (2), 134 (2), 135 (2), *136*, 183 (28), 187 (28), *192*, 225 (109), *231*, 262 (5), 277 (5), *288*
Barnes, R. P., 206 (6), 228 (6, 133), *229*, *232*
Barrett, E. K., 664 (21), *689*
Barrows, T. H., 267 (31), 274 (31), 284 (31), 285 (31), 286 (31), *288*
Barton, J. W., 401 (4), 402 (4), 405 (4), 406 (4), 407 (4), *408*
Baschang, G., 245 (15), 246 (15), *246*
Bashir, N., 33 (28), 52 (90), 60 (28, 90), 61 (28, 90), 62 (28), *73, 75,* 101 (156), *110*
Bass, K. C., 12 (53), *16*, 209 (19), 223 (19), *229*
Bass, R. G., 278 (99), *290*
Batkowski, T., 498 (24, 25), 540 (24, 25), *541, 542*
Battaglia, J., 752 (17), 758 (17), *759*
Batterham, T. J., 496 (11), *541*, 545 (11), *567*, 577 (8), *582*
Bauer, A., 95 (94), 105 (94), *108*
Bauer, C., 242 (5), 245 (5), *246*, 694 (20), *710*, 732 (16), 740 (16), *742*
Bauer, H.-H., 174 (97), *178*
Bauer, L., 376 (41), 377 (41), *397*, 665 (28), 684 (28), 686 (28), *689*

Bauer, S., 237 (12), *239*
Bauer, W., 125 (12), *128*, 701 (48), 707 (48), 708 (48), 709 (48), *711*
Baumann, H., 252 (24), 256 (24), *260*
Baumann, N., 254 (33), *260*
Baxter, I., 98 (128), 99 (128), *109*, 267 (46), 280 (46), *289*
Baxter, R. A., 89 (57), *107*
Bazir, N., 101 (171a), 102 (171a), 103 (171a), *110*
Beach, T. N., 14 (71), *16*, 250 (17), 255 (17), 256 (17), 257 (17), *259*, 267 (22), 272 (22), 280 (22), 281 (22), 282 (22), 283 (22), 284 (22), *288*
Beak, P., 275 (86), *290*
Beatty, W. A., 324 (109), *345*
Becherescu, P., 697 (39), 703 (39), *711*
Beck, K. M., 276 (89), 287 (89), *290*
Beck, W., 736 (38), *742*
Becker, B., 319 (83), 324 (83, 107), 335 (83), *345*
Becker, S., 216 (44), *230*
Beecham Group Ltd., 352 (9), 358 (9), *359*
Beesley, W. N., 184 (38), *192*
Beets, M. G., 23 (57), 24 (57), *27*
Begland, R. W., 98 (121), *109*, 153 (71), 158 (71), *160*, 185 (51), 187 (51), 190 (51), *192*
Begliomini, A., 184 (35), *192*
Bekarek, V., 210 (24), *229*
Beke, D., 238 (22), *239*
Beksha, V., 268 (59), 281 (59), *289*
Belcher, R., 224 (85), *231*
Bell, S. C., 86 (35), 102 (35), *106*, 284 (104), *290*
Belli, M. L., 172 (81), *178*
Bellini, A. M., 373 (32), 386 (32), *396*
Belozerova, K. A., 66 (15), *73*
BeMiller, J. N., 700 (46), *711*
Benedek-Vamos, M., 12 (50), *16*, 20 (20), 23 (20), *26*
Benko, P., 47 (75), 63 (75), *74*
Benkovic, P. A., 267 (29, 30, 31), 274 (31), 284 (30, 31), 285 (29, 31), 286 (31), *288*, 659 (14), 661 (14, 19), 662 (14, 19), 676 (14, 19), *689*
Benkovic, S. J., 267 (29, 30, 31), 274 (31), 284 (30, 31), 285 (29, 31), 286 (31), *288*, 659 (14), 661 (14), 662 (14), 676 (14), *689*
Bennett, G. M., 214 (35), 224 (35), *230*
Bentley, R., 448 (10), 453 (10), *454*
Benz, J., 168 (38), *176*
Berde, B., 327 (124), *346*
Berends, W., 12 (45), 13 (65), *16*, 94 (85), 104 (188), 105 (85, 188), *108, 111*, 149 (43), *160*, 168 (46), 169 (46, 50), 174 (46), *177*, 201 (11), 203 (11), 204 (11), *204*, 242 (3), 244 (3), 245 (3), *246*
Berezovskii, V. M., 39 (44), 56 (44), 57 (44), 58 (44), 60 (44), *74*, 155 (109), *161*, 238 (24), *239*
Berg, S. S., 255 (35), 256 (35), *260*
Bergmann, M., 312 (58), 319 (58), 321 (58), 336 (58), 337 (58), *344*
Bergstrom, F. W., 227 (117), *232*
Berkovitch, D., 84 (28), *106*, 247 (6), 248 (7), 250 (6), 255 (6), 256 (6), *259*
Bernal, I., 118 (32, 33), *122*
Bernanek, V., 210 (24, 25, 26, 27), 224 (25), 227 (27), *229*
Bernardi, L., 448 (12), 453 (12), *454*
Bernardi, R., 12 (60), *16*
Bernhard, G., *359* (17)
Bernstein, J., 498 (28), 525 (28), *542*
Bertini, F., 12 (58), 13 (58), *16*, 209 (17), *229*
Berzovskii, V. M., 143 (16), 144 (20), 153 (16, 68), 155 (16, 20), *159, 160*
Beska, E., 97 (103), *109*
Bestmann, H. J., 154 (79, 83), 156 (79), *161*
Bethune, V. G., 151 (50), 153 (50), 158 (50), *160*
Bettinetti, G. F., 267 (38, 39, 40), 276 (38), 277 (40), 280 (39), 281 (39), 287 (38), *288*, 629 (69), 646 (69), *652*
Beuchel, K. H., 186 (50), *192*
Beutel, R. H., 23 (54), 24 (54), *27*, 79 (4), *106*, 187 (62), *193*, 716 (21), 722 (21), 723 (21), *728*
Beuttenmueller, M., 473 (29), 478 (29), 489 (29), 490 (29), 492 (29), *493*
Bhaduri, A. P., 21 (29), *26*, 733 (42), 738 (42), 740 (42), 741 (42), *742*

Bhandarkar, R. M., 623 (43, 46), 643 (46), 645 (46), 646 (43), *652*
Bianchi, G., 227 (128), 228 (128), *232*
Bickert, E., 101 (172), *110*
Bicking, J. B., 153 (78), *161*
Bieksa, V., 284 (96), *290*, 672 (60), *690* (53)
Biemann, K., 272 (73), *289*, 324 (102), *345*
Biffin, M. E. C., 400 (2), 402 (2, 6), 407 (2), *408*
Biggerstaff, G., 247 (3), 255 (3), 257 (3), *259*
Bigotto, A., 292 (3), *342*, 361 (1), *395*
Bilancio, D. E., 247 (3), 255 (3), 257 (3), *259*
Bindra, A. P., 371 (28), 372 (28, 60, 61, 64, 66), 373 (28, 64), 374 (64), 375 (66), 385 (28, 60, 64, 66), 386 (60, 61, 64, 66), 387 (61, 66), *396, 397*
Biniecki, S., 188 (69), 190 (69), *193*
Binte, H. J., 125 (12), *128*, 701 (48), 707 (48), 708 (48), 709 (48), *711*
Birch, A. J., 323 (147), 324 (147), *347*
Bird, C. W., 22 (41), *26*, 100 (142), 104 (142), *110*, 218 (54), *230*
Birkhofer, L., 23 (61), *27*
Bitterlich, W., 224 (92), *231*
Black, P. J., 9 (24), *15*
Blackburn, W., 100 (137), *109*
Blaha, K., 305 (135), 307 (137, 138), 335 (137), 336 (135), *346*
Blanc, G. W., 247 (3), 255 (3), 257 (3), *259*
Blanz, E. J., 126 (19), 127 (19), 128 (19), *129*
Blaschke, W., 216 (44), *230*
Blears, D. J., 9 (26), *15*
Bloch, A., 376 (36), 377 (36), 379 (36), 382 (36), 394 (36), *396*
Blomquist, A. T., 133 (9), 136 (9), *136*, 156 (98), *161*, 237 (20), *239*
Bluhm, Th., 713 (42, 43, 44), 718 (45), 723 (45), 726 (45), *728, 729*
Boaz, H. E., 327 (154), *347*
Bobek, M., 376 (36), 377 (36), 379 (36), 382 (36), 394 (36), *396*
Bocharnikova, A. V., 294 (10), *342*
Bochkarev, V. N., 308 (45), *343*
Bodforss, S., 103 (173), *110*, 258 (44), 259 (44), *260*, 263 (13), 278 (13), *288*, 505 (39), 523 (39), 538 (39), 540 (39), *542*
Boekelheide, V., 296 (15), 300 (15), 328 (15), 330 (15), *342*
Bognar, R., 228 (139), *232*
Bohlmann, F., 9 (18), *15*, 266 (19), 279 (19), *288*
Böhme, H., 156 (99), *161*
Bohnsack, G., 324 (118), *346*
Boissonnas, R. A., 315 (64), 335 (64), *344*
Boldyrev, B. G., 732 (20), 737 (20), 736 (37), 738 (20), *742*
Bolhofer, W. A., 171 (65), *177*
Bolliger, G., *347* (145)
Bonse, G., 174 (97), *178*
Booth, G., 275 (83), 279 (83), *290*
Borkovec, J., 207 (12, 13), 222 (12), *229*
Bortnick, N. M., 299 (25), 328 (25), *343*
Bose, J. L., 10 (30), *15*, 41 (53), 60 (53), 61 (53), *74*, 234 (3), 238 (38), *238*, *239*
Bost, R. W., 225 (108), *231*, 238 (29), *239*
Bowie, R. A., 37 (38), *73*
Boyd, G. V.,599 (1), 616 (1), *650*
Boyer, J. H., 20 (23, 24), 22 (24), 23 (23), *26*, 82 (21), 101 (21), *106*, 224 (91), *231*, 608 (28), 634 (28), 636 (28), *651*
Brach, P. J., 353 (12), 357 (12), *359*
Bradsher, C. K., 467 (9), 478 (9), 479 (9), 482 (9), 483 (9), 487 (9), 488 (9), 489 (9), *493*
Brady, T. E., 62 (119), *76*, 128 (8), *128*, 131 (7), 134 (7), 136 (7), *136*, 214 (36), 228 (36), *230*
Braude, E. A., 226 (115), *232*
Braun, D., 237 (42), *240*
Bravo, L., 151 (48), *160*
Bredereck, H., 140 (8), 141 (8), 153 (8), 154 (8), *159*
Bremer, O., 548 (25), 565 (25), 566 (25), *567*
Brennan, J. A., 12 (49), *16*, 18 (3), 19 (3), 22 (3), 23 (3), 24 (3), 25 (3), *25*
Breska, V., 256 (40), *260*
Bressan, G. B., 164 (21), 172 (21), *176*
Brewer, A. D., 277 (9), *288*

Brignell, P. J., 9 (25), *15*
Brigsne, H., *359* (17)
Brill, E., 275 (87), 276 (87, 88), 287 (87, 88), *290*
Brimage, D. R. G., 237 (13), *239*
Brockmann, H., 324 (118), *346*
Brodskii, A. I., 63 (152), 64 (152), *77*
Brooker, A. C., 188 (70), *193*, 656 (8), 675 (8), *689*
Brooker, L. G. S., 184 (43), *192*, 665 (29), 667 (31), 668 (29, 31), 669 (29, 31), 670 (29, 31), 677 (29, 31), 678 (29, 31), 679 (29, 31), 680 (29), 681 (31), 682 (31), 683 (31), 685 (29, 31), 686 (31), 687 (31), *689*, 723 (39), *728*
Brose, E., 246 (20), *246*
Brown, C. W., 79 (2), 97 (2), 98 (2), 100 (2), *106*, 171 (72), 172 (72), *177*, 180 (13), 181 (13), 183 (13), 187 (13), *191*, 199 (4), 200 (4), 203 (4), *204*, 247 (4), 255 (4), *259*
Brown, D. J., 9 (19), *15*, 400 (2), 402 (2, 6), 407 (2), *408*
Brown, E. V., 25 (74), *27*, 497 (62), 531 (62), *543*
Brown, P., *347* (145)
Brown, R. F. C., 431 (2), 433 (2), *439*
Brown, W. V., 181 (17), 188 (17), *191*
Broyles, M. H., 24 (65), *27*
Bruch, E., 328 (122), *346*
Brugel, E., 411 (4), 417 (4), *429*
Bruice, T. C., 321 (143), 322 (144), *347*
Brunner, R., 324 (150), 326 (121), 340 (121), *346, 347*
Brutane, D., 100 (138), *110*
Bryant, H., 225 (110), *231*, 731 (4), 740 (4), *741*
Buchardt, O., 45 (72), *74*
Buchmann, G., 470 (16), 491 (16), *493*
Buda, L., 47 (75), 63 (75), *74*
Buddenberg, O., 648 (57), *652*
Budesinsky, M., 305 (135), 336 (135), *346*
Budesinsky, Z., 155 (89), 156 (89), 157 (89), *161, 178* (102), 188 (68), *193*
Buechel, K. E., 665 (58), 671 (58), *690*
Buechel, K. H., 664 (23, 24), 665 (23), 684 (23), *689*
Bueren, H., 547 (16), 564 (16), *567*
Buisson, P. J. C., 474 (38, 39), 481 (38), 484 (38, 39), *494*
Bulhak, B., 322 (84), 324 (110), *345*
Bullard, W. P., 661 (19), 662 (19), 676 (19), *689*
Bumpus, F. M., 325 (152), 336 (152), *347*
Bunting, J. W., 250 (16), *259*, 266 (17), *288*, 545 (10, 13), 546 (13), 567
Buratti, W., 12 (58), 13 (58), *16*
Burchall, G. R., 168 (41), *176*
Burdon, J., 268 (50), 278 (50), *289*
Burger, K., 100 (146), 102 (146), 101 (170), 102 (170), *110*
Buriks, R. S., 20 (23, 24), 22 (24), 23 (23), *26*, 224 (91), *231*
Burke, R. W., 118 (34), *122*
Burmistrov, S. I., 34 (29), *73*, 80 (14), 92 (66), 102 (66), *106, 107,* 127 (20), *129*, 135 (16), *136*, 145 (45), 152 (53), 157 (45), 158 (45), *160*, 163 (5), 172 (5), 173 (5), *175*, 626 (62), 647 (62), *652*, 696 (35), 697 (35), 704 (35), *711*, 724 (50), *729*
Burrell, R. A., 29 (6), 30 (6), 49 (6), 56 (6), 57 (6), *73,* 98 (116), *109*, 183 (29), 187 (29), *192*
Burton, D. E., 20 (22), 22 (22), 23 (22), 24 (22), *26*, 97 (106), 98 (106), 100 (106), 104 (106), 105 (191), *109, 111,* 171 (68), 174 (68), *177*, 202 (14), *204*, 221 (65), 222 (65), 223 (65), 226 (65), *230,* 237 (17), 238 (17), *239*
Butenandt, A., 324 (99), *345*
Buu-Hoi, N. P., 237 (18), *239*, 244 (12), *246*, 498 (20), 539 (20), 540 (20), *541*, 693 (7), 701 (7, 47), 703 (7, 47), 704 (7, 47), 705 (7, 47), 706 (47), 709 (47), *710,* 713 (12), 725 (49), 726 (12, 40), *728, 729*

Cain, C. K., 579 (14), 580 (14), 581 (14), *582*
Cameron, D. W., 98 (128), 99 (128), *109*, 267 (46), 280 (46), *289*
Campbell, E. C., 444 (4), 452 (4), 453 (4), *454*, 471 (21), 474 (21), 484 (36), 489 (21), *493*

Canevari, R., 476 (42), 480 (42), 485 (42), 486 (42), *494*
Caradonna, C., 503 (36), 535 (36), *542*
Carlisle, G. O., 118 (32), *122*
Carmack, M., 165 (25), 168 (25), 174 (25), *176*, 184 (42), *192*, 450 (18), 454 (18), *455*
Caronna, T., 8 (72), 12 (55, 60), *16*, 25 (76), *27*, 209 (17), *229*
Carr, A., 225 (110), *231*, 731 (4), 740 (4), *741*
Carter, S. D., 165 (28), 170 (28), *176*, 180 (44), 184 (44), 189 (44), 191 (44), *192*
Casagrande, C., 294 (12), 301 (12), 302 (12), 303 (12), 332 (12), *342*, 480 (43), 486 (43), *494*
Case, F. H., 12 (49), *16*, 18 (3), 19 (3), 22 (3), 23 (3), 24 (3), 25 (3), *25*, 156 (96), *161*, 173 (92), *178*, 237 (16), *239*
Caserio, M. C., 131 (3), 135 (3), *136*, 153 (63), *160*
Casini, G., 732 (19, 41), 736 (19, 41), 737 (41), 738 (41), 740 (19), *742*
Castellani-Bisi, C., 579 (18), *582*
Castellano, A., 13 (63), *16*
Castle, R. N., 166 (32), *176*, 207 (16), *229*
Catteau, J. P., 13 (63), *16*
Caubere, P., 269 (62), 281 (62), 282 (62), *289*
Cava, M. P., 151 (48), *160*
Cavagna, G., 102 (161), *110*
Cavagnol, J. C., 9 (20), 10 (32), *15*, 22 (42), 23 (42), 24 (42), *26*, 270 (70), 272 (74), 279 (74), 280 (74), 281 (74), 282 (74), 283 (74), 284 (74), *289*
Cavett, J. W., 184 (39), *192*
Cerletti, A., 327 (124), *346*
Cesari, A., 151 (49), *160*, 625 (61), 647 (61), *652*
Chabassier, S., 749 (23), 755 (23), 756 (23), *759*
Chai, S. Y., 601 (74), 634 (74), 635 (74), 636 (74), 638 (74), 639 (74), 640 (74), 642 (74), *652*
Chambers, R. D., 22 (39), *26*, 95 (89), 96 (89), 104 (89), *108*, 168 (48), 170 (48), 172 (48), 174 (48), *177*, 203 (19), *204*
Chang, C. J., 327 (154), *347*
Chang, Y. C., 364 (10), 365 (10), 384 (10), *396*
Chang, Y.-W., 98 (125), 99 (125), 105 (125, 192), *109, 111,* 172 (82), 174 (82), 175 (82), *178*
Chao, C. I. P., 293 (126), 302 (126), 329 (126), *346*
Chapman, D. D., 80 (9), 87 (9), 93 (68), 98 (68), 99 (9), 100 (68), 102 (68), 103 (68), *106, 107,* 173 (95), *178*, 189 (72), *193*, 624 (55), 643 (55), *652*, 668 (34, 40), *690*
Chapman, N. B., 164 (20), *176*, 180 (9), *191*
Charney, W., 323 (92), 335 (92), *345*
Charnyak, S. A., 21 (77), *27*
Charushin, V. N., 512 (64), 516 (64), *543*
Chasan, D. E., 41 (50), *74*
Chastain, B., 731 (4), 740 (4), *741*
Chattaway, F. D., 625 (59), 646 (59), 647 (59), *652*
Chatterjea, J. N., 103 (179), *111*
Chatterjee, R. K., 571 (8, 9, 10), 573 (8, 9), *574*
Cheeseman, G. W. H., 6 (4, 5), *6*, 8 (15), 9 (19, 21, 25), *15*, 22 (41), *26*, 30 (11), 31 (11), 32 (18), 46 (74), 56 (11, 74, 100), 58 (11), *73, 74, 75,* 79 (3, 7), 86 (38, 40), 89 (7), 90 (59), 95 (38, 88), 96 (7, 59, 88, 98), 97 (3, 7, 88), 98 (3, 7, 59, 88), 99 (59, 88), 101 (7), 102 (7, 88), 103 (59), 104 (7, 88), 105 (7, 88, 98), *106, 107, 108,* 112 (3), 114 (3), 115 (3), 119 (3), *121*, 162 (2), 165 (28, 30), 168 (44), 170 (28), 171 (69), 172 (2, 44), 174 (44, 98), *175, 176, 177, 178*, 180 (6, 14, 4), 183 (27), 184 (44), 188 (6, 14), 189 (6, 44), 191 (44), *191, 192*, 194 (4), 195 (4), 196 (4), *197*, 199 (2, 3), 200 (2, 3, 6), 201 (2, 6), 202 (2, 3, 15), 203 (2, 6), 204 (6), *204*, 213 (30), 218 (30, 54, 55), 220 (58), 227 (30), *229, 230*, 238 (27), *239*, 599 (1, 2, 3, 4), 600 (2, 4, 6, 8), 601 (2, 3, 9), 603 (9), 605 (21, 24), 607 (24, 26), 608 (4), 610 (4, 24), 615 (3), 616 (1, 11, 26), 617 (11, 21, 26), 618 (9, 26), 619 (3, 21, 24), 620 (21), 621 (4, 21, 24), 622

(9, 11), 633 (3, 11, 21, 24, 26), 634 (2, 3, 9, 11, 21, 24), 635 (2, 3, 4, 11, 26), 636 (11, 24), 637 (4), 638 (2, 24), 639 (4, 9, 24), 640 (4, 9, 24), 641 (4, 9, 24), 642 (24), *650, 651*
Chekir, K., 497 (61), 521 (61), 534 (61), 538 (61), 539 (61), 540 (61), *543, 567* (35)
Chelhot, N. C., 630 (80), 648 (80), *653*
Chellemi, D., 93 (75), *108*
Chelot, N. C., 39 (47), 48 (47), 62 (47), *74*, 134 (11), 136 (11), *136*
Chen, T. T., 13 (62), *16*, 206 (11), 223 (11), *229*, 262 (7), 277 (7), 278 (7), 283 (7), 285 (7), *288*
Chen, Y. S., 323 (90), 324 (90), *345*
Cheng, C. C., 148 (35), 153 (35), *159*
Chernov, V. A., 593 (9), 594 (9), 595 (9), 596 (9), *597*
Chernyak, S. A., 21 (27), 23 (27), *26*
Chernyshkova, L. A., 368 (25), 383 (25), *396*
Chichibabin, A. E., 498 (18), 502 (18), 540 (18), *541*
Childress, S. J., 86 (35), 102 (35), *106*, 284 (104), *290*
Chilton, W. S., 241 (2), 242 (2), 243 (2, 8), 244 (2, 8), 245 (2), *246*, 693 (11), 705 (11), *710*
Chirgadze, Y. N., 307 (40), *343*
Chittenden, M. L., 115 (17), *121*, 166 (35), *176*
Chitwood, H. C., 580 (19, 21), 581 (21), *582*
Chmielewska, I., 322 (84), 324 (110), *345*
Christie, W., 276 (90), 279 (90), 287 (90), *290*
Chrzanowski, R., 267 (30), 284 (30), *288*
Chrzanowski, R. L., 267 (31), 274 (31), 284 (31), 285 (31), 286 (31), *288*
Chugai Parmaceutical Co., 656 (7, 11), 658 (7, 11), 675 (7, 11), *689*
Chu Lesher, E., 382 (57), *397*
Chun-Shan Wang, 739 (33), 740 (36), *742*
Chupakhin, O. N., 90 (61, 62), *107*, 512 (64), 516 (64), *543*
Ciba Ltd., Brit. Pat. 1,021,195,81 (19), 101 (19), *106*
Ciba Ltd., Fr. Pat. 1,021,195, 404 (16), 406 (16), 407 (16), *409*
Ciba Ltd., Fr. Pat. 1,350,002, 404 (14), 406 (14), 407 (14), *408*
Ciba Ltd., Fr. Pat. 1,350,003, 404 (15), 406 (15), 407 (15), *409*
Ciba Ltd., Fr. Pat. 1,361,010, 113 (10), *121*
Ciba Ltd., Neth. Appl. 6,515,056, *728* (36)
Ciba Ltd., Neth. Appl. 6,613,937, 466 (8), 475 (8), 480 (8), 481 (8), 484 (8), 490 (8), 491 (8), *493*
Cimanis, A., 101 (150), *110*, 237 (39), *239*
Citterio, A., 8 (72), 12 (59), 13 (59), *16*, 18 (4), 25 (76), *25, 27*
Clark, B. A. J., 348 (1, 19), 349 (1, 3, 19), 350 (3), 356 (1, 3, 19), 357 (1, 3, 19), *359*
Clark, J. W., 223 (82), *231*
Clark, N. G., 22 (47), 23 (47), 24 (70), *26, 27*
Clarke, K., 180 (9), *191*
Clarke, P., 268 (58), 279 (58), 280 (58), 281 (58), *289*
Clark-Lewis, J. W., 32 (19), 53 (19), 58 (19), *73*, 91 (64), 98 (64), 99 (64), 102 (165), 103 (181), *107, 110, 111,* 141 (10, 12, 13), 147 (30), 153 (10, 30), 154 (10, 30), 155 (13, 88), 156 (10), *159, 161,* 267 (25), 283 (25), *288*, 504 (37), 506 (37), 507 (37), 512 (37), 518 (37), 521 (37), 522 (37), 531 (37), 532 (37), 533 (37), 534 (37), 535 (37), 536 (37), 537 (37), 538 (37), 539 (37), *542*, 545 (6), 547 (6), 548 (6), 552 (6, 28), 553 (6), 554 (6), 557 (6), 559 (6), 564 (6, 28), 565 (6, 28), 566 (6, 28), *567*, 659 (15, 17), 660 (17, 18), 661 (15), 675 (17), 676 (15), *689*
Claudi, F., 732 (19, 41), 736 (19, 41), 737 (41), 739 (41), 740 (19), *742*
Claypool, D. P., 72 (148), *77*
Clement, M. G., 184 (33), *192*
Clerin, D., 270 (65), 281 (65), 283 (65), *289*
Close, M., 247 (3), 255 (3), 257 (3), *259*
Close, V. A., 306 (33), 307 (33), *343*

Clough, F. W., 442 (15, 16), 443 (15), 451 (16, 17), *455*
Coffey, R. G., 423 (18), *429*
Cohen, L. A., 600 (13), 604 (13), 622 (13), 634 (13), *651*
Cohen, S. S., 93 (71), *108*
Cole, F. E., 376 (36), 377 (36), 379 (36), 382 (36), 394 (36), *396*
Cole, J. E., 168 (39), 172 (39), *176*
Colton, D. F., 237 (8), 238 (8, 26), *239*
Comfort, D. R., 267 (29), 285 (29), *288*, 659 (14), 661 (14), 662 (14), 676 (14), 689
Compagnon, O., 217 (51), *230*
Compagnon, P. L., 217 (51), *230*
Cook, A. H., 448 (10), 453 (10), *454*
Cooks, R. G., 693 (10), *710*
Cookson, G. H., 751 (13), 757 (13), *759*
Cookson, R. F., 600 (8), *651*, 655 (3), 658 (4), 674 (3), 675 (3), *689*
Copet, A., 579 (18), *582*
Cormack, M., 96 (96), *108*
Cort, L. A., 579 (16), 581 (16), *582*
Cotrel, C., 352 (8), 355 (8), 358 (8), *359*
Courtois, J. E., 695 (24), 706 (24), 707 (24), 708 (24), *710*
Coutts, R. T., 531 (50), 532 (50), *542*
Cox, J. M., 29 (6), 30 (6), 49 (6), 56 (6), 57 (6), *73*, 98 (116), *109*, 183 (29), 187 (29), *192*
Crabtree, A., 475 (40), 489 (40), 490 (40), *494*
Cragg, A. M., 227 (129), *232*
Cragoe, E. J., Jr., 143 (15), 151 (78), 153 (15), *159*, *161*, 187 (61), *193*, 378 (47, 48, 49), 380 (47, 48, 49), 389 (47, 48, 49), 390 (47, 49), 391 (47, 49), 392 (47, 49), 393 (47, 49), *397*
Crawford, H., 238 (35), *239*
Crawford, H. T., 21 (79), *27*
Credner, H. H., 698 (43, 44), 703 (44), 706 (43), *711*
Creech, H. J., 752 (16), 758 (16), *759*
Cresswell, R. M., 24 (68), *27*, 147 (31), 153 (31), 154 (31), 155 (31), 157 (31), *159*
Crichton, D. D., 278 (99), *290*
Crippa, A., 373 (31), 386 (31), *396*
Crippa, B., 625 (60), 647 (60), *652*
Crippa, G. B., 373 (31), 386 (31), *396*, 404 (12), 407 (12), 408 (12), *408*, 716 (19), 723 (19), 724 (19), *728*
Croce, P. D., 13 (70), *16*
Crolla, T., 8 (72), *16*, 25 (76), *27*
Cronin, T. H., 64 (134), 65 (134), *76*
Crow, W. D., 425 (24), *430*, 431 (2), 433 (2), *439*
Crowther, A. F., 171 (75), *177*
Crychtol, K., 252 (24), 256 (24), *260*
Cue, B. W., 49 (85), 56 (85), *75*
Cuiban, F., 81 (16), 97 (16), 98 (16), *106*, 114 (13), 119 (13), *121*, 166 (31), *176*, 556 (32), *567*
Cullen, E., 267 (42), *289*
Culvenor, C. C. J., 227 (118), *232*
Cunningham, H. C., 281 (100), 282 (100), *290*
Curd, F. H. S., 163 (19), 168 (19), 171 (75), 174 (19), *176*, *177*
Cuvigny, T., 216 (48), 226 (48), *230*
Cygler, M., *111* (199)
Cymerman-Craig, J., 267 (28), 279 (28), *288*
Czuba, W., 19 (17, 18), *26*

Dahn, H., 80 (13), 103 (178), *106*, *111*, 131 (5, 8), 134 (5), 135 (5, 8, 17), 136 (5), *136*, 139 (6), 140 (7, 7a), 153 (7), 154 (81, 82), 156 (6, 7, 7a), 157 (6, 7, 7a, 102), *159*, *161*, 173 (96), *178*, 223 (81, 83), 224 (83, 86), 226 (83), 227 (122), 228 (131), *231*, *232*, 695 (30), 696 (33, 34), 705 (30), 706 (33), *711*
Dalalian, H., 128 (29), *129*, 172 (84), *178*, 352 (10), 357 (10), *359*
Dalziel, J. A. W., 116 (24), 118 (35), *122*
Damagk, G., 112 (4), 119 (4), *121*
Damodaran, V. A., 268 (50), 278 (50), *289*
Dandegaonker, S. H., 104 (187), *111*, 169 (54), 173 (54), *177*
Danek, J., 148 (36), *159*
Danieli, R., 578 (10), *582*
Danilewicz, J. C., 22 (45), *26*
Danyluk, S. S., 9 (26), *15*
Danzig, M., 100 (137), *109*
Das, K. G., 10 (30), *15*

Datta, P. C., 432 (7), 437 (7), *440*
Daun, R. J., 184 (36), *192*
Davey, D. G., 163 (19), 168 (19), 171 (75), 174 (19), *176, 177*
Davidson, R. S., 237 (13), *239*
Davies, G. E., 423 (14, 16, 17), *429*
Davis, M. L., 55 (96), 56 (96), *75*
Davis, P. M. H., 22 (47), 23 (47), 24 (70), *26, 27*
Davis, R. B., 267 (42), *289*
Davydova, S. I., 80 (11), 99 (11), *106*
Day, A. R., 84 (26), 98 (26), *106*, 144 (18), 153 (18), *159*, 184 (47), 186 (47), 190 (47), *192*, 268 (53), 281 (100), 282 (100), *289, 290*, 293 (5), 294 (5), 300 (5), 301 (5), 302 (5), 327 (5), 328 (5), 329 (5), 330 (5), *342*, 376 (39, 40), 377, (39, 40), 389 (39), 394 (39, 40), 395 (39), *397*, 464 (4, 5), 469 (13, 14), 470 (4, 5), 471 (13, 14), 476 (4, 5, 13, 14), 480 (4, 13), 481 (5, 13, 14), 482 (4, 5, 13, 14), 483 (5, 13, 14), 484 (4, 5, 13, 14), 485 (13, 14), 486 (13, 14), *493*, 498 (21), 502 (21), 514 (21), 524 (21), 530 (21), 531 (21), 532 (21), 533 (21), 535 (21), 536 (21), 540 (21), *541*, 548 (23), 551 (23), 561 (23), 565 (23), *567*, 664 (22), 665 (22), 667 (22), 684 (22), 685 (22), 686 (22), *689*, 733 (21), 736 (21), 737 (21), *742*
De, S. C., 432 (7), 437 (7), *440*
De Alti, G., 292 (3), *342*, 361 (1), *395*
Dean, F. M., 227 (129), *232*
Deardorff, E. R., 118 (34), *122*
De Groot, Ae., 206 (7, 9), 228 (7), *229*, 278 (11), 285 (11), *288*
Degutis, J., 256 (40), *260*, 284 (96), *290*, 672 (60), *690* (53)
Degutis, Yu., 268 (59), 281 (59), *289*
Dekker, W. H., 156 (94), *161*
De Lajudie, P., 749 (23), 755 (23), 756 (23), *759*
de la Llosa, P., 93 (76), *108*, 604 (15), 622 (15), 635 (15), *651*
Dell, A., 148 (32a), *159*
Demagk, G., 163 (13), *176*
De Martino, G., 605 (18, 19), 618 (18, 19), 621 (18, 19, 75), 622 (75), 634 (18, 75), 635 (18, 19, 75), 636 (19, 75), 638 (19, 75), 639 (19, 75), 640 (19), 641 (75), *651, 652*
De Micheli, C., 227 (128), 228 (128), *232*
Denison, Y. V., 308 (45), *343*
Denzel, T., 560 (37), *567*
Depompei, M. F., 361 (62), 365 (62), 370 (62), 382 (62), 383 (62), *397*
DerKosch, J., 86 (39), *107*
Desai, R. B., 307 (139), 335 (139), *347*, 469 (10), 493
De Selms, R. C., 19 (14), 23 (14), 24 (14), 25 (14), *26*, 266 (20), 267 (45), 273 (45), 279 (45), 280 (45), 281 (45), 282 (20, 45), *288, 289*, 521 (55), 534 (55), *542*, 565 (33), 567
Desimoni, G., 435 (17), 439 (17), *440*
Deuschel, W., 163 (14), 169 (52), 172 (14, 52), 173 (14), *176, 177,* 180 (4), 184 (4, 57), 187 (4, 56, 57), *191, 193,* 200 (8), 203 (8), *204*, 735 (26), 739 (26), *742*
Dewar, M. J. S., 8 (9, 11), 12 (48), *15*, 18 (2), 22 (2), 24 (2), *25*
Dickore, K., 459 (6), 461 (6), *462*
di Dio, P., 125 (11), *128*
Dieckmann, K. P., 89 (52), 93 (52), *107*
Diel, P. J., 37 (39), *73*
Diels, O., 648 (57), *652*
Dikciuviene, A., 672 (60), *690* (53)
Dilcher, H., 400 (3), 402 (3), 405 (3), 407 (3), *408*
Dimroth, P., 624 (49, 50, 51), 625 (51), 643 (49, 50), *652*
Dirlam, J. P., 39 (153), *77*
Diskus, A., 736 (38), *742*
Dittrich, K., 98 (123), 109, 695 (29), 696 (32), 701 (35), 705 (35), 707 (35), 708 (35), *711*, 714 (13), *728*
Dixon, J. E., 322 (144), *347*
Dobas, I., 98 (120), *109*, 127 (24), *129*
Doerr, F., 713 (1, 2), *727*
Doerscheln, W., 262 (7), 277 (7), 278 (7), 283 (7), 285 (7), *288*
Dolecek, M., 224 (142), *232*
Dolejs, L., 693 (9), 701 (9, 49), *710, 711*

Domaschke, L., 579 (15), 580 (15), 581 (15), *582*
Dominy, B. W., 748 (8), 755 (8), *758*
Donahue, W. E., 222 (80), *231*
Dorgan, R. J., 348 (19), 349 (3, 19), 350 (3), 356 (3, 19), 357 (3, 19), *359*
Dorn, H., 400 (3, 200), 402 (3), 405 (3), 407 (3, 20), *408, 409*
Dornauer, H., 622 (76), *653*
Dornow, A., 153 (73), *160*, 256 (42), 257 (42), *260*, 548 (22), 555 (22), 565 (22), 566 (22), *567*, 665 (26), 684 (26), *689*
Dorofeenko, G. N., 744 (22), 754 (22), *759*
Dörr, F., 9 (23), *15*
Dorscheln, W., 13 (62), *16*, 206 (11), 223 (11), 229
Dose, W., 695 (29), *711*
Doty, M., 693 (8), *710*
Dou, H. J. M., 12 (52), *16*
Douglas, M. L., 56 (106), *75*
Douwe Egberts Koninklijke, 456 (3), 459 (3), 460 (3), *462*
Draber, W., 664 (24), *689*
Drefahl, G., 238 (33), *239*
Druey, J., 237 (15), *239*, 248 (10), 249 (10), 256 (10), 257 (10), 258 (10), *259*, 267 (26), 285 (26), 286 (26), *288*
Drumheller, J., 165 (29), *176*
Drummond, P. E., 324 (115), 337 (115), *346*
Drury, J., 79 (5), 101 (5), 102 (5), *106*
Druta, I., 247 (5), 252 (25), 253 (5), 256 (5, 25), *259, 260*, 610 (33), 641 (33), *651*, 656 (9), 675 (9), *689*, 744 (5), 754 (5), *758*
Duda, L., 404 (13), 407 (13), *408*
Dudley, K. H., 155 (93), 156 (93), *161*
Dukai, J., 56 (101), *75*
Dunham, D. E., 292 (2), 293 (2), 302 (2), 327 (2), *342*
Dunlop, A. P., 299 (18), *342*
Dunn, G., 89 (57), *107*
Duplan, J. C., 271 (72), 279 (72), 282 (72), 283 (72), *289*
Dury, K., 127 (25), *129*, 624 (49, 51), 625 (51), 643 (49), *652*, 732 (14), 739 (14), *742*
Dutcher, J. D., 706 (50), *711*
Dutt, N. K., 145 (26), 148 (37, 38), 152 (37), 153 (37), *159*, 245 (16), *246*
Dvoryantseva, G. G., 8 (5), *15*, 41 (49), *74*, 200 (9), *204*

Earley, J. V., 264 (16), 277 (16), 278 (16), *288*
Easley, W. K., 24 (65, 69), *27*, 247 (2), 255 (2), 256 (2), 257 (2), 258 (2), *259*, 255 (36), 256 (36), *260*
Eble, T. E., 307 (37), 319 (37), 323 (27), *343*
Edgar, J. A., 103 (181), *111*, 141 (13), 155 (13), *159*, 267 (25), 283 (25), *288*
Edwards, M. L., 38 (41), 55 (97), 58 (41), 59 (41), 65 (41, 97), 66 (97), 67 (41, 140), 68 (97), 69 (41), 70 (41), 71 (41), *73, 75, 76*
Efimova, G. A., 731 (6), 732 (18), 734 (18, 23, 24), 736 (18, 23, 24), 737 (24), 738 (18, 24), 739 (18), *741, 742*
Efros, L. S., 731 (6), 732 (18), 734 (18, 23, 24), 736 (18, 23, 24), 737 (24), 738 (18, 24), 739 (18), *741, 742*
Ege, G., 730 (39), *742*
Egli, C., 221 (59, 64), *230*
Eholzer, U., 49 (82), 63 (127, 129), 64 (82, 127, 129, 135, 137, 138), 65 (127, 129, 138, 142, 144), 66 (129, 137), 67 (129), 68 (129), 69 (129), 70 (82, 127, 129), 72 (129), *75, 76, 77*, 150 (46), 158 (46), *160*, 228 (140), *232*, 626 (78), 647 (78), 648 (78), *653*
Eichenberger, K., 89 (58), *107*
Eiden, F., 89 (54), *107*, 131 (15), 134 (20), 135 (15), *136*, 152 (54), 156 (54), *159*, 225 (130), *232*, 473 (28, 29), 478 (29), 488 (28), 489 (29), 490 (29), 491 (28), 492 (29), *493*, 613 (36), 637 (36), 639 (36), 641 (36), *651*, 744 (4), 751 (14, 15), 754 (4), 755 (4), 757 (14, 15), *758, 759*
Eisch, J., 13 (66), *16*, 268 (48), 284 (48), *289*
Eistert, B., 101 (157), 102 (157), *110*
Eldredge, C. H., 669 (46, 47), *690*
El-Hawehi, Z., 116 (22), 120 (22), 121 (22), *122*

Elina, A. S., 8 (5), *15*, 30 (9), 39 (45, 46), 41 (49), 43 (45, 64), 45 (73), 46 (45, 64), 47 (73 77, 78), 48 (80), 49 (83, 84), 56 (9, 45, 64, 83, 84), 57 (9, 46, 64, 80, 83, 84, 107, 108, 109), 58 (9, 45, 73, 80, 110), 59 (45, 111, 112, 116), 60 (9, 109), 61 (64, 83, 112), 62 (73), 63 (9, 46, 78, 83, 107, 111, 128), 64 (9, 78, 109, 110), 65 (9, 78, 107, 128), 66 (15, 76, 83, 112, 116, 128), 67 (111, 116, 128), 68 (76, 109), 69 (76, 109, 146), 70 (76, 77, 107, 109, 146), 71 (77, 111), 72 (77), *73, 74 75, 76, 77,* 95 (92), 98 (115, 115a), *108, 109*, 113 (19), *121*, 124 (7), 125 (7, 10, 14), 127 (7, 10, 14), 128 (7), *128*, 145 (21), 149 (44), 152 (21), 153 (58, 59, 153), 154 (59), 157 (44, 58), *159, 160*, 169 (51), 174 (100), *178*, 180 (11), 181 (19), 183 (19), 187 (19, 59), 189 (71), *191, 193*, 200 (9), *204,* 213 (29), 214 (34), 215 (40), 221 (34), 222 (29, 75), 224 (96), 225 (29), 226 (29), *229, 230, 231,* 256 (38), 257 (38), *260*, 273 (76), 279 (76), 282 (76), 283 (98), *289, 290*, 379 (68), 390 (68), 391 (68), 394 (67, 68), 395 (67), *397*, 667 (32), 684 (32), 686 (32), *689*
Ellingson, R. C., 578 (11), 581 (11), *582*
Elliott, R. D., 498 (26, 27, 30), 499 (30), 502 (26, 27, 32, 33, 34), 509 (27, 30, 32, 48, 49), 515 (48, 49), 524 (49), 525 (30, 32), 526 (27, 48, 49), 527 (26, 27, 30, 49), 528 (26, 27, 32, 34), 529 (26, 27), 530 (30), 531 (49), *542*, 545 (7), 551 (7), 555 (7, 29, 30), 558 (29, 30), 560 (7), 561 (7, 29, 30), 562 (7, 30), 563 (7, 30), *567*
Elvidge, J. A., 20 (22), 22 (22), 23 (22), 24 (22), *26*, 105 (191), *111*, 448 (10), 453 (10), *454*
Emerson, T. R., 42 (58, 59), *74*
Emmons, W. D., 226 (112), *232*
Enble, J. N., 568 (2), 569 (2), 570 (2), 572 (2), 573 (2), *573*
Engelhard, H., 103 (183), *111*
Engelman, M., 23 (62), 24 (62, 66), *27*, 225 (104, 105), *231*
Enslein, L., 101 (172), *110*
Eppstein, S. H., 324 (101), *345*
Erdelyi, I., 470 (17), 482 (17), 483 (17), 490 (17), 491 (17), *493*
Erlenmeyer, H., 156 (95), *161*
Ettlinger, L., 323 (93), *345*
Et'tsov, A. V., 105 (189), *111*
Euler, H., 24 (67), *27*
Evans, R. H., 324 (98), *345*
Evers, W. J., 23 (57), 24 (57), *27*
Evleth, E. M., 361 (1), *395*
Evodokimoff, V., 691 (1), 704 (1), *710*
Ewers, U., 545 (36), *567*
Exner, O., 210 (26), *229*

Fabio, P. E., 168 (40), *176*
Fabra, I., 353 (14), 358 (14), *359*
Facchetti, S., 579 (18), *582*
Fagano, J. F., 58 (149), *77*
Fahr, E., 85 (32), 102 (32), *106*
Fahrenholtz, K., 296 (15), 300 (15), 328 (15), 330 (15), *342*
Fakatsu, H., 364 (13, 14), 371 (13), *396*
Fallab, S., 352 (7), 357 (7), 358 (7), *359*, 437 (11), *440*
Fanta, P. E., 125 (15), 127 (15), 128 (30), *129*, 153 (61), *160*, 221 (69), *230*, 713 (9), 714 (9), 724 (9), *727*, 739 (33), 740 (36), *742*
Farben-Fabriken Bayer A-G., 64 (132), *76*, 171 (66), *177*, 275 (82), *290*
Farina, P. R., 267 (31), 274 (31), 284 (31), 285 (31), 286 (31), *288*
Farkas, J., 693 (9), 701 (9, 49), *710, 711*
Farkas, L., 103 (177), *111*
Farmer, E., 576 (4), *582*
Farooqi, M. A., 18 (6), 22 (6), *25*, 29 (8), 45 (8), 56 (8), 57 (8), 59 (8), *73*
Fattuta, S., 80 (12), 100 (12), 103 (184), *106, 111*, 172 (81), *178*, 237 (14), *239*
Favini, G., 22 (43), *26*, 292 (4), *342*, 496 (1), *541*, 545 (1), *566*, 576 (3), *582*
Fedoronko, M., 10 (31, 35), *15*, 86 (41), *107*, 127 (26), *129*, 224 (88), 226 (88), *231*, 235 (5), *239*
Feeney, J., 45 (72), *74*
Fegley, M. F., 299 (25), 328 (25), *343*
Fehn, J., 101 (170), 102 (170), *110*
Fehr, K., 702 (51), *711*

Feiden, K., 575 (1), 580 (1), *582*
Feil, D., 10 (38), *15*
Felder, E., 354 (15), 357 (15), *359*
Felici, M., 732 (19), 736 (19), 740 (19), *742*
Fellig, J., 382 (56), *397*
Felton, D. G. I., 716 (20), 722 (20), 723 (20), 724 (20), *728*
Ferappi, M., 732 (19), 736 (19), 740 (19), *742*
Ferrini, R., 294 (12), 301 (12), 302 (12), 303 (12), 332 (12), *342*, 480 (43), 486 (43), *494*
Feuer, H., 216 (45, 46), 221 (45, 46), *230*
Feuerbach, D. J., 217 (52), *230*
Feuerriegel, G., 328 (122), *346*
Field, G. F., 86 (34, 35a), 91 (63), 102 (34), 103 (34), *106, 107*, 264 (15, 16), 277 (15, 16), 278 (16), 284 (15), *288*
Fields, E. K., 433 (9), *440*
Fiesselman, H. F., 228 (137), *232*
Figueras, J., 237 (19), *239*, 263 (8), 277 (8), 283 (8), *288*
Filacchioni, G., 725 (54), *729*
Filipenko, T. Ya, 63 (128), 65 (128), 66 (128), 67 (128), *76*
Finzi, P. V., 227 (128), 228 (128), *232*
Fiorentini, G., 313 (59), 319 (59), 336 (59), *344*
Fioretti, A., 579 (13), 580 (13), 581 (13), *582*
Fischer, E., 315 (62, 63), 319 (63), *344*
Fischer, H., 245 (14), *246*
Fischer, O., 10 (34), 11 (34), *15*
Fischer, W., 85 (32), 102 (32), *106*
Fisher, G. H., 21 (35), 24 (35), 25 (35), *26*, 267 (24), 280 (24), *288*
Fisher, W. M., 25 (74), *27*
Fisons Pest Control Ltd., 22 (38), 23 (38), 24 (38), *26*, 94 (83), *108*
Fissekis, J. D., 101 (155), 102 (155), *110*
Flament, I., 456 (1), 457 (1), 459 (1, 7), 460 (1), 461 (1), *461, 462*
Flamini, V., 164 (22), *176*
Flanagan, K. J., 72 (148), *77*
Fleming, R. H., 626 (67), *652*
Flesher, A. M., 56 (105), *75*
Fleury, J. -P., 270 (65), 281 (65), 283 (65), *289*, 719 (47), 724 (47), *729*
Floodgate, G. D., 93 (72), *108*
Floss, H. G., 327 (154), *347*
Fodor, A., 324 (113), *346*
Fodor, G., 238 (22), *239*
Folkers, K., 21 (32), *26*
Fontanella, L., 296 (127), 310 (140), 330 (127), 332 (127), 335 (140), 336 (140), 337 (140), 338 (140), 339 (140), 340 (140), *346, 347*
Forchiassin, M., 172 (81), *178*
Ford, K. H., 21 (30), 22 (30), *26*
Foris, P., 47 (75), 63 (75), *74*
Fort, R. C., Jr., 600 (6), *651*
Fox, B. A., 498 (16), *541*
Foxton, M. W., 219 (56), *230*, 293 (8), 297 (8), 328 (8), 329 (8), *342*, 613 (37), 637 (37), *651*
Francis, J., 55 (98), 66 (98), *75*, 128 (28), *129*
Francis, N. R., 579 (16), 581 (16), *582*
Franzke, H., 470 (18), 471 (18), 478 (18), 482 (18), *493*
Frass, W., 237 (12), *239*
Frattesi, P., 314 (142), 318 (142), 336 (142), 340 (142), *347*
Fravolini, A., 184 (35), *192*
Freed, M. E., 125 (33), *129*, 293 (5), 294 (5, 13), 300 (5), 301 (5, 26), 302 (5, 13), 303 (26), 327 (5, 13), 328 (5), 329 (5), 330 (5, 13, 26), *342, 343*, 464 (4), 469 (11), 470 (4), 476 (4), 480 (4), 481 (11), 482 (4), 484 (4), 485 (11), *493*
Freedlander, B. L., 196 (16), *198*
Freeman, J. P., 226 (112), *231*
Freeman, S. K., 22 (49), *27*
Frei, A. J., 326 (121), 340 (121), *346*
French, D., 693 (14, 16, 17), 695 (14, 16, 17), 700 (16), 701 (14, 15), 702 (14), *710*
French, F. A., 126 (19), 127 (19), 128 (19), *129*
Frenzel, H., 117 (25), 119 (25), 120 (25), *122*
Frey, A. J., 315 (66), 316 (68), 317 (68, 78), 318 (66, 68), 321 (68), 326 (66, 78), 335 (66), 336 (68), 338 (68), 339 (66), *344*

Fric, I., 305 (135), 336 (135), *346*
Fried, K., 184 (34), *192*
Friesen, A. J. D., 322 (85), *345*
Fritsch, A. J., 124 (6), 125 (8), 127 (6), 128 (6), *128*, 215 (39), 223 (39), 224 (39), 225 (39), 226 (39), *230*
Fritche, D., 64 (131, 135, 137, 138), 65 (138, 142), 66 (138), *76, 77*
Fritsche, H., 154 (83), *161*
Fritz, H., 245 (15), 246 (15), *246*
Froehlich, J., 591 (1), *596*
Frye, J. L., 509 (49), 515 (49), 524 (49), 526 (49), 527 (49), 528 (49), 531 (49), *542*, 550 (30), 558 (30), 560 (30), 561 (30), 562 (30), 563 (30), *567*
Fryer, R. I., 91 (63), *107*, 264 (16), 270 (64), 277 (16), 278 (16), 279 (64), 280 (64), 281 (64), 284 (64), *288, 289*, 601 (12), 603 (12), 638 (12), *651*
Fuhr, K. H., 91 (63a), 98 (63a), *107*
Fujiki, S., 149 (39), 158 (39), *159*
Fuji Photo Film Co. Ltd., 414 (9), *429*
Fujisawa Pharmaceutical Co. Ltd., 436 (18), 439 (18), *440*
Fukuhara, T., 324 (103), *345*
Fumeaux, J. P., 695 (30), 705 (30), *711*
Fumia, A. J., Jr., 668 (33), 677 (33), 680 (33), *690*
Furst, A., 112 (2), 116 (2), 120 (2), *121*, 169 (55), 173 (55), *177*, 196 (16), *198*, 222 (74), *231*
Furuya, T., 414 (8), *429*
Fusco, R., 33 (25, 26), 34 (26), 56 (25), 57 (25, 26), 58 (25), 60 (25), 61 (25), *73*

Gabe, I., 252 (22, 23), 255 (22, 23), *260*, 720 (29, 30), 724 (29), *728*
Gabriel, S., 352 (10), 357 (10), *359*, 431 (3), 437 (3), *439*
Gaertner, B., 614 (40), 638 (40), *651*
Galasso, V., 292 (3), *342*, 361 (1), *395*
Gallagher, J. J., 89 (57), *107*
Galli, A., 294 (12), 301 (12), 302 (12), 303 (12), 332 (12), *342*, 480 (43), 486 (43), *494*
Galli, R., 12 (58, 60, 61), 13 (58, 61), *16*, 209 (17), *229*
Galosso, V., 464 (1), *493*
Gal'pern, M. G., 149 (41), 153 (41), 157 (41), *160*, 739 (34, 43), *742*
Galt, R. H. B., 19 (12), 23 (12), 24 (12), 25 (12), *26*, 29 (5), 56 (5), 63 (5, 122), 64 (5), 65 (5), 66 (5), 68 (5, 122), 70 (5), 71 (5), *73, 76*
Galzyrina, L. P., 131 (6), 134 (6), *136*
Gamba, A., 292 (4), *342*
Gandolfi, R., 227 (128), 228 (128), *232*
Ganguli, K. K., 118 (32, 33), *122*
Ganns, R., 174 (97), *178*
Garcia, E. E., 601 (12), 603 (12), 638 (12), *651*
Garcia, F., 244 (10), *246*
Gardini, G. P., 12 (54, 55, 56, 57, 58), 13 (58), *16*, 124 (32), *129*, 132 (18), 134 (18), 135 (18), *136*, 209 (18), 223 (18), *229*
Gardner, J. H., 97 (113), *109*, 172 (78), *177*, 189 (73), 190 (73), *193*
Gardner, J. N., 324 (115), 337 (115), *346*
Garkusha, G. A., 473 (27), 478 (27), 487 (27), *493*
Gasco, A., 436 (19), 439 (19), *440*
Gaugh, W. S., 625 (56), *652*, 668 (42), *690*
Gaumann, E., 323 (93), *345*
Gauss, W., 718 (25), 725 (25), *728*
Gawrilow, N., 324 (114), *346*
Geary, W. J., 237 (8), 238 (8, 25, 26, 30), *239*
Gehrlein, L., 718 (46), 726 (46), *729*
Geigy, A.-G., Jr., 184 (40), 188 (40), 190 (40), *192*
Geigy, J. R., 173 (91), *178*
Gentles, M. J., 323 (92), 335 (92), *345*
Gerchakov, S., 148 (33, 34), 153 (33, 34), *159*, 242 (4), 245 (4), *246*
Gerlach, K., 717 (23), 726 (23), *728*
Gerson, F., 578 (9), *582*
Ghirardini, M., 12 (59), 13 (59), *16*, 18 (4), 25 (76), *25, 27*
Giannella, M., 503 (35), 538 (35), *542*
Giardi, I., 164 (21), 172 (21), *176*
Gibson, C. S., 267 (21), 282 (21), *288*
Gier, D. W., 202 (13), *204*
Gierenz, G., 471 (23, 24), 476 (23), 480 (23, 24), 482 (23, 24), 483 (23, 24), 484 (23), 485 (23), 486 (23), *493*

Gilbert, E. E., 97 (102), *108*
Gillespie, H., 22 (40), 23 (40), *26*
Gillespie, H. B., 23 (62), 24 (62, 66), *27*, 225 (104, 105), *231*
Gilman, H., 13 (66), *16*, 224 (99, 103), *231*, 268 (48), 284 (48), *289*
Giori, P., 401 (5), 403 (5), 404 (11), 405 (5, 11), 407 (5), 408 (5), *408*, 749 (23), 755 (23), 756 (23), *759*
Gisin, B. F., 325 (151), *347*
Gist, L. A., 206 (6), 228 (6), *229*
Giuliano, R., 725 (54), *729*
Glebova, G. D., 155 (109), *161*
Gleinig, H., 224 (92), *231*
Glover, E. E., 361 (4), 368 (4), 370 (4), 382 (4), 383 (4), *395*, 444 (4), 452 (4), 453 (4), *454*, 464 (2, 3), 466 (2, 3), 467 (3), 470 (2), 471 (2, 21), 474 (36), 476 (2), 477 (2, 3), 478 (2, 3), 479 (2, 3), 480 (2), 482 (2), 484 (36), 485 (3), 487 (2), 488 (2, 3), 489 (2, 21), 492 (3), *493*
Glushkov, R. G., 153 (76), *160*, 367 (24), 370 (24), 383 (24), *396*, 425 (23), 429 (23), *430*
Göbel, W., 469 (12), 482 (12), 483 (12), 484 (12), *493*
Goeth, H., 262 (7), 277 (7), 278 (7), 283 (7), 285 (7), *288*
Gofman, S. M., 658 (12), *689*
Gogte, V. N., 10 (30), *15*
Golankiewicz, B., 210 (22), *229*
Golankiewicz, K., 210 (22), 222 (77), 224 (77), *229, 231*
Goldstein, H., 732 (9), 739 (9), 740 (9), *741*
Goldweber, M., 81 (17), 100 (17), 101 (17), 102 (17), *106*
Gómez-Sánchez, A., 101 (151), *110*, 244 (10, 11), *246*, 743 (2), *758*
Goncharova, I. N., 163 (16), 173 (16), *176*, 184 (45), 185 (45), 190 (45), 191 (45), *192*, 194 (7), 197 (7), *198*
Gonezi, C., 183 (31), *192*
Gonzalez, A., 55 (99), *75*
Gonzalez, F. G., 101 (151), *110*
Goodman, L., 371 (28), 372 (28, 60, 61, 66), 373 (28), 375 (65, 66), 385 (28, 60, 65, 66), 386 (60, 61, 65, 66), 387 (61, 65, 66), 388 (65), *396, 397*
Goodman, M., 310 (50), 334 (50), 336 (50), *344*
Goodwin, H. A., 237 (9), *239*
Gopalchari, R., 237 (11), *239*
Goralski, C. T., 216 (47), 221 (47), *230*
Gordienko, L. L., 63 (152), 64 (152), *77*, 150 (47), *160*
Gordon, B. S., 524 (59), 527 (59), *543*
Gore, P. H., 99 (152), *110*
Görlich, B., 99 (130), 101 (130), *109*
Gota, M., 115 (18), *122*, 187 (60), *193*
Göth, H., 13 (62), *16*, 206 (11), 223 (11), *229*
Goto, T., 238 (23), *239*, 250 (15), *259*, 362 (5, 59), 363 (6, 7, 8, 9), 364 (9, 11, 12, 13, 14), 365 (11, 16, 18), 366 (11, 19, 20, 21, 22), 367 (21), 370 (5), 371 (13), 383 (5, 11, 18, 22), 384 (5, 6, 18, 21, 22), *395, 396, 397*
Gowal, H., 154 (81), *161*
Grabosch, J., 507 (42), 539 (42), 540 (42), *542*, 626 (64), 647 (64), 648 (64), *652*
Grabowski, E. J. J., 377 (46), 378 (52), 380 (46, 52), 389 (46, 52), 390 (46, 52), 391 (46, 52), 392 (46, 52), 393 (46, 52), *397*
Grabowski, M. J., *111* (199)
Graff, S., 22 (40), 23 (40, 62), 24 (62, 66), *26, 27*, 225 (104, 105), *231*
Graham, J. H., 228 (133), *232*
Grammatica, S. J., 353 (12), 357 (12), *359*
Grammel, J., 101 (157), 102 (157), *110*
Grandberg, I. I., 402 (7, 9), 407 (9), 408 (9), *408*
Granik, V. G., 153 (76), *160*, 367 (24), 370 (24), 383 (24), *396*, 425 (23), 429 (23), *430*
Grant, D. M., 8 (12), 10 (12), *15*
Grantham, R. K., 604 (16), 641 (16), *651*
Grassmann, W., 324 (112), *346*
Greaves, R. J., 19 (14), 23 (14), 24 (14), 25 (14), *26*, 267 (45), 273 (45), 279 (45), 280 (45), 281 (45), 282 (45), *289*
Green, M., 308 (44), *343*
Grewe, F., 118 (36), *122*, 735 (28), *742*
Grifantini, M., 732 (19, 41), 736 (19, 41), 737 (41), 738 (41), 740 (19), *742*

Grimshaw, J., 19 (15), *26*
Grinev, A. N., 294 (9, 10), 327 (9), 328 (9), 329 (9), 330 (9), *342*, 368 (25), 383 (25), *396*, 614, (39), 618 (39), 638 (39), *651*, 673 (54), 688 (54), *690*
Griot, R., 317 (78), 326 (78), *344*
Grob, C. A., 311 (55), 312 (55), 319 (55), 336 (55), 338 (55), *344*
Gropper, H., 713 (1, 2), *727*
Gross, K. F., 156 (97), *161*, 182 (25), 188 (25), 190 (25), *192*
Grovenstein, E., 30 (14), 72 (14), *73*, 101 (159), *110*, 153 (62), *160*
Grünanger, P., 227 (128), 228 (128), *232*
Gualtieri, F., 238 (31), *239*, 503 (35, 36), 535 (36), 538 (35), *542*
Guarneri, M., 401 (5), 403 (5), 404 (11, 13), 405 (5, 11), 407 (5, 13), 408 (5), *408*
Guerret, P., 361 (2), 365 (2), 383 (2), *395*
Gum, W. F., 18 (7), 21 (28, 31), *25, 26*, 145 (25), *159*, 216 (47), 221 (47), *230*
Gum, W. F., Jr., 733 (22), 737 (22), *742*
Gumprecht, W. E., 168 (39), 172 (39), *176*
Gunther, H., 545 (36), *567*
Gupta, S. N. P., 103 (179), *111*
Guttmann, S., 315 (64), 316 (71, 73), 318 (71, 73, 80, 81), 326 (71, 73, 80, 81, 153), 335 (64, 73, 81), *344, 345, 347*

Habermalz, U., 591 (1), *596*
Habib, M. S., 18 (6), 22 (6), *25*, 29 (8), 33 (21, 27, 28), 44 (66), 45 (8), 46 (66), 47 (66), 51 (88), 52 (90, 91), 53 (94), 54 (95), 57 (8, 27, 94), 58 (27), 59 (8, 27), 60 (21, 28, 66, 90), 61 (28, 66, 90), 62 (28), 63 (94), 70 (66), *73, 74, 75*, 97 (108, 110), 98 (124), 101 (148, 156, 171a), 102 (163, 171a), 103 (163, 171a), 105 (148), *109, 110*, 138 (2), 146 (29), 152 (2), 153 (29, 64, 69, 75, 77), 154 (2, 29, 64, 77), 155 (29, 75, 77), 156 (77), *158, 159, 160*, 163 (8, 15), 172 (15), 173 (8, 15), *175, 176*, 184 (46), 188 (46), 189 (46), 190 (46), 191 (46), *192*, 224 (89), *231*
Hack, F., 235 (4), 238 (4), *239*
Hack, V., 221 (60), 227 (60), *230*
Hadacek, J., 207 (12), 222 (12), *229*
Haddadin, M. J., 35 (31, 33, 34), 38 (40), 39 (47), 40 (31, 48), 48 (47, 81), 49 (86), 61 (40), 62 (47, 81), 65 (48, 139, 145), 67 (139, 145), 68 (34, 145), 69 (34), 70 (33, 34, 48, 139, 145), 71 (48), 72 (139), *73, 74, 75, 76, 77*, 102 (160), *110*, 134 (11), 136 (11), *136*, 267 (32), *288*, 630 (80), 648 (80), *653*, 748 (8), 755 (8), *758*
Haefelinger, G., 235 (4), 238 (4), *239*
Hagaman, E. W., 327 (154), *347*
Hagendorn, I., 228 (140, 141), *232*
Hahn, W., *232* (143), 626 (79), *653*
Hahn, W. E., 626 (66), *652*
Hajek, M., 98 (120), *109*, 102 (168), 103 (168), *110*, 127 (24), *129*, 162 (4), 173 (4), *175*, 624 (53), 643 (53), *652*
Hajos, G., 739 (40), 740 (40), 741 (40), *742*
Halawi, S. S., 49 (86), *75*
Haley, C. A. C., 224 (95), *231*
Haley, T. J., 56 (105), *75*
Hall, G., 62 (120), 72 (120), *76*
Halpern, B., 306 (33), 307 (33), 315 (67), 335 (67), *343, 344*
Hamer, J., 11 (39), *16*, 261 (2), 276 (2), *287*
Hamilton, C. S., 97 (109), 99 (109), *109*, 171 (76), 172 (76), *177*, 180 (7), 188 (7), 189 (7), *191*, 214 (33), 223 (33), *230*
Hamilton, S. D., 655 (57), *690*, 746 (6), 747 (6), 749 (6), 755 (6), 757 (6), *758*
Hamlin, K. E., 276 (89), 287 (89), *290*
Hammann, I., 21 (33), *26*
Hampton, A., 8 (7), *15*, 19 (9), 21 (9), *22* (48), *26, 27*, 255 (34), *260*, 496 (13), 498 (31), 499 (31), 516 (31), 524 (58), 531 (31), *541, 542*, 546 (14), 548 (21), 560 (34), 561 (21), 564 (21), *567*
Hand, E. S., 210 (21), 213 (21), 218 (21, 53), 220 (53), 225 (21), *229, 230*, 605 (22, 23), 607 (22), 618 (22), 635 (22, 23), 636 (22, 23), *651*
Hansen, B., 352 (6), 358 (6), *359*
Hansen, H. V., 349 (4, 5), 350 (4, 5), 351 (5), 352 (4, 5), 356 (4, 5), 357 (4, 5), *359*

Hansen, S., 322 (85), *345*
Hardee, L. E., 442 (16), 451 (16), *455*
Harhash, A. H. E., 719 (26), 725 (26), *728*
Harley-Mason, J., 23 (59), 24 (59), *27*
Harnden, M. R., 306 (34), 307 (39), 308 (34, 47), 309 (34), 317 (34), 322 (34, 47), 334 (34, 47), 335 (34, 47), 336 (34, 47), 337 (34, 47), 338 (34), *343*
Harris, G., 22 (50), *27*
Harris, R. C., 181 (15), *191*
Harsanyi, K., 183 (31), *192*
Hartke, K., 201 (21), *204*
Hartman, K. E., 237 (10), *239*
Hartter, D. R., 98 (121), *109*, 153 (71), 158 (71), *160*, 185 (51), 187 (51), 190 (51), *192*
Harvey, J. A., 237 (36), *239*
Harzer, R., *359* (17)
Hasselquist, H., 24 (67), *27*, 59 (113), *76*
Hatano, M., 25 (73), *27*
Hatfield, W. E., 497 (62), 531 (62), *543*
Hattori, J., 171 (63, 64), *177*
Hattori, M., 115 (18), *121*, 198 (60), *193*
Hauth, H., 154 (82), *161*, 318 (80, 81), 326 (80, 81, 153), 335 (81), *344, 345, 347*
Hayashi, E., 29 (4), 30 (4, 12), 31 (4), 42 (55, 56), 44 (67, 68), 45 (69), 56 (4), 57 (4), 58 (4), 59 (4, 69, 114), 60 (4, 69, 114), 61 (4, 12, 69, 118), 62 (4, 12, 69, 114), 64 (4), 65 (4, 114), 68 (114), 70 (4, 12), 71 (4), *72, 73, 74, 76, 77*, 84 (30), 101 (30), 103 (175), *106, 111*, 119 (38), *122*, 127 (22), 128 (22, 31), *129* 146 (27), 153 (66, 67), 154 (66, 67), 155 (66), 156 (100), *159, 160, 161,* 163 (7), 166 (33), 172 (89), 173 (7, 89), *175, 178*, 187 (63), 188 (63), 190 (76), *193*, 199 (5), 203 (5), *204*, 207 (14), 221 (14), 222 (79), 223 (79), 227 (123), 228 (123), *229, 231, 232*, 237 (21), 238 (28), *239*, 720 (38), 723 (38), 724 (38), 725 (38), 726 (38), *728*
Hayashi, T., 940 (44), *742*
Hazlewood, S. J., 753 (18), 758 (18), *759*
Heese, M., 206 (11), 223 (11), *229*
Heffernan, M. L., 9 (24), *15*, 600 (72), *652*
Heilbron, I. M., 225 (113), 226 (113), *232*
Heilmeyer, L., 732 (11), 739 (11), *741*
Heine, H. W., 188 (70), *193*, 611 (34), 642 (34), *651*, 656 (8), 675 (8), *689*
Heininger, S. A., 254 (31), 258 (31), *260*
Heitzer, H., 49 (82), 64 (82), 70 (82), *75*, 150 (46), 158 (46), *160*, 626 (78), 647 (78), 648 (78), *653*
Helmick, L. S., 293 (126), 302 (126), 329 (126), *346*
Hemmerich, P., 155 (93), 156 (93, 95), *161*, 352 (7), 357 (7), 358 (7), *359*, 437 (11), *440*, 614 (40), 638 (40), *651*
Henry, D. W., 187 (58), *193*
Henry, R. L., 578 (11), 581 (11), *582*
Henseke, G., 98 (123), *109*, 124 (4, 5), 125 (12, 16), 127 (4, 5), 128 (4), *128, 129*, 131 (4), 134 (4), *136*, 154 (86), *161*, 172 (37, 83), *176, 178*, 214 (37), 215 (37), 221 (67), *230*, 242 (5), 244 (7), 245 (5, 7, 17, 18, 19), 246 (20), *246*, 280 (93), *290*, 405 (18), 407 (18), *409*, 506 (40), 523 (40), 539 (40), *542*, 693 (4, 5, 6), 694 (20), 695 (28, 29), 696 (32, 36), 698 (36), 699 (5, 6, 36), 700 (45), 701 (31, 48), 703 (4, 5), 704 (36), 705 (28, 31), 706 (53), 707 (31, 48), 708 (31, 48), 709 (48), *710, 711* (37), *711*, 713 (10), 714 (10, 13), 721 (10, 31), 722 (31), 724 (10), 725 (10), 727 (10), *728*, 732 (15, 16), 735 (15, 25), 739 (15), 740 (15, 16), *742*
Henzel, R. P., 611 (34), 642 (34), *651*
Hepworth, J. D., 550 (27), 558 (27), 564 (27), *567*
Hepworth, W., 168 (41), *176*
Hershenson, F. M., 376 (41), 377 (41), *397*
Herz, W., 294 (7), 327 (7), 328 (7), *342*
Herzog, H. L., 323 (92), 335 (92), *345*
Heseltine, D. W., 625 (56), *652*, 668 (33, 38, 39, 43, 44), 670 (48), 671 (49), 677 (33), 679 (43), 680 (33, 38), 681 (38), 682 (38), *690*
Hesse, M., 13 (62), *16*, 262 (7), 277 (7), 278 (7), 283 (7), 285 (7), *288*
Heubach, G., 224 (92), *231*
Heyneker, H., 102 (162), *110*
Hickson, J. L., 695 (26), 707 (26), *711*
Hielscher, M., 706 (54), *711*
Higgins, J., 94 (84), *108*

Hill, A. C., 147 (31), 153 (31), 154 (31), 155 (31), 157 (31), *159*
Hillers, S., 226 (114), *232*, 237 (39), *239*
Hiller, S. A., 101 (149, 150), *110*
Himmelspach, K., 693 (12), 703 (12, 52), *710, 711*
Hinchcliffe, A., 576 (4), *582*
Hinsberg, O., 713 (5), 723 (5), *727*
Hinsberg, S., 214 (31), 222 (31), *229*
Hinz, E., 153 (73), *160*, 65 (26), 684 (26), *689*
Hirako, T., 116 (20), *122*
Hirata, Y., 363 (6, 7, 8), 384 (6), *395, 396*
Hirobe, M., 399 (19), 406 (19), *409*
Hirt, R. C., 9 (20), *15*
Hisai, S., 196 (14), *198*
Hiskey, R. G., 228 (138), *232*
Hockenhull, D. J. D., 93 (72), *108*
Hoehn, H., 560 (37), *567*
Hoeppner, F., 216 (44), *230*
Hofer, F., 13 (67), *16*, 90 (60), 101 (60), *107*
Hofmann, A., 315 (66), 316 (68, 69, 72), 317 (68, 72, 78), 318 (66, 68, 69, 72, 80, 81), 319 (83), 321 (68), 324 (83, 108), 326 (66, 69, 72 78, 80, 81, 153), 327 (72, 120), 335 (66, 81, 83), 336 (68), 338 (68), 339 (66), *344, 345 347*
Hoffman, A. M., 168 (40), *176*
Hoffmann, J., 102 (168), 103 (168), *110*, 162 (4), 173 (4), *175*, 624 (53), 643 (53), *652*
Holland, A., 449 (14), *455*
Holland, D. O., 448 (11), 453 (11), *454*
Holley, A. D., 604 (14), 622 (14), 635 (14), *651*
Holley R. W. 604 (14), 622 (14), 635 (14), *651*
Holliday, R. E., 11 (39), *16*, 261 (2), 276 (2), *287*
Holtz, W. J., 378 (47), 380 (47), 389 (47), 390 (47), 391 (47), 392 (47), 393 (47), *397*
Homer, J., 170 (56), *177*
Honigberg, I. L., 148 (35), 153 (35), *159*
Hoover, J. R. E., 58 (149), *77*, 733 (21), 736 (21), 737 (21), *742*
Hoppe, H., 201 (21), *204*
Hormann, H., 324 (112), *346*
Horner, J. K., 187 (58), *193*
Horner, L., 95 (87), 97 (87), 98 (87), 99 (87), 104 (87), 105 (87), *108*, 170 (70), *177,* 202 (17), *204*, 238 (32), *239*
Horska, K., 701 (49), *711*
Horton, H., 242 (6), 243 (6), 245 (6), *246*
Hosi, H., 105 (195), *111*
Hosoi, H., 25 (75), *27*, 170 (58), 175 (58), *177*
Hotta, K., 105 (194), *111*
House, H. O., 228 (134, 135), *232*
Howe, R., 102 (162), *110*
Hoyer, G. A., 510 (51), 512 (60), 524 (60), 534 (60), 536 (60), *542, 543*
Hu, H. J., 118 (32), *122*
Huarte, V., 55 (100), *75*
Hubinger, H., 100 (137), *109*
Huffman, C. W., 221 (61), *230*
Hughes, D., 20 (22), 22 (22), 23 (22), 24 (22), *26*, 105 (191), *111*
Hughes, G. K., 99 (152), *110*, 753 (18), 758 (18), *759*
Huguenin, R., 316 (71, 73), 318 (71, 73), 326 (71, 73), 335 (73), *344*
Huguenin, R. L., 315 (64), 335 (64), *344*
Huisgen, R., 269 (61), 279 (61), 282 (61), *289*
Humphrey, W. D., 625 (59), 646 (59), 647 (59), *652*
Hüne, A., 79 (5), 101 (5), 102 (5), *106*
Hüni, A., 237 (15), *239*, 248 (10), 249 (10), 256 (10), 257 (10), 258 (10), *259*, 267 (26), 285 (26), 286 (26), *288*
Hunt, D., 324 (104), *345*
Hurd, C. D., 151 (50), 153 (50), 158 (50), *160*
Hurnaus, R., 100 (136), *109*
Huseby, R. M., 194 (3), 196 (3), 197 (3), *197*
Hutter, R., 323 (93), *345*

Ichikawa, K., 224 (102), 226 (102), *231*, 267 (44), 286 (44), *289*
Iida, E., 25 (73), *27*
Iijima, C., 29 (4), 30 (4, 12), 31 (4), 43 (63), 44 (63, 68), 45 (69), 56 (4), 57 (4), 58 (4), 59 (4, 69), 60 (40, 69), 61 (4, 12,

69, 118), 62 (4, 12, 69, 151), 64 (4), 65 (4), 70 (4, 12), 71 (4), *72, 73, 74, 76, 77*, 84 (33), 101 (30), 103 (175), *106, 111*, 127 (22), 128 (22, 31), *129*, 153 (67), 154 (67), 156 (100), *160, 161*, 163 (7), 166 (33), 173 (7), *175, 176*, 181 (20), 187 (63), 188 (20, 63), 190 (76), *192, 193*, 199 (5), 203 (5), *204*, 207 (14), 221 (14), 227 (123), 228 (123), *229, 232*, 237 (21), 238 (28), *239*, 665 (27), 667 (27), 686 (27), *689*

Iio, A., 86 (36), *107*

Ikeda, K., 207 (15), 228 (15), *229*

Ikeda, M., 235 (40, 41, 43), *239, 240*

Ikeda, T., 96 (100), 104 (100), 105 (100), *108*

Ilford Ltd., 382 (58), *397*

Illingsworth, B. D., 668 (35, 36, 37, 43), 678 (35, 36), 379 (35, 36, 37, 43), 682 (35, 36), *690*

Illuminati, G., 164 (21), 172 (21, 81), *176, 178*

Iltgen, A., 694 (19), 699 (19), 700 (19), 701 (19), 704 (19), 705 (19), *710*

Imada, K., 744 (21), 753 (21), 754 (21), *759*

Imaizumi, S., 105 (194), *111*

Imoto, E., 13 (64), *16*, 443 (1), 444 (2, 3), 452 (1, 2, 3), *454*

Inaba, S., 180 (10), 188 (10), *191*

Inagaki, K., *232* (144)

Inagaki, T., 87 (47), 103 (47), *107*, 740 (35), *742*

Ing, H. R., 141 (11), 153 (11), *159*

Ingeborg, I., 22 (53), *27*

Ingram, R. K., 224 (99), *231*

Inoue, N., 113 (7), 116 (7), 119 (7), *121*

Inoue, S., 238 (23), *239*, 362 (5, 59), 363 (9), 364 (9, 11, 12, 13), 365 (11, 18), 366 (11, 19, 20, 21, 22), 367 (21), 370 (5), 371 (13), 383 (5, 11, 18, 22), 384 (5, 18, 21, 22), *395, 396, 397*

Inoue, Y., 546 (12), *567*

Institoris, E., 470 (17), 482 (17), 483 (17), 490 (17), 491 (17), *493*

Ionescu, M., 81 (16), 97 (16), 98 (16), *106*, 114 (13), 119 (13), *121*, 166 (31), *176*, 556 (32), *567*

Iqbal, M., 33 (21), 51 (88), 54 (95), 60 (21), *73, 75*, 97 (110), 101 (148), 105 (148), *109, 110*, 138 (2), 152 (2), 153 (75, 77), 154 (2, 77), 155 (75, 77), 156 (77), *158, 160*

Irvine, G. M., 600 (72), *652*

Irving, H. M., 22 (50), *27*

Isenberg, M. M., 247 (3), 255 (3), 257 (3), *259*

Ishiguro, I., 105 (194), *111*

Ishikawa, M., 45 (71), *74*

Ismail, I. M., 165 (23), 169 (23), 170 (23), *176*

Isobe, M., 250 (15), *259*, 362 (59), 364 (12), *396, 397*

Isogai, Y., 411 (5), 414 (5), 426 (5), *429*

Isoyama, S., 715 (16), 724 (16), *728*

Israel, M., 498 (21), 502 (21), 507 (45), 514 (21), 522 (45), 524 (21), 530 (21), 531 (21, 45), 532 (21, 45), 533 (21, 45), 534 (45), 535 (21), 536 (21), 540 (21), *541, 542*, 548 (23), 551 (23), 561 (23), 565 (23), *567*

Issidorides, C. H., 35 (31, 33, 34), 38 (40), 39 (47), 40 (31, 48), 48 (47, 81), 61 (40), 62 (47, 81), 65 (48, 139, 145), 67 (139, 145), 68 (34, 145), 69 (34), 70 (33, 34, 48, 139, 145), 71 (48), 72 (139), *73, 74, 75, 76, 77*, 102 (160), *110*, 748 (8), 755 (8), *758*

Ito, H., 56 (103), *75*

Ivanovskaya, S. V., 658 (12), *689*

Iveronova, V. I., 304 (30), *343*

Iwanami, Y., 80 (10), 83 (10), 87 (47), 103 (47), *106, 107*, 715 (16), 724 (16), *728*, 732 (17), 739 (17), 740 (35), *742*

Iyer, R. N., 237 (11), *239*, 571 (7, 8, 10), 573 (7, 8), *574*

Jackman, D., 19 (13), *26*, 267 (33), 279 (33), 280 (33), *288*

Jackson, W. G., 97 (105), 105 (105), *109*, 307 (37), 319 (37), 323 (27), *343*

Jacobi, R., 124 (5), 127 (5), *128*, 165 (23), 169 (23), 170 (23), *176*, 221 (67), *230*, 245 (18), *246*, 280 (93), *290*

Jacobs, R. L., 352 (11), *359*

Jacquier, R., 361 (2), 365 (2), 383 (2), *395*

Jaenicke, L., 545 (36), *567*
James, D. S., 125 (15), 127 (15), *129*
James, W. J., 693 (16), 695 (16), 700 (16), *710*
Jankowski, W. C., 10 (27), *15*
Jann, K., 214 (38), *230*
Jantosovic, J., 184 (34), *192*
Januszewski, H., 10 (23), *15*
Jaquenoud, P. A., 315 (64), 335 (64), *344*
Jarrar, A. A., 40 (48), 49 (86), 65 (48), 70 (48), 71 (48), *74, 75*
Jeanmart, C., 352 (8), 355 (8), 358 (8), *359*
Jefferson, A. M., 238 (37), *239*
Jefford, C. W., 34 (30), *73*
Jenkins, P. W., 668 (39), 671 (49), *690*
Jensen, K. A., 23 (56), 25 (56), *27*
Jensen, R. E., 732 (10), 740 (10), *741*
Jezo, I., 10 (31, 35), *15*, 86 (41), *107*, 127 (26), *129*, 224 (88), 226 (88), *231*, 235 (5), *239*, 473 (34), *493*
Johnson, A. F., 278 (99), *290*
Johnson, A. W., 475 (40), 489 (40), 490 (40), *494*
Johnson, F. H., 363 (6), 384 (6), *395*
Johnson, F. H., 363 (7), *396*
Johnson, G. S., 693 (10), *710*
Johnson, J. L., 307 (37), 319 (37), 323 (37), *343*
Johnson, L. F., 10 (27), *15*
Johnston, J. D., 35 (34), 56 (102), 65 (136), 68 (34), 69 (34), 70 (34), *73, 75, 76*, 748 (8), 755 (8), *758*
Johnston, T. P., 423 (17), *429*
Jones, D. S., 368 (26), 370 (27), 384 (26, 27), *396*
Jones, E. R. H., 225 (113), 226 (113), *232*
Jones, G., 37 (38), *73*
Jones, J. H., 378 (47, 48, 49), 380 (47, 48, 49), 389 (47, 48, 49), 390 (47, 49), 391 (47, 49), 392 (47, 49), 393 (47, 49), *397*
Jones, L. C., 507 (45), 522 (45), 531 (45), 532 (45), 533 (45), 534 (45), *542*
Jones, R. G., 7 (1), *15*, 205 (2), 213 (2), 216 (42), 221 (2), *229, 230*
Jonsson, A., 222 (73), *231*, 267 (23), 283 (23), 284 (23), 286 (23), *288*
Joullie, M. M., 18 (7), 21 (28, 30, 31), 22 (30), *25, 26*, 145 (25), *159, 192* (53), 224 (101), 226 (101), *231*, 237 (10), *239*, 733 (22), 737 (22), *742*
Jucker, E., 373 (33), 387 (33), *396*, 435 (16), 438 (16), *440*
Jung, G., 547 (16), 564 (16), *567*
Jung, H., 168 (45), 174 (45), *176*
Junghanns, E., 95 (87), 97 (87), 98 (87), 99 (87), 104 (87), 105 (87), *108*, 171 (70), *177*, 202 (17), *204*, 238 (32), *239*
Jutisz, M., 93 (76), *108*, 604 (15), 622 (15), 635 (15), *651*

Kaganskii, M. M., 8 (5), *15*, 41 (49), *74*, 200 (9), *204*
Kaiser, E. M., *232* (145)
Kajiura, M., 715 (16), 724 (16), *728*
Kakoi, H., 366 (12), 383 (12), 384 (12), *396*
Kalinowska-Torz, J., 114 (14), *121*
Kallmayer, H. J., 724 (52), *729*
Kamiya, H., 307 (38), 324 (38, 97), 335 (97), *343, 345*
Kaneko, C., 42 (56), 45 (71), *74*
Kaneko, S., 45 (71), *74*
Kanematsu, K., 296 (16), 330 (16), *342*, 423 (22), 424 (22), 425 (22), 429 (22), *430*
Kano, H., 41 (52), *74*
Kappa, Th., 87 (46), *107*
Kappe, T., 275 (84), 285 (84), *290*
Kappe, Th., 210 (23), *229*
Karayannis, N. M., 41 (50), *74*
Karjalainen, A., 10 (29), *15*
Karle, I. L., 304 (133), *346*
Karlson, P., 324 (99), *345*
Karpetsky, T. P., 365 (15), *396*
Kartseva, T. V., 103 (176), *111*
Kashima, E., 196 (14), *198*
Kaspers, H., 186 (50), *192*
Kasuga, K., 399 (19), 406 (19), *409*
Katagiri, K., 148 (32), *159*
Kato, H., 301 (155), 327 (155), 329 (155), 331 (155), *347*
Kato, S., 317 (75), 339 (75), *344*
Katritzky, A. R., 8 (15), 9 (25), *15*, 43 (62), *74*, 86 (38), 95 (38), *107*
Katz, I., 23 (57), 24 (57), *27*
Kaul, L. L., 97 (104), *109*

Kavalek, J., 97 (127), 98 (127), *109*, 210 (24, 25, 26, 27), 224 (25, 90, 142), 227 (27), *229, 231, 232*
Kavarnos, S., 118 (33), *122*
Kaverina, N. V., 303 (129), 331 (129), 332 (129), *346*
Kawai, S., 25 (73, 75), *27*, 105 (195), *111*, 149 (39), 158 (39), *159*, 170 (58), 175 (58), *177*, 224 (102), 226 (102), *231*, 267 (44), 286 (44), *289*, 740 (44), *742*
Kawano, S., 377 (43, 44), 380 (43), 395 (43, 44), *397*, 593 (7, 10), 594 (7, 10), 595 (7, 10), *597*
Kaye, I. A., 498 (19), 515 (19), 525 (19), 527 (19), 533 (19), 540 (19), *541*
Kazitsyna, L. A., 431 (1), *439*
Keller-Schierlien, W., 153 (60), *160*, 323 (93), *345*
Kelly, M. G., 254 (28), *260*
Kenjo, Y., 715 (16), 724 (16), *728*
Kenner, G. W., 368 (26), 370 (27), 384 (26, 27), *396*
Keresztesy, J. C., 163 (18), 172 (18), 173 (18), *176*
Kerr, K. B., 184 (39), *192*
Kesler, E., 715 (14), 726 (14), *728*
Keuhle, E., 171 (61), 175 (61), *177*
Khoi, N. H., 237 (18), *239*, 726 (40), *728*
Kholodova, N. V., 744 (22), 754 (22), *759*
Khosla, M. C., 325 (152), 336 (152), *347*
Kidani, Y., *232* (144)
Kiel, W., 252 (26), *260*
Kieland, W., 610 (32), 638 (32), *651*
Kim, K. H., 67 (141), *76*
Kimura, T., 12 (44), *16*
King, F. E., 22 (47), 23 (47), 24 (70), *26*, *27*, 141 (10, 12), 153 (10), 154 (10), 156 (10), *159*, 659 (17), 660 (17, 18), 675 (17), 676 (15), *689*
King, K. F., 376 (41), 377 (41), *397*
King, T. F., 9 (20), *15*
Kinlin, T., 23 (60), 24 (60), *27*
Kinlin, T. E., 456 (2), *461*
Kinne, G., 547 (17), *567*
Kiprianov, A. I., 254 (32), *260*
Kirchner, G., 579 (15), 580 (15), 581 (15), *582*
Kirk, K. L., 600 (13), 604 (13), 622 (13), 634 (13), *651*
Kirsamov, A. V., 498 (18), 502 (18), 540 (18), *541*
Kirsch, W., 267 (42), *289*
Kiselev, M. I., 324 (116), 336 (116), *346*
Kishi, Y., 238 (23), *239*, 362 (59), 363 (6, 7), 366 (19, 20), 384 (6), *395, 396*, *397*
Kishimoto, H., 238 (23), *239*, 366 (19), *396*
Kissel, H. J., 732 (11), 739 (11), *741*
Klaudy, A., 21 (32), *26*
Klauke, E., 171 (61), 175 (61), *177*, 656 (10), 658 (10), *689*
Klebe, J. F., 702 (51), *711*
Klein, O., 154 (79), 156 (79), *161*
Klemenen, A., 56 (101), *75*
Klicnar, J., 98 (120), *109*, 102 (168), 103 (168), *110*, 127 (21, 24), *129*, 153 (56, 57), 157 (103), *160, 161*, 162 (4), 173 (4), *175*, 221 (70, 71), 222 (71, 76), 224 (76, 94), *230, 231*, 281 (94), *290*, 624 (53), 643 (53), *652*, 697 (41), 705 (41), *711*
Klutchko, S., 349 (4, 5), 350 (4, 5), 351 (5), 352 (4, 5), 356 (4, 5), 357 (4, 5), *359*
Klyuchko, G. V., 402 (9), 407 (9), 408 (9), *408*
Knobloch, W., 268 (57), 279 (57), *289*, 556 (31), 564 (31), *567*, 672 (52), *690*
Knoeppel, H., 579 (18), *582*
Knop, J. V., 713 (42, 43), *728*
Koaze, Y., 323 (87, 88, 89, 94), 324 (95), *345*
Kobayashi, H., 740 (32), *742*
Kobayashi, Y., 44 (65), 57 (65), 64 (65), *74* 293 (125), 296 (125), 328 (125), *346*
Kobayshi, T., 693 (13), 695 (13), *710*
Koblova, I. A., 184 (32), *192*
Kodaira, Y., 323 (91), 324 (91, 100), *345*
Kodak, Belg. Pat. 660, 253, 665 (30), 668 (30), 669 (30), 670 (30), 677 (30), 678 (30), 679 (30), 681 (30), 685 (30), *689*
Kohhler, M., *359* (17)
Koehnke, K., 701 (48), 707 (48), 708 (48), 709 (48), *711*
Koenigs, E., 547 (16, 17), 564 (16), *567*

Kohl, H., 135 (21), *136*
Kohlstock, B., 693 (5, 6), 699 (5, 6), 700 (45), 703 (5), *710, 711*
Köhnke, K., 125 (12), *128*
Koike, H., *232* (144)
Koike, S., 96 (100), 104 (100), 105 (100), *108*, 171 (63, 64), *177*
Kokoshko, Yu. Z., 42 (60), *74*
Koksharova, T. G., 152 (51), 158 (51), *160*
Kolesnikov, T., 732 (20), 737 (20), 738 (20), *742*
Kolesnikov, V. T., 736 (37), *742*
Kolinsky, J., 221 (60), 227 (60), *230*
Kollenz, G., 645 (77), 646 (77), *653*, 659 (16), 676 (16), *689*
Kolodynska, Z., 188 (69), 190 (69), *193*
Komescu, N., 56 (105), *75*
Komin, A. P., 96 (96), *108*, 165 (25), 168 (25), 174 (25), *176*, 184 (42), *192,* 450 (18), 454 (18), *455*
Komm, E., 324 (111), *346*
Konig, H., 269 (61), 279 (61), 282 (61), *289*
Konyukhov, V. N., 152 (51), 158 (51), *160*
Koppel, H. C., 148 (35), 153 (35), *159*
Korbonits, D., 183 (31), *192*
Korbukh, I. A., 400 (20), 407 (20), *409*
Kornfeld, E. C., 205 (2), 213 (2), 221 (2), *229*, 327 (154), *347*
Kornilov, M. Y., 358 (18), *359*
Korobitsyna, I. K., 431 (1), 432 (8), 437 (8), *439, 440*
Korte, F., 498 (29), 504 (29), 505 (38), *542*
Kosek, F., 127 (21), *129*, 153 (56), 157 (103), *160, 161*, 221 (70, 71), 222 (71), 224 (94), *230, 231*, 281 (94), *290*
Koshel, N. G., 83 (23), 99 (23, 133), 100 (23, 133), *106, 109*, 114 (15), 119 (15), 120 (15), *121*, 165 (26), 172 (26, 178), 173 (26), *176, 178*, 196 (10), 197 (10, 17), *198*
Kost, A. N., 402 (7), *408*
Kosuge, T., 307 (38), 324 (38, 97), 335 (97), *343, 345*
Kotz, P. J., 221 (61), *230*
Kovacik, V., 10 (31), *15*, 86 (41), *107, 127*, (26), *129*, 224 (88), 226 (88), *231*, 235 (5), *239*
Kovacs, J., 47 (75), 63 (75), *74*
Kovacs, O., 238 (22), *239*
Kovalev, E. G., 165 (26), 172 (26), 173 (26), *176*, 196 (10), 197 (10), *198*
Kovalev, G. V., 658 (12), *689*
Kozarenko, T. D., 304 (30), 310 (52), 315 (65), 319 (65), 334 (52, 65), *343, 344*
Kozhokina, L. F., 84 (29), 99 (29), 101 (29), *106*, 133 (10), 134 (14), *136*, 155 (90), *161*, 222 (78), *231*, 282 (101), *290*
Kradolfer, F., 323 (93), *345*
Krahn, R. C., 241 (2), 242 (2), 243 (2, 8), 244 (2, 8), 245 (2), *246*, 693 (11), 705 (11), *710*
Krahnert, L., 470 (16), 491 (16), *493*
Krajewski, J. J., 221 (61), *230*
Kram, T. C., 184 (37), *192*
Kranz, J., 184 (48), 190 (48), *192*
Kress, T. J., 8 (13), *15*
Krieger, H., 10 (29), *15*
Krimmel, C. P., 469 (15), 470 (15, 19), 482 (19), 488 (15, 19), 489 (15), 490 (15), *493*
Krönhke, F., 182 (24, 25), 183 (24), 188 (24, 25), 190 (25), *192*, 252 (26), *260*, 466 (7), 467 (7), 468 (7), 477 (7), 489 (7), 491 (7), 492 (7), *493*, 591, (1), *596*, 610 (32), 638 (32), *651*, 717 (23), 726 (23), *728*
Krönke, F., 84 (27), 101 (27, 153), *106*, *110*, 156 (97), *161*
Kruyff, J. J., 695 (22), 701 (22), 706 (22), 709 (22), *710*
Kubo, A., 42 (56), *74*
Kubota, K., 594 (12), *597*
Kubota, T., 41 (51), *74*
Kubota, Y., 323 (86), 324 (86), *345*
Kuehne, H., 556 (31), 564 (31), *567*
Kuhn, R., 127 (25), *129*, 245 (14), *246*, 732 (14), 739 (14), *742*
Kuhn-Kuhnenfelf, J., 474 (37), 490 (37), 491 (37), *494*
Kuk, S., 324 (113), *346*
Kul'chitskaya, N. E., 34 (29), *73*, 80 (14), *106*, 149 (45), 157 (45), 158 (45), *160*, 626 (62), 647 (62), *652*, 724 (50), *729*

Küllmar, K., 469 (12), 482 (12), 483 (12), 484 (12), *493*
Kumada, S., 196 (15), *198*
Kumadaki, I., 44 (65), 57 (65), 64 (65), *74*
Kumar, G., 21 (29), *26*, 733 (42), 738 (42), 740 (42), 741 (42), *742*
Kumashiro, I., 86 (43), *107*, 600 (7), 605 (20), 614 (38), 619 (20), 637 (38), 639 (20), 640 (20), 642 (20), *651*
Kummer, F., 730 (1), *741*
Kunstamm, M. P., 324 (98), *345*
Kushner, S., 128 (29), *129*, 172 (84), *178*, 352 (10), 357 (10), *359*
Kutsuma, T., 293 (125), 296 (125), 328 (125), *346*
Kwon, S., 235 (43), *240*
Kwong, S. F., 153 (78), *161*
Kyryacos, G., 100 (139), *110*

Lablache-Combier, A., 13 (63), *16*
Lacroix, A., 270 (65), 281 (65), 283 (65), *289*
Lagowski, J. M., 43 (62), *74*
Laird, A. H., 23 (59), 24 (59), *27*
LaLancette, E. A., 133 (9), 136 (9), *136*, 156 (98), *161*, 237 (20), *239*
Lalloz, L., 269 (62), 281 (62), 282 (62), *289*
Lambelin, G., 244 (12), *246*, 693 (7), 701 (7, 47), 703 (7, 47), 704 (7, 47), 705 (7, 47), 706 (47), 709 (47), *710, 711*, 725 (49), *729*
Lambie, A. J., 97 (106), 98 (106), 100 (106), 104 (106), *109*, 171 (68), 174 (68), *177*, 202 (14), *204*, 221 (65), 222 (65), 223 (65), 226 (65), *230*, 237 (17), 238 (17), *239*
Lampke, H., 470 (18), 471 (18), 478 (18), 482 (18), *493*
Landquist, J. K., 6 (3), *6*, 12 (42), *16*, 19 (10), 22 (10), 23 (10), 24 (10), *26*, 29 (3), 30 (13), 31 (13, 16), 32 (17), 45 (3), 46 (17), 48 (16), 55 (98), 56 (3, 17), 57 (3), 58 (3, 13, 16, 17), 59 (3, 13, 17), 60 (13, 16), 61 (13), 62 (13), 63 (3, 13, 16), 64 (3, 16), 65 (3, 13, 16), 66 (3, 13, 16, 98), 67 (13, 16), 68 (13, 16), 69 (13, 16), 70 (13, 16), 71 (13), 72 (13, 16), *72, 73, 75*, 95 (93), 96 (93), 99 (93, 147), 104 (93), 105 (93), *108, 110*, 120 (41), *122*, 123 (2), 125 (2), 127 (2), 128 (28), *128, 129*, 154 (80), *161*, 163 (6), 168 (6), 172 (6), 174 (6), *175*, 206 (4, 5), 209 (4), 214 (4), 215 (5), 221 (4), 222 (4, 5), 224 (4), 225 (4, 5), 226 (4, 5), 227 (5), 228 (5), *229*, 248 (8), 254 (30), 255 (8, 48), *259, 260*, 267 (35), 284 (35), 285 (35), *288*, 626 (65), 647 (65), *652*, 720 (27), 723 (27), 725 (48), 726 (27), *728, 729*, 732 (7), 739 (7), *741*
Lane, D. W. J., 21 (34), *26*, 97 (106), 98 (106), 100 (106), 104 (106), *109*, 171 (68), 174 (68), *177*, 202 (14), *204*, 221 (65), 222 (65), 223 (65), 226 (65), *230*, 237 (17), 238 (17), *239*
Lane, E. S., 21 (36), 22 (36), 23 (36), 25 (36, 71), *27*, 96 (101), 105 (193), *108, 111*, 174 (80), 175 (80), *178*, 268 (47), 282 (47), 286 (47), *289*
Lang, M., 719 (47), 724 (47), *729*
Langenbeck, U., 89 (52), 93 (52), *107*
Lanning, M. S., 93 (71), *108*
Lanyi, K., 470 (17), 482 (17), 483 (17), 490 (17), 491 (17), *493*
Lappin, G. R., 498 (23), 540 (23), *541*
Larini, G., 448 (12), 453 (12), *454*
Laseter, A. G., 548 (24), 551 (24), 557 (24), 559 (24), 560 (24), 561 (24), 562 (24), 563 (24), *567*
Lavrenova, G. I., 308 (46), *343*
Lavrenyuk, T. Ya., 63 (152), 64 (152), *77*, 150 (47), *160*
Lawrence, J. P., 216 (45, 46), 221 (45, 46), *230*
Lawrowsky, K., 324 (114), *346*
Le Bris, M. -T., 84 (28), *106*, 224 (147), *232*, 247 (6), 248 (7), 249 (46, 47), 250 (6), 251 (18, 19, 20), 255 (6, 18, 46, 47), 256 (6, 46, 47), 257 (18), 258 (18), *259*, 266 (18), 277 (18), *288*
Ledig, K. W., 697 (38), 703 (38), *711*
LeDouarec, J. C., 476 (42), 480 (42), 485 (42), 486 (42), *494*
Lee, D. Y., 372 (60, 64), 373 (64), 374 (64), 375 (65, 66), 385 (60, 64, 65), 386 (60,

64, 65), 387 (65), 388 (65), *397*
Leese, C. L., 98 (126), 99 (126), *109*, 123 (3), 127 (3), *128*, 205 (3), 206 (3), 221 (3), 224 (3), *229*, 496 (7), 498 (7), 502 (7), 504 (7), 508 (7), 515 (7), 516 (7), 518 (7), 525 (7), 526 (7), 532 (7), 533 (7), 534 (7), 540 (7), *541*
Lehmann, D., 506 (40), 523 (40), 539 (40), *542*, 701 (31), 705 (31), 707 (31), 708 (31), 714 (13), *711, 728*
Lehmann, F. E., 97 (81), *108*
Leigh, H. M., 324 (101), *345*
Leister, H., 84 (27), 101 (27), *106*, 182 (24), 183 (24), 188 (24), *192*
Leiter, J., 254 (27, 28), *260*
Lemke, W., 125 (16), *129*, 695 (28), 705 (28), *711*, 713 (10), 714 (10), 721 (10, 31), 722 (31), 724 (10), 725 (10), 727 (10), *728*, 732 (15), 735 (15, 25), 739 (15), 740 (15), *742*
Lempert, K., 249 (13), 154 (13), *259*
Leonard, N. J., 608 (28), 634 (28), 636 (28), *651*
Leone, A., 448 (12), 453 (12), *454*
Leresche, J. P., 80 (13), *106*
Lesiak, J., *232* (143), 626 (79), *653*
Lesiak, J. Z., 626 (66), *652*
Levene, P. A., 324 (109), *345*
Levey, M. R. W., 224 (101), 226 (101), *231*
Levi, A. A., 55 (98), 66 (98), *75*, 128 (28), *129*
Levi, E. M., 235 (6), *239*, 610 (31), 642 (31), *651*
Levshina, K. V., 131 (6), 134 (6), *136*
Levy, M. R. W., 21 (26), 23 (26), *26*
Lewis, I. M., 141 (11), 153 (11), *159*
Lewis, T. P., 10 (33), *15*, 497 (14), *541*
Ley, K., 35 (32), 36 (32), 49 (82), 63 (126, 127), 64 (82, 127, 135, 137, 138), 65 (32, 127, 138, 142, 143, 144), 66 (137), 70 (82, 127), *73, 75, 76, 77,* 150 (46), 158 (46), *160*, 626 (78), 647 (78), 648 (78), *653*
Lezenko, G. A., 254 (32), *260*
Lica, S., 252 (22), 255 (22), *260*, 720 (30), *728*
Liebenow, W., 245 (17), *246*
Liebig, R., 694 (21), 695 (21), 701 (21), 706 (21), *710*
Liepins, E., 237 (39), *239*
Lietz, G., 268 (57), 279 (57), *289*, 672 (52), *690*
Likhosherstov, A. M., 299 (21, 23), 303 (21, 128, 129), 328 (21, 128), 331 (129), 332 (128, 129), *343, 346*, 473 (30), 480 (30), *493*, 605 (17), 634 (17), *651*
Limbach, H. H., 730 (2), *741*
Lin, A. J., 125 (9), 127 (9), 128 (9), *128*, 267 (27), 273 (27), 280 (27), 281 (27), 286 (27), *288*, 373 (35), 375 (35), 388 (35), *396*, 661 (20), 662 (20), 676 (20), *689*
Linda, P., 89 (56), 97 (56), *107*, 164 (21, 22), 171 (77), 172 (21, 81), *176, 177, 178*
Linder, H. J., 228 (137), *232*
Lingappa, T. B., 324 (102), *345*
Lingappa, Y., 324 (102), *345*
Lions, F., 237 (9), *239*, 753 (18), 758 (18), *759*
Lippmann, E., 125 (11), *128*, 547 (19), 557 (19), 564 (19), 566 (19), *567*
Lisovskaya, N. N., 45 (70), 62 (70), *74*, 248 (9), 258 (9), *259*
L'Italien, Y. J., 99 (129), 100 (129), *109*, 172 (86), *178*, 180 (12), 189 (12), *191*
Litkei, G., 228 (139), *232*
Litzerman, R. A., 668 (38), 680 (38), 681 (38), 682 (38), *690*
Livingstone, D. B., 53 (92), *75*, 249 (14), 252 (14), 257 (14), 258 (14), *259*
Llamas, A., 55 (99), *75*
Loadman, M. J. R., 464 (2), 466 (2), 470 (2), 471 (2), 476 (2), 477 (2), 478 (2), 479 (2), 480 (2), 482 (2), 487 (2), 488 (2), 489 (2), *493*
Lockhart, D., 169 (53), *177*
Logemann, W., 102 (161), *110*
Lont, P. J., 163 (9), 165 (24), 172 (9), 173 (9), *175, 176*, 187 (64), 188 (64), *193*
Lorenzini, A., 93 (74), *108*
Lott, W. A., 498 (28), 525 (28), *542*
Lourie, A. D., 469 (13, 14), 471 (13, 14), 476 (13, 14), 480 (13), 481 (13, 14), 482 (13, 14), 483 (13, 14), 484 (13, 14),

485 (13, 14), 486 (13, 14), *493*
Lovell, C. H., 328 (130), 331 (130), 332 (130), *346*
Lown, J. W., 13 (68, 69), *16*, 220 (57), *230*, 293 (6), 297 (6, 17), 301 (6), 330 (6, 17), 331 (6), 333 (17), *342*, 599 (5), 612 (5, 35), 615 (5), 642 (5, 35), *651*, 717 (24), 726 (24), 727 (24), *728*
Lucente, G., 313 (59), 314 (142), 318 (142), 319 (59), 336 (59, 142), 340 (142), *344*, *347*
Lucken, E. A. C., 308 (44), *343*
Ludenscher, U., 305 (32), *343*
Luedemann, G., 323 (92), 335 (92), *345*
Luini, F., 376 (42), 394 (42), *397*
Lu Kashina, L. I., 80 (11), 99 (11), *106*
Luk'yanets, E. A., 149 (41), 153 (41), 157 (41), *160*, 739 (34, 43), *742*
Lumbroso, H., 8 (8), *15*
Lunazzi, L., 578 (10), *582*
Lunkenheimer, W., 186 (50), *192*, 665 (58), 671 (58), *690*
Lüttringhaus, A., 228 (140), *232*
Luzak, I., 473 (34), *493*
Lynch, B. M., 12 (52), *16*

McBride, J. A. H., 22 (39), *26*, 95 (89), 96 (89), 104 (89), *108*, 168 (48), 170 (48), 172 (48), 174 (48), *177*, 203 (19), *204*
McCapra, F., 364 (10), 365 (10), 384 (10), *396*
McFarland, J. W., 35 (34, 35), 39 (153), 66 (35), 68 (34, 35), 69 (34, 35), 70 (34), *73*, *77*
McKillop, A., 92 (65), 98 (65), 100 (65), 101 (65), 102 (65), *107*, 125 (13), *128*, 235 (7), *239*, 608 (27), 609 (30), 634 (27, 30), 635 (27, 30), 636 (27), *651*, 743 (3), 754 (3), *758*
McLaughlin, K. C., 7 (1), *15*, 205 (2), 213 (2), 216 (42), 221 (2), *229*, *230*
McMahon, D. M., 321 (143), *347*
McNair Scott, D. B., 382 (57), *397*
McNamee, R. W., 580 (21), 581 (21), *582*
McNutt, W. S., 143 (17), 155 (17), *159*
Madan, O. P., 138 (1a), 152 (1a), 153 (1a), 155 (1a), *158*, 172 (79), 173 (79), *178*
Mader, P. M., 104 (85), *111*, 263 (14), 278 (14), *288*
Maeda, K., 324 (115), 337 (115), *346*
Maffei, S., 11 (41), *16*, 267 (37), 276 (37), 280 (37), 282 (37), 286 (37), 287 (37), *288*, 354 (15), 357 (15), *359*, 716 (18), 720 (41), 724 (18), 726 (41), *728*
Magee, R. J., 94 (82), *108*, 202 (12), 203 (12), *204*
Mager, H. I. X., 12 (45), *16*, 94 (85), 104 (188), 105 (85, 188), *108*, *111*, 149 (43), *160*, 168 (46), 169 (46, 50), 174 (46), *177*, 201 (11), 203 (11), 204 (11), *204*, 242 (3), 244 (3), 245 (3), *246*
Magidson, O. Yu., 39 (46), 57 (46), 63 (46), *74*, 125 (10), 127 (10), *128*, 145 (21), 152 (21), 153 (21), *159*
Maguire, J., 415 (10), 418 (10), 419 (10), 420 (10), 422 (10), 427 (10), 428 (10), *429*
Maguire, J. A., 418 (12), 419 (12), 420 (12), 422 (12), 427 (12), 428 (12), *429*
Magyar, K., 47 (75), 63 (75), *74*
Mahajanshetti, C. S., 227 (121), 228 (121), *232*
Maibach, H. I., 21 (79), *27*
Maiorana, S., 33 (26), 34 (26), 57 (26), 73
Maitland, P., 224 (95), *231*
Maitlis, P. M., 12 (48), *16*, 18 (2), 22 (2), 24 (2), *25*
Maitte, P., 515 (65), 517 (65), 534 (65), 536 (65), 538 (65), 539 (65), 540 (65), 541 (65), *543*
Majer, O., 507 (43, 44), 531 (43), 533 (43), 535 (43), 536 (43), 537 (43, 44), 538 (44), *542*
Makovetskii, V. P., 358 (18), *359*
Malachowski, R., 457 (4), 460 (4), 461 (4), *462*
Malik, P. A., 623 (47), 644 (47), *652*
Mallett, S. E., 410 (2), 415 (2), 416 (2), 417 (2), 419 (2), 426 (2), 427 (2), 428 (2), *429*
Mamalis, P., 448 (11), 453 (11), *454*
Mandasescu, L., 252 (22, 23), 255 (22, 23), *260*, 720 (28, 29, 30), 724 (28, 29), *728*
Mandel, L. R., 380 (55), *397*
Manecke, G., 226 (111), *231*

Mania, D., 324 (115, 116), 337 (115, 116), *346*
Mann, F. G., 268 (49), 279 (49), 284 (49), *289*
Mann, R. K., 700 (46), *711*
Mansour, A. K. E., 719 (26), 726 (26), *728*
Marascia, F. J., 747 (7), 756 (7), *758*
Marchetti, L., 70 (147), *77*
Mariani, L., 310 (140), 335 (140), 336 (140), 337 (140), 338 (140), 339 (140), 340 (140), *347*
Marino, G., 89 (56), 97 (56), *107*, 171 (77), 172 (81), *177, 178*
Markova, G. A., 303 (129), 331 (129), 332 (129), *346*
Marks, B., 100 (140), *110*
Marrer, A., 13 (67), *16*
Marshall, H., 323 (92), 335 (92), *345*
Martelli, S., 732 (41), 736 (41), 737 (41), 738 (41), *742*
Martin, J. H., 324 (98), *345*
Martin, R. H., 367 (23), 376 (23), 377 (23), 384 (23), 393 (23), 395 (23), *396*
Martin, S. F., 115 (17), *121*, 166 (34, 35), *176*, 222 (72), 223 (72), *231*
Martinez, E., 55 (99), *75*
Martinotti, F., 713 (3), *727*
Maruyama, I., 93 (77), *108*
Marvel, C. S., 355 (16), *359*, 580 (20), *582*
Marwan, J., 64 (134), 65 (134), *76*
Marxer, A., 90 (60), 101 (60), *107*
Masahiro, H., 324 (103), *345*
Masaki, M., 571 (6), 573 (6), *574*
Mason, J. C., 36 (36), 51 (89), 53 (93), 60 (36), 61 (36, 93), 62 (93), 70 (36), *73, 75*, 101 (171), 102 (171), *110*
Mason, M. E., 23 (57), 24 (57), *27*
Mason, S. F., 9 (16), *15*, 376 (38), *397*, 496 (5, 8), *541*, 545 (3, 5), 564 (3), *567*, 577 (7), *582*
Massa, S., 621 (75), 622 (75), 634 (75), 635 (75), 636 (75), 637 (75), 638 (75), 639 (75), 641 (75), *652*
Massarani, E., 102 (169), *110*
Masters, B. R., 267 (41), 277 (41), *289*
Masuzawa, K., 571 (6), 573 (6), *574*
Mata, J. M., 28 (1), 63 (1, 123), *72, 76*
Matous, V., 97 (103), *109*
Matsumoto, K., 13 (68, 69), *16*, 293 (6), 297 (6, 17), 301 (6), 330 (6, 17), 331 (6), 333 (17), *342*, 599 (5), 612 (5, 35), 615 (5), 642 (5, 35), *650, 651*
Matsumoto, M., 86 (36), *107*
Matsumura, Y., 86 (36), *107*
Matsuo, J., 157 (104), *161*
Matsuo, K., 24 (64), *27*, 267 (36), 282 (36), 283 (36), *288*
Mauger, A. B., 307 (35, 139), 325 (119), 335 (139), *343, 346, 347*, 469 (10), 473 (35), *493*
Maury, G., 361 (2), 365 (2), 383 (2), *395*
Mazur, R. H., 314 (60), 335 (60), *344*
Meathrel, W. G., 250 (16), 266 (17), *259, 288*
Medne, A. Ya., 101 (148), 105 (148), *110*
Medvedeva, I., 215 (40), *230*
Medvedeva, M. I., 95 (92), *108*, 174 (100), *178*
Mee, J. D., 625 (56), *652*, 668 (38, 39, 41, 43, 44), 669 (46, 47), 670 (48), 671 (49), 679 (43), 680 (38), 681 (38), 682 (38), *690*
Meetz, H. K., 278 (99), *290*
Mehrotra, V. N., 103 (179), *111*
Mehta, S. R., 99 (132), *109*, 284 (102), *290*
Meienhofer, J., 311 (57), 324 (57), 335 (57), *344*
Meier, R., 254 (29), *260*
Meier, W., 311 (55), 312 (55), 319 (55), 336 (55), 338 (55), *344*
Meister, P. D., 324 (101), *345*
Melkonian, G. A., 694 (18), 697 (40), 700 (18, 40), 705 (18, 40), 709 (18, 40), *710, 711*
Meltzer, R. I., 349 (4, 5), 350 (4, 5), 351 (5), 352 (4, 5), 356 (4, 5), 357 (4, 5), *359*
Melvin, P., 254 (30), *260*
Menon, C. S., 94 (84), *108*
Menziani, E., 436 (19), 439 (19), *440*
Merchant, J. R., 623 (42, 43, 45, 46), 643 (46), 645 (45, 46), 646 (42, 43), *652*
Merlini, L., 87 (45), *107*, 229 (20), 731 (3), 739 (3), 740 (3), *741*
Merrifield, R. B., 325 (151), *347*
Mertes, M. P., 125 (9), 127 (9), 128 (9),

128, 267 (27), 273 (27), 280 (27), 281 (27), 286 (27), *288*, 373 (34, 35), 375 (34, 35), 387 (34), 388 (34, 35), *396*, 661 (20), 662 (20), 676 (20), *689*
Messer, M. N., 352 (8), 355 (8), 358 (8), *359*
Mesta, C. K., 104 (187), *111*, 169 (54), 173 (54), *177*
Meth-Cohn, O., 600 (73), 604 (16), 613 (73), 636 (73), 641 (16), *651, 652*
Metzger, K. G., 64 (131, 135, 137, 138), 65 (138, 142, 143), 66 (138), *76, 77*
Meyerson, S., 433 (9), *440*
Michalsky, J., 103 (182), *111*, 207 (12, 13), 222 (12), *229*
Micheel, F., 101 (157a), *110*
Middleton, E., Jr., 423 (18), *429*
Miescher, K., 447 (7, 8, 9), 448 (7), 453 (7, 8, 9), *454*
Miesel, J. L., 275 (86), *290*
Mika, N., 713 (1), *727*
Mikami, T., 56 (103), *75*
Mikhlina, E. E., 476 (41), 483 (41), *494*
Miki, Y., 235 (40, 41), *239, 240*
Milch, G., 86 (44), *107*
Miles, P., 227 (126), *232*
Millefiore, S., 10 (37), *15*, 713 (4), *727*
Miller, A. K., 63 (124), *76*
Miller, B., 274 (79), *289*
Miller, M. J., 242 (6), 243 (6), 245 (6), *246*
Miller, T. W., 28 (1), 63 (1, 123, 124), *72, 76*
Minamika, J., 235 (40), *239*
Minisci, F., 8 (72), 12 (54, 55, 56, 58, 60, 61), 13 (58, 61), *16*, 18 (4), 25 (7b), *25, 27*, 132 (18), 134 (18), 135 (18), *136*, 209 (17, 18), 223 (18), *229*
Minoli, G., 435 (17), 439 (17), *440*
Miragoli, G., 294 (12), 301 (12), 302 (12), 303 (12), 332 (12), *342*, 480 (43), 486 (43), *494*
Mirza, J., 277 (9), *288*
Misak, A., 180 (10), 188 (10), *191*
Misner, R. E., 62 (119), *76*, 131 (7), 134 (7), 136 (7), *136*, 214 (36), 228 (36), *230*
Mitscher, L. A., 324 (98), *345*
Miura, Y., 44 (67), 59 (114), 60 (114), 62 (114), 65 (114), 68 (114), *74, 76*, 172 (89), 173 (89), *178*, 222 (79), 223 (79), *231*
Miyagi, T., 171 (62), 174 (62), *177*
Miyagashima, T., 119 (38), *122*, 153 (66), 154 (66), 155 (66), *160*
Miyake, A., 196 (13), *198*
Miyazaki, H., 41 (51), *74*
Mizuno, H., 511 (52), 525 (52), *542*
Mizuno, Y., 207 (15), 228 (15), *229*
Mladineo, T. H., 274 (79), *289*
Modest, E. J., 507 (45), 522 (45), 531 (45), 532 (45), 533 (45), 534 (45), *542*
Moehring, H. U., 89 (52), 93 (52), *107*
Mohammady, A., 52 (91), *75*, 102 (163), 103 (163), *110*
Mohr, J., 402 (8), 408 (8), *408*
Moll, H., 133 (8), 135 (8, 17), *136*, 140 (7, 7a), 153 (7), 156 (7, 7a), 157 (7, 7a), *159*
Moller, J., 249 (13), 254 (13), *259*
Molnar, I., 672 (50, 51), 674 (50), 687 (50), 688 (50, 51), *690*
Mondelli, R., 87 (45), *107, 229* (20), 731 (3), 739 (3), 740 (3), *741*
Monge, A., 55 (100), *75*
Montgomery, J. A., 498 (26, 27, 30), 499 (30), 502 (26, 27, 32, 33, 34), 509 (27, 30, 32, 48, 49), 515 (48, 49), 524 (49), 525 (30, 32), 526 (27, 48, 49), 527 (26, 27, 30, 49), 528 (26, 27, 32, 34), 529 (26, 27), 530 (30), 531 (49), *542*, 545 (7, 8), 548 (24), 551 (7, 8, 24), 554 (8), 555 (7, 29, 30), 557 (8, 24), 558 (8, 29, 30), 559 (24), 560 (7, 24), 561 (7, 8, 24, 29, 30), 562 (7, 24, 30), 563 (7, 24, 30), *567*
Moody, K., 103 (181), *111*, 659 (15), 661 (15), 676 (15), *689*
Mookherjee, B. D., 23 (57), 24 (57), *27*
Moore, C., 166 (32), *176*, 207 (16), *229*
Moore, J. A., 747 (7), 756 (7), *758*
Moore, L. O., 223 (82), *231*
Moorhouse, A., 268 (58), 279 (58), 280 (58), 281 (58), *289*
Moreno, H., 97 (107), *109*, 221 (68), *230*

Moreno, H. R., 21 (35), 24 (35), 25 (35), *26*, 134 (12, 13), *136*, 145 (22, 23), 146 (22), 152 (23, 55), 153 (22), *159, 160*, 181 (18), 187 (18), *191*, 244 (9), 245 (13), *246*
Morgan, C. R. P., 659 (17), 660 (17), 675 (17), *689*
Mori, Y., 97 (111), *109*
Moriconi, E. J., 62 (119), *76*, 124 (6), 125 (8), 127 (6), 128 (6), *128*, 131 (7), 134 (7), 136 (7), *136*, 214 (36), 215 (39), 223 (39), 224 (39), 225 (39), 226 (39), 228 (36), *230*
Morikawa, T., 187 (63), 188 (63), *193*
Morinaga, K., 293 (125), 296 (125), 328 (125), *346*
Morita, T., 8 (11), *15*
Morley, J. S., 255 (39), 256 (39), *260*, 270 (71), 273 (71), 274 (71), 279 (71), 280 (71), *289*
Morris, H. R., 148 (32a), *159*
Morrison, D. C., 79 (6), 100 (6), 101 (6), *106*, 112 (2), 116 (2), 120 (2), *121*, 169 (55), 173 (55), *177*
Morrow, D. F., 85 (33), 102 (33, 180), 103 (180), *106, 111*
Mosby, W. L., 291 (1), 335 (1), *342*, 658 (13), *689*, 743 (1), *758*
Mosher, H. S., 266 (20), 270 (66), 271 (66), 282 (20, 66), 286 (66), *288, 289*, 521 (55, 56), 533 (56), 534 (55), *542*, 565 (33), *567*
Motoyama, R., 443 (1), 452 (1), *454*
Moureu, H., 227 (127), 228 (128), *232*
Muehlmann, F. L., 376 (40), 377 (40), 394 (40), *397*
Mueller, H., 225 (130), *232*, 751 (14, 15), 757 (14, 15), *759*
Mufarrij, N. A., 35 (34), 68 (34), 69 (34), 70 (34), *73*
Mühlstadt, M., 125 (11), *128*, 399 (1), 406 (1), *408*
Mukaiyama, T., 139 (4), 156 (4), *159*
Mulik, H., 56 (103), *75*
Müller, E., 101 (170), 102 (170), *110*
Müller, R., 9 (23), *15*
Munk, B., 100 (141), *110*
Munk, M., 172 (87), *178*, 267 (34), 280 (34), 282 (34), *288*
Murakoshi, T., 25 (75), *27*, 105 (195), *111*, 170 (58), 175 (58), *177*
Muralidhara, R., 23 (60), 24 (60), *27*, 456 (2), *461*
Murata, M., 423 (22), 424 (22), 425 (22), 429 (22), *430*
Muravich-Aleksandr Kh, L., 105 (189), *111*
Murray, H. C., 324 (101), *345*
Musante, C., 713 (7), 720 (7), 725 (7), *727*
Musatova, I. S., 41 (49), 47 (78), 57 (108), 58 (110), 59 (111), 63 (78, 111), 64 (110), 65 (78), 67 (111), 71 (111), *74, 75, 76*, 98 (115), *109*, 169 (51), *177*, 256 (38), 257 (38), *260*, 273 (76), 279 (76), 282 (76), *289*, 379 (68), 390 (68), 391 (68), 394 (67, 68), 395 (67), *397*, 667 (32), 684 (32), 686 (32), *689*
Musgrave, W. R. K., 22 (39), *26*, 95 (89), 96 (89), 104 (89), *108*, 168 (48), 170 (48), 172 (48), 174 (48), *177*, 203 (19), *204*
Mustafa, A., 719 (26), 726 (26), *728*
Myrbäck, K., 695 (27), 701 (27), 707 (27), *711*
Myshkina, L. A., 592 (2, 3, 4, 5, 6), 593 (2, 6, 8, 9, 13), 594 (9, 11, 13), 595 (2, 3, 6, 8, 9, 11, 13), 596 (2, 3, 4, 5, 6, 9, 13), *596, 597*

Nababsing, P., 12 (53), *16*, 209 (19), 223 (19), *229*
Nacci, V., 605 (18, 19), 618 (18, 19), 621 (18, 19), 634 (18), 635 (18, 19), 636 (19), 638 (19), 639 (19), 640 (19), *651*, 725 (54), *729*
Nachod, F. C., 157 (101), *161*
Nag, K., 145 (26), 148 (37), 152 (37), 153 (37), *159*, 245 (16), *246*
Nagasawa, Y., 127 (22), 128 (22), *129*, 153 (67), 154 (67), *160*, 207 (14), 221 (14), *229*
Nagarajan, K., 600 (10), 601 (10), 603 (10), 615 (10), 616 (10), 619 (10), 622 (10), 633 (10), 634 (10), 637 (10), 638 (10), 639 (10), 642 (10), *651*
Nagata, M., 470 (20), 471 (20), 480 (20), 482 (20), *493*
Nagoka, T., 12 (44), *16*

Nair, M. D., 99 (132), *109*, 284 (102), *290*, 623 (47), 644 (27), *652*
Nakagawa, Y., 464 (6), 469 (6), 475 (6), 477 (6), 478 (6), 480 (6), 482 (6), 484 (6), 485 (6), 491 (6), *493*
Nakajima, S., 12 (47), *16*, 19 (16), 23 (16), 24 (16), *26*, 216 (41), *230*
Nakamura, K., 196 (14), *198*
Nakamura, S., 324 (115, 116), 337 (115, 116), *346*
Nakao, M., 93 (77), *108*
Nakata, H., 42 (55, 56), *74*
Nano, G. M., 436 (19), 439 (19), *440*
Narayana, R., 571 (9), 573 (9), *574*
Nardi, A., 102 (169), *110*
Narita, S., 511 (63), 531 (63), 533 (63), 534 (63), 535 (63), 536 (63), 537 (63), 538 (63), 539 (63), *543*, 572 (11), *574*
Nasielski, R., 20 (20), 23 (20), *26*
Nasielski-Hinkens, R., 12 (50), *16*
Nast, R., 63 (127, 129), 64 (127, 129, 135, 138), 65 (127, 129, 138), 66 (129), 67 (129), 68 (129), 69 (129), 70 (127, 129), 72 (129), *76*
Nazarovo, L. M., 303 (128, 129), 328 (128), 331 (129), 332 (128, 129), *346*
Nazarova, L. S., 299 (21, 23), 303 (21, 129), 328 (21), 331 (129), 332 (129), *343*, *346*, 473 (30), 480 (30), *493*, 605 (17), 634 (17), *651*
Nefedova, O. I., 432 (8), 437 (8), *440*
Negishi, E., 144 (18), 153 (18), *159*, 268 (53), *289*
Nehter, R., 447 (7, 8), 448 (7), 453 (7, 8), *454*
Nei, K., 723 (37), *728*
Neip, L., 323 (93), *345*
Nelson, P. J., 410 (1), 411 (1), 415 (1), 416 (1), 417 (1), 418 (1, 11), 419 (11), 420 (11), 422 (11), 426 (1), 427 (11), 428 (1, 11), *429*
Nemeryuk, M. P., 592 (6), 593 (6), 594 (11), 595 (6, 11), 596 (6), *597*
Nemeth, P. E., 139 (5), 156 (5), *159*
Nesmeyanov, A. N., 227 (124), *232*
Neuberg, C., 713 (6), 724 (6), *727*
Neumann, R. E., 310 (54), *344*
Neumüller, G., 695 (25, 27), 700 (25), 701 (27), 706 (25), 707 (25, 27), *710*, *711*
Newbold, G. T., 20 (22), 21 (34), 22 (22), 23 (22), 24 (22), *26*, 30 (10), 33 (10), 54 (10), *73*, 89 (57), *107*, 97 (106), 98 (106), 100 (106), 104 (106), 105 (191), *109*, *111*, 171 (68), 174 (68), *177*, 202 (14), *204*, 221 (65), 222 (65), 223 (65), 226 (65), *230*, 237 (17), 238 (17), *239*
Nicholson, S., *429* (13)
Nicolaides, D. N., 715 (17), 723 (17), *728*
Nielsen, K. H., 93 (70), *107*
Ning, R. Y., 86 (34), 102 (34), 103 (34), *106*
Nishibe, K., 715 (16), 724 (16), *728*
Nishikawa, K., 180 (10), 188 (10), *191*, 364 (12), *396*
Nishimura, S., 444 (2, 3), 452 (2, 3), *454*
Nishimura, Y., 235 (43), *240*
Nitecki, D. E., 315 (67), 335 (67), *344*
Nitecki, D. N., 306 (33), 307 (33), *343*
Nivard, R. J. F., 135 (19), *136*
Noble, A. C., 267 (41), 274 (77), 279 (41), 280 (41), 282 (77), 283 (77), *289*
Noble, D., 531 (50), 532 (50), *542*
Nofre, C., 271 (72), 279 (72), 282 (72), 283 (72), *289*
Noguchi, S., 170 (57), 175 (57), *177*
Nohara, F., 470 (20), 471 (20), 480 (20), 482 (20), *493*
Nordin, P., 693 (8, 14), 695 (14, 23), 701 (23), *710*
Nordwig, A., 324 (112), *346*
Normant, H., 216 (48), 226 (48), *230*
Nottke, J. E., 723 (53), *729*
Nussbaum, J., 103 (178), *111*, 131 (5), 134 (5), 135 (5), 136 (5), *136*, 157 (102), *161*, 174 (96), *178*, 228 (131), *232*, 696 (34), *711*
Nyquist, E. B., *192* (53)

Oatis, J. E., 21 (35), 24 (35), 25 (35), *26*
Oberlinner, A., 252 (24), 256 (24), *260*
Occelli, E., 296 (127), 330 (127), 332 (127), *346*
Ochiai, E., 43 (61), *74*
Ochiai, S., 296 (16), 330 (16), *342*
Oda, R., 188 (67), *193*
Ogata, M., 41 (52), *74*

Ogawa, K., 470 (20), 471 (20), 480 (20), 482 (20), *493*
Ogg, R. A., 227 (117), *232*
Ogliaruso, M. A., 237 (36), *239*
Oguchi, S., 21 (37), 22 (37), 23 (37), 24 (37, 63), *26, 27*, 105 (196), *111*, 120 (39), 121 (39), *122*, 149 (39), 158 (39, 107, 108), *159, 161*, 168 (42), 174 (42), 175 (42), *176*
Ohashi, T., 324 (104, 149), *345, 347*
Ohkura, Y., 724 (51), *729*
Ohle, H., 694 (18, 19, 21), 695 (21, 22), 697 (40), 699 (19), 700 (18, 19, 40), 701 (19, 21, 22), 704 (19), 705 (18, 19, 40), 706 (21, 22, 54), 709 (18, 22, 40), *710, 711*
Ohloff, G., 456 (1), 457 (1), 459 (1), 460 (1), 461 (1), *461*
Ohmes, E., 730 (2), *741*
Ohmiya, S., 673 (55), 687 (55), 688 (55), *690*
Ohmomo, S., 324 (103, 104), *345*
Ohno, T., 740 (44), *742*
Ohta, K., 371 (29), 372 (29), 373 (29), 385 (29), *396*
Ohta, M., 571 (6), 573 (6), *574*
Ohtsura, M., 250 (15), *259*
Okamato, T., 399 (19), 406 (19), *409*, 411 (5, 6), 414 (5), 426 (5, 6), *429*
Okamoto, Y., 570 (5), 572 (5), *573*
Okawa, Y., 25 (75), *27*, 105 (195), *111*, 170 (58), 175 (58), *177*
Oksne, S., 86 (38), 95 (38), *107*
Okumura, K., 608 (29), 622 (29), 634 (29), *651*
Omiya, S., *176* (36), 624 (52), 644 (52), *652*
Omodei-Sale, A., 385 (63), 386 (63), 387 (63), *397*
Omran, S. M. A. E., 719 (26), 726 (26), *728*
Ooms, P. J. H., 135 (19), *136*
Osborn, A. R., 8 (6), *15*, 19 (8), *26*
Osdene, T. S., 144 (19), 153 (19), 155 (19), *159*, 182 (26), 187 (26), 188 (26), 189 (26, 74, 75), 190 (75), *192, 193*, 401 (4), 402 (4, 10), 405 (4), 406 (4, 10), 407 (4, 10), *408*, 496 (9), 508 (9, 46, 47), 518 (9), 524 (9, 46, 47), 525 (9), 526 (9, 46, 47), 532 (9), *541, 542*, 716 (21), 717 (22), 722 (21), 723 (21), 725 (32, 33), 726 (22), *728* (34, 35), *728*
Ossana, O. A., 353 (12), 357 (12), *359*
Otake, N., 324 (96), *345*
Otaki, A., 149 (39), 158 (39), *159*
Otomasu, H., 12 (47), *16*, 19 (16), 20 (21), 23 (16), 24 (16), *26*, 89 (55), *107*, 149 (42), 153 (42), 157 (42), 158 (42), *159*, 171 (71), 172 (85), *176* (36), *177, 178*, 216 (41), *230*, 624 (52), 644 (52), *652*, 656 (6), 673 (55, 56), 675 (6), 687 (55), 688 (55, 56), *689, 690*
Otsuji, Y., 13 (64), *16*
Ott, H., 293 (6), 297 (6), 301 (6), 316 (68), 317 (68, 78), 318 (68), 321 (68), 326 (78), 330 (6), 331 (6), 336 (68), 338 (68), *344*
Ourisson, G., 625 (58), 647 (58), *652*
Ovchinnikov, Y. A., 308 (45), *343*
Owens, C., 41 (50), *74*
Oxaal, Fr., 224 (97), *231*
Ozaki, T., 171 (63), *177*

Pachter, I. J., 139 (5), 156 (5), *159*
Padeiskaya, E. N., 58 (110), 64 (110), *75*
Padwa, A., 718 (46), 726 (46), *729*
Page, E., 247 (3), 255 (3), 257 (3), *259*
Pages, R. A., 432 (6), 437 (6), *440*
Pakoskey, A. M., 382 (57), *397*
Pailer, 93 (79), *108*, 103 (186), 104 (186), *111*, 715 (14), 726 (14), *728*, 732 (12), 740 (12), *741*
Pal, I. B., 664 (25), 665 (25), 685 (25), 687 (25), *689*
Palamidessi, G., 8 (8), *15*, 376 (42), 394 (42), *397*, 448 (12), 453 (12), *454*, 579 (13), 580 (13), 581 (13), *582*
Pallos, L., 47 (75), 63 (75), *74*
Palopoli, F. P., 601 (74), 634 (74), 635 (74), 636 (74), 638 (74), 639 (74), 640 (74), 642 (74), *652*
Panzica, R. P., 376 (37), 379 (37), 385 (37), *397*
Pan'shina, M. V., 658 (12), *689*
Panusova, S., 157 (103), *161*, 224 (94), *231*
Papke, E., 93 (69), 99 (69), 100 (69), 102 (69), *107*

Paquette, L. A., 294 (11), 299 (11), 300 (11), 302 (11), 327 (11), 328 (11), 329 (11), *342*, 473 (33), 482 (33), *493*
Paradis, A. D., 254 (27), *260*
Parker, E. D., 222 (74), *231*
Parnell, E. W., 255 (35), 256 (35), *260*
Parri, W., 749 (10), *758*
Parrick, J., 348 (19), 349 (3, 19), 350 (3), 356 (3, 19), 357 (3, 19), *359*
Parrini, V., 713 (7), 720 (7), 725 (7), *727*
Parsons, B. N., 275 (83), 279 (83), *290*
Patel, N. R., 373 (34), 375 (34), 387 (34), 388 (34), *396*
Paterson, J. W., 423, (18), *429*
Paton, D., 415 (10), 418 (10), 419 (10), 420 (10), 422 (10), 427 (10), 428 (10), *429*
Patterson, E. L., 324 (98), *345*
Patton, R., 201 (10), 203 (10), 204 (10), *204*
Paudler, W. W., 8 (13), *15*, 292 (2), 293 (2, 126), 302 (2, 126), 327 (2), 329 (126), *342, 346*, 361 (62), 365 (62), 370 (62), 382 (62), 383 (62), *397*, 405 (17), 407 (17), *409*
Paul, D. B., 353 (13), 355 (13), 357 (13), 358 (13), *359*
Paul, R., 474 (39), 484 (39), *494*
Paul, R. E., 474 (38), 481 (38), 484 (38), *494*
Pauli, O., 171 (61), 175 (61), *177*
Paulus, W., 171 (61), 175 (61), *177*
Pazur, J. H., 693 (15, 17), 695 (16), 701 (15, 17), *710*
Pearce, R. J., 63 (122), 68 (122), *76*
Peck, R. L., 464 (5), 470 (5), 476 (5), 481 (5), 482 (5), 483 (5), 484 (5), *493*
Peck, R. M., 752 (16), 758 (16), *759*
Peckham, W. D., 323 (92), 335 (92), *345*
Pedersen, C., 547 (18), 564 (18), *567*
Pendleton, L., 11 (40), *16*, 276 (91), 287 (91), *290*
"Peptides, Proc. European Symp., *5th* Oxford," 310 (49), *344*
Peratoner, E., *758* (9)
Perazzi, A., 296 (127), 330 (127), 332 (127), *346*
Perchinunno, M., 12 (60), *16*
Perchinunno, P., 12 (58), 13 (58), *16*
Percival, A., 97 (106), 98 (106), 100 (106), 104 (106), *109*, 171 (68), 174 (68), *177*, 202 (14), *204*, 221 (65), 222 (65), 223 (65), 226 (65), *230*, 237 (17), 238 (17), *239*
Perello, P. J., 268 (56), 281 (56), *289*
Pereslini, E. M., 49 (84), 56 (84, 107), 57 (84), 58 (110), 63 (107), 64 (110), 65 (107), 70 (110), *75*, 98 (115a), *109*, 187 (59), 189 (71), *193*
Perez Ossorio, R., 157 (105), 158 (105), *161*
Perkampus, H. H., 8 (14), 9 (22), *15*, 270 (68), *289*, 713 (42, 43), 718 (45), 723 (45), 726 (45), *728, 729*
Perrin, D. D., 545 (9, 13), 546 (13), *567*
Perroncito, G., 404 (12), 407 (12), 408 (12), *408*
Perry, T. L., 322 (85), *345*
Pershin, G. N., 66 (15), *73*, 103 (176), *111*
Pesson, M., 749 (23), 755 (23), 756 (23), *759*
Peter, P., 473 (28), 488 (28), 491 (28), *493*, 744 (4), 754 (4), 755 (4), *758*
Petering, H. G., 81 (15), 97 (15), *106*, 141 (9), 152 (9), *159*, 171 (74), *177*, 181 (22), 187 (22), *192*
Peterli, H. J., 170 (59), *177*
Peters, J. M., 254 (30), *260*
Petersen, S., 718 (25), 725 (25), *728*
Peterson, D. H., 324 (101), *345*
Petrov, V. I., 658 (12), *689*
Petrova, L. P., 228 (132), *232*
Petrovanu, M., 252 (25), 256 (25), *260*, 610 (33), 641 (33), *651*, 656 (9), 675 (9), *689*, 744 (5), 754 (5), *758*
Petrow, V., 227 (120), 228 (120), *232*, 498 (17), 512 (17), 515 (17), 527 (17), 530 (17), 531 (17), 533 (17), 540 (17), *541*, 750 (11, 12), 757 (11, 12), *758*
Petrzilka, T., 301 (27, 28), 302 (28), 307 (28), 322 (27, 28), 324 (28, 107), 328 (27, 28), 329 (28), 332 (28), 333 (28), 335 (28), 336 (28), 337 (28), 338 (28), *343, 345*
Pettersen, E., 323 (92), 335 (92), *345*
Pettit, G. R., *347* (145)
Petty, J. D., *232* (145)

Pfeiffer, F. R., 156 (96), *161*, 173 (92), *178*, 237 (16), *239*
Pfister, K., 79 (4), *106*, 181 (23), 187 (23), 188 (23), *192*, 276 (10), 279 (10), *288*
Pfizer Inc., 39 (43), *73*, 626 (63), 646 (63), 647 (63), *652*
Pflaum, R. T., 732 (10), 740 (10), *741*
Pflegel, P., 88 (48, 49), *107*
Pfleiderer, W., 88 (51), 89 (53), 98 (51), 100 (145), *107, 110*, 140 (8), 141 (8, 14), 146 (28), 153 (8, 14, 72), 154 (8), *159, 160*, 277 (9), *288*, 506 (41), 568 (1), *542, 573*
Pfliegel, T., 56 (101), *75*
Pflug, P., 82 (20), *106*
Phillips, J. N., 496 (12), *541*, 546 (15), *567*
Pieridou, M., 134 (11), 136 (11), *136*, 630 (80), 648 (80), *653*
Pietra, S., 11 (41), *16*, 267 (37), 276 (37), 280 (37), 282 (37), 286 (37), 287 (37), *288*, 354 (15), 357 (15), *359*
Pigini, M., 503 (35), 538 (35), *542*
Pinson, J., 262 (3), 276 (3), 277 (3), 278 (3), 283 (3), 284 (3), 286 (3), *288*, 497 (61), 521 (61), 534 (61), 538 (61), 539 (61), 540 (61), *543, 567* (35)
Piozzi, F., 101 (154), *110*
Pirkl, J., 97 (103), *109*
Pirola, O., 86 (37), 105 (37), *107*, 249 (12), 253 (12), 255 (12), *259*, 691 (2), 704 (2), *710*
Piskov, A. B., 184 (32), *192*
Pitre, D., 354 (15), 357 (15), *359*
Pittet, A. O., 23 (57, 60), 24 (57, 60), *27*, 456 (2), *461*
Placucci, G., 578 (10), *582*
Plakhina, G. D., 92 (67), *107*
Planckaert, B., 13 (63), *16*
Platt, B. C., 23 (55), 25 (55), *27*, 172 (88), *177*, 187 (65), 188 (65, 66), *193*
Plessen, H. V., 548 (22), 555 (22), 565 (22), 566 (22), *567*
Plevachuk, N. E., 47 (79), 62 (79), *75*, 126 (17), *129*, 214 (32), *229*
Plevachuk-Tarnaskaya, N. E., 98 (122), *109*, 127 (23), *129*
Poddubnaya, N. A., 308 (46), 324 (116), 336 (116), *343, 346*
Pohloudek-Fabini, R., 93 (69), 99 (69), 100 (69), 102 (69), *107*
Poisel, H., 312 (141), 321 (141), 322 (141), 335 (141), 336 (141), 337 (141), 338 (141), *347*
Polaks Frutal Works N. V., 456 (3), 459 (3), 460 (3), *462*
Polanc, S., 411 (27), 425 (27), 426 (27), 430
Pollak, P. I., 377 (45), 378 (50, 53), 380 (45, 50 53), 389 (45, 50, 53), 390 (45, 50, 53), 391 (45, 50, 53), 392 (45, 50, 53), 393 (45, 50, 53), *397*
Polyakova, N. A., 143 (16), 153 (16), 155 (16), *159*
Ponomarev, A. A., 299 (19, 20), 301 (19), 328 (19), 335 (20), *343*
Popp, F. D., 267 (41, 42), 274 (77), 279 (41), 280 (41), 282 (77), 283 (77), *289*
Poradowska, H., 19 (17, 18), *26*
Poroshin, K. T., 304 (30), 310 (52), 315 (65), 319 (65), 334 (52, 65), *343, 344*
Porta, O., 209 (17), *229*
Porter, Q. N., 400 (2), 402 (2, 6), 407 (2), *408*
Porter, T. H., 21 (32), *26*
Portlock, D. E., 217 (52), *230*
Postman, W., 30 (14), 72 (14), *73*, 101 (159), *110*, 153 (62), *160*
Postovskii, I. Ya., 83 (23), 90 (61, 62), 99 (23, 133), 100 (23, 133), *106, 107, 109*, 114 (15), 119 (15), 120 (15), *121*, 163 (16), 165 (26), 172 (26, 90), 173 (16, 26), *176, 178*, 184 (45), 185 (45), 190 (45), 191 (45), *192*, 194 (7), 196 (10), 197 (7, 10, 17), *198*
Postovskii, I. Y., 512 (64), 516 (64), *543*
Potoski, J. R., 125 (33), *129*
Potts, K. T., 194 (3), 196 (3), 197 (3), *197*, 410 (1), 411 (1, 4), 414 (7), 415 (1, 7), 416 (1), 417 (1, 4), 418 (1, 7, 11), 419 (7, 11), 420 (7, 11), 422 (11), 426 (1, 7), 427 (7, 11), 428 (1, 7, 11), *429*
Powalska, J., 114 (14), *121*
Pozharsky, A. F., 654 (1), 655 (1), 657 (1), 675 (1), *689*
Prasad, M., 324 (102), *345*
Pratt, E. F., 163 (18), 172 (18), 173 (18), *176*

Pratt, Y. T., 6 (2), *6*
Prebendowski, S., 457 (4), 460 (4), 461 (4), *462*
Predvoditeleva, G. S., 103 (176), *111*
Preister, M., 226 (146), *232*
Prelog, V., 153 (60), *160*, 323 (93), *345*
Preobrazhenskaya, M. N., 400 (20), 407 (20), *409*
Preston, J., 370 (27), 384 (27), *396*
Preston, H., Jr., 371 (28), 372 (28), 373 (28), 385 (28), *396*
Preston, R. K., 752 (16), 758 (16), *759*
Pretsch, E., 624 (54), 644 (54), 645 (54), *652*
Priester, M., 248 (11), 251 (45), 256 (11, 45), 257 (11), *259, 260*
Prijs, B., 156 (95), *161*
Pruckmayr, G., 93 (79), 103 (186), 104 (186), *108, 111*, 732 (12), 740 (12), *741*
Pryakhina, T. A., 152 (51), 158 (51), *160*
Puchkov, V. A., 308 (45), *343*
Pueschel, W., 698 (44), 703 (44), *711*
Pugmire, R. J., 8 (12), 10 (12), *15*
Pushhareva, Z. N., 152 (51), 158 (51), *160*
Pushkareva, Z. V., 42 (57, 60), *74*
Pytlewski, L. L., 41 (50), *74*

Quarg, G., 237 (42), *240*
Quilico, A., 12 (61), 13 (61), *16*
Qureshi, M. I., 29 (8), 33 (21), 45 (8), 54 (95), 56 (8), 57 (8), 59 (8), 60 (21), *73, 75*, 97 (110), 101 (148), 105 (148), *109, 110*, 138 (2), 152 (2), 153 (75, 77), 154 (2, 77), 155 (75, 77), 156 (77), *158, 160*
Qureshi, M. K., 18 (6), 22 (6), *25*

Rabouhans, M. L., 423 (15), *429*
Rabusic, E., 207 (12), 222 (12), *229*
Raevskii, K. S., 303 (128), 328 (128), 332 (128), *346*
Rafiq, M., 90 (59), 96 (59), 98 (59), 99 (59), 103 (59), *106*, 194 (4), 195 (4), 196 (4), *197*, 599 (1), 601 (9), 603 (9), 616 (1), 618 (9), 622 (9), 634 (9), 639 (9), 640 (9), 641 (9), *650*
Raines, S., 601 (74), 634 (74), 635 (74), 636 (74), 638 (74), 639 (74), 640 (74), 642 (74), *652*
Rajappa, S., 316 (132), 328 (132), 329 (132), *346*
Ralph, B. J., 102 (164), *110*
Ramage, G. R., 6 (3), *6*, 268 (55), 279 (55), *289*
Rampy, L. W., 568 (2), 569 (2), 570 (2), 572 (2), 573 (2), *573*
Rao, B. S., 102 (162), *110*
Rao, K. V., 19 (13), *26*, 267 (33), 279 (33), 280 (33), *288*
Rao, V. R., 600 (10), 601 (10), 603 (10), 615 (10), 616 (10), 619 (10), 622 (10), 633 (10), 634 (10), 637 (10), 638 (10), 639 (10), 642 (10), *651*
Rapos, P., 97 (103), *109*
Raulius, N. R., 613 (37), 637 (37), *651*
Ravel, J. M., 524 (59), 527 (59), *543*
Rawdah, T. N., 39 (47), 48 (47), 62 (47), *74*
Reavill, R. E., 9 (25), *15*
Rebel, W. J., 14 (71), *16*, 250 (17), 255 (17), 256 (17), 257 (17), *259*, 267 (22), 272 (22), 280 (22), 281 (22), 282 (22), 283 (22), 284 (22), *288*
Rees, C. W., 32 (20), 33 (20), 39 (20), 42 (58, 59), 53 (94), 57 (20, 94), 59 (20), 63 (20, 94), *73, 74, 75*, 88 (50), 98 (124), 100 (143), *107, 109, 110*, 146 (29), 153 (29, 64, 69), 154 (29, 64), 155 (29, 92), *159, 160, 161*
Regan, L. A., 85 (33), 102 (33), *106*, 102 (180), 103 (180), *111*
Regnier, G., 476 (42), 480 (42), 485 (42), 486 (42), *494*
Reich, F., 498 (22), 516 (22), 531 (22), *541*
Reicheneder, F., 624 (50), 643 (50), *652*
Reid, J., 706 (50), *711*
Reid, W., 103 (174), *111*, 135 (21), *136*, 214 (31), 222 (31), *229*, 263 (12), 288
Reidel, G., 163 (14), 172 (14), 173 (14), *176*, 735 (26), 739 (26), *742*
Reif, G., 315 (63), 319 (63), *344*
Reineke, L. M., 324 (101), *345*
Reinheckel, H., 153 (70), *160*, 163 (10), 172 (10), 173 (10), *175*, 194 (2), 196

(2), 197 (2), *197*, 715 (15), 724 (15), 725 (15), 727 (15), *728*, 732 (13), *741*
Reisser, F., 89 (53), *107*, 153 (72), *160*
Reusser, P., 323 (93), *345*
Rhone-Poulenc, S. A., 580 (22), *582*
Riccieri, F. M., 238 (31), *239*, 503 (36), 535 (36), *542*
Richards, A. J., 423 (18), *429*
Richards, C. G., 99 (131), 101 (131), *109*, 131 (2), 132 (2), 134 (2), 135 (2), *136*, 183 (28), *192*, 225 (109), *231*, 262 (5), 277 (5), *288*
Richardson, A., Jr., 630 (70), 649 (70), 650 (70), *652*, 753 (19, 20), 758 (19, 20), *759*
Richardson, H. W., 497 (62), 531 (62), *543*
Richardson, K. S. C., 322 (85), *345*
Rickett, R. M. W., 125 (15), 127 (15), *129*, 153 (61), *160*, 221 (69), *230*
Ridgewell, B. J., 8 (15), *15*
Ried, W., 507 (42), 539 (42), 540 (42), *542*, 626 (64), 647 (64), 648 (64), *652*
Riedel, G., 180 (4), 184 (4, 57), 187 (4, 56, 57), *191, 193*, 200 (8), 203 (8), *204*
Riley, J. G., 601 (12), 603 (12), 638 (12), *651*
Rink, M., 575 (1), 580 (1), *582*
Riolo, C. B., 713 (3), *727*
Rissi, E., 373 (33), 387 (33), *396*, 435 (16), 438 (16), *440*
Ristich, S. S., 221 (61), *230*
Ritter, H. W., 38 (41), 55 (97), 58 (41), 59 (41), 65 (41, 97), 66 (97), 67 (41), 68 (97), 69 (41), 70 (41), 71 (41), *73, 75*
Rittner, I., 307 (139), 335 (139), *347*, 469 (10), *493*
Robb, C. M., 153 (78), *161*
Roberts, J. D., 131 (3), 135 (3), *136*, 153 (63), *160*
Robertson, A., 102 (164), *110*
Robins, M. J., 8 (12), 10 (12), *15*
Robins, R. K., 8 (12), 10 (12), *15*
Robinson, C. A., 79 (4), *106*
Robinson, C. N., 227 (118), *232*
Robinson, G. E., 613 (37), 637 (37), *651*
Robstov, M. V., 476 (41), 483 (41), *494*
Roders, A., 8 (14), *15*, 270 (68), 289
Rodway, R. E., 655 (3), 658 (4), 674 (3), 675 (3), *689*
Roechling, H., 664 (23), 665 (23), 684 (23), *689*
Roedig, A., 174 (97), *178*
Rogers, W. P., 267 (28), 279 (28), *288*
Rogers Batcheler, M. L., 382 (57), *397*
Rohde, W., 276 (90), 279 (90), 287 (90), *290*
Roitshtein, L. M., 105 (189), *111*
Romanenko, V. D., 34 (29), *73*, 80 (14), 92 (66), 102 (66), *106, 107*, 127 (20), *129*, 135 (16), *136*, 149 (45), 152 (53), 157 (45), 158 (45), *160*, 163 (5), 172 (5), 173 (5), *175*, 626 (62), 647 (62), *652*, 696 (35), 697 (35), 704 (35), *711*, 724 (50), *729*
Romashkin, M. P., 732 (20), 737 (20), 738 (20), *742*
Romazanovich, N. P., 63 (152), 64 (152), *77*, 150 (47), *160*
Ronco, K., 221 (63), *230*
Rose, F. L., 410 (2, 3), 415 (2, 10), 416 (2, 3), 417 (2), 418 (10, 12), 419 (2, 10, 12), 420 (3, 10, 12), 422 (10, 12), 423 (16, 17), 426 (2), 427 (2, 10, 12), 428 (2, 10, 12), *429*
Rose, J. D., 498 (26, 27), 502 (26, 27, 34), 509 (27), 526 (27), 527 (26, 27), 528 (26, 27, 34), 529 (26, 27), *542*, 548 (24), 551 (24), 557 (24), 559 (24), 560 (24), 561 (24), 562 (24), 563 (24), *567*
Rosenberger, S., 447 (6), 448 (6), 453 (6), *454*
Rosenquist, U., 695 (27), 701 (27), 707 (27), *711*
Ross, R. E., 92 (65), 98 (65), 100 (65), 101 (65), 102 (65), *107*, 609 (30), 634 (30), 635 (30), *651*
Rossi, S., 33 (25, 26), 34 (26), 56 (25), 57 (25, 26), 58 (25), 60 (25), 61 (25), *73*, 86 (37), 105 (37), *107*, 249 (12), 253 (12), 255 (12), *259*, 313 (59), 319 (59), 336 (59), *344*, 691 (2), 704 (2), *710*
Rossotti, H. S., 22 (50), *27*
Roth, W., 22 (51), *27*
Rotter, U., 226 (111), *231*
Rotzler, G., 139 (6), 156 (6), 157 (6), *159*, 223 (81), *231*

Roy, P. D., 599 (1), 605 (21), 607 (26), 616 (1, 26), 617 (21, 26), 618 (26), 619 (21), 620 (21), 621 (21), 633 (21, 26), 634 (21), 635 (26), *650, 651*
Roy, S. K., 194 (3), 196 (3), 197 (3), *197*
Rozum, Yu. S., 45 (70), 62 (70), 63 (152), 64 (152), *74, 77*, 150 (47), *160*, 248 (9), 258 (9, 43), *259, 260*
Rozynov, B. V., 308 (45), *343*
Rua, G., 436 (19), 439 (19), *440*
Rudy, H., 507 (43, 44), 531 (43), 533 (43), 535 (43), 536 (43), 537 (43, 44), 538 (44), *542*
Ruehlmann, K., 376 (54), 379 (54), 395 (54), *395*
Ruggli, P., 227 (125), *232*, 228 (136), *232*
Ruhemann, S., 623 (41), 645 (41), *651*
Ruliffson, W. S., 693 (10), *710*
Rupe, H., 227 (125), *232*
Russell, D. M., 99 (131), 101 (131), *109*, 131 (2), 132 (2), 134 (2), 135 (2), *136*, 183 (28), 187 (28), *192*, 225 (109), *231*, 262 (5), 277 (5), *288*
Russell, R. A., 323 (147), 324 (147), *347*
Russell-Hill, D. Q., 164 (20), *176*
Rutner, H., 163 (11), 172 (11), *175*, 423 (20), 424 (20), 429 (20), *429*
Ryback, G., 99 (131), 101 (131), *109*, 131 (2), 132 (2), 134 (2), 135 (2), *136*, 183 (28), 187 (28), *192*
Rybin, L. V., 227 (124), *232*
Rybinskaya, M. I., 227 (124), *232*
Rydon, H. N., 23 (58), *27*, 98 (126), 99 (126, 134), 100 (134), *109*, 123 (3), 127 (3), *128*, 205 (3), 206 (3), 221 (3, 66), 224 (3, 66), *229, 230*, 309 (48), 310 (48), 311 (48), 315 (48), 334 (48), *344*, 496 (7), 498 (7), 502 (7), 504 (7), 508 (7), 515 (7), 516 (7), 518 (7), 525 (7), 526 (7), 532 (7), 533 (7), 534 (7), 540 (7), *541*
Rzeszotarski, W. J., 307 (139), 335 (139), *347*, 469 (10), *493*

Sacerdoti, S., 716 (19), 723 (19), 724 (19), *728*
Sachs, F., 697 (39), 703 (39), *711*
Sadilek, L., 103 (182), *111*
Sadtler Standard Spectra, 270 (67, 69), *289*
Safir, S. R., 128 (29), *129*, 172 (84), *178*, 352 (10), 357 (10), *359*
Safonova, T. S., 131 (6), 134 (6), *136*, 592 (2, 3, 4, 5, 6), 593 (2, 6, 8, 9, 13), 594 (9, 11, 13), 595 (2, 3, 6, 8, 9, 11, 13), 596 (2, 3, 4, 5, 6, 9, 13), *596, 597*
Saheb, S. E., 267 (32), *288*
Saikachi, H., 98 (114), *109*, 113 (5), 119 (5), *121*, 157 (104), *161*, 163 (12), 172 (12), *175*, 180 (3), 187 (3), *191*, 195 (9), 197 (9), *198*, 200 (7), 202 (7), *204*
St. Clair, R. L., 97 (158), *110*
Saint-Ruf, G., 244 (12), *246*, 498 (20), 539 (20), 540 (20), *541*, 693 (7), 701 (7, 47), 703 (7, 47), 704 (7, 47), 705 (7, 47), 706 (47), 709 (47), *710, 711*, 713 (12), 725 (49), 726 (12), *728, 729*
Saito, C., 180 (10), 188 (10), *191*
Saito, H., 720 (38), 723 (38), 724 (38), 725 (38), 726 (38), *726*
Saito, M., 693 (13), 695 (13), *710*
Sakai, S., 42 (56), *74*
Sakakibara, Y., 105 (198), *111*, 149 (39), 158 (39, 74), *159, 160*, 227 (119), *232*, 238 (34), *239*
Sakamoto, Y., 196 (14), *198*
Sakena, R., 571 (8, 9), *574*
Saldabols, N., 226 (114), *232*, 237 (39), *239*
Saldabols, N. O., 101 (149, 150), *110*
Sale, A. A., 88 (50), *107*
Salim Qureshi, M. A., 228 (133), *232*
Salzmann, U., 13 (67), *16*, 90 (60), 101 (60), *107*
Sam, D. J., 184 (49), 186 (49, 52), 187 (49, 52), *192*, 671 (59), *690*
Samek, Z., 701 (49), *711*
Samuelson, E., 224 (97), *231*
Sanayal, G. S., 245 (16), *246*
Sanderson, A., 456 (2), *461*
Sandoz Ltd., Meth. Appl. 6,609,202, 311 (56), 318 (56), 326 (56), 336 (56), 340 (56), *344*
Sandoz, Ltd., Neth. Appl. 6,511,933, 316 (70), 318 (70), 326 (70), 339 (70), *344*

Sandoz, Ltd., Belg. Pat. 617,451, 317 (77), 326 (77), *344*
Sandoz Ltd., Belg. Pat. 617,499, 317 (76), 326 (76), *344*
Sandoz Ltd., Neth. Appl. 6,412,779, 434 (15), 438 (15), *440*
Sandoz Ltd., Neth. Appl. 6,512,688, 316 (74), 326 (74), *344*
Sandrin, E., 315 (64), 335 (64), *344*
Sandris, C., 625 (58), 647 (58), *652*
Sanjurjo, J. L., 128 (29), *129*, 172 (84), *178*, 352 (10), 357 (10), *359*
Santelli, A. A., 725 (32, 33), *728* (35), *728*
Santilli, A. A., 189 (74), *193*, 508 (46, 47) 524 (46, 47), 526 (46, 47), *542*
Sanyal, G., 145 (26), *159*
Sanyal, G. S., 148 (37, 38), 152 (37), 153 (38), *159*
Sanz Burata, M., 157 (105, 106), 158 (105, 106), *161*
Saper, J., 498 (17), 512 (17), 515 (17), 527 (17), 530 (17), 531 (17), 533 (17), 540 (17), *541*, 750 (11, 12), 757 (11, 12), *758*
Sarava, R. F., 92 (67), *107*
Sarkis, G., 154 (85), *161*
Sarkis, G. Y., 61 (117), *76*, 233 (1), 237 (1), 238 (1), *238*
Sarsfield, A. A., 9 (25), *15*, 22 (41), *26*
Sasaki, T., 296 (16), 330 (16), *342*, 423 (22), 424 (22), 425 (22), 429 (22), *430*
Sasayama, K., 196 (12), *198*
Sasse, K., 112 (4), 113 (6), 116 (16), 117 (26, 28, 29), 118 (30, 36), 119 (4, 16), 120 (6, 28, 29), 121 (28), *121, 122*, 163 (13), 168 (45), 171 (61, 67), 174 (45, 67), 175 (61), *176, 177*, 181 (16), 188 (16), *191*, 735 (27, 28, 29, 30), *742*
Sassenberg, N., 256 (42), 257 (42), *260*
Sassi, K., 459 (6), 461 (6), *462*
Sato, H., 44 (65), 57 (65), 64 (65), *74*
Sato, K., 148 (32), *159*
Sato, M., 98 (200), *111*, 323 (86), 324 (86), *345*, 411 (5), 414 (5), 426 (5), *429*
Sato, S., 673 (55), 687 (55), 688 (55), *690*
Sato, T., 377 (43, 44), 380 (43), 395 (43, 44), *397*, 593 (7, 10), 594 (7, 10), 595 (7, 10), *597*
Sato, Y., 594 (12), *597*
Sauer, W., 165 (23), 169 (23), 170 (23), 172 (37, 83), *176, 178*, 405 (18), 407 (18), *409*, 696 (36), 698 (36), 699 (36), 704 (36), *711* (36), *711*
Savelli, G., 164 (22), *176*
Savins, E. G., 29 (6), 30 (6), 49 (6), 56 (6), 57 (6), *73*, 98 (116), *109*, 183 (29), 187 (29), *192*
Sawicki, E., 225 (110), *231*, 731 (4), 740 (4), *741*
Sax, K. J., Jr., 324 (98), *345*
Saxena, R., 571 (7, 10), 573 (7), *574*
Scalzo, M., 621 (75), 622 (75), 634 (75), 635 (75), 636 (75), 637 (75), 638 (75), 639 (75), 641 (75), *652*
Scarso, L., 93 (75), *108*
Schaaf, K. H., 216 (43), *230*
Schaefer, W., 225 (106), *231*
Schaffer, R. R., 221 (62), *230*
Scheeren, J. W., 135 (19), *136*
Scheinpflug, H., 168 (45), 174 (45), *176*
Schellenberg, M., 130 (1), 134 (1), 135 (1), 136 (1), *136*, 262 (4, 6), 277 (4, 6), *288*
Schelokov, V. I., 434 (13), 438 (13), *440*
Schelz, D., 226 (146), *232*, 248 (11), 251 (21, 45), 255 (21), 256 (11, 21, 45), 257 (11), *259, 260*, 725 (54), *729*, 739 (45), *742*
Schiemenz, G. P., 103 (183), *111*
Schindler, W., 170 (59), *177*
Schipper, E., 84 (26), 98 (26), *106*, 184 (47), 186 (47), 190 (47), *192*, 376 (39), 377 (39), 389 (39), 394 (39), 395 (39), *397*, 664 (22), 665 (22), 667 (22), 684 (22), 685 (22), 686 (22), *689*
Schlatter, J. M., 314 (60), 335 (60), *344*
Schleigh, W. R., 19 (14), 23 (14), 24 (14), 25 (14), *26*, 267 (45), 273 (45), 279 (45), 280 (45), 281 (45), 282 (45), *289*
Schlientz, W., 326 (121), 340 (121), *346*
Schlunke, H. P., 80 (13), *106*, 221 (63, 64), *230*, 254 (33), *260*
Schmid, H., 13 (62), *16*, 206 (11), 223 (11), *229*, 262 (7), 277 (7), 278 (7), 283 (7), 285 (7), *288*
Schmid, W., 37 (39), *73*
Schmidt, K. J., 21 (33), *26*, 22 (53), *27*

Schmidt, P., 89 (58), *107*
Schmidt, U., 312 (141), 321 (141), 322 (141), 335 (141), 336 (141), 337 (141), 338 (141), *347*
Schmidtberg, G., 713 (44), *728*
Schnegelberger, H., 466 (7), 467 (7), 468 (7), 477 (7), 489 (7), 491 (7), 492 (7), *493*
Schneider, H., 156 (99), *161*
Schneller, S. W., 194 (3), 196 (3), 197 (3), *197*, 414 (7), 415 (7), 417 (7), 418 (7), 419 (7), 420 (7), 426 (7), 427 (7), 428 (7), *429*, 442 (15, 16), 443 (15), 451 (16, 17), *455*
Schofield, K., 8 (6), *15*, 19 (8), *26*
Scholl, H. J., 656 (10), 658 (10), *689*
Schranner, M., 88 (51), 98 (51), *107*
Schroeder, W., 97 (105), 105 (105), *109*
Schroth, W., 269 (60), 283 (60), 284 (60), *289*
Schubart, R., 63 (127), 64 (127), 65 (127), 70 (127), *76*
Schuer, M., 713 (6), 724 (6), *727*
Schuler, W., 254 (29), *260*
Schultz, H. P., 21 (35), 22 (44), 24 (35), 25 (35), *26*, 81 (17), 83 (25), 97 (107), 98 (25), 100 (17, 137, 139, 140, 141), 101 (17), 102 (17), *106, 109, 110*, 134 (12, 13), *136*, 145 (22, 23), 146 (22), 148 (33, 34), 152 (23, 55), 153 (22, 33, 34), *159, 160*, 165 (29), 172 (87), *176, 178*, 181 (18), 187 (18), *191*, 201 (10), 203 (10), 204 (10), *204*, 221 (68), *230*, 242 (4), 244 (9), 245 (4, 13), *246*, 267 (24, 34), 275 (87), 276 (87, 88, 90), 279 (90), 280 (24, 34), 282 (34), 287 (87, 88, 90), *288, 290*
Schulze, W., 627 (68), 648 (68), *652*
Schunke, H. P., 154 (81), *161*
Schuyler, P., 267 (41), 274 (77), 279 (41), 280 (41), 282 (77), 283 (77), *289*
Schvedov, V. I., 368 (25), 383 (25), *396*
Schwab, A. W., 471 (25, 26), 487 (25, 26), *493*
Schwab, E., 328 (29), 329 (29), *343*
Schwaiger, W., 217 (49), 220 (49), 225 (49), 228 (49), *230*, 263 (103), 277 (103), 278 (103), *290*
Schweitzer, W., 125 (11), *128*
Schwenk, U., 95 (87), 97 (87), 98 (87), 99 (87), 104 (87), 105 (87), *108*, 171 (70), *177*, 202 (17), *204*, 238 (32), *239*
Schwyzer, R., 305 (32), *343*
Scientific Research Institute of Pharmacology and Chemotheraphy. Moscow, 299 (22), 327 (22), 328 (22), *343*, 473 (32), 482 (32), *493*
Scoffone, E., 93 (74, 75, 76), *108*, 604 (15), 622 (15), 635 (15), *651*
Scott, M. K., 294 (11), 299 (11), 300 (11), 302 (11), 327 (11), 328 (11), 329 (11), *342*, 473 (33), 482 (33), *493*
Sebesta, K., 701 (49), *711*
Seibl, J., 624 (54), 644 (54), 645 (54), *652*
Seiffert, W., 730 (2), *741*
Seki, T., 87 (47), 103 (47), *107*, 740 (35), *742*
Sekiguchi, T., *176* (36), 624 (52), 644 (52), *652*
Sellstedt, J. H., 633 (71), 649 (71), 650 (71), *652*
Selva, F., 86 (37), 105 (37), *107*, 249 (12), 253 (12), 255 (12), *259*, 691 (2), 704 (2), *710*
Semonsky, M., 327 (123), *346*
Sen, A. B., 138 (1a), 152 (1a), 153 (1a), 155 (1a), *158*, 172 (79), 173 (79), *178*, 571 (8, 9, 10), 573 (8, 9), *574*
Sen, A. K., 324 (115), 337 (115), *346*
Seng, F., 35 (32), 36 (32), 49 (82), 63 (126, 127, 128), 64 (82, 127, 129), 65 (32, 127, 129, 143, 144), 66 (129), 67 (129), 68 (129), 69 (129), 70 (82, 127, 129), 72 (129), *73, 75, 76, 77*, 150 (46), 158 (46), *160*, 626 (78), 647 (78), 648 (78), *653*
Senior, N., 254 (30), *260*
Senyavina, L. B., 307 (42), 317 (42), 318 (42), 335 (42), 337 (42), *343*
Senzavina, L. B., 434 (12), *440*
Sevcik, B., 148 (36), *159*
Seyhan, M., 127 (27), 128 (27), *129, 161* (84)
Shaffer, R. R., 56 (104), 64 (104), 65 (104), *75*
Shah, J., 623 (42, 43), 646 (42, 43), *652*

Shalaby, S. W., 105 (197), *111*, 226 (116), *232*
Shamsi, S. A., 102 (163), 103 (163), *110*
Shamsi, S. S., 52 (91), *75*
Shannon, J. S., 141 (13), 155 (13), *159*, 267 (25), 283 (25), *288*
Sharma, R. A., 376 (36), 377 (36), 379 (36), 382 (36), 394 (36), *396*
Sharma, S., 571 (7), 573 (7), *574*
Sharp, T. M., 172 (88), *178*, 23 (55), 25 (55), *27*, 187 (65), 188 (65, 66), *193*
Sharp, W., 89 (57), *107*
Sharpe, C. J., 12 (51), *16*, 18 (25), 25 (5, 72), *25, 27*
Shaw, E., 498 (28), 525 (28), *542*
Shchukina, M. N., 103 (176), *111*
Shea, P. J., 568 (2), 569 (2), 570 (2), 572 (2), 573 (2), *573*
Sheehan, J. C., 324 (115, 116), 337 (115, 116), *346*
Sheinker, Y. N., 307 (42), 317 (42), 318 (42), 335 (42), 337 (42), *343*
Sheinker, Yu. N., 49 (84), 56 (84, 107), 57 (84), 63 (107), 65 (107), *75*, 98 (115a), *109*, 187 (59), 189 (71), *193*, 434 (12), *440*
Shemyakin, M. M., 307 (41, 42, 43), 308 (45), 317 (42), 318 (42, 79), 335 (42, 43, 79), 336 (43), 337 (42, 43, 79), 338 (43), 339 (43, 79), 340 (43), *343, 344*, 434 (12, 13, 14), 438 (13, 14), *440*
Sheppard, R. C., 368 (26), 370 (27), 384 (26, 27), *396*
Shibata, K., 149 (39), 158 (39), *159*
Shibnev, V. A., 304 (30), 310 (52), 315 (65), 319 (65), 334 (52, 65), *343, 344*
Shigemitsu, K., 608 (29), 622 (29), 634 (29), *651*
Shiho, D., 95 (91), 98 (118), *107, 108*, 173 (94), *178*, 180 (5), 189 (5), 190 (78), *191, 193*, 194 (8), 195 (8), 196 (11, 12), 197 (8, 11), *198*
Shimada, T., 188 (67), *193*
Shimomura, O., 363 (6, 7), 384 (6), *395, 396*
Shiratori, O., 148 (32), *159*
Shive, W., 101 (155), 102 (155), *110*
Shkrob, A. M., 307 (41, 42), 317 (42), 318 (42, 79), 335 (42, 79), 337 (42, 79), 339 (79), 340 (79), *343, 344*, 434 (12, 13, 14), 438 (13, 14), *440*
Short, L. N., 8 (6), *15*, 19 (8), *26*
Shrive, W., 524 (59), 527 (59), *543*
Shriver, K. L., 568 (2), 569 (2), 570 (2), 572 (2), 573 (2), *573*
Shutskever, N. E., 304 (30), *343*
Shvedov, V. I., 294 (9, 10), 327 (9), 328 (9), 329 (9), 330 (9), *342*, 614 (39), 618 (39), *651*, 673 (54), 688 (54), *690*
Sickel, H., 310 (51), 315 (61), 319 (51, 61), 336 (51), *344*
Sidani, A. R., 72 (148), *77*
Siddappa, S., 227 (121), 228 (121), *232*
Sidorov, E. O., 90 (61, 62), *107*
Siegel, E., 171 (67), 174 (67), *177*
Siemion, I. Z., 304 (31, 134), 306 (136), 335 (136), 336 (136), *343, 346*
Sigiura, S., 238 (23), *239*
Sikora, H., 237 (12), *239*
Silk, J. A., 19 (11), 22 (52), 24 (52), *26, 27*, 29 (7), 31 (7, 16), 48 (16), 55 (98), 58 (16), 59 (7), 60 (16), 63 (7, 16), 64 (16), 65 (7, 16), 66 (7, 16, 98), 67 (7, 16), 68 (16), 69 (7, 16), 70 (16), 72 (16), *73, 75*, 120 (41), *122*, 123 (2), 125 (2), 127 (2), *128*, 154 (80), *161*, 206 (5), 215 (5), 222 (5), 224 (5, 100), 225 (5, 100), 226 (5, 100), 227 (5), 228 (5), *229, 231*, 254 (30), *260*, 626 (65), 647 (65), *652*
Silk, J. K., 128 (28), *129*
Silverman, G., 270 (64), 279 (64), 280 (64), 281 (64), 284 (64), *289*
Simek, A., *178* (102)
Siminetta, M., 545 (1), *566*
Simon, H., 224 (92), *231*
Simonek, I., 47 (75), 63 (75), *74*
Simonetta, M., 22 (43), *26*, 496 (1), *541*, 576 (3), *582*
Simonov, A. M., 635 (5), 654 (1, 2), 655 (1, 2), 657 (1, 5), 658 (12), 674 (2, 5), 675 (1, 5), *689*
Simpson, J. C. E., 6 (1), *6*, 17 (1), 22 (1), *25*, 78 (1), 79 (2), 97 (2), 98 (2), 100 (2), *106*, 123 (1), *128*, 137 (1), *158*, 162 (1), 171 (72), 172 (72), *175, 177*, 179

(1), 180 (13), 181 (13), 183 (13), 187 (1, 13), *191*, 199 (1, 4), 200 (4), 203 (4), *204*, 205 (1), *229*, 241 (1), *246*, 247 (1, 4), 255 (1, 4), *259*, 261 (1), *287*, 693 (3), 705 (3), *710*
Singh, R. P., 545 (6), 547 (6), 548 (6), 552 (6, 28), 553 (6), 554 (6), 557 (6), 559 (6), 564 (6, 28), 565 (6, 28), 566 (6, 28), *567*
Sircar, A. C., 664 (25), 665 (25), 685 (25), 687 (25), *689*
Skidmore, D., 21 (79), *27*
Skinner, C. G., 101 (155), 102 (155), *110*
Skinner, W. A., 21 (79), *27*
Skoldinov, A. P., 299 (21, 23), 303 (21, 128, 129), 328 (21, 128), 331 (129), 332 (128, 129), *343*, *346*, 473 (30), 480 (30), *493*, 605 (17), 634 (17), *651*
Skvortsov, I. M., 299 (19, 20), 301 (19), 328 (19), 335 (20), *343*
Slack, R., 449 (14), *455*
Slade, P., 471 (22), 473 (22), 487 (22), *493*
Slawinski, A. K., 116 (24), 118 (35), *122*
Slaymaker, S. C., 624 (48), 643 (48), *652*
Sleiter, G., 164 (21), 172 (21), *176*
Slezak, F. B., 498 (23), 540 (23), *541*
Smeby, R. R., 325 (152), 336 (152), *347*
Smith, A. A., 423 (15), *429*
Smith, A. E., 471 (22), 473 (22), 487 (22), *493*
Smith, A. I., 275 (85), *290*
Smith, B. B., 268 (49), 279 (49), 284 (49), *289*
Smith, D. L., 664 (21), *689*
Smith, E. L., 310 (54), *344*
Smith, J. G., 235 (6), *239*, 610 (31), 642 (31), *651*
Smith, N. H., 254 (28), *260*
Smith, P., 225 (113), 226 (113), *232*
Smith, P. W. G., 307 (36), 309 (48), 310 (48), 311 (48), 315 (48), 334 (48), *343*, *344*
Smith, R. F., 14 (71), *16*, 250 (17), 255 (17), 256 (17), 257 (17), *259*, 267 (22), 272 (22), 280 (22), 281 (22), 282 (22), 283 (22), 284 (22), *288*
Smith, S., 319 (82), 324 (82), 335 (82), *345*
Smith, S. C., 168 (41), *176*
Smith, V. K., 128 (29), *129*, 172 (84), *178*
Smith, V. K., Jr., 352 (10), 357 (10), *359*
Smith, W. R. D., 24 (68), *27*
Smith Broadent, H., 11 (40), 12 (46), *16*, 224 (103), *231*, 276 (91), 287 (91), *290*
Smolikova, J., 305 (135), 307 (137, 138), 335 (137), 336 (135), *346*
Smutney, E. J., 131 (3), 135 (3), *136*, 153 (63), *160*
Smyrniotis, P. Z., 95 (95), *108*
Snarey, M., 22 (45), *26*
Snatzke, G., 739 (40), 740 (40), 741 (40), *742*
Socha, J., 98 (127), 99 (127), *109*
Societa Farmaceutici Italia, 448 (13), 453 (13), *455*, 579 (12), 580 (12), 581 (12), *582*
Sociëtë des Usines Chimiques Rhône-Poulenc, 63 (130), *76*
Soddy, T., 13 (66), *16*, 268 (48), 284 (48), *289*
Soisson, D., 100 (137), *109*
Sokolov, S. V., 268 (51), 278 (51), *289*
Sokolova, A. S., 593 (9), 594 (9), 595 (9), 596 (9), *597*
Soldi, T. F., 713 (3), *727*
Solly, R. K., 431 (2), 433 (2), *439*
Solomons, J. A., 431 (4), 432 (4), 437 (4), *439*
Somerville, A. R., 423 (15, 16), *429*
Sonn, A., 352 (10), 357 (10), *359*, 431 (3), 437 (3), *439*
Sonnay, P., 456 (1), 457 (1), 459 (1), 460 (1), 461 (1), *461*
Soper, Q. F., 184 (41), 190 (77), *192*, *193*
Sörenson, N. A., 224 (97), *231*
Sorkin, E., 22 (51), *27*
Sorm, F., 701 (49), *711*
Sosnowski, C., 96 (97), *108*, 168 (49), 174 (49), *177*
Spano, F., 22 (40), 23 (40), 24 (66), *26*, *27* 225 (105), *231*
Spencer, H. E., 668 (43), 679 (43), *690*
Spikner, J. E., 93 (73), *108*
Spillmann, M., 447 (7), 448 (7), 453 (7), *454*

Spindler, J., 547 (19), 557 (19), 564 (19), 566 (19), *567*
Spinner, E., 496 (4), *541*, 545 (2), *567*, 577 (5), *582*
Spiro, V., 353 (14), 358 (14), *359*
Spoerri, P. E., 22 (49), *27*, 163 (11), 172 (11), *175*, 216 (43), *230*, 423 (20), 424 (20), 429 (20), *429*, 431 (4), 432 (4, 6), 437 (4, 6), *439, 440*
Spring, F. S., 30 (10), 33 (10), 54 (10), *73*, 89 (57), *107*
Springer, R. H., 148 (35), 153 (35), *159*
Srinivasan, V., 623 (45), 645 (45), *652*
Stacey, G. J., 30 (13), 31 (13), 58 (13), 59 (13), 60 (13), 61 (13), 62 (13), 63 (13), 65 (13), 66 (13), 67 (13), 68 (13), 69 (13), 70 (13), 71 (13), 72 (13), *73*, 163 (19), 168 (19), 171 (75), 174 (19), *176, 177*, 206 (4), 209 (4), 214 (4), 221 (4), 222 (4), 224 (4), 225 (4), 226 (4), *229*, 254 (30), 255 (48), *260*, 267 (35), 284 (35), 285 (35), *288*, 410 (3), 416 (3), 420 (3), *429* (13), 725 (48), *729*
Stachel, H.-D., 102 (166), *110*
Stadler, P., 318 (81), 326 (81), 335 (81), *345*
Stadler, P. A., 315 (62), 316 (69, 72), 317 (72, 78), 318 (69, 72, 80), 324 (150), 326 (69, 72, 78, 80, 121, 153), 327 (72), 340 (121), *344, 346, 347*
Stadtman, E. R., 95 (95), *108*
Stahlhofen, P., 103 (174), *111*, 263 (12), *288*
Stammer, C. H., 98 (200), *111*
Stanovnik, B., 410 (25), 411 (25, 27), 412 (26), 414 (26), 425 (25, 26, 27), 426 (25, 26, 27), *430*
Stapley, E. O., 63 (123, 124), *76*
Stapley, E. P., 28 (1), 63 (1), *72*
Starke, R., 693 (5), 699 (5), 703 (5), *710*
Stearns, B., 498 (28), 525 (28), *542*
Stefancich, G., 725 (54), *729*
Stefanescue, M., 252 (22), 255 (22), *260*, 720 (30), *728*
Stefaniak, L., 10 (28), *15*
Steglich, W., 82 (22), 99 (22), 100 (22, 136), 101 (22), 102 (22), *106, 109*
Stein, M. L., 238 (31), *239*
Stein, R. A., 153 (61), *160*, 221 (69), *230*
Steiner, P. R., 237 (13), *239*
Steinmetz, R., 130 (1), 134 (1), 135 (1), 136 (1), *136*, 262 (6), 277 (6), *288*
Stene, P., 267 (23), 283 (23), 284 (23), 286 (23), *288*
Stener, A., 80 (12), 100 (12), *106*, 713 (8), 726 (8), *727*
Stephen, W. I., 497 (15), *541*, 731 (5), 740 (5), *741*
Stephenson, O., 227 (120), 228 (120), *232*
Stepien, A., *111* (199)
Steresco, M., 81 (16), 97 (16), 98 (16), *106*, 114 (13), 119 (13), *121*, 166 (31), *176*, 556 (32), *567*
Sterk, H., 87 (46), *107*, 210 (23), *229*
Sterling Drug, Inc., 569 (4), 572 (4), *573*
Sternbach, L. H., 86 (34, 35a), 91 (63), 102 (34), 103 (34), *106, 107*, 264 (15, 16), 270 (64), 277 (15, 16), 278 (16), 279 (64), 280 (64), 281 (64), 284 (15, 64), *288, 289*
Stetter, H., 268 (52), 279 (52), *289*, 579 (17), 581 (17), *582*
Stevens, C. L., 228 (138), *232*
Stevens, J. R., 23 (54), 24 (54), *27*, 79 (4), 97 (113), *106, 109*, 172 (78), *177*, 187 (62), 189 (73), 190 (73), *193*, 716 (21), 722 (21), 723 (21), *728*
Stewart J. M., 94 (86), 100 (86), *108*
Steyn, P. S., 324 (105, 148), 337 (148), 339 (148), *345, 347*
Stingl, H. A., 227 (118), *232*
Stock, A. M., 222 (80), *231*
Stock, J. A., 324 (115), 337 (115), *346*
Stogryn, E. L., 282 (95), 286 (95), *290*
Stoicescu-Crivat, L., 252 (22, 23), 255 (22, 23), *260*, 720 (28, 29, 30), 724 (28, 29) *728*
Stoklosa, H. J., 497 (62), 531 (62), *543*
Stoll, A., 301 (27, 28), 301 (28), 307 (28), 319 (83), 322 (27, 28), 324 (28, 83, 106, 107, 108), 327 (120), 328 (27, 28), 329 (28), 332 (28), 333 (28), 335 (28, 83), 336 (28), 337 (28), 338)28), *343*, 345 *346*
Stone, P., 222 (73), *231*
Storr, R. C., 88 (50), *107*

Strakov, A. Ya., 100 (138), *110*
Straw, D., 82 (21), 101 (21), *106*
Streckenbach, B., 269 (60), 283 (60), 284 (60), *289*
Streef, J. W., 20 (78), *27*
Streier, M. P., 10 (32), *15*
Streuli, M., 732 (9), 739 (9), 740 (9), *741*
Striebeck, A., 101 (157a), *110*
Strier, M. P., 22 (42), 23 (42), 24 (42), *26*
Strugeon, B., 227 (120), 228 (120), *232*
Stuart, A., 277 (9), *288*
Stueben, K. C., 310 (50), 334 (50), 336 (50), *344*
Stuetz, P., 324 (150), 326 (153), *347*
Sturgeon, B., 750 (11), 757 (11), *758*
Sturm, B., 275 (81), *290*
Stutz, P., 318 (80, 81), 326 (80, 81), 335 (81), *344*, 345
Sucharda-Sobczyk, A., 304 (134), *346*
Sugimoto, N., 431 (5), 437 (5), *439*
Sugiura, S., 362 (5, 59), 363 (9), 364 (9, 11, 12, 13), 365 (11, 18), 366 (11, 19, 20, 21, 22), 367 (21), 370 (5), 371 (13), 383 (5, 11, 18, 22), 384 (5, 18, 21, 22), *395, 396, 397*
Sugiura, T., 740 (44), *742*
Sugiyama, H., 96 (100), 104 (100), 105 (100), *108*, 171 (64), *177*
Suling, C. H., 324 (118), *346*
Sullivan, A. P., 181 (23), 187 (23), 188 (23), *192*, 276 (10), 279 (10), *288*
Sullivan, M. X., 24 (69), *27*, 255 (36), 256 (36), *260*
Sumitomo Chemical Co. Ltd., 96 (99), 105 (99), *108*, 174 (99), *178*
Surrey, A. R., 157 (101), *161*
Sus, O., 447 (5, 6), 448 (5, 6), 453 (5, 6), *454*
Suschitzky, H., 83 (24), 99 (24), *106*, 227 (126), *232*, 630 (81), 649 (81), *653*
Suzuki, A., 324 (96), *345*
Suzuki, S., 146 (27), *159*
Suzuki, U., 315 (62), *344*
Swadesh, S., 299 (18), *342*
Sweeny, W., 89 (57), *107*
Switzer, J. L., 187 (55), *193*
Switzer, R. C., 187 (55), *193*
Sykes, A., 224 (85), *231*
Synak, J., 97 (103), *109*
Syono, K., 414 (8), *429*
Syrova, G. P., 48 (80), 57 (80), 58 (80), *75*, 124 (7), 125 (7), 127 (7), 128 (7), *128*, 149 (44), 157 (44), *160*
Szabo, Z., 470 (17), 482 (17), 483 (17), 490 (17), 491 (17), *493*
Szantho, V., 103 (177), *111*

Tabak, S. V., 402 (7), *408*
Tabaksfabrick-Koffiebranderijon Theehandle N. V., 456 (3), 459 (3), 460 (3), *462*
Tabner, B. J., 10 (36), *15*
Tabuchi, T., 324 (103, 104), *345*
Tabusa, F., 235 (43), *240*
Taeger, E., 116 (22), 120 (22), 121 (22), *122*
Taha, M. U., 40 (48), 65 (48), 70 (48), 71 (48), *74*
Tajana, A., 102 (169), *110*
Takada, A., 570 (5), 572 (5), *573*
Takada, T., 56 (103), *75*
Takahashi, H., 20 (21), *26, 176* (36), 324 (149), *347*, 624 (52), 644 (52), *652*, 656 (6), 675 (6), 673 (55, 56), 687 (55), 688 (55, 56), *689, 690*
Takayama, Y., 13 (64), *16*
Taki, H., 139 (4), 156 (4), *159*
Tagami, S., 95 (91), 98 (114, 118), *108*, *109*, 113 (5), 119 (5), *121*, 163 (12), 172 (12), 173 (95), *175, 178*, 180 (3, 5), 187 (3), 189 (5), 190 (78), *191, 193*, 194 (8), 195 (8, 9), 196 (11, 12), 197 (8, 9, 11), *198*, 200 (7), 202 (7), *204*
Talik, T., 548 (20), 566 (20), *567*
Talyor, E. C., 115 (17), *121*
Tamayo, M. L., 157 (105), 158 (105), *161*
Tamura, S., 324 (96), *345*
Tamura, Y., 235 (40, 41, 43), *239, 240*
Tanaka, S., 267 (44), 286 (44), *289*
Tanaka, T., 431 (5), 437 (5), *439*, 511 (63), 531 (63), 533 (63), 534 (63), 535 (63), 536 (63), 537 (63), 538 (63), 539 (63), *543*, 572 (11), *574*
Tanake, S., 224 (102), 226 (102), *231*
Tanner, H., 82 (22), 99 (22), 100 (22, 136), 101 (22), 102 (22), *106, 109*

Tappi, G., 436 (19), 439 (19), *440*
Tarasiejska, Z., 367 (23), 376 (23), 377 (23), 384 (23), 393 (23), 395 (23), *396*
Tardos, L., 470 (17), 482 (17), 483 (17), 490 (17), 491 (17), *493*
Tarnavskaya, N. E., 102 (167), 103 (167), 104 (167), *110*
Tate, M. E., 267 (28), 279 (28), *288*
Tatematsu, A., 42 (55, 56), *74*
Tatlow, J. C., 224 (85), *231*, 268 (50), 278 (50), *289*
Tatsuno, T., 323 (86), 324 (86), *345*
Taylor, E. C., 34 (30), *73*, 92 (65), 98 (65), 100 (65), 101 (65), 102 (65), *107*, 125 (13), *128*, 139 (3), 146 (28), 153 (3), *158, 159*, 166 (34, 35), *176*, 210 (21), 213 (21, 30), 218 (21, 30, 53), 220 (53), 222 (72), 223 (72), 225 (21), 228 (30), *229, 230, 231*, 235 (7), *239*, 401 (4), 402 (4, 10), 405 (4, 17), 406 (4, 10), 407 (4, 10, 17), *408, 409*, 506 (41), *542*, 600 (6), *651*, 605 (22, 23, 24), 607 (22, 24), 608 (27), 609 (30), 610 (24), 615 (24), 618 (22), 619 (24), 621 (24), 634 (24, 27, 30), 635 (22, 23, 27, 30), 636 (22, 23, 24, 27), 638 (24), 639 (24), 640 (24), 642 (24), *651*, 743 (3), 754 (3), *758*
Taylor, J. W., 30 (14), 72 (14), *73*, 101 (159), *110*, 153 (62), *160*
Taylor, P. J., 410 (3), 416 (3), 420 (3), *429* (3)
Teichmann, B., 693 (12), 703 (12), 703 (52), *710, 711*
Telang, S. A., 467 (9), 478 (9), 479 (9), 482 (9), 483 (9), 487 (9), 488 (9), 489 (9), *493*
Temple, C., Jr., 498 (26, 27, 30), 499 (30), 502 (26, 27, 32, 33, 34), 509 (27, 30, 32, 48, 49), 515 (48, 49), 524 (49), 525 (30, 32), 526 (27, 48, 49), 527 (26, 27, 30, 49), 528 (26, 27, 32, 34), 529 (26, 27), 530 (30), 531 (49), *542*, 545 (7), 548 (24), 551 (7, 24), 555 (7, 29, 30), 557 (24), 558 (29, 30), 559 (24), 560 (7, 24), 561 (7, 24, 29, 30), 562 (7, 24, 30), 563 (7, 24, 30), *567*
Tennant, G., 33 (22, 23, 24), 36 (36), 51 (23, 89), 53 (92, 93), 56 (22), 57 (22, 24), 58 (22, 24), 59 (24, 115), 60 (36, 115), 61 (22, 24, 36, 93, 115), 62 (93, 115), 70 (36), *73, 75, 76*, 84 (31), 97 (109), 99 (109, 135), 100 (31, 144), 101 (31, 171), 102 (31, 135, 171), *106, 109, 100*, 145 (24), 153 (24), 154 (24, 87), *159, 161*, 249 (14), 252 (14), 257 (14), 258 (14), *259*
Terada, H., 740 (44), *742*
Thakar, A. A., 41 (53), 60 (53), 61 (53), *74*, 238 (38), *239*
Theil, M., 225 (106), *231*
Theimer, E. T., 23 (57), 24 (57), *27*
Theriot, L. J., 118 (32, 33), *122*
Thetaz, C., 422 (28), 424 (28), *430*
Thibault, T. B., 97 (158), *110*
Thiele-Schleussner, 423 (19), 428 (19), *429*
Thomas, G. N., 22 (45), *26*
Thomas-Mamert, R., 458 (5), 460 (5), *462*, 732 (8), 740 (8), *741*
Thompson, M. J., 139 (3), 141 (13), 146 (28), 153 (3), 155 (13), *158, 159*, 267 (25), 283 (25), *288*, 504 (37), 506 (37, 41), 507 (37), 512 (37), 518 (37), 521 (37), 522 (37), 531 (37), 532 (37), 533 (37), 534 (37), 535 (37), 536 (37), 537 (37), 538 (37), 539 (37), *542*, 659 (15), 661 (15), 676 (15), *689*
Thompson, T. W., 410 (3), 416 (3), 420 (3), *429*
Thorp, J. M., 55 (98), 66 (98), *75*
Thorpe, J. M., 128 (28), *129*
Threlfall, T. L., 496 (10), 498 (10, 16), 501 (10), 502 (10), 504 (10), 505 (10), 508 (10), 510 (10), 512 (10), 513 (10), 514 (10), 515 (10), 517 (10), 518 (10), 519 (10), 520 (10), 524 (10), 525 (10), 526 (10), 530 (10), 531 (10), 532 (10), 533 (10), 534 (10), 535 (10), 537 (10), 538 (10), 539 (10), 540 (10), 541 (10), *541*, 550 (26), 552 (26), 553 (26), 558 (26), 559 (26), 564 (26), 565 (26), 566 (26), *567*
Tibbets, M. S., 23 (57), 24 (57), *27*
Tietzman, J. E., 312 (58), 319 (58), 321 (58), 336 (58), 337 (58), *344*
Timkova, P. M., 63 (128), 65 (128), 66 (128), 67 (128), *76*

Timmis, G. M., 144 (19), 153 (19), 155 (19), *159*, 182 (26), 187 (26), 188 (26), 189 (26), *192*, 319 (82), 324 (82), 335 (82), *345*, 496 (9), 508 (9), 518 (9), 524 (9), 525 (9), 526 (9), 532 (9), *541*, 716 (21), 717 (22), 722 (21), 723 (21), 726 (22), *728*
Timmons, C. J., 226 (115), *232*
Tinland, B., 496 (2), 502 (2), *541*, 545 (1), *566*
Tishler, M., 112 (1), 114 (1), 119 (1), *121*, 155 (91), 156 (91), *161*, 181 (21, 23), 187 (23), 188 (23), *192*, 276 (10), 279 (10), *288*, 410 (25), 411 (25, 27), 412 (26), 414 (26), 425 (25, 26, 27), 426 (25, 26, 27), *430*
Tisselli, E., 267 (40), 277 (40), *288*, 629 (69), 646 (69), *652*
Titov, V. V., 84 (29), 99 (29), 101 (29), *106*, 133 (10), 134 (14), *136*, 155 (90), *161*, 222 (78), *231*, 282 (101), *290*
Tittensor, E., 550 (27), 558 (27), 564 (27), *567*
Tocker, S., 294 (7), 327 (7), 328 (7), *342*
Toczko, K., 322 (84), 324 (110), *345*
Todd Miles, H., 95 (95), *108*
Toggweiler, U., 20 (23), 23 (23), *26*
Toja, E., 385 (63), 386 (63), 387 (63), *397*
Tokar, G., 47 (75), 63 (75), *74*
Tombacz, E., 86 (42), *107*
Tomcufcik, A. S., 168 (40), *176*
Tomimoto, M., 196 (13), *198*
Tong, H., 21 (79), *27*
Tong, Y. C., 382 (69), 394 (69), 395 (69), *397*, 450 (19), 451 (19), 454 (19), *455*
Tönjes, H., 228 (141), *232*
Tori, K., 41 (52), *74*, 250 (15), *259*
Torigoe, M., 149 (39), 158 (39), *159*
Torigoe, Y., 411 (5), 414 (5), 426 (5), *429*
Torikoshi, Y., 411 (6), 426 (6), *429*
Törz, E. S. G., 46 (74), 56 (74), *74*
Törzs, E. S. G., 238 (27), *239*
Tosi, G., 70 (147), *77*
Tosolini, G., 102 (161), *110*
Towell, E. E., 225 (108), *231*, 238 (29), *239*
Towle, J. L., 224 (98, 99), *231*
Towne, J. C., 93 (73), *108*
Townsend, L. B., 376 (37), 379 (37), 395 (37), *397*
Toyo Rayon Co. Ltd., 433 (10), *440*
Toyoshima, S., 317 (75), 339 (75), *344*
Tran Hong Thuy, 10 (34), 11 (34), *15*
Trappe, G., 268 (55), 279 (55), *289*
Traven, N. I., 592 (6), 593 (6), 595 (6), 596 (6), *597*
Traverso, J. J., 22 (46), *26*
Traxler, J. T., 221 (61), *230*
Traxler, P. M., *347* (145)
Trenholm, G., 474 (36), 484 (36), *494*
Trepanier, D. L., 568 (2, 3), 569 (2), 570 (2, 3), 572 (2, 3), 573 (2, 3), *573*
Treuner, U. D., 691 (55), *711*
Tripod, J., 254 (29), *260*
Tripp, R. C., 623 (44), *652*
Tristram, E. W., 377 (46), 378 (52), 380 (46, 52), 389 (46, 52), 390 (46, 52), 391 (46, 52), 392 (46, 52), 393 (46, 52), *397*
Troxler, F., 316 (72), 317 (72), 318 (72), 326 (72), 327 (72), *344*
Tserng, K. -Y., 665 (28), 684 (28), 686 (28), *689*
Tsizin, Y. S., 21 (27, 77), 23 (27), *26, 27*
Tsuji, J., 97 (111), *109*
Tsunoda, H., 323 (86), 324 (86), *345*
Tsutsumi, S., 116 (20), *122*
Tsutsumi, T., 196 (14), *198*
Tsyrul'nikova, L. G., 30 (9), 47 (76, 77, 78), 48 (80), 49 (84), 56 (9, 84), 57 (9, 80, 84, 107, 108, 109), 58 (9, 80), 59 (9, 111), 60 (9, 109), 63 (9, 78, 107, 111, 128), 64 (9, 109), 65 (9, 78, 107, 128), 66 (15, 76, 128), 67 (111, 128), 68 (76, 109), 69 (76, 109), 70 (76, 77, 107, 109), 71 (77, 111), 72 (77), *73, 74, 75, 76*, 95 (92), 98 (115, 115a), *108, 109*, 124 (7), 125 (7), 127 (7), 128 (7), *128*, 149 (44), 157 (44), *160*, 174 (100), *178*, 181 (19), 183 (19), 187 (19, 59), 189 (71), *191, 193*, 213 (29), 215 (40), 222 (29), 224 (96), 225 (29), 226 (29), *229, 230, 231*, 283 (98), *290*, 667 (32), 684 (32), 686 (32), *689*

Tuchtenhagen, G., 376 (54), 379 (54), 395 (54), *397*
Tuck, B., 218 (55), 220 (58), *230*, 599 (1, 2, 3, 4), 600 (2, 4), 601 (2, 3), 608 (4), 610 (4), 615 (3), 616 (1, 11), 617 (11), 619 (3), 621 (4), 622 (11), 633 (3, 11), 634 (2, 3), 635 (2, 3, 4, 11), 636 (11), 638 (2), 639 (4), 640 (4), 641 (4), *650*
Tulecki, J., 114 (14), *121*
Tull, R. J., 377 (45, 46), 378 (50, 52, 53), 380 (45, 46, 50, 52, 53), 389 (45, 46, 50, 52, 53), 390 (45, 46, 50, 52, 53), 391 (45, 46, 50, 52, 53), 392 (45, 46, 50, 52, 53), 393 (45, 46, 50, 52, 53), *397*
Turner, C. J., 599 (1), 616 (1), *650*
Turner, E. E., 169 (53), *177*
Tuszynska, M., 498 (25), 540 (25), *542*
Tyl'chinskaya, L. S., 143 (16), 153 (16), 155 (16), *159*
Tyurenkov, I. N., 658 (12), *689*

Uden, P. C., 497 (15), *541*, 731 (5), 740 (5), *741*
Ueda, T., 317 (75), 339 (75), *344*, 570 (5), 572 (5), *573*
Ueno, K., 740 (32), *742*
Undheim, K., 23 (58), *27*, 99 (134), 100 (134), *109*, 221 (66), 224 (66), *230*
Ungureanu, M., 247 (5), 252 (25), 253 (5), 256 (5, 25), *259, 260*, 610 (33), 641 (33), *651*, 656 (9), 675 (9), *689*, 744 (5), 754 (5), *758*
Unterstenhoefer, G., 113 (6), 116 (16), 117 (26, 28, 29), 118 (30, 36), 119 (16), 120 (6, 28, 29), 121 (28), *121, 122*, 181 (16), 188 (16), *191*, 735 (28, 29, 30), *742*
Urono, T., 594 (12), *597*
Uryukina, I. G., 635 (5), 654 (1, 2), 655 (1, 2), 657 (1, 5), 674 (2, 5), 675 (1, 5), *689*

Vajna de Pava, O., 227 (128), 228 (128), *232*
Valenta, A., 155 (89), 156 (89), 157 (89), *161*, 188 (68), *193*
Valenta, J. R., 58 (149), *77*
Valenta, V., *178* (102)
Valitova, F. G., 228 (132), *232*
Vallat, J. N., 244 (12), *246*, 693 (7), 701 (7, 47), 703 (7, 47), 704 (7, 47), 705 (7, 47), 706 (47), 709 (47), *710, 711*, 725 (49), *729*
van Beek, H. C. A., 13 (65), *16*
van der Meer, D., 10 (38), *15*
van der Plas, H. C., 20 (78), *27*, 163 (9), 165 (24), 172 (9), 173 (9), *175, 176*, 187 (64), 188 (64), *193*
Van Deusen, R. L., 580 (20), *582*
Vandoni, I., 496 (1), *541*, 545 (1), *566*, 576 (3), *582*
Van Dusen, J., 83 (25), 98 (25), *106*
Van Es, T., 234 (2), *238*
Van Geissen, G. J., 81 (15), 97 (15), *106*, 141 (9), 152 (9), *159*, 171 (74), *177*, 181 (22), 187 (22), *192*
Vanicek, V., 697 (42), 706 (42), *711*
van Koningsveld, H., 583 (1), *584*
Van Lare, E. J., 184 (43), *192*, 665 (29), 667 (31), 668 (29, 31, 35, 36, 37, 43), 669 (29, 31), 670 (29, 31), 677 (29, 31), 678 (29, 31, 35, 36), 679 (29, 31, 35, 36, 37, 43), 680 (29), 681 (31), 682 (29, 31, 35, 36), 683 (31), 685 (29, 31), 686 (31), 687 (31), *689, 690*, 723 (39), *728*
van Leeuwen, M., 7 (3), *15*
Van Soest, T. C., 217 (50), *230*
Varyukhina, L. V., 42 (57, 60), *74*
Vaughan, W. R., 163 (15), 172 (15), 173 (15), *176*, 184 (46), 188 (46), 189 (46), 190 (46), 191 (46), *192*, 623 (44), *652*
Vecera, M., 102 (168), 103 (168), *110*, 162 (4), 173 (4), *175*, 624 (53), 643 (53), *652*
Veisova, Z., 693 (9), 701 (9), *710*
Venkateswarlu, A., 600 (10), 601 (10), 603 (10), 615 (10), 616 (10), 619 (10), 622 (10), 633 (10), 634 (10), 637 (10), 638 (10), 639 (10), 642 (10), *651*
Verbeek, A. J., 165 (24), *176*
Verbeek, J., 13 (65), *16*
Vercek, B., 410 (25), 411 (25, 27), 412 (26), 414 (26), 425 (25, 26, 27), 426 (25, 26, 27), *430*
Verini, M. A., 579 (13), 580 (13), 581 (13), *582*

Verkade, P. E., 7 (3), *15*
Vetesnik, P., 210 (24, 25, 26, 27, 28), 222 (28, 76), 224 (25, 28, 76, 142), 227 (27), *229, 231, 232*
Vianello, E., 93 (74), *108*
Vicar, J., 305 (135), 307 (137, 138), 335 (137), 336 (135), *346*
Vilsmeier, W., 169 (52), 172 (52), *177*, 184 (57), 187 (57), *193*, 735 (26), 739 (26), *742*
Vinot, N., 497 (61), 515 (65), 517 (65), 521 (61), 534 (61, 65), 536 (65), 538 (61, 65), 539 (61, 65), 540 (61, 65), 541 (65), *543, 567* (35)
Visser, G. J., 206 (8, 9), *229*
Vitzthum, O. G., 456 (8), *462*
Vogler, H., 730 (39), *742*
Von Dobeneck, H., 352 (6), 358 (6), *359*
Von Dreele, R. B., *347* (145)
von Euler, H., 59 (113), *76*
Von Pechmann, H., *231* (107)
Von Saltza, M., 706 (50), *711*
Vos, A., 206 (8, 9), *229*
Vosatka, V., *178* (102)
Vouros, P., 272 (73), *289*
Vul'fson, N. S., 80 (11), 99 (11), *106*, 308 (45), *343*

Wadia, M. S., 41 (53), 60 (53), 61 (53), *74*, 238 (38), *239*
Wagner, A., 736 (38), *742*
Wagner, G., 88 (48, 49), *107*, 117 (25), 119 (25), 120 (25), *122*
Wagner-Jauregg, T., 624 (54), 644 (54), 645 (54), *652*
Wahl, H., 248 (7), *259*
Wait, S. C., Jr., 496 (3), *541*, 545 (1), *566*, 576 (2), *582*, 584 (2), *584*
Wajsman, E., *111* (199)
Wake, S., 13 (64), *16*
Wakefield, B. J., 83 (24), 99 (24), *106*, 630 (81), 649 (81), *653*
Waldradt, J. P., 23 (60), 24 (60), *27*
Walker, J. S., 9 (17), *15*
Walker, S. R., 423 (18), *429*
Wallace, R. A., 505 (38), *542*
Walradt, J. P., 456 (2), *461*
Walser, A., 270 (64), 279 (64), 280 (64), 281 (64), 284 (64), *289*
Wang, C.-S., 128 (30), *129*, 713 (9), 714 (9), 724 (9), *727*
Wang, S., 648 (57), *652*
Waravdekar, V. S., 254 (27), *260*
Ward, J. P., 217 (49), 220 (49), 225 (49), 228 (49), *230*, 263 (103), 277 (103), 278 (103), *290*
Ward, R. J., 170 (60), *177*
Ward, R. L., 203 (20), 204 (20), *204*
Warren, B. T., 62 (120), 72 (120), *76*
Warren, T. F., 449 (14), *455*
Wasleski, D. M., 202 (13), *204*
Wasson, J. R., 497 (62), 531 (62), *543*
Watanabe, H., 377 (43, 44), 380 (43), 395 (43, 44), *397*, 593 (7, 10), 594 (7, 10), 595 (7, 10), *597*
Watanabe, N., 196 (15), *198*
Wear, R. L., 97 (109), 99 (109), *109*, 171 (76), 172 (76), *177*, 180 (7), 188 (7), 189 (7), *191*, 214 (33), 223 (33), *230*
Webb, G. A., 10 (28), *15*
Weber, K. H., 95 (94), 105 (94), *108*
Webster, F. G., 669 (45), *690*
Weedon, B. C. L., 225 (113), 226 (113), *232*
Wegland, F., 154 (83), *161*
Wegler, R., 113 (6), 116 (16), 117 (26, 28, 29), 118 (30, 36), 119 (16), 120 (6, 28, 29), 121 (28), *121, 122*, 168 (45), 174 (45), *176*, 181 (16), 188 (16), *191*, 459 (6), 461 (6), *462*, 735 (28, 29, 30), *742*
Wegmann, J., 223 (83), 224 (83, 86), 226 (83), 227 (122), *231, 232*
Wehrli, F. W., 422 (28), 424 (28), *430*
Weidinger, H., 168 (43), *176*, 184 (48), 190 (48), *192*
Weijlard, J., 181 (21, 23), 187 (23), 188 (23), *192*, 276 (10), 279 (10), *288*
Weil, S., 458 (5), 460 (5), *462*, 732 (8), 740 (8), *741*
Weinberger, L., 353 (12), 357 (12), *359*
Weintraub, A., 324 (101), *345*
Weis, P., 227 (125), *232*
Weis, W., 466 (7), 467 (7), 468 (7), 477 (7), 489 (7), 491 (7), 492 (7), *493*
Weiser, D., 245 (14), *246*
Weiss, C., 216 (44), *230*

Weiss, W., 547 (17), *567*
Weissenhorn, R. G., 235 (4), 238 (4), *239*
Wellenreuther, G., 168 (43), *176*
Wellman, J. M., 155 (91), 156 (91), *161*
Wendt, G. R., 697 (38), 703 (38), *711*
Wentrup, C., 422 (28), 423 (21), 424 (21, 28), 425 (24), 429 (21), *429, 430*
Werbel, L. M., 365 (17), 384 (17), *396*, 752 (17), 758 (17), *759*
Werde, F., 328 (122), *346*
Werkhoff, P., 456 (8), *462*
Werner, L. H., 447 (7, 9), 448 (7), 453 (7, 9), *454*
Wersin, G., 318 (80, 81), 326 (80, 81, 153), 335 (81), *344, 345, 347*
Werstiuk, E. S. G., 6 (5), *6*, 56 (100), *75*
Wesley, J. W., 496 (3), *541*, 545 (1), *566*, 576 (2), *582*, 584 (2), *584*
West, D. E., 100 (143), *110*, 155 (92), *161*
Westermayer, G., 235 (4), 238 (4), *239*
Westley, J. W., 306 (33), 307 (33), 315 (67), 335 (67), *343, 344*
Weston, A. W., 276 (89), 287 (89), *290*
Weston Hurst, E., 254 (30), *260*
Westphal, G., 206 (10), 225 (10), 226 (10), *229*
Westphal, O., 214 (38), *230*, 693 (12), 703 (12, 52), *710, 711*
Westwood, R., 220 (57), *230*
Wettstein, A., 447 (7, 8, 9), 448 (7), 453 (7, 8, 9), *454*
Weygand, F., 82 (22), 99 (22), 100 (22, 146), 101 (22), 102 (22, 146), *106*, *100*, 702 (51), *711*
Whistler, R. L., 695 (26), 707 (26), *711*
White, E. H., 24 (64), *27*, 267 (36), 282 (36), 283 (36), *288*, 324 (146), *347*, 365 (15), *396*
Whitehead, C. W., 22 (46), *26*
Whitman, P. J., 148 (33), 153 (33), *159*, 242 (4), 245 (4), *246*, 267 (24), 280 (24), *288*
Whittaker, R. A., 83 (24), 99 (24), *106*, 630 (81), 649 (81), *653*
Whittle, C. W., 11 (40), *16*, 276 (91), 287 (91), *290*
Wibberley, D. G., 531 (50), 532 (50), *542*
Wiberg, K., 497 (14), *541*
Wiberg, K. B., 10 (33), *15*
Widmann, A., 23 (61), *27*
Wieczorek, J. S., 548 (20), 566 (20), *567*
Wild, G. M., 693 (16), 695 (16), 700 (16), *710*
Wildersohn, M., 516 (53), 517 (53), 530 (53), 531 (53), 533 (53), 536 (53), *542*, 734 (31), 735 (31), 739 (31), *742*
Wiley, P. F., 267 (43), 283 (43), 284 (43), 286 (43), *289*
Wiley, R. H., 624 (48), 643 (48), *652*
Willems, H., 318 (80, 81), 326 (80, 81, 153), 335 (81), *344, 345, 347*
Williams, C., 21 (36), 22 (36), 23 (36), 25 (36, 71), *26, 27*, 105 (193), *111*, 174 (80), 175 (80), *178*, 268 (47), 282 (47), 286 (47), *289*
Williams, D. H., 148 (32a), *159*
Williams, J. H., 128 (29), *129*, 172 (84), *178*, 352 (10), 357 (10), *359*
Williams, R. L., 105 (197), *111*, 226 (116), *232*
Willis, G. H., 214 (35), 224 (35), *230*
Willman, N., 222 (73), *231*, 267 (23), 283 (23), 284 (23), 286 (23), *288*
Wilson, F. H., 275 (80, 81), *289, 290*
Wilson, G., 270 (70), *289*
Wilson, K., 180 (9), *191*
Wilson, R. A., 23 (57), 24 (57), *27*
Wilson, R. M., 79 (4), *106*, 112 (1), 114 (1), 119 (1), *121*
Winfield, G., 227 (129), *232*
Winnefeld, W., 238 (33), *239*
Winter, M., 244 (7), 245 (7), *246*
Winterfeld, K., 469 (12), 470 (18), 471 (18, 23, 24), 476 (23), 478 (18), 480 (23, 24), 482 (12, 18, 23, 24), 483 (12, 23, 24), 484 (12, 23), 485 (23), 486 (23), *493*, 516 (53), 517 (53), 530 (53), 531 (53), 533 (53), 536 (53), *542*, 734 (31), 735 (31), 739 (31), *742*
Wintersteiner, O., 706 (50), *711*
Wiselogle, F. Y., 272 (74), 279 (74), 280 (74), 281 (74), 282 (74), 283 (74), 284 (74), *289*
Witanowski, M., 10 (28), *15*

Wittmann, H., 474 (37), 490 (37), 491 (37), *494*
Wojciechowski, L., 96 (97), 105 (190), *108, 111*, 116 (23), 118 (31), 120 (31), *122,* 168 (49), 174 (49), *177*
Wolf, F. J., 23 (54), 24 (54), *27*, 28 (1), 63 (1, 123), *72, 76*, 79 (4), *106*, 112 (1), 114 (1), 119 (1), *121*, 162 (3), *175*, 180 (2), 187 (2, 62), *191, 193*, 716 (21), 722 (21), 723 (21), *728*
Wolf, M., 633 (71), 649 (71), 650 (71), *652*
Wolfe, J. F., 217 (52), *230*
Wood, H. C. S., 24 (68), *27*, 147 (31), 153 (31), 154 (31), 155 (31), 157 (31), *159*, 277 (9), *288*
Wood, N. F., 545 (8), 551 (8), 554 (8), 557 (8), 558 (8), 561 (8), *567*
Woolley, D. W., 94 (86), 100 (86), *108*
Worley, S. D., 8 (9, 10), *15*
Woyciesjes, A., 323 (92), 335 (92), *345*
Wunsch, E., 310 (49), 324 (112), *344, 346*
Wuonola, M. A., 184 (49), 186 (49), 187 (49), *192*, 671 (59), *690*
Wynberg, H., 206 (7, 9), 228 (7), *229*, 278 (11), 285 (11), *288*

Yada, Y., 25 (75), *27*, 105 (195), *111*, 170 (58), 175 (58), *177*
Yajima, I., 25 (75), *27*, 105 (95), *111*, 170 (58), 175 (58), *177*
Yakhontov, L. N., 349 (2), 356 (2), *359*
Yakunina, N. G., 400 (20), 407 (20), *409*
Yamada, S., 12 (44), *16*, 42 (56), 45 (71), *74*
Yamamoto, H., 93 (77, 78), *108*, 171 (62), 174 (62), *177*
Yamamoto, K., 44 (68), *74*, 196 (15), *198*
Yamamoto, S., 723 (37), *728*
Yamatani, W., 723 (37), *728*
Yamazaki, T., 470 (20), 471 (20), 480 (20), 482 (20), *493*
Yandle, J. R., 10 (36), *15*
Yanina, A. D., 476 (41), 483 (41), *494*
Yankovskaya, I. S., 237 (39), *239*
Yano, T., 740 (32), *742*
Yarish, M. E., 736 (37), *742*
Yellin, T. O., 93 (80), 100 (80), *108*
Yokoe, I., 42 (56), 45 (71), *74*
Yoneda, M., 115 (18), *122*, 187 (60), *193*
Yonezawa, T., 86 (36), *107*
Yoshida, K., 89 (55), *107*, 171 (71), *177*, 656 (6), 675 (6), *689*, 673 (55, 56), 687 (55), 688 (55, 56), *690*
Yoshioka, I., 149 (42), 153 (42), 157 (42), 158 (42), *159*
Yoshioka, K., 12 (44), *16*, 171 (64), *177*
Yoshizumi, H., 42 (55, 56), *74*
Young, A. C., 255 (37), *260*
Young, J. W., 125 (8), *128*
Yruela Antinolo, M., 244 (10, 11), *246*, 743 (2), *758*
Yukimoto, Y., 296 (16), 330 (16), *342*
Yur'ev Yu, K., 431 (1), 432 (8), 437 (8), *439, 440*
Yurkevich, A. M., 144 (20), 155 (20), *159*, 238 (24), *239*
Yurugi, S., 196 (13), *198*

Zahner, H., 323 (93), *345*
Zahr, G. E., 39 (47), 48 (47), 62 (47), *74*
Zaitsu, K., 724 (51), *729*
Zakusov, V. V., 303 (128), 328 (128), 330 (131), 332 (128), *346*
Zally, W., 91 (63), *107*, 264 (16), 277 (16), 278 (16), *288*
Zamet, J. J., 38 (40), 61 (40), *73*, 102 (160), *110*
Zamora, M. L., 365 (17), 384 (17), *396*, 752 (17), 758 (17), *759*
Zeigler, E., 275 (84), 285 (84), *290*
Zeller, E. A., 614 (40), 638 (40), *651*
Zeller, P., 228 (136), *232*
Zellner, G., 103 (186), 104 (186), *111*, 256 (41), 258 (41), *260*, 272 (75), 279 (75), 282 (75), 283 (75), 285 (75), 286 (75), *289*, 732 (12), 740 (12), *741*
Zellner, H., 83 (18), 93 (79), *106, 108*, 103 (186), 104 (186), *111*, 256 (41), 258 (41), *260*, 272 (75), 279 (75), 282 (75), 283 (75), 285 (75), 286 (75), *289*, 732 (12), 740 (12), *741*
Zenno, H., 196 (15), *198*
Ziauddin, 33 (21, 27, 28), 44 (66), 46 (66), 47 (66), 51 (88), 52 (90), 54 (95), 56 (27), 57 (27), 58 (27), 59 (27), 60 (21, 28, 66, 90), 61 (28, 66, 90), 62 (28),

70 (66), *73, 74, 75,* 97 (110), 101 (148, 156, 171a), 102 (171a), 103 (171a), 105 (148), *109, 110,* 138 (2), 152 (2), 154 (2), *158,* 163 (8), 173 (8), *175,* 224 (89), *231*
Ziegler, E., 474 (37), 490 (37), 491 (37), *494*
Ziegler, J. B., 522 (57), *542*
Zillig, W., 324 (99), *345*
Zimmerman, H., 730 (1), *741*
Zinato, E., 172 (81), *178*
Zmujdzin, A., 7 (2), *15*
Zoga, S., 377 (43, 44), 380 (43), 395 (43, 44), *397,* 593 (10), 594 (10), 595 (10), *597*
Zoroastrova, V. M., 59 (150), *77,* 224 (87, 93), *231,* 285 (97), *290*
Zschiedrich, J., 399 (1), 406 (1), *408*
Zugravescu, I., 247 (5), 252 (25), 253 (5), 256 (5, 25), *259, 260,* 610 (33), 641 (33), *651,* 656 (9), 675 (9), *689,* 744 (5), 754 (5), *758*
Zumstein, O., 310 (53), 335 (53), *344*
Zunza, H., 268 (56), 281 (56), *289*
Zykova, T. N., 103 (176), *111*

Subject Index

Compounds appearing in the subject index are those which are referred to either by name or structure in the main text. Compounds appearing only in the tables at the end of individual chapters are not included.

3-Acetamido-6-methyl-8-propyl-*s*-triazolo-[4,3-*a*] pyrazine, 415, 422
2-Acetamidoquinoxaline, 30, 183
2-Acetoacetyl-3-ethoxyquinoxaline, 131
2-Acetonyl-7-bromo-3,4-dihydro-3-oxopyrido[2,3-*b*] pyrazine, 505
3-Acetonyl-7-bromo-1,2-dihydro-2-oxopyrido[2,3-*b*] pyrazine, 505
3-Acetonyl-1,2-dihydro-2-oxobenzo[*g*]-quinoxaline, 731
2-Acetonyl-3,4-dihydro-3-oxopyrido[2,3-*b*] pyrazine, 504, 520
3-Acetonyl-1,2-dihydro-2-oxopyrido[2,3-*b*] pyrazine, 520
3-Acetoxy-2,6-dimethylpyrazolo[1,5-*a*] pyrazine, 399
7-Acetyl-3-amino-2-carboxamido-5,6,7,8-tetrahydroimidazo[1,2-*a*] pyrazine, 367, 371
2-Acetyl-6-amino-3,4-dihydro-3-oxo-pyrido[2,3-*b*] pyrazine, 508
2-Acetyl-6-amino-3-methylpyrido[2,3-*b*] pyrazine, 508
2-Acetyl-3-benzylideneperhydro-1,4-dioxopyrrolo[1,2-*a*] pyrazine:
  hydrolysis of, 321
  preparation of, 312
  reduction of, 322
  ring opening of, 321
2-Acetyl-3-benzylperhydro-1,4,6-trioxo-pyrrolo[1,2-*a*] pyrazine:
  i.r. of, 307
  p.m.r. and stereochemistry of, 306
  preparation of, 309
4-Acetyl-1-(2-chloroethyl)-1,2,3,4-tetra-hydro-3-methyl-2-oxobenzo[*g*]-quinoxaline-5,10-dione, 732
2-Acetyl-1,4-dihydro-3-methylquinoxaline, 262
4-Acetyl-3,4-dihydro-3-oxoquinoxaline-2-carboxy-*N*-methylanilide, 33
Acetylenedicarboxylic acid esters:
  reaction of, with 1,2-diaminonaphthalene, 715
    with 2,3-diaminonaphthalene, 732
    with methylpyrazines, 297
    with methylquinoxalines, 219
    with *o*-phenylenediamines, 80, 83
    with pyrazinium ylids, 296, 297
    with quinoxalinium imines, 253, 254
    with quinoxalinium ylids, 252, 253
2-Acetyl-3-methylquinoxaline:
  i.r. of, 132
  isolation of, 130, 131
  oxidation of, 31
  reduction of, 262
7-Acetylperhydro-1,1-dimethyl-5,8-dioxo-thiazolo[3,4-*a*] pyrazine, 447
2-Acetylperhydro-1,4,6-trioxopyrrolo-[1,2-*a*] pyrazine:
  i.r. of, 307
  p.m.r. and stereochemistry of, 306
  preparation of, 309
2-Acetylquinoxaline, 131
7-Acetyl-5,6,7,8-tetrahydrotetrazolo-[1,5-*a*] pyrazine, 425
6-Acryloylamino-2,3-diphenylquinoxa-line, 237
4-Acyl-3,4-dihydroquinoxalin-2-ones:
  biological properties of, 93
3-Alkoxycarbonylmethylenequinox-alin-2-ones, 80
2-Alkoxy-3-chloroquinoxalines, 200
2-Alkoxyquinoxaline-4-oxides, 44

2-Alkoxyquinoxalines:
preparations of, 199
table of, 202
uses of, 202
1-Alkyl-2-amino-1,5,6,7-tetrahydroquinoxaline-3-carboxylates, 270
2-Alkyl-1,2-dihydro-3-methylquinoxalines, 217
2-Alkylperhydropyrrolo[1,2-*a*]pyrazines, 294
Alkylquinoxaline mono-*N*-oxides, 42
2-Alkylquinoxaline-4-oxides, 44
2-Alkylquinoxalines:
preparations of, 205, 209
reactions of, 211, 220
table of, 221
1-Alkylquinoxalinium salts, 250
1-Alkyltetrahydroquinoxalines, 272
Alloxan:
reaction of, with 4-chloro-*o*-phenylenediamine, 141, 182
with 1,2-diaminonaphthalene, 722
with 3,4-diaminopyridine, 553
with 2,3-diaminopyridines, 507
with *NN*-dimethyl-*o*-phenylenediamine, 141
with *o*-phenylenediamine, 140, 181
Alloxazine, 140
Alloxazines, 142, 143
Aminoacetaldehyde acetals:
reaction of, with 2-acylpyrroles, 294, 295
with 2-acylthiazoles, 444, 445
with 2-chloropyrazine, 365, 366
2-Aminoalkylquinoxalines, 207
2-Amino-3-arylquinoxalines, 182
2-Aminobenzo[*f*]quinoxaline, 722
3-Aminobenzo[*f*]quinoxaline, 722
3-Aminobenzo[*f*]quinoxaline-2-carboxylic acid, 716, 722
6-Amino-7-carboxamidopyrido[2,3-*b*]pyrazine, 511
7-Amino-6-carboxamidothieno[2,3-*b*]pyrazine, 442
6-Amino-2-carboxy-3,4-dihydro-3-oxopyrido[2,3-*b*]pyrazine, 508
5-Amino-6-chloro-2,3-dihydro-2-oxoimidazo[4,5-*b*]pyrazine, 378
2-Amino-3-chloroquinoxaline, 112, 180
3-Amino-7-chloroquinoxaline-2-carboxylic acid, 143
2-Amino-3-cyano-5,6-dihydrobenzo[*f*]quinoxaline, 181
3-Amino-2-cyano-5,6-dihydrobenzo[*f*]quinoxaline, 719
6-Amino-7-cyanopyrido[2,3-*b*]pyrazine, 511
2-Amino-3-cyanoquinoxaline, 185
3-Amino-1,2-dihydro-1-methyl-2-oxoquinoxaline, 181
1-Amino-3,4-dihydro-4-methyl-3-oxoquinoxalinium hydroxide inner salt:
cycloaddition reactions of, 254
isolation of, 86, 249
6-Amino-3,4-dihydro-3-oxopyrido[2,3-*b*]pyrazine, 512
6-Amino-2,3-dimethylquinoxaline, 249
6-Amino-3-methylpyrido[2,3-*b*]pyrazine, 513
2-Amino-3-methylquinoxaline, 183
cycloaddition reactions of, 254
isolation of, 86, 249
6-Amino-3,4-dihydro-3-oxoquinoxaline-2-carboxylic acid, 143
2-Amino-3,4-dihydroquinoxaline, 263, 264
3-Amino-1,2-dihydroquinoxalines, 264
8-Amino-2,3-dihydro-3-thioxoimidazo-[1,5-*a*]pyrazine:
desulfurisation of, 373
pka of, 371
preparation of, 372
5-Amino-2,3-dimethoxyquinoxaline, 201
2*H*-3-Amino-6,7-dimethylpyrazino[2,3-*b*]-[1,4]thiazine, 592, 593
8-Amino-2,3-dimethylpyrido[3,4-*b*]-pyrazine, 548
3-Amino-5,8-dimethyl-*s*-triazolo[4,3-*a*]pyrazine, 420
1*H*-3-Amino-5,6-diphenylpyrazolo[3,4-*b*]pyrazine, 405
8-Amino-2,3-diphenylpyrido[3,4-*b*]-pyrazine, 548
7-Amino-6-ethoxycarbonylthieno[2,3-*b*]pyrazine, 442
3-Aminoimidazo[1,2-*a*]pyrazine, 362, 366
8-Aminoimidazo[1,5-*a*]pyrazine, 371
4-Aminoimidazo[1,2-*a*]quinoxaline, 655, 657
8-Amino-3-mercaptoimidazo[1,5-*a*]-pyrazine, 371
3-Amino-5-methyl-8-propyl-*s*-triazolo[4,3-*a*]pyrazine, 420

3-Amino-6-methyl-8-propyl-*s*-triazolo[4,3-*a*]pyrazine:
  acylation of, 420
  acyl derivatives, isomerisation of, 415
  biological activity of, 423
  rearrangement of, 422
3-Amino-2-(2-oxazolin-2-yl)benzo[*f*]quinoxalines, 722
2-Amino-3-phenylbenzo[*f*]quinoxaline, 717
3-Amino-2-phenylbenzo[*f*]quinoxaline, 717
3-Aminopyrazolo[3,4-*b*]quinoxaline, 696
5-Aminopyrido[3,4-*b*]pyrazine, 499, 548
7-Aminopyrido[3,4-*b*]pyrazine, 559
8-Aminopyrido[2,3-*b*]pyrazine, 499, 516
8-Aminopyrido[3,4-*b*]pyrazine, 548
6-Aminopyrido[2,3-*b*]pyrazines:
  table of, 525
4-Aminopyrrolo[1,2-*a*]quinoxaline:
  5-oxide of, 603
  preparation of, 601, 618
2-Aminoquinoxaline:
  pka of, 183
  preparations of, 49, 179
  reactions of, 30, 163, 183
  u.v. and tautomerism of, 183
5-Aminoquinoxaline, 18
3-Aminoquinoxaline-2-carboxylic acid, 151
3-Aminoquinoxaline-2-carboxylic acids:
  unambiguous synthesis of, 144
5-Aminoquinoxaline-2,3-dione, 201
2-Aminoquinoxaline 1,4-dioxide, 41
2-Aminoquinoxaline 1-oxide:
  preparations of, 30, 49, 180
  protonation and pka of, 41
  reaction of, with benzenesulfonyl chloride, 49
3-Aminoquinoxaline 1-oxide, 41
2-Aminoquinoxaline 1-oxides, 183
2-Aminoquinoxalines:
  preparations of, 179-183
  reactions of, 183
  table of, 187
  uses of, 184
3-Aminoquinoxaline-2-sulfonamide, 115
3-Aminoquinoxaline-2-thione, 112, 115
3-Aminoquinoxaline-2-thiones, 113, 181
1-Aminoquinoxalin-2-one, 88
3-Aminoquinoxalin-2-one:
  preparation of, 84, 181
  reaction of, with phosphorus pentasulfide, 113
5-Amino-*s*-triazolo[1,5-*a*]pyrazine, 412
3-Amino-*s*-triazolo[4,3-*a*pyrazines:
  acylation of, 420
  biological activity of, 422
  preparation of, 418
  rearrangement of, 420
8-Amino-*s*-triazolo[1,5-*a*]pyrazine, 414
7-Aminothieno[2,3-*b*]pyrazine, 442
7-Aminothieno[2,3-*b*]pyrazine-6-carboxylic acid, 442
6- and 7-Amino-1,2,3-trimethylquinoxalinium chlorides, 249
4-Anilinopyrrolo[1,2-*a*]quinoxaline, 602
2-Anilinoquinoxaline, 181
2-Anilinoquinoxalines, 180
2-Anilinothiocarbonylperhydropyrrolo-[1,2-*a*]pyrazine, 293, 302
2-(D-Arabotetrahydroxybutyl)benzo-[*f*]quinoxaline:
  preparation of, 714
  reactions of, 720, 721
3-(D-Arabotetrahydroxybutyl)benzo-[*f*]quinoxaline:
  oxidation of, 721
  preparation of, 714
2-(D-Arabotetrahydroxybutyl)benzo-[*g*]quinoxaline, 735
2-(D-Arabotetrahydroxybutyl)quinoxaline, 242, 244
2-(D-Arabotetrahydroxybutyl)quinoxalines, 145
2-Aralkylquinoxalines, 209
2-Aralkyl-1,2,3,4-tetrahydroquinoxalines, 209
Aroylcyanide anils:
  condensation with *o*-phenylenediamines, 182
2-Aryl-1,2-dihydro-3-methylquinoxalines, 217
3-Aryl-1,2-dihydroquinoxalines, 263
4*H*-3-Aryl-6,7-dimethylpyrazino[2,3-*b*][1,4]thiazine, 592
2-Aryloxy-3-chloroquinoxalines, 200
2-Aryloxyquinoxalines:
  cyclisation to benzofuro[2,3-*b*]quinoxalines, 200
  preparation of, 199
  table of, 202
3-Arylpyrazino[2,3-*b*][1,4]thiazines, 592

2-Arylpyrido[2,3-*b*]pyrazines, 509
2-Arylpyrido[3,4-*b*]pyrazines, 551
3-Arylpyrido[2,3-*b*]pyrazines, 502
N-Aryl-2,3-quinoxalinedicarboximides, 149
2-Arylquinoxalines:
  preparation of, 233
  table of, 237
Azaindolizines:
  theoretical electronic structures, comparison of, 292
7-Azaindolizines, *see* Pyrrolo[1,2-*a*]-pyrazines
3-Azaphenazines, *see* Pyrido[3,4-*b*]quinoxalines
7-Azapyrrocolines, *see* Pyrrolo[1,2-*a*]pyrazines
2-Azaquinolizidines, *see* Pyrido[1,2-*a*]pyrazines
8-Azaquinolizidines, *see* Pyrido[1,2-*a*]pyrazines
5-Azaquinoxalines, *see* Pyrido[2,3-*b*]pyrazines
Azepino[1,2-*a*]quinoxalines, 219
6-Azido-7-nitroquinoxalines, 20
Azidopyrazine, 424
2-Azidoquinoxaline 1-oxide, 49
6-(1-Aziridinyl)benzo[*f*]quinoxaline, 718
2-(1-Aziridinyl)-3-chloro-6-nitroquinoxaline, 180
Azirino[1,2-*a*]quinoxalines, 611

Beirut reaction, 35, 37
Benzil:
  reaction of, with 3,4-diaminofurazan, 436
    with 4,5-diaminopyrazoles, 402
    with 2,3-diaminopyridine, 498
    with 2,3-diaminoquinoxaline, 186
    with 3,4-diaminothiophenes, 443, 444
    with *o*-phenylenediamine, 234
    with *N*-phenyl-*o*-phenylenediamine, 263, 265
Benzils:
  preparation of, 234
[1,4]-Benzodioxino[2,3-*b*]quinoxaline, 170
Benzofuro[2,3-*b*]quinoxalines, 200
Benzo[*f*]quinoxaline:
  N-oxidation of, 720
  physical properties of, 712, 713
  preparations of, 713, 718
Benzo[*g*]quinoxaline:
  attempted oxidation of, 734
  preparation of, 732
  u.v. and ^{1}H n.m.r. of, 730
Benzo[*f*]quinoxaline-2-carboxaldehyde, 720, 721
Benzo[*f*]quinoxaline-3-carboxaldehyde, 721, 722
Benzo[*g*]quinoxaline-2-carboxaldehyde, 735
Benzo[*f*]quinoxaline-2-carboxylic acid, 720
Benzo[*f*]quinoxaline-3-carboxylic acid, 721
Benzo[*g*]quinoxaline-2-carboxylic acid, 735
Benzo[*g*]quinoxaline-5,10-dione, 731, 734
Benzo[*g*]quinoxaline-5,10-diones:
  α-methyl derivatives, reactions of, 734
  polarographic reduction of, 731
  preparations of, 732, 733
  table of, 736
Benzo[*g*]quinoxaline-2,3-dithione, 735
Benzo[*f*]quinoxaline 4-oxide, 720
Benzo[*f*]quinoxalines:
  preparations of, 713-719
  reactions of, 719-723
  table of, 723
  uses of, 723
Benzo[*g*]quinoxalines:
  preparations of, 731, 732
  reactions of, 734
  table of, 739
Benzoylacetonitrile:
  reaction of, with benzofuroxan, 36, 37
    with quinoxaline *N*-oxides, 51
2-Benzoyl-3-benzylquinoxaline, 131
1-Benzoyl-7-chloro-1,2-dihydro-3-methylaminoquinoxaline, 264
4-Benzoyl-6-chloro-3,4-dihydroquinoxalin-2-one, 85
4-Benzoyl-6-chloro-1,2,3,4-tetrahydro-2-oxoquinoxaline, 264
4-Benzoyl-3,4-dihydro-6-methylthioquinoxalin-2-one, 85
4-Benzoyl-3,4-dihydroquinoxalin-2-one, 90
6-Benzoyl-7,8-dimethoxycarbonyl-1,4-dimethylpyrrolo[1,2-*a*]pyrazine, 296
6-Benzoyl-7,8-dimethoxycarbonyl-1,4-dimethylpyrrolo[1,2-*a*]pyrazine, 300

6-Benzoyl-7,8-dimethoxycarbonyl-1-methylpyrrolo[1,2-*a*]pyrazine, 296
6-Benzoyl-7,8-dimethoxycarbonyl-3-methylpyrrolo[1,2-*a*]pyrazine, 296
7-Benzoyl-4-methyl-1-phenylpyrazin[2,1-*c*][1,4]thiazine, 591
3-Benzoylquinoxaline-2-carboxylic acid, 140
2-Benzoylquinoxalines, 131, 132
1-Benzoyl-1,2,3,3a-tetrahydro-2-oxo-3-phenylimidazo[1,2-*a*]quinoxaline, 656
8-Benzylamino-2-methyl-*s*-triazolo[1,5-*a*]pyrazine, 411, 414
Benzyl cyanide:
  reaction of, with 1-amino-2-nitrosonaphthalene, 716
    with *o*-nitrosoaminobenzenes, 182
2*H*-2-Benzyl-1,3-dihydro-1,3-dioxopyrrolo-[3,4-*b*]quinoxaline, 625
6*H*-6-Benzyl-2,3-dimethyl-5,7-diphenylpyrrolo[3,4-*b*]pyrazine, 353
8*H*-2-Benzyl-1,2,3,4,6,7-hexahydro-1-oxopyrido[1,2-*a*]pyrazine, 479
3-Benzylideneperhydro-1,4-dioxopyrrolo-[1,2-*a*]pyrazine, 321
1-Benzylimidazo[1,2-*a*]pyrazine-7-oxide, 368
2-Benzyloxycarbonylperhydro-1,4-dioxopyrrolo[1,2-*a*]pyrazine, 310, 313
3-Benzylperhydro-1,4-dioxopyrrolo[1,2-*a*]pyrazine:
  biological activity of, 323
  epimerisation of, 318
  preparations of, 310, 313
  ring opening of, 319
3-Benzylperhydro-1,4-dioxo-2-pyruvoylpyrrolo[1,2-*a*]pyrazine:
  hydrolysis of, 319
  preparations of, 308, 312
  structural determination of, 308
3-Benzylperhydro-2-(2-methoxypropionyl)-1,4-dioxopyrrolo[1,2-*a*]pyrazine, 319
3-Benzylperhydro-8a-methyl-1,4-dioxopyrrolo[1,2-*a*]pyrazine, 311
2-Benzylperhydro-1-oxopyrido[1,2-*a*]pyrazine, 475, 479
3-Benzylperhydro-1,4,6-trioxopyrrolo[1,2-*a*]pyrazine:
  i.r. of, 307
  p.m.r. and stereochemistry of, 306
  preparation of, 308
2-Benzylquinoxaline, 209
2-Benzylquinoxalines, 131
3-Benzylquinoxalin-2-ones, 93
2,3-Bis(acetoxymethyl)quinoxaline 1,4-dioxide, 55
5,6-Bis(bromomethyl)-3-phenylisoxazolo-[4,5-*b*]pyrazine, 435
2,3-Bis(bromomethyl)quinoxaline:
  oxidation of, 124, 215
  preparation of, 214
  reaction of, with diethylamine, 214
2,3-Bis(carbethoxythio)quinoxaline, 117, 118
2,3-Bis(diethoxycarbonylmethyl)-1,2,3,4-tetrahydroquinoxaline, 269
2,3-Bis(3-dimethylaminopropyl)-1,2,3,4-tetrahydroquinoxaline, 13
Bis(6,7-diphenylpyrrolo[1,2-*a*]pyrazin-8-yl)disulfide, 297
5,7-Bis(ethoxycarbonylamino)-1,2-dihydro-3-methylpyrido[3,4-*b*]pyrazine, 554
5,7-Bis(ethoxycarbonylamino)-3-methylpyrido[3,4-*b*]pyrazine, 555, 557
2,3-Bis(1-*β*-glucosylmercapto)quinoxaline, 117
2,3-Bis(hydroxymethyl)quinoxaline 1,4-dioxide:
  antibacterial activity of, 55
  reactions of, 48, 149
2,3-Bis(methoxycarbonyl)-1,4-diphenylpyrrolo[1,2-*a*]quinoxaline, 611
2,3-Bis-(*p*-nitrophenoxy)quinoxaline, 201
2,3-Bis(2-pyridyl)benzo[*g*]quinoxalines, 731
2,3-Bis(2-pyridyl)pyrido[2,3-*b*]pyrazine, 497, 498
2,3-Bis(thiomethyl)quinoxalines, 221
2,3-Bis(*m*-toluidino)quinoxaline, 169
2,3-Bis-(*p*-tolylthio)quinoxaline, 31
2-Bromo-3-alkylquinoxalines, 163
2*H*-2-Bromo-3,4-dihydro-6,7-dimethyl-3-oxopyrazino[2,3-*b*][1,4]thiazine, 593
4-Bromo-1,2-dihydro-1-oxo-2,3-diphenylpyrido[1,2-*a*]pyrazinium bromide:
  preparations of, 467, 477
  reduction of, 479
7-Bromo-1,2-dihydro-2-oxopyrido[2,3-*b*]pyrazine, 505

6-Bromo-2,3-dimethylpyrazino[2,3-*b*]pyrazine, 579
6-Bromo-2,3-dimethylquinoxaline, 20
3-Bromoimidazo[1,2-*a*]pyrazine, 370
1-Bromo-3-methylimidazo[1,5-*a*]pyrazine, 374
7-Bromo-3-methylpyrido[2,3-*b*]pyrazine, 518
2-Bromomethylquinoxaline, 214
2-Bromo-3-methylquinoxaline 1,4-dioxide, 51
α-Bromomethylquinoxalines, 124
7-Bromo-3-methylquinoxalin-2-one, 83
6-Bromo-2-methylthiazolo[4,5-*b*]pyrazine, 448
3-*p*-Bromophenylbenzo[*f*]quinoxaline 1-oxide, 717
7-Bromopyrido[2,3-*b*]pyrazine, 514, 519
1-Bromopyrrolo[1,2-*a*]quinoxaline, 616
3-Bromopyrrolo[1,2-*a*]quinoxaline, 616, 617
2-Bromoquinoxaline, 163, 164
6-Bromoquinoxaline, 9, 19
6-Bromothiazolo[4,5-*b*]pyrazine, 448
5-Bromo-*s*-triazolo[1,5-*a*]pyrazine, 412
6-(4-*n*-Butylaminobutyl)quinoxaline, 18
2-Butylaminoquinoxaline, 181
1-Butyl-3-cyano-2,3-dihydropyrrolo[2,3-*b*]quinoxaline, 623
1-Butyl-1,2,3,4,5,6-hexahydro-6,7-dimethyl-2,3,5-trioxopyrido[3,4-*b*]pyrazine, 548
2-*t*-Butylquinoxaline:
  oxidation of, 29
  photochemical reactions of, 262
  preparation of, 206
1-Butyl-1,2,3,4-tetrahydro-2,3-dioxopyrido[3,4-*b*]pyrazine, 548
2-*t*-Butyl-1,2,3,4-tetrahydroquinoxaline, 206
Carbadox:
  antibacterial properties of, 55
3-Carbamoyl-1,2-dihydro-2-oxoquinoxaline, 145
3-(2-Carbamoylethyl)quinoxalin-2-one, 93
2-Carbamoyl-3-phenylquinoxaline, 139
2-Carboxy-3,4-dihydro-4-methyl-3-oxopyrido[2,3-*b*]pyrazine, 518
2-Carboxy-4,5-dihydro-4-oxopyrazolo[1,5-*a*]quinoxaline, 691
2-Carbethoxymethyl-3,4-dihydro-3-oxoquinoxaline, 210
3-(2-Carboxyethyl)perhydro-1,4-dioxopyrrolo[1,2-*a*]pyrazine, 322
2-(3-Carboxypropyl)quinoxalin-3-one, 92
3-Carboxypyrazolo[3,4-*b*]quinoxaline, 701, 703
2-Carboxyquinoxaline 1-oxide, 39
2-Chloroacetylquinoxaline, 146
8-Chloroalloxazine, 141
2-Chlorobenzo[*f*]quinoxaline, 720
3-Chlorobenzo[*f*]quinoxaline, 720
6-Chloro-2,3-bis(chloromethyl)quinoxaline, 221
2-Chloro-3-cycloalkylaminoquinoxalines, 163
Chlorodiazines:
  kinetic studies of halogen mobility in, 163
4-Chloro-1,2-dihydroimidazo[1,2-*a*]quinoxaline, 656
2*H*-2-Chloro-3,4-dihydro-6,7-dimethyl-3-oxopyrazino[2,3-*b*][1,4]thiazine, 593
5-Chloro-2,3-dihydro-6-mercapto-2-oxoimidazo[4,5-*b*]pyrazine, 378
5-Chloro-2,3-dihydro-2-oxoimidazo[4,5-*b*]-pyrazine, 378, 380
7-Chloro-3,4-dihydro-3-oxoquinoxaline-2-carboxyureide, 141
6-Chloro-2,3-dimethylquinoxaline 1,4-dioxide, 56
4-Chloro-1-ethoxypyrrolo[1,2-*a*]quinoxaline, 607
5-Chloro-6-ethylamino-2,3-dihydro-2-oxoimidazo[4,5-*b*]pyrazine, 377, 380
8-(2-Chloroethyl)perhydro-1-oxopyrazino-[2,1-*c*][1,4]oxazine, 588
2-Chloro-3-hydrazinoquinoxaline, 112
6-Chloro-5-methoxy-2,3-dimethylquinoxaline, 221
1-Chloro-4-methylpyrrolo[1,2-*a*]quinoxaline, 605
2-Chloro-3-methylquinoxaline, 31

2-Chloro-5-methylquinoxaline, 163
2-Chloro-3-methylquinoxaline 1,4-dioxide, 49
2-Chloro-3-methylquinoxalines, 207
8-Chloro-2-methyl-*s*-triazolo[1,5-*a*]pyrazine, 411
5-Chloro-2,3,6,7,8-pentafluoroquinoxaline, 170
2-Chloro-3-phenylquinoxaline, 31, 163
4-Chloropyrazolo[1,5-*a*]quinoxaline, 691
6-Chloropyrido[2,3-*b*]pyrazine, 516
7-Chloropyrido[3,4-*b*]pyrazine, 559
1-Chloropyrrolo[1,2-*a*]quinoxaline, 605, 616
3-Chloropyrrolo[1,2-*a*]quinoxaline, 616
2-Chloroquinoxaline:
 reactions of, 163-167
6-Chloroquinoxaline:
 reactions of, 19, 29
 u.v. of, 9
6-Chloroquinoxaline-2-carboxylic acid 1,4-dioxide, 29
2-Chloroquinoxaline 1-oxide, 49
6-Chloroquinoxaline 1-oxide, 18
7-Chloroquinoxaline 1-oxide, 29
2-Chloroquinoxalines:
 preparations of, 162-163
 reactions of, 163-167, 207
 table of, 171
 uses of, 168
3-Chloroquinoxaline-2-thione, 114
2-Chloro-3-substituted quinoxalines, 163, 167
5-Chloro-*s*-triazolo[1,5-*a*]pyrazine, 411
6-Chloro-*s*-triazolo[1,5-*a*]pyrazine, 411
COP:
 biological activity of, 586
Cyanine dyes based on imidazo[4,5-*b*]-quinoxalines:
 table of, 677
Cyanoacetic acid:
 reaction of, with *o*-nitrosoaminobenzenes, 144
  with *o*-nitrosoaminonaphthalenes, 716, 717
2-Cyanoalkylquinoxalines, 207
3-Cyano-1,2-dihydro-2-oxoquinoxaline, 51, 145
3-Cyano-1,2-dihydro-2-oxoquinoxaline 4-oxide, 51
6-Cyano-7,8-dimethoxycarbonyl-1-methoxypyrrolo[1,2-*a*]pyrazine, 297
6-Cyano-7,8-dimethoxycarbonylpyrrolo-[1,2-*a*]pyrazine, 297
2-Cyanoquinoxaline, 146
Cyclopentapyrazines:
 table of, 460
Cypridina luciferin:
 bioluminescence of, 363
 structure of, 364
 synthesis of, 366

cis-Decahydroquinoxaline, 275
dl-trans-Decahydroquinoxaline, 276
Decahydroquinoxalines:
 table of, 287
cis-Decahydroquinoxalin-2-one, 276
2,3-Dehydroquinoxaline, 151
Diacetyl:
 reaction of, with 3-amino-2-anilinopyridine, 507
  with cyclopentane-1,2-diamine, 456, 457
  with 4,5-diaminopyrazoles, 402, 403
  with 2,3-diaminopyridine, 498
  with 3,4-diaminopyrroles, 353, 354
  with 3,4-diamino[1,2,5]thiadiazole, 450
  with *N*-substituted-*o*-phenylenediamines, 248
2,3-Dialkoxyquinoxalines, 201
 table of, 203
2,3-Dialkylquinoxalines:
 preparations of, 205-209
 reactions of, 31, 211-220
 table of, 223
 uses of, 221
2,3-Diallyl-1,2,3,4-tetrahydroquinoxaline, 13
2,3-Diaminobenzo[*g*]quinoxaline, 735
5,7-Diamino-2-methylpyrido[3,4-*b*]pyrazine, 499, 551
6,8-Diamino-3-methylpyrido[2,3-*b*]pyrazine, 499, 502
2,3-Diaminopyridine:
 reaction of, with alloxan, 507

with symmetrical α-dicarbonyl compounds, 498
synthesis of, 498
with unsymmetrical α-dicarbonyl compounds, 501
3,4-Diaminopyridine:
reaction of, with alloxan, 553
with α-ketoaldehydes, 550
with α-ketoesters, 552
with symmetrical α-dicarbonyl compounds, 547
3,6-Diaminopyrido[2,3-*b*]pyrazines, 508, 524
6,7-Diaminoquinoxaline-2,3-diones, 97
2,3-Diaminoquinoxalines:
acylation of, 186
cyclic derivatives of, 186
preparation of, 184
table of, 190
2,3-Dianilinoquinoxaline, 169
2,3-Diaryloxyquinoxalines:
preparation and properties, 201, 202
table of, 203
2,3-Diarylquinoxalines:
preparation of, 234
table of, 237
uses of, 237
4,7-Diaza-5,6-benzindoles, *see* Pyrrolo-[2,3-*b*]quinoxalines
1,4-Diazabicyclo[4.4.0]decanes, *see* Pyrido[1,2-*a*]pyrazines
1,4-Diazabicyclo[4.3.0]nonanes, *see* Pyrrolo[1,2-*a*]pyrazines
4,7-Diazaindoles, *see* Pyrrolo[2,3-*b*]pyrazines
1,4-Diaza-4-methyl-8-oxa-9-oxobicyclo-[4.3.0]nonane, 435
7,10-Diazaperhydronaphthalenes, *see* Pyrido[1,2-*a*]pyrazines
5,8-Diazaquinoxalines, *see* Pyrazino[2,3-*b*]pyrazines
Dibenzo[*a,c*]phenazine, 235
2,3-Dibenzoyl-1,4-diphenylpyrrolo[1,2-*a*]quinoxaline, 611
2,3-Dibenzoylquinoxaline, 134
2,3-Dibenzylquinoxaline, 214
1,4-Dibromo-2,3-butanedione, 214
2-Dibromomethylquinoxaline, 214
3-Dibromomethylquinoxaline-2-carboxaldehyde, 215
6,8-Dibromopyrrolo[1,2-*a*]pyrazine, 302
1,3-Dibromopyrrolo[1,2-*a*]quinoxaline, 616
2,3-Dibromopyrrolo[1,2-*a*]quinoxaline, 616
2,3-Dibromoquinoxaline, 96, 168
6,7-Dibromoquinoxaline-2,3-dione, 96
5,6-Dibromo-1,2,5-thiadiazolo[3,4-*b*]pyrazine, 450
2,3-Di-*t*-butyl-1,2-dihydroquinoxaline, 264
2,3-Di-*t*-butylquinoxaline, 206
2;3-Dicarboxyquinoxaline 1,4-dioxide, 150
2,3-Dichlorobenzo[*g*]quinoxaline, 735
6,7-Dichloro-2,3-dimethylquinoxaline, 31
2,3-Dichloro-6-[2-hydroxyethylcarbamoyl]-quinoxaline, 171
2,3-Dichloro-6-methoxyquinoxaline, 168
2-Dichloronitromethylquinoxaline, 216
2,3-Dichloro-5-nitroquinoxaline, 168
2,3-Dichloro-6-nitroquinoxaline, 163, 180
2,3-Dichloropyrido[2,3-*b*]pyrazine, 516, 517
1,3-Dichloropyrrolo[1,2-*a*]quinoxaline, 616
2,3-Dichloroquinoxaline:
amination of, 180, 184
preparation of, 168
reactions of, 163, 168-170
6,7-Dichloroquinoxaline, 18
2,3-Dichloroquinoxalines:
preparation of, 168
reactions of, 163, 169, 170
table of, 174
uses of, 171
2,3-Dichloroquinoxaline-6-sulfonates, 171
6,7-Dichloroquinoxalin-2-one, 80
2,3-Dichloro-6-sulfonamidoquinoxalines, 171
5,6-Dichloro-1,2,5-thiadiazolo[3,4-*b*]pyrazine, 450, 451
1*H*-5,7-Dichloro-2-trifluoromethylimidazo-[4,5-*b*]quinoxaline, 664
2,3-Dicyanoquinoxaline, 185
2,3-Dicycloalkylaminoquinoxalines, 163, 185
1,4-Diethoxycarbonyl-1,4-dihydro-2,3-diphenylquinoxaline, 235
2-(*p*-Diethylaminoacetamidophenyl)-3-methylquinoxaline, 221
2-Diethylaminoquinoxalines, 184
Diethyl mesoxalate:
reaction of, with 3,4-diaminopyridine, 552
with 2,3-diaminopyridines, 504

with methylamino aminopyridines, 506, 507
with *o*-phenylenediamines, 80, 138
1,3-Diethyl-2-methylimidazo[4,5-*b*]-quinoxalinium chloride:
preparation of, 667
use as intermediate for dyestuffs, 668-671
2,3-Diethylquinoxaline, 209
5,7-Diethyl-2,3,5,7-tetrahydro-5,7-dimethylfuro[3,4-*b*]pyrazine, 431, 432
8*H*-2,2-Diethyl-1,2,3,4-tetrahydro-7-methoxy-8-oxopyrido[1,2-*a*]-pyrazinium chloride, 469, 470
2,3-Difluoroquinoxaline, 168, 170
1,4-Diformyl-1,2,3,4-tetrahydroquinoxalines, 267
2,3-Dihalogenoquinoxalines:
table of, 176
2-(Dihalonitromethyl)quinoxalines, 221
2,3-Dihydrazinoquinoxaline, 194
5*H*-6,7-Dihydrocyclopentapyrazines:
flavor and aroma properties of, 456, 459
preparation of, 457
spectroscopic properties of, 456
6*H*-5,7-Dihydro-5,7-diiminopyrrolo[3,4-*b*]pyrazine, 353
4,5-Dihydro-2,3-dimethoxycarbonyl-5-methyl-4-oxopyrazolo[1,5-*a*]quinoxaline, 253
5*H*-6,7-Dihydro-2,3-dimethylcyclopentapyrazine, 456, 457
5*H*-6,7-Dihydro-5,7-dimethylcyclopentapyrazine, 457
1,2-Dihydro-1,3-dimethyl-2-methylenequinoxaline, 266
5,6-Dihydro-6,7-dimethyl-5-oxo-2,3-diphenylpyrido[3,4-*b*]pyrazine, 548
2,6-Dihydro-1,3-dimethyl-6-oxo-7,8-diphenylpyrrolo[1,2-*a*]pyrazine, 293
3,7-Dihydro-2,8-dimethyl-3-oxo-6-phenylimidazo[1,2-*a*]pyrazine, 363
2*H*-3,4-Dihydro-6,7-dimethyl-3-oxopyrazino[2,3-*b*][1,4]thiazine, 593, 594
1,2-Dihydro-1,3-dimethyl-2-oxopyrido[3,4-*b*]pyrazine, 552
3,4-Dihydro-2,4-dimethyl-3-oxopyrido[3,4-*b*]pyrazine, 552
3,4-Dihydro-2,5-dimethyl-3-oxopyrrolo-[1,2,3-*de*]quinoxaline, 630
1,2-Dihydro-1,3-dimethyl-2-oxoquinoxaline, 32
2,3-Dihydro-5,6-dimethyl-3-oxo-*s*-triazolo[4,3-*a*]pyrazine, 415, 418
2,3-Dihydro-5,8-dimethyl-3-oxo-*s*-triazolo[4,3-*a*]pyrazine, 415
1,2-Dihydro-2,3-dimethyl-2-(2-tetrahydrofuryl)quinoxaline, 262
2,3-Dihydro-5,8-dimethyl-3-thioxo-*s*-triazolo[4,3-*a*]pyrazine, 415
5,7-Dihydro-5,7-dioxofuro[3,4-*b*]pyrazine, 431-433
5,6-Dihydro-5,6-dioxo[1,2,5]oxadiazolo-[3,4-*b*]pyrazine, 436
3-(5,6-Dihydro-1,6-dioxo[1,2,3]oxadiazolo[3,4-*a*]pyrazinium-8-yl]propionic acid, 435, 436
6*H*-5,7-Dihydro-5,7-dioxopyrrolo[3,4-*b*]pyrazine, 352, 354
5,7-Dihydro-5,7-dioxopyrrolo[3,4-*b*]pyrazines:
preparation of, 352
reduction of, 354, 355
2*H*-1,3-Dihydro-1,3-dioxopyrrolo[3,4-*b*]quinoxaline, 625, 629
5,6-Dihydro-3,6-diphenylbenzo[*f*]quinoxaline, 718
2,3-Dihydro-5,6-diphenylimidazo[1,2-*a*]pyrazine, 367
1,5-Dihydro-2,3-diphenyl-1-oxopyrrolo[1,2-*a*]quinoxaline, 599
3,4-Dihydro-2,3-diphenylpyrrolo[1,2,3-*de*]quinoxaline, 630, 631
3,4-Dihydro-2-ethoxycarbonyl-3-oxopyrido[2,3-*b*]pyrazine, 504
1,2-Dihydro-1-hydroxy-7-methyl-2-oxoquinoxaline 4-oxides, 53
1,2-Dihydro-2-hydroxy-1-methylquinoxaline, 266
5,6-Dihydro-6-hydroxy-8-methylthiazolo-[3,2-*a*]pyrazinium bromide 7-oxide, 445
1,2-Dihydro-3-hydroxy-2-oxo-1-methylquinoxaline, 51
3,4-Dihydro-1-hydroxy-3-oxopyrido[1,2-*a*]pyrazinium betaine, 477
8*H*-5,6-Dihydro-9-hydroxy-8-oxopyrido-[1,2-*a*]quinoxaline, 744
2*H*-3,4-Dihydro-3-hydroxy-3,6,7-trimethylpyrazino[2,3-*b*][1,4]thiazine, 592

1,2-Dihydro-3-hydroxy-2,5,10-trioxo-pyrido[2,3-*b*]quinoxaline, 747
2,3-Dihydroimidazo[1,2-*a*]pyrazine, 361, 367
1,2-Dihydroimidazo[1,2-*a*]quinoxaline, 656
4,5-Dihydroimidazo[1,2-*a*]quinoxaline, 657
4*H*-5,6-Dihydroimidazo[1,5,4-*de*]quinoxaline, 672
6,7-Dihydro-5-mercapto-6-oxo-1-phenyl-pyrazolo[3,4-*b*]pyrazine, 404, 406
1,2-Dihydro-4-methoxyimidazo[1,2-*a*]quinoxaline, 656
1,2-Dihydro-4-methylimidazo[1,2-*a*]quinoxaline, 656
4*H*-5,6-Dihydro-2-methylimidazo[1,5,4-*de*]quinoxaline, 672, 674
3,4-Dihydro-2-methyl-3-methylene-4-phenylpyrido[2,3-*b*]pyrazine, 507
1,2-Dihydro-1-methyl-3-(1-methylhydrazino)-2-oxoquinoxaline, 195
1,2-Dihydro-3-methyl-2-oxobenzo[*f*]-quinoxaline, 720
1,2-Dihydro-3-methyl-2-oxobenzo[*g*]-quinoxaline-5,10-dione, 734
3,7-Dihydro-2-methyl-3-oxoimidazo[1,2-*a*]pyrazine:
  preparation of, 366
  protonation of, 362
  reactions of, 363, 370
4*H*-1,2-Dihydro-4-methyl-2-oxoimidazo-[4,5-*b*]quinoxaline, 665
1,2-Dihydro-1-methyl-2-oxo-3-phenoxy-quinoxaline, 180
3,7-Dihydro-2-methyl-3-oxo-6-phenyl-imidazo[1,2-*a*]pyrazine, 366
4,5-Dihydro-3-methyl-4-oxo-1-phenylimidazo[1,5-*a*]quinoxaline, 659
6,7-Dihydro-3-methyl-6-oxo-1-phenyl-pyrazolo[3,4-*b*]pyrazine, 403, 405
1,2-Dihydro-4-methyl-2-oxo-3-phenyl-pyrido[2,3-*b*]quinoxaline, 747
1,5-Dihydro-2-methyl-1-oxo-4-phenyl-pyrrolo[1,2-*a*]quinoxaline, 615
1,2-Dihydro-1-methyl-2-oxo-3-phenyl-quinoxaline 4-oxide, 52
1,2-Dihydro-1-methyl-2-oxopyrido[2,3-*b*]pyrazine, 507
1,2-Dihydro-3-methyl-2-oxopyrido[2,3-*b*]pyrazine:
  preparation of, 504
  reactions of, 513, 517
1,2-Dihydro-3-methyl-2-oxopyrido[3,4-*b*]pyrazine, 552
3,4-Dihydro-2-methyl-3-oxopyrido[2,3-*b*]pyrazine, 504, 517
3,4-Dihydro-2-methyl-3-oxopyrido[3,4-*b*]pyrazine:
  preparation of, 552
  reactions of, 558, 560
3,4-Dihydro-4-methyl-3-oxopyrido[2,3-*b*]pyrazine, 504, 507
1,5-Dihydro-4-methyl-1-oxopyrrolo[1,2-*a*]quinoxaline, 607
4,5-Dihydro-5-methyl-4-oxopyrrolo[1,2-*a*]quinoxaline, 615
1,2-Dihydro-1-methyl-2-oxoquinoxaline, 32
3,4-Dihydro-4-methyl-3-oxoquinoxaline-2-carboxy-*N*-methylanilide, 32
3,4-Dihydro-4-methyl-3-oxoquinoxaline-2-carboxyureide, 147
1,2-Dihydro-1-methyl-2-oxoquinoxaline 4-oxide, 51, 145
1,2-Dihydro-3-(*N*-methyl-*N*-phenylcarbamoyl)-1-methyl-2-oxoquinoxaline 4-oxide, 53
3,4-Dihydro-1-methylpyrrolo[1,2-*a*]pyrazine, 299
2*H*-1,3-Dihydro-2-methylpyrrolo[3,4-*b*]quinoxaline, 626
1,2-Dihydro-3-methylquinoxaline, 261
1,2-Dihydro-3-methyl-2-thioxopyrido[2,3-*b*]pyrazine, 517, 518
3,4-Dihydro-2-methyl-3-thioxopyrido[2,3-*b*]pyrazine, 517
3,4-Dihydro-2-methyl-3-thioxopyrido[3,4-*b*]pyrazine, 558, 559
1,2-Dihydro-1-methyl-2-thioxoquinoxaline, 114
1,2-Dihydro-4-methyl-1-thioxo-*s*-triazolo-[4,3-*a*]quinoxaline, 415
1,2-Dihydro-2-methyl-3-*p*-tolylquinoxaline, 263
3,4-Dihydro-2-methyl-3-*p*-tolylquinoxaline, 263
2*H*-1,3-Dihydro-2-methyl-1,4,9-trioxo-pyrrolo[3,4-*b*]quinoxaline, 626

3,4-Dihydro-3-methylene-2,4-diphenyl-pyrido[2,3-*b*] pyrazine, 507
6,7-Dihydro-5-morpholin-4-yl-6-oxo-1-phenylpyrazolo[3,4-*b*] pyrazine, 404, 406
5,10-Dihydro-4-nitro-10-phenylpyrido-[3,4-*b*] quinoxaline, 750
1,2-Dihydro-2-oxobenzo[*g*] quinoxalines, 731
1,2-Dihydro-1-oxo-2,3-diphenylpyrido-[1,2-*a*] pyrazinium bromide:
preparation of, 466
reactions of, 477, 479
5,6-Dihydro-6-oxo-8,10-diphenylpyrido-[1,2-*a*] quinoxalinium perchlorate, 744
2,3-Dihydro-2-oxo-1,3-diphenylpyrrolo-[2,3-*b*] quinoxaline, 623, 624
2,3-Dihydro-2-oxoimidazo[4,5-*b*]-pyrazine, 376, 377
3,7-Dihydro-3-oxoimidazo[1,2-*a*] pyrazine, 362, 365
2,3-Dihydro-2-oxoimidazo[4,5-*b*] pyrazines:
biological activity of, 380
preparation of, 377, 378
reactions of, 380
1*H*-2,3-Dihydro-2-oxoimidazo[4,5-*b*] quinoxaline, 665
3,7-Dihydro-3-oxo-2-phenylimidazo[1,2-*a*] pyrazine, 365
1,2-Dihydro-1-oxo-3-phenylpyrido[1,2-*a*] pyrazinium bromide, 466, 467
1,5-Dihydro-1-oxo-4-phenylpyrrolo[1,2-*a*] quinoxaline, 607
1*H*-2,4-Dihydro-2-oxo-3-phenylpyrrolo-[2,3-*b*] quinoxaline, 624
1,2-Dihydro-2-oxo-3-phenylquinoxaline 4-oxide, 51, 52
3,4-Dihydro-3-oxo-2-phenylquinoxaline 1-oxides, 252
3,4-Dihydro-3-oxo-1-phenylquinoxaline-2-thione, 113
1,2-Dihydro-2-oxopyrazino[2,3-*b*]-pyrazine, 577, 578
2*H*-3,4-Dihydro-3-oxopyrazino[2,3-*b*]-[1,4] thiazines, 594
4,5-Dihydro-4-oxopyrazolo[1,5-*a*]-quinoxaline, 691
1,2-Dihydro-2-oxopyrido[2,3-*b*] pyrazine:
preparation of, 505, 509
reactions of, 519, 523
1,2-Dihydro-2-oxopyrido[3,4-*b*] pyrazine, 553, 557
3,4-Dihydro-3-oxopyrido[2,3-*b*] pyrazine:
preparation of, 505
reactions of, 512, 519
3,4-Dihydro-3-oxopyrido[3,4-*b*] pyrazine:
preparation of, 553, 558
reactions of, 546, 558
5,8-Dihydro-8-oxopyrido[2,3-*b*] pyrazine, 496
1,2-Dihydro-1-oxopyrido[1,2-*a*] pyrazinium salts:
preparation of, 470, 471
u.v. and i.r. of, 464
1,4-Dihydro-4-oxopyrido[2,3-*b*] quinoxaline, 747
1,5-Dihydro-1-oxopyrrolo[1,2-*a*] quinoxaline, 605
4,5-Dihydro-4-oxopyrrolo[1,2-*a*] quinoxaline, 601, 615
1,5-Dihydro-1-oxopyrrolo[1,2-*a*] quinoxalines:
preparation of, 605
reaction of, with phosphoryl chloride, 619
u.v. and tautomerism of, 599
1*H*-2,3-Dihydro-2-oxopyrrolo[2,3-*b*] quinoxalines, 624
3,4-Dihydro-3-oxoquinoxaline-2-carboxylic acid, 146
3,4-Dihydro-3-oxoquinoxaline-2-carboxyureide, 141, 147
1,2-Dihydro-2-oxoquinoxaline 4-oxide, 51, 145
1,2-Dihydro-2-oxoquinoxaline 4-oxides, 33, 51
2,3-Dihydro-3-oxo-*s*-triazolo[4,3-*a*] pyrazines, 415, 418
1,2-Dihydro-2-oxo-3-ureidopyrido[2,3-*b*] pyrazine, 521
1,2-Dihydro-2-oxo-3-ureidopyrido[3,4-*b*] pyrazine, 553
3,4-Dihydro-3-oxo-2-ureidopyrido[2,3-*b*] pyrazine, 521, 522
1,2-Dihydro-4-phenylimidazo[1,2-*a*] quinoxaline, 655, 656
4,5-Dihydro-4-phenylpyrrolo[1,2-*a*] quinoxaline, 601, 618
1,2-Dihydro-3-phenylquinoxaline, 262, 263
1,4-Dihydro-2-phenylquinoxaline, 262

2*H*-1,3-Dihydro-2-propylpyrrolo[3,4-*b*]quinoxaline, 626
1,2-Dihydropyrido[3,4-*b*]pyrazines, 558
3,4-Dihydropyrido[2,3-*b*]pyrazines, 515
5,10-Dihydropyrido[3,4-*b*]quinoxaline, 750
1,4-Dihydroquinoxaline, 261
1,2-Dihydroquinoxalines:
  preparation of, 261
  table of, 276
1,4-Dihydroquinoxalines:
  table of, 276
6,9-Dihydro-2,3,7,8-tetramethylbenzo[*g*]-quinoxaline-5,10-dione, 733
2,3-Dihydro-3-thioxoimidazo[1,5-*a*]pyrazine, 372, 375
4,5-Dihydro-4-thioxopyrrolo[1,2-*a*]quinoxaline:
  preparation of, 601
  reactions of, 615, 621
5*H*-6,7-Dihydro-2,5,7-trimethylcyclopentapyrazine, 457
3,4-Dihydro-1,2,4-trimethyl-2-oxoquinoxalinium salts, 252
1,2-Dihydro-2,2,3-trimethylquinoxaline, 264
1,2-Dihydro-1,2,3-triphenyl-2-quinoxalinol, 263
5,6-Dihydroxy[1,2,5]oxadiazolo[3,4-*b*]pyrazine, 436
2,3-Dihydroxythieno[3,4-*b*]pyrazine, 443
Diiminosuccindinitrile, 184
1,3-Diiodopyrrolo[1,2-*a*]quinoxaline, 616
2,3-Diisopropylquinoxaline, 30
7,8-Dimethoxycarbonyl-1,4-dimethylpyrrolo-[1,2-*a*]pyrazine, 296, 301
7,8-Dimethoxycarbonylpyrrolo[1,2-*a*]pyrazines, 296
2,3-Dimethoxyquinoxaline, 31, 201
5,8-Dimethoxyquinoxalines, 21
1,3-Dimethylalloxan, 141
5-(2-Dimethylaminoethyl)amino-6-chloro-2,3-dihydro-2-oxoimidazo[4,5-*b*]-pyrazine, 380
1-[2-(Dimethylamino)ethyl]-3-phenylquinoxalinium chloride hydrochloride, 249
2*H*-2-(4-Dimethylanilino)pyrrolo[3,4-*b*]-quinoxaline, 627
2,3-Dimethyl-5,7-dihydro-5,7-dioxofuro-[3,4-*b*]pyrazine, 431
2,3-Dimethyl-6-methylaminoquinoxaline, 249
2,3-Dimethyl-5-nitroquinoxaline, 31
5,6-Dimethyl-3-phenylisoxazolo[4,5-*b*]pyrazine, 435
5,6-Dimethyl-1-phenylpyrazolo[3,4-*b*]pyrazine, 402
5,6-Dimethyl-2-phenyl-*s*-triazolo[1,5-*a*]pyrazine, 410
5,8-Dimethyl-3-phenyl-*s*-triazolo[4,3-*a*]pyrazine, 416, 417
2,3-Dimethylpyrido[2,3-*b*]pyrazine:
  oxidation of, 515
  reduction of, 521
  styryl derivatives from, 514
2,3-Dimethylpyrido[3,4-*b*]pyrazine, 547, 560
2,4-Dimethylpyrrolo[1,2-*a*]quinoxaline, 617, 618
2,3-Dimethylquinoxaline:
  preparation of, 206
  protonation of, 211
  reactions of, 30, 212-220
1,4-Dimethylquinoxaline-2,3-dione, 96
6,7-Dimethylquinoxaline-2,3-dione, 95
2,3-Dimethylquinoxaline 1,4-dioxide:
  biological properties of, 55, 56
  preparation of, 30, 36
  reactions of, 47, 48
2,3-Dimethylquinoxaline 1-oxide:
  preparation of, 30
  reactions of, 46, 47
1,3-Dimethylquinoxalinium perchlorate, 247
1,3-Dimethylquinoxalinium salts, 248, 250
5,6-Dimethyl-1,2,5-thiadiazolo[3,4-*b*]pyrazine, 450
5,8-Dimethyl-*s*-triazolo[1,5-*a*]pyrazine, 414
5,8-Dimethyl-*s*-triazolo[4,3-*a*]pyrazine, 414, 420
5,6-Dinitroquinoxaline, 18
6,7-Dinitroquinoxaline-2,3-dione, 96, 216
Dioxidine, *see* 2,3-Bis(hydroxymethyl)-quinoxaline 1,4-dioxide
2,3-Diphenoxyquinoxaline, 201
3,6-Diphenylbenzo[*f*]quinoxaline, 718
2,3-Diphenylbenzo[*g*]quinoxaline, 731

1,3-Diphenyl-2-methylpyrrolo[3,4-*b*]quinoxaline, 134
5,6-Diphenyl[1,2,5]oxadiazolo[3,4-*b*]pyrazine, 436
1,3-Diphenylpyrazolo[3,4-*b*]quinoxaline, 696, 697
2,3-Diphenylpyrido[3,4-*b*]pyrazine, 547
1,3-Diphenylpyrido[1,2-*a*]pyrazinium bromide, 467
1,3-Diphenylpyrrolo[1,2-*a*]pyrazine, 294
7,8-Diphenylpyrrolo[1,2-*a*]pyrazines, 297, 298
2,3-Diphenylpyrrolo[2,3-*b*]quinoxaline, 622
2,3-Diphenylquinoxaline:
  formation of, 235
  mass spectrum of, 235
  as photosensitiser, 237
  reactions of, 30, 235
2,3-Diphenylquinoxaline 1,4-dioxide, 30
2,3-Diphenylquinoxaline 1-oxide, 30, 45
1,3-Diphenylquinoxalin-2-one, 85
5,6-Diphenyltetrazolo[1,5-*a*]pyrazine, 424, 425
2,3-Diphenylthieno[3,4-*b*]pyrazine, 443
4-(2,3-Diphenylthieno[3,4-*b*]pyrazin-5-yl)butyric acid, 440
1,3-Diphenylthieno[3,4-*b*]quinoxaline, 134
2,3-Di-(2-pyridyl)quinoxaline, 237
5,8- and 6,7-Disubstituted 2,3-dimethylquinoxalines:
  u.v. and ^1H n.m.r. spectra of, 211
6,7-Disubstituted 2,3-dimethylquinoxalines:
  oxidation of, 31
2,3-Disubstituted quinoxaline 1,4-dioxides, 267
2,3-Disubstituted quinoxalines:
  oxidation of, 31
2,3-Disubstituted-1,2,3,4-tetrahydrobenzo-[*g*]quinoxaline-5,10-diones, 732
1,4-Dithia-1,4-dihydrophenazino[2,3-*b*]-quinoxaline, 169
1,3-Dithiolo[4,5-*b*]pyrazines:
  table of, 461
1,3-Dithiolo[4,5-*b*]quinoxaline, 118

Echinomycin, 148
Eradex, *see* Quinoxaline-2,3-dithione cyclic trithiocarbonate
Ergot alkaloids, 324
Ergotamine, 326
Ergotaminine, 326
3-(D-Erythrotrihydroxypropyl)6,7-dimethylpyrazolo[3,4-*b*]quinoxaline, 244
2-(L-Erythrotrihydroxypropyl)quinoxaline, 241
1-Ethoxycarbonylaminoquinoxalinium hydroxide:
  inner salt, 249, 254
2-Ethoxycarbonyl-3,4-dihydro-4-methyl-3-oxopyrido[3,4-*b*]pyrazine, 553
3-Ethoxycarbonyl-1,2-dihydro-1-methyl-2-oxopyrido[3,4-*b*]pyrazine, 553, 557
3-Ethoxycarbonyl-1,4-dihydro-1-methyl-4-oxopyrido[2,3-*b*]quinoxaline, 747, 749
2-Ethoxycarbonyl-3,4-dihydro-3-oxopyrido-[2,3-*b*]pyrazine, 519
3-Ethoxycarbonyl-1,2-dihydro-2-oxopyrido-[2,3-*b*]pyrazine, 519
3-Ethoxycarbonyl-1,2-dihydro-2-oxopyrido-[3,4-*b*]pyrazine, 552, 557
3a-Ethoxycarbonyl-1,2,3,3a,4,5-hexahydro-1,4-dioxopyrrolo[1,2-*a*]quinoxaline, 605
7-Ethoxycarbonyl-8-hydroxy-3-methoxypyrido[2,3-*b*]pyrazine, 510
3-Ethoxycarbonylmethyl-1,2-dihydro-2-oxobenzo[*f*]quinoxaline, 715
2-Ethoxycarbonylmethyl-1,2-dihydro-2-oxobenzo[*g*]quinoxaline, 732
2-Ethoxycarbonylmethyl-3-methylquinoxaline, 206
3-Ethoxycarbonylmethylquinoxalin-2-one:
  preparation of, 80
  tautomerism of, 87, 210
  tricyclic derivatives from, 92
4*H*-3-Ethoxycarbonyl-4-oxopyrazino[1,2-*a*]pyrimidine, 569
5-Ethoxycarbonylquinoxaline, 18
6-Ethoxycarbonylquinoxaline, 18
3-Ethoxycarbonyl-1,2,3,4-tetrahydro-1-methyl-4-oxopyrido[2,3-*b*]quinoxaline, 747
6-Ethoxycarbonylthieno[2,3-*b*]pyrazine, 442
1-Ethoxy-2,3-diformyl-4,5-dihydro-4-oxopyrrolo[1,2-*a*]quinoxaline, 614
1-Ethoxy-3,4,6,7,8,8a-hexahydro-4-oxopyrrolo[1,2-*a*]pyrazine, 316

7(or 6)-Ethoxy-3-methylquinoxalin-2-one, 93
9a-Ethoxyperhydro-8-methyl-4,7-dioxopyrazino[2,1-*b*] [1,3] oxazine, 585
Ethyl acetoacetate:
reaction of, with *o*-aminoazobenzenes, 144, 145
2-amino-3-nitrosopyridines, 508
quinoxaline *N*-oxides, 51
Ethyl 3-acetonylquinoxaline-2-carboxylate, 139
5-Ethylamino-2,3-dihydro-2-oxoimidazo-[4,5-*b*] pyrazine, 380
Ethyl benzo[*f*] quinoxaline-3-carboxylate, 716
Ethyl 3-chloroquinoxaline-2-carboxylate, 138
Ethyl cyanoacetate:
reaction of, with benzofuroxans, 36
Ethyl 2-(3,4-dihydro-3-oxo-2-quinoxalinyl)propionate, 93
1,2-Ethylenediamine:
reaction of, with 2-acetylfuran, 299, 300
with chelidonic acid, 471, 473, 474
with cyclopentane-1,2-diones, 457, 458
with α,ω-dihalocarboxylic acids, 473, 474
with 2-oxo-4,5-diphenylimidazole, 379
with sucrose, 473
Ethyl ethoxalylacetate:
reaction of, with benzofuroxan, 38
with 2,3-diaminopyridines, 503
with *o*-phenylenediamine, 80
Ethyl 3-ethoxyquinoxaline-2-carboxylate, 131
Ethyl 3-methylquinoxaline-2-carboxylate, 125
Ethyl 3-methylquinoxalin-2-ylpyruvate, 210
7-Ethylperhydro-2-methyl-6-oxopyrazino-[1,2-*c*] pyrimidine, 571
2-Ethyl-1-(2-phenylethyl)imidazo[1,5-*a*] pyrazine, 377
2-Ethyl-3-phenylquinoxaline, 31
1-Ethylpyrido[1,2-*a*] pyrazinium bromide:
2-oxide, preparation of, 467
2-Ethylquinoxaline, 29
3-Ethylquinoxalin-2-one, 209
2-Ethyl-1,2,3,4-tetrahydro-1,4-dimethylquinoxaline, 269
8*H*-2-Ethyl-1,2,3,4-tetrahydro-7-methoxy-8-oxopyrido[1,2-*a*] pyrazine, 469
1*H*-7-Ethyl-2,3,5,6-tetrahydro-1-methylcyclopentapyrazine, 459
4-Ethyl-1,2,3,4-tetrahydro-2-oxobenzo[*g*]-quinoxaline-5,10-dione, 732
Ethyl 3,6,7-trimethylquinoxaline-2-carboxylate, 144

Flavazoles, *see* Pyrazolo[3,4-*b*] quinoxalines (flavazoles)
6-Fluoroalkylquinoxaline-2,3-diones, 97
2-Fluoroquinoxaline, 163, 164
6-Fluoroquinoxaline, 19
Fluorubin:
preparation of, 170, 186
uses of, 171, 187
2-Formyl-3,4-dihydro-3-oxobenzo[*f*] quinoxaline:
phenylhydrazone of, 714
3-Formyl-2-hydroxymethylquinoxaline 1-oxide:
internal hemiacetal of, 48
7-Formyl-5-methylpyrrolo[2,3-*b*] pyrazine, 349
7-Formylpyrrolo[2,3-*b*] pyrazine, 349, 350
2-Formylquinoxaline 1,4-dioxide, 47
4-Formyl-1,2,3,4-tetrahydro-2-oxoquinoxalines, 267
Furazono[3,4-*b*] pyrazines, *see* [1,2,5]Oxadiazolo[3,4-*b*] pyrazines
Furo[3,4-*b*] pyrazines:
table of, 437
2-(2-Furyl)quinoxalines, 244

Glucazidones:
reformulation of, 244
Glycyl-L-proline anhydride:
crystal structure of, 304, 307
i.r. of, 307
preparation of, 310, 311, 315
ring opening of, 319
Glyoxal:
reaction of, with 4,5-diaminopyrazoles, 402, 403
with 2,3-diaminopyridine, 498
with 3,4-diamino[1,2,5] thiadiazole, 450
with *o*-phenylenediamines, 7, 17
with *o*-quinone dioximes, 38, 39

with 2,3,4-triaminopyridine, 498, 499
Glyoxalino(1′:5′-1:2)quinoxalines, *see* Imidazo[1,5-*a*]quinoxalines
Glyoxylic acid esters:
reaction of, with diaminoisothiazoles, 449
with 2,3-diaminopyridines, 505, 506
with *o*-phenylenediamines, 79

α-Halogenoacetaldehyde acetals:
reaction of, with 2-acylpyrroles, 294, 295
with 2-aminopyrazine, 365, 366
with 2-benzoylimidazole, 368, 369
Halogenopyrrolo[1,2-*a*]quinoxalines:
nucleophilic displacements of, 619, 620
preparation of, 616
rearrangements of, 617
2-Halogenoquinoxalines:
reaction of, with potassium amide in liquid ammonia, 164
table of, 171
2-Halogeno-3-substituted quinoxalines:
reaction of, with potassium amide in liquid ammonia, 164
1,4,5,6,7,8-Hexaazanaphthalenes, *see* Pyrazino[2,3-*e*]-*v*-tetrazines
Hexachloroquinoxaline, 168, 170
Hexafluoroquinoxaline, 170
1,2,3,4,5,6-Hexahydro-6,7-dimethyl-3,5-dioxopyrido[3,4-*b*]pyrazine, 555
1,2,3,4,5,6-Hexahydro-6,7-dimethyl-5-oxo-pyrido[3,4-*b*]pyrazine, 555
10*H*-5,6,6a,7,8,9-Hexahydro-6,10-dioxo-pyrido[1,2-*a*]quinoxaline, 743
1,2,3,3a,4,5-Hexahydro-1,4-dioxopyrrolo-[1,2-*a*]quinoxaline, 605, 609
1,2,3,4,5,6-Hexahydro-2,3-dioxopyrrolo-[1,2,3-*de*]quinoxaline, 630
trans-4a,5,6,7,8,8a-Hexahydro-2,3-diphenylquinoxaline, 275
1,2,3,3a,4,5-Hexahydro-2-hydroxy-1-oxopyrrolo[1,2-*a*]quinoxaline, 608, 622
Hexahydroimidazo[1,5-*a*]quinoxalines, 658
1,2,3,3a,4,5-Hexahydro-5-methyl-1,3-dioxoimidazo[1,5-*a*]quinoxaline, 660, 661
1,2,3,4,5,6-Hexahydro-2-methylene-5-methyl-3-oxopyrrolo[1,2,3-*de*]quinoxaline, 630
1,2,3,3a,4,5-Hexahydro-2-(4-methylphenyl)-imidazo[1,5-*a*]quinoxaline, 662
6*H*-1,2,7,8,9,9a-Hexahydro-1-oxo-2-phenylpyrido[1,2-*a*]pyrazine, 479
6*H*-1,2,7,8,9,9a-Hexahydro-1-oxo-3-phenyl-pyrido[1,2-*a*]pyrazine, 466, 481
8*H*-1,2,3,4,6,7-Hexahydro-1-oxopyrido[1,2-*a*]pyrazine:
preparation of, 477
reactions of, 475, 478
1,2,3,3a,4,5-Hexahydro-1-oxopyrrolo-[1,2-*a*]quinoxaline, 608
1,2,3,3a,4,5-Hexahydro-4-oxopyrrolo-[1,2-*a*]quinoxaline, 604
L-Histidyl-L-proline anhydride, 322
spiro-Hydantoins:
formation of, from pyrido[2,3-*b*]pyrazines, 507
from pyrido[3,4-*b*]pyrazines, 554
2-Hydrazino-3-alkoxyquinoxalines, 195
2-Hydrazino-3-methoxyquinoxaline, 195
2-Hydrazinoquinoxaline, 114
2-Hydrazinoquinoxalines:
preparations of, 194
reactions of, 195, 196
table of, 196
uses of, 196
3-Hydrazinoquinoxaline-2-thione:
hydrolysis of, 114
preparation of, 113
use of, 116
3-Hydrazinoquinoxaline-2-thiones, 112, 116
5-Hydrazino-*s*-triazolo[1,5-*a*]pyrazine, 412
8-Hydrazino-*s*-triazolo[1,5-*a*]pyrazine, 412, 414
3-(*o*-Hydroxyaryl)quinoxaline 1-oxides, 37
6-Hydroxy-1,3-dimethyl-7,8-diphenyl-pyrrolo[1,2-*a*]pyrazine:
alkylation of, 301
preparation of, 297
spectra of, 293
3-Hydroxy-5,6-dimethyl-2-phenylpyrazolo-[3,4-*b*]pyrazine, 402
3-Hydroxy-5,6-dimethyl-*s*-triazolo[4,3-*a*]pyrazine, 415
3-Hydroxy-5,8-dimethyl-*s*-triazolo[4,3-*a*]pyrazine, 415
6-Hydroxy-7,8-diphenylpyrrolo[1,2-*a*]pyrazine, 297

3-Hydroxyimidazo[1,2-*a*]pyrazine, 362
2-Hydroxyimidazo[4,5-*b*]pyrazine, 376
2-Hydroxyimino-1,2-dihydroquinoxaline, 183
4*H*-2-Hydroxy-4-imino-3-phenylpyrazino-[1,2-*a*]pyrimidine, 569
6-Hydroxy-3-methyl-7,8-diphenylpyrrolo-[1,2-*a*]pyrazine, 301
1-Hydroxy-4-methyl-2,3-diphenylpyrrolo-[1,2-*a*]quinoxaline, 612, 615
6-Hydroxy-3-methylisothiazolo[4,5-*b*]pyrazine, 449
1*H*-3-Hydroxy-5-methylpyrazolo[3,4-*b*]pyrazine, 402
1*H*-3-Hydroxy-6-methylpyrazolo[3,4-*b*]pyrazine, 402
3-Hydroxy-1-methylpyrazolo[3,4-*b*]quinoxaline, 696
2-Hydroxymethylquinoxaline, 125
3-Hydroxymethylquinoxaline-2-carboxaldehyde, 125
2-Hydroxymethylquinoxaline 1,4-dioxide, 48
3-Hydroxy-2-phenylpyrazolo[3,4-*b*]pyrazine, 402
1-Hydroxy-3-phenylpyrido[1,2-*a*]pyrazinium betaine, 477
Hydroxypyrazino[2,3-*b*]pyrazines, 577
1*H*-3-Hydroxypyrazolo[3,4-*b*]pyrazine, 401, 405
1*H*-6-Hydroxypyrazolo[3,4-*b*]pyrazine, 401
8-Hydroxypyrido[2,3-*b*]pyrazine, 496
Hydroxypyrido[2,3-*b*]pyrazines, 496, 503
Hydroxypyrido[3,4-*b*]pyrazines, 545
5-Hydroxyquinoxaline, 8, 13
5-Hydroxyquinoxaline 1,4-dioxide, 19
2-Hydroxyquinoxalines, *see* Quinoxalin-2-ones
1-Hydroxyquinoxalin-2-one 4-oxides, 39
3-Hydroxy-*s*-triazolo[4,3-*a*]pyrazines, 415, 418
1*H*-3-Hydroxy-1,5,6-trimethylpyrazolo-[3,4-*b*]pyrazine, 406

Imidazolino[1,2-*a*]pyrazines, *see* Imidazo-[1,2-*a*]pyrazines
Imidazolo-5′,4′:2,3-quinoxalines, *see* Imidazo[4,5-*b*]quinoxalines
Imidazo[1,2-*a*]piperazines, *see* Imidazo-[1,2-*a*]pyrazines
Imidazo[1,2-*a*]pyrazine:
  electrophilic substitution of, 370
  physical properties of, 360, 361
  preparation of, 365
Imidazo[1,5-*a*]pyrazine:
  preparation of, 372
  reactions of, 373, 375
  u.v. and p.m.r. of, 370, 371
Imidazo[4,5-*b*]pyrazine:
  biological activity of, 382
  N-oxidation of, 379
  preparation of, 376
Imidazo[1,2-*a*]pyrazine 7-oxide, 368
Imidazo[1,5-*a*]pyrazine 7-oxide:
  p.m.r. of, 371
  preparation of, 373
1-*H*-Imidazo[4,5-*b*]pyrazine 4-oxide, 379, 382
1*H*-Imidazo[4,5-*b*]pyrazine 4-oxides, 377, 379
Imidazo[1,2-*a*]pyrazines:
  preparation of, 365
  protonation of, 361, 362
  quaternary salts of, 368
  table of, 382
  u.v. and p.m.r. of, 361
Imidazo[1,5-*a*]pyrazines:
  preparation of, 372
  spectra of, 371
  table of, 385
Imidazo[4,5-*b*]pyrazines:
  biological activity of, 380-382
  preparation of, 376
  table of, 394
  u.v. and p.m.r. of, 375, 376
Imidazo[1,2-*a*]quinoxaline:
  hypotensive activity of, 658
  preparation of, 654, 655
  reactions of, 657
  u.v. and pka of, 654
1*H*-Imidazo[4,5-*b*]quinoxaline:
  amphoteric nature of, 664
  preparation of, 665
  ring opening of, 667
Imidazo[1,2-*a*]quinoxalines:
  biological activity of, 658
  preparation of, 655
  table of, 674
Imidazo[1,5-*a*]quinoxalines:
  table of, 675

Imidazo[1,5,4-*de*]quinoxalines:
  preparation of, 672
  table of, 687
Imidazo[4,5-*b*]quinoxalines:
  acidity of, 664
  cyanine dyestuffs from, 663
  preparation of, 664-667
  table of, 684
  uses of, 671
  u.v. of, 664
Imidazo[4,5-*b*]quinoxalinocarbocyanines:
  structure of, 663
  table of, 677
Imidazo[4,5-*b*]quinoxalinoindocarbocyanines:
  structure of, 663
  table of, 677
Imidazo[4,5-*b*]quinoxalinothiacarbocyanines:
  structure of, 663
  table of, 677
1-Iodopyrrolo[1,2-*a*]quinoxaline, 616, 618
3-Iodopyrrolo[1,2-*a*]quinoxaline, 616, 618
2-Iodoquinoxaline, 163, 164
6-Iodoquinoxaline, 19
Isoalloxazines, 143
3-Isobutylperhydro-1,4-dioxopyrrolo[1,2-*a*]pyrazine, *see* L-Leucyl-L-proline anhydride
3-Isobutylperhydropyrrolo[1,2-*a*]pyrazine, 301
Isonitrosoketones:
  reaction of, with 1,2-diaminonaphthalene, 716
    with *o*-phenylenediamines, 205, 249
2-Isopropyl-3-methylquinoxaline, 30
2-Isopropyl-3-phenylquinoxaline, 31
2-Isopropylquinoxaline, 29
Isothiazolo[4,5-*b*]pyrazines, 449, 454
Isoxazolo[4,5-*b*]pyrazines: 435
  table of, 439

L-Leucyl-L-proline anhydride:
  biological activity of, 323
  conformation of, 304
  occurrence of, 324
  preparation of, 310, 315
  ring opening of, 319
  X-ray diffraction of, 304

Malononitrile:
  reaction of, with 2-aminopyrazine-3-carboxaldehyde, 511
    with benzofuroxan, 36
6-Mercapto-1,3-dimethyl-7,8-diphenylpyrrolo[1,2-*a*]pyrazine, 298
3-Mercapto-5,8-dimethyl-*s*-triazolo[4,3-*a*]pyrazine:
  i.r. and tautomerism of, 415
  methylation, 420
6-Mercapto-7,8-diphenylpyrrolo[1,2-*a*]pyrazine, 297
1-Mercapto-2,3-diphenylpyrrolo[1,2-*a*]quinoxaline, 612
6-Mercapto-1-methyl-7,8-diphenylpyrrolo-[1,2-*a*]pyrazine, 298
6-Mercapto-3-methyl-7,8-diphenylpyrrolo-[1,2-*a*]pyrazine, 298
2-Mercaptoquinoxaline 1-oxide:
  zinc salt, bactericidal and fungicidal properties of, 55
S-2-(3-Mercaptoquinoxalinyl)thiouronium chloride, 116, 118
5-Mercapto-*s*-triazolo[1,5-*a*]pyrazine, 412
3-Mercapto-*s*-triazolo[4,3-*a*]pyrazines, 419
2-Mesitylquinoxaline, 235
Mesoxalic acid:
  reaction of, with *o*-phenylenediamine, 138
3-Methoxycarbonyl-4,7-dimethylpyrazolo-[1,5-*a*]pyrazine, 398
3-Methoxycarbonyl-4,7-dimethylpyrazolo-[1,5-*a*]pyrazine 5-oxide, 398
4-Methoxycarbonylpyrrolo[1,2-*a*]quinoxaline, 601
2-Methoxy-3-methylquinoxaline, 31
6-Methoxy-2-methylthiazolo[4,5-*b*]pyrazine, 448
2-Methoxy-3-phenylquinoxaline, 31
2-Methoxyquinoxaline, 86, 200
5-Methoxyquinoxaline, 19
6-Methoxyquinoxaline, 19
6-Methoxyquinoxaline-2,3-dione, 168
6-Methoxyquinoxaline-5,8-dione, 21
5-Methoxyquinoxaline 1,4-dioxide, 29
2-Methoxyquinoxalines, 199
3-Methylaminoquinoxaline-2-carboxylic acid, 141
3-Methylaminoquinoxaline-2-thione, 113
2-Methylbenzo[*f*]quinoxaline, 716

3-Methylbenzo[*f*]quinoxaline, 716
Methyl 2,3-dibromopropionate:
  reaction of, with *NN'*-dibenzenesulfonyl-*o*-phenylenediamine, 144
Methyl 3,4-dihydro-3-oxoquinoxaline-2-carboxylate:
  spiro derivatives from, 147
2-Methyl-1,3-diphenylimidazo[4,5-*b*]-quinoxalinium tosylate, 670
2*H*-2-Methyl-1,3-diphenylpyrrolo[3,4-*b*]quinoxaline, 629, 630
5-Methyl-2,3-diphenylthieno[3,4-*b*]pyrazine, 444
Methylhydrazine:
  reaction of, with chloroquinoxalines, 194
2-(1-Methylhydrazino)-3-phenylquinoxaline, 194
2-Methylimidazo[1,2-*a*]pyrazine, 365
3-Methylimidazo[1,2-*a*]pyrazine, 365
3-Methylimidazo[1,5-*a*]pyrazine, 372-375
1-Methylimidazo[4,5-*b*]pyrazine, 375, 376
2-Methylimidazo[4,5-*b*]pyrazine, 376
1-Methylimidazo[1,2-*a*]pyrazine 7-oxide, 368
1-Methylimidazo[4,5-*b*]pyrazine 4-oxide, 376, 377
1*H*-2-Methylimidazo[4,5-*b*]quinoxaline, 665, 667
4*H*-4-Methylimidazo[4,5-*b*]quinoxaline, 665
2-Methyl-3-methylthioquinoxaline 1,4-dioxide, 49
4*H*-2-Methyl-4-oxopyrazino[2,3-*d*][1,3]-oxazine, 587
8-Methyl-2-phenylimidazo[1,2-*a*]-pyrazine, 365
3-Methyl-1-phenylpyrazolo[3,4-*b*]quinoxaline, 697
1-Methyl-3-phenylpyrido[1,2-*a*]pyrazinium bromide, 467
2-Methyl-4-phenylpyrrolo[1,2-*a*]quinoxaline, 610
4-Methyl-2-phenylpyrrolo[1,2-*a*]quinoxaline, 610
2-Methyl-3-phenylquinoxaline, 31, 213
8-Methyl-6-phenylthiazolo[3,2-*a*]pyrazinium bromide, 445
1-Methylpyrazolo[3,4-*b*]pyrazine, 400
2-Methylpyrazolo[3,4-*b*]pyrazine, 400
3-Methylpyrazolo[3,4-*b*]quinoxaline, 693
2-Methylpyrido[3,4-*b*]pyrazine, 550, 557
3-Methylpyrido[2,3-*b*]pyrazine, 502
3-Methylpyrido[3,4-*b*]pyrazine, 555
1-Methylpyrido[1,2-*a*]pyrazinium bromide:
  2-oxide, 470, 478
1-Methylpyrrolo[1,2-*a*]pyrazine, 294
7-Methylpyrrolo[2,3-*b*]pyrazine, 350
1-Methylpyrrolo[1,2-*a*]quinoxaline, 616
4-Methylpyrrolo[1,2-*a*]quinoxaline, 601, 615
2-Methylquinoxaline:
  oxidation of, 29
  preparations of, 13, 205, 207
  quaternisation of, 247
  reactions of, 213-220, 263
  reduction of, 261, 267
6-Methylquinoxaline, 19
3-Methylquinoxaline-2-carboxaldehyde:
  arylhydrazones of, 126, 214
  preparations of, 125, 215
3-Methylquinoxaline-2-carboxaldehyde 1,4-dioxide, 55
Methyl quinoxaline-2-carboxylate, 144
1-Methylquinoxaline-2,3-dione, 96
6-Methylquinoxaline-5,8-diones, 21
2-Methylquinoxaline 1,4-dioxide, 47
6-Methylquinoxaline-2,3-dithione cyclic dithiocarbonate, 118
Methylquinoxaline orange, 220
2-Methylquinoxaline 1-oxide, 39, 46
5-Methylquinoxaline 1-oxide, 163
2-Methylquinoxalines:
  conversion into quinoxaline-2-carboxylic acids, 145
1-Methylquinoxalinium iodide, 250
1-Methylquinoxalin-2-one:
  hydroxylation and phenylation of, 89
  preparation of, 88
  u.v. and pka of, 86
3-Methylquinoxalin-2-one, 80, 83
3-Methylquinoxalin-2-ones, 83
2-Methylsulfinylquinoxaline, 181
2-Methylsulfonylquinoxaline, 114
2-Methylsulfonylquinoxaline 4-oxide, 114, 115
6-Methylthiazolo[3,2-*a*]pyrazinium bromide:
  7-oxide, 445

8-Methylthiazolo[3,2-*a*]pyrazinium bromide:
preparation of, 444
7-oxide, preparation of, 445
3-Methylthioimidazo[1,5-*a*]pyrazine, 375
4-Methylthiopyrrolo[1,2-*a*]quinoxaline, 616
2-Methylthioquinoxaline, 31, 114
2-Methyl-*s*-triazolo[1,5-*a*]pyrazine, 411
2-Methyl-*s*-triazolo[1,5-*a*]pyrazine 7-oxide, 411, 412
Morestan, *see* 6-Methylquinoxaline-2,3-dithione cyclic dithiocarbonate
2-Morpholino-3-pyrrolidinylquinoxaline, 185
4*H*-2-(2-Morpholin-4-ylethoxy)-4-oxo-3-phenylpyrazino[1,2-*a*]pyrimidine, 570
1*H*-3-Morpholin-4-ylpyrrolo[3,4-*b*]quinoxaline, 626

Naphthopiperazines, *see* Pyrazino[2,3-*b*]pyrazines
6-Nitro-2,3-diphenylquinoxaline, 20
2-Nitromethylquinoxaline, 216
3-Nitro-2-phenylpyrrolo[1,2-*a*]quinoxaline, 610
p-Nitrophenyl-2-quinoxalinysulfide, 114
1-Nitropyrrolo[1,2-*a*]quinoxaline, 616
3-Nitropyrrolo[1,2-*a*]quinoxaline, 616
5-Nitroquinoxaline, 18
5-Nitroquinoxaline-2,3-dione, 168
6-Nitroquinoxaline-2,3-dione:
insecticidal properties of, 97
isolation of, 216
preparation of, 94, 96
5- and 6-Nitroquinoxalines, 19, 267
'6-Nitroquinoxalines:
amination of, 20

1-Oxa-7,10-diazabicyclo[4.4.0]decanes, *see* Pyrazino[2,1-*b*][1,3]oxazines
1-Oxa-6,9-diazabicyclo[4.3.0]nonanes, *see* Oxazolo[3,2-*a*]pyrazines
[1,2,3]Oxadiazolo[3,4-*a*]pyrazines:
table of, 439
[1,2,5]Oxadiazolo[3,4-*b*]pyrazines:
table of, 439
Oxalic acid esters:
reaction of, with 2-aminomethylpiperidine, 470
with 3,4-diaminofurazan, 436
with 2,3-diaminopyridine, 498
with 3,4-diaminothiophene, 443
with 2- or 3-methylpyrido[2,3-*b*]-pyrazines, 513
with *o*-phenylenediamines, 94, 95
Oxazino[3,4-*a*]pyrazines, *see* Pyrazino[1,2-*c*][1,3]oxazines
Oxazolo[3,2-*a*]pyrazines:
table of, 438
Oxazolo[3,4-*a*]pyrazines:
table of, 438
Oxazolo[3,2-*a*]pyrrolo[2,1-*c*]pyrazines:
hydrolysis of, 319
i.r. and tautomerism of, 307
occurrence of, 303, 324
preparation of, 317
2-Oxo-1,3-dithiolo[4,5-*b*]pyrazines, 459
4*H*-4-Oxopyrazino[2,3-*d*][1,3]oxazines, 587
4*H*-4-Oxopyrazino[1,2-*a*]pyrimidine, 569
2-Oxopyrrolo[2,3-*b*]quinoxalines:
table of, 643

1,4,5,7,8-Pentaazanaphthalenes, *see* Pyrazino[2,3-*e*]-*as*-triazines
1,4,6,7,8-Pentaazanaphthalenes, *see* Pyrazino[2,3-*d*]-*v*-triazines
Pentane 2,3,4-trione:
reaction of, with *o*-phenylenediamine, 130
Perchloroimidazo[1,2-*a*]pyrazine, 370
Perhydro-2,8-dimethyl-4,6-dioxopyrazino-[1,2-*a*]pyrazine, 575
Perhydro-1,1-dimethyl-5,8-dioxothiazolo-[3,4-*a*]pyrazine, 447, 448
Perhydro-2,8-dimethylpyrazino[1,2-*a*]pyrazine, 575
Perhydro-1,4-dioxo-3,3-diphenylpyrido-[1,2-*a*]pyrazine, 469
Perhydro-5,8-dioxo-1-phenylthiazolo[3,4-*a*]-pyrazine, 448
Perhydro-1,4-dioxopyrido[1,2-*a*]pyrazine, 304. *See also* Glycyl-L-proline anhydride
Perhydro-3,4-dioxopyrido[1,2-*a*]pyrazine, 470
Perhydro-1,4-dioxopyrrolo[1,2-*a*]pyrazines:
biological activity of, 322
conformation of, 304
epimerisation of, 314, 318

occurrence of, 323
preparation of, 308-316
reactions of, 316, 317, 319, 322
resolution of, 307
spectra of, 304, 307, 308
stereochemistry of, 305, 306
table of, 334
Perhydro-5,8-dioxothiazolo[3,4-*a*]pyrazines:
preparation of, 447, 448
table of, 453
Perhydro-2-hydroxyacetyl-1,4-dioxopyrrolo[1,2-*a*]pyrazines, 307
Perhydro-3-(4-hydroxybenzyl)-1,4-dioxopyrrolo[1,2-*a*]pyrazine, 323
Perhydro-9-hydroxy-1-tetrahydrofur-2-ylpyrido[1,2-*a*]pyrazine, 474
Perhydroimidazo[1,5-*a*]pyrazines, 373, 375
Perhydro-3-isopropyl-1,4-dioxopyrrolo-[1,2-*a*]pyrazine, *see D*-Valyl-*L*-proline anhydride
Perhydro-2-lauroyl-1,4,6-trioxopyrrolo-[1,2-*a*]pyrazine, 317, 322
Perhydro-7-methyl-1,3-dioxo-2-phenylimidazo[1,5-*a*]pyrazine, 373
Perhydro-3-methyl-1;4-dioxo-2-phenylpyrrolo[1,2-*a*]pyrazine, 310
Perhydro-2-methyl-1,4-dioxopyrido[1,2-*a*]pyrazine, 469
Perhydro-2-methyl-3,4-dioxopyrido[1,2-*a*]pyrazine, 471, 478
Perhydro-3-methyl-1,4-dioxopyrrolo[1,2-*a*]pyrazine:
p.m.r. and stereochemistry of, 305
preparations of, 310, 315
Perhydro-7-methyl-3-oxo-oxazolo[3,4-*a*]pyrazine, 435
Perhydro-6-methyl-1-oxopyrido[1,2-*a*]pyrazine, 470
Perhydro-5-methyl-1-oxopyrido[1,2-*a*]pyrazinium iodide, 475, 480
Perhydro-2-(1-methyl-2-phenyl)ethylpyrrolo[1,2-*a*]pyrazine, 300
Perhydro-6-methyl-2-phenylpyrido[1,2-*a*]pyrazine, 469
Perhydro-oxazolo[3,2-*a*]pyrazines, 434
Perhydro-2-oxo-3a,7a-diphenylimidazo-[4,5-*b*]pyrazine, 379
Perhydro-1-oxo-2,3-diphenylpyrido[1,2-*a*]pyrazine, 479
Perhydro-1-oxo-3,3-diphenylpyrido[1,2-*a*]pyrazine, 481
Perhydro-1-oxo-3-phenylpyrido[1,2-*a*]pyrazine, 475
Perhydro-1-oxopyrido[1,2-*a*]pyrazine:
alkylation of, 475
preparation of, 470, 479
Perhydro-3-oxopyrido[1,2-*a*]pyrazine, 471, 473
Perhydro-1-oxopyrrolo[1,2-*a*]pyrazine, 294, 302
Perhydro-3-oxopyrrolo[1,2-*a*]pyrazine, 299, 302
Perhydro-1-oxopyrrolo[1,2-*a*]pyrazines, 299, 300
Perhydro-2-phenylpyrido[1,2-*a*]pyrazine, 469
Perhydro-2-propylpyrrolo[1,2-*a*]pyrazine, 300
Perhydropyrazino[2,1-*b*][1,3]oxazines:
table of, 589
Perhydropyrazino[2,3-*b*]pyrazine, 579, 580
Perhydropyrazino[1,2-*c*]pyrimidines:
table of, 573
Perhydropyrido[1,2-*a*]pyrazines:
biological activity of, 480
i.r. and conformational analysis of, 464
preparation of, 469
reactions of, 476
table of, 482
Perhydropyrrolo[1,2-*a*]pyrazine:
acylation and alkylation of, 301
preparations of, 299, 302
Perhydropyrrolo[1,2-*a*]pyrazines:
biological activity of, 303
preparation of, 294
quaternisation of, 300
Perhydropyrrolo[1,2-*a*]quinoxaline, 608
Perhydro-7,7,8a-trimethyl-1,6-dioxopyrrolo[1,2-*a*]pyrazine, 299
Perhydro-1,6,8-trioxo-7-phenylpyrazino[1,2-*c*]pyrimidine, 571
Perhydro-1,6,8-trioxopyrazino[1,2-*c*]pyrimidine, 571
Perhydro-1,4,6-trioxopyrrolo[1,2-*a*]pyrazine, 317
Perhydro-1,4,6-trioxopyrrolo[1,2-*a*]pyrazines, 322
2-Phenacetyl-3-phenylquinoxaline, 133
Phenacyl bromide:

reaction of, with 2-acylpyridine oximes, 467, 468
with 2-acylpyrroles, 294, 295
with 2-acylthiazole oximes, 445, 446
with 2-amino-3-mercaptopyrazines, 592, 593
with 2-amino-3-methylpyrazine, 365, 366
with 2-carboxamidopyridines, 466
with imidazole-2-carboxaldehyde oxines, 368
with *o*-phenylenediamine, 263
1-Phenacylquinoxalinium bromide, 252
3-Phenacylquinoxalin-2-one, 87
2-Phenoxyquinoxaline, 200
2-Phenoxyquinoxalines, 199
3-Phenylbenzo[*f*]quinoxaline, 717, 719
3-Phenylbenzo[*f*]quinoxaline-5,6-quinone, 719
2-(*N*-Phenylcarbamoyl)quinoxaline 1-oxide, 39
Phenylglyoxylic acid:
reaction of, with *o*-phenylenediamine, 79
8-Phenylimidazo[1,2-*a*]pyrazine, 368
1-Phenylimidazo[1,2-*a*]pyrazine 7-oxide, 368
4-Phenylimidazo[1,2-*a*]quinoxaline, 654
3-Phenylisoxazolo[4,5-*b*]pyrazine, 435
α-Phenyl-3-methyl-2-quinoxalineacetonitrile, 207
2-Phenyl-3-piperidinoquinoxalines, 184
1-Phenylpyrazolo[3,4-*b*]quinoxaline, 693, 695
1-Phenylpyrazolo[3,4-*b*]quinoxalines:
preparation of, 693-696
use of, for characterising sugars, 693
2-Phenylpyrido[3,4-*b*]pyrazine, 550
1-Phenylpyrido[1,2-*a*]pyrazinium bromide, 471
1-Phenylpyrrolo[1,2-*a*]pyrazine:
p.m.r. of, 293
preparations of, 294, 302
4-Phenylpyrrolo[1,2-*a*]quinoxaline, 601
2-Phenylquinoxaline, 18, 261
5-Phenylquinoxaline, 18
6-Phenylquinoxaline, 18
α-Phenyl-2-quinoxalineacetonitrile, 146
3-Phenyl-2-quinoxalinecarboxamide, 31
2-Phenylquinoxaline 1,4-dioxide, 30
2-Phenylquinoxaline 1-oxide:
irradiation of, 45
isolation of, 43, 44
2-Phenylquinoxaline 4-oxide:
preparation of, 30
reactions of, 44, 45, 163
2-Phenylquinoxalines, 234, 235
3-Phenylquinoxalin-2-one:
preparations of, 82, 84
reaction of, with 3-dimethylaminopropylmagnesium chloride, 90
3-Phenyl-2-quinoxalinylacrylic acid, 213
6-Phenylthiazolo[3,2-*a*]pyrazinium bromide, 445
8-Phenylthiazolo[3,2-*a*]pyrazinium bromide, 445
1-Phenyl-*s*-triazolo[4,3-*a*]quinoxaline, 416
1-Phenyl-7-trifluoromethylquinoxaline-2,3-dione, 95
Piperazino[*hi*]benzimidazoles, *see* Imidazo[1,5,4-*de*]quinoxalines
Piperazinopiperazines, *see* Pyrazino[2,3-*b*]pyrazines
Piperazino[1,2-*a*]piperazines, *see* Pyrazino[1,2-*a*]pyrazines
Piperazino[9:10]pyridines, *see* Pyrido[1,2-*a*]pyrazines
2-Polyhydroxyalkylquinoxalines:
o.r.d. of, 241, 243
preparation of, 241
reactions of, 241, 244
table of, 245
Pyrazine-2,3-dicarboxylic anhydride, *see* 5,7-Dihydro-5,7-dioxofuro[3,4-*b*]pyrazine
Pyrazine-2,3-dicarboxylic imide, *see* 6*H*-5,7-Dihydro-5,7-dioxopyrrolo[3,4-*b*]pyrazine
Pyrazino[1,2-*c*][1,3]oxazines, 586
Pyrazino[2,1-*b*][1,3]oxazines, 585
Pyrazino[2,1-*c*][1,4]oxazines, 588
table of, 590
Pyrazino[2,3-*d*][1,3]oxazines, 587
Pyrazino[2,3-*d*][1,3]oxazin-4-ones:
table of, 589
Pyrazino[2,3-*b*]pyrazine, 576-579
Pyrazino[1,2-*a*]pyrazines, 575
table of, 580
Pyrazino[2,3-*b*]pyrazines:
biological activity of, 580
covalent hydration of, 577

preparation of, 578
table of, 581
Pyrazino[1,2-*a*]pyrimidines, 568
table of, 572
Pyrazino[1,2-*c*]pyrimidines, 570
Pyrazino[2,3-*e*]-*v*-tetrazines, 584
Pyrazino[2,1-*c*][1,4]thiazines, 591
table of, 595
Pyrazino[2,3-*b*][1,4]thiazines, 592
table of, 595
Pyrazino[1,2-*a*]-*s*-triazines, 583
Pyrazino[2,3-*d*]-*v*-triazines, 584
Pyrazino[2,3-*e*]-*as*-triazines, 583
3-(2-Pyrazinyl)imidazo[1,5-*a*]pyrazine, 372
1*H*-Pyrazolo[3,4-*b*]pyrazine, 400, 405
Pyrazolo[3,4-*b*]pyrazine-5,6-dicarboxylic acid, 405
Pyrazolo[1,5-*a*]pyrazines, 398
table of, 406
Pyrazolo[3,4-*b*]pyrazines:
p.m.r. of, 400
preparation of, 402
protonation of, 402
table of, 407
Pyrazolo[1,5-*a*]quinoxaline, 691
Pyrazolo[3,4-*b*]quinoxaline:
preparation of, 696
reactions of, 405, 698, 699
u.v. of, 693
Pyrazolo[*a*]quinoxalines, *see* Pyrazolo[1,5-*a*]quinoxalines
Pyrazolo[1,5-*a*]quinoxalines, 691
table of, 704
Pyrazolo[2,3-*a*]quinoxalines, *see* Pyrazolo-[1,5-*a*]quinoxalines
Pyrazolo[2,3-*b*]quinoxalines, *see* Pyrazolo-[3,4-*b*]quinoxaline (flavazoles)
Pyrazolo[3,4-*b*]quinoxalines (flavazoles):
preparations of, 244, 693-695
table of, 704
u.v. of, 693
Pyrazolo-4′,5′:2,3-quinoxalines, *see* Pyrazolo[3,4-*b*]quinoxalines (flavazoles)
Pyrazo[3,4-*b*]quinoxalines, *see* Pyrazolo-[3,4-*b*]quinoxalines (flavazoles)
Pyridazino[4,5-*b*]quinoxalines, 152
Pyridino[2′,3′:2,3]pyrazines, *see* Pyrido[2,3-*b*]pyrazines
Pyridino[3′,4′:2,3]pyrazines, *see* Pyrido[3,4-*b*]pyrazines
Pyridino[1,2-*a*]quinoxalines, *see* Pyrido[1,2-*a*]quinoxalines
Pyridino-2′,3′:2,3-quinoxalines, *see* Pyrido[2,3-*b*]quinoxalines
Pyridopiperazines, *see* Pyrido[1,2-*a*]pyrazines
Pyrido[2,3-*b*]pyrazine:
physical properties of, 496, 497
preparation of, 498
reactions of, 512, 515, 517
Pyrido[3,4-*b*]pyrazine:
covalent hydration and pka of, 546
oxidation of, 557
preparation of, 547
spectra of, 545
Pyrido[1,2-*a*]pyrazines:
table of, 487
Pyrido[2,3-*b*]pyrazines:
biological properties of, 524
physical properties of, 496, 497
preparations of, 497-509
reactions of, 512-524
table of, 530
Pyrido[3,4-*b*]pyrazines:
biological properties of, 560
physical properties of, 545, 546
preparations of, 547-556
reactions of, 556-560
table of, 564
Pyrido[1,2-*a*]pyrazinium betaines, 474, 477
Pyrido[1,2-*a*]pyrazinium bromide:
2-oxide, preparation of, 467, 470
reactions of, 476, 479
reduction of, 479
u.v. of, 464
Pyrido[1,2-*a*]pyrazinium oxides, 479
Pyrido[1,2-*a*]pyrazinium salts, 479
Pyrido[2,3-*b*]quinoxaline, 746, 747
Pyrido[3,4-*b*]quinoxaline:
preparation of, 750
spectra of, 747, 749
Pyrido[1,2-*a*]quinoxalines, 743
table of, 753
Pyrido[2,3-*b*]quinoxalines:
bond fixation in, 746
table of, 755
Pyrido[3,4-*b*]quinoxalines:
bond fixation in, 749

table of, 757
Pyrido[1,2,3-*de*]quinoxalines, 752
table of, 758
Pyrrolidino(1′,2′:1,2)piperazines, *see* Pyrrolo[1,2-*a*]pyrazines
Pyrrolo[1,2-*a*]pyrazine, 292-294, 302
Pyrrolo[2,3-*b*]pyrazine, 348, 349
Pyrrolo[1,2-*a*]pyrazines:
biological activity of, 303
i.r. and p.m.r. of, 292, 293
preparation of, 294-300
reactions of, 300-302
table of, 327
Pyrrolo[2,3-*b*]pyrazines:
preparation of, 348, 349
reactions of, 349-352
spectra of, 348
table of, 356
Pyrrolo[3,4-*b*]pyrazines, 355
table of, 357
2,5-Pyrrolopyrazines, *see* Pyrrolo[1,2-*a*]pyrazines
Pyrrolo[2,3-*c*]pyridines:
formation of, 294
Pyrrolo[1,2-*a*]quinoxaline:
amination of, 618
electrophilic substitution of, 598, 616-618
physical properties of, 598-600
preparation of, 601
quaternisation of, 615
Pyrrolo[1,2-*a*]quinoxalines:
physical properties of, 599-600
preparations of, 601-614
reactions of, 615-622
table of, 633
uses of, 622
Pyrrolo[2,3-*b*]quinoxalines, 622
table of, 644
Pyrrolo[3,4-*b*]quinoxalines, 625
table of, 646
Pyrrolo[1,2,3-*de*]quinoxalines, 630
table of, 649
Pyrrolo[1,2-*a*]quinoxaline 2-sulfonic acid, 617
Pyrrolo[1,2-*a*]quinoxaline 3-sulfonic acid, 616, 617
Pyruvaldehyde:
reaction of, with aminopyrazines, 366, 367
with 4,5-diamino-3-hydroxypyrazole, 402, 403
with 2,3-diaminopyridines, 502
with *o*-phenylenediamine, 205
with *o*-quinone dioximes, 38, 39
with 1,2,3,5-tetraaminopyridine, 499
Pyruvic acid esters:
reaction of, with 2-amino-1-phenylazo-naphthalene, 716
with 3,4-diaminopyridine, 552
with 2,3-diaminopyridines, 504

Quinolo-1,4-diazines, *see* Pyrido[1,2,3-*de*]quinoxalines
Quinoxaline:
addition reactions of, 13, 268
oxidation of, 12
physical properties of, 7-11
preparation of, 7
reduction of, 11, 261, 267, 276
substitution reactions of, 12, 13, 18, 262
Quinoxaline antibiotics, 148
Quinoxaline-2-carboxaldehyde:
preparations of, 123, 215
reactions of, 125, 131
Quinoxaline-2-carboxaldehydes:
preparations of, 123, 124
table of, 127
Quinoxaline-2-carboxaldehyde thiosemicarbazones, 126
Quinoxaline-2-carboxylic acid:
acid chloride of, 146
pka of, 145
preparation of, 125
uses of, 148
Quinoxaline-2-carboxylic acids:
pka of, 145
preparations of, 127-145
reactions of, 146
table of, 152
Quinoxaline-2,3-dicarboxaldehyde, 124, 125
Quinoxaline-2,3-dicarboxylic acid:
derivatives of, 150, 151
isolation of, 215
preparation of, 149
Quinoxaline-2,3-dicarboxylic acids:
preparation of, 149, 150
table of, 157

Quinoxaline-2,3-dicarboxylic imide, *see* 2*H*-1,3-Dihydro-1,3-dioxopyrrolo[3,4-*b*]quinoxaline
Quinoxaline-2,3-dione:
preparation of, 94
reactions of, 96, 168, 209
tautomerism of, 95
Quinoxaline-2,3-diones:
biological properties of, 97
preparations of, 94, 95
reactions of, 96
table of, 104
Quinoxaline-5,8-diones, 20, 21
Quinoxaline-2,3-dione 6-sulfonic acid, 96
Quinoxaline 1,4-dioxide:
antibacterial properties of, 55
physical properties of, 41, 42
polarography of, 42
preparation of, 29
reactions of, 45, 46
Quinoxaline di-*N*-oxides:
biological properties of, 55, 66
preparations of, 29-40
reactions of, 45-51
spectra of, 40-42
table of, 63
Quinoxaline-2,3-dithione, 116, 117
Quinoxaline-2,3-dithione cyclic trithiocarbonate:
insecticidal and fungicidal properties of, 118
Quinoxaline-2,3-dithiones:
acylation of, 117
preparation of, 116
table of, 120
uses of, 118
Quinoxaline 2(or 3)-ketones:
table of, 134
Quinoxaline mono-*N*-oxides:
i.r. of, 40
mass spectra of, 42
preparations of, 29-34, 38-40
quaternisation of, 45, 248
reactions of, 42-51
table of, 56
Quinoxaline 1-oxide:
physical properties of, 40-42
polarography of, 42
preparation of, 29
reactions of, 43-45
Quinoxalines:
ionisation properties of, 7,8
i.r. spectra of, 8, 9
n.m.r. spectra of, 9, 10
N-oxidation of, 29-33
reactions of, 2-5
reviews on, 6
synthesis of, 1
u.v. spectra of, 9
Quinoxaline-2-thione, 112, 114
Quinoxaline-2-thiones:
acylation of, 114
preparation of, 112
table of, 119
uses of, 116
Quinoxalinium salts:
cyanine dyes from, 252
preparations of, 247-249
reactions of, 250-254
table of, 255
uses of, 254
Quinoxalino[2,3-*b*]benzothiazines, 170
4,5-Quinoxalinoimidazoles, *see* Imidazo-[4,5-*b*]quinoxalines
Quinoxalin-2-one:
pka and tautomerism of, 86
preparations of, 79, 81, 82
reactions of, 88-90
u.v. and mass spectra of, 86
Quinoxalin-2-ones:
physical properties of, 86, 87
preparations of, 79-86
reactions of, 88-91
reduction of, 267
table of, 97
tricyclic derivatives from, 91-93
uses of, 93, 94
Quinoxalino[2,3-*c*]pyrroles, *see* Pyrrolo-[3,4-*b*]quinoxalines
2-Quinoxalinoyl chloride, 146
2-(2-Quinoxalinyl)acetophenone, 146
Quinoxalin-2-ylcarbamate-*N*-oxides, 49
2-Quinoxalinyl ketones, 131
table of, 134

Sodium mesoxalate:
reaction of, with *o*-phenylenediamine, 138
Spiro-furo[3,4-*b*]pyrazines, 432
table of, 437

2-Styrylquinoxalines, 209, 213
2-Substituted aminoquinoxalines:
preparation of, 180
p-Substituted anilinomethyl-1,2,3,4-tetrahydroquinoxalines, 267, 274
5- and 6-Substituted 2,3-dimethylquinoxalines:
pka values of, 212
u.v. and ^{1}H n.m.r. spectra of, 210
6-Substituted 2,3-dimethylquinoxalines:
oxidation of, 31
1-Substituted 2,3-dimethylquinoxalinium perchlorates, 248, 251
1-Substituted and 1,4-disubstituted quinoxaline-2,3-diones, 95
2-Substituted quinoxalines:
oxidation of, 29
5-Substituted quinoxalines:
oxidation of, 19, 29
reduction of, 19
6-Substituted quinoxalines:
oxidation of, 19, 29
reduction of, 19, 267
2-Substituted-3-trifluoromethylquinoxalines:
^{1}H n.m.r. spectra of, 41
1-oxides and 1,4-dioxides, ^{1}H n.m.r. spectra of, 42
2-Sulfonamidoquinoxaline, 184

5,6,11,12-Tetraazanaphthacenes, 170
1,4,5,8-Tetraazanaphthalenes, *see* Pyrazino[2,3-*b*]pyrazines
2,3,5,10-Tetrachlorobenzo[*g*]quinoxaline, 734, 735
5,6,7,8-Tetrachloroquinoxaline, 20, 21
5,6,7,8-Tetrachloroquinoxalin-2-one, 94
1,1′,3,3′-Tetraethylimidazo[4,5-*b*]quinoxalinocyanine iodide, 664
5,6,7,8-Tetrafluoroquinoxaline-2,3-dione, 95, 96
1*H*-2,3,5,6-Tetrahydro-1,5-dimethylcyclopentapyrazine, 457, 458
1*H*-2,3,5,6-Tetrahydro-1,7-dimethylcyclopentapyrazine, 457, 458
1*H*-2,3,5,6-Tetrahydro-5,7-dimethylcyclopentapyrazine, 457
5*H*-4a,6,7,7a-tetrahydro-2,3-dimethylcyclopentapyrazine, 456
1,2,3,4-Tetrahydro-1,4-dimethyl-2,3-dioxoquinoxaline, 209
4,5,6,7-Tetrahydro-2,5-dimethyl-4-oxopyrazolo[1,5-*a*]pyrazine, 399
1,2,3,4-Tetrahydro-1,4-dimethylquinoxaline, 269
cis-1,2,3,4-Tetrahydro-2,3-dimethylquinoxaline:
preparation of, 266
u.v. and n.m.r. spectra of, 270, 271
dl-trans-1,2,3,4-Tetrahydro-2,3-dimethylquinoxaline:
n.m.r. spectrum and conformation of, 271
preparation of, 266
resolution of, 267
1,2,3,4-Tetrahydro-1,4-dimethyl-2-vinylquinoxaline, 269
1,2,3,4-Tetrahydro-2,3-dioxobenzo[*g*]-quinoxaline-5,10-dione, 734
1,2,3,4-Tetrahydro-2,3-dioxobenzo[*f*]-quinoxaline dioxime, 715
1,2,3,4-Tetrahydro-2,3-dioxo-1-phenylpyrido[3,4-*b*]pyrazine, 548
1,2,3,4-Tetrahydro-2,3-dioxopyrazino[2,3-*b*]pyrazine, 577, 578
8*H*-1,2,3,4-Tetrahydro-1,8-dioxopyrido[1,2-*a*]pyrazine, 478
1,2,3,4-Tetrahydro-2,3-dioxopyrido[2,3-*b*]pyrazine, 519
1,2,3,4-Tetrahydro-2,3-dioxopyrido[3,4-*b*]pyrazine, 547, 558
1,2,3,4-Tetrahydro-1,3-dioxopyrido[1,2-*a*]pyrazinium bromide:
preparation of, 466
reduction of, 480
u.v. and n.m.r. spectra of, 464
8*H*-1,2,3,4-Tetrahydro-1,8-dioxopyrido-[1,2-*a*]pyrazin-6-ylcarboxylic acid, 473, 478
1,2,3,3a-Tetrahydro-3a,4-diphenylpyrrolo-[1,2-*a*]quinoxaline, 610
3,4,5,6-Tetrahydro-2,3-diphenylpyrrolo-[1,2,3-*de*]quinoxaline, 630, 631
1,2,3,4-Tetrahydro-2,3-diphenylquinoxaline, 268
1,2,3,4-Tetrahydro-1-ethylquinoxaline, 272
5*H*-2,3,6,7-Tetrahydro-5-hydroxy-6,7-dioxocyclopentapyrazine, 457
1,2,3,4-Tetrahydro-2-hydroxymethyl-1,4-ditosylquinoxaline, 268, 273

1,2,3,4-Tetrahydro-1-hydroxy-2-oxopyrido-[2,3-*b*] pyrazine, 510
1*H*-2,3,5,6-Tetrahydro-7-methylcyclopentapyrazine, 457, 458
1,2,3,5-Tetrahydro-5-methyl-1,3-dioxoimidazo[1,5-*a*] quinoxaline, 659, 660
1,2,3,4-Tetrahydro-1-methyl-2,3-dioxoquinoxaline, 183
5,6,7,8-Tetrahydro-7-methyl-1-methylamino-6,8-dioxoimidazo[1,5-*a*] pyrazine, 373
1,2,3,4-Tetrahydro-2-methyl-3-oxopyrido-[3,4-*b*] pyrazine, 555
1,2,3,4-Tetrahydro-3-methylpyrido[3,4-*b*] pyrazine, 555
1,2,3,4-Tetrahydro-1-methylquinoxaline, 272
1,2,3,4-Tetrahydro-2-methylquinoxaline, 262, 267
1,2,3,4-Tetrahydro-6-nitroquinoxaline, 273
5*H*-2,3,6,7-Tetrahydro-6-oxocyclopentapyrazin-5-yl carboxylic acid, 458
1,2,3,4-Tetrahydro-2-oxopyrazino[1,2-*a*] pyrimidinium chloride, 569
1,2,3,4-Tetrahydro-2-oxopyrido[2,3-*b*] pyrazine, 515, 519
1,2,3,4-Tetrahydro-2-oxopyrido[3,4-*b*] pyrazine, 556
1,2,3,4-Tetrahydro-3-oxopyrido[3,4-*b*] pyrazine, 555
1,2,3,4-Tetrahydro-1-oxopyrido[1,2-*a*] pyrazinium bromide, 479
1,2,3,4-Tetrahydro-3-oxopyrido[1,2-*a*] pyrazinium chloride, 471
1,2,3,4-Tetrahydro-2-oxopyrrolo[1,2,3-*de*] quinoxalines, 632
1,2,3,4-Tetrahydro-2-phenylpyrido[1,2-*a*] pyrazinium bromide, 469
6,7,8,9-Tetrahydro-4-phenylpyrrolo[1,2-*a*] quinoxaline, 614
1,2,3,4-Tetrahydropyrazino[2,3-*b*] quinoxaline, 655
1,2,3,4-Tetrahydroquinoxaline:
  acylation of, 273
  physical properties of, 270, 272
  preparations of, 266-268
  reduction of, 276
1,2,3,4-Tetrahydroquinoxalines:
  preparations of, 266-270
  reactions of, 272-274
  table of, 278
  uses of, 274, 275
5,6,7,8-Tetrahydroquinoxalin-2-one, 275
2,3,5,7-Tetrahydro-5,5,7,7-tetramethylfuro[3,4-*b*] pyrazine, 431, 432
1*H*-2,3,5,6-Tetrahydro-1,5,5-trimethylcyclopentapyrazine, 459
1*H*-2,3,5,6-Tetrahydro-3,5,7-trimethylcyclopentapyrazine, 457
2-Tetrahydroxybutylquinoxalines, 242, 244
Tetrazolo[1,5-*a*] piperazines, *see* Tetrazolo[1,5-*a*] pyrazines
Tetrazolo[1,5-*a*] pyrazine, 423, 424
Tetrazolo[1,5-*a*] pyrazines:
  table of, 429
Tetrazolo[1,5-*c*] pyrazines, *see* Tetrazolo-[1,5-*a*] pyrazines
Tetrazolo[1,5-*a*] quinoxalines, 196
1,2,5-Thiadiazolo[3,4-*b*] pyrazine, 450
1,2,5-Thiadiazolo[3,4-*b*] pyrazines:
  table of, 454
Thiazolo[4,5-*b*] pyrazines:
  table of, 453
Thiazolo[3,2-*a*] pyrazin-4-ium bromide, 444-446
Thiazolo[3,2-*a*] pyrazinium salts, 444
  table of, 452
Thieno[2,3-*b*] pyrazine, 441, 442
Thieno[3,4-*b*] pyrazine, 443
Thieno[2,3-*b*] pyrazines, 442
  table of, 451
Thieno[3,4-*b*] pyrazines, 443
  table of, 452
2-Thioxo-1,3-dithiolo[4,5-*b*] pyrazines, 459
6-*p*-Toluenesulfonamidoquinoxaline, 19
p-Tolyl-D-isoglucosamine:
  reaction of, with *o*-phenylenediamine, 242
2,3,4-Triaminopyridine:
  formation of pyridopyrazines from, 498-500
1,3,6-Triazaacenaphthylenes, *see* Imidazo-[1,5,4-*de*] quinoxaline
2,9,10-Triazaanthracenes, *see* Pyrido[3,4-*b*] quinoxalines
1,3,8-Triazabicyclo[4.4.0] decanes, *see* Pyrido[3,4-*b*] pyrazines
1,4,8-Triazabicyclo[4.3.0] nonanes, *see* Imidazo[1,5-*a*] pyrazines

1,3a,6-Triazaindenes, *see* Imidazo[1,2-*a*]pyrazines
1,4,7-Triazaindenes, *see* Pyrrolo[3,4-*b*]pyrazines
1,4,5-Triazanaphthalenes, *see* Pyrido[2,3-*b*]pyrazines
1,4,6-Triazanaphthalenes, *see* Pyrido[3,4-*b*]pyrazines
s-Triazolo[1,5-*a*]pyrazine, 410-412, 414
s-Triazolo[4,3-*a*]pyrazine, 415, 417, 418
s-Triazolo[1,5-*a*]pyrazines:
  preparation of, 411
  table of, 425
  uses of, 414
  u.v. and mass spectra of, 410
s-Triazolo[2,3-*a*]pyrazines, *see* *s*-Triazolo[1,5-*a*]pyrazines
s-Triazolo[4,3-*a*]pyrazines:
  mass spectra of, 417
  preparation of, 418
  table of, 426
  u.v. of, 415
v-Triazolo[4,5-*b*]pyrazines:
  table of, 428
s-Triazolo[4,3-*a*]quinoxalines, 196
2-Tribromomethylquinoxaline, 214
1,2,3-Tribromopyrrolo[1,2-*a*]quinoxaline, 616, 618
2,3,7-Trichloro-6-methylaminosulfonylquinoxaline, 171
5,6,8-Trichloro-7-methylquinoxaline, 20
1,2,3-Trichloropyrrolo[1,2-*a*]quinoxaline, 616
2,3,6-Trichloroquinoxaline, 163
6,7,8-Trimethoxycarbonyl-1,3-dimethylpyrrolo[1,2-*a*]pyrazine, 297
6,7,8-Trimethoxycarbonyl-3-methylpyrrolo[1,2-*a*]pyrazine, 293, 297
1,2-Trimethylenepiperazines, *see* Pyrrolo-[1,2-*a*]pyrazines
3,6,7-Trimethylpyrazolo[3,4-*b*]quinoxaline, 703
1,8-Trimethylenequinoxalines, *see* Pyrrolo-[1,2-*a*]pyrazines
1,1,4-Trimethylquinoxalinium iodide, 273
1,2,3-Trimethylquinoxalinium salts, 250, 251
3,6,7-Trimethylquinoxalin-2-one, 83
2-Trimethylsilyloxyquinoxaline, 88
2,5,6-Trimethyl-*s*-triazolo[1,5-*a*]pyrazine, 411
2-Trioxanylquinoxaline, 124
3,5,6-Triphenylisoxazolo[4,5-*b*]pyrazine, 435
3,5,6-Triphenyl-*s*-triazolo[4,3-*a*]pyrazine, 418
2,5,6-Triphenyl-*v*-triazolo[4,5-*b*]pyrazines, 423

D-Valyl-L-proline anhydride:
  biological activity of, 323
  formation of, 324
  occurrence of, 323
  preparations of, 311, 315

2-(L-Xylotetrahydroxybutyl)quinoxaline, 242